21ST CENTURY ASTRONOMY

Stars and Galaxies

SIXTH EDITION

21ST CENTURY ASTRONOMY

Stars and Galaxies

SIXTH EDITION

Laura Kay
Barnard College

Stacy Palen
Weber State University

George Blumenthal
University of California—Santa Cruz

W. W. NORTON & COMPANY
NEW YORK • LONDON

W. W. Norton & Company has been independent since its founding in 1923, when William Warder Norton and Mary D. Herter Norton first published lectures delivered at the People's Institute, the adult education division of New York City's Cooper Union. The firm soon expanded its program beyond the Institute, publishing books by celebrated academics from America and abroad. By midcentury, the two major pillars of Norton's publishing program—trade books and college texts—were firmly established. In the 1950s, the Norton family transferred control of the company to its employees, and today—with a staff of four hundred and a comparable number of trade, college, and professional titles published each year—W. W. Norton & Company stands as the largest and oldest publishing house owned wholly by its employees.

Printed in Canada

Editor: Erik Fahlgren
Project Editor: Diane Cipollone
Editorial Assistant: Sara Bonacum
Managing Editor, College: Marian Johnson
Managing Editor, College Digital Media: Kim Yi
Production Manager: Ashley Horna
Media Editor: Rob Bellinger
Associate Media Editor: Arielle Holstein
Media Project Editor: Danielle Belfiore
Media Editorial Assistant: Kelly Smith
Marketing Manager: Katie Sweeney
Design Director: Rubina Yeh
Photo Editor: Trish Marx
Director of College Permissions: Megan Schindel
Permissions Associate: Elizabeth Trammell
Composition: Graphic World
Manufacturing: Transcontinental

ISBN 978-0-393-64471-5 (pbk)

W. W. Norton & Company, Inc., 500 Fifth Avenue, New York, NY 10110-0017
wwnorton.com
W. W. Norton & Company Ltd., 15 Carlisle Street, London W1D 3BS
1 2 3 4 5 6 7 8 9 0

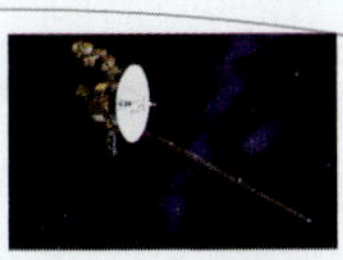

Aloha and Bon Voyage to our late co-author Brad Smith, who showed the world the outer solar system through his cameras on the Voyager spacecraft.

LAURA KAY thanks her wife, M.P.M. She dedicates this book to her late uncle, Lee Jacobi, for an early introduction to physics, and to her late colleagues at Barnard College, Tally Kampen and Sally Chapman.

STACY PALEN thanks her husband, John Armstrong, for his patient support during this project.

GEORGE BLUMENTHAL gratefully thanks his wife, Kelly Weisberg, and his children, Aaron and Sarah Blumenthal, for their support during this project. He also wants to thank Professor Robert Greenler for stimulating his interest in all things related to physics.

BRIEF CONTENTS

Chapters 8–12 are not included in this edition.

CONTENTS

Chapters 8–12 are not included in this edition.

PART II THE SOLAR SYSTEM

PART III STARS AND STELLAR EVOLUTION

PART IV GALAXIES, THE UNIVERSE, AND COSMOLOGY

Working It Out

Chapters 8–12 are not included in this edition.

AstroTours

Chapters 8–12 are not included in this edition.

AstroTour animations, Interactive Simulations, and Astronomy in Action Videos are all available from the free Student Site at the Digital Resources Site, and they are also integrated into assignable Smartwork5 exercises. **digital.wwnorton.com/astro6**

Interactive Simulations

Chapters 8–12 are not included in this edition.

Astronomy in Action Videos

Chapters 8–12 are not included in this edition.

Preface

Dear Student

Why is it a good idea to take a science course, and in particular, why is astronomy a course worth taking? Many people choose to learn about astronomy because they are curious about the universe. Your instructor likely has two basic goals in mind for you as you take this course. The first is to understand some basic physical concepts and how they apply to the universe around us. The second is to think like a scientist and learn to use the scientific method not only to answer questions in this course but also to be able to evaluate and understand new information you encounter in your life. We have written the sixth edition of *21st Century Astronomy* with these two goals in mind.

Throughout this book, we emphasize not only the content of astronomy (for example, the differences among the planets, the formation of chemical elements) but also *how* we know what we know. The scientific method is a valuable tool that you can carry with you and use for the rest of your life. One way we highlight the process of science is the **Process of Science Figures.** In each chapter, we have chosen one discovery and provided a visual representation illustrating the discovery or a principle of the process of science. In these figures, we try to illustrate that science is not a tidy process, and that discoveries are sometimes made by different groups, sometimes by accident, but always because people are trying to answer a question and show why or how we think something is the way it is.

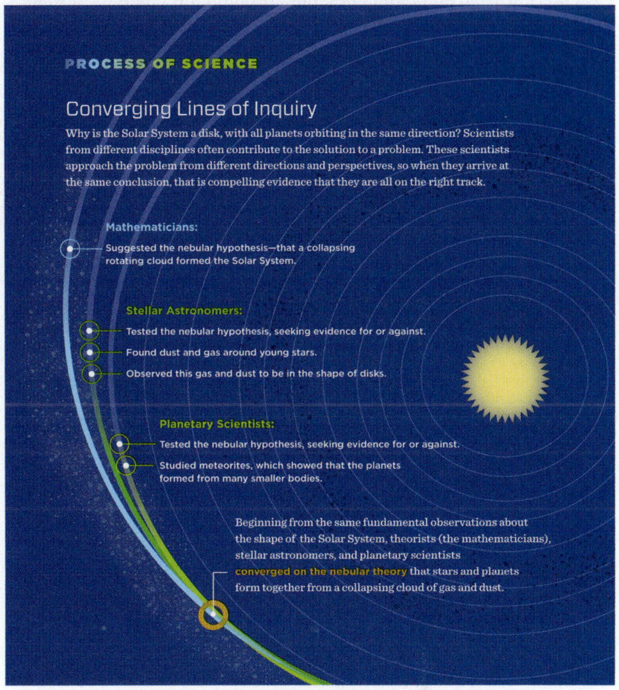

The most effective way to learn something is to "do" it. Whether playing an instrument or a sport or becoming a good cook, reading "how" can only take you so far. The same is true of learning astronomy. We have written this book to help you "do" as you learn. Sometimes this means doing an activity to apply a concept and sometimes we provide a tool to make reading a more active process.

- At the beginning of each chapter, we have provided a set of Learning Goals to guide you as you read. There is a lot of information in every chapter, and the Learning Goals should help you focus on the most important points.
- We present a big-picture question in association with the chapter-opening photo at the beginning of each chapter. For each of these, we have tried to pose a question that is not only relevant to its chapter but also something you may have wondered about. We hope that these questions, plus the photographs that accompany them, capture your attention as well as your imagination.

LEARNING GOALS

By the end of this chapter, you should be able to:

LG 1 Describe how our understanding of planetary system formation developed from the work of both planetary and stellar scientists.

LG 2 Discuss the role of gravity and angular momentum in explaining why planets orbit the Sun in a plane and why they revolve in the same direction that the Sun rotates.

LG 3 Explain how temperature at different locations in the protoplanetary disk affects the composition of planets, moons, and other bodies.

LG 4 Discuss the processes that resulted in the formation of planets and other objects in our Solar System.

LG 5 Describe how astronomers find planets around other stars and derive exoplanet properties.

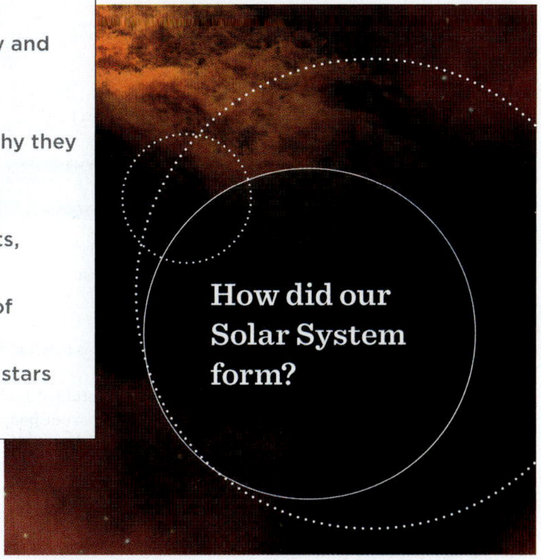

CHECK YOUR UNDERSTANDING 7.1

Which of the following pieces of evidence support the nebular hypothesis? (Choose all that apply.) (a) Planets orbit the Sun in the same direction. (b) The Solar System is relatively flat. (c) Earth has a large Moon. (d) We observe disks of gas and dust around other stars.

- There are **Check Your Understanding** questions at the end of each chapter section. These questions are designed to be answered quickly if you have understood the previous section. The answers are provided in the back of the book so you can check your answer and decide if further review is necessary.

READING ASTRONOMY NEWS

ARTICLES QUESTIONS

A multiplanet system discovered around a relatively nearby star

Four Earth-sized planets detected orbiting the nearest Sun-like star

August 9, 2017, by TIM STEPHENS

A new study by an international team of astronomers reveals that four Earth-sized planets orbit the nearest Sun-like star, Tau Ceti, which is about 12 light years away and visible to the naked eye. These planets have masses as low as 1.7 Earth mass, making them among the smallest planets ever detected around nearby Sun-like stars. Two of them are super-Earths located in the habitable zone of the star, meaning they could support liquid surface water.

According to lead author Fabo Feng of the University of Hertfordshire, UK, the researchers are getting tantalizingly close to the 10-centimeter-per-second limit required for detecting Earth analogs. "Our detection of such weak wobbles is a milestone in the search for Earth analogs and the understanding of the Earth's habitability through comparison with these analogs," Feng said. "We have introduced new methods to remove the noise in the data in order to reveal the weak planetary signals."

The outer two planets around Tau Ceti

see how star's activity differed at different wavelengths and use that information to separate this activity from signals of planets," Tuomi said.

The researchers painstakingly improved the sensitivity of their techniques and were able to rule out two of the signals the team had identified in 2013 as planets. "But no matter how we look at the star, there seem to be at least four rocky planets orbiting it," Tuomi said. "We are slowly learning to tell the difference between wobbles caused by planets and those caused by stellar active

ARTICLES QUESTIONS

1. How do planets cause their stars to "wobble"?
2. With the spectroscopic method, why is finding Earth-sized planets harder than finding giant planets?
3. How did the astronomers improve their ability to find small planets?
4. What is a debris disk?
5. What would be the disadvantage to a planet's being tidally locked to its star? (Hint: Refer to Chapter 2.)

- As a citizen of the world, you make judgments about science, distinguishing between good science and pseudoscience. You use these judgments to make decisions in the grocery store, pharmacy, car dealership, and voting booth. You may base these decisions on the presentation of information you receive through the media, which is very different from the presentation in class. One important skill is the ability to recognize what is credible and to question what is not. To help you hone this skill, we have provided **Reading Astronomy News** sections at the end of every chapter. These features include a news article with questions to help you make sense of how science is presented to you. It is important that you learn to be critical of the information you receive, and these features will help you do that.
- While we know a lot about the universe, science is an ongoing process, and we continue to search for new answers. To give you a glimpse of what we don't know, we provide an **Unanswered Questions** feature near the end of each chapter. Most of these questions represent topics that scientists are currently studying.

? Unanswered Questions

- How typical is the Solar System? Only within the past decade have astronomers found other systems containing four or more planets, and so far the observed distributions of large and small planets in those multiplanet systems have looked different from those of the Solar System. Computer simulations of planetary system formation suggest that a system with stable orbits and a planetary distribution like those of the Solar System may develop only rarely. Improved supercomputers can run more complex simulations, which can be compared with the observations to better understand how solar systems are configured.
- How Earth-like must a planet be before scientists declare it to be "another Earth"? An editorial in the science journal *Nature* cautioned that scientists should define "Earth-like" in advance—before multiple discoveries of planets "similar" to Earth are announced and a media frenzy ensues. Must a planet be of similar size and mass, be located in the habitable zone, and have spectroscopic evidence of liquid water before we call it "Earth 2.0"?

- The language of science is mathematics, and it can be as challenging to learn as any other language. The choice to use mathematics as the language of science is not arbitrary; nature "speaks" math. To learn about nature, you will need to speak its language. We don't want the language of math to obscure the concepts, so we have placed this book's mathematics in **Working It Out** boxes to make it clear when we are beginning and ending a mathematical argument, so that you can spend time with the concepts in the chapter text and then revisit the mathematics of the concept to study the formal language of the argument. You will learn to work with data and identify when data aren't quite right. We want you to be comfortable reading, hearing, and speaking the language of science, and we will provide you with tools to make it easier.

7.1 Working It Out Angular Momentum

In its simplest form, the angular momentum (L) of a system is given by

$$L = m \times v \times r$$

where m is the mass, v is the speed at which the mass is moving, and r represents how spread out the mass is.

Let's apply that relationship to the angular momentum of Jupiter in its orbit about the Sun. The angular momentum from one body orbiting another is called *orbital* angular momentum, $L_{orbital}$. The mass (m) of Jupiter is 1.90×10^{27} kilograms (kg), the speed of Jupiter in orbit (v) is 1.31×10^4 meters per second (m/s), and the radius of Jupiter's orbit (r) is 7.79×10^{11} meters. Putting all that together gives

$$L_{orbital} = (1.90 \times 10^{27}\ \text{kg}) \times (1.31 \times 10^4\ \text{m/s}) \times (7.79 \times 10^{11}\ \text{m})$$

$$L_{orbital} = 1.94 \times 10^{43}\ \text{kg m}^2/\text{s}$$

where R is the radius of the sphere and P is the rotation period of its spin.

Let's compare Jupiter's orbital angular momentum with the Sun's spin angular momentum to investigate the distribution of angular momentum in the Solar System. Appendix 2 provides the Sun's radius (6.96×10^8 meters) and mass (1.99×10^{30} kg), and its rotation period is 24.5 days = 2.12×10^6 seconds. If we assume that the Sun is a uniform sphere, the spin angular momentum of the Sun is

$$L_{spin} = \frac{4 \times \pi \times (1.99 \times 10^{30}\ \text{kg}) \times (6.96 \times 10^8\ \text{m})^2}{5 \times (2.12 \times 10^6\ \text{s})}$$

$$L_{spin} = 1.14 \times 10^{42}\ \text{kg m}^2/\text{s}$$

$L_{orbital}$ of Jupiter is about 17 times greater than L_{spin} of the Sun. Thus, most of the angular momentum of the Solar System now

- Each chapter concludes with an **Origins** section, which relates material or subjects found in the chapter to questions about the origin of the universe and the origin of life in the universe and on Earth. Astrobiologists have made much progress in recent years on understanding how conditions in the universe may have helped or hindered the origin of life, and in each Origins we explore an example from its chapter that relates to how the universe and life formed and evolved.
- At the end of each chapter, we have provided several types of questions, problems, and activities for you to practice your skills. The **Test Your Understanding** questions focus on more detailed facts and concepts from the chapter. **Thinking about the Concepts** questions ask you to synthesize information and explain the "how" or "why" of a situation. **Applying the Concepts** problems give you a chance to practice the quantitative skills you learned in the chapter and to work through a situation mathematically. The **Using the Web** questions represent other opportunities to "learn by doing." Using the Web sends you to websites of space missions, observatories, experiments, or archives to access recent observations, results, or press releases. Other sites are for "citizen science" projects in which you can contribute to the analysis of new data.
- **Explorations** show you how to use the concepts and skills you learned in an interactive way. Some of the book's Explorations ask you to use animations and simulations on the Student Site, while the others are hands-on, paper-and-pencil activities that use everyday objects such as ice cubes or balloons.

The resources outside of the book (at the Student Site) can help you understand and visualize many of the physical concepts described in the book. **AstroTours and Interactive Simulations** are represented by icons in the margins of the book. There is also a series of short **Astronomy in Action** videos that are represented by icons in the margins and available at the Student Site. These videos feature one of the authors (and several students) demonstrating physical concepts at work. Your instructor might assign these videos to you or you might choose to watch them on your own to create a better picture of each concept in your mind.

Astronomy gives you a sense of perspective that no other field of study offers. The universe is vast, fascinating, and beautiful, filled with a wealth of objects that, surprisingly, can be understood using only a handful of principles. By the end of this book, you will have gained a sense of your place in the universe.

USING THE WEB

46. Using the exoplanet catalogs:
    a. Go to the "Catalog" Web page (http://exoplanet.eu/catalog) of the Extrasolar Planets Encyclopedia. Look for a star that has multiple planets. Make a graph showing the distances of the planets from that star, and note the masses and sizes of the planets. Put the Solar System planets on the same axis. How does that extrasolar planet system compare with the Solar System?
    b. Go to the "Exoplanets Data Explorer" website (http://exoplanets.org) and click on "Table." That website lists planets that have detailed orbital data published in scientific journals, and it may have a smaller total count than the website in part (a). Pick a planet discovered recently, as specified in the "First Reference" column. What is the planet's minimum mass? What is its semimajor axis and the period of its orbit? What is the eccentricity of its orbit? Click on the star name in the first column to get more information. Does that planet have a radial velocity curve? Was it observed in transit, and if so, what is the planet's radius and density? Is it more like Jupiter or more like Earth?

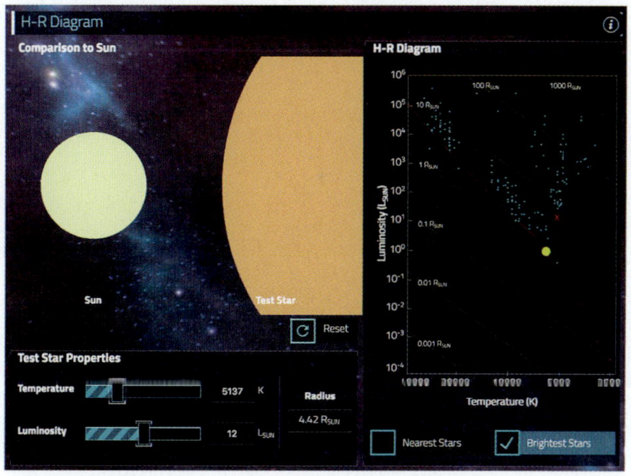

Dear Instructor

We wrote this book with a few overarching goals: to inspire students, to make the material interactive, and to create a useful and flexible tool that can support multiple learning styles.

As scientists and as teachers, we are passionate about the work we do. We hope to share that passion with students and inspire them to engage in science on their own. Through our own experience, familiarity with education research, and surveys of instructors, we have come to know a great deal about how students learn and what goals teachers have for their students. We have explicitly addressed many of these goals and learning styles in this book, sometimes in large, immediately visible ways such as the inclusion of features but also through less obvious efforts such as questions and problems that relate astronomical concepts to everyday situations or a fresh approach to organizing material.

For example, many teachers state that they would like their students to become "educated scientific consumers" and "critical thinkers" or that their students should "be able to read a news story about science and understand its significance." We have specifically addressed these goals in our **Reading Astronomy News** feature, which presents a news article and a series of questions that guide a student's critical thinking about the article, the data presented, and the sources.

Education research shows that the most effective way to learn is by doing. **Exploration** activities at the end of each chapter are hands-on, asking students to take the concepts they've learned in the chapter and apply them as they interact with animations and simulations on the Student Site or work through pencil-and-paper activities. Many of these Explorations incorporate everyday objects and can be used either in your classroom or as activities at home. The **Using the Web** problems direct students to "citizen science" projects, where they can contribute to the analysis of new astronomical data. Other problems send students to websites of space missions, observatories, collaborative projects, and catalogs to access the most current observations, results, and news releases. These Web problems can be used for homework, lab exercises, recitations, or "writing across the curriculum" projects.

We also believe students should be exposed to the more formal language of science—mathematics. We have placed the math in **Working It Out** boxes, so it does not interrupt the flow of the text or get in the way of students' understanding of conceptual material. But we've gone further by beginning with fundamental ideas in early Working It Out boxes and slowly building in complexity through the book. We've also worked to remove some of the stumbling blocks that affect student confidence by providing calculator hints, references to earlier Working It Out boxes, and detailed, fully worked examples. Many chapters include problems on reading and interpreting graphs. Appendix 1, "Mathematical Tools," summarizes some math concepts for students.

Discussion of basic physics is contained in Part I to accommodate courses that use the *Solar System* or *Stars and Galaxies* volumes. A "just-in-time" approach to introducing the physics is still possible by bringing in material from Chapters 2–6 as needed. For example, the sections on tidal forces in Chapter 4 can be taught along with the moons of the Solar System in Part II, or with mass transfer in binary stars in Part III, or with galaxy interactions in Part IV. Spectral lines in Chapter 5 can be taught with planetary atmospheres in Part II or with stellar spectral types in Part III, and so on.

In our overall organization, we have made several efforts to encourage students to engage with the material and build confidence in their scientific skills as they proceed through the book. For planets, stars and galaxies, we have organized the material to cover the general case first and then delve into more details with specific examples. Thus, you will find "planetary systems" before our own Solar System, "stars" before the Sun, and "galaxies" before the Milky Way. This allows us to avoid frustrating students by making assumptions about what they know about stars or galaxies or forward-referencing to basic definitions and overarching concepts. This organization also implicitly helps students understand their place in the universe: our galaxy and our star are each one of many. They are specific examples of a physical universe in which the same laws apply everywhere. Planets have been organized comparatively to emphasize that science is a process of studying individual examples that lead to collective conclusions. All of these organizational choices were made with the student perspective in mind and a clear sense of the logical hierarchy of the material.

We begin in Chapter 19 by introducing galaxies as a whole and our measurements of them, including recession velocities. Then we address the Milky Way in Chapter 20—a specific example of a galaxy that we can discuss in detail. This follows the repeating motif of moving from the general to the specific that exists throughout the text and gives students a basic grounding in the concepts of spiral galaxies, supermassive black holes, and dark matter before they need to apply those concepts to the specific example of our own galaxy. Chapter 21, "The Expanding Universe," covers the cosmological principle, the Hubble expansion, and the observational evidence for the Big Bang.

The **Origins** sections illustrate how astrobiologists and other scientists approach the study of a scientific question from the chapter related to the origin of the universe and of life. Material about exoplanets is introduced in Chapter 7, incorporated into other chapters when appropriate, and continued in Chapter 24. We revised each chapter, streamlining some topics, and updating the science to reflect the progress in the field. This includes:

- Updates on new telescopes and space missions
- New images and results from Solar System exploration missions to the Moon, Mars, Jupiter, Saturn, and Pluto
- Updated graphical data on climate change
- Results from Kepler and other missions on populations of exoplanets
- Results from the GAIA mission on stellar distances, the HR diagram, and the Milky Way
- Results from LIGO on the discovery of gravitational waves
- Updated cosmology from the Planck mission
- Many new articles for Reading Astronomy News, such as "2,500 Miles of Citizen Scientists", "NASA killed Cassini to Avoid Contaminating Saturn's Moons", and "LIGO Just Detected the Oldest Gravitational Waves Ever Discovered"
- Revised Process of Science figures for each chapter
- Revised simulations for some Explorations exercises

Many professors find themselves under pressure from accrediting bodies or internal assessment offices to assess their courses in terms of learning goals. Each chapter has Learning Goals and end-of-chapter Summary to correspond to the chapter's Learning Goals. In Smartwork5, questions and problems are tagged and can be sorted by Learning Goal. Smartwork5 contains more than 2,000 questions and problems that are tied directly to this text, including the Check Your Understanding questions and versions of the Reading Astronomy News and Exploration questions. Any of these could be used as a reading quiz to be completed before class or as homework. Every question in Smartwork5 has hints and answer-specific feedback so that students are coached to work toward the correct answer. An instructor can easily modify any of the provided questions, answers, and feedback or can create his or her own questions.

We've also created a series of 23 videos explaining and demonstrating concepts from the text, accompanied by questions in Smartwork5. You might assign these videos prior to lecture—either as part of a flipped modality or as a "reading quiz." In either case, you can use the diagnostic feedback from the questions in Smartwork5 to tailor your in-class discussions. Or you might show them in class, to stimulate discussion. Or you might simply use them as a jumping-off point—to get ideas for activities to do with your own students.

We continue to look for better ways to engage students, so please let us know how these features work for your students.

Digital Media for Students

digital.wwnorton.com/astro6

smartwork5 Online Homework

Steven Desch, Guilford Technical Community College

Diane Friend, University of Montana

Ana M. Larson, University of Washington

Violet Mager, Penn State Wilkes-Barre

Ryan Oelkers, Vanderbilt University

Todd Young, Wayne State College

William Younger, Tidewater Community College

More than 2,000 questions support *21st Century Astronomy, Sixth Edition*—all with answer-specific feedback, hints, and ebook links. Questions include Summary Self-Tests, versions of the Explorations (based on AstroTours and new Interactive Simulations), and questions based on the Reading Astronomy News features. Astronomy in Action video questions focus on getting students to come to class prepared and on overcoming common misconceptions. Process of Science Guided Inquiry Assignments help students apply the scientific method to important questions in astronomy, challenging them to think like scientists. Rounding out the Smartwork5 course are ranking, sorting, and labeling exercises—all classified according to the same Learning Goals used in the textbook and test bank.

Smartwork5 can be set up to work right in your LMS, with student scores flowing directly to your LMS gradebook. Your local Norton representative can help you set up LMS integration.

At Play in the Cosmos: The Videogame

Jeff Bary (Colgate University), Adam Frank (University of Rochester), Learning Games Network (University of Wisconsin and Massachusetts Institute of Technology), Gear Learning at UW-Madison's Wisconsin Center for Education Research

This ground-breaking and award-winning educational videogame is designed to engage introductory astronomy students by asking them to confront challenges using concepts that span the scope of the course. Using the process of science, students complete missions by flying their ship and choosing the right scientific tools for each task. The videogame can be downloaded to PC, Mac, and tablets and can be included with new copies of the Sixth Edition at no additional cost.

A complete videogame instructor's manual provides suggestions for meaningfully integrating the game into your course—for example, as a lab activity or as homework. The manual, written by lead game author Jeff Bary, provides mission overviews, plus multiple-choice and discussion questions for every mission.

Norton Ebook

The 21st Century Astronomy, Sixth Edition ebook provides students an enhanced reading experience at a fraction of the cost of a print textbook. Students are able to have an active reading experience and can take notes, bookmark, search, highlight, and even read offline. Instructors can add notes for students to see as they read the text. Norton Ebooks can be viewed on—and synced among—all computers and mobile devices. Every question in Smartwork5 provides a reference link to the ebook so that students have easy access to their textbook when completing homework assignments.

Student Site

W. W. Norton's student website features the following:

- Thirty AstroTour animations. These animations, some of which are interactive, use art from the text to help students visualize important physical and astronomical concepts. All are now tablet-compatible.
- Seven new Interactive Simulations, authored by Stacy Palen, pair with the Exploration activities in the text, allowing students to explore topics such as Moon phases, Kepler's laws, and the Hertzsprung-Russell diagram.
- Twenty-three Astronomy in Action videos that feature author Stacy Palen demonstrating the most important concepts in a visual, easy to understand, and memorable way.

Learning Astronomy by Doing Astronomy: Collaborative Lecture Activities

Stacy Palen, Weber State University

Ana Larson, University of Washington

Students learn best by doing. Devising, writing, testing, and revising suitable in-class activities that use real astronomical data, illuminate astronomical concepts, and pose probing questions that ask students to confront misconceptions can be challenging and time consuming. In this workbook, the authors draw on their experience teaching thousands of students in many different types of courses (large in-class, small in-class, hybrid, online, flipped, and so forth) to bring 30 field-tested activities that can be used in any classroom today. The activities have been designed to require no special software, materials, or equipment and to take no more than 50 minutes to do.

Starry Night Planetarium Software (College Version) and Workbook

Steven Desch, Guilford Technical Community College

Michael Marks, Bristol Community College

Starry Night is a realistic, user-friendly planetarium simulation program designed to allow students in urban areas to perform observational activities on a computer screen. Norton's unique accompanying workbook offers observation assignments that guide students' virtual explorations and help them apply what they've learned from the text reading assignments.

For Instructors

Instructor's Manual

Ana Larson, University of Washington

This resource includes brief chapter overviews, suggested discussion points, notes on the Process of Science figures, AstroTour animations, Astronomy in Action videos, Interactive Simulations, Reading Astronomy News, Explorations, and worked solutions to Check Your Understanding and end-of-chapter questions and problems. Also included are notes on teaching with *Learning Astronomy by Doing Astronomy: Collaborative Lecture Activities* and answers to the *Starry Night Workbook* exercises.

Norton Interactive Instructor's Guide (IIG)

This new and searchable online resource is designed to help instructors prepare for lecture in real time. It contains the following resources, each tagged by topic and chapter:

- Discussion Points
- Notes on the Process of Science figures
- AstroTour animations
- Astronomy in Action videos
- Interactive Simulations
- Notes on Reading Astronomy News and Explorations
- Notes on teaching with Learning Astronomy by Doing Astronomy: Collaborative Lecture Activites
- Solutions to *Check Your Understanding and end-of-chapter* questions and problems
- Answers to *Starry Night Workbook* exercises
- PowerPoint Lecture Slides
- All art and tables in JPEG and PowerPoint formats
- LMS Coursepacks, available in Blackboard, Canvas, Desire2Learn, and Moodle formats

All of the above content is also available on the traditional Instructor's Resource Site.

Teaching Astronomy by Doing Astronomy blog tada101.com

This new resource features posts by Stacy Palen on teaching ideas, classroom activities, and current events. Instructors looking for additional ways to incorporate active learning will find this blog to be an invaluable resource.

PowerPoint Lecture Slides

Christa Speights, Northern Kentucky University

These ready-made lecture slides integrate selected textbook art, all Check Your Understanding and Working It Out questions from the text, class questions, and links to the AstroTour animations. These lecture slides are fully editable and are available in Microsoft PowerPoint format.

Test Bank

Matthew Newby, Temple University

The Test Bank has been revised using Bloom's Taxonomy and provides over 2,400 multiple-choice and short-answer problems. Each chapter of the Test Bank consists of five question levels classified according to Bloom's Taxonomy:

Remembering
Understanding
Applying
Analyzing
Evaluating

Problems are further classified by section and difficulty level, making it easy to construct tests and quizzes that are meaningful and diagnostic. The Test Bank assesses a common set of Learning Objectives consistent with the textbook and Smartwork5 online homework.

LMS Coursepacks

Norton's Coursepacks, available for use in various Learning Management Systems (LMSs), feature all Test Bank questions, links to the AstroTours, worksheets based on the Explorations and Astronomy in Action videos, and automatically graded versions of selected end-of-chapter Test Your Understanding multiple-choice questions. Coursepacks are available in BlackBoard, Canvas, Desire-2Learn, and Moodle formats.

Acknowledgments

The authors would like to acknowledge the extraordinary efforts of the staff at W. W. Norton: Sara Bonacum who kept things flowing smoothly; Jane Miller and Trish Marx for managing the numerous photos; and the copy editor, Gabe Waggoner, who made sure that all the grammar and punctuation survived the multiple rounds of the editing process. We would especially like to thank John Murdzek for the developmental editing process, and Diane Cipollone, who shepherded the manuscript through production.

Erik Fahlgren was the editor. Ashley Horna managed the production and Rubina Yeh was the design director. Rob Bellinger, Arielle Holstein, and Kelly Smith worked on the media and supplements, and Katie Sweeney will help get this book into the hands of people who can use it.

We gratefully acknowledge the contributions of the authors who worked on previous editions of *21st Century Astronomy*: Jeff Hester, Gary Wegner and the late Dave Burstein, Ron Greeley, Brad Smith, and Howard Voss, with special thanks to Dave for starting the project, to Jeff for leading the original authors through the first edition, and to Brad for leading the second and third editions.

Laura Kay
Stacy Palen
George Blumenthal

And we would like to thank the reviewers, whose input at every stage improved the book:

Sixth Edition Reviewers

Manuel Alvarado, El Paso Community College
Gagandeep Anand, Boston University
Simon Balm, Santa Monica College
Raymond Benge, Tarrant College
Ryan Bennett, University of North Texas
Brett Bochner, Hofstra University
Jack Brockway, Radford University
Shea Brown, University of Iowa
Spencer Buckner, Austin Peay State University
C. Austin Campbell, El Paso Community College
Christina Cavalli, Austin Community College
Gerald Cecil, University of North Carolina- Chapel Hill
Scott Cochran, Central Michigan University
Tara Cotton, University of Georgia
Robert Coyne, Texas Tech University
Steve Desche, Arizona State University
Michael Endl, Austin Community College
Duncan Farrah, University of Hawaii
Sunil Fernandes, University of Texas- San Antonio
Ken Gayley, University of Iowa
Christopher Gerardy, University of North Carolina- Charlotte
Guillermo Gonzalez, Ball State University
Ioannis Haranas, Wilfrid Laurier College
Javier Hasbun, University of West Georgia
Jim Higdon, Georgia Southern University
Scott Hildreth, Chabot College
Christopher Hodges, Sacramento State University
Mike Hood, Mt. San Antonio College
James Imamura, University of Oregon
Jennifer Jones, Arapahoe Community College
Bruno Jungweirt, North Carolina State University
Patrick Kelly, Dalhousie University
Bethuel Khamala, El Paso Community College
Keigo Fukumura, James Madison University
Rafael Lang, Purdue University
Kenneth Lanzetta, Stony Brook University
Lee LaRue, Paris Junior College
Denis Leahy, University of Calgary
Hector Leal, University of Texas- Rio Grande Valley
Hyun-chul Lee, University of Texas- Rio Grande Valley
Ludwik Lembryk, University of Toledo
Isaac Lopez, Boston University
Petrus Martens, Georgia State University
Justin Mason, Old Dominion University
Douglas McElroy, Chaffey College
Robert Morehead, Texas Tech University
Hon-kie Ng, Florida State University
Merav Opher, Boston University
Robert Parks, Louisiana State University
Jon Pedicino, College of the Redwoods
Nicolas Pereyra, University of Texas- Rio Grande Valley
Bob Powell, University of West Georgia
Barry Rice, Sierra College
Ruben Sandapen, Acadia University
Eric Schlegel, University of Texas- San Antonio
Michael Schwartz, Santa Monica College
Parampreet Singh, Louisiana State University
James Sy, El Paso Community College
Catherine Tabor, El Paso Community College
Benjamin Team, University of West Georgia
Fiorella Terenzi, Florida International University
Lennart Van Haaften, Texas Tech University
Jasper Wall, University of British Columbia
Colin Wallace, University of North Carolina- Chapel Hill
G. Scott Watson, Syracuse University
James Webb, Florida International University
Julia Wickett, Collin College
Kurt Williams, Texas A&M University
Fred Wilson, Collin College
David Wood, San Antonio College

Reviewers of Previous Editions

Scott Atkins, University of South Dakota
Timothy Barker, Wheaton College
Peter A. Becker, George Mason University
Timothy C. Beers, Michigan State University
David Bennum, University of Nevada–Reno
Edwin Bergin, University of Pittsburgh
William Blass, University of Tennessee
Steve Bloom, Hampden Sydney College
Daniel Boice, University of Texas at San Antonio
Bram Boroson, Clayton State University
David Branning, Trinity College
Julie Bray-Ali, Mt. San Antonio College
Suzanne Bushnell, McNeese State University
Paul Butterworth, George Washington University
Juan E. Cabanela, Minnesota State University–Moorhead
Amy Campbell, Louisiana State University
Michael Carini, West Kentucky University
Supriya Chakrabarti, Boston University
Robert Cicerone, Bridgewater State College
David Cinabro, Wayne State University
Judith Cohen, California Institute of Technology
Eric M. Collins, California State University–Northridge
John Cowan, University of Oklahoma–Norman
Debashis Dasgupta, University of Wisconsin–Milwaukee
Robert Dick, Carleton University
Gregory Dolise, Harrisburg Area Community College
Tom English, Guilford Technical Community College
David Ennis, The Ohio State University
John Finley, Purdue University
Matthew Francis, Lambuth University
Kevin Gannon, College of Saint Rose
Todd Gary, O'More College of Design
Christopher Gay, Santa Fe College
Parviz Ghavamian, Towson University
Martha Gilmore, Wesleyan University
Bill Gutsch, St. Peter's College
Karl Haisch, Utah Valley University
Charles Hawkins, Northern Kentucky University
Sebastian Heinz, University of Wisconsin–Madison
Barry Hillard, Baldwin Wallace College
Paul Hintzen, California State University–Long Beach
Paul Hodge, University of Washington
William A. Hollerman, University of Louisiana at Lafayette
Hal Hollingsworth, Florida International University
Olencka Hubickyj-Cabot, San Jose State University
Kevin M. Huffenberger, University of Miami
Adam Johnston, Weber State University
Steven Kawaler, Iowa State University
Monika Kress, San Jose State University
Jessica Lair, Eastern Kentucky University
Alex Lazarian, University of Wisconsin–Madison
Kevin Lee, University of Nebraska–Lincoln
Matthew Lister, Purdue University
M. A. K. Lodhi, Texas Tech University
Leslie Looney, University of Illinois at Urbana–Champaign
Jack MacConnell, Case Western Reserve University
Kevin Mackay, University of South Florida
Dale Mais, Indiana University–South Bend
Michael Marks, Bristol Community College
Norm Markworth, Stephen F. Austin State University
Kevin Marshall, Bucknell University
Stephan Martin, Bristol Community College
Amanda Maxham, University of Nevada–Las Vegas
Chris McCarthy, San Francisco State University
Ben McGimsey, Georgia State University
Charles McGruder, West Kentucky University
Janet E. McLarty-Schroeder, Cerritos College
Stanimir Metchev, Stony Brook University
Chris Mihos, Case Western University
Milan Mijic, California State University–Los Angeles
J. Scott Miller, University of Louisville
Scott Miller, Sam Houston State University
Kent Montgomery, Texas A&M University–Commerce
Andrew Morrison, Illinois Wesleyan University
Edward M. Murphy, University of Virginia
Kentaro Nagamine, University of Nevada–Las Vegas
Ylva Pihlström, University of New Mexico
Jascha Polet, California State Polytechnic University
Dora Preminger, California State University–Northridge
Daniel Proga, University of Nevada–Las Vegas
Laurie Reed, Saginaw Valley State University
Judit Györgyey Ries, University of Texas
Allen Rogel, Bowling Green State University
Kenneth Rumstay, Valdosta State University
Masao Sako, University of Pennsylvania
Samir Salim, Indiana University–Bloomington
Ata Sarajedini, University of Florida
Paul Schmidtke, Arizona State University
Ann Schmiedekamp, Pennsylvania State University
Jonathan Secaur, Kent State University
Ohad Shemmer, University of North Texas
Caroline Simpson, Florida International University
Paul P. Sipiera, William Rainey Harper College
Ian Skilling, University of Pittsburgh
Tammy Smecker-Hane, University of California–Irvine
Allyn Smith, Austin Peay State University
Jason Smolinski, State University of New York at Oneonta
Roger Stanley, San Antonio College
Ben Sugerman, Goucher College
Neal Sumerlin, Lynchburg College
Angelle Tanner, Mississippi State University
Christopher Taylor, California State University–Sacramento
Donald Terndrup, The Ohio State University
Todd Thompson, The Ohio State University
Glenn Tiede, Bowling Green State University
Frances Timmes, Arizona State University
Trina Van Ausdal, Salt Lake Community College
Walter Van Hamme, Florida International University
Karen Vanlandingham, West Chester University
Nilakshi Veerabathina, University of Texas at Arlington
Paul Voytas, Wittenberg University
Ezekiel Walker, University of North Texas
James Webb, Florida International University
Paul Wiita, Georgia State University
Richard Williamon, Emory University
David Wittman, University of California–Davis

About the Authors

Laura Kay is an award-winning professor and Chair of Physics and Astronomy at Barnard College, where she has taught since 1991. She received a BS degree in physics and an AB degree in feminist studies from Stanford University, and MS and PhD degrees in astronomy and astrophysics from the University of California, Santa Cruz. As a graduate student she spent 13 months at the Amundsen Scott station at the South Pole in Antarctica. She studies active galactic nuclei, using ground-based and space telescopes. She teaches courses in astronomy, astrobiology, women and science, and polar exploration. At Barnard she has served as chair of the Physics & Astronomy Department, chair of the Women's Studies Department, chair of Faculty Governance, and interim associate dean for Curriculum and Governance.

Stacy Palen is an award-winning professor in the physics department and the director of the Ott Planetarium at Weber State University. She received her BS in physics from Rutgers University and her PhD in physics from the University of Iowa. As a lecturer and postdoc at the University of Washington, she taught Introductory Astronomy more than 20 times over 4 years. Since joining Weber State, she has been very active in science outreach activities ranging from star parties to running the state Science Olympiad. Stacy does research in formal and informal astronomy education and the death of Sun-like stars. She spends much of her time thinking, teaching, and writing about the applications of science in everyday life. She then puts that science to use on her small farm in Ogden, Utah.

George Blumenthal has been a professor of astronomy and astrophysics at the University of California, Santa Cruz since 1972, and served as the university's Chancellor from 2006 to 2019. He received his BS degree from the University of Wisconsin–Milwaukee and his PhD in physics from the University of California, San Diego. As a theoretical astrophysicist, George's research encompasses several broad areas, including the nature of the dark matter that constitutes most of the mass in the universe, the origin of galaxies and other large structures in the universe, the earliest moments in the universe, astrophysical radiation processes, and the structure of active galactic nuclei such as quasars. Besides teaching and conducting research, he has served as Chair of the UC Santa Cruz Astronomy and Astrophysics Department, has chaired the Academic Senate for both the UC Santa Cruz campus and the entire University of California system, and has served as the faculty representative to the UC Board of Regents.

21ST CENTURY ASTRONOMY

Stars and Galaxies

SIXTH EDITION

1 Thinking Like an Astronomer

Now is a fascinating time to be studying this most ancient of the sciences. Loosely translated, the word **astronomy** means "patterns among the stars." But modern astronomy—the astronomy we talk about in this book—is about far more than looking at the sky and cataloging the visible stars. The contents of the universe, its origin and fate, and the nature of space and time have become the subjects of rigorous scientific investigation. Humans have long speculated about our *origins*. How and when did the Sun, Earth, and Moon form? Are other galaxies, stars, planets, and moons similar to our own? The answers that scientists are finding to those questions are changing our view of not only the cosmos but also ourselves.

LEARNING GOALS

In this chapter, we will begin studying astronomy by exploring our place in the universe and the methods of science. By the end of this chapter, you should be able to:

LG 1 Describe the size and age of the universe and Earth's place in it.

LG 2 Explain how astronomers use the scientific method to study the universe.

LG 3 Show how scientists use mathematics, including graphs, to find patterns in nature.

LG 4 Describe our astronomical origins.

The first view of Earth seen by humans from deep space. In December 1968, *Apollo 8* astronauts photographed Earth above the Moon's limb. ▶▶▶

Why is this considered one of the most famous photos ever taken?

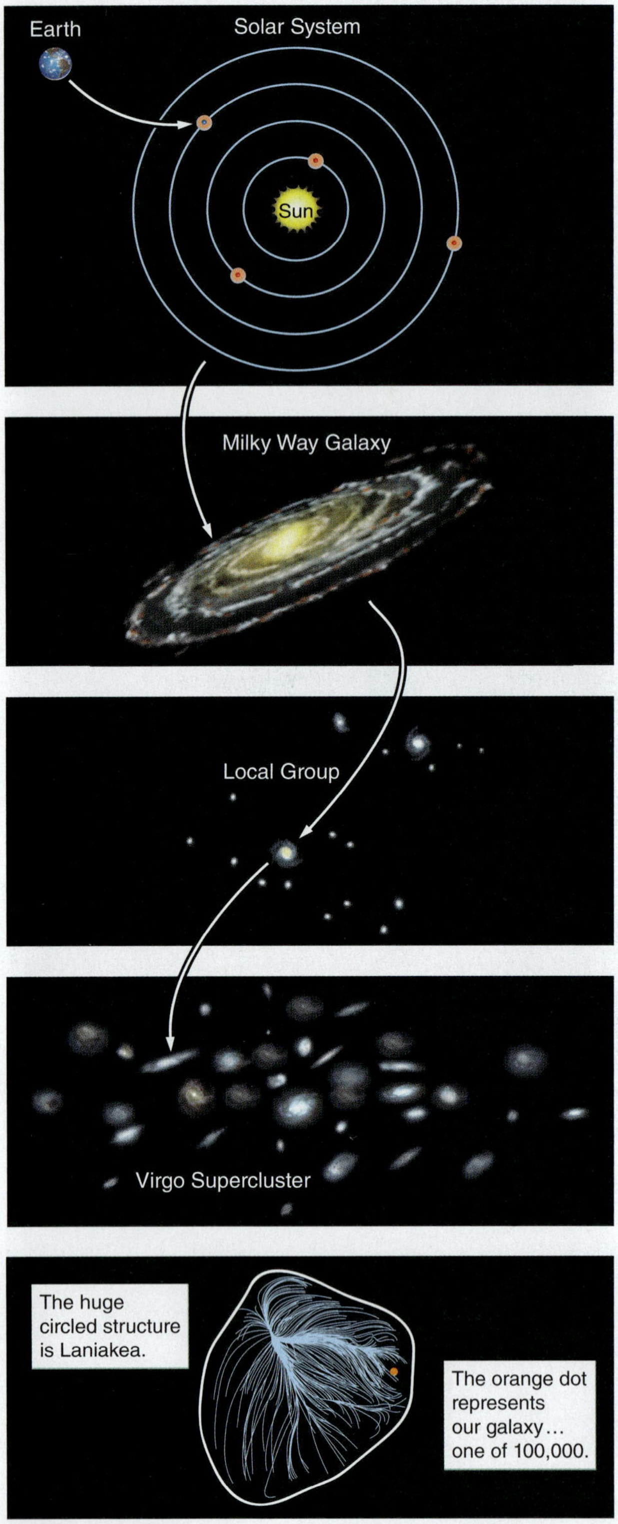

Figure 1.1 Our cosmic address is Earth, Solar System, Milky Way Galaxy, Local Group, Virgo Supercluster, Laniakea Supercluster. We live on Earth, a planet in our Solar System, orbiting the Sun, a star in the Milky Way Galaxy. The Milky Way is a large galaxy within the Local Group of galaxies, located in the Virgo Supercluster, one of four superclusters in the Laniakea Supercluster.

1.1 Earth Occupies a Small Place in the Universe

Astronomers contemplate our place in the universe by studying Earth's position in space and time. Locating Earth in the larger universe is the first step in learning the science of astronomy. In this section, you will get a feel for the neighborhood in which Earth is located. You will also begin to explore the scale of the universe in space and time.

Our Place in the Universe

Most people receive their postal mail at an address—house or building number, street, city, state, and country. If we expand our view to include the enormously vast universe, however, our "cosmic address" might include our planet, star, galaxy, galaxy group or cluster, and galaxy supercluster.

We reside on a planet called Earth, which is orbiting under the influence of gravity around a star called the **Sun**. The Sun is an ordinary middle-aged star more massive and luminous than some stars but less massive and luminous than others. The Sun is extraordinary only because of its importance to us within our own **Solar System**. Our Solar System consists of eight planets (listed in order of distance from the Sun): Mercury, Venus, Earth, Mars, Jupiter, Saturn, Uranus, and Neptune. It also contains many smaller bodies, such as dwarf planets (for example, Pluto, Ceres, and Eris), asteroids (for example, Ida and Eros), and comets (for example, Halley). All those objects are gravitationally bound to the Sun.

The Sun is located about halfway out from the center of the **Milky Way Galaxy**, a flattened collection of stars, gas, and dust. Our Sun is just one among several hundred billion stars scattered throughout our galaxy, and planets orbit many of those stars.

The Milky Way is a member of a collection of a few dozen galaxies called the **Local Group**. Most galaxies in that group are much smaller than the Milky Way. As we look farther outward, the Local Group is part of a vastly larger collection of thousands of galaxies—a **supercluster**—called the Virgo Supercluster, which is part of an even larger grouping called the Laniakea Supercluster. The observable universe has millions of superclusters.

We can now define our cosmic address—Earth, Solar System, Milky Way Galaxy, Local Group, Virgo Supercluster, Laniakea Supercluster—as illustrated in **Figure 1.1**. Yet even that address is not complete because it encompasses only the *local universe*. The part of the universe that we can see—the *observable universe*—extends to many times the size of Laniakea in every direction. That volume contains about 1 trillion (thousand billion) to 2 trillion galaxies. The entire universe is much larger than the local universe and contains much more than the observed planets, stars, and galaxies. Astronomers estimate that about 95 percent of the universe is made up of matter that does not interact with light, known as *dark matter*, and a form of energy that permeates all space, known as *dark energy*. Dark matter and dark energy aren't well understood, and they are among the many exciting areas of research in astronomy.

The Scale of the Universe

As you saw in Figure 1.1, the size of the universe completely dwarfs our human experience. We can start by comparing astronomical sizes and distances to something more familiar. For example, the diameter of our Moon (3,474 kilometers [km]) is slightly greater than the distance between New York, New York, and Ogden, Utah (**Figure 1.2a**), where two of the authors of this book work. The distance from Earth to the Moon is about 100 times the Moon's diameter, and the planet Saturn with its majestic rings would fill much of that distance (**Figure 1.2b**). The distance from Earth to the Sun is about 400 times the Earth–Moon distance, and the distance to the planet Neptune is about 30 times the Earth–Sun distance.

As we move out from the Solar System to the stars, however, the enormous distances become difficult to comprehend. The nearest star is about 9,000 times farther from the Sun than the Sun's distance to Neptune. The diameter of our Milky Way Galaxy is 30,000 times the distance to that nearest star. The Andromeda Galaxy, the nearest large galaxy similar to the Milky Way, is about 600,000 times farther than that nearest star. The diameter of the Local Group of galaxies is about 4 times the distance to Andromeda, and the diameter of the Laniakea Supercluster, which includes the Local Group and many other galaxy groups and clusters, is 50 times larger than the Local Group. As noted earlier, that supercluster is just one of millions in the observable universe.

To get a better sense of those distances, imagine a model in which the objects and distances in the universe are 1 billion times smaller than they really are. In that model, Earth is about the size of a marble or a peanut M&M (about 1.3 centimeters [cm], or half an inch), the Moon is 38 cm away, and the Sun is 150 meters away. Neptune is 4.5 km from the Sun, and the nearest star to the Sun is about 40,000 km away (or about the length of the real Earth's circumference). The model Milky Way Galaxy would fill the real Solar System nearly to the orbit of Saturn. The distance between the model Milky Way and Andromeda galaxies would fill the real Solar System 20 times farther, beyond humanity's most distant space probe. The model Laniakea Supercluster would fill the real Solar System and go about one-eighth of the way to the nearest star.

When thinking about distances in the universe, discussing the time it takes to travel to various places can be helpful. If someone asks you how far it is to the nearest city, you might say 100 km or you might say 1 hour. In either case, you will have given that person an idea of how far the city is. In astronomy, the speed of a car on the highway is far too slow to be useful. Instead, we use the fastest speed in the universe—the speed of light. Light travels at 300,000 kilometers per second (km/s). **Figure 1.3** shows how distances that range from the size of Earth to the size of the observable universe can be measured in the time it takes light to travel that far. Light can circle Earth, for example—a distance of 40,000 km—in just under 1/7 of a second. The circumference of Earth, then, is 1/7 of a light-second. Even relatively small distances in astronomy are so vast that they are measured in units of **light-years (ly)**: the distance light travels in 1 year, about 9.5 trillion km, or 6 trillion miles.

Because light takes time to reach us, we see astronomical objects as they were in the past: how far back in time depends on the object's distance from us. Because light from the Moon takes 1 1/4 seconds to reach us, we see the Moon as it was 1 1/4 seconds ago. Because light from the Sun takes 8 1/3 minutes to reach us, we see the Sun as it was 8 1/3 minutes ago. We see the nearest star as it was more than 4 years ago and objects across the Milky Way as they were tens of

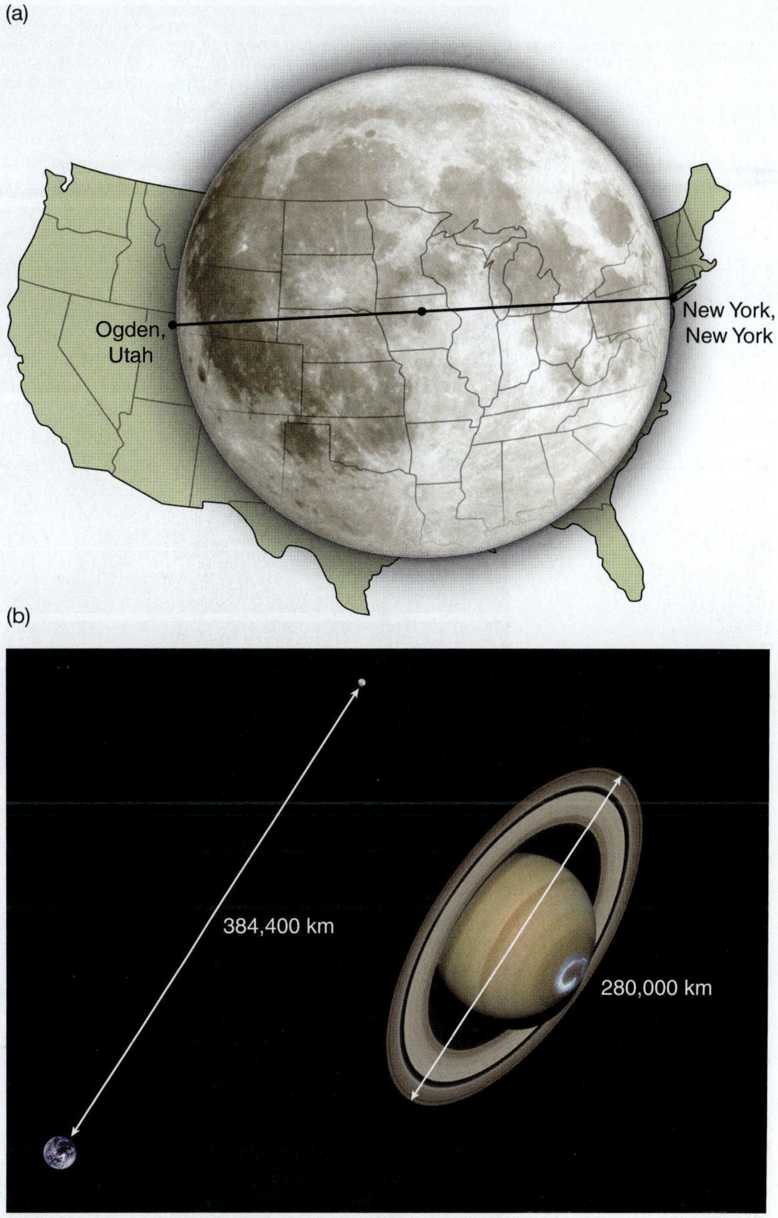

Figure 1.2 (a) The diameter of the Moon is about the same as the distance between New York, New York, and Ogden, Utah. (b) The size of Saturn, including the rings, is about 70 percent of the distance between Earth and the Moon.

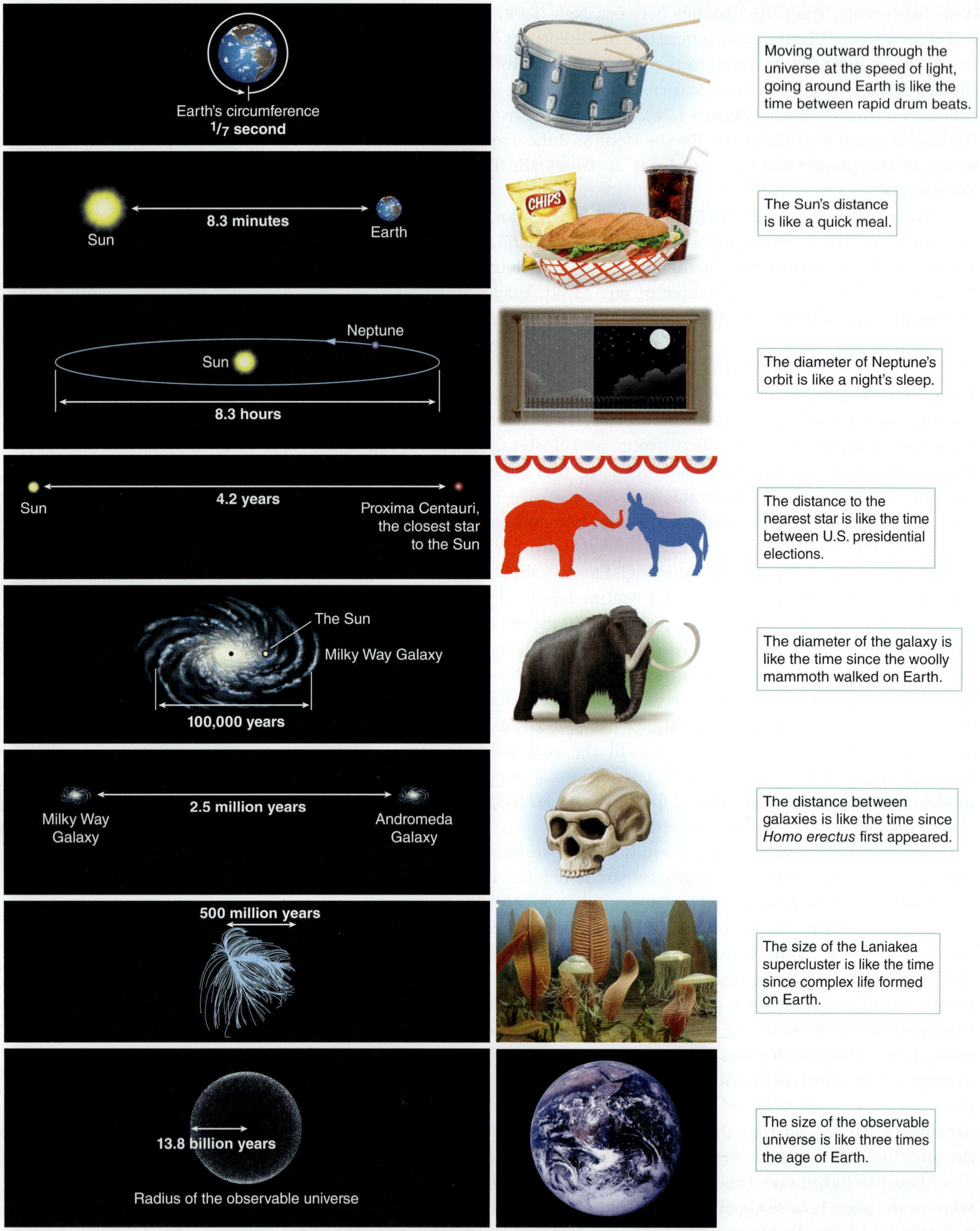

Figure 1.3 Thinking about the time light takes to travel between objects helps us comprehend the vast distances in the observable universe.

thousands of years ago. The light from the Virgo Cluster of galaxies has travelled for 50 million years to reach us. The light from the most distant observable objects has been traveling for almost the age of the universe—nearly 13.8 billion years.

The vast distances from Earth to other objects in the universe tell us that we occupy a very small part of the space in the universe and a very small part of time. Earth and the Solar System are only about one-third as old as the universe. Animals have existed on Earth for even less time. Imagine the age of the universe and the important events in it as though they took place within a single day, as illustrated in **Figure 1.4**. In that timeline, the Big Bang begins the cosmic day at midnight, and the original light chemical elements are created within the first 2 seconds. The first stars and galaxies appear within the first 10 minutes. Our Solar System formed from recycled gas and dust left over from previous generations of stars, at about 4 P.M. on that cosmic clock. The first bacterial life appears on Earth at 5:20 P.M., the first land animals at 11:20 P.M., and modern humans at 11:59:59.8 P.M.—that is, with just 1/5 of a second left in the cosmic day. We humans appeared quite recently in the history of the universe.

CHECK YOUR UNDERSTANDING 1.1

Rank the following from smallest to largest: (a) a light-minute, (b) a light-year, (c) a light-hour, (d) the radius of Earth, (e) the distance from Earth to the Sun, (f) the radius of the Solar System.

1.2 Science Is a Way of Viewing the Universe

Humans have long paid attention to the sky and the stars, making new discoveries and contributing new ideas about the universe. Those discoveries and new ideas still occur often and evolve rapidly, making astronomy a dynamic science. To view the universe through the eyes of an astronomer, you need to understand how science itself works. Throughout this book, we emphasize not only scientific discoveries but also the *process* of science. We begin with this section, which examines the scientific method.

The Scientific Method

The **scientific method** is a systematic way of exploring the world on the basis of developing and then testing new ideas or explanations. Often, the method begins with a fact—an observation or a measurement. For example, you might observe that the weather changes predictably each year and wonder why that happens. You then create a **hypothesis**, a testable explanation of the observation: "I think that it is cold in the winter and warm in the summer because Earth is closer to the Sun in the summer." You come up with a test: if it is cold in the winter and warm in the summer because Earth is closer to the Sun in the summer, then everywhere on the planet will be cold in the winter—Australia should have winter at the same time of year as the United States. You can use that test to check your hypothesis. In January, you travel from the United States to Australia and find that it is summer in Australia. Your hypothesis has just been proved incorrect, so we say that it has been **falsified**. That is different from the word's meaning in common usage, in which "falsified" evidence has been manipulated to misrepresent the truth. Your test has two important elements that all scientific tests share. Your observation is reproducible: anyone who goes to Australia will find the same result. And

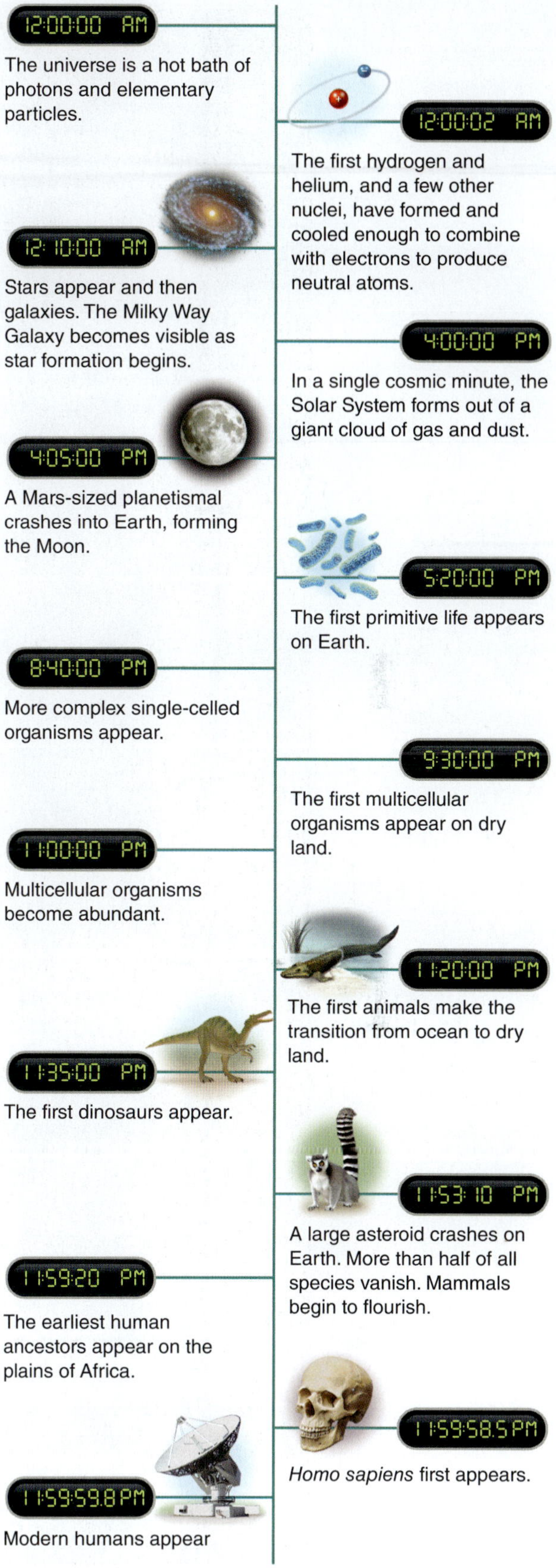

Figure 1.4 This cosmic timeline presents the history of the universe as a 24-hour day.

your result is repeatable: if you conducted a similar test next year or the year after, you would get the same result. Because you have falsified your hypothesis, you must revise or replace it to be consistent with the new data.

Any idea that is not testable—that is, not falsifiable—must be accepted or rejected based on intuition alone, so it is not a scientific idea. A falsifiable hypothesis or idea does not have to be testable using current technology, but we must be able to imagine an experiment or observation that could prove the idea wrong if we could carry it out. As continuing tests support a hypothesis by failing to disprove it, scientists come to accept the hypothesis as a theory. A **theory** is a well-developed idea or group of ideas that must agree with known physical laws and make testable predictions of a given phenomenon or system. As with "falsified," the scientific meaning of "theory" is different from the meaning in common usage. In everyday language, theory may mean a guess: "Do you have a theory about who did it?" In everyday language, a theory can be something we don't take too seriously. "After all," people say, "it's only a theory."

In stark contrast, scientists use the word *theory* to mean a carefully constructed proposition that takes into account every piece of relevant data as well as our entire understanding of how the world works. A theory has been or will be used to make testable predictions, and all those predictions must come true. Every attempt to prove it false fails. A classic example is Einstein's theory of relativity, which Chapter 18 covers in some depth. For more than a century, scientists have tested the predictions of the theory of relativity and have not been able to falsify it. Even after 100 years of verification, if a prediction of the theory of relativity failed tomorrow, the theory would require revision or replacement. As Einstein himself noted, a theory that fails only one test is proved false. In that sense, all scientific knowledge is subject to challenge. In addition, scientific results should be reproducible so that other people get the same results.

In the loosely defined hierarchy of scientific knowledge, an *idea* is a notion about how something might be. Moving up the hierarchy, we come to a *fact*—an observation or measurement. The measured value of Earth's radius is a fact. A *hypothesis* is an idea that leads to testable predictions. A hypothesis may be the forerunner of a scientific theory, may be based on an existing theory, or both. At the top of the hierarchy is a *theory*: an idea that has been examined carefully, is consistent with all existing theoretical and observational knowledge, and makes testable predictions. Ultimately, the success of the predictions is the deciding factor between competing theories. A scientific *law* is a series of observations that enables making predictions but has no underlying explanation of why the phenomenon occurs. A *law* of daytime might say that the Sun rises and sets once each day, whereas a *theory* of daytime might say that the Sun rises and sets once each day because Earth spins on its axis. Scientists themselves can be sloppy about how they use those words, so you may sometimes see them used differently from how we have defined them here.

As the **Process of Science Figure** shows, the scientific method follows a specific sequence. Scientists begin with an observation or idea, followed by analysis, followed by a hypothesis, followed by prediction, followed by further observations or experiments to test the prediction. A hypothesis may lead to a scientific theory, may be based on an existing theory, or both. Ultimately, the basis for deciding among competing theories is the success of their predictions. Scientists can use theories to take their knowledge a step further by building theoretical models. A **theoretical model** describes the properties of a particular object or system in terms of known physical laws or theories, which are used to connect the theoretical model to the behavior of a complex system.

PROCESS OF SCIENCE

The Scientific Method

The scientific method is a formal procedure used to test the validity of scientific hypotheses and theories.

An idea or observation leads to a falsifiable hypothesis. This hypothesis is tested by observation or experiment, and is either accepted as a tested theory or rejected based on the results. The green loop repeats indefinitely as scientists continue to test the theory.

Scientific Principles

Scientific **principles** are general ideas or a sense about the universe that often guides the construction of new theories. Two principles used in studying astronomy are the *cosmological principle* and *Occam's razor*.

The **cosmological principle** is the testable assumption that the physical laws that apply here and now also apply everywhere and at all times. That principle also implies that the universe has no special locations or directions and that matter and energy obey the same physical laws throughout space and time as they do today on Earth. The principle means that the same physical laws that we observe and apply in laboratories on Earth can be used to understand what goes on in the centers of stars or in distant galaxies. Each new theory that comes from applying the cosmological principle to observations of the universe around us adds to our confidence in the validity of that principle, which we discuss in more detail in Chapter 21.

Occam's razor states that when facing two hypotheses that explain all the observations equally well, we should favor the one that requires the fewest assumptions until we have evidence to the contrary. For example, a hypothesis that atoms are constructed differently in the Andromeda Galaxy and in the Milky Way Galaxy would violate the cosmological principle. That hypothesis also would require many assumptions to explain how atoms in the Andromeda Galaxy are constructed differently and yet still appear to behave in the same way as atoms in the Milky Way. For example, suppose you assume that the center of an atom in Andromeda is negatively charged, whereas the center of an atom in the Milky Way is positively charged. That assumption would require you to make additional assumptions about the location of the boundary between Andromeda-like matter and Milky Way–like matter and about why atoms on the boundary between the two regions did not destroy each other. You also would need an assumption about why atoms in the two regions are constructed so differently. If reasonable experimental evidence ever challenges the validity of the cosmological principle, scientists will construct a new description of the universe that takes those new data into account. Until then, the cosmological principle is the hypothesis that has the fewest assumptions, thereby satisfying Occam's razor. To date, the cosmological principle has been repeatedly tested and remains valid.

In many sciences, researchers can conduct controlled experiments to test hypotheses. That experimental method is often unavailable to astronomers: we cannot change the tilt of Earth or the temperature of a star to see what happens. Instead, astronomers work from observations or models. Astronomers make multiple observations by using various methods and then create mathematical and physical models based on established science to explain the observations.

For example, when astronomers first discovered planets orbiting other stars, those new extrasolar planets, or *exoplanets*, were most often giant gaseous planets (similar to Jupiter) in short-period orbits very close to their star. Those planets are the easiest to discover because they have a strong pull on their star, as we discuss in Chapter 7. However, their proximity to their star is completely unlike the situation in our own Solar System, where the giant planets are far from the Sun. Those discoveries challenged existing ideas about how our Solar System formed. As different observers using multiple telescopes found more and more of those planets, astronomers realized they needed new explanations of planet formation to explain how such large planets could wind up so close to their star. Astronomers could not build different solar systems—but they could use the known laws of physics to create computer simulations of planetary systems. When researchers did that, they found that planets can migrate, moving to orbits closer to or farther from their star. Planetary scientists are searching for evidence to test that idea in our own Solar System, where that process may have occurred early in the life of the Sun. In that example, the theory was motivated by the earlier observations.

Alternatively, an astronomer might make predictions from an existing successful mathematical or physical model and then conduct observations and analyze data to test the predictions. One example is the discovery of black holes. In the late 18th century, two scientists hypothesized the existence of "dark stars": massive objects having such strong gravity that light could not escape. At that time, the scientists had no way to test that hypothesis. More than 100 years later, in the early 20th century, Karl Schwarzschild (1873–1916) studied Einstein's relativity equations and calculated that those collapsed dark stars would be very small, with a radius of only a few kilometers. Fifty years later, those objects were named *black holes*. No evidence of their existence was available until the 1970s and 1980s, when the new technology of space-based X-ray telescopes made possible the observations needed to test the hypothesis. Einstein's existing theory made the discovery of black holes possible.

The scientific method provides the rules for testing whether an idea is false, but it offers no insight into where the idea came from in the first place or how an experiment was designed. Scientists discussing their work use words such as *insight*, *intuition*, and *creativity*. Scientists speak of a beautiful theory in the same way that an artist speaks of a beautiful painting or a musician speaks of a beautiful song. Science has an aesthetic that is as human and as profound as any found in the arts.

In one important respect, though, science is different from art or music. Whereas art and music are judged by a human jury alone, the validity of a scientific hypothesis or theory is subject to the natural world. Nature alone is the judge of which theories can be kept and which theories must be discarded. What we want to be true does not matter. In the history of science, many a beautiful and beloved theory has been abandoned.

Scientific Revolutions

Scientific inquiry is necessarily dynamic. Scientists do not have all the answers and must constantly refine their ideas in response to new data and new insights. The vulnerability of knowledge may seem like a weakness. "Gee, you really don't know anything," the cynical person might say. But that vulnerability is actually the scientific process's greatest strength: it means that science self-corrects. New information eventually overturns incorrect ideas. In science, even our most cherished ideas about the nature of the physical world remain fair game, subject to challenge by new evidence. Many of history's best scientists earned their status by falsifying a universally accepted idea. That is a powerful motivation for scientists to challenge old ideas constantly—to formulate and test new explanations for their observations.

For example, the classical physics that Sir Isaac Newton developed in the 17th century to explain motion, forces, and gravity withstood the scrutiny of scientists for more than 200 years. During the late 19th and early 20th centuries, however, a series of scientific revolutions completely changed our understanding of the nature of reality. The work of Albert Einstein (**Figure 1.5**) is representative of those scientific revolutions. Einstein's special and general theories of relativity replaced Newton's mechanics. Einstein did not prove Newton wrong but rather

Figure 1.5 Albert Einstein is perhaps the most famous scientist of the 20th century, and he was *Time* magazine's selection for Person of the Century. Einstein helped usher in two scientific revolutions.

showed that Newton's theories were a special case of a far more general and powerful set of physical theories. Einstein's new ideas unified the concepts of mass and energy and destroyed the conventional notion of space and time as separate concepts.

Throughout this text, you will encounter many other discoveries that forced scientists to abandon accepted theories. Einstein himself never embraced the view of the world offered by *quantum mechanics*—a second revolution he helped start. Yet quantum mechanics, a statistical description of the behavior of particles smaller than atoms, has held up for more than 100 years. In science, all authorities are subject to challenge, even Einstein.

Science is a way of thinking about the world. It is a search for the relationships that make our world what it is. Scientific knowledge is an accumulated collection of ideas about how the universe works, yet scientists are always aware that what is known today may be superseded tomorrow. A scientist assumes that order exists in the universe and that the human mind can grasp the essence of the rules underlying that order. Scientists build on those assumptions to make and then test predictions, finding the underlying rules that allow humanity to solve problems, invent new technologies, or find a new appreciation for the natural world. In the final analysis, science has found such a central place in our civilization because *science works*.

Figure 1.6 Since ancient times, people recognized that patterns in the sky change with the seasons. Those and other patterns shape our lives. These star maps show the sky in the Northern Hemisphere during each season.

CHECK YOUR UNDERSTANDING 1.2

The scientific method is a process by which scientists: (a) prove theories to be known facts; (b) gain confidence in theories by failing to prove them wrong; (c) show all theories to be wrong; (d) survey what most scientists think about a theory.

1.3 Astronomers Use Mathematics to Find Patterns

Scientific thinking allows scientists to make predictions. Once a pattern has been observed, such as the daily rising and setting of the Sun, scientists can predict what will happen next.

Imagine that the patterns in your life became disrupted, so that the world became entirely unpredictable. For example, what would life be like if you could not predict whether an object you dropped would fall up or down? Or what if one morning the Sun rose in the east and set in the west, and the next day it rose in the west and set in the east? In fact, objects do fall toward the ground. The Sun rises, sets, and then rises again at predictable times and in predictable locations. Spring turns into summer, summer turns into autumn, autumn turns into winter, and winter turns into spring again. The rhythms of nature produce patterns in our lives, and those patterns give us clues about the nature of the physical world. Astronomers identify and characterize those patterns and use them to understand the world around us. Some of the most easily identified patterns in nature are those we see in the sky. What in the sky will look different or the same a week from now? A month from now? A year from now? As you can see in **Figure 1.6**, patterns in the sky mark the changing of the seasons. Because the seasons determine planting and harvesting times, astronomy—which studies those patterns—is the oldest of all sciences. We describe many other examples of patterns in the sky in Chapter 2.

Astronomers use mathematics to analyze patterns and to communicate complex material compactly and accurately. Because studying patterns in nature is so important to science, it should come as no surprise that mathematics is the language of science. Many people find mathematics to be a major obstacle that prevents them from appreciating the beauty and elegance of the world as seen through the eyes of a scientist. In this book, we strive to explain any necessary math in everyday language. We will describe what equations mean and help you use them in a way that allows you to connect scientific concepts to the world. Your responsibility is to accept the challenge and make an honest effort to understand the material. **Working It Out 1.1** and **Working It Out 1.2** review some basics of mathematical tools and graphs. At the back of the book, Appendix 1 explains some essential mathematical tools, and Appendix 2 contains physical constants of nature. Other appendixes contain data tables with key information about planets, moons, and stars.

CHECK YOUR UNDERSTANDING 1.3

When you see a pattern in nature, it is usually evidence of: (a) a theory's being displayed; (b) a breakdown of random clustering; (c) an underlying physical law; (d) a coincidence.

1.1 Working It Out Mathematical Tools

Scientists use mathematics to understand the patterns they observe and to communicate that understanding to others. The following mathematical tools will be useful in our study of astronomy:

Scientific notation. Scientific notation is how we handle numbers of vastly different sizes. Writing out 7,540,000,000,000,000,000,000 in standard notation is inefficient. Scientific notation uses the first few digits (the *significant* ones) and counts the number of decimal places to create the condensed form 7.54×10^{21}. Similarly, instead of writing out 0.000000000005, we write 5×10^{-12}. The exponent on the 10 is positive or negative depending on whether the decimal point moves right or left, respectively. For example, the average distance to the Sun is 149,600,000 km, but astronomers usually express that value as 1.496×10^{8} km.

Ratios. Ratios are a useful way to compare objects. A star may be "10 times as massive as the Sun" or "10,000 times as luminous as the Sun." Those expressions are ratios.

Proportionality. Often, understanding a concept amounts to understanding the *sense* of the relationships that it predicts or describes. "If you have twice as far to go, getting there will take you twice as long." "If you have half as much money, you will be able to buy only half as much gas." Those are examples of proportionalities.

Appendix 1 further explains the mathematical tools used in this book.

1.2 Working It Out Reading a Graph

Scientists often convey complex information and mathematical patterns in graphical form. Reading graphs is an important skill not only in astronomy but also in life. Economists, social scientists, mortgage brokers, financial analysts, urban planners, historians, medical professionals, and scientists all use graphs to evaluate and communicate important information.

Graphs typically have two axes: a horizontal axis (the x-axis) and a vertical axis (the y-axis). The x-axis usually shows an independent variable, the one a researcher might have control over in an experiment, whereas the y-axis shows the dependent variable, which—in many experiments—is the variable a researcher is studying.

Graphs can take different shapes. Suppose we plot the distance a car travels over time, as shown in **Figure 1.7a**. In a linear graph, each interval on an axis represents the same-sized step. Each step on the horizontal axis of the graph in Figure 1.7a represents 5 minutes. Each step on the vertical axis represents a distance of 5 km traveled by the car. Data are plotted on the graph, with one dot for each observation; for example, after 20 minutes, the car has traveled 20 km.

Drawing a line through those data indicates the trend (or relationship) of the data. To understand what the trend means, scientists often find the slope of the line, which is the relationship of the line's rise along the y-axis to its movement along the x-axis. To find the slope, we look at the change between two points on the vertical axis divided by the change between the same two points on the horizontal axis. For example, finding the slope of the line gives

$$\text{Slope} = \frac{\text{Change in vertical axis}}{\text{Change in horizontal axis}}$$

$$\text{Slope} = \frac{(15 - 10)\text{ km}}{(15 - 10)\text{ min}}$$

$$= 1\text{ km/min}$$

There, the trend tells us that the car is traveling at 1 kilometer per minute (km/min), or 60 kilometers per hour (km/h).

Many observations of natural processes do not yield a straight line on a graph—for example, an exponential process. Think about what happens when you catch a cold. You feel fine when you get up in the morning at 7 A.M. At 9 A.M. you feel a little tired. By 11 A.M. you have a bit of a sore throat or a sniffle and think, "I wonder whether I'm getting sick," and by 1 P.M., you have a runny nose, congestion, fever, and chills. That process is exponential because the virus that has infected you reproduces exponentially.

For the sake of this discussion, suppose that the virus produces one copy of itself each time it invades a cell. (Viruses actually produce 1,000–10,000 copies each time they invade a cell, so the exponential curve is much steeper.) One virus infects a cell and multiplies, so now two viruses exist—the original and a copy. Those viruses invade two new cells, and each one produces a copy. Now four viruses are there. After the next cell invasion, we have eight. Then 16, 32, 64, 128, 256, 512, 1,024, 2,048, and so on. That behavior is plotted in **Figure 1.7b**.

Seeing what's happening in the early stages of an exponential curve can be difficult because the later numbers are so much larger than the earlier ones. That's why we sometimes plot that type of data *logarithmically*, by putting the logarithm (the power of 10) of the data on the vertical axis, as shown in **Figure 1.7c**. Now each step on the axis represents 10 times as many viruses as the previous step. Even though we draw all the steps the same size on the page, they represent different-sized steps in the data—the number of viruses, for example. We often use that technique in astronomy because it has a second, related advantage: very large variations in the data can easily fit on the same graph.

Each time you see a graph, you should first understand the axes—what data are plotted on the graph? Then you should check whether the axes are linear or logarithmic. Finally, you can look at the actual data or lines in the graph to understand how the system behaves.

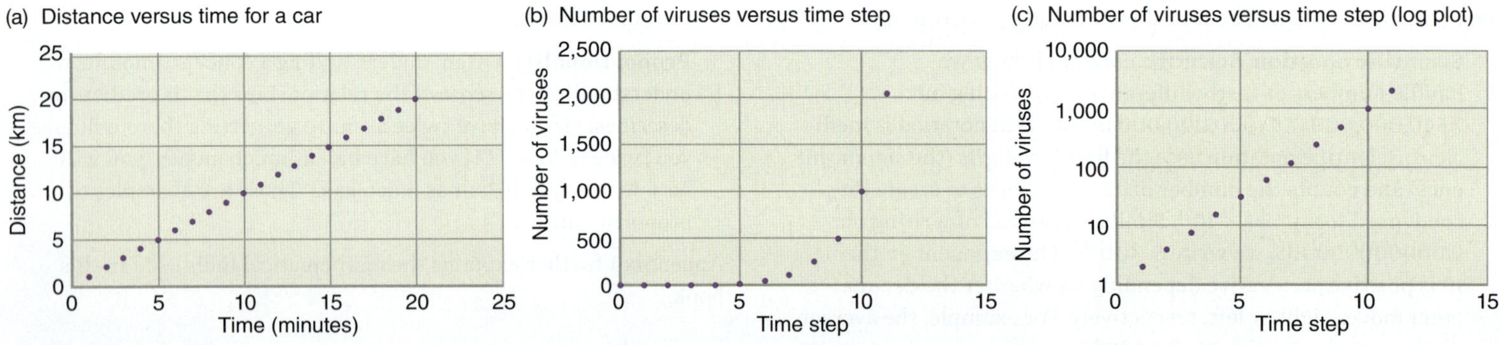

Figure 1.7 Graphs such as these show relationships between quantities. (a) The relationship between time and distance traveled. (b) The relationship between time and number of viruses. (c) The data in part (b) plotted logarithmically.

Origins An Introduction

How and when did the universe begin? What combination of events led to the existence of humans as sentient beings living on a small, rocky planet orbiting a typical middle-aged star? Are others like us scattered throughout the galaxy?

Earlier in the chapter we mentioned the theme of origins. Throughout this book, that theme involves much more than how humans came to be on Earth. In these Origins sections, which conclude each chapter, we look into the origin of the universe and the origin of life on Earth. We also examine the possibilities of life elsewhere in the Solar System and beyond—a subject called **astrobiology**. Our origins theme includes the discovery of planets around other stars and how they compare with the planets of our own Solar System.

Later in the book, we present observational evidence that supports the **Big Bang** theory, which states that the universe started expanding from an infinitesimal size about 13.8 billion years ago. In the early universe, only the lightest chemical elements were found in substantial amounts: hydrogen and helium, and tiny amounts of lithium and beryllium. However, we live on a planet with a central core consisting mostly of heavy elements such as iron and nickel, surrounded by outer layers made up of rocks containing large amounts of silicon and various other elements—all heavier than the original elements. The human body contains carbon, nitrogen, oxygen, calcium, phosphorus, and a host of other chemical elements—except for hydrogen itself, all are heavier than hydrogen and helium. If those heavier elements that make up Earth and our bodies were not present in the early universe, where did they come from?

The answer to that question lies within the stars themselves (**Figure 1.8**). In the core of a star, light elements, such as hydrogen, combine to form more massive atoms, which eventually leads to atoms such as carbon. When a star nears the end of its life, it often loses much of its material—including some of the new atoms formed in its interior—by blasting it back into interstellar space. That material combines with material lost from other stars—some of which produced even more massive atoms as they exploded—to form large clouds of dust and gas. Those clouds go on to make new stars and planets, similar to our Sun and Solar System. Earlier "generations" of stars supplied the building blocks for the chemical processes that we see in the universe, including life. The atoms that make up much of what we see were formed in the cores of stars. The phrase "we are stardust" is not just poetry. We are actually made of recycled stardust.

Studying origins also gives examples of the process of science. Many physical processes in chemistry, planetary science, physics, and astronomy that are seen on Earth or in the Solar System are observed across the Milky Way and throughout the universe. As of this writing, though, the only biology we know about is that which exists on Earth. At this point in human history, then, much of what scientists can say about the origin of life on Earth and the possibility of life elsewhere is reasoned extrapolation and educated speculation. In these Origins sections, we address some of those hypotheses and try to be clear about which are speculative and which have been tested.

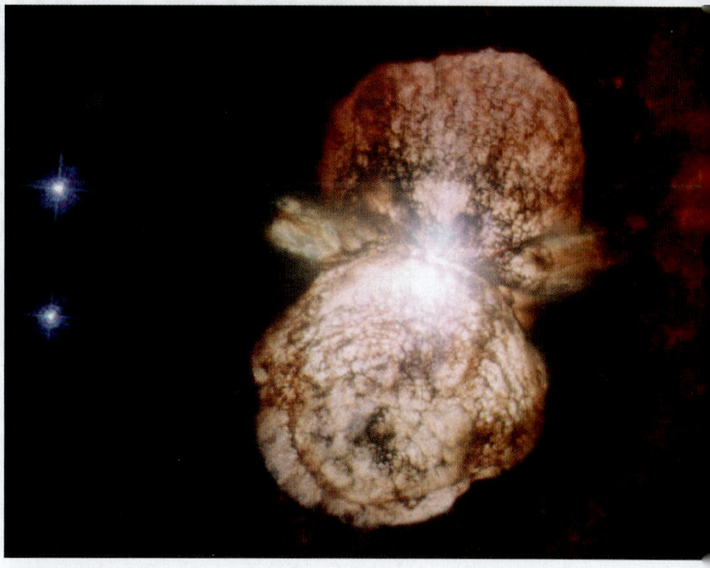

Figure 1.8 You and everything around you are composed of atoms forged in the interiors of stars that lived and died before the Sun and Earth formed. The supermassive star Eta Carinae, shown here, is ejecting a cloud of chemically enriched material just as earlier generations of stars once did to enrich the gas that would become our Solar System.

READING ASTRONOMY NEWS

ARTICLES QUESTIONS

Astronomers think there are 10 times more galaxies than previously estimated.

A Universe of 2 Trillion Galaxies

Royal Astronomical Society

An international team of astronomers, led by Christopher Conselice, professor of astrophysics at the University of Nottingham, have found that the universe contains at least 2 trillion galaxies, 10 times more than previously thought. The team's work, which began with seedcorn funding from the Royal Astronomical Society, appears in the *Astrophysical Journal.*

Astronomers have long sought to determine how many galaxies there are in the observable universe, the part of the cosmos where light from distant objects has had time to reach us. Over the last 20 years, scientists have used images from the Hubble Space Telescope to estimate that the universe we can see contains around 100–200 billion galaxies. Current astronomical technology allows us to study just 10 percent of these galaxies, and the remaining 90 percent will be only seen once bigger and better telescopes are developed.

Professor Conselice's research is the culmination of 15 years' work, part-funded by a research grant from the Royal Astronomical Society awarded to Aaron Wilkinson, an undergraduate student at the time. Aaron, now a PhD student at the University of Nottingham, began by performing the initial galaxy-counting analysis, work that was crucial for establishing the feasibility of the larger-scale study.

Professor Conselice's team then converted pencil beam images of deep space from telescopes around the world, and especially from the Hubble telescope, into three-dimensional maps. These allowed them to calculate the density of galaxies as well as the volume of one small region of space after another. This painstaking research enabled the team to establish how many galaxies we have missed—much like an intergalactic archaeological dig.

The results of this study are based on the measurements of the number of observed galaxies at different epochs—different instances in time—through the universe's history. When Professor Conselice and his team at Nottingham, in collaboration with scientists from the Leiden Observatory at Leiden University in the Netherlands and the Institute for Astronomy at the University of Edinburgh, examined how many galaxies there were at a given epoch, they found that there were significantly more at earlier times.

It appears that when the universe was only a few billion years old, there were 10 times as many galaxies in a given volume of space as there are within a similar volume today. Most of these galaxies were low-mass systems with masses similar to those of the satellite galaxies surrounding the Milky Way.

Professor Conselice said, "This is very surprising as we know that, over the 13.7 billion years of cosmic evolution since the Big Bang, galaxies have been growing through star formation and mergers with other galaxies. Finding more galaxies in the past implies that significant evolution must have occurred to reduce their number through extensive merging of systems."

He continued, "We are missing the vast majority of galaxies because they are very faint and far away. The number of galaxies in the universe is a fundamental question in astronomy, and it boggles the mind that over 90 percent of the galaxies in the cosmos have yet to be studied. Who knows what interesting properties we will find when we study these galaxies with the next generation of telescopes?"

ARTICLES QUESTIONS

1. Explain how observing galaxies at different distances tells us about how galaxies existed in different periods.
2. Why does that result suggest that galaxies started small?
3. What is meant by "observable universe"?
4. Watch Hubblecast 96 based on this news story (http://spacetelescope.org/videos/heic1620a/). What else did you learn from the video?
5. Do a search to see if there is a new estimate of the number of observable galaxies.

Source: "A universe of two trillion galaxies," Royal Astronomical Society News Archive, October 13, 2016. Reprinted with permission.

Summary

Astronomy seeks answers to many compelling questions about the universe. Astronomy uses all available tools to follow the scientific method. The process of science is based on objective reality, physical evidence, and testable hypotheses. Scientists continually strive to improve their understanding of the natural world and must be willing to challenge accepted truths as new information becomes available.

LG 1 **Describe the size and age of the universe and Earth's place in it.** We reside on a planet orbiting a star at the center of a solar system in a vast galaxy that is one of many in the universe. We occupy a very tiny part of the universe in space and time.

LG 2 **Explain how astronomers use the scientific method to study the universe.** The scientific method is an approach to learning about the physical universe. The method includes observations, forming hypotheses, making predictions to test and refine those hypotheses, and repeated testing of theories. All scientific knowledge is provisional. Like art, literature, and music, science is a creative human activity; it is also a remarkably powerful, successful, and aesthetically beautiful way of viewing the world.

LG 3 **Show how scientists use mathematics, including graphs, to find patterns in nature.** Mathematics provides many of the tools that astronomers need to understand the patterns we see and to communicate that understanding to others.

LG 4 **Describe our astronomical origins.** We are a product of the universe: The very atoms we're made of formed in stars that died long before the Sun and Earth formed. The atoms of the chemical elements that make up matter come from particles that have been around since the Big Bang, processed through stars, and collected into the Sun and the Solar System.

? Unanswered Questions

- What makes up the universe? We have listed planets, stars, and galaxies as making up the cosmos, but astronomers now have evidence that 95 percent of the universe is composed of dark matter and dark energy, which we do not understand. Scientists are using the largest telescopes and particle colliders on Earth, as well as telescopes and experiments in space, to explore what makes up dark matter and what constitutes dark energy.
- Does life as we know it exist elsewhere in the universe? At the time of this writing, no scientific evidence indicates that life exists on any other planet. Our universe is enormously large and has existed for a great length of time. What if life is too far away or existed too long ago for us ever to "meet"?

Questions and Problems

Test Your Understanding

1. Rank the following in order from smallest to largest.
   a. Local Group
   b. Milky Way
   c. Solar System
   d. universe
   e. Sun
   f. Earth
   g. Laniakea Supercluster
   h. Virgo Supercluster
2. If an event were to take place on the Sun, approximately how long would the light it generates take to reach us?
   a. 8 minutes
   b. 11 hours
   c. 1 second
   d. 1 day
   e. It would reach us instantaneously.
3. *Understanding* in science means that
   a. we have accumulated lots of facts.
   b. we can connect facts through an underlying idea.
   c. we can predict events on the basis of accumulated facts.
   d. we can predict events on the basis of an underlying idea.
4. The cosmological principle states that
   a. on a large scale, the universe is the same everywhere at a given time.
   b. the universe is the same at all times.
   c. our location is special.
   d. all of the above
5. The Sun is part of
   a. the Solar System.
   b. the Milky Way Galaxy.
   c. the universe.
   d. all of the above

6. A light-year is a measure of
   a. distance.
   b. time.
   c. speed.
   d. mass.
7. Occam's razor states that
   a. the universe is expanding in all directions.
   b. the laws of nature are the same everywhere in the universe.
   c. if two hypotheses fit the facts equally well, the simpler one is more likely to apply.
   d. patterns in nature are really manifestations of random occurrences.
8. The circumference of Earth is 1/7 of a light-second. Therefore,
   a. if you were traveling at the speed of light, you would travel around Earth 7 times in 1 second.
   b. light travels a distance equal to Earth's circumference in 1/7 of a second.
   c. neither a nor b
   d. both a and b
9. According to the graphs in Figures 1.7b and c, by how much did the number of viruses increase in four time steps?
   a. It doubled.
   b. It tripled.
   c. It quadrupled.
   d. It went up more than 10 times.
10. Any explanation of a phenomenon that includes a supernatural influence is not scientific because
    a. it is not based on a hypothesis.
    b. it is wrong.
    c. people who believe in the supernatural are not credible.
    d. science is the study of the natural world.
11. "All scientific knowledge is provisional." In that context, *provisional* means
    a. "wrong."
    b. "relative."
    c. "temporary."
    d. "incomplete."
12. When we observe a star that is 10 light-years away, we are seeing that star
    a. as it is today.
    b. as it was 10 days ago.
    c. as it was 10 years ago.
    d. as it was 20 years ago.
13. Which of the following was *not* made in the Big Bang?
    a. hydrogen
    b. lithium
    c. beryllium
    d. carbon
14. "We are stardust" means that
    a. Earth exists because two stars collided.
    b. the atoms in our bodies have passed through (and often formed in) stars.
    c. Earth is formed primarily of material that used to be in the Sun.
    d. Earth and the other planets will eventually form a star.
15. According to our current understanding of the universe, the following astronomical events led to the formation of you. Place them in order of their occurrence over astronomical time.
    a. Stars die and distribute heavy elements into the space between the stars.
    b. Hydrogen and helium are made in the Big Bang.
    c. Enriched dust and gas gather into clouds in interstellar space.
    d. Stars are born and process light elements into heavier ones.
    e. The Sun and planets form from a cloud of interstellar dust and gas.

Thinking about the Concepts

16. Suppose you lived on the planet named "Tau Ceti e" that orbits Tau Ceti, a nearby star in our galaxy. How would you write your cosmic address?
17. Imagine yourself living on a planet orbiting a star in a very distant galaxy. What does the cosmological principle tell you about the physical laws at that distant location?
18. If the Sun suddenly exploded, how soon after the explosion would we know about it?
19. If a star exploded in the Andromeda Galaxy, how long would that information take to reach Earth?
20. Give an example of a scientific theory that a newer theory has superseded. As scientists developed that new theory, where on the Process of Science Figure did a change occur so that the old theory became invalid and the new theory was accepted?
21. Some people have proposed the theory that extraterrestrials (aliens) visited Earth in the remote past. Can you think of any tests that could support or refute that theory? Is it falsifiable?
22. What does the word *falsifiable* mean? Give an example of an idea that is not falsifiable. Give an example of an idea that is falsifiable.
23. Explain how the word *theory* is used differently in the context of science than it is in everyday language.
24. What is the difference between a *hypothesis* and a *theory* in science?

25. Suppose the tabloid newspaper at your local supermarket claimed that children born under a full Moon become better students than children born at other times.
  a. Is that theory falsifiable?
  b. If so, how could it be tested?

26. A textbook published in 1945 stated that light from the Andromeda Galaxy takes 800,000 years to reach Earth. In this book, we assert that it takes 2,500,000 years. What does that difference tell you about a scientific "fact" and how our knowledge evolves?

27. Astrology makes testable predictions. For example, it predicts that the horoscope for your star sign on any day should fit you better than horoscopes for other star signs. Read the daily horoscopes for all the astrological signs in a newspaper or online. How many of them might fit the day you had yesterday? Repeat the experiment every day for a week and record which horoscopes fit your day each day. Did your horoscope sign consistently best describe your experiences?

28. A scientist on television states that it is a known fact that life does not exist beyond Earth. Would you consider that scientist's statement reputable? Explain your answer.

29. Some astrologers use elaborate mathematical formulas and procedures to predict the future. Does doing so show that astrology is a science? Why or why not?

30. Why can it be said that we are made of stardust? Explain why current theory supports that statement.

Applying the Concepts

31. Review Working It Out 1.1., and then convert the following numbers to scientific notation:
  a. 7,000,000,000
  b. 0.00346
  c. 1,238

32. Review Working It Out 1.1, and then convert the following numbers to standard notation:
  a. 5.34×10^8
  b. 4.1×10^3
  c. 6.24×10^{-5}

33. If a car is traveling at 35 km/h, how far does it travel in
  a. 1 hour?
  b. half an hour?
  c. 1 minute?

34. Review Appendix 1. The surface area of a sphere is proportional to the square of its radius. How many times larger is the surface area if the radius is
  a. doubled?
  b. tripled?
  c. halved (divided by 2)?
  d. divided by 3?

35. The average distance from Earth to the Moon is 384,400 km. How many days would it take you, traveling at 800 km/h—the typical speed of jet aircraft—to reach the Moon?

36. The average distance from Earth to the Moon is 384,400 km. In the late 1960s, astronauts reached the Moon in about 3 days. How fast (on average) must they have been traveling (in kilometers per hour) to cover that distance in that time? Compare that speed with the speed of a jet aircraft (800 km/h).

37. (a) If light takes about 8 1/3 minutes to travel from the Sun to Earth, and Neptune is 30 times farther from Earth than the Sun is, how long does light from Neptune take to reach Earth? (b) Radio waves travel at the speed of light. What does that fact imply about the problems you would have if you tried to conduct a two-way conversation between Earth and a spacecraft orbiting Neptune?

38. The distance from Earth to Mars varies from 56 million km to 400 million km. How long does a radio signal traveling at the speed of light take to reach a spacecraft on Mars when Mars is closest and when Mars is farthest away?

39. The surface area of a sphere is proportional to the square of its radius. The radius of the Moon is only about one-quarter that of Earth. How does the surface area of the Moon compare with that of Earth?

40. A remote Web page may sometimes reach your computer by going through a satellite orbiting approximately 3.6×10^4 km above Earth's surface. What is the minimum delay, in seconds, before the Web page shows up on your computer?

41. Imagine that you have become a biologist, studying rats in Indonesia. Usually, Indonesian rats maintain a constant population. Every half century, however, those rats suddenly begin to multiply exponentially. Then the population crashes back to the constant level. Sketch a graph that shows the rat population over two of those episodes.

42. New York is 2,444 miles from Los Angeles. What is that distance in car-hours? In car-days? (Assume a travel speed of 60 mph.)

43. Some people theorize that a tray of hot water will freeze more quickly than a tray of cold water when both are placed in a freezer.
  a. Does that theory make sense to you?
  b. Is the theory falsifiable?
  c. Do the experiment yourself. Note the results. Was your intuition borne out?

44. A pizzeria offers a 9-inch-diameter pizza for $12 and an 18-inch-diameter pizza for $24. Are both offerings equally economical? If not, which is the better deal? Explain your reasoning.

45. The circumference of a circle is given by $C = 2\pi r$, where r is the radius of the circle.
   a. Calculate the approximate circumference of Earth's orbit around the Sun, assuming that the orbit is a circle with a radius of 1.5×10^8 km.
   b. A year has 8,766 hours, so how fast, in kilometers per hour, does Earth move in its orbit?
   c. How far along in its orbit does Earth move in 1 day?

USING THE WEB

46. Go to the interactive "The Scale of the Universe 2" (Web page at https://apod.nasa.gov/apod/ap140112.html, or download the app or find on YouTube). Start at 10° (human size) and scale upward; clicking on an object gives you its exact size. What astronomical bodies are about the size of the United States? What objects are about the size of Earth? What stars are larger than the distance from Earth to the Sun? How many light-days is the distance of *Voyager 1* to Earth? What objects are about the size of the distance from the Sun to the nearest star? How many times larger is the Milky Way than the Solar System? How many times larger is the Local Group than the Milky Way? How many times larger is the observable universe than the Local Group?

47. a. For a video representation of the scale of the universe, view the short video *The Known Universe* at the Hayden Planetarium website (https://amnh.org/our-research/hayden-planetarium/digital-universe), which takes the viewer on a journey from the Himalayan mountains to the most distant galaxies. How far have broadcast radio programs from Earth traveled? Is the Sun a particularly luminous star in comparison with others? Do you think the video is effective for showing the size and scale of the universe?
   b. A similar film produced in 1996 in IMAX format, *Cosmic Voyage*, can be found online on YouTube. Watch the "powers of 10" zoom out to the cosmos, starting at the 7-minute mark, for about 5 minutes. Do the "powers of 10" circles add to your understanding of the size and scale of the universe?

48. Go to the Astronomy Picture of the Day (APOD) app or website (https://apod.nasa.gov/apod) and click on "Archive" to look at the recent pictures and videos. Submissions to the website come from all around the world. Pick one and read the explanation. Was the image or video taken from Earth or from space? Is it a combination of several images? Does it show Earth, our Solar System, objects in our Milky Way Galaxy, more distant galaxies, or something else? Would someone who has not studied astronomy understand the explanation? Do you think the website promotes a general interest in astronomy?

49. Throughout this book, we examine how the media cover discoveries in astronomy and space. On your favorite news website (or one assigned by your instructor), find a recent article about astronomy or space. Does the website have a separate section for science? Is the article you selected based on a press release, on interviews with scientists, or on an article in a scientific journal? Use Google News or the equivalent to see how widespread the coverage of the story is. Have many newspapers carried it? Has it been picked up internationally? Has it been discussed in blogs? Do you think the story was interesting enough to be covered?

50. Go to a blog about astronomy or space. Is the blogger a scientist, a science writer, a student, or an enthusiastic amateur astronomer? What is the current topic of interest? Is it controversial? Are readers making many comments? Is the blog something you would want to read again?

EXPLORATION

digital.wwnorton/astro6

Logic is fundamental to studying science and to scientific thinking. A logical fallacy is an error in reasoning, which good scientific thinking avoids. For example, "because Einstein said so" is not an adequate argument. No matter how famous the scientist is (even if he is Einstein), he or she must still supply a logical argument and evidence to support a claim. Anyone who claims that something must be true because Einstein said it has committed the logical fallacy known as an *appeal to authority.* Many types of logical fallacies exist, but a few crop up often enough in discussions about science that you should be aware of them.

Ad hominem. In that fallacy, you attack the person making the argument instead of the argument itself. For example, "A well-known actor says dogs are better pets than cats, but I think that actor is an idiot, so cats must be better."

Appeal to belief. That fallacy has the general pattern "Most people believe X is true; therefore, X is true." For example, "Most people believe Earth orbits the Sun. Therefore, Earth orbits the Sun." Even if your conclusion is correct, you may still have committed a logical fallacy in your argument.

Begging the question. In that fallacy, also known as circular reasoning, you assume that the claim is true and then use that assumption as evidence to prove the claim is true. For example, "I am trustworthy; therefore, I must be telling the truth." No real evidence is presented for the conclusion.

Biased sample. If a sample drawn from a smaller pool has a bias, conclusions about the sample cannot be applied to a larger pool. For example, imagine you poll students at your university and find that 30 percent of them visit the library at least once per week. Then you conclude that 30 percent of Americans visit the library at least once per week. You have committed the biased sample fallacy because university students are not a representative sample of the American public.

Post hoc ergo propter hoc. That term is Latin for "after this, therefore because of this." Just because one thing follows another doesn't mean that one caused the other. For example, "An eclipse occurred, and then the king died. Therefore, the eclipse killed the king." That fallacy is often connected to related inverse reasoning: "If we can prevent an eclipse, the king won't die."

Slippery slope. In that fallacy, you claim that a chain reaction of events will take place, inevitably leading to a conclusion that no one could want. For example, "If I don't get A's in all my classes, I will not ever be able to get into graduate school, and then I won't ever be able to get a good job, and then I will be living in a van down by the river until I'm old and starve to death." None of those steps actually follows inevitably from the one before.

The following are examples of logical fallacies. Identify the fallacy represented. Each fallacy just discussed is represented once.

1 You get a chain email threatening terrible consequences if you break the chain. You move the message to your spam box. Later that day you get in a car accident. The next morning, you retrieve the chain email and send it along.

____

____

____

2 If I get question 1 on the assignment wrong, then I'll get question 2 wrong as well, and before you know it, I will never catch up in the class.

____

____

____

3 All my friends love the band Degenerate Electrons. Therefore, all people my age love that band.

____

____

____

4 Eighty percent of Americans believe in the tooth fairy. Therefore, the tooth fairy exists.

____

____

5 My professor says that the universe is expanding. But my professor is a geek, and I don't like geeks, so the universe can't be expanding.

____

____

6 When applying for a job, you use a friend as a reference. Your prospective employer asks you how she can be sure your friend is trustworthy, and you say, "I can vouch for him."

____

____

2 Patterns in the Sky—Motions of Earth and the Moon

Ancient peoples learned to use the patterns they observed in the sky to predict how the length of day, the seasons, and the appearance of the Moon would change. Some people understood those patterns well enough to create complicated calendars and predict rare eclipses. But now we can see those patterns with the perspective of centuries of modern science, and we can explain those changes as being a consequence of the motions of Earth and the Moon. Discovering what causes those patterns has shown us the way outward into the universe.

LEARNING GOALS

Here in Chapter 2, we examine the patterns in the sky and on Earth and the underlying motions that cause them. By the end of this chapter, you should be able to:

LG 1 **Describe how Earth's rotation about its axis and its revolution around the Sun affect how we perceive celestial motions from different places on Earth.**

LG 2 **Explain why seasons change throughout the year.**

LG 3 **Describe the factors that create the phases of the Moon.**

LG 4 **Sketch the alignment of Earth, the Moon, and the Sun during eclipses of the Sun and the Moon.**

Total eclipse of the Sun on August 21, 2017. ▶▶▶

What causes a solar eclipse?

Figure 2.1 (a) One suspected use of Stonehenge 4,000 years ago was to keep track of celestial events. (b) The Mayan El Caracol at Chichén Itzá in Mexico (906 CE) is believed to have been designed to align with Venus. (c) The Beijing Ancient Observatory in China (1442 CE) includes ancient astronomical instruments as well as some brought by European Jesuits in the 17th and 18th centuries.

2.1 Earth Spins on Its Axis

Ancient humans may not have known that they were "stardust," but they did sense a connection between their lives on Earth and the sky above. Before our modern technological civilization, people's lives were more attuned to the ebb and flow of nature, which includes the patterns in the sky. By watching the repeating patterns of the Sun, Moon, and stars in the sky, people could predict when the seasons would change and when the rains would come and the crops would grow. Ancient astronomers with knowledge of the sky—priests and priestesses, natural philosophers, and explorers—all had knowledge of the world that others did not, and knowledge was power. Some of those early observations and ideas live on today in the names of stars and the apparent grouping of stars we call **constellations**, in calendars based on the Moon and Sun that many cultures still use, and in the astronomical names of the days of the week.

From Mesopotamia to Africa, from Europe to Asia, from the Americas to the British Isles, the archaeological record holds evidence that ancient cultures built structures to study astronomical positions and events. **Figure 2.1** shows some examples. From the 8th through 17th centuries, people used pre-telescopic astronomical observatories to study the sky for timekeeping and navigation. Many of those structures and observatories are now national historical or UNESCO World Heritage sites.

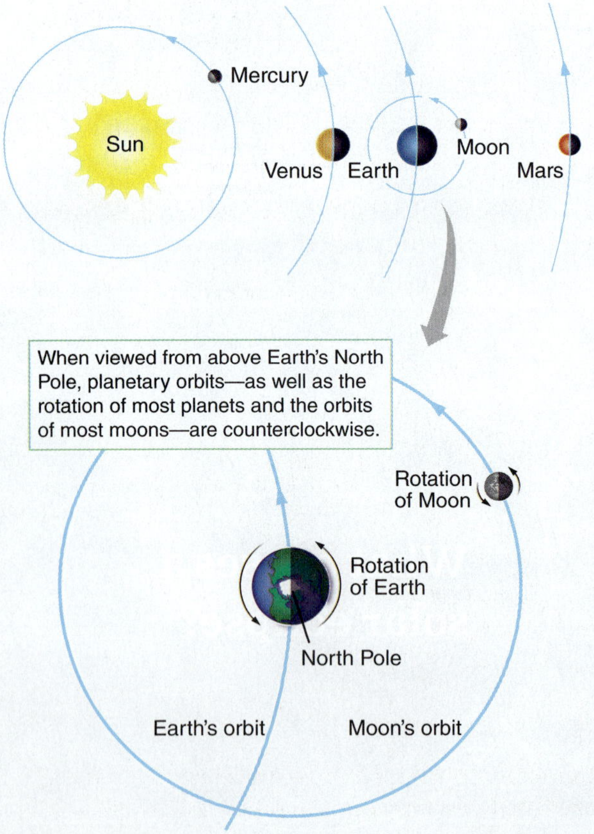

Figure 2.2 Motions in the Solar System, as viewed from above Earth's North Pole. (Not drawn to scale.)

The Celestial Sphere

Long before Christopher Columbus sailed west from Spain, Aristotle and other Greek philosophers knew that Earth was a sphere. However, because Earth seems stationary, they did not realize that its motion caused the changes they observed in the sky from day to day and year to year. As we explain here, Earth's rotation on its axis determines the rising and setting of the Sun, Moon, and stars—one of the rhythms that governs life on Earth (**Figure 2.2**).

One reason the ancients did not suspect that Earth rotates was that they could not perceive its spinning motion. As Earth rotates about its axis, the planet's surface is moving quite fast—about 1,674 kilometers per hour (km/h) at the equator. We do not feel that motion any more than we would feel the motion of a car with a perfectly smooth ride cruising down a straight highway. Nor do we feel the direction of Earth's spin, although the hourly motion of the Sun, Moon, and stars across the sky reveals that direction. Earth's **North Pole** is at the north end of Earth's rotation axis. Imagine that you are in space far above Earth's North Pole. From there you would see Earth complete a counterclockwise rotation, once each 24-hour period, as Figure 2.2 shows. As the rotating Earth carries an observer on the surface from west to east, objects in the sky appear to move in the opposite direction, from east to west. As seen from Earth's surface, each object seems to move across the sky in a path called its **apparent daily motion**.

To help visualize the apparent daily motions of the Sun and stars, think of the sky as a huge sphere with the stars painted on its surface and Earth at its center. From ancient Greek times to the Renaissance, many people believed that to be a true representation of the heavens. Astronomers call that imaginary sphere the **celestial sphere** (**Figure 2.3**). The celestial sphere is a useful concept because it is easy to visualize, but never forget that it is, in fact, imaginary.

Each point on the celestial sphere indicates a direction in space. Directly above Earth's North Pole is the **north celestial pole** (**NCP**). Directly above Earth's **South Pole**, which is at the south end of Earth's rotation axis, is the **south celestial pole** (**SCP**). Directly above Earth's **equator** is the **celestial equator**, an imaginary circle that divides the sky into a northern half and a southern half. Just as the north celestial pole is the projection of the direction of Earth's North Pole into the sky, the celestial equator is the projection of the plane of Earth's equator into the sky. Just as Earth's North Pole is 90° away from Earth's equator, the north celestial pole is 90° away from the celestial equator. If you are in the Northern Hemisphere and you point one arm toward the celestial equator and one arm toward the north celestial pole, your arms will always form a right angle, so the north celestial pole is 90° away from the celestial equator. The same holds true for the Southern Hemisphere: the angle between the celestial equator and the south celestial pole is always 90° as well.

Between the celestial poles and the equator, objects have positions on the celestial sphere with coordinates analogous to latitude and longitude on Earth. **Latitude** indicates distance north or south from Earth's equator. On the celestial sphere, **declination** similarly indicates the distance of an object north or south of the celestial equator (from 0° to ±90°). On Earth, **longitude** measures how far east or west you are from the Royal Observatory in Greenwich, England. On the celestial sphere, **right ascension** is similar to longitude on Earth. Right ascension measures a celestial body's angular distance eastward along the celestial equator from the point where the Sun's path crosses the celestial equator from south to north. Those coordinates are used to quickly locate objects in the sky. The **ecliptic** is the Sun's path in the sky throughout the year (Figure 2.3). In Appendix 7, we discuss detailed descriptions and illustrations of latitude and longitude as well as coordinates used with the celestial sphere.

The **zenith** is the point in the sky directly above you, wherever you are, as **Figure 2.4a** shows. You can find the **horizon** by standing up and pointing your right hand at the zenith and your left hand straight out from your side. Turn in a complete circle. Your left hand has traced out the entire horizon. You can divide the sky into an eastern half and a western half with a line that runs from the horizon at due north through the zenith to the horizon at due south. That imaginary north–south line is the **meridian**, shown as a dashed line in Figure 2.4a. **Figure 2.4b** shows those locations on the celestial sphere. The meridian line continues around the far side of the celestial sphere, through the **nadir** (the point directly below you), and back to the starting point due north.

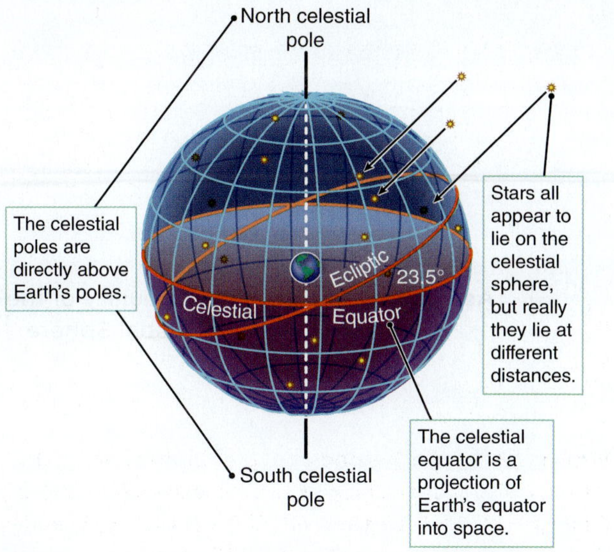

Figure 2.3 The celestial sphere is a useful fiction in thinking about the stars' appearance and apparent motion in the sky.

AstroTour: The Celestial Sphere and the Ecliptic

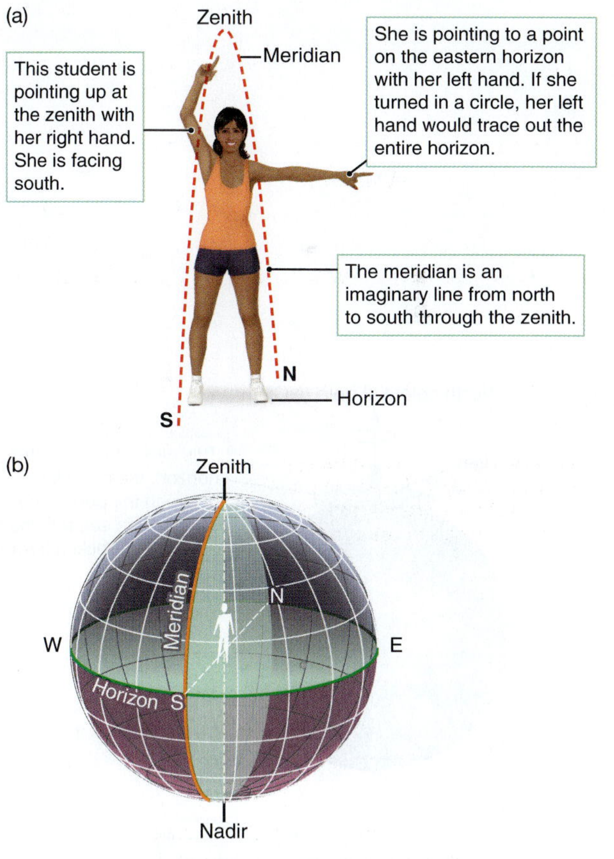

Figure 2.4 (a) The meridian is a line on the celestial sphere that runs from north to south, dividing the sky into an east half and a west half. (b) At any location on Earth, the sky is divided into an east half and a west half by an imaginary meridian projected onto the celestial sphere.

Astronomy in Action: Vocabulary of the Celestial Sphere

Take a moment to visualize all those locations in space. To see how to use the celestial sphere, consider the Sun at noon and at midnight. Local noon occurs when the Sun crosses the meridian at your location. That is the highest point above the horizon that the Sun will reach on any given day. The highest point is almost never the zenith. You have to be in a specific place on a specific day for the Sun to be directly overhead at noon. One such place and time is at latitude 23.5° north of the equator on June 20. From our perspective on Earth, the celestial sphere appears to rotate, carrying the Sun across the sky to its highest point at noon, over toward the west to set in the evening. In *reality*, the Sun remains in the same place in space through the entire 24-hour period, and Earth rotates so that any given location on Earth faces a different direction at every moment. When it is noon where you live, Earth has rotated so that you face most directly toward the Sun. Half a day later, at midnight, your location on Earth has rotated to face most directly away from the Sun. Local midnight occurs when the Sun is precisely opposite from its position at local noon.

The View from the Poles

AstroTour: The View from the Poles

The apparent daily motions of the stars and the Sun depend on where you live. For example, the apparent daily motions of celestial objects in a northern place such as Alaska are different from the apparent daily motions seen from a tropical island such as Hawaii. To understand why your location matters, let's examine the view of the stars from the poles—and then use them to guide our thinking about the view of the stars from other latitudes.

Imagine that you are the person in **Figure 2.5a** who is standing on the North Pole watching the sky. There, the north celestial pole is directly overhead at the zenith. Ignore the Sun for the moment and pretend that you can always see stars in the sky. You are standing where Earth's axis of rotation intersects its surface, which is like standing at the center of a rotating carousel. As Earth rotates, the spot directly above you remains fixed over your head while everything else in the sky appears to revolve counterclockwise around that spot. **Figure 2.5b** depicts that overhead view.

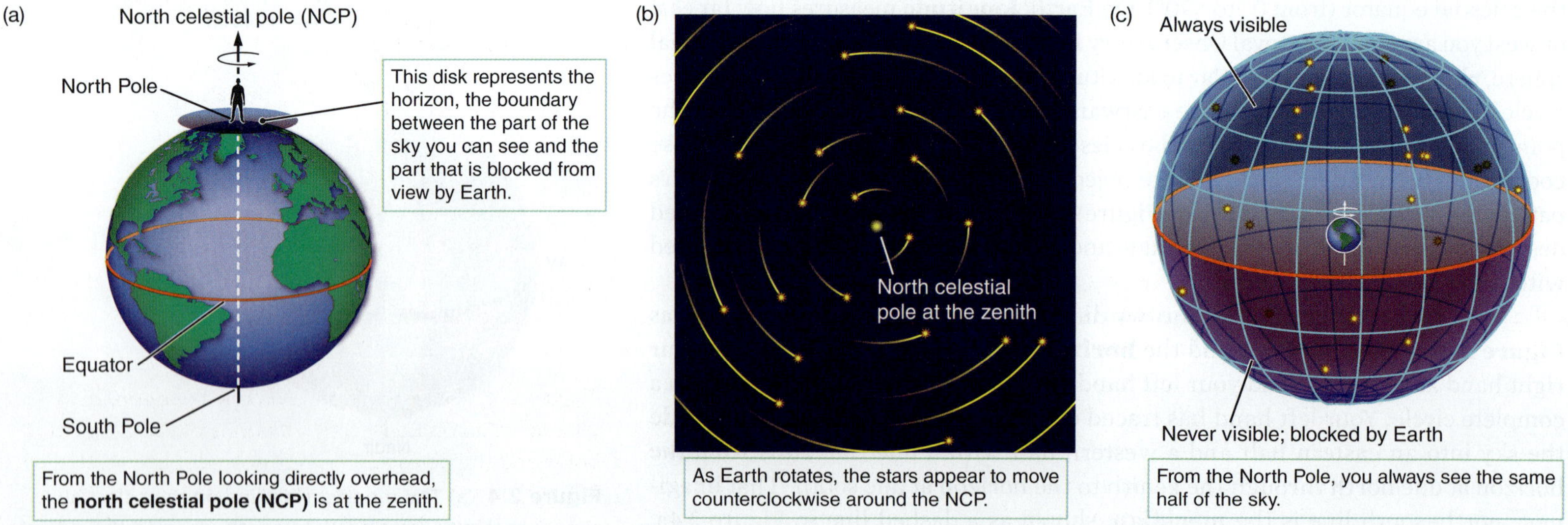

Figure 2.5 As viewed from (a) Earth's North Pole, (b) stars move throughout the night in counterclockwise, circular paths about the zenith. (c) The same half of the sky is always visible from the North Pole.

Wherever you are on Earth, you can see only half the sky. The horizon is the boundary between the part of the sky you can see and the other half of the sky that is blocked by Earth. Except at the poles, the visible half of the sky changes constantly as Earth rotates because the zenith points to different locations in the sky as Earth carries you around. If you are standing at the North Pole, however, the zenith is always in the same location in space, so objects visible from the North Pole follow circular paths that always have the same **altitude**, or angle above the horizon. Objects close to the zenith appear to follow small circles, whereas objects near the horizon follow the largest circles (Figure 2.5b). The view from the North Pole is special: you will always see the *same* half of the celestial sphere from there because nothing rises or sets each day as Earth turns (**Figure 2.5c**).

The view from Earth's South Pole is much the same—with two major differences. First, the South Pole is on the opposite side of Earth from the North Pole, so the visible half of the sky at the South Pole is precisely the half hidden from view at the North Pole. Second, stars appear to move clockwise around the south celestial pole rather than counterclockwise as they do at the north celestial pole. To visualize why those motions are different, stand up and spin around from right to left. As you look at the ceiling, things appear to move counterclockwise, but as you look at the floor, they appear to be moving clockwise.

CHECK YOUR UNDERSTANDING 2.1A

No matter where you are on Earth, stars appear to rotate about a point called the: (a) zenith; (b) celestial pole; (c) nadir; (d) meridian.

The View Away from the Poles

Suppose that you leave the North Pole to travel south to lower latitudes. Imagine a line from the center of Earth to your location on the surface, as in **Figure 2.6**. Now imagine a second line from Earth's center to the point on the equator closest to you. The angle between those two lines is your latitude. At the North Pole, for example, those two imaginary lines form a 90° angle. At the equator, they form a 0° angle. So the latitude of the North Pole is 90° north, and the latitude of the equator is 0°. The South Pole is at latitude 90° south.

Your latitude determines the part of the sky that you can see throughout the year. As you move south from the North Pole, your zenith moves away from the north celestial pole, and so the horizon moves as well. At the North Pole, the horizon makes a 90° angle with the north celestial pole, which is at the zenith. At a latitude of 60° north, as in Figure 2.6, your horizon is tilted 60° from the north celestial pole. No matter where you are on Earth, the angle between your horizon and the north celestial pole is equal to your latitude. The situation is the same in the Southern Hemisphere—your latitude is the altitude of the south celestial pole. At the equator, at a latitude of 0°, the north and south celestial poles would be at the northern and southern horizons, respectively.

One way to solidify your understanding of the view of the sky at different latitudes is to draw pictures like the one in Figure 2.6. If you can draw a picture like that for any latitude—filling in the values for each angle in the drawing and imagining what the sky looks like from there—you are developing a working knowledge of the appearance of the sky. That knowledge will prove useful later when we discuss a variety of phenomena, such as the changing of the seasons.

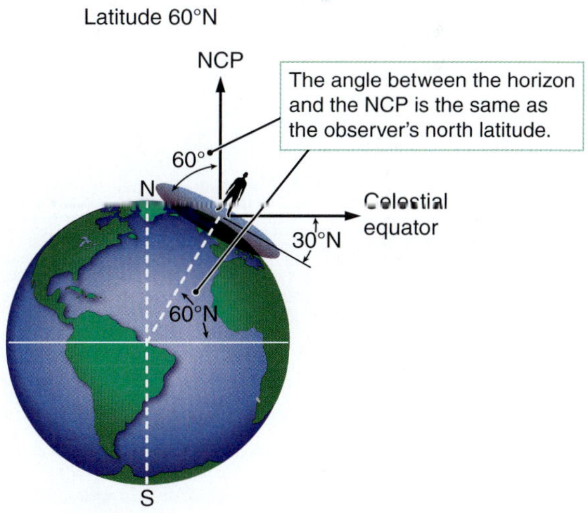

Figure 2.6 Your perspective on the sky depends on your location on Earth. The locations of the celestial poles and the celestial equator in an observer's sky depend on the observer's latitude. Here, an observer at latitude 60° north sees the north celestial pole at an altitude of 60° above the northern horizon and the celestial equator 30° above the southern horizon.

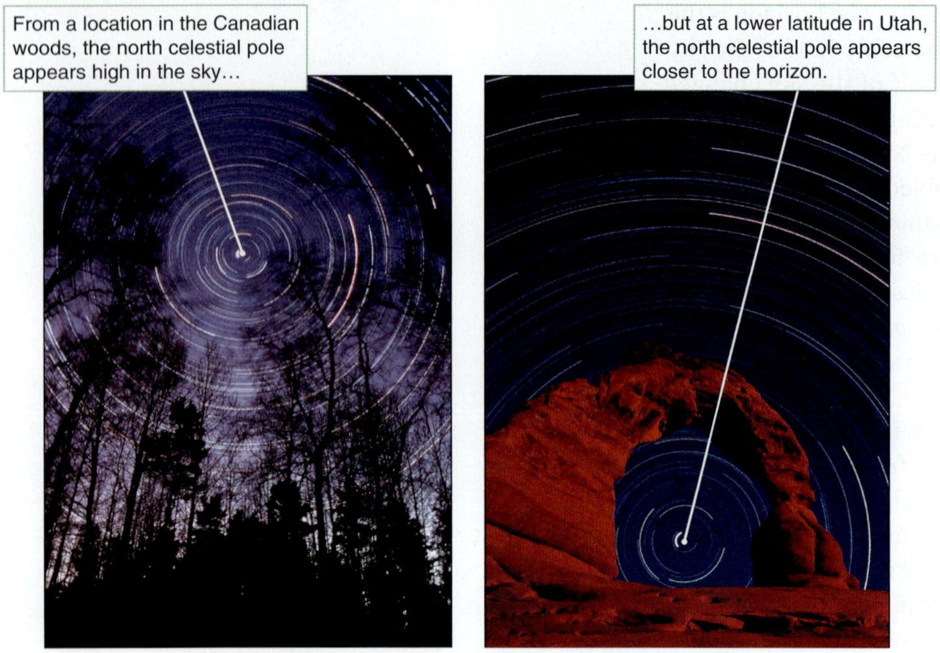

Figure 2.7 Time exposures of the sky showing the apparent motions of stars through the night. Note the difference in the circumpolar portion of the sky as seen from the two northern latitudes.

MOTIONS OF THE STARS AND THE CELESTIAL POLES **Figure 2.7** shows two time-lapse views of the sky from different latitudes. The stars' apparent motions about the celestial poles also differ from latitude to latitude. The visible part of the sky constantly changes as stars rise and set with Earth's rotation. From that perspective the horizon appears fixed, whereas the stars appear to move. From those latitudes, if we focus our attention on the north celestial pole, we see much the same thing we saw from Earth's North Pole. The north celestial pole remains fixed in the sky, and all the stars appear to move throughout the night in counterclockwise, circular paths around that point. The north celestial pole is no longer directly overhead as it was at the North Pole, however, so the apparent circular paths of the stars are now tipped with respect to the horizon. (More correctly, your horizon is now tipped with respect to the apparent circular paths of the stars.)

From the vantage point of an observer in the Northern Hemisphere, stars located near the north celestial pole in the sky never dip below the horizon and thus stay visible all night. Recall from Figure 2.6 that the latitude is equal to the altitude of the north celestial pole. Stars closer to the north celestial pole than that angle never dip below the horizon as they complete their apparent paths around the pole. Those stars are **circumpolar** ("around the pole"). Another group of stars, near the south celestial pole, never rises above the horizon in the Northern Hemisphere. Stars between those that never rise and those in the circumpolar region (that is, the ones that never set) can be seen for *only part* of each 24-hour day. Those stars appear to rise and set as Earth turns. The equator is the only place on Earth where you can see the entire sky over the course of 24 hours. From there, the north and south celestial poles sit on the northern and southern horizons, respectively, and all the stars move through the sky each 24-hour day. (Even though the Sun lights the sky for roughly half that time, the stars are still there.)

Figure 2.8 shows the orientation of the sky as seen by observers at four latitudes. For an observer at the North Pole (Figure 2.8a), the celestial equator lies exactly along the horizon. The north celestial pole is at the zenith, and the southern half of the sky is never visible. Stars neither rise nor set; their paths form circles parallel to the horizon. For an observer at the equator (Figure 2.8b), the celestial poles are both at the horizon, and all the stars are visible in a 24-hour period, rising straight up and setting straight down each day.

At other latitudes, the celestial equator intersects the horizon due east and due west. Therefore, a star on the celestial equator rises due east and sets due west. Stars located north of the celestial equator rise north of east and set north of west. Stars located south of the celestial equator rise south of east and set south of west.

From everywhere else on Earth (except at the poles), half of the celestial equator is always visible above the horizon. Therefore, any object located on the celestial equator is visible half the time—above the horizon for 12 hours each day. Objects not on the celestial equator are above the horizon for different amounts of time. Figures 2.8c and d show that stars in the observer's hemisphere are visible

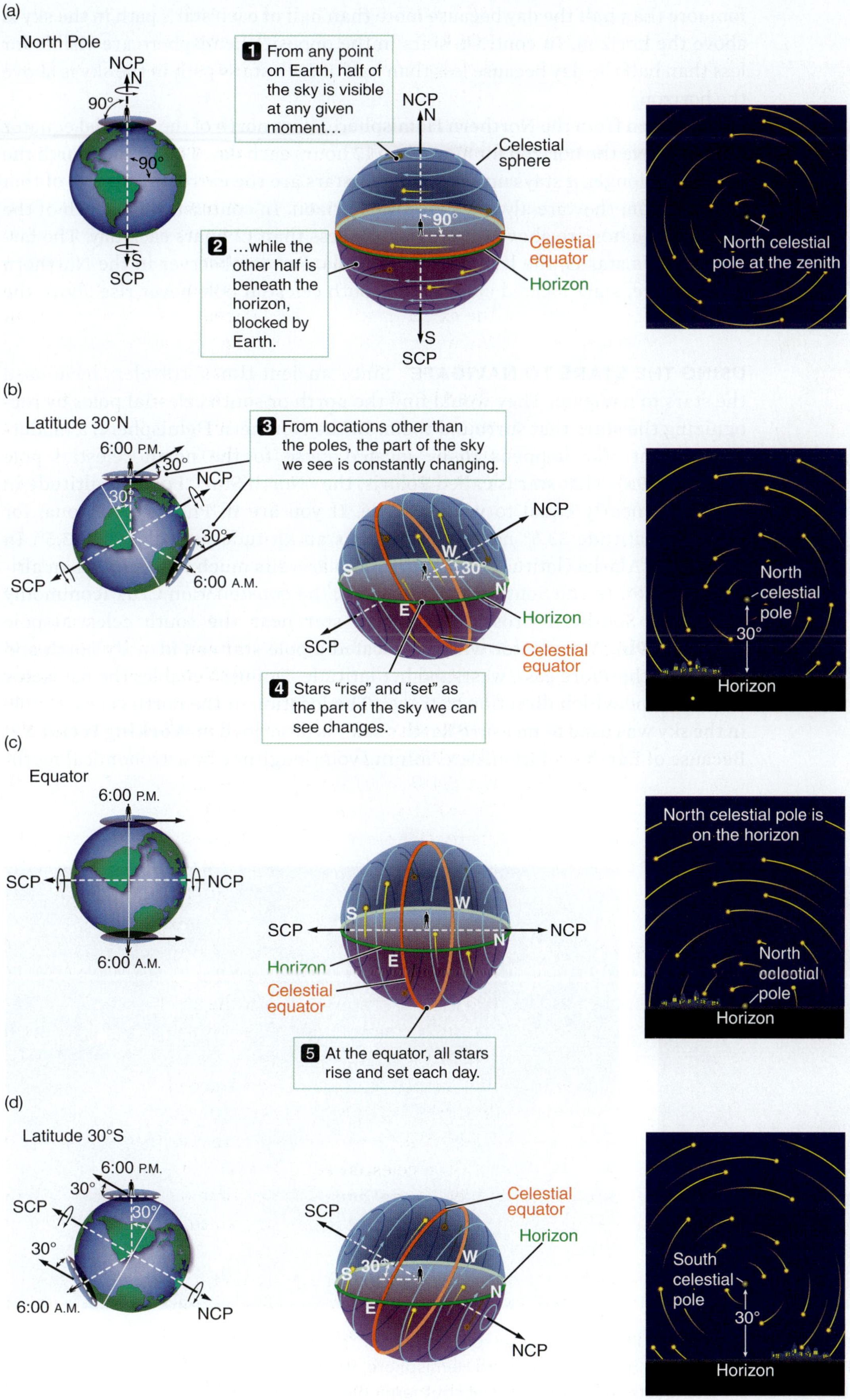

Figure 2.8 The celestial sphere is shown here as viewed by observers at four latitudes. At all locations other than the poles, stars rise and set as the part of the celestial sphere that we see changes during the day.

for more than half the day because more than half of each star's path in the sky is above the horizon. In contrast, stars in the opposite hemisphere are visible for less than half the day because less than half of each star's path in the sky is above the horizon.

As seen from the Northern Hemisphere, stars north of the celestial equator remain above the horizon for more than 12 hours each day. The farther north the star is, the longer it stays up. Circumpolar stars are the extreme example of that phenomenon; they are always above the horizon. In contrast, stars south of the celestial equator are above the horizon for less than 12 hours each day. The farther south a star is, the less time it is visible. For an observer in the Northern Hemisphere, stars located close to the south celestial pole never rise above the horizon.

USING THE STARS TO NAVIGATE Since ancient times, travelers have used the stars to navigate. They would find the north or south celestial poles by recognizing the stars that surround them. In the Northern Hemisphere, a moderately bright star happens to be located close to the north celestial pole (**Figure 2.9a**). That star is called Polaris, the "North Star." Polaris's altitude in the sky is nearly equal to your latitude. If you are in Phoenix, Arizona, for example (latitude 33.5° north), Polaris has an altitude very close to 33.5°. In Fairbanks, Alaska (latitude 64.6° north), Polaris sits much higher, with an altitude of 64.6°. In the Southern Hemisphere, the constellation Crux (commonly called the Southern Cross) points to a star near the south celestial pole (**Figure 2.9b**). A navigator who has located a pole star can identify north and south and therefore east, west, and her latitude. Doing so enables the navigator to determine which direction to travel. The location of the north celestial pole in the sky was used to measure Earth's size, as described in **Working It Out 2.1**. Because of Earth's rotation, determining your longitude by astronomical meth-

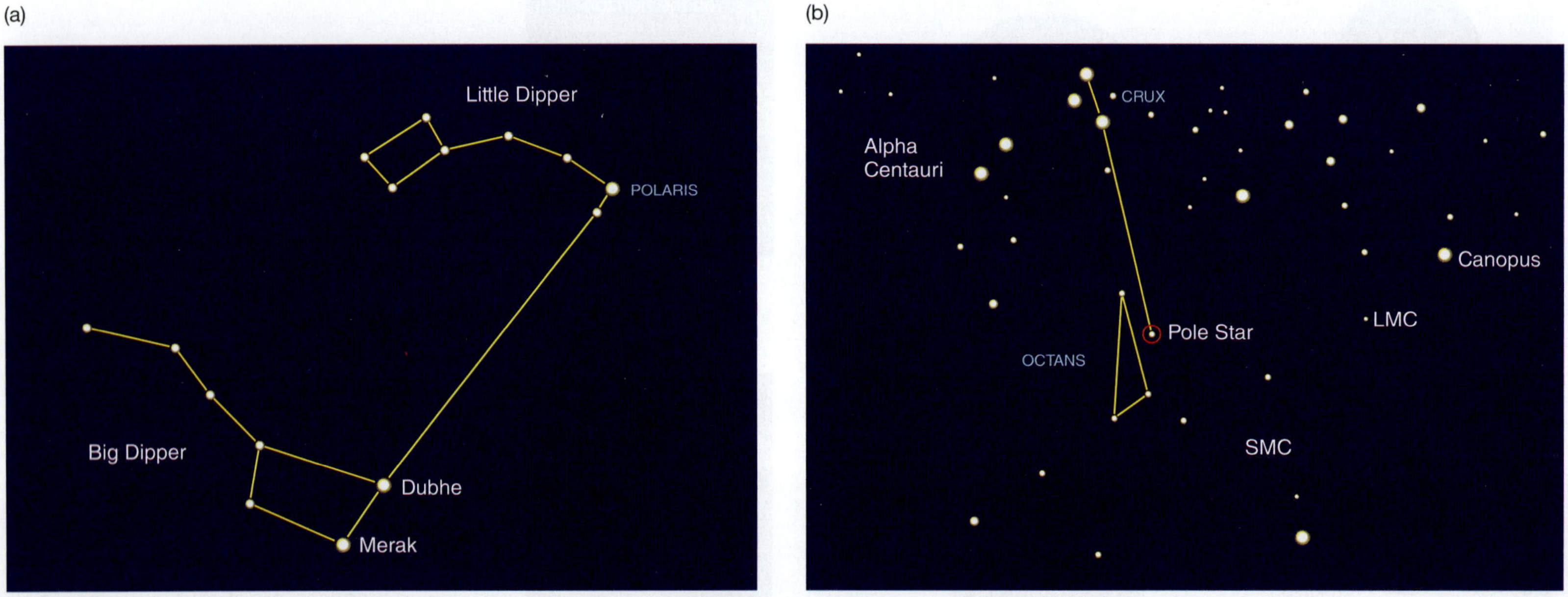

Figure 2.9 Groups of stars near the pole stars in the sky can be used to locate a pole star. (a) Two bright "pointer stars" in the cup of the Big Dipper point toward Polaris, the "North Star." (b) In the Southern Hemisphere, the constellation Crux and two of its bright pointer stars can be used to locate the relatively faint southern pole star.

2.1 Working It Out How to Estimate Earth's Size

We can use the location of the north celestial pole in the sky to estimate Earth's size. Suppose we start out in Phoenix, Arizona, and we observe the north celestial pole to be 33.5° above the horizon. If we head north, by the time we reach the Grand Canyon, about 290 km from Phoenix, we notice that the north celestial pole has risen to about 36° above the horizon. That difference between 33.5° and 36° (2.5°) is 1/144 of the way around a circle. (A circle is 360°, and 2.5°/360° = 1/144.)

That means we must have traveled 1/144 of the way around Earth's circumference, C, by traveling the 290 km between Phoenix and the Grand Canyon. In other words,

$$\frac{1}{144} \times C = 290 \text{ km}$$

When we rearrange the expression, Earth's circumference is given by

$$C = 144 \times 290 \text{ km} \approx 42{,}000 \text{ km}$$

Earth's actual circumference is just over 40,000 km, so our simple calculation was close. The circumference of a circle is equal to $2\pi r$, where r is its radius. Earth's radius, then, is given by

$$\text{Radius} = \frac{C}{2\pi} = \frac{40{,}000 \text{ km}}{2\pi} = 6{,}400 \text{ km}$$

In about 230 BCE, using much the same method, the Greek astronomer Eratosthenes (276–194 BCE) was the first to accurately measure Earth's size. As **Figure 2.10** illustrates, Eratosthenes used the distance between his home city of Alexandria, Egypt, and the city of Syene (now Aswân, also in Egypt), which was 5,000 "stadia." He noticed that on the first day of summer in Syene, the sunlight reflected directly off the water in a deep well, so the Sun must have been nearly at the zenith. By measuring the shadow of the Sun from an upright stick in Alexandria, he saw that the Sun was about 7.2° south of the zenith on the same date. Assuming that Earth was spherical and that Syene was directly south of Alexandria, he determined the distance between the two cities to be 7.2° divided by 360, or 1/50 of Earth's circumference.

Accurately estimating distances in those days was hard. Although historians know Eratosthenes concluded that the cities were 5,000 stadia apart, they are still not at all sure of the value of his stadion unit. If the stadion was 185 meters, the size of Greek stadia, Eratosthenes would have worked the math similarly:

$$\frac{1}{50} \times C = 5{,}000 \text{ stadia} \times 185 \text{ meters/stadion}$$
$$= 925{,}000 \text{ meters} = 925 \text{ km}$$

Eratosthenes would have found Earth's circumference to be

$$C = 50 \times 925 \text{ km} = 46{,}250 \text{ km}$$

That result is only about 16 percent higher than the modern value.

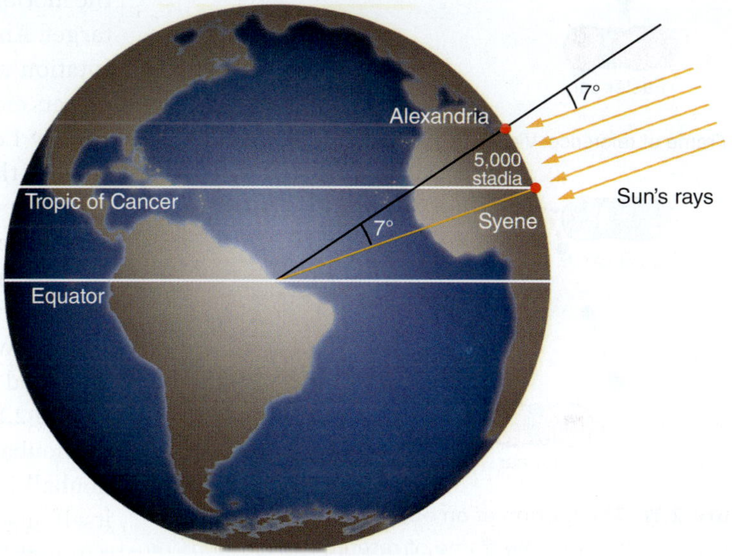

Figure 2.10 Eratosthenes used observations and basic calculations to estimate Earth's size.

ods is much more complicated. Longitude cannot easily be determined from astronomical observation alone.

Relative Motions and Frame of Reference

Why don't we feel motion as Earth spins on its axis and moves through space in its orbit around the Sun? The answer lies in the concept of a **frame of reference**, a coordinate system within which an observer measures positions and motions. The difference in motion between two individual frames of reference is called the

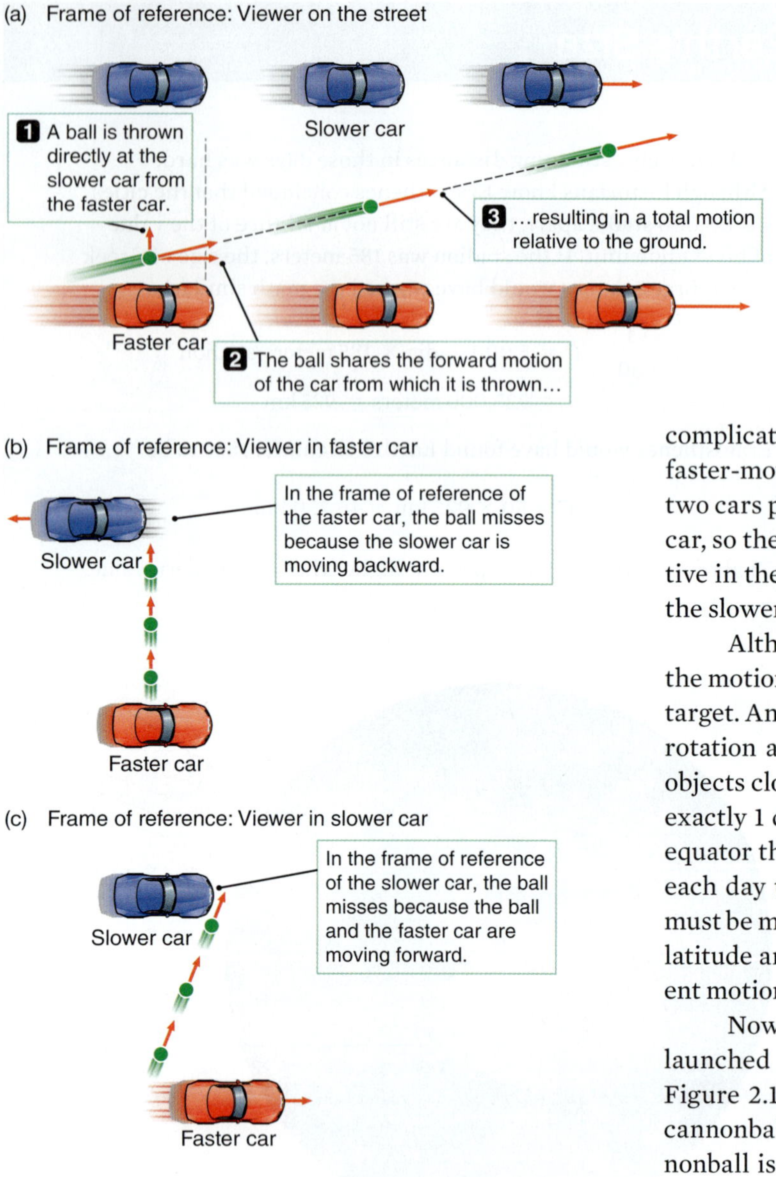

Figure 2.11 The motion of an object is always measured relative to the frame of reference of the observer.

relative motion. For example, imagine that you are riding in a car traveling down a straight section of highway at a constant speed. Under those conditions, without looking out the window or feeling road vibrations, you cannot easily tell the difference between being in the moving car and sitting in a parked car. Because everything in the car is moving together, you can measure only the relative motions between objects in the car, and they are all zero—no motion is observed. Similarly, the resulting relative motions between objects that are near each other on Earth are zero because they have the same rotational motion. That is why we do not notice Earth's motion.

Now imagine two cars driving down the road at different speeds, as **Figure 2.11a** shows. For the moment, ignore any real-world complications, such as wind resistance. If you were to throw a ball from the faster-moving car directly out the side window at the slower-moving car as the two cars passed, you would miss. The ball shares the forward motion of the faster car, so the ball outruns the forward motion of the slower car. From your perspective in the faster car (**Figure 2.11b**), the slower car lagged behind the ball. From the slower car's perspective (**Figure 2.11c**), your car and the ball sped on ahead.

Although we cannot feel Earth's rotation, it influences things as diverse as the motion of weather patterns and how an artillery gunner must aim at a distant target. An object on Earth's surface moves in a circle each day around the planet's rotation axis. That circle is larger for objects near the equator and smaller for objects closer to one of the poles, but all objects complete their circular motion in exactly 1 day. As you can see in **Figure 2.12**, Earth's surface moves faster at the equator than at higher latitudes. An object closer to the equator has farther to go each day than one nearer a pole does. Therefore, the object nearer the equator must be moving faster than the one at higher latitude. If an object starts out at one latitude and then moves to another, that difference in speed influences its apparent motion over the surface.

Now consider two locations at different latitudes. Imagine a cannonball launched directly north from a point in the Northern Hemisphere, as in Figure 2.12a. Because the cannon is closer to the equator than its target is, the cannonball is moving toward the east faster than its target. Even though the cannonball is fired toward the north, it shares the eastward velocity of the cannon itself, just like the ball thrown from the faster car in Figure 2.11. Therefore, the cannonball is *also* moving toward the east faster than its target. As the cannonball flies north, it too moves toward the east faster than the ground underneath it does. To an observer on the ground, the cannonball appears to curve toward the east as it outruns the eastward motion of the ground it is crossing. The farther north the cannonball flies, the greater the difference between its eastward velocity and the ground's. Thus, the cannonball follows a path that appears to curve more and more to the east the farther north it goes. If you are located in the Northern Hemisphere and fire a cannonball *south* toward the equator, as in Figure 2.12b, the opposite effect will occur. Now the cannon is moving toward the east more slowly than its target. As the cannonball flies toward the south, its eastward motion lags behind that of the ground underneath it, and the cannonball appears to curve toward the west.

That curving motion of objects as a result of the difference in Earth's rotation speeds at different latitudes is called the **Coriolis effect**. In the Northern Hemisphere, the Coriolis effect causes a cannonball fired north to drift to the east as seen from the ground. In other words, the cannonball appears to curve to the

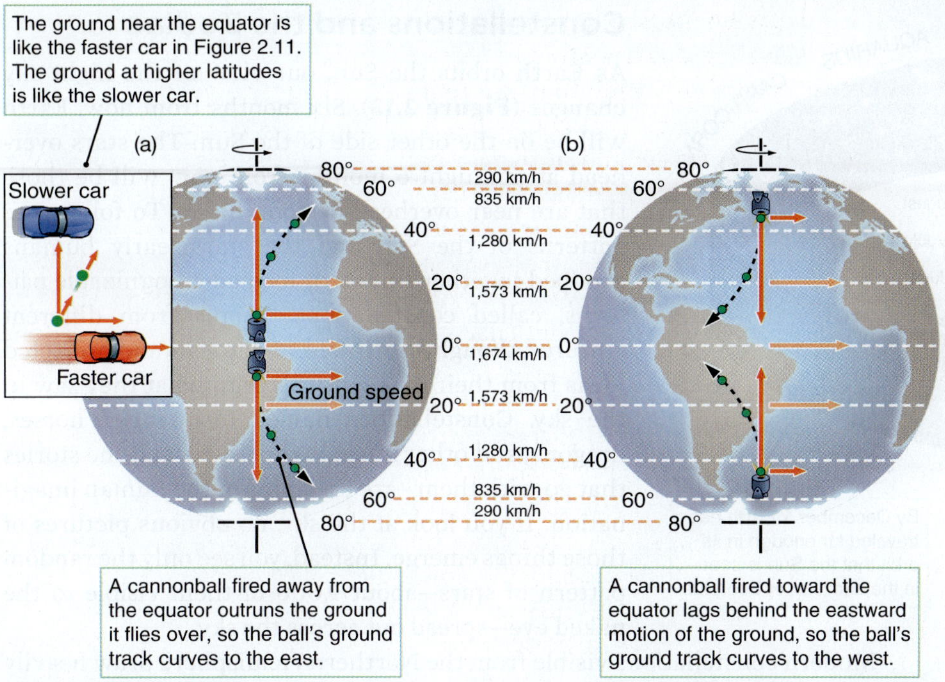

Figure 2.12 The Coriolis effect causes objects to be deflected as they move across Earth's surface. The green dashed curve shows the cannonball's path as measured by a local observer.

right. A cannonball fired south appears to curve to the west, which also gives it the appearance of curving to the right. In the Northern Hemisphere, the Coriolis effect seems to deflect things to the *right*. In the Southern Hemisphere, it seems to deflect things to the *left*. In between, at the equator itself, the Coriolis effect vanishes.

On Earth, the effect is enough to deflect a fly ball hit north or south into deep left field in a stadium in the northern United States by about a half a centimeter. At some time or other, the Coriolis effect has probably determined the outcome of a baseball game.

CHECK YOUR UNDERSTANDING 2.1B

If the star Polaris has an altitude of 35°, we know that: (a) our longitude is 55° east; (b) our latitude is 55° north; (c) our longitude is 35° west; (d) our latitude is 35° north.

2.2 Revolution about the Sun Leads to Changes during the Year

Earth orbits (or **revolves**) around the Sun in the same direction that Earth spins about its axis—counterclockwise as viewed from above Earth's North Pole (see Figure 2.2). A **year** is the time Earth takes to complete one revolution around the Sun. Earth's motion around the Sun is responsible for many of the patterns of change we see in the sky and on Earth, including changes in the stars we see overhead. Because of that motion, the stars in the night sky change throughout the year, and Earth experiences seasons.

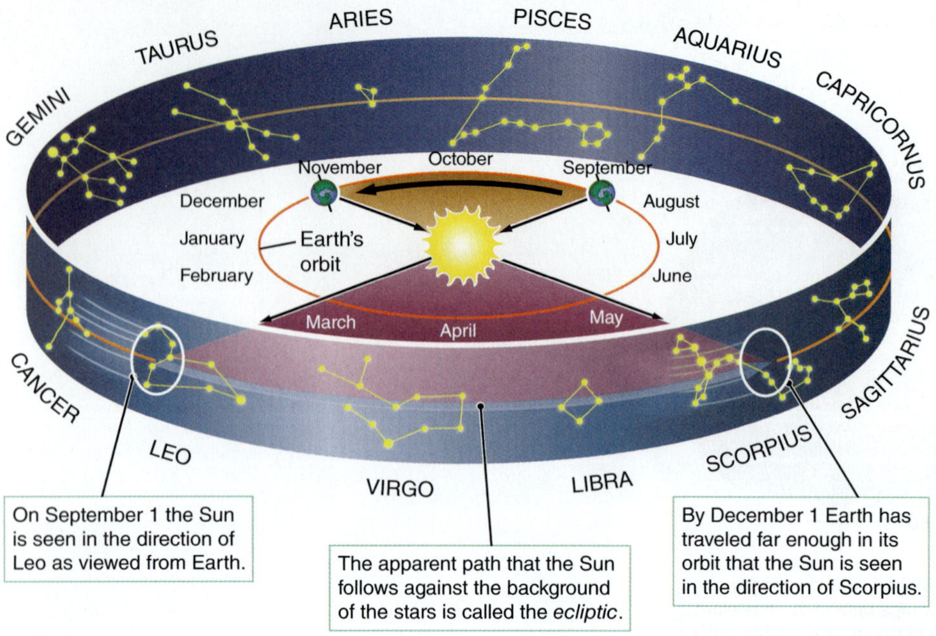

Figure 2.13 As Earth orbits the Sun, different stars appear in the night sky, and the Sun's apparent position against the background of stars changes. The imaginary circle that the Sun's annual path traces is called the ecliptic. Constellations along the ecliptic form the zodiac.

Constellations and the Zodiac

As Earth orbits the Sun, our view of the night sky changes (**Figure 2.13**). Six months from now, Earth will be on the other side of the Sun. The stars overhead at midnight 6 months from now will be those that are near overhead at noon today. To follow the patterns of the Sun and the stars, early humans grouped together stars that formed recognizable patterns, called constellations. People from different cultures, though, saw different patterns and projected ideas from their own cultures onto what they saw in the sky. Constellations named for winged horses, dragons, and other imaginary things—and the stories that go with them—are creations of the human imagination. If you look at the sky, no obvious pictures of those things emerge. Instead, you see only the random pattern of stars—about 9,000 of them visible to the naked eye—spread out across the sky.

Modern constellations visible from the Northern Hemisphere draw heavily from the list of constellations that the Alexandrian astronomer Ptolemy compiled 2,000 years ago. Modern constellation names in the southern sky come from European explorers who visited the Southern Hemisphere during the 17th and 18th centuries. Today, astronomers use an officially sanctioned set of 88 constellations to serve as a road map of the sky. Every star in the sky lies within the borders of a single constellation, and the names of constellations are used in naming their component stars. For example, Sirius, the brightest star in the sky, lies within the constellation Canis Major (meaning "big dog"). Following the Greek alphabet, astronomers call the brightest star in a constellation *alpha*; the second brightest, *beta*; and so forth. The official name of Sirius is Alpha Canis Majoris, indicating that it is the brightest star in Canis Major. Appendix 7 includes sky maps showing the constellations.

If you noted the position of the Sun in relation to the stars each day for a year, you would find that the Sun traces out a great circle against the background of the stars. On September 1, the Sun appears to be in the direction of the constellation Leo. Six months later, on March 1, Earth is on the other side of the Sun, and the Sun appears from our perspective on Earth to be in the direction of the constellation Aquarius. Recall that the Sun's apparent path against the background of the stars is called the ecliptic (the blue band in Figure 2.13). The constellations that lie along the ecliptic through which the Sun appears to move are called the constellations of the **zodiac**.

Earth's Axial Tilt and the Seasons

To understand why the seasons change, we need to consider the combined effects of Earth's rotation on its axis and revolution around the Sun. Many people believe that seasons change because Earth is closer to the Sun in summer and farther away in winter. How can we falsify that hypothesis? We can predict that if the distance between Earth and the Sun caused the seasons, then all of Earth should experience summer at the same time of year. But the United States has summer in June, whereas Australia has summer in December. In modern times, we can directly measure the distance: Earth is actually closest to the Sun at the beginning of January. We have just falsified the hypothesis tying seasonal change to distance, and we need to look for another explanation that explains all the available facts.

As Earth orbits the Sun, the Sun appears to move along the ecliptic, which is tilted 23.5° with respect to the celestial equator. That tilt occurs because Earth's axis of rotation is tilted 23.5° from the perpendicular to Earth's orbital plane. The days are longer in summer than in winter, and the Sun is higher in the sky as it crosses the meridian in summer than in winter.

Figure 2.14 shows that as Earth moves around the Sun, its axis always points in the same direction in space (toward Polaris). As it orbits, Earth is sometimes on one side of the Sun and sometimes on the other. As a result, Earth's North Pole is sometimes tilted more toward the Sun, and other times the South Pole is tilted more toward the Sun. When Earth's North Pole is tilted toward the Sun, an observer on Earth views the Sun north of the celestial equator. For observers in the Northern Hemisphere, the Sun is above the horizon more than 12 hours each day, thus making the days longer than 12 hours. Six months later, when Earth's North Pole is tilted away from the Sun, an observer in the same place views the Sun south of the celestial equator and the days are shorter.

AstroTour: The Earth Spins and Revolves

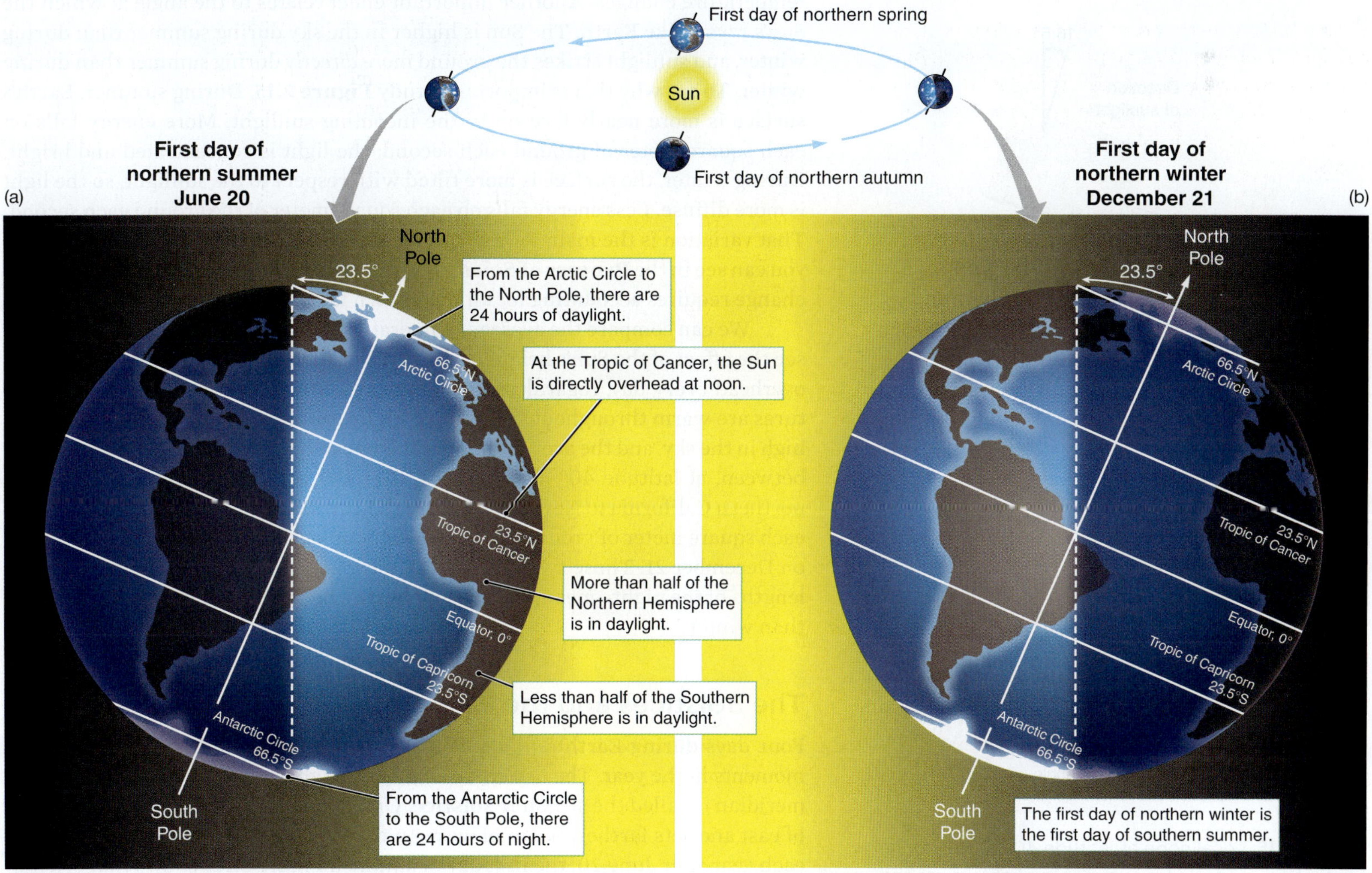

Figure 2.14 (a) On the first day of the northern summer (around June 20, the summer solstice), the northern end of Earth's axis is tilted most nearly toward the Sun, whereas the Southern Hemisphere is tipped away. (b) Six months later, on the first day of the northern winter (around December 21, the winter solstice), the situation is reversed. Seasons are opposite in the two hemispheres.

Astronomy in Action: The Cause of Earth's Seasons

In the preceding paragraph, we specified the *Northern* Hemisphere because seasons are opposite in the Southern Hemisphere. Look again at Figure 2.14. Around June 20, while the Northern Hemisphere is enjoying the long days and short nights of summer, Earth's South Pole is tilted away from the Sun. It is winter in the Southern Hemisphere; less than half of the Southern Hemisphere is illuminated by the Sun, and the days are shorter than 12 hours. On December 21, Earth's South Pole is tilted toward the Sun. It is summer in the Southern Hemisphere, so its days are long and the nights are short.

To understand how the combination of Earth's axial tilt and its path around the Sun creates seasons, consider a limiting case. If Earth's spin axis were exactly perpendicular to the plane of Earth's orbit (the **ecliptic plane**), the Sun would always be on the celestial equator. At every latitude (except the poles), the Sun would follow the same path through the sky every day, rising due east each morning and setting due west each evening. The Sun would be above the horizon exactly half the time, and days and nights would always be exactly 12 hours long everywhere on Earth. In short, each day would be just like the last, and Earth would have no seasons.

The changing length of days through the year only partly explains seasonal temperature changes. Another important effect relates to the angle at which the Sun's rays strike Earth. The Sun is higher in the sky during summer than during winter, and sunlight strikes the ground *more directly* during summer than during winter. To see why that is important, study **Figure 2.15**. During summer, Earth's surface is more nearly face-on to the incoming sunlight. More energy falls on each square meter of ground each second; the light is concentrated and bright. During winter, the surface is more tilted with respect to the sunlight, so the light is more diffuse. Less energy falls on each square meter of the ground each second. That variation is the main reason why summer is hotter and winter is colder. As you can see in the **Process of Science Figure**, determining the causes of seasonal change requires accounting for all the known facts.

We can compare the average temperatures at different latitudes on Earth to see the effect of the Sun's height in the sky. Near the equator, the Sun passes high overhead every day, regardless of the season. As a result, the average temperatures are warm throughout the year. At high latitudes, however, the Sun is *never* high in the sky, and the average temperatures can be cold even during summer. In between, at latitude 40° north, which stretches across the United States from northern California to New Jersey, more than twice as much solar energy falls on each square meter of ground per second at noon on June 20 as falls there at noon on December 21. Those two effects—the directness of sunlight and the different lengths of the night—mean that the Sun heats a hemisphere more during summer than winter.

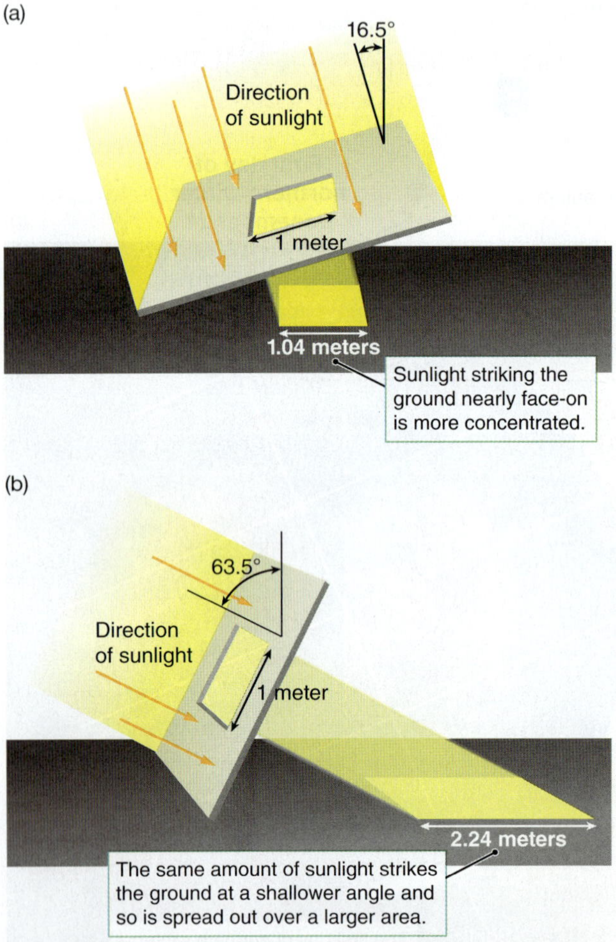

Figure 2.15 Local noon at latitude 40° north. (a) On the first day of northern summer, sunlight strikes the ground almost face-on. (b) On the first day of northern winter, sunlight strikes the ground more obliquely, and less than half as much sunlight falls on each square meter of ground each second.

The Solstices and the Equinoxes

Four days during Earth's orbit (two solstices and two equinoxes) mark unique moments in the year. The day when the Sun is highest in the sky as it crosses the meridian is called the **summer solstice**. On that day, the Sun rises farthest north of east and sets farthest north of west. In the Northern Hemisphere, that occurs each year near June 20, the first day of summer. Figure 2.14a shows that orientation of Earth and Sun.

Six months after the summer solstice, around December 21, the North Pole is tilted away from the Sun (Figure 2.14b). That day is the **winter solstice** in the Northern Hemisphere—the shortest day of the year and the first day of winter in

PROCESS OF SCIENCE

Theories Must Fit All the Known Facts

Many people misunderstand the phenomenon of changing seasons because they do not account for all the relevant facts.

HYPOTHESIS 1

We have seasons because Earth is closer to the Sun in the summer and farther away in winter.

THE TEST If this were true, both the Northern and Southern Hemispheres would have summer in July. However, the Northern and Southern Hemispheres experience opposite seasons.

THE CONCLUSION The hypothesis is falsified.

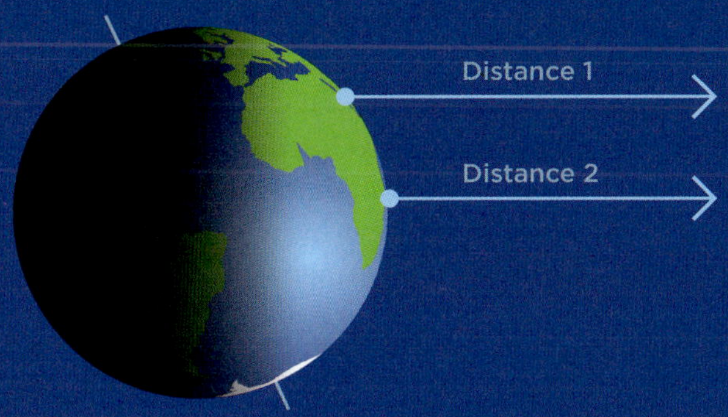

HYPOTHESIS 2

We have seasons because the tilt of Earth's axis causes one hemisphere to be significantly closer to the Sun than the other.

THE TEST If this were true, the distances would have to be very different to cause such a large effect. Earth is tiny compared to its distance from the Sun: The distances between each hemisphere and the Sun differ by less than 0.004 percent.

THE CONCLUSION The hypothesis is falsified.

HYPOTHESIS 3

We have seasons because Earth's tilt changes the distribution of energy—one hemisphere receives more intense light than the other.

THE TEST If this were true, the amount of sunlight striking the ground in the summer would have to be more than in the winter, and the days would have to be longer in summer.

THE CONCLUSION Seasons are caused primarily by a change in illumination due to Earth's tilt. During winter, less energy falls on each square meter of ground per second.

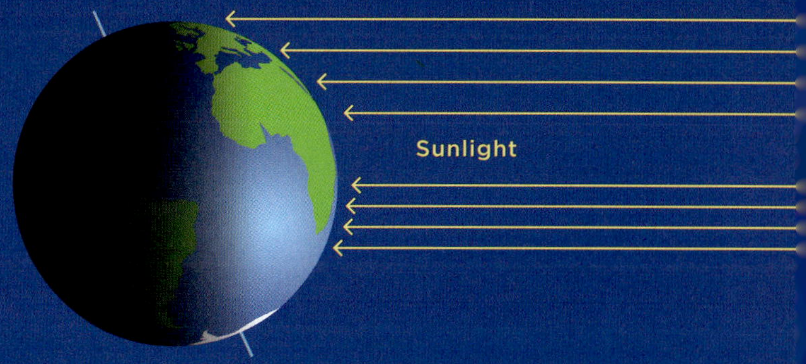

New information often challenges misconceptions. Incorporating this new information may alter current theories to improve the explanation of observed phenomena.

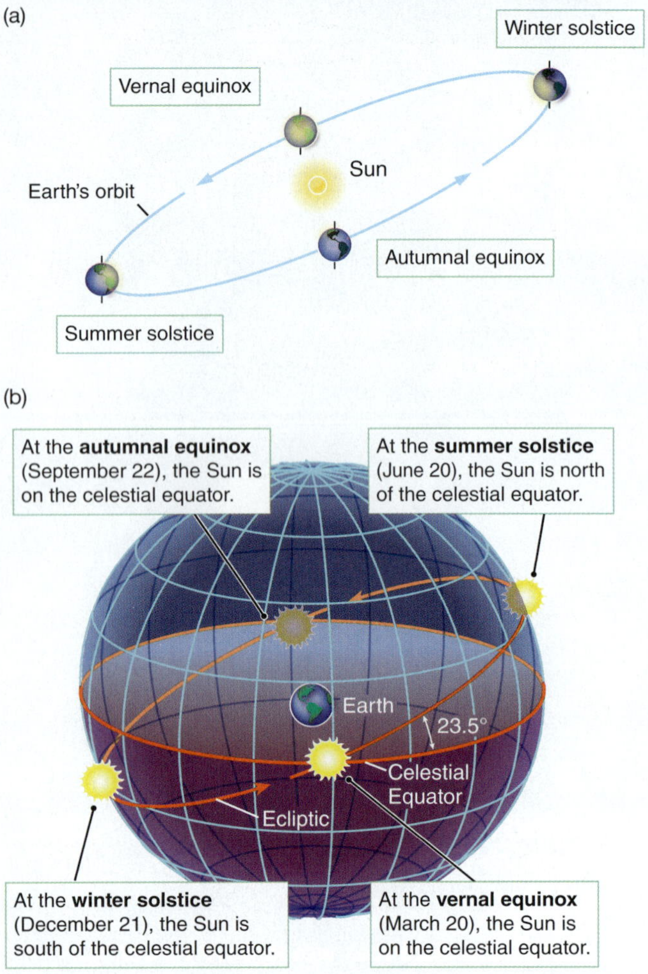

Figure 2.16 Earth's motion around the Sun from the frame of reference of (a) the Sun and (b) Earth.

Astronomy in Action: The Earth-Moon-Sun System

the Northern Hemisphere. Almost all cultural traditions in the Northern Hemisphere include some sort of major celebration in late December. Those winter festivals celebrate the return of the source of Earth's light and warmth. The days have stopped growing shorter and are beginning to get longer, and spring will come again.

Between the two solstices, the ecliptic crosses the celestial equator on two days. On those days, the Sun lies directly above Earth's equator. We call those days *equinoxes*, which means "equal night," because the entire Earth experiences 12 hours of daylight and 12 hours of darkness (except at the poles). Halfway between summer solstice and winter solstice, the **autumnal equinox** marks the beginning of fall in the Northern Hemisphere; it occurs around September 22. Halfway between winter solstice and summer solstice, the **vernal equinox** marks the beginning of spring in the Northern Hemisphere; it occurs around March 20. In the Southern Hemisphere, the dates for the summer and winter solstices, and the autumnal and vernal equinoxes, are reversed from those in the Northern Hemisphere.

Figure 2.16 shows the solstices and equinoxes from two perspectives. Figure 2.16a shows Earth in orbit around a stationary Sun, whereas Figure 2.16b shows the Sun's apparent motion along the celestial sphere, which is how it appears to observers on Earth. In both cases, we are looking at the plane of Earth's orbit from the side, so that it is shown in perspective and looks quite flattened. We also have tilted the images so that Earth's North Pole points straight up. The equinoxes correspond to points in the sky where the celestial equator meets the ecliptic. Practice shifting between those perspectives. You will know that you understand them when you can look at a position in either panel and predict the corresponding positions of the Sun and Earth in the other panel.

Just as a pot of water on a stove takes time to heat up when the burner is turned on and to cool when the burner is turned off, Earth takes time to respond to changes in heating from the Sun. The hottest months of northern summer are usually July and August, which come after the summer solstice, when days are growing shorter. Similarly, the coldest months of northern winter are usually January and February, which occur after the winter solstice, when days are growing longer. Temperature changes on Earth lag behind changes in the amount of heating we receive from the Sun.

That picture of the seasons must be modified somewhat near Earth's poles. At latitudes north of 66.5° north and south of 66.5° south, the Sun is circumpolar

Figure 2.17 This composite photo shows the midnight Sun, which is visible in latitudes above 66.5° north (or south). In the 360-degree panoramic view, the Sun moves 15° each hour.

for a part of the year surrounding the first day of summer. Those lines of latitude are the **Arctic Circle** and the **Antarctic Circle**, respectively (see Figure 2.14). When the Sun is circumpolar, it is above the horizon 24 hours per day, earning the polar regions the nickname "land of the midnight Sun" (**Figure 2.17**). An equally long period surrounds the first day of winter, when the Sun never rises and the nights are 24 hours long. The Sun never rises high in the Arctic or Antarctic sky, so the sunlight is never very direct. Even with the long days at the height of summer, the Arctic and Antarctic regions remain relatively cool.

In contrast, on the equator, *all* stars, including the Sun, are above the horizon approximately 12 hours per day. Days and nights there are 12 hours long throughout the year. The Sun passes directly overhead on the first day of spring and the first day of autumn because on those days the Sun is on the celestial equator. Sunlight is most direct, perpendicular to the ground, at the equator on those days. On the summer solstice, the Sun is at its northernmost point along the ecliptic. On that day, and on the winter solstice, the Sun is farthest from the zenith at noon, and therefore sunlight is least direct at the equator.

Latitude 23.5° north is called the Tropic of Cancer, and latitude 23.5° south is called the Tropic of Capricorn (Figure 2.14). The band between those latitudes is called the **Tropics**. If you live in the tropics—in Rio de Janeiro or Honolulu, for example—the Sun will be directly overhead at noon twice during the year.

Precession of the Equinoxes

Two thousand years ago, when Ptolemy and his associates were formalizing their knowledge of the positions and motions of objects in the sky, the Sun appeared in the constellation Cancer on the first day of northern summer and in the constellation Capricornus on the first day of northern winter. Today, the Sun is in Taurus on the first day of northern summer and in Sagittarius on the first day of northern winter. Why have the constellations in which solstices appear changed? Two motions are actually associated with Earth and its axis: Earth spins on its axis, but its axis also wobbles like the axis of a spinning top slowing down (**Figure 2.18a**). The wobble is very slow: the north celestial pole takes about 26,000 years to complete one trip around a large circle centered on the north ecliptic pole. Polaris is the star we now see near the north celestial pole. However, if you could travel several thousand years into the past or the future, the point about which the northern sky appears to rotate would no longer be near Polaris. Instead, the stars would rotate about another point on the path shown in Figure 2.18b. That figure shows the path of the north celestial pole through the sky during one cycle of that wobble.

The celestial equator is perpendicular to Earth's axis. Therefore, as Earth's axis wobbles, the celestial equator also must wobble. As it does so, the locations where the celestial equator crosses the ecliptic—the equinoxes—change as well. During each 26,000-year wobble of Earth's axis, the locations of the equinoxes make one complete circuit around the celestial equator. That change of the position of the equinox, due to the wobble of Earth's axis, is called the **precession of the equinoxes**.

CHECK YOUR UNDERSTANDING 2.2

If Earth's axis were tilted by 45°, instead of its actual tilt of 23.5°, how (if at all) would the seasons be different from what they are now? (a) The seasons would remain the same. (b) Summers would be colder. (c) Winters would be shorter. (d) Winters would be colder.

(a)

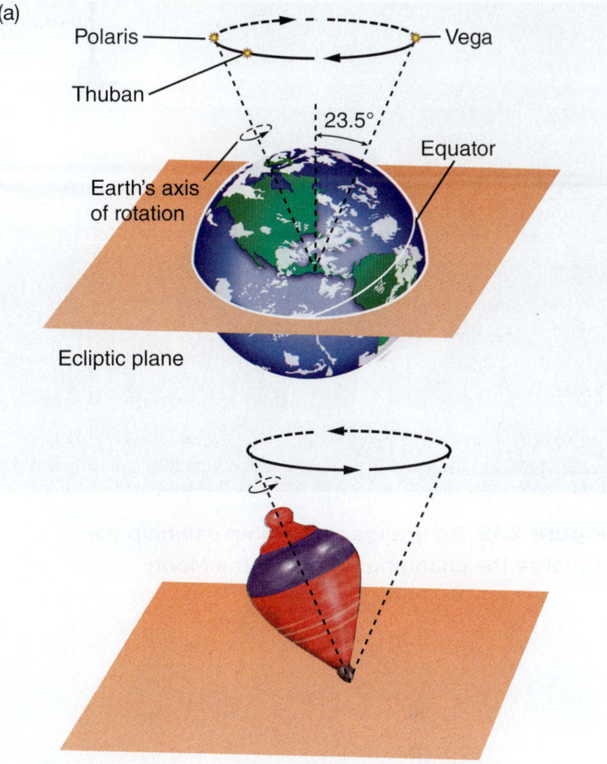

(b)

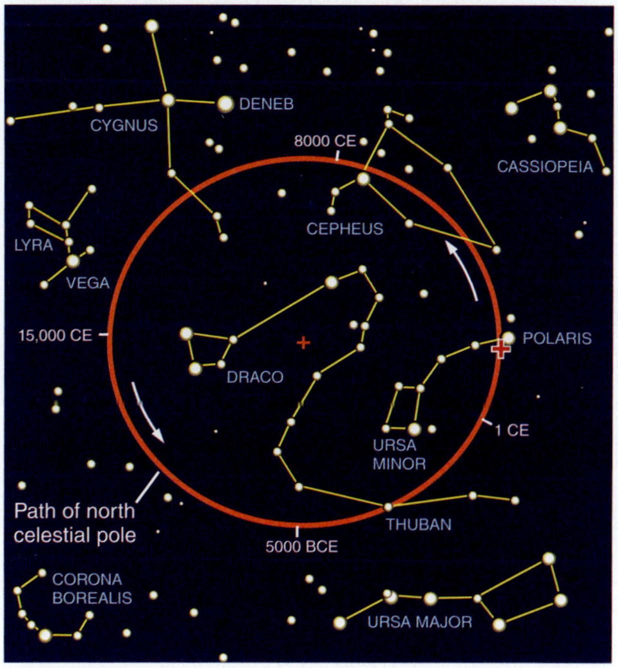

Figure 2.18 (a) Earth's axis of rotation changes orientation in the same way that the axis of a spinning top changes orientation. Polaris is the current North Pole star, Thuban was the North Pole star in 3000 BCE, and Vega will be the Pole star in 14,000 CE. (b) That precession causes the projection of Earth's rotation axis to move in a circle, centered on the north ecliptic pole (orange cross in the center). The red cross on the circle marks the projection of Earth's axis on the sky in the early 21st century.

Figure 2.19 An orange and a lamp can help you visualize the changing phases of the Moon.

Astronomy in Action: Phases of the Moon

AstroTour: The Moon's Orbit: Eclipses and Phases

Simulation: Phases of the Moon

2.3 The Moon's Appearance Changes as It Orbits Earth

After the Sun, the most prominent object in our sky is the Moon. Just as Earth orbits around the Sun, the Moon orbits around Earth once every 27.32 days. In this section, we discuss the phases of the Moon as seen from Earth.

The Changing Phases of the Moon

The Moon and its changing aspects have been the frequent subject of mythology, art, literature, and music. In mythology, the Moon was the Greek goddess Artemis, the Roman goddess Diana, and the Inuit god Igaluk. We speak of the "man in the Moon," the "harvest Moon," and sometimes a "blue Moon."

Unlike the Sun, the Moon has no light source of its own; instead, the Moon shines by reflected sunlight. As the Moon orbits Earth, our view of the illuminated portion of the Moon constantly changes. Those different appearances of the Moon are called **phases**. During a new Moon, when the Moon is between Earth and the Sun, the side facing away from us is illuminated, and during a full Moon, when Earth is between the Sun and the Moon, the side facing toward us is illuminated. The rest of the time, we can see only part of the illuminated portion from Earth. Sometimes the Moon appears as a circular disk in the sky. Other times it is nothing more than a thin sliver, or its face appears dark.

To help you visualize the changing phases of the Moon, use an orange, a lamp, and your head (**Figure 2.19**). Your head is Earth, the orange is the Moon, and the lamp is the Sun. Turn off all the other lights in the room, and step back as far from the lamp as you can. Hold up the orange slightly above your head so that the lamp illuminates it from one side. Move the orange clockwise around your head and watch how the appearance of the orange changes. When you are between the orange and the lamp, the face of the orange that is toward you is fully illuminated. The orange appears to be a bright, circular disk. As the orange moves around its circle, you will see a progression of lighted shapes, depending on how much of the bright side and how much of the dark side of the orange you can see. That progression of shapes mimics the changing phases of the Moon.

Figure 2.20 shows the changing phases of the Moon. The **new Moon** occurs when the Moon is between Earth and the Sun. The far side is illuminated, but the near side is in darkness and we cannot see it. The new Moon appears close to the Sun in the sky, so it is up in the daytime with the Sun: it rises in the east at sunrise, crosses the meridian near noon, and sets in the west near sunset. A new Moon is never above the horizon in the nighttime sky.

A few days after a new Moon, as the Moon orbits Earth, a sliver of its illuminated half, called a **waxing crescent Moon**, becomes visible. *Waxing* here means "growing in size and brilliance"; the name refers to the fact that the Moon appears to be "filling out" from night to night at that time. From our perspective, the Moon has also moved away from the Sun in the sky. Because the Moon travels around Earth in the same direction in which Earth rotates, we now see the Moon trailing the Sun, so the Moon is east of the Sun in the sky. A waxing crescent Moon is visible in the western sky in the evening, near the setting Sun but remaining above the horizon after the Sun sets. The "horns" of the crescent always point directly away from the Sun.

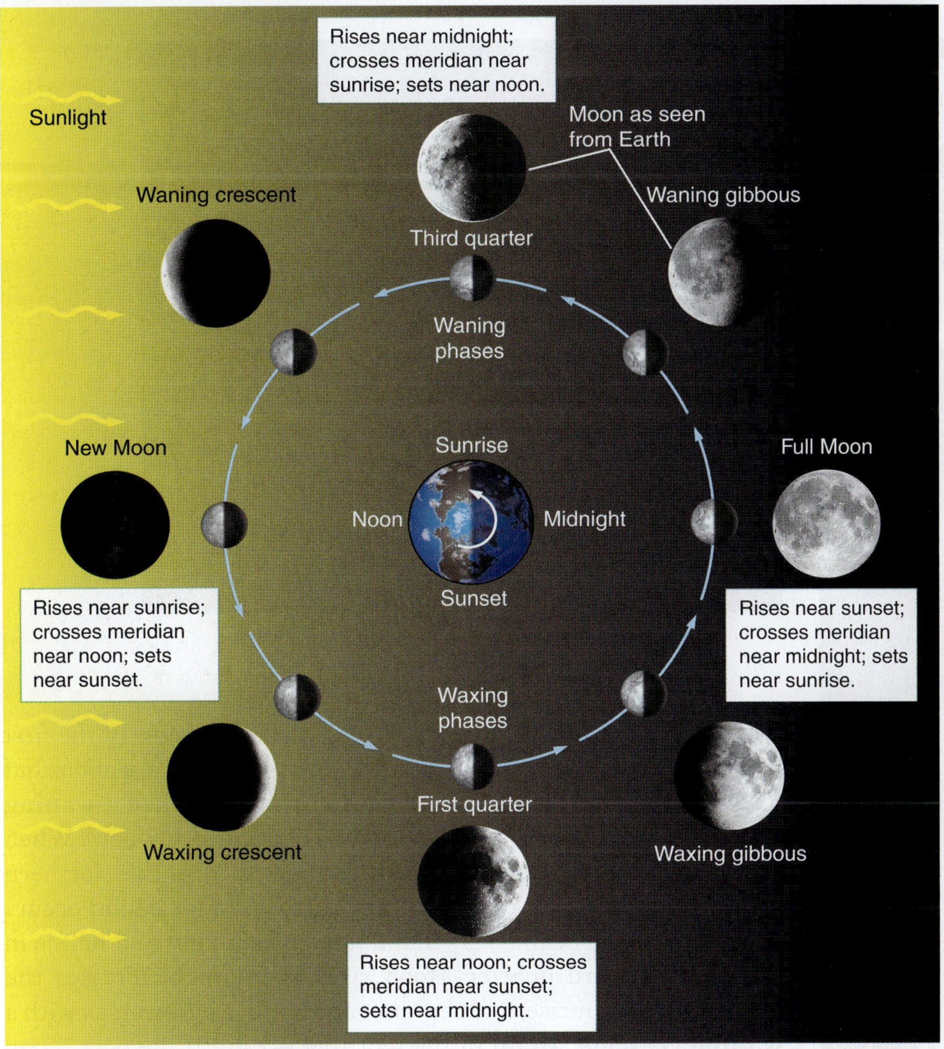

Figure 2.20 The inner circle of images (connected by blue arrows) shows the Moon as it orbits Earth, as seen by an observer far above Earth's North Pole. The Sun is on the left. The outer ring of images shows the corresponding phases of the Moon as seen from the northern hemisphere of Earth.

As the Moon moves farther along in its orbit, and the angle between the Sun and Moon grows larger, more and more of its near side becomes illuminated. About a week after the new Moon, half of the near side of the Moon is illuminated and half is in darkness. That phase is called a **first quarter Moon** because the Moon has moved a quarter of the way around Earth and has completed the first quarter of its cycle from new Moon to new Moon (Figure 2.20). The first quarter Moon rises at noon, crosses the meridian at sunset, and sets at midnight.

As the Moon moves beyond first quarter, more than half of its near side is illuminated. That phase is called a **waxing gibbous Moon**, from the Latin *gibbus*, meaning "hump." The waxing gibbous Moon continues nightly to "grow" until finally we see the entire near side of the Moon illuminated—a **full Moon**. Earth is now between the Sun and the Moon, which appear opposite each other in the sky when viewed from Earth. The full Moon rises as the Sun sets, crosses the meridian at midnight, and sets in the morning as the Sun rises.

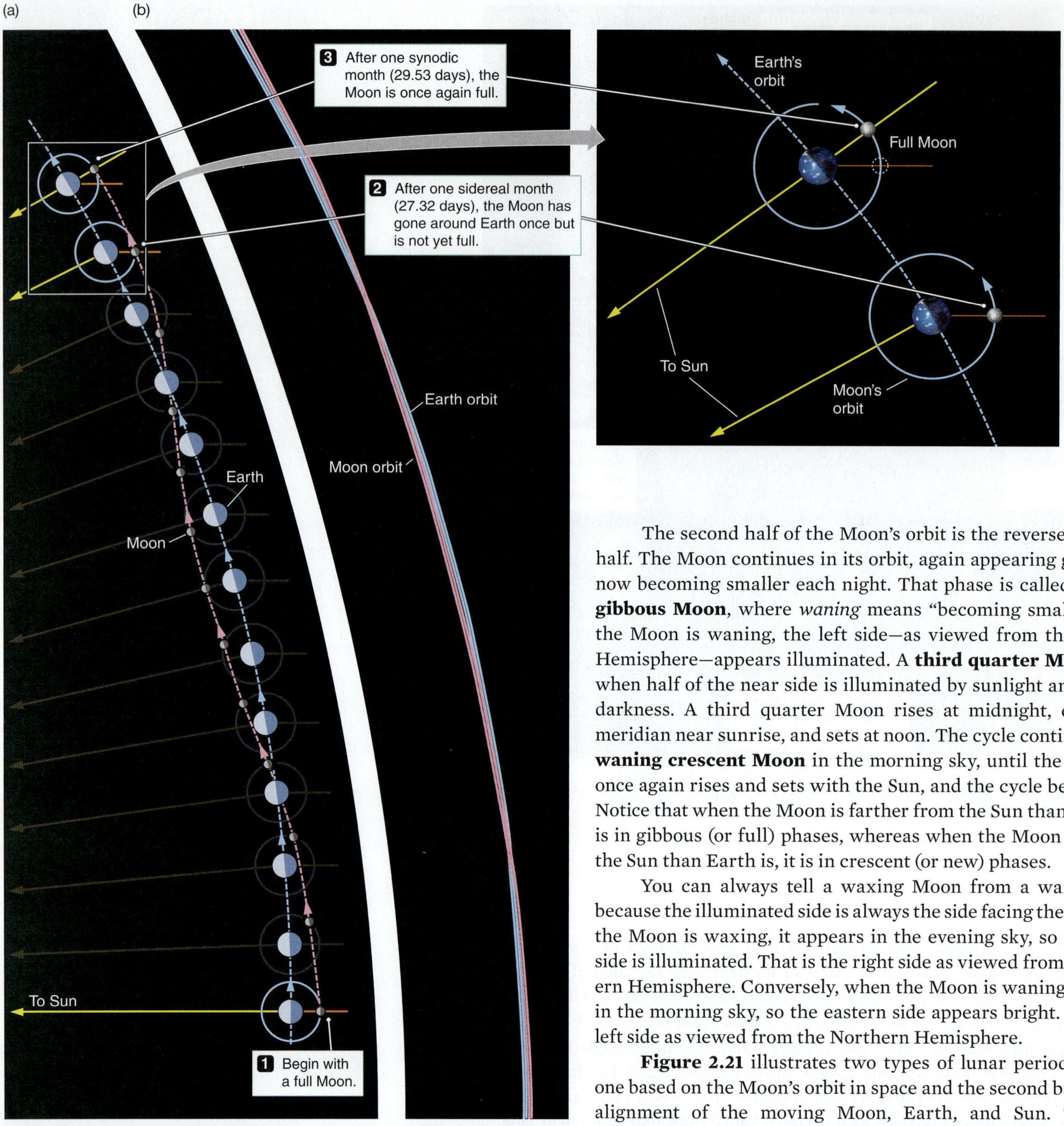

Figure 2.21 (a) The Moon completes one sidereal orbit in 27.32 days, but the synodic period (the period between phases seen from Earth) from one full Moon to the next is 29.53 days. The horizontal orange line to the right of the Moon indicates a fixed direction in space. (b) Here the orbits of Earth and the Moon are shown to scale, but the sizes of Earth and the Moon are not.

The second half of the Moon's orbit is the reverse of the first half. The Moon continues in its orbit, again appearing gibbous but now becoming smaller each night. That phase is called a **waning gibbous Moon**, where *waning* means "becoming smaller." When the Moon is waning, the left side—as viewed from the Northern Hemisphere—appears illuminated. A **third quarter Moon** occurs when half of the near side is illuminated by sunlight and half is in darkness. A third quarter Moon rises at midnight, crosses the meridian near sunrise, and sets at noon. The cycle continues with a **waning crescent Moon** in the morning sky, until the new Moon once again rises and sets with the Sun, and the cycle begins again. Notice that when the Moon is farther from the Sun than Earth is, it is in gibbous (or full) phases, whereas when the Moon is closer to the Sun than Earth is, it is in crescent (or new) phases.

You can always tell a waxing Moon from a waning Moon because the illuminated side is always the side facing the Sun. When the Moon is waxing, it appears in the evening sky, so its western side is illuminated. That is the right side as viewed from the Northern Hemisphere. Conversely, when the Moon is waning, it appears in the morning sky, so the eastern side appears bright. That is the left side as viewed from the Northern Hemisphere.

Figure 2.21 illustrates two types of lunar periods: the first one based on the Moon's orbit in space and the second based on the alignment of the moving Moon, Earth, and Sun. The Moon completes one orbit around Earth in 27.32 days: that **sidereal period** is how long the Moon takes to return to the same location in its orbit. However, because the relationships among Earth, the Moon, and the Sun change as a result of Earth's orbital motion, going from one full Moon to the next takes 29.53 days. That cycle is known as the Moon's **synodic period** and is the basis for our "month," because it is what we can easily observe from Earth.

Do not try to memorize all possible combinations of where the Moon is in the sky at what phase and at what time of day. Instead, work on understanding the motion and phases of the Moon, and then use your understanding to figure out the specifics of any given case. To study the phases of the Moon, draw a picture like Figure 2.20, and use it to follow the Moon around its orbit. From your drawing, figure out what phase you would see and where it would appear in the sky at a given time of day. You might also try the simulations described in "Exploration: The Phases of the Moon" at the end of the chapter.

The Moon's Visible Face

Although the Moon's illumination varies as it orbits, one aspect of the Moon's appearance does *not* change: the face that we see. If we were to go outside next week, next month, 20 years from now, or 200 centuries from now, we would still see the same side of the Moon that we see tonight. That is true because the Moon rotates on its axis exactly once for each revolution that it makes around Earth.

Imagine walking around the Washington Monument while always keeping your face toward the monument. By the time you complete one circle around the monument, your head has turned completely around once. When you were south of the monument, you were facing north; when you were east of the monument, you were facing west; and so on. But someone looking at you from the monument would always see your face. The Moon does the same thing, rotating on its axis once per revolution around Earth, always keeping the same face toward Earth (**Figure 2.22**). Objects whose revolution and rotation are synchronized (or "in sync") with each other are said to be in **synchronous rotation**. That effect occurs as a result of the gravitational and tidal forces between Earth and the Moon (Section 4.4). In later chapters we discuss other examples of synchronous rotation.

The Moon's *far side*, facing away from Earth, is often called the "dark side of the Moon." In fact, no side of the Moon is always dark. At any given time, half of the Moon is in sunlight and half is in darkness—just as for Earth. The side of the Moon that faces away from Earth, the "far side," spends just as much time in sunlight as the side of the Moon that faces Earth.

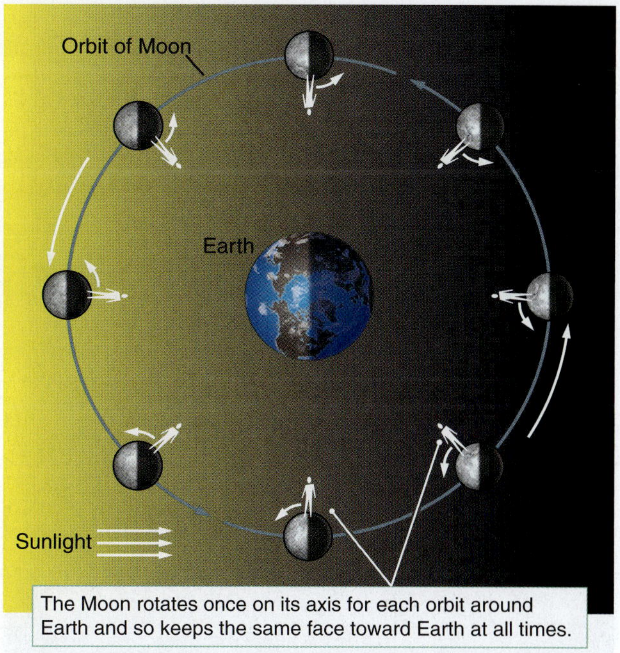

Figure 2.22 The Moon rotates once on its axis for each orbit around Earth—an effect called synchronous rotation. Here, the Sun is far to the left of the Earth-Moon system.

CHECK YOUR UNDERSTANDING 2.3

If you see the Moon rising just as the Sun is setting, what phase is the Moon in? What is the phase if the Moon is setting at noon?

2.4 Calendars Are Based on the Day, Month, and Year

Archaeologists tell us that developing agriculture was crucial for the rise of human civilization, and keeping track of the seasons and best times of the year to plant and harvest was critical to successful farming. Records going back to the dawn of humanity suggest that people kept track of time by following the patterns in the sky, especially those of the Sun, Moon, and stars. Some anthropologists have speculated that notches on fragments of bone found in southern France represent a 33,000-year-old lunar calendar. In this section, we examine some different calendars.

Figure 2.23 The ancient Egyptian calendar used a system of 12 months, plus festival days.

Lunar and Lunisolar Calendars

As civilizations developed around the globe, different cultures tried to solve the "problem" of the calendar. The rates of rotation of Earth and the revolutions of the Moon around Earth and of Earth around the Sun are not multiples of one another. A lunar cycle (from full Moon to full Moon) is 29.5 days, and a solar cycle—the time the Sun takes to appear to move from and return to its highest possible point in the sky at noon on the summer solstice—is 365.24 days. One solar cycle, then, has 12.38 lunar cycles. Those fractions of days and months are what make calendars complicated.

Some of the oldest known calendars come from the Egyptians, the Babylonians, and the Chinese. The ancient Egyptians used a system of 12 months of 30 days each—which added up to 360 days—and then added five "festival days" to the end of the year (**Figure 2.23**). Without leap years, the seasons in that 365-day year drifted, so an extra month was added when necessary. When we consider how some Western countries celebrate the days between the modern December holidays and the new year, an end-of-year calendar break for festivals seems like a good solution!

The Babylonians started the 24-hour day and 7-day week—7 for the Sun, the Moon, and the 5 planets visible with the naked eye. They created the first **lunisolar calendar**, in which a month began with the first sighting of the lunar crescent, and a 13th month was added when needed to catch up to the solar year. As the Babylonians developed mathematics, they discovered that 235 lunar months equals 19 solar years (and 6,940 days). Then they created a calendar cycle that consisted of 19 years, 12 of which have 12 months and 7 of which have 13 months, and then the cycle repeats. The ancient Hebrew calendar adopted that lunisolar cycle, and the Jewish calendar still uses it today. That type of calendar keeps holidays in the same season from year to year, even though the dates change.

The ancient Chinese calendar, dating back several thousand years, also is lunisolar. By about 500 BCE, the Chinese were using a year of 365.25 days and a system similar to that of the Babylonians of adding a 13th month to some years. To mark out a year, a few other cultures used stellar calendars, which followed the position of a bright star such as Sirius or certain prominent groups of stars in their sky, such as the Pleiades or the Big Dipper.

The Islamic calendar is purely lunar, with no added 13th lunar month. Their 12 months of 29 or 30 days each add up to 354 days—11.24 days short of the solar year. For that reason, the Islamic New Year and all other holidays drift earlier in each solar year. In the Islamic calendar, a holiday may fall in the winter in some years, and then a few years later it will have moved back to autumn.

The Modern Civil Calendar

The international civil calendar used today is a solar calendar known as the **Gregorian calendar**. It is based on the **tropical year**, which measures the 365.242 solar days between vernal (spring) equinoxes. A **solar day** is the 24-hour period of Earth's rotation that brings the Sun back to the same local meridian. That measure is in contrast to the **sidereal day**, the time Earth takes to make one rotation and face the same star on the meridian. The sidereal day is about 23 hours 56 minutes and is slightly shorter than the solar day because of Earth's motion around the Sun.

The Gregorian calendar includes a system of **leap years**—years in which a 29th day is added to February—decreed by Julius Caesar in 45 BCE to make up for the extra fraction of a day. Leap years prevent the seasons from slowly sliding through the year to become increasingly out of sync with the months, so that we don't end up experiencing winter in December one year and in August in later years.

The Gregorian calendar is named for Pope Gregory XIII. He was concerned that the Easter holiday—the first Sunday after the first full Moon after March 21—was drifting away from the vernal equinox. Julius Caesar's rule of one leap year every 4 years resulted in an average year of 365.25 days, but the actual year is 365.242 days. That difference of 0.008 day is about 11.5 minutes per year, or about 3 days every 400 years, and by Gregory's time it had caused the date of the vernal equinox to drift in the Julian calendar by about 10 days. In 1582, therefore, Pope Gregory decreed that 10 days would be deleted from the calendar to move the vernal equinox back to March 21. To make that system work out better in the future, he declared that only century years divisible by 400 are leap years, thereby deleting 3 leap years (and the 3 days) every 400 years. Catholic countries followed that system immediately, but Protestant countries did not adopt it until the 1700s. Eastern Orthodox countries, including Russia, did not switch from the Julian to the Gregorian calendar until the 1900s. One slight further revision—making years divisible by 4,000 into common 365-day years—has been proposed so the modern Gregorian calendar will slip by about only 1 day in 20,000 years.

Despite international adoption of the Gregorian calendar, billions of people still celebrate holidays and festivals according to a lunisolar or lunar calendar. Chinese New Year, Passover, Easter, Ramadan, Rosh Hashanah, and Diwali, among others, have dates that change from one year to the next because they are based on lunar months from lunisolar or lunar calendars. The astronomy of people from long ago is still in use today.

CHECK YOUR UNDERSTANDING 2.4

Suppose that the astronomical cycles were even multiples of each other, so that a cycle of lunar phases was precisely 30 days and Earth's orbital period was precisely 12 months. In such a system, how would the dates of "wandering" holidays, such as Chinese New Year or Ramadan, change?

2.5 Eclipses Result from the Alignment of Earth, the Moon, and the Sun

Ancient peoples attuned to the patterns of the sky must have been terrified to watch the Sun being eaten away as if by a giant dragon or to see the full Moon turning the color of blood. An **eclipse** is the total or partial obscuration of one celestial body, or the light from that body, by another celestial body. Archaeological evidence suggests that ancient peoples tried hard to figure out the pattern of eclipses and thereby bring them into the orderly scheme of the heavens. Stonehenge, pictured in Figure 2.1a, may have enabled its builders to predict when eclipses might occur. Ancient Chinese, Babylonian, and Greek astronomers had figured out that eclipses occur in cycles, and they used that knowledge to predict when and where eclipses would occur. In this section, we describe the types of eclipses and their frequency.

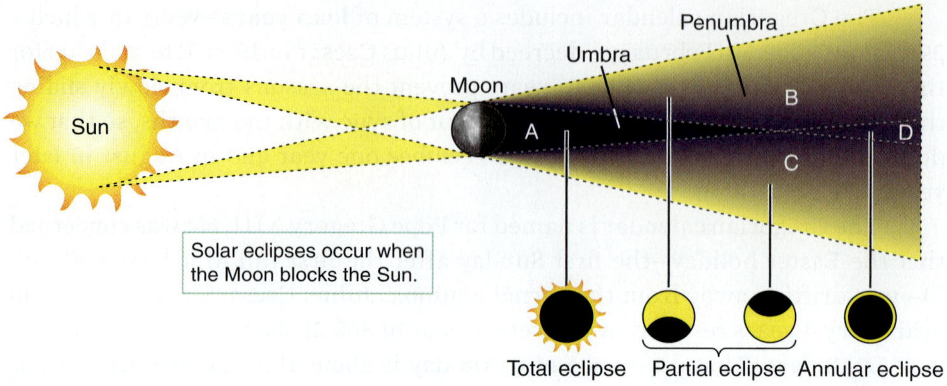

Figure 2.24 Different parts of the Sun are blocked at different places within the Moon's shadow. An observer on Earth in the umbra (point A) sees a total solar eclipse, observers in the penumbra (points B and C) see a partially eclipsed Sun, and observers at point D see an annular solar eclipse. Not to scale.

Solar Eclipses

A **solar eclipse** occurs when the Moon passes between Earth and the Sun; observers on Earth in the shadow of the Moon will see the eclipse. Three types of solar eclipses exist: *total*, *partial*, and *annular*. Consider the structure of the shadow of the Sun cast by a round object such as the Moon, as shown in **Figure 2.24**. An observer at point A would see the Sun blocked. That darkest, inner part of the shadow is called the **umbra**. If a location on Earth passes through the Moon's umbra, the Moon blocks all the Sun's light, and a **total solar eclipse** will be observed (Chapter Opening Figure and **Figures 2.25** and **2.26a**). At points B and C in Figure 2.24, an observer can see one side of the disk of the Sun but not the other. Only partially in shadow, that outer region is the **penumbra**. If a location on Earth passes through the Moon's penumbra, viewers there will observe a **partial solar eclipse**, in which the Moon's disk blocks the light from a portion of the Sun's disk.

In the third type of solar eclipse, called an **annular solar eclipse**, the Sun appears as a bright ring surrounding the dark disk of the Moon (**Figure 2.26b**). An observer at point D in Figure 2.24 is far enough from the Moon that it appears smaller than the Sun. An object's apparent size in the sky depends on the object's actual size and its distance from us. The Sun is about 400 times the diameter of the Moon, and the distance between the Sun and Earth is about 400 times more than the distance between the Moon and Earth. As a result, the Moon and Sun have almost the same apparent size in the sky. In addition, the Moon's orbit is not a perfect circle. When the Moon and Earth are a bit closer together than average, the Moon appears slightly larger in the sky than the Sun. An eclipse occurring then will be total for some observers. When the Moon and Earth are farther apart than average, the Moon appears smaller than the Sun, so eclipses occurring then will be annular for some observers. Among all solar eclipses, one-third are total at some location on Earth, one-third are annular, and one-third are only partial.

Figure 2.25 The full spectacle of a total solar eclipse.

Figure 2.27 shows the geometry of a solar eclipse when the Moon's shadow falls on Earth's surface. Figures like that usually show Earth and the Moon much closer together than they really are. The page is too small to draw them to scale and still show the critical details. The relative sizes and distances between Earth

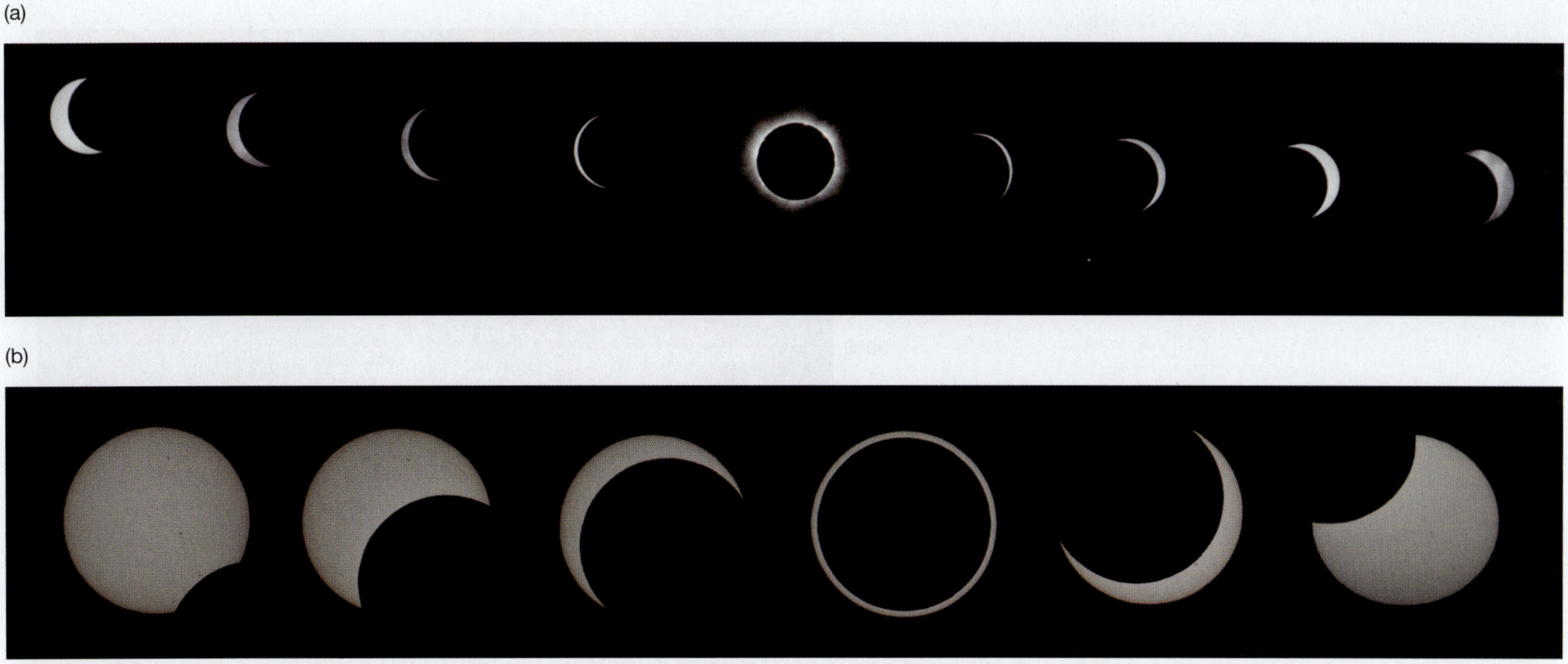

Figure 2.26 Time-lapse images of the Sun taken (a) during a total solar eclipse and (b) during the annular solar eclipse of May 20, 2012. The Sun set during the ending phases.

and the Moon are roughly equivalent to the difference between a basketball and a tennis ball, respectively, placed 7 meters apart. Figure 2.27b shows the geometry of a solar eclipse with Earth, the Moon, and the separation between them drawn to scale. Compare that drawing with Figure 2.27a to see why drawings of Earth and the Moon are rarely drawn to scale. If the Sun were drawn to scale in Figure 2.27a, it would be bigger than your head and located almost 64 meters off the left side of the page.

From any particular location, you are more likely to observe a partial solar eclipse than a total solar eclipse. Where the Moon's penumbra touches Earth, it has a diameter of almost 7,000 km—large enough to cover a substantial fraction of Earth. Thus, a partial solar eclipse is often visible from many locations on Earth.

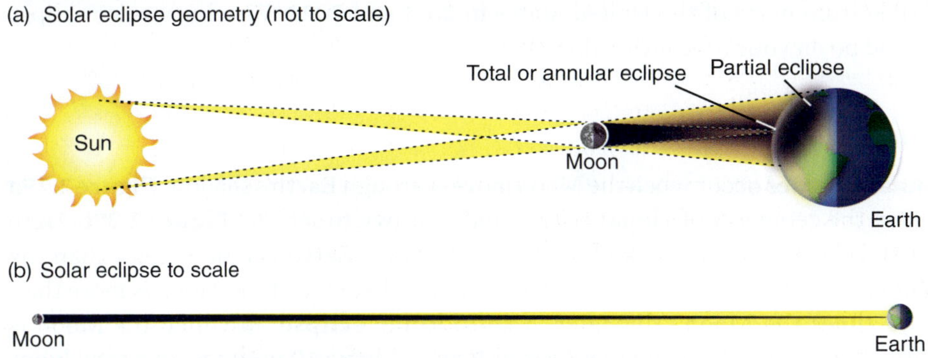

Figure 2.27 (a, b) A solar eclipse occurs when the shadow of the Moon falls on the surface of Earth. Note that (b) is drawn to scale.

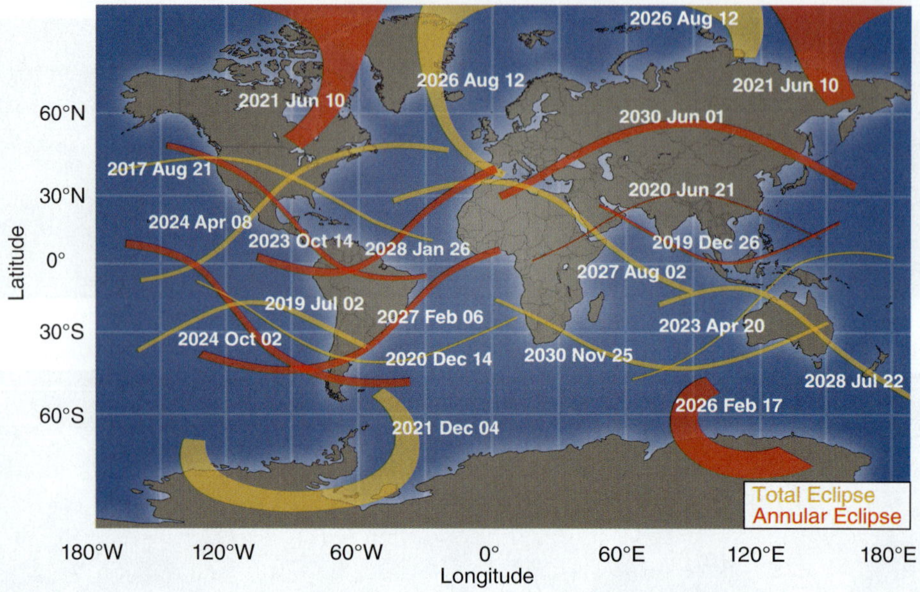

Figure 2.28 The paths of total solar eclipses through 2030. Solar eclipses occurring in Earth's polar regions cover more territory because the Moon's shadow hits the ground obliquely.

In contrast, the path along which a total solar eclipse is visible, shown in **Figure 2.28**, covers only a tiny fraction of Earth's surface. Even when the distance between Earth and the Moon is at a minimum, the umbra is only 270 km wide on Earth. As the Moon moves along in its orbit, that tiny shadow sweeps across Earth at a few thousand kilometers per hour. In addition, the Moon's shadow falls on Earth's curved surface, causing the region shaded by the Moon during a solar eclipse to be elongated by various amounts. The curvature can even cause an eclipse that started out as annular to become total.

As a result of those factors, a total solar eclipse can never last longer than 7 1/2 minutes and is usually significantly shorter. Even so, it is one of the most amazing sights in nature. People all over the world flock to the most remote corners of Earth to witness the fleeting spectacle of the bright disk of the Sun blotted out of the daytime sky. Perhaps you saw some of the total eclipse visible from much of the United States in August 2017. The next total solar eclipse visible in the continental United States will take place in 2024. Annular eclipses will be visible from parts of the United States in 2021 and 2023. Viewing a solar eclipse should be on your lifetime to-do list!

Lunar Eclipses

Lunar eclipses occur when the Moon moves through Earth's shadow. **Figure 2.29a** shows the geometry of a lunar eclipse and is drawn to scale in **Figure 2.29b**. Here Earth is between the Sun and the Moon. Because Earth is much larger than the Moon, the dark umbra of Earth's shadow at the distance of the Moon is more than 2 1/2 times the Moon's diameter. A **total lunar eclipse**, in which the Moon is entirely within Earth's shadow, lasts as long as 1 hour 40 minutes. In a total lunar eclipse, the Moon often appears red (**Figure 2.30a**). That "blood-red Moon," as literature and poetry have called it, occurs because red light from the Sun is bent

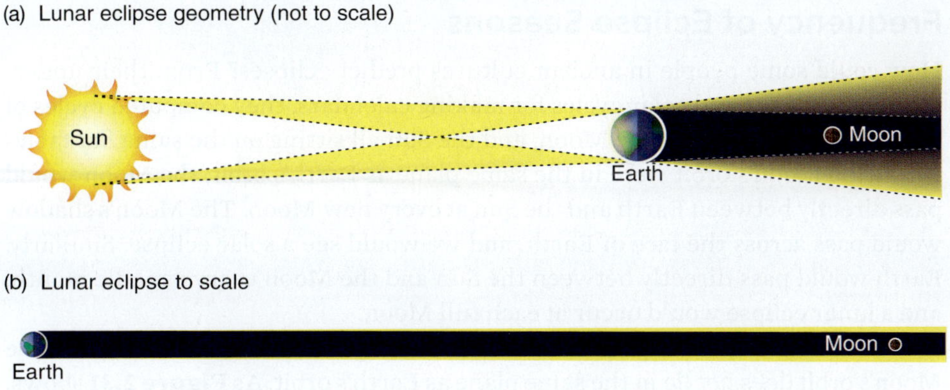

Figure 2.29 (a, b) A lunar eclipse occurs when the Moon passes through Earth's shadow. Note that (b) is drawn to scale.

as it travels through Earth's atmosphere and then illuminates the Moon. Earth's atmosphere absorbs or scatters other colors of sunlight away from the Moon and therefore does not illuminate it.

A **penumbral lunar eclipse** occurs when the Moon passes through the penumbra of Earth's shadow; those are noticeable only from a very dark location or when the Moon passes within about 1,000 km of the umbra. If Earth's shadow incompletely covers the Moon, some of the Moon's disk remains bright and some of it is in shadow. The result is a **partial lunar eclipse**. **Figure 2.30b** shows a composite of images taken at different times during a partial lunar eclipse. In the middle frame, Earth's shadow nearly completely eclipses the Moon.

Many more people have observed a total lunar eclipse than have observed a total solar eclipse. To see a total solar eclipse, you must be located within that very narrow band of the Moon's shadow as it moves across Earth's surface. In contrast, when the Moon is immersed in Earth's shadow, anyone located in the hemisphere facing the Moon can see it. As a result, from any location, total eclipses of the Moon are relatively common, and you may have seen at least one.

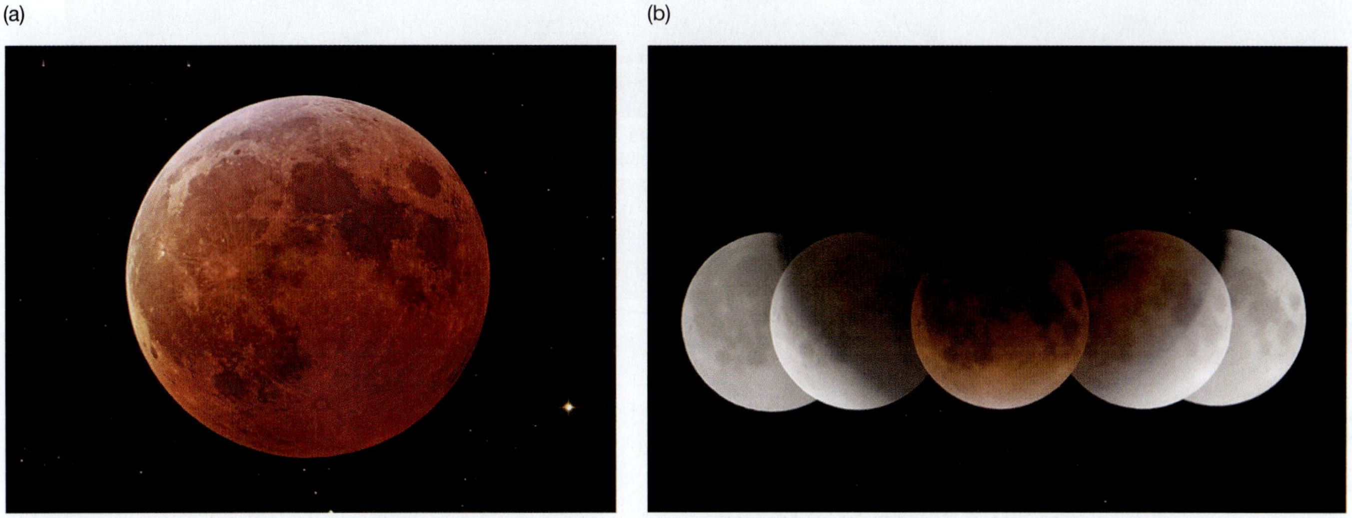

Figure 2.30 (a) During a total lunar eclipse, the Moon often appears blood red. (b) A time-lapse series of photographs of a partial lunar eclipse clearly shows Earth's shadow. Note the size of Earth's shadow in comparison with the size of the Moon.

Frequency of Eclipse Seasons

How could some people in ancient cultures predict eclipses? From their understanding of lunar and solar cycles for making calendars, they computed cycles of eclipses. Imagine Earth, the Moon, and the Sun all sitting on the same flat tabletop. If the Moon's orbit were in the same plane as Earth's orbit, the Moon would pass directly between Earth and the Sun at every new Moon. The Moon's shadow would pass across the face of Earth, and we would see a solar eclipse. Similarly, Earth would pass directly between the Sun and the Moon every synodic month, and a lunar eclipse would occur at each full Moon.

Solar and lunar eclipses do *not* happen every month, however, because the Moon's orbit does *not* lie in the same plane as Earth's orbit. As **Figure 2.31** shows, the plane of the Moon's orbit around Earth is inclined by about 5.2° with respect to the plane of Earth's orbit around the Sun. The line along which the orbital planes of the Sun and the Moon intersect is called the **line of nodes**. For part of the year, the line of nodes points generally toward the Sun. During those times, called **eclipse seasons**, a new Moon passes directly between the Sun and Earth, casting its shadow on Earth's surface and causing a solar eclipse. Similarly, a full Moon occurring during an eclipse season passes through Earth's shadow, causing a lunar eclipse. An eclipse season lasts only 38 days. That's how long the Sun is close enough to the line of nodes for eclipses to occur. Usually, the line of nodes

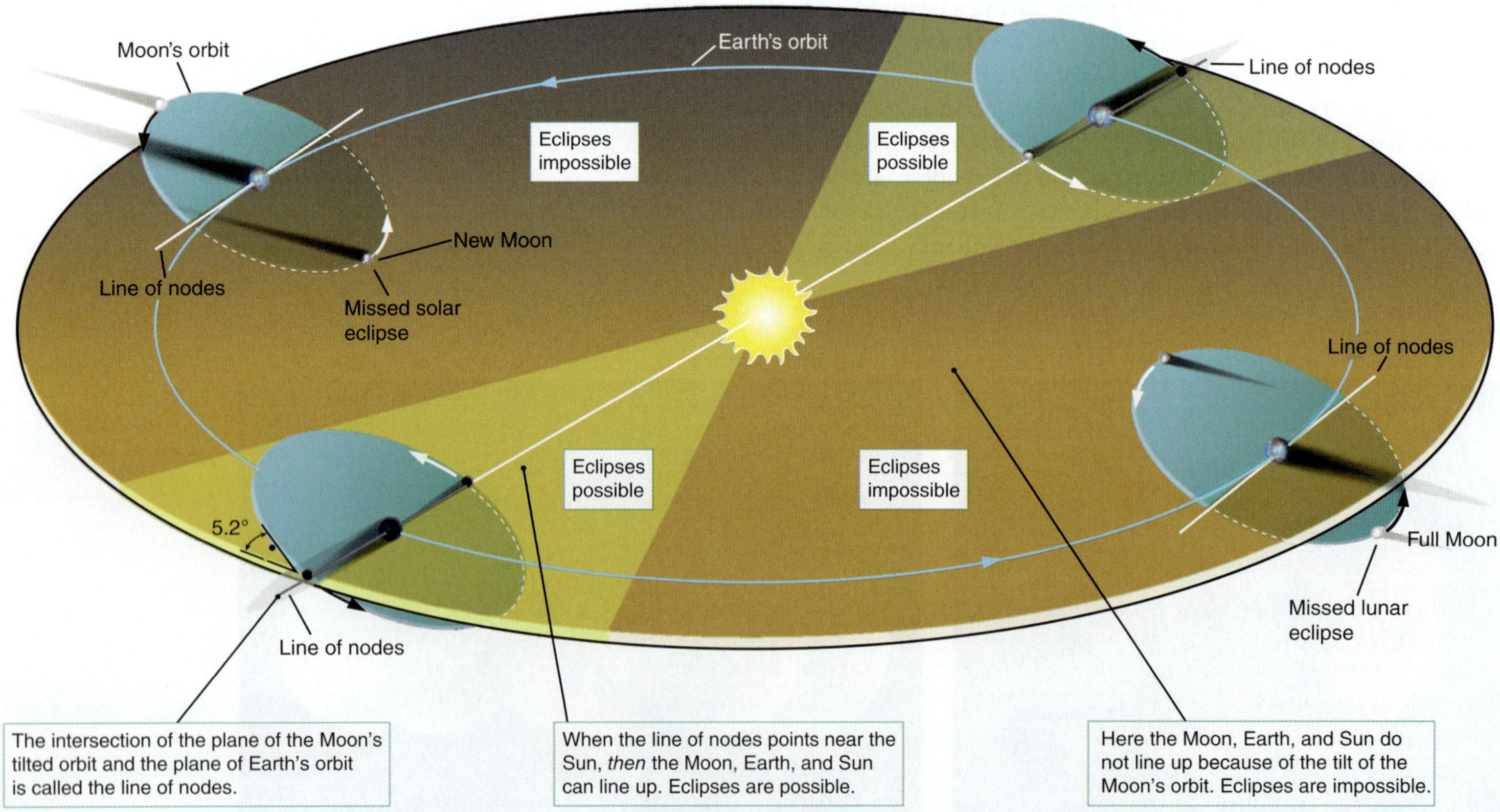

Figure 2.31 Eclipses are possible only when the Sun, Moon, and Earth lie along (or very close to) an imaginary line known as the line of nodes. When the Sun does not lie along the line of nodes, Earth passes under or over the shadow of a new Moon, and a full Moon passes under or over the shadow of Earth.

points farther away from the Sun, and Earth, Moon, and Sun cannot line up closely enough for an eclipse to occur. At those times, a solar eclipse cannot take place because the shadow of a new Moon passes "above" or "below" Earth. Similarly, no lunar eclipse can occur because a full Moon passes "above" or "below" Earth's shadow.

If the plane of the Moon's orbit always had the same orientation, eclipse seasons would occur twice a year, as suggested in Figure 2.31. In actuality, eclipse seasons occur about every 5 months 20 days. The roughly 10-day difference occurs because the plane of the Moon's orbit slowly wobbles, much like the wobble of a spinning plate balanced on the end of a circus performer's stick. As the Moon's orbital plane wobbles, the line of nodes changes direction. That wobble rotates in the direction opposite the direction of the Moon's motion in its orbit. That is, the line of nodes moves clockwise as viewed from above Earth's orbital plane. One wobble of the Moon's orbit takes 18.6 years, so we say that the line of nodes regresses by 360° every 18.6 years, or 19.4° per year. That rate amounts to about a 20-day regression each year. If January 1 marks the middle of an eclipse season, the next eclipse season will be centered around June 20, and the one after that around December 10.

CHECK YOUR UNDERSTANDING 2.5

If the Moon were in its same orbital plane but twice as far from Earth, which of the following would happen? (a) The Moon would not go through phases. (b) Total eclipses of the Sun would not be possible. (c) Annular eclipses of the Sun would not be possible. (d) Total eclipses of the Moon would not be possible.

Origins The Obliquity of Earth

Earth's various motions give rise to the most basic of patterns faced by life on Earth. Earth's rotation is responsible for the cycle of night and day. Earth's axial tilt and its passage around the Sun bring the change of the seasons. As life evolved on Earth, it had to adapt to those patterns.

Earth's range of climate, based on distance from the equator, probably has contributed to the broad diversity of life on our planet. Earth's biodiversity includes life that adapted to the long, cold polar nights and to the much higher temperatures of the tropics. Earth's life adapted to seasonal patterns in rain and drought, leading to acquired seasonal patterns of migration and reproduction. If Earth had no axial tilt, the poles would continuously be in winter and probably too cold for humans. Latitudes near the equator would be consistently even warmer than they are now.

Periodic changes in Earth's axial tilt might have affected life. If Earth's tilt were larger than 23.5°, the seasonal variation would be even stronger. If the tilt were smaller, the seasonal variation would be weaker. Chinese, Indian, Greek, and Arabic records going back 3,000 years indicate that people estimated the tilt by measuring the length of the shadow from a vertical pole on the day of the solstice. Earth's axial tilt actually varies from 22.1° to 24.5° over a 41,000-year cycle. Currently, the tilt is about midway between the two extremes and getting smaller. It will reach its minimum value of 22.1° in about 10,000 years. Scientists are studying whether that variation in tilt correlates with periods of temperature change on Earth, especially the times of ice ages.

READING ASTRONOMY NEWS

ARTICLES QUESTIONS

Citizen scientists across the nation captured images of the Sun's corona to create the largest dataset of its kind.

2,500 Miles of Citizen Scientists

LAURA REED, sciencenode.org

On August 21 many of us stood outside and marveled at the solar eclipse.

But for some, the eclipse was more than just a show. It was an opportunity to participate in the National Solar Observatory (NSO) Citizen Continental-America Telescopic Eclipse (CATE) Experiment.

The Citizen CATE Experiment is funded by federal, corporate, and private sources, including several companies that donated equipment to the project. Volunteer observers and schools get to keep the equipment used in the study.

A grand experiment

Scientists, students, and volunteers assembled along the 2,500-mile path of eclipse totality tracked the Sun using 68 identical telescopes.

Using specialized software and instrument packages, they produced more than 1,000 images from the start of the partial solar eclipse until totality.

The goal was to capture 90 minutes of continuous, high-resolution, rapid-cadence images detailing a difficult-to-capture region of the solar atmosphere: the Sun's inner corona.

This is the first time scientists have collected research-quality observations of the corona during the eclipse's entire transit across the U.S.

"This dataset is extraordinary," says Matt Penn, principal investigator for the project. "Normally during a solar eclipse, we get about 2 minutes of data in the region closest to the photosphere. But Citizen CATE allows us to get an hour and a half of data."

My corona

The corona is the Sun's outer atmosphere. It is difficult for scientists to study because the photosphere, or solar surface, is so bright that it overpowers the faint corona. We can only see the corona when something obscures the photosphere.

(Think of a light so bright that it makes it difficult to see an object close to you. Blocking the light with your hand allows a better view of the object.)

Scientists can create artificial eclipses using an instrument called a coronagraph that covers the Sun's bright disk. The proximity of the instrument to the observer distorts the Sun's edge, making precise observation and measurement difficult.

During a real eclipse, the Moon blocks the Sun. The Moon's great distance lets scientists measure and study the corona in greater detail.

The process

On the day of the event, skies were clear for 58 of the 68 observation sites.

Observers, spaced about 50 miles from each other, started observations when the Moon's shadow appeared on their horizon. They captured images every 10 seconds during totality.

The rapid cadence of imagery along with a 2-arcsecond pixel resolution should help scientists understand the intensity of the corona over an extended period as well as the motions of prominences, coronal inflows, coronal mass ejections, and other active regions.

The next total solar eclipse in the United States will be April 8, 2024. The path of totality will begin in Texas, moving north from Mexico, and will exit the U.S. via Maine.

Start making plans to join the next team of citizen scientists.

ARTICLES QUESTIONS

1. Why does a solar eclipse move across the surface of the Earth?
2. Why did scientists want data from along the whole track of the eclipse?
3. Why can the corona be seen only during an eclipse?
4. Would you have expected a lunar eclipse near the date of the solar eclipse? Explain.
5. See whether any results are available on the project website (http://citizencate.org/). What did participants learn from this experiment?

Summary

The motions of Earth and the Moon are responsible for many of the repeating patterns that can be observed in the sky. Calendars keep track of time by using those patterns. Earth's rotation about its axis causes daily patterns of rising and setting, and Earth's revolution around the Sun causes yearly patterns of the stars in the sky and marks the passage of the seasons. Earth's tilt on its axis changes both the length of daytime and the intensity of sunlight, causing the seasons. Life on Earth adapted to those seasonal variations. Earth's axial tilt varies slightly over tens of thousands of years. The Moon's revolution around Earth causes the monthlong pattern of the phases of the Moon. Occasionally, alignments of Earth, the Moon, and the Sun cause eclipses.

LG 1 **Describe how Earth's rotation about its axis and its revolution around the Sun affect how we perceive celestial motions from different places on Earth.** Earth's daily rotation on its axis causes the apparent daily motion of the Sun, Moon, and stars. Our location on Earth and Earth's location in its orbit around the Sun determine which stars we see at night. You can determine your latitude from the altitude of the pole star. When observing objects in our sky, we need to consider how Earth moves in relation to other objects. The ecliptic is the path that the Sun appears to take through the stars.

LG 2 **Explain why seasons change throughout the year.** A year is the time Earth takes to complete one revolution around the Sun. Constellations are patterns of stars that reappear in the same place in the sky at the same time of night each year. The tilt of Earth's axis determines the seasons by changing the angle at which sunlight strikes Earth's surface in different locations. The changing angle of sunlight and the differing length of the day cause the seasonal variations on Earth. Equinoxes and solstices mark the changing seasons.

LG 3 **Describe the factors that create the phases of the Moon.** The relative locations of the Sun, Earth, and Moon determine the phases of the Moon. The Moon takes one sidereal month to complete one revolution around Earth and one synodic month to go through a cycle of phases. The Moon's motion around Earth causes it to be illuminated differently at different times. When farther from the Sun than Earth is, the Moon is in gibbous phases. When closer to the Sun than Earth is, the Moon is in crescent phases.

LG 4 **Sketch the alignment of Earth, the Moon, and the Sun during eclipses of the Sun and the Moon.** A solar eclipse occurs when the new Moon is in the plane of Earth and the Sun, and the shadow of the Moon falls on Earth. A lunar eclipse occurs when the full Moon is in the plane of Earth and the Sun, and the shadow of Earth falls on the Moon. Twice a year, at new or at full Moon, the Moon is exactly in line between Earth and the Sun. At those times, eclipses occur.

? Unanswered Questions

- How long will Earth continue to have total solar eclipses? Those occur because the Moon and the Sun are coincidentally the same size in our sky, but will that always be the case? An object's observed size in the sky depends on its actual diameter and its distance from us. One or both of those can change. The Moon is slowly moving away from Earth by about 4 meters per century. Over time, the Moon will appear smaller in the sky, and it won't be able to cover the full disk of the Sun. Although we can measure how fast the Moon is moving away from Earth now, we are less certain of how that rate may change with time. A lesser and more uncertain effect comes from the Sun—which will continue to brighten slowly, as it has throughout its history. With that brightening, the Sun's actual diameter will slightly increase, and it will appear larger in our sky. A more distant Moon and a larger Sun will eventually (in hundreds of millions of years) end total eclipses on Earth.

Questions and Problems

Test Your Understanding

1. Constellations are groups of stars that
   a. are close to one another in space.
   b. are bound to each other by gravity.
   c. are relatively close to one another in Earth's sky.
   d. all have the same composition.
2. Where on Earth can you stand and, over the course of a year, see the entire sky?
   a. only at the North Pole
   b. at either pole
   c. at the equator
   d. anywhere
3. Day and night are caused by
   a. the tilt of Earth on its axis.
   b. the rotation of Earth on its axis.
   c. the revolution of Earth around the Sun.
   d. the revolution of the Sun around Earth.
4. Polaris, the North Star, is unique because
   a. it is the brightest star in the night sky.
   b. it is the only star in the sky that doesn't move throughout the night.
   c. it is always located at the zenith, for any observer.
   d. it has a longer path above the horizon than any other star.
5. An angle exists between the ecliptic and the celestial equator because
   a. Earth's axis is tilted with respect to its orbit.
   b. Earth's orbit is tilted with respect to the orbits of other planets.
   c. the Sun follows a rising and falling path through space.
   d. the Sun's orbit is tilted with respect to Earth's.
6. Earth's axial tilt causes the seasons because
   a. one hemisphere of Earth is closer to the Sun in summer.
   b. the days are longer in summer.
   c. the rays of light strike the ground more directly in summer.
   d. both a and b
   e. both b and c
7. Which is *not* true on the vernal and autumnal equinoxes?
   a. Every nonpolar place on Earth has 12 hours of daylight and 12 hours of darkness.
   b. The Sun rises due east and sets due west.
   c. The Sun is located on the celestial equator.
   d. The motion of the stars in the sky is different from that on other days.
8. We always see the same side of the Moon because
   a. the Moon does not rotate on its axis.
   b. the Moon rotates on its axis once for each revolution around Earth.
   c. the other side of the Moon is unlit when it is facing Earth.
   d. the other side of the Moon is on the opposite side of Earth when it is facing Earth.
9. If you see the Moon on the meridian at sunrise, the phase of the Moon is
   a. waxing gibbous.
   b. full.
   c. first quarter.
   d. third quarter.
10. A lunar eclipse occurs when ______________ shadow falls on ______________.
    a. Earth's; the Moon
    b. the Moon's; Earth
    c. the Sun's; the Moon
    d. the Sun's; Earth
11. Different cultures created different calendars because
    a. they had measured different lengths of the day, month, and year.
    b. they used different definitions of the day, month, and year.
    c. the number of days in a month and the number of days and months in a year are not integers.
    d. calendars are completely arbitrary.
12. Which stars we see at night depends on
    a. our location on Earth.
    b. Earth's location in its orbit.
    c. the time of the observation.
    d. all of the above.
13. On the summer solstice in June, the Sun will be directly above ______________ and all locations north of ______________ will experience daylight all day.
    a. the Tropic of Cancer; the Antarctic Circle
    b. the Tropic of Capricorn; the Arctic Circle
    c. the Tropic of Cancer; the Arctic Circle
    d. the Tropic of Capricorn; the Antarctic Circle
14. The Sun, Moon, and stars
    a. appear to move each day because the celestial sphere rotates about Earth.
    b. change their relative positions over time.
    c. rise north or south of west and set north or south of east, depending on their location on the celestial sphere.
    d. always remain in the same position with respect to one another.

15. If you see the first quarter Moon on the meridian, the Sun must be
    a. on the western horizon.
    b. on the eastern horizon.
    c. below the horizon.
    d. on the meridian.

Thinking about the Concepts

16. Seafarers such as Columbus used Polaris to navigate as they sailed from Europe to North America. When Magellan sailed the South Seas, he could not use Polaris. Explain why.
17. If you were standing at Earth's North Pole, where would you see the north celestial pole in relation to your zenith?
18. Observers in the Northern Hemisphere see the zodiacal constellation Gemini in the winter. Why don't they see it in the summer?
19. Imagine that you are flying along in a jetliner.
    a. Describe ways to tell that you are moving.
    b. If you look down at a building, which way is it moving relative to you?
20. Astronomers are sometimes asked to serve as expert witnesses in court cases. Suppose you are called as an expert witness, and the defendant states that he could not see the pedestrian because the full Moon was casting long shadows across the street at midnight. Is that claim credible? Why or why not?
21. Imagine that one person was developing a theory of seasons as described in the three hypotheses in this chapter's Process of Science Figure. Compare that process with the flowchart of Chapter 1's Process of Science Figure. How would the development of that theory look on the diagram?
22. Why is the winter solstice *not* the coldest time of year?
23. Earth spins on its axis and wobbles like a top.
    a. How long does it take to complete one spin?
    b. How long does it take to complete one wobble?
24. At what approximate time of day do you see the full Moon near the meridian? At what time is the first quarter (waxing) Moon on the eastern horizon? Use a sketch to help explain your answers.
25. Assume that the Moon's orbit is circular, and imagine that you are standing on the side of the Moon that faces Earth.
    a. How would Earth appear to move in the sky as the Moon made one revolution around Earth?
    b. How would the "phases of Earth" appear to you compared with the phases of the Moon as seen from Earth?
26. If people on Earth were observing a lunar eclipse, what would you see from the Moon?
27. From any given location, why are you more likely to witness a partial eclipse of the Sun than a total eclipse?
28. Why don't we see a lunar eclipse each time the Moon is full or witness a solar eclipse each time the Moon is new?
29. How would Earth's temperature variation be different if it were tilted 90° on its axis (like the planet Uranus)?
30. Explain how a cyclic change in Earth's tilt could affect the planet's seasonal temperatures.

Applying the Concepts

31. Earth spins along at 1,674 km/h at the equator. Use that fact, along with the length of the day, to calculate Earth's equatorial diameter.
32. Determine the latitude where you live. Draw and label a diagram showing that your latitude is the same as (a) the altitude of the north celestial pole and (b) the angle (along the meridian) between the celestial equator and your local zenith. What is the Sun's altitude at noon as seen from your home at the winter solstice and at the summer solstice?
33. If the angle between your zenith and Polaris is 40°, what is the altitude of Polaris? What is your latitude?
34. The southernmost star in a group of stars known as the Southern Cross lies approximately 65° south of the celestial equator. What is the farthest-north latitude from which the entire Southern Cross is visible? Can it be seen in any U.S. states? If so, which ones?
35. Imagine that you are standing on the South Pole at the time of the southern summer solstice.
    a. How far above the horizon will the Sun be at noon?
    b. How far above (or below) the horizon will the Sun be at midnight?
36. Suppose the tilt of Earth's equator with respect to its orbit were 10° instead of 23.5°. At what latitudes would the Arctic and Antarctic circles and the Tropics of Cancer and Capricorn be located?
37. The Moon's orbit is tilted by about 5° with respect to Earth's orbit around the Sun. What is the highest altitude in the sky that the Moon can reach, as seen in Philadelphia (latitude 40° north)?
38. Suppose you would like to witness the midnight Sun (when the Sun appears just above the northern horizon at midnight), but you don't want to travel any farther north than necessary.
    a. How far north (that is, to which latitude) would you have to go?
    b. At what time of year would you make that trip?

39. If, as some historians believe, Eratosthenes used the Egyptian stadion, about 157.5 meters, what would he have computed for the size of Earth? How close is that result to the modern value?

40. a. The vernal equinox is now in the zodiacal constellation Pisces. The precession of Earth's axis will eventually cause the vernal equinox to move into Aquarius. How long, on average, does the vernal equinox spend in each zodiacal constellation?
    b. Stonehenge was erected roughly 4,000 years ago. Referring to the zodiacal constellations shown in Figure 2.13, identify the constellation in which those ancient builders saw the vernal equinox.

41. Referring to Figure 2.18, estimate when Vega, the fifth-brightest star in our sky (excluding the Sun), will once again be the northern pole star.

42. The Moon's apparent diameter in the sky is approximately ½°. How long does the Moon take to move 360°? About how long does the Moon take to move a distance equal to its own diameter across the sky?

43. An object's apparent size in the sky is proportional to its actual diameter divided by its distance. The Moon has a radius of 1,737 km, with an average distance of 3.780×10^5 km from Earth. The Sun has a radius of 696,000 km, with an average distance of 1.496×10^8 km from Earth. Show that the apparent sizes of the Moon and Sun in our sky are approximately the same.

44. Earth has an average radius of 6,371 km. If you were standing on the Moon, how much larger would Earth appear in the lunar sky than the Moon appears in our sky?

45. How would the length of the eclipse season change if the plane of the Moon's orbit were inclined less than its current 5.2° to the plane of Earth's orbit? Explain your answer.

USING THE WEB

46. Go to the U.S. Naval Observatory website (USNO "Data Services," at http://aa.usno.navy.mil/data). Look up the times for sunrise and sunset for your location for the current week. (You can change the dates one at a time or bring up a table for the entire month.) How are the times changing from one day to the next? Bring up the "Duration of Daylight/Darkness Table for One Year" page for your location. Are the days getting longer or shorter? When do the shortest and the longest days occur? Look up a location in the opposite hemisphere (Northern or Southern). When are the days shortest and longest?

47. Go to the "Earth and Moon Viewer" website (http://fourmilab.ch/earthview). Under "Viewing the Earth," click on "latitude, longitude and altitude" and enter your approximate latitude and longitude, and 40,000 for altitude; then select "View Earth." Are you in daytime or nighttime? Now play with the locations; keep the same latitude but change to the opposite hemisphere (Northern or Southern). Is it still night or day? Go back to your latitude, and this time enter 180° minus your longitude, and change from west to east, or from east to west, so that you are looking at the opposite side of Earth. Is it night or day there? What do you see at the North Pole (latitude 90° north) and the South Pole (latitude 90° south)? At the bottom of your screen you can play with the time. Move back 12 hours. What do you observe at your location and at the poles?

48. Go to the U.S. Naval Observatory website (USNO "Data Services," at http://aa.usno.navy.mil/data). Look up the Moon data for today. When will it rise and set? What is the phase? How will it change over the next 4 weeks? Enter one day at a time or look at the yearlong tables for moonrise and moonset and for the dates of primary phases. What time of day does a third quarter Moon rise? When (and in what phases) can you see the Moon in the daytime?

49. Using the times of moonrise and moonset that you found in question 48, make a plan to observe the Moon directly for a few weeks. Take a picture of the Moon (or make a sketch) every other day. How is the brightness of the Moon changing? When it's daytime, estimate how far the Moon is from the Sun in the sky. When it's nighttime, use your pictures to see whether the stars near the Moon in the sky are the same every night.

50. Go to the "NASA Eclipse" website (https://eclipse.gsfc.nasa.gov/eclipse.html). When is the next lunar eclipse? Will it be visible at your location if the skies are clear? Is it a total or partial eclipse? When is the next solar eclipse? Will it be visible at your location? Compare the fraction of Earth that the solar eclipse will affect with the fraction for the lunar eclipse. Why are lunar eclipses visible in so many more locations?

EXPLORATION

digital.wwnorton.com/astro6

Visit the Digital Resources Page and on the Student Site open the "Phases of the Moon" Interactive Simulation in Chapter 2.

Study the diagrams shown in the simulator. The largest window shows a view of the Earth-Moon system as seen from above Earth's North Pole. The Sun is far off the screen to the left. An observer stands on Earth. The small window at upper right shows how the Moon looks from the Northern Hemisphere. The small window at lower right shows the observer's location, with the Sun and Moon pictured as flat disks in the sky.

1 Given the relative positions of the observer and the Sun, approximately what time is it for this observer: 6:00 A.M., noon, 6:00 P.M., or midnight?

2 Where is the Moon in the observer's sky: on the eastern horizon, on the western horizon, below the horizon, or crossing the meridian?

3 What is the phase of the Moon?

4 Imagine yourself on the Moon in the image shown in the larger window. If you looked toward Earth, what phase of Earth would you see?

Now select "start animation." Allow the animation to run until the Moon is 20 percent illuminated (as shown in the upper small window on the right).

5 In which direction does the Moon orbit Earth: clockwise or counterclockwise?

6 For observers in the Northern Hemisphere, which side of the Moon is illuminated first after a new Moon: right or left?

7 If you observe a crescent Moon with the horns of the crescent pointing right, is the Moon waxing or waning?

Grab the Moon with your mouse and drag it to first quarter. Drag the observer so that her local time is approximately midnight.

8 Where is the first quarter Moon in the observer's sky: on the eastern horizon, on the western horizon, below the horizon, or crossing the meridian?

The time, the phase of the Moon, and the Moon's location in the sky are related.

9 Arrange the observer and the Moon so that the Moon is full and crossing the meridian. What time is it for the observer: 6:00 A.M., noon, 6:00 P.M., or midnight?

10 Arrange the observer and the Moon so that the Moon is in third quarter and the time for the observer is approximately noon (the Sun is on the meridian). Where is the Moon in the observer's sky: on the eastern horizon, on the western horizon, below the horizon, or crossing the meridian?

11 Arrange the observer and the Moon so that it is approximately 6:00 P.M. (sunset) for the observer, and the Moon is just rising on the eastern horizon. What is the phase of the Moon?

Many other combinations of the time, phase of the Moon, and Moon's location are available to play with. Challenge yourself to be able to set up any two of those and find the third. When you can do that without the simulator, just by making the picture in your head, you will really understand the phases of the Moon.

3 Motion of Astronomical Bodies

The birth of modern astronomy dates back to when astronomers and mathematicians discovered regular patterns in the motions of the planets. A successful theory of how Earth moved and how it fit in with its neighbors in the Solar System was the first step toward understanding our planet's place in the universe.

LEARNING GOALS

Here in Chapter 3, we examine how astronomers came to understand that Earth and other planets orbit the Sun. By the end of this chapter, you should be able to:

LG 1 **Compare the geocentric and heliocentric models of the Solar System.**

LG 2 **Use Kepler's laws to describe how objects in the Solar System move.**

LG 3 **Explain how Galileo's astronomical discoveries convinced him that the heliocentric model was correct.**

LG 4 **Describe the physical laws of motion discovered by Galileo and Newton.**

The shadows of a few of Jupiter's Galilean moons fall on the planet. ►►►

Who discovered Jupiter's moons?

3.1 The Motions of Planets in the Sky

When people in ancient times looked up at the sky, they saw the Sun, Moon, and stars rise in the east and set in the west and *appear* to move around Earth. The ancient peoples were aware of five planets (*planet* means "wandering star") because from one night to the next they moved in a generally eastward direction among the stars, whose positions were fixed on the celestial sphere. The ancients did *not* know, however, that Earth was similar to those planets. A successful theory of how Earth and the planets move and how Earth fits in with its neighbors in the Solar System was the first step to understanding Earth's place in the universe. The history of how those ideas evolved—from Earth at the center of all things to Earth as just another planet—is a prime example of science's self-correcting nature.

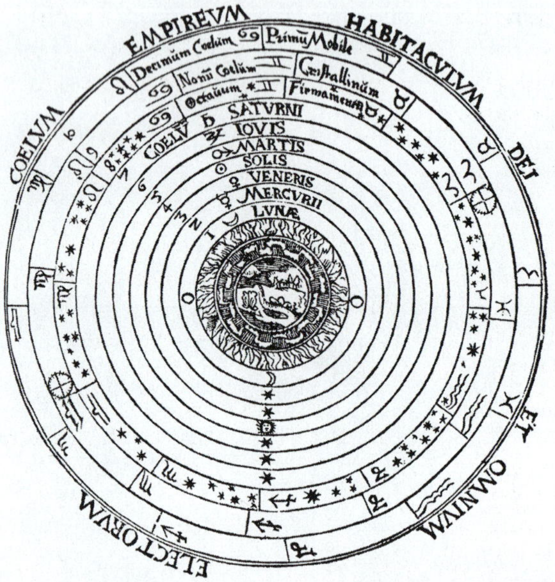

Figure 3.1 In the Ptolemaic view of the heavens, Earth is at the center, orbited by the Moon, Mercury, Venus, the Sun, Mars, Jupiter, and Saturn.

The Geocentric Model

As discussed in Chapter 2, hard evidence of the motions of Earth was remarkably difficult to come by. Largely because we do not feel Earth moving through space, early astronomers developed a **geocentric** (Earth-centered) **model** of the Solar System to explain what they observed in the sky. Looking up at the sky, people saw that the Sun, Moon, planets, and stars appeared to move around Earth. Before the 17th century, Western educated people believed that the Sun, Moon, and known planets all moved in circles around a stationary Earth. **Figure 3.1** illustrates the geocentric model of the Alexandrian astronomer Ptolemy (Claudius Ptolemaeus, 90–168 CE).

The geocentric model, though, did not account for all observations. Ancient astronomers knew that the planets usually have an eastward **prograde motion**, in which each night they move a little eastward with respect to the background stars. Sometimes, however, the ancients observed apparent **retrograde motion**, in which the planets appear to move westward for a time before resuming their normal eastward travel. **Figure 3.2** shows retrograde motion for Mars.

That odd behavior of the five "naked eye" planets—Mercury, Venus, Mars, Jupiter, and Saturn—created a puzzling problem for the geocentric model as

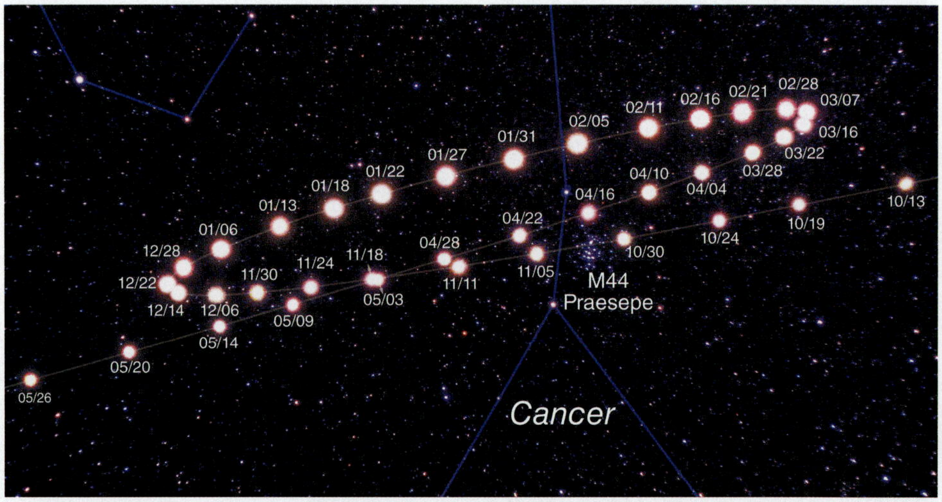

Figure 3.2 This time-lapse photographic series shows Mars moving in apparent retrograde motion.

Ptolemy summarized it in 150 CE. Some astronomers and philosophers in ancient times, such as Aristarchus of Samos (310–230 BCE), hypothesized that the Sun might be the center of the Solar System, but they lacked the tools to test that hypothesis or the mathematical insight to formulate a more complete and testable model. But most other astronomers of the time were skeptical because they thought that if Earth moved around the Sun, they should feel Earth's motion. Therefore, they preferred the geocentric model, in which the Sun, Moon, and planets all moved in perfect circles around a stationary Earth, with the "fixed stars" located somewhere way out beyond the planets.

Because the geocentric model in its simplest form failed to explain the retrograde motion of the planets, Ptolemy added an embellishment called an *epicycle*—a small circle superimposed on each planet's larger circle (**Figure 3.3**). Thus, as the planet travels along its larger circle around Earth, it also moves periodically along its epicycle. When moving along the smaller circle in the opposite direction of the forward motion of the larger circle, the planet's forward motion would be reversed and it would appear to move backward in the sky. Ptolemy's model had many epicycles, but they made the model *work*, in the sense that this model was reasonably successful at predicting the positions of planets in the sky. For nearly 1,500 years, Ptolemy's model of the heavens was the accepted paradigm in the Western world.

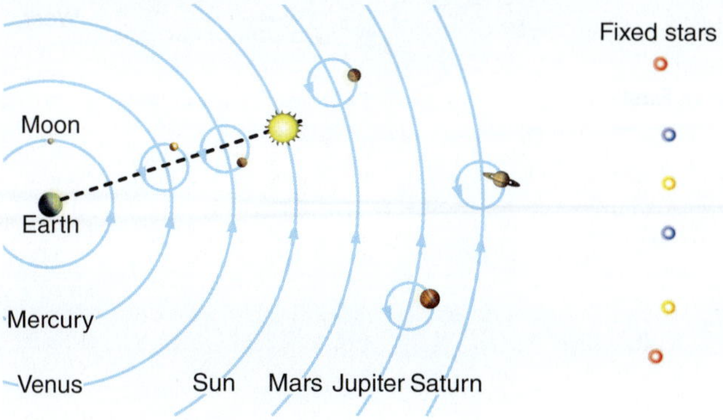

Figure 3.3 To reconcile retrograde motion with the geocentric model of the Solar System, Ptolemy added loops called epicycles to each planet's circular orbit around Earth.

CHECK YOUR UNDERSTANDING 3.1A

How did the ancients know the planets were different from the stars?

Copernicus Proposes a Heliocentric Model

Nicolaus Copernicus (1473–1543—**Figure 3.4**) is famous for placing the Sun, rather than Earth, at the center of the Solar System. He was not the first person to suggest that Earth orbited the Sun; a few ancient Greek and medieval Arab astronomers had briefly considered the idea when they saw problems with the Earth-centered model. However, Copernicus was the first to develop a comprehensive mathematical model with the Sun at the center that later astronomers could test. That model was the beginning of the Copernican Revolution. Through the work of 16th- and 17th-century scientists such as Tycho Brahe, Galileo Galilei, Johannes Kepler, and Isaac Newton, the **heliocentric** (Sun centered) **model** of the Solar System became one of the best-corroborated theories in all of science.

Figure 3.4 Nicolaus Copernicus rejected the ancient Greek model of an Earth-centered universe and replaced it with a model that centered on the Sun.

Copernicus was multilingual and highly educated: he studied philosophy, canon (Catholic) law, medicine, economics, mathematics, and astronomy in his native Poland and in Italy. He conducted astronomical observations from a small tower, and sometime around 1514 he started writing about heliocentricity. Eighteen years later, he completed his manuscript. He did not publish the book because he knew his ideas would be controversial: philosophical and religious views of the time held that humanity, and thus Earth, must be the center of the universe. Late in his life, Copernicus was finally persuaded to publish his ideas, and his great work *De revolutionibus orbium coelestium* ("On the Revolutions of the Heavenly Spheres") appeared in 1543, the year he died.

Figure 3.5 shows Copernicus's model with the planets orbiting around the Sun. That model explained the observed motions of Earth, the Moon, and the planets, including retrograde motion, much more simply than the geocentric model did. It's like when you are in a car or train and you pass a slower-moving

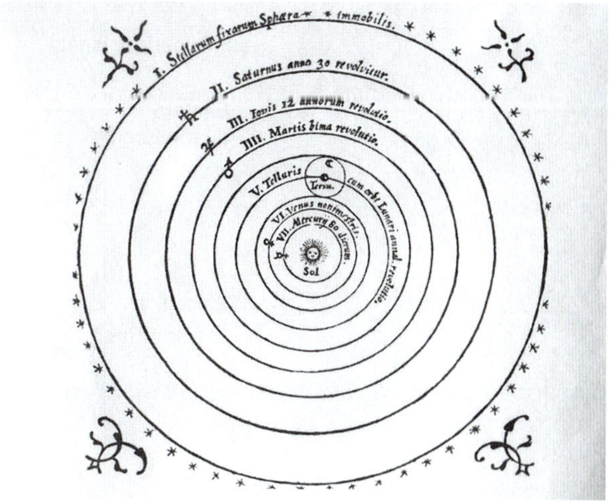

Figure 3.5 The Copernican heliocentric view of the Solar System (II–VII) and the fixed stars (I). The Sun is at the center, orbited by Mercury, Venus, Earth, Mars, Jupiter, and Saturn. The Moon orbits Earth.

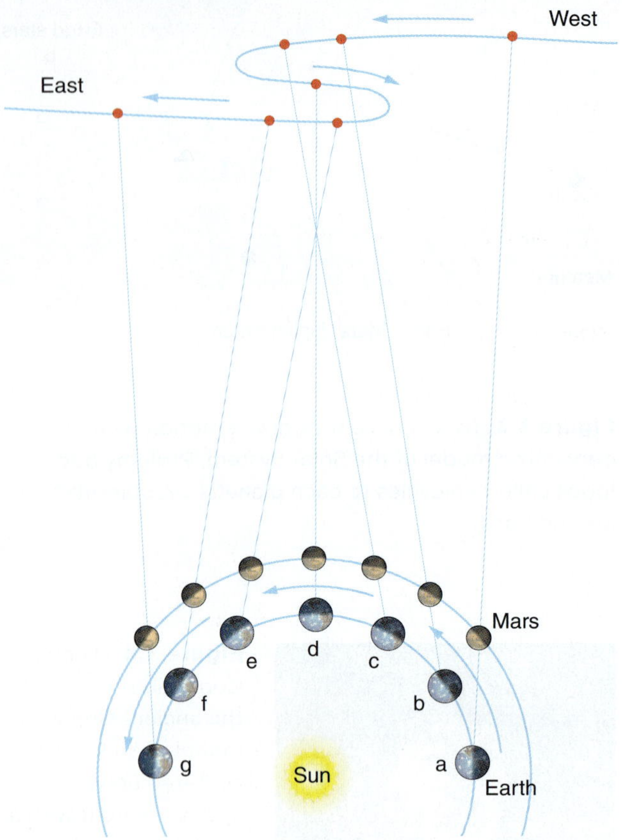

Figure 3.6 The Copernican model explains the apparent retrograde motion of Mars (see Figure 3.2) as seen in Earth's sky when Earth passes Mars in its orbit. (Not drawn to scale.)

car or train: the other vehicle seems as though it is moving backward. Without an external frame of reference, telling which vehicle is moving and in which direction can be hard. Through his model, Copernicus provided that frame of reference for the Sun and its planets.

In the Copernican model, planets farther from the Sun undergo apparent retrograde motion when Earth overtakes them in their orbits. **Figure 3.6** illustrates that effect for Mars. Conversely, planets closer to the Sun than Earth is—Mercury and Venus—move in apparent retrograde motion when overtaking Earth. Except for the Sun, all Solar System objects exhibit apparent retrograde motion. The effect diminishes with increasing distance from Earth. Retrograde motion is an illusion caused by the relative motion between Earth and the other planets.

Copernicus still conceived of the planets as moving in perfectly circular orbits at constant speeds, and so he needed to use some epicycles to match the observations. His model made testable predictions of the location of each planet on a given night. Those predictions were at least as accurate as those of the geocentric model. But the heliocentric model was, overall, simpler than the geocentric model and became the basis for further refinements in understanding how Earth moved. As copies of *De Revolutionibus* and Copernicus's ideas slowly spread across Europe, other scientists began to consider and then accept the heliocentric model, and a scientific revolution began.

Scaling the Solar System

From his observations, Copernicus deduced the correct order of the planets and concluded that planets closer to the Sun traveled faster than planets farther out. He also realized he needed to consider two categories of planets: **inferior** planets, which are closer to the Sun than Earth is, and **superior** planets, which are farther from the Sun than Earth is.

In Copernicus's model, Earth, another planet, and the Sun periodically align in space to form either a line or a right triangle. **Figure 3.7a** shows a superior planet in line with the Sun and Earth but on the other side of the Sun from Earth—a configuration called a *conjunction*. A superior planet in conjunction rises

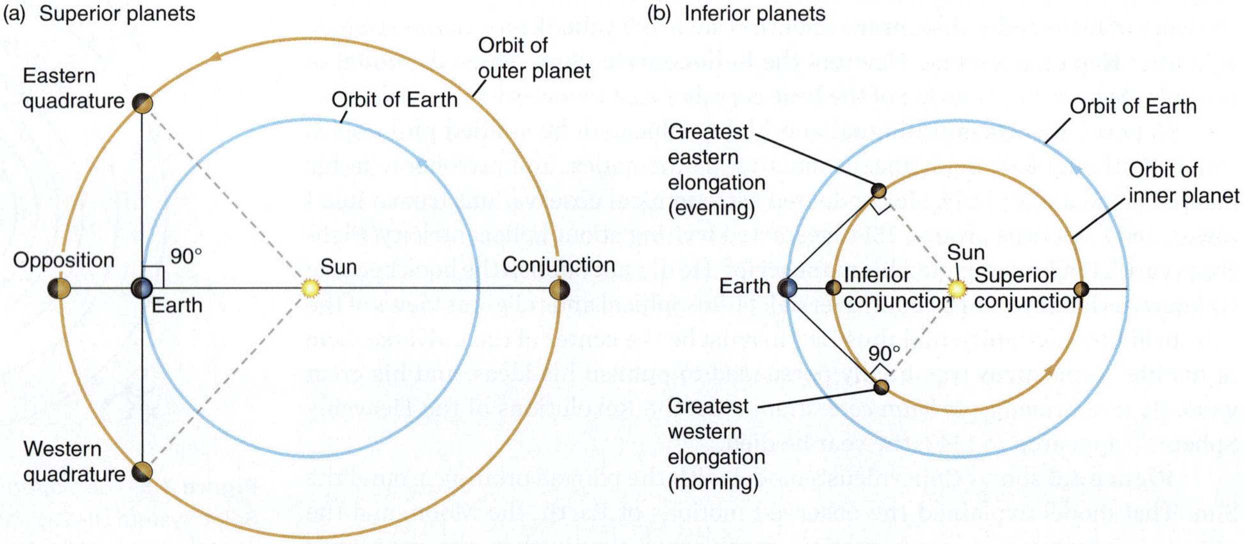

Figure 3.7 Planetary configurations for (a) superior (outer) planets and (b) inferior (inner) planets.

and sets in the sky with the Sun. When a superior planet is in conjunction, it is at its farthest from Earth, so it is at its faintest, too. You won't see the planet at all exactly at conjunction because it's on the far side of the Sun.

In contrast, when a superior planet is in line with the Sun and Earth on the same side of the Sun as Earth, the configuration is an *opposition* (Figure 3.7a). At opposition, the superior planet is "opposite" the Sun in the sky: like a full Moon, it rises when the Sun sets and sets when the Sun rises. When the superior planet is in opposition, it is at its closest to Earth and thus at its brightest; therefore, opposition is the best time to observe the planet in the sky. Opposition is also the time when the planet exhibits retrograde motion because that is exactly when Earth is overtaking the planet in its orbit. *Quadrature* is said to occur when Earth, the Sun, and a superior planet form a right triangle in space (Figure 3.7a).

For an inferior planet, the configurations are slightly different (**Figure 3.7b**). If the inferior planet is between Earth and the Sun, we call the configuration an *inferior conjunction*—when the planet is closest to Earth. If the inferior planet is on the other side of the Sun from Earth, we call the configuration a *superior conjunction*—when the planet is farthest from Earth. When the inferior planet forms a right triangle with Earth and the Sun, and thus is the farthest it gets from the Sun in the sky, we call the configuration the *greatest elongation*. When you are standing on Earth and looking toward the inner planets, you always see them close to the Sun in the sky, so you see Mercury and Venus only within a few hours of sunrise or sunset. The best time to observe those planets is at greatest elongation because they will have the greatest separation from the Sun in the sky.

Copernicus used the geometry of those alignments along with his observations of the positions of the planets in the sky, including their altitudes and the times they rose and set, to estimate the planet–Sun distances in multiples of the Earth–Sun distance. He realized that two types of orbital periods existed. Similar to the terms used for lunar orbits, a planet's *sidereal period* is how long the planet takes to make one orbit around the Sun with respect to the stars and return to the same point in space. A planet's *synodic period* is how long the planet takes to return to the same configuration with the Sun and Earth, such as inferior conjunction to inferior conjunction or opposition to opposition. The synodic period is what can be observed from Earth.

In **Figure 3.8a**, Earth and the superior planet are in opposition at point A. Superior planets move around the Sun more slowly than Earth does, so Earth orbits the Sun once and then catches up to the superior planet to form the next opposition at point B. In **Figure 3.8b**, Earth and the inferior planet are in inferior conjunction at point A. An inferior planet moves around the Sun faster than Earth does, so it completes one sidereal period and then must continue in its orbit to catch up to Earth for the next inferior conjunction at point B. **Table 3.1** shows that those relative distances Copernicus calculated are remarkably close to the distances modern methods yield. **Working It Out 3.1** shows the numerical details. Copernicus's model not only predicted planetary positions in the sky but also could be used to compare the distances between the planets and the Sun accurately and thus to set the scale of the Solar System.

CHECK YOUR UNDERSTANDING 3.1B

The planet Uranus will be observed in retrograde motion when: (a) Uranus is closest to the Sun; (b) Uranus is farthest from the Sun; (c) Earth overtakes Uranus in its orbit; (d) Uranus overtakes Earth in its orbit.

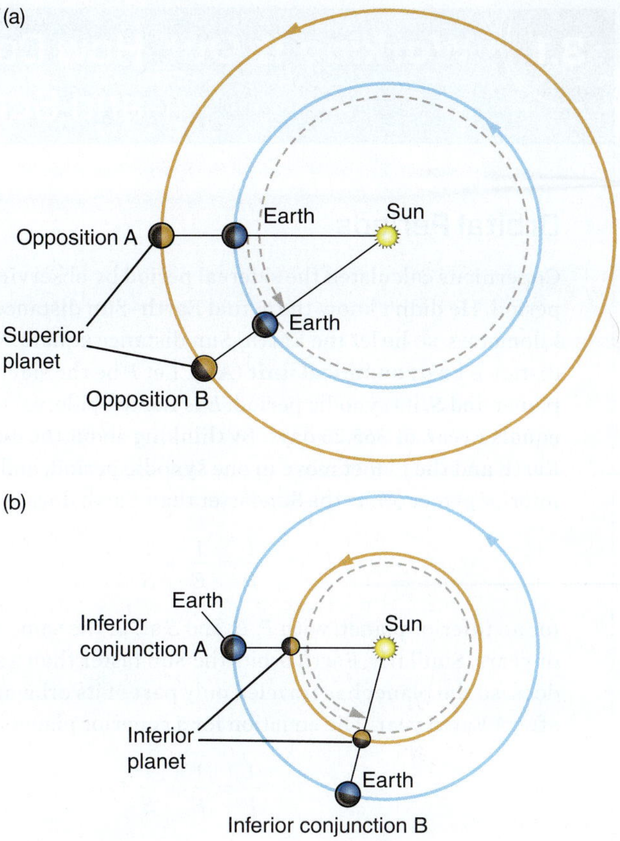

Figure 3.8 The synodic periods of planets indicate how long they take to return to the same configuration with Earth and the Sun. (a) Earth completes one orbit around the Sun first and then catches up to the superior planet. (b) Inferior planets complete a full orbit around the Sun first and then catch up to Earth.

TABLE 3.1 Copernicus's Scale of the Solar System

Planet	Copernicus's Value (AU)	Modern Value (AU)
Mercury	0.38	0.39
Venus	0.72	0.72
Earth	1.00	1.00
Mars	1.52	1.52
Jupiter	5.22	5.20
Saturn	9.17	9.58

AU = astronomical unit.

3.1 Working It Out How Copernicus Computed Orbital Periods and Scaled the Solar System

Orbital Periods

Copernicus calculated the sidereal period by observing the synodic period. He didn't know the actual Earth–Sun distance in miles or kilometers, so he let the Earth–Sun distance equal 1. We call that distance 1 **astronomical unit** (**AU**). Let P be the sidereal period of a planet and S, its synodic period. E is Earth's sidereal period, which equals 1 year, or 365.25 days. By thinking about the distance that Earth and the planet move in one synodic period, and noting that an inferior planet orbits the Sun faster than Earth does, we can show that

$$\frac{1}{P} = \frac{1}{E} + \frac{1}{S}$$

for an inferior planet, with P, E, and S all in the same units of days or years. Similarly, Earth orbits the Sun faster than a superior planet does, so the planet has traveled only part of its orbit around the Sun after 1 Earth year. The equation for a superior planet is

$$\frac{1}{P} = \frac{1}{E} - \frac{1}{S}$$

For Saturn, the time that passes between oppositions—the date of maximum brightness—shows that Saturn's synodic period (S) is 378 days, or 378 ÷ 365.25 = 1.035 years. Then, to compute Saturn's sidereal period (P) in years, we use $S = 1.035$ yr and $E = 1$ yr in the equation for a superior planet:

$$\frac{1}{P} = \frac{1}{1\,\text{yr}} - \frac{1}{1.035\,\text{yr}} = 1 - 0.966\,\text{yr}^{-1} = 0.034\,\text{yr}^{-1}$$

Thus,

$$P = \frac{1}{0.034\,\text{yr}^{-1}} = 29.4\,\text{yr}$$

Saturn's sidereal period is 29.4 years, which means that Saturn takes 29.4 years to travel around the Sun and return to where it started in space.

Scaling the Solar System

Copernicus used the configurations of the planets shown in Figure 3.7 along with their sidereal periods to compute the relative distances of the planets. For the superior planets, he measured the fraction of the circular orbit that the planet completed in the time between opposition and quadrature, and then he used trigonometry to solve for the planet–Sun distance in astronomical units (see Figure 3.7a). For the inferior planets, he had a right triangle at the point of greatest elongation, and then he used right-triangle trigonometry to solve for the planet–Sun distance in astronomical units (see Figure 3.7b). Copernicus's values are impressively similar to modern values (see Table 3.1). Copernicus still did not know the actual value of the astronomical unit in miles, but he was the first to accurately compute the relative distances of the planets from the Sun.

3.2 Kepler's Laws Describe Planetary Motion

Copernicus did not understand *why* the planets move about the Sun, but he realized that his heliocentric picture offered a way to compute the planets' relative distances. His theory is an example of **empirical science**, which seeks to describe patterns in nature as accurately as possible. Copernicus's work was revolutionary because he challenged the accepted geocentric model and proposed that Earth is one planet among many. His conclusions paved the way for other great empiricists, including Tycho Brahe and Johannes Kepler.

Figure 3.9 Tycho Brahe, known commonly as Tycho, was one of the greatest astronomical observers before the invention of the telescope.

Tycho Brahe's Observations

Tycho Brahe (1546–1601—**Figure 3.9**) was a Danish astronomer of noble birth who entered university at age 13 to study philosophy and law. After seeing a partial solar eclipse in 1560, Tycho (conventionally referred to by his first name) became interested in astronomy. A few years later, he observed Jupiter and Saturn near each other in the sky, though not exactly where one would expect from the astronomical tables based on Ptolemy's model. Tycho gave up studying law and devoted himself to making better tables of the planets' positions in the sky.

The king of Denmark granted Tycho the island of Hven, between Sweden and Denmark, to build an observatory. Tycho designed and built new instruments, operated a printing press, and taught students and others how to conduct observations. With the help of his sister Sophie, Tycho carefully measured the precise positions of planets in the sky over several decades, developing the most comprehensive set of planetary data then available. He created his own geo-heliocentric model, shown in **Figure 3.10**. In Tycho's model, the planets orbit the Sun, and the Sun and planets orbit Earth. His model gained some acceptance among people who preferred to keep Earth at the center for philosophical or religious reasons. Tycho lost his financial support when the king died, and in 1600 he relocated to Prague.

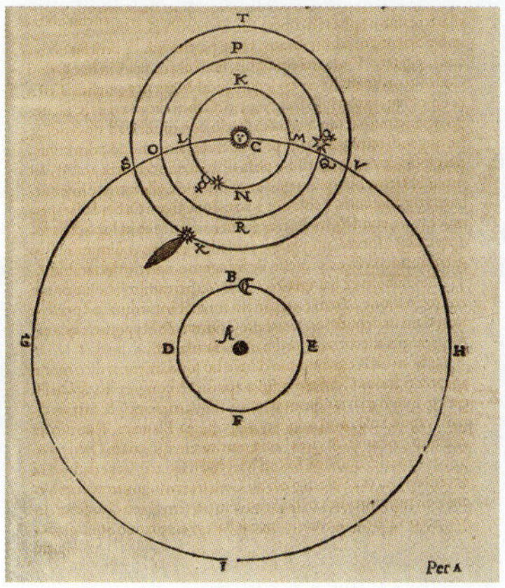

Figure 3.10 Tycho's geo-heliocentric model, showing the Moon and Sun orbiting Earth, with the other planets orbiting the Sun.

Kepler's Laws

In 1600, Tycho hired a more mathematically inclined astronomer, Johannes Kepler (1571–1630—**Figure 3.11**), as his assistant. Kepler, who had studied Copernicus's ideas, was responsible for the next major step toward understanding the planets' motions. Kepler inherited the records of Tycho's observations when he died. Working first with Tycho's observations of Mars, Kepler deduced three rules, now generally referred to as **Kepler's laws**, which accurately describe how the planets move. Kepler's laws are empirical: they use existing data to make predictions about future behavior but do not include an underlying theory of why the objects behave as they do.

Figure 3.11 Johannes Kepler explained the motions of the planets with three empirically determined laws.

KEPLER'S FIRST LAW Comparing Tycho's extensive planetary observations with predictions from Copernicus's heliocentric model, Kepler expected the data to confirm circular orbits for planets orbiting the Sun. Instead, he found disagreements between his predictions and Tycho's observations. He was not the first to notice such discrepancies. Rather than discard Copernicus's model, Kepler revised it.

By replacing Copernicus's circular orbits with *elliptical* orbits, Kepler could predict the positions of planets for any day, and his predictions fit Tycho's observations almost perfectly. Looking like an elongated circle, an **ellipse** is symmetric from right to left and from top to bottom. As shown in **Figure 3.12a**, you can draw an ellipse by attaching the two ends of a piece of string to a piece of paper, stretching the string tight with the tip of a pencil, and then drawing around those two points while keeping the string tight. Each point at which the string is attached is a **focus** (plural: **foci**) of the ellipse. Notice in **Figure 3.12b** that the ellipse becomes more circular as the two foci move closer together. As the two foci move farther apart, the ellipse becomes more and more elongated. The **eccentricity** (*e*) of an ellipse measures that elongation; it is determined by the separation between the

Figure 3.12 (a) We can draw an ellipse by attaching a length of string to a piece of paper at two points (called foci) and then pulling the string around as shown. The long axis is called the major axis. (b) Ellipses range from circles ($e = 0$) to elongated eccentric shapes. e = eccentricity.

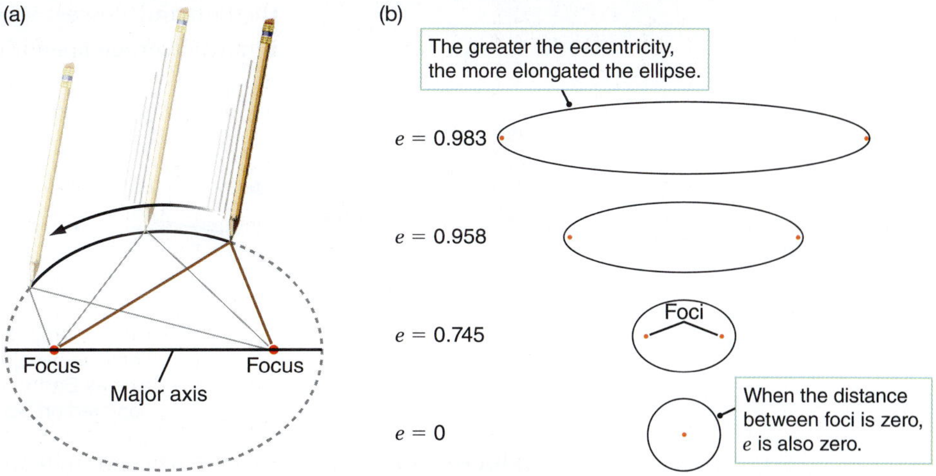

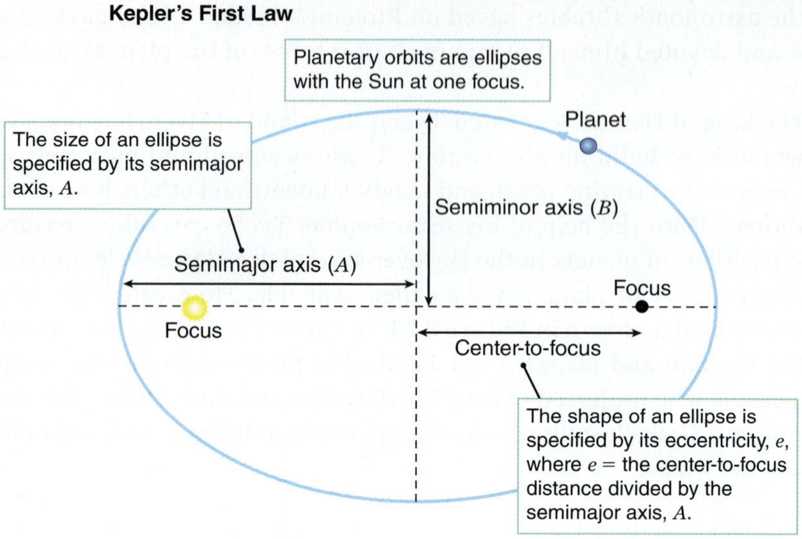

Figure 3.13 According to Kepler's first law, planets move in elliptical orbits with the Sun at one focus. (Nothing is at the other focus.) The orbit's eccentricity is the center-to-focus distance divided by the semimajor axis.

two foci divided by the length of the long axis, called the **major axis**. A circle has an eccentricity of 0 because the two foci coincide at the center. The more elongated the ellipse becomes, the closer its eccentricity gets to 1.

AstroTour: Kepler's Laws

Kepler's first law of planetary motion states that the orbit of each planet is an ellipse with the Sun located at one focus (see the **Process of Science Figure**). Nothing but empty space is at the other focus. **Figure 3.13** illustrates Kepler's first law and shows the features of an ellipse that can be measured. The dashed lines in Figure 3.13 represent the major (long) and minor (short) axes of the ellipse. Half the length of the major axis of the ellipse is called the **semimajor axis**, A. The average distance between the Sun and a planet equals the semimajor axis of the planet's orbit.

Simulation: Planetary Orbits

The eccentricities of planetary orbits vary widely, but most planetary objects in our Solar System have nearly circular orbits. As **Figure 3.14a** shows, Earth's orbit, with an eccentricity of 0.017, is very nearly a circle centered on the Sun; therefore the distance variation is small. In contrast, dwarf planet Pluto's orbit, depicted in **Figure 3.14b**, has an eccentricity of 0.249. The orbit is noticeably elongated, with the Sun offset from center.

KEPLER'S SECOND LAW By analyzing Tycho's observational data of changes in the planets' positions, Kepler found that a planet moves fastest when closest to the Sun and slowest when farthest from the Sun. For example, we now measure Earth's average speed in its orbit around the Sun at 29.8 kilometers per second

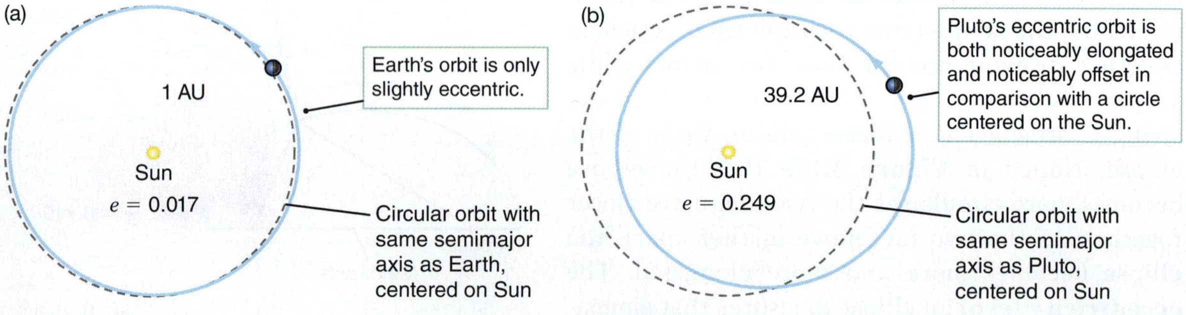

Figure 3.14 The orbits of (a) Earth and (b) Pluto in comparison with circles around the Sun. e = eccentricity.

PROCESS OF SCIENCE

Theories Are Falsifiable

Early astronomers studied the motions of the planets, but did not understand why the planets behaved as they do.

Copernicus
1473–1543

The Hypothesis

Copernicus proposed that planets moved in circular orbits (with epicycles) around the Sun.

Tycho
1546–1601

The Observation

Tycho, born three years after Copernicus died, observed and collected enormous amounts of data about planet positions.

Kepler
1571–1630

The Prediction

Kepler, 25 years younger than Tycho, used Copernicus's model and Tycho's data to predict where the planets should be.

The Test

When Kepler compared the predictions with actual data, they disagreed. Copernicus's idea was falsified!

The New Hypothesis

Planet orbits are not circular. They are elliptical.

Newton
1643–1726

This new hypothesis gained acceptance a century later when Newton showed that planetary orbits are an inevitable consequence of gravitational attraction.

In order for a theory to be scientific, it must be falsifiable, even if the test can't be carried out until decades or centuries later. Disproving an old theory is part of the self-correcting nature of science: It always leads to deeper understanding.

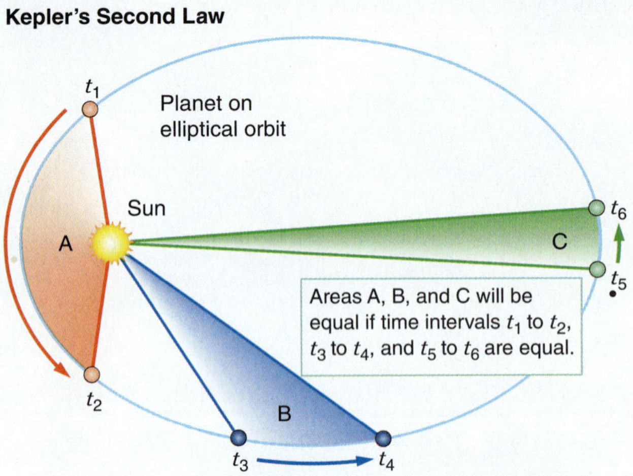

Figure 3.15 An imaginary line between a planet and the Sun sweeps out an area as the planet orbits. According to Kepler's second law, if the three time intervals shown are equal, the three areas A, B, and C will be the same.

(km/s). When closest to the Sun, Earth travels at 30.3 km/s. At its farthest from the Sun, Earth travels at 29.3 km/s.

Kepler found a way to elegantly describe the changing speed of a planet in its orbit around the Sun. **Figure 3.15** shows a planet at six points along the ellipse of its orbit (t_1 to t_6). Imagine a straight line connecting the Sun with that planet. We can think of the line as "sweeping out" an area as it moves with the planet from one point to another. Area A (in orange) is swept out between times t_1 and t_2, area B (in blue) is swept out between times t_3 and t_4, and area C (in green) is swept out between times t_5 and t_6. When the planet is closest to the Sun (area A), it is moving the fastest, but the distance between the planet and the Sun is small. Kepler realized that changes in the distance from the Sun to a planet and changes in a planet's speed work together so that the area a planet sweeps out in the same amount of time is always the same, regardless of where the planet is in its orbit. Therefore, if the three time intervals in the figure are equal (that is, $t_1 \rightarrow t_2 = t_3 \rightarrow t_4 = t_5 \rightarrow t_6$), the three areas A, B, and C must be equal as well.

Kepler's second law, also called Kepler's **law of equal areas**, states that the imaginary line connecting a planet to the Sun sweeps out equal areas in equal times, regardless of where the planet is along the ellipse of its orbit. That law applies to only one planet at a time. The area swept out by Earth in a given time interval is always the same. Likewise, the area swept out by Mars in a given time is always the same. But the area swept out by Earth and the area swept out by Mars in a given time are *not* the same. That law can be used to find the speed of a planet anywhere in its orbit.

KEPLER'S THIRD LAW Kepler looked for patterns in the planets' orbital periods. Compared with planets closer to the Sun, he found that planets farther from the Sun have longer orbits and orbit the Sun more slowly. Kepler discovered a mathematical relationship between a planet's sidereal period—how many years the planet takes to go around the Sun and return to the same position in the Solar System—and its average distance from the Sun in astronomical units. **Kepler's third law** states that in those units, the square of the sidereal

3.2 Working It Out Kepler's Third Law

Kepler's third law states that the square of the period of a planet's orbit measured in years, P_{years}, is equal to the cube of the semimajor axis of the planet's orbit measured in astronomical units, A_{AU}. As an equation, the law says

$$(P_{\text{years}})^2 = (A_{\text{AU}})^3$$

For mathematical convenience, Kepler used units based on Earth's orbit—astronomical units and years. If other units were used, such as kilometers and days, P^2 would still be proportional to A^3, but the constant of proportionality would not be 1.

To calculate the average size of Neptune's orbit in astronomical units, you first need to find out how long Neptune's period is in Earth years, which you can determine by observing the synodic period and from that computing its sidereal period (using Working It Out 3.1). Neptune's sidereal period is 165 years. Plugging that number into Kepler's third law gives the following result:

$$(P_{\text{years}})^2 = (165)^2 = 27{,}225 = (A_{\text{AU}})^3$$

To solve that equation, you must first square 165 to get 27,225 and then take its cube root (see Appendix 1 for calculator hints). Then

$$A_{\text{AU}} = \sqrt[3]{27{,}225} = 30.1$$

The semimajor axis of Neptune's orbit—that is, the average distance between Neptune and the Sun—is 30.1 AU.

TABLE 3.2 Kepler's Third Law: $P^2 = A^3$

Planet	Period P (years)	Semimajor Axis A (AU)	$\frac{P^2}{A^3}$
Mercury	0.241	0.387	$\frac{0.241^2}{0.387^3} = 1.00$
Venus	0.615	0.723	$\frac{0.615^2}{0.723^3} = 1.00$
Earth	1.000	1.000	$\frac{1.000^2}{1.000^3} = 1.00$
Mars	1.881	1.524	$\frac{1.881^2}{1.524^3} = 1.00$
Ceres	4.599	2.765	$\frac{4.599^2}{2.765^3} = 1.00$
Jupiter	11.86	5.204	$\frac{11.86^2}{5.204^3} = 1.00$
Saturn	29.46	9.582	$\frac{29.46^2}{9.582^3} = 0.99$*
Uranus	84.01	19.201	$\frac{84.01^2}{19.201^3} = 1.00$
Neptune	164.79	30.047	$\frac{164.79^2}{30.047^3} = 1.00$
Pluto	247.92	39.482	$\frac{247.92^2}{39.482^3} = 1.00$
Eris	557.00	67.696	$\frac{557.00^2}{67.696^3} = 1.00$

*Ratios are not exactly 1.00 because of slight perturbations from the gravity of other planets.

period (P) of a planet's orbit is equal to the cube of the semimajor axis (A) of the planet's orbit.

Table 3.2 lists the periods and semimajor axes of the orbits of the eight classical planets and three of the dwarf planets, along with the values of the ratio P^2 divided by A^3. Those data are also plotted in **Figure 3.16**. Kepler referred to that relationship as his **harmonic law** or, more poetically, as the "Harmony of the Worlds." Kepler's third law is explored further in **Working It Out 3.2**. Kepler's laws enhanced the heliocentric mathematical model of Copernicus and led to its greater acceptance.

CHECK YOUR UNDERSTANDING 3.2

Order the following from largest to smallest semimajor axis: (a) a planet with a period of 84 Earth days; (b) a planet with a period of 1 Earth year; (c) a planet with a period of 2 Earth years; (d) a planet with a period of 0.5 Earth years.

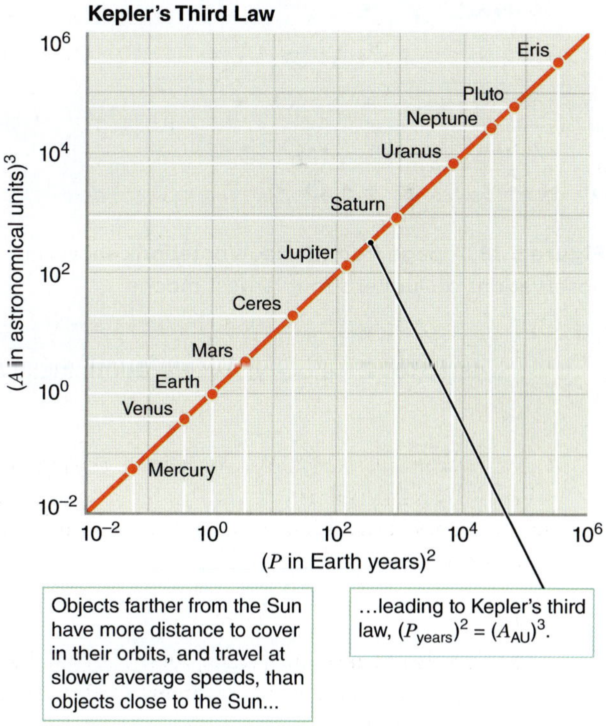

Figure 3.16 A plot of A^3 versus P^2 for objects in our Solar System shows that they obey Kepler's third law. (By plotting powers of 10 on each axis, we can fit both large and small values on the same plot. We will use that approach often.)

Figure 3.17 Galileo Galilei laid the physical framework for Newton's laws.

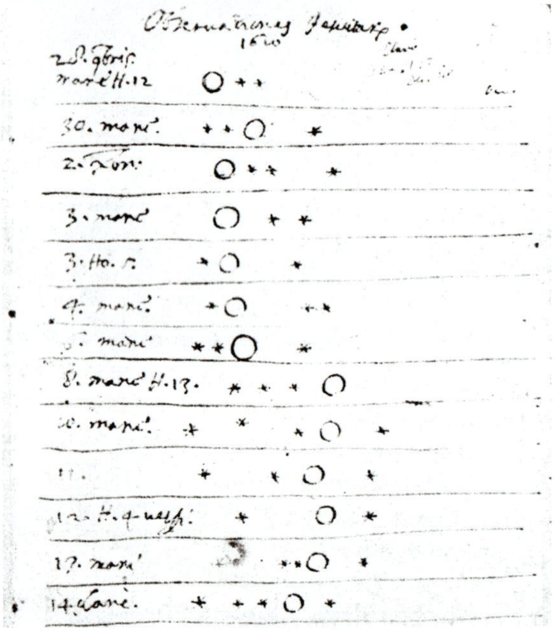

Figure 3.18 A page from Galileo's notebook shows his observations of Jupiter's four largest moons.

AstroTour: Velocity, Acceleration, Inertia

3.3 Galileo's Observations Supported the Heliocentric Model

Galileo Galilei (1564–1642—**Figure 3.17**) was the first to use a telescope to make and report significant discoveries about astronomical objects. Galileo's telescopes were relatively small, yet sufficient for him to observe spots on the Sun, the uneven surface and craters of the Moon, and the many stars in the band of light in the sky called the Milky Way.

Galileo's Observations

Galileo provided the first observational evidence that some objects in the sky do not orbit Earth. When Galileo turned his telescope on Jupiter, he observed four "stars" in a line near the planet. Over time he observed that the objects changed position from night to night (**Figure 3.18**). Galileo hypothesized that those objects were moons orbiting Jupiter. Those four moons are the largest of Jupiter's many moons and are still called the Galilean moons. Galileo also estimated the relative distance of each moon from Jupiter and the periods of their orbits, and he showed that they followed Kepler's third law.

Galileo also observed that Venus went through phases like the Moon. He noticed that the phases of Venus were correlated with how large Venus appeared in his telescope. In a geocentric model in which Venus orbits Earth like the Moon does, the apparent size of Venus would change only slightly, when it looped an epicycle. In the heliocentric model, however, the Earth–Venus distance varies by a lot, and Venus's size changes accordingly. When Venus is in its gibbous to full phases, it is farther away, on the other side of the Sun from Earth, and smaller in the sky. When Venus is in its crescent to new phases, it is closer, on the same side of the Sun as Earth, and larger in the sky (**Figure 3.19**). His observations of Jupiter's moons and the phases of Venus in particular convinced Galileo that Copernicus was correct to place the Sun at the center of the Solar System.

In addition to his astronomical observations, Galileo did important work on the motion of objects. Unlike natural philosophers, who thought about objects in motion but did not actually experiment with them, Galileo conducted experiments with falling and rolling objects. As with his telescopes, Galileo improved or developed new technology to enable him to conduct those experiments. For example, by carefully rolling balls down an inclined plane and by dropping various objects from a height, he found that a falling object travels a distance proportional to the square of the time it has been falling. If he simultaneously dropped two objects of different masses, they reached the ground at the same time, showing that all objects falling to Earth accelerate at the same rate, independent of their mass.

Galileo's observations and experiments with many types of moving objects, such as carts and balls, led him to disagree with the Greek philosophers about when and why objects continue to move or come to rest. Before Galileo, it was thought that an object's natural state was to be at rest. But he found that an object naturally does what it was doing until a force acts on it. That is, an object in motion continues moving along a straight line with a constant speed until a force acts on it to change that state of motion. That idea of *inertia*, which Newton later adopted as his first law of motion, has implications for not only the motion of carts and balls but also the orbits of planets.

Dialogue Concerning the Two Chief World Systems

Galileo faced considerable danger because of his work. His later life was consumed by conflict with the Catholic Church over his support of the Copernican system. In 1632, Galileo published his best-selling book, *Dialogo sopra i due massimi sistemi del mondo* ("Dialogue Concerning the Two Chief World Systems"). The *Dialogo* presents a brilliant philosopher named Salviati as the champion of the Copernican heliocentric view of the universe. The defender of an Earth-centered universe, Simplicio—who uses arguments made by the classical Greek philosophers and the pope—sounds silly and ignorant.

Galileo, a religious man with two daughters in a convent, thought he had the Catholic Church's tacit approval for his book. But when he placed several of the pope's geocentric arguments in the unflattering mouth of Simplicio, the perceived attack on the pope got the church's attention. Galileo was put on trial for heresy, sentenced to prison, and eventually placed under house arrest. To escape a harsher sentence, Galileo was forced to publicly recant his belief in the Copernican theory that Earth moves around the Sun. According to one story, as he left the courtroom, Galileo stamped his foot on the ground and muttered, "And yet it moves!"

The *Dialogo* was placed on the pope's Index of Prohibited Books, along with Copernicus's *De Revolutionibus*. Yet Galileo's work traveled across Europe, was translated into other languages, and was read by other scientists. Galileo spent his final years compiling his research on inertia and other ideas into the book *Discourses and Mathematical Demonstrations Relating to Two New Sciences*, which was published in 1638 in Holland, outside the Catholic Church's jurisdiction. (In 1992, Pope John Paul II apologized for the "Galileo Case.")

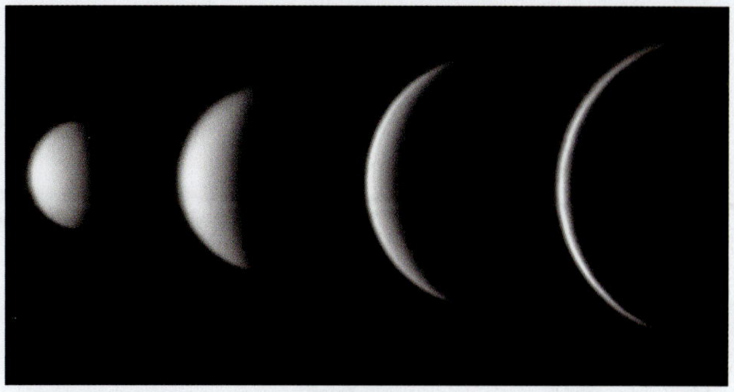

Figure 3.19 Modern photographs of the phases of Venus show that when we see Venus more illuminated, it also appears smaller, implying that Venus is farther away then.

CHECK YOUR UNDERSTANDING 3.3

Which of Galileo's astronomical observations did Copernicus's model explain better than Ptolemy's? (a) sunspots; (b) craters on the Moon; (c) moons of Jupiter; (d) apparent size and phases of Venus

3.4 Newton's Three Laws Help Explain How Celestial Bodies Move

Empirical laws, such as Kepler's laws, describe what happens, but they do not explain why. Kepler described the orbits of planets as ellipses, but he did not explain why they should be so. To take that next step in the scientific process, scientists use basic physical principles and the tools of mathematics to derive the empirically determined laws. Alternatively, a scientist might start with physical laws and predict relationships, which are then verified or falsified through experiment and observation. If those predictions are verified, the scientist may have determined something fundamental about how the universe works.

Sir Isaac Newton (1642–1727—**Figure 3.20**) took that next step in explaining the nature of motion. Newton was a student of mathematics at Cambridge University when it closed because of the Great Plague and students were sent home to the safer countryside. Over the next 2 years, he studied on his own, and at the age of 23 he invented calculus, which would become crucial to his development of the physics of motion. (The German mathematician Gottfried Leibniz independently developed calculus around the same time.)

Figure 3.20 Sir Isaac Newton formulated three laws of motion.

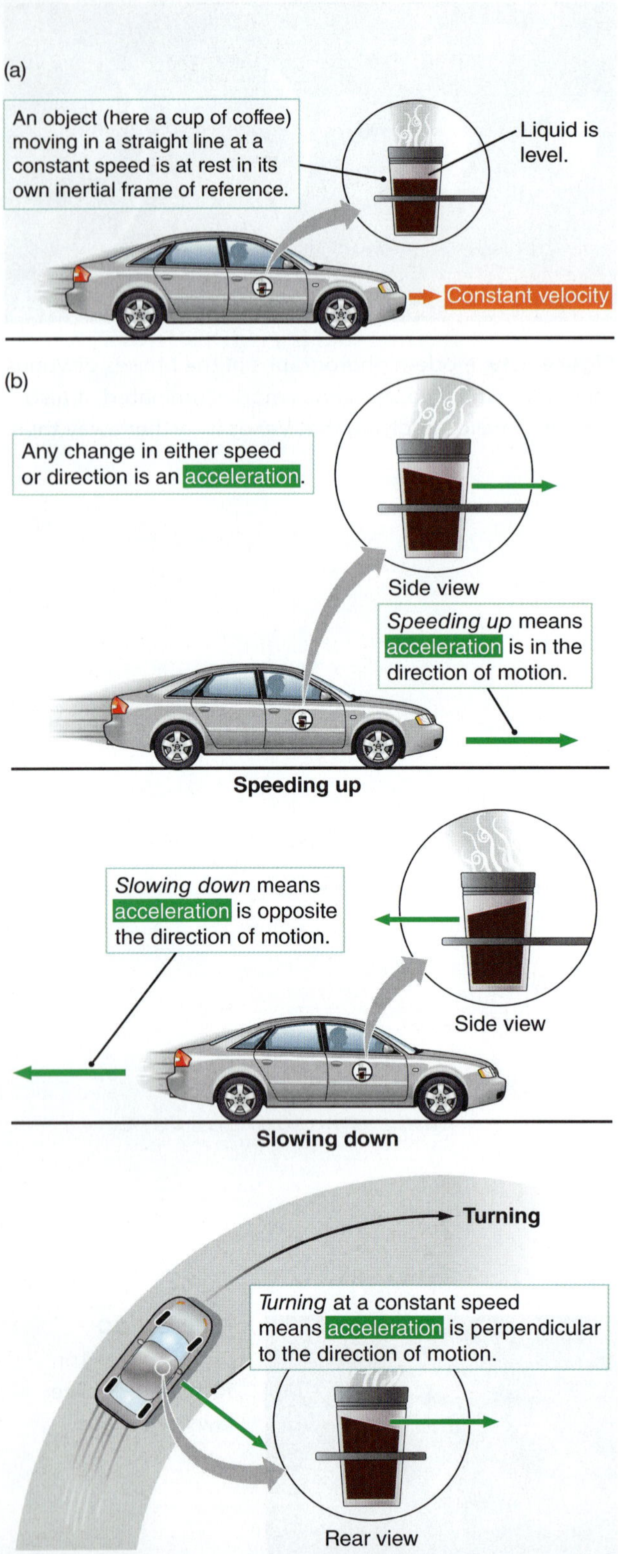

Figure 3.21 (a) An object moving in a straight line at a constant speed is at rest in its own inertial frame of reference. (b) Any change in an object's velocity is an acceleration. When you are driving, for example, any time your speed changes or you follow a curve in the road, you are accelerating. (Throughout the text, velocity arrows are red and acceleration arrows are green.)

Building on the work of Kepler, Galileo, and others, Newton proposed three physical laws that govern the motions of all objects in the sky and on Earth. To understand how the planets and all other celestial bodies move, you must understand these three laws.

Newton's First Law: Objects at Rest Stay at Rest; Objects in Motion Stay in Motion

A **force** (***F***) is a push or a pull on an object. Two or more forces can oppose one another such that they are perfectly balanced and cancel out. For example, gravity pulls down on you as you sit in your chair, but the chair pushes up on you with an exactly equal and opposite force. As a result, you stay motionless. Forces that cancel out do not affect an object's motion. When forces combine to produce an effect, we often use the term *net force*, or sometimes just *force*.

Imagine that you are driving a car, and your book is on the seat next to you. A rabbit runs across the road in front of you, and you hit the brakes hard. You feel the seat belt tighten to restrain you. At the same time, your book flies off the seat and hits the dashboard. You have just experienced what Newton describes in his first law of motion. **Inertia** is an object's tendency to maintain its state—either of uniform motion or of rest—until a net force pushes or pulls it. In the stopping car, you did not hit the dashboard because the force of the seat belt slowed you down. The book, however, hit the dashboard because no such force acted upon it.

Newton's first law of motion describes inertia and states that an object in motion tends to stay in motion, in the same direction, until a net force acts upon it; and an object at rest tends to stay at rest until a net force acts upon it. Galileo's law of inertia became the cornerstone of physics as Newton's first law.

Recall from Section 2.1 the concept of a frame of reference. Within a frame of reference, only the relative motions between objects have any meaning. Without external clues, you cannot tell the difference between sitting still and traveling at constant speed in a straight line. For example, if you close your eyes while riding in the passenger seat of a quiet car on a smooth road, you would feel as though you were sitting still. In the earlier example in this chapter, your book was "at rest" beside you on the front seat of your car, but a person standing by the side of the road would see the book moving past at the same speed as the car. People in a car approaching you would see the book moving quite fast—at the speed they are traveling plus the speed you are traveling! All those perspectives are equally valid, and all those speeds of the book are correct when measured in the appropriate reference frame.

A reference frame moving in a straight line at a constant speed is called an **inertial frame of reference**. Any inertial frame of reference is as good as another. In the inertial frame of reference of a cup of coffee, for example, **Figure 3.21a** shows that the cup is at rest in its own frame even if the car is moving quickly down the road.

Newton's Second Law: Motion Is Changed by Forces

What if a net force does act? In the earlier example, you were traveling in the car, and your motion slowed when the force of the seat belt acted upon you. Forces change an object's motion—by changing either the speed or the direction. That effect reflects **Newton's second law of motion**: if a net force acts on an object, the object's motion changes.

In the driver's seat of a car, you have several controls, including an accelerator and a brake pedal, which you use to make the car speed up or slow down. A *change in speed* is one way the car's motion can change. But you also have the steering wheel in your hands. When you are moving down the road and you turn the wheel, your speed does not necessarily change, but the direction of your motion does. A *change in direction* also is a kind of change in motion.

Together, an object's speed and direction are called **velocity** (***v***). "Traveling at 50 kilometers per hour (km/h)" indicates speed; "traveling north at 50 km/h" indicates velocity. **Acceleration** (***a***) describes the changes in an object's velocity. For example, if you go from 0 to 100 km/h in 4 seconds, you feel a strong push from the seat back as it shoves your body forward, causing you to accelerate along with the car. However, if you take 2 minutes to go from 0 to 100 km/h, the acceleration is so slight that you hardly notice it.

Partly because a car's gas pedal is often called the accelerator, some people think *acceleration* always means that an object is speeding up. But in physics, *any change in speed or direction is an acceleration.* **Figure 3.21b** illustrates that point by showing what happens to the coffee in a cup as the car speeds up, slows down, or turns. Slamming on your brakes and going from 100 to 0 km/h in 4 seconds is just as much an acceleration as going from 0 to 100 km/h in 4 seconds. Similarly, the acceleration you experience as you go through a fast, tight turn at a constant speed is every bit as real as the acceleration you feel when you slam your foot on the accelerator or the brake. Whether speeding up, slowing down, turning left, or turning right—if you are not moving in a straight line at a constant speed, you are accelerating.

Newton's second law of motion says that a net force causes acceleration. An object's acceleration depends on two things. First, as shown in **Figure 3.22**, the acceleration depends on the strength of the net force acting on the object to change its motion. If the forces acting on the object do *not* add up to zero, a net force is present and the object accelerates (Figure 3.22a). The stronger the net force, the greater the acceleration (Figure 3.22b). Push on something twice as hard and it experiences twice as much acceleration. Push on something 3 times as hard and its acceleration is 3 times as great. The resulting change in motion occurs in the direction the net force points. Push an object away from you, and it will accelerate away from you.

An object's acceleration also depends on its inertia. You can push some objects easily, such as the empty box from a new refrigerator. But you can't easily shove the actual refrigerator around, however, even though it is about the same size as the box. Figure 3.22c shows that the greater the mass, the greater the inertia, and the *less* acceleration that will occur in response to the same net force. That relationship among acceleration, force, and mass is expressed mathematically in **Working It Out 3.3**.

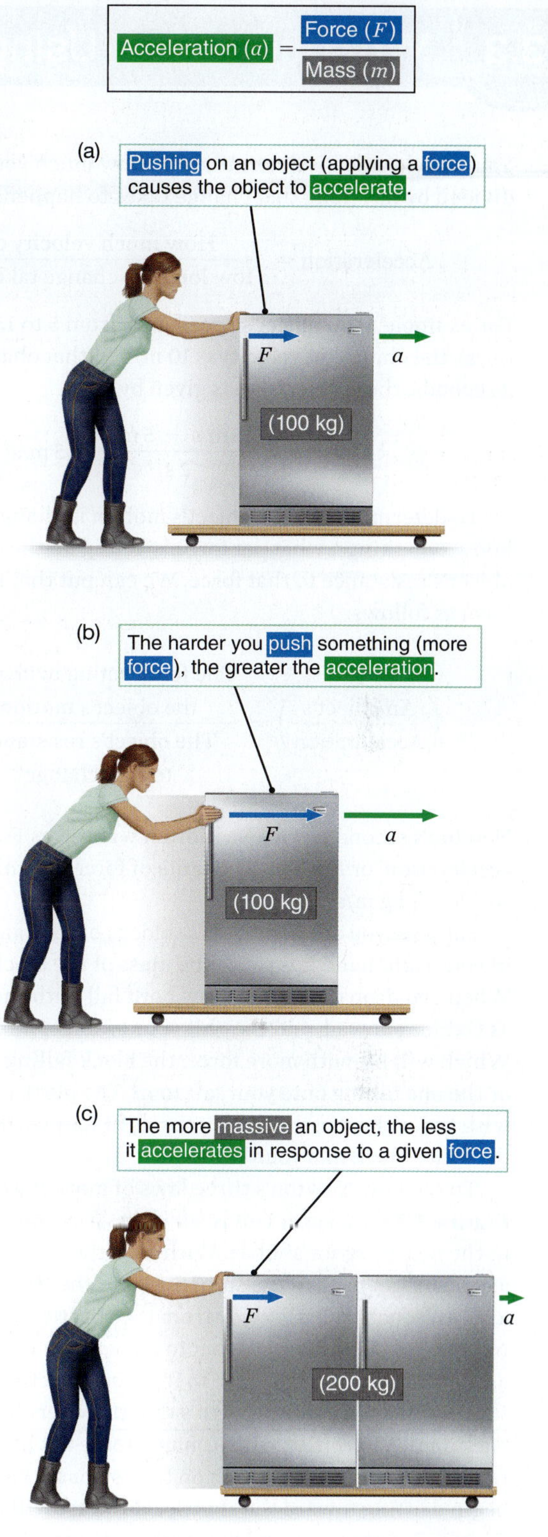

Figure 3.22 According to Newton's second law of motion, an object's acceleration is the force acting on the object divided by the object's mass. (Throughout the text, force arrows are blue.)

Newton's Third Law: Whatever Gets Pushed Pushes Back

Imagine that you are standing on a skateboard and pushing yourself along with your foot. Each shove of your foot against the ground sends you faster along your way. But why? You accelerate because as you push on the ground, the ground pushes back on you.

Part of Newton's genius was his ability to see patterns in such everyday events. Newton realized that *every* time one object exerts a force on another, the second object exerts a matching force on the first. That second force is as strong as the first force but is in the opposite direction. When you accelerate yourself on the skateboard, you push backward on Earth, and Earth pushes you forward. As

3.3 Working It Out Using Newton's Laws

Your acceleration is determined by how much your velocity changes, divided by how long that change takes to happen:

$$\text{Acceleration} = \frac{\text{How much velocity changes}}{\text{How long the change takes to happen}}$$

For example, if an object's speed goes from 5 to 15 meters per second (m/s), the change in velocity is 10 m/s. If that change happens in 2 seconds, the acceleration is given by

$$a = \frac{15\ \text{m/s} - 5\ \text{m/s}}{2\ \text{s}} = 5\ \text{m/s}^2$$

To determine how an object's motion is changing, we need to know two things: what net force is acting on the object and the object's resistance to that force. We can put that idea into equation form as follows:

$$\left(\frac{\text{An object's}}{\text{Acceleration}}\right) = \frac{\text{The force acting to change the object's motion}}{\text{The object's resistance to that change}} = \frac{\text{Force}}{\text{Mass}}$$

Newton's second law above is often written as Force = mass × acceleration, or $F = ma$. The units of force are called **newtons (N)**, so $1\ \text{N} = 1\ \text{kg m/s}^2$.

Suppose you are holding two blocks of the same size, the block in your right hand has twice the mass of the block in your left hand. When you drop the blocks, they both fall with the same acceleration, as Galileo showed, and they hit your two feet at the same time. Which will hit with more force: the block falling onto your right foot or the one falling onto your left foot? The block in your right hand, with twice the mass, will hit your right foot with twice the force that the other block hits your left foot.

To see how Newton's three laws of motion work together, study **Figure 3.23**. An astronaut is adrift in space, motionless with respect to the nearby space shuttle. With no tether to pull on, how can the astronaut get back to the ship? Suppose the 100-kg astronaut throws a 1-kg wrench directly away from the shuttle at a speed of 10 m/s. Newton's second law says that to change the wrench's motion, the astronaut must apply a force to it in the direction away from the shuttle. Newton's third law says that the wrench must therefore push back on the astronaut with as much force but in the opposite direction. The force of the wrench on the astronaut causes the astronaut to begin drifting toward the shuttle. How fast will the astronaut move? Turn to Newton's second law again. Because the astronaut has more mass, he will accelerate less than the wrench will. A force that causes the 1-kg wrench to accelerate to 10 m/s will have much less effect on the 100-kg astronaut. Because acceleration equals force divided by mass, the 100-kg astronaut will experience only 1/100 as much acceleration as the 1-kg wrench. The astronaut will drift toward the shuttle, but only at the leisurely rate of 1/100 × 10 m/s, or 0.1 m/s.

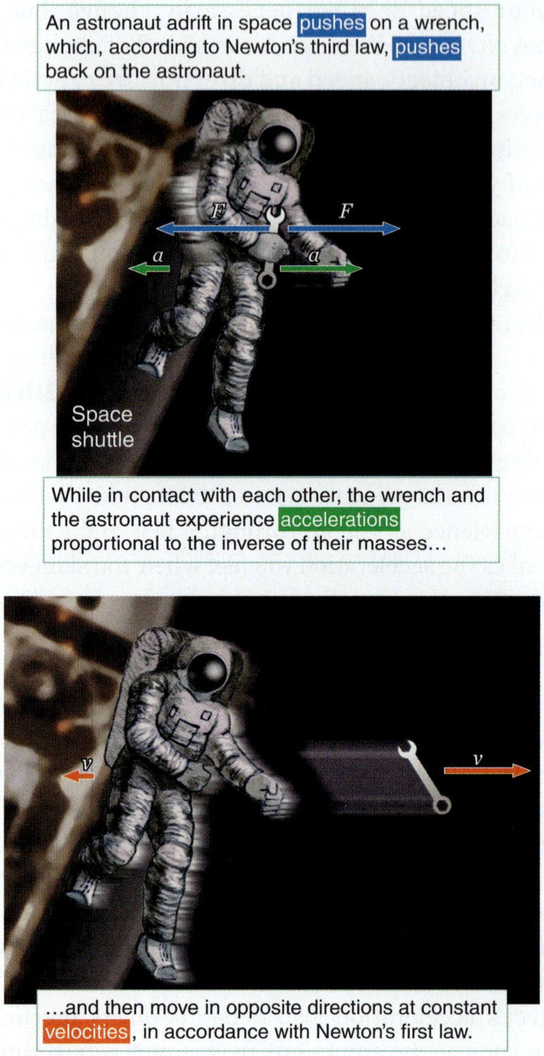

Figure 3.23 According to Newton's laws, if an astronaut adrift in space throws a wrench, the two will move in opposite directions. Their speeds will depend on their masses: the same force will produce a smaller acceleration of a more massive object than of a less massive object. (Acceleration and velocity arrows not drawn to scale.)

shown in **Figure 3.24**, when a woman moves a load on a cart by pulling a rope, the rope pulls back, and when a car tire pushes back on the road, the road pushes forward on the tire. Similarly, when Earth pulls on the Moon, the Moon pulls on Earth, and when a rocket engine pushes hot gases out of its nozzle, those hot gases push back on the rocket, propelling it into space.

All those force pairs are examples of **Newton's third law of motion**, which says that forces always come in pairs, with those two forces always equal in strength but opposite in direction. The forces in those pairs always act on two objects. Your weight pushes down on the floor, and the floor pushes back up on your feet with the same amount of force. For every force, an equal force in the opposite direction is *always* there.

CHECK YOUR UNDERSTANDING 3.4

Imagine a planet moving in a perfectly circular orbit around the Sun. Is that planet accelerating? (a) Yes, because it is constantly changing its speed. (b) Yes, because it is constantly changing its direction. (c) No, because its speed is not constantly changing. (d) No, because planets do not accelerate.

Origins Planets and Orbits

In addition to the planets in our own Solar System, researchers have detected thousands of planets orbiting stars other than our own Sun. Those planets' orbits can be calculated and understood by applying the same three Kepler's laws that we have discussed here. Astrobiologists think that a planet's orbit around its star affects the chances of life developing.

Consider the average distance of the planet from its star. A planet close to its star receives more energy than a planet farther out. If Earth were closer to the Sun, it would be hotter throughout the year—probably so hot that water would evaporate and no longer exist as a liquid. If Earth were farther from the Sun, it would be colder and all surface water would probably freeze. Liquid water was crucial for life to form on Earth, so some astronomers look for planets at a distance from their star such that liquid water can exist (that distance varies depending on the star's temperature and size).

Next consider the eccentricity of a planet's orbit. Recall from Figure 3.14a that Earth's orbit differs from a circle by less than 2 percent. Thus, Earth's distance from the Sun varies by only about 5 million km throughout the year; and as we saw in Chapter 2, seasons change on Earth because of the tilt of Earth's axis, not because of slight changes in distance from the Sun. Mars, however, has about the same axial tilt as Earth. But because the orbit of Mars is more eccentric, Mars has greater seasonal variation. The distance between Mars and the Sun varies from 1.38 AU (207 million km) to 1.67 AU (249 million km)—an eccentricity of 9 percent (or ~40 million km). As a result, the seasons on Mars are unequal. They are shorter when Mars is closer to the Sun and moving faster, and longer when Mars is farther from the Sun and moving slower. The inequality of the martian seasons affects the overall stability of the planet's temperature and climate. When we look at planets orbiting other stars, we see that many have orbital eccentricities even higher than that of Mars and therefore large variations in temperature.

Earth is at the right distance from the Sun to have temperatures that permit water to be liquid, and its orbital eccentricity is low enough that the average planetary temperature does not change much during its orbit. Those orbital characteristics have contributed to making the conditions on Earth suitable for life to develop.

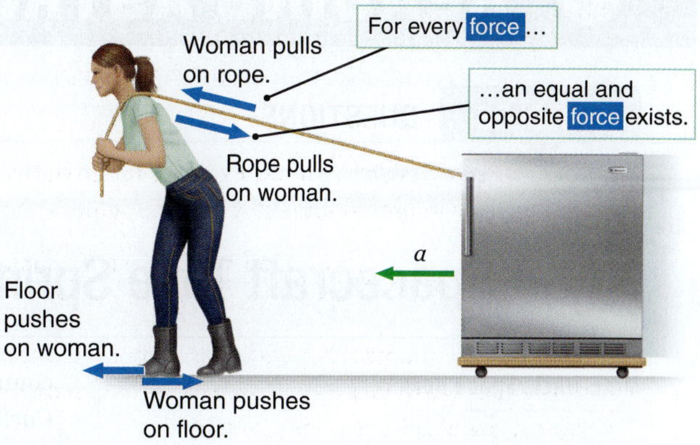

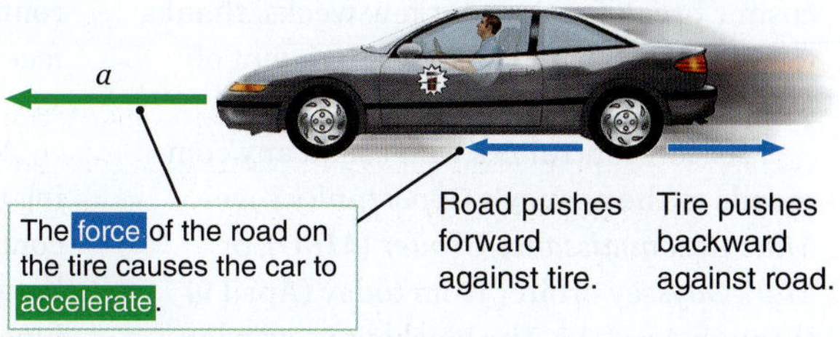

Figure 3.24 Newton's third law states that for every force, an equal and opposite force is always present. Those opposing forces always act on the two objects in the same pair.

Astronomy in Action: Velocity, Force, and Acceleration

AstroTour: Velocity, Acceleration, Inertia

READING ASTRONOMY NEWS

ARTICLES QUESTIONS

In this article we see that planetary alignments have practical implications. A spacecraft on the other side of the Sun will lose radio contact with Earth.

NASA Spacecraft Take Spring Break at Mars

MIKE WALL, Space.com

NASA's robotic Mars explorers are taking a cosmic break for the next few weeks, thanks to an unfavorable planetary alignment of Mars, the Earth, and the Sun.

Mission controllers won't send any commands to the agency's *Opportunity* rover, *Mars Reconnaissance Orbiter* (*MRO*), or *Mars Odyssey* orbiter from today (April 9) through April 26. The blackout is even longer for NASA's car-size *Curiosity* rover, which is slated to go solo from April 4 through May 1.

The cause of the communications moratorium is a phenomenon called a Mars solar conjunction, during which the Sun comes between Earth and the Red Planet (**Figure 3.25**). Our star can disrupt and degrade interplanetary signals in this formation, so mission teams won't be taking any chances.

"Receiving a partial command could confuse the spacecraft, putting them in grave danger," NASA officials explain in a video posted last month by the agency's Jet Propulsion Laboratory (JPL) in Pasadena, California.

Opportunity and *Curiosity* will continue performing stationary science work, using commands already beamed to the rovers. *Curiosity* will focus on gathering weather data, assessing the martian radiation environment, and searching for signs of subsurface water and hydrated minerals, officials said Monday (April 8).

MRO and *Odyssey* will also keep studying the Red Planet from above, and they'll continue to serve as communications links between the rovers and Earth. The conjunction will also affect the European Space Agency's *Mars Express* orbiter, officials have said.

Odyssey will send rover data home as usual during conjunction, though the orbiter may have to relay information multiple times due to dropouts. *MRO*, on the other hand, entered record-only mode on April 4. The spacecraft will probably have about 52 gigabits of data to relay when it's ready to start transmitting again on May 1, *MRO* officials have said.

Mars solar conjunctions occur every 26 months, so NASA's Red Planet veterans have dealt with them before. This is the fifth conjunction for *Opportunity*, in fact, and the sixth for *Odyssey*, which began orbiting Mars in 2001.

But it'll be the first for *Curiosity*, which touched down on August 5, kicking off a 2-year surface mission to determine if the Red Planet has ever been capable of supporting microbial life.

"The biggest difference for this 2013 conjunction is having *Curiosity* on Mars," *Odyssey* mission manager Chris Potts, of NASA's Jet Propulsion Laboratory in Pasadena, California, said in a statement last month.

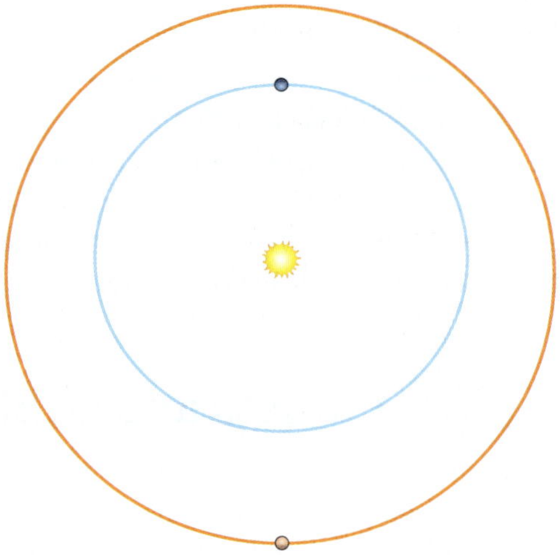

Figure 3.25 Every 26 months, Mars and Earth are on opposite sides of the Sun, preventing communication between the planets.

ARTICLES QUESTIONS

1. How often do Mars solar conjunctions occur? Does that interval correspond to the sidereal period or to the synodic period of Mars?
2. Give two reasons that conjunction is a bad time to view Mars in the sky.
3. Make a sketch of the Mars solar conjunction, showing Mars, Earth, and the Sun. Using the orbital period data in Table 3.2, add the positions of Mars and Earth at the next Mars conjunction. Will it take place at the same location in space?
4. View this short video from NASA: https://www.jpl.nasa.gov/video/details.php?id=1204. Do you think it explains the concepts well to someone who is not studying astronomy?
5. If people were on a mission on Mars, losing contact would be troubling. How might NASA plan to avoid losing contact with astronauts on Mars during conjunctions?

Summary

Early astronomers hypothesized that Earth was stationary at the center of the Solar System. Later astronomers realized that a Sun-centered Solar System was much simpler and could explain what people observed in the sky. Planets, such as Jupiter, orbit the Sun, not Earth. Kepler's laws describe the planets' elliptical orbits around the Sun, including details about how fast a planet travels at various points in its orbit. Those laws helped Newton develop his laws of motion, which govern how all objects move (not just orbiting ones). Orbital semimajor axis, eccentricity, and stability may affect a planet's suitability to foster life.

LG 1 **Compare the geocentric and heliocentric models of the Solar System.** Earth's motion is hard to detect, so before the Copernican Revolution, most people accepted a geocentric model of the Solar System, in which all objects orbit Earth. But in that model, the planets' apparent retrograde motion was hard to understand. Copernicus created the first comprehensive mathematical model of the Solar System with the Sun at the center, called a heliocentric model. His model explained apparent retrograde motion as a visual illusion seen when an inner planet passes an outer planet in their orbits.

LG 2 **Use Kepler's laws to describe how objects in the Solar System move.** Using Tycho's observational data, Kepler developed empirical rules to describe the motions of the planets. Kepler's three laws state that (1) planets move in elliptical orbits around the Sun; (2) planets move fastest when closest to the Sun and slowest when farthest from the Sun, so that the planets sweep out equal areas in equal times; and (3) the orbital period (P) of a planet squared equals the semimajor axis (A) of its orbit cubed: $P^2 = A^3$.

LG 3 **Explain how Galileo's astronomical discoveries convinced him that the heliocentric model was correct.** Galileo invented astronomical telescopes and used them to observe sunspots, craters on the Moon, and moons orbiting Jupiter. He also saw Venus going through phases, like the Moon, but changing its apparent size in each phase. The observations of Venus were hard to explain with Ptolemy's model.

LG 4 **Describe the physical laws of motion discovered by Galileo and Newton.** Galileo studied the physics of falling objects and discovered the principle of inertia. Newton's laws state that (1) objects do not change their motion unless they experience a net force, (2) force = mass × acceleration, and (3) every force has an equal and opposite force. Net forces cause accelerations (changes in motion). Inertia resists changes in motion.

? Unanswered Questions

- Would the history of scientific discoveries in physics and astronomy have been different if the Inquisition had not prosecuted Galileo? Galileo wrote the *Dialogo* after the Catholic Church in 1616 ordered him not to "hold or defend" the idea that Earth moves and the Sun is still. And he wrote his equally famous *Discorsi e Dimostrazioni Matematiche* (often shortened in English to "Two New Sciences") while under house arrest after his trial. However undeterred Galileo appeared to be, the effects of the decrees, prohibitions, and prosecutions might have dissuaded other scientists in Catholic countries from pursuing that type of work. Indeed, after Galileo's experiences, the center of the scientific revolution moved north to Protestant Europe.
- What percentage of planets are in unstable orbits? In younger planetary systems, planets might migrate in their orbits because of the presence of other massive planets nearby. We explain in Chapter 7 that Uranus and Neptune might have migrated in that way. Some planets have been discovered moving through the galaxy without any obvious orbit around a star—and therefore are not in the stable orbits we see in our own Solar System.

Questions and Problems

Test Your Understanding

1. An *empirical science* is one based on
   a. hypothesis.
   b. calculus.
   c. computer models.
   d. observed data.
2. When Earth catches up to a slower-moving outer planet and passes it in its orbit, the outer planet
   a. exhibits retrograde motion.
   b. slows down because it feels Earth's gravitational pull.
   c. becomes dimmer as it passes through Earth's shadow.
   d. moves into a more elliptical orbit.

3. Copernicus's model of the Solar System was superior to Ptolemy's because
   a. it had a mathematical basis and computed the spacing of the planets.
   b. it was much more accurate.
   c. it did not require epicycles.
   d. it fit the telescopic data better.
4. An inferior planet is one that is
   a. smaller than Earth.
   b. larger than Earth.
   c. closer to the Sun than Earth.
   d. farther from the Sun than Earth.
5. The time a planet takes to come back to the same position in space in relation to the Sun is called its ______________ period.
   a. synodic
   b. sidereal
   c. heliocentric
   d. geocentric
6. Suppose a planet is discovered orbiting a star in a highly elliptical orbit. While the planet is close to the star, it moves ______________, but while it is far away, it moves ______________.
   a. faster; slower
   b. slower; faster
   c. retrograde; prograde
   d. prograde; retrograde
7. If a superior planet is observed from Earth to have a synodic period of 1.2 years, what is its sidereal period?
   a. 0.54 years
   b. 1.8 years
   c. 4.0 years
   d. 6.0 years
8. A net force must be acting when an object
   a. accelerates.
   b. changes direction but not speed.
   c. changes speed but not direction.
   d. all of the above
9. For Earth, $P^2/A^3 = 1.0$ (in appropriate units). Suppose a new dwarf planet is discovered that is 14 times farther from the Sun than Earth is. For that planet,
   a. $P^2/A^3 = 1.0$.
   b. $P^2/A^3 > 1.0$.
   c. $P^2/A^3 < 1.0$.
   d. we can't know the value of P^2/A^3 without more information.
10. Galileo observed that Venus had phases that correlated with the size of its image in his telescope. From that information, you may conclude that Venus
    a. is the center of the Solar System.
    b. orbits the Sun.
    c. orbits Earth.
    d. orbits the Moon.
11. Kepler's second law says that
    a. planetary orbits are ellipses with the Sun at one focus.
    b. the square of a planet's orbital period equals the cube of its semimajor axis.
    c. net forces cause changes in motion.
    d. planets move fastest when closest to the Sun.
12. The average speed of a new planet in orbit has been reported to be 33 km/s. When it is closest to its star it moves at 31 km/s, and when it is farthest from its star it moves at 35 km/s. Those data must be in error because
    a. the average speed is far too fast.
    b. Kepler's third law says the planet has to sweep out equal areas in equal times, so the speed of the planet cannot change.
    c. Kepler's second law says the planet must move fastest when closest to its star, not when farthest away.
    d. with those numbers, the square of the orbital period will not be equal to the cube of the semimajor axis.
13. Galileo observed that Jupiter has moons. From that information, you may conclude that
    a. Jupiter is the center of the Solar System.
    b. Jupiter orbits the Sun.
    c. Jupiter orbits Earth.
    d. some things do not orbit Earth.
14. If you start from rest and accelerate at 10 mph/s and end up traveling at 60 mph, how long did it take?
    a. 1 second
    b. 6 seconds
    c. 60 seconds
    d. 0.6 seconds
15. Planets with high eccentricity may be unlikely candidates for life because
    a. the speed varies too much.
    b. the period varies too much.
    c. the temperature varies too much.
    d. the orbit varies too much.

Thinking about the Concepts

16. Study Figure 3.2. During normal motion, does Mars move toward the east or west? Which direction does it travel when moving retrogradely? For how many days did Mars move retrogradely? If one of the martian missions were photographing *Earth* in the sky during those days, what would it have observed?

17. Copernicus and Kepler engaged in what is called empirical science. What do we mean by *empirical*?

18. Explain why Saturn's synodic period is very close to 1 Earth year (a sketch may help).

19. Make a sketch of Earth, Venus, and the Sun in a geocentric model and in a heliocentric model. Label the Sun and Earth, and then show the changing phases of Venus during one of its orbits. What would we observe in each model? Why was the invention of the telescope necessary to distinguish between those models?

20. Experiment with falling objects as Galileo did. When you drop pairs of objects with different masses, do they reach the ground at the same time? Do they hit the ground with the same force? Does that approach work with a sheet of paper or a tissue—why or why not?

21. A planet's orbital speed around the Sun varies. When is the planet moving fastest in its orbit? Slowest?

22. The Moon's orbit around Earth is elliptical, too, with an eccentricity of 0.05. How does that value compare with the eccentricity of Earth's orbit? How do those elliptical orbits explain the types of solar eclipses discussed in Chapter 2?

23. Galileo came up with the concept of inertia. What do we mean by *inertia*?

24. If Kepler had lived on Mars, would he have deduced the same empirical laws for the motion of the planets? Explain.

25. What is the difference between speed and velocity? Between velocity and acceleration?

26. When involved in an automobile collision, a person not wearing a seat belt will move through the car and often strike the windshield directly. Which of Newton's laws explains why the person continues forward, even though the car stopped?

27. When riding in a car, we can sense changes in speed or direction through the forces that the car applies on us. Do we wear seat belts in cars and airplanes to protect us from speed or from acceleration? Explain your answer.

28. An astronaut standing on Earth can easily lift a wrench having a mass of 1 kg, but not a scientific instrument with a mass of 100 kg. On the International Space Station, he can manipulate both, although the scientific instrument responds much more slowly than the wrench. Explain why.

29. The Process of Science Figure illustrates that scientific ideas are always open to challenge. Construct an argument that the constant process of challenging and falsifying ideas is a strength of science, rather than a weakness.

30. How might you expect conditions on Earth to be different if the eccentricity of its orbit were 0.17 instead of 0.017?

Applying the Concepts

31. Study the graph in Figure 3.16. Is that graph linear or logarithmic? From the data on the graph, find the approximate semimajor axis and period of Saturn. Show your work.

32. Study Figure 3.19, which shows that Venus's apparent size changes as it goes through phases. Approximately how many times larger is Venus in the sky at the thinnest crescent than at the gibbous phase shown? Approximately how many times closer is Venus to us at the phase of that thinnest crescent than at the gibbous phase?

33. Suppose a dwarf planet is discovered orbiting the Sun with a semimajor axis of 50 AU. What would be the planet's orbital period?

34. The orbital period of Uranus is 84 years. Compute the semimajor axis of its orbit. How much time passes between oppositions of Uranus?

35. Dwarf planet Ceres is 2.77 AU from the Sun. Its synodic period is 1.278 years.
    a. Use Working It Out 3.1 to find the sidereal period in years.
    b. Use Kepler's law to find the sidereal period in years.
    c. Compare your results for (a) and (b).

36. Suppose you read online that "experts have discovered a new planet with a distance from the Sun of 2 AU and a period of 3 years." Use Kepler's third law to argue that this is impossible.

37. Show, as Galileo did, that Kepler's third law applies to the four Galilean moons of Jupiter by calculating P^2 divided by A^3 for each moon. (Data on the moons can be found in Appendix 4.)

38. In a period of 3 months, a planet travels 30,000 km with an average speed of 3.8 m/s. Some time later, the same planet travels 65,000 km in 3 months. How fast is the planet traveling at that later time? During which period is the planet closer to the Sun?

39. If you were on Mars, how often would you see retrograde motion of *Earth* in the martian night sky? (You can view a picture at https://mars.jpl.nasa.gov/allaboutmars/nightsky/retrograde.)

40. The elliptical orbit of a comet that a spacecraft recently visited is 1.24 AU from the Sun at its closest approach and 5.68 AU from the Sun at its farthest.
    a. Sketch the comet's orbit. When is it moving fastest? Slowest?
    b. What is the semimajor axis of its orbit? How long does the comet take to go around the Sun?
    c. What is the distance from the Sun to the "center" of the ellipse? What is the eccentricity of the comet's orbit?

41. You are driving down a straight road at a speed of 90 km/h, and you see another car approaching you at a speed of 110 km/h along the road.
    a. With respect to your own frame of reference, how fast is the other car approaching you?
    b. With respect to the other driver's frame of reference, how fast are you approaching the other driver's car?

42. During the latter half of the 19th century, a few astronomers thought a planet might be circling the Sun inside Mercury's orbit. They even gave it a name: Vulcan. We now know that Vulcan does not exist. If a planet with an orbit one-fourth the size of Mercury's actually existed, what would its orbital period be in comparison with that of Mercury?

43. Suppose you are pushing a small refrigerator of mass 50 kg on wheels. You push with a force of 100 N.
    a. What is the refrigerator's acceleration?
    b. Assume the refrigerator starts at rest. How long will the refrigerator accelerate at that rate before it gets away from you (that is, before it is moving faster than you can run—on the order of 10 m/s)?

44. If a 100-kg astronaut pushes on a 5,000-kg satellite and the satellite accelerates at 0.1 m/s^2, what acceleration does the astronaut experience in the opposite direction?

45. Sketch the orbit of Mars by using the information provided in the "Origins: Planets and Orbits" section of the chapter for the closest and farthest distances of Mars from the Sun.
    a. What is the orbit's major axis? Semimajor axis?
    b. What is the distance from the "center" of the orbit to the Sun? Compute the orbit's eccentricity. Compare that value with the eccentricity of Earth's orbit.

USING THE WEB

46. Go to the Web page "This Week's Sky at a Glance" (https://www.skyandtelescope.com/observing/sky-at-a-glance/) at the *Sky & Telescope* magazine website. Which planets are visible in your sky this week? Why are Mercury and Venus visible only in the morning before sunrise or in the evening just after sunset? Before telescopes, how did people know the planets were different from the stars?

47. Look up the dates for the next opposition of Mars, Jupiter, or Saturn. One source is the NASA "Sky Events Calendar" (https://eclipse.gsfc.nasa.gov/SKYCAL/SKYCAL.html). Check only the "Planet Events" box in "Section 2: Sky Events"; and in Section 3, generate a Sky Events Calendar or Table for the year. If you are coming up on an opposition (Opp), take pictures of the planet over the next few weeks. Can you see its position move in retrograde motion with respect to the background stars?

48. Refer to the Web page from question 47 to find the current observational positions of all the planets.
    a. Which ones are in or near to conjunction, opposition, or greatest elongation?
    b. Which are visible in the morning sky? In the evening sky?
    c. To connect the time we can see the planets on Earth with their physical alignments with Earth and the Sun in space, sketch the Solar System with Earth, the Sun, and the planets as it looks today from "above." Check your result with NASA's "Solar System Simulator" (https://space.jpl.nasa.gov): set it for "Show Me Solar System" as seen from above, and set the field of view at 2°, then 20°, and then 45° to see the inner and then the outer planets. Does your sketch agree with the simulator?

49. Go to the Museo Galileo website and view the Room VII video and exhibit at https://catalogue.museogalileo.it/room/RoomVII.html. What did his telescope look like? What other instruments did he use? From the museum page you can link to short videos (in English) on his science and his trial (http://catalogue.museogalileo.it/index/VideoIndexByThematicArea.html#s7). For example, click on "Galileo's micrometer." How did he measure the separation of the moons from Jupiter? How did that measurement allow him to show that the moons obeyed Kepler's third law? Why is Galileo often considered the first modern scientist? Why is his middle finger on display in the museum?

50. Go to the online "Extrasolar Planets Encyclopedia" (http://exoplanet.eu/catalog).
    a. Find a planet with an orbital period similar to that of Earth (click on "Period" twice and it will show decreasing values in days). What is the semimajor axis of the exoplanet's orbit? If it is very different from 1 AU, the star's mass is different from the Sun's. Click on the star name in the first column to see the star's mass. What is the planet's orbital eccentricity?
    b. Click on "Planet" to sort by name, and select a star with multiple planets. Verify that Kepler's third law applies by showing that the value of P^2/A^3 is about the same for each planet of that star. How eccentric are the orbits of the multiple planets?

EXPLORATION

Kepler's Laws

digital.wwnorton.com/astro6

In this Exploration, we examine how Kepler's laws apply to the orbit of Mercury. Visit the Digital Resources Page and on the Student Site open the "Planetary Orbits" Interactive Simulation in Chapter 3. That simulator animates the orbits of the planets, enabling you to control the simulation speed, as well as several other parameters. Here we focus on exploring Mercury's orbit, but you may wish to spend some time examining the orbits of other planets as well.

Kepler's First Law

In the "Orbit Settings" panel, use the drop-down menu next to "set parameters for" to select "Mercury" and then click "OK." Click the "Kepler's 1st Law" tab at the bottom of the control panel. Use the radio buttons to select "show empty focus" and "show center."

1 How would you describe the shape of Mercury's orbit?

____

____

Deselect "show empty focus" and "show center," and select "show semiminor axis" and "show semimajor axis." Under "Visualization Options," select "show grid."

2 Use the grid markings to estimate the ratio of the semiminor axis to the semimajor axis.

____

____

3 Calculate the eccentricity of Mercury's orbit from that ratio by using $e = [1 - (\text{Ratio})^2]^{1/2}$.

____

____

Kepler's Second Law

Click on "reset" near the top of the control panel, set parameters for Mercury, and click "OK." Then click on the "Kepler's 2nd Law" tab at the bottom of the control panel. Slide the "adjust size" slider to the right until the fractional sweep size is 1/8.

Click on "start sweeping." The planet moves around its orbit, and the simulation fills in area until one-eighth of the ellipse is filled. Click on "start sweeping" again as the planet arrives at the rightmost point in its orbit (that is, at the point in its orbit farthest from the Sun). You may need to slow the animation rate by using the slider under "Animation Controls." Click on "show grid" under the visualization options. (If the moving planet annoys you, pause the animation.) One easy way to estimate an area is to count the squares.

4 Count the squares in the yellow area and in the red area. You will need to decide what to do with fractional squares. Are the areas the same? Should they be?

____

____

Kepler's Third Law

Click on "reset" near the top of the control panel, set parameters for Mercury, and then click on the "Kepler's 3rd Law" tab at the bottom of the control panel. Select "show solar system orbits" in the "Visualization Options" panel. Study the graph. Use the eccentricity slider to change the simulated planet's eccentricity. Make the eccentricity first smaller and then larger.

5 Did anything in the graph change?

____

6 What do your observations of the graph tell you about the dependence of the period on the eccentricity?

____

Set the parameters back to those for Mercury. Now use the semimajor axis slider to change the semimajor axis of the simulated planet.

7 What happens to the period when you make the semimajor axis smaller?

____

____

8 What happens when you make it larger?

____

____

9 What do those results tell you about the dependence of the period on the semimajor axis?

____

____

4 Gravity and Orbits

Here in Chapter 4, we explore the physical laws that explain the regular patterns in the motions of the planets. Because the Sun is far more massive than all the other parts of the Solar System combined, its gravity shapes the motions of every object in its vicinity, from the almost circular orbits of some planets to the extremely elongated orbits of comets.

LEARNING GOALS

By the end of this chapter, you should be able to:

LG 1 Explain the elements of Newton's universal law of gravitation.

LG 2 Use the laws of motion and gravitation to explain planetary orbits.

LG 3 Explain how tidal forces from the Sun and Moon create Earth's tides.

LG 4 Describe how tidal forces affect solid bodies.

The International Space Station in orbit around Earth. ▶▶▶

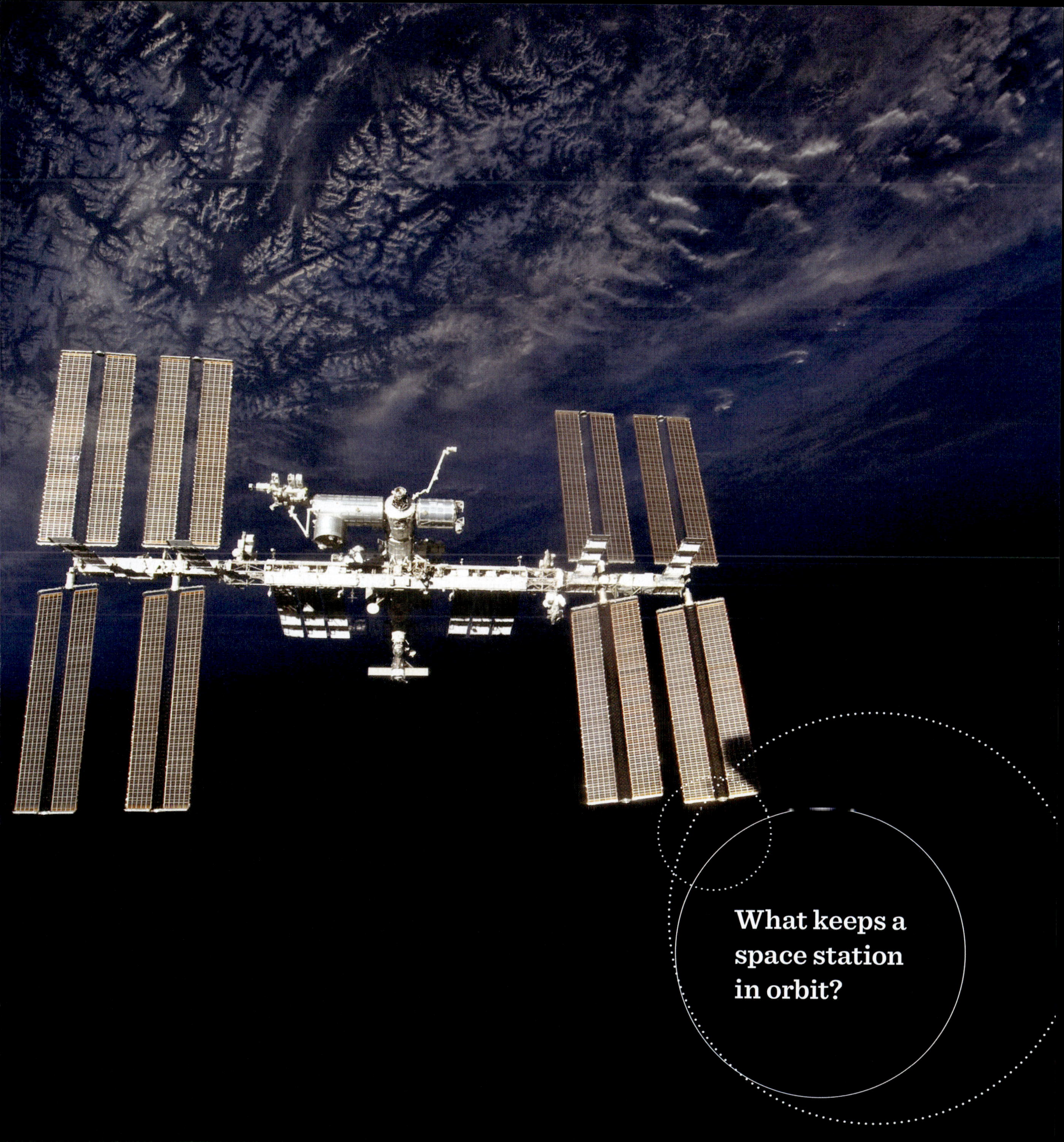

What keeps a space station in orbit?

4.1 Gravity Is an Attractive Force between Objects

In Chapter 3, we explored Kepler's work on the movement of the planets around the Sun and Newton's laws of motion. Here in Chapter 4, we build on those concepts to understand Newton's universal law of gravitation. Although some properties of gravity were observed before Newton, his work connected the everyday phenomenon of falling objects to the motion of the planets around the Sun. Newton's law of gravity combined Kepler's empirical laws and Newton's own laws of motion.

Gravity, Mass, and Weight

Many forces that we see in everyday life involve direct contact between objects. The cue ball on a pool table slams into the eight ball, knocking it into the pocket. The shoe of the child pushing a scooter presses directly against the pavement. When physical contact occurs between two objects, the source of the forces between them is easy to see.

If you drop a ball, it falls toward the ground. The ball picks up speed as it falls, accelerating downward toward Earth. According to Newton's second law, where there is acceleration, there is force. But where is the force that causes the ball to accelerate? The ball falling toward Earth is an example of a different kind of force, one that acts at a distance across the intervening void of space. The ball falling toward Earth is accelerating in response to the force of gravity. **Gravity**, one of the fundamental forces of nature, is the mutually attractive force between objects with mass.

Recall from Chapter 3 that Galileo discovered that all freely falling objects accelerate toward Earth at the same rate, regardless of their mass. Drop a marble and a book at the same time and from the same height, and they will hit the ground together. Note that air resistance becomes a factor at higher speeds, but the effect is negligible for dense, slow objects. The acceleration of falling objects due to gravity near the surface of Earth, which Galileo also measured experimentally, is usually written as g (lowercase) and has an average value across the surface of Earth of 9.8 meters per second squared (m/s^2). The value of g varies slightly across Earth's surface, ranging from 9.78 m/s^2 at the equator to 9.83 m/s^2 at the poles. That variation exists because Earth is not a perfect sphere: its rotation makes it flatter at the poles, so the radius of Earth is smaller there.

After working out the laws governing the motion of objects, Newton realized that if all objects, regardless of mass, fall with the same acceleration, the gravitational *force* on an object must be determined by the object's *mass*.

Newton's second law (Section 3.4) states that acceleration equals force divided by mass, or $a = F/m$. The acceleration due to gravity can be the same for all objects only if the value of the force divided by the mass is the same for all objects. In other words, an object twice as massive has double the gravitational force acting on it; an object three times as massive has triple the gravitational force acting on it, and so on.

The gravitational force acting on an object attracted by a planet is called the object's **weight**. On the surface of Earth, weight equals mass multiplied by the acceleration of gravity at Earth's surface, g. In common language, we often use weight and mass interchangeably. To be more scientifically precise, astronomers use *mass* to refer to the amount of matter in an object and *weight* to refer to the force that the planet's gravitational pull exerts on that object. Your mass is the same no matter what planet or moon you are on, but your weight will be different:

$$F_{\text{weight}} = m \times g$$

where F_{weight} is an object's weight in newtons (N), the metric unit of force; m is the object's mass in kilograms (kg); and g is Earth's constant for acceleration due to gravity, 9.8 m/s^2. On Earth, then, an object with a *mass* of 1 kg has a *weight* of 9.8 N (**Figure 4.1**). On the Moon, the acceleration due to gravity is 1.6 m/s^2, which is about one-sixth of its value on Earth, so a 1-kg mass has a weight of 1.6 N on the Moon. Similarly, your weight on the Moon would be about one-sixth of your weight on Earth.

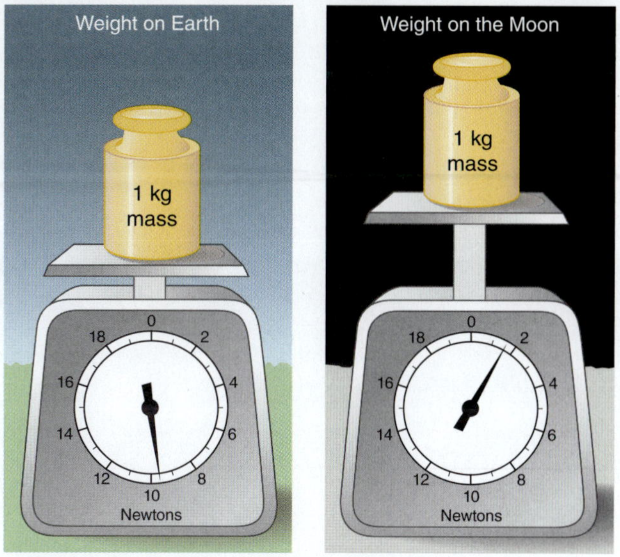

Figure 4.1 On the Moon, a mass of 1 kg has 1/6 the weight (displayed in newtons) that it has on Earth.

Newton's Law of Gravity

As Newton told the story, he saw an apple fall from a tree to the ground. He reasoned that if gravity is a force that depends on mass, a gravitational force should exist between *any* two masses, including between a falling apple and Earth. That great insight came from applying his third law of motion to gravity. Recall that Newton's third law states that for every force, an equal and opposite force is present. Therefore, if Earth exerts a force of 9.8 N on a 1-kg mass sitting on its surface, that 1-kg mass must also exert a force of 9.8 N on Earth. Drop a 7-kg bowling ball and it falls toward Earth, but at the same time Earth falls toward the 7-kg bowling ball. We don't notice the planet's motion because Earth is very massive, so it has a lot of resistance to changes in its motion. In the time a 7-kg bowling ball takes to fall to the ground from a height of 1 kilometer (km), Earth has "fallen" toward the bowling ball by only a tiny fraction of the size of an atom.

Newton reasoned that if doubling the mass of any object doubles the gravitational force between the object and Earth, doubling the mass of Earth ought to do the same thing. In short, the gravitational force between Earth and an object equals the product of the two masses multiplied by *something*:

$$\text{Gravitational force} = \text{Something} \times \text{Mass of Earth} \times \text{Mass of object}$$

If the mass of the object is 2 times greater, the force of gravity will be 2 times greater. Likewise, if the mass of Earth happened to be 3 times what it is, the force of gravity also would have to be 3 times greater. If the mass of both Earth and the object were greater by those amounts, the gravitational force would increase by a factor of 2×3, or 6 times. Because objects fall toward the center of Earth, we know that gravity is an attractive force acting along a line between the two masses.

If gravity is a force that depends on mass, a gravitational force should exist between *any* two masses. Suppose we have two masses—call them mass 1 and mass 2, or m_1 and m_2 for short. The gravitational force between them is *something* multiplied by the product of the masses:

$$\text{Gravitational force between two objects} = \text{Something} \times m_1 \times m_2$$

We have gotten this far just by combining Galileo's observations of falling objects with (1) Newton's laws of motion and (2) Newton's belief that Earth is a mass just like any other mass. But what about that "something" in the previous expression?

Kepler had already thought about that question. He reasoned that because the Sun is the focal point for planetary orbits, the Sun must exert an influence over the motions of the planets. Kepler speculated that the influence must grow weaker with distance from the Sun because the planets closer to the Sun moved much faster than the farther ones. Kepler did not know about forces or inertia or gravity as the cause of celestial motion, but he thought that geometry alone suggested how that solar "influence" might change for planets progressively farther from the Sun.

To see why the influence must become weaker, imagine you have a certain amount of paint to spread over the surface of a sphere. If the sphere is small, you can apply a thick coat of paint. If the sphere is larger, however, the paint has to be spread farther, so the coat will be thinner. The surface area of a sphere depends on the square of the sphere's radius. Double the radius of a sphere, and the sphere's surface area becomes 4 times what it was. If you paint that new, larger sphere, the paint must cover 4 times as much area, and the thickness of the paint will be only a fourth of what it was on the smaller sphere. If you triple the radius of the sphere, the sphere's surface will be 9 times larger and the coat of paint will be only one-ninth as thick.

Kepler reasoned that as the influence of the Sun extended farther and farther into space, it would have to spread out to cover the surface of a larger and larger imaginary sphere centered on the Sun. The Sun's influence should diminish with the square of the distance from the Sun—a relationship known as an **inverse square law**.

Kepler had an interesting idea, but not a scientific hypothesis with testable predictions. He couldn't explain how the Sun influences the planets, and he lacked the mathematical tools to calculate how an object would move under such an influence. Newton, however, had both. If gravity is a force between *any* two objects, a gravitational force should then exist between the Sun and each of the planets. If that gravitational force were the same as Kepler's "influence," gravity might behave according to an inverse square law.

Newton's expression for gravity came to look like this:

$$\text{Gravitational force between two objects} = \text{Something} \times \frac{m_1 \times m_2}{(\text{Distance between objects})^2}$$

The expression still has a "something," and that something is a constant of proportionality. That constant determines the strength of gravity between objects, and it is the same for all pairs of objects. Newton named it the **universal gravitational constant**. Newton estimated that gravitational constant, written as G (uppercase), by using Galileo's measurement of g, estimates of Earth's radius, and a guess at the mass of Earth by assuming it had about the same density as typical rocks. Not until many years later was the actual value of G first measured. Today the value of G is accepted as 6.67×10^{-11} N m^2/kg^2 or its equivalents: 6.67×10^{-11} m^3/(kg s^2) or 6.67×10^{-20} km^3/(kg s^2).

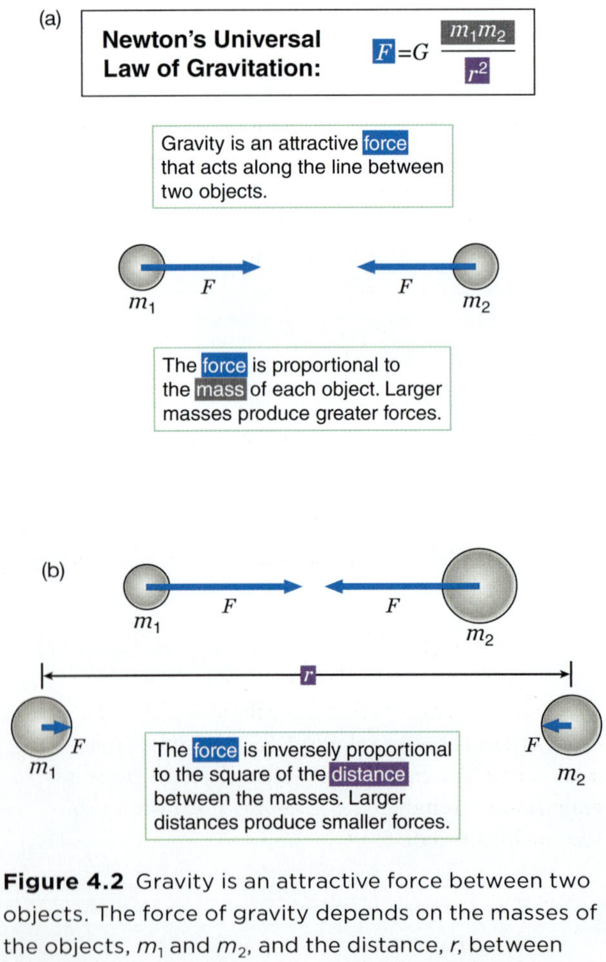

Figure 4.2 Gravity is an attractive force between two objects. The force of gravity depends on the masses of the objects, m_1 and m_2, and the distance, r, between them.

A Universal Law of Gravitation

Newton's **universal law of gravitation**, illustrated in **Figure 4.2**, states that gravity is a force between any two objects having mass. The force due to gravity has the following properties:

1. It is an attractive force acting along a straight line between the two objects.
2. It is proportional to the mass of one object (m_1) multiplied by the mass of the other object (m_2). If we double m_1, the force (F) increases by a factor of 2. Likewise, if we double m_2, F increases by a factor of 2.

3. It is inversely proportional to the square of the distance r between the centers of the two objects. The graph in **Figure 4.3** shows that if we double r, F decreases by a factor of 4. If we triple r, F falls by a factor of 9.

Written as a mathematical formula, the universal law of gravitation is

$$F_{\text{grav}} = G \times \frac{m_1 \times m_2}{r^2}$$

where F_{grav} is the force of gravity between two objects; m_1 and m_2 are the masses of objects 1 and 2, respectively; r is the distance between the centers of mass of the two objects; and G is the universal gravitational constant. The relationship between the force of gravity and the masses of and distance between two objects is further explored in **Working It Out 4.1**.

Gravity pulls you toward the center of Earth. Gravity holds the planets and stars together and keeps the thin blanket of air we breathe close to Earth's surface. The planets, including Earth, orbit the Sun, and gravity holds them in orbit. Gravity caused a vast

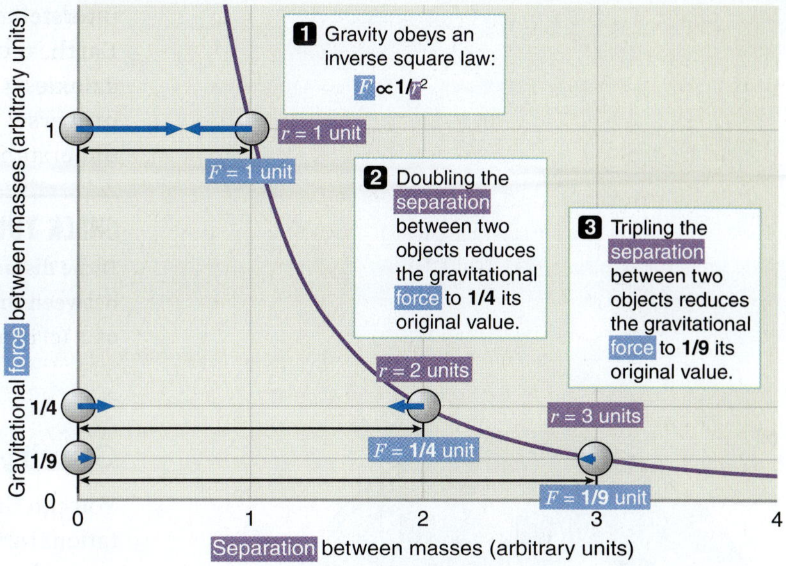

Figure 4.3 As two objects move apart, the gravitational force between them decreases by the inverse square of the distance between them.

4.1 Working It Out Playing with Newton's Laws of Motion and Gravitation

For any two objects, the force of gravity is directly proportional to the masses and inversely proportional to the *square* of the distance between them. Let's look at a few examples of how to use this equation:

Changing the Distance

How would the gravitational force between Earth and the Moon change if the distance between them were doubled? In that example, the masses of Earth and the Moon stay the same, and r becomes $2r$. We can calculate how the force changes by writing the equation for distance r and again for distance $2r$, and then taking a ratio to compare them:

$$\frac{F_{\text{grav at distance }2r}}{F_{\text{grav at distance }r}} = \frac{G \times \dfrac{M_{\text{Earth}} M_{\text{Moon}}}{(2r)^2}}{G \times \dfrac{M_{\text{Earth}} M_{\text{Moon}}}{r^2}}$$

We can cancel out the constant G and the masses of Earth and the Moon, which do not change. Then we need to multiply both the numerator and denominator of the fraction by r^2, remembering that both the 2 and the r get squared in $(2r)^2 = 4r^2$. The equation becomes

$$\frac{F_{\text{grav at distance }2r}}{F_{\text{grav at distance }r}} = \frac{\cancel{G} \times \cancel{M_{\text{Earth}} M_{\text{Moon}}}}{\cancel{G} \times \cancel{M_{\text{Earth}} M_{\text{Moon}}}} \times \frac{r^2}{(2r)^2}$$

$$\frac{F_{\text{grav at distance }2r}}{F_{\text{grav at distance }r}} = \frac{\cancel{r^2}}{4\cancel{r^2}} = \frac{1}{4}$$

Doubling the distance reduced the force by a factor of 4; that is, the force is 1/4 as strong.

Gravitational Acceleration

We have two ways to think about the gravitational force that Earth exerts on an object with mass m located on the surface of Earth. Recall Newton's second law of motion: $F = m \times a$. Here we are considering the gravitational force and the acceleration due to gravity, or $F_{\text{grav}} = m \times g$. The other way to think about the force is from the perspective of the universal law of gravitation:

$$F_{\text{grav}} = G \times \frac{M_{\text{Earth}} \times m}{R^2_{\text{Earth}}}$$

Here, M_{Earth} is the mass of Earth, and R_{Earth} is the radius of Earth. The two expressions describing that force must be equal. Therefore,

$$m \times g = G \times \frac{M_{\text{Earth}} \times m}{R^2_{\text{Earth}}}$$

The mass m is on both sides of the equation, so we can cancel it out. The equation then becomes

$$g = G \times \frac{M_{\text{Earth}}}{R^2_{\text{Earth}}}$$

The expression shows that the gravitational acceleration (g) experienced by an object of mass m on the surface of Earth is determined by Earth's mass and radius. The mass of the object itself (m) appears nowhere in the expression, so changing m does not affect the gravitational *acceleration* of an object on Earth.

interstellar cloud of gas and dust to collapse 4.5 billion years ago to form the Sun, Earth, and the rest of the Solar System. Gravity binds colossal groups of stars into galaxies. Gravity shapes space and time, and it can affect the ultimate fate of the universe. We return often to the concept of gravity because it is central to an understanding of the universe.

CHECK YOUR UNDERSTANDING 4.1A

If the distance between Earth and the Sun were cut in half, the gravitational force between those two objects would: (a) decrease by a factor of 4; (b) decrease by a factor of 2; (c) increase by a factor of 2; (d) increase by a factor of 4.

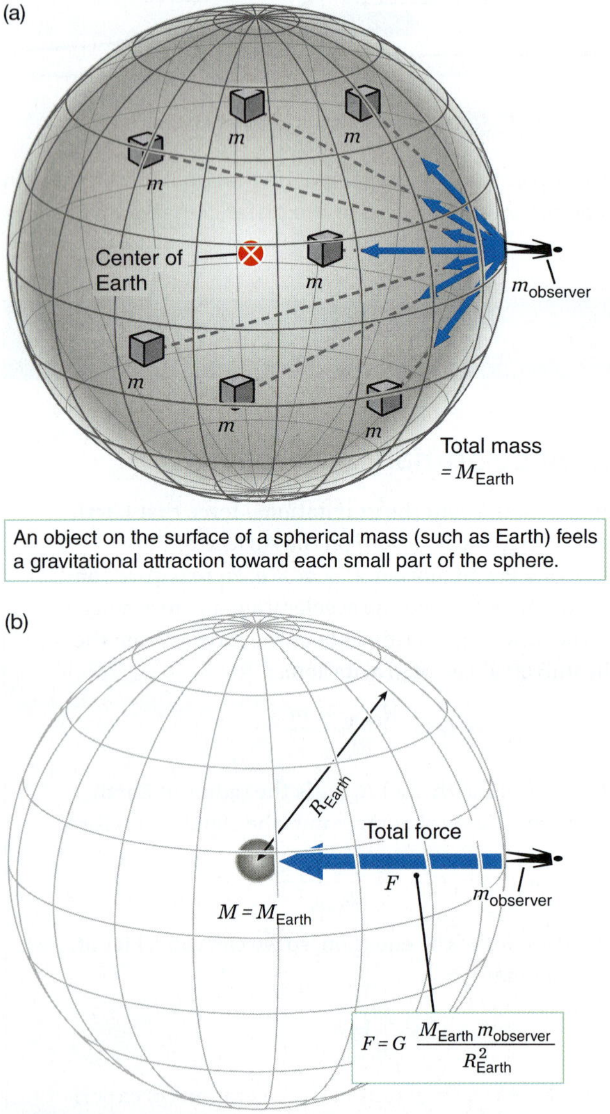

An object on the surface of a spherical mass (such as Earth) feels a gravitational attraction toward each small part of the sphere.

The net force is the same as if we scooped up the mass of the entire sphere and concentrated it at a point at the center.

Figure 4.4 Outside a sphere, the net gravitational force due to a spherical mass is the same as the gravitational force from the same mass concentrated at a point at the center of the sphere.

Gravity Differs from Place to Place within an Object

You can think of Earth as a collection of small masses, each of which feels a gravitational attraction toward every other small part of Earth. The mutual gravitational attraction that occurs among all parts of the same object is called *self-gravity*.

As you sit reading this book, you are exerting a gravitational attraction on every other fragment of Earth, and every other fragment of Earth is exerting a gravitational attraction on you. Your gravitational interaction is strongest with the parts of Earth closest to you. The parts of Earth on the other side of our planet are much farther from you, so their pull on you is correspondingly less.

The net effect of all those forces is to pull you (or any other object) toward Earth's center. If you drop a hammer, it falls directly toward the ground. Because Earth is nearly spherical, for every piece of Earth pulling you toward your right, a corresponding piece of Earth is pulling you toward your left with just as much force. For every piece of Earth pulling you forward, a corresponding piece of Earth is pulling you backward. Because Earth is almost spherically symmetric, all those "sideways" forces (see the blue arrows in **Figure 4.4a**) cancel out, leaving behind an overall force that points toward Earth's center (**Figure 4.4b**).

Some parts of Earth are closer to you and others are farther away, but an average distance exists between you and all the small fragments of Earth pulling on you. That average distance is the distance between you and the center of Earth. As illustrated in Figure 4.4b, the overall pull that you feel is the same as it would be if all the mass of Earth were concentrated at a single point located at the very center of the planet.

That relationship is true for any spherically symmetric object. Outside the object, its gravity behaves as though all the mass of that object were concentrated at a point at its center. That relationship is important in many applications. For example, when you estimate your weight on another planet, you are calculating the force of gravity between you and the planet. The "distance" in the gravitational equation will be the distance between you and the center of the planet, which is just the radius of the planet plus your altitude.

CHECK YOUR UNDERSTANDING 4.1B

If Earth shrank to a smaller radius but kept the same mass, would the gravitational force between Earth and the Moon: (a) decrease; (b) increase; or (c) stay the same? Would everyone's weight at Earth's surface: (a) increase; (b) decrease; or (c) stay the same?

4.2 An Orbit Is One Body "Falling Around" Another

Kepler's laws on the motions of planets enabled astronomers to predict the positions of the planets accurately, but those laws did not explain why the planets behave as they do. Newton's work explained why planets orbit the Sun.

Newton used his laws of motion and his proposed law of gravity to predict the paths of planetary orbits. His calculations showed that those orbits should be ellipses with the Sun at one focus, that planets should move faster when closer to the Sun, and that the square of the period of a planet's orbit should vary as the cube of the semimajor axis of that elliptical orbit. Newton's universal law of gravitation *predicted* that planets should orbit the Sun in just the way that Kepler's empirical laws described. By explaining Kepler's laws, Newton found important corroboration for his law of gravitation.

Gravity and Orbits

Newton's laws tell us how forces change an object's motion and how objects interact with one another through gravity. To know where an object will be at any given time, we have to "add up" the object's motion over time. Newton invented calculus to do that, but here we aim just for a conceptual understanding.

In Newton's time, the closest thing to making a heavy object fly was shooting cannonballs out of a cannon, so he used cannonballs in "thought experiments" about planetary motions. If you drop a cannonball, it falls directly to the ground, like any other mass does. A cannonball fired out of a cannon that is level with the ground behaves differently, however, as shown in **Figure 4.5a**. The cannonball still falls to the ground in the same amount of time as it does when it is dropped, but while falling it also travels over the ground, following a curved path that carries it a horizontal distance before it finally lands. The faster the cannonball moves when it is fired from the cannon, the farther it will go before it hits the ground (**Figure 4.5b**).

In the real world, that experiment reaches a natural limit. To travel through air, the cannonball must push air out of its way—an effect normally referred to as *air resistance*—which slows it down. But we can ignore such real-world complications in this thought experiment. Instead imagine that, having inertia, the cannonball continues along its course until it runs into something. The faster the cannonball moves when it is fired, the farther it goes before hitting the ground. If the cannonball flies far enough, Earth's surface curves out from under it, as shown in **Figure 4.5c**. Eventually a point is reached (**Figure 4.5d**) at which the cannonball is flying so fast that the surface of Earth curves away from the cannonball at the same rate at which the cannonball is falling toward Earth. When that occurs, the cannonball, which always falls toward the center of Earth, is, in a sense, falling around the world. An **orbit** is the path of one object that freely falls around another.

Why do astronauts appear to float freely about the cabin of a spacecraft? It is not because they have escaped Earth's gravity; Earth's gravity is what holds them in their orbit. Instead, the answer lies in Galileo's early observation that all objects fall with the same acceleration, regardless of their mass. The astronauts and the spacecraft are both in orbit around Earth, moving in the same direction, at the same speed, and experiencing the same gravitational acceleration, so they fall around Earth together. **Figure 4.6** demonstrates that point. The astronaut is orbiting Earth just as the spacecraft is orbiting Earth. On the surface of Earth,

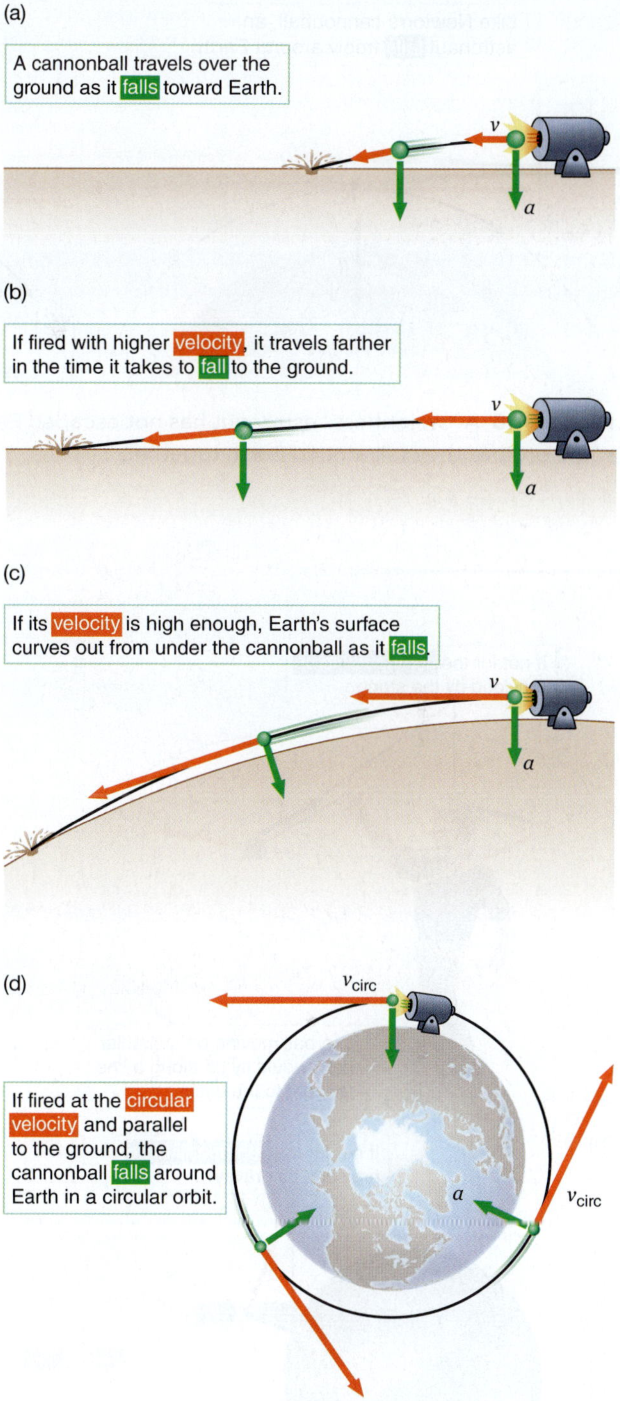

Figure 4.5 Newton realized that a cannonball fired at the right speed would fall around Earth in a circle. Velocity (*v*) is indicated by a red arrow and acceleration (*a*), by a green arrow.

AstroTour: Newton's Laws and Universal Gravitation

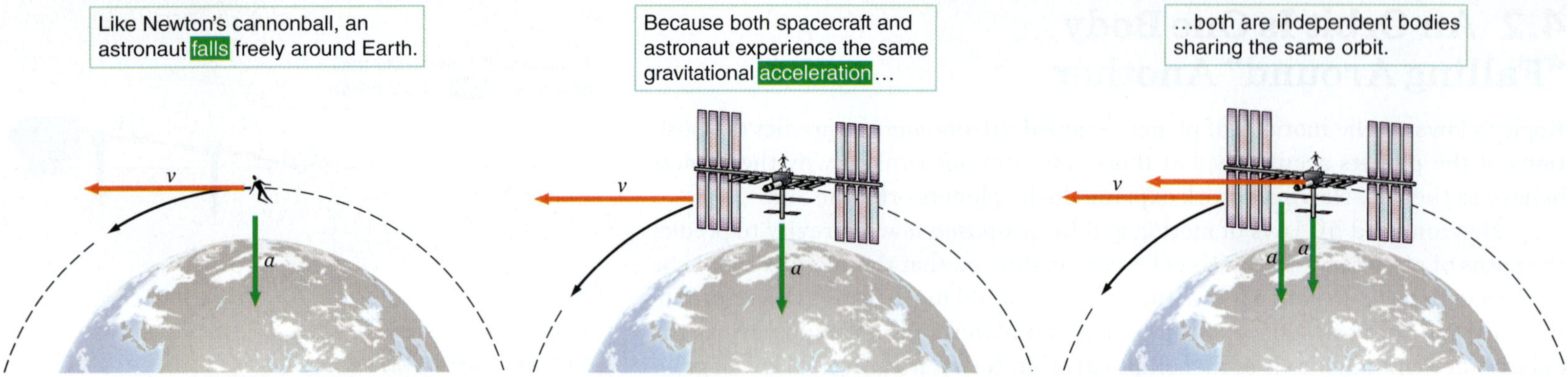

Figure 4.6 A "weightless" astronaut has not escaped Earth's gravity. Rather, an astronaut and a spacecraft share the same orbit as they fall around Earth together.

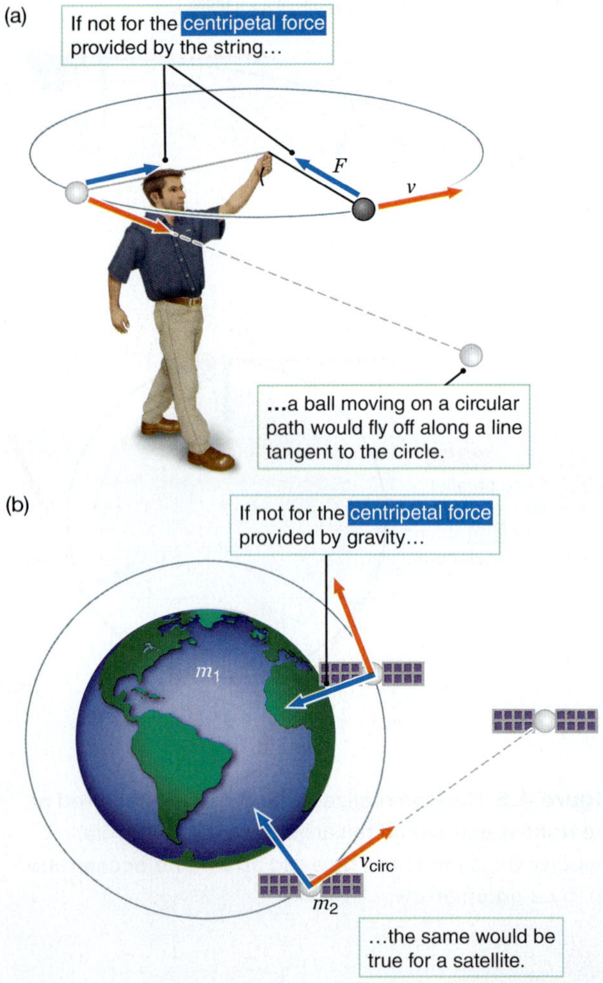

Figure 4.7 (a) A string provides the centripetal force that keeps a ball moving in a circle. (We are ignoring the smaller force of gravity that also acts on the ball.) (b) Gravity provides the centripetal force that holds a satellite in a circular orbit about Earth.

your body tries to fall toward the center of Earth, but the ground gets in the way. You feel your weight when you are standing on Earth because the ground pushes on you to oppose the force of gravity, which pulls you downward. In the spacecraft, however, nothing interrupts the astronaut's fall because the spacecraft is falling around Earth in the same orbit as the astronaut. The astronaut is in **free fall**, falling freely in Earth's gravity. The **Process of Science Figure** illustrates the universality of Newton's law of gravitation.

What Velocity Is Needed to Reach Orbit?

How fast must Newton's cannonball be fired for it to fall around the world? The cannonball would be in **uniform circular motion**, meaning it moves along a circular path at constant speed. That type of motion is discussed in more depth in Appendix 8. Another example of uniform circular motion is a ball whirling around your head on a string, as shown in **Figure 4.7a**. If you let go of the string, the ball will fly off in a straight line in whatever direction it is traveling at the time, just as Newton's first law predicts for an object in motion. The string prevents the ball from flying off by constantly changing the direction the ball is traveling. The central force of the string on the ball is called a **centripetal force**: a force toward the center of the circle. Using a more massive ball, speeding up its motion, and making the string shorter so that the turn is tighter all are ways to increase the force needed to keep a ball moving in a circle.

For Newton's cannonball (or a satellite), no string holds the ball in its circular motion. Instead, gravity supplies the force, as illustrated in **Figure 4.7b**. The force of gravity must be just enough to keep the satellite moving on its circular path. Because that force has a specific strength, the satellite must therefore be moving at a particular speed around the circle, which we call its **circular velocity** (v_{circ}). If the satellite were moving at any other velocity, it would not be moving in a circular orbit. Remember the cannonball: If it is moving too slowly, it will drop below the circular path and hit the ground. Similarly, if the cannonball is moving too fast, its motion will carry it above the circular orbit. Only a cannonball moving at just the right velocity—the circular velocity—will fall around Earth on a circular path (see Figure 4.5d).

Newton's thought experiment became a reality in 1957, when the Soviet Union launched the first artificial object to orbit Earth. They used a rocket to lift Sputnik 1, an object about the size of a basketball, high enough above Earth's upper atmosphere that air resistance wasn't an issue. Sputnik 1 achieved a high enough speed that it fell around Earth, just like Newton's imaginary cannonball.

PROCESS OF SCIENCE

Universality

The laws of physics are the same everywhere and at all times.
The principle of universality underlies our understanding of the natural world.

1604

Galileo

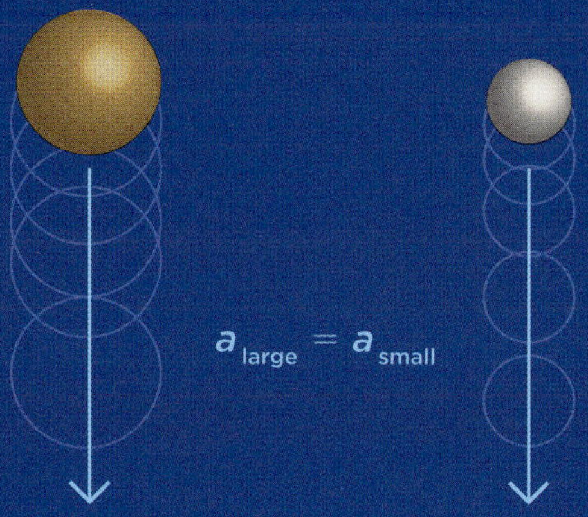

In 1604, Galileo determined that all objects on Earth have the same gravitational acceleration, *a*. Universality means that all objects on any world share the same gravitational acceleration with each other.

1971

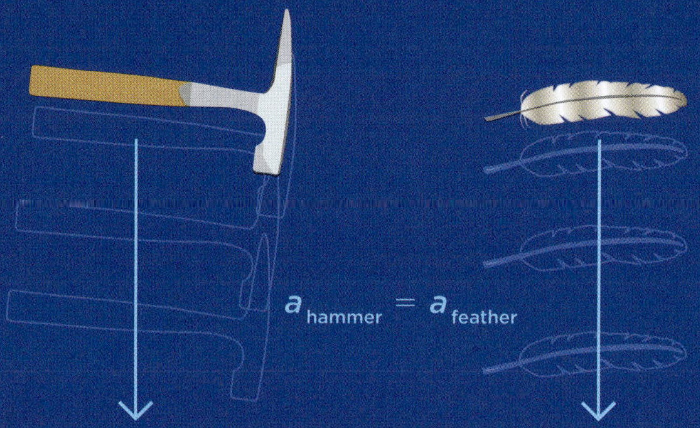

David Scott and *Apollo 15*

In 1971, *Apollo 15* commander David Scott tested the gravitational law with a hammer and a falcon feather on the Moon. With no air resistance, the objects had the same gravitational acceleration.

The physical laws that apply to falling objects also apply to planets orbiting the Sun, to stars orbiting within the galaxy, and to galaxies orbiting each other. This universality allows us to draw conclusions about very distant objects.

When one object is falling around a much more massive object, we say that the less massive object is a **satellite** of the more massive object. Planets are satellites of the Sun, and moons are natural satellites of planets. Newton's imaginary cannonball and Sputnik 1 were satellites (*sputnik* means "satellite" in Russian). An Earth-orbiting spacecraft and the astronauts inside it are independent satellites of Earth that conveniently happen to share the same orbit.

The Shape of Orbits

Some Earth satellites travel a circular path at constant speed. Just like the ball on the string, satellites traveling at the circular velocity always stay the same distance from Earth, neither speeding up nor slowing down in orbit. But what if the satellite then fired its rockets and started traveling *faster* than the circular velocity? Earth's pull is as strong as ever, but because the satellite has greater speed, Earth's gravity does not bend the satellite's path sharply enough to hold it in a circle. When that happens, the satellite begins to climb above a circular orbit.

AstroTour: Elliptical Orbit

As the distance between the satellite and Earth increases, the satellite slows down. Think about a ball thrown upward into the air, as shown in **Figure 4.8a**. As the ball climbs higher, the pull of Earth's gravity opposes its motion, slowing the ball down. The ball climbs more and more slowly until its vertical motion stops for an instant and then is reversed; the ball then begins to fall back toward Earth, speeding up along the way. A satellite does the same thing. As the satellite climbs above a circular orbit and begins to move away from Earth, Earth's gravity opposes the satellite's outward motion, slowing the satellite down. The farther the satellite is from Earth, the more slowly the satellite moves—just like the ball thrown into the air. Also just like the ball, the satellite reaches a maximum height on its curving path and then begins falling back toward Earth. When it does, Earth's gravity speeds the satellite up as it gets closer and closer to Earth. The satellite's orbit has changed from circular to elliptical.

Any object in an elliptical orbit, including a planet orbiting the Sun (**Figure 4.8b**), will therefore lose speed as it pulls away from what it is orbiting

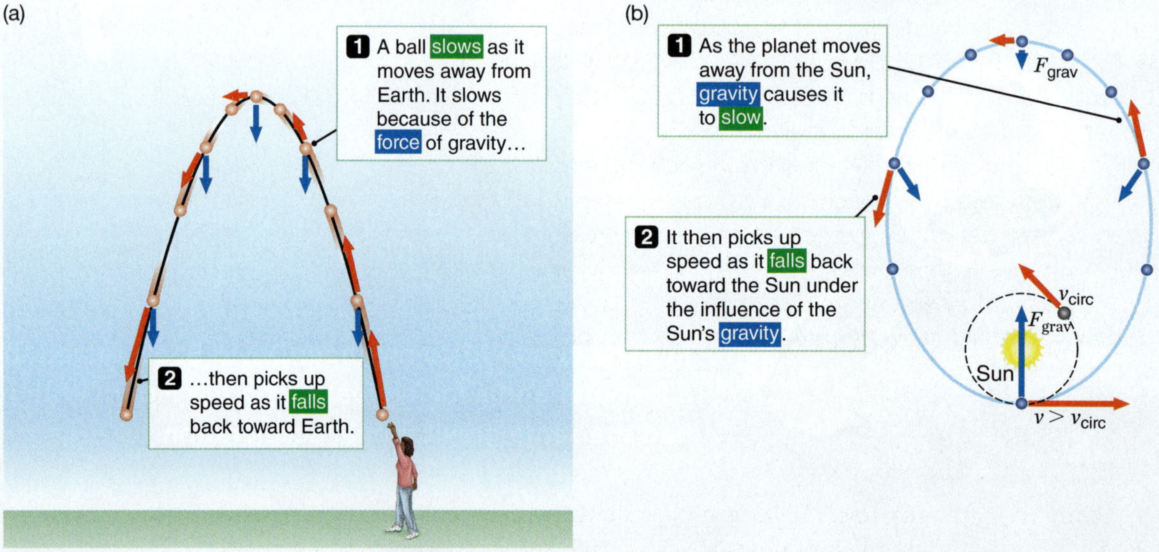

Figure 4.8 (a) A ball thrown into the air slows as it climbs away from Earth and then speeds up as it heads back toward Earth. (b) A planet in an elliptical orbit around the Sun does the same thing. (Although no planet has an orbit as eccentric as the one shown here, the orbits of comets can be far more eccentric.)

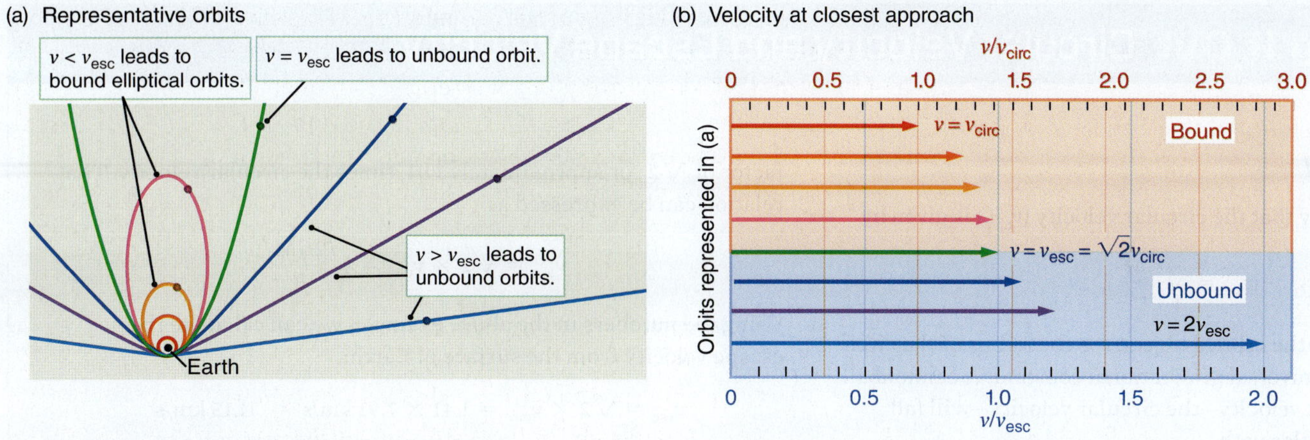

Figure 4.9 (a) Various orbits that share the same point of closest approach but differ in velocity at that point. (b) Closest-approach velocities for the orbits in (a). An object's velocity at closest approach determines the orbit's shape and whether the orbit is bound or unbound. v_{circ} = circular velocity; v_{esc} = escape velocity.

and then gain that speed back again as gravity causes it to fall inward toward what it is orbiting. Gravity, therefore, helps explain Kepler's second law from Section 3.2—namely, that a planet moves fastest when closest to the Sun and slowest when farthest from the Sun.

Newton's laws do more than explain Kepler's laws: they predict different types of orbits beyond Kepler's empirical experience. **Figure 4.9a** shows a series of satellite orbits, each with the same point of closest approach to Earth but with different velocities at that point, as indicated in **Figure 4.9b**. The more speed a satellite has at its closest approach to Earth, the farther the satellite can pull away from Earth, and the more eccentric its orbit becomes. As long as it remains elliptical, no matter how eccentric, the orbit is called a **bound orbit** because the satellite is gravitationally bound to the object it is orbiting.

In that sequence of faster and faster satellites, a point comes at which the satellite is moving so fast that gravity cannot reverse its outward motion, so the object travels away from Earth, never to return. The lowest speed at which an object can permanently leave the gravitational grasp of another mass is called the **escape velocity**, v_{esc}. Once a satellite's velocity at closest approach equals or exceeds v_{esc}, and it is no longer gravitationally bound to the object it was orbiting, we say it is in an **unbound orbit**.

As Figure 4.9a shows, an object with a velocity *less* than the escape velocity (v_{esc}) has an elliptically shaped orbit and follows the same path over and over again, whereas unbound orbits do not close like an ellipse. An object such as a comet on an unbound orbit makes only a single pass around the Sun and then continues away from the Sun into deep space, never to return. Circular velocity and escape velocity are further explored in **Working It Out 4.2**.

Estimating Mass by Using Newton's Version of Kepler's Laws

Newton's calculations opened up a new way to investigate the universe. He showed that the same physical laws that describe the flight of a cannonball on Earth—or the legendary apple falling from a tree—also describe the motions of the planets through the heavens. His laws of motion and gravitation predict all three of Kepler's empirical laws of planetary motion. Newton's version of Kepler's

4.2 Working It Out Circular Velocity and Escape Velocity

Circular Velocity

In Appendix 8, we show that the circular velocity (v_{circ}) is given by

$$v_{circ} = \sqrt{\frac{GM}{r}}$$

where M is the mass of the orbited object, r is the radius of the circular orbit, and G is the universal gravitational constant. A cannonball moving at just the right velocity—the circular velocity—will fall around Earth on a circular path.

We can use that equation to show how fast Newton's cannonball would have to travel to stay in its circular orbit. The average radius of Earth is 6,370 km, the mass of Earth is 5.97×10^{24} kg, and the gravitational constant is 6.67×10^{-20} km^3/(kg s^2). Inserting those values into the expression for v_{circ}, we get

$$v_{circ} = \sqrt{\frac{[6.67 \times 10^{-20}\,\text{km}^3/(\text{kg s}^2)] \times (5.97 \times 10^{24}\,\text{kg})}{6{,}370\,\text{km}}} = 7.91\,\text{km/s}$$

To stay in its circular orbit, Newton's cannonball would have to be traveling about 8 kilometers per second (km/s)—more than 28,000 kilometers per hour (km/h). That's well beyond the range of a typical cannon, but rockets routinely reach those speeds.

Now let's compare that speed with that needed to launch a satellite from the Moon into orbit just above the lunar surface. The radius of the Moon is 1,740 km, and its mass is 7.35×10^{22} kg. Those values give a considerably lower circular velocity than the one for Earth:

$$v_{circ} = \sqrt{\frac{(6.67 \times 10^{-20}\,\text{km}^3/\text{kg s}^2) \times (7.35 \times 10^{22}\,\text{kg})}{1{,}740\,\text{km}}} = 1.68\,\text{km/s}$$

Escape Velocity

Sending a spacecraft to another planet requires launching it with a velocity greater than Earth's escape velocity. The escape velocity is a factor of $\sqrt{2}$, or approximately 1.41, times the circular velocity. That relation can be expressed as

$$v_{esc} = \sqrt{2} \times v_{circ} = \sqrt{\frac{2GM}{R}}$$

Using the numbers in the above example, we can calculate the escape velocity from the surface of Earth:

$$v_{esc} = \sqrt{2} \times v_{circ} = 1.41 \times 7.91\,\text{km/s} = 11.15\,\text{km/s}$$

To leave Earth, a rocket must have a speed of at least 11.15 km/s, or 40,140 km/h.

As with weight, the escape velocity from other astronomical objects is different from the escape velocity from Earth. Ida is a small asteroid orbiting the Sun between the orbits of Mars and Jupiter. Ida has an average radius of 15.7 km and a mass of 4.2×10^{16} kg. Therefore,

$$v_{esc} = \sqrt{\frac{2 \times [6.67 \times 10^{-20}\,\text{km}^3/(\text{kg s}^2)] \times (4.2 \times 10^{16}\,\text{kg})}{15.7\,\text{km}}}$$

$$v_{esc} = 0.019\,\text{km/s} = 68\,\text{km/h}$$

A baseball thrown at about 130 km/h (81 mph) would easily escape from Ida's surface and fly off into interplanetary space.

When two objects are closer to having the same mass, however—such as dwarf planet Pluto and its moon Charon, or a large planet and a star, or two stars—both objects experience significant accelerations in response to their mutual gravitational attraction. The two objects are both orbiting about a common point located between them, called the **center of mass**, so we now must think of them as falling around each other. Each mass is moving on its own elliptical orbit around the two objects' mutual center of mass. From measuring the size and period of any orbit, we can calculate the *sum* of the masses of the orbiting objects. Almost all knowledge about the masses of astronomical objects comes directly from applying Newton's version of Kepler's third law.

laws can be used to estimate the mass of the Sun from the orbit of Earth, as seen in **Working It Out 4.3**.

When Earth, a much less massive object, is orbiting the Sun, a much more massive object, the Sun's gravity has a strong influence on Earth, but Earth's gravity has little effect on the Sun. Therefore, the Sun remains nearly motionless from Earth's orbit around it.

Astronomy in Action: Center of Mass

CHECK YOUR UNDERSTANDING 4.2

Rank the circular velocities of the following objects from smallest to largest: (a) a 5-kg object orbiting Earth halfway to the Moon; (b) a 10-kg object orbiting Earth just above Earth's surface; (c) a 15-kg object orbiting Earth at the same distance as the Moon.

4.3 Working It Out Calculating Mass from Orbital Periods

Newton's Version of Kepler's Third Law

The time a planet takes to complete one orbit around the Sun (the orbital period, P) equals the distance traveled divided by the planet's speed. For simplicity, let's assume the orbit is circular. Thus, the time an object takes to make one trip around the Sun is the circumference of the circle ($2\pi r$) divided by the object's speed (v). The speed of the planet must be equal to the circular velocity discussed in Working It Out 4.2. Putting those relationships together, we have

$$\text{Orbital period } (P) = \frac{\text{Circumference of orbit}}{\text{Circular velocity}} = \frac{2\pi r}{\sqrt{\dfrac{GM_{\text{Sun}}}{r}}}$$

Squaring both sides of the equation gives

$$P^2 = \frac{4\pi^2 r^2}{\dfrac{GM_{\text{Sun}}}{r}} = \frac{4\pi^2}{GM_{\text{Sun}}} \times r^3$$

The square of the period of an orbit is equal to a constant ($4\pi^2/GM_{\text{Sun}}$) multiplied by the cube of the radius of the orbit. That is Kepler's third law (P^2 = constants $\times A^3$) applied to circular orbits. When Kepler used Earth units of years and astronomical units for the planets, he was taking a ratio of their periods and orbital radii with those for Earth, so the constants canceled out. Using calculus, Newton similarly could derive Kepler's third law for elliptical orbits with semimajor axis A instead of radius r:

$$P^2 = \frac{4\pi^2}{GM_{\text{Sun}}} \times A^3$$

Mass of the Sun

If we can measure the size and period of any orbit, we can use Newton's universal law of gravitation to calculate the mass of the object being orbited. To do so, we rearrange Newton's form of Kepler's third law above to read

$$M = \frac{4\pi^2}{G} \times \frac{A^3}{P^2}$$

Everything on the right side of that equation is either a constant (4, π, and G) or a measurable quantity (the semimajor axis A and period P of an orbit). The left side of the equation is the mass of the object at the focus of the ellipse. For example, we can find the mass of the Sun by noting the period and semimajor axis of the orbit of a planet around the Sun. Let's use the numbers for Earth. Whenever we have an equation with G, putting everything else into the same units as G (km, kg, s) is best. So first we must compute the number of seconds in 1 year: $P = 1$ yr $= 365.24$ days/yr $\times$ 24 h/day $\times$ 60 min/h $\times$ 60 s/min $= 3.16 \times 10^7$ s. The semimajor axis $A = 1$ AU $= 1.5 \times 10^8$ km. Then the mass of the Sun can be computed:

$$M_{\text{Sun}} = \frac{4\pi^2}{G} \times \frac{A^3}{P^2} = \frac{4\pi^2}{6.67 \times 10^{-20}\ \text{km}^3/(\text{kg s}^2)} \times \frac{(1.5 \times 10^8\ \text{km})^3}{(3.16 \times 10^7\ \text{s})^2}$$

$$M_{\text{Sun}} = 2.0 \times 10^{30}\ \text{kg}$$

We could have used the period and semimajor axis of another Solar System planet to get the same result for the mass of the Sun.

4.3 Tidal Forces Are Caused by Gravity

The rise and the fall of the oceans are called Earth's **tides**. Coastal dwellers long ago noted that the strength of the tides varies with the phase of the Moon. Tides are strongest during a new or a full Moon and are weakest during first quarter or third quarter Moon. Here in Section 4.3, we see how tides result from differences between the strength of the gravitational pull of the Moon and Sun on one part of Earth in comparison with their pull on other parts of Earth.

Astronomy in Action: Tides

Tides and the Moon

Recall from Figure 4.4 that each small part of an object feels a gravitational attraction toward every other small part of the object, and that self-gravity differs from place to place. In addition, each small part of an object feels a gravitational attraction toward every other mass in the universe, and those external forces also differ from place to place within the object.

Simulation: Tides on Earth

AstroTour: Tides and the Moon

The Moon's gravity pulls on Earth as though the Moon's mass were concentrated at its center. The side of Earth that faces the Moon is closer to the Moon

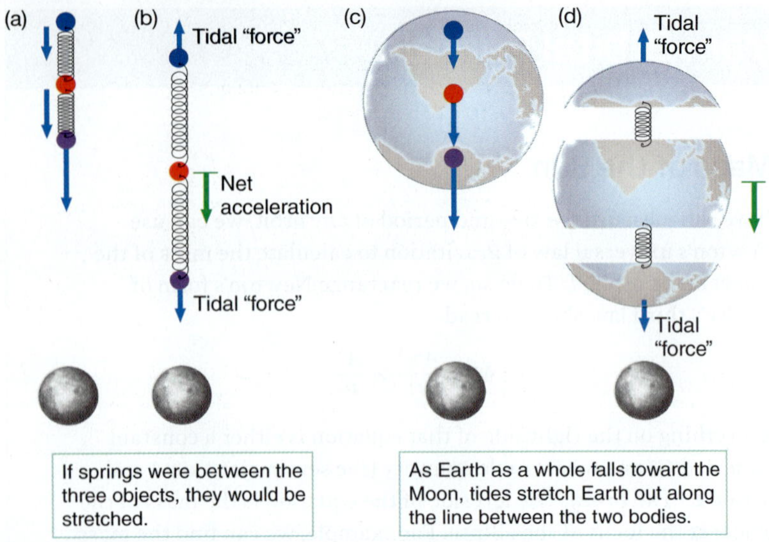

Figure 4.10 (a) Imagine three objects connected by springs. (b) The springs are stretched as though forces were pulling outward on each end of the chain. (c) Similarly, three locations on Earth experience different gravitational attractions toward the Moon. (d) The difference in the Moon's gravitational attraction across Earth causes Earth's tides.

than the rest of Earth is, so it feels a stronger-than-average gravitational attraction toward the Moon. In contrast, the side of Earth facing away from the Moon is farther than average from the Moon, so it feels a weaker-than-average attraction toward the Moon. The pull of the Moon on the near side of Earth is about 7 percent greater than its pull on the far side of Earth.

To understand the consequence of that variation in the pull of the Moon, imagine three rocks being pulled by gravity toward the Moon. A rock closer to the Moon feels a stronger force than a rock farther from the Moon. Now suppose the three rocks are connected by springs (**Figure 4.10a**). As the rocks are pulled toward the Moon, the purple rock pulls away from the red rock, and the red rock pulls away from the blue rock. Therefore, the differences in the gravitational forces they feel will stretch *both* springs (**Figure 4.10b**). Now, instead of springs, imagine that the rocks are at different places on Earth (**Figure 4.10c**). On the side of Earth away from the Moon, the force is smaller (as indicated by the shorter arrow), so that part gets left behind (**Figure 4.10d**). Those differences in the Moon's gravitational attraction on different parts of Earth are called **tidal forces**.

Figure 4.11 shows how Earth is stretched, causing a tidal bulge. The Moon is not pushing the far side of Earth away; rather, the Moon simply is not pulling on the far side of Earth as hard as it is pulling on the planet as a whole. The far side of Earth is "left behind" as the rest of the planet is pulled more strongly toward the Moon. Figure 4.11 shows that a net force also is squeezing inward on Earth in the direction perpendicular to the line between Earth and the Moon. Together, the stretching by tidal forces along the line between Earth and the Moon and the squeezing by tidal forces perpendicular to that line distort Earth's shape, like a rubber ball caught in the middle of a tug-of-war.

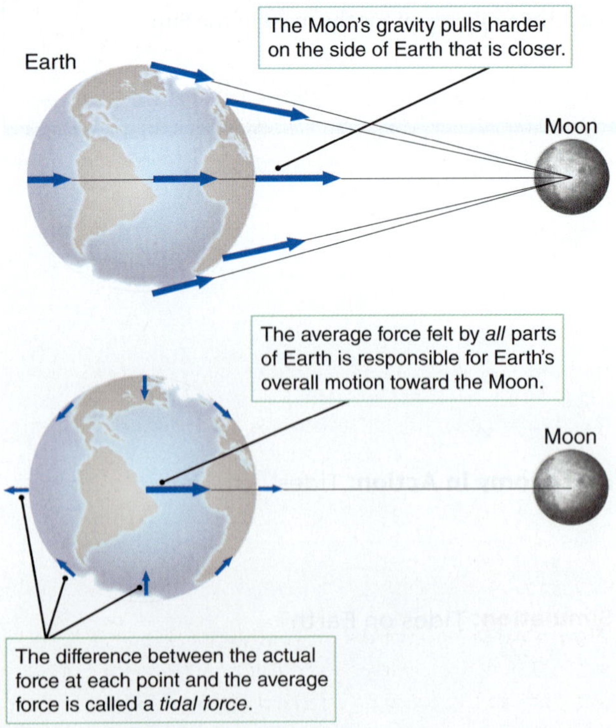

Figure 4.11 Tidal forces stretch Earth along the line between Earth and the Moon but compress Earth perpendicular to that line.

If the surface of Earth were perfectly smooth and covered with an ocean of uniform depth and Earth did not rotate, the Moon would pull our oceans into an elongated **tidal bulge** like that shown in **Figure 4.12a**. The water would be at its deepest on the side toward the Moon and on the side away from the Moon, and the water would be at its shallowest midway between. But Earth is *not* covered with perfectly uniform oceans, and Earth *does* rotate. As any point on Earth rotates through the ocean's tidal bulges, that point experiences the ebb and flow of the tides. In addition, friction between the spinning Earth and its tidal bulge drags the oceanic tidal bulge around in the direction of Earth's rotation, as illustrated in **Figure 4.12b**.

Figure 4.12c shows the tides you would experience throughout a day. Begin as the rotating Earth carries you through the tidal bulge on the Moonward side of the planet. Because Earth's rotation drags the tidal bulge, the Moon is not exactly overhead but is instead high in the western sky. When you are at the high point in the tidal bulge, the ocean around you has risen higher than average—called a *high tide*. The Moon rises 50 minutes later than it did on the previous day, so the time between the Moon's return to the same position in the sky from the previous day is 24h 50m. Each high and low tide is 1/4 of that, or about 6 1/4 hours later. So, 6 1/4 hours after the high tide, somewhat after the Moon has settled beneath the western horizon, the rotation of Earth carries you through a point where the ocean is lower than average—called a *low tide*. If you wait another 6 1/4 hours, it is again high tide. You are now passing through the region where ocean water is "left behind" (with respect to Earth as a whole) in the tidal bulge on the side of Earth that is away from the Moon. The Moon at that time is hidden from view on

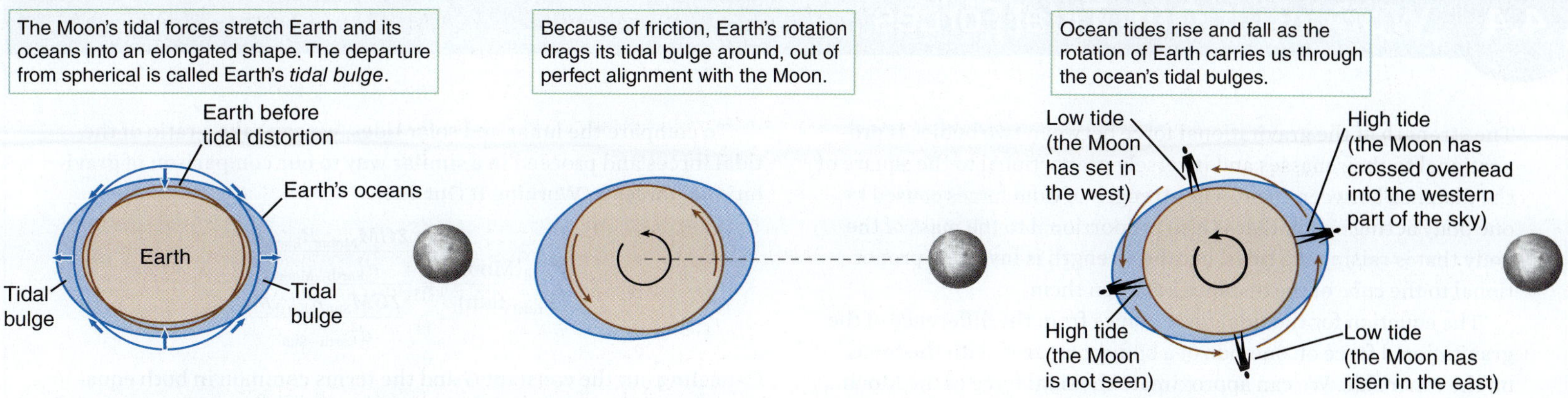

Figure 4.12 (a) Tidal forces pull Earth and its oceans into a tidal bulge. (b) Earth's rotation pulls its tidal bulge slightly out of alignment with the Moon. (c) As Earth's rotation carries us through those bulges, we experience the ocean tides. The magnitude of the tides has been exaggerated here for clarity. In these figures, the observer is looking down from above Earth's North Pole. Sizes and distances are not to scale.

the far side of Earth. About 6 1/4 hours later, sometime after the Moon has risen above the eastern horizon, it is low tide. About 25 hours after you started this journey—the amount of time the Moon takes to return to the same point in the sky from which it started—you again pass through the tidal bulge on the near side of the planet. That is the age-old pattern by which mariners have lived their lives for millennia: the twice-daily coming and going of high tide, shifting through the day in lockstep with the passing of the Moon.

Ocean currents and local geography, such as the shapes of Earth's shorelines and ocean basins, complicates the simple picture of tides. In addition, oceanwide oscillations occur, similar to water sloshing around in a basin. As the oceans respond to the tidal forces from the Moon, they flow around the various landmasses that break up the water covering our planet. Some places, such as the Mediterranean Sea and the Baltic Sea, are protected from tides by their relatively small sizes and the narrow passages connecting those bodies of water to the larger ocean. In other places, the shape of the land funnels the tidal surge from a large region of ocean into a relatively small area, concentrating its effect, as at eastern Canada's Bay of Fundy (**Figure 4.13**).

Solar Tides

Tides resulting from the pull of the Moon are called **lunar tides**, but the Sun also influences Earth's tides. The gravitational pull of the Sun causes Earth to stretch along a line pointing approximately in the direction of the Sun. The side of Earth closer to the Sun is pulled toward the Sun more strongly than the side of Earth away from the Sun, just as the side of Earth closest to the Moon is pulled more strongly toward the Moon. Tides on Earth resulting from differences in the gravitational pull of the Sun are called **solar tides**. The absolute strength of the Sun's pull on Earth is nearly 200 times greater than the strength of the Moon's pull on Earth. However, the Sun's gravitational attraction does not change by much from one side of Earth to the other because the Sun is much farther away than the Moon. As a result, solar tides are only about half as strong as lunar tides (**Working It Out 4.4**).

(a)

(b)

Figure 4.13 The world's most extreme tides are found in the Bay of Fundy in eastern Canada. Water rocks back and forth in the bay with a period of about 13 hours, close to the 12.5-hour period of the tides. The shape of the basin amplifies the tides so that the difference in water depth between low tide (a) and high tide (b) is extreme—typically about 14.5 meters and as much as 16.6 meters.

4.4 Working It Out Tidal Forces

The strength of the gravitational force between two bodies is proportional to their masses and inversely proportional to the square of the distance between them. The strength of tidal forces caused by one body acting on another is also proportional to the mass of the body that is raising the tides, but the strength is inversely proportional to the *cube* of the distance between them.

The equation for the tidal force comes from the difference of the gravitational force on one side of a body compared with the force on the other side. We can approximate the tidal force of the Moon acting on a mass m on the surface of Earth with

$$F_{\text{tidal}}(\text{Moon}) = \frac{2GM_{\text{Moon}}R_{\text{Earth}}m}{d^3_{\text{Earth–Moon}}}$$

where R_{Earth} is Earth's radius and $d_{\text{Earth-Moon}}$ is the distance between Earth and the Moon.

Let's compare the Moon's tidal force acting on Earth with the Sun's tidal force acting on Earth. F_{tidal} from the Moon is given in the preceding equation; F_{tidal} from the Sun is given by

$$F_{\text{tidal}}(\text{Sun}) = \frac{2GM_{\text{Sun}}R_{\text{Earth}}m}{d^3_{\text{Earth–Sun}}}$$

To compare the lunar and solar tides, we can take a ratio of the tidal forces and proceed in a similar way to our comparison of gravitational forces in Working It Out 4.1:

$$\frac{F_{\text{tidal}}(\text{Moon})}{F_{\text{tidal}}(\text{Sun})} = \frac{\dfrac{2GM_{\text{Moon}}R_{\text{Earth}}m}{d^3_{\text{Earth–Moon}}}}{\dfrac{2GM_{\text{Sun}}R_{\text{Earth}}m}{d^3_{\text{Earth–Sun}}}}$$

Canceling out the constant G and the terms common in both equations (m and R_{Earth}) gives

$$\frac{F_{\text{tidal}}(\text{Moon})}{F_{\text{tidal}}(\text{Sun})} = \frac{M_{\text{Moon}}}{M_{\text{Sun}}} \times \frac{d^3_{\text{Earth–Sun}}}{d^3_{\text{Earth–Moon}}} = \frac{M_{\text{Moon}}}{M_{\text{Sun}}} \times \left(\frac{d_{\text{Earth–Sun}}}{d_{\text{Earth–Moon}}}\right)^3$$

Using the values from Appendix 4, $M_{\text{Moon}} = 7.35 \times 10^{22}$ kg, $M_{\text{Sun}} = 2 \times 10^{30}$ kg, $d_{\text{Earth-Moon}} = 384{,}400$ km, and $d_{\text{Earth-Sun}} = 1.5 \times 10^8$ km, gives

$$\frac{F_{\text{tidal}}(\text{Moon})}{F_{\text{tidal}}(\text{Sun})} = \frac{7.35 \times 10^{22}\ \text{kg}}{2 \times 10^{30}\ \text{kg}} \times \left(\frac{1.5 \times 10^8\ \text{km}}{384{,}400\ \text{km}}\right)^3 = 2.2$$

Thus, the tidal force between Earth and the Moon is 2.2 times stronger than the tidal force between Earth and the Sun. The Sun is an important factor, too, though, and that's why the tides change depending on the alignment of the Moon and the Sun with Earth.

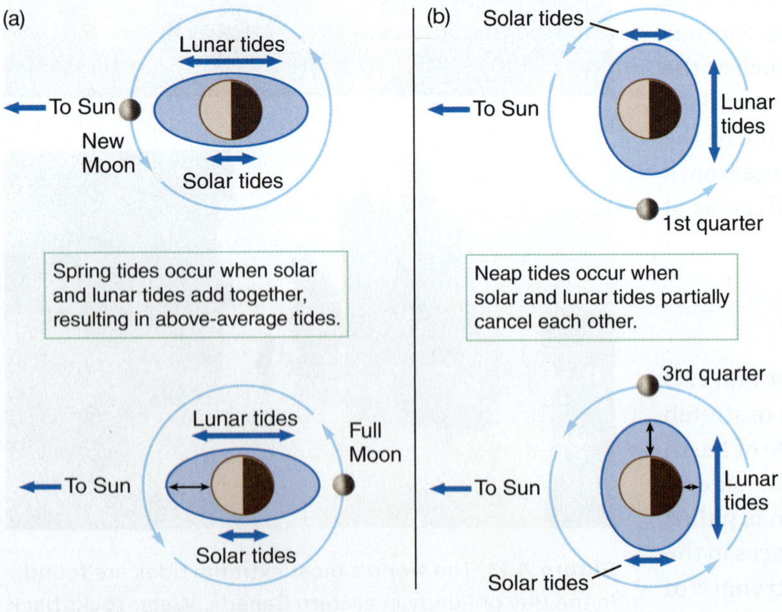

Figure 4.14 Solar tides are about half as strong as lunar tides. The interactions of solar and lunar tides result in either (a) spring tides when they are added together or (b) neap tides when they partially cancel each other.

Solar and lunar tides interact. When the Moon and the Sun are lined up with Earth (**Figure 4.14a**), at either new or full Moon, the lunar and solar tides on Earth overlap. That alignment creates more extreme tides that range from higher high tides to lower low tides. The extreme tides near the new or full Moon are called **spring tides**—not because of the season but because the water appears to spring out of the sea. Conversely, when the Moon, Earth, and Sun make a right angle (**Figure 4.14b**), at the Moon's first and third quarters, the lunar and solar tidal forces stretch Earth in different directions, creating less extreme tides known as **neap tides**. The word *neap* is derived from the Saxon word *neafte*, which means "scarcity": at those times of the month, shellfish and other food gathered in the tidal region are less accessible because the low tide is higher than at other times. Neap tides are only about half as strong as average tides and only a third as strong as spring tides.

CHECK YOUR UNDERSTANDING 4.3

Rank the following in order of weakest to strongest total tides: (a) new Moon in July; (b) first quarter Moon in July; (c) full Moon in January; (d) third quarter Moon in January.

4.4 Tidal Forces Affect Solid Bodies

In Section 4.3, we focused on how the liquid of Earth's oceans moves in response to the tidal forces from the Moon and Sun. But those tidal forces also affect the solid body of Earth. As Earth rotates through its tidal bulge, the solid body of the planet is constantly being deformed by tidal forces. Earth is somewhat elastic (like a rubber ball), and tidal stresses cause a vertical displacement of about 30 centimeters (cm) between high tide and low tide, or roughly a third of the displacement of the oceans. Deforming the shape of a solid object requires energy. (For a practical demonstration of that fact, hold a rubber ball in your hand and squeeze and release it a few dozen times.) That energy from the deformation is converted into thermal energy by friction in Earth's interior. That friction opposes and takes energy from the rotation of Earth, causing Earth to gradually slow. Earth's internal friction adds to the slowing caused by friction between Earth and its oceans as the planet rotates through the tidal bulge of the oceans. As a result, Earth's days are currently getting about 1.7 milliseconds (ms) longer every century. That sounds insignificant, but it adds up: when dinosaurs ruled 200 million years ago, the day was closer to 23 hours long, and 200 million years into the future, the day will be close to 25 hours.

Tidal Locking

Other solid bodies besides Earth experience tidal forces. For example, the Moon has no bodies of liquid to make tidal forces obvious, but its shape is distorted in the same manner as Earth's. Because of Earth's much greater mass and the Moon's smaller radius, Earth's tidal effects on the Moon are stronger than the Moon's tidal effects on Earth. Given that the average tidal deformation of Earth is about 30 cm, the average tidal deformation of the Moon should be about 6 meters. However, what we actually observe on the Moon is a tidal bulge of about 20 meters. That unexpectedly large displacement exists because the Moon's tidal bulge was "frozen" into its relatively rigid crust at an earlier time, when the Moon was closer to Earth and tidal forces were much stronger than they are today. Planetary scientists sometimes call that deformation the Moon's *fossil tidal bulge*.

Recall that the Moon's rotation period equals its orbital period (see Figure 2.22). That synchronous rotation of the Moon is a result of **tidal locking**. Early in the Moon's history, the period of its rotation was almost certainly different from its orbital period. As the Moon rotated through its extreme tidal bulge, however, friction within the Moon's crust was tremendous, rapidly slowing the Moon's rotation. After a fairly short time, the period of the Moon's rotation equaled the period of its orbit. When its orbital and rotation periods became equalized, the Moon no longer rotated with respect to its tidal bulge. Instead, the Moon and its tidal bulge rotated *together*, in lockstep with the Moon's orbit around Earth. As illustrated in **Figure 4.15**, that scenario continues today as the tidally distorted Moon orbits Earth, always keeping the same face and the long axis of its tidal bulge toward Earth.

Tidal forces affect not only the rotations of the Moon and Earth but also their orbits. Because of its tidal bulge, Earth is not perfectly spherical. Therefore, the material in Earth's tidal bulge on the side nearer the Moon pulls on the Moon more strongly than does material in the tidal bulge on the back side of Earth. Because the tidal bulge on the Moonward side of Earth "leads" the Moon

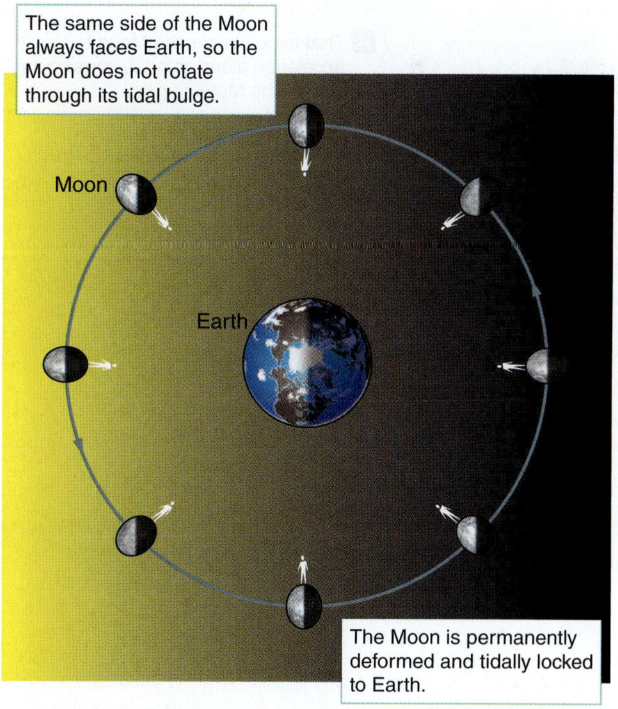

Figure 4.15 Tidal forces due to Earth's gravity lock the Moon's rotation to its orbital period.

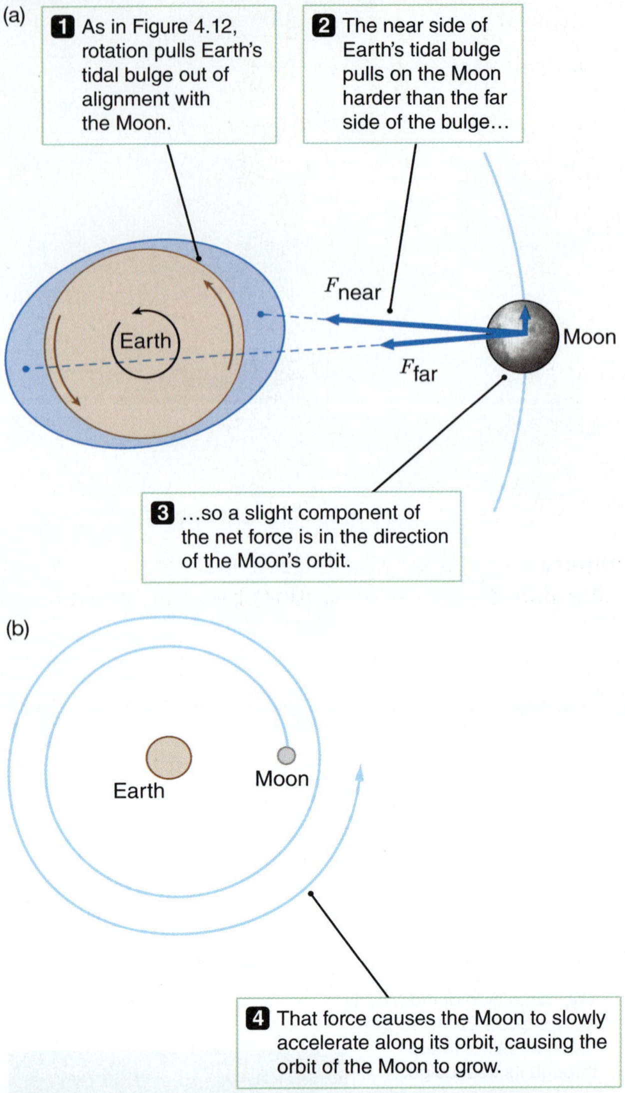

Figure 4.16 Interaction between Earth's tidal bulge and the Moon causes the Moon to accelerate in its orbit and the Moon's orbit to grow.

somewhat, as shown in **Figure 4.16**, the gravitational attraction of the bulge causes the Moon to accelerate slightly along the direction of its orbit around Earth. The rotation of Earth is dragging the Moon along with it. The acceleration of the Moon in the direction of its orbital motion causes the orbit of the Moon to grow larger. At present, the Moon is drifting away from Earth at a rate of 3.83 cm per year.

As the Moon grows more distant, the lunar month gets about 0.014 second longer each century. If that increase in the radius of the Moon's orbit were to continue long enough (about 50 billion years), Earth would become tidally locked to the Moon, just as the Moon is now tidally locked to Earth. At that point, the Earth's period of rotation, the Moon's period of rotation, and the Moon's orbital period would all be the same—about 47 of our current days—and the Moon would be about 43 percent farther from Earth than it is today. That situation will never actually occur, however, because the Sun will have burned itself out long before then.

The effects of tidal forces are apparent throughout the Solar System. Most of the moons in the Solar System are tidally locked to their parent planets, and for dwarf planet Pluto and its largest moon, Charon, each is tidally locked to the other.

Tidal locking is only one way that orbits and rotations can be coupled. Tidal forces have coupled the planet Mercury's rotation to its very elliptical orbit around the Sun. Unlike the Moon's synchronous rotation, however, Mercury spins on its axis three times for every two trips around the Sun. The period of Mercury's orbit—87.97 Earth days—is exactly 1 1/2 times the 58.64 days that Mercury takes to spin once on its axis. When Mercury comes to the point in its orbit that is closest to the Sun, one hemisphere faces the Sun, and then in the next orbit, the other hemisphere faces the Sun.

Tidal Forces on Many Scales

We normally think of the effects of tidal forces as small in comparison with the force of gravity holding an object together, yet tidal effects can be extremely destructive. Consider for a moment the fate of a small moon, asteroid, or comet that wanders too close to a massive planet such as Jupiter or Saturn. All objects in the Solar System larger than about a kilometer in diameter are held together by their self-gravity. However, the self-gravity of a small object such as an asteroid, a comet, or a small moon is feeble. In contrast, the tidal forces close to a massive object such as Jupiter can be very strong. If the tidal forces trying to tear an object apart become greater than the self-gravity trying to hold the object together, the object will break into pieces.

The **Roche limit** is the distance at which a planet's tidal forces become greater than the self-gravity of a smaller object—such as a moon, asteroid, or comet—causing the object to break apart. For a smaller object having the same density as the planet, the Roche limit is about 2.45 times the planet's radius. Such an object bound together solely by its own gravity can remain intact when it is outside a planet's Roche limit, but not when it is inside the limit. Objects such as the International Space Station and other Earth satellites are not torn apart, even though they orbit well within Earth's Roche limit, because chemical and physical bonds hold them together, not just self-gravity.

We have concentrated on the role that tidal forces play on Earth and the Moon, but tidal forces exist throughout the Solar System and the universe.

Anytime two objects of significant size or two collections of objects interact gravitationally, the gravitational forces will differ from one place to another within the objects, giving rise to tidal effects. Tidal disruption of small bodies is the source of the particles that make up the rings of the giant planets. Tidal interactions can cause material from one star in a binary pair to be pulled onto the other star. Tidal effects can strip stars from clusters consisting of thousands of stars. Galaxies can pass close enough together to strongly interact gravitationally. When that happens, as in **Figure 4.17**, tidal effects can grossly distort both galaxies. Tidal forces even play a role in shaping huge collections of galaxies—the largest known structures in the universe.

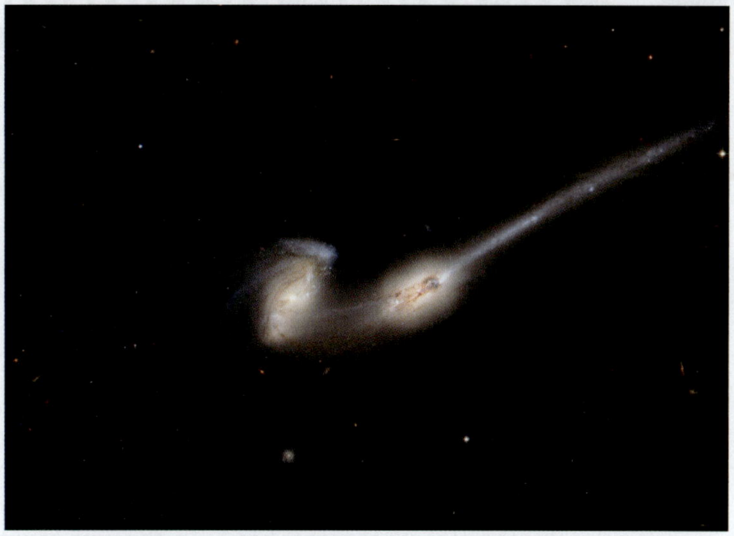

Figure 4.17 The tidal "tails" seen here are characteristic of tidal interactions between galaxies.

CHECK YOUR UNDERSTANDING 4.4

The Moon always keeps the same face toward Earth because of: (a) tidal locking; (b) tidal forces from the Sun; (c) tidal forces from the Earth and Sun.

Origins Tidal Forces and Life

Earth's rotation is slowing down as the Moon slowly moves away into a larger orbit, so in the distant past the Moon was closer and Earth rotated faster. In that time, tides would have been stronger and the interval between high tides would have been shorter. Tides are also affected by the configuration of the continents and oceans on Earth, so precisely how much faster Earth rotated billions of years ago is unknown. What is known, though, is that the stronger and more frequent tides would have provided additional energy to the oceans of the young Earth.

Scientists debate whether life on Earth originated deep in the ocean, on the surface of the ocean, or in hot springs on land (see Chapter 24). The tides shaped the regions in the margins between land and ocean, such as tide pools and coastal flats. Some researchers think that those border regions, which alternate between wet and dry with the tides, could have been places where concentrations of biochemicals periodically became high enough for more complex reactions to take place. Those complex reactions were important to the earliest life. Later, those border regions may have been important as advanced life moved from the sea to the land.

Elsewhere in the Solar System, the giant planets Jupiter and Saturn are far from the Sun and thus very cold. Both Jupiter and Saturn have many moons, and the closest moons would experience strong tidal forces from their respective planets. Several of those moons are thought to have a liquid ocean underneath an icy surface. Tidal forces from Jupiter or Saturn supply the heat to keep the water in a liquid state. Astrobiologists think that those subsurface liquid oceans are perhaps the most probable location for life elsewhere in the Solar System.

On Earth, the solar tides are about half as strong as lunar tides. A planet with a closer orbit would experience much stronger tidal forces from its star. Many planets detected outside the Solar System have orbits very close to their stars (see Chapter 7), so those exoplanets experience strong tidal forces. The planets might be tidally locked, too, so they have a synchronous rotation like the Moon does around Earth, with one side of the planet always facing the star and one side facing away. How might life on Earth have evolved differently if half the planet was in perpetual night and half in perpetual day?

Astronomers and physicists have wondered whether the value of Newton's gravitational constant, G, has changed over time. This press release reports on one experiment to measure the value of G.

ARTICLES QUESTIONS

Exploding Stars Prove Newton's Law of Gravity Unchanged over Cosmic Time

By Swinburne Media Centre

Australian astronomers have combined all observations of supernovae ever made to determine that the strength of gravity has remained unchanged over the last 9 billion years.

Newton's gravitational constant, known as G, describes the attractive force between two objects, together with the separation between them and their masses. It had previously been suggested that G could have been slowly changing over the 13.8 billion years since the Big Bang.

If G had been decreasing over time, for example, this would mean the Earth's distance to the Sun was slightly smaller in the past, meaning that we would experience longer seasons now compared to much earlier points in the Earth's history.

But researchers at Swinburne University of Technology in Melbourne have now analysed the light given off by 580 supernova explosions in the nearby and far universe and have shown that the strength of gravity has not changed.

"Looking back in cosmic time to find out how the laws of physics may have changed is not new," said Professor Jeremy Mould. "But supernova cosmology now allows us to do this with gravity."

A Type Ia supernova marks the violent death of a star called a white dwarf, which is as massive as our Sun but packed into a ball the size of our Earth.

Telescopes can detect the light from this explosion and use its brightness as a "standard candle" to measure distances in the universe, a tool that helped Australian astronomer Professor Brian Schmidt in his 2011 Nobel Prize–winning work discovering the mysterious force Dark Energy.

Professor Mould and his PhD student Syed Uddin at the Swinburne Centre for Astrophysics and Supercomputing and the ARC Centre of Excellence for All-sky Astrophysics (CAASTRO) assumed that these supernova explosions happen when a white dwarf reaches a critical mass or after colliding with other stars to "tip it over the edge."

"This critical mass depends on Newton's gravitational constant G and allows us to monitor it over billions of years of cosmic time—instead of only decades as was the case in previous studies," Professor Mould said.

Despite these vastly different time spans, their results agree with findings from the Lunar Laser Ranging Experiment that has been measuring the distance between the Earth and the Moon since NASA's *Apollo* missions in the 1960s and has been able to monitor possible variations in G at very high precision.

"Our cosmological analysis complements experimental efforts to describe and constrain the laws of physics in a new way and over cosmic time," Mr. Uddin said.

In their current publication, the Swinburne researchers were able to set an upper limit on the change in Newton's gravitational constant of 1 part in 10 billion per year over the past 9 billion years.

ARTICLES QUESTIONS

1. Explain how observing distant supernovae means observing back in time.
2. Why are physicists still measuring G, centuries after Newton described that constant?
3. What quantities that you studied here in Chapter 4 would change if the value of G changed over time?
4. Why is it important to the process of science that this result agrees with a result from a totally different experiment?
5. If the average distance to the Sun were slightly less in the past, would solar tides have been weaker or stronger? Would you expect that change to have been as important to tides on Earth as changes in the distance to the Moon?

Source: "Exploding stars prove Newton's law of gravity unchanged over cosmic time," Swinburne University of Technology Media Center News Release, March 24, 2014. Reprinted by permission of Swinburne University of Technology.

Summary

Objects stay in orbit because of gravity. Newton's laws of motion and his proposed law of gravity predict the paths of planetary orbits and explain Kepler's laws. Newton's calculations showed that those orbits should be ellipses with the Sun at one focus, that planets should move faster when closer to the Sun, and that the square of the period of a planet's orbit should vary as the cube of the semimajor axis of that elliptical orbit. Newton also mathematically confirmed Galileo's conclusion that falling objects have an accelerated motion independent of their mass. Tidal forces provide energy to Earth's oceans. Tide pools on Earth may have been a site of early biochemical reactions. Some moons in the outer Solar System might have liquid water because of tidal heating from their respective planets.

LG 1 **Explain the elements of Newton's universal law of gravitation.** Gravity is a force between any two objects with mass. Gravity is one of the fundamental forces of nature—it binds the universe together. The force of gravity is proportional to the mass of each object and inversely proportional to the square of the distance between them.

LG 2 **Use the laws of motion and gravitation to explain planetary orbits.** An orbit is one body "falling around" another. Planets orbit the Sun in elliptical orbits. All objects affected by gravity move either in bound elliptical orbits or unbound paths. Orbits are ultimately given their shape by the gravitational attraction of the objects involved, which in turn reflects the masses of those objects.

LG 3 **Explain how tidal forces from the Sun and Moon create Earth's tides.** Tides on Earth are the result of differences between how hard the Moon and Sun pull on one part of Earth in comparison with their pull on other parts of Earth. The primary cause of tides is the Moon, which stretches out Earth. The tides are the strongest when the Sun, Moon, and Earth are aligned. As Earth rotates, tides rise and fall twice each day.

LG 4 **Describe how tidal forces affect solid bodies.** Tidal forces lock the Moon's rotation to its orbit around Earth. Tidal forces can break up an object if its gets too close to a more massive object. Tidal forces are observed throughout the universe—in planets and moons, pairs of stars, and interacting galaxies.

? Unanswered Question

- What range of gravities will support human life? Humans have evolved to live on Earth's surface, but what happens when humans go elsewhere? What are the limits for our hearts, lungs, eyes, and bones? At the higher end of human tolerance, fighter pilots have been trained to experience about 10 times the normal surface gravity on Earth for very short periods (too long and they black out). Astronauts who spend several months in near-weightless conditions experience medical problems such as bone loss. On the Moon or Mars, humans will weigh much less than on Earth. Many science fiction tales have been written about what happens to children born on a space station or on another planet or moon with low surface gravity: would their hearts and bodies ever be able to adjust to the higher surface gravity of Earth, or must they stay in space forever?

Questions and Problems

Test Your Understanding

1. In Newton's universal law of gravitation, the force is
   a. proportional to both masses.
   b. proportional to the radius.
   c. proportional to the radius squared.
   d. inversely proportional to the orbiting mass.
2. If we wanted to increase a satellite's altitude above Earth and keep it in a stable orbit, we also would need to
   a. increase its orbital speed.
   b. increase its weight.
   c. decrease its weight.
   d. decrease its orbital speed.

3. An object in a(n) __________ orbit in the Solar System will remain in its orbit forever. An object in a(n) __________ orbit will escape from the Solar System.
   a. unbound; bound
   b. circular; elliptical
   c. bound; unbound
   d. elliptical; circular

4. Compared with your mass on Earth, your mass on the Moon would be
   a. lower because the Moon is smaller than Earth.
   b. lower because the Moon has less mass than Earth.
   c. higher because of the combination of the Moon's mass and size.
   d. the same—mass doesn't change.

5. If you went to Mars, your weight would be
   a. higher because you are closer to the center of the planet.
   b. lower because Mars has two small moons instead of one big moon, so less tidal force is present.
   c. lower because the lower mass and smaller radius of Mars combine to a lower surface gravity.
   d. the same as on Earth.

6. Venus has about 80 percent of Earth's mass and about 95 percent of Earth's radius. Your weight on Venus will be
   a. 20 percent more than on Earth.
   b. 20 percent less than on Earth.
   c. 10 percent more than on Earth.
   d. 10 percent less than on Earth.

7. The connection between gravity and orbits enables astronomers to measure the __________ of stars and planets.
   a. distances
   b. sizes
   c. masses
   d. compositions

8. If the Moon were twice as massive, how would the strength of lunar tides change?
   a. The highs would be higher, and the lows would be lower.
   b. Both the highs and the lows would be higher.
   c. The highs would be lower, and the lows would be higher.
   d. Nothing would change.

9. If Earth had half its current radius, how would the strength of lunar tides change?
   a. The highs would be higher, and the lows would be lower.
   b. Both the highs and the lows would be higher.
   c. The highs would be lower, and the lows would be higher.
   d. Both the highs and the lows would be lower.

10. If the Moon were 2 times closer to Earth than it is now, the gravitational force between Earth and the Moon would be
    a. 2 times stronger.
    b. 4 times stronger.
    c. 8 times stronger.
    d. 16 times stronger.

11. If the Moon were 2 times closer to Earth than it is now, the tides would be
    a. 2 times stronger.
    b. 4 times stronger.
    c. 8 times stronger.
    d. 16 times stronger.

12. If two objects are tidally locked to each other,
    a. the tides always stay on the same place on each object.
    b. the objects always remain in the same place in each other's sky.
    c. the objects are falling together.
    d. both a and b

13. Spring tides occur only when
    a. the Sun is near the vernal equinox.
    b. the Moon's phase is new or full.
    c. the Moon's phase is first quarter or third quarter.
    d. it is either spring or fall.

14. If an object crosses from farther to closer than the Roche limit, it
    a. can no longer be seen.
    b. begins to accelerate very quickly.
    c. slows down.
    d. may be torn apart.

15. Self-gravity is
    a. the gravitational pull of a person.
    b. the force that holds objects such as people and lamps together.
    c. the gravitational interaction of all the parts of a body.
    d. the force that holds objects on Earth.

Thinking about the Concepts

16. Both Kepler's laws and Newton's laws tell us something about the motion of the planets, but they have a fundamental difference. What is that difference?

17. Explain the difference between circular velocity and escape velocity. Which of those must be larger? Why?

18. Explain the difference between weight and mass.

19. Weight on Earth is proportional to mass. On the Moon, too, weight is proportional to mass, but the constant of proportionality is different on the Moon from what it is on Earth. Why? Explain why that difference does not violate the universality of physical law, as described in the Process of Science Figure.

20. Two comets are leaving the vicinity of the Sun, one traveling in an elliptical orbit and the other in an unbound orbit. What can you say about the future of those two comets? Would you expect either of them to return eventually?

21. What is the advantage of launching satellites from spaceports located near the equator? Would you expect satellites to be launched to the east or to the west? Why?

22. Explain how to use celestial orbits to estimate an object's mass. What observational quantities do you need to make that estimation?

23. Suppose astronomers discovered an object approaching the Sun in an unbound orbit. What would that say about the origin of the object?

24. What determines the strength of gravity at various radii between Earth's center and its surface?

25. The best time to dig for clams along the seashore is when the ocean tide is at its lowest. What phases of the Moon would be best for clam digging? What would be the best times of day during those phases?

26. The Moon is on the meridian at your seaside home, but your tide calendar does not show that it is high tide. What might explain that apparent discrepancy?

27. We may have an intuitive feeling for why lunar tides raise sea level on the side of Earth facing the Moon, but why is sea level also raised on the side facing away from the Moon?

28. Tides raise and lower the level of Earth's oceans. Can they do the same for Earth's landmasses? Explain your answer.

29. Lunar tides raise the ocean surface by less than 1 meter. How can tides as large as 5–10 meters occur?

30. Most commercial satellites are well inside the Roche limit as they orbit Earth. Why are they not torn apart?

Applying the Concepts

31. Mars has about one-tenth the mass of Earth and about half of Earth's radius. What is the value of gravitational acceleration on the surface of Mars in comparison with that on Earth? Compare your estimated mass and weight on Mars with your mass and weight on Earth. Do Hollywood movies showing people on Mars accurately portray that difference in weight?

32. Earth speeds along at 29.8 km/s in its orbit. Neptune's nearly circular orbit has a radius of 4.5×10^9 km, and the planet takes 164.8 years to make one trip around the Sun. Calculate how fast Neptune moves along in its orbit.

33. Venus's circular velocity is 35.03 km/s, and its orbital radius is 1.082×10^8 km. Use that information to calculate the mass of the Sun.

34. At the surface of Earth, the escape velocity is 11.2 km/s. What would be the escape velocity at the surface of a very small asteroid having a radius 10^{-4} that of Earth and a mass 10^{-12} that of Earth?

35. How long does Newton's cannonball, moving at 7.9 km/s just above Earth's surface, take to complete one orbit around Earth?

36. When a spacecraft is sent to Mars, it is first launched into an Earth orbit with circular velocity.
    a. Describe the shape of that orbit.
    b. What minimum velocity must we give the spacecraft to send it on its way to Mars?

37. Earth's average radius is 6,370 km and its mass is 5.97×10^{24} kg. Show that the acceleration of gravity at the surface of Earth is 9.81 m/s^2.

38. Using 6,370 km for Earth's radius, compare the gravitational force acting on a NASA rocket when it is sitting on its launchpad with the gravitational force acting on it when it is orbiting 350 km above Earth's surface.

39. The International Space Station travels on a nearly circular orbit 350 km above Earth's surface. What is its orbital speed?

40. Using the last equation in Working It Out 4.1, calculate the value of g on Earth. Compare that result with the measured value.

41. As described in Working It Out 4.4, tidal force is proportional to the masses of the two objects and is inversely proportional to the cube of the distance between them. Some astrologers claim that your destiny is determined by the "influence" of the planets that are rising above the horizon at the moment of your birth. Compare the tidal force of Jupiter (mass, 1.9×10^{27} kg; distance, 7.8×10^8 km) with that of the doctor in attendance at your birth (mass, 80 kg; distance, 1 meter).

42. The asteroid Ida (mass, 4.2×10^{16} kg) is attended by a tiny asteroidal moon, Dactyl, which orbits Ida at an average distance of 90 km. If you ignore the mass of the tiny moon, what is Dactyl's orbital period in hours?

43. Suppose you go skydiving.
    a. Just as you fall out of the airplane, what is your gravitational acceleration?
    b. Would that acceleration be bigger, smaller, or the same if you were strapped to a flight instructor and so had twice the mass?
    c. Just as you fall out of the airplane, what is the gravitational force on you? (Assume that your mass is 70 kg.)
    d. Would that force be bigger, smaller, or the same if you were strapped to a flight instructor and so had twice the mass?

44. Assume that a planet just like Earth is orbiting the bright star Vega at a distance of 1 astronomical unit (AU). Vega has twice the mass of the Sun.
    a. How long in Earth years will that planet take to complete one orbit around Vega?
    b. How fast is the Earth-like planet traveling in its orbit around Vega?

45. Suppose that in the past the Moon was 80 percent of the distance from Earth that it is now. Calculate how much stronger the lunar tides would have been. How would the neap and spring tides be different from now?

USING THE WEB

46. Go to NASA's "Apollo 15 Hammer-Feather Drop" Web page (https://nssdc.gsfc.nasa.gov/planetary/lunar/apollo_15_feather_drop.html or https://youtube.com/watch?v=oYEgdZ3iEKA) and watch the video from *Apollo 15* of astronaut David Scott dropping the hammer and falcon feather on the Moon. What did that experiment show? What would happen if you tried that on Earth with a feather and a hammer? Would it work? Suppose instead you dropped the hammer and a big nail. How would they fall? How does the acceleration of falling objects on the Moon compare with the acceleration of falling objects on Earth?

47. Go to the Exploratorium's "Your Weight on Other Worlds" Web page (http://exploratorium.edu/ronh/weight), which will calculate your weight on other planets and moons in our Solar System. On which objects would your weight be higher than it is on Earth? What difficulties would human bodies have in a higher-gravity environment? For example, could you easily get up out of bed and walk? What are the possible short-term and long-term effects of lower gravity on the human body? What types of life on Earth might adapt well to a different gravity?

48. a. Watch the first 12 minutes of the episode "Gravity" in season 2 of the series *The Universe* (http://history.com/shows/the-universe/season-2/episode-17 or http://dailymotion.com/video/x1p90qs or on YouTube) to see several illustrated examples of gravity. Why does the tennis ball appear to float when dropped at the top of the roller coaster? How could Newton imagine a satellite orbiting Earth centuries before it was possible? Why does the speed of the cannonball determine whether it goes into orbit? What was the technical difficulty in launching a satellite?
    b. In the same video as in part (a), watch the trip on the zero-G plane (the "vomit comet"), at 23–29 minutes. How does the plane simulate zero-G? Why does it last for only 20–30 seconds? How is that similar to the roller coaster in part (a)?

49. Go to a website that will show you the times for high and low tides—for example, http://saltwatertides.com. Pick a location and bring up the tide table for today and the next 14 days. Why do two high tides and two low tides occur every day? What is the difference in the height of the water between high and low tides? In the last few columns of the table, the times of moonrise and moonset are indicated, as well as the percentage of lunar illumination. Does the time of the high tide lead or follow the highest position of the Moon in the sky? Compare with Figure 4.12c: what phases of the Moon have the greatest differences in the height of high and low tides?

50. Figure 4.17 shows two galaxies pulling at each other, most likely after they have already collided. Watch "The Mice—Interacting Galaxies" on YouTube (https://youtube.com/watch?v=pR4pzthwqkA) to see a simulation of that interaction. Galaxies are not solid objects: they contain stars and gas and dust and a lot of empty space. Explain how those tidal tails can result from that type of interaction.

EXPLORATION

digital.wwnorton.com/astro6

In the Exploration of Chapter 3, we used the "Planetary Orbits" Interactive Simulation to explore Kepler's laws for Mercury. Now that we know how Newton's laws explain why Kepler's laws describe orbits, we will revisit the simulator to explore the Newtonian features of Mercury's orbit. Visit the Digital Resources Page and on the Student Site open the "Orbits" Interactive Simulation in Chapter 3.

Acceleration

To begin exploring the simulation, set parameters for "Mercury" in the "Orbit Settings" panel and then click "OK." Click the "Newtonian Features" tab at the bottom of the control panel. Select "show solar system orbits" and "show grid" under "Visualization Options." Change the animation rate to 0.01, and select the "start animation" button.

Examine the graph at the bottom of the panel.

1 Where is Mercury in its orbit when the acceleration is smallest?

2 Where is Mercury in its orbit when the acceleration is largest?

3 What are the values of the largest and smallest accelerations?

In the "Newtonian Features" graph, mark the boxes for "vector" and "line" that correspond to the acceleration. Specifying those parameters will insert an arrow that shows the direction of the acceleration and a line that extends the arrow.

4 To what Solar System object does the arrow point?

5 In what direction is the force on the planet?

Velocity

Examine the graph at the bottom of the panel again.

6 Where is Mercury in its orbit when the velocity is smallest?

7 Where is Mercury in its orbit when the velocity is largest?

8 What are the values of the largest and smallest velocities?

Add the velocity vector and line to the simulation by clicking on the boxes in the graph window. Study the resulting arrows carefully.

9 Are the velocity and the acceleration always perpendicular (is the angle between them always 90°)?

10 If the orbit were a perfect circle, what would be the angle between the velocity and the acceleration?

Hypothetical Planet

In the "Orbit Settings" panel, change the semimajor axis to 0.8 AU.

11 How does that imaginary planet's orbital period now compare with Mercury's?

Now change the semimajor axis to 0.1 AU.

12 How does the planet's orbital period now compare with Mercury's?

13 Summarize your observations of the relationship between the speed of an orbiting object and the semimajor axis.

5 Light

Our knowledge of the universe beyond Earth comes from light emitted, absorbed, or reflected by astronomical objects. Light carries information about the temperature, composition, and speed of the objects. Light also tells us about the nature of the material that the light passed through on its way to Earth. Yet light plays a far larger role in astronomy than that of being a messenger. Light is one of the primary means by which energy is transported throughout the universe. Stars, planets, and vast clouds of gas and dust filling the space between the stars heat up as they absorb light and cool off as they emit light. Light carries energy generated in the heart of a star outward through the star and off into space. Light transports energy from the Sun outward through the Solar System, heating the planets; and light carries energy away from each planet, allowing each one to cool. The balance between those two processes establishes each planet's temperature and therefore a planet's possible suitability for life.

LEARNING GOALS

An astronomer must try to understand the universe by studying the light and other particles that reach Earth from distant objects. By the end of this chapter, you should be able to:

LG 1 Describe the wave and particle properties of light, and describe the electromagnetic spectrum.

LG 2 Describe how to measure the chemical composition of distant objects by using the unique spectral lines of different types of atoms.

LG 3 Describe the Doppler effect and how it can be used to measure the motion of distant objects.

LG 4 Explain how the spectrum of light that an object emits depends on its temperature.

LG 5 Differentiate luminosity from brightness, and illustrate how distance affects each.

The visible part of the electromagnetic spectrum is laid out in the colors of this rainbow. ▶▶▶

What is light?

5.1 Light Brings Us the News of the Universe

Is light a **wave**—a disturbance that travels from one point to another—or is it made up of particles? Scientists have come to understand that light sometimes acts like a wave and sometimes acts like a particle. We begin with a discussion of how fast light travels. We then discuss its wavelike and particle-like properties.

The Speed of Light

In the 1670s, Danish astronomer Ole Rømer (1644–1710) studied the movement of the moons of Jupiter, measuring the times when each moon disappeared behind the planet. To his amazement, the observed times did not follow the regular schedule that he predicted from Kepler's laws. Sometimes the moons disappeared behind Jupiter sooner than expected, whereas at other times they disappeared behind Jupiter later than expected. Rømer realized that the difference depended on where Earth was in its orbit. If he began tracking the moons when Earth was closest to Jupiter, the moons were almost 17 minutes "late" by the time Earth was farthest from Jupiter. When Earth was once again closest to Jupiter, the moons again passed behind Jupiter at the predicted times.

Rømer correctly concluded that his observations did not represent a failure of Kepler's laws. Instead, he was seeing the first clear evidence that light travels at a finite speed. As shown in **Figure 5.1**, the moons appeared "late" when Earth was farther from Jupiter because of the time needed for light to travel the extra distance between the two planets. Over the course of Earth's yearly trip around the Sun, the distance between Earth and Jupiter changes by 2 astronomical units (AU). The speed of light equals that distance divided by Rømer's 16.7-minute delay, or about 3×10^5 kilometers per second (km/s). The value that Rømer actually announced in 1676 was a bit on the low side—2.25×10^5 km/s—because the size of Earth's orbit was not well known. Modern measurements of the speed of light give a value of 2.99792458×10^5 km/s in a **vacuum** (a region of space devoid of matter). The speed of light in a vacuum is one of nature's fundamental constants, usually written as c (lowercase). The speed of light through any medium, such as air or glass, is always less than c.

The International Space Station moves around Earth at a speed of about 28,000 kilometers per hour (km/h), taking 91 minutes to complete one orbit. Light travels almost 40,000 times faster than that and can circle Earth in only

Figure 5.1 Danish astronomer Ole Rømer realized that apparent differences between the predicted and observed orbital motions of Jupiter's moons depend on the distance between Earth and Jupiter. He used those observations to measure the speed of light. (The superscript letters in "16^m 40^s" stand for minutes and seconds of time, respectively.)

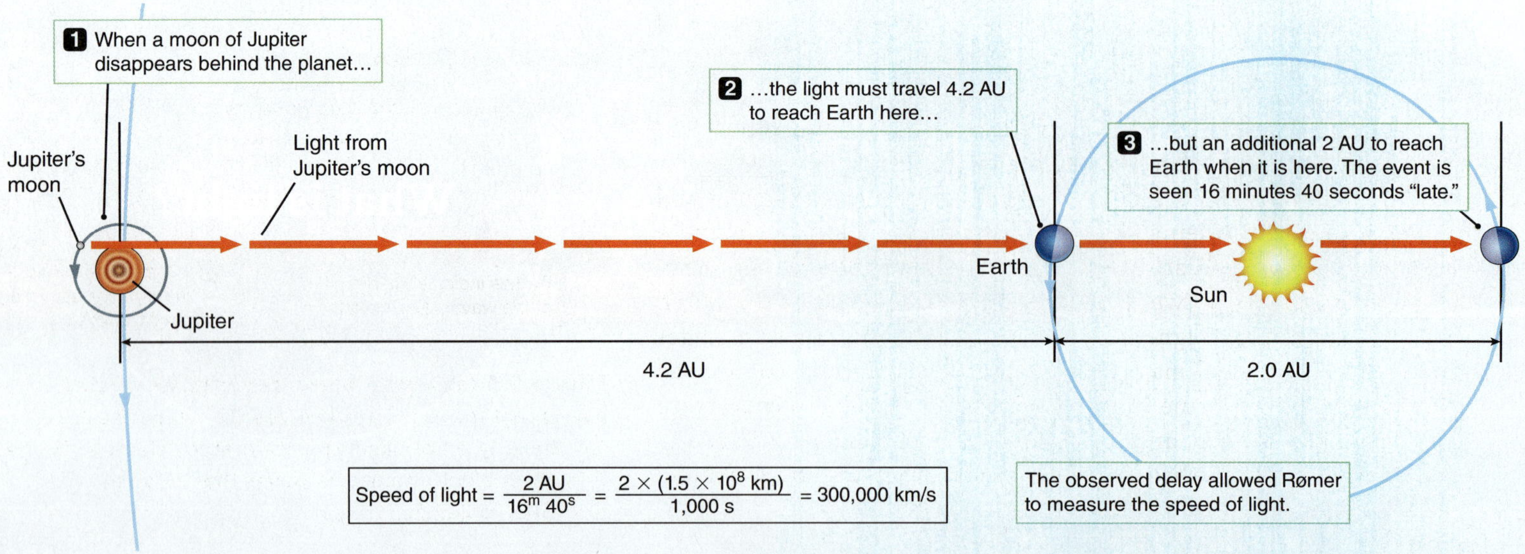

$$\text{Speed of light} = \frac{2\ \text{AU}}{16^m\ 40^s} = \frac{2 \times (1.5 \times 10^8\ \text{km})}{1{,}000\ \text{s}} = 300{,}000\ \text{km/s}$$

1/7 of a second. Because light is so fast, its travel time is a convenient way to express cosmic distances. Light takes 1 1/4 seconds to travel between Earth and the Moon, so we say that the Moon is 1 1/4 light-*seconds* from Earth. The Sun is 8 1/3 light-*minutes* away, and the next-nearest star is 4 1/3 light-*years* distant. Thus, a *light-year* is defined as the distance light travels in 1 year, or about 9.5 trillion km. The light-year is a measure of distance.

While traveling at that high speed, light carries energy from place to place. **Energy** is the ability to do work, and it comes in many forms. **Kinetic energy** is the energy of moving objects. **Thermal energy** is closely related to kinetic energy and is the sum of all the random motion of atoms, molecules, and particles, by which we measure their temperature. For example, when light from the Sun strikes the pavement, the pavement heats up. Light carried that energy from the Sun to the pavement. Rømer knew how long light took to travel a given distance, but physicists would take more than 200 years to figure out what light actually is.

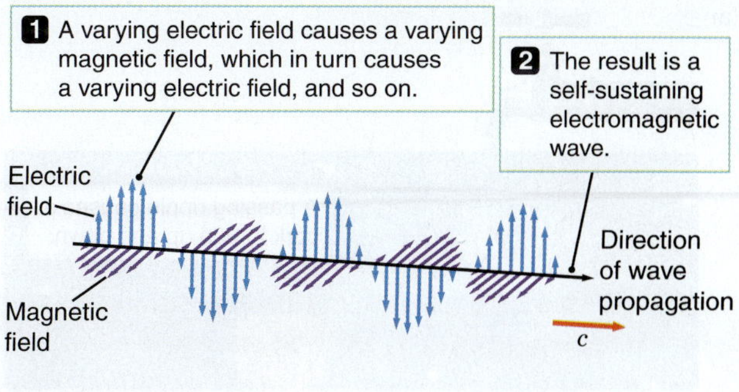

Figure 5.2 An electromagnetic wave consists of oscillating electric and magnetic fields that are perpendicular both to each other and to the direction in which the wave travels.

Light as an Electromagnetic Wave

In the late 19th century, the Scottish physicist James Clerk Maxwell (1831–1879) introduced the concept that electricity and magnetism are two components of the same physical phenomenon. An **electric force** is the push or pull between electrically charged particles that make up atoms, such as protons and electrons, arising from their electric charges. Particles with opposite charges attract, and those with like charges repel. A **magnetic force** is a force between electrically charged particles arising from their motion.

To describe those electric and magnetic forces, Maxwell considered what happens when charged particles move in *electric fields* and *magnetic fields*. An **electric field** is a measure of the electric force on a charge at any point in space. Similarly, a **magnetic field** is a measure of the magnetic force acting on a small magnet at any point in space.

Maxwell summarized the behavior of electric fields and magnetic fields in four elegant equations. Among other things, those equations say that a changing electric field causes a magnetic field and that a changing magnetic field causes an electric field. A change in the motion of a charged particle causes a changing electric field, which causes a changing magnetic field, which causes a changing electric field, and so on. You can see that interaction in **Figure 5.2**. Once the process starts, a self-sustaining procession of oscillating electric and magnetic fields moves out in all directions through space. In other words, an accelerating charged particle gives rise to an **electromagnetic wave**. Those electromagnetic waves, and the accelerating charges that generate them, are the sources of electromagnetic radiation. Maxwell's equations also predict the speed at which an electromagnetic wave should travel, which agrees with the measured speed of light (c).

Maxwell's wave description of light also suggests how light originates and how it interacts with matter. When a drop of water falls from the faucet into a sink full of water, it causes a disturbance, or wave, like the one shown in **Figure 5.3a**. The wave moves outward as a ripple on the surface of the water. **Figure 5.3b** shows that electromagnetic waves resulting from periodic changes in the strength of the electric and magnetic fields move out through space away from their source in much the same way. The ripples in the sink, however, are distortions of the water's surface, and they require a **medium**: a substance to travel through. Light waves move through empty space—what we call a vacuum—in the absence of a medium.

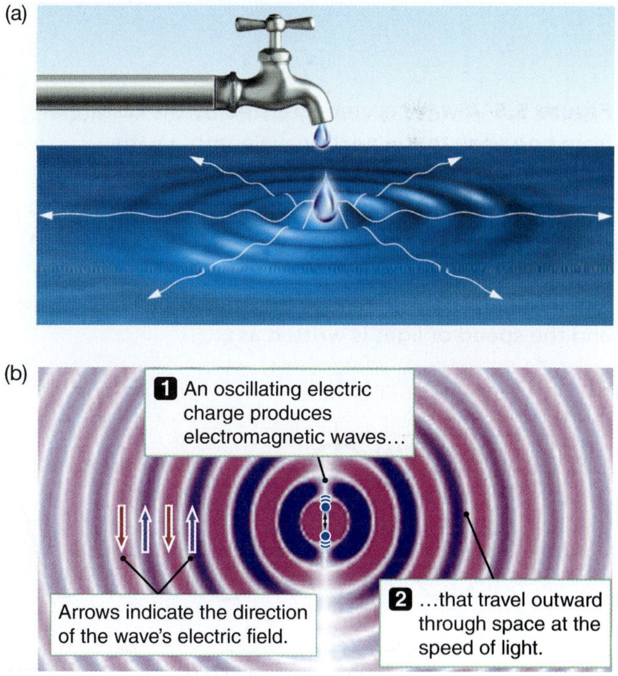

Figure 5.3 (a) A drop falling into water generates waves that move outward across the water's surface. (b) Similarly, an oscillating (accelerated) electric charge generates electromagnetic waves that move away at the speed of light.

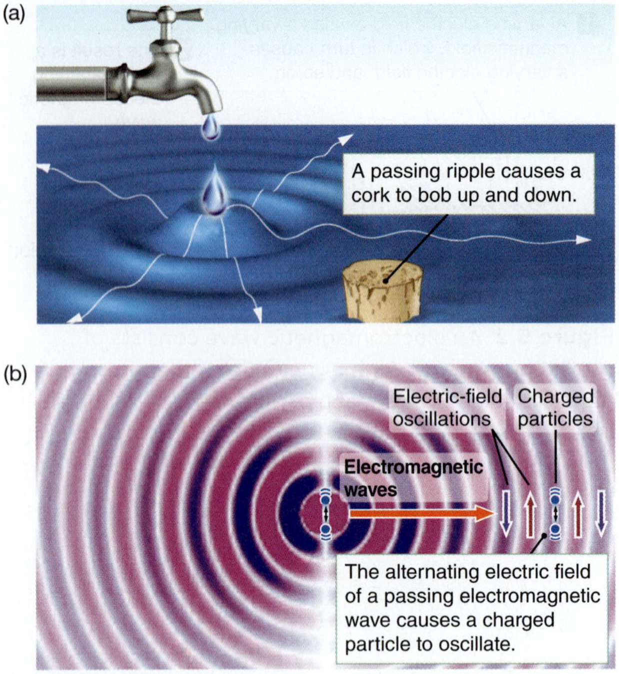

Figure 5.4 (a) When waves moving across the surface of water reach a cork, they cause the cork to bob up and down. (b) Similarly, a passing electromagnetic wave causes an electric charge to oscillate in response to the wave.

Now imagine that a cork is floating in the sink (**Figure 5.4a**). The cork remains stationary until the ripple from the dripping faucet reaches it. As the ripple passes by, the rising and falling water causes the cork to rise and fall. That can happen only if the wave is carrying energy—a conserved quantity that gives objects and particles the ability to do work. Light waves similarly carry energy through space and cause electrically charged particles to vibrate, as in **Figure 5.4b**.

Characterizing Waves

We discuss several kinds of waves in this book, including electromagnetic waves crossing the vast expanse of the universe and seismic waves traveling through Earth. Waves are generally characterized by four quantities: *amplitude*, *speed*, *frequency*, and *wavelength*. Each of those quantities is illustrated in **Figure 5.5**. The **amplitude** of a wave is the height of the wave above the undisturbed position (Figure 5.5a). For water waves, the amplitude is how far the water is lifted up by the wave. For light waves, the amplitude is related to the brightness of the light. A water wave travels at a particular speed, v (Figure 5.5b), through the water. The water itself doesn't travel; it just moves up and down at the same location. For waves such as those in water, that speed varies and depends on the density of the substance the wave moves through, among other things. Light, in contrast, always moves through a vacuum at the same speed, $c \approx 300{,}000$ km/s.

The distance from one crest of a wave to the next is the **wavelength**, usually denoted by the Greek letter lambda, λ (Figure 5.5c). The number of wave crests passing a point in space each second is called the wave's **frequency**, f. The unit of frequency is cycles per second, which is called **hertz (Hz)** after the 19th-century physicist Heinrich Hertz (1857–1894), who was the first to experimentally confirm Maxwell's predictions about electromagnetic radiation.

Figures 5.5c and 5.5d show that waves with longer (larger) wavelengths have lower (smaller) frequencies, whereas waves with shorter (smaller) wavelengths have higher (larger) frequencies. Higher-frequency waves carry more energy. Imagine standing on an ocean beach with the waves lapping up against you: the amount of energy you feel from the waves increases as the frequency of the ocean waves increases.

Waves travel a distance of one wavelength each cycle, so the speed of a wave can be found by multiplying the frequency and the wavelength. Translating that idea into math, we have $v = \lambda f$. The speed of light in a vacuum is always c, so once the wavelength of a wave of light is known, its frequency is known, and vice versa. Because light travels at constant speed, its wavelength and frequency are inversely proportional to each other: if the wavelength increases, the frequency decreases. Waves can carry a tremendous amount of information, such as complex and beautiful music, which travels by sound waves. As you continue your study of the universe, time and time again you will find that the information we receive, whether about the interior of Earth or about a distant star or galaxy, rides in on a wave.

Figure 5.5 A wave is characterized by the distance from one peak to the next (wavelength, λ), the frequency of the peaks (f), the maximum deviations from the medium's undisturbed state (amplitude), and the velocity (v) at which the wave pattern travels from one place to another. In an electromagnetic wave, the amplitude is the maximum strength of the electric field, and the speed of light is written as c.

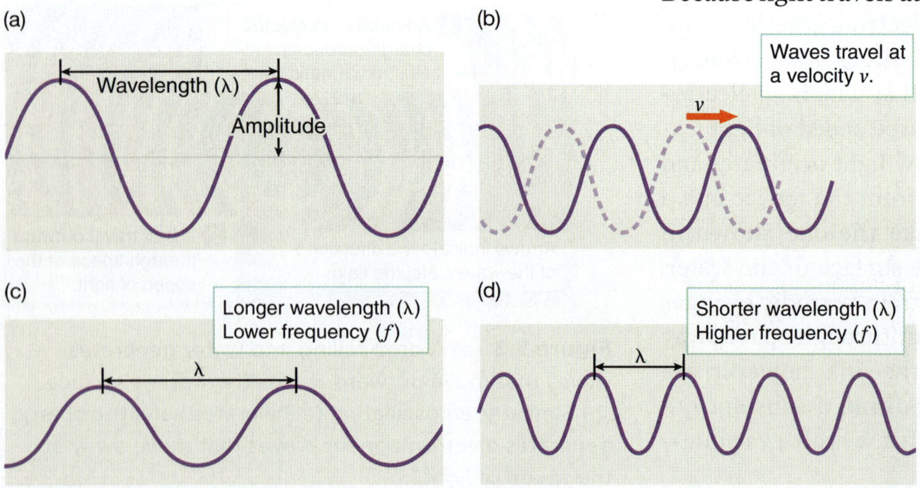

The Electromagnetic Spectrum

Most light signals are made up of many wavelengths. A rainbow, such as the one shown in the chapter-opening figure, is created when white light interacts with water droplets and is spread out into its component colors. Light spread out by wavelength is called a **spectrum**. At the long-wavelength (and therefore low-frequency) end of the visible spectrum is red light. At the short-wavelength (high-frequency) end is violet light. A commonly used unit for the wavelength of visible light is the **nanometer** (**nm**). A nanometer is one-billionth (10^{-9}) of a meter. Human eyes can see light between violet (about 380 nm) and red (750 nm). Stretched out between the two, in a rainbow, is the rest of the visible spectrum.

The light-sensitive cells in our eyes respond to visible light, but that is only a small sample of the range of possible wavelengths for electromagnetic radiation. Light can have wavelengths much shorter or much longer than your eyes can perceive. The whole range of wavelengths of light—collectively called the **electromagnetic spectrum**—is shown in **Figure 5.6**. Most of the electromagnetic spectrum—and therefore most of the information in the universe—is invisible to the human eye. To detect light outside the visible range, we must use specialized detectors of various kinds, as we explain in Chapter 6.

Refer to Figure 5.6 as we tour the electromagnetic spectrum, beginning with the shortest wavelengths and working our way to the longest ones. The very shortest wavelengths of light are called **gamma rays**, or sometimes gamma radiation. Because that light has the shortest wavelengths, it has the highest frequency and the highest energy, so it penetrates matter easily. Wavelengths between 0.1 and 40 nm are called **X-rays**. You have probably encountered X-rays at the dentist's office or in a hospital's emergency room—X-ray light has enough energy to penetrate skin and muscle but is stopped by denser bone. **Ultraviolet** (**UV**) **radiation** has wavelengths between 40 and about 380 nm—longer than X-rays but shorter than visible light. You are familiar with that type of light from sunburns: UV light has enough energy to penetrate your skin, but not much deeper.

Infrared (**IR**) **radiation** has longer wavelengths than the reddest wavelengths in the visible range. You are familiar with infrared radiation because you often feel it as heat. When you hold your hand next to a hot stove, some of the heat you feel is carried to your hand by infrared radiation emitted from the stove. In that sense, you could think of your skin as being a giant infrared eyeball—it is

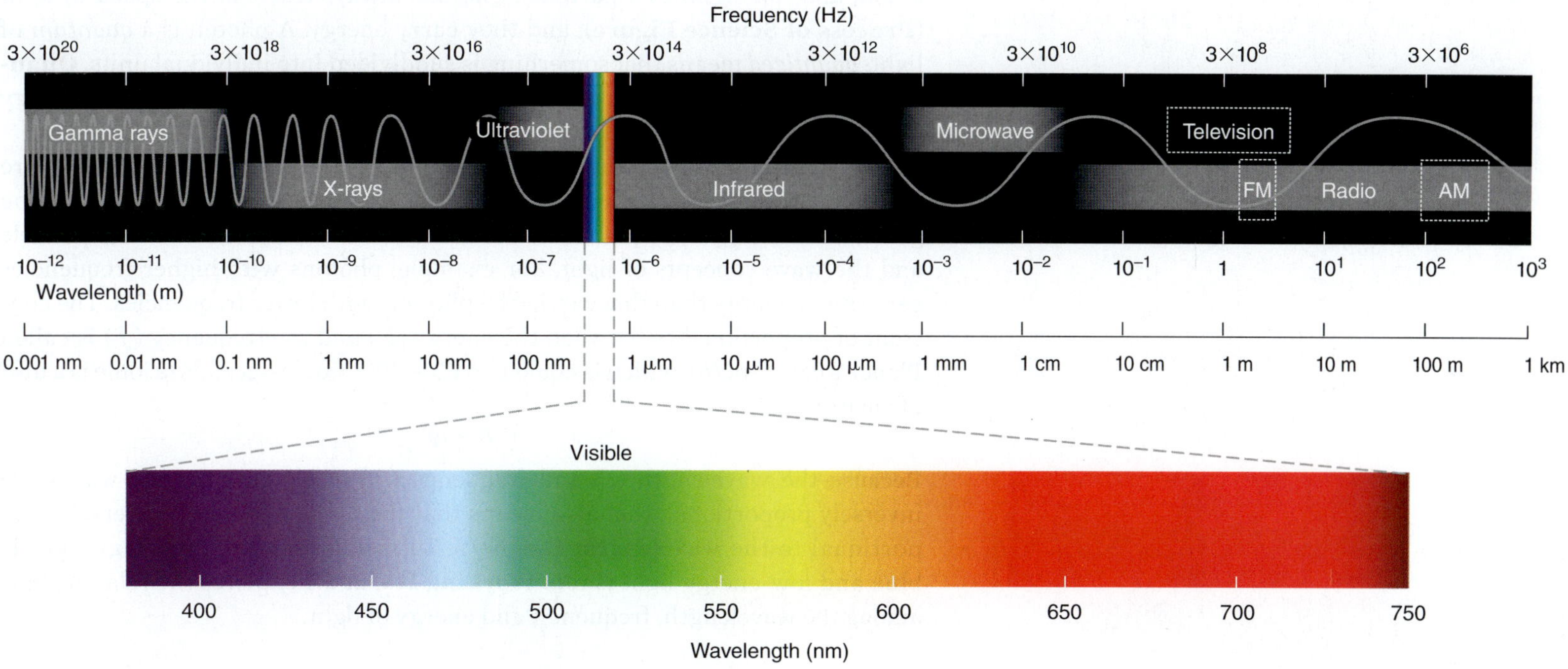

Figure 5.6 By convention, the electromagnetic spectrum is divided into loosely defined regions ranging from gamma rays to radio waves. Throughout the book, we use the following labels to indicate the form of radiation used to produce astronomical images, with an icon to remind you: G = gamma rays; X = X-rays; U = ultraviolet; V = visible; I = infrared; R = radio. If more than one region is represented, multiple labels are highlighted.

sensitive to infrared wavelengths. Infrared radiation also is used in television remote controls, and night vision goggles detect infrared radiation from warm objects such as animals. A useful unit for infrared light is the **micron** (**μm**, where μ is the Greek letter mu). One micron is 1,000 nm, or one-millionth (10^{-6}) of a meter. Infrared wavelengths are longer than red light and shorter than 500 microns.

Microwave radiation has even longer wavelengths than those of infrared radiation. The microwave in your kitchen heats the water in food by using light of those wavelengths. The longest-wavelength light, which has wavelengths longer than a few centimeters, is called **radio waves**. Light of those wavelengths in the form of FM, AM, television, and cell phone signals is used to transmit information.

CHECK YOUR UNDERSTANDING 5.1A

Rank the following in order from longest wavelength to shortest wavelength: (a) gamma rays; (b) visible light; (c) infrared light; (d) ultraviolet light; (e) radio waves

Light as a Particle

Although the wave theory of light describes many observations, it does not provide a complete picture of the properties of light. Many of the difficulties with the wave model of light have to do with how light interacts with small particles such as atoms and molecules.

The work of Albert Einstein and other scientists modified our understanding of light to show that light both acts like a wave and acts like a particle. In 1905, Einstein explained the *photoelectric effect*, in which subatomic particles called electrons are emitted when surfaces are illuminated by electromagnetic radiation greater than a certain frequency. He showed that the rate at which electrons are emitted depends only on the amount of incoming light and that the speed of the electrons depends only on the frequency of the incoming radiation. That work earned Einstein the Nobel Prize in Physics in 1921. In the particle model, light is made up of massless particles called **photons** (*phot-* means "light," as in *photograph*; and *-on* signifies a particle). Photons always travel at the speed of light (**Process of Science Figure**), and they carry energy. A photon is a *quantum* of light; *quantized* means that something is subdivided into individual units. **Quantum mechanics** is a branch of physics that deals with the quantization of energy and of other properties of matter.

The energy of a photon and the frequency of the electromagnetic wave are directly proportional to each other: the higher the frequency of the light, the greater the energy each photon carries. That relationship connects the particle and the wave concepts of light. For example, photons with higher frequencies carry more energy than that carried by photons with lower frequencies. The constant of proportionality between the energy (E) and the frequency (f) is called Planck's constant (h), which is equal to 6.63×10^{-34} joule-seconds (a joule is a unit of energy):

$$E = hf$$

Because the wavelength (λ) and frequency (f) of electromagnetic waves are inversely proportional, that also means that the photon energy is inversely proportional to the wavelength: $E = hc/\lambda$. In visible light, high-energy light is blue and low-energy light is red. **Working It Out 5.1** explores the relationships among the wavelength, frequency, and energy of light.

PROCESS OF SCIENCE

Agreement between Fields

Scientists working on very different problems in different fields all found the same results: light has a speed that can be measured.

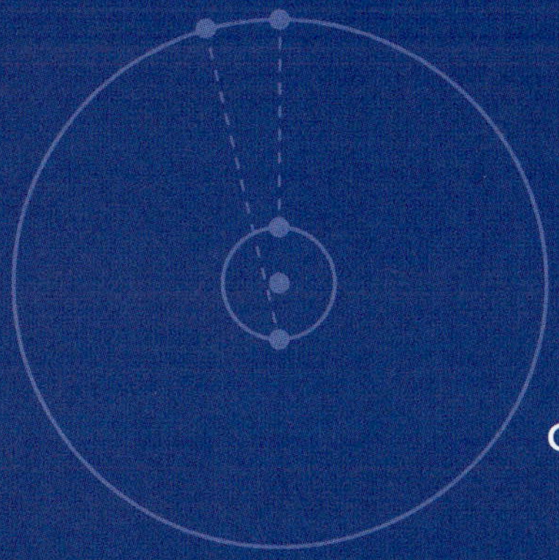

Ole Rømer (1644–1710) studied eclipses of Jupiter's moons.

Rømer calculated the speed of light from the eclipse delays of Jupiter's moons.

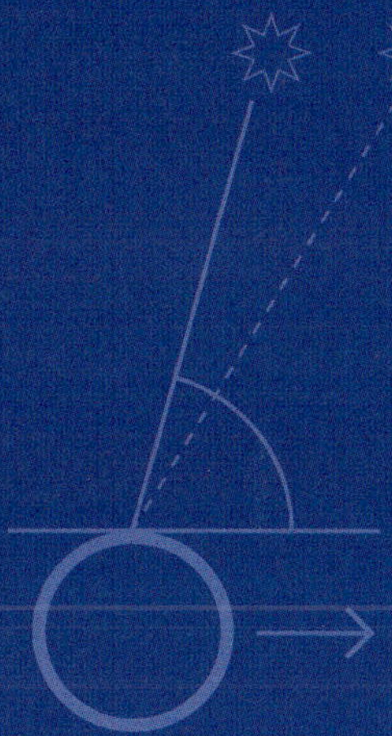

In 1727, James Bradley (1693–1762) studied the apparent motions of stars, which appeared to make small circles because of the relative motion of Earth.

A half century after Rømer's measurement, Bradley's motion studies led to a more accurate measurement of the speed of light.

$$c = 1/\left(\sqrt{\mu_0 \varepsilon_0}\right)$$

James Clerk Maxwell (1831–1879) studied electricity and magnetism.

Maxwell determined that the speed of electromagnetic waves must be a constant, and is equal to the speed of light.

$$t' = \frac{t}{\sqrt{1 - \frac{v^2}{c^2}}}$$

Einstein (1879–1955) explored the consequences of the fact that the speed of light is the same for all observers.

Predictions from this theory have been tested repeatedly since 1905.

Astronomers and physicists converged on an understanding that photons travel at the "speed of light"—a fundamental constant of the universe that is the same for all observers.

5.1 Working It Out Working with Electromagnetic Radiation

Wavelength and Frequency

When you tune to a radio station to, say, 770 AM, you are receiving an electromagnetic signal that travels at the speed of light and is broadcast at a frequency of 770 kilohertz (kHz), or 7.7×10^5 Hz. We can use the relationship between wavelength and frequency, $c = \lambda f$, to calculate the wavelength of the AM signal:

$$\lambda = \frac{c}{f} = \left(\frac{3 \times 10^8\ \text{m/s}}{7.7 \times 10^5/\text{s}}\right) = 390\ \text{m}$$

FM wavelengths are much shorter than AM wavelengths. For example, a station at 89.5 FM broadcasts signals with a wavelength of 3.4 m.

The human eye is most sensitive to light in green and yellow wavelengths, about 500–590 nm. If we examine green light with a wavelength of 530 nm, we can compute its frequency:

$$f = \frac{c}{\lambda} = \left(\frac{3 \times 10^8\ \text{m/s}}{530 \times 10^{-9}\ \text{m}}\right) = 5.66 \times 10^{14}/\text{s} = 5.66 \times 10^{14}\ \text{Hz}$$

That frequency corresponds to 566 *trillion* wave crests passing by each second.

Photon Energy

How does the energy of an X-ray photon with a wavelength of 1 nm compare with the energy of a visible-light photon with a wavelength of 530 nm (the same green light used in the previous calculation)? The equation for the energy of a photon is $E = hf$. Because $f = c/\lambda$, substituting c/λ for f yields the inverse relationship, $E = hc$. Because we are making a *comparison*, we can take a ratio, and then the constants h and c cancel out:

$$\frac{E_{\text{X-ray photon}}}{E_{\text{visible photon}}} = \frac{hc/\lambda_{1\ \text{nm}}}{hc/\lambda_{530\ \text{nm}}} = \frac{\cancel{hc}}{\cancel{hc}} \times \frac{\lambda_{500\ \text{nm}}}{\lambda_{1\ \text{nm}}} = \frac{530\ \text{nm}}{1\ \text{nm}} = 530$$

The X-ray photon has 530 times the energy of the visible-light photon.

In the particle description of light, the electromagnetic spectrum is a spectrum of photon energies. The higher the frequency of the electromagnetic wave, the greater the energy carried by each photon. Photons of shorter wavelength (higher frequency) carry more energy than that carried by photons of longer wavelength (lower frequency). For example, photons of blue light carry more energy than that carried by photons of longer-wavelength red light. Ultraviolet photons carry more energy than that carried by photons of visible light, and X-ray photons carry more energy than that carried by ultraviolet photons. The lowest-energy photons are radio wave photons. A beam of red light can carry just as much energy as a beam of blue light, but the red beam will have more photons than the blue beam.

AstroTour: Light as a Wave, Light as a Photon

CHECK YOUR UNDERSTANDING 5.1B

As wavelength increases, the energy of a photon ________ and its frequency ________. (a) increases; decreases (b) increases; increases (c) decreases; decreases (d) decreases; increases

5.2 The Quantum View of Matter Explains Spectral Lines

Light and matter interact, and that interaction allows us to detect matter even at great distances in space. To understand that interaction, we must first understand the building blocks of matter. Here in Section 5.2, we review atomic structure and the process by which astronomers identify the chemical elements in astronomical objects.

Atomic Structure

Matter is anything that occupies space and has mass. Atoms are composed of a central massive **nucleus**, which contains **protons** with a positive charge and **neutrons**, which have no charge. A cloud of negatively charged **electrons** surrounds the nucleus. Atoms with the same number of protons are all of the same type of chemical **element**. For example, an atom with two protons, shown in **Figure 5.7a**, is the element helium. An atom with six protons is the element carbon, one with eight protons is the element oxygen, and so forth. An element may have many **isotopes**: atoms with the same number of protons but different numbers of neutrons. **Molecules** are groups of atoms bound together by shared electrons. A single teaspoon of water contains about 10^{23} atoms—about as many atoms as the number of stars in the observable universe.

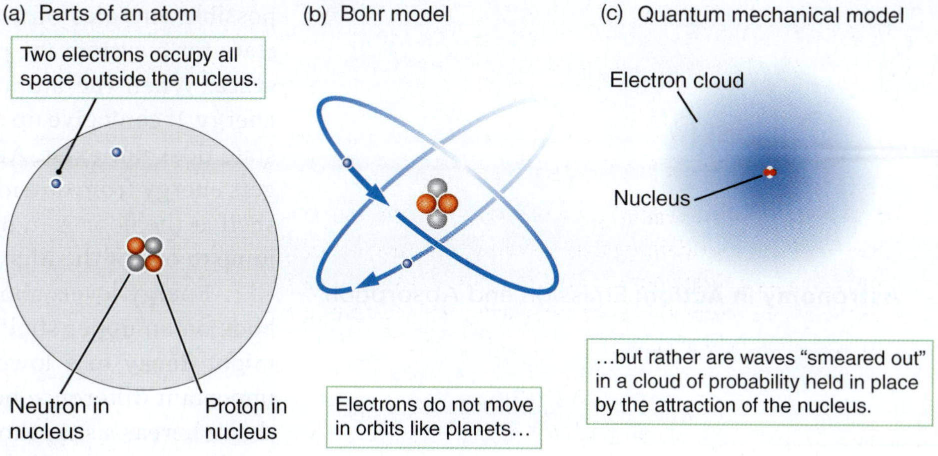

Figure 5.7 (a) An atom (here, helium) is made up of a nucleus consisting of positively charged protons and electrically neutral neutrons and is surrounded by much less massive negatively charged electrons. (b) Atoms are often drawn as miniature "solar systems," but this Bohr model is incorrect. (c) Electrons are actually smeared out around the nucleus in quantum mechanical clouds of probability.

For an atom to be electrically neutral, it must have the same number of electrons as protons. Electrons have much less mass than protons or neutrons, so almost all the mass of an atom is found in its nucleus. The Danish physicist Niels Bohr (1885–1962) proposed a model of the atom in 1913 in which a massive nucleus sits in the center and the smaller electrons orbit around it, much as planets orbit around the Sun (**Figure 5.7b**).

AstroTour: Atomic Energy Levels and the Bohr Model

The **Bohr model**, however, is an incomplete description of the atom. Just as waves of light have particle-like properties, particles of matter also have wavelike properties. Once that was understood, the Bohr model was modified so that the positively charged nucleus is surrounded by electron "clouds" or "waves," as shown in **Figure 5.7c**. In that quantum mechanical model, it is not possible to know precisely where the electron is in its orbit. The wave characteristics of particles make it impossible to pin down simultaneously their exact location and their exact velocity; some uncertainty will always exist. That is why a featureless cloud is used to represent electrons in orbit around an atomic nucleus.

Figure 5.8 (a) Energy states of an atom are analogous to shelves in a bookcase. You can move a book from one shelf to another, but books can never be placed between shelves. (b) Atoms exist in one allowed energy state or another but never in between. The ground state is the lowest possible level.

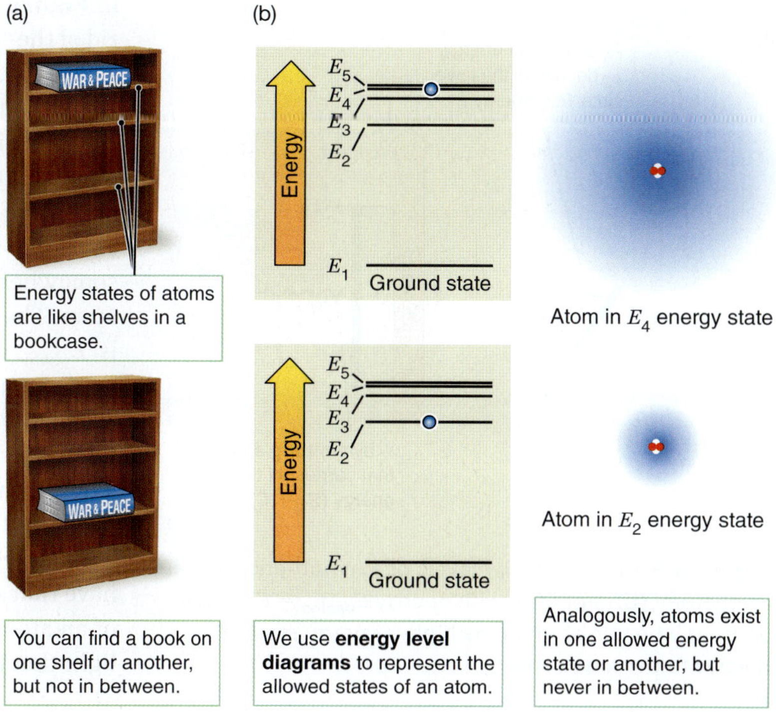

Atomic Energy Levels

Because of their wavelike properties, electrons in an atom can take on only certain specific energies that depend on the energy states of the atom. The form that the electron waves take depends on the possible energy states of atoms. The energy states of atoms are like the shelves of the bookcase depicted in **Figure 5.8a**. The energy of an atom might correspond to the energy of one state or to the energy of the next state, but the energy of the atom is never found between the two states, just as a book can be on only one shelf at a time and cannot be partly on one shelf and partly on another. A given atom may have many energy states available to it, but those stable states are *discrete*.

Astronomers keep track of the allowed states of an atom by using energy level diagrams, as shown in **Figure 5.8b**. Both the bookcase and the energy level diagram are simplifications of the possible energies of a three-dimensional system. The lowest possible energy state for a system (or part of a system) such as an atom is called the **ground state**. When the atom is in the ground state, the electron has its minimum energy. It can't give up any more energy to move to a lower state because a lower state does not exist. An atom will remain in its ground state forever unless it gets energy from outside. In the bookcase analogy, a book sitting on the bottom shelf at the floor is in its ground state. It has nowhere left to fall, and it cannot jump to one of the higher shelves on its own.

Astronomy in Action: Emission and Absorption

Energy levels above the ground state are called **excited states**. Just as a book on an upper shelf might fall to a lower shelf, an atom in an excited state might **decay** to a lower state by getting rid of some of its extra energy. An important difference between the atom and the book on the shelf, however, is that whereas a snapshot might catch the book falling between the two shelves, the atom will never be caught between two energy states. When the transition occurs from a higher state to a lower one, the difference in energy between the two states is carried off all at once. A common way for an atom to do that is to emit a photon. The photon emitted by the atom carries away exactly the amount of energy that atom loses as it goes from the higher energy state to the lower energy state. Similarly, atoms moving from a lower energy state to a higher energy state can absorb only certain specific energies.

To make an analogy with money, suppose you have a penny (1 cent), a nickel (5 cents), and a dime (10 cents), totaling 16 cents. Now imagine that you give away the nickel and are left with 11 cents. You never had exactly 13 cents or 13.6 cents. You had 16 cents and then 11 cents. Atoms don't accept and give away money to change energy states, but they do accept and give away photons with well-defined energies.

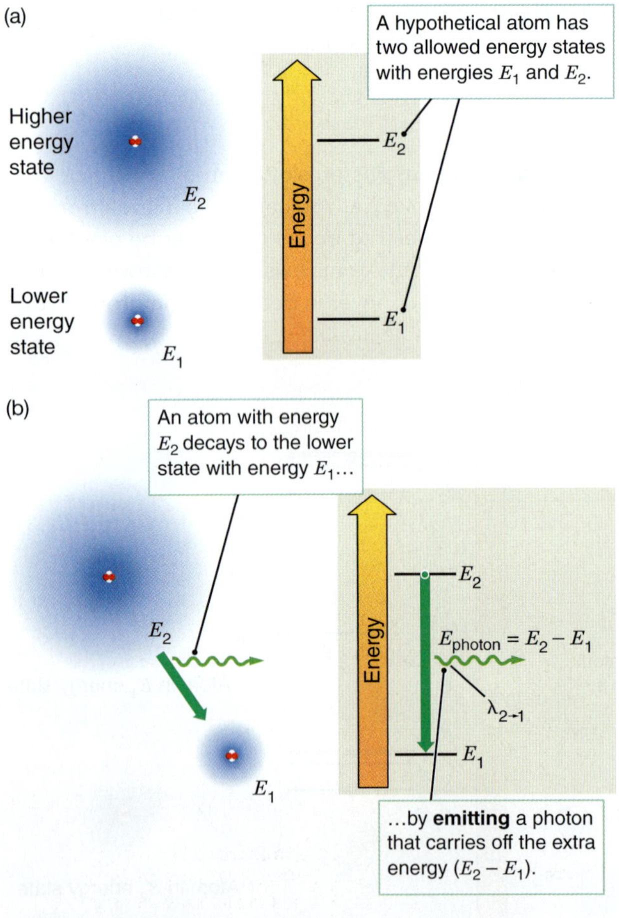

Figure 5.9 (a) The energy levels of a hypothetical two-level atom. (b) A photon with energy $E_{photon} = E_2 - E_1$ is emitted when an atom in the higher energy state (E_2) decays to the lower energy state (E_1).

EMISSION SPECTRA Imagine a hypothetical atom that has only two available energy states. The energy of the lower energy state (the ground state) is E_1, and the energy of the higher energy state (the excited state) is E_2. The energy levels of that atom can be represented in an energy level diagram such as the one in **Figure 5.9a**. An atom in the excited state moves to the ground state by getting rid of the "extra" energy all at once. In that **emission** process, a photon is released. The atom goes from one energy state to another, but it never has an amount of energy in between.

In **Figure 5.9b**, the green arrow pointing straight down indicates that the atom went from the higher state (E_2) to the lower state (E_1). The atom lost an amount of energy equal to $E_2 - E_1$, the difference between the two states. Because energy is never truly lost or created, the energy the atom lost has to show up somewhere. Here, the energy shows up in the form of a photon emitted by the atom. The energy of the emitted photon matches the energy lost by the atom; that is, $E_{photon} = E_2 - E_1$.

An atom can emit photons with energies corresponding *only* to the difference between two of its allowed energy states. Because the energy of a photon is related to the frequency or wavelength of electromagnetic radiation, in the transition from level 2 to level 1, a photon of energy $E_{photon} = E_2 - E_1$ has a specific wavelength $\lambda_{2\to1}$ and a specific frequency $f_{2\to1}$. Therefore, those emitted photons have a specific color, and every photon emitted in any transition from E_2 to E_1 will have that same color defined by a specific wavelength. The energy level structure of an atom determines the wavelengths of the photons it emits—the color of the light that the atom gives off.

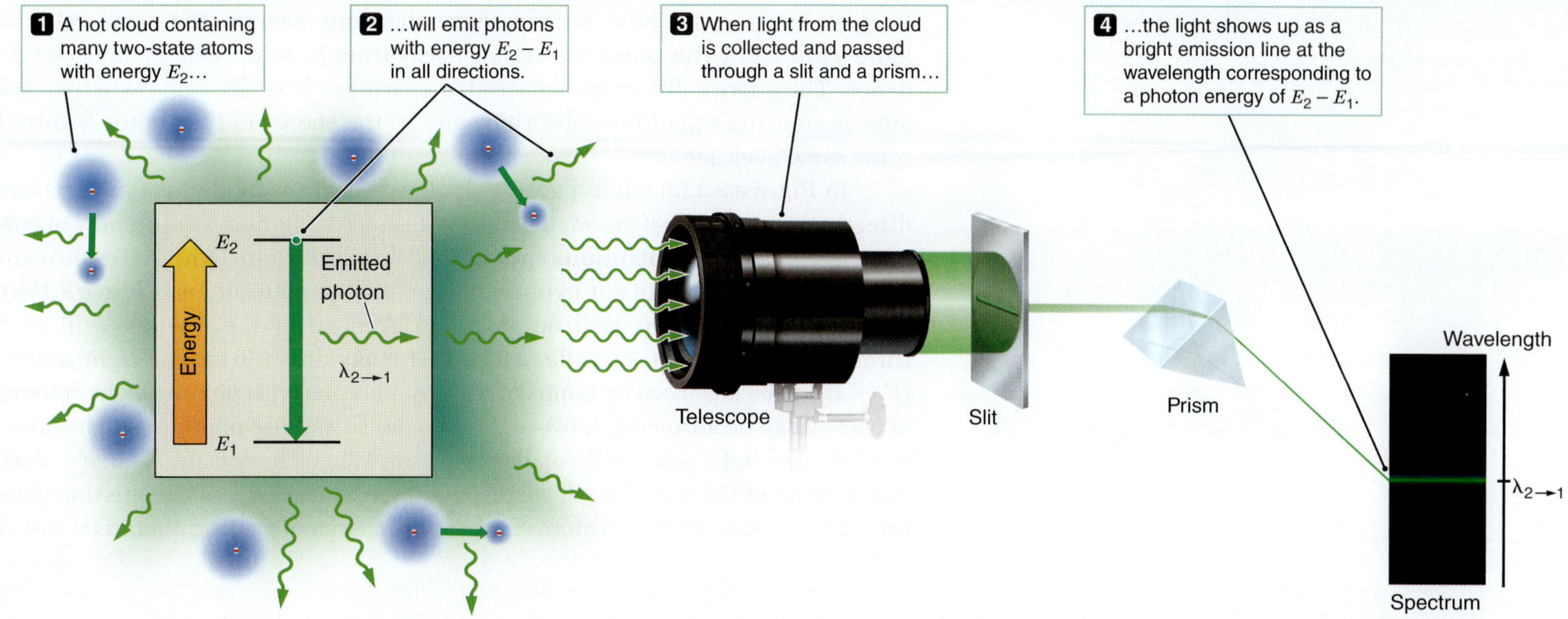

Figure 5.10 A cloud of gas containing atoms with two energy states, E_1 and E_2, emits photons with an energy $E_{photon} = E_2 - E_1$, which appear in the spectrum (right) as a single bright *emission line*.

Figure 5.10 illustrates the light coming from a cloud of gas consisting of the hypothetical two-state atoms in Figure 5.9. Any atom in the higher energy state (E_2) quickly decays and emits a photon in a random direction, and an enormous number of photons come pouring out of the cloud of gas. Instead of containing photons of all different energies—light of all different colors—that light contains only photons with the specific energy $E_2 - E_1$ and wavelength $\lambda_{2\to1}$. In other words, all the light coming from those atoms in the cloud is the same color. If you spread the light out into its component colors, only one color would be present—a single bright line called an **emission line**.

Why was the atom in the excited state E_2 in the first place? An atom sitting in its ground state will remain there unless it absorbs just the right amount of energy to kick it up to an excited state. In general, the atom either absorbs the energy of a photon or it collides with another atom or an unattached electron and absorbs some of the other particle's energy. In a neon sign, an alternating electric field inside the glass tube pushes electrons in the gas back and forth through the neon gas inside the tube. Some of those electrons crash into atoms of the gas, knocking them into excited states. The atoms then drop back down to their ground states by emitting photons, causing the gas inside the tube to glow a color characteristic of the element neon.

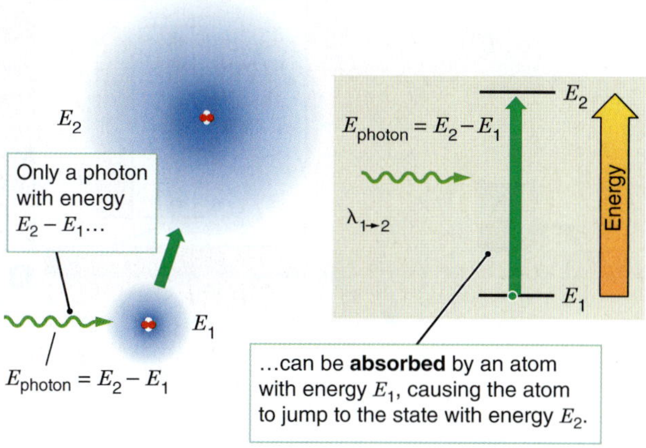

Figure 5.11 An atom in a lower energy state (E_1) may absorb a photon of energy $E_{photon} = E_2 - E_1$, leaving the atom in a higher energy state (E_2).

ABSORPTION SPECTRA An atom in a low energy state can absorb the energy of a passing photon and jump up to a higher energy state, as shown in **Figure 5.11**. Once again, the energy required to go from E_1 to E_2 is the difference in energy between the two states, $E_2 - E_1$. For a photon to cause an atom to jump from E_1 to E_2, it must provide exactly that much energy. The only photons that can excite atoms from E_1 to E_2 are photons with $E_{photon} = E_2 - E_1$. As with emission, those photons have a corresponding frequency ($f_{1\to2} = E_{photon}/h$) and wavelength

($\lambda_{1\rightarrow 2} = hc/E_{photon}$). Those photons have the same energy—the same color of light—emitted by the atoms when they decay from E_2 to E_1. That is not a coincidence. The energy difference between the two levels is the same whether the atom is emitting a photon or absorbing one, so the energy of the photon involved is the same in either case.

In **Figure 5.12a**, white light (with all wavelengths of photons in it) passes directly through a glass prism, which spreads out the light into a rainbow of colors. That is called a **continuous spectrum**. When the white light passes through a cool cloud composed of our hypothetical gas of two-state atoms (**Figure 5.12b**), however, some photons will be absorbed. Almost all the photons will pass through the cloud of gas unaffected because they do not have the right energy ($E_2 - E_1$) to be absorbed by atoms of the gas. Photons with just the right amount of energy can be absorbed, however, and as a result, those photons will be missing from the light passing through the prism. When that occurs, a sharp, dark line appears at the wavelength corresponding to the energy of the missing photons. That process by which atoms capture the energy of passing photons is called

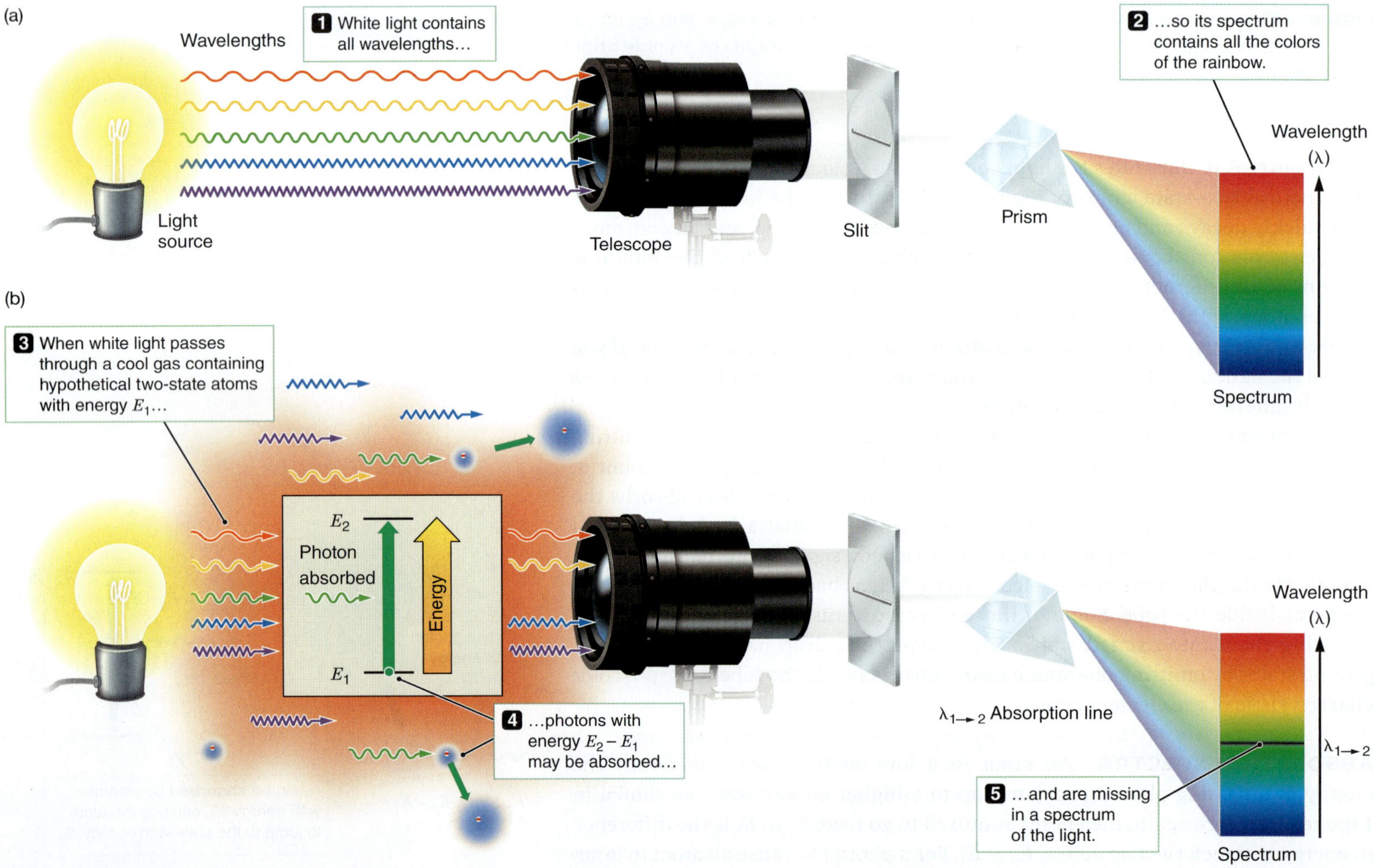

Figure 5.12 (a) When passed through a prism, white light produces a spectrum containing all colors. (b) When light of all colors passes through a cloud of hypothetical two-state atoms, photons with energy $E_{photon} = E_2 - E_1$ may be absorbed, leading to the dark absorption line in the spectrum.

absorption, and the dark line seen in the spectrum is called an **absorption line**. **Figure 5.13a** shows the absorption lines in the spectrum of a star. The same spectrum is shown again in **Figure 5.13b** as a graph of the brightness at every wavelength. The brightness in Figure 5.13b drops abruptly at the wavelengths corresponding to the dark lines in Figure 5.13a. Places between the dark lines are brighter and therefore higher on the graph than the absorption lines.

When an atom absorbs a photon, it may quickly decay to its previous lower energy state, emitting a photon with the same energy as the photon it just absorbed. If the atom reemits a photon just like the one it absorbed, why does the absorption matter? The photon that was taken out of the passing light was replaced, but all the absorbed photons were originally traveling in the same direction, whereas the emitted photons are traveling in random directions. In other words, some photons with energies equal to $E_2 - E_1$ are diverted from their original paths by their interaction with atoms. If you look at a white light through the cloud in Figure 5.12b, you will observe an absorption line at a wavelength of $\lambda_{2\rightarrow1}$, but if you look at the cloud from another direction, you will observe an emission line at that same wavelength.

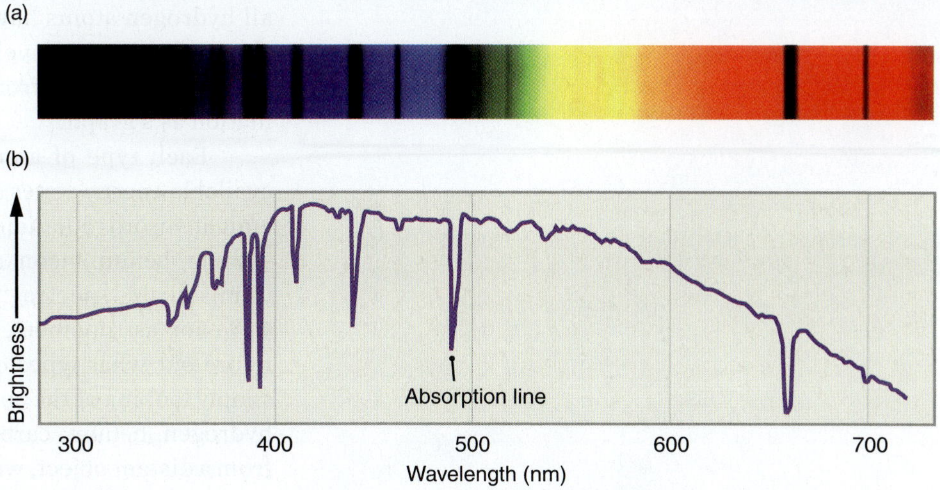

Figure 5.13 Absorption lines in the spectrum of a star as an image (a) and a graph (b).

AstroTour: Atomic Energy Levels and Light Emission and Absorption

Spectral Fingerprints of Atoms

In the mid-19th century, Gustav Kirchoff (1824–1887) first observed emission, continuous, and absorption spectra from the three types of sources shown in Figures 5.10 and 5.12. He summarized his findings as three empirical laws—but he did *not* know about energy levels in atoms. Kirchoff, together with Robert Bunsen (1811–1899), concluded that the dark-line (absorption) spectrum of the Sun was the "reverse" of the bright-line (emission) spectrum that would be produced by the Sun's atmosphere alone and identified some of the elements on the Sun. (A few years later the element helium was discovered in the solar spectrum.) Others observed the stars. Noteworthy among them were William (1824–1910) and Margaret Lindsay (1848–1915) Huggins, who published an atlas of stellar spectra in 1899 and who showed that the types of atoms seen in the stars are the same as those found on Earth.

The study of spectra of astronomical objects is fundamental to our understanding of the universe. Astronomers who study spectra will say that "a spectrum is worth a thousand pictures" because of the wealth of information that can come from it. We refer to spectra in every chapter in this book.

Real atoms can occupy many more than just two possible energy states; therefore, any given type of atom can emit and absorb photons at many wavelengths. An atom with three energy states, for example, might jump from state 3 to state 2, or from state 3 to state 1, or from state 2 to state 1. The three distinct emission lines in the spectrum from a gas made up of those atoms would have wavelengths of $hc/(E_3 - E_2)$, $hc/(E_3 - E_1)$, and $hc/(E_2 - E_1)$, respectively.

An atom's allowed energy states are determined by the interactions among the electrons and the nucleus. Every neutral hydrogen atom consists of a nucleus containing one proton, plus a single electron in a cloud surrounding the nucleus. Therefore, every hydrogen atom has the same energy states available to it, and all hydrogen atoms have the same emission and absorption lines. **Figure 5.14a** shows the energy level diagram of hydrogen. **Figure 5.14b** shows the visible emission spectrum from hydrogen, and **Figure 5.14c** displays that same information as a graph.

Each type of atom—that is, each chemical element—has a unique set of available energy states and therefore a unique set of wavelengths at which it can emit or absorb radiation. **Figure 5.14d** shows the emission spectra of the elements sodium, helium, neon, and mercury. Those unique sets of wavelengths serve as unmistakable spectral "fingerprints" for each chemical element.

For astronomers, spectral fingerprints are crucial. They let astronomers figure out what types of atoms (or molecules) are present in distant objects by simply looking at the spectrum of light from those objects. If the spectral lines of hydrogen, helium, carbon, oxygen, or any other element are visible in the light from a distant object, we know that element is present in that object. The strength of a line is determined in part by how many atoms of that type are present in the source. By measuring the strength of the lines from different types of atoms in the spectrum of a distant object, astronomers can often infer the relative amounts of elements that make up the object. In addition, by looking at the relative strength of lines from the same element, astronomers often can determine the temperature, density, and pressure of the material.

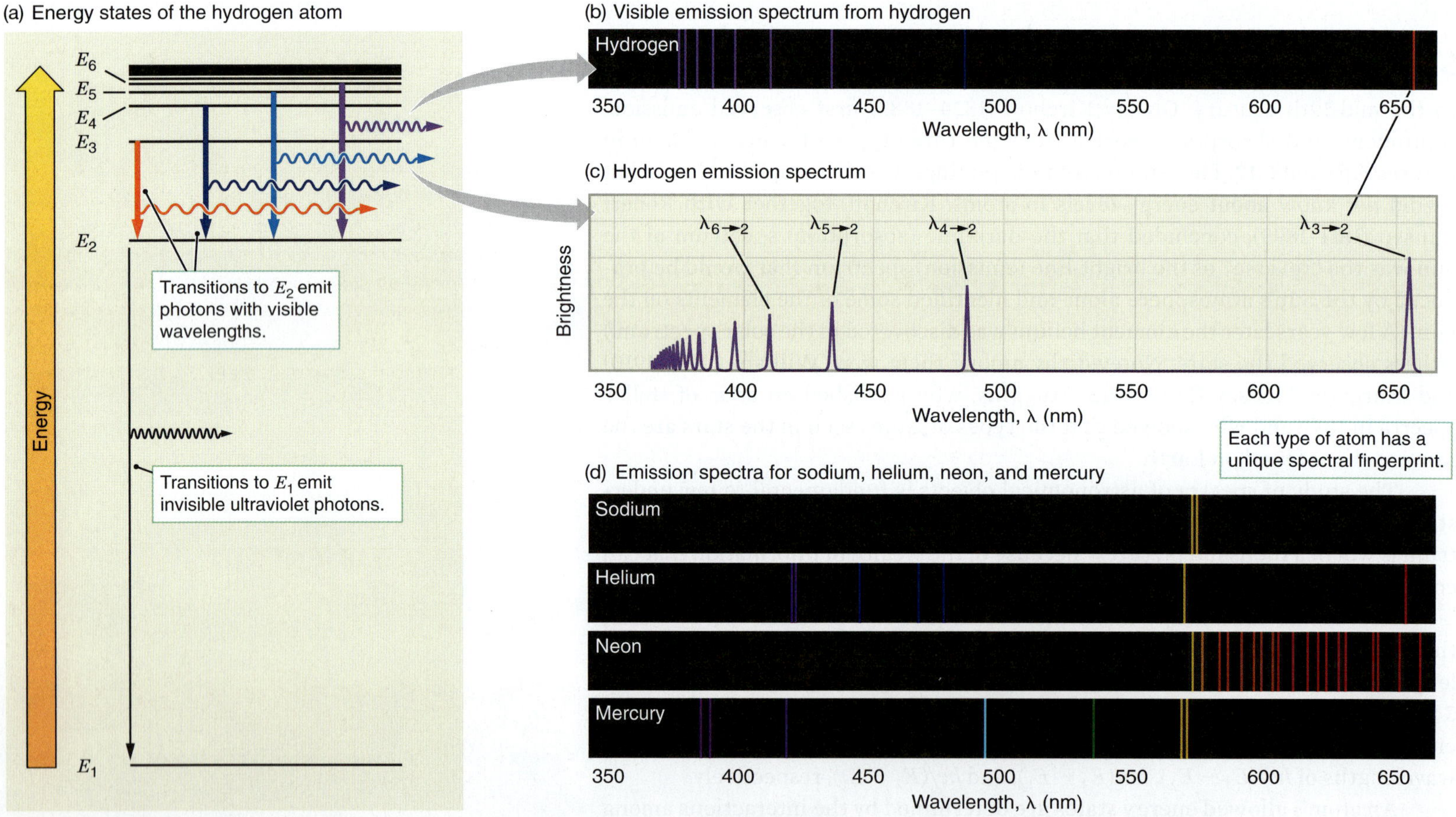

Figure 5.14 (a) The energy states of the hydrogen atom. Decays to level E_2 emit photons in the visible part of the spectrum. (b) This visible emission spectrum is what you might see if you looked at the light from a hydrogen lamp projected through a prism onto a screen. (c) The brightness at every wavelength can be measured to produce a graph of the brightness of spectral lines versus their wavelength. (d) Emission spectra from gaseous sodium, helium, neon, and mercury.

Spectra are how we know what makes up the stars and planets, and that we on Earth are composed of the same elements. Astronomers use the relative abundance of the elements in the Sun (called the **solar abundance**) as a standard reference. Hydrogen (H) is the most abundant element in the Sun (**Figure 5.15**), followed by helium (He) and 13 others. Those 15 elements make up 99.99 percent of the mass of the Sun. Most of the elements on the regular periodic table (in the lower right) make up less than 0.01 percent of the mass of the Sun.

Excitement and Decay

If a book on a level shelf is not disturbed, it will sit there forever; something must *cause* the book to fall off the shelf. Similarly, physicists have wondered what causes an atom in an excited state to jump down to a lower energy state and emit a photon. Sometimes an atom in a higher energy state can be "stimulated" into emitting a photon—but usually nothing causes the atom to jump to the lower energy state. Instead, the atom decays *spontaneously*. Although scientists can determine on average how long the atom is likely to remain in the excited state, they can't know exactly when a particular atom will decay until after the decay has happened. An atom decays at a random moment that is not influenced by anything in the universe and cannot be predicted.

Toys that glow in the dark are an example of that phenomenon. Photons in sunlight or from a lightbulb are absorbed by certain phosphorescent atoms in the toy, knocking those atoms into excited energy states. The excited states of the atoms in the toy last for many seconds, unlike the excited energy states of many atoms that tend to decay in a small fraction of a second. If on average those atoms tend to

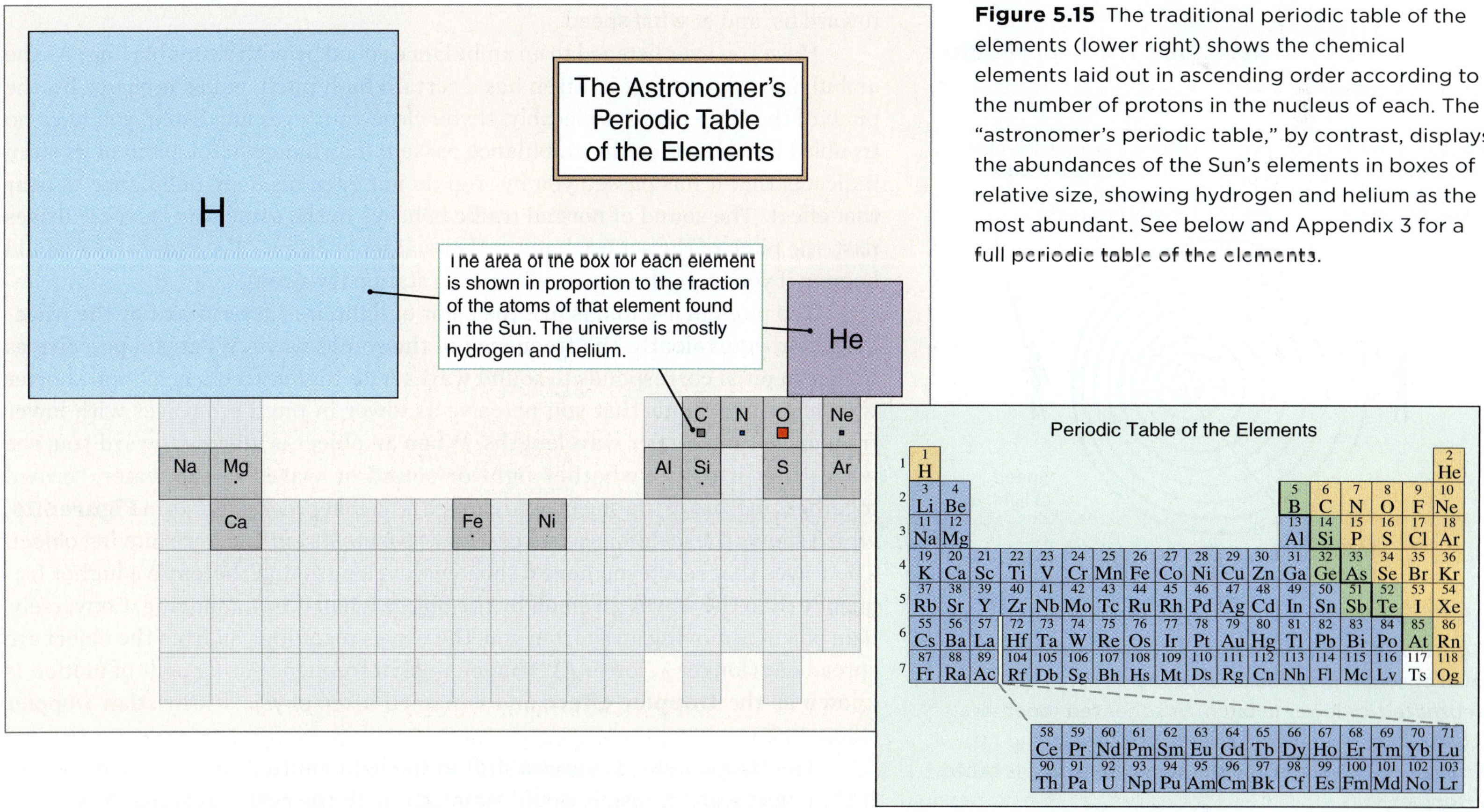

Figure 5.15 The traditional periodic table of the elements (lower right) shows the chemical elements laid out in ascending order according to the number of protons in the nucleus of each. The "astronomer's periodic table," by contrast, displays the abundances of the Sun's elements in boxes of relative size, showing hydrogen and helium as the most abundant. See below and Appendix 3 for a full periodic table of the elements.

remain in their excited state for 1 minute before decaying and emitting a photon, then after 1 minute a 50-50 chance exists that any particular atom in the toy will have decayed and a 50-50 chance that the atom will remain in its excited state. Although it is impossible to say exactly which atoms will decay, about half of the trillions and trillions of atoms in the toy will decay within 1 minute, and the brightness of the glow from the toy will have dropped to half of what it was. After each minute, half of the remaining excited atoms decay, and the glow from the toy drops to half of what it was 1 minute earlier. Thus, the glow from the toy slowly fades away.

In deep space, where atoms can remain undisturbed for long periods, certain excited states of atoms last, on average, for tens of millions of years or even longer. An atom may have been in such an excited energy state for a few seconds, a few hours, or 50 million years when, in an instant, it decays to the lower energy state without anything having caused it to do so. Physicists can only calculate the *probabilities* that certain events will take place.

CHECK YOUR UNDERSTANDING 5.2

How can spectra tell us the chemical composition of a distant star?

5.3 The Doppler Shift Indicates Motion Toward or Away from Us

Astronomy in Action: Doppler Shift

AstroTour: The Doppler Effect

You have already seen that light is a tightly packed bundle of information that can reveal a wealth of information about the physical state of material located tremendous distances away. Here in Section 5.3, we explain how light can be used to determine whether a distant astronomical object is moving away from us or toward us, and at what speed.

Have you ever listened to an ambulance speed by with sirens blaring? As the ambulance approaches, its siren has a certain high pitch, but as it passes by, the pitch of the siren drops noticeably. If you close your eyes and listen, you have no trouble knowing when the ambulance passed; the change in the pitch of its siren indicates that it has passed you by. You do not even need an ambulance to hear that effect. The sound of normal traffic behaves in the same way. As a car drives past, the pitch of the sound that it makes suddenly drops. (The same effect would happen if you were the one driving past a stationary siren).

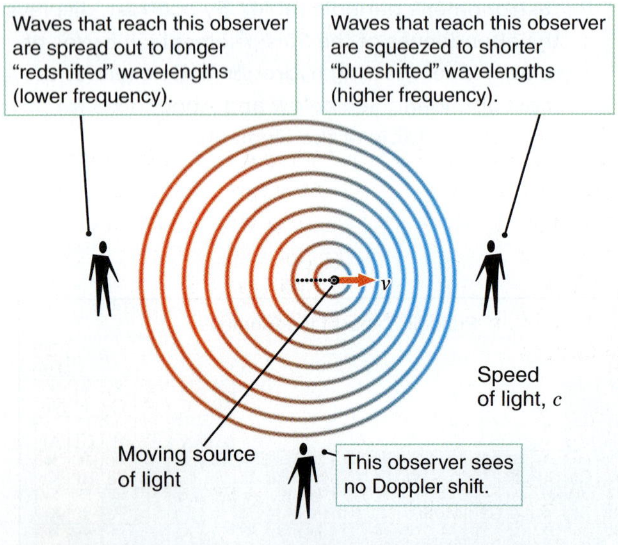

Figure 5.16 Motion of a light or sound source relative to an observer may cause waves to be spread out (*redshifted*, or lower in pitch) or squeezed together (*blueshifted*, or higher in pitch). A change in the wavelength of light or the frequency of sound is called a *Doppler shift.*

The pitch of a sound is like the color of light: it is determined by the wavelength or, equivalently, the frequency of the sound wave. What you perceive as higher in pitch corresponds to sound waves with higher frequencies and shorter wavelengths. Sounds that you perceive as lower in pitch are waves with lower frequencies and longer wavelengths. When an object is moving toward you, the waves that it emits, whether light or sound or waves in the water, "crowd together" in front of the object. You can see how that process works in **Figure 5.16**, which shows the locations of successive wave crests emitted by a moving object. The waves that reach you have a shorter wavelength and therefore a higher frequency than the waves given off by the object when it is not moving. Conversely, if an object is moving away from you, the waves reaching you from the object are spread out (longer λ, lower f). That change in frequency as a result of motion is known as the **Doppler effect** and is named after physicist Christian Doppler (1803–1853).

The Doppler effect causes a shift in the light emitted from a moving object. If the object were at rest, it would emit light with the **rest wavelength** (λ_{rest}), as shown in **Figure 5.17a**. If an object such as a star is moving *toward* you, the light reaching you from the object has a shorter wavelength than its rest wavelength—the light is "bluer" than the rest wavelength, and the light is described as **blueshifted**, as shown by the blue waves in **Figure 5.17b**. In contrast, light from a source moving *away* from you is shifted to longer wavelengths. The light that you see is "redder" than if the source were not moving away from you and is described as **redshifted**, as shown by the red waves in **Figure 5.17c**. The faster the object is moving with respect to you, the larger the shift. The amount by which the wavelength of light is shifted by the Doppler effect is called the light's **Doppler shift**, which depends on the speed of the object emitting the light.

The Doppler shift provides information only about the **radial velocity** (v_r) of the object, which is the part of the motion that is toward you or away from you. The radial velocity is the rate at which the distance between you and the object is changing: if v_r is positive, the object is getting farther away from you; if v_r is negative, the object is getting closer. At the moment the ambulance is passing you, it is getting neither closer nor farther away, so the pitch you hear is the same as the pitch heard by the crew riding on the truck. Similarly, a distant object moving slowly across the sky does not move toward or away from you, and so its light will not be Doppler shifted from your point of view.

Doppler shifts become especially useful when you are looking at an object that has emission or absorption lines in its spectrum. Those spectral lines enable astronomers to determine how rapidly the object is moving toward or away from Earth. To determine that velocity, astronomers first identify the spectral line as being from a certain chemical element, which has a unique rest wavelength (λ_{rest}) measured in a lab on Earth. They then measure the observed wavelength (λ_{obs}) in the spectrum of the distant object. The difference between the rest wavelength and the observed wavelength indicates the object's radial velocity. See **Working It Out 5.2**.

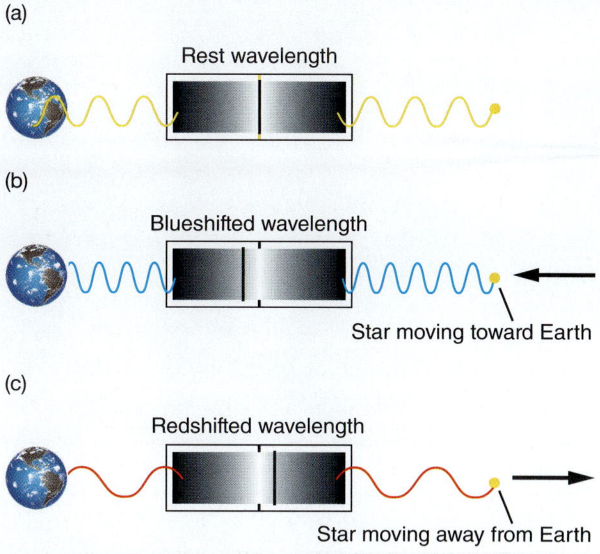

Figure 5.17 From their rest wavelength (a), spectral lines of astronomical objects are blueshifted if they are moving toward the observer (b) and redshifted if they are moving away from the observer (c).

5.2 Working It Out Making Use of the Doppler Effect

The Doppler formula for objects moving at a radial velocity (v_r) much less than the speed of light is given by

$$v_r = \frac{\lambda_{obs} - \lambda_{rest}}{\lambda_{rest}} \times c$$

A prominent spectral line of hydrogen atoms has a rest wavelength, λ_{rest}, of 656.3 nm (see Figure 5.14b). Suppose that you measure the wavelength of that line in the spectrum of a distant object and find that instead of seeing the line at 656.3 nm, you see the line at a wavelength, λ_{obs}, of 659.0 nm. What is its radial velocity? Using the above equation,

$$v_r = \frac{659.0\,\text{nm} - 656.3\,\text{nm}}{656.3\,\text{nm}} \times (3 \times 10^5\,\text{km/s})$$

$$v_r = 1{,}200\,\text{km/s}$$

Thus, the object is moving away from you with a radial velocity of 1,200 km/s.

Suppose instead that you know the velocity and want to compute the wavelength at which you would observe the spectral line. Earth's nearest stellar neighbor, Proxima Centauri, is moving toward us at a radial velocity of −21.6 km/s. What is the observed wavelength, λ_{obs}, of a magnesium line in Proxima Centauri's spectrum that has a rest wavelength, λ_{rest}, of 517.27 nm? We can rearrange the Doppler formula to solve for λ_{obs}:

$$\lambda_{obs} = \left(1 + \frac{v_r}{c}\right) \times \lambda_{rest}$$

$$\lambda_{obs} = \left(1 + \frac{-21.6\,\text{km/s}}{3 \times 10^5\,\text{km/s}}\right) \times 517.27\,\text{nm} = 517.23\,\text{nm}$$

Although the observed Doppler blueshift ($\lambda_{obs} - \lambda_{rest} = 517.23 - 517.27$) is only −0.04 nm, it is easily measured with modern instrumentation.

CHECK YOUR UNDERSTANDING 5.3

Which of the following Doppler shifts indicates the fastest-approaching object? (a) 0.04 nm; (b) 0.06 nm; (c) −0.04 nm; (d) −0.06 nm

5.4 Temperature Affects the Spectrum of Light That an Object Emits

The balance between heating and cooling in an object determines its temperature. If an object's temperature is constant, those two must be in balance. Here in Section 5.4, we examine that balance and see how we can use it to predict the temperatures of planets and stars.

Astronomy in Action: Changing Equilibrium

Equilibrium and Balance

Your body is heated by the release of chemical energy inside it. Sometimes your body is also heated by energy from your surroundings. If you are standing in sunshine on a hot day, the hot air around you and the sunlight falling on you both heat you. In response to that heating, your body cools itself off by perspiring: water seeps from the pores in your skin and evaporates. The energy to evaporate the water comes from your body. As the perspiration evaporates, it cools you down because it removes heat energy from your body. For your body temperature to remain stable, the heating must be balanced by the cooling. If more heating than cooling occurs, your body temperature increases; if more cooling than heating, your body temperature decreases.

In a tug-of-war contest between two perfectly matched teams, each team pulls on the rope, but the force of one team's pull is only enough to match, not overcome, the force exerted by the other team. A picture taken now and another taken 5 minutes from now would not differ in any significant way. In that static equilibrium, opposing forces balance each other exactly. Static equilibrium can be stable, unstable, or neutral. A marble in a bowl is in a stable equilibrium: if it moves, it will return to its original position at the bottom of the bowl. A book standing on its edge, unsupported on either side, is in an unstable equilibrium: if you nudge it, it will fall over, not settle back into its original position. When an unstable equilibrium is disturbed, it moves further away from equilibrium rather than back toward it.

Equilibrium also can be dynamic, which means the system is constantly changing so that one source of change is exactly balanced by another source of change, and the configuration of the system remains the same. In **Figure 5.18**, placing a can with a hole cut in the bottom under an open water faucet provides a simple example of dynamic equilibrium. The depth of the water in the can determines how fast water pours out through the hole in the bottom of the can. When the water reaches just the right depth (Figure 5.18a), water pours out of the hole in the bottom of the can at the same rate it pours into the top of the can from the faucet. The water leaving the can balances the water entering, and equilibrium is established. If you took a picture now and another picture in a few minutes, little of the water in the can would be the same, but the pictures would be indistinguishable.

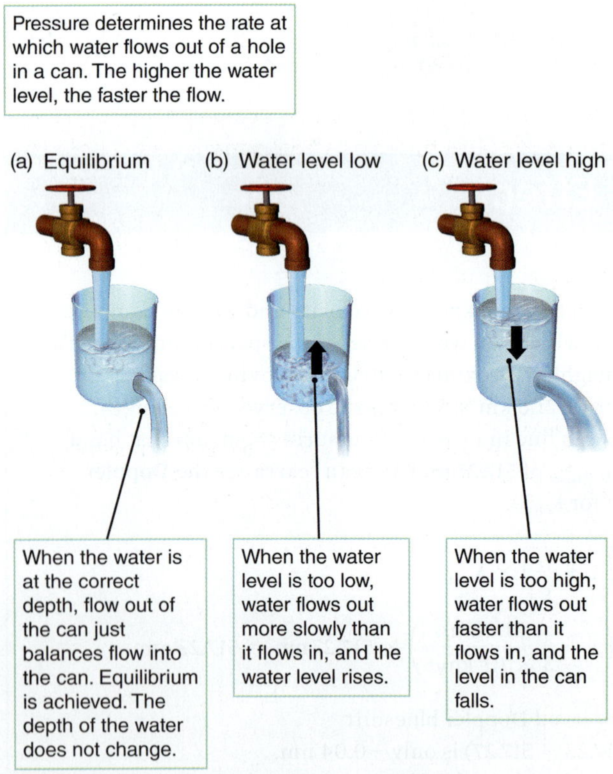

Figure 5.18 The relative rates at which water flows into and out of a can determine the water level in the can. This is an example of dynamic equilibrium.

If a system is not in equilibrium, its configuration will change. If the level of the water in the can is too low (Figure 5.18b), water will not flow out of the bottom of the can fast enough to balance the water flowing in. The water level will begin to rise until the amount coming in equals the amount going out. A picture taken now and another taken a short time later would not look the same. Conversely, if the water level in the can is too high (Figure 5.18c), water will flow out of the can faster than it flows in. The water level will begin to fall until the amount coming in equals the amount going out. Once again, if the system is not in equilibrium, its configuration will change.

When heating is balanced by cooling, we call it **thermal equilibrium**. Planets have a dynamic but stable thermal equilibrium, and electromagnetic radiation plays a crucial role in maintaining that balance. Energy from sunlight heats the surface of a planet, driving its temperature up, and the planet emits thermal radiation into space, cooling it down. For a planet to remain at the same average temperature over time, the energy it radiates into space must exactly balance the energy it absorbs from the Sun. **Figure 5.19** shows how the equilibrium temperature of a planet is analogous to the water level in Figure 5.18. We return to planetary equilibrium later in the chapter. Many kinds of equilibrium exist besides thermal equilibrium, some of which we encounter later in the book.

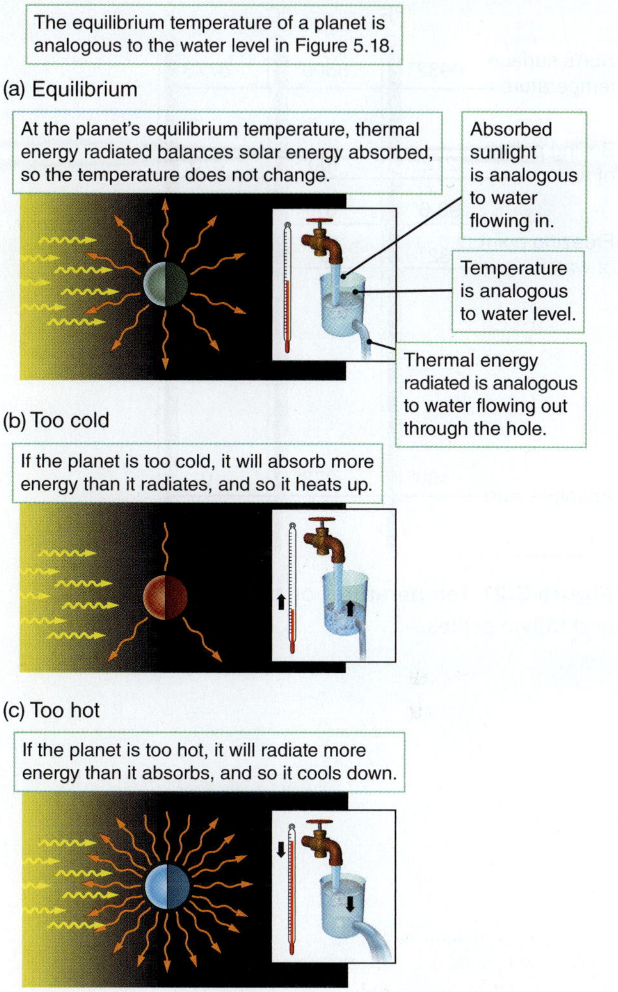

Figure 5.19 Planets are heated by absorbing sunlight (and sometimes by internal heat sources) and cooled by emitting thermal radiation into space. In the absence of other sources of heating or means of cooling, the equilibrium between those processes determines the temperature of the planet.

Temperature

In everyday life, we define hot and cold subjectively: something is hot or cold when it feels hot or cold. When we measure *temperature*, we specify degrees on a thermometer, but the way we define a degree is arbitrary. If you grew up in the United States, you probably think of temperatures in degrees Fahrenheit (°F), whereas if you grew up almost anywhere else in the world, you think of temperatures in degrees Celsius (°C).

Temperature is a measurement of how energetically the atoms that make up an object are moving about. The air around us is composed of vast numbers of atoms and molecules. Those molecules are moving about every which way. Some move slowly; some move more rapidly. All atoms and molecules are constantly in motion. The average kinetic energy (E_K) is given by $E_K = 1/2\,mv^2$, where m is the mass of an atom or molecule and v is its velocity. The more energetically the atoms or molecules are bouncing about, the higher the object's temperature. In fact, the random motions of atoms and molecules are often called their **thermal motions** to emphasize the connection between those motions and temperature. **Figure 5.20** shows that when the temperature of a gas is increased, the kinetic energy is increased, in which case the atoms move faster.

The atoms and molecules in a solid body (like you) cannot move about freely like a gas, but they still move back and forth around their average location, and temperature measures the amount of that movement. If an object is hotter than you are, thermal energy flows from that object into you. At the atomic level, that means the object's atoms are bouncing more energetically than the atoms in your body are, so if you touch the object, its atoms collide with your atoms, causing the atoms in your body to move faster. Your body gets hotter as thermal energy flows from the object to you. At the same time, those collisions rob the particles in the object of some of their energy. Their motions slow down, and the hotter object cools. Heating processes increase the average thermal energy of an object's particles, whereas cooling processes decrease the average thermal energy of those particles.

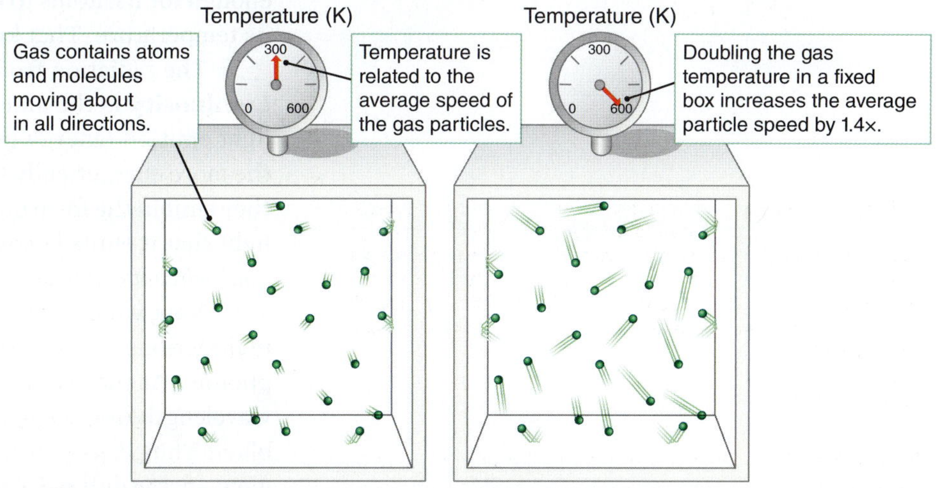

Figure 5.20 Higher gas temperatures correspond to faster-moving atoms.

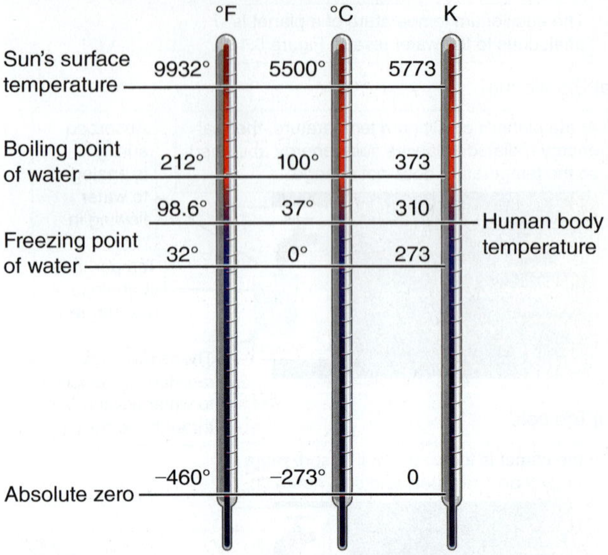

Figure 5.21 Temperatures on the Fahrenheit, Celsius, and Kelvin scales.

The three most common temperature scales are shown in **Figure 5.21**. On the Fahrenheit scale, water at sea level has a freezing point of 32°F and a boiling point of 212°F. On the Celsius scale, water freezes at 0°C and boils at 100°C. Because the number of degrees between freezing and boiling on those two scales is different (180°F vs. 100°C), a 1-degree change measured in °F is not the same as a 1-degree change measured in °C.

As the motions of the particles in an object slow down, the temperature decreases more and more. The lowest possible temperature, at which all thermal motions would stop, is called **absolute zero**. Absolute zero corresponds to −273.15°C and −459.57°F. It also is the starting point of the **Kelvin temperature scale**. The size of one unit on the Kelvin scale, called a **kelvin (K)**, is the same as the Celsius degree. To convert between °C and K, just add 273.15 to the temperature in °C. Thus, water freezes at 273.15 K and water boils at 373.15 K. The Kelvin scale has no negative temperatures.

When temperatures are measured in kelvins, the average thermal energy of particles is proportional to the measured temperature. Thus, the average thermal energy of the atoms in an object with a temperature of 200 K is twice the average thermal energy of the atoms in an object with a temperature of 100 K.

Temperature, Luminosity, and Color

We have seen the way discrete atoms emit and absorb radiation, which leads to a useful understanding of emission lines and absorption lines that tell us about the physical state and motion of distant objects. But not all objects have spectra dominated by discrete spectral lines. As you saw in Figure 5.12a, if you pass the light from a lightbulb through a prism, instead of discrete bright and dark bands you will see light spread out smoothly from the blue end of the spectrum to the red. Similarly, if you look closely at the spectrum of the Sun, you will see absorption lines, and you will see light smoothly spread out across all colors of the spectrum—the continuous spectrum noted earlier.

We can think of a dense material as being composed of a collection of charged particles being jostled as their thermal motions cause them to run into their neighbors. The hotter the material is, the more violently its particles are being jostled. A charged particle radiates anytime it accelerates, so the jostling of particles that results from their thermal motions causes them to give off a continuous spectrum of electromagnetic radiation. That is why any material dense enough for its atoms to be jostled by their neighbors emits light simply because of its temperature. That kind of radiation is called **thermal radiation**.

The radiation from an object changes as the object heats up or cools down. **Luminosity** is the amount of light *leaving* a source—that is, the total amount of light emitted (energy per second, measured in watts, W). The hotter the object, the more energetically the charged particles within it move, and the more energy they emit in the form of electromagnetic radiation. So, as an object gets hotter, the light that it emits becomes more intense. In other words, *an object is more luminous when it is hotter.*

Now, what color light does an object emit? As the object gets hotter, the thermal motions of its particles become more energetic, producing more energetic photons. As the average energy of the photons that it emits increases, the average wavelength of the emitted photons decreases, and the light from the object gets bluer. Thus, *hotter objects are bluer.* If you heat a piece of metal, the metal will glow—first a dull red, then orange, and then yellow. The hotter the metal becomes, the more the highly energetic blue photons become mixed with the less energetic red photons, and the color of the light shifts from red toward blue. The light becomes more intense and bluer as the metal becomes hotter.

Blackbodies are objects that emit electromagnetic radiation only because of their temperature, not their composition. Blackbodies emit just as much thermal radiation as they absorb from their surroundings. Physicist Max Planck (1858–1947) graphed the intensity of the emitted radiation across all wavelengths and obtained the characteristic curves that we now call **Planck spectra** or **blackbody spectra**. **Figure 5.22** shows blackbody spectra for objects at several temperatures.

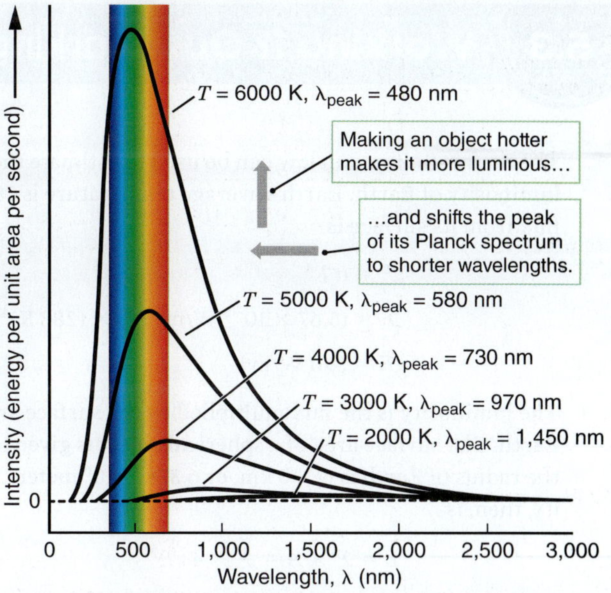

Figure 5.22 This graph shows blackbody spectra emitted by sources with temperatures of 2000, 3000, 4000, 5000, and 6000 K. At higher temperatures, the peak of the spectrum shifts toward shorter wavelengths, and the amount of energy radiated per second from each square meter of the source increases.

Blackbody Laws

In the real world, the light from stars such as the Sun and the thermal radiation from a planet often come close to having blackbody spectra. Those objects follow the Stefan-Boltzmann law, which relates luminosity with temperature, and Wien's law, which relates temperature with color.

STEFAN-BOLTZMANN LAW As the temperature of an object increases, the object gives off more radiation at every wavelength, so the luminosity of the object should increase. Adding up all the energy in a blackbody spectrum shows that the increase in luminosity is proportional to the fourth power of the temperature. That relationship, luminosity $\propto T^4$, is known as the **Stefan-Boltzmann law**. It was discovered in the laboratory by physicist Josef Stefan (1835–1893) and derived mathematically by his student Ludwig Boltzmann (1844–1906).

Measuring all the photons emitted by Earth or the Sun in all possible directions is hard, but measuring the *flux* is easier. The **flux** ($\mathcal{F}$) is the amount of energy radiated by each square meter of the surface of an object each second. Flux is proportional to luminosity. You can find the luminosity by multiplying the flux by the total surface area. According to the Stefan-Boltzmann law, the flux is given by $\mathcal{F} = \sigma T^4$, where σ (the Greek letter sigma) is the **Stefan-Boltzmann constant**. It equals 5.67×10^{-8} W/(m^2 K^4), where 1 watt = 1 joule per second (J/s).

The Stefan-Boltzmann law says that an object rapidly becomes more luminous as its temperature increases. If the temperature of an object doubles, the amount of energy being radiated each second increases by a factor of $2^4 = 16$. If the temperature of an object increases by a factor of 3, the energy being radiated by the object each second increases by a factor of $3^4 = 81$. A lightbulb with a filament temperature of 3000 K radiates 16 times as much light as it would if the filament temperature were 1500 K. Even modest changes in temperature can result in large changes in the luminosity of an object.

Astronomy in Action: Wien's Law

WIEN'S LAW Look again at Figure 5.22. The wavelength where the blackbody spectrum is at its peak, λ_{peak}, corresponds to where the electromagnetic radiation from an object is greatest. As the temperature, T, increases, the peak of the spectrum shifts toward shorter wavelengths. For example, λ_{peak} = 970 nm for a 3000 K object, but only 480 nm for a 6000 K object. Photon energy and wavelength are inversely related; thus, as the peak wavelength decreases, the average photon energy increases. The object becomes bluer. The physicist Wilhelm Wien (1864–1928) found that the peak wavelength in the spectrum is inversely proportional to the temperature of the object ($\lambda_{peak} \propto 1/T$). **Wien's law** states that if you double the temperature, the peak wavelength becomes half of what it was. If you increase the temperature by a factor of 3, the peak wavelength becomes a third

5.3 Working It Out Working with the Stefan-Boltzmann Law and Wien's Law

The Stefan-Boltzmann law can be used to estimate the flux and luminosity of Earth. Earth's average temperature is 288 K, so the flux from its surface is

$$\mathcal{F} = \sigma T^4$$

$$\mathcal{F} = (5.67 \times 10^{-8}\ \text{W/m}^2\ \text{K}^4) \times (288\ \text{K})^4$$

$$\mathcal{F} = 390\ \text{W/m}^2$$

The luminosity is the flux multiplied by the surface area (A) of Earth. The surface area of a spherical object is given by $4\pi R^2$, and the radius of Earth is 6,378 km, or 6.378×10^6 meters. The luminosity, then, is

$$L = \mathcal{F} \times A = \mathcal{F} \times 4\pi R^2$$

$$L = (390\ \text{W/m}^2) \times [4\pi(6.378 \times 10^6\ \text{m})^2]$$

$$L \approx 2 \times 10^{17}\ \text{W}$$

Earth emits the equivalent of the energy used by 2,000,000,000,000,000 (2 million billion) hundred-watt lightbulbs. That value is still not anywhere close to the amount emitted by the Sun.

If astronomers measure the spectrum of an object emitting thermal radiation and find where the peak in the spectrum is, Wien's law can be used to calculate the temperature of the object. Wien's law can be written as

$$T = \frac{2{,}900{,}000\ \text{nm K}}{\lambda_{\text{peak}}}$$

What, for example, is the surface temperature of the Sun? The spectrum of the light coming from the Sun peaks at a wavelength of $\lambda_{\text{peak}} = 500$ nm, so

$$T = \frac{2{,}900{,}000\ \text{nm K}}{500\ \text{nm}} = 5800\ \text{K}$$

What is the peak wavelength at which Earth radiates? Using Earth's average temperature of 288 K in Wien's law gives

$$\lambda_{\text{peak}} = \frac{2{,}900{,}000\ \text{nm K}}{288\ \text{K}} = 10{,}100\ \text{nm} = 10.1\ \mu\text{m}$$

Thus, Earth's radiation peaks in the infrared region of the spectrum.

of what it was. Stefan-Boltzmann's law and Wien's law are further explored in **Working It Out 5.3**. We use those laws later in the chapter to estimate the temperatures of the planets.

CHECK YOUR UNDERSTANDING 5.4

When you look at the sky on a dark night you see stars of different colors. Rank them from coolest to hottest. (a) orange; (b) red-orange; (c) yellow; (d) red; (e) blue

5.5 The Brightness of Light Depends on the Luminosity and Distance of the Source

Whereas luminosity is the amount of light *leaving* a source, the **brightness** of electromagnetic radiation is the amount of light *arriving* at a particular location. Therefore, the observed brightness depends on the luminosity and the distance of the light source. For example, replacing a 50-W lightbulb with a 100-W bulb makes a room twice as bright because it doubles the light reaching any point in the room. But brightness also depends on the distance from a source of electromagnetic radiation. For example, if you needed more light to read this book, you could replace the bulb in your lamp with a more luminous bulb or you could move the book closer to the light. Conversely, if a light were too bright, you could move away from it. Our everyday experience teaches us that as we move away from a light, its brightness decreases.

The particle description of light offers another way to think about the brightness of radiation and how brightness depends on distance. Suppose you had a piece of cardboard that measured 1 meter by 1 meter. To make the light falling on the cardboard twice as bright, you would need to double the number of photons that hit the cardboard each second. Tripling the brightness of the light would mean increasing the number of photons hitting the cardboard each second by a factor of 3, and so on. Brightness depends on the number of photons falling on each square meter of a surface each second.

Now imagine a lightbulb sitting at the center of a spherical shell (**Figure 5.23**). Photons from the bulb travel in all directions and land on the inside of the shell. To find the number of photons landing on each square meter of the shell during each second—that is, to determine the brightness of the light—take the *total* number of photons given off by the lightbulb each second and divide by the number of square meters over which those photons have to be spread. The surface area of a spherical shell is given by the formula $A = 4\pi d^2$, where d is the distance between the bulb and the surface of the sphere. The number of photons striking one square meter each second is equal to the total number of photons emitted each second divided by the surface area $4\pi d^2$.

Now increase the size of the spherical shell while keeping the total number of photons given off by the lightbulb each second the same. As the shell becomes larger, the photons from the lightbulb must spread out to cover a larger surface area. Each square meter of the shell receives fewer photons each second, so the brightness of the light decreases. If the shell's surface is moved twice as far from the light, the area over which the light must spread increases by a factor of $2^2 = 2 \times 2 = 4$. The photons from the bulb spread out over 4 times as much area, so the number of photons falling on each square meter each second becomes 1/4 of what it was. If the surface of the sphere is 3 times as far from the light, the area over which the light must spread increases by a factor of $3^2 = 3 \times 3 = 9$, and the number of photons per second falling on each square meter becomes 1/9 of what it was originally. That is the same kind of inverse square relationship you saw for gravity in Chapter 4. The brightness of the light from an object is inversely proportional to the square of the distance from the object. Twice as far means one-fourth as bright.

The idea of photons streaming and spreading onto a surface from a light explains why brightness follows an inverse square law. In practice, however, talking about the average *energy* falling on a surface each second, rather than the number of photons, is usually more more useful to an astronomer.

The luminosity of an object is the total number of photons given off by the object multiplied by the energy of each photon. So, instead of thinking about how the number of photons must spread out to cover the surface of a sphere, we can think about how the energy carried by the photons must spread out to cover the surface of a sphere. The brightness of the light is the amount of energy falling on a square meter in a second, and it equals the luminosity L divided by the area of the sphere, which depends on the radius squared. That tells us, for example, that the brightness of the Sun depends on the inverse square of the planet's distance from the Sun. That relationship factors in when we estimate the equilibrium temperatures of the planets in **Working It Out 5.4**.

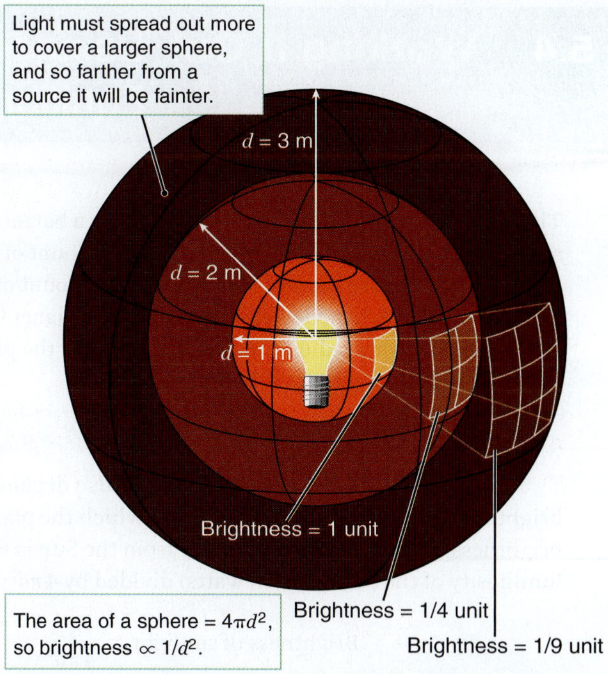

Figure 5.23 Light obeys an inverse square law as it spreads away from a source. Twice as far means one-fourth as bright, three times as far means one-ninth as bright, and so on.

Astronomy in Action: Inverse Square Law

CHECK YOUR UNDERSTANDING 5.5

The average distance of Mars from the Sun is 1.4 AU. How bright is the Sun on Mars compared with its brightness on Earth? (a) 1.4 times brighter; (b) about 2 times brighter; (c) about 2 times fainter; (d) 1.4 times fainter

5.4 Working It Out Using Radiation Laws to Calculate Equilibrium Temperatures of Planets

The temperature of a planet is determined by a balance between the amount of sunlight being absorbed and the amount of energy being radiated back into space. We begin with the amount of sunlight being absorbed. When viewed from the Sun, a planet looks like a circular disk with a radius equal to the radius of the planet, R_{planet}. The area of the planet that is lit by the Sun is

$$\text{Absorbing area of planet} = \pi R^2_{planet}$$

The amount of energy striking a planet also depends on the brightness of sunlight at the distance at which the planet orbits. The brightness of sunlight at a distance d from the Sun is equal to the luminosity of the Sun (L_{Sun}, in watts) divided by $4\pi d^2$:

$$\text{Brightness of sunlight} = \frac{L_{Sun}}{4\pi d^2}$$

A planet does not absorb *all* the sunlight that falls on it. **Albedo**, a, is the fraction of the sunlight that reflects from a planet. The corresponding fraction of the sunlight absorbed by the planet is 1 minus the albedo $(1 - a)$. A planet covered in snow would have a high albedo (close to 1), whereas a planet covered by black rocks would have a low albedo, close to 0:

$$\text{Fraction of sunlight absorbed} = 1 - a$$

We can now calculate the energy absorbed by the planet each second. Writing that relationship as an equation, we say that

$$\begin{pmatrix}\text{Energy absorbed}\\ \text{by the planet}\\ \text{each second}\end{pmatrix} = \begin{pmatrix}\text{Absorbing}\\ \text{area of}\\ \text{the planet}\end{pmatrix} \times \begin{pmatrix}\text{Brightness}\\ \text{of sunlight}\end{pmatrix} \times \begin{pmatrix}\text{Fraction}\\ \text{of sunlight}\\ \text{absorbed}\end{pmatrix}$$

$$= \pi R^2_{planet} \times \frac{L_{Sun}}{4\pi d^2} \times (1 - a)$$

Now let's turn to the other piece of the equilibrium: the amount of energy that the planet radiates away into space each second. We can calculate that amount by multiplying the number of square meters of the planet's total surface area by the energy radiated by each square meter each second. The surface area for the planet is given by $4\pi R^2_{planet}$. According to the Stefan-Boltzmann law, the energy radiated by each square meter each second is given by σT^4. Thus,

$$\begin{pmatrix}\text{Energy radiated}\\ \text{by the planet}\\ \text{each second}\end{pmatrix} = \begin{pmatrix}\text{Surface}\\ \text{area of}\\ \text{the planet}\end{pmatrix} \times \begin{pmatrix}\text{Energy radiated}\\ \text{per square meter}\\ \text{per second}\end{pmatrix}$$

$$= 4\pi R^2_{planet} \times \sigma T^4$$

If the planet's temperature is to remain stable—not heating up or cooling down—then each second the "energy radiated" must be equal to "energy absorbed":

$$\begin{pmatrix}\text{Energy radiated}\\ \text{by the planet}\\ \text{each second}\end{pmatrix} = \begin{pmatrix}\text{Energy absorbed}\\ \text{by the planet}\\ \text{each second}\end{pmatrix}$$

When we set those two quantities equal to each other, we arrive at the following expression:

$$4\pi R^2_{planet}\sigma T^4 = \pi R^2_{planet}\frac{L_{Sun}}{4\pi d^2}(1 - a)$$

Canceling out πR^2_{planet} on both sides, and rearranging the equation to put T on one side and everything else on the other, gives

$$T^4 = \frac{L_{Sun}(1 - a)}{16\sigma\pi d^2}$$

If we take the fourth root of each side, we get

$$T = \left(\frac{L_{Sun}(1 - a)}{16\sigma\pi d^2}\right)^{1/4}$$

Putting in the appropriate numbers for the known luminosity of the Sun, L_{Sun}, and the constants π and σ yields a simpler equation:

$$T = 279\,\text{K} \times \left(\frac{1 - a}{d^2_{AU}}\right)^{1/4}$$

where d_{AU} is the distance of the planet from the Sun in astronomical units.

To use that equation, we need to know a planet's distance from the Sun and its average albedo. For a blackbody ($a = 0$) at 1 AU from the Sun, the temperature is 279 K. For Earth, with an albedo of 0.3 and a distance from the Sun of 1 AU, the temperature is

$$T = 279\,\text{K} \times \left(\frac{1 - 0.3}{1^2}\right)^{1/4} = 255\,\text{K}$$

(Calculator hint: To take a fourth root, you can take the square root twice, or use the x^y button with $y = 0.25$.)

Earth is cooler than a blackbody at 1 AU from the Sun because its average albedo is greater than zero. If Earth's albedo changed or the Sun's luminosity changed, that would affect the result. When we examine planets around other stars, we must use the luminosity of the particular star in the equation, instead of the Sun's luminosity, so the temperature at 1 AU will be different from what it is for Earth.

Origins Temperatures of Planets

In the previous chapters, we discussed how a planet's axial tilt and its orbital shape affect its temperature and thus its prospects for life. Now let's get more specific about the temperatures of planets, using what you learned in this chapter about thermal radiation. For a planet at an equilibrium temperature, the energy radiated by a planet exactly balances the energy absorbed by the planet. If the planet is hotter than that equilibrium temperature, it will radiate energy faster than it absorbs sunlight, and its temperature will decrease. If the planet is cooler than that temperature, it will radiate energy slower than it absorbs sunlight, and its temperature will increase.

Planets at different distances from the Sun will have different temperatures, and the temperature should be inversely proportional to the square root of the distance, as you saw in Working It Out 5.4. **Figure 5.24** plots the actual and predicted temperatures of nine solar system objects. Each vertical orange bar shows the range of temperatures found on the surface of the planet or, for the giant planets, at the top of the planet's clouds. The black dots show the predictions made using the equation in Working It Out 5.4. For most planets, the predictions are not too far off, indicating that our basic understanding of *why* planets have the temperatures they have is probably pretty good. The data for Mercury, Mars, and Pluto agree particularly well.

Sometimes, however, the predictions are wrong. For Earth, the actual measured temperature is a bit higher than the predicted temperature, and for Venus the actual surface temperature is much higher than the prediction.

The predicted values assume that the temperature of the planet is the same everywhere, but planets are likely to be hotter on the day side than on the night side. The predictions also assume that a planet's only source of energy is sunlight and that the fraction of sunlight reflected is constant over the surface of each planet. The values also incorporate the assumption that the planets absorb and radiate energy into space as blackbodies.

The discrepancies between the calculated and the measured temperatures of some of the planets indicate that for those planets, some or all of those assumptions are incorrect. For example, the planet may have its own source of energy besides sunlight, or it may have an atmosphere that traps more solar energy and increases the temperature of the planet. Understanding the temperatures of planets makes it possible to hypothesize which planets are more suitable for life.

Figure 5.24 Predicted temperatures for the planets and dwarf planet Pluto are based on the equilibrium between absorbed sunlight and thermal radiation into space. These temperatures are compared with ranges of observed surface temperatures.

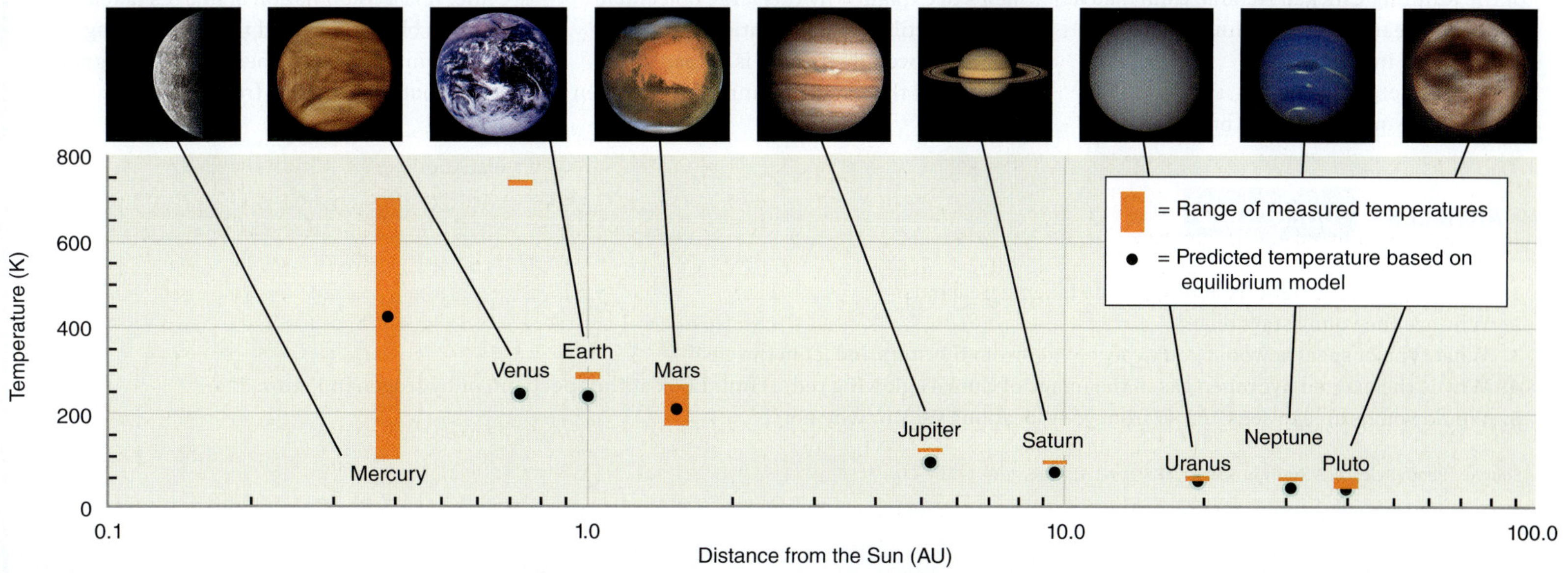

ARTICLES QUESTIONS

This press release from the European Southern Observatory (ESO) in Chile was widely reprinted in news sites around the world.

A Study in Scarlet

ESO

This new image from ESO's La Silla Observatory in Chile reveals a cloud of hydrogen called Gum 41 (**Figure 5.25**). In the middle of this little-known nebula, brilliant hot young stars are giving off energetic radiation that causes the surrounding hydrogen to glow with a characteristic red hue.

This area of the southern sky, in the constellation of Centaurus (The Centaur), is home to many bright nebulae, each associated with hot newborn stars that formed out of the clouds of hydrogen gas. The intense radiation from the stellar newborns excites the remaining hydrogen around them, making the gas glow in the distinctive shade of red typical of star-forming regions. Another famous example of this phenomenon is the Lagoon Nebula, a vast cloud that glows in similar bright shades of scarlet.

The nebula in this picture is located some 7,300 light-years from Earth. Australian astronomer Colin Gum discovered it on photographs taken at the Mount Stromlo Observatory near Canberra, and included it in his catalog of 84 emission nebulae, published in 1955. Gum 41 is actually one small part of a bigger structure called the Lambda Centauri Nebula, also known by the more exotic name of the Running Chicken Nebula. Gum died at a tragically early age in a skiing accident in Switzerland in 1960.

Figure 5.25 The Gum 41 Nebula.

In this picture of Gum 41, the clouds appear to be quite thick and bright, but this is actually misleading. If a hypothetical human space traveler could pass through this nebula, it is likely that they would not notice it as—even at close quarters—it would be too faint for the human eye to see. This helps to explain why this large object had to wait until the mid-twentieth century to be discovered—its light is spread very thinly and the red glow cannot be well seen visually.

This new portrait of Gum 41—likely one of the best so far of this elusive object—has been created using data from the Wide Field Imager (WFI) on the MPG/ESO 2.2-meter telescope at the La Silla Observatory in Chile. It is a combination of images taken through blue, green, and red filters, along with an image using a special filter designed to pick out the red glow from hydrogen.

ARTICLES QUESTIONS

1. How long has the light from that nebula taken to reach us?
2. Why are the young stars blue?
3. What type of spectra would you expect to get from the stars and from the gas?
4. Why is the excited hydrogen gas in the image of Gum 41 glowing red? (Hint: Look at the spectrum of hydrogen in Figure 5.14.)
5. Would you be able to see Gum 41 from your location? Why or why not?

Source: "Photo Release: A Study in Scarlet," ESO.org, April 16, 2014.

Summary

Light carries both information and energy throughout the universe. The speed of light in a vacuum is 300,000 km/s; nothing can travel faster. Visible light is only a tiny portion of the entire electromagnetic spectrum. Atoms absorb and emit radiation at unique wavelengths, giving them spectral fingerprints. A planet's temperature depends on its distance from its star, its albedo, and the luminosity of its star.

LG 1 **Describe the wave and particle properties of light, and describe the electromagnetic spectrum.** Light is simultaneously a stream of particles called photons and an electromagnetic wave. Different types of electromagnetic radiation, from gamma rays to visible light to radio waves, are electromagnetic waves that differ in frequency and wavelength.

LG 2 **Describe how to measure the chemical composition of distant objects by using the unique spectral lines of different types of atoms.** Nearly all matter is composed of atoms, and light can reveal the identity of the chemical elements present in matter. The electron energy levels of each element have different (unique) spacings, and the wavelengths of the photons each element emits correspond to the differences in those levels. As a result, we can identify different chemical elements and molecules in distant objects.

LG 3 **Describe the Doppler effect and how it can be used to measure the motion of distant objects.** Because of the Doppler effect, light from receding objects is redshifted to longer wavelengths, and light from approaching objects is blueshifted to shorter wavelengths. The wavelength shifts of the spectral lines indicate how fast an astronomical object is moving toward or away from Earth.

LG 4 **Explain how the spectrum of light that an object emits depends on its temperature.** Temperature is a measure of how energetically particles are moving in an object. A light source that emits electromagnetic radiation because of its temperature is called a blackbody. A blackbody emits a continuous spectrum. The total amount of energy emitted is proportional to the temperature to the fourth power, and the peak wavelength, which determines its color, is inversely proportional to the temperature.

LG 5 **Differentiate luminosity from brightness, and illustrate how distance affects each.** The luminosity of an object is the amount of light it emits. The brightness of an object is proportional to its luminosity divided by its distance squared. Thus, the brightness of the Sun is different when measured from each Solar System planet, but the luminosity is the same.

? Unanswered Questions

- Has the speed of light always been 300,000 km/s? Some theoretical physicists have questioned whether light traveled much faster earlier in the history of our universe. The observational evidence that may test that idea comes from studying the spectra of the most distant objects—whose light has been traveling for billions of years—and determining whether billions of years ago chemical elements absorbed light differently from how they do today. So far, no evidence exists that the speed of light changes.
- Will traveling faster than the speed of light ever be possible? Our current understanding of the science says no. A staple of science fiction films and stories is spaceships that go into "warp speed" or "hyperdrive"—moving faster than light—to traverse the huge distances of space (and visit a different planetary system every week). If that premise is simply fictional and the speed of light is a true universal limit, travel between the stars will take many years. Because all electromagnetic radiation travels at the speed of light, even an electromagnetic signal sent to another planetary system would take many years to get there. Interstellar visits (and interstellar conversations) will be quite prolonged.

Questions and Problems

Test Your Understanding

1. If the Sun instantaneously stopped giving off light, what would happen on Earth?
   a. Earth would immediately get dark.
   b. Earth would get dark 8 minutes later.
   c. Earth would get dark 27 minutes later.
   d. Earth would get dark 1 hour later.

2. Why is an iron atom a different element from a sodium atom?
   a. A sodium atom has fewer neutrons in its nucleus than an iron atom has.
   b. An iron atom has more protons in its nucleus than a sodium atom has.
   c. A sodium atom is bigger than an iron atom.
   d. A sodium atom has more electrons.

3. Suppose an atom has three energy levels, specified in arbitrary units as 10, 7, and 5. In those units, which of the following energies might an emitted photon have? (Select all that apply.)
   a. 3
   b. 2
   c. 5
   d. 4

4. When a boat moves through the water, the waves in front of the boat bunch up, whereas the waves behind the boat spread out. That is an example of
   a. the Bohr model.
   b. the wave nature of light.
   c. emission and absorption.
   d. the Doppler effect.

5. As a blackbody becomes hotter, it also becomes ________ and ________.
   a. more luminous; redder
   b. more luminous; bluer
   c. less luminous; redder
   d. less luminous; bluer

6. Which of the following factors does *not* directly influence the temperature of a planet?
   a. the luminosity of the Sun
   b. the distance from the planet to the Sun
   c. the albedo of the planet
   d. the size of the planet

7. Two stars are of equal luminosity. Star A is 3 times as far from you as star B. Star A appears ________ star B.
   a. 9 times brighter than
   b. 3 times brighter than
   c. the same brightness as
   d. 1/3 as bright as
   e. 1/9 as bright as

8. When less energy radiates from a planet, its ________ increases until a new ________ is achieved.
   a. temperature; equilibrium
   b. size; temperature
   c. equilibrium; size
   d. temperature; size

9. How does the speed of light in a medium compare with the speed in a vacuum?
   a. The speed is the same in both a medium and a vacuum because the speed of light is a constant.
   b. The speed in the medium is always faster than the speed in a vacuum.
   c. The speed in the medium is always slower than the speed in a vacuum.
   d. The speed in the medium may be faster or slower, depending on the medium.

10. When an electron moves from a higher energy level in an atom to a lower energy level,
    a. a continuous spectrum is emitted.
    b. a photon is emitted.
    c. a photon is absorbed.
    d. a redshifted spectrum is emitted.

11. In Figure 5.14, the red photons come from the transition from E_3 to E_2. Those photons will have the ________ wavelengths because they have the ________ energy compared with that of the other photons.
    a. shortest; least
    b. shortest; most
    c. longest; least
    d. longest; most

12. Star A and star B appear equally bright in the sky. Star A is twice as far away from Earth as star B. How do the luminosities of stars A and B compare?
    a. Star A is 4 times as luminous as star B.
    b. Star A is 2 times as luminous as star B.
    c. Star B is 2 times as luminous as star A.
    d. Star B is 4 times as luminous as star A.

13. What is the surface temperature of a star that has a peak wavelength of 290 nm?
    a. 1000 K
    b. 2000 K
    c. 5000 K
    d. 10,000 K
    e. 100,000 K
14. If a planet is in thermal equilibrium,
    a. no energy is leaving the planet.
    b. no energy is arriving on the planet.
    c. the amount of energy leaving equals the amount of energy arriving.
    d. the temperature is very low.
15. The temperature of an object has a specific meaning as it relates to the object's atoms. A high temperature means that the atoms
    a. are very large.
    b. are moving very fast.
    c. are all moving together.
    d. have a lot of energy.

Thinking about the Concepts

16. The speed of light in a vacuum is 3×10^5 km/s. Can light travel at a lower speed? Explain your answer.
17. Is light a wave or a particle or both? Explain your answer.
18. If any of the experiments mentioned in the Process of Science Figure had *not* agreed with the others, what would that mean for the conclusion that light has a finite, constant speed?
19. If photons of blue light have more energy than photons of red light, how can a beam of red light carry as much energy as a beam of blue light?
20. Patterns of emission or absorption lines in spectra can uniquely identify individual atomic elements. How can the positive identification of atomic elements be used to test the validity of the cosmological principle discussed in Chapter 1?
21. An atom in an excited state can drop to a lower energy state by emitting a photon. Can we predict exactly how long the atom will remain in the higher energy state? Explain your answer.
22. Spectra of astronomical objects show both bright and dark lines. Describe what those lines indicate about the atoms responsible for the spectral lines.
23. Astronomers describe certain celestial objects as being *redshifted* or *blueshifted*. What do those terms indicate about the objects?
24. An object somewhere near you is emitting a pure tone at middle C on the octave scale (262 Hz). You, having perfect pitch, hear the tone as A above middle C (440 Hz). Describe the motion of that object relative to where you are standing.
25. During a popular art exhibition, the museum staff finds it necessary to protect the artwork by limiting the total number of viewers in the museum at any particular time. New viewers are admitted at the same rate that others leave. Is that system an example of static equilibrium or of dynamic equilibrium? Explain.
26. A favorite object for amateur astronomers is the double star Albireo, with one of its components a golden yellow and the other a bright blue. What do those colors tell you about the relative temperatures of the two stars?
27. The stars you see in the night sky cover a large range of brightness. What does that range tell you about the distances of the various stars? Explain your answer.
28. Why is it not surprising that sunlight peaks in the "visible"?
29. In Figure 5.24, why is the range of temperatures so much greater for Mercury than for the other planets?
30. Suppose you want to find a planet with the same temperature as Earth. What could you say about the size of the orbit of such a planet if it is orbiting a cooler red star? A yellow star like the Sun? A hotter blue star?

Applying the Concepts

31. You are tuned to 790 on AM radio. That station is broadcasting at a frequency of 790 kHz (7.90×10^5 Hz). You switch to 98.3 on FM radio. That station is broadcasting at a frequency of 98.3 MHz (9.83×10^7 Hz).
    a. What are the wavelengths of the AM and FM radio signals?
    b. Which broadcasts at higher frequencies, AM or FM?
    c. What are the photon energies of the two broadcasts?
32. Your microwave oven cooks by vibrating water molecules at a frequency of 2.45 gigahertz (GHz), or 2.45×10^9 Hz. What is the wavelength, in centimeters, of the microwave's electromagnetic radiation?
33. Assume that an object emitting a pure tone of 440 Hz is on a vehicle approaching you at a speed of 25 m/s. If the speed of sound at this particular atmospheric temperature and pressure is 340 m/s, what will be the frequency of the sound that you hear? (Hint: Keep in mind that frequency is inversely proportional to wavelength.)
34. You observe a spectral line of hydrogen at a wavelength of 502.3 nm in a distant galaxy. The rest wavelength of that line is 486.1 nm. What is the radial velocity of that galaxy? Is it moving toward you or away from you?

35. If half of the phosphorescent atoms in a glow-in-the-dark toy give up a photon every 30 minutes, how bright (relative to its original brightness) will the toy be after 2 hours?

36. How bright would the Sun appear from Neptune, 30 AU from the Sun, compared with its brightness as seen from Earth? The spacecraft *Voyager 1* is now about 140 AU from the Sun and heading out of the Solar System. Compare the brightness of the Sun as seen from *Voyager 1* with that seen from Earth.

37. On a dark night you notice that a distant lightbulb happens to have the same brightness as a firefly 5 meters away from you. If the lightbulb is a million times more luminous than the firefly, how far away is the lightbulb?

38. Two stars appear to have the same brightness, but one star is 3 times more distant than the other. How much more luminous is the more distant star?

39. A panel with an area of 1 square meter (m^2) is heated to a temperature of 500 K. How many watts is the panel radiating into its surroundings?

40. The Sun has a radius of 6.96×10^5 km and a blackbody temperature of 5780 K. Calculate the Sun's luminosity.

41. Some of the hottest stars known have a blackbody temperature of 100,000 K. What is the peak wavelength of their radiation? What type of radiation is it?

42. Your body emits radiation at a temperature of about 98.6°F.
    a. What is that temperature in kelvins? What is the peak wavelength, in microns, of your emitted radiation? What region of the spectrum is this?
    b. If you assume an exposed body surface area of $0.25\ m^2$, how many watts of power do you radiate?

43. A planet with no atmosphere at 1 AU from the Sun would have an average blackbody surface temperature of 279 K if it absorbed all the Sun's electromagnetic energy falling on it (albedo = 0).
    a. What would the average temperature on that planet be if its albedo were 0.1, typical of a rock-covered surface?
    b. What would the average temperature be if its albedo were 0.9, typical of a snow-covered surface?

44. The orbit of Eris, a dwarf planet, carries it out to a maximum distance of 97.7 AU from the Sun. If you assume an albedo of 0.8, what is the average temperature of Eris when it is farthest from the Sun?

45. Suppose our Sun had 10 times its current luminosity. What would the average blackbody surface temperature of Earth be if Earth had the same albedo?

USING THE WEB

46. a. Go to the website for NASA's Astronomy Picture of the Day (https://apod.nasa.gov/apod/ap101027.html) and study the picture of the Andromeda Galaxy in visible light and in ultraviolet light (slide the mouse on and off to flip between the two pictures). Which light represents a hotter temperature? What differences do you see in the two images?
    b. Go to the APOD archive (https://apod.nasa.gov/cgi-bin/apod/apod_search) and enter "false color" in the search box. Examine a few images that come up in the search. What does *false color* mean in that context? What wavelength(s) were the pictures exposed in? What is the color coding; that is, what wavelength does each color in the image represent? You can read more about false color at http://chandra.harvard.edu/photo/false_color.html.

47. Crime scene investigators may use different types of light to examine a crime scene. Search for "forensic lighting" in your browser. What wavelengths of light are used to search for blood and saliva? For fingerprints? Why is having access to different kinds of light useful for an investigator? Search on "forensic spectroscopy" and select a recent report. How is spectroscopy being used in crime scene investigations?

48. Using Google Images or an equivalent website, search for "night vision imaging" and "thermal imaging." How do night vision goggles and thermal-imaging devices work differently from regular binoculars or cameras? When are night vision or thermal-imaging tools useful?

49. The Transportation Security Administration (TSA) uses several types of imaging devices to screen passengers in airports. Search for "TSA imaging" in your browser. What wavelengths of light are being used in those devices? What concerns do passengers have about some of those imaging devices?

50. Go to the NASA Earth Observations website (https://neo.sci.gsfc.nasa.gov) and look at a map of Earth's albedo (click on "albedo" in the menu for "energy" or "land" if it didn't come up). Compare that map with those of 2, 4, 6, 8, and 10 months earlier. Which parts of Earth have the lowest and highest albedos? In which parts do the albedos seem to change the most with the time of the year? Would you expect ice, snow, oceans, clouds, forests, and deserts to add or subtract in each case from the total Earth albedo? Which parts of Earth are not showing up on this map?

EXPLORATION

digital.wwnorton.com/astro6

Visit the Digital Resources Page and on the Student Site open the "Light as a Wave, Light as a Photon" AstroTour in Chapter 5. Watch the first section and then click through, using the "Play" button, until you reach "Section 2 of 3."

Here we explore the following questions: How many properties does a wave have? Are any of those properties related to each other?

Work your way to the experimental section, where you can adjust the properties of the wave. Watch the simulation for a moment to see how fast the frequency counter increases.

1 Increase the wavelength by pressing the arrow key. What happens to the rate of the frequency counter?

2 Reset the simulation and then decrease the wavelength. What happens to the rate of the frequency counter?

3 How are the wavelength and frequency related to each other?

4 Imagine that you increase the frequency instead of the wavelength. How should the wavelength change when you increase the frequency?

5 Reset the simulation, and increase the frequency. Did the wavelength change in the way you expected?

6 Reset the simulation, and increase the amplitude. What happens to the wavelength and the frequency counter?

7 Decrease the amplitude. What happens to the wavelength and the frequency counter?

8 Is the amplitude related to the wavelength or frequency?

9 Why can't you change the speed of this wave?

6 The Tools of the Astronomer

In Chapter 5, we explained that astronomers learn about the physical and chemical properties of distant planets, stars, and galaxies by studying the light they emit. That electromagnetic radiation, though, must first be collected and processed before it can be analyzed and converted to useful knowledge. Here in Chapter 6, we describe the tools that astronomers use to capture and scrutinize that information.

LEARNING GOALS

By the end of this chapter, you should be able to:

LG 1 Compare how the two main types of optical telescopes gather and focus light.

LG 2 Summarize the main types of detectors that are used on telescopes.

LG 3 Explain why some wavelengths of radiation must be observed from high, dry, and remote observatories on Earth, or from space.

LG 4 Explain the benefits of sending spacecraft to study the planets and moons of our Solar System.

LG 5 Describe other astronomical tools that contribute to the study of the universe.

The twin 10-meter Keck reflectors on Mauna Kea, Hawaii, have a multiple-mirror, compact design. ►►►

Why are most telescopes on remote mountaintops?

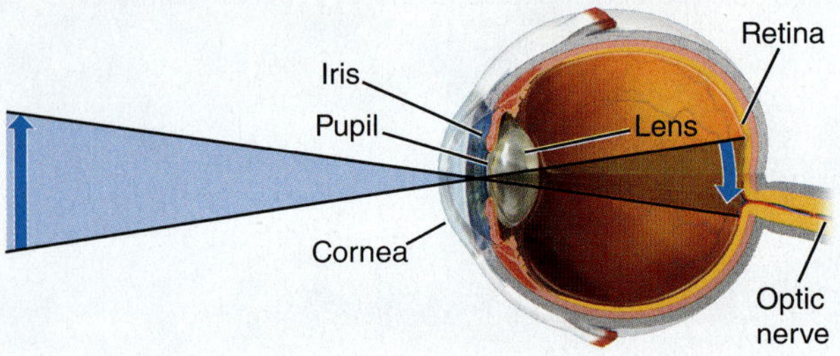

Figure 6.1 A schematic view of the human eye creating an image of an object (the blue arrow).

6.1 The Optical Telescope Revolutionized Astronomy

Astronomical observations began with the human eye—information about the overall colors of stars and their brightness in the night sky is apparent even to the naked eye—that is, unassisted by binoculars or telescopes or filters. The development of **telescopes**—devices for collecting and focusing light—in the 17th century greatly increased the amount of light that can be collected from astronomical objects. With modern telescopes, astronomers can detect light that has been traveling across space for billions of years—even electromagnetic radiation from soon after the Big Bang, the beginning of the universe itself.

The Eye

Human eyes are sensitive to light with wavelengths ranging from about 380 nanometers (nm) (deep violet) to 750 nm (far red). A simplified schematic of the human eye is shown in **Figure 6.1**. The part of the human eye that detects light is called the retina, and the individual receptor cells that respond to light falling on the retina are called rods and cones. The center of the human retina consists solely of cones, which detect color and provide the greatest visual acuity. Away from the center, rods and cones intermingle, with rods dominating far from the center, where they are responsible for peripheral vision.

Our vision is limited by the eye's angular **resolution**, which refers to how close two points of light can be to each other before we can no longer distinguish them. Unaided, the best human eyes can resolve objects separated by 1 arcminute (1/60 of a degree), an angular distance of about 1/30 the diameter of the full Moon. (A more in-depth description of angular units—radians, degrees, arcminutes, and arcseconds—can be found in Appendix 1.4.) That may seem small, but thousands of stars and galaxies may reside within a patch of sky with that diameter.

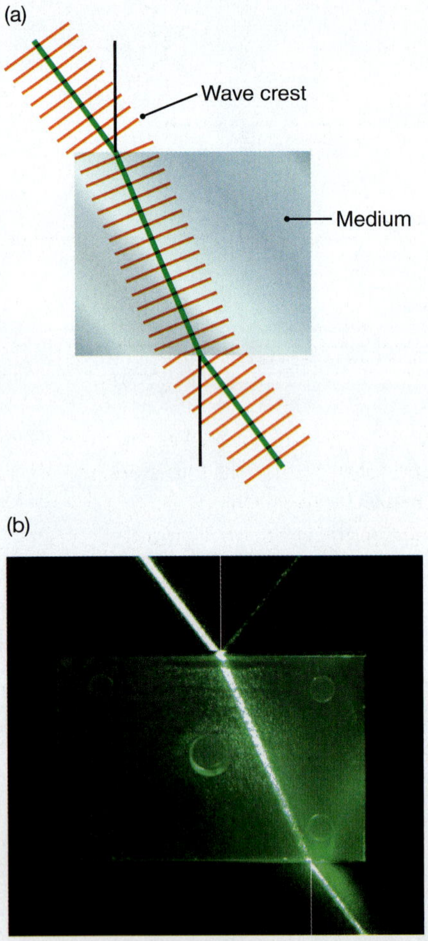

Figure 6.2 (a) When wavefronts enter a new medium, they bend in a new direction relative to a line perpendicular to the surface (black lines). (b) An actual light ray entering and leaving a medium.

Refracting Telescopes

Optical telescopes come in two primary types: **refracting telescopes**, which use lenses, and **reflecting telescopes**, which use mirrors. For all telescopes, the "size" of the telescope refers to the diameter of the largest mirror or lens, which determines the light-collecting area. That diameter is called the **aperture**. The light-gathering power of a telescope is proportional to its opening—that is, to the square of its diameter. The larger the aperture, the more light the telescope can collect. A "1-meter telescope" has a primary mirror (or lens) that is 1 meter in diameter. The aperture of the human eye is 6–7 millimeters (mm).

In the late 13th century, craftsmen in Venice were making small lentil-shaped disks of glass that could be mounted in frames and worn over the eyes to improve vision. More than 300 years later, Hans Lippershey (1570–1619), a spectacle maker living in the Netherlands, put two of his lenses together in a tube. With that new instrument, he saw distant objects magnified and could see farther. Galileo Galilei heard news of that invention and constructed one of his own. Recall from Chapter 3 that by 1610, Galileo had become the first to see the phases of Venus and the moons of Jupiter and was among the first to see craters on the Moon. He also was the first to realize that the Milky Way is made up of many individual stars. The refracting telescope—one that uses lenses—quickly revolutionized the science of astronomy.

Refraction is the basis for the refracting telescope. Recall from Chapter 5 that the speed of light is constant in a vacuum, but through a medium such as air or glass, the speed of light is always lower. As light enters a new medium, its speed changes. If the light strikes the surface at an angle, some of the crest of the wave arrives at the surface earlier and some arrives later. You can see that happen in **Figure 6.2a**, a schematic diagram of wave crests (red lines) striking a medium at an angle. **Figure 6.2b** shows an actual light ray passing into and out of a medium (here, glass). The ray bends each time the medium changes. That bending of light when it changes the medium through which it travels is called **refraction**.

The amount of refraction depends on the properties of the medium. A medium's index of refraction (n) is equal to the ratio of the speed of light in a vacuum (c) to its speed in the medium (v). That relationship can be expressed by the equation $n = c/v$. For example, most glass has an index of refraction of approximately 1.5, so the speed of light in glass is 300,000 kilometers per second (km/s) divided by 1.5, or 200,000 km/s. The light bends by an amount that depends on the index of refraction of the materials involved and the angle at which the light strikes.

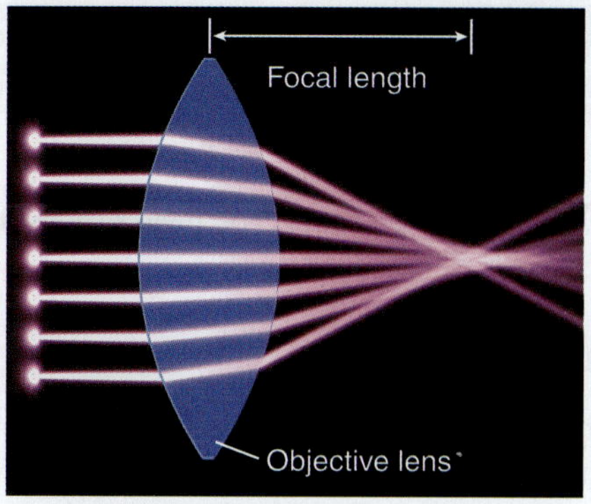

Figure 6.3 For a curved lens such as the one shown, refraction causes the light to focus to a point. That point is in a slightly different location for different wavelengths (colors) of light.

The primary lens in a refracting telescope is a simple convex lens, called the **objective lens** (**Figure 6.3**), whose curved surfaces refract the light from a distant object. That refracted light forms an image on the telescope's **focal plane**, which is perpendicular to the *optical axis*—the path that light takes through the center of the lens. Because the telescope's glass lens is curved, light at the outer edges of the lens strikes the surface more obliquely than light near the center. As a result, light at the outer edges of the lens is refracted more than light near its center. The lens concentrates the light rays entering the telescope, bringing them to a sharp focus at a distance called the **focal length**. Sometimes focal length is specified on a telescope as *focal ratio*, which equals focal length divided by aperture; that term may be familiar to you from the lenses used in photography.

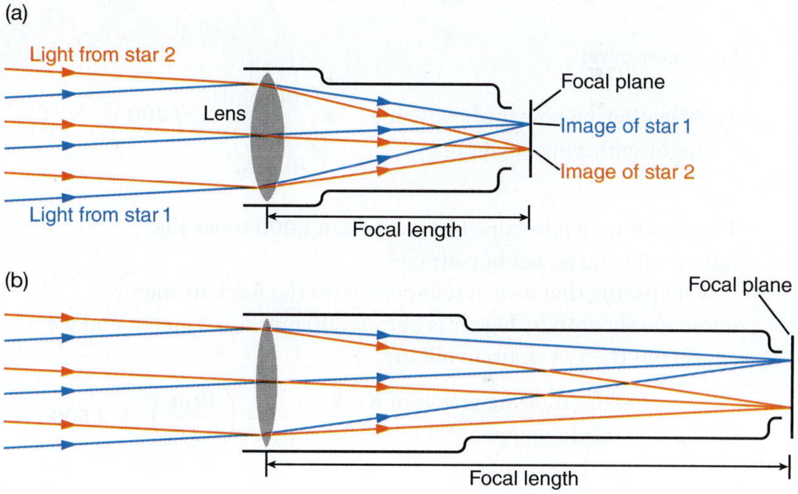

Figure 6.4 (a) A refracting telescope uses a lens to collect and focus light from two stars, forming images of the stars on its focal plane. (b) A telescope with a longer focal length produces larger images.

Figure 6.4 illustrates how a telescope uses the light that passes through its lenses. Figure 6.4a shows the light from two stars passing through a lens and converging at the focal plane of the lens. Figure 6.4b shows the same situation for a lens with a longer focal length. Longer focal lengths increase the size and separation of objects in the focal plane. Aperture and focal length are the two most important parameters of a telescope. The image can be viewed with an **eyepiece**—a changeable lens whose focal length determines the magnification (**Working It Out 6.1**). In modern research, however, the images are sent directly to a camera or other detector.

Refracting telescopes have two major shortcomings. First, physical limits constrain the size of refracting telescopes. The larger the area of the objective lens, the more light-gathering power it has and the fainter the stars we can observe. As objective lenses get larger, however, they get heavier, and a massive piece of glass at the end of a very long tube sags too much under the force of gravity. Refracting telescopes grew in size until the 1897 completion of the Yerkes 1-meter refractor (**Figure 6.5**), the world's largest operational refracting telescope. Located in Williams Bay, Wisconsin, the Yerkes telescope consists of a 450-kilogram (kg) objective lens mounted at the end of a 19.2-meter tube.

Figure 6.5 (a) The Yerkes 1-meter refractor is the world's largest refracting telescope. (b) The main parts of a refractor.

(a)

(b)

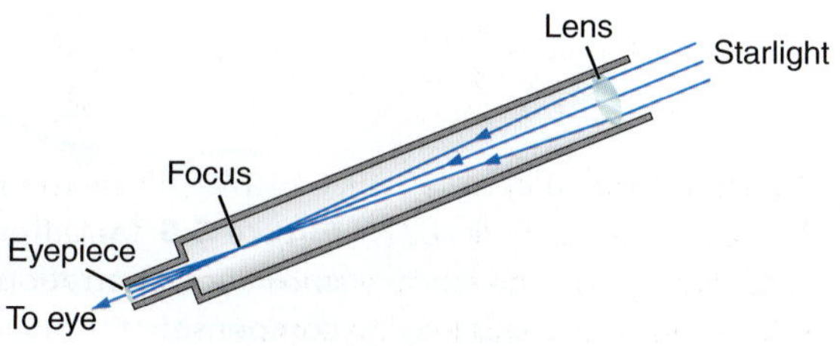

6.1 Working It Out Telescope Aperture and Magnification

If you are shopping for a telescope, you need to consider its aperture and magnification.

Aperture

The light-gathering power of a telescope is proportional to the area of its lens or mirror, $\pi \times (D/2)^2$, and thus to the square of the aperture (D). The amount of light a telescope collects increases as the aperture increases. How does the light-gathering power of a telescope with a diameter of 200 mm, or 8 inches, compare with that of the pupil of your eye, which is about 6 mm in the dark?

$$\text{Light-gathering power of telescope} = \frac{\pi}{4} \times (200\text{ mm})^2$$

and

$$\text{Light-gathering power of eye} = \frac{\pi}{4} \times (6\text{ mm})^2$$

So, to compare:

$$\frac{\text{Light-gathering power of telescope}}{\text{Light-gathering power of eye}} = \frac{\frac{\not\pi}{\not4}(200\text{ mm})^2}{\frac{\not\pi}{\not4}(6\text{ mm})^2} = \left(\frac{200}{6}\right)^2 = 1{,}111$$

Thus, an 8-inch telescope has more than 1,000 times the light-gathering power of your eye.

Comparing that 8-inch telescope with the Keck 10-meter telescope shows why bigger is better: 200 mm = 0.2 meter, and we cancel out the $\pi/4$ again to obtain

$$\frac{\text{Light-gathering power of Keck}}{\text{Light-gathering power of 8-inch telescope}} = \left(\frac{10\text{ m}}{0.2\text{ m}}\right)^2 = 2{,}500$$

Even larger telescopes, 25–40 meters in diameter, are under construction.

Magnification

Most telescopes have a set focal length and come with a collection of eyepieces. The magnification of the image in the telescope is given by

$$\text{Magnification} = \frac{\text{Telescope focal length}}{\text{Eyepiece focal length}}$$

Suppose the focal length of the 200-mm telescope in the preceding example is 2,000 mm. Combined with the focal length of a standard eyepiece, 25 mm, that telescope will give the following magnification:

$$\text{Magnification} = \frac{2{,}000\text{ mm}}{25\text{ mm}} = 80$$

That telescope and eyepiece combination has a magnifying power of 80, meaning that a crater on the Moon will appear 80 times larger in the telescope's eyepiece than it does when viewed by the naked eye. An eyepiece that has a focal length of 8 mm will have about 3 times more magnifying power, or 250 times.

A higher magnification will not necessarily let you see an object better. A faint and fuzzy image will not look clearer when magnified.

The second major shortcoming of refracting telescopes is **chromatic aberration** (**Figure 6.6**). Starlight is made up of all the colors of the rainbow, and each color refracts at a slightly different angle because the index of refraction depends on the wavelength of the light. Shorter (bluer) wavelengths are refracted more than longer (redder) wavelengths (Figure 6.6a). That wavelength-dependent difference in refraction, which spreads the white light out into its spectral colors, is

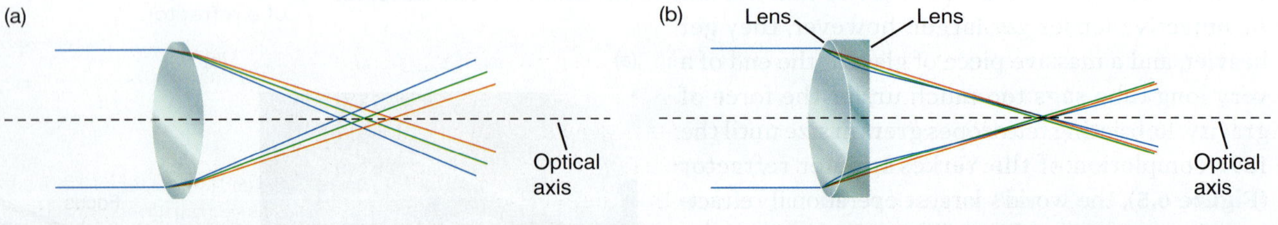

Figure 6.6 (a) Different wavelengths of light come to different foci along the optical axis of a simple lens, causing chromatic aberration. (b) A compound lens using two types of glass with different indices of refraction can compensate for much of the chromatic aberration, so different colors of light all come to a focus at the same point.

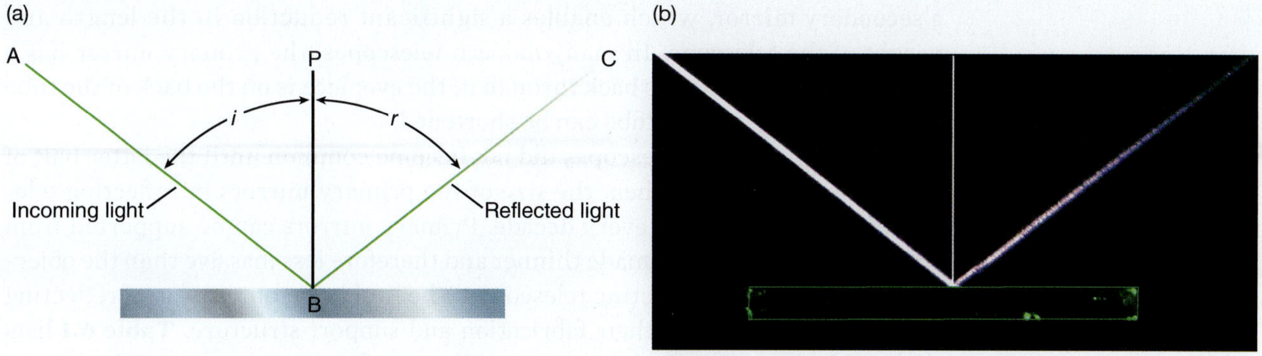

Figure 6.7 (a) When a ray of incoming light (AB) shines on a flat surface, it reflects from the surface, becoming the reflected ray BC. The angle between AB and PB, the perpendicular to the surface, is the angle of incidence (*i*). The angle between BC and PB is the angle of reflection (*r*). The angles of incidence and reflection are always equal. (b) Light from a laser beam is reflected from a flat glass surface.

called **dispersion**. Dispersion causes bluer light to come to a shorter focus than that of the longer visible wavelengths, creating chromatic aberration. In a refracting telescope with a simple convex lens, chromatic aberration produces haloed images around the star. That issue was addressed in early telescopes by increasing the focal length, but then the telescope requires a longer tube. Now manufacturers of quality cameras and telescopes use a **compound lens** composed of two types of glass to correct for chromatic aberration (Figure 6.6b).

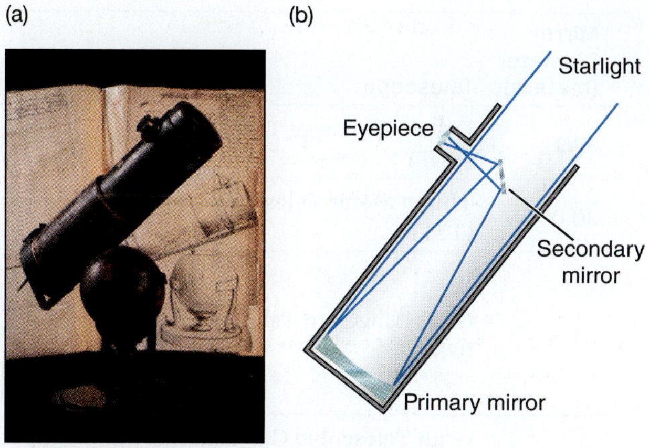

Figure 6.8 (a) Newton's reflecting telescope. (b) A Newtonian focus telescope has a curved primary and a flat secondary mirror, as shown in the sketch. The eyepiece is on the side of the tube.

Reflecting Telescopes

Another property of light, called **reflection**, is the basis for reflecting telescopes. When light encounters a different medium—here, going from air to glass—some of it will be reflected from the surface of the new medium. In other words, some of the light will reverse its direction of travel. The most common example occurs when light encounters an ordinary flat mirror (**Figure 6.7**). The angle of the incoming light and the angle of outgoing light are always equal. A reflected image in a mirror looks like the original image, although left and right are interchanged.

In 1668, Isaac Newton designed a reflecting telescope, which uses mirrors instead of lenses (**Figure 6.8a**). The angle of reflection does not depend on the wavelength of light, so chromatic aberration is *not* a problem in reflecting telescopes. The important parts of Newton's reflecting telescope are shown schematically in **Figure 6.8b**. To make that reflecting telescope, Newton cast a 2-inch primary mirror made of copper and tin and polished it to a special curvature. He then placed the primary mirror at the bottom of a tube with a secondary flat mirror mounted above it at a 45° angle. The second mirror directed the focused light to an eyepiece on the outside of the tube.

Astronomers use mirrors with a surface that curves inward toward the incoming light, called *concave* mirrors. The same rules of incidence and reflection for flat mirrors hold here for each ray of light, but for concave mirrors the reflected rays do not maintain the same angle with respect to each other as they do with a flat mirror. Concave mirrors will reflect the rays so that they converge to form an image (**Figure 6.9**). If the incoming light rays are parallel, as from a distant source such as a star, the reflected light rays cross at the focal length of the mirror. The light path from the primary mirror to the focal plane can be "folded" by using

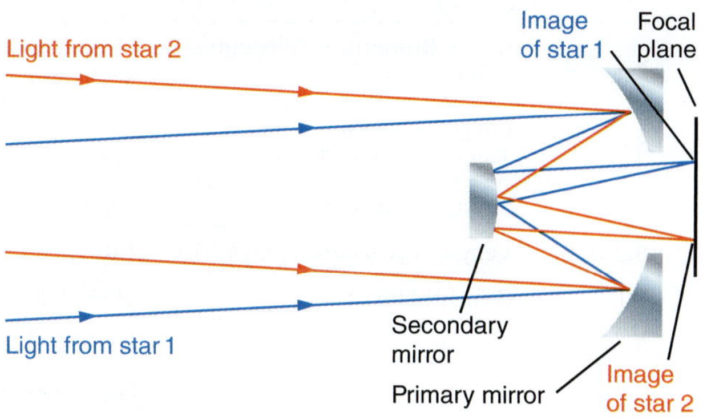

Figure 6.9 Large reflecting telescopes often use Cassegrain focus, in which a secondary mirror directs the light back through a hole in the concave parabolic primary mirror to an accessible focal plane behind the primary mirror.

a secondary mirror, which enables a significant reduction in the length and weight of the telescope. In many modern telescopes, the primary mirror has a hole so that light can pass back through it; the eyepiece is on the back of the tube of the telescope, and the tube can be shortened.

AstroTour: Geometric Optics and Lenses

Large reflecting telescopes did not become common until the latter half of the 18th century. Since then, the size of the primary mirrors in reflecting telescopes has grown larger every decade. Primary mirrors can be supported from the back, and they can be made thinner and therefore less massive than the objective lenses found in refracting telescopes. The limitation on the size of reflecting telescopes is the cost of their fabrication and support structure. **Table 6.1** lists the world's largest optical telescopes. All are reflecting telescopes. The largest

TABLE 6.1 The World's Largest Optical Telescopes

Mirror Diameter (meters)	Telescope	Sponsor(s)	Location	Operational Date
39.3	Extremely Large Telescope (ELT)	European Southern Observatory (Europe, Chile, Brazil)	Cerro Armazones, Chile	Under construction
30.0	Thirty Meter Telescope (TMT)	International collaboration led by Caltech, U. of California, U. of Hawaii, China, Japan, India, and Canada	Mauna Kea, Hawaii, or Canary Islands, Spain	TBD
24.5	Giant Magellan Telescope (GMT)	Carnegie Institution, Arizona State U., Harvard U., Smithsonian Institution, U. of Arizona, U. of Texas, Texas A&M U., U. of Chicago, São Paulo Research Foundation (FAPESP), Australian National U., Astronomy Australia Ltd., and Korea Astronomy and Space Science Institute	Cerro Las Campanas, Chile	Under construction
10.4	Gran Telescopio Canarias (GTC)	Spain, Mexico, U. of Florida	Canary Islands	2007
10	Keck I	Caltech, U. of California, NASA	Mauna Kea, Hawaii	1993
10	Keck II	Caltech, U. of California, NASA	Mauna Kea, Hawaii	1996
10	South African Large Telescope (SALT)	South Africa, USA, UK, Germany, Poland, New Zealand, India	Sutherland, South Africa	2005
10	Hobby-Eberly Telescope (HET)	U. of Texas, Penn State U., Stanford U., Germany	Mount Fowlkes, Texas	1999
8.4 × 2	Large Binocular Telescope (LBT)	U. of Arizona, Ohio State U., Italy, Germany, Arizona State, and others	Mount Graham, Arizona	2008
8.4	Large Synoptic Survey Telescope (LSST)	National Science Foundation, Dept. of Energy, and other partners through the private LSST corporation	Cerro Pachón, Chile	Under construction
8.3	Subaru Telescope	Japan	Mauna Kea, Hawaii	1999
8.2 × 4	Very Large Telescope (VLT)	European Southern Observatory	Cerro Paranal, Chile	2000
8.1	Gemini North	USA, UK, Canada, Chile, Brazil, Argentina, Australia	Mauna Kea, Hawaii	1999
8.1	Gemini South	USA, UK, Canada, Chile, Brazil, Argentina, Australia	Cerro Pachón, Chile	2000
6.5	MMT	Smithsonian Institution, U. of Arizona	Tucson, Arizona	2000
6.5	Magellan I	Carnegie Institution, U. of Arizona, Harvard U., U. of Michigan, MIT	Cerro Las Campanas, Chile	2000
6.5	Magellan II	Carnegie Institution, U. of Arizona, Harvard U., U. of Michigan, MIT	Cerro Las Campanas, Chile	2002

single mirrors constructed today are 8 meters in diameter, but reflecting telescopes even bigger than that are designed to make use of an array of smaller segments. The primary mirror of each of the 10-meter twin Keck telescopes is made up of 36 hexagon-shaped segments that are 1.8 meters in diameter (**Figure 6.10**). Located on 4,100-meter-high Mauna Kea in Hawaii, the Keck telescopes are among the world's largest reflecting telescopes. Each one has 4 million times the light-gathering power of the human eye.

Figure 6.10 Each of the Keck 10-meter reflectors uses an aligned group of 36 hexagonal mirrors to collect light.

CHECK YOUR UNDERSTANDING 6.1A

Why are all large astronomical telescopes reflectors (choose all reasons that apply)? (a) chromatic aberration is minimized; (b) they are not as heavy; (c) they can be shorter; (d) the glass doesn't need to be curved.

Optical and Atmospheric Limitations

The *resolution* of a telescope determines how close two points of light can be to each other before they are indistinguishable (**Figure 6.11**). Review Figure 6.4a to see the path followed by rays of light from two distant stars as they pass through the lens of a refracting telescope. Figure 6.4b illustrated that increasing the focal length increases the size of and separation between the images that a telescope produces. That is one reason why telescopes provide a much clearer view of the stars than that obtained with the naked eye. The focal length of a human eye is typically about 20 mm, whereas telescopes used by professional astronomers often have focal lengths of tens or even hundreds of meters. Such telescopes make images that are far larger than those formed by the human eye, and consequently they contain far more detail.

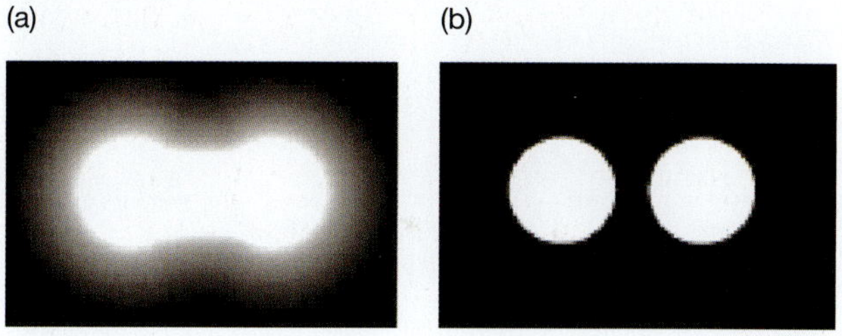

Figure 6.11 Resolution is the ability to separate two images that appear close together. When resolution is lower (a), the two images blend together. When resolution is higher (b), individual images can be distinguished.

Focal length explains only one difference between the resolution of telescopes and that of the unaided eye. The other difference results from the wave nature of light. **Figure 6.12** shows that light waves spread out from the edges of the lens or mirror when they pass through the aperture of a telescope. The distortion that occurs as light passes the edge of an opaque object is called **diffraction**. Diffraction "diverts" some of the light from its path, slightly blurring the image

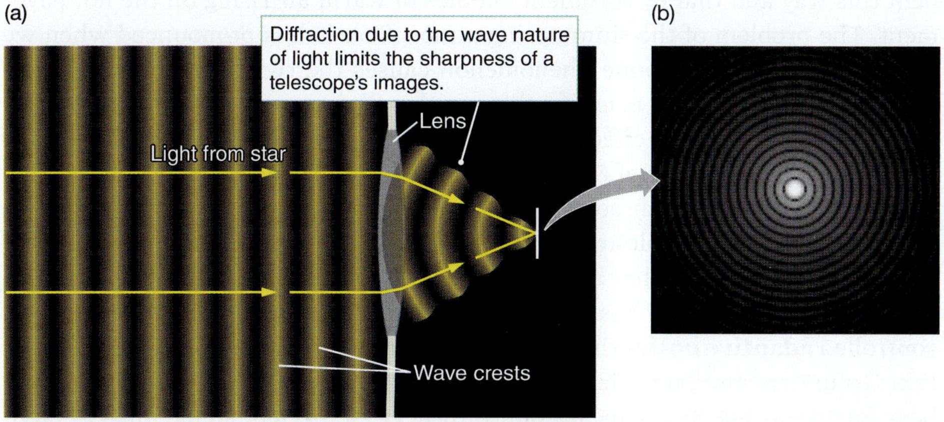

Figure 6.12 (a) Light waves from a star are diffracted by the edges of a telescope's lens or mirror. (b) That diffraction causes the stellar image to be blurred, limiting a telescope's ability to resolve objects.

6.2 Working It Out Diffraction Limit

The practical limit on the angular resolution of a telescope (θ) is called the diffraction limit. That limit is determined by the ratio of the wavelength of light passing through the telescope (λ) to the diameter of the aperture (D):

$$\theta = 2.06 \times 10^5 \left(\frac{\lambda}{D}\right) \text{arcsec}$$

The constant, 2.06×10^5, has units of arcseconds (arcsec). An **arcsecond** is a tiny angular measure found by first dividing the sky into 360 degrees, then dividing each degree by 60 to get arcminutes, and then dividing each arcminute by 60 to get arcseconds. An arcsecond is 1/1,800 of the size of the Moon in the sky, or about the size of a tennis ball if you could see it from 13 km (8 miles) away.

Both λ and D must be expressed in the same units—usually meters. The smaller the ratio of λ/D, the better the resolution. For example, the size of the human pupil (see Figure 6.1) ranges from about 2 mm in bright light to 8 mm (0.008 meter) in the dark. Visible (green) light has a wavelength (λ) of 550 nm—that is, 550×10^{-9} meter, or 5.5×10^{-7} meter. Using those values for the aperture and the wavelength gives

$$\theta = 2.06 \times 10^5 \left(\frac{5.5 \times 10^{-7}\ \text{m}}{0.008\ \text{m}}\right) \text{arcsec} = 14\ \text{arcsec}$$

or about 1/4 arcmin for the theoretical resolution. The typical actual resolution of the human eye is 1 arcmin. We do not achieve the theoretical resolution with our eyes because the physical properties of our eyes are not perfect.

How does the resolution of the human eye compare with that of the Hubble Space Telescope when it is operating in the visible part of the spectrum? Its primary mirror has a diameter of 2.4 meters. Substituting that value for D and again using visible (green) light gives

$$\theta = 2.06 \times 10^5 \left(\frac{5.5 \times 10^{-7}\ \text{m}}{2.4\ \text{m}}\right) \text{arcsec} = 0.047\ \text{arcsec}$$

or about 300 times better than the theoretical resolving power of the human eye.

made by the telescope. The degree of blurring depends on the wavelength of the light and the telescope's aperture. The larger the aperture, the smaller the problem diffraction poses. The best resolution that a given telescope can achieve is known as the **diffraction limit** (**Working It Out 6.2**).

Larger telescopes have better resolution and can distinguish objects that appear closer together. Theoretically, the 10-meter Keck telescopes have a diffraction-limited resolution of 0.0113 arcsec in visible light, which would be good enough for you to read newspaper headlines 60 km away. But for telescopes with apertures larger than about a meter, Earth's atmosphere stands in the way of better resolution. If you have ever looked out across a large asphalt parking lot on a summer day, you have probably seen the distant horizon shimmer as light is bent this way and that by turbulent bubbles of warm air rising off the hot pavement. The problem of the shimmering atmosphere is less pronounced when we look overhead, but the same phenomenon causes the twinkling of stars in the night sky. As telescopes magnify the angular diameter of an object, they also magnify the shimmering effects of the atmosphere. The limit on the resolution of a telescope on the surface of Earth caused by that atmospheric distortion is called **astronomical seeing**. One advantage of launching telescopes such as the Hubble Space Telescope into orbit around Earth is that they are not hindered by astronomical seeing because they are above the atmosphere.

Modern technology has improved ground-based telescopes with computer-controlled **adaptive optics** that compensate for much of the atmosphere's distortion. To understand how adaptive optics work, we need to look more closely at how Earth's atmosphere smears out an otherwise perfect stellar image. Light from a distant star arrives at the top of Earth's atmosphere with flat, parallel wave crests. If Earth's atmosphere were perfectly uniform, the crests would remain flat as they reached the objective lens or primary mirror of a ground-based telescope. After making their way through the telescope's optical system, the crests would produce a tiny diffraction disk in the focal plane, as shown in Figure 6.12b. But Earth's atmosphere is not uniform. It is filled with bubbles of air that have slightly different temperatures from those of their surroundings. Different temperatures mean different densities, and different densities mean different refractive properties, so each bubble bends light differently.

Those air bubbles act as weak lenses, and by the time the waves reach the telescope they are far from flat, as shown in **Figure 6.13**. Instead of a tiny diffraction disk, the image in the telescope's focal plane is distorted and swollen, degrading the resolution. Adaptive optics flatten out that distortion. First, an optical device within the telescope constantly measures the wave crests. Then, before reaching the telescope's focal plane, the light is reflected off yet another mirror, which has a flexible surface. A computer analyzes the light and bends the flexible mirror so that it accurately corrects for the distortion caused by the air bubbles. **Figure 6.14** shows an example of an image corrected by adaptive optics. To calibrate those systems, a laser beam is shone into the atmosphere to create an artificial "star" in the same direction as the telescope is pointing. The widespread use of adaptive optics has made the image quality of ground-based telescopes competitive with the quality of Hubble images from space at some wavelengths. Telescopes also can have active optics, in which actuators on the rear of the mirror correct for factors such as wind, temperature, and sagging from gravity.

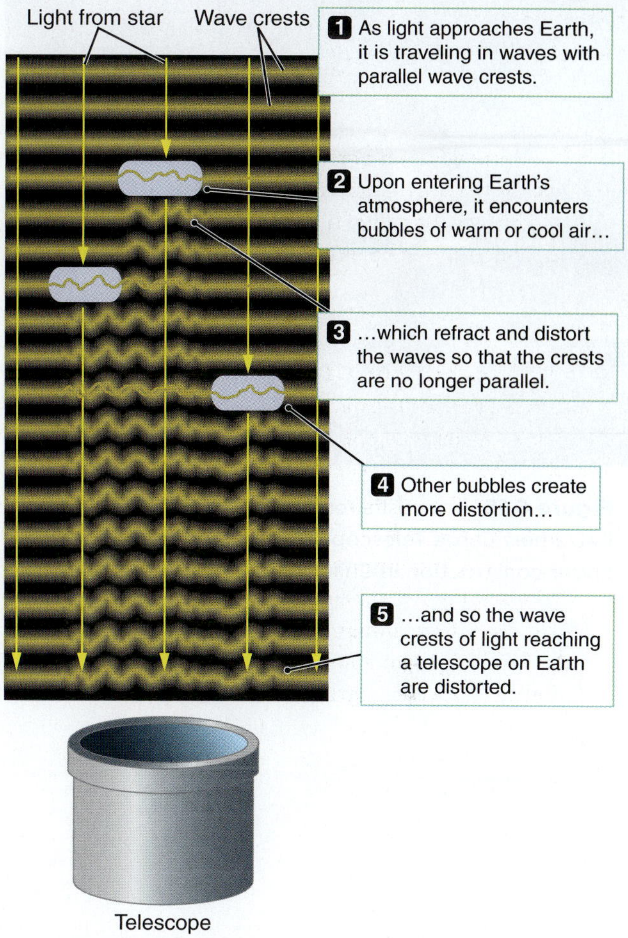

Figure 6.13 Bubbles of warmer or cooler air in Earth's atmosphere distort the wavefront of light from a distant object.

Observatory Locations

What makes a good location for a telescope on Earth? According to Table 6.1, astronomers look for sites that are high, dry, and dark. The best sites are far away from the lights of cities; in locations with little moisture, humidity, or rain; and where the atmosphere is relatively still. Telescopes are located as high as possible so that they get above a significant part of Earth's atmosphere, which distorts images and blocks infrared and microwave light. Many telescopes are situated on remote, high mountaintops surrounded by desert or ocean. Recall from Chapter 2 that the stars that can be seen throughout the year depend on latitude, and only at the equator would a telescope have access to all the stars in the sky. But because equatorial latitudes have tropical weather—wet, humid, and stormy—they are poor locations for telescopes. To cover the entire sky, astronomers have built telescopes in both northern and southern locations. In the United States, large telescopes are located in California, Arizona, New Mexico, Texas, and Hawaii. The largest southern-sky observatories are found in Chile, South Africa, and Australia. The twin Gemini telescopes, designed to be a matched pair, are located in Hawaii in the Northern Hemisphere and in Chile in the Southern Hemisphere.

Newer and larger telescopes are planned for many of the same locations listed in Table 6.1. The 8-meter Large Synoptic Survey Telescope (LSST) is headed for Cerro Pachón in Chile, current site of the Gemini South telescope. The Giant Magellan Telescope (GMT), consisting of seven 8-meter mirrors in a pattern equivalent to a 24.5-meter mirror, will be constructed at Cerro Las Campanas in Chile. The Thirty Meter Telescope (TMT) is planned for Mauna Kea in Hawaii or the Canary Islands, and the European Southern Observatory (ESO) is building the 39-meter Extremely Large Telescope (ELT) at

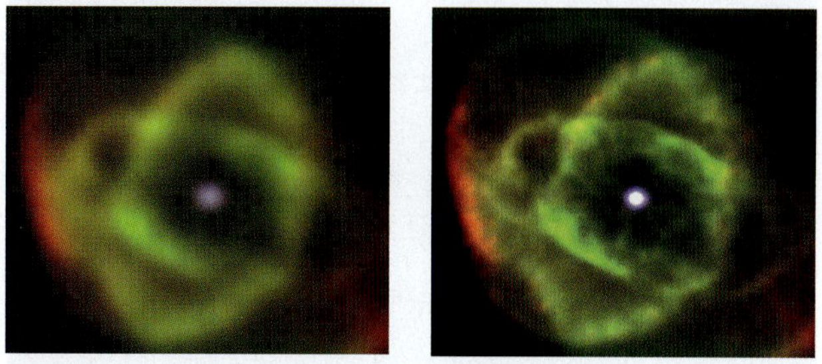

Figure 6.14 These images of the Cat's Eye Nebula from the Palomar Observatory telescope without (left) and with (right) adaptive optics show the benefit of the technique.

Figure 6.15 An artist's rendering of the European Extremely Large Telescope, a reflecting telescope under construction in Chile.

Cerro Armazones in Chile (**Figure 6.15**). As telescopes get larger—and more expensive—international collaboration becomes even more important.

Today's professional astronomers rarely look through the eyepiece of a telescope because they learn much more and make better use of observing time by permanently recording an object's image at a variety of wavelengths or seeing its light spread out into a revealing spectrum. Some astronomers no longer travel to telescopes at all, instead observing remotely from the base of the mountain or far away at their own institutions.

Professional and amateur astronomers alike are concerned about loss of the dark sky. As cities and suburbs around the world grow and expand, the use of outdoor artificial light becomes more widespread. Pictures from space show how bright many areas of Earth are at night (**Figure 6.16**). In the United States, two-thirds of the population resides in an area too bright to see the Milky Way in the sky at night, and it has been estimated that by 2025 the continental United States will have almost no dark skies. Increased air pollution also dims the view of the night sky in many locations. The U.S. National Park Service now advertises evening astronomy programs in natural, unpolluted dark skies as one of the reasons to visit some parks. Several international astronomy associations are working with UNESCO (the United Nations Educational, Scientific and Cultural Organization) to promote the "right to starlight," arguing that for historical, cultural, and scientific reasons, it would be a huge loss if humanity could no longer view the stars. Those organizations are encouraging countries to create starlight reserves and starlight parks where people can experience increasingly rare dark skies and a natural nocturnal environment.

CHECK YOUR UNDERSTANDING 6.1B

In practice, the smallest angular size that you can resolve with a 10-inch telescope is governed by the: (a) blurring caused by Earth's atmosphere; (b) diffraction limit of the telescope; (c) size of the primary mirror; (d) magnification of the telescope.

6.2 Optical Detectors and Instruments Used with Telescopes

Beginning in the 1800s, the development of film photography, and later digital photography, revolutionized astronomy, allowing astronomers to detect fainter and more distant objects than possible to detect with the eye alone. Here in Section 6.2, we examine some of the more common types of detectors.

Integration Time and Quantum Efficiency

Originally, the retina of the human eye was the only astronomical detector. The limit of the faintest stars we can see with our unaided eyes is determined in part by two factors that are characteristic of all detectors: *integration time* and *quantum efficiency*.

Integration time is the limited time interval during which the eye can add up photons—that is analogous to leaving the shutter open on a camera. The brain "reads out" the information gathered by the eye about every 100 milliseconds (ms). Anything that happens faster than that appears to happen all at once. If two images on a computer screen appear 30 ms apart, you will see them as a single image

Figure 6.16 This satellite image of Earth at night shows that few populated areas are free from light pollution.

because your eyes will add up (or integrate) whatever they see over an interval of 100 ms or less. If the images occur 200 ms apart, however, you will see them as separate images. That relatively brief integration time is the most important factor limiting our nighttime vision. Stars too faint to be seen with the unaided eye are those from which you receive too few photons for your eyes to process in 100 ms.

Quantum efficiency determines how many responses occur for each photon received. For the human eye, 10 photons must strike a cone within 100 ms to activate a single response. So the quantum efficiency of our eyes is about 10 percent: for every 10 events, the eye sends one signal to the brain. Together, integration time and quantum efficiency determine the rate at which photons must arrive at the retina before the brain says, "I see something." Astronomers seek to use detectors with longer integration times and higher quantum efficiency than our eyes can achieve.

Figure 6.17 A photograph of the Moon taken by John W. Draper in 1840.

From Photographic Plates to Charge-Coupled Devices

For more than two centuries after the invention of the telescope, astronomers struggled with the problem of surface brightness. Only *point sources* such as stars appear brighter in a telescope. Extended astronomical objects, such as the Moon, appear bigger in the eyepiece, but the light is spread over that larger image, so their surfaces are no brighter than they appear to the unaided eye. Even when astronomers built larger telescopes, nebulae and galaxies appeared larger, but the details of these faint objects remained elusive. The problem was not with the telescopes but with the limitations of optics and the human eye. Only with the longer exposure times made possible by the invention of photography and the later development of electronic cameras could astronomers finally discern intricate details in faint objects.

In 1840, John W. Draper (1811–1882), a New York chemistry professor, created the earliest known astronomical photograph (**Figure 6.17**). By the late 1800s, astronomers had created thousands of photographic plates with permanent images of planets, nebulae, and galaxies. The quantum efficiency of most photographic systems used in astronomy was poorer than that of the human eye—typically 1–3 percent. But unlike the eye, photography can overcome poor quantum efficiency by leaving the shutter open on the camera, increasing the integration time to many hours of exposure. Photography enabled astronomers to record and study objects invisible to the human eye. However, one problem is that the response of photography to light is not linear, especially at long exposures, so if you doubled the exposure time, you did not get twice as much light on your image. By the middle of the 20th century, the search was on for electronic detectors that would overcome photography's problems in sensitivity, spectral range, and nonlinearity.

In 1969, scientists at Bell Laboratories invented a detector called a **charge-coupled device**, or **CCD**. By the late 1970s, the CCD had become the detector of choice in almost all astronomical-imaging applications. CCDs are linear, so doubling the exposure means you record twice as much light. As a result, they are good for measuring objects that vary in brightness, as well as for faint objects that require long exposures. CCDs have a quantum efficiency far superior to that of photography or the eye—up to 80 percent at some wavelengths. That improvement dramatically increases the ability to view faint objects with short exposure times.

A CCD is an ultrathin wafer of silicon—less than the thickness of a human hair—that is divided into a two-dimensional array of picture elements, or **pixels** (**Figure 6.18a**). When a photon strikes a pixel, it creates a small electric charge

(a)

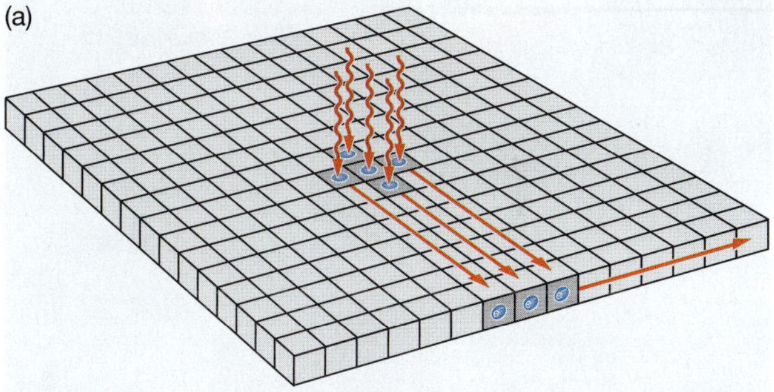

(b)

Figure 6.18 (a) In this simplified diagram of a charge-coupled device (CCD), photons from a star land on pixels (represented by gray squares) and produce free electrons within the silicon. The electron charges are electronically moved sequentially to the collecting register at the bottom. Each row is then moved out to the right to an electronic amplifier, which converts the electric charge of each pixel into a digital signal. (b) This large CCD (about 6 inches across) contains 12,288 × 8,192 pixels.

(a)

(b)

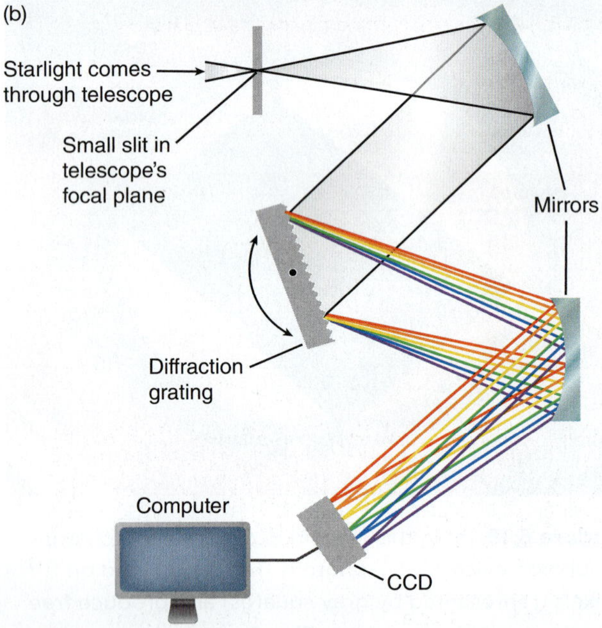

Figure 6.19 (a) A spectrum is created by the reflection of light from the closely spaced tracks of a CD. (b) In a grating spectrograph, light goes through the telescope and then a slit, where it is reflected to the diffraction grating and split into components. The spectrum is recorded on the CCD.

within the silicon. As each CCD pixel is read out, the digital signal that flows to the computer is nearly proportional to the accumulated charge. That is what we mean when we say that the CCD is a linear device. If a CCD is exposed to too much light, however, it can lose its linearity.

Liquid nitrogen or helium is used to cool the CCDs down to very low temperatures to reduce noise caused by the movement of the charge-carrying electrons within the silicon wafer. The first astronomical CCDs were small arrays containing a few hundred thousand pixels. The larger CCDs used in astronomy today may contain more than 100 million pixels (**Figure 6.18b**). Still larger arrays are under development as ever-faster computing power keeps up with image-processing demands.

The output from a CCD is a digital signal that can be sent directly from the telescope to image-processing software or stored electronically for later analysis. Nearly every spectacular astronomical image in ultraviolet, visible, or infrared wavelength that you find online was recorded by a CCD in a telescope either on the ground or in space. CCDs are found in many everyday devices, too, such as digital cameras, digital video cameras, and camera phones.

Your cell phone takes color pictures by using a grid of CCD pixels arranged in groups of three. Each pixel in a group is constructed to respond to only a particular range of colors—to only red light, for example. That also is true for digital image displays. You can see for yourself if you place a small drop of water on the screen of your smartphone or tablet and turn it on. The water magnifies the grid of pixels so that you can see them individually. The grid degrades the angular resolution of the camera because each spot in the final image requires three pixels of information. Astronomers instead use all the pixels on the camera to determine how many photons fall on each pixel, without regard to color. They put filters in front of the camera to allow light of only particular wavelengths to pass through, such as the light of a specific spectral line. Color pictures such as those from the Hubble Space Telescope are constructed by taking multiple pictures, coloring each one, and then carefully aligning and overlapping them to produce beautiful and informative images. Sometimes the colors are "true"; that is, they are close to the colors you would see if you were actually looking at the object with your eyes. Other times "false" colors represent different portions of the electromagnetic spectrum and tell you the temperature or composition of different parts of the object. Using changeable filters instead of designated color pixels gives astronomers greater flexibility and greater resolution.

Spectrographs

Spectroscopy is the study of an object's *spectrum* (plural: *spectra*)—its electromagnetic radiation split into component wavelengths. **Spectrographs** (sometimes called **spectrometers**) are instruments that take the spectrum of an object and then record it. The first spectrographs used prisms to disperse the light. Modern spectrographs use a **diffraction grating**, which is made by engraving closely spaced lines on glass to disperse incoming light into its constituent wavelengths. **Figure 6.19a** shows the light reflected from a CD or DVD; the closely spaced tracks act as a grating and create a spectrum. **Figure 6.19b** shows a grating spectrograph: Light from an astronomical object enters a telescope and passes through a slit. The light is reflected onto the diffraction grating, which creates a spectrum like the one shown in Figure 5.13. The spectra are recorded and then analyzed. Some modern spectrographs use bundles of optical fibers, or masks with multiple slits, to obtain spectra simultaneously from multiple objects in the field of view of the telescope.

CHECK YOUR UNDERSTANDING 6.2

CCD cameras have much higher quantum efficiency than that of other detectors. Therefore, CCD cameras: (a) can collect photons for longer times; (b) can collect photons of different energies; (c) can generate a signal from fewer photons; (d) can split light into different colors.

6.3 Astronomers Observe in Wavelengths Beyond the Visible

Recall from Section 5.4 (and Wien's law) that an object's temperature can be found from the peak wavelength of its continuous spectrum. Extending beyond visible light, radio or infrared telescopes are used to study cool objects, such as clouds of dust, whereas X-ray or gamma-ray telescopes are used to study violently hot gas. Thus, astronomers use telescopes that observe at all the wavelengths of the electromagnetic spectrum. **Figure 6.20** shows Earth's **atmospheric windows** and which parts of the spectrum they let in. The largest window is in radio wavelengths, including microwaves at the short-wavelength end of the radio window. Those telescopes can be built on the ground. However, gamma-ray,

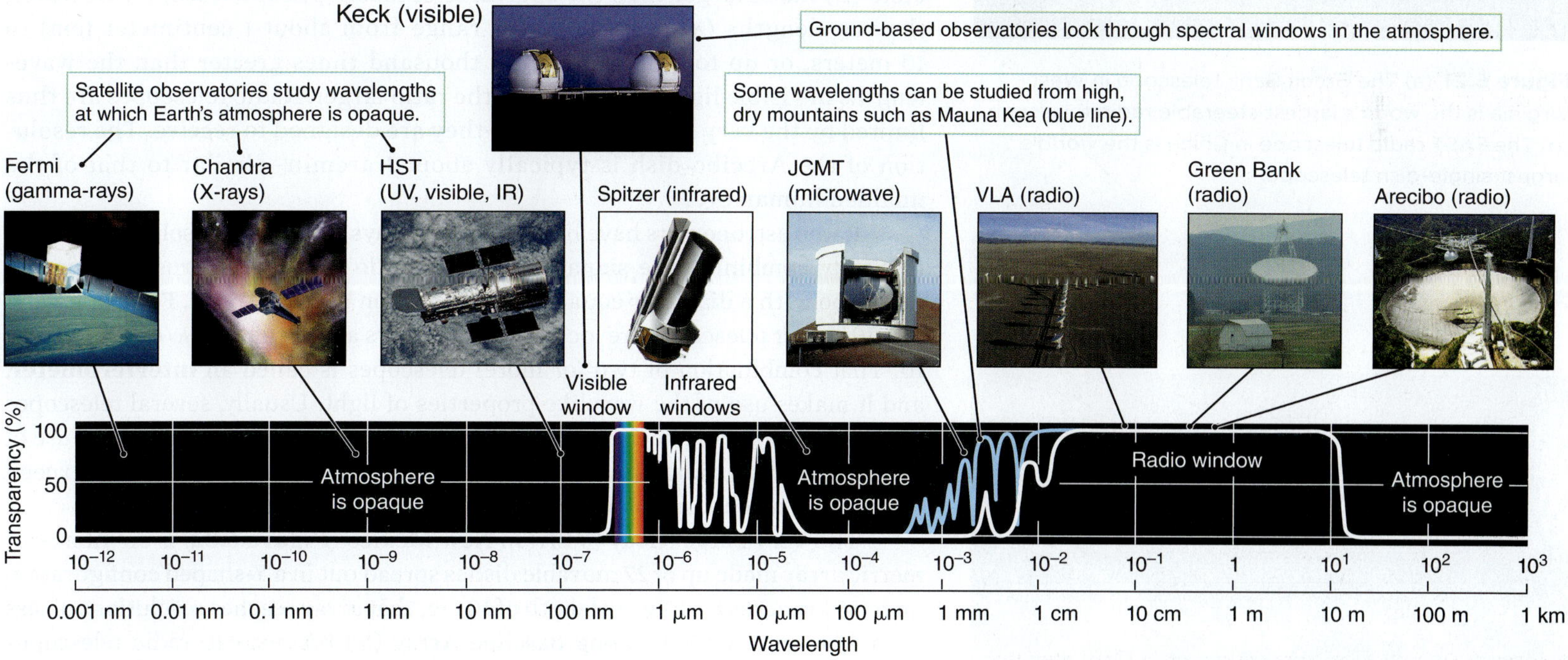

Figure 6.20 Earth's atmosphere blocks most electromagnetic radiation. Fermi = Fermi Gamma-ray Space Telescope (orbiting); Chandra = Chandra X-ray Observatory (orbiting); HST = Hubble Space Telescope (orbiting); Keck = Keck Observatory (Hawaii); Spitzer = Spitzer Space Telescope (orbiting); JCMT = James Clerk Maxwell Telescope (Hawaii); VLA = Very Large Array (New Mexico); Green Bank = Robert C. Byrd Green Bank Telescope (West Virginia); Arecibo = Arecibo Observatory (Puerto Rico).

(a)

(b)

Figure 6.21 (a) The Green Bank telescope in West Virginia is the world's largest steerable radio telescope. (b) The FAST radio telescope in China is the world's largest single-dish telescope.

X-ray, ultraviolet, and most of the infrared light arriving at Earth fails to reach the ground because it is partially or completely absorbed by ozone, water vapor, carbon dioxide, and other molecules in Earth's atmosphere. Light at those wavelengths has to be observed from space.

Radio Telescopes

Karl Jansky (1905–1950), a young physicist working for Bell Laboratories in the early 1930s, identified a radio source in the Milky Way in the direction of the galactic center, in the constellation Sagittarius. Jansky's discovery marked the birth of radio astronomy, and in his honor, the basic unit for the strength of a radio source is called the **jansky (Jy)**. A few years later, Grote Reber (1911–2002), a radio engineer and ham radio operator, built his own radio telescope and conducted the first survey of the sky at radio frequencies. Reber was largely responsible for the rapid advancement of radio astronomy in the post–World War II era.

Most radio telescopes are large, steerable dishes, typically tens of meters in diameter, such as the one shown in **Figure 6.21a**. The world's largest single-dish radio telescopes are the new FAST (Five-hundred-meter Aperture Spherical Telescope) dish in China (**Figure 6.21b**) and the 305-meter Arecibo dish built into a natural bowl-shaped depression in Puerto Rico. Those large single-dish telescopes are not steerable, so they can observe only sources that pass within 20° of the zenith as Earth's rotation carries them overhead.

As large as radio telescopes are, they have relatively poor angular resolution. Recall that a telescope's angular resolution is determined by the ratio λ/D, so the resolution decreases as that ratio increases. Radio telescopes have diameters (D) much larger than the apertures of most optical telescopes. However, the wavelengths (λ) of radio waves range from about 1 centimeter (cm) to 10 meters, or up to several hundred thousand times greater than the wavelengths of visible light, which makes the ratio larger. Radio telescopes are thus limited by the very long wavelengths they are designed to receive. The resolution of the Arecibo dish is typically about 1 arcmin—similar to that of the unaided human eye.

Radio astronomers have had to develop ways to improve resolution. Mathematically combining the signals from two radio telescopes turns them into a telescope with a diameter equal to the separation between them. For example, if two 10-meter telescopes are located 1,000 meters apart, the D in λ/D is 1,000, not 10. That combination of two (or more) telescopes is called an **interferometer**, and it makes use of the wavelike properties of light. Usually, several telescopes are used in an arrangement called an **interferometric array**. Through the use of very large arrays, radio astronomers can better observe bright sources and exceed the angular resolution possible with optical telescopes.

Figure 6.22 The VLA in New Mexico combines signals from 27 telescopes so that they act as one "very large" telescope.

The Very Large Array (VLA) in New Mexico (**Figure 6.22**) is an interferometric array made up of 27 movable dishes spread out in a Y-shaped configuration up to 36 km across. At a wavelength of 10 cm, that array reaches resolutions of less than 1 arcsec. The Very Long Baseline Array (VLBA) uses 10 radio telescopes spread out over more than 8,000 km from the Virgin Islands in the Caribbean to Hawaii in the Pacific. At a wavelength of 10 cm, that array can attain resolutions of better than 0.003 arcsec. A radio telescope put into near-Earth orbit as part of a Space Very Long Baseline Interferometer (SVLBI) overcomes even that limit. The new Event Horizon Telescope combines many of the most advanced existing radio telescopes, from Greenland to the South Pole, to make an *Earth-sized* interferometer. For a few nights each year, all the telescopes observe the same object, with a combined resolution that may be good enough to image objects near the center of the Milky Way.

Some radio telescopes use many small dishes. The Atacama Large Millimeter/submillimeter Array (ALMA; **Figure 6.23**), located at an elevation of 5,000 meters in the Atacama Desert in Chile, was completed in 2013. That project, an international collaboration of astronomers from Europe, North America, East Asia, and Chile, consists of 66 dishes with diameters of 12 and 7 meters for observations in the 0.3- to 9.6-mm-wavelength range. The Square Kilometre Array (SKA) is designed to have *thousands* of small radio dishes, which together will act as one dish with a collecting area of 1 square kilometer (km^2). Twenty countries are supporting that telescope, which will be located in Australia and South Africa.

Optical telescopes also can be combined in an array to yield resolutions greater than those of single telescopes, although for technical reasons the individual units cannot be spread as far apart as radio telescopes. The Very Large Telescope Interferometer (VLTI) in Chile, operated by ESO, combines four 8-meter telescopes with four movable 1.8-meter auxiliary telescopes. It has a baseline of up to 200 meters, yielding angular resolution of about 0.001 arcsec. The six-telescope Center for High Angular Resolution Astronomy (CHARA) array in California works in the visible and near-infrared regions. It has a baseline of 330 meters with resolution of 0.0003 arcsec.

Figure 6.23 The new Atacama Large Millimeter/submillimeter Array (ALMA) telescope in the Atacama Desert in northern Chile has many international partners.

Infrared Telescopes

Molecules such as water vapor in Earth's atmosphere block infrared (IR) photons from reaching astronomical telescopes on the ground, so telescopes that observe in the infrared [0.75–30 microns (μm)] are at the highest locations. Mauna Kea, a dormant volcano and home of the Mauna Kea Observatories (MKO), rises 4,200 meters above the Pacific Ocean. At that altitude, the MKO telescopes sit above 40 percent of Earth's atmosphere; but more important, 90 percent of Earth's atmospheric water vapor lies below. Still, for the infrared astronomer, the remaining 10 percent is troublesome.

Airborne observatories overcome atmospheric absorption of infrared light by placing telescopes above most of the water vapor in the atmosphere. NASA's Stratospheric Observatory for Infrared Astronomy (SOFIA) (**Figure 6.24**), a joint project with the German Aerospace Center (DLR), is a modified 747 airplane that carries a 2.5-meter telescope and works in the far-infrared region of the spectrum, from 1 to 650 μm. It flies in the stratosphere at an altitude of about 12 km, above 99 percent of the water vapor in Earth's lower atmosphere. Because airplanes are highly mobile, SOFIA can observe in both the Northern and Southern Hemispheres. Other infrared wavelengths must be observed from space.

Figure 6.24 SOFIA is a 2.5-meter infrared telescope mounted in a Boeing 747 aircraft.

Orbiting Observatories

Gaining full access to the complete electromagnetic spectrum requires getting completely above Earth's atmosphere. The first astronomical satellite was the British Ariel 1, launched in 1962 to study solar UV and X-ray radiation. Today, a

TABLE 6.2 Selected Space Observatories

Telescope	Sponsor(s)	Description	Year
Hubble Space Telescope (HST)	NASA, ESA	Optical, infrared, ultraviolet observations	1990–
Chandra X-ray Observatory	NASA	X-ray imaging and spectroscopy	1999–
X-ray Multi-Mirror Mission (XMM-Newton)	ESA	X-ray spectroscopy	1999–
Galaxy Evolution Explorer (GALEX)	NASA	Ultraviolet observations	2003–2012
Spitzer Space Telescope	NASA	Infrared observations	2004–
Neil Gehrels Swift Observatory	NASA	Gamma-ray bursts	2004–
Convection Rotation and Planetary Transits (COROT) space telescope	CNES (France)	Planet finder	2006–2013
Fermi Gamma-ray Space Telescope	NASA, European partners	Gamma-ray imaging and gamma-ray bursts	2008–
Planck telescope	ESA	Cosmic microwave background radiation	2009–2013
Herschel Space Observatory	ESA	Far-infrared and submillimeter observations	2009–2013
Kepler telescope	NASA	Planet finder	2009–2018
Solar Dynamics Observatory (SDO)	NASA	Sun, solar weather	2010–
RadioAstron	Russia	Very-long-baseline interferometry in space	2011–
Nuclear Spectroscopic Telescopic Array (NuSTAR)	NASA	High-energy X-ray	2012–
Gaia	ESA	Optical, digital 3D space camera	2013–
Transiting Exoplanet Survey Satellite (TESS)	NASA, MIT, SAO	Survey entire sky for transiting planets around nearby and bright stars	2018–
CHEOPS (CHaracterising ExOPlanet Satellite)	ESA	High-precision brightness measurements of stars known to have planets	2019 (scheduled)
James Webb Space Telescope (JWST)	NASA, ESA, Canadian Space Agency	Optical and infrared; replacement for HST	2021 (scheduled)

multitude of orbiting astronomical telescopes cover the electromagnetic spectrum from gamma rays to microwaves, with more in the planning stage (**Table 6.2**). Optical telescopes, such as the 2.4-meter Hubble Space Telescope (HST; **Figure 6.25a**), operate successfully at low Earth orbit, 600 km above Earth's surface. Launched in 1990, HST has been the workhorse for UV, visible, and IR space astronomy for more than 25 years. Low Earth orbit is also the region where the International Space Station and many scientific satellites orbit.

For certain other satellites and space telescopes, however, 600 km is not high enough. The Chandra X-ray Observatory (**Figure 6.25b**), NASA's X-ray telescope, cannot see through even the tiniest traces of atmosphere and therefore orbits more than 16,000 km above Earth's surface. NASA's Spitzer Space Telescope, an infrared telescope, is so sensitive that it needs to be completely free from Earth's own infrared radiation. The solution was to put it into a *solar* orbit, trailing tens of millions of kilometers behind Earth. The James Webb Space Telescope, scheduled to replace the HST, will observe primarily in infrared wavelengths. It will be located 1.5 million miles away from Earth, orbiting the Sun at a fixed distance from the Sun and Earth.

Orbiting telescopes located above the atmosphere are unaffected by atmospheric image distortions, weather, or brightening night skies. But space observatories are much more expensive than ground-based observatories and can be difficult or impossible to repair. The HST required several servicing missions, but such missions are impossible for the observatories in more distant Earth orbits. Ground-based telescopes at even the most remote mountaintop locations can receive shipments of replacement parts in a few days; space telescopes cannot. Some wavelengths can be observed from space only, but issues of cost and repair are why ground-based telescopes are much more prevalent.

(a)

(b)

Figure 6.25 (a) The Hubble Space Telescope (HST) observes in the ultraviolet, visible, and infrared regions of the electromagnetic spectrum. (b) NASA's Chandra X-ray Observatory, shown in this artist's conception, collects X-rays. Different telescopes in space allow astronomers to observe at all wavelengths of the electromagnetic spectrum.

CHECK YOUR UNDERSTANDING 6.3

Which of the following is the biggest disadvantage of putting a telescope in space?
(a) Astronomers don't have as much control in choosing what to observe.
(b) Astronomers have to wait until the telescopes come back to Earth to get their images.
(c) Space telescopes can observe only in certain parts of the electromagnetic spectrum.
(d) Space telescopes are much more expensive and riskier than similar ground-based telescopes.

6.4 Planetary Spacecraft Explore the Solar System

Recall from Chapter 2 that everyone always sees the same face of the Moon from Earth because the Moon's orbital and rotational periods are equal. The first view of the "far" side was in 1959, when the Soviet flyby mission *Luna 3* sent back pictures showing that the far side of the Moon was very different from its Earth-facing half. No matter how powerful our ground-based or Earth-orbiting telescopes, sometimes we need to send a spacecraft for a different view.

Only in the past half century has the technology existed to explore the Solar System. Spacecraft have now visited all the planets and some of their moons, as well as a few comets and asteroids, providing the first close-up views of those distant worlds. The study of the Solar System from space is an international collaboration involving NASA, the European Space Agency (ESA), the Russian Federal Space Agency (Roscosmos), the Japan Aerospace Exploration Agency (JAXA), the China National Space Administration (CNSA), and the Indian Space Research Organisation (ISRO). Other countries may soon join the endeavor. In this section, we look at the types of spacecraft used to explore our Solar System.

Flybys and Orbiters

Exploration of the Solar System began with a reconnaissance phase, using spacecraft to fly by or orbit a planet or other body. A **flyby** is a spacecraft that first approaches and then continues flying past the target. As those spacecraft speed by, instruments aboard them briefly probe the physical and chemical properties of the target and its environment.

Flyby missions are the most common first phase of exploration. They cost less than orbiters or landers and are easier to design and execute. Flyby spacecraft,

(a)

(b)

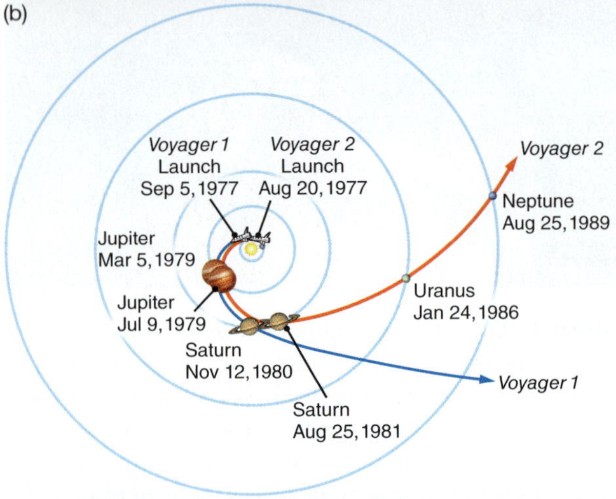

Figure 6.26 The two *Voyager* spacecraft (a) flew past the outer planets and (b) have recently passed the boundary of our Solar System.

Figure 6.27 The robotic rover *Curiosity* took this selfie on the surface of Mars.

such as *Voyager 1* and *Voyager 2*, can sometimes visit several worlds during their travels (**Figure 6.26**). The downside of flyby missions is that because of the physics of orbits, the spacecraft must move by very swiftly. They are limited to just a few hours or at most a few days in which to conduct close-up studies of their targets. Flyby spacecraft give astronomers their first close-up views of Solar System objects, and sometimes the data obtained are then used to plan follow-up studies.

More detailed reconnaissance work is done by spacecraft known as **orbiters** because they orbit around their target. Such missions are intrinsically more difficult than flyby missions because they have to make risky maneuvers and use their limited fuel to change their speed to enter an orbit. But orbiters can linger, looking in detail at more of the surfaces of the objects they are orbiting and studying things that change with time, such as planetary weather.

Orbiters use remote-sensing instrumentation like that used by Earth-orbiting satellites to study our own planet. Those instruments include tools such as cameras that take images at different wavelength ranges, radar that can map surfaces hidden beneath obscuring layers of clouds, and spectrographs that analyze the electromagnetic spectrum. Those instruments enable planetary scientists to map other worlds, measure the heights of mountains, identify geological features and rock types, watch weather patterns develop, measure the composition of atmospheres, and get a general sense of the place. Additional instruments make measurements of the extended atmospheres and space environment through which they travel.

Landers, Rovers, and Atmospheric Probes

Reconnaissance spacecraft provide a wealth of information about a planet, but no better way exists to explore a planet than doing so within a planet's atmosphere or on solid ground. Spacecraft have landed on the Moon, Mars, Venus, Saturn's large moon Titan, several asteroids, and a comet. Those spacecraft have taken pictures of planetary surfaces, measured surface chemistry, and conducted experiments to determine the physical properties of the surface rocks and soils.

Using **landers**—spacecraft that touch down and remain on the surface—has several disadvantages. Because of the expense, only a few landings in limited areas are practical. With that limitation, the results may apply only to the small area around the landing site. Imagine, for example, what a different picture of Earth you might get from a spacecraft that landed in Antarctica, as opposed to a spacecraft that landed in a volcano or on the floor of a dry riverbed. Sites to be explored with landed spacecraft must be carefully chosen on the basis of reconnaissance data. Some landers have wheels and can explore the vicinity of the landing site. Such remote-controlled vehicles, called **rovers**, were used first by the Soviet Union on the Moon four decades ago and more recently by the United States on Mars. **Figure 6.27** shows a self-portrait of the *Curiosity* rover on Mars.

Atmospheric probes descend into the atmospheres of planets and continually measure and send back data on temperature, pressure, and wind speed, along with other properties, such as chemical composition. Atmospheric probes have survived all the way to the solid surfaces of Venus and of Saturn's moon Titan (**Figure 6.28**), sending back streams of data during their descent. An atmospheric probe sent into Jupiter's atmosphere never reached that planet's surface because, as we discuss later in the book, Jupiter does not have a solid surface in the same sense that terrestrial planets and moons do. After sending back its data, the Jupiter probe eventually melted and vaporized as it descended into the hotter layers of the planet's atmosphere.

Sample Returns

If you pick up a rock from the side of a road, you might learn a lot from the rock by using tools that you could carry in your pocket or in your car. Much better, though, would be to pick up a few samples and carry them back to a laboratory equipped with a full range of state-of-the-art instruments that can measure chemical composition, mineral type, age, and other information needed to reconstruct the story of your rock sample's origin and evolution. The same is true of Solar System exploration. One of the most powerful methods for investigating remote objects is to collect samples of the objects and bring them back to Earth for study. So far, only samples of the Moon, a comet, and streams of charged particles from the Sun have been collected and returned to Earth. Scientists have found meteorites on Earth that are pieces of Mars that were blasted loose when objects crashed into that world. In the next 10 years, robotic "sample and return" missions may go to Mars.

Figure 6.28 Artist's illustration showing the Huygens space probe descending and landing on Saturn's moon Titan in 2005.

The missions discussed so far in this section have all been conducted with robotic spacecraft. The only spacecraft that took people to another world were the *Apollo* missions to the Moon. That program ran from 1961 to 1972 and included several missions before the actual Moon landings. The *Apollo 8* astronauts brought back the famous picture of Earth viewed over the surface of the Moon (see the opening figure of Chapter 1). Each mission from *Apollo 11* through *Apollo 17* had three astronauts—two to land on the Moon and one to remain in orbit. *Apollo 13* did not reach the Moon but returned to Earth safely. Twelve American astronauts walked on the Moon between 1969 and 1972 and brought back a total of 382 kg of rocks and other material (**Figure 6.29**).

The return of extraterrestrial samples to Earth is governed by international treaties and standards to ensure that the samples do not contaminate our world. For example, before the lunar samples that the *Apollo* missions brought back could be studied, they (and the astronauts) had to be quarantined and tested for the presence of alien life-forms. The same international standards apply to spacecraft landing on other planets. The goal of those standards is to avoid transporting life-forms from Earth to another planet; we do not want to "discover" life that we, in fact, introduced. In addition, we do not want to potentially harm life that may exist on other planets.

With many missions under way and others on the horizon, robotic exploration of the Solar System is an ongoing, dynamic activity. Appendix 5 summarizes some recent and current missions. We will discuss some of those missions in the relevant chapter. Information on the latest discoveries can be found on mission websites and in science news sources.

Figure 6.29 *Apollo 11* astronaut Buzz Aldrin on the moon, as photographed by mission commander Neil Armstrong, in 1969.

CHECK YOUR UNDERSTANDING 6.4

Spacecraft are the most effective way to study planets in our Solar System because: (a) planets move too fast across the sky for us to image them well from Earth; (b) planets cannot be imaged from Earth; (c) spacecraft can collect more information than is available just from images from Earth; (d) space missions are easier than long observing campaigns.

6.5 Other Tools Contribute to the Study of the Universe

High-profile space missions have sent back stunning images and data from across the electromagnetic spectrum. But astronomers use other tools as well, including particle accelerators and colliders, neutrino and gravitational-wave detectors, and super-computers.

Figure 6.30 The ATLAS particle detector at CERN's Large Hadron Collider near Geneva, Switzerland. The instrument's enormous size is evident from the person standing near the bottom center of the picture.

Particle Accelerators

Ever since the early years of the 20th century, physicists have been peering into the structure of the atom by observing what happens when small particles collide. By the 1930s, physicists had developed the technology to accelerate charged subatomic particles such as protons to very high speeds and then observe what happens when they slam into a target. From such experiments, physicists have discovered many kinds of subatomic particles and learned about their physical properties. High-energy particle colliders have proved to be an essential tool for physicists studying the basic building blocks of matter.

Astronomers have realized that to understand the very largest structures seen in the universe, it is important to understand the physics that took place during the earliest moments in the universe, when everything was extremely hot and dense. High-energy particle colliders that physicists use today are designed to approach the energies of the early universe. The effectiveness of particle accelerators is determined by the energy they can achieve and the number of particles they can accelerate. Modern particle colliders such as the Large Hadron Collider near Geneva, Switzerland (**Figure 6.30**), reach very high energies. Particles also can be studied from space. The Alpha Magnetic Spectrometer, installed on the International Space Station in 2011, searches for some of the most exotic forms of matter, such as dark matter, antimatter, and high-energy particles called cosmic rays.

Neutrinos and Gravitational Waves

The **neutrino** is an elusive elementary particle that plays a major role in the physics of the interiors of stars. Neutrinos are extremely difficult to detect. In less time than you take to read this sentence, a thousand trillion (10^{15}) solar neutrinos from the Sun are passing through your body, even during the night. Neutrinos are so nonreactive with matter that they can pass right through Earth (and you) as though it (or you) weren't there at all. To be observed, a neutrino has to interact with a detector. Neutrino detectors typically record only one of every 10^{22} (10 billion trillion) neutrinos passing through them, but that's enough to reveal processes deep within the Sun or the violent death of a star 160,000 light-years away.

Experiments designed to look for neutrinos originating outside Earth are buried deep underground in mines or caverns or under the ocean or ice to ensure that only neutrinos are detected. For example, the ANTARES experiment uses the Mediterranean Sea as a neutrino telescope. Detectors located 2.5 km under the sea, off the coast of France, observe neutrinos that originated in objects visible in southern skies and passed through Earth. In the IceCube neutrino observatory located at the South Pole in Antarctica, the neutrino detectors are 1.5–2.5 km under the ice, and they observe neutrinos that originated in objects visible in northern skies (**Figure 6.31**).

Figure 6.31 The IceCube neutrino telescope at the South Pole, Antarctica.

Another elusive phenomenon is the **gravitational wave**. Gravitational waves are disturbances in a gravitational field, similar to the waves that spread out from the disturbance you create when you toss a pebble onto the quiet surface of a pond (**Process of Science Figure**). Several facilities, including the Laser Interferometer Gravitational-Wave Observatory (LIGO) and the European VIRGO interferometer, have been constructed to detect gravitational waves. Recently, LIGO has detected several pairs of coalescing binary black holes and neutron stars, as we explore in Chapter 18.

PROCESS OF SCIENCE

Technology and Science Are Symbiotic

Scientists have been searching for waves that carry gravitational information for nearly 100 years, but the accuracy of their measurements has been limited by the available technology.

1916

Einstein predicts that changing gravitational fields should produce gravitational waves. He calculates the size of the effect, and concludes that it will be impossible to detect them.

c. 1970

Joseph Weber constructs precision-machined bars of metal that should "ring" as a gravitational wave passes by. His results were never confirmed.

1990s

Construction begins on LIGO, an observatory which uses lasers to precisely measure the change in the length of long tubes as gravitational waves pass by.

2015

LIGO detects gravitational waves for the first time, announcing the discovery in early 2016, 100 years after Einstein's prediction. This first detection was followed by several more in the next two years.

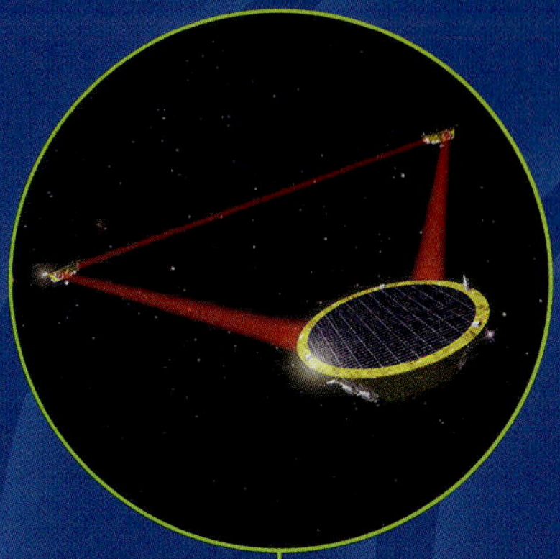

Future (2034?)

LISA, a gravitational wave observatory in space, will be more sensitive; it will be able to detect much weaker events than LIGO.

Technology and science develop together. New technologies enable humans to ask new scientific questions. Asking new scientific questions pushes the development of better instrumentation and new technologies.

Figure 6.32 These images show supercomputer simulations of the collision of two galaxies. Astronomers compare simulations such as those with telescopic observations.

Computers

Astronomers use powerful computers to gather, analyze, and interpret data. A single CCD image may contain millions of pixels, with each pixel displaying roughly 30,000 levels of brightness. That adds up to several trillion pieces of information in each image. To analyze their data, astronomers typically do calculations for *every single pixel* of an image to remove unwanted contributions from Earth's atmosphere or to correct for instrumental effects. Astronomers conduct many types of sky surveys—in which one or more telescopes survey a specific part of the sky—yielding thousands of images that need to be analyzed.

High-performance computers also play an essential role in generating and testing theoretical models of astronomical objects. Even when we completely understand the underlying physical laws that govern the behavior of a particular object, often the object is so complex that calculating its properties and behavior would be impossible without the assistance of high-performance computers. As discussed in Chapter 4, for example, you can use Newton's laws to compute the orbits of two stars that are gravitationally bound to each other because their orbits take the form of simple ellipses. However, understanding the orbits of the several hundred billion stars that make up the Milky Way Galaxy is not so easy, even though the underlying physical laws are the same.

Computer modeling is used to determine the interior properties of stars and planets, including Earth. Although astronomers cannot see beneath the surfaces of those bodies, they have a surprisingly good understanding of their interiors, which we describe in later chapters. Astronomers start a model by assigning well-understood physical properties to tiny volumes within a planet or star. The computer assembles an enormous number of those individual elements into an overall representation. The result is a rather good picture of what the interior of the star or planet is like.

Astronomers also use high-performance computers to study how astronomical objects, systems of objects, and the universe as a whole evolve. For example, astronomers create models of galaxies and then run computer simulations to study how those galaxies might change over billions of years. **Figure 6.32** shows a simulation of the collision of two galaxies. The results of the computer simulations are then compared with telescopic observations. If the simulations do not match the observations, the model is adjusted and the simulations are run again until general agreement exists between them.

CHECK YOUR UNDERSTANDING 6.5

High-performance computers have become one of an astronomer's most important tools. Which of the following require the use of that type of computer? (Choose all that apply.) (a) analyzing images taken with very large CCDs; (b) generating and testing theoretical models; (c) pointing a telescope from object to object; (d) studying how astronomical objects or systems evolve.

Origins Microwave Telescopes Detect Radiation from the Big Bang

In this chapter, we explored the tools of the astronomer, from basic optical telescopes to instruments that observe in different wavelengths. Now let's examine in more detail one type of telescope that has aided the study of the origin of the universe. Recall from Chapter 1 that astronomers think the universe originated with a hot Big Bang. The multiple strands of evidence for that conclusion are discussed in Chapter 21. Here, we look at one piece: the observation of faint microwave radiation left over from the early hot universe. Two Bell Laboratories physicists, Arno Penzias (1933–) and Robert Wilson (1936–), were working on satellite communications when they first detected that radiation in 1964 with a microwave dish antenna in New Jersey. Today, we routinely use cell phones and handheld GPS devices that communicate directly with satellites, but at the time, that capability was at the limit of technology.

Penzias and Wilson needed a very sensitive microwave telescope for the work they were doing for Bell Labs because any spurious signals coming from the telescope itself might wash out the faint signals bounced off a satellite. To that end, they were working hard to eliminate all possible sources of interference originating from within their instrument, including keeping the telescope free of bird droppings. No matter how carefully they tried to eliminate sources of extraneous noise, they always still detected a faint signal at microwave wavelengths. That faint signal was the same in every direction and turned out to be from the Big Bang. Penzias and Wilson shared the 1978 Nobel Prize in Physics for discovering the **cosmic microwave background radiation (CMB)** left over from the Big Bang itself.

Since 1964, astronomers from around the world have designed increasingly precise instruments to measure that radiation from the ground, from high-altitude balloons, from rockets, and from satellites. The Russian experiment RELIKT-1, launched in 1983, found some limits on the variation of the CMB. The COBE (Cosmic Background Explorer) satellite, launched in 1989, showed that the spectrum of that radiation precisely matched that of a blackbody with a temperature of 2.73 K—exactly what was predicted for the radiation left over from the Big Bang. (Compare **Figure 6.33** with the curves in Figure 5.22.) The data also showed some slight differences in temperature—small fractions of a degree—over the map of the sky. Those slight variations tell us about how the universe evolved from one dominated by radiation to one that contains structures such as galaxies, stars, planets, and us. John Mather and George Smoot shared the 2006 Nobel Prize in Physics for that work.

In 1998 and 2003, a high-altitude balloon experiment called BOOMERANG (short for "balloon observations of millimetric extragalactic radiation and geophysics") flew over Antarctica at an altitude of 42 km to study CMB variations and estimate the overall geometry of the universe. The *WMAP* (Wilkinson Microwave Anisotropy Probe) satellite, launched in 2001, created an even more detailed map of the temperature variations in that radiation, yielding more precise values for the age and shape of the universe and the presence of dark matter and dark energy. The Planck space telescope, operated from 2009 to 2013 by the European Space Agency, was much more sensitive than *WMAP* and studied those CMB variations in even more precise detail. The Atacama Cosmology Telescope and Simons Array (Chile) and the South Pole Telescope (Antarctica) study that radiation to look for evidence of when galaxy clusters formed. Those experiments and observations have opened up the current era of precision cosmology, in which astronomers can make detailed models of how the universe was born, eventually leading to stars, planets, and us.

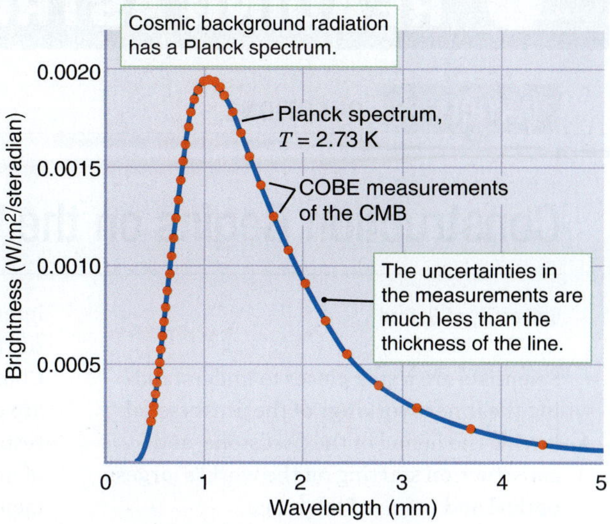

Figure 6.33 This graph shows the spectrum of the cosmic microwave background radiation (CMB) as measured by the COBE satellite (red dots). A steradian is a unit of solid angle. The uncertainty in the measurement at each wavelength is much less than the size of a dot. The line running through the data is a Planck blackbody spectrum with a temperature of 2.73 K.

READING ASTRONOMY NEWS

ARTICLES QUESTIONS

The Extremely Large Telescope will be located high on a mountain in Chile.

Construction Begins on the World's First Super Telescope

University of Oxford

Scientists are a step closer to understanding the inner-workings of the universe following the laying of the first stone, and construction starting on the world's largest optical and infrared telescope.

With a main mirror 39 metres in diameter, the Extremely Large Telescope (ELT) is going to be, as its name suggests, enormous. Unlike any other before it, ELT is also designed to be an adaptive telescope and has the ability to correct atmospheric turbulence, taking telescope engineering to another level.

To mark the construction's milestone, a ceremony was held at ESO's Paranal residencia in northern Chile, close to the site of the future giant telescope which will be on top of Cerro Armazones, a 3,046-metre peak mountain.

Among many other representatives from industry, the significance of the project was highlighted by the attendance of the Director General of ESO, Tim de Zeeuw, and President of the Republic of Chile, Michelle Bachelet Jeria.

The ELT is being built by the European Southern Observatory (ESO), an international collaboration supported by the UK's Science and Technology Facilities Council (STFC). Oxford University scientists are playing a key role in the project, and are responsible for the design and construction of its spectrograph, 'HARMONI', an instrument designed to simultaneously take 4,000 images, each in a slightly different colour. The visible and near-infrared instrument will harness the telescope's adaptive optics to provide extremely sharp images.

'HARMONI' will enable scientists to form a more detailed picture of the formation and evolution of objects in the Universe. Supporting researchers to view everything from the planets in our own solar system and stars in our own and nearby galaxies with unprecedented depth and precision, to the formation and evolution of distant galaxies that have never been observed before.

Niranjan Thatte, Principal Investigator for 'HARMONI' and Professor of Astrophysics at Oxford's Department of Physics, said: 'For me, the ELT represents a big leap forward in capability, and that means that we will use it to find many interesting things about the Universe that we have no knowledge of today.

'It is the element of "exploring the unknown" that most excites me about the ELT. It will be an engineering feat, and its sheer size and light grasp will dwarf all other telescopes that we have built to date.'

A time capsule, created by members of the ESO team and sealed at the event, will serve as a lasting memory of the research and the scale of ambition and commitment behind it. Contents include a copy of a book describing the original scientific aims of the telescope, images of the staff that have and will play a role in its construction and a poster of an ELT visualisation. The cover of the time capsule is engraved with a hexagon made of Zerodur, a one-fifth scale model of one of the ELT's primary mirror segments.

Tim de Zeeuw, Director General of ESO, said: 'The ELT will produce discoveries that we simply cannot imagine today, and it will surely inspire numerous people around the world to think about science, technology and our place in the Universe. This will bring great benefit to the ESO member states, to Chile, and to the rest of the world.'

ARTICLES QUESTIONS

1. Why do astronomers want to build at this particular location?
2. What are the advantages of larger telescopes?
3. How can adaptive optics make images sharper?
4. Why will this telescope observe only in optical and infrared wavelengths?
5. Do a Web search to see the status of this project. What countries are partners? Does any local opposition to the project exist in the host country?

Summary

Earth's atmosphere blocks many spectral regions and distorts telescopic images. Telescopes are sited to be above as much of the atmosphere as possible. Telescopes are matched to the wavelengths of observation, with different technologies required for each region of the spectrum. The aperture of a telescope both determines its light-gathering power and limits its resolution; larger telescopes are better in both measures. Modern CCD cameras have better quantum efficiency and longer integration times, allowing astronomers to study fainter and more distant objects than were observable with earlier detectors. Telescopes observing at microwave wavelengths have detected radiation left over from the Big Bang.

LG 1 **Compare how the two main types of optical telescopes gather and focus light.** The telescope is the astronomer's most important tool. Ground-based telescopes that observe in visible wavelengths are either refractors (lenses) or reflectors (mirrors). All large astronomical telescopes are reflectors. Large telescopes collect more light and have greater resolution. The diffraction limit is the limiting resolution of a telescope.

LG 2 **Summarize the main types of detectors that are used on telescopes.** Photography improved the ability of astronomers to record details of faint objects seen in telescopes. CCDs are today's astronomical detector of choice because they are much more linear, have a broader spectral response, and can send electronic images directly to a computer. Spectrographs are specialized instruments that take the spectrum of an object to reveal what the object is made of and many other physical properties.

LG 3 **Explain why some wavelengths of radiation must be observed from high, dry, and remote observatories on Earth, or from space.** Radio, near-infrared, and optical telescopes can see through our atmosphere. Those types of telescopes can be arrayed to greatly increase angular resolution. Putting telescopes in space solves problems created by Earth's atmosphere.

LG 4 **Explain the benefits of sending spacecraft to study the planets and moons of our Solar System.** Most of what is known about the planets and their moons comes from observations by spacecraft. Flyby and orbiting missions obtain data from space, and landers and rovers collect data from the ground.

LG 5 **Describe other astronomical tools that contribute to the study of the universe.** Astronomers also use particle accelerators, neutrino detectors, and gravitational-wave detectors to study the universe. High-performance computers are essential to acquiring, analyzing, and interpreting astronomical data.

? Unanswered Questions

- Will telescopes be placed on the Moon? The Moon has no atmosphere to make stars twinkle, cause weather, or block certain wavelengths of light from reaching its surface. All parts of the Moon have days and nights that last for 2 Earth weeks each. China had a small (15 cm) ultraviolet telescope operated from Earth on their *Chang'e 3* lander. One proposal is for a small array of radio telescopes on the side of the Moon facing Earth that would study the Sun and the solar wind. Eventually, telescopes would go on the far side of the Moon, which faces away from the light and radio radiation of Earth. On the far side one might put an array of hundreds of radio telescopes that would be deployed to study the earliest formation of stars and galaxies. Another proposal is for the Lunar Liquid Mirror Telescope (LLMT), with a diameter of 20–100 meters, to be located at one of the Moon's poles. Gravity would settle the rotating liquid into the necessary parabolic shape, and those liquid mirror telescopes are much simpler than are arrays of telescopes with large glass mirrors. Astronomers debate whether telescopes on the Moon would be easier to service and repair than those in space and whether problems caused by lunar dust would outweigh any advantages.

- Will human exploration of the Solar System occur within your lifetime? Since the *Apollo* program, humans have not returned to the Moon or traveled to other planets or moons in the Solar System. Sending humans to the worlds of the Solar System is much more complicated, risky, and expensive than sending robotic spacecraft. Humans need life support such as air, water, and food. Radiation in space can be dangerous. Furthermore, human explorers would expect to return to Earth, whereas most spacecraft do not come back. Astronomers and space scientists have heated debates about human spaceflight versus robotic exploration. Some argue that true exploration requires that human eyes and brains actually go there; others argue that the costs and risks are too high for the potential additional scientific knowledge. Beyond basic exploration, we also do not know whether humans will ever permanently colonize space.

Questions and Problems

Test Your Understanding

1. If one telescope has an aperture of 20 cm, and another has an aperture of 30 cm, and if aperture size is the only difference, then which should you choose, and why?
   a. The 20 cm, because the light-gathering power will be better.
   b. The 20 cm, because the image size will be larger.
   c. The 30 cm, because the light-gathering power will be better.
   d. The 30 cm, because the image size will be larger.
2. Which of the following can be observed from Earth's surface? (Choose all that apply.)
   a. radio waves
   b. gamma radiation
   c. far UV light
   d. X-ray light
   e. visible light
3. Match the following properties of telescopes (lettered) with their corresponding definitions (numbered).

a. aperture	(1) two or more telescopes connected to act as one
b. resolution	(2) distance from lens to focal plane
c. focal length	(3) diameter
d. chromatic aberration	(4) ability to distinguish close objects
e. diffraction	(5) computer-controlled correction for atmospheric distortion
f. interferometer	(6) color-separating effect
g. adaptive optics	(7) smearing effect due to sharp edge

4. Two 10-meter telescopes, separated by 85 meters, can operate as an interferometer. What is its resolution when it observes in the infrared at a wavelength of 2 microns?
   a. 0.01 arcsec
   b. 0.005 arcsec
   c. 0.2 arcsec
   d. 0.05 arcsec
5. Arrays of radio telescopes can produce much better resolution than single-dish telescopes can because they work based on the principle of
   a. reflection.
   b. refraction.
   c. diffraction.
   d. interference.
6. Refraction is caused by
   a. light bouncing off a surface.
   b. light changing colors as it enters a new medium.
   c. light changing speed as it enters a new medium.
   d. two light beams interfering.
7. The light-gathering power of a 4-meter telescope is ___________ than that of a 2-meter telescope.
   a. 4 times larger
   b. 8 times larger
   c. 16 times smaller
   d. 2 times smaller
8. Improved resolution is helpful to astronomers because
   a. they often want to look in detail at small features of an object.
   b. they often want to look at very distant objects.
   c. they often want to look at many objects close together.
   d. all of the above
9. The part of the human eye that acts as the detector is the
   a. retina.
   b. pupil.
   c. lens.
   d. iris.
10. Cameras that use adaptive optics provide higher-spatial-resolution images primarily because
    a. they operate above Earth's atmosphere.
    b. deformable mirrors are used to correct the blurring due to Earth's atmosphere.
    c. composite lenses correct for chromatic aberration.
    d. they simulate a much larger telescope.
11. The advantage of an interferometer is that
    a. the resolution is dramatically improved.
    b. the focal length is dramatically increased.
    c. the light-gathering power is dramatically increased.
    d. diffraction effects are dramatically decreased.
    e. chromatic aberration is dramatically decreased.
12. The angular resolution of a ground-based telescope is usually determined by
    a. diffraction.
    b. the focal length.
    c. refraction.
    d. atmospheric seeing.
13. A grating can spread white light out into a spectrum of colors because of the property of
    a. reflection.
    b. interference.
    c. dispersion.
    d. diffraction.

14. Why would astronomers put telescopes in airplanes?
    a. to get the telescopes closer to the stars
    b. to get the telescopes above most of the water vapor in Earth's atmosphere
    c. to be able to observe one object for more than 24 hours without stopping
    d. to allow the telescopes to observe the full spectrum of light

15. If we could increase the quantum efficiency of the human eye, doing so would
    a. allow humans to see a larger range of wavelengths.
    b. allow humans to see better at night or in other low-light conditions.
    c. increase the resolution of the human eye.
    d. decrease the resolution of the human eye.

Thinking about the Concepts

16. Galileo's telescope used simple lenses. What is the primary disadvantage of using a simple lens in a refracting telescope?

17. The largest astronomical refractor has an aperture of 1 meter. List several reasons why building a larger refractor with twice that aperture would be impractical.

18. Your camera may have a zoom lens, ranging between wide angle (short focal length) and telephoto (long focal length). How does the size of an object in the camera's focal plane differ between wide angle and telephoto?

19. Optical telescopes reveal much about the nature of astronomical objects. Why do astronomers also need information provided by gamma-ray, X-ray, infrared, and radio telescopes?

20. For light reflecting from a flat surface, the angles of incidence and reflection are the same. That is also true for light reflecting from the curved surface of a reflecting telescope's primary mirror. Sketch a curved mirror and several of those reflecting rays.

21. Consider two optically perfect telescopes having different diameters but the same focal length. Is the image of a star larger or smaller in the focal plane of the larger telescope? Explain your answer.

22. Study the Process of Science Figure. Make a flowchart for the symbiosis between technology and science that led to the detection of gravitational waves.

23. Explain adaptive optics and how they improve a telescope's image quality.

24. Explain integration time and quantum efficiency and how each contributes to the detection of faint astronomical objects.

25. Some people believe that we put astronomical telescopes on high mountaintops or in orbit because doing so gets them closer to the objects they are observing. Explain what is wrong with that popular misconception, and give the actual reason telescopes are located in those places.

26. Humans have sent various kinds of spacecraft—including flybys, orbiters, and landers—to all the planets in our Solar System. Explain the advantages and disadvantages of each of those types of spacecraft.

27. If Earth has meteorites that are pieces of Mars, why is going to Mars and bringing back samples of the martian surface so important?

28. Humans had a first look at the far side of the Moon as recently as 1959. Why had we not seen it earlier—when Galileo first observed the Moon with his telescope in 1610?

29. Where are neutrino detectors located? Why are neutrinos so difficult to detect?

30. Why do telescopes in space give a better picture of the leftover radiation from the Big Bang?

Applying the Concepts

31. Compare the light-gathering power of the Thirty Meter Telescope with that of the dark-adapted human eye (aperture, 8 mm) and with that of one of the 10-meter Keck telescopes.

32. Study the photograph of light entering and leaving a block of refractive material in Figure 6.2b. Use a protractor to measure the angles of the green light as it enters the block and as it leaves the block. How are those angles related?

33. Many amateur astronomers start out with a 4-inch (aperture) telescope and then graduate to a 16-inch telescope. By what factor does the light-gathering power of the telescope increase with that upgrade? How much fainter are the faintest stars that can be seen in the larger telescope?

34. The resolution of the human eye is about 1.5 arcmin. What would the aperture of a radio telescope (observing at 21 cm) have to be to have that resolution? Even though the atmosphere is transparent at radio wavelengths, humans do not see light in the radio range. Using your calculations and logic, explain why.

35. Assume that you have a telescope with an aperture of 1 meter. Compare the telescope's theoretical resolution when you are observing in the near-infrared region of the spectrum ($\lambda = 1{,}000$ nm) with that when you are observing in the violet region of the spectrum ($\lambda = 400$ nm).

36. Assume that the maximum aperture of the human eye, D, is approximately 8 mm and the average wavelength of visible light, λ, is 5.5×10^{-4} mm.
    a. Calculate the diffraction limit of the human eye in visible light.
    b. How does the diffraction limit compare with the actual resolution of 1–2 arcmin (60–120 arcsec)?
    c. To what do you attribute the difference?

37. The diameter of the full Moon in the focal plane of an average amateur's telescope (focal length, 1.5 meters) is 13.8 mm. How big would the Moon be in the focal plane of a very large astronomical telescope (focal length, 250 meters)?

38. One of the earliest astronomical CCDs had 160,000 pixels, each recording 8 bits (256 levels of brightness). A new generation of astronomical CCDs may contain a billion pixels, each recording 15 bits (32,768 levels of brightness). Compare the number of bits of data that each of those two CCD types produces in a single image.

39. Consider a CCD with a quantum efficiency of 80 percent and a photographic plate with a quantum efficiency of 1 percent. If an exposure time of 1 hour is required to photograph a celestial object with a given telescope, how much observing time would be saved by substituting a CCD for the photographic plate?

40. The VLBA uses an array of radio telescopes ranging across 8,000 km of Earth's surface from the Virgin Islands to Hawaii.
    a. Calculate the angular resolution of the array when radio astronomers are observing interstellar water molecules at a microwave wavelength of 1.35 cm.
    b. How does that resolution compare with the angular resolution of two large optical telescopes separated by 100 meters and operating as an interferometer at a visible wavelength of 550 nm?

41. When operational, the Space VLBI may have a baseline of 100,000 km. What will be the angular resolution when studying interstellar molecules emitting at a wavelength of 17 mm from a distant galaxy?

42. The *Mars Reconnaissance Orbiter* (*MRO*) flies at an average altitude of 280 km above the martian surface. If its cameras have an angular resolution of 0.2 arcsec, what is the size of the smallest objects that the *MRO* can detect on the martian surface?

43. At this writing, *Voyager 1* is about 140 astronomical units (AU) from Earth, continuing to record its environment as it departed the boundary of our Solar System.
    a. How far away is *Voyager 1*, in kilometers?
    b. How long do observational data take to come back to us from *Voyager 1*?
    c. How does *Voyager 1*'s distance from Earth compare with that of the nearest star (other than the Sun)?

44. Gravitational waves travel at the speed of light. Their speed, wavelength, and frequency are related as $c = \lambda \times f$. If we were to observe a gravitational wave from a distant cosmic event with a frequency of 10 hertz (Hz), what would be the wavelength of the gravitational wave?

45. Compute the peak of the blackbody spectrum with a temperature of 2.73 K. What region of the spectrum is this?

USING THE WEB

46. A webcast for the International Year of Astronomy 2009 called "Around the World in 80 Telescopes" can be accessed at http://eso.org/public/events/special-evt/100ha.html. The 80 telescopes are situated all over, including Antarctica and space. Pick two of the telescopes and watch the videos. Do you think those videos are effective for public outreach for the observatory in question? For astronomy in general? For each telescope you choose, answer the following questions: Does the telescope observe in the Northern Hemisphere or the Southern Hemisphere? What wavelengths does the telescope observe? What are some of the key science projects at the telescope?

47. Most major observatories have their own websites. Use the link in Question 46 to find a master list of telescopes, and click on a telescope name to link to an observatory website (or run a search on names from Tables 6.1 and 6.2). For the telescope you choose, answer the following questions: What is that telescope's "claim to fame"—is it the largest? At the highest altitude? At the driest location? With the darkest skies? The newest? Does the observatory website have news releases? What is a recent discovery from that telescope?

48. Go to the website for the International Dark Sky Association (http://darksky.org/). Watch the short "Losing the Dark" film (available under the "Resources" tab). Is a "dark sky place" located near you? What are the ecological arguments against too much light at night?

49. What is the current status of the James Webb Space Telescope (https://jwst.nasa.gov)? How is it different from the Hubble Space Telescope? What are some of the instruments for the JWST and its planned projects? What is the current estimated cost of the JWST?

50. Pick a mission from Appendix 5, go to its website, and see what's new. For the mission you choose, answer the following questions: Is the spacecraft still active? Is it sending images? What new science is coming from this mission?

EXPLORATION

Geometric Optics and Lenses

digital.wwnorton.com/astro6

Visit the Digital Resources Page and on the Student Site open the "Geometric Optics and Lenses" AstroTour in Chapter 6. Read through the animation until you reach the optics simulation, pictured in **Figure 6.34**. The simulator shows a converging lens and a pencil. Rays come from the pencil on the left of the converging lens, pass through the lens, and make an image to the right of the lens. The view that would be seen by an observer at the position of the eye is shown in the circle at upper right. Initially, when the pencil is at position 2.3 and the eye is at position 2.0, the pencil is out of focus and blurry.

1 Is the eraser at the top or the bottom of the actual pencil? (That answer becomes important later.)

Using the red slider in the upper left of the window, try moving the pencil to the right. Pause when the observer's eye sees a recognizable pencil (even if it's still blurry).

2 Does the eye see the pencil right side up or upside down?

3 This is somewhat analogous to the view through a telescope. The objects are very far from the lenses, and the observer sees things upside down in the telescope. If an object in your field of view is at the top of the field and you want it in the center, should you move the telescope up or down?

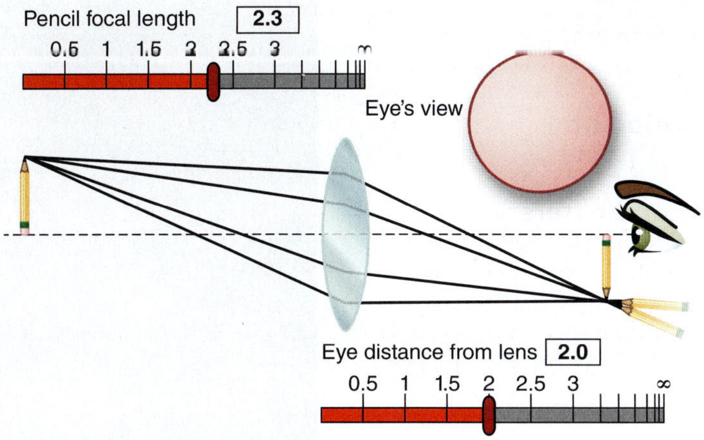

Figure 6.34 Use this simulation to change the position of the object of the eye to explore what will be seen for various configurations.

Now return the pencil to position 2.3. Use the red slider in the lower right of the window to move the eye closer to the lens (to the left).

4 At what distance does the image of the pencil first become crisp and clear?

5 Is the pencil right side up or upside down?

6 In practice at the telescope, we do not move the observer back (away from the eyepiece) to bring the image into focus. Why not?

7 Instead of moving the observer, we use a focusing knob to move the lens in the eyepiece, which brings the image into focus. Imagine that you are looking through the eyepiece of a telescope and the image is blurry. You turn the focusing knob and things get blurrier! What should you try next?

8 Now imagine that you get the image focused just right, so it is crisp and sharp. The next person to use the telescope wears glasses and insists that the image is blurry. When you look through the telescope again, though, the image is still crisp. Explain why your experiences differ.

Step through the animation to the next picture. Carefully study the two telescopes shown and the path the light takes through them.

9 Which telescope has a longer focal length: the top one or the bottom one?

10 Which telescope produces an image with the red and the blue stars more separated: the top one or the bottom one?

11 A longer focal length is an advantage in one sense, but it's not the entire story. What are some disadvantages of a telescope with a very long focal length?

7 The Formation of Planetary Systems

The planetary system containing Earth—our Solar System—is a by-product of the birth of the Sun, but the physical processes that shaped the formation of the Solar System are not unique to it. The same processes have formed many other multiplanet systems. Here in Chapter 7, we examine how planetary systems are born and evolve.

LEARNING GOALS

By the end of this chapter, you should be able to:

LG 1 Describe how our understanding of planetary system formation developed from the work of both planetary and stellar scientists.

LG 2 Discuss the role of gravity and angular momentum in explaining why planets orbit the Sun in a plane and why they revolve in the same direction that the Sun rotates.

LG 3 Explain how temperature at different locations in the protoplanetary disk affects the composition of planets, moons, and other bodies.

LG 4 Discuss the processes that resulted in the formation of planets and other objects in our Solar System.

LG 5 Describe how astronomers find planets around other stars and derive exoplanet properties.

From clouds of gas and dust, planetary systems are born. ▶▶▶

How did our Solar System form?

7.1 Planetary Systems Form around a Star

Earth is part of a collection of **planets**—large, round isolated bodies that orbit a star. Astronomers call a system of planets surrounding a star a **planetary system**. The Solar System, shown in **Figure 7.1**, is the planetary system that includes Earth, seven other planets, and the Sun. The system also includes moons that orbit planets and small bodies that occupy particular regions of the Solar System, such as the asteroid belt or the Kuiper Belt. Our Solar System is a tiny part of our galaxy, which is a tiny part of the universe. Review Figure 1.3 to remind yourself of the size scales involved. Light takes about 4 hours to travel to Earth from Neptune, the outermost known planet in the Solar System, but light from the most distant galaxies has taken more than 13 *billion* years to reach Earth.

Until the latter part of the 20th century, the origin of the Solar System remained speculative. Over the past century, with the aid of spectroscopy, astronomers have determined that the Sun is an ordinary star, one of hundreds of billions in its galaxy, the Milky Way, and that the Milky Way is an ordinary galaxy, one of hundreds of billions in the universe. In the past few decades, stellar astronomers studying the formation of stars and planetary scientists analyzing clues about the history of the Solar System have arrived at the same picture of the early Solar System—but from two very different directions. That unified understanding provides the foundation for the way astronomers now think about the Sun and the objects that orbit it. In this section, we look at how the work of stellar and planetary scientists converged to inform our understanding of planetary system formation.

AstroTour: Solar System Formation

The Nebular Hypothesis

The first plausible theory for the formation of the Solar System, the **nebular hypothesis**, was proposed in 1755 by the German philosopher Immanuel Kant (1724–1804) and conceived independently in 1796 by the French astronomer Pierre-Simon Laplace (1749–1827). Kant and Laplace argued that a rotating cloud of interstellar gas, or **nebula** (Latin for "cloud"), gradually collapsed and flattened to form a disk with the Sun at its center. Surrounding the Sun were rings of material from which the planets formed. That configuration would explain why the planets orbit the Sun in the same direction in the same plane. The nebular hypothesis remained popular throughout the 19th century, and those basic principles of the hypothesis are still retained today.

Our modern theory of planetary system formation calculates the conditions required for a cloud of interstellar gas to collapse under the force of their own self-gravity to form stars. Recall from Chapter 4 that self-gravity is the gravitational attraction between the parts of an object, such as a planet or star, that pulls all the parts toward the object's center. That inward force is opposed by either structural strength (for rocks that make up terrestrial planets) or the outward force resulting from gas pressure and radiation pressure within a star. If the outward force is less than self-gravity, the object contracts; if it is greater,

Figure 7.1 Our Solar System includes planets, moons, and other small bodies. Sizes and distances shown are not to scale.

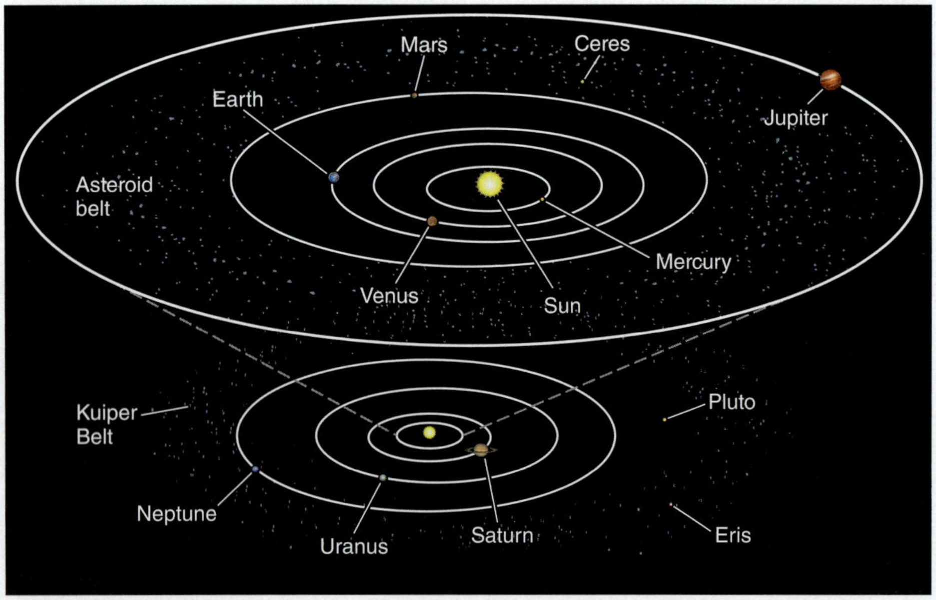

the object expands. In a stable object, the inward and outward forces are balanced.

In support of the nebular hypothesis, disks of gas and dust have been observed surrounding young stellar objects (**Figure 7.2**). From that observational evidence, stellar astronomers have shown that, much like a spinning ball of pizza dough spreads out to form a flat crust, the cloud that produces a star—the Sun, for example—collapses first into a rotating disk. Material in the disk eventually suffers one of three fates: it travels inward onto the forming star at its center, it remains in the disk itself to form planets and other objects, or it is ejected back into interstellar space.

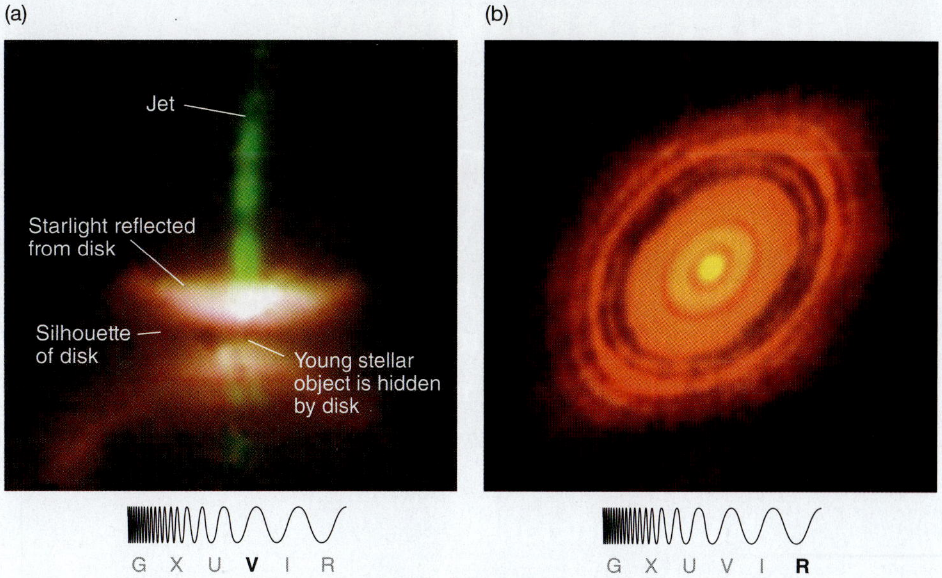

Figure 7.2 (a) Hubble Space Telescope image of a disk around newly formed stars. The dark band is the silhouette of the disk seen edge on. Bright regions are dust illuminated by the star's light. The jets are shown in green. (b) The Atacama Large Millimeter/submillimeter Array (ALMA) obtained this image of a protoplanetary disk around the star HL Tau. This image shows substructures and possibly planets in the system's dark patches.

Planetary Scientists and the Convergence of Evidence

While astronomers were working to understand star formation, other groups of scientists with very different backgrounds were piecing together the history of the Solar System. Planetary scientists, geochemists, and geologists looking at the current structure of the Solar System inferred what some of its early characteristics must have been. The orbits of all the planets lie very close to a single plane, so the early Solar System must have been flat. In addition, all the planets orbit the Sun in the same direction, so the material from which the planets formed must have been orbiting the Sun in the same direction as well.

To find out more, scientists study samples of the very early Solar System. Rocks that fall to Earth from space, known as **meteorites**, include pieces of material left over from the Solar System's youth. Many meteorites, such as the one in **Figure 7.3**, resemble a piece of concrete in which pebbles and sand are mixed with a much finer filler, suggesting that the larger bodies in the Solar System must have grown from the aggregation of smaller bodies. That finding suggests an early Solar System in which the young Sun was surrounded by a flattened disk of both gaseous and solid material. Our Solar System formed from that swirling disk of gas and dust.

As astronomers and planetary scientists compared notes, they realized they had arrived at the same picture of the early Solar System from two completely different directions. The rotating disk from which the planets formed was the remains of the disk that had accompanied the formation of the Sun. Earth, along with all the other orbiting bodies that make up the Solar System, formed from the remnants of an *interstellar cloud* that collapsed to form the local star, the Sun. The connection between the formation of stars and the origin and later evolution of the Solar System is one of the cornerstones of both astronomy and planetary science—a central theme of our understanding of our Solar System (see the **Process of Science Figure**).

Figure 7.3 Meteorites are the surviving pieces of Solar System fragments that land on planets. This meteorite formed from many smaller components that stuck together.

CHECK YOUR UNDERSTANDING 7.1

Which of the following pieces of evidence support the nebular hypothesis? (Choose all that apply.) (a) Planets orbit the Sun in the same direction. (b) The Solar System is relatively flat. (c) Earth has a large Moon. (d) We observe disks of gas and dust around other stars.

PROCESS OF SCIENCE

Converging Lines of Inquiry

Why is the Solar System a disk, with all planets orbiting in the same direction? Scientists from different disciplines often contribute to the solution to a problem. These scientists approach the problem from different directions and perspectives, so when they arrive at the same conclusion, that is compelling evidence that they are all on the right track.

Mathematicians:

- Suggested the nebular hypothesis—that a collapsing rotating cloud formed the Solar System.

Stellar Astronomers:

- Tested the nebular hypothesis, seeking evidence for or against.
- Found dust and gas around young stars.
- Observed this gas and dust to be in the shape of disks.

Planetary Scientists:

- Tested the nebular hypothesis, seeking evidence for or against.
- Studied meteorites, which showed that the planets formed from many smaller bodies.

Beginning from the same fundamental observations about the shape of the Solar System, theorists (the mathematicians), stellar astronomers, and planetary scientists **converged on the nebular theory** that stars and planets form together from a collapsing cloud of gas and dust.

7.2 The Solar System Began with a Disk

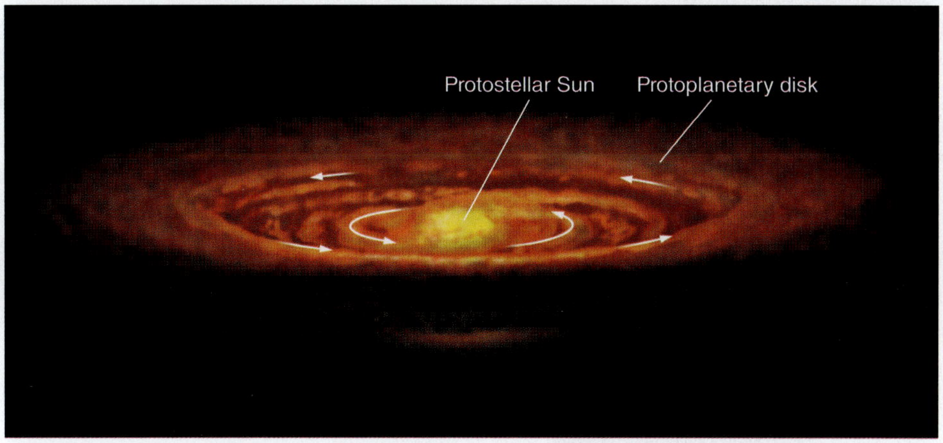

Figure 7.4 Think of the young Sun as being surrounded by a flat, rotating disk of gas and dust that was flared at its outer edge.

Planets form in a disk around young stars, but what are some of the specifics of the process? **Figure 7.4** illustrates the young Solar System as it appeared roughly 5 billion years ago. At that time, the Sun was still a **protostar**—a large ball of gas but not yet hot enough in its center to be a star. As the cloud of interstellar gas collapsed to form the protostar, its gravitational energy was converted into heat energy and radiation. Surrounding the protostellar Sun was a flat, orbiting disk of gas and dust. Each bit of the material in that thin disk orbited the Sun in accordance with the same laws of motion and gravitation that govern the orbits of the planets. The disk around the Sun—like the disks that astronomers see today surrounding protostars elsewhere in our galaxy—is called a **protoplanetary disk**. The disk probably contained less than 1 percent of the mass of the star forming at its center, but that amount was more than enough to account for the bodies that make up the Solar System today.

The Collapsing Cloud and Angular Momentum

Astronomy in Action: Angular Momentum

Conservation of *angular momentum* causes protoplanetary disks to form. **Angular momentum** is a conserved quantity associated with a revolving or rotating system and depends on both the velocity and distribution of the system's mass. The angular momentum of an isolated system is always conserved; that is, it remains unchanged unless acted on by an external force. A figure-skater spinning on the ice (**Figure 7.5**), like any other rotating object, has some amount of angular momentum. Unless frictional forces act to reduce her angular momentum, she will always have the same amount of angular momentum.

The amount of angular momentum depends on three factors:

1. How fast the object is rotating. The faster an object is rotating, the more angular momentum it has.
2. The mass of the object. If a similarly sized bowling ball and a basketball are spinning at the same speed, the bowling ball has more angular momentum because it has more mass.
3. How the mass of the object is distributed relative to the spin axis—that is, how spread out the object is. For an object of a given mass and rate of rotation, the more spread out it is, the more angular momentum it has. A spread-out object rotating slowly might have the same angular momentum as a compact object rotating rapidly.

Figure 7.5 A figure-skater relies on the principle of conservation of angular momentum to change the speed at which she spins.

Both an ice-skater and a collapsing interstellar cloud are subject to the **conservation of angular momentum**: the angular momentum must remain the same in the absence of an external force. For angular momentum to be conserved, a change in one of the above quantities (the rate of spin, mass, or distribution of mass) must be accompanied by a change in another quantity. Because an ice-skater's mass doesn't change, for example, she can control how rapidly she spins

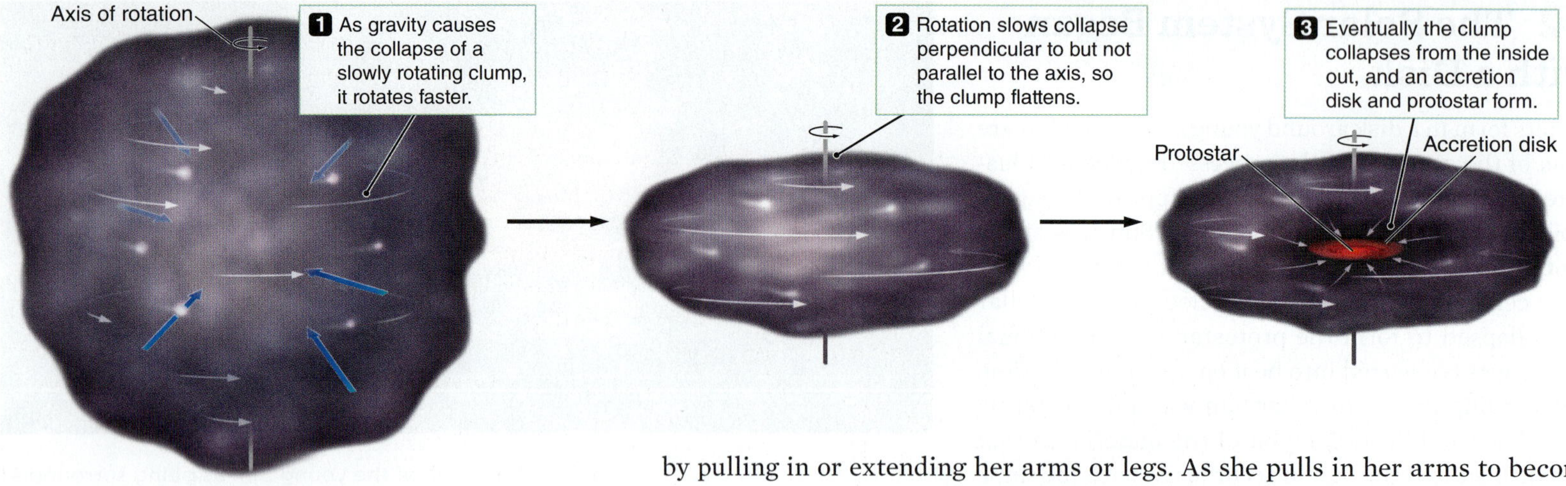

Figure 7.6 A rotating interstellar cloud collapses in a direction parallel to its axis of rotation, thus forming an accretion disk.

by pulling in or extending her arms or legs. As she pulls in her arms to become more compact, she changes her distribution of mass and must spin faster to maintain the same angular momentum. When her arms are held tightly in front of her and one leg is wrapped around the other, the skater's spin becomes a blur. She finishes with a flourish by throwing her arms and leg out—an action that abruptly slows her spin by spreading out her mass. The skater's angular momentum remains constant throughout the maneuver. Similarly—as shown in **Figure 7.6**—the cloud that formed our Sun rotated faster and faster as it collapsed, just as the ice-skater speeds up when she pulls in her arms.

That description, however, presents a puzzle. Suppose the Sun formed from a typical cloud—one about a light-year across and rotating so slowly that completing one rotation took a million years. By the time the cloud collapsed to the size of the Sun today, it would have been spinning so fast that one rotation would occur every 0.6 second. That rate is more than 3 million times faster than our Sun actually spins. At that rate of rotation, the Sun would tear itself apart. It *appears* that angular momentum was *not* conserved in the actual formation of the Sun—but that can't be right because angular momentum *must* be conserved. We must be missing something. Where did the angular momentum go?

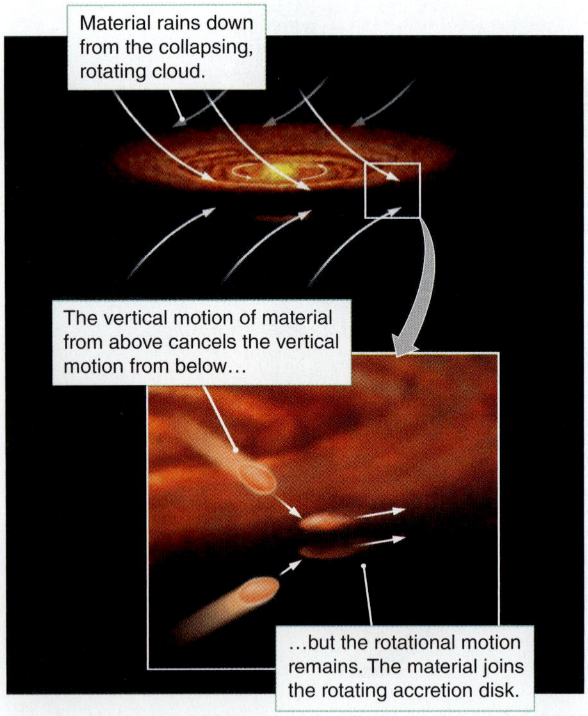

Figure 7.7 Gas from a rotating cloud falls inward from opposite sides, piling up onto a rotating disk.

The Formation of an Accretion Disk

To understand how angular momentum is conserved in disk formation, we must think in three dimensions. Imagine that the ice-skater bends her knees, compressing herself downward instead of bringing her arms toward her body. As she does so, she again makes herself less spread out, but her rate of spin does not change because no part of her body has become any closer to the axis of spin. Similarly, as shown in Figure 7.6, a clump of a molecular cloud can flatten out without speeding up by collapsing parallel to its axis of rotation. Instead of collapsing into a ball, the interstellar cloud flattens into a disk. As the cloud collapses, its self-gravity increases and the inner parts begin to fall freely inward, raining down on the growing object at the center. The outer portions of the cloud lose the support of the collapsed inner portion, and they start falling inward, too. As that material makes its final inward plunge, it lands on a thin, rotating disk—called an **accretion disk**—that forms from the accretion of material around a massive object.

The formation of accretion disks is common in the universe. The process is shown in **Figure 7.7**. As material falls onto the disk, its motion perpendicular to the disk stops abruptly, but the motion of its mass *parallel* to the surface of the

7.1 Working It Out Angular Momentum

In its simplest form, the angular momentum (L) of a system is given by

$$L = m \times v \times r$$

where m is the mass, v is the speed at which the mass is moving, and r represents how spread out the mass is.

Let's apply that relationship to the angular momentum of Jupiter in its orbit about the Sun. The angular momentum from one body orbiting another is called *orbital* angular momentum, $L_{orbital}$. The mass (m) of Jupiter is 1.90×10^{27} kilograms (kg), the speed of Jupiter in orbit (v) is 1.31×10^4 meters per second (m/s), and the radius of Jupiter's orbit (r) is 7.79×10^{11} meters. Putting all that together gives

$$L_{orbital} = (1.90 \times 10^{27}\ \text{kg}) \times (1.31 \times 10^4\ \text{m/s}) \times (7.79 \times 10^{11}\ \text{m})$$

$$L_{orbital} = 1.94 \times 10^{43}\ \text{kg m}^2/\text{s}$$

Calculating the *spin* angular momentum of a spinning object, such as a skater, a planet, a star, or an interstellar cloud, is more complicated. Here, we must add up the individual angular momenta of *every tiny mass element* within the object. For a uniform sphere, the spin angular momentum is

$$L_{spin} = \frac{4\pi mR^2}{5P}$$

where R is the radius of the sphere and P is the rotation period of its spin.

Let's compare Jupiter's orbital angular momentum with the Sun's spin angular momentum to investigate the distribution of angular momentum in the Solar System. Appendix 2 provides the Sun's radius (6.96×10^8 meters) and mass (1.99×10^{30} kg), and its rotation period is 24.5 days = 2.12×10^6 seconds. If we assume that the Sun is a uniform sphere, the spin angular momentum of the Sun is

$$L_{spin} = \frac{4 \times \pi \times (1.99 \times 10^{30}\ \text{kg}) \times (6.96 \times 10^8\ \text{m})^2}{5 \times (2.12 \times 10^6\ \text{s})}$$

$$L_{spin} = 1.14 \times 10^{42}\ \text{kg m}^2/\text{s}$$

$L_{orbital}$ of Jupiter is about 17 times greater than L_{spin} of the Sun. Thus, most of the angular momentum of the Solar System now resides in the orbits of its major planets.

For a collapsing sphere to conserve L_{spin}, its rotation period P must be proportional to R^2. As with the skater, when a sphere decreases in radius, its rotation period decreases; that is, it spins faster.

disk adds to the disk's total angular momentum. Thus, the angular momentum of the infalling material is transferred to the accretion disk. The rotating accretion disk has a radius of hundreds of astronomical units, and that is *thousands* of times greater than the radius of the star that will eventually form at its center. Therefore, most of the angular momentum in the original interstellar cloud ends up in the accretion disk rather than in the central protostar (see **Working It Out 7.1** for an example of the relevant calculation).

AstroTour Traffic Circle Analogy

Most of the matter that lands on the accretion disk either becomes part of the star or is ejected back into interstellar space in the form of jets or other outflows (Figure 7.2a). Those jets are bipolar—they come in pairs aligned along an axis. Material swirling in the bipolar jets carries angular momentum away from the accretion disk in the general direction of the poles of the rotation axis. However, a small amount of material is left behind in the disk. The objects in that leftover disk—the dregs of the process of star formation—form planets and other objects that orbit the star.

Formation of Large Objects

Random motions of the gas within the protoplanetary disk eventually push the smaller grains of solid material toward larger grains. As that happens, the smaller grains stick to the larger grains. The "sticking" process among smaller grains is due to the same static electricity that causes dust bunnies to grow under your bed.

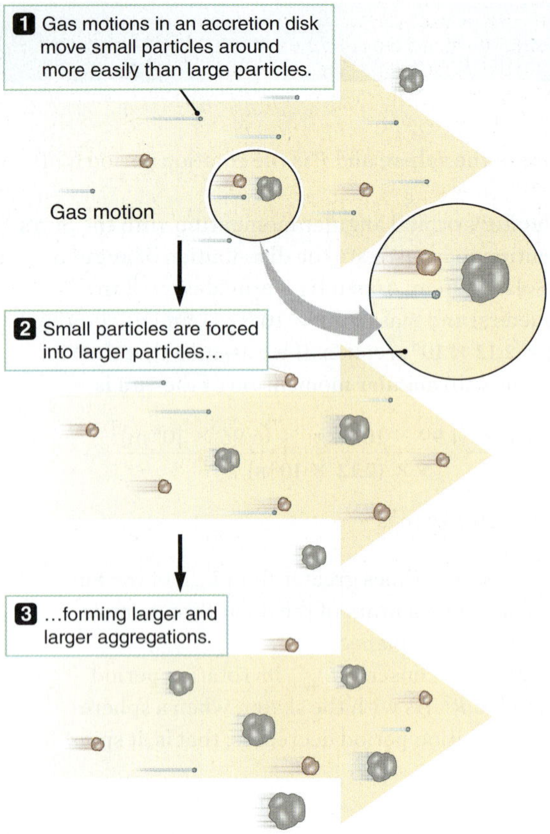

Figure 7.8 Motions of gas in a protoplanetary disk blow smaller particles of dust into larger particles, making the larger particles larger still. That process continues, eventually creating objects many meters in size.

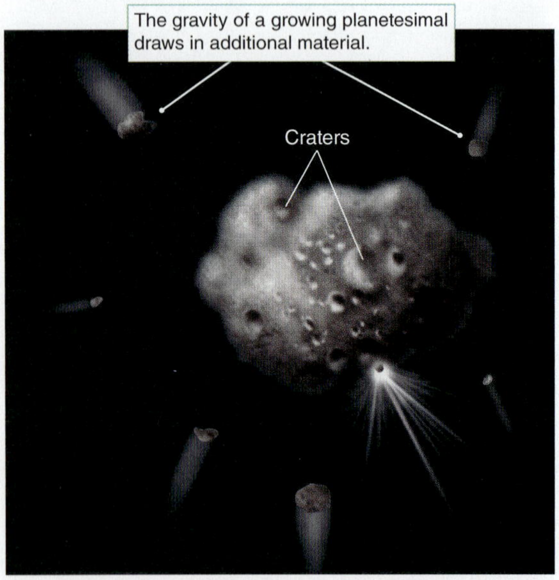

Figure 7.9 The gravity of a planetesimal is strong enough to attract surrounding material, which causes the planetesimal to grow more rapidly.

Starting out at only a few microns (μm) across—about the size of particles in smoke—the slightly larger bits of dust grow to the size of pebbles and then to clumps the size of boulders, which are not as easily pushed around by gas (**Figure 7.8**). When clumps grow to about 100 meters across, the objects are so far apart that they collide less often, and their growth rate slows down but does not stop. Within a protoplanetary disk, the larger dust grains become larger at the expense of the smaller grains.

For two large clumps to stick together rather than explode into many small pieces, they must bump into each other very gently: collision speeds must be about 0.1 m/s or less. If your stride is about a meter long, a collision speed of 0.1 m/s would correspond to only one step every 10 seconds. The process does not yield larger and larger bodies with every collision. When violent collisions occur in an accretion disk, larger clumps break back into smaller pieces. But over a long time, large bodies do form.

Objects continue to grow by "sweeping up" smaller objects that get in their way. Those objects can eventually measure up to several hundred meters across. As the clumps reach the size of about a kilometer, a different process becomes important. Those kilometer-sized objects are massive enough that their gravity pulls on nearby bodies, as shown in **Figure 7.9**. Those bodies of rock and ice are known as **planetesimals** ("tiny planets") and eventually combine with one another to form planets. Besides chance collisions with other objects, a planetesimal's gravity can now pull in and capture small objects outside its direct path. That process speeds up the growth of planetesimals, so larger planetesimals quickly consume most of the remaining bodies near their orbits. The final survivors of that process are large enough to be called planets. As with the major bodies in orbit around the Sun, some of the planets may be relatively small and others quite large.

CHECK YOUR UNDERSTANDING 7.2

Where does most of the angular momentum of the original cloud go? (a) into the orbital angular momentum of planets; (b) into the star; (c) into the spin of the planets; (d) lost along the jets from the star

7.3 The Inner Disk and Outer Disk Formed at Different Temperatures

In the Solar System, the inner planets are small and mostly rocky, whereas the outer planets are very large and mostly gaseous. That distinct difference between the inner and outer Solar System can be explained by how the local disk environment affects the formation process. In this section, we examine these differences.

Energy in the Disk

The accretion disks surrounding young stars form from interstellar material that may have a temperature of only a few kelvins, but the disks themselves reach temperatures of hundreds of kelvins or more. Astronomers want to understand what heats up the disk around a forming star so that we can calculate how hot those disks get.

The law of **conservation of energy** states that unless energy is added to or taken away from a system from the outside, the total amount of energy in the system must remain constant. The form the energy takes, however, can change. With that concept in mind, why does gas falling on the disk make the disk hot? Imagine you are working against gravity by lifting a heavy object, such as a bowling ball. Lifting the bowling ball takes energy, and the law of conservation of energy states that energy is never lost. Where does that energy go? The energy is stored and changed into a form called **gravitational potential energy**. If you drop the bowling ball, it falls, and as it falls, it speeds up. The gravitational potential energy that was stored is converted to energy of motion, which is called **kinetic energy**. When the bowling ball hits the floor, it stops suddenly. The bowling ball loses its energy of motion, so what form does that energy take now? Some of the energy is converted into the sound the bowling ball makes when it hits the floor, too, and some goes into heating and distorting the floor. But much of the energy is converted into thermal energy. The atoms and molecules that make up the bowling ball are moving around within the bowling ball a bit faster than they were before the bowling ball hit, so the bowling ball and its surroundings, including the floor, grow a tiny bit warmer. Similarly, as gas falls toward the disk surrounding a protostar, gravitational potential energy is converted first to kinetic energy, causing the gas to pick up speed. When the gas hits the disk and stops suddenly, that kinetic energy is turned into thermal energy.

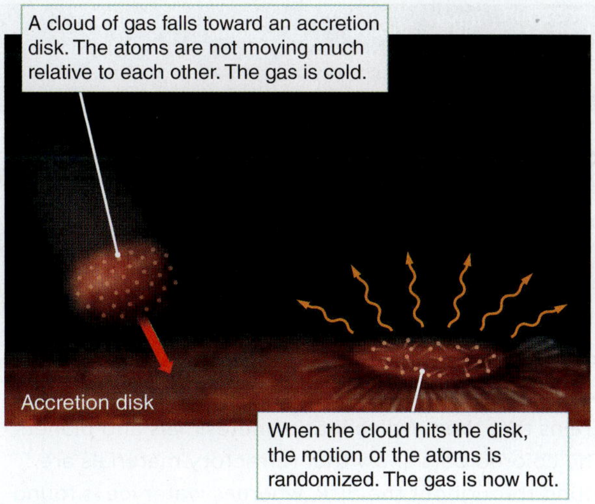

Figure 7.10 Atoms in a gas fall together until they hit the accretion disk, at which point their motions become randomized, raising the temperature of the gas.

Material falling onto the accretion disk around a forming star causes the disk to heat up too (**Figure 7.10**). The amount of heating depends on *where* the material hits the disk. Material hitting the inner part of the disk (the *inner disk*) has fallen farther and picked up greater speed within the gravitational field of the forming star than material hitting the disk farther out. Like a brick dropped from a tall building, material striking the inner disk is moving quite rapidly when it hits, so it heats the inner disk to high temperatures. In contrast, material falling onto the outer part of the disk (the *outer disk*) is moving much more slowly, like a brick dropped from just a foot or so above the ground. As a result, the temperature at the outermost parts of the disk is not much higher than that of the original interstellar cloud. Stated another way: material falling onto the inner disk converts more gravitational potential energy into thermal energy than does material falling onto the outer disk.

The energy released as material falls onto the disk is not the only source of thermal energy in the disk. Even before the nuclear reactions that will one day power the new star have ignited, the conversion of gravitational energy into thermal energy drives the temperature at the surface of the protostar to several thousand kelvins, and it drives the luminosity of the huge ball of glowing gas to many times the luminosity of the present-day Sun. For the same reasons that Mercury is hot whereas Pluto is not (see Chapter 5), the radiation streaming outward from the protostar at the center of the disk drives the temperature in the inner parts of the disk even higher, increasing the difference in temperature between the inner and outer parts of the disk.

The Compositions of Planets

Temperature affects whether a material exists in a solid form. On a hot summer day, ice melts and water quickly evaporates; on a cold winter night, water in your breath freezes into tiny ice crystals. Metals and rocky materials, such as iron, **silicates** (minerals containing silicon and oxygen), and carbon, remain solid even at high temperatures. Substances that can withstand high temperatures without melting or being vaporized are called **refractory materials**. Other materials,

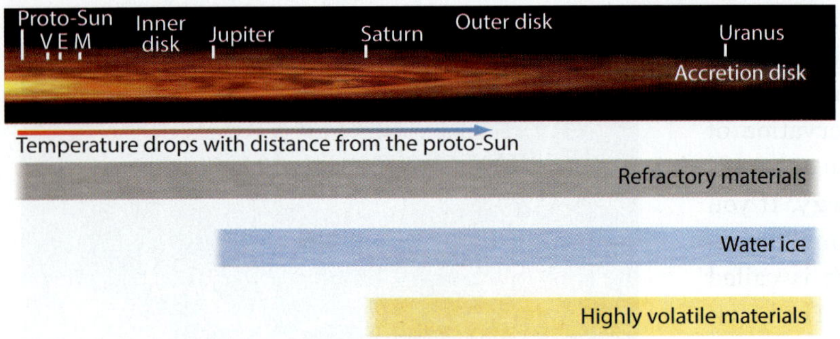

Figure 7.11 Differences in temperature within a protoplanetary disk determine the composition of dust grains that then evolve into planetesimals and planets. The colored bars show that refractory materials are found throughout the disk, whereas water ice is found only outside Jupiter's orbit, and highly volatile materials are found only outside Saturn's orbit. Shown here are the proto-Sun and the orbits of Venus (V), Earth (E), Mars (M), Jupiter, Saturn, and Uranus.

such as water, ammonia, and methane, remain in a solid form only if their temperature is very low. Those materials, which become gases at moderate temperatures, are called **volatile materials** (or *volatiles* for short). Astronomers generally call the solid form of any volatile material an **ice**.

Differences in temperature from place to place within the protoplanetary disk significantly affect the makeup of the dust grains in the disk. **Figure 7.11** illustrates that only refractory substances exist in the hottest parts of the disk—the area closest to the protostar. In the inner disk, dust grains are composed almost entirely of refractory materials. Some substances can survive in solid form somewhat farther out, including some hardier volatiles, such as water ice and certain chemical compounds that are **organic** (meaning that they contain molecules with a carbon-hydrogen bond). Those solids add to the materials that make up dust grains. In the coldest, outermost parts of the accretion disk, far from the central protostar, highly volatile components such as methane, ammonia, and carbon monoxide ices and other organic molecules survive only in solid form. The differences in composition of dust grains within the disk are reflected in the composition of the planets formed from that dust. Planets that form closer to the central star tend to be made up mostly of refractory materials, such as rock and metals, but are deficient in volatiles. Planets that form farther from the central star contain not only refractory materials but also large quantities of ices and organic materials.

In the Solar System, the inner planets are composed of rocky material surrounding metallic cores of iron and nickel. Objects in the outer Solar System, including moons, giant planets, and comets, are composed largely of ices of various types. But not all planetary systems are so neatly organized as our Solar System. When planets around other stars were first discovered, they appeared to be very different, with large planets close to their respective stars. Astronomers now think that **chaotic** encounters—in which a small change in the initial state of a system can lead to a large change in the final state of the system—may change the organization of planetary compositions. In a process called **planet migration**, the force of gravity from all the nearby objects can move some planets so that they end up far from the place of their birth. In our Solar System, for example, Uranus and Neptune originally may have formed nearer to the orbits of Jupiter and Saturn but were then driven outward to their current locations by gravitational encounters with Jupiter and Saturn. A planet also can migrate when it gives up some of its orbital angular momentum to the disk material that surrounds it. Such a loss of angular momentum causes the planet to slowly spiral inward toward the central star. Thus, the order of planets in a system can change.

Formation of an Atmosphere

Once a solid planet has formed, it may continue to grow by capturing gas from the protoplanetary disk. To do so, it must act quickly. Young stars and protostars emit fast-moving particles and intense radiation that can quickly disperse the gaseous remains of the accretion disk. Gaseous planets such as Jupiter probably have only about 10 million years to form and to grab whatever gas they can. Because of their strong gravitational fields, more massive young planets can capture more of the hydrogen and helium gas that makes up the bulk of the disk. What follows is much like the formation of a star and protoplanetary disk, but on a smaller scale—namely, gas from a mini accretion disk moves inward and falls onto the planet.

The gas that a planet captures when it forms—primarily hydrogen and helium—is called the planet's **primary atmosphere**. The primary atmosphere of a large planet can be more massive than the solid body, as with Jupiter. Some of the solid material in the mini accretion disk might stay behind to coalesce into larger bodies in much the same way that particles of dust in the protoplanetary disk came together to form planets. The result is a mini "solar system"—a group of moons that orbit about the planet.

A less massive planet also may capture some gas from the protoplanetary disk, only to lose it later. The gravity of small planets may be too weak to hold low-mass gases such as hydrogen or helium. Even if a small planet can gather some hydrogen and helium from its surroundings, that primary atmosphere will not last long. In the inner solar system, the temperatures are higher, so the hydrogen and helium atoms are moving faster than in the outer solar system and will escape from a small planet. The atmosphere that remains around a small planet such as Earth is a **secondary atmosphere**, which forms later in the life of a planet. Volcanism is one important source of a secondary atmosphere because it releases heavier and thus slower-moving gases, such as carbon dioxide, water vapor, and other gases trapped in the planet's interior. In addition, volatile-rich comets that formed in the outer parts of the disk fall inward toward the new star long after its planets have formed, and they sometimes collide with planets. **Comets** are icy planetesimals that survive planetary accretion. They may serve as a significant source of water, organic compounds, and other volatile materials on planets close to the central star.

CHECK YOUR UNDERSTANDING 7.3

In our Solar System, the inner planets are rocky because: (a) the original cloud had more rocky material near the center; (b) warm temperatures in the inner disk caused the inner planetesimals to be formed of only rocky material; (c) the inner disk filled a smaller volume, and so it was denser; (d) the hydrogen and helium atoms were too low mass to remain in the inner disk.

7.4 The Formation of Our Solar System

Nearly 5 billion years ago, the Sun was still a protostar surrounded by a protoplanetary disk of gas and dust. During the next few hundred thousand years, much of the dust in the disk had collected into planetesimals—clumps of rock and metal near the emerging Sun, and aggregates of rock, metal, ice, and organic materials farther from the Sun. In this section, we look at the formation of the types of planets in our own Solar System.

The Terrestrial Planets

Within the inner 5 astronomical units (AU) of the disk, several rock and metal planetesimals quickly grew larger to become the dominant masses in their orbits. With their ever-strengthening gravitational fields, they either captured most of the remaining planetesimals or ejected them from the inner part of the disk. **Figure 7.12** shows some results from a computer simulation of how that might have happened. The dominant planetesimals became planet-sized bodies with masses ranging between 5 percent and 100 percent of Earth's mass. Those dominant planetesimals evolved into the **terrestrial planets**, which are rocky, Earth-like planets. Today, the surviving terrestrial planets are Mercury, Venus,

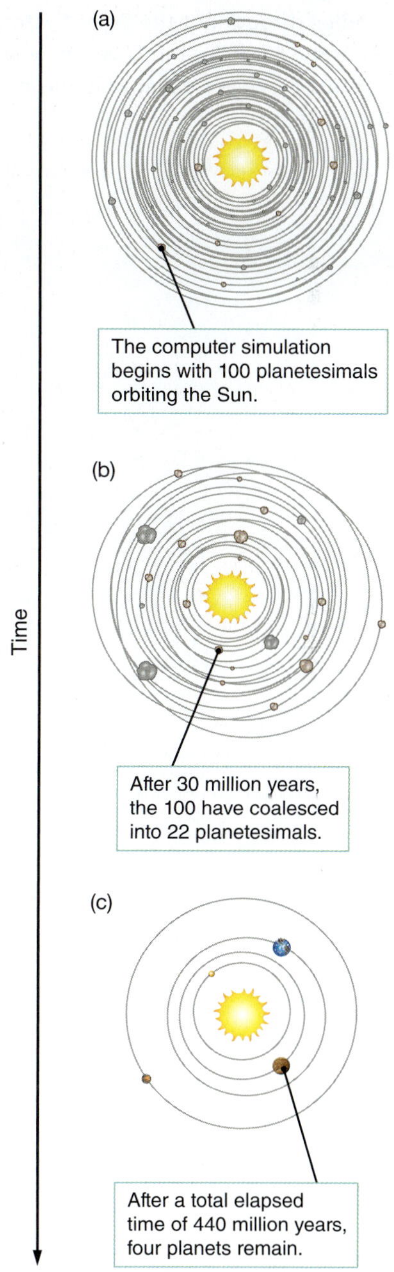

Figure 7.12 Computer models simulate how material in the protoplanetary disk became clumped into the planets over time. Only a few planets remain at the end.

Figure 7.13 Large impact craters on Mercury (and on other solid bodies throughout the Solar System) record the final days of the Solar System's youth, when planets and planetesimals grew as smaller planetesimals rained down on their surfaces.

Earth, and Mars. Earth's Moon is often grouped with those terrestrial planets because of its similar physical and geological properties, even though it is not a planet itself and formed differently. It is possible that one or two other planets or large moons formed in the young Solar System but were later destroyed.

For several hundred million years after the four surviving terrestrial planets formed, leftover pieces of debris still in orbit around the Sun continued to rain down on the surfaces of those planets. Today, we can still see the scars of those early impacts on the cratered surfaces of all the terrestrial planets (**Figure 7.13**). That rain of debris continues even today, but at a much lower rate.

Before the proto-Sun became a true star, gas in the inner part of the protoplanetary disk was still plentiful. During that early period the two larger terrestrial planets, Earth and Venus, may have held on to weak primary atmospheres of hydrogen and helium, but those thin atmospheres were soon lost to space. The terrestrial planets did not develop thick atmospheres until the formation of the secondary atmospheres that now surround Venus, Earth, and Mars. Mercury's size and proximity to the Sun and the Moon's small mass prevented those bodies from retaining significant secondary atmospheres.

The Giant Planets

Beyond 5 AU from the Sun, in a much colder part of the accretion disk, planetesimals combined to form several bodies with masses about 5–20 times that of Earth (5–20 M_{Earth}). Those planet-sized objects formed from planetesimals containing volatile ices and organic compounds in addition to rock and metal. In a process astronomers call *core accretion–gas capture*, mini accretion disks formed around those planetary cores, capturing massive amounts of hydrogen and helium and funneling that material onto the planets. Four such massive bodies became the cores of the **giant planets**—Jupiter, Saturn, Uranus, and Neptune. Those giant planets are many times the mass of any terrestrial planet.

Jupiter's massive solid core captured and retained the most gas—roughly 300 M_{Earth}. The other outer planetary cores captured less hydrogen and helium, perhaps because their cores were less massive or because less gas was available to them. Saturn ended up with less than 100 M_{Earth} of gas, and Uranus and Neptune grabbed less than 20 M_{Earth} of gas.

The core accretion model indicates that a Jupiter-like planet could take up to 10 million years to accumulate. Some planetary scientists think that our protoplanetary disk could not have survived long enough to form gas giants such as Jupiter through the general process of core accretion. All the gas may have dispersed in roughly half that time, cutting off Jupiter's supply of hydrogen and helium. An alternative explanation is a process called *disk instability,* in which the protoplanetary disk suddenly and quickly fragments into massive clumps equivalent to those of a large planet. Both core accretion and disk instability may have played a role in the formation of our own and other planetary systems.

During the formation of the planets, gravitational energy was converted into thermal energy as the individual atoms and molecules moved faster. That conversion warmed the gas surrounding the cores of the giant planets. Proto-Jupiter and proto-Saturn probably became so hot that they glowed a deep red color, similar to the heating element on an electric stove. Their internal temperatures may have been even higher.

In the mini accretion disks surrounding the giant planets, some of the remaining material combined into small bodies, which became moons. A **moon** is any natural satellite in orbit about a planet or asteroid. The composition of the moons that formed around the giant planets followed the same trend as that of the planets that formed around the Sun: the innermost moons formed under the hottest conditions and therefore contained the smallest amounts of volatile material. For example, the closest of Jupiter's many moons may have experienced high temperatures from nearby Jupiter's glowing so intensely that it would have evaporated most of the volatile substances in the inner part of its mini accretion disk.

Remaining Planetesimals

Not all planetesimals in the disk became planets. For example, dwarf planets orbit the Sun but have not cleared other, smaller bodies from their orbits. Ceres and Pluto (Figure 7.1) are dwarf planets. More dwarf planets, along with many smaller bodies, are found in the Kuiper Belt, beyond Pluto's orbit. Asteroids are small bodies found inside Jupiter's orbit around the Sun; most are located in the main asteroid belt between the orbits of Mars and Jupiter. Jupiter's gravity kept the region between Jupiter and Mars so stirred up that most planetesimals there never formed a large planet.

Planetesimals persist to this day in the outermost part of the Solar System as well. Formed in a deep freeze, those objects have retained most of the highly volatile materials found in the grains present when the accretion disk formed. Unlike the crowded inner part of the disk, the outermost parts had planetesimals too sparsely distributed for large planets to grow. Icy planetesimals in the outer Solar System that survived planetary accretion remain today as **comet nuclei**. The frozen, distant dwarf planets Pluto and Eris are especially large examples of those residents of the outer Solar System.

Many Solar System objects show evidence of cataclysmic impacts that reshaped worlds, suggesting that the early Solar System must have been a remarkably violent and chaotic place. The dramatic difference in the terrain of the northern and southern hemispheres on Mars, for example, has been interpreted as the result of one or more colossal collisions. The leading theory for the origin of our Moon is that it resulted from the collision of an object with Earth. Mercury has a crater on its surface from an impact so devastating that it caused the crust to buckle on the opposite side of the planet. In the outer Solar System, one of Saturn's moons, Mimas, has a crater roughly one-third the diameter of the moon itself. Uranus suffered one or more collisions violent enough to knock it on its side. As a result, its equatorial plane is tilted at almost a right angle to its orbital plane. Other examples are discussed in later chapters.

CHECK YOUR UNDERSTANDING 7.4

Suppose that astronomers found a rocky, terrestrial planet beyond the orbit of Neptune. What is the most likely explanation for its origin? (a) It formed close to the Sun and migrated outward. (b) It formed in that location and was not disturbed by migration. (c) It formed later in the Sun's history than other planets. (d) It is a captured planet that formed around another star.

7.5 Planetary Systems Are Common

When astronomers turn their telescopes to young nearby stars, they see disks of the same type from which the Solar System formed. When the light from the central star is blocked (**Figure 7.14**), evidence of the planetary disk is observed. The

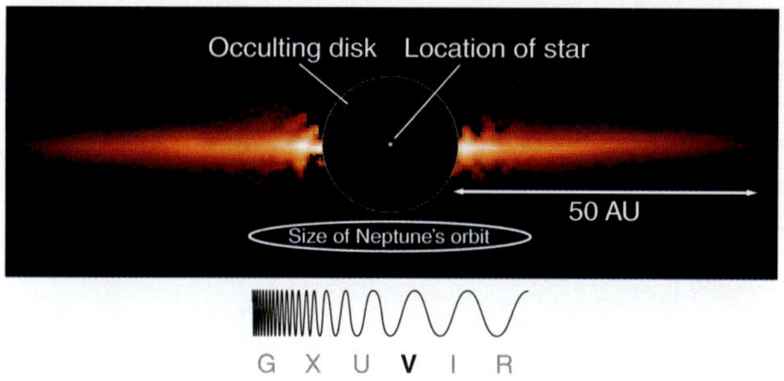

Figure 7.14 An edge-on circumstellar dust disk is seen extending outward to 60 AU from the young (12-million-year-old) star AU Microscopii. The star itself, whose brilliance would otherwise overpower the circumstellar disk, is hidden behind an occulting disk (opaque mask) placed in the telescope's focal plane. The star's position is represented by the dot.

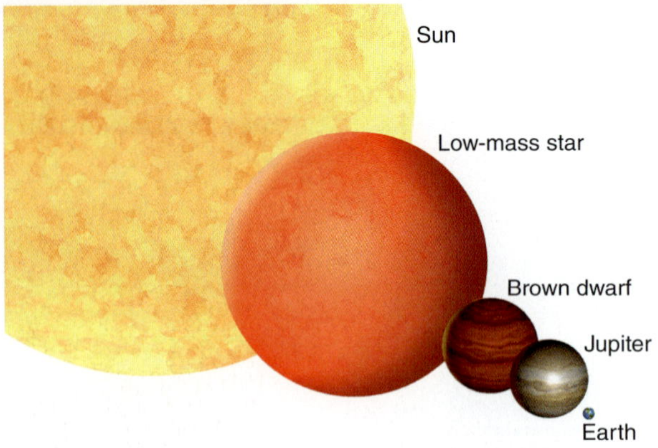

Figure 7.15 A comparison of the diameters of the Sun, a low-mass star, a brown dwarf, Jupiter, and Earth.

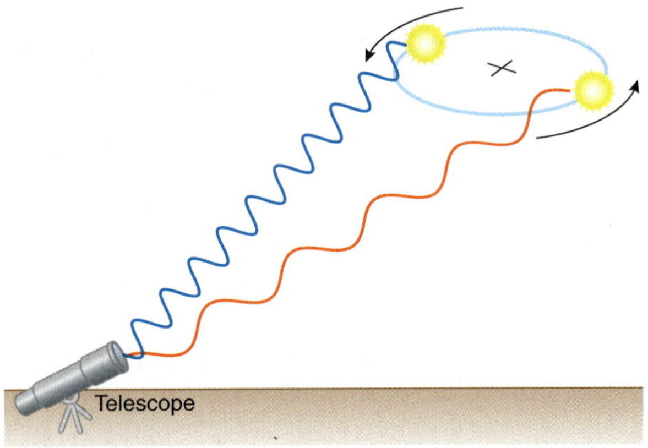

Figure 7.16 Doppler shifts observed in the spectrum of a star as it moves around its common center of gravity with its planet. The spectral lines are blueshifted as the star moves toward us and redshifted as it moves away from us.

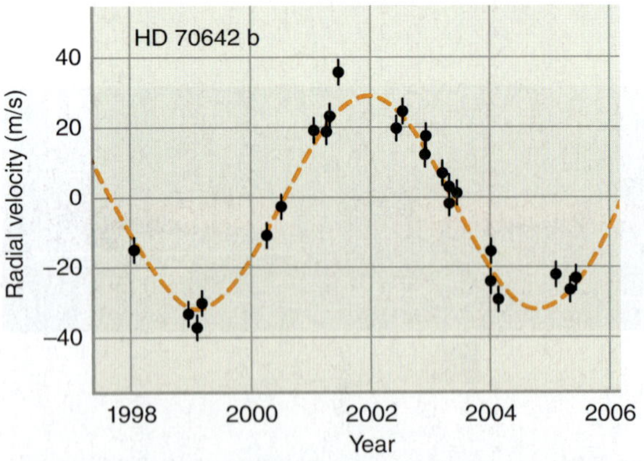

Figure 7.17 Radial velocity data for a star with a planet. A positive velocity is motion away from the observer, whereas a negative velocity is motion toward the observer. The plot repeats as the planet completes another orbit around the star.

physical processes that led to the formation of the Solar System should be commonplace wherever new stars are being born. Compared with stars, however, planets are small and dim objects. They shine primarily by reflection and therefore are millions to billions of times fainter than their host stars. Thus, they were difficult to find until advances in telescope detector technology in the 1990s enabled astronomers to discover them through indirect methods. In 1995, astronomers announced the first confirmed **extrasolar planet**, usually shortened to **exoplanet**—a planet orbiting around a star other than the Sun. Today, the number of known exoplanets has grown to the thousands, and new discoveries occur almost daily.

The IAU defines an exoplanet as an object that orbits a star other than the Sun and has a mass less than 10–13 Jupiter masses (10–13 M_{Jup}). Objects more massive than 10–13 M_{Jup} but less massive than 0.08 solar masses (0.8 M_{Sun}; about 80 M_{Jup}) are **brown dwarfs**. Objects more massive than 0.08 M_{Sun} are defined as stars. **Figure 7.15** compares the diameters of typical objects of each class.

The Search for Exoplanets

The first planets were discovered indirectly, by observing their gravitational tug on the central star. As technology has improved, other methods have become more productive. Astronomers now have direct images of planets orbiting stars and have taken spectra of some exoplanets to observe the composition of their atmospheres. Almost certainly, between the time we write this and the time you read it, new discoveries will have been made. The field is advancing extremely quickly: more than 100 projects are searching for exoplanets from the ground and from space. We will now look at each discovery method.

THE RADIAL VELOCITY METHOD As a planet orbits a star, the planet's gravity tugs the star around ever so slightly. If the star has a radial velocity (Chapter 5), an observable Doppler shift may appear in the spectrum of the star. **Figure 7.16** illustrates that motion. When the star is moving toward us (negative radial velocity), the light is blueshifted; when the star is moving away from us (positive radial velocity), the light is redshifted. That pattern of radial velocity repeats. After detecting those changes in the radial velocity (**Figure 7.17**), astronomers can infer the existence of the planet and find the planet's mass and distance from the star.

The smallest planet that can be found depends on the precision of the observations. For example, in our Solar System, Jupiter's mass is greater than the mass of all the other planets, asteroids, and comets combined, so Jupiter is the planet with the largest effect on the Sun. Both the Sun and Jupiter orbit a common center of gravity (sometimes called center of mass—the location where the effect of one mass balances the other) that lies just outside the surface of the Sun, as shown in **Figure 7.18**. Jupiter tugs the Sun around in a circle with a speed of 12 m/s. Alien astronomers would find that the Sun's radial velocity varies by ±12 m/s, with a period equal to Jupiter's orbital period of 11.86 years. From that information, the astronomers would rightly conclude that the Sun has at least one planet with a mass comparable to Jupiter's. Without greater precision, the observers would be unaware of the other, less massive planets. To detect Saturn, they would need to improve the precision of their measurements to 2.7 m/s. Earth would not be detectable unless the aliens could detect motions as small as 0.09 m/s.

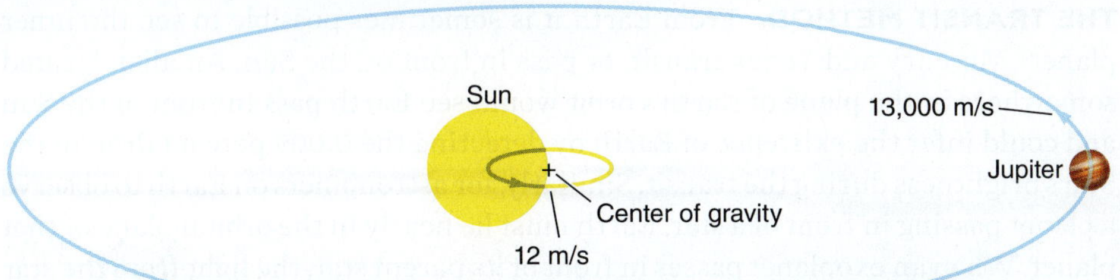

Figure 7.18 The Sun and Jupiter orbit around a common center of gravity (+), which lies just outside the Sun's surface. Spectroscopic measurements made by an extrasolar astronomer would reveal the Sun's radial velocity varying by ±12 m/s over 11.86 years, which is Jupiter's orbital period. Jupiter travels around its orbit at a speed of 13,000 m/s.

The precision of radial velocity instruments has been about 0.3 m/s. The technique enabled astronomers to detect giant planets around solar-type stars, but not yet to find planets with masses similar to Earth's. Finding the signal of the Doppler shift in the noise of the observation requires the star to be quite bright in our sky. A new instrument at the Very Large Telescope is expected to improve precision and be able to detect smaller planets. **Working It Out 7.2** explains more about the spectroscopic radial velocity method.

Astronomy in Action: Doppler Shift

Simulation: Radial Velocity

7.2 Working It Out Estimating the Size of the Orbit of a Planet

In the spectroscopic radial velocity method, the star is moving about its center of mass, and its spectral lines are Doppler-shifted accordingly. Recall from Figure 7.18 that an alien astronomer looking toward the Solar System would observe a shift in the wavelengths of the Sun's spectral lines—caused by the presence of Jupiter—of ~12 m/s.

Figure 7.17 showed the radial velocity data for a star with a planet discovered with that method. How do astronomers use that method to estimate the distance (A) of the planet from the star? Recall from Chapter 4 that Newton generalized Kepler's law relating the period of an object's orbit to the orbital semimajor axis:

$$P^2 = \frac{4\pi^2}{G} \times \frac{A^3}{M}$$

where A is the semimajor axis of the orbit, P is its period, and M is the combined mass of the two objects. To find A, we rearrange the equation as follows:

$$A^3 = \frac{G}{4\pi^2} \times M \times P^2$$

According to the graph of radial velocity observations in Figure 7.17, the period of the orbit is 5.7 years. A year has 3.16×10^7 seconds, so

$$P = 5.7\ \text{yr} \times (3.16 \times 10^7\ \text{s/yr}) = 1.8 \times 10^8\ \text{s}$$

The mass of the star is much greater than the mass of the planet, so the combined masses of the star and the planet can be approximated as the mass of the star, which here is about equal to the mass of the Sun, 2×10^{30} kg. (Stellar masses can be estimated from their spectra.) The gravitational constant G is $6.67 \times 10^{-20}\ \text{km}^3/(\text{kg s}^2)$. Plugging in the numbers gives

$$A^3 = \frac{6.67 \times 10^{-20} \dfrac{\text{km}^3}{\text{kg s}^2}}{4\pi^2} \times (2 \times 10^{30}\ \text{kg}) \times (1.8 \times 10^8\ \text{s})^2$$

$$A^3 = 1.1 \times 10^{26}\ \text{km}^3$$

taking the cube root,

$$A = 4.8 \times 10^8\ \text{km}$$

If we convert that number into astronomical units (where 1 AU = 1.5×10^8 km), the semimajor axis of the orbit of this planet is

$$A = \frac{4.8 \times 10^8\ \text{km}}{1.5 \times 10^8\ \text{km/AU}} = 3.2\ \text{AU}$$

The planet is 3.2 times farther from its star than Earth is from the Sun.

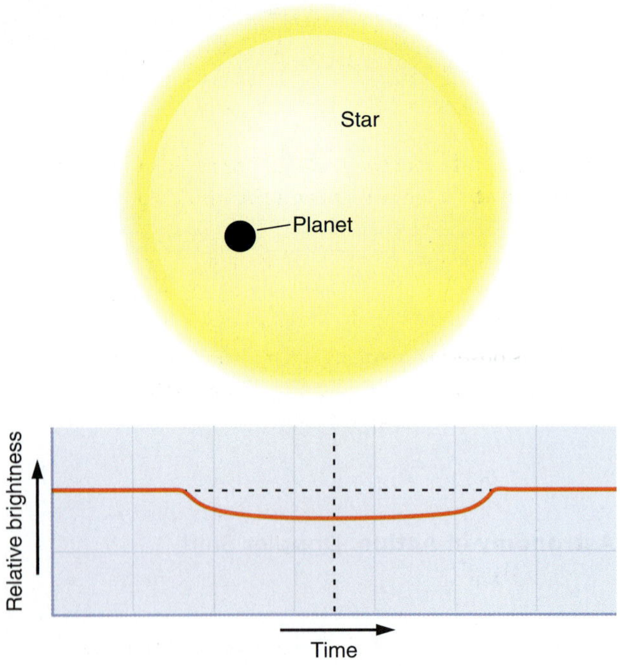

Figure 7.19 A planet that passes in front of a star blocks some of the light coming from the star's surface, causing the brightness of the star to decrease slightly. (The decrease in brightness is exaggerated here.)

THE TRANSIT METHOD From Earth it is sometimes possible to see the inner planets Mercury and Venus transit, or pass in front of, the Sun. An alien located somewhere in the plane of Earth's orbit would see Earth pass in front of the Sun and could infer the existence of Earth by detecting the 0.009 percent drop in the Sun's brightness during the transit. Similarly, for astronomers on Earth to observe a planet passing in front of a star, Earth must lie nearly in the orbital plane of that planet. When an exoplanet passes in front of its parent star, the light from the star diminishes by a tiny amount (**Figure 7.19**). This is the **transit method** of detecting exoplanets, in which we observe the dimming of starlight as a planet passes in front of its parent star. Whereas the radial velocity method gives us the mass of the planet and its orbital distance from a star, the transit method provides the radius of a planet. **Working It Out 7.3** shows how the radii are estimated.

Current ground-based technology limits the sensitivity of the transit method to about 0.1 percent of a star's brightness. Telescopes in space improve the sensitivity because smaller dips in brightness can be measured. The small Fench COROT telescope (27 cm) discovered 32 planets during its 6 years of operation (2007–2013). NASA's 0.95-meter Kepler telescope has discovered many planets and has found thousands more candidates that are being investigated further. **Figure 7.20** illustrates how systems of multiple planets are identified with that method: if one planet is found, then multiple sets of transits can indicate that other planets are orbiting the same star. Several thousand exoplanets have been detected from ground-based and space telescopes by using the transit method.

OTHER METHODS The gravitational field of an unseen planet can act like a lens, bending the light from a distant star in such a way that it causes the star to brighten temporarily while the planet is passing in front of it. Because the effect is small, it is usually called *microlensing*. Like the radial velocity method, microlensing provides an estimate of the mass of the planet. To date, about 60 exoplanets have been found with that technique.

Planets also may be detected by *astrometry*—precisely measuring the position of a star in the sky. If the system is viewed from "above," the star moves in a mini-orbit as the planet pulls it around. That motion is generally tiny and therefore very difficult to measure. For systems viewed from above the plane of the planet's orbit, however, none of the prior methods will work because the planet neither passes in front of the star nor causes a shift in its speed along the line of sight. Space missions such as the Gaia observatory, launched in 2013 by the European Space Agency, conduct such observations.

Direct imaging involves taking a picture of the planet directly. The technique is conceptually straightforward but is technically difficult because it involves searching for a relatively faint planet in the overpowering glare of a bright star—a challenge far more difficult than looking for a star in a clear, bright daytime sky. Even when an object is detected by direct imaging, an astronomer must still determine whether the observed object is actually a planet. Suppose we detect a faint object near a bright star. Could it be a more distant star that just happens to be in the line of sight? Future observations could tell whether the object shares the bright star's motion through space, but it also could be a brown dwarf rather than a true planet. An astronomer would need to make further observations to determine the object's mass.

Figure 7.20 Multiple planets can be detected through multiple transits with different changes in brightness. The arrows point to the changes in the total light as the three planets transit the star.

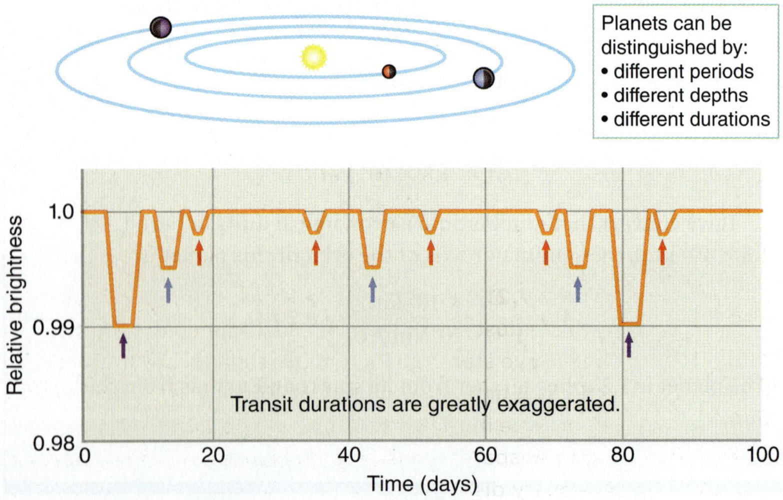

7.3 Working It Out Estimating the Radius of an Exoplanet

The masses of exoplanets can often be estimated using Kepler's laws and the conservation of angular momentum. When planets are detected with the transit method, astronomers can estimate the radius of an exoplanet. In that method, astronomers look for planets that eclipse their stars and then observe how much the star's light decreases during that eclipse (see Figure 7.19). When Venus or Mercury transits the Sun, a black circular disk is visible on the face of the circular Sun. During the transit, the amount of light from the transited star is reduced by the area of the circular disk of the planet divided by the area of the circular disk of the star:

$$\% \text{ reduction in light} = \frac{\text{Area of disk of planet}}{\text{Area of disk of star}} = \frac{\pi R^2_{\text{planet}}}{\pi R^2_{\text{star}}} = \frac{R^2_{\text{planet}}}{R^2_{\text{star}}}$$

Then, to solve for the radius of the planet, astronomers need an estimate of the radius of the star and a measurement of the percentage of reduction in light during the transit. The radius of a star is estimated from the surface temperature and the luminosity of the star.

Kepler-11, for example, is a system of at least six planets that transit a star. The radius of the star, R_{star}, is estimated to be 1.1 times the radius of the Sun, or $1.1 \times (7.0 \times 10^5 \text{ km}) = 7.7 \times 10^5$ km. The light from the Kepler-11 star is observed to decrease by 0.077 percent, or 0.00077 (see Figure 7.19), from planet Kepler-11c. What is the radius of Kepler-11c?

$$0.00077 = \frac{R^2_{\text{Kepler-11c}}}{R^2_{\text{star}}} = \frac{R^2_{\text{Kepler-11c}}}{(7.7 \times 10^5 \text{ km})^2}$$

$$R^2_{\text{Kepler-11c}} = 4.5 \times 10^8 \text{ km}^2 \quad \text{thus} \quad R_{\text{Kepler-11c}} = 2.1 \times 10^4 \text{ km}$$

Dividing $R_{\text{Kepler-11c}}$ by R_{Earth} (6,400 km) shows that $R_{\text{Kepler-11c}} = 3.3\ R_{\text{Earth}}$.

Some planets have been discovered through that method with large ground-based telescopes operating in the infrared region of the spectrum, using adaptive optics. **Figure 7.21** is an infrared image of Beta Pictoris b. A related form of direct observation involves separating the spectrum of a planet from the spectrum of its star to obtain information about the planet directly. Large ground-based telescopes have obtained spectra of the atmospheres of some exoplanets and have found, for example, carbon monoxide and water in those atmospheres.

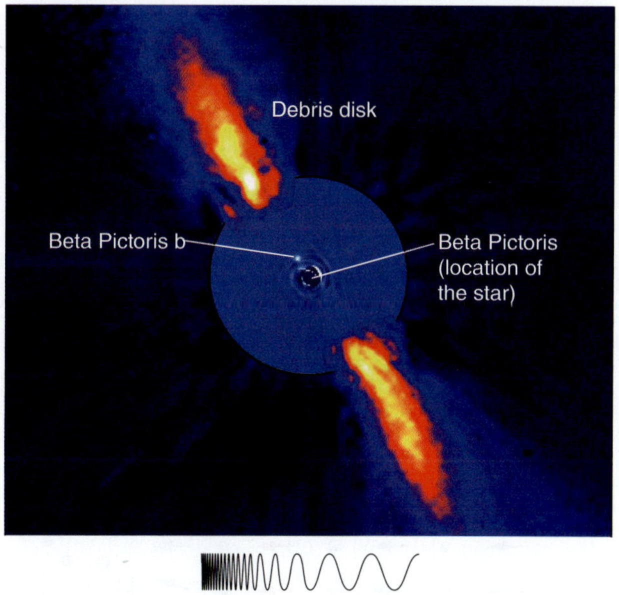

Figure 7.21 The planet Beta Pictoris b is seen orbiting within a dusty debris disk that surrounds the bright naked-eye star Beta Pictoris. The planet's estimated mass is 8 times that of Jupiter. The star is hidden behind an opaque mask, and the planet appears through a semitransparent mask used to subdue the brightness of the dusty disk.

Types of Exoplanets

Searches for exoplanets have been remarkably successful. Between the discovery of the first (in 1995) and this writing, nearly 4,000 more have been confirmed, and thousands more candidates are under investigation. As the number of observed systems with single and multiple planets increases, astronomers can compare those worlds with Solar System planets, finding more variation than they expected. The field is changing so fast that the most up-to-date information can be found only online.

The first discoveries included many **hot Jupiters**—Jupiter-sized planets orbiting solar-type stars in circular or highly eccentric orbits that bring them closer to their parent stars than Mercury is to our own Sun. Those planets were among the first to be detected because they are relatively easy targets for the spectroscopic radial velocity method. The large mass of a nearby hot Jupiter tugs the star very hard, creating large radial velocity variations in the star. In addition, those large planets orbiting close to their parent stars are more likely to pass in front of the star periodically and reveal themselves via the transit method.

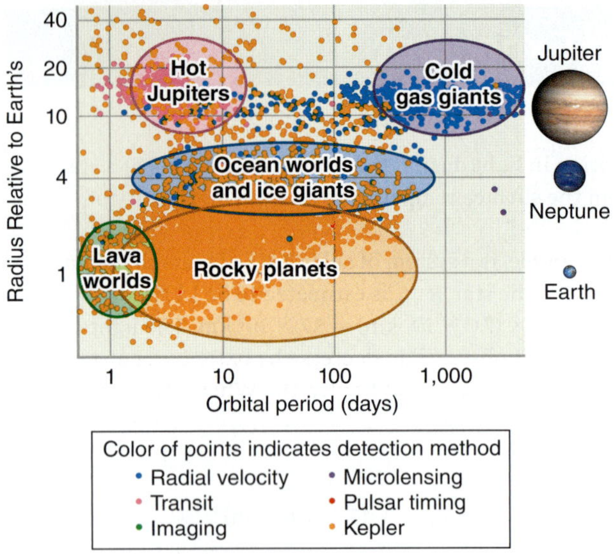

Figure 7.22 Different populations of exoplanets. Color of points indicates detection method. Jupiter-sized planets can be close to their star and hot, or they can be far from their star and cold. Neptune-sized planets are probably composed of water and ice. Earth-sized planets have higher density, indicating that they are made of rock.

Therefore, those hot Jupiter systems are easier to find than smaller, more distant planets. Astronomers realized that those hot Jupiter systems are not representative of most planetary systems; they were just easier to find. Scientists call that type of bias a *selection effect*. **Figure 7.22** illustrates the types of exoplanet populations.

Astronomers were surprised by the hot Jupiters because, according to the theory of planet formation described earlier in the chapter, those giant, volatile-rich planets should not have been able to form so close to their parent stars. From theories based on the Solar System, astronomers expected that Jupiter-type planets should form in the more distant, cooler regions of the protoplanetary disk, where the volatiles that make up much of their composition can survive. Hot Jupiters may form much farther from their parent stars and later migrate inward to a closer orbit. That migration may be caused by an interaction with gas or planetesimals in which orbital angular momentum is transferred from the planet to its surroundings, allowing it to spiral inward.

The radial velocity method yields an estimate of the mass of a planet (see Working It Out 7.2), and the transit method yields an estimate of the size of a planet (see Working It Out 7.3). With both pieces of information, the density (mass divided by the volume) of the planet can be computed to determine whether the planet is mostly gaseous (low density) or mostly rock (high density). Many of the new planets being discovered by Kepler are mini-Neptunes (gaseous planets with masses of 2–10 M_{Earth}) or super-Earths (rocky planets more massive than Earth). **Figure 7.23** illustrates the formation of those two types.

Planets with longer orbital periods, and therefore larger orbits, can be discovered only when the observations have gone on long enough to observe more than one complete orbit. Some of the exoplanets have highly elliptical orbits compared with those in the Solar System. Planets have been found with orbits that are highly tilted with respect to the plane of the rotation of their star, and some planets move in orbits whose direction is opposite that of their star's rotation. Multiple-planet systems have been observed in which the larger mini-Neptunes alternate with smaller super-Earths. The multiple-planet systems that have been found with the transit method reside in flat systems like our own, offering further evidence that the planets formed in a flat protoplanetary disk around a young star. But the current hypothesis to explain the Solar System's inner, small rocky planets and outer, large gaseous planets may not apply to all other planetary systems.

In addition, some planets detected through microlensing seem to be wandering freely through the Milky Way. Those planets may have been ejected from their solar systems after they formed and are no longer in gravitationally bound orbits around their stars.

We return to exoplanets at several points in this book—namely, when we review the planets in our own Solar System, when we consider the types of stars, and when we discuss the search for Earth-like planets in Chapter 24. The frequent new discoveries requiring revisions of existing theories make exoplanets one of the most exciting topics in astronomy today.

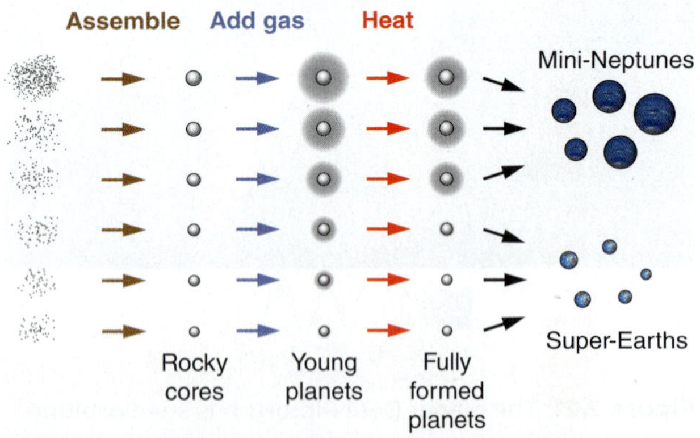

Figure 7.23 Many exoplanets are between the sizes of Earth and Neptune (~4 R_E). "Super-Earths" began with smaller rocky cores and less material, whereas "mini-Neptunes" had larger cores and collected more gas while forming.

CHECK YOUR UNDERSTANDING 7.5

What is the most common method for discovering an Earth-mass planet around a star? (a) Doppler spectroscopy; (b) direct imaging; (c) transit; (d) astrometric

Origins The Search for Earth-Sized Planets

The discovery of planetary systems, many different from the Solar System, shows us that the formation of planets often, and perhaps always, accompanies the formation of stars. The implications of that conclusion are profound. Planets are a common by-product of star formation. In a galaxy of 200 billion stars and a universe of hundreds of billions of galaxies, how many planets (and moons) might exist? And with all those planets in the universe, how many might have conditions suitable for the particular category of chemical reactions that we refer to as "life"? (We return to that point in Chapter 24.)

The Kepler Mission was developed by NASA to find Earth-sized and larger planets in orbit about a variety of stars. Kepler is a 0.95-meter telescope with 42 CCD detectors that observes approximately 150,000 stars in 100 square degrees of sky to look for planetary transits. To confirm a planetary detection, transits need to be observed three times with repeatable changes in brightness, duration of transit times, and computed orbital period. Kepler could detect a dip of 0.01 percent in the brightness of a star—sensitive enough to detect an Earth-sized planet. Kepler identified the first Earth-sized planets in 2011. Stars with transiting planets detected by Kepler also are observed spectroscopically to obtain radial velocity measurements that can lead to an estimate of a planet's mass. If a planet's radius and mass are known, the planet's density (mass per volume) can be estimated, too. From the density, astronomers can get a sense of whether the planet is composed primarily of gas, rock, ice, water, or some mixture of those.

On Earth, liquid water was essential for life to form and evolve. Because life on Earth is the only example of life for which we have evidence, we do not know whether liquid water is a cosmic requirement, but it is a place to start. The Kepler Mission searched for rocky planets at the right distance from their stars to permit the existence of liquid water, a distance known as the **habitable zone**. The location and width of the zone will vary depending on the temperature of the star. If a planet is too close to its star, water will exist only as a vapor; if too far, water will be frozen as ice. In the Solar System, Earth is the only planet currently in the habitable zone. Although announcements of new planets often state whether the planet is in the habitable zone, just being in the zone doesn't guarantee that the planet actually has liquid water—or that the planet is inhabited! An example of an Earth-sized planet in a habitable zone is shown in **Figure 7.24**.

Kepler identified thousands of planet candidates, some in the habitable zones of their respective stars. The candidates must be confirmed by follow-up observations of more transits or of radial velocities before they are officially announced as planet detections. Citizen scientists contribute to that search through the Exoplanet Explorers project; recently, the project announced the discovery of a system with five planets.

Astronomers are looking forward to results from new missions such as the Transiting Exoplanet Survey Satellite (TESS), launched in 2018, and the James Webb Space Telescope (JWST), scheduled for launch in 2020. TESS will monitor 200,000 nearby stars for dimming caused by planetary transits, whereas JWST will observe in the infrared and should be able to detect gases in the atmospheres of exoplanets.

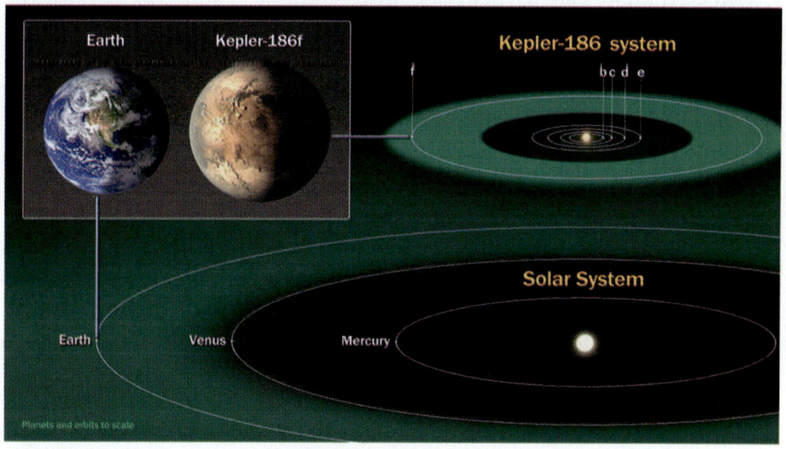

Figure 7.24 Artist's conception of the Kepler-186 system. Located about 500 light-years from Earth, that system has a planet in its habitable zone. The Solar System is shown for comparison. The Kepler-186 star is cooler than the Sun; thus, its habitable zone is closer to the star.

ARTICLES QUESTIONS

A multiplanet system discovered around a relatively nearby star

Four Earth-Sized Planets Detected Orbiting the Nearest Sun-like Star

August 9, 2017, by TIM STEPHENS

A new study by an international team of astronomers reveals that four Earth-sized planets orbit the nearest Sun-like star, Tau Ceti, which is about 12 light years away and visible to the naked eye. These planets have masses as low as 1.7 Earth mass, making them among the smallest planets ever detected around nearby Sun-like stars. Two of them are super-Earths located in the habitable zone of the star, meaning they could support liquid surface water.

The planets were detected by observing the wobbles in the movement of Tau Ceti. This required techniques sensitive enough to detect variations in the movement of the star as small as 30 centimeters per second.

"We are now finally crossing a threshold where, through very sophisticated modeling of large combined data sets from multiple independent observers, we can disentangle the noise due to stellar surface activity from the very tiny signals generated by the gravitational tugs from Earth-sized orbiting planets," said coauthor Steven Vogt, professor of astronomy and astrophysics at UC Santa Cruz.

According to lead author Fabo Feng of the University of Hertfordshire, UK, the researchers are getting tantalizingly close to the 10-centimeter-per-second limit required for detecting Earth analogs. "Our detection of such weak wobbles is a milestone in the search for Earth analogs and the understanding of the Earth's habitability through comparison with these analogs," Feng said. "We have introduced new methods to remove the noise in the data in order to reveal the weak planetary signals."

The outer two planets around Tau Ceti are likely to be candidate habitable worlds, although a massive debris disk around the star probably reduces their habitability due to intensive bombardment by asteroids and comets.

The same team also investigated Tau Ceti four years ago in 2013, when coauthor Mikko Tuomi of the University of Hertfordshire led an effort in developing data analysis techniques and using the star as a benchmark case. "We came up with an ingenious way of telling the difference between signals caused by planets and those caused by star's activity. We realized that we could see how star's activity differed at different wavelengths and use that information to separate this activity from signals of planets," Tuomi said.

The researchers painstakingly improved the sensitivity of their techniques and were able to rule out two of the signals the team had identified in 2013 as planets. "But no matter how we look at the star, there seem to be at least four rocky planets orbiting it," Tuomi said. "We are slowly learning to tell the difference between wobbles caused by planets and those caused by stellar active surface. This enabled us to essentially verify the existence of the two outer, potentially habitable planets in the system."

Sun-like stars are thought to be the best targets in the search for habitable Earth-like planets due to their similarity to the Sun. Unlike more common smaller stars, such as the red dwarf stars Proxima Centauri and Trappist-1, they are not so faint that planets would be tidally locked, showing the same side to the star at all times. Tau Ceti is very similar to the sun in its size and brightness, and both stars host multi-planet systems.

ARTICLES QUESTIONS

1. How do planets cause their stars to "wobble"?
2. With the spectroscopic method, why is finding Earth-sized planets harder than finding giant planets?
3. How did the astronomers improve their ability to find small planets?
4. What is a debris disk?
5. What would be the disadvantage to a planet's being tidally locked to its star? (Hint: Refer to Chapter 2.)

Summary

Stars and their planetary systems form from collapsing interstellar clouds of gas and dust, following the laws of gravity and conservation of angular momentum. Conservation of angular momentum produces an accretion disk around a protostar that often fragments to form multiple planets, as well as smaller objects such as asteroids and dwarf planets, through the gradual accumulation of material into larger and larger objects. Multiple methods exist for finding planets around other stars, and those planets are now thought to be very common. This field of study is evolving very quickly as technology advances.

LG 1 **Describe how our understanding of planetary system formation developed from the work of both planetary and stellar scientists.** Planets are a common by-product of star formation, and many stars are surrounded by planetary systems. Gravity pulls clumps of gas and dust together, causing them to shrink and heat up. Angular momentum must be conserved, leading to both a spinning central star and an accretion disk that rotates and revolves in the same direction as the central star. Solar System meteorites show that larger objects build up from smaller objects.

LG 2 **Discuss the role of gravity and angular momentum in explaining why planets orbit the Sun in a plane and why they revolve in the same direction that the Sun rotates.** As particles orbit the forming star, the cloud of dust and gas flattens into a plane. Conservation of angular momentum determines both the speed and the direction of the revolution of the objects in the forming system. Dust grains in the protoplanetary disk first stick together because of collisions and static electricity. As those objects grow, they eventually have enough mass to attract other objects gravitationally. Once that occurs, they begin emptying the space around them. Collisions of planetesimals lead to the formation of planets.

LG 3 **Explain how temperature at different locations in the protoplanetary disk affects the composition of planets, moons, and other bodies.** The temperature is higher near the central protostar, forcing volatile elements, such as water, to evaporate and leave the inner part of the disk. Planets in the inner part of the disk will have fewer volatiles than those in the outer part of the disk. The gas that a planet captures when it forms is the planet's primary atmosphere. Less massive planets lose their primary atmospheres and then form secondary atmospheres.

LG 4 **Discuss the processes that resulted in the formation of planets and other objects in our Solar System.** In the current model of the formation of the Solar System, solid terrestrial planets formed in the inner disk, where temperatures were high, whereas giant gaseous planets formed in the outer disk, where temperatures were low. Dwarf planets such as Pluto formed in the asteroid belt and in the region beyond the orbit of Neptune. Asteroids and comet nuclei remain today as leftover debris.

LG 5 **Describe how astronomers find planets around other stars and derive exoplanet properties.** Astronomers find planets around other stars by using the radial velocity method, the transit method, microlensing, astrometry, and direct imaging. As technology has improved, the number and variety of known exoplanets has increased dramatically, with thousands of planets and planet candidates discovered orbiting other stars near the Sun within the Milky Way Galaxy in just the past few years.

? Unanswered Questions

- How typical is the Solar System? Only within the past decade have astronomers found other systems containing four or more planets, and so far the observed distributions of large and small planets in those multiplanet systems have looked different from those of the Solar System. Computer simulations of planetary system formation suggest that a system with stable orbits and a planetary distribution like those of the Solar System may develop only rarely. Improved supercomputers can run more complex simulations, which can be compared with the observations to better understand how solar systems are configured.
- How Earth-like must a planet be before scientists declare it to be "another Earth"? An editorial in the science journal *Nature* cautioned that scientists should define "Earth-like" in advance—before multiple discoveries of planets "similar" to Earth are announced and a media frenzy ensues. Must a planet be of similar size and mass, be located in the habitable zone, and have spectroscopic evidence of liquid water before we call it "Earth 2.0"?

Questions and Problems

Test Your Understanding

1. Place the following events in the order that corresponds to the formation of a planetary system.
   a. Gravity collapses a cloud of interstellar gas.
   b. A rotating disk forms.
   c. Small bodies collide to form larger bodies.
   d. A stellar wind "turns on" and sweeps away gas and dust.
   e. Primary atmospheres form.
   f. Primary atmospheres are lost.
   g. Secondary atmospheres form.
   h. Dust grains stick together by static electricity.
2. If the radius of an object's orbit is halved, and angular momentum is conserved, what must happen to the object's speed?
   a. It must be halved.
   b. It must stay the same.
   c. It must be doubled.
   d. It must be squared.
3. Unlike the giant planets, the terrestrial planets formed when
   a. the inner Solar System was richer in heavy elements than the outer Solar System.
   b. the inner Solar System was hotter than the outer Solar System.
   c. the outer Solar System took up more volume than the inner Solar System, so more material was available to form planets.
   d. the inner Solar System was moving faster than the outer Solar System.
4. The terrestrial planets and the giant planets have different compositions because
   a. the giant planets are much larger.
   b. the terrestrial planets formed closer to the Sun.
   c. the giant planets are made mostly of solids.
   d. the terrestrial planets have few moons.
5. The spectroscopic radial velocity method preferentially detects
   a. large planets close to the central star.
   b. small planets close to the central star.
   c. large planets far from the central star.
   d. small planets far from the central star.
   e. The method detects all those planets equally well.
6. The concept of disk instability was developed to solve the problem that
   a. Jupiter-like planets migrate after formation.
   b. not enough gas was in the Solar System to form Jupiter.
   c. the early solar nebula probably dispersed too soon to form Jupiter.
   d. Jupiter consists mostly of volatiles.
7. Because angular momentum is conserved, an ice-skater who throws her arms out will
   a. rotate more slowly.
   b. rotate more quickly.
   c. rotate at the same rate.
   d. stop rotating.
8. Clumps grow into planetesimals by
   a. gravitationally pulling in other clumps.
   b. colliding with other clumps.
   c. attracting other clumps with opposite charge.
   d. conserving angular momentum.
9. The transit method preferentially detects
   a. large planets close to the central star.
   b. small planets close to the central star.
   c. large planets far from the central star.
   d. small planets far from the central star.
   e. The method detects all those planets equally well.
10. If the radius of a spherical object is halved, what must happen to the period so that the spin angular momentum is conserved?
    a. It must be divided by 4.
    b. It must be halved.
    c. It must stay the same.
    d. It must double.
    e. It must be multiplied by 4.
11. The amount of angular momentum in a spherical object does *not* depend on
    a. its radius.
    b. its mass.
    c. its rotation speed.
    d. its temperature.
12. The planets in the inner part of the Solar System are made primarily of refractory materials; the planets in the outer Solar System are made primarily of volatiles. That difference occurs because
    a. refractory materials are heavier than volatiles, so they sank farther into the nebula.
    b. no volatiles were in the inner part of the accretion disk.
    c. the volatiles on the inner planets were lost soon after the planets formed.
    d. the outer Solar System has gained more volatiles from space since formation.
13. If scientists want to find out about the composition of the early Solar System, the best objects to study are
    a. the terrestrial planets.
    b. the giant planets.
    c. the Sun.
    d. asteroids and comets.
14. The direction of revolution in the plane of the Solar System was determined by
    a. the plane of the galaxy in which the Solar System sits.
    b. the direction of the gravitational force within the original cloud.
    c. the direction of rotation of the original cloud.
    d. the amount of material in the original cloud.
15. A planet in the "habitable zone"
    a. is close to the central star.
    b. is far from the central star.
    c. is the same distance from its star as Earth is from the Sun.
    d. is at a distance where liquid water can exist on the surface.

Thinking about the Concepts

16. What is the source of the material that now makes up the Sun and the rest of the Solar System?

17. Describe the different ways by which stellar astronomers and planetary scientists each came to the same conclusion about how planetary systems form.

18. What is a protoplanetary disk? What are two reasons that the inner part of the disk is hotter than the outer part?

19. Physicists describe certain properties, such as angular momentum and energy, as being *conserved*. What does that mean? Do conservation laws imply that an individual object can never lose or gain angular momentum or energy? Explain your reasoning.

20. The Process of Science Figure in this chapter makes the point that different areas of science must agree with one another. Suppose that a few new exoplanets are discovered that appear not to have formed from the collapse of a stellar nebula (for example, the planetary orbits might be in random orientations). What will scientists do with that new information?

21. How does the law of conservation of angular momentum control a figure-skater's rate of spin?

22. What is an accretion disk?

23. Describe how tiny grains of dust grow to become massive planets.

24. Look under your bed, the refrigerator, or any similar place for dust bunnies. Once you find them, blow one toward another. Watch carefully and describe what happens as they meet. What happens if you repeat that action with additional dust bunnies? Will those dust bunnies ever have enough gravity to begin pulling themselves together? If they were in space instead of on the floor, might that happen? What force prevents their mutual gravity from drawing them together into a "bunny-tesimal" under your bed?

25. Why do we find rocky material everywhere in the Solar System but find large amounts of volatile material only in the outer regions?

26. Why could the four giant planets collect massive gaseous atmospheres, whereas the terrestrial planets could not? Explain the source of the secondary atmospheres surrounding the terrestrial planets.

27. Describe four methods that astronomers use to search for exoplanets. What are the limitations of each method; that is, what circumstances are necessary to detect a planet by each method?

28. Why is it difficult to obtain an image of an exoplanet?

29. Many of the first exoplanets that astronomers found orbiting other stars were giant planets with Jupiter-like masses and with orbits located very close to their parent stars. Explain why those characteristics are a selection effect of the discovery method.

30. How has the Kepler telescope found Earth-like planets, and what do astronomers mean by "Earth-like"?

Applying the Concepts

31. In Figure 7.17, what is the maximum radial velocity of HD 70642 b in meters per second? Convert that number to miles per hour (mph). How does that value compare with the speed at which Earth orbits the Sun (67,000 mph)?

32. Use Appendix 4 to answer the following:
    a. What is the total mass of all the planets in the Solar System, expressed in Earth masses (M_{Earth})?
    b. What fraction of that total planetary mass is Jupiter?
    c. What fraction does Earth represent?

33. Compare Earth's orbital angular momentum with its spin angular momentum by using the following values: $m = 5.97 \times 10^{24}$ kg, $v = 29.8$ kilometers per second (km/s), $r = 1$ AU, $R = 6{,}378$ km, and $P = 1$ day. Assume that Earth is a uniform body. What fraction does each component (orbital and spin) contribute to Earth's total angular momentum? Refer to Working It Out 7.1.

34. Venus has a radius 0.949 times that of Earth and a mass 0.815 times that of Earth. Venus's rotation period is 243 days. What is the ratio of Venus's spin angular momentum to that of Earth? Assume that Venus and Earth are uniform spheres.

35. Jupiter has a mass equal to 318 times Earth's mass, an orbital radius of 5.2 AU, and an orbital velocity of 13.1 km/s. Earth's orbital velocity is 29.8 km/s. What is the ratio of Jupiter's orbital angular momentum to that of Earth?

36. The asteroid Vesta has a diameter of 530 km and a mass of 2.7×10^{20} kg.
    a. Calculate the density (mass/volume) of Vesta.
    b. The density of water is 1,000 kg/m^3, whereas that of rock is about 2,500 kg/m^3. What does that difference tell you about the composition of Vesta?

37. Using data in Appendix 4, compute the densities of Venus, Jupiter, and Neptune. Compare with the densities of rock, water, and gas.

38. Recalling Kepler's laws, put the three planets in Figure 7.20 in order from fastest to slowest. Compare the duration of the transits in Figure 7.20. Why does the outermost planet have the longest duration?

39. Suppose you can measure radial velocities of about 0.3 m/s. Suppose you are observing a spectral line with a wavelength of 575 nanometers (nm). How large a shift in wavelength would a radial velocity of 0.3 m/s produce?

40. Earth tugs the Sun around as it orbits, but that effect (only 0.09 m/s) is much smaller than that of any known exoplanet. How large a shift in wavelength does that effect cause in the Sun's spectrum at 500 nm?

41. If an alien astronomer observed a plot of the light curve as Jupiter passed in front of the Sun, by how much would the Sun's brightness drop during the transit?

42. A planet has been found to orbit a 1-M_{Sun} star in 200 days.
    a. What is the orbital radius of that exoplanet?
    b. Compare its orbit with that of the planets around our own Sun. What temperatures must that planet experience?

43. One of the planets orbiting the star Kepler-11 has an orbital radius of 1.1 R_{Sun}. If the planet has a radius of 4.5 R_{Earth}, by how much does the brightness of Kepler-11 decrease when that planet transits the star?

44. Kepler detected a planet with a diameter of 1.7 Earth (D_{Earth}).
    a. How much larger is the volume of that planet than Earth's?
    b. Assume that the density of the planet is the same as Earth's. How much more massive is it than Earth?

45. The planet COROT-11b was discovered using the transit method, and astronomers have followed up with radial velocity measurements, so both its radius (1.43 R_{Jup}) and its mass (2.33 M_{Jup}) are known. The density provides a clue about whether the object is gaseous or rocky.
    a. What is the mass of the planet in kilograms?
    b. What is the planet's radius in meters?
    c. What is the planet's volume?
    d. What is the planet's density? How does that density compare with the density of water (1,000 kg/m^3)? Is the planet likely to be rocky or gaseous?

USING THE WEB

46. Using the exoplanet catalogs:
    a. Go to the "Catalog" Web page (http://exoplanet.eu/catalog) of the Extrasolar Planets Encyclopedia. Look for a star that has multiple planets. Make a graph showing the distances of the planets from that star, and note the masses and sizes of the planets. Put the Solar System planets on the same axis. How does that extrasolar planet system compare with the Solar System?
    b. Go to the "Exoplanets Data Explorer" website (http://exoplanets.org) and click on "Table." That website lists planets that have detailed orbital data published in scientific journals, and it may have a smaller total count than the website in part (a). Pick a planet discovered recently, as specified in the "First Reference" column. What is the planet's minimum mass? What is its semimajor axis and the period of its orbit? What is the eccentricity of its orbit? Click on the star name in the first column to get more information. Does that planet have a radial velocity curve? Was it observed in transit, and if so, what is the planet's radius and density? Is it more like Jupiter or more like Earth?

47. Space missions:
    a. Go to the website for the Kepler Mission (https://nasa.gov/mission_pages/kepler/main/). How many confirmed planets has Kepler discovered? How many planet candidates are there? When did the mission end? Search for the latest version of the "Kepler Orrery," an animation that shows multiplanet systems discovered by Kepler. Do most of those systems look like our own?
    b. Go to the website for the TESS mission (https://tess.gsfc.nasa.gov/). Are there any results?

48. Citizen science projects I:
    a. Go to the "Exoplanet Explorers" website (https://zooniverse.org/projects/ianc2/exoplanet-explorers) Exoplanet Explorers is part of the Zooniverse, a citizen science project that invites individuals to participate in a major science project by using their own computers. To participate in that or any of the other Zooniverse projects mentioned in later chapters, you will need to sign up for an account. Read through the sections under "About," including the FAQ. What are some of the advantages to crowdsourcing Kepler data analysis? Back on the Exoplanet Explorers home page, click on "Classify" and then "Show the project tutorial" under the example on the right. Also click on the "Field Guide" to see more examples. Then look at the data presented and classify.
    b. Go to the "Planet Hunters" website (http://planethunters.org), an older Zooniverse project. Read through the sections under "Science," including the FAQ. Back on the Planet Hunters home page, click on "Start Classifying" and go through the tutorial. Classify a few images.

49. Citizen science projects II: Go to the "Disk Detective" website at http://diskdetective.org/, another Zooniverse project for which you will need to open an account as in problem 48. In that project, you will look at observations of young stars to see whether evidence exists for a planetary disk. Under "Menu," read "Science," "About," "FAQ," and then "Classify." Work through an example and then classify a few images.

50. Go to the "Super Planet Crash" Web page (http://stefanom.org/spc/) or the App. Read "Help" to see the rules. First build a system like ours with four Earth-sized planets in the inner 2 AU. Is that system stable? What happens if you add in super-Earths or "ice giants"? Build up a few completely different planetary systems and see what happens. What types of situations cause instability in the inner 2 AU of those systems?

EXPLORATION

Exploring Exoplanets

digital.wwnorton.com/astro6

Visit the Digital Resources Page and on the Student Site open the "Radial Velocity" Interactive Simulation in Chapter 7. That applet has several panels that allow you to experiment with the variables important for measuring radial velocities. First, in the window labeled "Visualization Controls," check the box to show multiple views. Compare the views shown in panels 1–3 with the colored arrows in the last panel to see where an observer would stand to see the view shown. Start the animation (in the "Animation Controls" panel), and allow it to run while you watch the planet orbit its star from each of the views shown. Stop the animation, and in the "Presets" panel, select "Option A" and then click "set."

1 Is Earth's view of that system most nearly like the "side view" or most nearly like the "orbit view"?

2 Is the orbit of that planet circular or elongated?

3 Study the radial velocity graph in the upper right panel. The blue curve shows the radial velocity of the star over a full period. What is the maximum radial velocity of the star?

4 The horizontal axis of the graph shows the "phase," or fraction of the period. A phase of 0.5 is halfway through a period. The vertical red line indicates the phase shown in views in the upper left panel. Start the animation to see how the red line sweeps across the graph as the planet orbits the star. The period of the planet is 365 days. How many days pass between the minimum radial velocity and the maximum radial velocity?

5 When the planet moves away from Earth, the star moves toward Earth. The sign of the radial velocity tells the direction of the motion (toward or away). Is the radial velocity of the star positive or negative at that time in the orbit? If you could graph the radial velocity of the planet at that point in the orbit, would it be positive or negative?

In the "Presets" window, select "Option B" and then click "set."

6 What has changed about the orbit of the planet as shown in the views in the upper left panel?

7 When is the planet moving fastest: when it is close to the star or when it is far from the star?

8 When is the star moving fastest: when the planet is close to it or when the planet is far away?

9 Explain how an astronomer would determine, from a radial velocity graph of the star's motion, whether the orbit of the planet was circular or elongated.

10 Study the "Earth view" panel at the top of the window. Would that planet be a good candidate for a transit observation? Why or why not?

In the "System Orientation" panel, change the inclination to 0.0.

11 Now is Earth's view of the system most nearly like the "side view" or most nearly like the "orbit view"?

12 How does the radial velocity of the star change as the planet orbits?

13 Click the box that says "show simulated measurements," and change the "noise" to 1.0 m/s. The gray dots are simulated data, and the blue line is the theoretical curve. Use the slider bar to change the inclination. What happens to the radial velocity as the inclination increases? (Hint: Pay attention to the vertical axis as you move the slider, not just the blue line.)

14 What is the smallest inclination for which you would find the data convincing? That is, what is the smallest inclination for which the theoretical curve is in good agreement with the data?

13

Taking the Measure of Stars

To all but the largest telescopes, even nearby stars are just points of light in the night sky. Astronomers study the stars by observing their light, by using the laws of physics discussed in earlier chapters, and by finding patterns in subgroups of stars that are then applied to other stars. Astronomers use knowledge of geometry, radiation, and orbits to begin to answer basic questions about stars, such as how they are similar to or different from the Sun, and whether they might have planets orbiting around them as the Sun does.

LEARNING GOALS

By the end of this chapter, you should be able to:

LG 1 Explain how astronomers use the brightness of nearby stars and their distances from Earth to determine how luminous they are.

LG 2 Explain how astronomers obtain the temperatures, sizes, and composition of stars.

LG 3 Describe how astronomers estimate the masses of stars.

LG 4 Classify stars and organize that information on a Hertzsprung-Russell (H-R) diagram.

LG 5 Explain how the mass and composition of a main-sequence star determine its luminosity, temperature, and size.

The constellation Orion. In addition to stars we can see numerous nebula of gas and dust. ►►►

Why do stars have different colors?

13.1 Astronomers Measure the Distance, Brightness, and Luminosity of Stars

The stars in the sky differ in brightness and color, but does one star appear brighter than another because it has a higher luminosity and is emitting more light or because it is closer to us? In this section, we explain how to find the distances to nearby stars and how to use distance and apparent brightness to find the luminosity of a star.

Stereoscopic Vision

Your two eyes have different views that depend on the distance to the object you are viewing. Hold up your finger in front of you, close to your nose. View it with your right eye only and then with your left eye only. Each eye sends a slightly different image to your brain, so your finger *appears* to move back and forth relative to the background behind it. Now hold up your finger at arm's length, and blink your right eye and then your left. Your finger appears to move much less. The way your brain combines the different information from your eyes to perceive the distances to objects around you is called **stereoscopic vision**. **Figure 13.1a** shows an overhead view of the experiment you just performed with your finger. The left eye sees the blue pencil almost directly between the green balls on the bookshelf. But the right eye sees the blue pencil to the left of both balls. Similarly, the position of the pink pencil appears to vary. Because the pink pencil is closer to the observer, its position appears to change more than the position of the blue pencil—seeming to move from the right of the blue pencil to the left of the blue pencil.

Stereoscopic vision enables you to judge the distances of objects as far away as a few hundred meters, but beyond that it is of little use. Your right eye's view of a mountain several kilometers away is indistinguishable from the view seen by your left eye—all you can determine is that the mountain is too far away for you to judge its distance stereoscopically. The distance over which your stereoscopic vision works is limited by the separation between your eyes—about 6 centimeters (cm). If you could increase the distance between your eyes by several meters, the view from each eye would be different enough for you to determine the distances to objects kilometers away.

You cannot separate your eyes, but you can compare pictures taken with a camera from two widely separated locations. The greatest separation possible without leaving Earth is to let Earth's orbital motion carry us from one side of the Sun to the other. If you take a picture of the sky tonight and then wait 6 months and take another picture, the distance between the two locations is the diameter of Earth's orbit (2 astronomical units [AU]), which gives us more powerful stereoscopic vision.

Astronomy in Action: Parallax

Distances to Nearby Stars

Figure 13.1b, which depicts Earth's orbit from far above the Solar System, shows how astronomers apply the concept of stereoscopic vision to measure the distances to stars. The change in position of Earth over 6 months is like the distance between the right eye and the left eye in Figure 13.1a. The nearby (pink and blue) stars are like the pink and blue pencils, whereas the distant yellow stars are like the green balls on the bookcase. Because of the shift in perspective as Earth orbits the Sun, nearby stars appear to shift their positions. The pink star, which is closer,

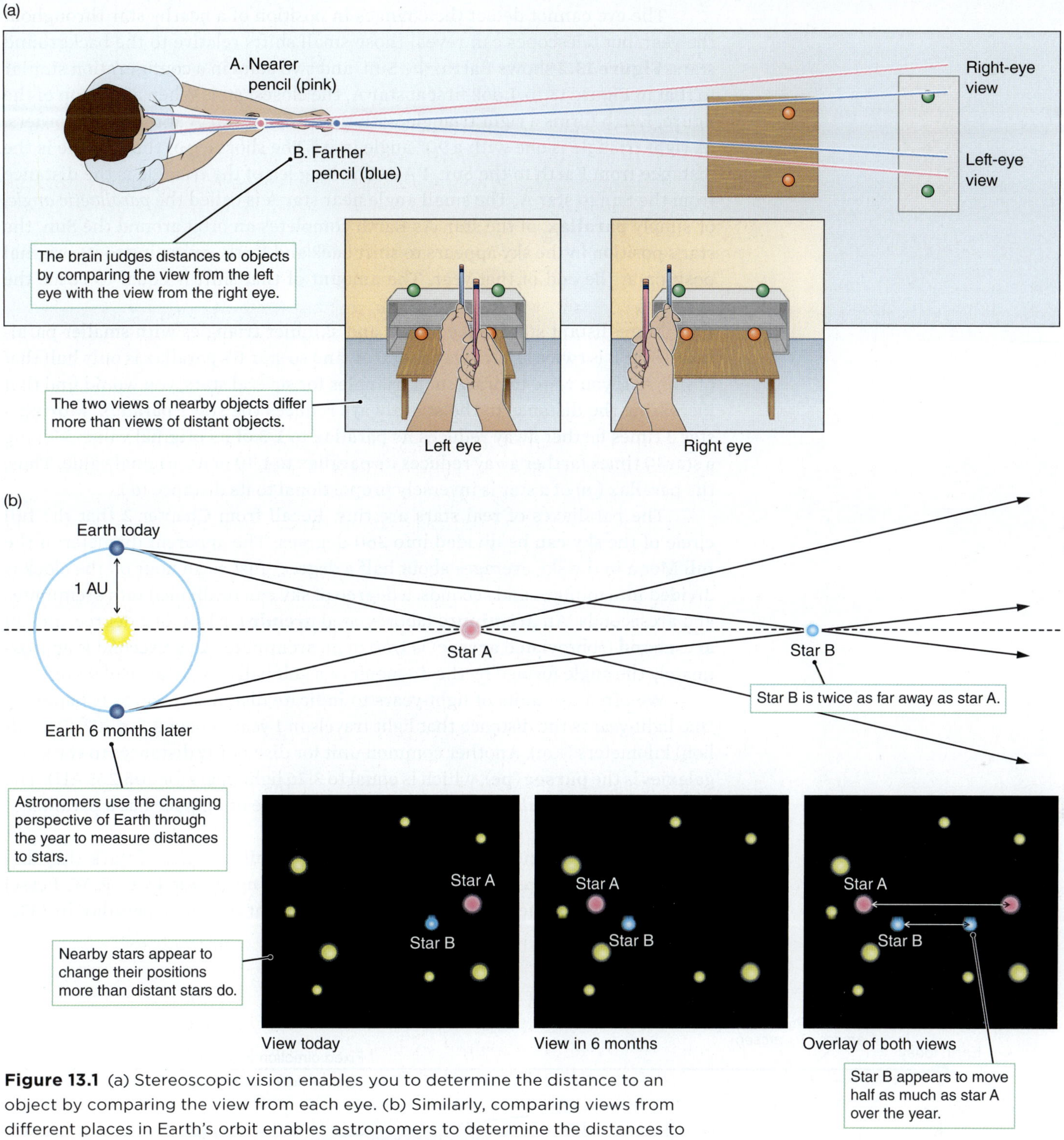

Figure 13.1 (a) Stereoscopic vision enables you to determine the distance to an object by comparing the view from each eye. (b) Similarly, comparing views from different places in Earth's orbit enables astronomers to determine the distances to stars. As Earth moves around the Sun, the apparent positions of nearby stars change more than the apparent positions of more distant stars. (Diagram not to scale.)

appears to move farther than the more distant blue star. Over 1 full year, the nearby star appears to move one way and then back again with respect to distant background stars, returning to its original position at the end of that year. We can determine the distance to the star by applying geometry to the amount of that apparent shift.

The eye cannot detect the changes in position of a nearby star throughout the year, but telescopes can reveal those small shifts relative to the background stars. **Figure 13.2** shows Earth, the Sun, and two stars in a configuration similar to that in Figure 13.1b. Look first at star A, the closest star. When at the top of the figure, Earth forms a right triangle with the Sun and star A at the other corners. (A right triangle is one with a 90° angle in it.) The short leg of the triangle is the distance from Earth to the Sun, 1 AU. The long leg of the triangle is the distance from the Sun to star A. The small angle near star A is called the *parallactic angle*, or simply **parallax**, of the star. As Earth completes an orbit around the Sun, the star's position in the sky appears to shift back and forth, returning to its original position at the end of that year. The amount of that shift is equal to twice the parallax.

More distant stars make longer and skinnier triangles with smaller parallaxes. Star B is twice as far away as star A, and so star B's parallax is only half that of star A. If you were to draw such triangles for several stars, you would find that increasing the distance to the star always reduces the star's parallax. Moving a star 3 times farther away reduces its parallax to 1/3 of its original value. Moving a star 10 times farther away reduces its parallax to 1/10 of its original value. Thus, the parallax (p) of a star is inversely proportional to its distance (d).

The parallaxes of real stars are tiny. Recall from Chapter 2 that the full circle of the sky can be divided into 360 degrees. The apparent diameter of the full Moon in the sky averages about half a degree. Just as an hour on the clock is divided into minutes and seconds, a degree of sky can be divided into arcminutes and arcseconds. An **arcminute** (abbreviated **arcmin**) is 1/60 of a degree, and an **arcsecond** (abbreviated **arcsec**) is 1/60 of an arcminute. An arcsecond is approximately the angle formed by the diameter of a golf ball at a distance of 9 km.

We often use units of light-years to indicate distances to stars (Chapter 1). One light-year is the distance that light travels in 1 year—about 9.5×10^{12} (9.5 trillion) kilometers (km). Another common unit for discussing distances to stars and galaxies is the **parsec (pc)**, which is equal to 3.26 light-years (or 206,265 AU). The term is short for *parallax second*—a star at a distance of 1 parsec has a parallax of 1 arcsecond.

When astronomers began to measure the parallax angles of stars, they discovered that stars are very distant objects (**Working It Out 13.1**). F. W. Bessel (1784–1846) made the first successful measurement of stellar parallax in 1838,

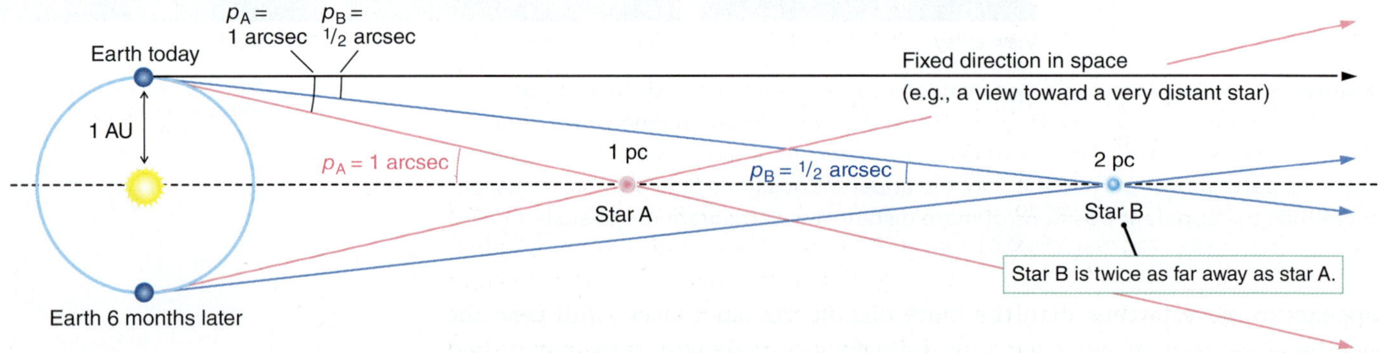

Figure 13.2 The parallax (p) of a star is inversely proportional to its distance. More distant stars have smaller parallaxes. (Diagram not to scale.)

13.1 Working It Out Parallax and Distance

If star B in Figure 13.2 is twice as far away from us as star A, star B will have half the parallax of star A. The parallax of a star (p) is inversely proportional to its distance (d):

$$p \propto \frac{1}{d} \quad \text{or} \quad d \propto \frac{1}{p}$$

If the angle at the apex of a triangle is 1 arcsec, and the base of the triangle is 1 AU, the length of the triangle is 1 parsec. One parsec equals 206,265 AU (see Appendix 1), or 3.09×10^{13} km, or 3.26 light-years. Astronomers use the unit of parsecs because doing so makes the relationship between distance and parallax easier to manage than when the distance is measured in light-years.

A star with a parallax of 1 arcsec is at a distance of 1 pc (Figure 13.2). The inverse proportionality between distance and parallax becomes

$$\left(\begin{array}{c}\text{Distance measured}\\ \text{in parsecs}\end{array}\right) = \frac{1}{\left(\begin{array}{c}\text{Parallax measured}\\ \text{in arcseconds}\end{array}\right)}$$

or

$$d(\text{pc}) = \frac{1}{p(\text{arcsec})}$$

Suppose that the parallax of a star is measured to be 0.5 arcsec. The distance can be found by

$$d(\text{pc}) = \frac{1}{0.5} = 2 \text{ parsecs}$$

Similarly, a star with a measured parallax of 0.01 arcsec is located at a distance of $1/0.01 = 100$ pc.

After the Sun, the next-closest star to Earth is Proxima Centauri. Located at a distance of 4.24 light-years, Proxima Centauri is a faint member of a system of three stars called Alpha Centauri. What is that star's parallax? First, we convert the distance to parsecs:

$$d = 4.24 \text{ light-years} \times \frac{1 \text{ parsec}}{3.26 \text{ light-years}} = 1.30 \text{ parsecs}$$

Then,

$$p(\text{arcsec}) = \frac{1}{1.30 \text{ pc}} = 0.77 \text{ arcsec}$$

Even the closest star to the Sun has a parallax of only about 3/4 arcsec.

reporting a parallax of 0.314 arcsec for the star 61 Cygni. That finding implied that 61 Cygni was 3.2 pc away, or 660,000 times as far away as the Sun. With that one measurement, Bessel increased the known volume of the universe by a factor of 10,000. Today, astronomers know of at least 60 stars, some in multiple-star systems, within 5 pc (16.3 light-years) of the Sun. In the neighborhood of the Sun, each star or star system has on average a volume of about 260 cubic light-years of space to itself.

Most stars are so far away that the parallax angle is too small to measure using ground-based telescopes, which are limited by Earth's atmosphere. In the 1990s, the European Space Agency's *Hipparcos* satellite measured the positions and parallaxes of over 100,000 stars with an uncertainty of about ±0.001 arcsec. *Hipparcos'* successor is *Gaia*, a space mission that has been studying stellar parallaxes since the end of 2013. *Gaia*'s second data release in 2018 included parallaxes of more than 1 billion stars, with uncertainties of ±0.00004 to ±0.0007 arcsec, depending on the star's brightness.

Note that since the distance depends on 1 over the parallax, an uncertainty in the parallax will mean a corresponding range of distances. For example, a star with a measured parallax of 0.004 ± 0.001 arcsec really has a parallax between 0.003 and 0.005 arcsec. That range of parallaxes gives a corresponding distance range of 200–333 pc from Earth. If the measured parallax is 0.004 ± 0.0001 arcsec, then the distance range is 244–256 parsecs. Thus astronomers are very excited about this recent data from *Gaia*. Other methods of measuring distance to more remote stars are discussed later.

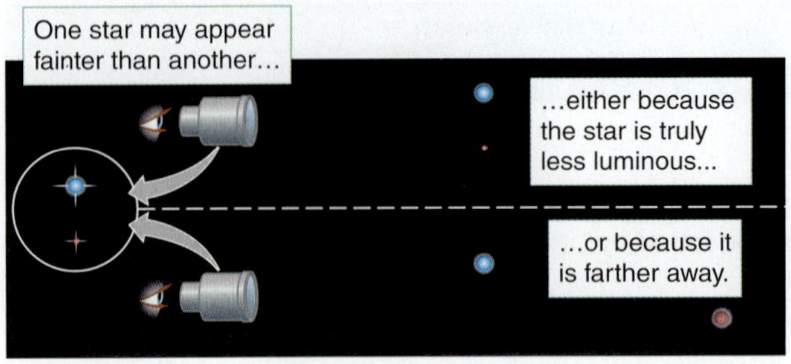

Figure 13.3 The brightness of a star visible in our sky depends on both its luminosity—how much light it emits—and its distance.

Luminosity, Brightness, and Distance

The stars in Earth's sky are of different brightnesses, where brightness corresponds to the amount of energy falling on a square meter of area each second in the form of electromagnetic radiation (Chapter 5). Although the brightness of a star can be measured directly, it does not immediately give much information about the star itself. As illustrated in **Figure 13.3**, a star in the night sky may in fact appear bright only because it is nearby. Conversely, a faint star may be a powerful beacon, still visible despite its tremendous distance.

Astronomers measure the brightness of stars by comparing them with one another. The system they use dates back 2,100 years, when the Greek astronomer Hipparchus classified stars according to their brightness. The details of his system, still in use today, are discussed in **Working It Out 13.2** and **Appendix 7**.

To learn about the actual properties of a star, astronomers need to know the total energy radiated by a star each second—the star's luminosity. Recall from

13.2 Working It Out The Magnitude System

The **magnitude** system of brightness for celestial objects can be traced back 2,100 years to the ancient Greek astronomer Hipparchus, who classified the brightest stars he could see as being "of the first magnitude" and the faintest as being "of the sixth magnitude." Later, astronomers defined Hipparchus's 1st magnitude stars as being exactly 100 times brighter than his 6th magnitude stars. Hipparchus must have had typical eyesight because an average person under dark skies can see stars only as faint as 6th magnitude. Today, telescopes extend our vision far into space. The Hubble Space Telescope (HST) can integrate for long exposures and detect stars as faint as 30th magnitude. The limits also have been extended to stars brighter than 1st magnitude by using zero and negative numbers. A negative magnitude signifies that an object is *brighter* than an object at zero magnitude. For example, Sirius, the brightest star in the sky, has a magnitude of −1.46. Venus can be as bright as magnitude −4.4, or about 15 times brighter than Sirius and bright enough to cast a shadow. The magnitude of the full Moon is −12.6, and that of the Sun is −26.7 (**Figure 13.4**).

Looking at that system mathematically, with five steps between the 1st and 6th magnitudes, you see that each step is equal to the fifth root of 100, or $100^{1/5}$, approximately 2.512. That system is logarithmic, but instead of the usual base 10, it is base 2.512. Thus, a 5-magnitude difference in brightness equals $(2.512)^5$, or 100, times the difference in brightness. Fifth magnitude stars are 2.512 times brighter than 6th magnitude stars, and 4th magnitude stars are $2.512 \times 2.512 = 6.310$ times brighter than 6th magnitude stars. A 2.5-magnitude difference equals $(2.512)^{2.5}$, or 10, times difference in brightness. The brightness ratio between any two stars is equal to $(2.512)^N$, where N is the magnitude difference between them.

Because the limit of the Hubble Space Telescope (HST) is 30, and $30 - 6 = 24$, HST can detect stars that are $(2.512)^{24} = 4 \times 10^9$, or 4 billion, times fainter than the magnitude 6 that the naked eye can see. Or if we compare the Sun and the Moon, the Sun is 14 magnitudes, or $(2.512)^{14} = 4 \times 10^5$ (400,000) times brighter than the full Moon. (For more detailed calculations and a table of magnitudes and brightness differences, see Appendix 7.)

The magnitude of a star, as we have discussed it so far, is called the star's **apparent magnitude** because it is the brightness of the star as it *appears* in Earth's sky. If all stars were located exactly 10 pc (32.6 light-years) away from Earth, the brightness of each star would reflect its luminosity. If the distance from Earth to a star is known, astronomers compute how bright the star would appear if it were located at 10 pc. The **absolute magnitude** of a star—its apparent magnitude at a distance of 10 pc—measures the star's luminosity.

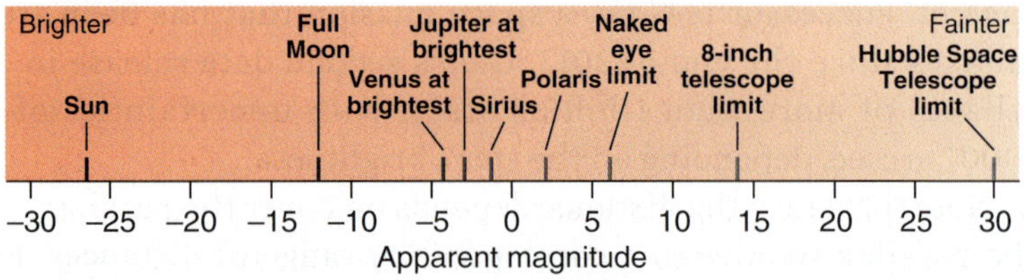

Figure 13.4 Apparent magnitude indicates the apparent brightness of an object in our sky. The brightest objects have a negative apparent magnitude, while telescopes have extended the observable range to fainter objects with higher magnitudes.

Chapter 5 that the brightness of an object that has a known luminosity L and is located at a distance d is given by

$$\text{Brightness} = \frac{\text{Total light emitted per second}}{\text{Area of sphere of radius } d} = \frac{\text{Luminosity}}{4\pi d^2}$$

You can rearrange that equation, moving the quantities you know how to measure (distance and brightness) to the right-hand side and the quantity you would like to know (luminosity) to the left, to get

$$\text{Luminosity} = 4\pi d^2 \times \text{Brightness}$$

That equation is used to find how much total light a star must be giving off to appear as bright as it does when seen from Earth.

Different stars have different luminosities. The Sun provides a convenient comparison when measuring the properties of stars, including their luminosity. The measured luminosity of the Sun is $L_{\text{Sun}} = 3.8 \times 10^{26}$ watts (W). The most luminous stars exceed the luminosity of the Sun by a million times ($10^6\ L_{\text{Sun}}$), whereas the least luminous stars have luminosities less than 1/10,000 that of the Sun ($10^{-4}\ L_{\text{Sun}}$). The most luminous stars are therefore more than 10 billion (10^{10}) times more luminous than the least luminous stars. Only a very small fraction of stars are near the upper end of that range of luminosities. Most stars are at the faint end of that distribution, less luminous than the Sun. **Figure 13.5** shows the relative number of stars compared with their luminosities in solar units. (Distances for the nearest stars are obtained from their parallaxes; other methods—discussed later—are used for the more distant stars.)

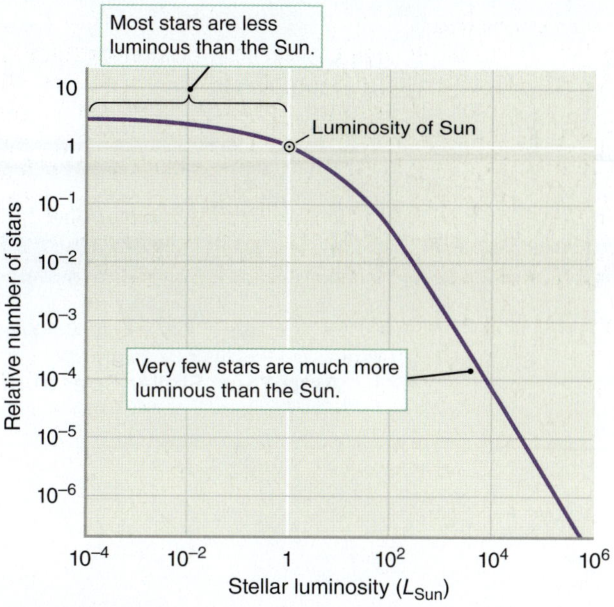

Figure 13.5 The distribution of the luminosities of stars is plotted logarithmically, with increments in powers of 10.

CHECK YOUR UNDERSTANDING 13.1

Stars A and B appear equally bright, but star A is twice as far away from us as star B. Which of the following is true? (a) Star A is twice as luminous as star B. (b) Star A is 4 times as luminous as star B. (c) Star B is twice as luminous as star A. (d) Star B is 4 times as luminous as star A. (e) Star A and star B have the same luminosity because they have the same brightness.

13.2 Astronomers Can Determine the Temperature, Size, and Composition of Stars

Two everyday concepts—stereoscopic vision and the fact that objects appear brighter when closer—have provided the tools needed to measure the distance and luminosity of the closest stars. Stars that appear to be faint points of light in the night sky are in fact luminous beacons located at great distances. The laws of radiation described in Chapter 5 reveal still more about stars.

The gas in the outer layers of stars is dense enough that the radiation from a star comes close to obeying the same laws as the radiation from solid objects, such as the heating element on an electric stove. We can therefore use what we already know about Planck blackbody radiation (Chapter 5) to understand the radiation from stars. According to the Stefan-Boltzmann law, the hotter of two same-sized objects is more luminous, and according to Wien's law, hotter objects are bluer. In this section, we use those two laws to measure the temperatures and sizes of stars. We also develop a more detailed understanding of the line emission mentioned in Chapter 5 to obtain information about the composition of stars.

Wien's Law Revisited: The Color and Surface Temperature of Stars

Wien's law (see Working It Out 5.3) shows that the temperature of an object determines the peak wavelength of its spectrum. As the temperature of the surface of an object increases, the light that it emits gets bluer. Stars with especially hot surfaces are blue, stars with especially cool surfaces are red, and yellow-white stars such as the Sun are in between. If you obtain a spectrum of a star and measure the wavelength at which the spectrum peaks, Wien's law can tell you the temperature of the star's surface. The color of a star tells you about the temperature only at the surface because that layer is giving off most of the radiation that we see. (Stellar interiors are far hotter, as we discuss in Chapter 14.)

AstroTour: Stellar Spectrum

Astronomers can determine the temperature of a star without a spectrum. They often measure the colors of stars by comparing the brightness at two specific wavelengths. The brightness of a star is usually measured through an optical **filter**—sometimes just a piece of colored glass—that lets through only a small range of wavelengths. Two of the most common filters are a blue filter, which allows light with wavelengths of about 440 nanometers (nm) to pass through, and a "visual" (yellow-green) filter, which allows light with wavelengths of about 550 nm to pass through. The ratio of brightness between the blue and visual filters is called the *color index* of the star (see Appendix 7 for more details). From a pair of pictures of a group of stars, each taken through a different filter, we can find an approximate value of the surface temperature of every star in the picture—perhaps hundreds or even thousands—all at once. That type of analysis shows that many more cool stars exist than hot stars and that most stars have surface temperatures lower than that of the Sun.

Classifying Stars by Surface Temperature

Although the hot "surface" of a star emits radiation with a spectrum very close to a smooth Planck blackbody curve, that light must then escape through the outer layers of the star's atmosphere. The atoms and molecules in the cooler layers of the star's atmosphere leave their absorption line fingerprints in the escaping light (**Figure 13.6**). Under some circumstances, the atoms and molecules in the star's atmosphere, along with any gas that might be found near the star, can produce emission lines in stellar spectra. Absorption and emission lines complicate how astronomers use the laws of Planck blackbody radiation to interpret light from stars, but spectral lines provide a wealth of information about the state and composition of the gas in a star's atmosphere.

The spectra of stars were first classified during the late 1800s, long before stars, atoms, and radiation were well understood. Stars were classified by the appearance of the dark bands (now known as absorption lines) seen in their spectra. The original ordering of that classification was arbitrarily based on the prominence of particular absorption lines known to be associated with the element hydrogen. Stars with the strongest hydrogen lines were denoted *A stars*, stars with weaker hydrogen lines were denoted *B stars*, and so on.

Annie Jump Cannon (1863–1941) led an effort at the Harvard College Observatory to examine and classify the spectra of hundreds of thousands of stars systematically. She dropped many of the earlier spectral types, keeping only seven that were later reordered on the basis of surface temperatures. Spectra of stars of several spectral types are shown in **Figure 13.7**. The hottest stars, with surface temperatures above 30,000 K, are denoted *O stars*. O stars have only weak absorption lines from hydrogen and helium. The coolest stars—*M stars*—have temperatures as low

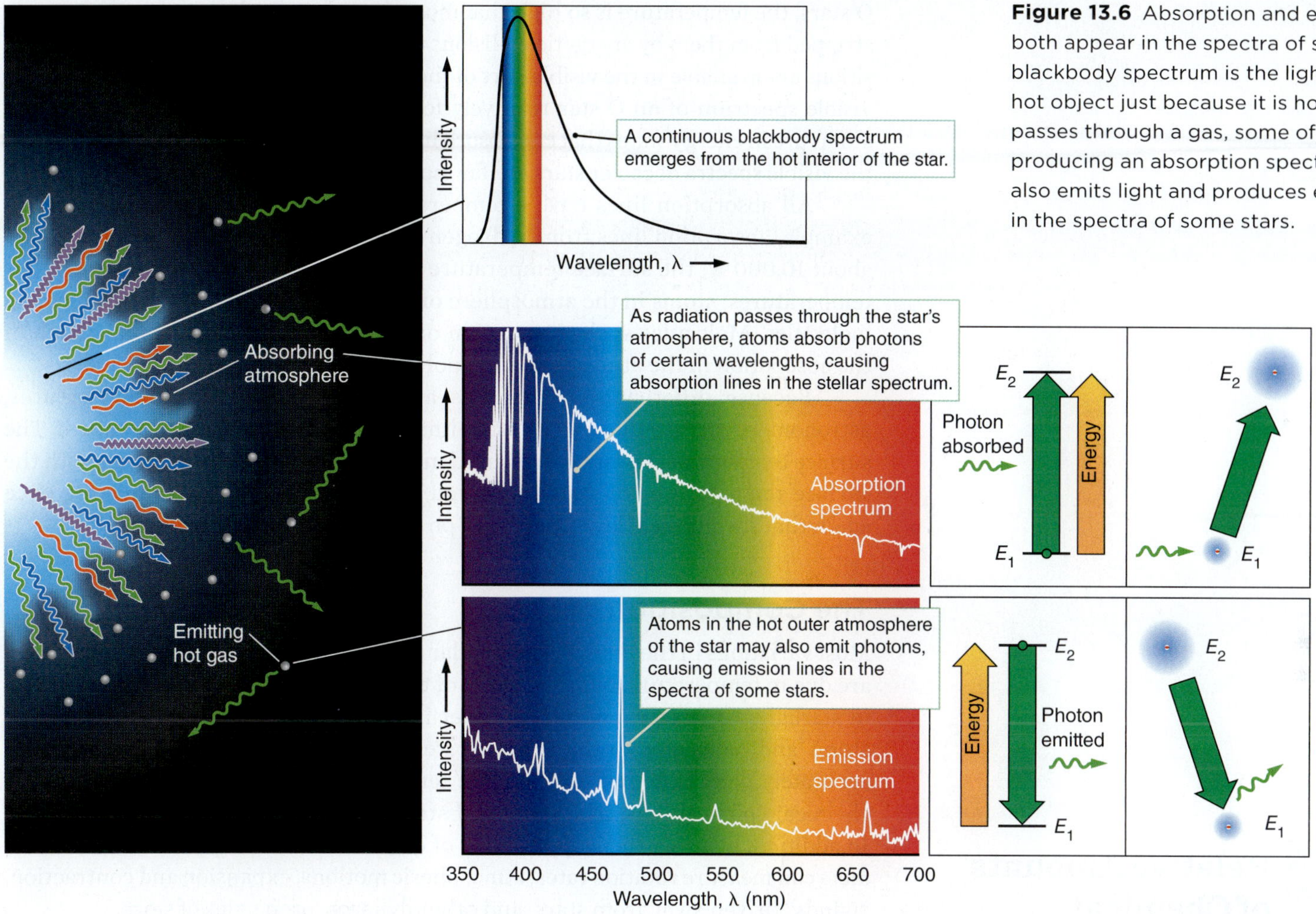

Figure 13.6 Absorption and emission lines both appear in the spectra of stars. The Planck blackbody spectrum is the light emitted from a hot object just because it is hot. As that light passes through a gas, some of it is absorbed, producing an absorption spectrum. Hot gas also emits light and produces emission lines in the spectra of some stars.

as about 2800 K. M stars have absorption lines from many types of atoms and molecules. The sequence of **spectral types** of stars, from hottest to coolest, is O, B, A, F, G, K, M. That sequence has undergone several modifications, most recently to add cooler objects known as brown dwarfs with spectral types L, T, and Y.

Astronomers divide the main spectral types into a finer sequence of subclasses by adding numbers to the letter designations. For example, the hottest B stars are B0 stars, slightly cooler B stars are B1 stars, and so on. The coolest B stars are B9 stars, only slightly hotter than A0 stars. The boundaries between spectral types are not always easy to determine. A hotter-than-average G star is very similar to a cooler-than-average F star. The Sun is a G2 star.

In Figure 13.7, notice that not only are hot stars bluer than cool stars but also the absorption lines in their spectra are different. The temperature of the gas in the atmosphere of a star affects the state of the atoms in that gas, affecting in turn the energy level transitions available to absorb radiation (see Section 5.2 to review atomic energy levels). In

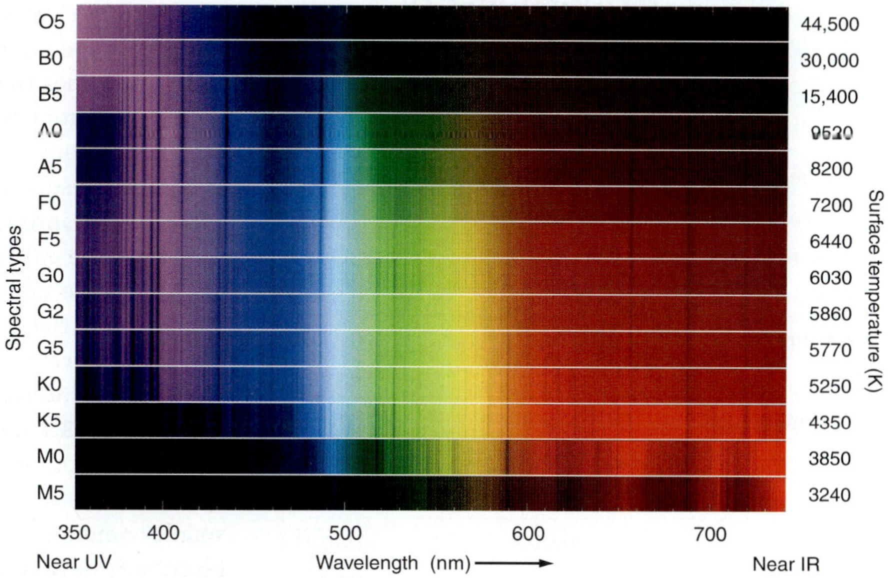

Figure 13.7 Visible light spectra of stars with different spectral types, ranging from hot, blue O stars to cool, red M stars. Hotter stars are more luminous at shorter wavelengths. The dark lines are absorption lines. These spectra have been adjusted to show only the brightest part of each spectrum.

O stars, the temperature is so high that most atoms have had one or more electrons stripped from them by energetic collisions within the hot gas. As a result, few transitions are available in the visible part of the electromagnetic spectrum, making the visible spectrum of an O star relatively featureless. At lower temperatures, more atoms are in energy states that can absorb light in the visible part of the spectrum, so the visible spectra of cooler stars are far more complex than the spectra of O stars.

All absorption lines have a temperature at which they are strongest. For example, absorption lines from hydrogen are most prominent at temperatures of about 10,000 K, the surface temperature of an A star. At the very lowest stellar temperatures, atoms in the atmosphere of a star react with one another, forming molecules. Molecules such as titanium oxide (TiO) are responsible for much of the absorption in the atmospheres of cool M stars.

Because different spectral lines are formed at different temperatures, astronomers can use those absorption lines to measure a star's temperature. The surface temperatures of stars measured in that way agree extremely well with the surface temperatures of stars measured by using Wien's law, confirming once again that the physical laws that apply on Earth apply to stars as well.

The Composition of Stars

Most variations in the lines of a particular chemical element seen in stellar spectra are due to temperature, but the details of the absorption and emission lines found in starlight also carry a wealth of other information. By applying the physics of atoms and molecules to the study of stellar absorption lines, astronomers can accurately determine not only surface temperatures of stars but also pressures, chemical compositions, magnetic-field strengths, and other physical properties. In addition, by using the Doppler shift of emission and absorption lines, astronomers can measure rotation rates, atmospheric motions, expansion and contraction, "winds" driven away from stars, and other dynamic properties of stars.

The chemical composition of a star also is found from its spectra. Each type of atom has different energy levels and thus different spectral lines (Chapter 5). The patterns of spectral lines are measured in laboratories on Earth and then used to identify the atoms (or molecules) present in stars. For example, if a star has absorption lines that correspond to the energy difference between two levels in the calcium atom, we know that calcium is present in the atmosphere of the star.

The strengths of various absorption lines tell us not only what kinds of atoms are present in the gas but also the amount of each. However, we must take great care in interpreting spectra to account properly for the temperature and density of the gas in the atmosphere of a star. Recall Figure 5.15—the "astronomer's periodic table of the elements." Typically, hydrogen composes more than 90 percent of the atoms in the atmosphere of a star, whereas helium accounts for most of what remains. All the other chemical elements, collectively called **heavy elements** or **massive elements**, are present in only very small amounts.

Table 13.1 shows the chemical composition of the atmosphere of the Sun. The Sun's composition is fairly typical for stars nearby, but the percentages of specific heavy elements can vary tremendously from star to star. Some stars have smaller amounts of heavier elements than the Sun. The existence of such stars, essentially devoid of more massive elements, provides important clues about the origin of chemical elements and the chemical evolution of the universe. Many of the atoms that make up Earth and its atmosphere (such as iron, silicon, nitrogen, oxygen, and carbon) exist as only a small percentage of the Sun.

TABLE 13.1 Relative Amounts of Chemical Elements in the Atmosphere of the Sun

Element	Percentage of Atoms in the Sun	Percentage of Sun's Mass
Hydrogen	92.5	74.5
Helium	7.4	23.7
Oxygen	0.064	0.82
Carbon	0.039	0.37
Neon	0.012	0.19
Nitrogen	0.008	0.09
Silicon	0.004	0.09
Magnesium	0.003	0.06
Iron	0.003	0.16
Sulfur	0.001	0.04
Total of others	0.001	0.03

The Stefan-Boltzmann Law and Finding the Sizes of Stars

Stars are so far away that most cannot be imaged as more than point sources. Finding the size of a star, then, involves putting together other observable measurements, such as temperature and luminosity.

The temperature of a star can be found directly, either from its color through Wien's law (**Figure 13.8a**) or from the strength of its spectral lines. The temperature of the surface of a star is one factor that influences its luminosity. If a large star and a small star are the same temperature, they will emit the same energy from every patch of surface, but the large star has more patches, so it will be more luminous altogether. Conversely, if two stars are the same size, the hotter one will be more luminous than the cooler one. That relationship is an application of the Stefan-Boltzmann law (**Figure 13.8b**). A small, hot star might even be more luminous than a larger cool star.

The luminosity of a star can be found from its brightness and its distance. Because the luminosity depends on both the temperature and the size of the star, we can use the luminosity and the temperature to determine the radius of the star (**Figure 13.8c** and **Working It Out 13.3**). Astronomers have used the luminosity-temperature-radius relationship to estimate the radii of many thousands of stars.

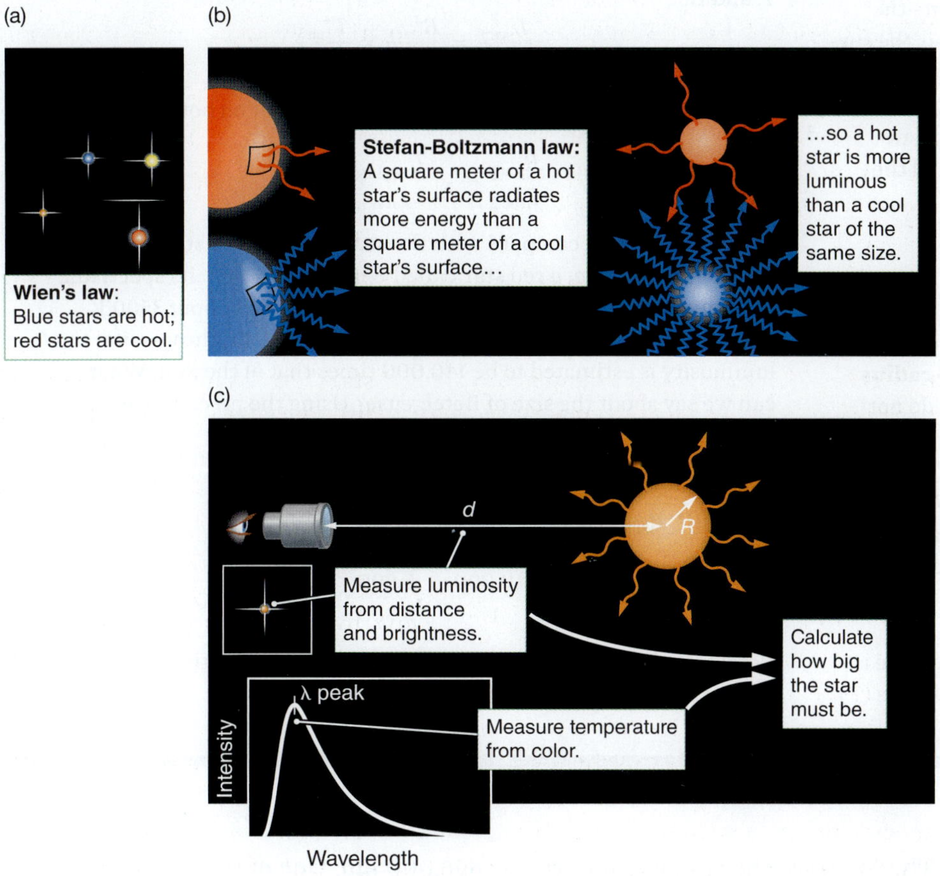

Figure 13.8 (a) The temperature of a star can be found from its color through Wien's law. (b) The luminosity depends on both the temperature and the size of the star. Blue stars are hotter; red stars are cooler. (c) Once the temperature and the luminosity are known, the size of the star can be calculated.

13.3 Working It Out Estimating the Sizes of Stars

According to the Stefan-Boltzmann law (Chapter 5), the amount of energy radiated each second by each square meter of the surface of a star is equal to the constant σ multiplied by the surface temperature of the star raised to the fourth power:

$$\begin{pmatrix}\text{Energy radiated each}\\ \text{second by 1 m}^2\text{ of surface}\end{pmatrix} = \sigma T^4$$

To find the total amount of light radiated each second by the star, we need to multiply the radiation emitted per second from each square meter by the number of square meters of the star's surface:

$$\begin{pmatrix}\text{Energy radiated}\\ \text{each second}\end{pmatrix} = \begin{pmatrix}\text{Energy radiated each}\\ \text{second by 1 m}^2\text{ of surface}\end{pmatrix} \times \begin{pmatrix}\text{Surface}\\ \text{area}\end{pmatrix}$$

The left-hand term in that equation—the total energy emitted by the star per second (in units of joules per second [J/s] = watts [W])—is the star's luminosity, L. The middle term—the energy radiated by each square meter of the star per second (in units of joules per square meter per second [J/m²/s])—can be replaced with the σT^4 factor from the Stefan-Boltzmann law. The remaining term—the number of square meters covering the surface of the star—is the surface area of a sphere, $A_{sphere} = 4\pi R^2$ (in units of square meters [m²]), where R is the radius of the star.

If we replace the words in the equation with the appropriate mathematical expressions for the Stefan-Boltzmann law and the area of a sphere, the equation for the luminosity becomes

$$L = \sigma T^4 \times 4\pi R^2$$

Combining gives

$$L = 4\pi R^2 \sigma T^4 \text{ J/s (W)}$$

That last equation is called the **luminosity-temperature-radius relationship** for stars. Because the constants (4, π, and σ) do not change, the luminosity of a star is proportional only to R^2T^4. Make a star 3 times as large, and its surface area becomes $3^2 = 9$ times as large. Because 9 times as much area is present to radiate, 9 times as much radiation is emitted. Make a star twice as hot, and each square meter of the star's surface radiates $2^4 = 16$ times as much energy. Thus, larger, hotter stars are more luminous than smaller, cooler stars.

Now, how large must a star of a given temperature be to have a total luminosity of L? The star's luminosity (L) and temperature (T) are measurable quantities, and the star's radius (R) is what we want to know. We can rearrange the previous equation, moving the properties that we know how to measure (temperature and luminosity) to the right-hand side of the equation and the property that we would like to know (the radius of the star) to the left-hand side. After doing some algebra, we find that

$$R = \sqrt{\frac{L}{4\pi\sigma T^4}} = \frac{1}{T^2} \times \sqrt{\frac{L}{4\pi\sigma}}$$

Again, the right-hand side of the equation contains only things that we know or can measure. The constants 4, π, and σ are always the same. We can find L, the luminosity of the star, from the measurements of the star's brightness and parallax (although only for nearby stars with known parallax). T is the surface temperature of the star, which can be measured from its color. From the relationship of those measurements, we now know something new: the size of the star.

If we compare two stars, the constants all cancel out, leaving L, T, and R:

$$\frac{L_{star\,1}}{L_{star\,2}} = \frac{R^2_{star\,1}}{R^2_{star\,2}} \times \frac{T^4_{star\,1}}{T^4_{star\,2}}$$

rearranging the terms, and taking the square root on both sides:

$$\frac{R_{star\,1}}{R_{star\,2}} = \sqrt{\frac{L_{star\,1}}{L_{star\,2}}} \times \frac{T^2_{star\,2}}{T^2_{star\,1}}$$

How does the Sun compare with the second-brightest star in the constellation Orion, a red star called Betelgeuse? From its spectrum, we know that Betelgeuse's surface temperature T is about 3500 K. Its distance is about 200 pc, and from that and its brightness, its luminosity is estimated to be 140,000 times that of the Sun. What can we say about the size of Betelgeuse? Using the preceding equation, we can determine the following:

$$\frac{R_{Betelgeuse}}{R_{Sun}} = \sqrt{\frac{L_{Betelgeuse}}{L_{Sun}}} \times \frac{T^2_{Sun}}{T^2_{Betelgeuse}}$$

$$\frac{R_{Betelgeuse}}{R_{Sun}} = \sqrt{\frac{140{,}000}{1}} \times \frac{5{,}800^2}{3{,}500^2} = {\sim}370 \times 2.7 \sim 1{,}000$$

Betelgeuse has a radius about 1,000 times larger than that of the Sun, making it a *supergiant*.

The radius of the Sun, written R_{Sun}, is 696,000 km. One of the smallest types of stars, called white dwarfs, have radii only about 1 percent of the Sun's radius—about the size of Earth. The largest stars, called red supergiants, can have radii more than 1,000 times that of the Sun. Many more stars are toward the small end of that range—smaller than the Sun—than giant stars.

CHECK YOUR UNDERSTANDING 13.2

If star A has twice the surface temperature of the Sun but has the same luminosity, the radius of star A must be ______________ the radius of the Sun. (a) 16 times; (b) 4 times; (c) 1/2; (d) 1/4

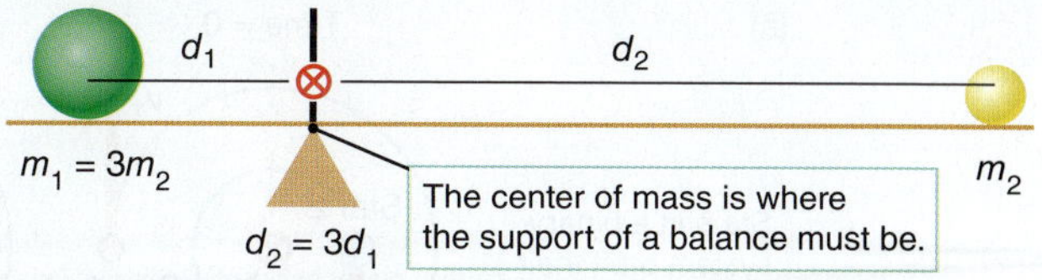

Figure 13.9 The center of mass of two objects is the "balance" point on a line joining the centers of two masses.

13.3 Measuring the Masses of Stars in Binary Systems

Determining the mass of a star is difficult. Astronomers cannot use the amount of light from a star or the star's size as a measure of its mass. Stars can be large or small, faint or luminous. However, more massive stars *always* have stronger gravity. When astronomers are trying to determine the masses of astronomical objects, they almost always wind up looking for the effects of gravity. In Chapter 4, you learned that Kepler's laws of planetary motion are the result of gravity and that the properties of the orbit of a planet can be used to measure the mass of the Sun. Similarly, astronomers can study two stars that orbit each other to determine their masses.

About half of the higher-mass stars in the sky are actually systems consisting of several stars moving about under the influence of their mutual gravity. Most of those are **binary stars**, in which two stars orbit each other in elliptical orbits as predicted by Newton's version of Kepler's laws. That version of Kepler's laws can be used to find the mass of a star, as we will demonstrate. However, most low-mass stars are single, and low-mass stars far outnumber higher-mass stars, so most stars are single and their mass cannot be found that way. In this section, we look at how astronomers measure the masses of stars in binary systems.

Binary Star Orbits

The **center of mass** is the balance point of a system. If the two objects were sitting on a seesaw in a gravitational field, the support of the seesaw would have to be directly under the center of mass for the objects to balance (**Figure 13.9**). When Newton applied his laws of motion to the problem of orbits, he found that two objects must move in elliptical orbits around each other and that their common center of mass lies at one focus shared by both ellipses (**Figure 13.10**). The center of mass, which lies along the line between the two objects, remains stationary. The two objects will always be found on exactly opposite sides of the center of mass.

Imagine that you are watching a binary star, as shown in **Figure 13.11a**. As seen from above, two stars orbit the common center of mass. Star 1, which is less massive, must complete its orbit in the same time as star 2, which is more massive. Because the less massive star has farther to go around the center of mass, it must be moving *faster* than the more massive star. In that face-on view, no determination of the Doppler shift (Chapter 5) can be made because all the motion is in the plane of the sky, and none is toward or away from the observer.

When a system is edge-on to the observer, however, the observer can take advantage of the Doppler shift to find out about the motion. **Figure 13.11b**

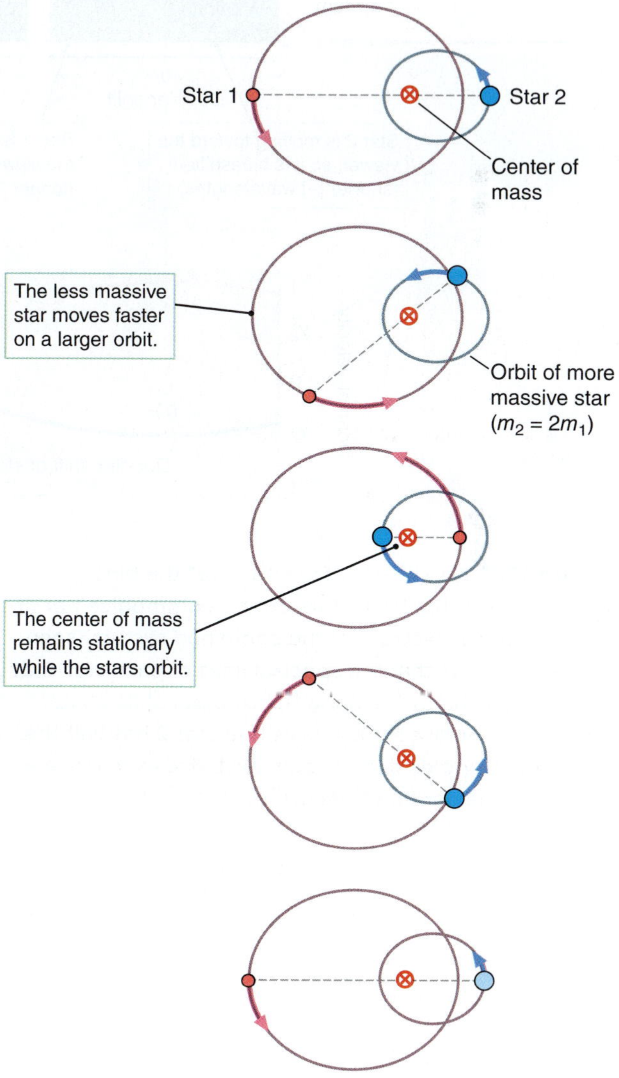

Figure 13.10 In a binary star system, the two stars orbit on elliptical paths about their common center of mass. Here, the blue star has twice the mass of the red one. The eccentricity of the orbits is 0.5. The time steps between the frames are equal.

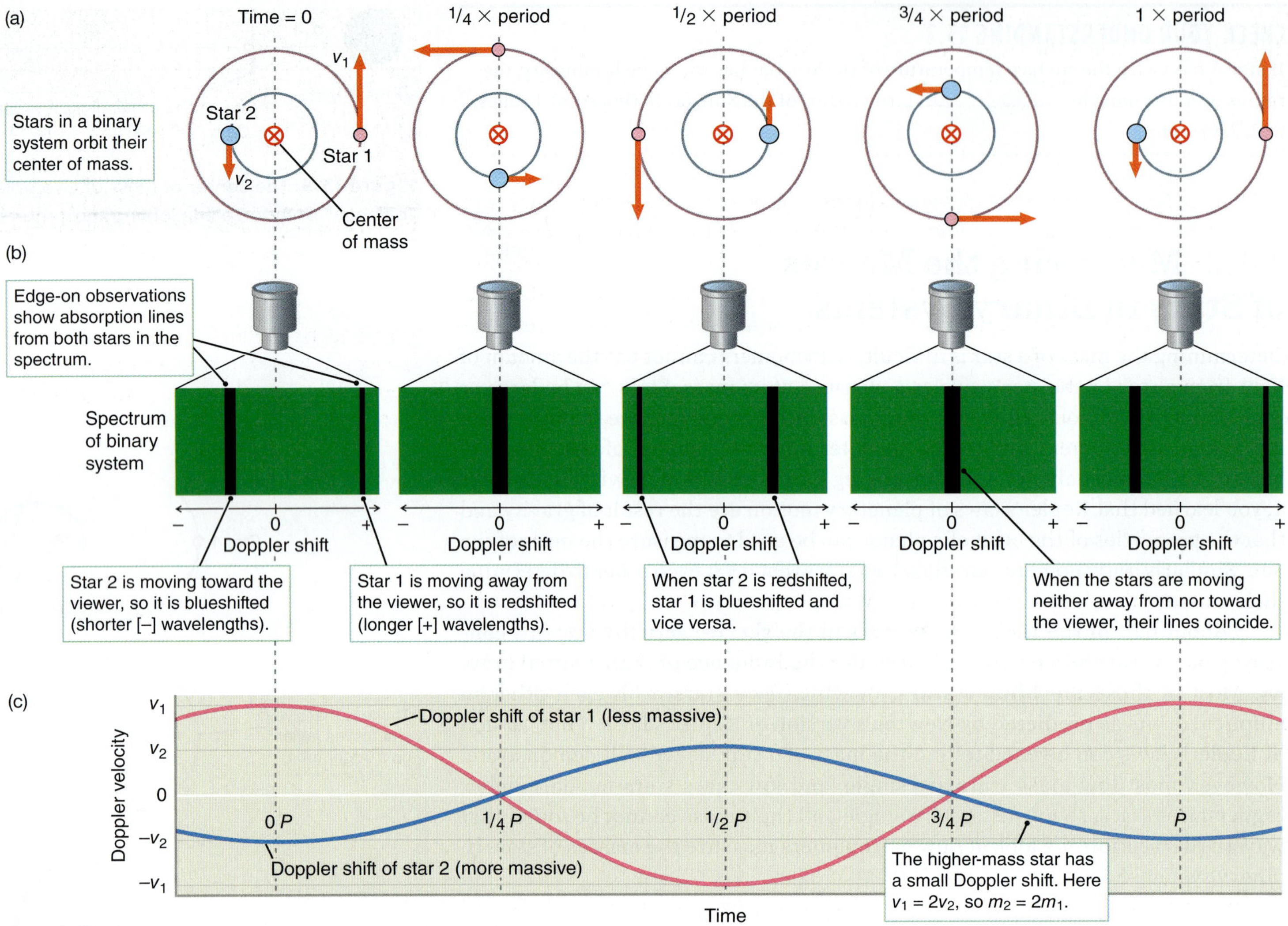

Figure 13.11 (a) The view from "above" the binary system shows that both stars orbit a common center of mass. (b) The spectrum of the combined system (seen edge-on) shows that the spectral lines of each star shift back and forth. (c) Graphing the Doppler shift of star 1 with star 2 versus time reveals that star 2 has half the maximum Doppler shift, so star 2 is twice as massive as star 1. *P* is the period of the orbit.

shows observations of the spectrum of the combined system associated with each position in Figure 13.11a, if it were edge-on instead of face-on. The spectral lines of the stars shift back and forth as they move toward and away from the observer. Because the two stars are always on exactly opposite sides of their center of mass, they always move in opposite directions. Thus, when star 2 approaches, star 1 recedes. The light coming from star 2 will be shifted to *shorter* wavelengths because of the Doppler effect as it approaches, so the light will be *blue*shifted, and the light coming from star 1 will be shifted to *longer* wavelengths as it recedes, so the light will be *red*shifted. Half an orbital period later, the situation is reversed: lines from star 1 are blueshifted, and lines from star 2 are redshifted.

The less massive star has a larger orbit—and consequently moves more quickly—than the more massive star. **Figure 13.11c** compares the velocity obtained from the maximum Doppler shift for star 1 with the velocity obtained from the maximum Doppler shift for star 2. That comparison gives the ratio of the masses of the two stars:

$$\frac{v_1}{v_2} = \frac{m_2}{m_1}$$

By observing the spectrum of the system, we can find from that equation the relative masses of the two stars—star 2 is 2 times as massive as star 1—but we can't find the actual mass of either star from those observations alone.

The Masses of Binary Stars

In Chapter 4, we ignored the complexity of the motion of two objects around their common center of mass. Now, however, that very complexity enables us to measure the masses of the two stars in a binary system. If we can measure the period of the binary system and the average separation between the two stars, Kepler's third law gives us the total mass in the system: the sum of the two masses. Because the analysis in the previous subsection gives us the ratio of the two masses, we now have two relationships between two unknowns. We have all we need to determine the mass of each star separately. In other words, if we know that star 2 is 2 times as massive as star 1, and we know that star 1 and star 2 together are 3 times as massive as the Sun, we can calculate separate values for the masses of star 1 and star 2.

Depending on the type of system, several ways are available to measure the average separation and the orbital period. In a **visual binary** system, the system is close enough to Earth, and the stars are far enough from each other, that we can take pictures that show the two stars separately (**Figure 13.12**). Then, astronomers can directly measure the shapes and periods of the orbits of the two stars just by watching them as they orbit each other. Those values can be used with Doppler measurements of the radial (line-of-sight) velocities of the stars to solve for the ratio of the two masses.

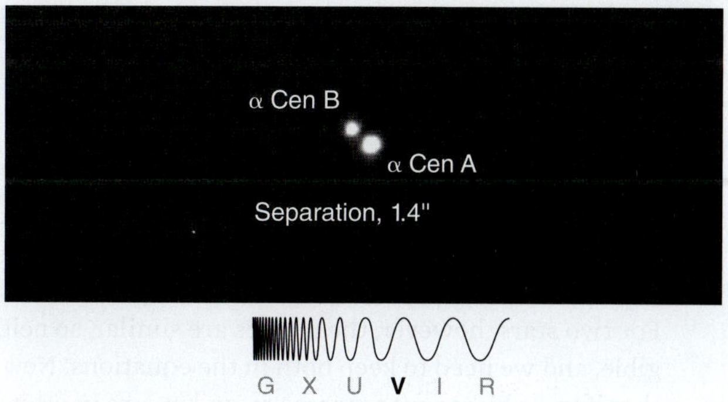

Figure 13.12 The two stars of a visual binary are resolved. These stars are two components of the Alpha Centauri system, the nearest star system to the Solar System.

In most binary systems, however, the two stars are so close together and so far away from us that we cannot see the stars separately. The identification of those stars as binary systems is indirect and comes from observing periodic variations in the *light* from the star or from observing periodic changes in the *spectrum* of the star. If we view a binary system nearly edge-on, so that one star passes in front of the other, it is called an **eclipsing binary**. An observer will see a repeating dip in brightness as one star passes in front of (eclipses) the other (**Figure 13.13**). If the stars have different temperatures, a repeating pattern will occur of a smaller dip in brightness when the hotter star eclipses the cooler one, followed by a larger dip in brightness when the cooler star eclipses the hotter one. The pattern of those dips also gives an estimate of the relative sizes (radii) of the two stars. That procedure for identifying binary systems—similar to the transit method for finding extrasolar planets discussed in Chapter 7—works only when the system is viewed nearly edge-on. The Kepler space telescope has discovered thousands of eclipsing binaries in addition to finding new extrasolar planets.

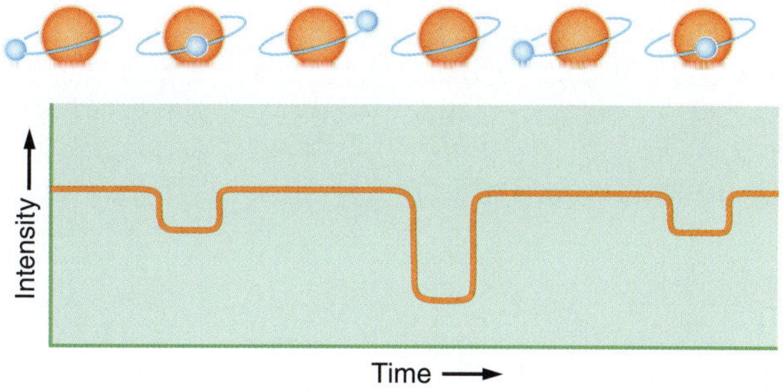

Figure 13.13 In an eclipsing binary system, the system is viewed nearly edge-on, so that the stars repeatedly pass behind each other, blocking some of the light. When the blue star passes in front of the larger, cooler star, less light is blocked than when the smaller star passes behind the red star. The shape of the dips in the light curve of an eclipsing binary can reveal information about the relative size and surface brightness of the two stars.

In a **spectroscopic binary** system, the spectral lines of the two stars show periodic changes as they are Doppler-shifted away from each other, first in one direction and then in the other (Figure 13.11). The period of the orbit is determined from the time for a set of spectral lines to go from approaching to receding and back again. The orbital velocities of the stars and the period of the orbit give the size of the orbit because distance equals velocity multiplied by time. Consequently, astronomers can estimate the combined masses of the two stars. To calculate the individual masses, an estimate of the tilt of the orbit is needed. Thus, spectroscopic binary masses are more approximate than those in eclipsing binary systems.

13.4 Working It Out Measuring the Mass of an Eclipsing Binary Pair

In Working It Out 4.3, we used Newton's version of Kepler's third law to calculate the mass of the Sun by observing the orbital period of one of its planets. In that special case, the Sun's mass is so much greater than the mass of a planet that the planet's mass is negligible. For two stars, however, the masses are similar, so neither is negligible, and we need to keep both in the equations. Newton showed that if two objects with masses m_1 and m_2 are in orbit about each other, the period of the orbit, P, is related to the average distance between the two masses, the semimajor axis A, by the equation

$$P^2 = \frac{4\pi^2 A^3}{G(m_1 + m_2)}$$

Rearranging that equation turns it into an expression for the sum of the masses of the two objects:

$$m_1 + m_2 = \frac{4\pi^2}{G} \times \frac{A^3}{P^2}$$

We could use the equation that way, with the masses of the two stars in kilograms (kg), the distance between them in kilometers (km), the period of their orbit in seconds (s), and the gravitational constant G in $km^3/(kg\ s)$. However, astronomers often think about stellar masses in units of the Sun's mass. If we divide that equation by the "mass of the Sun" equation in Working It Out 4.3,

$$M_{Sun} = \frac{4\pi^2}{G} \times \frac{A^3}{P^2}$$

(where M_{Sun} = mass of the Sun, $A = 1$ AU, and $P = 1$ year), the constants cancel out and the equation simplifies to

$$\frac{m_1}{M_{Sun}} + \frac{m_2}{M_{Sun}} = \frac{A^3_{AU}}{P^2_{years}}$$

Therefore, if we know both m_1/m_2 from measuring velocities by Doppler shifts and $m_1 + m_2$ from the observed orbital properties, we can solve for the separate values of m_1 and m_2.

Suppose you are an astronomer studying a binary star system. After observing the star for several years, you accumulate the following information about the system:

1. The star is an eclipsing binary.
2. The period of the orbit is 2.63 years.
3. Star 1 has a Doppler velocity that varies between +20.4 and −20.4 km/s.
4. Star 2 has a Doppler velocity that varies between +6.8 and −6.8 km/s.
5. The stars are in circular orbits. You know that because the Doppler velocities about the star are symmetric; the approach and recession speeds of the star are equal.

Those data are summarized in **Figure 13.14.** You begin your analysis by noting that the star is an eclipsing binary, which tells you that the orbit of the star is edge-on to your line of sight. The Doppler velocities tell you the total orbital velocity of each star, and you determine the size of the orbits by using the relationship

$$\text{Distance} = \text{Speed} \times \text{Time}$$

In one orbital period, star 1 travels around a circle—a distance of

$$d = (20.4\ \text{km/s}) \times (2.63\ \text{yr}) = 53.7\ \text{km} \times \text{yr/s}$$

A binary system can fall into more than one of those three categories, regardless of how it was originally discovered. If a spectroscopic binary system is also a visual or eclipsing binary, the orbit and masses of the stars can be completely solved (**Working It Out 13.4**). Historically, most stellar masses were measured for stars in eclipsing binary systems rather than for those in visual or spectroscopic binaries. New observational capabilities, however, have increased the number of known visual binaries by greatly improving the ability to see the stars in a binary directly. Accurate measurements of masses have been obtained for several hundred binary stars, about half of which are eclipsing binaries. The range of stellar masses found in that way is not nearly as great as the range of stellar luminosities. The least massive stars have masses of about 0.08 M_{Sun}, whereas the most massive stars appear to have masses up to about 200 M_{Sun}.

CHECK YOUR UNDERSTANDING 13.3

Which of the following properties must be measured to determine the masses of stars in a typical binary system? (Choose all that apply.) (a) the period of the orbits of the two stars; (b) the average separation between the two stars; (c) the radii of the two stars; (d) the velocities of the two stars.

Multiply by the number of seconds in a year:

$$d = 53.7 \frac{\text{km} \times \text{yr}}{\text{s}} \times \frac{3.16 \times 10^7\,\text{s}}{\text{yr}} = 1.70 \times 10^9\,\text{km}$$

That distance is the circumference of the star's orbit, or 2π times the radius of the star's orbit, A_1. Thus, star 1 is following an orbit with a radius of

$$A_1 = \frac{d}{2\pi} = \frac{1.70 \times 10^9\,\text{km}}{2\pi} = 2.7 \times 10^8\,\text{km}$$

To convert that result to astronomical units, use the relation $1\,\text{AU} = 1.50 \times 10^8$ km:

$$A_1 = 2.7 \times 10^8\,\text{km} \times \frac{1\,\text{AU}}{1.50 \times 10^8\,\text{km}} = 1.8\,\text{AU}$$

A similar analysis of star 2 shows that its orbit has a radius of $A_2 = 0.6$ AU.

Next, apply Newton's version of Kepler's third law. Because the stars are always on opposite sides of the center of mass, the semi-major axis $A_{AU} = 1.8\,\text{AU} + 0.6\,\text{AU} = 2.4\,\text{AU}$. Because you know A and the period P (measured as 2.63 years), you can calculate the total mass of the two stars:

$$\frac{m_1}{M_{Sun}} + \frac{m_2}{M_{Sun}} = \frac{(A_{AU})^3}{(P_{years})^2} = \frac{(2.4)^3}{(2.63)^2} = 2.0$$

Thus, the combined mass of the two stars is twice the mass of the Sun. To sort out the individual masses of the stars, use the measured velocities and the fact that the mass and velocity are inversely proportional:

$$\frac{m_2}{m_1} = \frac{v_1}{v_2} = \frac{20.4\,\text{km/s}}{6.8\,\text{km/s}} = 3.0$$

Star 2 is 3 times as massive as star 1. In mathematical terms, $m_2 = 3 \times m_1$. Substituting into the equation

$$m_1 + m_2 = 2.0\,M_{Sun}$$

gives

$$m_1 + 3m_1 = 2.0\,M_{Sun}$$

or $4m_1 = 2.0\,M_{Sun}$, so $m_1 = 0.5\,M_{Sun}$. Because $m_2 = 3 \times m_1$, $m_2 = 1.5\,M_{Sun}$.

Star 1 has a mass of $0.5\,M_{Sun}$, and star 2 has a mass of $1.5\,M_{Sun}$. You have just found the masses of two distant stars.

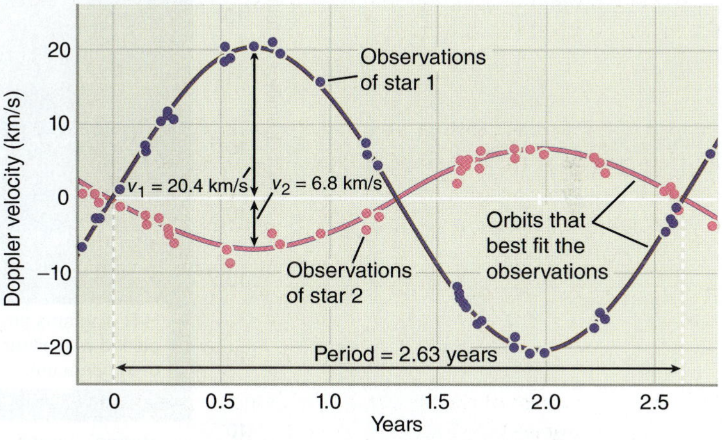

Figure 13.14 Doppler velocities of the stars in an eclipsing binary are used to measure the masses of the stars.

13.4 The Hertzsprung-Russell Diagram Is the Key to Understanding Stars

Measuring basic properties of stars is only the first step in understanding; the next step is to look for patterns in their properties. In the early 20th century, Ejnar Hertzsprung (1873–1967) and Henry Norris Russell (1877–1957) independently studied the properties of stars. Both astronomers plotted the luminosities of stars versus their surface temperatures—a diagram that came to be known as the *Hertzsprung-Russell diagram*, or simply **H-R diagram**. We use that diagram often for the study of stars. In this section, we take a first look at that important diagram and how stars are organized within it.

The H-R Diagram

The H-R diagram (**Figure 13.15**) is a graph of luminosity versus temperature that can be used to study how stars change over time. The spectral type is plotted on the horizontal axis (the *x*-axis), along with the surface temperature plotted backward: temperature is higher on the left and lower on the right. Hot, blue stars are on the

Figure 13.15 The Hertzsprung-Russell (H-R) diagram is used to plot the properties of stars. More luminous stars are at the top of the diagram. Hotter stars are on the left. Stars of the same radius (*R*) lie along the dotted lines moving from upper left to lower right. Absolute magnitudes are discussed in Working It Out 13.2 and Appendix 7.

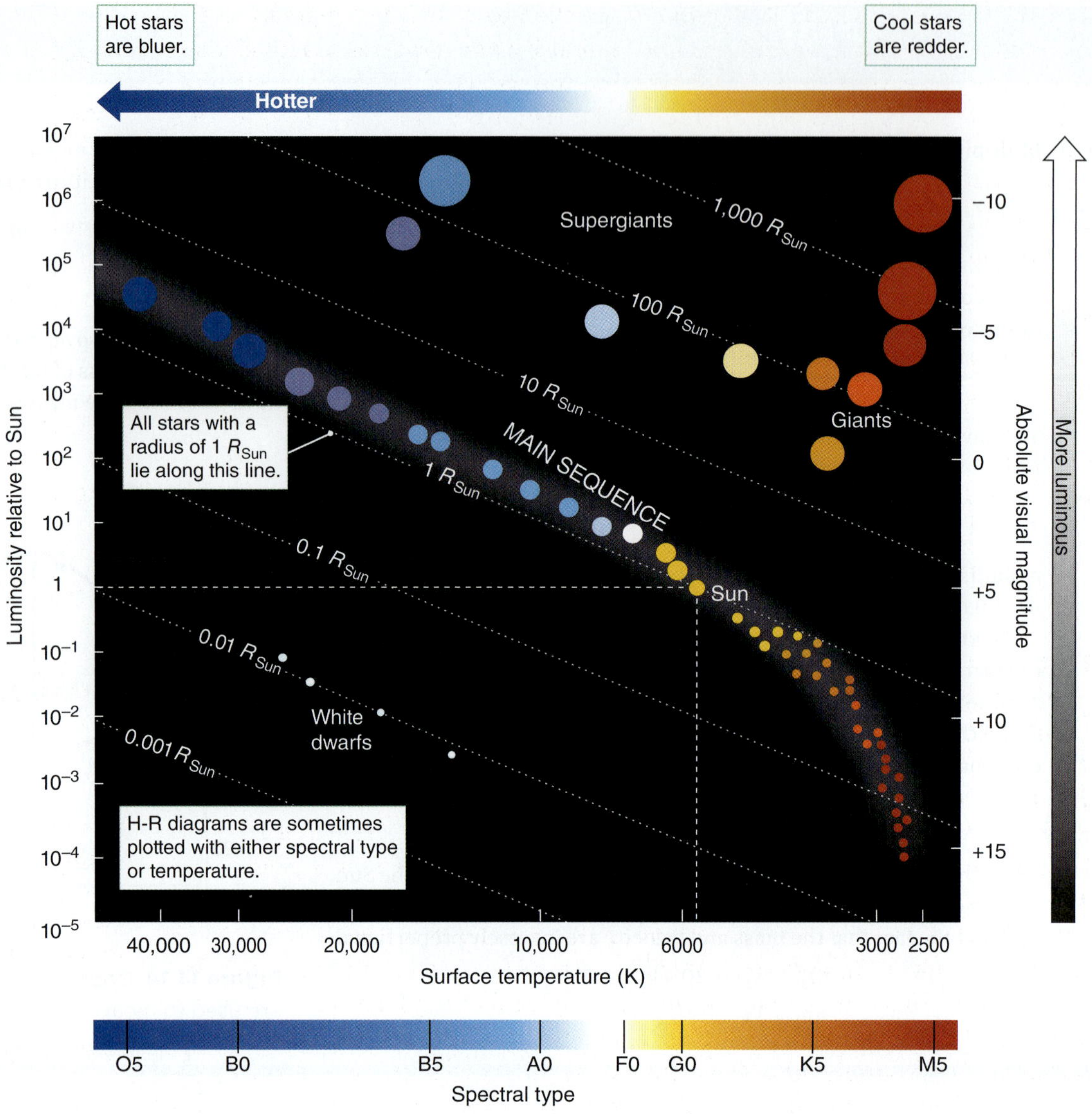

left side of the H-R diagram; cool, red stars are on the right. Temperature is plotted logarithmically. For example, consider two points on the H-R diagram. One represents a star with a surface temperature of 40,000 K, and the other a star with a surface temperature of 20,000 K—a temperature change of a factor of 2. The interval along the axis between those points is therefore the same size as the interval between points representing stars with surface temperatures of 10,000 and 5000 K—with that same change by a factor of 2. The horizontal axis is sometimes labeled with another characteristic that corresponds to temperature, such as the color.

AstroTour: H-R Diagram

Simulation: H-R Diagram

The luminosity of stars is plotted along the vertical axis (the *y*-axis). More luminous stars are toward the top of the diagram; less luminous stars are toward the bottom. Sometimes the luminosity axis is labeled with the absolute visual magnitude instead of luminosity, as shown on the right-hand *y*-axis. As with the temperature axis, luminosities are plotted logarithmically. Here, each step along the left-hand *y*-axis corresponds to a multiplicative factor of 10 in the luminosity. To understand why the plotting is done that way, recall that the most luminous stars are 10 billion times more luminous than the least luminous stars, yet all those stars must fit on the same plot.

Each point on the H-R diagram is specified by a surface temperature and luminosity. Therefore, we can use the luminosity-temperature-radius relationship described earlier in the chapter to find the radius of a star at that point as well. A star in the upper right corner of the H-R diagram is very cool, so each square meter of its surface radiates only a small amount of energy. That star also is extremely luminous. It must be huge, then, to account for its high luminosity, despite the feeble radiation coming from each square meter of its surface. Conversely, a star in the lower left corner of the H-R diagram is very hot, which means that a large amount of energy is coming from each square meter of its surface. That star has a very low overall luminosity, however, so it must be very small. Moving up and to the right takes you to larger and larger stars; moving down and to the left takes you to smaller and smaller stars. All stars of the same radius lie along slanted lines across the H-R diagram. Astronomers can note the properties of a star—its temperature, color, size, and luminosity—from a glance at its position on the H-R diagram. The discovery and study of those patterns led to an understanding of the astrophysics of stars (see the **Process of Science Figure**).

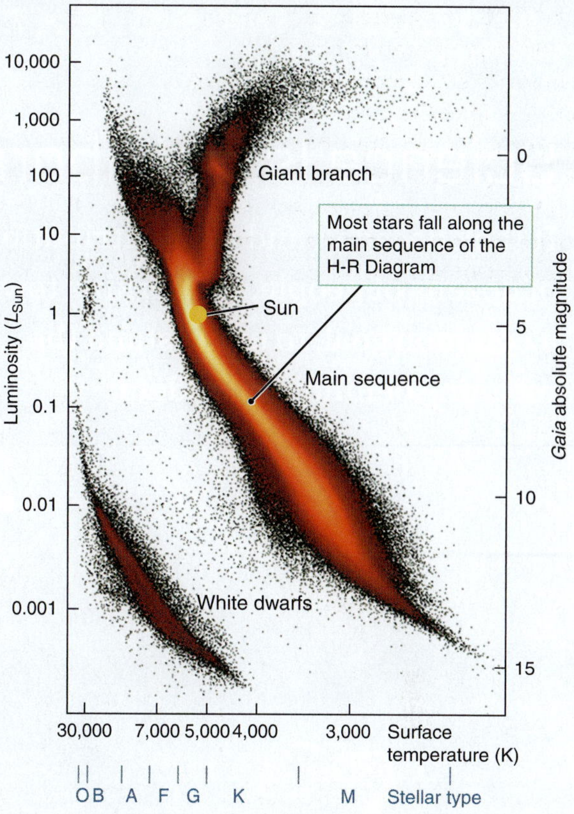

Figure 13.16 An H-R diagram for 4 million stars within 5,000 light-years of the Sun plotted from data obtained by the *Gaia* satellite clearly shows the main sequence. Most of the stars lie along this band running from the lower right of the diagram toward the upper left.

The Main Sequence

Figure 13.16 shows 4 million stars plotted on an H-R diagram. The data are based on observations of stars within 5,000 light-years of the Sun, near enough for parallax measurements obtained by the *Gaia* satellite. A quick look at the diagram immediately shows what was first discovered in the original diagrams of Hertzsprung and Russell. About 90 percent of the stars in the sky lie in a well-defined region running across the H-R diagram from lower right to upper left, known as the **main sequence**. On the left end of the main sequence are the O stars: hotter, larger, and more luminous than the Sun. On the right end of the main sequence are the M stars: cooler, smaller, and fainter than the Sun. If you know where a star lies on the main sequence, you know its approximate luminosity, surface temperature, and size.

The H-R diagram offers a useful method for finding the distance to main-sequence stars. Astronomers can determine whether a star is on the main sequence by looking at the absorption lines in its spectrum. The spectral type also is determined from the spectral lines, and that spectral type indicates the star's temperature. Once that value on the *x*-axis is known, we can then read up to the main sequence and then across to the *y*-axis to find the star's luminosity. Recall that the luminosity, brightness, and distance are all connected. We can think about how far away a star of a particular luminosity must be to have the observed brightness. So we can find the star's distance by comparing a star's luminosity, obtained from the H-R diagram, with its apparent brightness. That method of determining distances to main-sequence stars from the spectra, luminosity, and brightness of stars is called **spectroscopic parallax**. Details of that method are discussed in Appendix 7. Despite the similarity between the names, the method is very different from the parallax method using trigonometry. Spectroscopic parallax is useful to much larger distances than trigonometric parallax, although it is less precise.

From a combination of observations of binary star masses, parallax, luminosity measurements, and mathematical models, astronomers have determined that stars of different masses lie on different parts of the main sequence. Stellar mass increases smoothly from the lower right to the upper left along the main sequence (**Figure 13.17**). If a main-sequence star is *less* massive than the Sun, it is also smaller, cooler, redder, and less luminous than the Sun and is located to the lower right of the Sun on the main sequence. Conversely, if a main-sequence star is *more* massive than the Sun, it is also larger, hotter, bluer, and more luminous

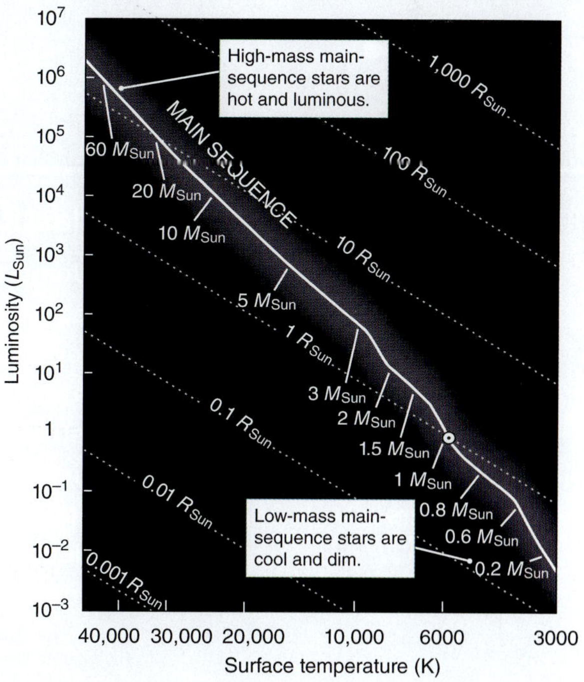

Figure 13.17 Mass determines the location of a star along the main sequence.

PROCESS OF SCIENCE

Science Is Collaborative

It took decades, and the contributions of dozens of people, all working toward a common goal, to understand the meaning behind stellar data.

Annie Jump Cannon

Ejnar Hertzsprung

Henry Norris Russell

Meghnad Saha

Cecilia Payne-Gaposchkin

1800s
The Observations

About 500,000 photographs of stellar spectra were obtained by many astronomers at many telescopes.

1900s
The Classification

Annie Jump Cannon led a team that classified all the spectra according to the strengths of particular absorption lines at particular wavelengths.

1910s
The Graph

Hertzsprung and Russell independently developed what will later be called the H-R diagram. They did not understand why the *x*-axis, ordered O-B-A-F-G-K-M, revealed a nice band across the middle of the diagram. Russell hypothesized this result must be caused by a single stellar characteristic.

1920s
The Understanding

Meghnad Saha showed that the stellar characteristic in question was temperature. Cecilia Payne-Gaposchkin showed that stars were mostly composed of hydrogen and helium. Modern astrophysics was born; others went on to develop the under-standing of stellar atmospheres.

Scientific discoveries sometimes seem to occur suddenly. Instead, new scientific knowledge usually results from the effort of many people working for many years to solve a problem.

than the Sun and is located to the upper left of the Sun on the main sequence. The mass of a star determines where on the main sequence the star will lie.

Table 13.2 summarizes the properties of the spectral classes of main-sequence stars. *All* main-sequence stars with a mass of 1 M_{Sun} are G2 stars like the Sun and have the *same* surface temperature, size, and luminosity as the Sun. Similarly, if a main-sequence star is classified as B0, it has a surface temperature of about 30,000 K, a luminosity about 32,500 times that of the Sun, a mass of about 17.5 M_{Sun}, and a radius of about 6.7 R_{Sun}. If a different main-sequence star is classified as M5, it has a surface temperature of 3170 K, a luminosity of about 0.008 L_{Sun}, a mass of about 0.21 M_{Sun}, and a radius of about 0.29 R_{Sun}.

The relationship between the mass and the luminosity of stars is very sensitive. Relatively small differences in the masses of stars result in large differences in their main-sequence luminosities. From determining the luminosities of binary stars with measured mass, a relationship between the mass and luminosity was found. That **mass-luminosity relationship**, usually expressed as $L \propto M^{3.5}$, is shown in **Figure 13.18a**. The exact exponent varies from about 3 to 4 for different ranges of stellar masses, but the method is useful for estimating masses of single stars. The mass also correlates to the size of a star (**Figure 13.18b**) and to the temperature of a star (**Figure 13.18c**).

The mass and chemical composition of a main-sequence star determine how large it is, what its surface temperature is, how luminous it is, what its internal structure is, how long it will live, how it will evolve, and what its final fate will be. A star must have a balance between gravity trying to hold the star together and the energy released by nuclear reactions within the star trying to blow it apart. The mass of the star determines the strength of its gravity, which in turn determines how much energy must be generated in its interior to prevent it from collapsing under its own weight. The mass of a star determines where that balance is struck.

TABLE 13.2 Properties of Main-Sequence Stars

Spectral Type	Temperature (K)	Mass (M_{Sun})	Radius (R_{Sun})	Luminosity (L_{Sun})
O5	42,000	60	13	500,000
B0	30,000	17.5	6.7	32,500
B5	15,200	5.9	3.2	480
A0	9800	2.9	2.0	39
A5	8200	2.0	1.8	12.3
F0	7300	1.6	1.4	5.2
F5	6650	1.4	1.2	2.6
G0	5940	1.05	1.06	1.25
G2 (Sun)	5780	1.00	1.00	1.0
G5	5560	0.92	0.93	0.8
K0	5150	0.79	0.93	0.55
K5	4410	0.67	0.80	0.32
M0	3840	0.51	0.63	0.08
M5	3170	0.21	0.29	0.008

Stars Not on the Main Sequence

Although 90 percent of stars are main-sequence stars, some stars are found in the upper right portion of the H-R diagram, well above the main sequence (see Figure 13.15). Those stars, which must be luminous, cool, and large, with radii hundreds or thousands of times the radius of the Sun, are called giants. At the other extreme are stars found in the far lower left corner of the H-R diagram. Those stars are the tiny white dwarfs, comparable to the size of Earth. Their small surface areas explain why they have such low luminosities, despite having high temperatures.

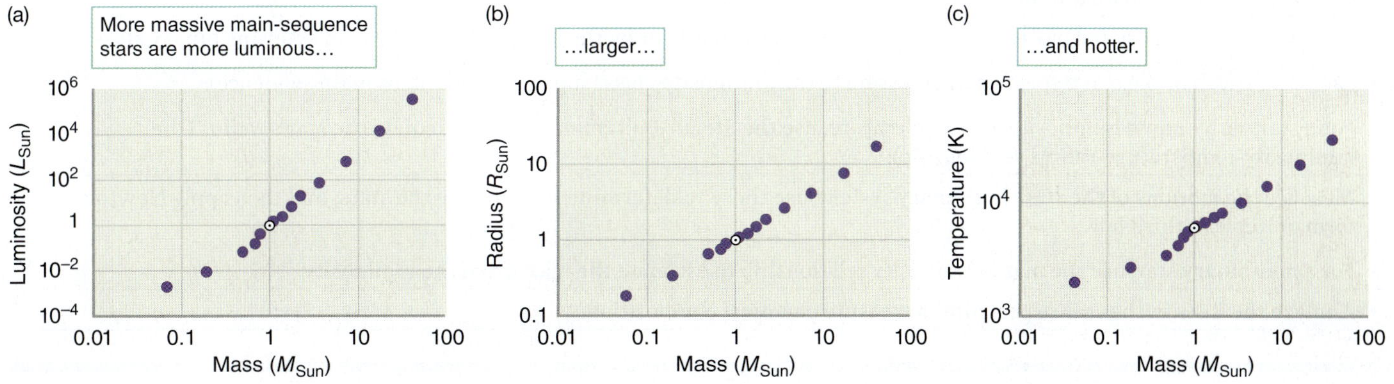

Figure 13.18 These graphs plot luminosity (a), radius (b), and temperature (c) versus mass for stars along the main sequence. The mass of a main-sequence star determines its other properties.

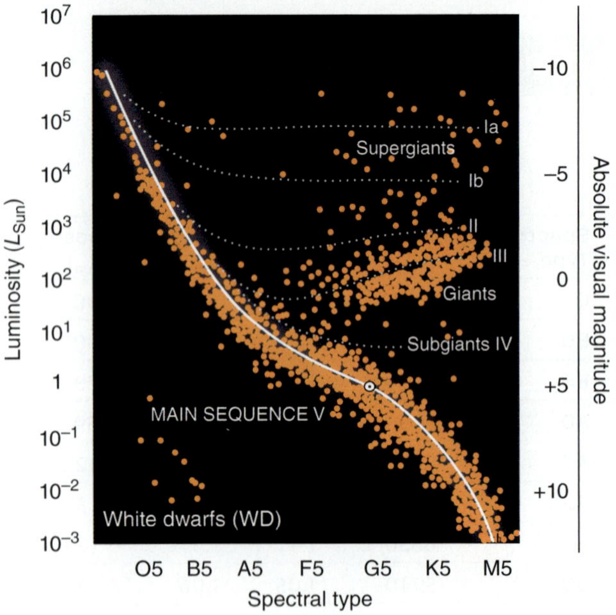

Figure 13.19 Stellar luminosity classes indicate the size (radius) of a star at each spectral type.

Stars that lie off the main sequence on the H-R diagram can be identified by their luminosities (determined by their distance) or by slight differences in their spectral lines. The width of a star's spectral lines indicates the density and surface pressure of gas in the star's atmosphere. In general, denser stars have broader lines. Puffed-up stars above the main sequence have lower densities and lower surface pressure and narrower absorption lines than main-sequence stars.

When using the H-R diagram to estimate the distance to a star by the spectroscopic parallax method, astronomers must know whether the star is on, above, or below the main sequence to find the star's luminosity. The spectral line widths of stars both on and off the main sequence indicate **luminosity class**, which tells us the relative *size* of the star within each spectral class. Supergiant stars, the largest stars that we see, are luminosity class I, bright giants are class II, giants are class III, subgiants are class IV, main-sequence stars are class V, and white dwarfs are class WD. Luminosity classes I–IV lie above the main sequence, whereas class WD falls below and to the left of the main sequence (**Figure 13.19**). Thus, the complete spectral classification of a star includes both its spectral type (indicating temperature and color) and its luminosity class (indicating relative size).

The existence of the main sequence, together with the fact that the mass of a main-sequence star determines where on the sequence it will lie, creates a grand pattern that makes it possible to understand stars in fundamental ways. The existence of stars that do *not* follow that pattern raises yet more questions. In the coming chapters, you will learn that the main sequence tells us what stars are and how they work, and that stars off the main sequence reveal how stars form, evolve, and die. **Table 13.3** summarizes the techniques that astronomers use to determine some of the basic properties of stars. Of the properties listed in the table,

TABLE 13.3 Taking the Measure of Stars

Property	Methods
Luminosity	• For a star with a known distance, measure the brightness and then apply the inverse square law of radiation: $\text{Luminosity} = 4\pi \times \text{Distance}^2 \times \text{Brightness}$ • For a star without a known distance, take a spectrum of the star to determine its spectral and luminosity classes, plot them on an H-R diagram, and read the luminosity from the diagram.
Temperature	• Measure the color index of the star by using blue and visual filters. Use Wien's law to relate the color to a temperature. • Take a spectrum of the star, and estimate the temperature from its spectral class by noting which spectral lines are present.
Distance	• For a relatively nearby star (within a few hundred parsecs), measure the parallax shift of the star over the year. • For a more distant star, find the luminosity by using the H-R diagram as noted earlier, and then use the spectroscopic parallax method to relate luminosity, distance, and brightness.
Size	• For a few of the largest and closest stars, measure the size directly or by the length of the eclipse in eclipsing binary stars. • From the width of the star's spectral lines, estimate the luminosity class (supergiant, giant, or main sequence). • For a star with known luminosity and temperature, use the Stefan-Boltzmann law to calculate the star's radius (the luminosity-temperature-radius relationship).
Mass	• Measure the motions of the stars in a binary system, use those to determine the orbits of the stars, and then apply Newton's form of Kepler's third law. • For a non-binary star, use the mass-luminosity relationship to estimate the mass from the luminosity.
Composition	• Analyze the lines in the star's spectrum to measure chemical composition.

only temperature, distance, and composition can be *measured*. Luminosity must be *inferred* from the H-R diagram or calculated from distance and brightness, and size and mass must be *calculated*. Other measurable properties include brightness, color, spectral type, and parallax shift.

CHECK YOUR UNDERSTANDING 13.4

Choose the two qualities that describe a star located in the lower right of the H-R diagram: (a) hot; (b) cool; (c) high luminosity; (d) low luminosity.

Origins Habitable Zones

How might a basic property of a star, such as its luminosity, color, mass, or surface temperature, affect the chance of a planet with life being in orbit around that star? The only known life is that on planet Earth, where liquid water was essential for life to form and evolve. Whether liquid water is an absolute requirement for life elsewhere is not known, but the presence of water is a good starting point for determining where to look. So astronomers look for planets at the right distance from their stars to have a planetary temperature that permits water to exist in a liquid state on their surfaces—a range of distances known as the **habitable zone**. On planets whose orbits are closer to their star than the habitable zone, water would exist only as a vapor—if at all. On planets that have orbits beyond the habitable zone, water would be permanently frozen as ice.

An important factor for estimating the temperature of a planet is the brightness of the sunlight that falls on that planet (see the Chapter 5 "Origins" section and Working It Out 5.4). That factor depends on the luminosity of the star and the planet's distance from the star. In the Solar System, the habitable zone ranges from ~0.9 to ~1.4 AU, which includes Earth but just misses Venus and Mars. Main-sequence stars less luminous than the Sun are cooler and have narrower habitable zones, minimizing the chance that a planet will form within that slender zone. Main-sequence stars more massive than the Sun are hotter and have larger habitable zones. **Figure 13.20** illustrates those zones around Sun-like, hotter, and cooler stars.

Astronomers are now finding planets in the habitable zones of their respective stars. Methods of planet detection, as discussed in Chapter 7, work best when the planet is close to its star. As of this writing, the Kepler Mission has identified and confirmed a few dozen planets in habitable zones by using the transit method and has found more candidate planets that need to be confirmed.

The distance from a star at a certain temperature is not the only consideration for whether a planet has water. The presence of a planetary atmosphere also is a factor. More massive planets can retain their atmospheres, which can trap heat and raise the planet's temperature, as we saw in Chapter 5 for Venus and Earth. Smaller planets have a lower gravitational pull and may not be able to retain an atmosphere. In addition, some habitable zones may be near planets, not stars. Some of the giant planets in the cold, outer part of our own Solar System have moons with liquid water (Chapters 11 and 24). The heat keeping the water liquid comes from the nearby planet, not from the Sun.

Finally, *habitable* does not mean *inhabited*; it means only that the planet is at the right distance from its star that it could have liquid water. Identifying planets in their habitable zone is a first step to selecting which planets are most interesting for further study.

Figure 13.20 The distance and extent of a habitable zone (green) surrounding a star depends on the star's temperature. Regions too close to the star are too hot (red) and those too far away are too cold (blue) to permit the existence of liquid water. The orbits of Mercury, Venus, Earth, and Mars have been drawn around these stars for scale.

ARTICLES QUESTIONS

NASA reports on a new technique for using the Hubble Space Telescope to measure parallax angles.

NASA's Hubble Extends Stellar Tape Measure 10 Times Farther into Space

NASA Press Release

Using NASA's Hubble Space Telescope, astronomers now can precisely measure the distance of stars up to 10,000 light-years away—10 times farther than previously possible.

Astronomers have developed yet another novel way to use the 24-year-old space telescope by employing a technique called spatial scanning, which dramatically improves Hubble's accuracy for making angular measurements. The technique, when applied to the age-old method for gauging distances called astronomical parallax, extends Hubble's tape measure 10 times farther into space (**Figure 13.21**).

"This new capability is expected to yield new insight into the nature of dark energy, a mysterious component of space that is pushing the universe apart at an ever-faster rate," said Nobel laureate Adam Riess of the Space Telescope Science Institute (STScI) in Baltimore, Maryland.

Parallax, a trigonometric technique, is the most reliable method for making astronomical distance measurements, and a practice long employed by land surveyors here on Earth. The diameter of Earth's orbit is the base of a triangle, and the star is the apex where the triangle's sides meet. The lengths of the sides are calculated by accurately measuring the three angles of the resulting triangle.

Astronomical parallax works reliably well for stars within a few hundred light-years of Earth. For example, measurements of the distance to Alpha Centauri, the star system closest to our Sun, vary only by 1 arcsec. This variance in distance is equal to the apparent width of a dime seen from 2 miles away.

Stars farther out have much smaller angles of apparent back-and-forth motion that are extremely difficult to measure. Astronomers have pushed to extend the parallax yardstick ever deeper into our galaxy by measuring smaller angles more accurately.

This new long-range precision was proven when scientists successfully used Hubble to measure the distance of a special class of bright stars called Cepheid variables, approximately 7,500 light-years away in the northern constellation Auriga. The technique worked so well, they are now using Hubble to measure the distances of other far-flung Cepheids.

Such measurements will be used to provide firmer footing for the so-called cosmic "distance ladder." This ladder's "bottom rung" is built on measurements to Cepheid variable stars that, because of their known brightness, have been used for more than a century to gauge the size of the observable universe. They are the first step in calibrating far more distant extra-galactic milepost markers such as Type Ia supernovae.

Riess and the Johns Hopkins University in Baltimore, Maryland, in collaboration with Stefano Casertano of STScI, developed a technique to use Hubble to make measurements as small as five-billionths of a degree.

To make a distance measurement, two exposures of the target Cepheid star were taken 6 months apart, when Earth was on opposite sides of the Sun. A very subtle shift in the star's position was measured to an accuracy of 1/1,000 the width of a single image pixel in Hubble's Wide Field Camera 3, which has 16.8 megapixels total. A third exposure was taken after another 6 months to allow for the team to subtract the effects of the subtle space motion of stars, with additional exposures used to remove other sources of error.

Riess shares the 2011 Nobel Prize in Physics with another team for his leadership in the 1998 discovery that the expansion rate of the universe is accelerating—a phenomenon widely attributed to a mysterious, unexplained dark energy filling the universe. This new high-precision distance measurement technique is enabling Riess to gauge just how much the universe is stretching. His goal is to refine estimates of the universe's expansion rate to the point where dark energy can be better characterized.

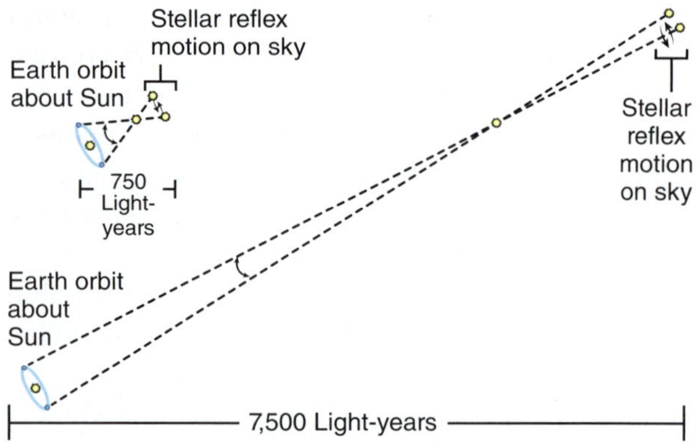

Figure 13.21 By applying a technique called spatial scanning to an age-old method for gauging distances called astronomical parallax, scientists now can use NASA's Hubble Space Telescope to make precision distance measurements 10 times farther into our galaxy than was previously possible.

ARTICLES QUESTIONS

1. How many parsecs are in 7,500 light-years? In 10,000 light-years?
2. What is the parallax angle of a star 7,500 light-years away? Ten thousand light-years away?
3. Why were the exposures taken 6 months apart?
4. Why is making parallax measurements from space better than doing so from the ground on Earth?
5. How does improving the accuracy of the distances to nearby stars from trigonometric parallax affect astronomers' estimates of farther stars' distances, measured using spectroscopic parallax?

Summary

Finding the distances to stars is a difficult but important task for astronomers. Parallax and spectroscopic parallax are two methods that astronomers use to determine distances to stars. Brightness and distance can be used to obtain the luminosity. Careful study of the light from a star, including its spectral lines, gives the temperature, size, and composition of the star. Study of binary systems gives the masses of stars of various spectral types, which we can extend to all stars of the same spectral type. The H-R diagram shows the relationship among the various physical properties of stars. The mass of a star is the major determining factor in the changes it undergoes. The habitable zone is the distance from a star in which a planet could have the right temperature for liquid water to exist on its surface. Stars of different luminosities and temperatures have habitable zones of different widths at different distances from the star.

LG 1 **Explain how astronomers use the brightness of nearby stars and their distances from Earth to determine how luminous they are.** The distance to a nearby star is measured by finding its parallax—by measuring how the star's apparent position changes in the sky over a year. The nearest star (other than the Sun) is about 4 light-years (1.3 parsecs) away. The brightness of a star in the sky can be measured directly, and brightness and distance can be used to obtain the star's luminosity—how much light the star emits.

LG 2 **Explain how astronomers obtain the temperatures, sizes, and composition of stars.** The color of a star depends on its temperature; blue stars are hotter and red stars are cooler. The radius can be computed from the temperature and luminosity of the star. Small, cool stars greatly outnumber large, hot stars. Spectral lines carry a great deal of information about a star, including what chemical elements and molecules are present in the star.

LG 3 **Describe how astronomers estimate the masses of stars.** Masses of stars are measured in binary star systems by observing the effects of the gravitational pull between the stars. Newton's universal law of gravitation and Kepler's laws connect the motion of the star to the forces they experience and thus to their masses.

LG 4 **Classify stars and organize that information on a Hertzsprung-Russell (H-R) diagram.** The H-R diagram shows the relationship among the various physical properties of stars. Temperature increases to the left, so that hotter stars lie on the left side of the diagram, whereas cooler stars lie on the right. Luminosity increases vertically, so that the most luminous stars lie near the top of the diagram. A star's luminosity class and temperature indicate its size. The mass and composition of a main-sequence star determine its luminosity, temperature, and size. Ninety percent of stars lie along the main sequence.

LG 5 **Explain how the mass and composition of a main-sequence star determine its luminosity, temperature, and size.** The mass and composition of a main-sequence star determine its original position on the H-R diagram. The main sequence on the H-R diagram is actually a sequence of masses. That position connects its other properties such as its luminosity, temperature, and size.

? Unanswered Questions

- What is the *upper* limit for stellar mass? Both theory and observation have shown that the *lower* limit for stellar mass is approximately 0.08 M_{Sun} with temperature about 2000 K. However, neither theory nor observation has yielded a definitive value for the upper limit. Many astronomers believe that the upper limit lies somewhere around 150–200 M_{Sun}. The very first stars that formed in the universe might have been even larger.
- Are planets with life likely to be orbiting around main-sequence stars of type M? Those low-luminosity stars are the most common type in the Milky Way, and the Kepler telescope has detected many planets orbiting such stars. However, the habitable zone of an M star is very close to the star, so that the planet may be strongly affected by streams of charged particles blowing off the star. Unless the planet has a strong protective magnetic field, that radiation may decrease or completely strip a planet of its atmosphere. Another complication is that a close-in planet is likely to be tidally locked to the star, so that one hemisphere of the planet receives light and the other hemisphere is permanently dark. That imbalance of light and heat might make the planet uninhabitable.

Questions and Problems

Test Your Understanding

1. Star A and star B are nearly the same distance from Earth. If star A is half as bright as star B, which of the following statements must be true?
   a. Star B is farther away than star A.
   b. Star B is twice as luminous as star A.
   c. Star B is hotter than star A.
   d. Star B is larger than star A.
2. Star A and star B are two nearby stars. If star A is blue and star B is red, which of the following statements must be true?
   a. Star A is hotter than star B.
   b. Star A is cooler than star B.
   c. Star A is farther away than star B.
   d. Star A is more luminous than star B.
3. Star A and star B are two stars nearly the same distance from Earth. If star A is blue and star B is red, but they have equal brightness, which of the following statements is true?
   a. Star A is more luminous than star B.
   b. Star A is larger than star B.
   c. Star A is smaller than star B.
   d. Star A is less luminous than star B.
4. What does it most likely mean when a star has very weak hydrogen lines and is blue?
   a. The star is too hot for hydrogen lines to form.
   b. The star has no hydrogen.
   c. The star is too cold for hydrogen lines to form.
   d. The star is moving too fast to measure the lines.
5. Star A and star B are a binary system. If the Doppler shift of star A's absorption lines is 3 times the Doppler shift of star B's absorption lines, which of the following statements is true?
   a. Star A is 3 times as massive as star B.
   b. Star A is one-third as massive as star B.
   c. Star A is closer than star B.
   d. The binary pair is moving toward Earth, but star A is farther away.
6. Star A and star B are two red stars at nearly the same distance from Earth. If star A is many times brighter than star B, which of the following statements is true?
   a. Star A is a main-sequence star, whereas star B is a red giant.
   b. Star A is a red giant, whereas star B is a main-sequence star.
   c. Star A is hotter than star B.
   d. Star A is a white dwarf, whereas star B is a red giant.
7. Star A and star B are two blue stars at nearly the same distance from Earth. If star A is many times brighter than star B, which of the following statements is true?
   a. Star A is a main-sequence star, whereas star B is a red giant.
   b. Star A is a main-sequence star, whereas star B is a blue giant.
   c. Star A is a white dwarf, whereas star B is a blue giant.
   d. Star A is a blue giant, whereas star B is a white dwarf.
8. In which region of an H-R diagram would you find the main-sequence stars with the widest habitable zones?
   a. upper left
   b. upper right
   c. center
   d. lower left
   e. lower right
9. If star A is more massive than star B, and both are main-sequence stars, star A is ___________ than star B. (Choose all that apply.)
   a. more luminous
   b. less luminous
   c. hotter
   d. colder
   e. larger
   f. smaller
10. A telescope on Mars could measure the distances to more stars than can be measured from Earth because
    a. the resolution of the telescope would be better.
    b. Mars has a thin atmosphere.
    c. it would be closer to the stars.
    d. the parallax "baseline" would be longer.
11. Star A and star B are two nearby stars. If star A has a parallactic angle 4 times as large as star B's, which of the following statements is true?
    a. Star A is one-quarter as far away as star B.
    b. Star A is 4 times as far away as star B.
    c. Star A has moved through space one-quarter as far as star B.
    d. Star A has moved through space 4 times as far as star B.
12. If star A appears twice as bright as star B but is also twice as far away, star A is ___________ as luminous as star B.
    a. 8 times
    b. 4 times
    c. twice
    d. half
13. In Table 13.1, the percentage of hydrogen in the Sun decreases when changing from percentage by number of atoms to percentage by mass, but the percentage of helium increases. Why?
    a. Hydrogen is more massive than helium.
    b. Helium is more massive than hydrogen.
    c. Hydrogen is located in a different part of the Sun.
    d. The mass of hydrogen is hard to measure.
14. Capella (in the constellation Auriga) is the sixth-brightest star in the sky. When viewed with a high-power telescope, it turns out that Capella is actually two pairs of binary stars: the first pair are G-type giants, whereas the second pair are M-type main-sequence stars. What color does Capella appear to be?
    a. red
    b. yellow
    c. blue
    d. The color cannot be determined from that information.

15. An eclipsing binary system has a primary eclipse (star A is eclipsed by star B) that is deeper (more light is removed from the light curve) than the secondary eclipse (star B is eclipsed by star A). What does that information tell you about stars A and B?
    a. Star A is hotter than star B.
    b. Star B is hotter than star A.
    c. Star B is larger than star A.
    d. Star B is moving faster than star A.

Thinking about the Concepts

16. The distances of nearby stars are determined by their parallaxes. Why are the distances of stars farther from Earth more uncertain?

17. To know certain properties of a star, you must first determine the star's distance. For other properties, knowing its distance is unnecessary. Explain why an astronomer does or does not need to know a star's distance to determine each of the following properties: size, mass, temperature, color, spectral type, and chemical composition.

18. Albireo, in the constellation Cygnus, is a visual binary system whose two components can easily be seen with even a small, amateur telescope. Viewers describe the brighter star as "golden" and the fainter one as "sapphire blue."
    a. What does that description tell you about the relative temperatures of the two stars?
    b. What does that description tell you about their respective sizes?

19. Very cool stars have temperatures around 2500 K and emit Planck spectra with peak wavelengths in the red part of the spectrum. Do those stars emit any blue light? Explain your answer.

20. The stars Betelgeuse and Rigel are both in the constellation Orion. Betelgeuse appears red, whereas Rigel is bluish white. To the eye, the two stars seem equally bright. If you can compare the temperature, luminosity, or size from just that information, do so. If not, explain why.

21. Explain why the stellar spectral types (O, B, A, F, G, K, M) are not in alphabetical order. What sequence of temperatures is defined by those spectral types?

22. Other than the Sun, the only stars whose mass astronomers can measure *directly* are those in eclipsing or visual binary systems. Why? How do astronomers estimate the masses of stars that are not in eclipsing or visual binary systems?

23. Once the mass of a certain spectral type of star located in a binary system has been determined, it can be assumed that all other stars of the same spectral type and luminosity class have the same mass. Why is that a reasonable assumption?

24. Explain why the Kepler Mission is finding eclipsing binary stars while it is searching for extrasolar planets by using the transit method.

25. Scientific advances often require the participation of scientists from all over the world, working on the same problem over many decades, even centuries. Compare that mode of "collaboration" with collaborations in your courses (perhaps on final projects or papers). What mechanisms must be in place to allow scientists to collaborate across space and time in that way?

26. What would happen to our ability to measure stellar parallax if we were on Mars? What if we were on Venus or Jupiter?

27. Figure 13.7 has an absorption line at about 410 nm that is weak for O stars and weak for G stars but very strong in A stars. That particular line comes from the transition from the second excited state of hydrogen up to the sixth excited state. Why is that line weak in O stars? Why is it weak in G stars? Why is it strongest in the middle of the range of spectral types?

28. Which kinds of binary systems are best observed edge-on? Which kinds are best observed face-on?

29. In Figure 13.10, two stars orbit a common center of mass.
    a. Explain why star 2 has a smaller orbit than star 1.
    b. Re-sketch that picture for the case in which star 1 has a very low mass, perhaps close to that of a planet.
    c. Re-sketch that picture for the case in which star 1 and star 2 have the same mass.

30. If our Sun were a blue main-sequence star, and Earth was still 1 AU from the Sun, would you expect Earth to be in the habitable zone? What about if our Sun were a red main-sequence star?

Applying the Concepts

31. Suppose that Figure 13.1b included a third star, located 4 times as far away as star A. How much less than star A would the third star appear to move each year? How much less than star B?

32. Suppose you see an object jump from side to side by half a degree as you blink back and forth between your eyes. How much farther away is an object that moves only one-third of a degree?

33. Logarithmic (log) plots show major steps along an axis scaled to represent equal factors, most often factors of 10. Why do astronomers sometimes use a log plot instead of the more conventional linear plot? Is the horizontal axis of the H-R diagram in Figure 13.15 logarithmic or linear?

34. Figure 13.5 is plotted logarithmically on both axes. The luminosities are in units of solar luminosities.
    a. How much more luminous than the Sun is a star on the far right side of the plot?
    b. How much less luminous than the Sun is a star on the far left side of the plot?

35. Compared with the Sun, how luminous, large, and hot is a star that has 10 times the mass of the Sun? Use Figure 13.17 to answer that question.

36. Sirius, the brightest star in the sky, has a parallax of 0.379 arcsec. What is its distance in parsecs? In light-years? How long does the light take to reach Earth?

37. Sirius is actually a binary pair of two A-type stars. The brighter star is called the "Dog Star" and the fainter is called the "Pup Star" because Sirius is in the constellation Canis Major (meaning "big dog"). The Dog Star appears about 6,800 times brighter than the Pup Star, even though both stars are at the same distance from Earth. Compare the temperatures, luminosities, and sizes of those two stars.

38. Sirius and its companion orbit around a common center of mass with a period of 50 years. The mass of Sirius is 2 times the mass of the Sun.
    a. If the orbital velocity of the companion is 2.35 times greater than that of Sirius, what is the mass of the companion?
    b. What is the semimajor axis of the orbit?

39. Sirius is 25 times more luminous than the Sun, and Polaris (the "North Pole Star") is 2,500 times more luminous than the Sun. Sirius appears 24 times brighter than Polaris. How much farther away is Polaris than Sirius? Use your answer from Problem 36 to find the distance of Polaris in light-years.

40. Betelgeuse (in Orion) has a parallax of 0.00451 ± 0.00080 arcsec, as measured by the Hipparcos satellite. What is the distance to Betelgeuse, and what is the uncertainty in that measurement?

41. The star Achernar has a Hipparchus parallax of 0.02339 arcsec and appears about as bright as Betelgeuese (previous problem) in the sky. Which star is actually more luminous? Knowing that Betelgeuse appears reddish, whereas Achenar appears bluish, which star would you say is hotter, and why?

42. The Sun is about 16 trillion (1.6×10^{13}) times brighter than the faintest stars visible to the naked eye.
    a. How far away (in astronomical units) would an identical solar-type star be if it were just barely visible to the naked eye?
    b. What would be its distance in light-years?

43. If $m_1 = m_2$ in Figure 13.9, where would the center of mass be located? If $m_1 = 2m_2$, where would the center of mass be located?

44. Find the peak wavelength of blackbody emission for a star with a temperature of about 10,000 K. In what region of the spectrum does that wavelength fall? What color is that star?

45. About 1,470 watts (W) of solar energy hits each square meter of Earth's surface. Use that value and the distance to the Sun to calculate the Sun's luminosity.

USING THE WEB

46. Go to the European Space Agency's *Gaia* mission website (https://www.esa.int/Our_Activities/Space_Science/Gaia). How has it helped astronomers determine the distances to more stars? Why is making parallax measurements from space better than measuring from the ground on Earth? Do a search for *Gaia* news (for example, at http://sci.esa.int/gaia/)—what has been learned from the released data?

47. Go to the Kepler Eclipsing Binary Catalog (http://keplerebs.villanova.edu) to see what new observations look like. Pick a few stars to study. Look at the last two columns (figures, labeled "LC Figs" and "SC Figs"). The "raw" and "dtr" figures are rough, but the "pf" figure shows a familiar light curve. How deep is the eclipse; that is, how much lower is the "normalized flux" during maximum eclipse?

48. Do a search for a photograph of your favorite constellation (or go outside and take a picture yourself). Can you see different colors in the stars? What do the colors tell you about the surface temperatures of the stars? From your photograph, can you tell which are the three brightest stars in the constellation? Those stars will be named "alpha" (α), "beta" (β), and "gamma" (γ) for that constellation. Look up the constellation online and see whether you chose the right stars. What are their temperatures and luminosities? What are their distances?

49. Citizen science: Go to the website for the Stellar Classification Online Public Exploration (SCOPE) project (http://scope.pari.edu/takepart.php). The project uses crowdsourcing to classify stars seen on old photographic plates of pictures taken in the Southern Hemisphere. Create an account, review the science and the FAQs, and then click on "To Take Part" to see some practice examples. Then go to "Classify," choose a photographic plate, and classify a few stars.

50. Watch the short NASA video about planets in binary star systems: https://nasaviz.gsfc.nasa.gov/11106. Why do astronomers think these systems are not very common? Do a search on "circumbinary planets"—have many more been found?

EXPLORATION

The H-R Diagram

digital.wwnorton.com/astro6

Visit the Digital Resources Page and on the Student Site open the "H-R Diagram" Interactive Simulation for Chapter 13. This simulation enables you to compare stars on the H-R diagram in two ways. You can compare an individual star (marked by a red *X*) to the Sun by varying its properties in the box in the left half of the window. Or you can compare groups of the nearest and brightest stars. Play around with the controls for a few minutes to familiarize yourself with the simulation.

Begin by exploring how changes to the properties of the individual star change its location on the H-R diagram. First, press the "Reset" button at the top right of the window.

Decrease the temperature of the star by dragging the temperature slider to the left. Notice that the luminosity remains the same. Because the temperature has decreased, each square meter of star surface must be emitting less light. What other property of the star changes to keep the total luminosity of the star constant?

Predict what will happen when you move the temperature slider all the way to the right. Now do it. Did the star behave as you expected?

1 As you move to the left across the H-R diagram, what happens to the radius?

____

____

____

2 What happens as you move to the right?

____

____

Press "Reset" and experiment with the luminosity slider.

3 As you move up on the H-R diagram, what happens to the radius?

____

____

4 What happens as you move down?

____

____

____

Press "Reset" again and predict how you would have to move the slider bars to move your star into the red giant portion of the H-R diagram (upper right). Adjust the slider bars until the star is in that area. Were you correct?

5 How would you adjust the slider bars to move the star into the white dwarf area of the H-R diagram?

____

____

Press the "Reset" button and explore the right-hand side of the window. Add the nearest stars to the graph by clicking their radio button under "Plotted Stars." Using what you learned above, compare the temperatures and luminosities of those stars with the Sun (marked by the *X*).

6 Are the nearest stars generally hotter or cooler than the Sun?

____

7 Are the nearest stars generally more or less luminous than the Sun?

____

Press the radio button for the brightest stars. That action will remove the nearest stars and add the brightest stars in the sky to the plot. Compare those stars with the Sun.

8 Are the brightest stars generally hotter or cooler than the Sun?

____

9 Are the brightest stars generally more or less luminous than the Sun?

____

10 How do the temperatures and luminosities of the brightest stars in the sky compare with the temperatures and luminosities of the nearest stars? Does that information support the claim in the chapter that more low-luminosity stars than high-luminosity stars exist? Explain.

____

____

____

____

14 Our Star—The Sun

Because the Sun is the only star close to Earth, much of our detailed knowledge about stars has come from studying the Sun. In Chapter 13, we looked at the physical properties of distant stars, including their mass, luminosity, size, temperature, and chemical composition. Here in Chapter 14, we ask fundamental questions about Earth's local star: How does the Sun work? Where does it get its energy? How has its luminosity changed over the billions of years since the Solar System formed?

LEARNING GOALS

By the end of this chapter, you should be able to:

LG 1 Describe the balance between the forces that determine the structure of the Sun.

LG 2 Explain how mass is converted to energy in the Sun's core and how long the Sun will take to use up its fuel.

LG 3 Sketch a physical model of the Sun's interior, and list the ways that energy moves outward from the Sun's core toward its surface.

LG 4 Describe how observations of solar neutrinos and seismic vibrations on the surface of the Sun test astronomers' models of the Sun.

LG 5 Describe the solar activity cycles of 11 and 22 years, and explain how those cycles are related to the Sun's changing magnetic field.

This image is a combination of several extreme ultraviolet images of the Sun from the Solar Dynamics Observatory. ▶▶▶

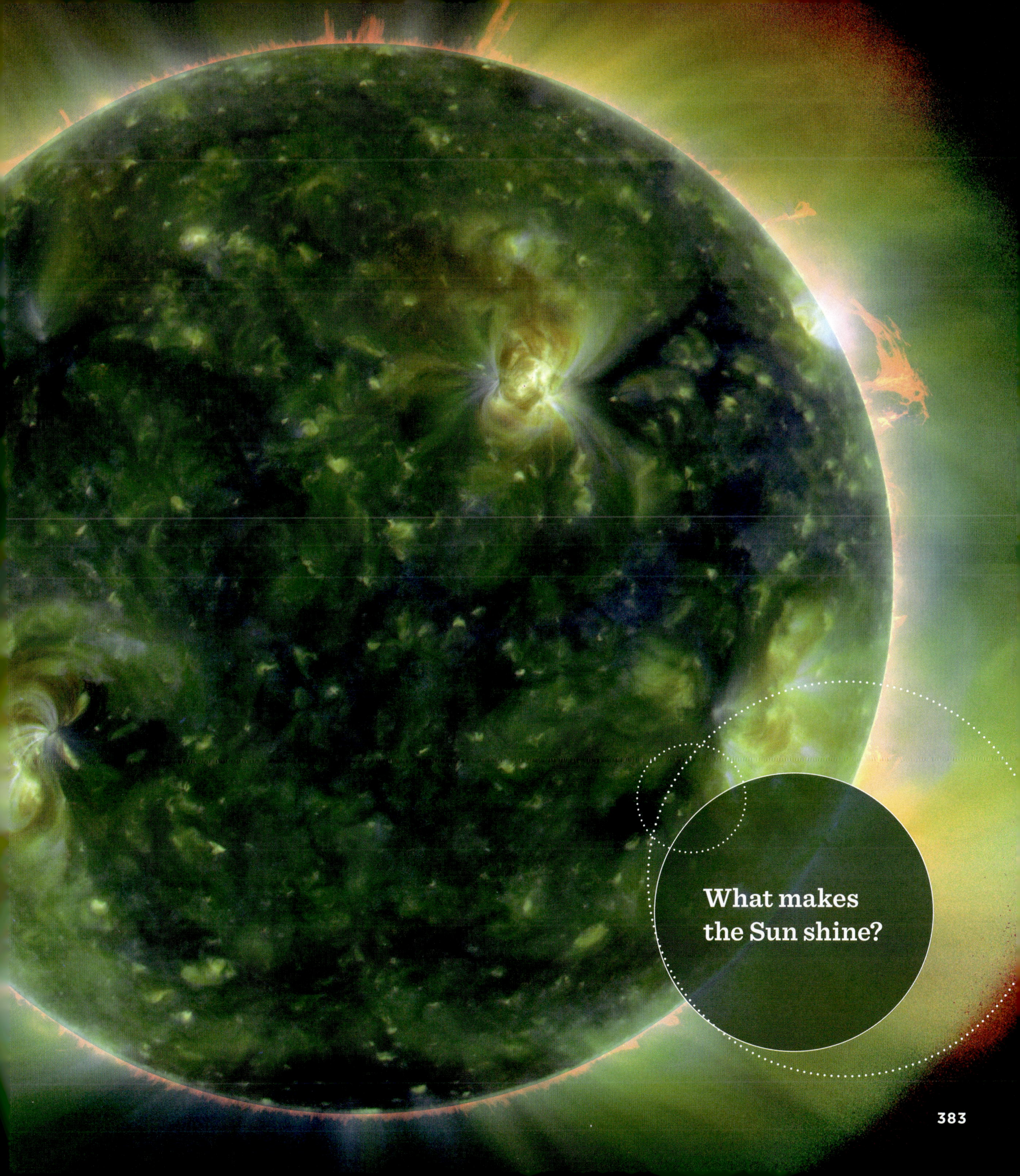

What makes the Sun shine?

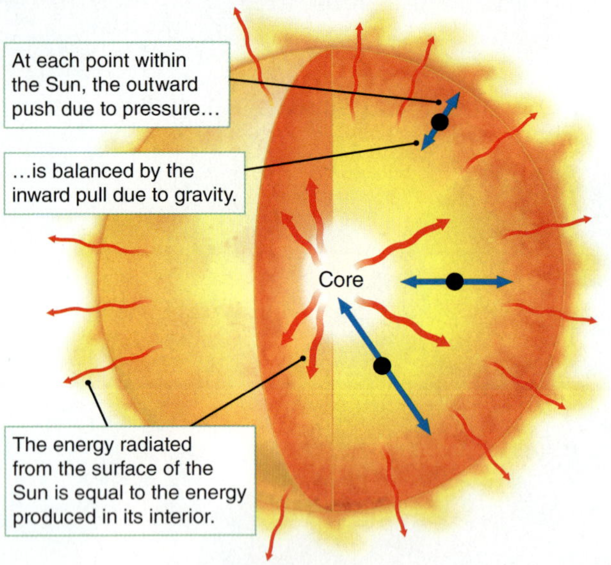

Figure 14.1 The structure of the Sun is determined by the balance between the forces due to pressure and gravity and the balance between the energy generated in its core and the energy radiated from its surface.

Figure 14.2 (a) This cutaway figure shows how the fraction of radius given in the *x*-axis of the graphs in (b) is measured. The energy produced by the Sun is generated in the Sun's core. (b) Pressure, density, and temperature all increase toward the center of the Sun.

(a)

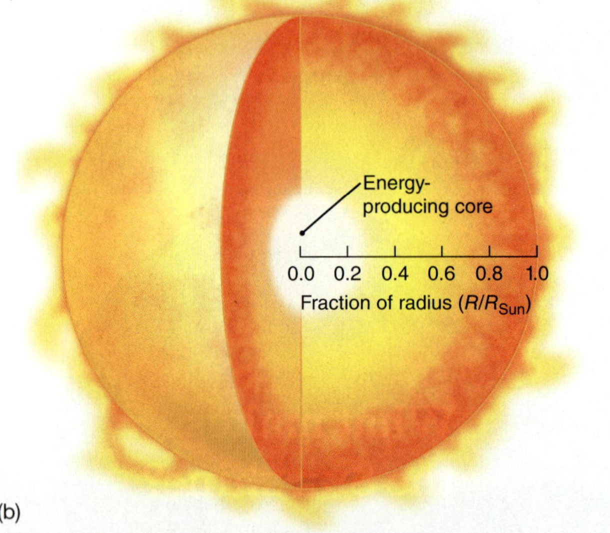

(b)

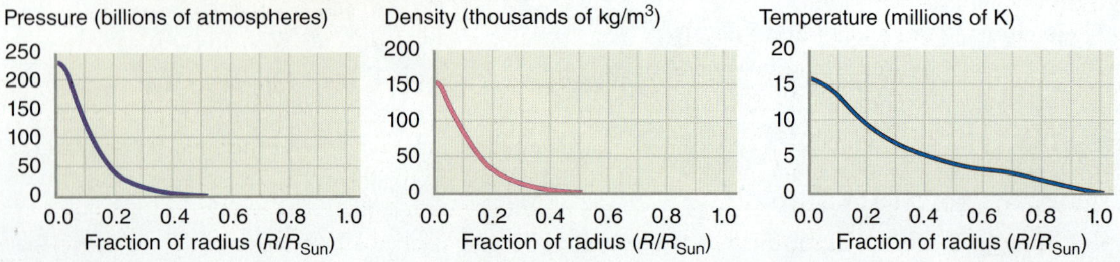

14.1 The Sun Is Powered by Nuclear Fusion

Energy from the Sun is responsible for daylight, for Earth's weather and seasons, and for terrestrial life itself. At a luminosity of 3.85×10^{26} watts (W), the Sun produces more energy in a second than all the electric power plants on Earth could generate in a half-million years. In this section, we look at energy production in the Sun.

Hydrostatic Equilibrium

Geologists learn about the interior of Earth by using a combination of physics, detailed computer models, and experiments that test the predictions of those models. The task of exploring the interior of the Sun is much the same. Like Earth's structure, the structure of the Sun is governed by several physical processes and relationships. Using physics, chemistry, and the properties of matter and radiation, astronomers can express those processes and relationships as mathematical equations. They then solve those equations and create a model of the Sun. One of the great successes of 20th century astronomy is the construction of a theoretical model of the Sun that agrees with observations of the mass, composition, size, temperature, and luminosity of the real thing.

The structure of the Sun results from a balance between the outward force due to the pressure of radiation produced inside and the inward force of gravity. That balance is known as **hydrostatic equilibrium**. The outward pressure is a result of local energy in the process of finding its way to the surface of the Sun from deep in its interior. To understand how hydrostatic equilibrium affects the Sun, we need to know how those forces are produced and how they continually change to balance each other.

The balance between the forces due to radiation pressure and gravity is illustrated in **Figure 14.1**. The Sun is a huge ball of hot gas. Deep in the Sun's interior, the outer layers press downward because of gravity, producing a large inward force. To maintain balance, the outward force due to pressure must be equally large. If gravity were not balanced by radiation pressure, the Sun would collapse. If radiation pressure were greater than gravity, the Sun would blow itself apart. At every point within the Sun's interior, the radiation pressure must be just enough to hold up the weight of all the layers above that point. If the Sun were not in a stable hydrostatic equilibrium, forces within it would not be in balance, and the size of the Sun would change.

Hydrostatic equilibrium becomes an even more powerful concept when combined with the way gases behave. Deeper in the interior of the Sun, the weight of the material above becomes greater, and hence the radiation pressure increases. In a gas, higher pressure means higher density, higher temperature, or both. **Figure 14.2** shows how conditions vary inside the Sun. The graphs in Figure 14.2b, which are based on calculations, show that as the radiation pressure increases toward the center of the Sun, the density and temperature of the gas increase as well.

CHECK YOUR UNDERSTANDING 14.1A

Hydrostatic equilibrium in the Sun means that: (a) the Sun does not change; (b) the Sun absorbs and emits equal amounts of energy; (c) the outward force from radiation pressure balances the weight of overlying layers; (d) energy produced in the core per unit time equals energy emitted at the surface per unit time.

Nuclear Fusion

A second fundamental balance within the Sun is the balance of energy (see Figure 14.1). Stars like the Sun are remarkably stable objects. To remain in balance, the Sun must produce just enough energy in its interior to replace the energy radiated away from its surface. That energy balance tells us how much energy must be produced in the interior of the Sun and how that energy finds its way from the interior to the Sun's surface, where it is radiated away. Theoretical models of stellar evolution indicate that the luminosity of the Sun is increasing, but very, very slowly. From geological records and the models, astronomers estimate that the Sun's luminosity 4.5 billion years ago was about 70 percent of its current luminosity.

One of the most basic questions facing the pioneers of stellar astrophysics was how the Sun and other stars get their energy. In the 19th century, physicists proposed that the Sun was slowly shrinking and that the core was heating up as a result of that gravitational contraction. Calculations soon showed, however, that gravitational contraction could power the Sun for only millions of years. Geological and biological evidence available at the time suggested that Earth was tens of millions or hundreds of millions of years old. By the early 20th century, radiometric dating suggested that Earth was more than a billion years old, and therefore gravitational contraction could not be the source of the Sun's energy. In the 1930s, using theoretical and laboratory physics, nuclear physicists concluded that the Sun's energy comes from nuclear reactions at its core, capable of powering the star for billions of years.

The nucleus of most hydrogen atoms consists of a single proton (Chapter 5). Nuclei of all other atoms are built from a mixture of protons and neutrons. Most helium nuclei, for example, consist of two protons and two neutrons. Protons have a positive electric charge, whereas neutrons have no electric charge. Because like charges repel, and the closer they are the stronger the force, all the protons in an atomic nucleus are continually repelling each other with a tremendous force. The nuclei of atoms should fly apart because of electric repulsion. They don't, though, because they are held together by the **strong nuclear force**, which overcomes that repulsion. However, the strong nuclear force acts only over very short distances, on the order of 10^{-15} meters—about the size of the atomic nucleus, or about a hundred-thousandth the size of an atom.

Compared with the energy required to free an electron from an atom, the amount of energy required to tear a nucleus apart is enormous. Conversely, when an atomic nucleus (with a mass up to the nucleus of iron) is formed from component parts, energy is released. **Nuclear fusion**—the process of combining two less massive atomic nuclei into a single, more massive atomic nucleus—occurs when atomic nuclei are brought close enough together for the strong nuclear force to overcome the force of electric repulsion (**Figure 14.3**). Many kinds of nuclear fusion can occur in stars. In main-sequence stars like the Sun, the primary energy generation process is the fusion of hydrogen into helium. That process is sometimes called "hydrogen burning" even though it has nothing to do with fire or other chemical combustion. The fusion of hydrogen into helium always takes several steps, but the net result is that four hydrogen nuclei become one helium nucleus plus energy.

The energy produced in nuclear reactions comes from converting mass to energy. The exchange rate between mass and energy is given by Einstein's equation, $E = mc^2$, in which E is energy, m is mass, and c^2 is the speed of light squared. For any nuclear reaction, we can determine the mass turned into energy by calculating the mass lost. To find that lost mass, we subtract the mass of the outputs from the mass of the inputs. In hydrogen fusion, the inputs are four hydrogen nuclei, and the output is a helium nucleus plus energy. The mass of four separate

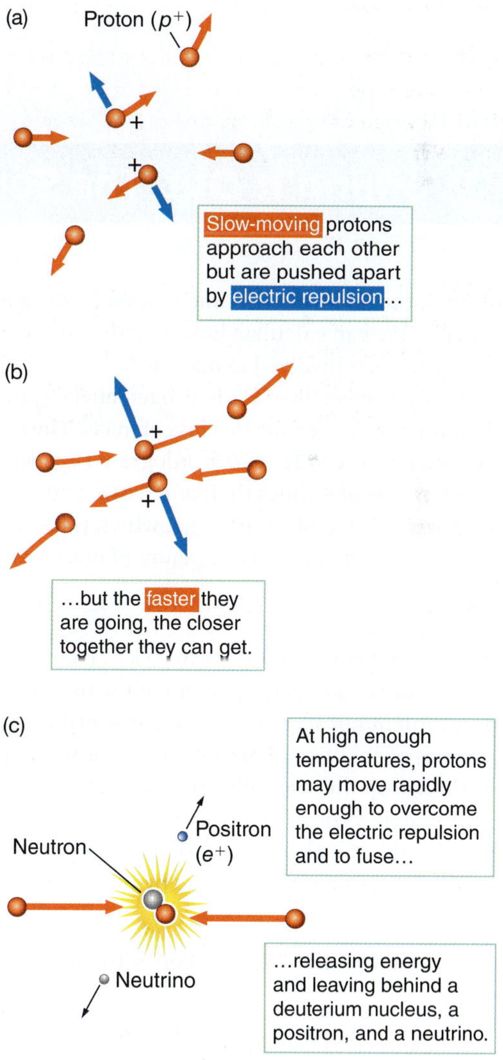

Figure 14.3 (a) Atomic nuclei are positively charged and electrically repel each other. (b) The faster that two nuclei are moving toward each other, the closer they will get before veering away. (c) At the temperatures and densities found in the centers of stars, nuclei can overcome that electric repulsion, so fusion takes place.

hydrogen nuclei is 1.007 times greater than the mass of a single helium nucleus (with two protons and two neutrons); so when hydrogen fuses to make helium, 0.7 percent of the mass of the hydrogen is converted to energy.

Although each fusion reaction produces a small amount of energy, the total mass of the Sun is very large, so a great deal of hydrogen is available to fuse. When the amount of energy produced by nuclear fusion is compared with the luminosity of the Sun, we see that those reactions can power the Sun for 10 billion years—a time frame longer than the 4.6-billion-year age of the Solar System (measured from radioactive dating of meteorites from space and the Moon). Details of that calculation are provided in **Working It Out 14.1**.

Energy is produced in the Sun's innermost region, the **core**, where the conditions are the most extreme. The density of matter in the core is about 150 times the density of water, and the temperature is about 15 million kelvins (K). Under those conditions, the atomic nuclei have tens of thousands of times more kinetic energy than that of atoms at room temperature and can slam into each other hard enough to overcome the electric repulsion, allowing the strong nuclear force to act (Figure 14.3c). In hotter and denser gases, such collisions happen more often. For that reason, the rate of nuclear fusion reactions is extremely sensitive to the

14.1 Working It Out The Source of the Sun's Energy

Like all stars, the Sun has a lifetime limited by the amount of fuel available. We can calculate how long the Sun will live by comparing the mass involved in nuclear fusion with the amount of mass available. Converting four hydrogen nuclei (protons) into a single helium nucleus results in a loss of mass. The mass of a single hydrogen nucleus is 1.6726×10^{-27} kilogram (kg). So, four hydrogen nuclei have a mass of 4 times that, or 6.6904×10^{-27} kg. The mass of a helium nucleus is 6.6447×10^{-27} kg, which is less than the mass of the four hydrogen nuclei. The amount of mass lost, m, is

$$m = 6.6904 \times 10^{-27}\ \text{kg} - 6.6447 \times 10^{-27}\ \text{kg} = 0.0457 \times 10^{-27}\ \text{kg}$$

We can write that result as 4.57×10^{-29} kg—a mass loss of about 0.7 percent. Conversion of 0.7 percent of the mass of the hydrogen into energy might not seem very efficient—until we compare it with other sources of energy and discover that it is millions of times more efficient than even the most efficient chemical reactions.

Using Einstein's equation $E = mc^2$, where c is the speed of light (3×10^8 m/s), along with the definition of a joule ($1\ \text{J} = 1\ \text{kg m}^2/\text{s}^2$), we can calculate the energy released by that mass-to-energy conversion:

$$E = mc^2 = (4.57 \times 10^{-29}\ \text{kg}) \times (3.00 \times 10^8\ \text{m/s})^2 = 4.11 \times 10^{-12}\ \text{J}$$

Each reaction that takes four hydrogen nuclei and turns them into a helium nucleus releases 4.11×10^{-12} J of energy, which doesn't seem like very much. But atoms are very small. Fusing a single kilogram of hydrogen into helium releases about 6.3×10^{14} J of energy—about the equivalent of the chemical energy released in burning 100,000 barrels of oil. To see how much the Sun must be fusing per second to produce its current luminosity, we divide the luminosity of the Sun by that amount of energy per kilogram:

$$\frac{\text{Luminosity of Sun}}{\text{Energy per kilogram}} = \frac{3.9 \times 10^{26}\ \text{J/s}}{6.3 \times 10^{14}\ \text{J/kg}} = 6.2 \times 10^{11}\ \text{kg/s}$$

For the Sun to produce as much energy as it does, it must convert roughly 620 billion kg of hydrogen into helium every second (and about 4 billion kg of matter—0.7 percent—is converted to energy in the process). The Sun has been fusing hydrogen at that rate for at least the age of Earth and the Solar System—4.6 billion years. How much longer will the Sun last?

Astronomers estimate that only 10 percent of the Sun's total mass will ever be involved in fusion because the other 90 percent will never get hot enough or dense enough for the strong nuclear force to make fusion happen. Ten percent of the mass of the Sun is $(0.1) \times (2 \times 10^{30})$ kg, or 2×10^{29} kg. That is the amount of "fuel" the Sun has available. The Sun consumes hydrogen at a rate of 620 billion kg/s, so each year the Sun consumes about:

$$M_{\text{year}} = (6.2 \times 10^{11}\ \text{kg/s}) \times (3.16 \times 10^7\ \text{s/yr}) \sim 2 \times 10^{19}\ \text{kg/yr}$$

If we know how much fuel the Sun has (2×10^{29} kg), and we know how much the Sun fuses each year (2×10^{19} kg/yr), we can divide the amount by the rate to find the lifetime of the Sun:

$$\text{Lifetime} = \frac{M_{\text{fuel}}}{M_{\text{year}}} = \frac{2 \times 10^{29}\ \text{kg}}{2 \times 10^{19}\ \text{kg/yr}} = 10^{10}\ \text{yr}$$

When the Sun was formed, it had enough fuel to power it for about 10 billion years. The Sun is nearly halfway through its lifetime of hydrogen fusion.

temperature and the density of the gas, which is why those energy-producing collisions are concentrated in the Sun's core. Half the energy produced by the Sun is generated within the inner part of the core: the inner 9 percent of the Sun's radius, or less than 0.1 percent of the volume of the Sun.

The conversion of four hydrogen nuclei to one helium nucleus is the most significant source of energy in main-sequence stars. Hydrogen is the most abundant element in the universe, so it is the most abundant source of nuclear fuel at the beginning of a star's lifetime. Hydrogen also is the easiest atom to fuse. Hydrogen nuclei—protons—have an electric charge of +1. The electric barrier that must be overcome to fuse protons is the repulsion of one proton against another. To fuse two carbon nuclei, for example, the repulsion of the six protons in one carbon nucleus pushing against the six protons in another carbon nucleus must be overcome. The repulsion between two carbon nuclei is 36 times stronger than that between two hydrogen nuclei. Therefore, hydrogen fusion occurs at a much lower temperature than any other type of nuclear fusion.

AstroTour: The Solar Core

The Proton-Proton Chain

Simulation: Proton-Proton Chain

To test the theory that the Sun shines because of nuclear fusion, astronomers can analyze the predicted by-products of the nuclear reactions. In the core of the Sun and in other low-mass stars, hydrogen fusion takes place in a series of nuclear reactions called the **proton-proton chain**, which has three branches. The most important branch, responsible for about 85 percent of the energy generated in the Sun, consists of the three steps illustrated in **Figure 14.4**. Each step produces particles and/or energy in the form of light. We begin by following the creation of the helium nucleus and then go back to find out what happens to the other products of the reaction.

Follow along in Figure 14.4 as we step through the proton-proton chain. The nucleus of hydrogen consists of one proton. In the first step, two protons fuse. During that process, one proton is transformed into a neutron. To conserve spin and charge, two particles are emitted: a positively charged particle called a **positron** and a neutral particle called a **neutrino**. Energy also is emitted in the form of gamma ray photons. The new atomic nucleus formed by the first step in the chain consists of a proton and a neutron. Recall from Chapter 5 that an *isotope* of an element has the same number of protons but a different number of neutrons. Thus, the new atomic nucleus is still hydrogen because it has only one proton. That particular isotope of hydrogen is common enough that it has its own name—deuterium, written as ^2H (H is the element symbol for hydrogen, which always has one proton, and 2 is the *atomic mass number*—the total number of protons and neutrons in the nucleus).

In the second step of the proton-proton chain, another proton slams into the deuterium nucleus, forming the nucleus of an isotope of helium, ^3He, consisting of two protons and a neutron. The energy released in that step is carried away as a highly energetic gamma-ray photon. Those first two steps are shown twice in Figure 14.4a because they must occur twice to produce a single ^4He nucleus.

Figure 14.4 The Sun and all other main-sequence stars get their energy by fusing the nuclei of four hydrogen atoms to make a single helium atom. In the Sun, about 85 percent of the energy produced comes from the branch of the proton-proton chain illustrated here.

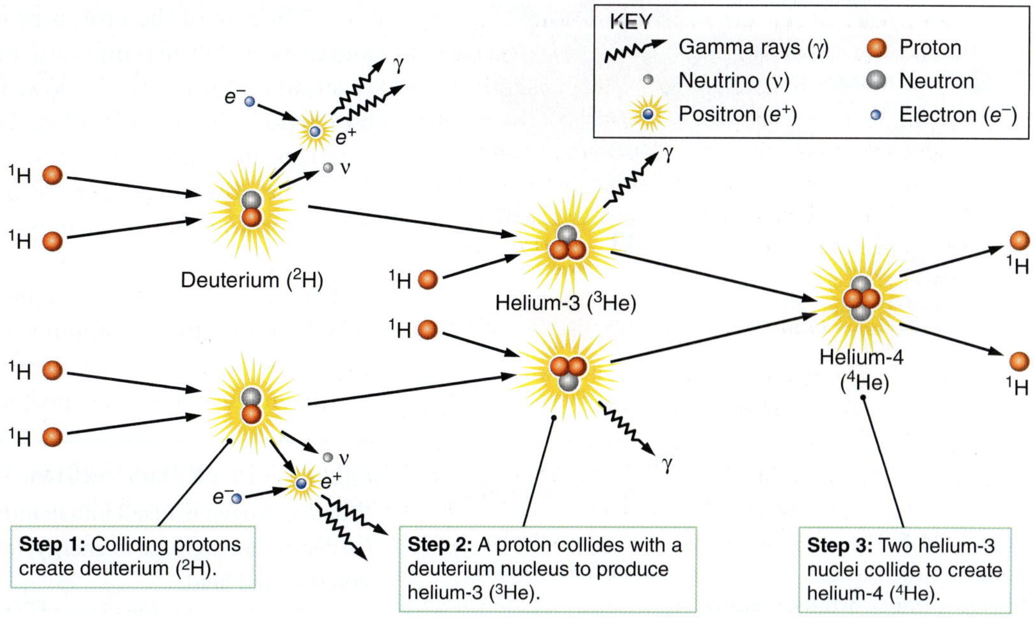

In the third and final step of the proton-proton chain, two ^{3}He nuclei collide and fuse, producing an ordinary ^{4}He nucleus and ejecting two protons in the process. The energy released in that step is the kinetic energy of the helium nucleus and two ejected protons. Overall, four hydrogen nuclei have combined to form one helium nucleus.

Now let's go back and look at what happens to the other products of the reaction. In step 1, a positron—a particle of antimatter—is produced. **Antimatter** particles have the same mass as a corresponding matter particle but have opposite values of other properties, such as charge. The positron ($e+$) is the antimatter counterpart of an electron (e^-). When matter (electrons) and antimatter (positrons) meet, they annihilate each other, and their total mass is converted to energy in the form of gamma-ray photons (γ). That happens to the emitted positrons inside the Sun, and the emitted photons from the annihilation carry away part of the energy released when the two protons fused. Those photons heat the surrounding gas. The gamma rays emitted in step 2 similarly heat the gas. The thermal energy produced in the core of the Sun takes one hundred thousand years or more to find its way to the Sun's surface, and so the light we see from the Sun indicates what the Sun's core was doing a very long time ago.

The neutrino emitted in step 1 has a very different fate. Neutrinos are particles that have no charge, have very little mass, and travel at nearly the speed of light. They interact so weakly with ordinary matter that the neutrino escapes from the Sun without further interactions with any other particles. The core of the Sun is buried beneath 700,000 kilometers (km) of dense, hot matter, yet the Sun is transparent to neutrinos—essentially all of them travel into space as though the outer layers of the Sun were not there. Because they travel at nearly the speed of light, neutrinos from the center of the Sun arrive at Earth after 8.3 minutes. Therefore, we can use them to probe what the Sun is doing today.

The dominant branch of the proton-proton chain can be written symbolically as follows:

Step 1: $^1\text{H} + {^1\text{H}} \rightarrow {^2\text{H}} + e^+ + \nu$ and then $e^+ + e^- \rightarrow \gamma + \gamma$

Step 2: $^2\text{H} + {^1\text{H}} \rightarrow {^3\text{He}} + \gamma$

Step 3: $^3\text{He} + {^3\text{He}} \rightarrow {^4\text{He}} + {^1\text{H}} + {^1\text{H}}$

The rate of the proton-proton chain reaction depends on both temperature and density. At the temperature and pressure that exist within the Sun's core, the reaction rate is relatively slow. The Sun's slow nuclear fusion is fortunate for life on Earth: if the hydrogen fused quickly, the Sun would have exhausted its supply long ago, and life might not have had time to evolve.

The other 15 percent of the Sun's energy is generated by variations of the proton-proton chain. The most common variation happens in step 3, in which ^{3}He fuses with an existing ^{4}He to create beryllium (^{7}Be), which decays to lithium (^{7}Li) and energy, and then the ^{7}Li plus one ^{1}H become two ^{4}He. In a less common variation, the beryllium combines with hydrogen to become boron (^{8}B), which then decays to beryllium and then to two ^{4}He. In both variations, ultimately four hydrogen nuclei become one helium nucleus.

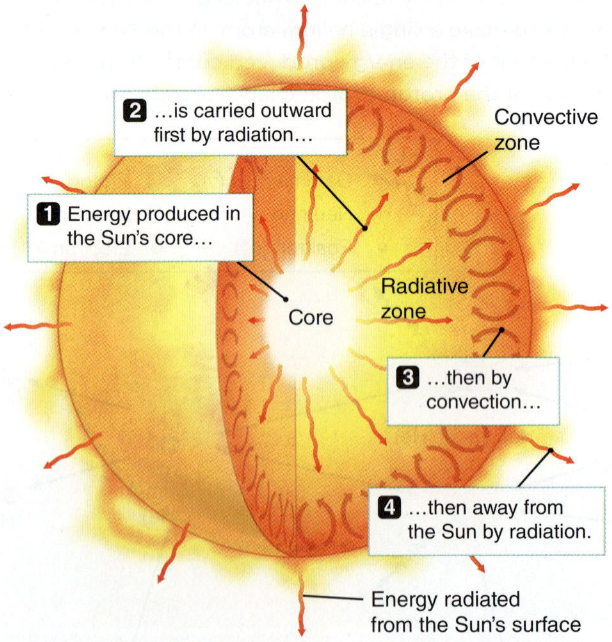

Figure 14.5 The interior of the Sun is divided into zones on the basis of where energy is produced and how it is transported outward.

CHECK YOUR UNDERSTANDING 14.1B

When hydrogen is fused into helium, energy is released from: (a) gravitational collapse; (b) conversion of mass to energy; (c) the increase in pressure; (d) the decrease in the gravitational field.

14.2 Energy Is Transferred from the Interior of the Sun

Although geologists cannot travel deep inside Earth to find out how it is structured, they can build a model of its interior by using data on how seismic waves travel during earthquakes. Similarly, astronomers can create a model of the Sun's interior by using their knowledge of the balance of forces and energy within the Sun and an understanding of how energy moves from one place to another. Those models can be tested by observations of waves traveling through the Sun and by studying neutrinos from the Sun.

Astronomy in Action: Random Walk

Energy Transport

Some of the energy released by hydrogen fusion in the core of the Sun escapes directly into space in the form of neutrinos. However, most of the energy heats the solar interior and then moves outward through the Sun to the surface, a process known as **energy transport**. Energy transport, a key determinant of the Sun's structure, can occur by *conduction*, *convection*, or *radiation*.

Conduction is important primarily in solids. When you pick up a hot object, for example, your fingers are heated by conduction. That happens because energetic thermal vibrations of atoms and molecules cause neighboring atoms and molecules to vibrate more rapidly as well. Conduction is typically ineffective in a gas because the atoms and molecules are too far apart to transmit vibrations to one another efficiently. Conduction does not play a key role in transporting energy from the core of the Sun to its surface, but conduction will be relevant later when we discuss dying stars.

In the Sun, energy is transported by convection and radiation through zones, as shown in **Figure 14.5**. The mechanism of energy transport from the center of the Sun outward depends on the decreasing temperature and density as the radius increases. First, energy moves outward through the inner layers of the Sun as radiation in the form of photons. Next, energy moves by convection in parcels of gas. Finally, energy radiates from the Sun's surface as light. We look at each process in turn.

Near the core, **radiation** transfers energy from hotter to cooler regions via photons (**Figure 14.6**), which carry the energy with them. Recall from your study of radiation in Chapter 5 that the hotter region contains more (and more energetic) photons than the cooler region. More photons move from the hotter, very crowded region to the cooler, less crowded region than in the reverse direction. A net transfer of photons and photon energy occurs from the hotter region to the cooler region, and radiation carries energy outward from the Sun's core.

The transfer of energy from one point to another by radiation also depends on how freely radiation can move from one point to another within a star. The degree to which matter blocks the flow of photons through it is called **opacity**. The opacity of a material depends on many things, including the density of the material, its composition, its temperature, and the wavelength of the photons moving through it.

Energy transfer by radiation is most efficient in regions with low opacity. The radiative zone (see Figure 14.5) is the region in the inner part of the Sun in which the opacity is relatively low, and radiation carries the energy produced in the

Figure 14.6 Higher-temperature regions deep within the Sun produce more radiation than do lower-temperature regions farther out. Although radiation flows in both directions, more radiation flows from the hotter regions to the cooler regions than from the cooler regions to the hotter regions. Therefore, radiation carries energy outward from the inner parts of the Sun. For simplicity, this illustration includes only the most common photons (those with wavelengths at the peak of the blackbody curve). Photons of wavelengths representing all colors are present in all regions, with more of all kinds in the hotter regions and fewer of all kinds in the cooler regions.

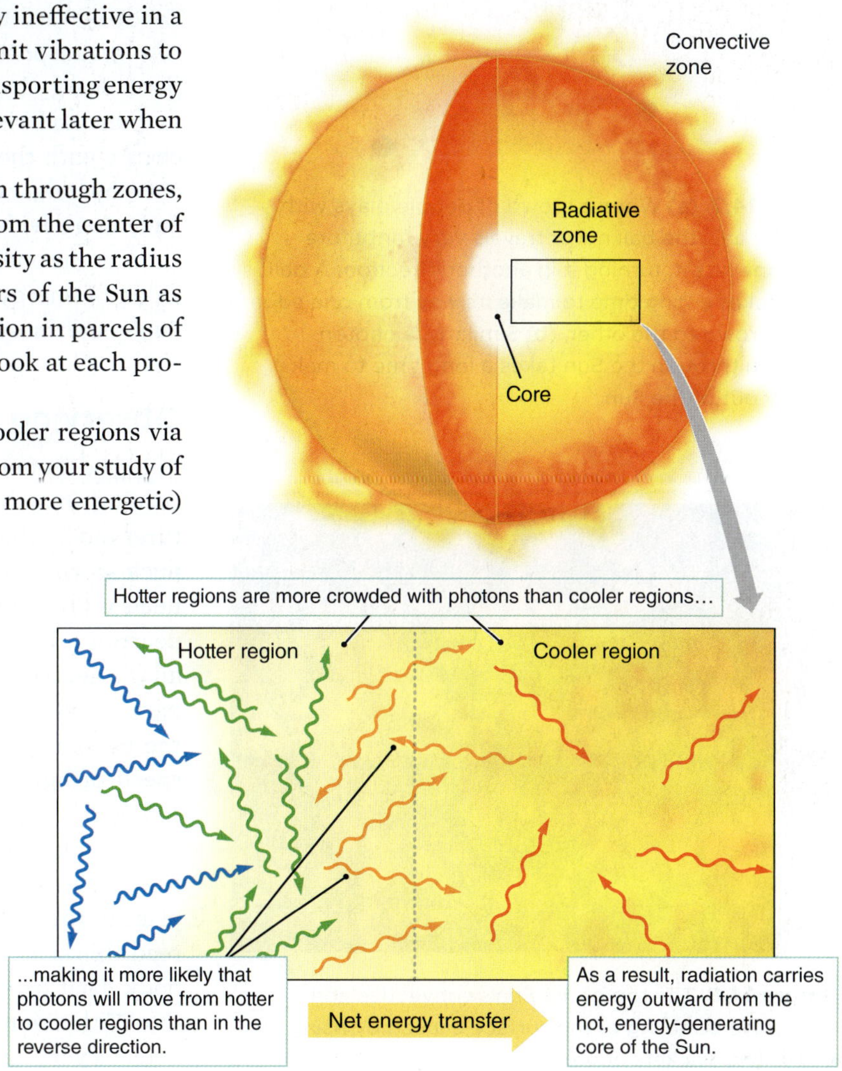

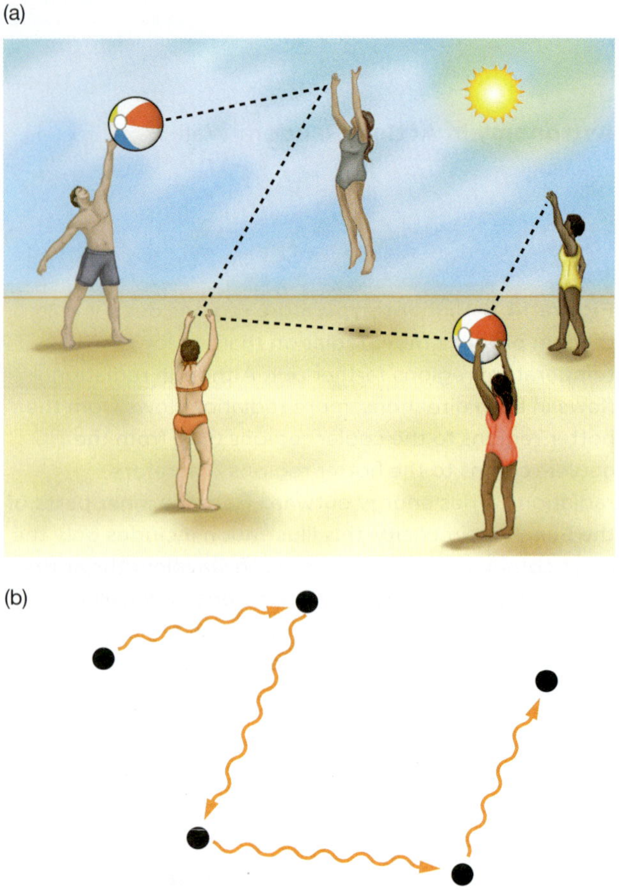

Figure 14.7 (a) When a crowd of people plays with a beach ball, the ball never travels very far before someone hits it, turning it in another direction. A ball often takes a long time to make its way from one edge of the crowd to the other. (b) Similarly, a photon traveling through the Sun takes a long time to make its way out of the Sun.

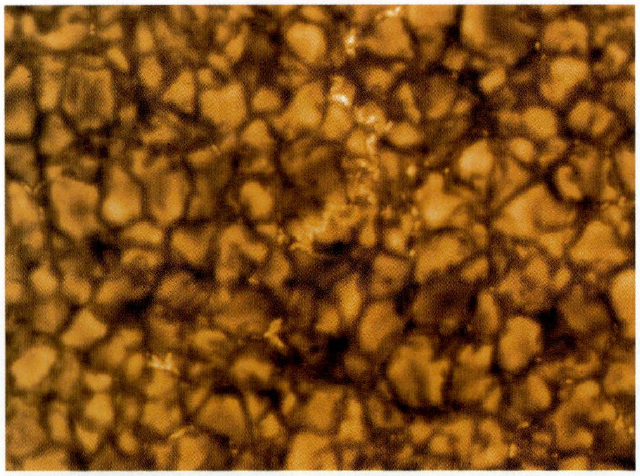

Figure 14.8 The top of the Sun's convective zone shows the bubbling of the surface caused by rising and falling packets of gas.

core outward through the star. That radiative zone extends about 70 percent of the way out toward the surface of the Sun. Even though the region's opacity is low enough for radiation to dominate convection as an energy transport mechanism, photons still travel only a short distance within the region before being absorbed and then reemitted, or scattered by matter, much like a beach ball being batted about by a crowd of people (**Figure 14.7**). Each interaction sends the photon in an unpredictable direction—not necessarily toward the surface of the star. The distances between interactions are so short that, on average, the energy of a gamma-ray photon produced in the interior of the Sun takes about one hundred thousand or more years to find its way to the outer layers of the Sun. Opacity holds energy inside the Sun and lets it seep away only slowly. As it travels, the gamma-ray photon gradually is converted to lower-energy photons, emerging mostly as optical and infrared radiation from the surface.

From a peak of 15 million K at the center of the Sun, the temperature falls to about 2 million to 3 million K at the outer margin of the radiative zone. At that cooler temperature, the opacity is higher, so radiation is less efficient at carrying energy from one place to another. The energy flowing outward through the Sun "piles up" against that edge of the radiative zone.

Nearer the surface of the Sun, transfer by radiation becomes inefficient and the temperature changes quickly. Instead, **convection** takes over. Convection carries energy from inside a planet to its surface or from the Sun-heated surface of Earth upward through Earth's atmosphere. Convection also plays an important role in transporting energy outward from the interior of the Sun. Convection transports energy by moving packets of hot gas, like hot-air balloons, which become buoyant and rise up through the lower-temperature gas above them, carrying energy with them. The solar convective zone (see Figure 14.5) extends from the outer boundary of the radiative zone outward to just below the visible surface of the Sun, where evidence of convection can be seen in the bubbling surface (**Figure 14.8**).

In the outermost layers of the Sun, radiation again takes over as the primary mode of energy transport, and it is radiation that transports energy from the Sun's outermost layers off into space.

Observing Neutrinos from the Core of the Sun

The model of energy production and energy transport in the Sun just presented correctly matches observed global properties of the Sun such as its size, temperature, and luminosity. The nuclear fusion model of the Sun predicts exactly which nuclear reactions should be occurring in the core of the Sun and at what rate. The nuclear reactions that make up the proton-proton chain produce a vast number of neutrinos. Because neutrinos barely interact with other ordinary matter, almost all the neutrinos produced in the heart of the Sun travel freely through the outer parts of the Sun and on into space as though the outer layers of the Sun were not there. Solar neutrinos produced in the core of the Sun, traveling at nearly the speed of light, take only 8.3 minutes to reach Earth—much quicker than the 100,000-year journey of photons leaving the core.

Neutrinos interact so weakly with matter that they are extremely difficult to observe. Nevertheless, an extremely large number of nuclear reactions take place in the Sun, so the Sun produces an enormous number of neutrinos. As you read this sentence, about 400 trillion solar neutrinos are passing through your body. That happens even at night because neutrinos easily pass through Earth, too. With that many neutrinos moving about, a neutrino detector does not have to detect a very large percentage of them to be effective.

A neutrino telescope looks very different from other telescopes. The first apparatus designed to detect solar neutrinos was built 1,500 meters underground, within the Homestake Mine in Lead, South Dakota. Astronomers filled a tank with 100,000 gallons of a dry-cleaning fluid called perchloroethylene (C_2Cl_4). Astronomers predicted that, over 2 days, roughly 10^{22} solar neutrinos passed through the Homestake detector. Of those, on average only one neutrino interacted with a chlorine atom within the fluid to form a radioactive isotope of argon. Over time, a measurable amount of argon was produced.

The Homestake experiment operated from the late 1960s to the early 1990s and detected that argon isotope—evidence of neutrinos from the Sun, confirming that nuclear fusion powers the Sun. Astronomers noticed, however, that they were measuring only about one-third as many solar neutrinos as predicted by solar models. The difference between the predicted and measured number of solar neutrinos was called the **solar neutrino problem**.

One possible explanation of the solar neutrino problem was that the working model of the structure of the Sun was somehow wrong. That possibility seemed unlikely, however, because of the many other successful predictions of the solar model. A second possibility was that an understanding of the neutrino itself was incomplete. The neutrino was long thought to have zero mass, like photons, and to travel at the speed of light. But if neutrinos actually do have a tiny amount of mass, particle physics suggests that solar neutrinos should oscillate—alternate back and forth—among three kinds of neutrinos: the electron, muon, and tau neutrinos. According to that explanation, early neutrino experiments could detect only the electron neutrino and, consequently, observed only about a third of the expected number of neutrinos. Since then, many other neutrino detectors have been built, each using different reactions to detect neutrinos of different energies or different types. Experiments at high-energy physics labs, nuclear reactors, and neutrino telescopes around the world have shown that neutrinos do have a nonzero mass and do oscillate among neutrino types.

Solving the solar neutrino problem is a good example of how science works—how a better model of the neutrino showed that the solar neutrino problem was real and not merely an experimental mistake, and how a single set of anomalous observations was later confirmed by other, more sophisticated experiments. All that effort led to a better understanding of basic physics (see the **Process of Science Figure**).

Probing the Sun's Interior

Models of Earth's interior predict how density and temperature change from place to place. Those differences affect the seismic waves traveling through Earth, bending the paths that they travel. Geologists test models of Earth's interior by comparing measurements of seismic waves from earthquakes with model predictions of how seismic waves should travel through the planet.

Just as geologists use seismic waves from earthquakes to probe the interior of Earth, solar physicists use the surface oscillations of the Sun to probe the solar interior. The science that uses solar oscillations to study the Sun is called **helioseismology**. Detailed observations of motions of material from place to place across the surface of the Sun show that the Sun vibrates or rings, something like a bell that has been struck. Unlike a well-tuned bell—which vibrates primarily at one frequency—the vibrations of the Sun are very complex. In the Sun, many frequencies of vibrations occur simultaneously, causing some parts of the Sun to bulge outward and some to draw inward. Those motions help

PROCESS OF SCIENCE

Learning from Failure

The first detection of solar neutrinos raised more questions than it answered.

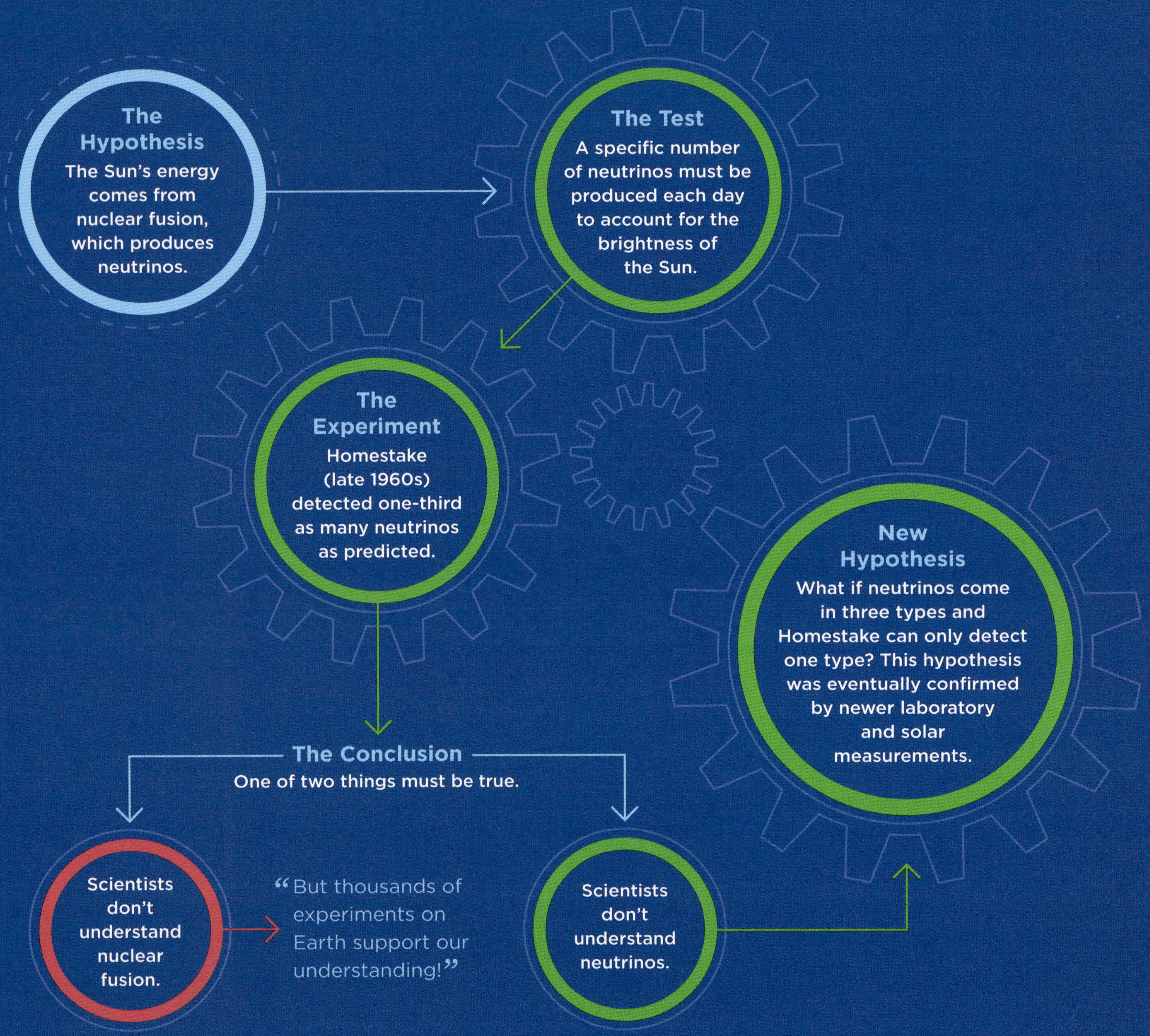

Part of the "scientific attitude" is to find failure exciting. When experiments do not turn out as expected, good scientists get excited because there is something new to understand!

researchers probe what lies below. **Figure 14.9** illustrates the motions of the different parts of the Sun, with red and blue areas moving in opposite directions. Some waves are amplified and some are suppressed, depending on how they overlap as they travel through the Sun. Astronomers study those waves by using the Doppler effect (see Chapter 5), which distinguishes between parts of the Sun that move toward the observer and those that move away.

To detect the disturbances of helioseismic waves on the surface of the Sun, astronomers must measure Doppler shifts of less than 0.1 m/s while detecting changes in brightness of only a few parts per million at any given location on the Sun. Tens of millions of wave motions are possible within the Sun. Some waves travel around the circumference of the Sun, yielding information about the density of the upper convection zone. Other waves travel through the interior of the Sun, revealing the density structure of the Sun close to its core. Still others travel inward toward the center of the Sun, until they are bent by the changing solar density and return to the surface.

All those wave motions are going on at the same time, so sorting them out requires computer analysis of long, unbroken strings of solar observations from several sources. The Global Oscillation Network Group (GONG) is a network of six solar observation stations spread around the world that enables astronomers to observe the surface of the Sun approximately 90 percent of the time.

To interpret helioseismology data, scientists compare the measurements of the strength, frequency, and wavelengths of the waves against predicted vibrations calculated from models of the solar interior. That technique serves as a powerful test of models of the solar interior and has led both to some surprises and to improvements in the models. For example, some scientists proposed that the solar neutrino problem might be solved if the models had overestimated the amount of helium in the Sun. That explanation was ruled out by analyzing the waves that penetrate to the core of the Sun. Helioseismology showed that the value for opacity used in early solar models was too low. That realization led astronomers to recalculate the location of the bottom of the convective zone. Both theory and observation now put the base of the convective zone at 70 percent of the way out from the center of the Sun, with an uncertainty in that number of less than 0.5 percent.

Working back and forth between observation and theory has enabled astronomers to probe the otherwise inaccessible interior of the Sun. We now know that the energy is produced by nuclear fusion deep in the core and that it moves outward by radiation to a point about 70 percent of the radius of the Sun. Then it travels outward by convection to the surface. We also know how the temperature, density, and pressure change with radius and how those factors change the opacity at different distances from the center. Even though researchers usually cannot sample directly or set up controlled experiments, collaboration between theory and observation is essential to observational sciences such as astronomy.

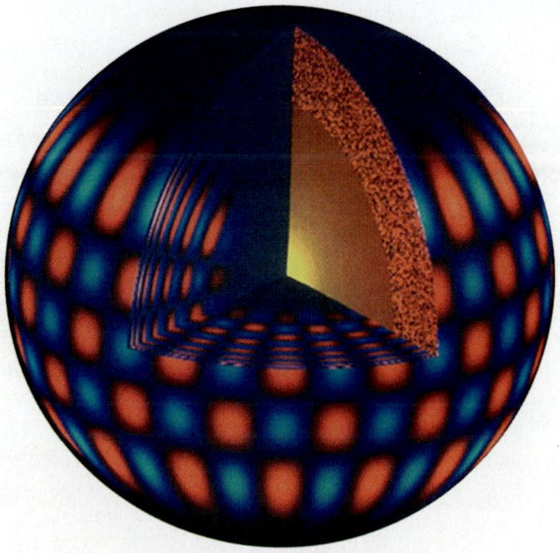

Figure 14.9 The interior of the Sun rings like a bell as helioseismic waves move through it. This figure shows one particular mode of the Sun's vibration. Red indicates regions where gas is traveling inward, whereas blue indicates regions where gas is traveling outward. Astronomers observe those motions via Doppler shifts

CHECK YOUR UNDERSTANDING 14.2

How do neutrinos help us understand what is going on in the core of the Sun? (a) Neutrinos from distant objects pass through the Sun, probing the interior. (b) Neutrinos from the Sun pass easily through Earth. (c) Neutrinos created in fusion reactions at the core of the Sun easily escape.

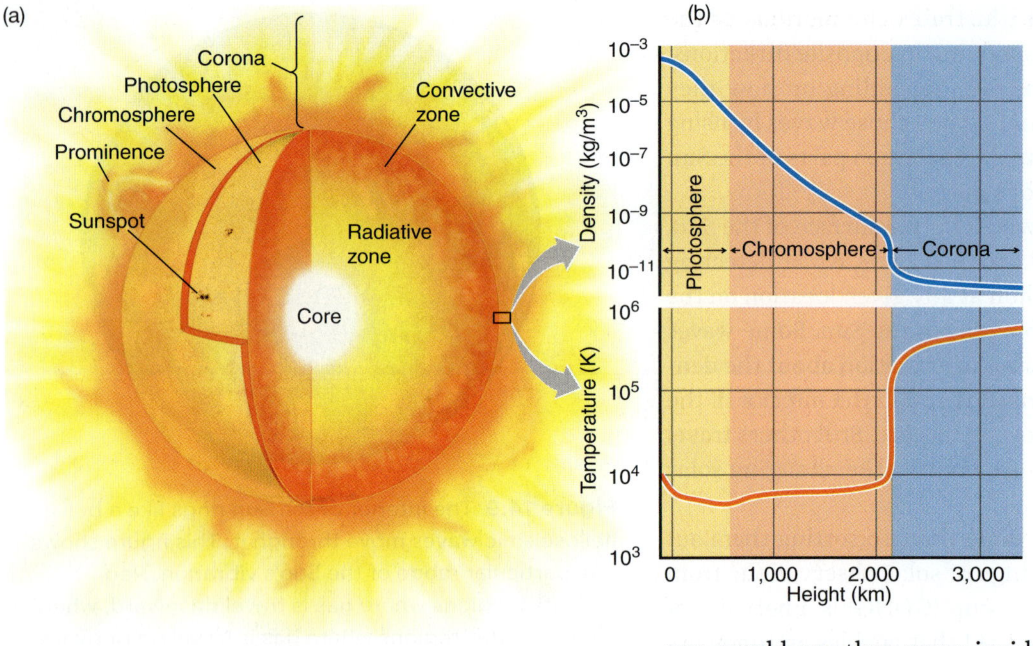

Figure 14.10 (a) The components of the Sun's atmosphere are located above the convective zone. (b) The density and temperature of the Sun's atmosphere change abruptly at the boundary between the chromosphere and corona. The *y*-axes are logarithmic.

14.3 The Atmosphere of the Sun

Beyond the convective zone lie the outer layers of the Sun, collectively known as the Sun's atmosphere. Those layers, shown in **Figure 14.10a**, include the *photosphere*, the *chromosphere*, and the *corona*. We can observe those layers of the Sun directly by using telescopes and satellites. Observations of the Sun's atmosphere are important because activity in the Sun's atmosphere has consequences for human infrastructure such as power grids and satellites in orbit around Earth.

The Sun is a large ball of gas and so, unlike Earth, has no solid surface. Its apparent surface is like a fog bank on Earth. After people walk into a fog bank and disappear from view, you would say they were inside the fog bank, even though they never passed through a definite boundary. The apparent surface of the Sun is similar. Light from the Sun's surface can escape directly into space, so we can see it. Light from below the Sun's surface cannot escape directly into space, so we cannot see it.

At the base of the atmosphere is the **photosphere**: the apparent surface of the Sun or its visible boundary. That is where features such as sunspots can be seen. Above that photosphere is the **chromosphere**, a region of strong emission lines. The top layer is the **corona**, which can be viewed during a solar eclipse as a halo around the Sun. In the Sun's atmosphere, the density of the gas drops very rapidly with increasing altitude. The graphs in **Figure 14.10b** show how density and temperature change across the atmosphere of the Sun. The temperature increases sharply across the boundary between the chromosphere and the corona, whereas the density falls sharply across the same boundary. In this section, we explore each layer, beginning at the bottom with the photosphere.

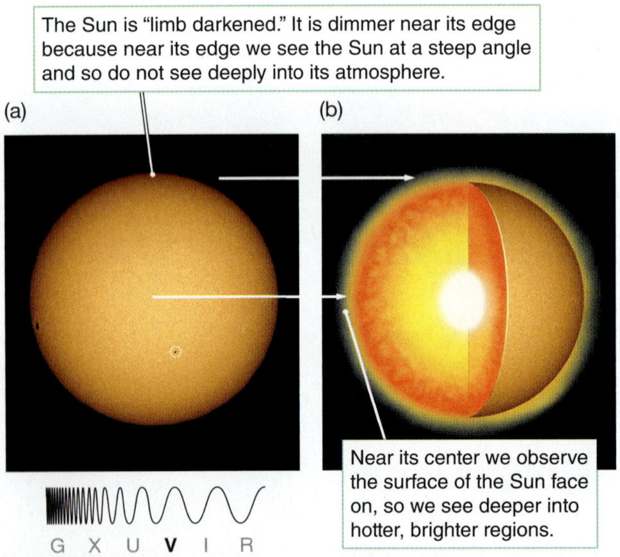

Figure 14.11 (a) When viewed in visible light, the Sun appears to have a sharp outline, even though it has no true surface. The center of the Sun appears brighter, whereas the limb of the Sun is darker—an effect known as limb darkening. (b) Looking at the center of the Sun allows us to see deeper into the Sun's interior than we do when looking at the edge of the Sun. Because higher temperature means more luminous radiation, the center of the Sun appears brighter than its limb.

The Photosphere

The **effective temperature** of the photosphere is calculated from the Sun's luminosity and radius by using the Stefan-Boltzmann law (see Chapter 5). The photosphere has an effective temperature of 5780 K, ranging from 6600 K to 4500 K over a 500-km-thick zone. The Sun appears to have a well-defined surface and a sharp outline when viewed from Earth because 500 km does not look very thick when viewed from a distance of 150 million km.

In **Figure 14.11a**, the Sun appears fainter near its edges than near its center, an effect called **limb darkening**. That effect is an artifact of the structure of the Sun's photosphere. When looking near the edge of the Sun, you are looking through the photosphere at a steep angle. As a result, you do not see as deeply into the Sun as when you are looking directly down through the photosphere near the center of the Sun's disk. The light from the limb of the Sun comes from a shallower layer that is cooler and fainter, as shown in **Figure 14.11b**.

In the Sun's atmosphere, the density of the gas drops very rapidly with increasing altitude. All visible solar phenomena take place in the Sun's atmosphere. Most of the radiation from below the Sun's photosphere is absorbed by matter and reemitted at the photosphere as a blackbody spectrum.

As we examine the structure of the Sun in more detail, however, that simple description of the spectra of stars turns out to be incomplete. Light from the solar photosphere must escape through the upper layers of the Sun's atmosphere, which affects the spectrum we observe. In Chapter 13, we discussed the presence of absorption lines in the spectra of stars. Now we can take a closer look at how those absorption lines form. As photospheric light travels upward, atoms in the solar atmosphere absorb the light at discrete wavelengths, forming absorption lines. Because the Sun appears so much brighter than any other star, its spectrum can be studied in far more detail, so specially designed telescopes and high-resolution spectrometers have been built specifically to study the Sun's light. The solar spectrum is shown in **Figure 14.12**. Absorption lines from more than 70 elements have been identified. Analysis of those lines forms the basis for much of astronomers' knowledge of the solar atmosphere, including the composition of the Sun. That is also the starting point for understanding the atmospheres and spectra of other stars.

Figure 14.12 This high-resolution spectrum of the Sun stretches from 400 nanometers (nm) in the lower left corner to 700 nm in the upper right corner and shows dark absorption lines. The spectrum was produced by passing the Sun's light through a prism-like device and then cutting and folding the single long spectrum (from blue to red) into rows so that it will fit in a single image taken by a camera.

The Chromosphere and Corona

The temperature falls from 6600 K at the photosphere's bottom to 4400 K at its top. At that point, the trend reverses and the temperature slowly begins to climb, rising to about 6000 K at a height of 1,500 km above the top of the photosphere (see Figure 14.10b). This region of increasing temperature is called the chromosphere (**Figure 14.13a**). The reason for the chromosphere's temperature reversal with increasing height is not well understood, but it may be caused by magnetic waves propagating through the region and depositing their energy at the top of the chromosphere.

The chromosphere was discovered in the 19th century during observations of total solar eclipses (**Figure 14.13b**). The chromosphere is seen most strongly at the solar limb as a source of emission lines, especially a particular hydrogen line produced when the electron falls from the third energy state to the second energy state. That line is known as the Hα line (the "hydrogen alpha line"). The deep red color of the Hα line is what gives the chromosphere its name; that word means "the place where color comes from." The element helium was discovered in 1868 from a spectrum of the chromosphere of the Sun nearly 30 years before it was found on Earth. Helium is named after *helios*, the Greek word for "Sun."

At the top of the chromosphere, across a transition region only about 100 km thick, the temperature suddenly soars, whereas the density abruptly drops (see Figure 14.10b). Above that transition lies the outermost region of the Sun's atmosphere, the corona, where temperatures reach 1 million to 2 million K. The corona is thought to be heated by magnetic fields and micro solar flares.

The Sun's corona has been known since ancient times: it is visible during total solar eclipses as an eerie outer glow stretching several solar radii beyond the Sun's surface (**Figure 14.13c**). Because it is so hot, the solar corona also is a strong source of X-rays. Those X-ray photons have so much energy that many

Figure 14.13 (a) This spacecraft image of the Sun shows fine structure in the chromosphere extending outward from the photosphere. (b) The chromosphere is visible during a total eclipse. (c) This eclipse image shows the Sun's corona, consisting of million-kelvin gas that extends for millions of kilometers beyond the surface of the Sun.

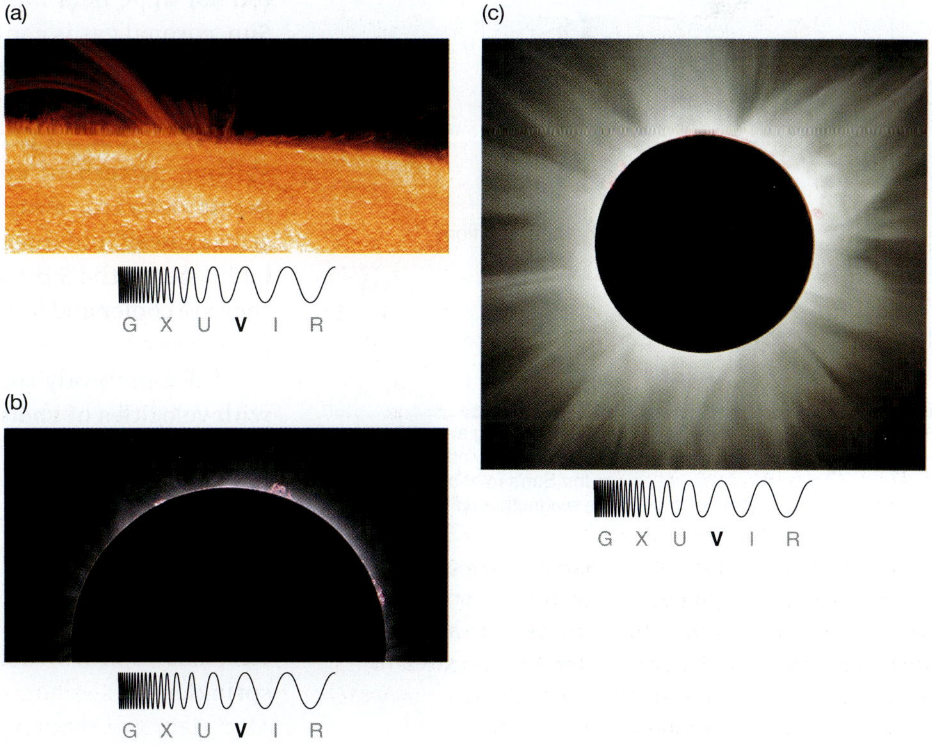

Figure 14.14 This close-up image of the Sun shows the tangled structure of coronal loops.

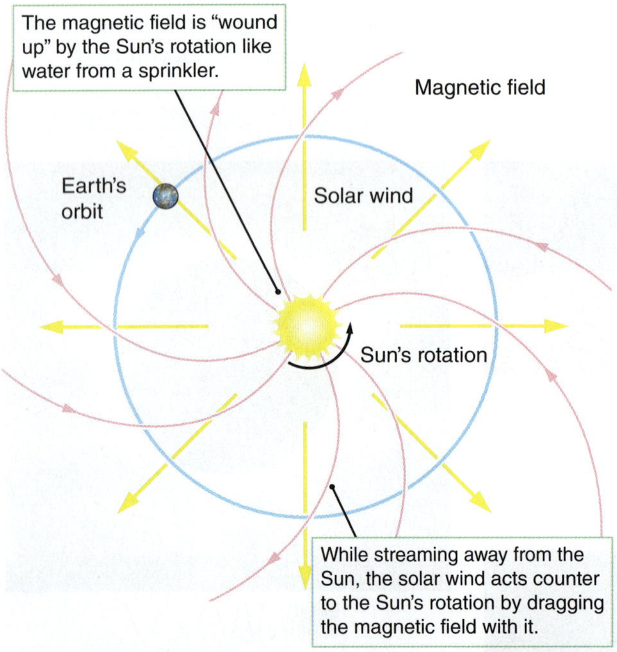

Figure 14.15 The solar wind (yellow arrows) streams away from active areas and coronal holes on the Sun. As the Sun rotates, the solar wind takes on a spiral structure, much like the spiral of water that streams away from a rotating lawn sprinkler.

electrons are stripped away from nuclei, leaving atoms in the corona highly ionized.

CHECK YOUR UNDERSTANDING 14.3

The surface of the Sun appears sharp in visible light because: (a) the photosphere is cooler than the layers below it; (b) the photosphere is thinner than the other layers in the Sun; (c) the photosphere is less dense than the convection zone; (d) the Sun has a distinct surface.

14.4 The Atmosphere of the Sun Is Very Active

The atmosphere of the Sun is a very turbulent place. The best-known features on the surface of the Sun are relatively dark blemishes in the solar photosphere, called **sunspots**. Sunspots come and go over time, though they remain long enough for us to determine the rotation rate of the Sun. Those spots are associated with active regions: loops of material and explosions that fling particles far out into the Solar System. Long-term patterns have been observed in the variations of sunspots and active regions, revealing that the magnetic field of the Sun is constantly changing.

Solar Activity Is Caused by Magnetic Effects

The magnetic field (see Chapter 5) of the Sun causes virtually all the structure seen in the Sun's atmosphere. High-resolution images of the Sun show *coronal loops* that make up much of the Sun's lower corona (**Figure 14.14**). That texture is the result of magnetic structures called flux tubes. Magnetic fields are responsible for much of the structure of the corona as well. The corona is far too hot to be held in by the Sun's gravity, but over most of the surface of the Sun, coronal gas is confined by magnetic loops with both ends firmly anchored deep within the Sun. The magnetic field in the corona acts almost like a network of rubber bands that coronal gas is free to slide along but cannot cross. In contrast, about 20 percent of the surface of the Sun is covered by an ever-shifting pattern of **coronal holes**, large regions where the magnetic field points outward, away from the Sun, and where coronal material is free to stream away into interplanetary space as the solar wind. In extreme ultraviolet images of the Sun we see coronal holes as dark regions, which indicates that they are cooler and less dense than their surroundings (see the chapter-opening photograph).

The relatively steady part of the solar wind consists of lower-speed flows with velocities of about 350 km/s and higher-speed flows with velocities up to about 700 km/s. The higher-speed flows originate in coronal holes. Depending on their speed, particles in the solar wind take 2–5 days to reach Earth. Often, 2–5 days after a coronal hole passes across the center of the face of the Sun, the speed and density of the solar wind reaching Earth increases. The solar wind drags the Sun's magnetic field along with it (**Figure 14.15**), so the magnetic field in the solar wind gets "wound up" by the Sun's rotation. Consequently, the solar wind has a spiral structure resembling the stream of water from a rotating lawn sprinkler.

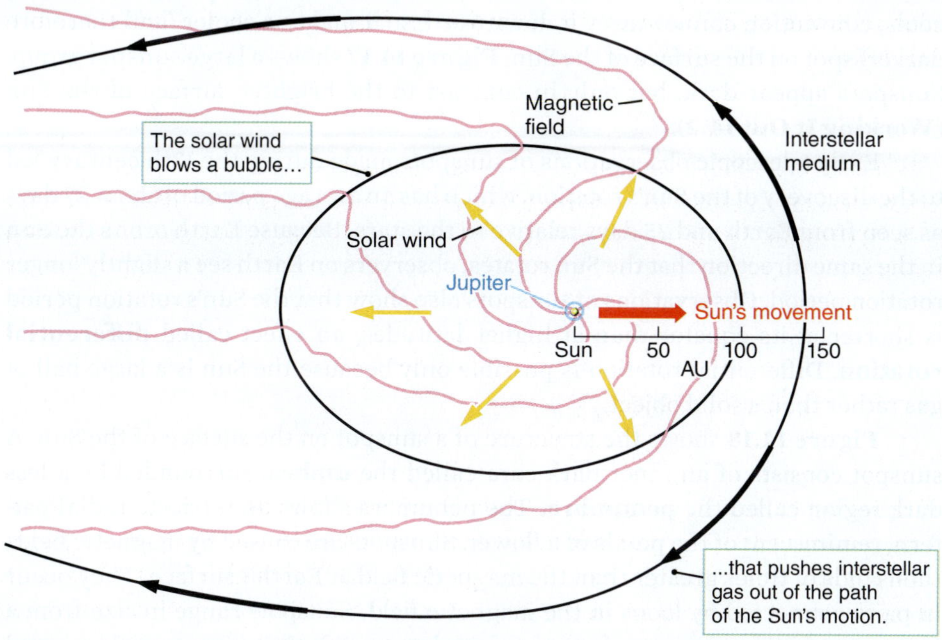

Figure 14.16 The solar wind (yellow arrows) streams away from the Sun for about 100 AU until it finally piles up against the pressure of the interstellar medium through which the Sun is traveling. The *Voyager 1* spacecraft crossed that boundary in 2012.

The effects of the solar wind are felt throughout the Solar System. The solar wind blows the tails of comets away from the Sun, shapes the magnetospheres of the planets, and supplies the energetic particles that power Earth's spectacular auroral displays. Using space probes, astronomers have observed the solar wind extending out to 100 astronomical units (AU) from the Sun. But the solar wind does not go on forever. The farther it gets from the Sun, the more it spreads out. Just like radiation, the density of the solar wind follows an inverse square law. At a distance of about 100 AU from the Sun, the solar wind stops abruptly. Here it piles up against the pressure of the **interstellar medium**, the gas and dust that lie between stars in a galaxy. **Figure 14.16** shows the region of space over which the solar wind is measured. The *Voyager 1* spacecraft has crossed the outer edge of that boundary and sent back the first direct measurements of true interstellar space. The *Interstellar Boundary Explorer* spacecraft, launched in 2008, also is exploring the region.

Sunspots and Changes in the Sun

Sunspots have been noted since antiquity. Telescopic observations of sunspots date back almost 400 years, and records exist of naked-eye observations by Chinese, Greek, and medieval astronomers centuries before that. *But remember: never look directly at the Sun!* Direct viewing through a commercial solar filter is safe, as is projecting the image through a telescope or binoculars onto a surface such as paper and looking only at the projection. Many websites have live images of the Sun viewed through ground and space telescopes (see the "Using the Web" problems at the end of the chapter). Sunspots are places where material is trapped at the surface of the Sun by magnetic-field lines. When that material

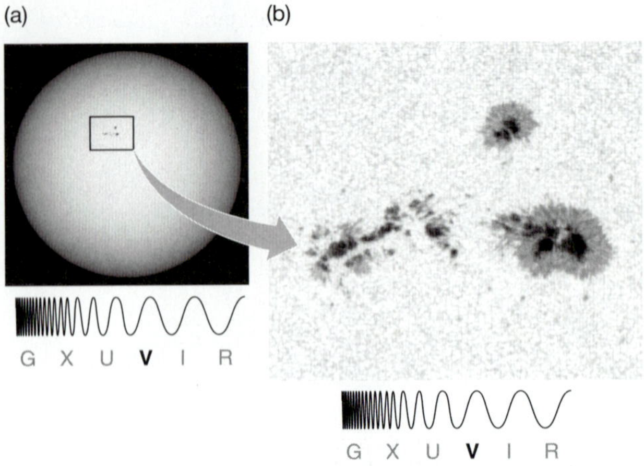

Figure 14.17 (a) This image from the Solar Dynamics Observatory (SDO), taken in 2010, shows a large sunspot group. Sunspots are magnetically active regions cooler than the surrounding surface of the Sun. (b) This high-resolution view shows the sunspots in this group.

cools, convection cannot carry it downward, so it makes a cooler (and therefore darker) spot on the surface of the Sun. **Figure 14.17** shows a large sunspot group. Sunspots appear dark, but only in contrast to the brighter surface of the Sun (**Working It Out 14.2**).

Early telescopic observations of sunspots made during the 17th century led to the discovery of the Sun's rotation, which has an average period of about 27 days as seen from Earth and 25 days relative to the stars. Because Earth orbits the Sun in the same direction that the Sun rotates, observers on Earth see a slightly longer rotation period. Observations of sunspots also show that the Sun's rotation period is shorter at its equator than at higher latitudes, an effect called **differential rotation**. Differential rotation is possible only because the Sun is a large ball of gas rather than a solid object.

Figure 14.18 shows the structure of a sunspot on the surface of the Sun. A sunspot consists of an inner dark core called the **umbra**, surrounded by a less dark region called the **penumbra**. The penumbra shows an intricate radial pattern, reminiscent of the petals of a flower. Sunspots are caused by magnetic fields thousands of times greater than the magnetic field at Earth's surface. They occur in pairs connected by loops in the magnetic field. Sunspots range in size from a few tens of kilometers across up to complex groups that may contain several dozen individual spots and span as much as 150,000 km. The largest sunspot groups can be seen without a telescope.

Although sunspots occasionally last 100 days or longer, half of all sunspots come and go in about 2 days, and 90 percent are gone within 11 days. The number and distribution of sunspots change in a pattern averaging 11 years called the **sunspot cycle**. **Figure 14.19a** shows data for several recent cycles. At the beginning of a cycle, sunspots appear at solar latitudes of about 30° north and south of

14.2 Working It Out Sunspots and Temperature

Sunspots are about 1500 K cooler than their surroundings. What does that lower temperature tell us about their luminosity? Think back to the Stefan-Boltzmann law in Chapter 5. The flux, $\mathcal{F}$, from a blackbody is proportional to the fourth power of the temperature, T:

$$\mathcal{F} = \sigma T^4$$

The constant of proportionality is the Stefan-Boltzmann constant, σ, which has a value of 5.67×10^{-8} W/(m^2 K^4).

The flux is the amount of energy coming from a square meter of surface every second. How much less energy comes out of a sunspot than out of the rest of the Sun? Using 4500 and 6000 K for the temperatures of a typical sunspot and the surrounding photosphere, respectively, we can set up two equations:

$$\mathcal{F}_{spot} = \sigma T^4_{spot} \quad \text{and} \quad \mathcal{F}_{surface} = \sigma T^4_{surface}$$

We could solve each of those separately and then divide the value of $\mathcal{F}_{spot}$ by $\mathcal{F}_{surface}$ to find out how much fainter the sunspot is, but solving for the ratio of the fluxes eliminates σ:

$$\frac{\mathcal{F}_{spot}}{\mathcal{F}_{surface}} = \frac{\sigma T^4_{spot}}{\sigma T^4_{surface}} = \frac{T^4_{spot}}{T^4_{surface}} = \left(\frac{T_{spot}}{T_{surface}}\right)^4$$

Plugging in our values for T_{spot} and $T_{surface}$ gives

$$\frac{\mathcal{F}_{spot}}{\mathcal{F}_{surface}} = \left(\frac{4500\ \text{K}}{6000\ \text{K}}\right)^4 = 0.32$$

and multiplying both sides by $F_{surface}$ gives

$$\mathcal{F}_{spot} = 0.32\mathcal{F}_{surface}$$

So, the amount of energy coming from a square meter of sunspot every second is about one-third as much as the amount of energy coming from a square meter of surrounding surface every second. In other words, the sunspot is about one-third as bright as the surrounding photosphere. If you could cut out the sunspot and place it elsewhere in the sky, it would be brighter than the full Moon.

the solar equator. Over the following years, sunspots are found closer to the equator as their number increases to a maximum and then declines. As the last few sunspots approach the equator, new sunspots again begin appearing at middle latitudes, and the next cycle begins. **Figure 14.19b** shows the number of sunspots at a given latitude plotted against time: that diagram of opposing diagonal bands is called the sunspot "butterfly diagram."

In the early 20th century, solar astronomer George Ellery Hale (1868–1938) was the first to show that the 11-year sunspot cycle is actually half of a 22-year magnetic cycle during which the direction of the Sun's magnetic field reverses after each 11-year sunspot cycle. **Figure 14.19c** shows how the average strength of the magnetic field at every latitude has changed over more than 35 years. The direction of the Sun's magnetic field flips at the maximum of each sunspot cycle. Sunspots come in pairs, with one spot (the leading sunspot) in front of the other with respect to the Sun's rotation. In one 11-year sunspot cycle, the leading sunspot in each pair tends to be a north magnetic pole, whereas the trailing sunspot tends to be a south magnetic pole. In the next 11-year sunspot cycle, that polarity is reversed: the leading sunspot in each pair is a south magnetic pole, whereas the trailing sunspot tends to be a north magnetic pole. The

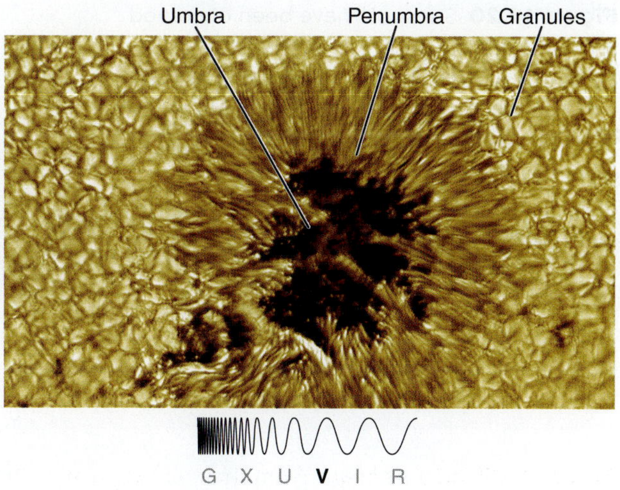

Figure 14.18 This very high-resolution view of a sunspot shows the dark umbra surrounded by the lighter penumbra. The solar surface around the sunspot bubbles with separate cells of hot gas called *granules*. The smallest features are about 100 km across.

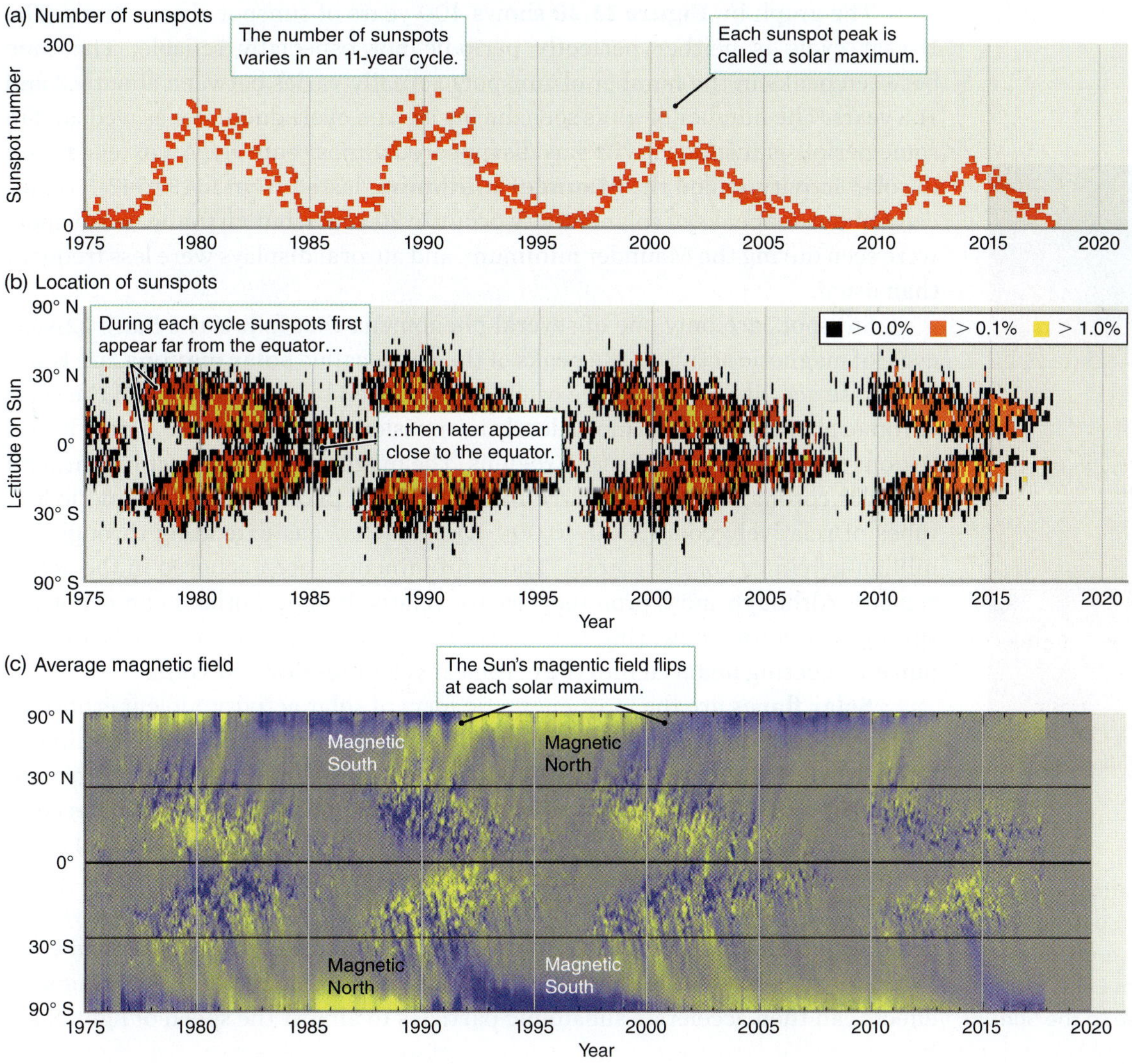

Figure 14.19 (a) The number of sunspots varies, as shown in this graph of the past few solar cycles. (b) The "solar butterfly" diagram shows the fraction of the Sun covered by sunspots at each latitude. The data are color coded to show the percentage of the strip at that latitude that is covered in sunspots at that time: black, 0 to less than 0.1 percent; red, 0.1–1.0 percent; yellow, greater than 1.0 percent. (c) The Sun's magnetic poles flip every 11 years. Yellow indicates magnetic north, whereas blue indicates magnetic south.

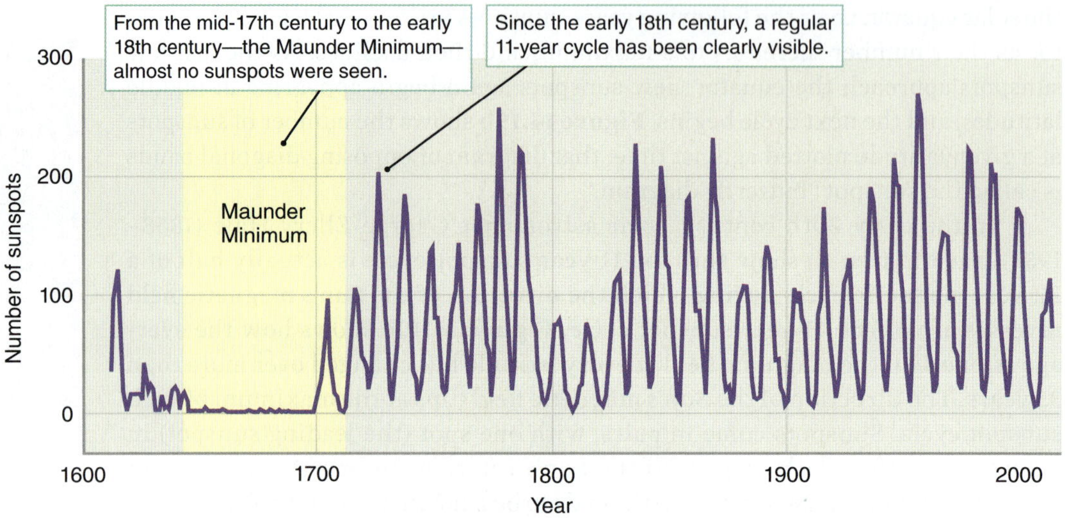

Figure 14.20 Sunspots have been observed for hundreds of years. In this plot, the 11-year cycle in the number of sunspots (half of the 22-year solar magnetic cycle) is clearly visible. Sunspot activity varies greatly. The period from the middle of the 17th century to the early 18th century, when almost no sunspots were seen, is called the *Maunder minimum*.

transition between those two magnetic polarities occurs near the peak of each sunspot cycle. Magnetic activity on the Sun affects the photosphere, chromosphere, and corona.

The graph in **Figure 14.20** shows 400 years of sunspot observations. The 11-year cycle is neither perfectly periodic nor especially reliable. The time between peaks in the number of sunspots actually varies between about 9.7 and 11.8 years. The number of spots seen during a given cycle fluctuates as well, and in some periods sunspot activity has disappeared almost entirely. An extended lull in solar activity, called the **Maunder minimum**, lasted from 1645 to 1715. Typically, about six peaks of solar activity occur in 70 years, but virtually no sunspots were seen during the Maunder minimum, and auroral displays were less frequent than usual.

Sunspots are only one of several phenomena that follow the Sun's 22-year cycle of magnetic activity. The peaks of the cycle, called **solar maxima**, are times of intense activity. Sunspots are often accompanied by a brightening of the solar chromosphere that is seen most clearly in emission lines such as Hα. Those bright regions are known as solar active regions. The magnificent loops arching through the solar corona, shown in **Figure 14.21**, are solar **prominences**, magnetic flux tubes of relatively cool (5000–10,000 K) but dense gas extending through the million-kelvin gas of the corona. Those prominences are anchored in the active regions. Although most prominences are relatively quiet, others can erupt out through the corona, towering a million kilometers or more over the surface of the Sun and ejecting material into the corona at velocities of 1,000 km/s.

Solar flares are the most energetic form of solar activity, violent eruptions in which enormous amounts of magnetic energy are released over a few minutes to a few hours. **Figure 14.22** shows solar flares erupting from two sunspot groups. The images in Figures 14.22a and b, taken in ultraviolet light, show material at very high temperatures. The spots in the visible-light image (Figure 14.22c) are at the base of the activity seen in Figures 14.22a and b. Solar flares can heat gas to temperatures of 20 million K, and they are the source of intense X-rays and gamma rays. Hot **plasma** (consisting of atoms stripped of some or all of their electrons) moves outward from flares at speeds that can reach 1,500 km/s. Magnetic effects can then accelerate subatomic particles to almost the speed of light. Such

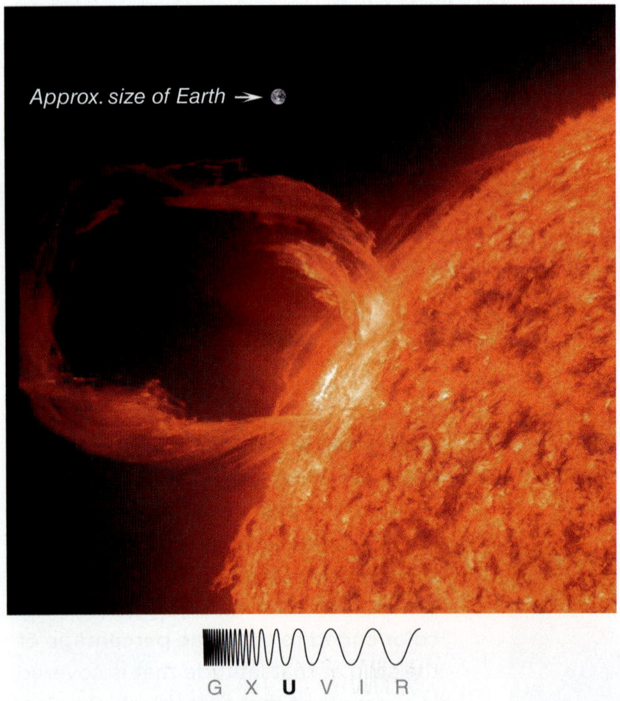

Figure 14.21 Solar prominences are magnetically supported arches of hot gas that rise high above active regions on the Sun. Here, you can see a close-up view at the base of a large prominence. An image of Earth is included for scale (it is not actually that close to the Sun).

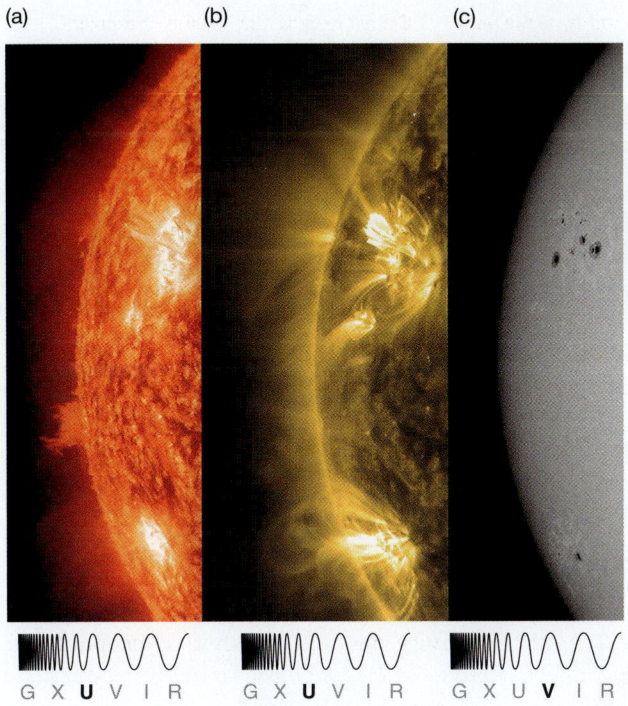

Figure 14.22 The Solar Dynamics Observatory (SDO) observed these active regions of the Sun that produced solar flares in August 2011. (a) Activity near the surface at 60,000 K is visible in extreme ultraviolet light (along with a prominence rising up from the Sun's edge). (b) Viewed at other ultraviolet wavelengths, many looping arcs and plasma heated to about 1 million K become visible. (c) The dark spots in this image are the magnetically intense sunspots that are the sources of all the activity.

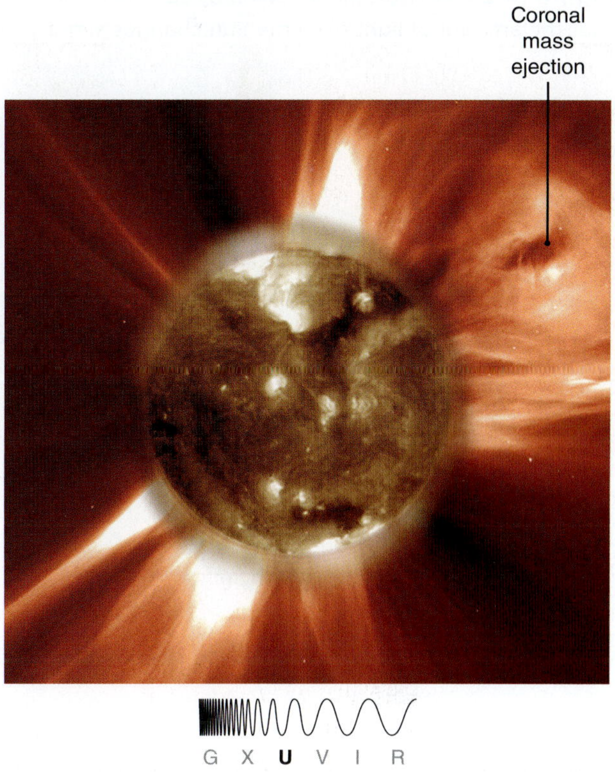

Figure 14.23 This Solar and Heliospheric Observatory (*SOHO*) image shows a coronal mass ejection (upper right); a simultaneously recorded ultraviolet image of the solar disk is superimposed.

events, called **coronal mass ejections (CMEs)** (**Figure 14.23**), send powerful bursts of energetic particles outward through the Solar System. CMEs occur about once per week during the minimum of the sunspot cycle and as often as several times per day near the maximum of the cycle.

Solar Activity Affects Earth

The amount of solar radiation received at the distance of Earth from the Sun has been measured to be, on average, 1,361 watts per square meter (W/m^2). Satellite measurements of the amount of radiation coming from the Sun (**Figure 14.24**) show that this value varies by as much as 0.2 percent over periods of a few weeks, as dark sunspots in the photosphere and bright spots in the chromosphere move across the disk. Overall, however, the increased radiation from active regions on the Sun more than makes up for the reduction in radiation from sunspots. On average, the Sun seems to be about 0.1 percent brighter during the peak of a solar cycle than it is at its minimum.

Solar activity affects Earth in many ways. Solar active regions are the source of most of the Sun's extreme ultraviolet and X-ray emissions—energetic radiation that heats Earth's upper atmosphere and, during periods of increased solar activity, causes Earth's upper atmosphere to expand. When that happens, the swollen upper atmosphere can significantly increase the atmospheric drag on spacecraft

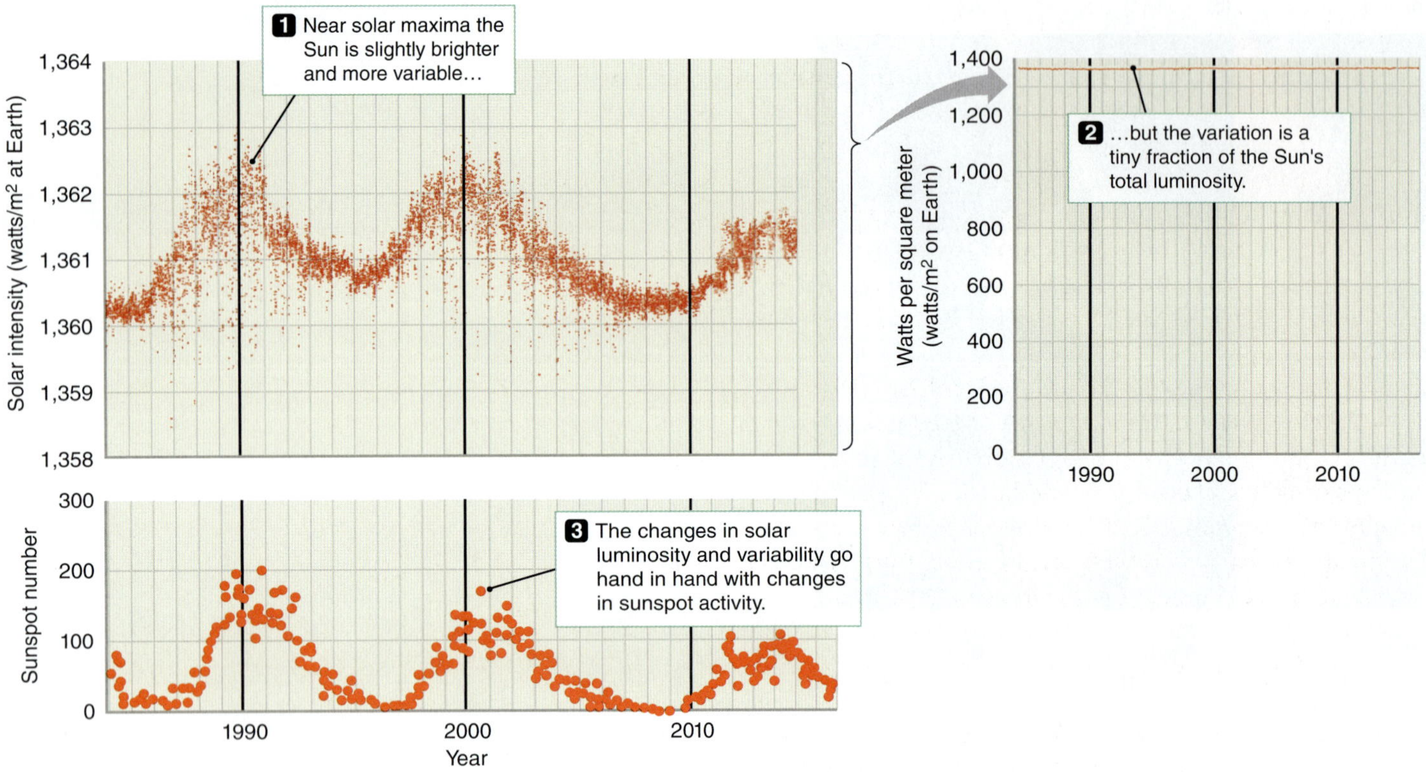

Figure 14.24 Measurements taken by satellites show that the amount of light from the Sun changes slightly.

orbiting at relatively low altitudes, such as that of the Hubble Space Telescope, causing their orbits to decay. Periodic boosts have been necessary to keep the Hubble Space Telescope in its orbit.

Earth's magnetosphere is the result of the interaction between Earth's magnetic field and the solar wind. Increases in the solar wind accompanying solar activity, especially CMEs directed at Earth, can disrupt Earth's magnetosphere. Spectacular auroras can accompany such events, as can magnetic storms that have disrupted electric power grids and caused blackouts across large regions. CMEs emitted in the direction of Earth also hinder radio communication and navigation, and they can damage sensitive satellite electronics, including communication satellites. In addition, energetic particles accelerated in solar flares pose one of the greatest dangers to human exploration of space.

Detailed observations from the ground and from space help astronomers understand the complex nature of the solar atmosphere. The Solar and Heliospheric Observatory (*SOHO*) spacecraft is a joint mission between NASA and the European Space Agency (ESA). *SOHO* moves in lockstep with Earth at a location approximately 1,500,000 km from Earth that is almost directly in line between Earth and the Sun. *SOHO* carries 12 scientific instruments that monitor the Sun and measure the solar wind upstream of Earth. In addition, NASA's Solar Dynamics Observatory (SDO) studies the solar magnetic field to predict when major solar events will occur, instead of simply responding after they happen.

CHECK YOUR UNDERSTANDING 14.4

Sunspots appear dark because: (a) they have very low density; (b) magnetic fields absorb most of the light that falls on them; (c) they are regions of very high pressure; (d) they are cooler than their surroundings.

Origins The Solar Wind and Life

Energetic particles accelerated in solar flares need to be considered when astronauts are orbiting Earth in a space station or, someday, traveling to the Moon or farther. Earth's magnetic field protects life on the surface from those energetic particles, which travel along the magnetic-field lines to Earth's poles, creating the auroras. But the Moon does not have that protection because its magnetic field is very weak. Astronauts on the lunar surface would be exposed to as much radiation as astronauts traveling in space. The strength of the solar wind varies with the solar cycle, as noted in Section 14.4, so the exposure danger varies as well.

The Solar System is surrounded by the **heliosphere** (**Figure 14.25**), in which the solar wind blows against the interstellar medium and clears out an area like the inside of a bubble. As the Sun and Solar System move through the Milky Way Galaxy, passing in and out of interstellar clouds, that heliosphere protects the entire Solar System from galactic high-energy particles known as cosmic rays that originate primarily in high-energy explosions of massive dying stars. When the Sun is in its lower-activity state, the heliosphere is weaker, so more galactic cosmic rays enter the Solar System. In addition, the intensity of those cosmic rays depends on where the Sun and Solar System are located in their orbit about the center of the Milky Way Galaxy.

Some scientists have theorized that at times when the Sun was quiet and the heliosphere was weaker than average, and the Solar System was passing through a particular part of the galaxy, the cosmic-ray flux in the Solar System—and on Earth—increased. That increased flux possibly led to a disruption in Earth's ozone layer and possibly contributed to a mass extinction in which many species died out on Earth.

Thus, in addition to heat and light on Earth, the extension of the Sun through the solar wind may have affected the evolution of life on Earth—and may affect the ability of humans to live and work in space.

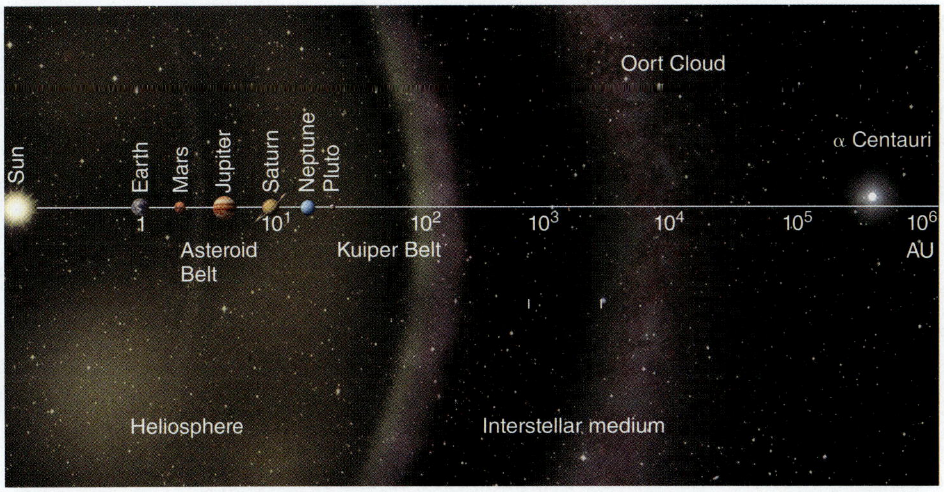

Figure 14.25 The heliosphere of the Sun is a bubble of charged particles covering the Solar System. The heliosphere is formed by the solar wind blowing against the interstellar medium. Both the *Voyager* spacecraft are now past 100 AU. The scale, indicating how far objects are from the Sun, is logarithmic.

READING ASTRONOMY NEWS

ARTICLES QUESTIONS

Scientists discuss major solar storms that affect Earth.

Carrington-Class CME Narrowly Misses Earth

By DR. TONY PHILLIPS, Science@NASA

Last month (April 8–11, 2014), scientists, government officials, emergency planners, and others converged on Boulder, Colorado, for NOAA's Space Weather Workshop—an annual gathering to discuss the perils and probabilities of solar storms.

The current solar cycle is weaker than usual, so you might expect a correspondingly low-key meeting. On the contrary, the halls and meeting rooms were abuzz with excitement about an intense solar storm that narrowly missed Earth.

"If it had hit, we would still be picking up the pieces," says Daniel Baker of the University of Colorado, who presented a talk entitled *The Major Solar Eruptive Event in July 2012: Defining Extreme Space Weather Scenarios.*

The close shave happened almost two years ago. On July 23, 2012, a plasma cloud or "CME" rocketed away from the Sun as fast as 3,000 km/s, more than 4 times faster than a typical eruption. The storm tore through the Earth's orbit, but fortunately Earth wasn't there. Instead it hit the *STEREO-A* spacecraft. Researchers have been analyzing the data ever since, and they have concluded that the storm was one of the strongest in recorded history. "It might have been stronger than the Carrington Event itself," says Baker.

The Carrington Event of September 1859 was a series of powerful CMEs that hit Earth head-on, sparking Northern Lights as far south as Tahiti. Intense geomagnetic storms caused global telegraph lines to spark, setting fire to some telegraph offices and disabling the "Victorian Internet." A similar storm today could have a catastrophic effect on modern power grids and telecommunication networks. According to a study by the National Academy of Sciences, the total economic impact could exceed $2 trillion or 20 times greater than the costs of a Hurricane Katrina. Multi-ton transformers fried by such a storm could take years to repair and impact national security.

A recent paper in *Nature Communications* authored by UC Berkeley space physicist Janet G. Luhmann and former postdoc Ying D. Liu describes what gave the July 2012 storm Carrington-like potency. For one thing, the CME was actually *two* CMEs separated by only 10 to 15 minutes. This double storm cloud traveled through a region of space that had been cleared out by another CME only four days earlier. As a result, the CMEs were not decelerated as much as usual by their transit through the interplanetary medium.

Had the eruption occurred just one week earlier, the blast site would have been facing Earth, rather than off to the side, so it was a relatively narrow escape.

When the Carrington Event enveloped Earth in the 19th century, technologies of the day were hardly sensitive to electromagnetic disturbances. Modern society, on the other hand, is deeply dependent on Sun-sensitive technologies such as GPS, satellite communications, and the Internet.

"The effect of such a storm on our modern technologies would be tremendous," says Luhmann.

During informal discussions at the workshop, Nat Gopalswamy of the Goddard Space Flight Center noted that "without NASA's *STEREO* probes, we might never have known the severity of the 2012 superstorm. This shows the value of having 'space weather buoys' located all around the Sun."

It also highlights the potency of the Sun even during so-called "quiet times." Many observers have noted that the current solar cycle is weak, perhaps the weakest in 100 years. Clearly, even a weak solar cycle can produce a very strong storm. Says Baker, "We need to be prepared."

ARTICLES QUESTIONS

1. What is a coronal mass ejection (CME)?
2. Why would a CME cause disruptions on Earth?
3. Explain how the 2012 storm missed Earth by 1 week.
4. More sensationalistic headlines for this story claimed that Earth almost "was sent back to the Dark Ages." What did they mean by that exaggeration?
5. Go to the NASA press release for this event (https://science.nasa.gov/science-news/science-at-nasa/2014/23jul_superstorm/), and click to watch the 4-minute "ScienceCast" video. What happened during the Carrington CME in 1859? Is that video effective at communicating the science information to the nonspecialist?

Summary

The forces due to pressure and gravity balance each other in hydrostatic equilibrium, maintaining the Sun's structure. Nuclear reactions converting hydrogen to helium are the source of the Sun's energy. Energy created in the Sun's core moves outward to the surface, first by radiation and then by convection. The solar wind may adversely affect astronauts located in space or on planets that lack a protective magnetic field, but it also has protected the Solar System from galactic, high-energy, cosmic-ray particles.

LG 1 **Describe the balance between the forces that determine the structure of the Sun.** The outward radiation pressure of the hot gas inside the Sun balances the inward pull of gravity at every point. That balance is dynamically maintained. An energy balance is also maintained, with the energy produced in the core of the Sun balancing the energy lost from the surface.

LG 2 **Explain how mass is converted to energy in the Sun's core and how long the Sun will take to use up its fuel.** In the core of the Sun, mass is converted to energy via the proton-proton chain. When four hydrogen atoms fix to one helium atom, some mass is lost. That mass is released as energy, nearly all of which leaves the Sun either as photons or as neutrinos. Neutrinos are elusive, almost massless particles that interact only very weakly with other matter. Observations of neutrinos confirm that nuclear fusion is the Sun's primary energy source.

LG 3 **Sketch a physical model of the Sun's interior, and list the ways that energy moves outward from the Sun's core toward its surface.** The interior of the Sun is divided into zones defined by how energy is transported in that region. Energy moves outward through the Sun by radiation and by convection.

LG 4 **Describe how observations of solar neutrinos and seismic vibrations on the surface of the Sun test astronomers' models of the Sun.** The Sun has multiple layers, each with a characteristic pressure, density, and temperature. Neutrinos directly probe the interior of the Sun. That model of the interior of the Sun has been tested by helioseismology, in much the same way that the model of Earth's interior has been tested by seismology.

LG 5 **Describe the solar activity cycles of 11 and 22 years, and explain how those cycles are related to the Sun's changing magnetic field.** Activity on the Sun follows a cycle that peaks every 11 years but takes 22 full years for the magnetic field to reverse. Sunspots are photospheric regions cooler than their surroundings, and they reveal the cycles in solar activity. Material streaming away from the Sun's corona creates the solar wind, which moves outward through the Solar System until it meets the interstellar medium. Solar storms, including ejections of mass from the corona, produce auroras and can disrupt power grids and damage satellites.

? Unanswered Questions

- Will nuclear fusion become a major source of energy production on Earth? Scientists have been working on controlled nuclear fusion for more than 60 years, since the first hydrogen bombs were developed. So far, however, replicating the conditions inside the Sun has posed too many difficulties. Nuclear fusion requires that we have hydrogen isotopes at very high temperature, density, and pressure, just as when a hydrogen bomb explodes. However, controlled nuclear fusion requires that we confine that material long enough to get more energy out than we put in. Several major experiments have attempted to fuse isotopes of hydrogen. An alternative approach is to fuse an isotope of helium, ^{3}He, which has only three particles in the nucleus (two protons and one neutron). On Earth, ^{3}He is found in very limited supply. But ^{3}He is in much greater abundance on the Moon, so some people propose setting up mining colonies on the Moon to extract ^{3}He for use in fusion reactions on Earth or possibly even on the Moon (see question 49b at the end of the chapter).
- Are large timescale variations in Earth's climate—ice ages—related to solar activity? Solar activity affects Earth's upper atmosphere and may affect weather patterns as well. Variations in the amount of radiation from the Sun might be responsible for past variations in Earth's climate. Current models indicate that observed variations in the Sun's luminosity could account for only about 0.1-K differences in Earth's average temperature—much less than the effects due to the ongoing buildup of carbon dioxide in Earth's atmosphere. Triggering the onset of an ice age may require a sustained drop in global temperatures of only 0.2–0.5 K, so astronomers are continuing to investigate a possible link between solar variability and long-timescale changes in Earth's climate.

Questions and Problems

Test Your Understanding

1. The physical model of the Sun's interior has been confirmed by observations of
   a. neutrinos and seismic vibrations.
   b. sunspots and solar flares.
   c. neutrinos and positrons.
   d. sample returns from spacecraft.
   e. sunspots and seismic vibrations.

2. Place in order the following steps in the fusion of hydrogen into helium. If two or more steps happen simultaneously, use an equals sign (=).
   a. A positron is emitted.
   b. One gamma ray is emitted.
   c. Two hydrogen nuclei are emitted.
   d. Two ^{3}He collide and become ^{4}He.
   e. Two hydrogen nuclei collide and become ^{2}H.
   f. Two gamma rays are emitted.
   g. A neutrino is emitted.
   h. One deuterium nucleus and one hydrogen nucleus collide and become ^{3}He.

3. Sunspots, flares, prominences, and coronal mass ejections are all caused by
   a. magnetic activity on the Sun.
   b. electrical activity on the Sun.
   c. the interaction of the Sun's magnetic field and the interstellar medium.
   d. the interaction of the solar wind and Earth's magnetic field.
   e. the interaction of the solar wind and the Sun's magnetic field.

4. The structure of the Sun is determined by not only the balance between the forces due to ________ and gravity but also the balance between energy generation and energy ________.
   a. pressure; production
   b. pressure; loss
   c. ions; loss
   d. solar wind; production

5. In the proton-proton chain, four hydrogen nuclei are converted to a helium nucleus. That does not happen spontaneously on Earth because the process requires
   a. vast amounts of hydrogen.
   b. very high temperatures and densities.
   c. hydrostatic equilibrium.
   d. very strong magnetic fields.

6. The solar neutrino problem pointed to a fundamental gap in our knowledge of
   a. nuclear fusion.
   b. neutrinos.
   c. hydrostatic equilibrium.
   d. magnetic fields.

7. Sunspots change in number and location during the solar cycle. That phenomenon is connected to
   a. the rotation rate of the Sun.
   b. the temperature of the Sun.
   c. the magnetic field of the Sun.
   d. the tilt of the axis of the Sun.

8. Suppose an abnormally large amount of hydrogen suddenly fused in the core of the Sun. Which of the following would be observed first?
   a. The Sun would become brighter.
   b. The Sun would swell and become larger.
   c. The Sun would become bluer.
   d. The Sun would emit more neutrinos.

9. The solar corona has a temperature of 1 million to 2 million K; the photosphere has a temperature of only about 6000 K. Why isn't the corona much, much brighter than the photosphere?
   a. The magnetic field traps the light.
   b. The corona emits only X-rays.
   c. The photosphere is closer to us.
   d. The corona has a much lower density.

10. The Sun rotates once every 25 days relative to the stars. The Sun rotates once every 27 days as seen from Earth. Why are those two numbers different?
    a. The stars are farther away.
    b. Earth is smaller.
    c. Earth moves in its orbit as the Sun rotates.
    d. The Sun moves relative to the stars.

11. Place the following regions of the Sun in order of increasing radius.
    a. corona
    b. core
    c. radiative zone
    d. convective zone
    e. chromosphere
    f. photosphere
    g. a sunspot

12. Coronal mass ejections
    a. carry away 1 percent of the mass of the Sun each year.
    b. are caused by breaking magnetic fields.
    c. are always emitted in the direction of Earth.
    d. are unimportant to life on Earth.

13. As energy moves out from the Sun's core toward its surface, it first travels by ________, then by ________, and then by ________.
    a. radiation; conduction; radiation
    b. conduction; radiation; convection
    c. radiation; convection; radiation
    d. radiation; convection; conduction

14. Energy is produced primarily in the center of the Sun because
    a. the strong nuclear force is too weak elsewhere.
    b. that's where neutrinos are created.
    c. that's where most of the helium is.
    d. the temperature and density are high enough in the core.
15. The solar wind pushes on the magnetosphere of Earth, changing its shape, because
    a. the solar wind is so dense.
    b. the magnetosphere is so weak.
    c. the solar wind contains charged particles.
    d. the solar wind is so fast.

Thinking about the Concepts

16. Explain how hydrostatic equilibrium acts as a safety valve to keep the Sun at its constant size, temperature, and luminosity.
17. Two of the three atoms in a molecule of water (H_2O) are hydrogen. Why are Earth's oceans not fusing hydrogen into helium and setting Earth ablaze?
18. Why are neutrinos so difficult to detect?
19. Explain the proton-proton chain through which the Sun generates energy by converting hydrogen to helium.
20. On Earth, nuclear power plants use *fission* to generate electricity. In fission, a heavy element such as uranium is broken into many atoms, where the total mass of the fragments is less than that of the original atom. Explain why fission could not be powering the Sun today.
21. If an abnormally large amount of hydrogen suddenly fused in the core of the Sun, what would happen to the rest of the Sun? Would the Sun change as seen from Earth?
22. Study the Process of Science Figure. If the follow-up experiments did not detect the other types of neutrinos, what would have been the next step for scientists at that point?
23. What is the solar neutrino problem, and how was it solved?
24. The Sun's visible "surface" is not a true surface, but a feature called the photosphere. Explain why the photosphere is not a true surface.
25. How are orbiting satellites and telescopes affected by the Sun?
26. Describe the solar corona. Under what circumstances can it be seen without special instruments?
27. In the proton-proton chain, the mass of four protons is slightly greater than the mass of a helium nucleus. Explain what happens to that "lost" mass.
28. What have sunspots revealed about the Sun's rotation?
29. Why are different parts of the Sun best studied at different wavelengths? Which parts are best studied from space?
30. Why is studying the interaction of the solar wind with the interstellar medium important?

Applying the Concepts

31. In Figure 14.10, density and temperature are both graphed versus height.
    a. Is the height axis linear or logarithmic? How do you know?
    b. Is the density axis linear or logarithmic? How do you know?
    c. Is the temperature axis linear or logarithmic? How do you know?
32. Using the data in Figures 14.19b and c, present an argument that sunspots occur in regions of strong magnetic field.
33. Study Figure 14.17a and Figure 14.20.
    a. Estimate the fraction of the Sun's surface covered by the large sunspot group in Figure 14.17a. (Remember that you are seeing only one hemisphere of the Sun.)
    b. From the graph in Figure 14.20, estimate the average number of sunspots that occurs at solar maximum.
    c. On average, what fraction of the Sun could be covered by sunspots at solar maximum? Is that a large fraction?
    d. Compare your conclusion with the graph of intensity in Figure 14.24. Does that graph make sense to you?
34. The Sun has a radius equal to about 2.3 light-seconds. Explain why a gamma ray produced in the Sun's core does not emerge from the Sun's surface 2.3 seconds later.
35. Assume that the Sun's mass is about 300,000 Earth masses and that its radius is about 100 times that of Earth. The density of Earth is about 5,500 kg/m^3.
    a. What is the average density of the Sun?
    b. How does that value compare with the density of Earth? With the density of water?
36. The Sun shines by converting mass into energy according to $E = mc^2$. Show that if the Sun produces 3.85×10^{26} J of energy per second, it must convert 4.3 million metric tons (4.3×10^9 kg) of mass per second into energy.
37. Assume that the Sun has been producing energy at a constant rate over its lifetime of 4.6 billion years (1.4×10^{17} seconds).
    a. How much mass has it lost creating energy over its lifetime?
    b. The current mass of the Sun is 2×10^{30} kg. What fraction of its current mass has been converted into energy over the lifetime of the Sun?
38. Suppose our Sun was an A5 main-sequence star, with twice the mass and 12 times the luminosity of the Sun, a G2 star. How long would that A5 star fuse hydrogen to helium? What would that mean for Earth?
39. Imagine that the source of energy inside the Sun changed abruptly.
    a. How long would it take before a neutrino telescope detected the event?
    b. When would a visible-light telescope see evidence of the change?

40. On average, how long do particles in the solar wind take to reach Earth from the Sun if they are traveling at an average speed of 400 km/s?

41. If a sunspot appears only 70 percent as bright as the surrounding photosphere, and the photosphere has a temperature of approximately 5780 K, what is the temperature of the sunspot?

42. The hydrogen bomb represents an effort to create a process similar to what takes place in the core of the Sun. The energy released by a 5-megaton hydrogen bomb is 2×10^{16} J.
   a. This textbook, *21st Century Astronomy*, has a mass of about 1.6 kg. If all its mass were converted to energy, how many 5-megaton bombs would it take to equal that energy?
   b. How much mass did Earth lose each time a 5-megaton hydrogen bomb was exploded?

43. Verify the claim made at the start of this chapter that the Sun produces more energy per second than all the electric power plants on Earth could generate in a half-million years. Estimate or look up how many power plants are on the planet and how much energy an average power plant produces. Be sure to account for different kinds of power, such as coal, nuclear, and wind.

44. Let's examine the reason that the Sun cannot power itself by chemical reactions. Using Working It Out 14.1 and the fact that an average chemical reaction between two atoms releases 1.6×10^{-19} J, estimate how long the Sun could emit energy at its current luminosity. Compare that estimate with the known age of Earth.

45. The Sun could get energy from gravitational contraction for a time period of ($GM^2_{Sun}/R_{Sun}L_{Sun}$). How long would the Sun last at its current luminosity? (Be careful with units!)

USING THE WEB

46. a. Go to the SDO website (http://sdo.gsfc.nasa.gov). Under "Data," select "The Sun Now" and view the Sun at many wavelengths. What activity do you observe in the images at the location of any sunspots seen in the "HMI Intensitygram" images? (You can download a free SDO app by Astra to get real-time images on your mobile device.) Look at a recent news story from the SDO website. What was observed, and why is it newsworthy?
   b. Go to the *STEREO* mission's website (https://stereo.gsfc.nasa.gov). What is *STEREO*? Where are the spacecraft located? How does that configuration enable observations of the entire Sun at once? (You can download the app "3-D Sun" to get the latest images on your mobile device.)

47. a. What are the science goals of NASA's *Interface Region Imaging Spectrograph* (*IRIS*) mission (https://www.nasa.gov/mission_pages/iris/)? What has it discovered?
   b. An older space mission, *SOHO* (Solar and Heliospheric Observatory; https://sohowww.nascom.nasa.gov), was launched in 1995 by NASA and ESA. Click on "The Sun Now" to see today's images. The Extreme Ultraviolet imaging Telescope (EIT) images are in the far ultraviolet and show violent activity. How do those images differ from the ones of SDO in question 46b?
   c. Go to the Daniel K. Inouye Solar Telescope (DKIST) website (https://atst.nso.edu). That adaptive-optics telescope under construction on Haleakala, Maui, will be the largest solar telescope. Why is studying the magnetic field of the Sun important? What are some advantages of studying the Sun from a ground-based telescope instead of a space-based telescope? What wavelengths does the DKIST observe? Why is Maui a good location? When is the telescope scheduled to be completed?

48. a. Go to the Space Weather website (http://spaceweather.com). Are any solar flares happening today? What is the sunspot number? Is it about what you would expect for this year? (In the column at left, click on "What is the sunspot number?" to see a current graph.) Are any coronal holes present today?
   b. Citizen science: Go to the Solar Stormwatch II website (https://www.zooniverse.org/projects/shannon-/solar-stormwatch-ii), a Zooniverse project from the Royal Observatory in Greenwich, England. Create an account for Zooniverse if you don't already have one. Log in and click on "Get Started" and "Project Tutorial." Mark up two sets of images (and save a screen shot for your homework).

49. a. Go to the National Ignition Facility (NIF) & Photon Science website (https://lasers.llnl.gov/about/). Under "Science," click on "How to Make a Star." How are lasers used in experiments to develop controlled nuclear fusion on Earth? How does the fusion reaction here differ from that in the Sun?
   b. An alternative approach is to fuse ^{3}He + ^{3}He instead of the hydrogen isotopes. On Earth, however, ^{3}He is in limited supply. Helium-3 is in much greater abundance on the Moon, so some people propose setting up mining colonies on the Moon to extract ^{3}He for fusion reactions on Earth. Do an Internet search on "helium 3 Moon." Which countries are talking about going to the Moon for that purpose? What is the timeline for when that might happen? What are the difficulties?

50. a. Go to https://voyager.jpl.nasa.gov/mission/status/. Where are the *Voyager* spacecraft now? Has *Voyager 2* crossed into interstellar space?
   b. Go to the website for the Interstellar Boundary Explorer (*IBEX*) (https://www.nasa.gov/mission_pages/ibex/). What has *IBEX* learned about the solar wind and the interstellar medium?

EXPLORATION

The Proton-Proton Chain

digital.wwnorton.com/astro6

The proton-proton chain powers the Sun by fusing hydrogen into helium. That fusion process produces several types of particles as by-products, as well as energy. In this Exploration, we study the steps of the proton-proton chain in detail, with the intent of helping you keep them straight.

Visit the Digital Resources Page on the Student Site and open the "Proton-Proton Chain" Interactive Simulation in Chapter 14.

Watch the animation all the way through once.

Play the animation again, pausing after the first collision. Two hydrogen nuclei (both positively charged) have collided to produce a new nucleus with only one positive charge.

1 Which particle carried away the other positive charge?

2 What is a neutrino? Did the neutrino enter the reaction, or was the neutrino produced in the reaction?

Compare the interaction on the top with the interaction on the bottom.

3 Did the same reaction occur in each instance?

Resume playing the animation, pausing it after the second collision.

4 What two types of nuclei entered the collision? What type of nucleus resulted?

5 Was charge conserved in that reaction, or did a particle have to carry charge away?

6 What is a gamma ray? Did the gamma ray enter the reaction, or was it produced by the reaction?

Resume the animation again, and allow it to run to the end.

7 What nuclei enter the final collision? What nuclei are produced?

8 In chemistry, a catalyst is a species that facilitates a reaction but is not used up in the process. Do any nuclei act like catalysts in the proton-proton chain?

Make a table of inputs and outputs. Which particles in the final frame of the animation were inputs to the reaction? Which were outputs? Fill in your table with those inputs and outputs.

9 Which outputs are converted into energy that leaves the Sun as light?

10 Which outputs could become involved in another reaction immediately?

11 Which output is likely to stay in that form for a very long time?

15 The Interstellar Medium and Star Formation

Stars form from clouds of dust and gas, which collapse until nuclear fusion begins. The formation process is faster for more massive stars. The most massive stars take tens of thousands of years to form, whereas the least massive stars take hundreds of millions of years. To understand star formation, astronomers have observed many stars at various stages of development. In this chapter, we will describe the interstellar environment from which stars both large and small form. Then we will focus on the forming star—the *protostar*—and explain how it becomes a star.

LEARNING GOALS

By the end of the chapter, you should be able to:

LG 1 Describe the types and states of material that exist in the space between the stars and how that material is detected.

LG 2 Explain the conditions under which a cloud of gas can contract into a stellar system and the role that gravity and angular momentum play in forming stars and planets.

LG 3 List the steps in the evolution of a protostar, and explain how the mass of the protostar affects its evolution.

LG 4 Describe the track of a protostar as it evolves to a main-sequence star on the Hertzsprung-Russell (H-R) diagram.

Stars are forming in this giant molecular cloud. ▶▶▶

How do stars form?

15.1 The Interstellar Medium Fills the Space between the Stars

The space between the stars contains thin gas, interrupted by giant clouds of cool gas and dust. Those clouds can be observed in the visible part of the spectrum in three ways: as they absorb light from objects behind them, emit light from excited atoms, or reflect starlight from nearby stars. Together, the clouds and the thin gas are known as the **interstellar medium**.

Stars interact with the interstellar medium. Stars form from the interstellar medium and return much of their material back to it when they die. Outflows from dying stars pile up material in their path like snow in front of a snowplow, clearing out vast hot bubbles. Those hot bubbles of high-pressure gas compress clouds, driving up their densities and triggering the formation of new stars. Interstellar clouds are destroyed by those violent events. In turn, new clouds are formed from the swept-up gas. Those clouds go on to produce new stars. In addition, energy from stars heats and stirs the interstellar medium. For example, ultraviolet radiation from massive, hot stars warms gas that then pushes outward into its surroundings. In this section, we survey the varied states of the interstellar medium.

(a)

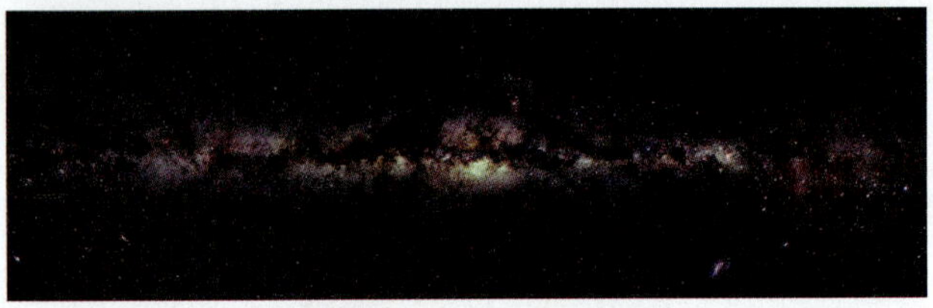

(b)

Figure 15.1 (a) This all-sky picture of the Milky Way was taken in visible light. The dark splotches blocking the view are dusty interstellar clouds. The center of this and all the other all-sky images in this chapter is the center of the Milky Way. (b) This all-sky picture was taken in the near infrared. Infrared radiation penetrates the interstellar dust, providing a clearer view of the stars in the disk of the Milky Way.

The Composition and Density of the Interstellar Medium

The Sun formed from the interstellar medium, so the chemical composition of the interstellar medium near the Sun is similar to the chemical composition of the Sun (see Table 13.1). In the interstellar medium, hydrogen accounts for about 90 percent of the number of atomic nuclei, and the remaining 10 percent is almost all helium. The more massive elements account for only 0.1 percent of the atomic nuclei, or about 2 percent of the mass in the interstellar medium (because those atoms are more massive than hydrogen or helium, they account for a larger fraction of the mass in the interstellar medium than they do of the number of individual atoms). Roughly 99 percent of interstellar matter is a gas, consisting of individual atoms or molecules moving about freely, as the molecules in the air do.

The air that you breathe has a density 100 billion billion times greater than the average density of the interstellar medium. Each cubic centimeter (cm^3) of the air around you contains about 2.7×10^{19} molecules. The local interstellar medium has an average density of about 0.1 atom/cm^3. Stated another way, imagine a tube with the diameter of your fist that stretches from the Solar System to the center of our galaxy, 26,000 light-years away. Now imagine a tube with the same diameter that stretches between your eye and the floor on Earth. The amount of material in both tubes is about the same.

Interstellar Dust and Its Effects on Light

About 1 percent of the mass in the interstellar medium is in the form of solid grains, called **interstellar dust**. Ranging in size from little more than large molecules up to particles about 1 micron (μm) across, those solid grains are more like the particles of soot from a candle flame than the dust that collects on a windowsill. It would take several hundred average interstellar grains to span the thickness of a single human hair. Interstellar dust begins to form when larger atoms such as iron, silicon, and carbon stick together to form grains in dense, relatively cool environments such as the outer atmospheres and "stellar winds" of cool, red giant stars or in dense material thrown into space by stellar explosions. Once those grains are in the interstellar medium, other atoms and molecules stick to them. About half of the interstellar matter that is more massive than helium is found in interstellar grains.

Interstellar dust is extremely effective at absorbing and diverting light, so the view of distant objects is affected even by the low-density interstellar medium. That effect is called **interstellar extinction**. Go out on a dark summer night in the Northern Hemisphere (or on a dark winter night in the Southern Hemisphere) and look closely at the Milky Way, visible as a band of diffuse light running through the constellation Sagittarius. A dark "lane" is running roughly down the middle of that bright band, splitting it in two, as shown in **Figure 15.1a**. That dark band is caused by interstellar extinction, dust that dims the light from distant stars. Recall the comparison of the air between you and the floor with the interstellar medium stretching to the center of the galaxy. If the interstellar medium were compressed to the same density as air, it would be so full of dust that you could not easily see your hand 10 centimeters (cm) in front of your face. When spread out between Earth and the distant stars, the interstellar dust dims light just as effectively. Although interstellar extinction is most noticeable in the dark lanes in the Milky Way, it has a lesser but still important effect on starlight coming from all directions in the galaxy.

Figure 15.1 shows two images of the Milky Way Galaxy: one taken in visible light, the other taken in the infrared (IR). Those two images are *all-sky images*: they portray the entire sky surrounding Earth. They have been oriented so that the disk of the Milky Way, in which the Solar System is embedded, runs horizontally across the center of the image. The dark clouds that block the shorter-wavelength visible light (Figure 15.1a) seem to have vanished in the longer-wavelength infrared image (**Figure 15.1b**). That observation shows that interstellar extinction affects different regions of the electromagnetic spectrum differently. When astronomers observe in long-wavelength regions of the spectrum, such as infrared or radio wavelengths, they can make observations through the clouds to the center of the galaxy and beyond.

To understand why dust obscures short-wavelength radiation but not long-wavelength radiation, think about waves on the ocean, as shown in **Figure 15.2a**. Imagine you are on the ocean in a boat. If the ocean waves have a wavelength much longer than your boat, the waves cause you to move slowly up and down. But that is all that happens: no energy is transferred from the wave to the boat, so the wave travels on unaffected by the interaction. The situation is different if the wavelength of the waves is close in size to the length of your boat. For example, if the waves are roughly half the size of your boat, the front of the boat may be on a wave crest while the back of the boat is in a trough or vice versa. The boat will tip wildly back and forth as the waves pass by, and energy will be transferred from the wave to the boat; the wave loses energy in the interaction. If the size of the boat and the wavelength of the waves are the right match, even low waves will rock the boat. Comparing those two situations shows that the wave loses very little energy when it is much bigger than the boat. But the wave loses a lot of energy when it is of similar size or smaller than the boat, and the wave can be blocked by the boat if the boat is large enough in comparison.

The interaction of light with matter is more involved than that of a boat rocking on the ocean, but the same basic idea applies, as shown in **Figure 15.2b**. Tiny interstellar dust grains effectively block the transmission of short-wavelength ultraviolet light and blue light because they have wavelengths comparable to or smaller than the typical size of dust grains. Our view of the Milky Way in those wavelengths is limited. In contrast, longer-wavelength infrared and radio radiation interact less strongly with the tiny interstellar dust grains, so in those portions of the spectrum we get a more complete view of the Milky Way.

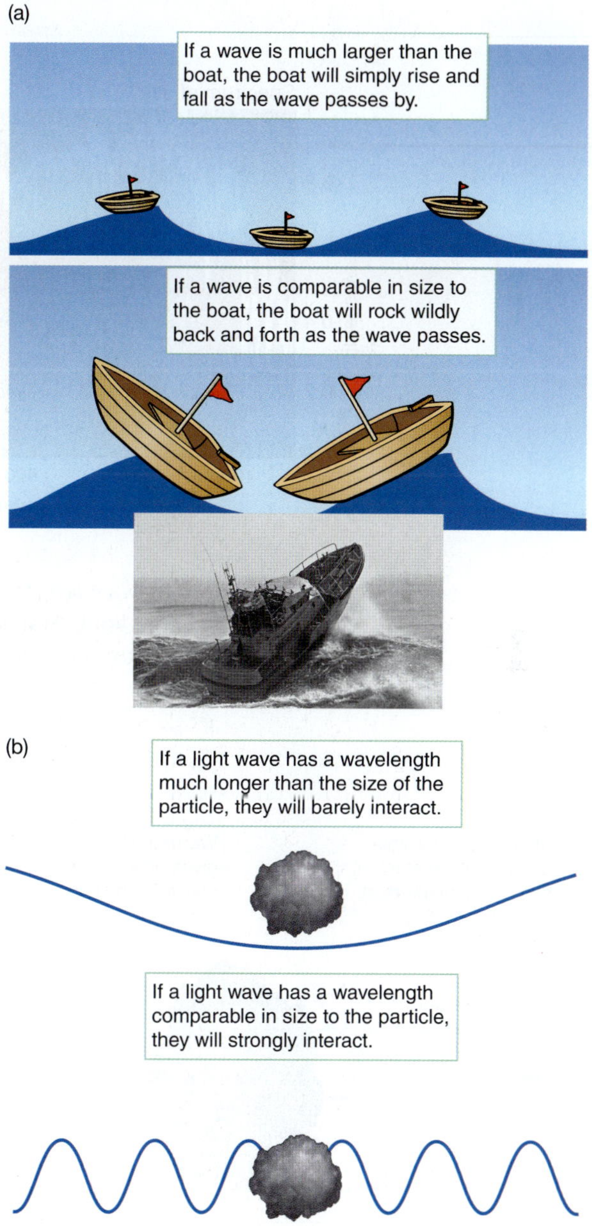

Figure 15.2 (a) Just as boats interact most strongly with ocean waves that have a wavelength similar to the length of the boat, (b) particles interact most strongly with light that has wavelengths near the same size as the particle.

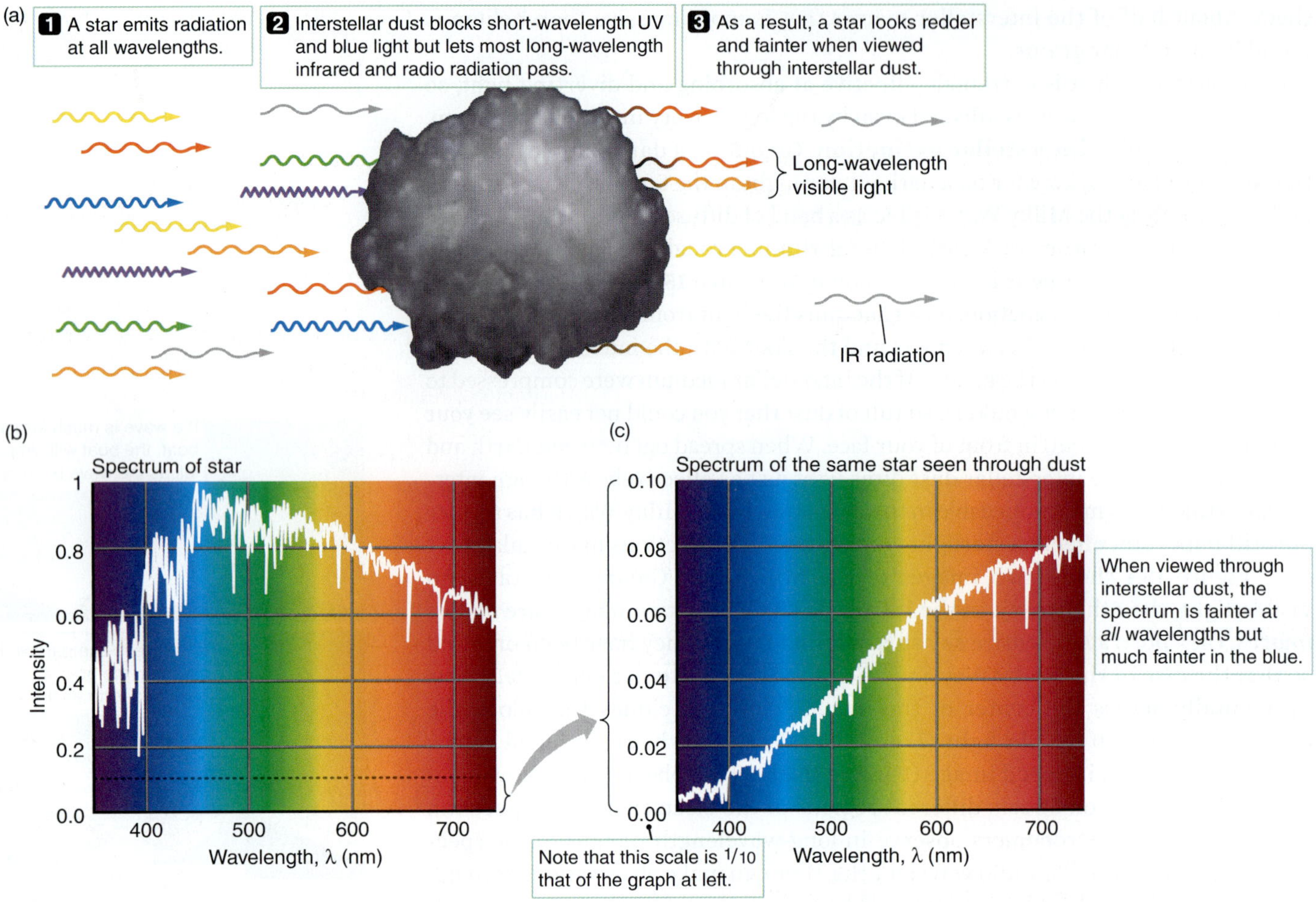

Figure 15.3 (a) The wavelengths of ultraviolet and blue light are close to the size of interstellar grains, so the grains effectively block that light. Grains are less effective at blocking longer-wavelength light. Therefore, the spectrum of a star (b) when seen through an interstellar cloud (c) appears fainter and redder.

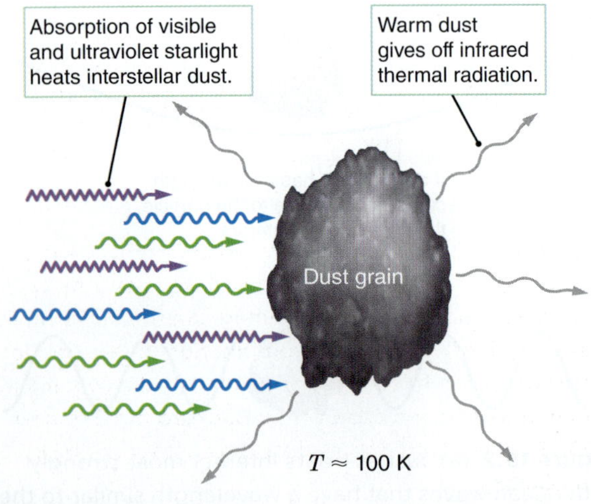

Figure 15.4 The temperature of interstellar grains depends on the equilibrium between absorbed and emitted radiation.

The presence of dust significantly affects the spectrum of an object, as shown in **Figure 15.3**. Extinction occurs at all wavelengths (Figure 15.3a), causing an object viewed through dust to be fainter at all wavelengths than it would be otherwise. Extinction also affects short-wavelength blue light more than it affects long-wavelength red light, so the object appears less blue than it really is (Figure 15.3b and c). Removing the blue light causes an object to appear more red, so that effect is called **reddening**. Correcting for the reddening effect of dust can be one of the most difficult parts of interpreting astronomical observations, often adding to uncertainty in the measurement of an object's properties.

At longer infrared wavelengths, interstellar extinction from small gas particles is less effective, but larger dust particles still play an important role. In Chapter 5, we discussed how Wein's law can be used to relate the temperature of an object to the wavelength at which it shines most brightly, and how the equilibrium between absorbed sunlight and emitted thermal radiation determines the temperatures of the terrestrial planets (see Figure 5.19). As **Figure 15.4** illustrates, a similar equilibrium is at work in interstellar space, where dust is heated by starlight and surrounding gas to temperatures of tens to hundreds of kelvins. According to Wien's law, dust with a temperature of 100 K shines most strongly at a wavelength of 29 μm, whereas cooler dust—at 10 K—shines most strongly at a

15.1 Working It Out Dust Glows in the Infrared

The temperature of interstellar dust can be found from its spectrum. Wien's law, discussed in Chapter 5, relates the temperature of an object to the peak wavelength (λ_{peak}) of its emitted radiation. For warm dust at a temperature of 100 K (recall that 1 μm = 10^{-6} meter = 1,000 nanometers [nm]):

$$\lambda_{peak} = \frac{2{,}900\ \mu\text{m K}}{T} = \frac{2{,}900\ \mu\text{m K}}{100\ \text{K}} = 29\ \mu\text{m}$$

For cooler dust, at 10 K:

$$\lambda_{peak} = \frac{2{,}900\ \mu\text{m K}}{T} = \frac{2{,}900\ \mu\text{m K}}{10\ \text{K}} = 290\ \mu\text{m}$$

The temperature and the peak wavelength are inversely proportional, so if the temperature drops, the peak wavelength gets longer. For the temperatures common for dust in the interstellar medium, the peak wavelength is in the far-infrared region of the electromagnetic spectrum.

longer wavelength (**Working It Out 15.1**). Such warm dust produces much of the diffuse light present in infrared observations. **Figure 15.5a**, from NASA's Wide-field Infrared Survey Explorer (WISE), shows the sky at combined wavelengths of 3.4, 12, and 22 μm. That light is produced by dust hotter than 100 K. In **Figure 15.5b**, a far-infrared image of the sky shows the Milky Way's dark, dusty, cooler clouds shining at a wavelength of 100 μm.

(a)

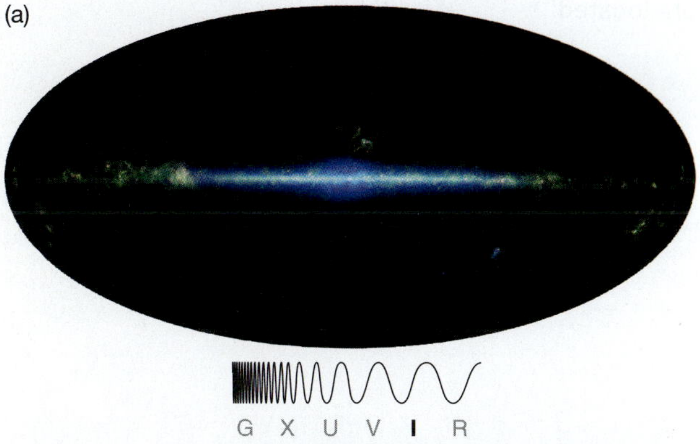

(b)

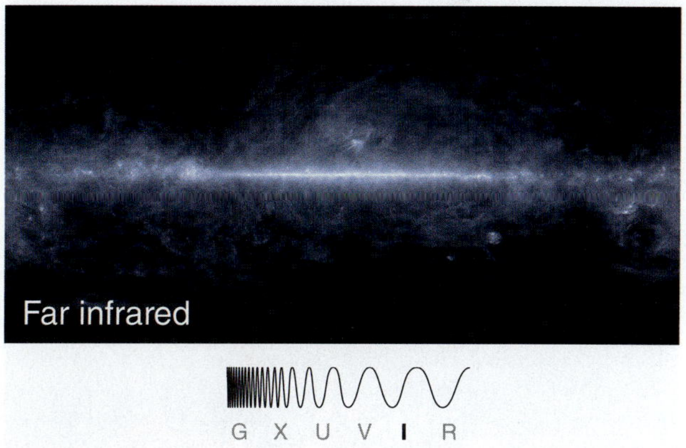

Figure 15.5 (a) This WISE infrared image shows the plane of the Milky Way at combined wavelengths of 3.4, 12, and 22 μm. (b) This all-sky image, in the far-infrared wavelength of 100 μm, is a combination of two images from two space telescopes—the Infrared Astronomical Satellite (IRAS) and the Cosmic Background Explorer (COBE). The image shows dust throughout the Milky Way Galaxy.

Temperatures and Densities of Interstellar Gas

The gas and dust in interstellar space are roughly evenly divided between dense regions called **interstellar clouds** and the **intercloud gas** spread between them. The clouds fill about 2 percent of the volume of interstellar space, whereas the remaining 98 percent is filled with thin gas.

The properties of intercloud gas vary from place to place (**Table 15.1**). About half of the volume of interstellar space is filled with an intercloud gas

TABLE 15.1 Typical Properties of Components of the Interstellar Medium

Component	Temperature (K)	Number Density (atoms/cm^3)	Size of Cube per Gram* (km)	State of Hydrogen
Hot intercloud gas	~1 million	~0.005	~8,000	Ionized
Warm intercloud gas	~8000	0.01–1	~800	Ionized or neutral
Cold intercloud gas	~100	1–100	~80	Neutral
Interstellar clouds	~10	100–1,000	~8	Molecular or neutral

*This is the length of one side of the cube of space you would need to search to find 1 gram of the material. It is another way of thinking about density.

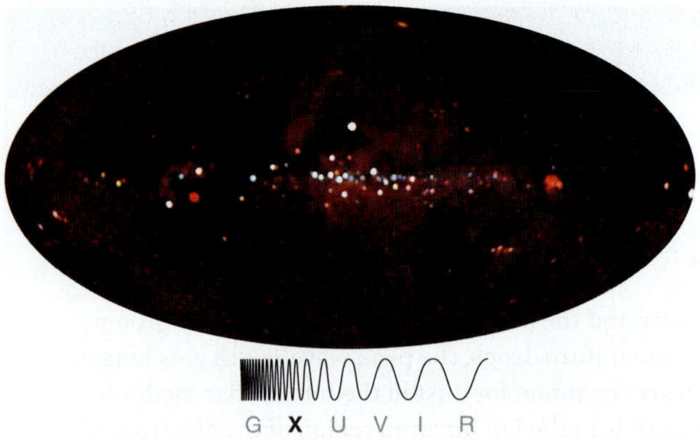

Figure 15.6 Bright spots in this image are distant X-ray sources, including objects such as bubbles of very hot, high-pressure gas surrounding the sites of recent explosions of supernovae. Supernovae are more common in the disk of the Milky Way, where most stars are located.

that is extremely hot—millions of kelvins—heated primarily by the energy of tremendous stellar explosions called *supernovae* (**Figure 15.6**). Because the temperature is so high, the atoms in the gas are moving very rapidly. The gas has an extremely low density; typically, you would have to search a liter (1,000 cm^3) or more of hot intercloud gas to find a single atom. Therefore, if you were floating in that million-kelvin intercloud gas, it would do little to keep you warm. Because so few atoms are present, you would radiate energy away and cool off much faster than the very hot gas around you could replace the lost energy.

Hot intercloud gas glows faintly in the X-ray portion of the electromagnetic spectrum, and orbiting X-ray telescopes observe the entire sky aglow with faint X-rays. That observation indicates that the Solar System is passing through a bubble of hot intercloud gas that may be the remnant of a supernova explosion 300,000 years ago. The bubble has a density of about 0.005 hydrogen atom/cm^3 and is at least 650 light-years across.

Not all intercloud gas is as hot as that around our Solar System. Most other intercloud gas is "warm," with a temperature of about 8000 K and a density ranging from about 0.01 to 1 atom/cm^3. Ultraviolet starlight with wavelengths shorter than 91.2 nm has enough energy to **ionize** hydrogen, meaning to strip the electron away from the nucleus. About half of the volume of warm intercloud gas is kept ionized by ultraviolet starlight. That ionized gas "uses up" the ultraviolet light from stars, so the remaining half of the volume of warm intercloud gas is protected and remains in an un-ionized state. Similarly, the ozone layer in Earth's upper atmosphere shields the planet surface from ultraviolet light from the Sun.

One way to look for both warm and hot interstellar gas is to study the spectra of distant stars. Most commonly, that gas produces absorption lines in the spectra of distant stars when atoms in the gas absorb starlight at particular wavelengths. Those absorption lines can be used to find the temperature, density, and chemical composition of the gas. Regions of warm, ionized intercloud gas, as in supernova remnants, also can produce emission lines when protons and electrons constantly recombine into hydrogen atoms. When a proton and an electron combine to form a neutral hydrogen atom, energy is emitted in the form of electromagnetic radiation. Typically, the resulting hydrogen atom is left in an excited state (see Chapter 5). The atom then drops down to lower and lower energy states, emitting a photon at each step. Therefore, warm, ionized interstellar gas glows in emission lines characteristic of hydrogen, with the Hα (hydrogen alpha) line often the strongest. That line occurs in the red part of the spectrum at a wavelength of 656.3 nm. Other elements undergo a similar process.

The faint, diffuse emission in **Figure 15.7** comes mostly from warm (about 8000 K), ionized intercloud gas glowing in Hα. The bright spots are called **H II regions** ("H two regions") because the hydrogen atoms are ionized; H I atoms are neutral, and H II atoms are ionized. H II regions form a rough sphere around hot luminous O and B stars because those stars produce enough ultraviolet radiation to ionize even relatively dense interstellar clouds around them. H II regions indicate areas of active star formation because O stars do not live long enough to move very far from where they formed. H II regions around O stars are the very clouds from which those stars were born.

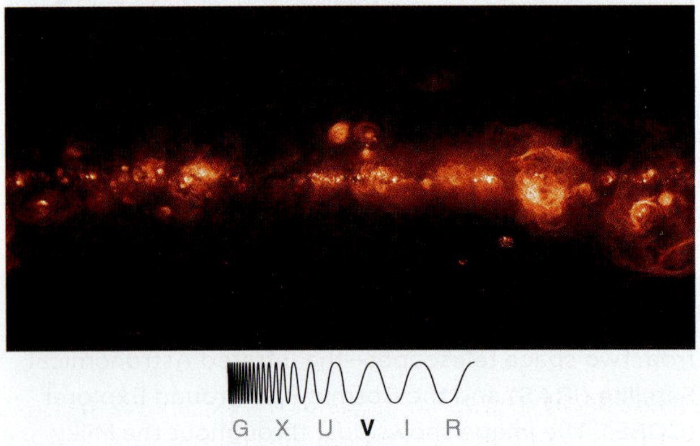

Figure 15.7 Warm interstellar gas (about 8000 K) glows in the Hα line of hydrogen. This image of the Hα emission from much of the northern sky reveals the complex structure of the interstellar medium.

One of the closest H II regions to the Sun is the Orion Nebula, located 1,340 light-years from the Sun in the constellation Orion (**Figure 15.8**). Almost all the ultraviolet light that powers that nebula comes from a single hot star, and only a few hundred stars are forming in its immediate vicinity. In contrast, a dense star

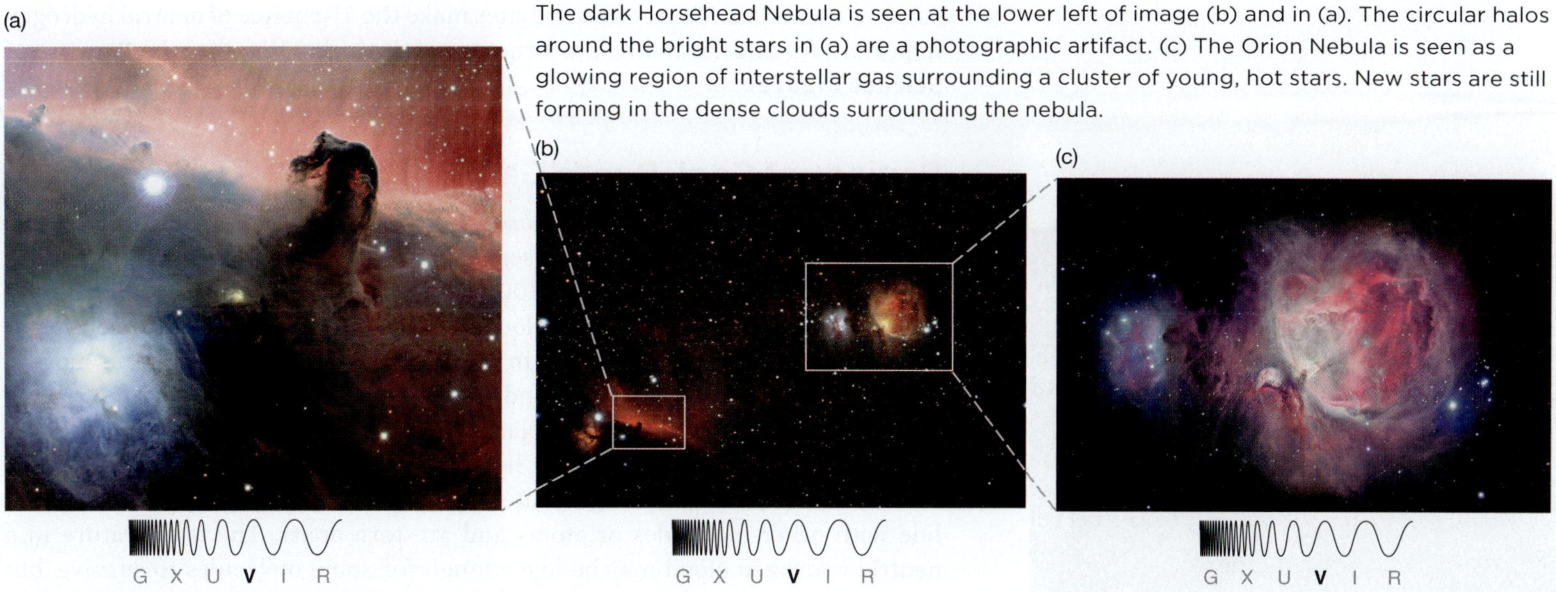

Figure 15.8 The Orion Nebula is only a small part of the larger Orion star-forming region. The dark Horsehead Nebula is seen at the lower left of image (b) and in (a). The circular halos around the bright stars in (a) are a photographic artifact. (c) The Orion Nebula is seen as a glowing region of interstellar gas surrounding a cluster of young, hot stars. New stars are still forming in the dense clouds surrounding the nebula.

cluster containing thousands of hot, luminous stars powers a giant H II region called 30 Doradus, located in the Large Magellanic Cloud, a small companion galaxy to the Milky Way located 160,000 light-years away. If 30 Doradus were as close as the Orion Nebula, it would be bright enough in the nighttime sky to cast shadows.

Warm, *neutral* hydrogen gas also produces an emission line, although it gives off radiation in a way different from that of warm, *ionized* hydrogen. Many subatomic particles, including protons and electrons, have a property called *spin* that causes them to behave as though each particle has a bar magnet, with a north and a south pole, built into it. As demonstrated in **Figure 15.9**, a hydrogen atom can exist in only two configurations: either the magnetic "poles" of the proton and electron point in opposite directions or they are aligned. Those configurations have different energies. When the two "magnets" point in the same direction, the atom has slightly less energy than when they point in the opposite direction. If left undisturbed long enough, a hydrogen atom in the higher energy unaligned state will spontaneously jump to the lower energy state, emitting a photon in the process. The energy difference between the two magnetic spin states of a hydrogen atom is extremely small, so the emitted photon has a wavelength of 21 cm, in the radio region of the spectrum. Interactions between atoms in the gas will later bump the hydrogen atoms back to the higher energy state, refreshing the supply of atoms that can produce the 21-cm line.

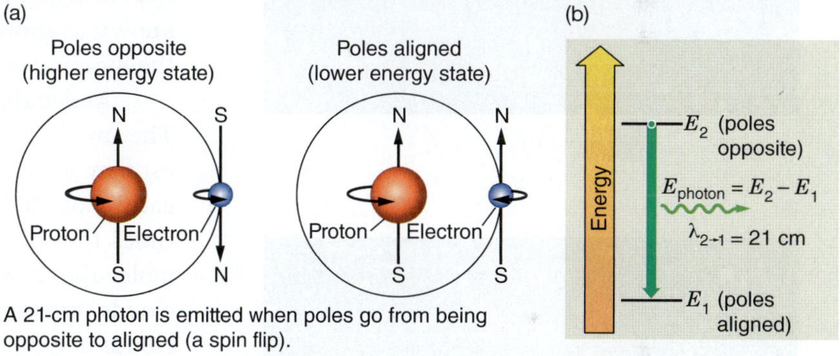

Figure 15.9 (a) A slight difference in energy occurs when the poles of the proton and electron are aligned compared with when they are opposite. (b) This energy difference corresponds to a photon of 21 cm.

The tendency for hydrogen atoms to emit 21-cm radiation is extremely weak. On average, you would have to wait about 11 million years for an individual hydrogen atom in the higher energy state to jump spontaneously to the lower energy state and give off a photon. But the universe has a lot of hydrogen, so at any given time, many atoms are making that transition. In **Figure 15.10**, the sky is aglow with 21-cm radiation from neutral hydrogen. Because of its long wavelength, 21-cm radiation freely penetrates dust in the interstellar medium, enabling astronomers to observe neutral hydrogen throughout the galaxy. Measurements

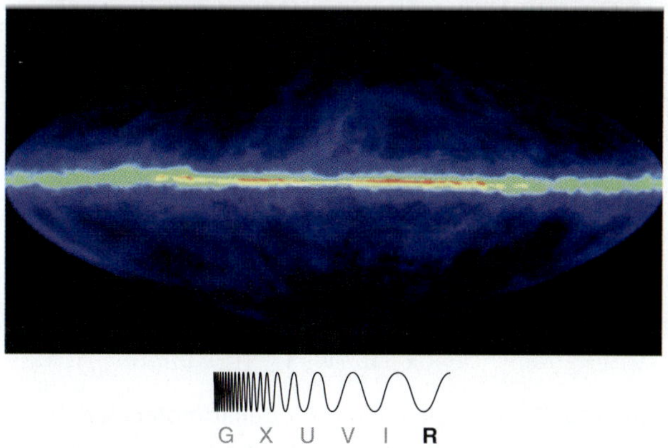

Figure 15.10 This radio image of the sky shows the distribution of neutral hydrogen gas throughout the galaxy. Red indicates directions of the highest hydrogen density, and blue and black show areas with little hydrogen. Radio waves penetrate interstellar dust, allowing astronomers to probe the structure of the galaxy.

of the Doppler shift of the line indicate how fast the emitting gas is moving toward us or away from us. Those two attributes make the 21-cm line of neutral hydrogen important for understanding the structure of the Milky Way (see the **Process of Science Figure**).

Regions of Cool, Dense Gas

Most interstellar clouds are composed primarily of isolated neutral hydrogen atoms and are much cooler and denser than the warm intercloud gas. Interstellar clouds have temperatures of about 100 K and densities of about 1–100 atoms/cm^3. On average, atoms in interstellar clouds are moving at about 20 kilometers per second (km/s). In contrast, hot gas in the interstellar medium has atoms moving at thousands of kilometers per second.

On Earth, finding atoms in isolation is uncommon; most atoms are in molecules. In most of interstellar space, however, including most interstellar clouds, molecules do not survive long. If interstellar gas is too hot, molecules soon collide with other molecules or atoms and are torn apart. The temperature in a neutral hydrogen cloud may be low enough for some molecules to survive, but ultraviolet photons with enough energy to break molecules apart can penetrate neutral hydrogen clouds. The inner regions of the densest interstellar clouds are known as **molecular clouds** because in those regions, dust blocks almost all of the ionizing radiation, and so molecules persist.

Molecular clouds range from a few solar masses to 10 million solar masses. The smallest molecular clouds may be less than half a light-year across; the largest may be more than a thousand light-years. **Giant molecular clouds** are typically about 100–200 light-years across and have masses a few hundred thousand times that of the Sun. The Milky Way Galaxy contains several thousand giant molecular clouds and many more smaller ones. Despite that large number of clouds, molecular clouds fill only about 0.1 percent of interstellar space. Such clouds may be rare, but they are extremely important because they are the cradles of star formation.

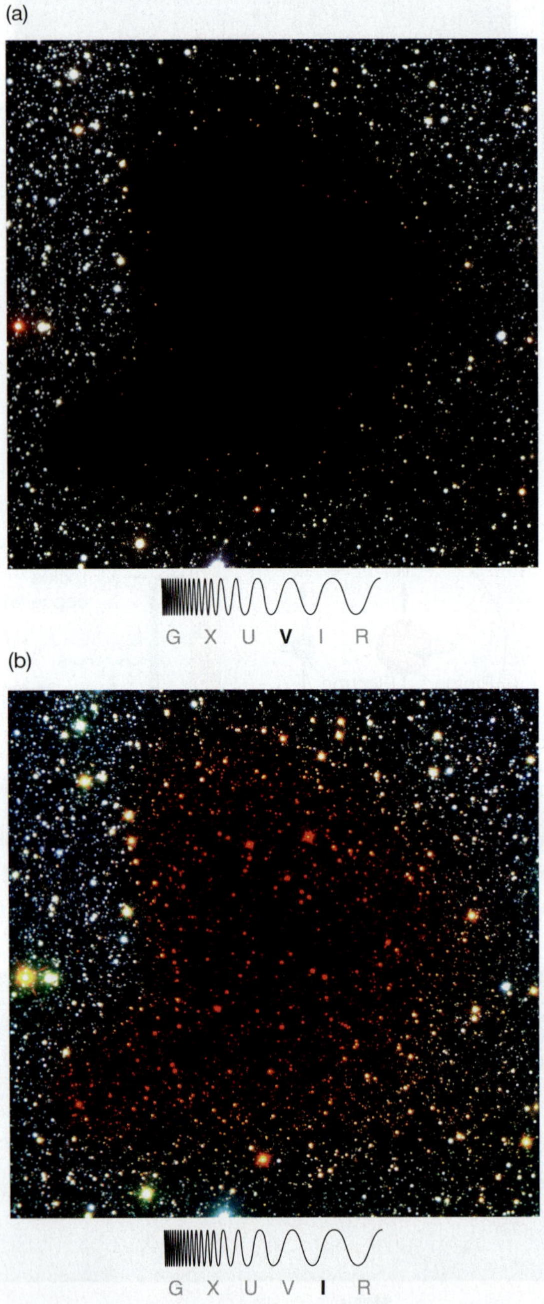

Figure 15.11 In visible light, interstellar molecular clouds are seen in silhouette against a background of stars and glowing gas. (a) Light from background stars is blocked by dust and gas in nearby Barnard 68, a dense, dark molecular cloud. (b) Infrared wavelengths can penetrate much of that gas and dust, as seen in this false-color image of Barnard 68.

In images such as those of **Figure 15.11**, the dust in molecular clouds causes a silhouette in visible light against a background of stars. Long-wavelength infrared radiation passes through that dust, revealing sources inside and behind the cloud. Inside such clouds it is dark and usually very cold, typically only about 10 K. Most of those clouds have densities of about 100–1,000 molecules/cm^3, but densities as high as 10^{10} molecules/cm^3 have been observed. Even at 10^{10} molecules/cm^3, that gas is still less than a billionth as dense as the air around you. In that cold, relatively dense environment, atoms combine to form a wide variety of molecules.

Molecules radiate as they make transitions between energy states. Molecular energy states are determined by the way the molecules rotate or vibrate, for example. Each type of molecule is unique in its properties, and thus unique in its energy states, and so molecular emission lines are useful in the same way that atomic emission lines are useful. The wavelengths of emission lines from molecules are an unmistakable fingerprint of the kinds of molecules responsible for them. Because some of the transitions are in the radio or infrared portion of the spectrum, those molecules can be detected even deep inside a molecular cloud. Observations of molecular lines reveal the innermost workings of the densest and most opaque interstellar clouds. By far the most common component of molecular clouds is molecular hydrogen (H_2), which consists of two hydrogen atoms and is the

PROCESS OF SCIENCE

All Branches of Science Are Interconnected

Studies of the natural world on the smallest scales and the largest scales inform one another.

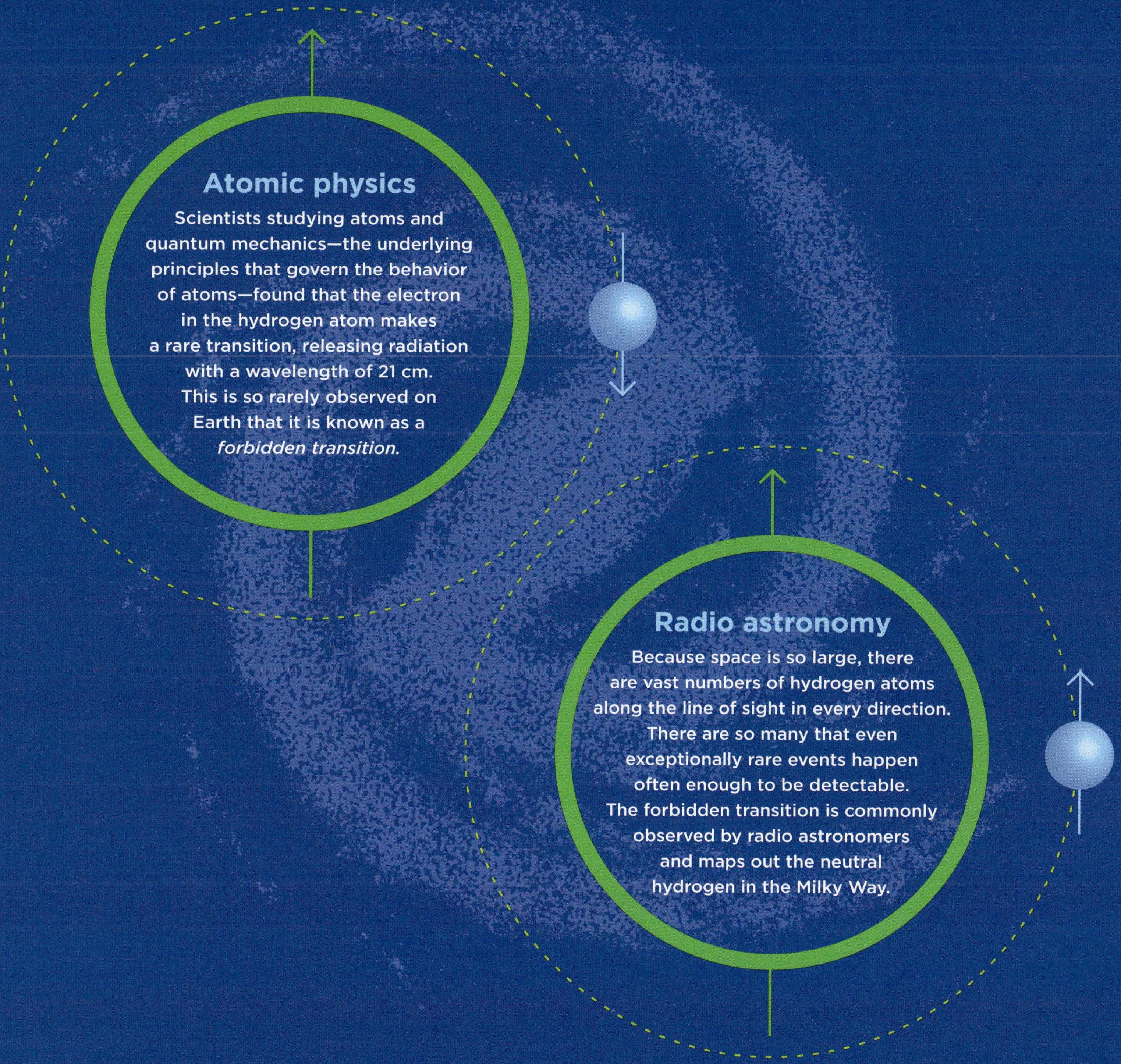

Radio astronomers use the physics of tiny atoms to understand the behavior of an enormous galaxy, which is 10^{30} times larger than an atom.

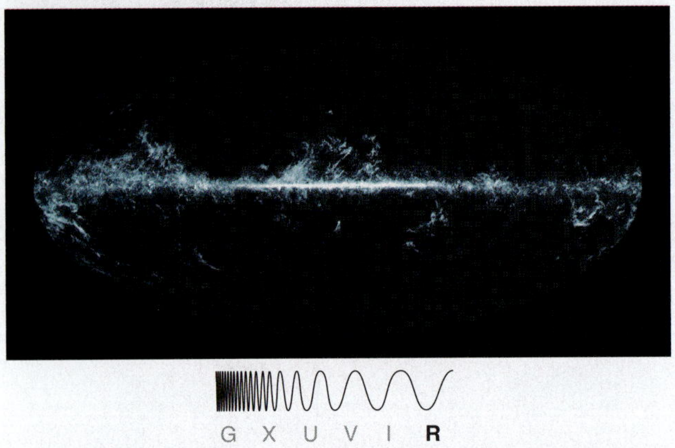

Figure 15.12 This all-sky image from the Planck observatory, a spacecraft that observed in the microwave region of the spectrum, shows the distribution of carbon monoxide (CO), which traces molecular clouds where stars are born.

smallest possible molecule. Cold molecular hydrogen is very difficult to observe directly because it has no emission lines in the long-wavelength regions of the spectrum.

In addition to molecular hydrogen, approximately 150 other molecules have been observed in interstellar space. They range from very simple structures such as carbon monoxide (CO), to complex organic compounds such as methanol (CH_3OH), to some molecules with large carbon chains. Very large carbon molecules, made of hundreds of individual atoms, have sizes between those of large interstellar molecules and small interstellar grains. Among the more important molecules is CO. The ratio of CO to H_2 is relatively constant, as far as has been tested, and interstellar CO (**Figure 15.12**) is often used to estimate the amounts and distribution of interstellar H_2, which is more difficult to observe directly.

CHECK YOUR UNDERSTANDING 15.1

When visible light from an object passes through the interstellar medium (choose all that apply): (a) the object appears dimmer; (b) the object appears bluer; (c) the object appears brighter; (d) the object appears redder.

15.2 Molecular Clouds Are the Cradles of Star Formation

Stars and planets form from large clouds of dust and gas in the interstellar medium. In this section, we explore the first steps of that process as a cloud begins to contract and fragment to form stars.

Self-Gravity in the Molecular Cloud

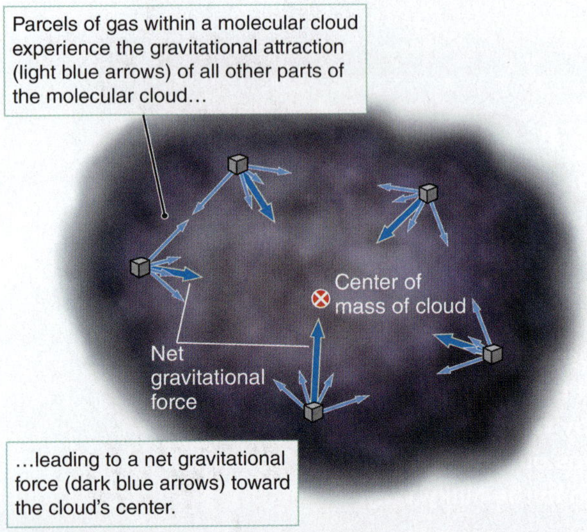

Figure 15.13 Self-gravity causes a molecular cloud to collapse, drawing parcels of gas toward a single point inside the cloud. The lighter blue arrows are examples of forces on the parcel due to other parcels of gas. The darker blue arrows show the sum of all those forces. That net force always points toward the center of mass of the cloud.

As shown in **Figure 15.13**, each part of an interstellar cloud experiences a gravitational attraction from every other part of the cloud. The sum of all those forces acting on a particular parcel of gas will always point toward the center of mass of the cloud. That sum is the *net force* of gravity and indicates the direction in which the parcel will begin to move. The gravitational attraction between all the parts of a cloud is called *self-gravity*, which acts to hold the cloud together.

Recall from our discussion of the Sun in Chapter 14 that hydrostatic equilibrium is the balance between gravity and pressure in a stable object. Interstellar clouds are not always in hydrostatic equilibrium—most interstellar clouds are large and very thin, so their self-gravity is weak. The internal pressure pushing out is much stronger than the self-gravity, so the cloud should expand. But the much hotter gas surrounding the clouds also exerts a pressure inward on a cloud. That external hot gas helps hold a cloud together and often serves as a trigger for collapse. If a cloud is massive enough and dense enough (or becomes so after a triggering event), self-gravity becomes stronger than the internal pressure, so the clouds collapse under their own weight, beginning a chain of events that will form a new generation of stars.

If self-gravity in a molecular cloud is much greater than internal pressure, gravity should win outright, and the cloud should rapidly collapse toward its center. In practice, the process goes very slowly because several other effects stand in the way of the collapse. One effect that slows the collapse of a cloud is conservation of angular momentum, described in Chapter 7. Others are turbulence, the effects of magnetic fields, and thermal pressure. Even though those effects may slow the collapse of a molecular cloud, in the end gravity will dominate. One part of the cloud can lose angular momentum to another part of the cloud, allowing the part of the cloud with less angular momentum to collapse further. Neutral matter crosses magnetic field lines, gradually increasing the gravitational pull toward the center until the force on the charged particles is large enough to drag the magnetic field toward the center as well. Turbulence fades away. The details of those processes are complex and are the subject of much current research. The important point is that the effects that slow the collapse of a molecular cloud are temporary, and gravity is persistent. As the forces that oppose the cloud's self-gravity gradually fade, the cloud shrinks.

Astronomy in Action: Angular Momentum

AstroTour: Star Formation

Molecular Clouds Fragment as They Collapse

Some regions within a molecular cloud are denser and collapse more rapidly than surrounding regions. **Figure 15.14** shows the process of collapse in a molecular cloud. Slight variations in the density of the cloud become very dense, localized concentrations of gas. Instead of collapsing into a single object, the molecular cloud fragments into very dense **molecular-cloud cores**. A single molecular cloud may form hundreds or thousands of molecular-cloud cores, each of which is typically a few light-months in size. Some of those dense cores will eventually form stars.

As a molecular-cloud core collapses, the cloud's self-gravity grows stronger because the force of gravity is inversely proportional to the square of the radius. Suppose a cloud is 4 light-years across. When the cloud has collapsed to 2 light-years across, the different parts of the cloud are, on average, only half as far apart as when the collapse started. As a result, the self-gravity is 4 times stronger. When the cloud is one-fourth as large as it was at the beginning of the

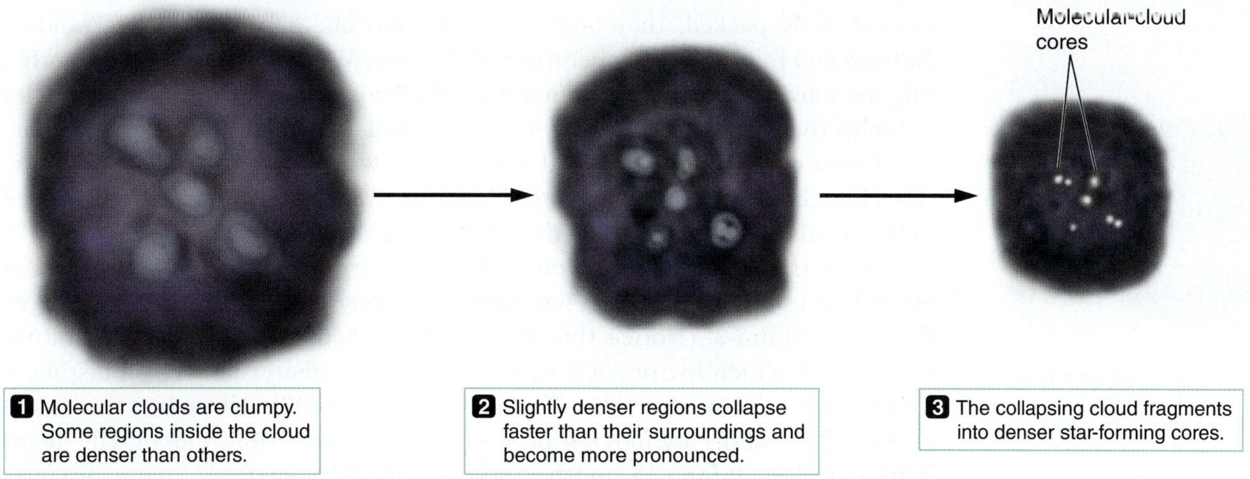

Figure 15.14 When a molecular cloud collapses, denser regions within the cloud collapse more rapidly than less dense regions. As that process continues, the cloud fragments into several very dense molecular-cloud cores embedded within the large cloud. Those cloud cores may go on to form stars.

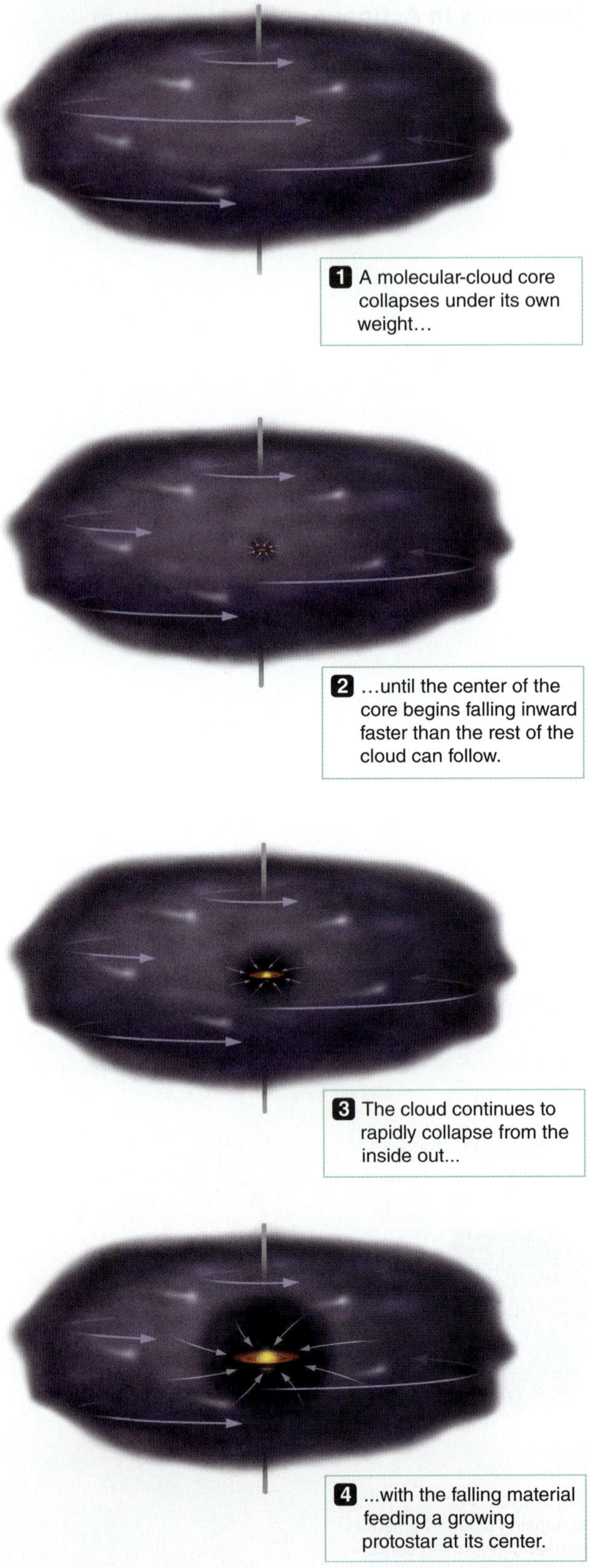

Figure 15.15 When a molecular-cloud core gets very dense, it collapses from the inside out. Conservation of angular momentum causes the infalling material to form an accretion disk that feeds the growing protostar.

collapse, the self-gravity is 16 times stronger. As a core collapses, the self-gravity increases; as self-gravity increases, the collapse speeds up; as the collapse speeds up, the self-gravity increases even faster.

Eventually, gravity overwhelms the opposing forces due to pressure, magnetic fields, and turbulence. That happens first near the center of the molecular-cloud core where the cloud material is densest. The inner parts of the molecular-cloud core start to fall rapidly inward. Those inner layers supported the weight of the layers farther out. Without the support of that inner material, the more distant material begins to fall freely toward the center. The molecular-cloud core collapses from the inside out, as shown in **Figure 15.15**.

CHECK YOUR UNDERSTANDING 15.2

Molecular clouds fragment as they collapse because: (a) the rotation of the cloud throws some mass to the outer regions; (b) the density increases fastest in the center of the cloud; (c) density variations from place to place grow larger as the cloud collapses; (d) the interstellar wind is stronger in some places than others.

15.3 Formation and Evolution of Protostars

Because of conservation of angular momentum, material that falls inward in a collapsing molecular-cloud core accumulates in a flat, rotating accretion disk. Most of that material eventually finds its way inward to the center of the disk. The object forming there, which will eventually become a star, is called a **protostar**. Just as an ice skater spins faster when she draws her arms in (recall Figure 7.5), the protostar spins faster than the original cloud. In this section, we follow the evolution of the protostar as it becomes a star.

A Protostar Forms

As particles fall toward the center, they move faster and faster. As they become more densely packed, they begin to crash into one another, causing random motions and raising the temperature of the core. When the particles are hotter, they are moving faster, in random directions. Those random motions of particles are collectively known as the *thermal energy*. Thus, the collapse of the molecular-cloud core converts gravitational energy that was stored in the large, diffuse cloud into thermal energy, and the gas in the outer layers of the protostar is heated to thousands of kelvins, causing the protostar to shine.

Because of the accumulation of thermal energy, the core gradually reaches more than 1 million K. At that temperature, deuterium can fuse with hydrogen to create helium-3. (Notice that this reaction occurs at a lower temperature than that at which hydrogen fusion occurs.) Once deuterium fusion begins, it drives convection in the core. Recall from Chapter 14 that convection is the transport of energy by moving packets of gas. Convection temporarily keeps more gas from falling in and creates an apparent "surface"—more properly called a photosphere. The photosphere radiates away energy from the protostar. The hotter it gets, the more energy it radiates and the bluer that radiation becomes (see Working It Out 5.3). Because the protostar has not yet finished collapsing, the photosphere of a protostar is tens of thousands of times larger than the photosphere of the Sun today. Each square meter radiates away energy, so despite being thousands of times more luminous than the Sun, the protostar has a lower temperature.

Although the protostar is extremely luminous, astronomers often cannot observe it in visible light for two reasons: First, the photosphere of the protostar is relatively cool, so most of its radiation is in the infrared part of the spectrum. Second, and even more important, the protostar is buried deep in the heart of a dense and dusty molecular cloud. Instead, astronomers view protostars in the infrared part of the spectrum because much of the longer-wavelength infrared light from a protostar can escape through the cloud. Sometimes, as the dust absorbs the visible light, it warms up, and that heated dust also glows in the infrared. Even when astronomers cannot view a protostar directly, they can sometimes view that heated dust and know that a collapsing molecular-cloud core is hidden inside.

Sensitive infrared instruments developed since the 1980s have revolutionized the study of protostars and other young stellar objects. Clouds that appeared dark in the visible region of the spectrum, when viewed in the infrared, have revealed themselves to be clusters of dense cloud cores, young stellar objects, and glowing dust. Nearby hot stars sometimes blow away concealing dust and gas, so that molecular-cloud cores can be viewed more directly. For example, stars are forming in nodules attached to the tops of the columns of dust and gas in the Eagle Nebula, shown in **Figure 15.16**.

The Evolving Protostar

At any instant, the protostar is in balance: the forces from hot gas pushing outward and the force of gravity pulling inward exactly oppose each other. However, that balance is constantly changing. Once the core switches from convection to radiation, the deuterium in the core becomes depleted because it is no longer being replaced by material from the outer layers brought in by convection. The

Figure 15.16 The Eagle Nebula contains dense columns of molecular gas and dust at the edge of an H II region. The yellow box in the left image identifies the region magnified in the second and third images.

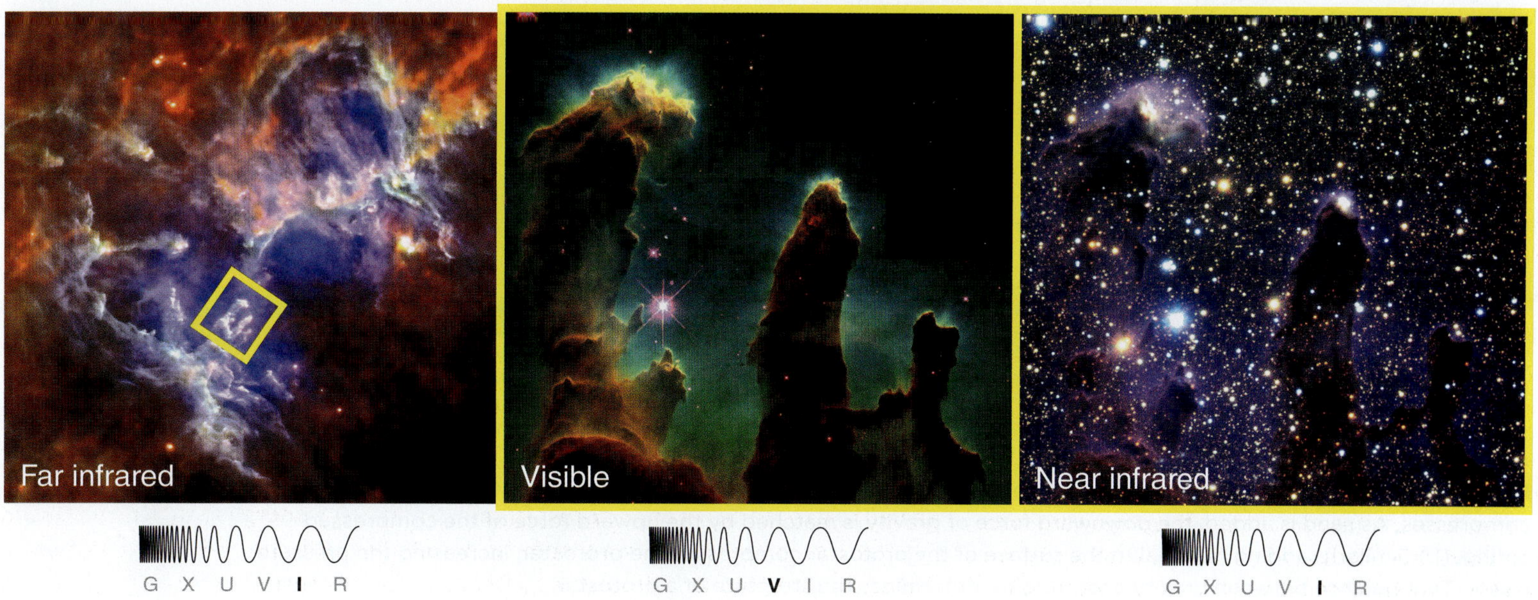

nuclear reactions slow down, reducing the pressure from the hot gas and allowing material to resume falling onto the protostar. That infalling material adds to the mass and self-gravity of the protostar and therefore increases the weight that inner layers of the protostar must support. The protostar also slowly loses its internal thermal energy by radiating it away.

How can an object be in perfect balance and yet be changing at the same time? Consider an Earth-bound example. **Figure 15.17a** shows a simple spring balance, which works on the principle that the more a spring is compressed, the harder it pushes back. You can measure the weight of an object by determining the point at which the pull of gravity and the push of the spring are equal.

When sand is poured slowly onto the spring balance, at any instant the downward weight of the sand is balanced by the upward force of the spring. As the weight of the sand increases, the spring is slowly compressed. The spring and the weight of the sand are always in balance, but that balance is changing as more sand is added. Turning to **Figure 15.17b**, you can see that the situation is analogous to that of the protostar, in which the outward pressure of the gas behaves like the spring. The self-gravity is always matched by the internal pressure pushing out.

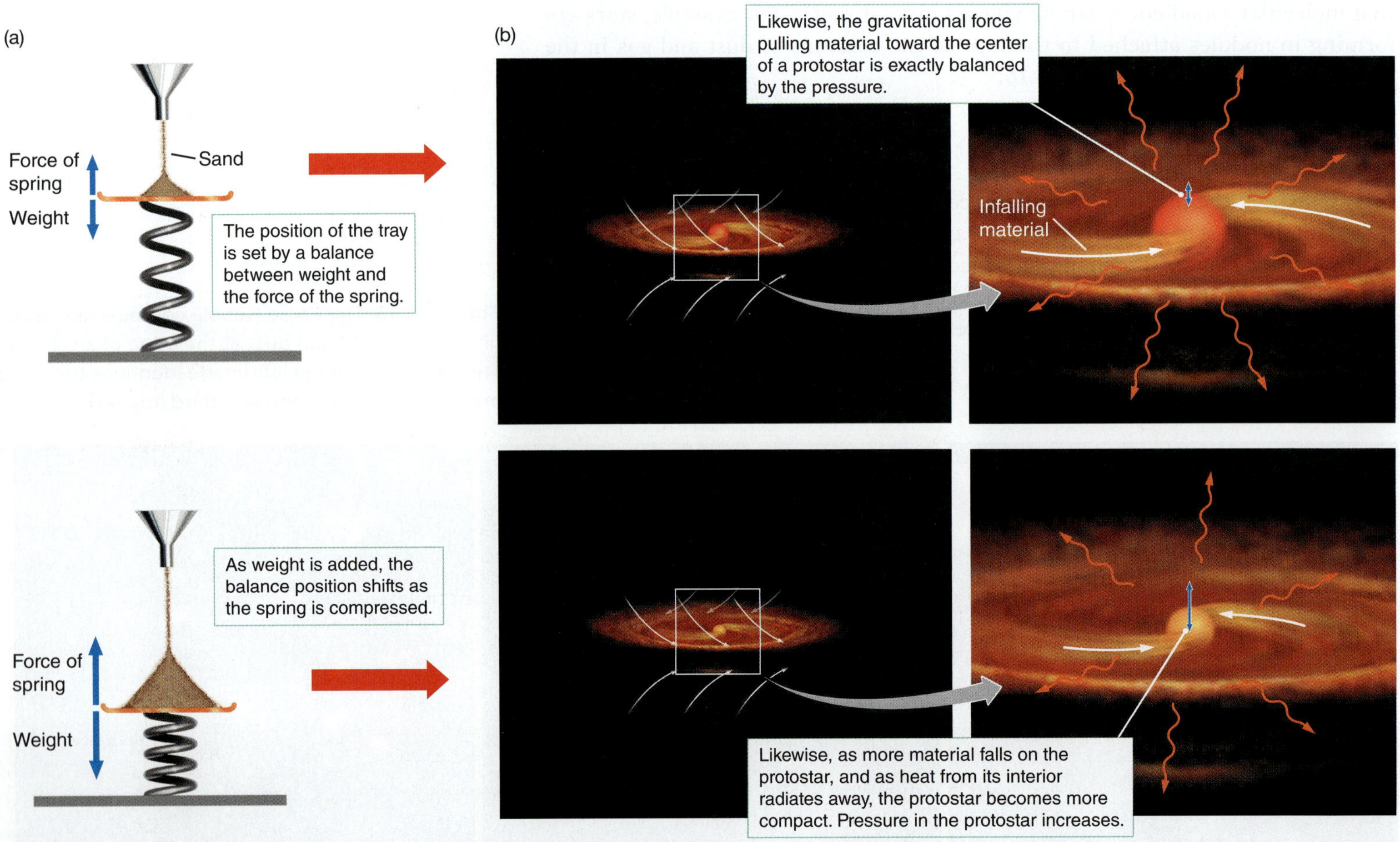

Figure 15.17 (a) As weight is added to a pan on top of a spring, the pressure on the spring increases, and the spring compresses. As sand is added, the downward force of gravity is matched by the upward force of the compressed spring. (b) Similarly, adding material to the surface of the protostar compresses the protostar, increasing the pressure inside. That balance between gravity and pressure determines the structure of a protostar.

Material falls onto the protostar, adding to its self-gravity. While the protostar slowly loses internal thermal energy by radiating it away, the material that has fallen onto the protostar compresses the protostar and heats it up. The interior becomes denser and hotter, and the pressure rises—just enough to balance the increased weight of the material above it. Dynamic balance is always maintained as the protostar slowly contracts.

Figure 15.18 illustrates that chain of events as the protostar shrinks. Gravitational energy is converted to thermal energy, which heats the core, raising the pressure to oppose gravity. That process continues, with the protostar becoming smaller and smaller and its interior growing hotter and hotter. If the protostar is massive enough, its interior will eventually become so hot that nuclear fusion of hydrogen to helium can begin. That is when the transition from protostar to star takes place. The distinction between the two is that a protostar draws its energy from gravitational collapse, whereas a star draws its energy from thermonuclear reactions in its interior.

Whether the protostar will actually become a star depends on its mass. As the protostar slowly collapses, the temperature at its center rises. If the protostar's mass is greater than about 0.08 times the mass of the Sun (0.08 M_{Sun}), the

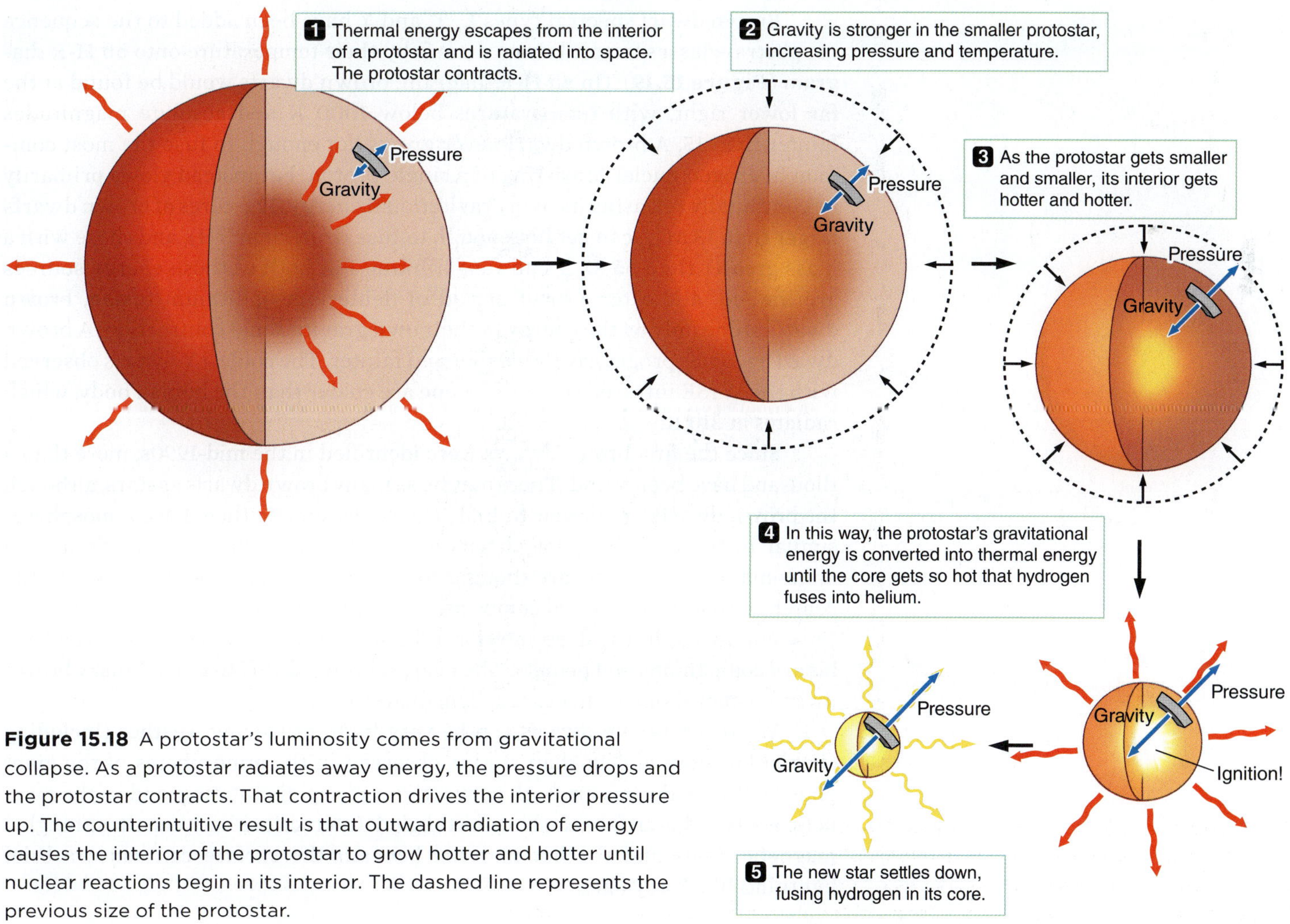

Figure 15.18 A protostar's luminosity comes from gravitational collapse. As a protostar radiates away energy, the pressure drops and the protostar contracts. That contraction drives the interior pressure up. The counterintuitive result is that outward radiation of energy causes the interior of the protostar to grow hotter and hotter until nuclear reactions begin in its interior. The dashed line represents the previous size of the protostar.

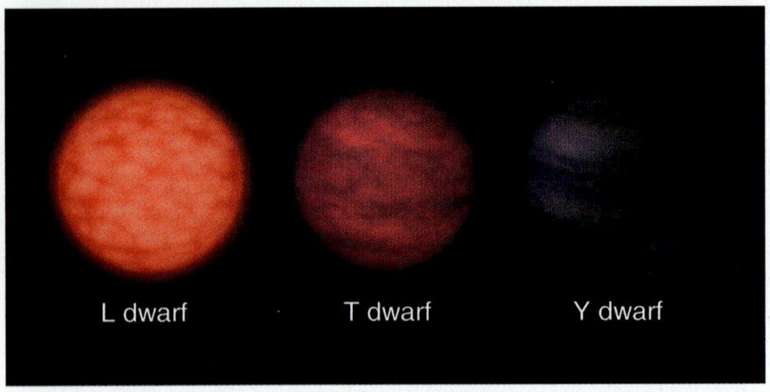

Figure 15.19 This artist's conception shows the three types of brown dwarf stars: L dwarfs (T ~1700 K); T dwarfs (T ~1200 K); and Y dwarfs (T ~500 K).

temperature in its core will eventually reach 10 million K and fusion of hydrogen into helium will begin. The newly born star will once again adjust its structure until it is radiating energy away from its surface at just the rate that energy is being liberated in its interior. As it does so, it achieves hydrostatic and thermal equilibrium and "settles" onto the main sequence of the Hertzsprung-Russell (H-R) diagram, where it will spend most of its life.

Brown Dwarfs

If the mass of the protostar is less than 0.08 M_{Sun}, it will never reach the point at which sustained nuclear fusion takes place. An object of roughly that mass is called a **brown dwarf** or sometimes a *substellar object*. Though forming in the same way that a star forms, in many respects a brown dwarf is more like a giant planet Jupiter than like a star. The International Astronomical Union (IAU) has set the smallest mass for a brown dwarf at 13 Jupiter masses (13 M_{Jup}), although some astronomers think that 10 M_{Jup} is more likely the lower limit at which brief nuclear fusion can occur. If the mass of the object is greater than 13 M_{Jup}, it must be a brown dwarf, not a supermassive planet. The upper limit of brown dwarf masses is about 70 M_{Jup}. Despite that range of masses, brown dwarfs all have radii about the same as Jupiter's radius because more massive brown dwarfs are denser (Figure 7.15).

Brown dwarf spectral types L, T, and Y have been added to the sequence of spectral classes to sort brown dwarfs by their temperature onto an H-R diagram (**Figure 15.19**). On an H-R diagram, brown dwarfs would be found at the far lower right, with temperatures below 1000 K and absolute magnitudes fainter than 15. A brown dwarf never grows hot enough to fuse the most common hydrogen nuclei consisting of a single proton, but instead glows primarily by continually releasing its own gravitational energy. The cores of brown dwarfs larger than 13 M_{Jup} can get hot enough to fuse deuterium (^2H), and those with a mass greater than 65 M_{Jup} can fuse lithium. But both of those energy sources are limited, and after a brief period of deuterium or lithium fusion, brown dwarfs shine only by the energy of their own gravitational contraction. A brown dwarf becomes progressively smaller and fainter. The coldest Y dwarfs observed with the WISE infrared space telescope are colder than the human body, which radiates at 310 K.

Since the first brown dwarfs were identified in the mid-1990s, more than a thousand have been found. There may be as many brown dwarfs as stars, although the brown dwarfs are harder to find. The cooler among them have atmospheres similar to those of the giant planets of the Solar System, full of methane and ammonia. Brown dwarfs are thought to be convective, all the way down to the center, so that the elemental composition of a brown dwarf's atmosphere reflects the composition in the deep interior. Like stars, brown dwarfs sometimes have binary companions and planets. The nearest brown dwarf to us is a binary brown dwarf system about 2 parsecs (6.5 light-years) away.

Winds on brown dwarfs can be very high, producing weather (including clouds) far more violent than storms observed in the atmospheres of the giant planets. Brown dwarfs are both convective and rapidly rotating, and so the magnetic fields at the surface can become tangled. Interactions between those tangled magnetic fields and the hot material beneath the surface can cause a brief lightning-like X-ray flare.

CHECK YOUR UNDERSTANDING 15.3

The energy required to begin nuclear fusion in a protostar originally came from: (a) the gravitational potential energy of the protostar; (b) the kinetic energy of the protostar; (c) the wind from nearby stars; (d) the pressure from the interstellar medium.

15.4 Evolution before the Main Sequence

Protostars and young evolving stars change their location on the H-R diagram as they settle into the main sequence, where they will spend the bulk of their lives. In this section, we examine some of the early stages in the life of the new stars.

The Evolutionary Track of an Evolving Star

In Chapter 13, we introduced the H-R diagram and used it to help explain how the properties of stars differ. As stars evolve through their lifetimes, their position on the H-R diagram changes, creating a path known as an **evolutionary track**. Because a protostar is so large, it is more luminous than a star of the same temperature on the main sequence, so a protostar's evolutionary track is located above the main sequence on the H-R diagram. **Figure 15.20** shows the evolutionary tracks of protostars of several masses. In a higher-mass protostar, a radiative zone develops around the core, causing the surface temperature to increase. Because of that increase in surface temperature, the luminosity of a higher-mass protostar stays roughly constant even as it contracts; on the H-R diagram, its track moves left horizontally to the main sequence. Lower-mass protostars are convective all the way to the core, so the surface temperature stays about the same as they collapse. As those protostars collapse, the luminosity decreases. On the H-R diagram, the track moves vertically down to the main sequence; that path is called a **Hayashi track** (Figure 15.20b).

In the 1960s, the theoretical physicist Chushiro Hayashi (1920–2010) explained the difference between the surface temperature of a star or protostar and the temperature deep in its interior. Hayashi showed that the atmospheres of stars and protostars contain a natural thermostat: the H^- ion. (An H^-, or "H minus," ion is a hydrogen atom that has acquired an extra electron and therefore has a negative charge.) The amount of H^- in the atmosphere of a protostar is highly sensitive to the temperature at the protostar's surface. The cooler the atmosphere of a star, the more slowly atoms and electrons are moving and the easier it is for a hydrogen atom to hold on to an extra electron. As a result, the cooler the atmosphere of the star, the more H^- is present.

The H^- ion, in turn, helps control how much energy a star or protostar radiates away. The more H^- the atmosphere of the star or protostar has, the more opaque the atmosphere is and the more effectively the thermal energy of the protostar is trapped in its interior. Imagine that the surface of the protostar is "too cool," meaning that extra H^- forms in the atmosphere and makes the atmosphere of the protostar more opaque. The atmosphere thus traps more of the radiation trying to escape, and the trapped energy heats up the star. As the temperature climbs, the H^- ions are changed to neutral H atoms. Now imagine the

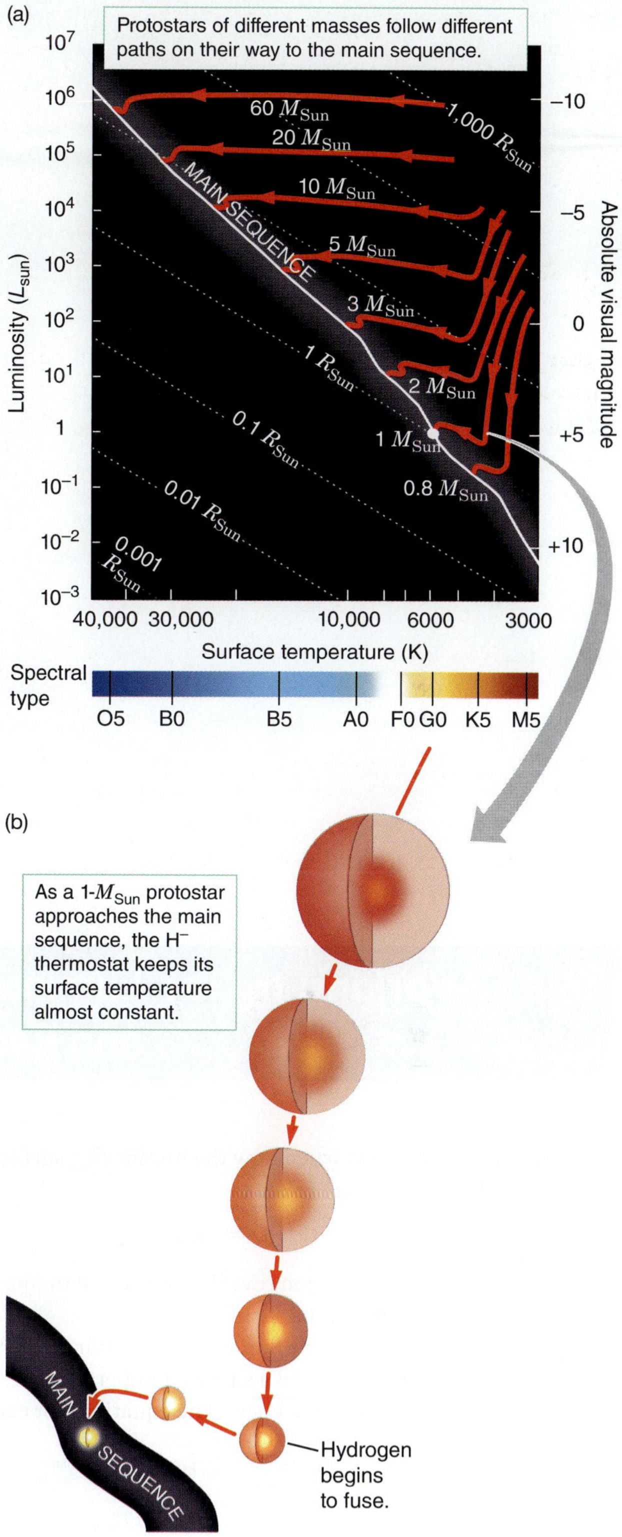

Figure 15.20 (a) The evolution of pre-main-sequence stars can be followed on the H-R diagram. More massive protostars in the upper right portion of the diagram follow horizontal tracks. (b) The roughly vertical, constant-temperature part of the evolutionary track of a low-mass protostar is called the Hayashi track.

other possibility—that the protostar is too hot. Then H^- in the protostar's atmosphere is destroyed, so the atmosphere becomes more transparent, allowing radiation to escape more freely from the interior. Because the protostar cannot hold on to enough of its energy to stay warm, the surface cools. In either case—too cold or too hot—H^- is formed or destroyed, respectively, until the star's atmosphere once again traps just the right amount of escaping radiation. The H^- ion is basically doing the same thing that you do with your bedcovers at night. If you get too cold, you pile on extra covers to trap your body's thermal energy and keep warm (corresponding to more H^- ions with the star). If you get too hot, you kick off some covers to cool off (fewer H^- ions).

The amount of H^- in the atmosphere keeps the surface temperature of the lower-mass protostar somewhere between about 3000 and 5000 K, depending on the protostar's mass and age. Because the surface temperature of the protostar is not changing much, the amount of energy per unit time (power) radiated away by each square meter of the surface of the protostar does not change much, either. Recall the Stefan-Boltzmann law from Chapter 5, which says that the amount radiated by each square meter of an object's surface depends on its temperature. As the protostar shrinks, the area of its surface shrinks as well. With fewer square meters of surface from which to radiate, the luminosity of the protostar drops. As viewed from outside, the protostar stays at nearly the same temperature and color but gradually gets fainter as it evolves toward its eventual life as a main-sequence star. The relationship among surface temperature, luminosity, and radius of protostars is further explored in **Working It Out 15.2**.

15.2 Working It Out Luminosity, Surface Temperature, and Radius of Protostars

In Chapter 13, you learned how the luminosity, surface temperature, and radius of a star are related:

$$L = 4\pi R^2 \sigma T^4$$

What can that equation reveal about the changing properties of the protostar as its radius shrinks?

Suppose that when the Sun was a protostar, it had a radius 10 times what it is now and a surface temperature of 3300 K. What would its luminosity have been? The equations for each are

$$L_{\text{protostar}} = 4\pi R^2_{\text{protostar}} \sigma T^4_{\text{protostar}}$$

and

$$L_{\text{Sun}} = 4\pi R^2_{\text{Sun}} \sigma T^4_{\text{Sun}}$$

We can set that relationship up as a ratio, comparing the luminosity of the protostar Sun with its luminosity now, L_{Sun}:

$$\frac{L_{\text{protostar}}}{L_{\text{Sun}}} = \frac{4\pi R^2_{\text{protostar}} \sigma T^4_{\text{protostar}}}{4\pi R^2_{\text{Sun}} \sigma T^4_{\text{Sun}}}$$

We rewrite that expression as follows, grouping like terms together:

$$\frac{L_{\text{protostar}}}{L_{\text{Sun}}} = \frac{4\pi\sigma}{4\pi\sigma} \times \left(\frac{R_{\text{protostar}}}{R_{\text{Sun}}}\right)^2 \times \left(\frac{T_{\text{protostar}}}{T_{\text{Sun}}}\right)^4$$

Then we cancel out the constants, $4\pi\sigma$, and use the value for $T_{\text{Sun}} = 5780$ K from Chapter 14. We know that the protostar's radius is 10 times that of the Sun, so $R_{\text{protostar}}/R_{\text{Sun}} = 10$. Then the equation becomes

$$\frac{L_{\text{protostar}}}{L_{\text{Sun}}} = \left(\frac{10}{1}\right)^2 \times \left(\frac{3{,}300}{5{,}780}\right)^4 = 10^2 \times (0.57)^4 = 10.6$$

So the Sun was about 10.6 times more luminous as a protostar than it is now. We see that on the H-R diagram of protostars (see Figure 15.20). As a 1-M_{Sun} star approaches the main sequence on the diagram, it moves down (toward lower luminosity) and to the left (toward higher surface temperature).

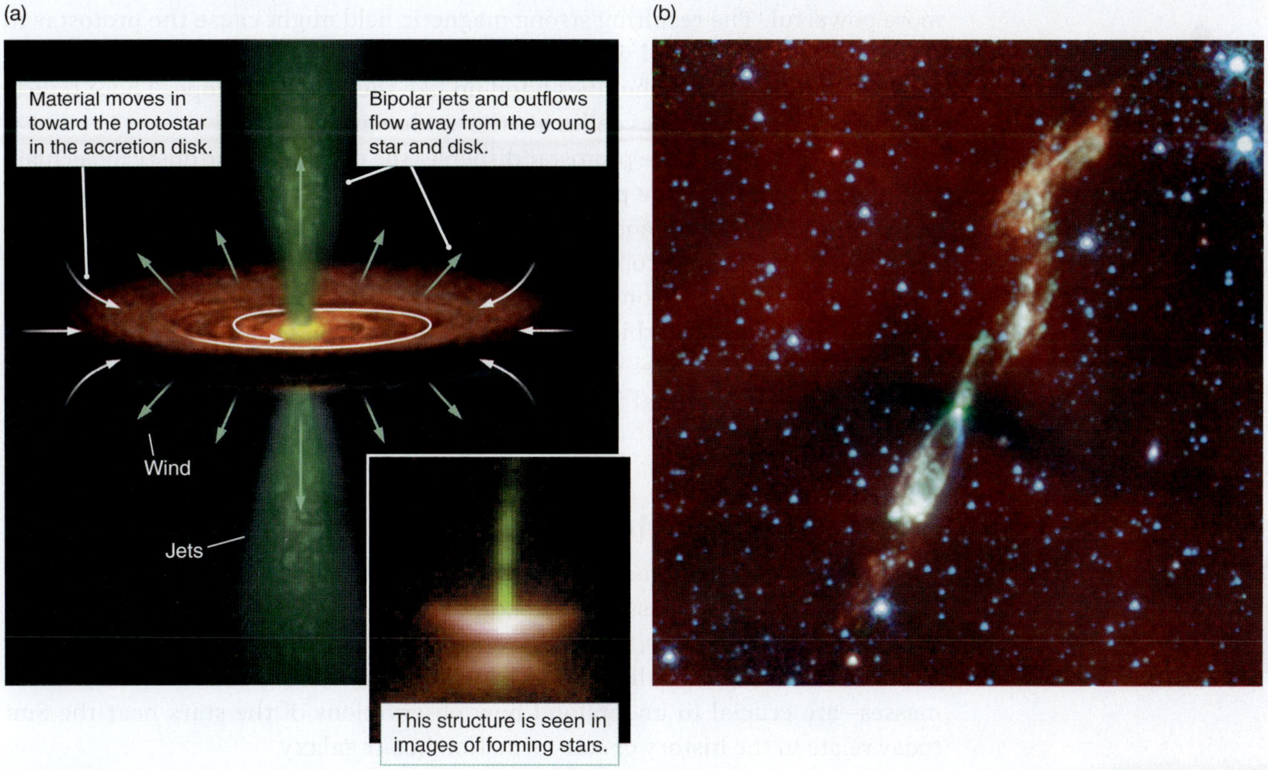

Figure 15.21 (a) Material falls onto an accretion disk around a protostar and then moves inward, eventually falling onto the protostar. In the process, some of that material is driven away in powerful jets that stream perpendicular to the disk. (b) This infrared Spitzer Space Telescope image shows jets streaming outward from a young, developing star. Note the nearly edge-on, dark accretion disk surrounding the young star.

Bipolar Outflow

As shown in **Figure 15.21a**, material falls onto the accretion disk around a young stellar object and moves inward toward the equator of the star. Meanwhile, other material is blown away from the protostar and disk in two opposite directions from the plane of the disk. The resulting stream of material away from the protostar is called a **bipolar outflow**. Powerful outflows can disrupt the molecular-cloud core and accretion disk from which the protostar formed, stopping material from falling onto the protostar.

Some bipolar outflows from young stellar objects are slow and fairly disordered, but others produce remarkable **jets** of material moving at hundreds of kilometers per second (**Figure 15.21b**). The material in those jets flows out into the interstellar medium, where it heats, compresses, and pushes away surrounding interstellar gas. Knots of glowing gas accelerated by jets are called **Herbig-Haro objects** (or **HH objects** for short), named after the two astronomers who first identified them and associated them with star formation. An example is shown in **Figure 15.22**.

The origin of outflows from protostars is not well understood, but current models suggest that outflows are the result of magnetic interactions between the protostar and the disk. The interior of a lower-mass protostar on its Hayashi track is convective. That convection, coupled with the protostar's rapid rotation, forms a dynamo, similar to the dynamo that drives the Sun's magnetic field but much

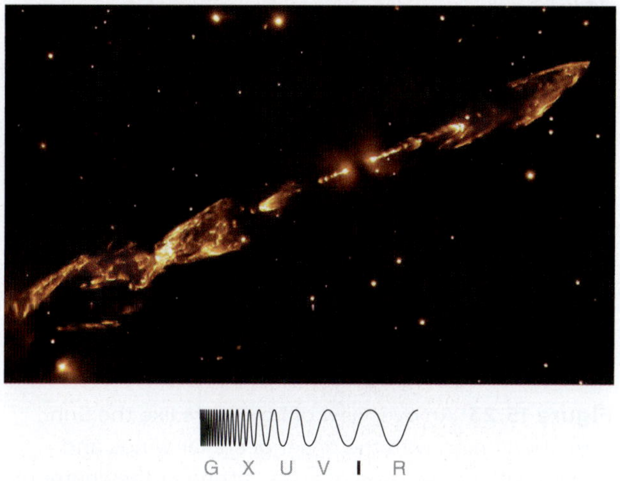

Figure 15.22 In this image of Herbig-Haro object 212, taken by ESO's Infrared Spectrometer And Array Camera (ISAAC), the accretion disk of the protostar is visible as a fuzzy dark band cutting between the two jets. The protostar is concealed inside that accretion disk and is only a few thousand years old. The jets are remarkably symmetric, with several knots appearing at intervals. That finding suggests that the jet pulses regularly and over a short timescale—possibly as short as 30 years. At the ends of the jets, ejected gas collides with interstellar dust and gas at several hundred kilometers per second, creating bow shocks.

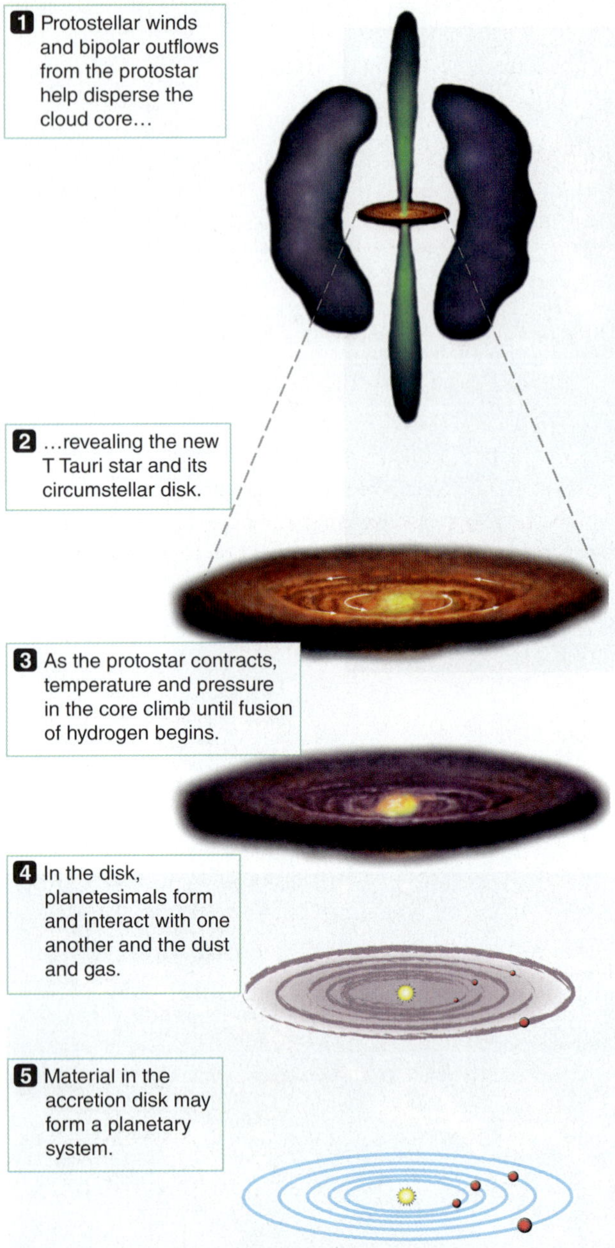

Figure 15.23 An overview of how stars like the Sun form, beginning with the onset of stellar winds and ending with the ignition of a star sitting at the center of a revolving system of planets.

more powerful. The resulting strong magnetic field might cause the protostar to begin blowing a powerful wind. As the rotation of the protostar drags the magnetic field lines around, the lines wind up like the fibers in a rope. Those tightly wound magnetic field lines collimate the outflow into jets.

As the wind from the protostar disperses the remains of the dusty molecular-cloud core from which the protostar formed, the first direct, visible-light view of the protostar emerges. Some of those protostars of lower mass are called **T Tauri stars**. The name comes from the first recognized member of that class of objects, the star labeled T in the constellation Taurus. Higher-mass protostars of spectral type B or A are called Herbig Be or Ae stars. **Figure 15.23** summarizes the evolution of a protostar from the onset of winds to the formation of planets (recall that steps 4 and 5 were covered in Chapter 7).

The Influence of Mass

Astronomers are interested in how and why molecular clouds fragment to form stars with a range of masses. Astronomers do not understand why some cloud cores become 1-M_{Sun} stars, whereas others become 5- or 10-M_{Sun} stars. The details of that division—specifically, what fraction of newly formed stars will have which masses—are crucial to understand how observations of the stars near the Sun today relate to the history of star formation in our galaxy.

A look around the disk of our galaxy reveals a variety of stars—some very old and others very young. If those were the only stars available to study, scientists could not easily learn much about how stars evolve. But astronomers have long known that stars are often found close together in collections called **star clusters**. Star clusters are collections of stars that all formed in the same place, from the same material, and at about the same time. **Figure 15.24** shows one such star cluster: a group called the Pleiades, or "Seven Sisters." Clusters such as that one serve as extremely useful samples for studying star formation. Even though the few brightest and most massive stars in a cluster dominate any observation of a cluster, most of the stars in a cluster are less massive than the Sun. In fact, some star-forming regions seem to form no high-mass stars at all.

After a molecular-cloud core collapses, how a protostar evolves depends almost entirely on its mass. A star like the Sun takes about 10 million years or so to descend its Hayashi track and become a star on the main sequence. The total time for such a star to form—from initial fragmentation up to the ignition of hydrogen fusion—might be more like 30 million years. Because the self-gravity of a more massive core is stronger, more massive cores collapse to form stars more quickly. A 10-M_{Sun} star will form in only 100,000 years. A 100-M_{Sun} star might take less than 10,000 years. By comparison, a 0.1-M_{Sun} star might take 100 million years finally to reach the main sequence.

The 30 million years that the Sun took to form is a long time, yet still just a tiny fraction of the 10 billion years during which the Sun will steadily fuse hydrogen into helium as a main-sequence star. Because stars spend so much longer on the main sequence than they do as protostars, it is no wonder that so few among the many stars visible in the sky are protostars. But every star was once a protostar, including the Sun.

CHECK YOUR UNDERSTANDING 15.4

What causes the wind from a protostar to form jets? (a) The accretion disk can supply material to the jet only along the equator of the star. (b) Magnetic fields wind up and direct the outflow. (c) The star is larger at the equator, so material leaves the poles more easily. (d) The star spins faster at the equator, so material leaves the poles more easily.

Origins Star Formation, Planets, and Life

When astronomers consider the possibility of other life in the universe, one of the first things they think about is the formation of stars and planets. Life probably needs planets, and planets form along with stars. The conditions under which a star is born, and the mass and chemical composition that it has when it begins its nuclear fusion, set the stage for the rest of its life. In Chapter 7, we noted that most stars have planets that form at about the same time as the star. If the star is going to have rocky planets with hard surfaces (such as the planets of the inner Solar System) or gaseous planets with molten rocky cores and rocky moons (such as the planets of the outer Solar System), the material from which the star and planets form must be "enriched" with the heavy elements that make up those rocky surfaces.

Those enriched clouds would also provide elements essential to life on Earth. In addition to the presence of organic molecules mentioned earlier in the chapter, astronomers have detected water in star-forming regions such as W3 IRS5 (**Figure 15.25**). Water exists as ice mixed with dust grains in the cool molecular clouds or exists as vapor when it is closer to a protostar and the dust grains and ice evaporate. In 2011, the Herschel Space Observatory detected oxygen molecules (O_2, the type we breathe) in a star-forming complex in Orion. Oxygen is the third-most-common element in the universe, yet it had not been decisively observed before in molecular form. That oxygen also may have come from the melting and evaporation of water ice on the tiny dust grains.

As noted in Chapter 13, astronomers had doubted that planets could exist in stable orbits in binary star systems, but now a few such systems have been found. Planets that form within associations of O and B stars may be too unstable to last very long. Isolated planets unattached to any star are moving through the Milky Way. Perhaps those rogue planets were gravitationally ejected soon after they formed in a multiple system. But those planets do not have a source of energy like Earth's Sun. Astronomers theorize that only planets that orbit stars can support life. So when trying to estimate the possibility of life in the galaxy, astronomers include estimations of the rate of formation of stars in the galaxy, along with the fraction of stars that have planets. Advances in the study of star formation and planet detection help astronomers understand better the conditions under which life might develop elsewhere.

Figure 15.24 The Pleiades, or "Seven Sisters," is a cluster of hundreds of young stars, about seven of which are easily visible to the naked eye. The diffuse blue light around the stars is starlight scattered by interstellar dust.

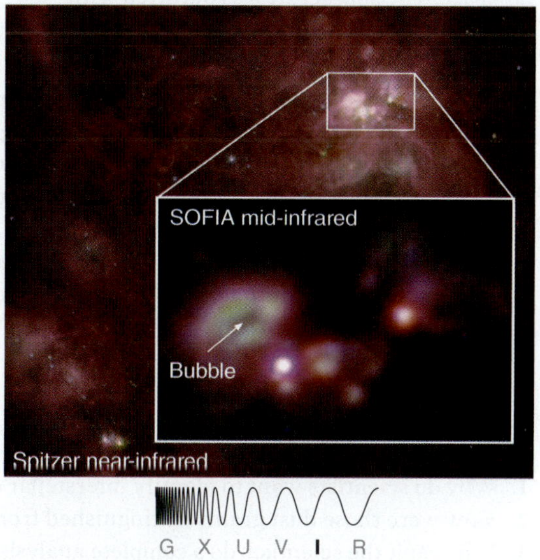

Figure 15.25 Water was detected in the W3 IRS5 star-forming complex. Here, W3 IRS5 is observed with the Spitzer Space Telescope in near infrared and with the Stratospheric Observatory for Infrared Astronomy (SOFIA) telescope in mid-infrared. A massive star has cleared the dust and gas from a small bubble, sweeping it into a dense shell (green).

ARTICLES QUESTIONS

Astronomers report on their search for dust from outside the Solar System.

Interstellar Dust Discovered inside NASA Spacecraft

By IRENE KLOTZ, *Discovery News*

Thanks to a massive effort by 30,716 volunteers, scientists have pinpointed what appear to be seven precious specks of dust from outside the Solar System, each bearing unique stories of exploded stars, cold interstellar clouds, and other past cosmic lives.

The Herculean effort began eight years ago after NASA's *Stardust* robotic probe flew by Earth to deposit a capsule containing samples from a comet and dust grains from what scientists hoped would be interstellar space. The spacecraft was outfitted with panels containing a smoke-like substance called aerogel that could trap and preserve fast-moving particles.

Stardust twice put itself into position to fish for interstellar grains, which are so small that a trillion of them would fit in a teaspoon. The only way scientists back on Earth would be able to find them was by the microscopic trails the grains made as they plowed into the aerogel.

"When we did the math we realized it would take us decades to do the search ourselves," physicist Andrew Westphal, with the University of California, Berkeley, told *Discovery News*.

The team used an automated microscope to scan the collector and put out a call for volunteers.

"This whole approach was treated with pretty justifiable criticism by people in my community. They said, 'How can you trust total strangers to take on this project?'" Westphal said.

"We really didn't know how else to do it. We still don't," he added.

Recruits were trained and had to pass a test before they were given digital scans to peruse. Scientists sometimes inserted images with known trails just to see if the volunteers, known as "dusters," would spot them.

"We were very pleased to see that people are really good at finding these tracks, even really, really difficult things to find," Westphal said.

More than 50 candidate dust motes turned out to be bits of the spacecraft itself, but scientists found seven specks that bear chemical signs of interstellar origin and travel.

The grains are surprisingly diverse in shape, size, and chemical composition. The larger ones, for example, have a fluffy, snowflake-like structure.

Additional tests are needed to verify the grains' interstellar origins and ferret out their histories. But the grains are so tiny that with currently available technology, additional analysis would mean their demise.

"It'll probably be years before we can do a lot more with these samples," said space scientist Mike Zolensky, who oversees NASA's collection of cosmic dust, moon rocks, and other extraterrestrial samples at the Johnson Space Center in Houston.

"But we've got them safely tucked away and we can hang on to them until those techniques come along," Zolensky said.

ARTICLES QUESTIONS

1. Why do scientists want to identify interstellar dust grains?
2. How were those dust grains distinguished from Solar System grains?
3. Why can't the scientists do a complete analysis on those particles?
4. How did volunteer "citizen scientists" assist with that project?
5. Visit the website noted in question 49 in the "Questions and Problems" section later in this chapter. Is that project continuing? What else has been discovered?

Source: Irene Klotz, "Interstellar dust discovered inside NASA spacecraft." Originally published by Seeker, August 14, 2014. Reprinted by permission of Group Nine Media.

Summary

Astronomers have never watched the full process of a single star forming from beginning to end. Instead, they have observed many stars at different stages in their formation and evolution at different wavelengths, and they have used their knowledge of physical laws to tie those observations together into a coherent, consistent description of how, why, and where stars form. Stars form from material in the interstellar medium, under the influence of gravity, often triggered by external factors such as the winds of hot stars nearby. The conditions under which a star is born determine whether it will have planets along with the chemical elements required by life as it exists on Earth.

LG 1 Describe the types and states of material that exist in the space between the stars and how that material is detected. The interstellar medium ranges from cold, relatively dense molecular clouds to hot, tenuous intercloud gas heated and ionized by energy from stars and stellar explosions. Dust and gas in the interstellar medium blocks much visible light but becomes more transparent at longer, infrared wavelengths. Different phases of the interstellar medium emit various types of radiation and can be observed at different wavelengths, ranging from radio waves to X-rays. Neutral hydrogen cannot be detected at visible and infrared wavelengths, but its presence is revealed by its 21-cm emission.

LG 2 Explain the conditions under which a cloud of gas can contract into a stellar system and the role that gravity and angular momentum play in forming stars and planets. Star formation begins when the self-gravity of dense clouds exceeds outward pressure. The clouds collapse, heat up, and fragment to form stars. The conservation of angular momentum is important to the formation of disks during the collapse. Forming stars are detected directly from their infrared emission and indirectly from the dust around them, which is heated by the forming star.

LG 3 List the steps in the evolution of a protostar, and explain how the mass of the protostar affects its evolution. Protostars collapse, radiating away their gravitational energy until fusion starts in their cores. When protostars reach hydrostatic and thermal equilibrium, they settle onto the main sequence. Stars form in clusters from dense cores buried within giant molecular clouds. A protostar must have a mass of at least 0.08 M_{Sun} to become a true star. Brown dwarfs are neither stars nor planets but have masses in between.

LG 4 Describe the track of a protostar as it evolves to a main-sequence star on the Hertzsprung-Russell (H-R) diagram. Because star formation takes tens of thousands to millions of years, astronomers learn about the formation of stars from observations of many protostars at various stages of development. Higher-mass protostars move horizontally across the H-R diagram, at roughly constant luminosity, to the main sequence. Lower-mass protostars drop nearly straight down, at constant temperature, toward the main sequence. Once hydrogen fusion begins, the star moves nearly horizontally until it reaches the main sequence.

? Unanswered Questions

- Many questions about star formation remain. For example, how must theories be modified to explain how binary stars or other multiple-star systems form? At what point during star formation is it determined that a collapsing cloud core will form several stars instead of just one? Some models suggest that this split may happen early in the process, during the fragmentation and collapse of the molecular cloud. The advantage of those ideas is that they provide a natural way of dealing with much of the cloud core's angular momentum: it goes into the orbital angular momentum of the stars around each other. Other models suggest that additional stars may form from the accretion disk around an initially single protostar.
- Do high-mass and low-mass stars form very differently? The smallest stars, spectral type M, are most likely to form as single stars, but a high fraction of medium-mass stars are formed in binary pairs. According to one theory, those binaries start out as triple systems, from which the smallest star is ejected, leading to a remaining pair and a single star. The highest-mass stars are less likely to form alone; many form in larger groups of massive stars in which the formation of one large star may trigger the formation of another nearby in the molecular cloud.

Questions and Problems

Test Your Understanding

1. Phases of the interstellar medium include (choose all that apply)
   a. hot, low-density gas.
   b. cold, high-density gas.
   c. hot, high-density gas.
   d. cold, low-density gas.

2. Dust in the interstellar medium can be observed in
   a. visible light.
   b. infrared radiation.
   c. radio waves.
   d. X-rays.

3. The interstellar medium in the Sun's region of the galaxy is closest in composition to that of
   a. the Sun.
   b. Jupiter.
   c. Earth.
   d. comets in the Oort Cloud.

4. Interstellar dust is effective at blocking visible light because
   a. the dust is so dense.
   b. dust grains are so few.
   c. dust grains are so small.
   d. dust grains are so large.

5. Hot intercloud gas is heated primarily by
   a. starlight.
   b. protostars.
   c. neutrinos.
   d. supernova explosions.

6. Astronomers determined the composition of the interstellar medium from
   a. observing its emission and absorption lines.
   b. measuring the composition of the planets.
   c. return samples from spacecraft.
   d. composition of meteorites.

7. In astronomy, the term *bipolar* refers to outflows that
   a. point in opposite directions.
   b. alternate between expanding and collapsing.
   c. rotate about a polar axis.
   d. show spiral structure.

8. Which of the following has contributed most to our understanding of star formation?
   a. Astronomers have observed star formation as it happens for a few stars.
   b. Astronomers have observed star formation as it happens for many stars.
   c. Astronomers have observed many stars at various steps of the formation process.
   d. Theoretical models predict how stars form.

9. Cold neutral hydrogen can be detected because
   a. it emits light when electrons drop through energy levels.
   b. it blocks the light from more distant stars.
   c. it is always hot enough to glow in the radio and infrared wavelengths.
   d. the atoms in the gas change spin states.

10. The Hayashi track is a nearly vertical evolutionary track on the H-R diagram for low-mass protostars. Which of the following would you expect from a protostar moving along a vertical track?
    a. The star remains the same brightness.
    b. The star remains the same luminosity.
    c. The star remains the same color.
    d. The star remains the same size.

11. Which two forces establish hydrostatic equilibrium in an evolving protostar?
    a. the force from pressure and gravity
    b. the force from pressure and the strong nuclear force
    c. gravity and the strong nuclear force
    d. energy emitted and energy produced

12. Suppose you are studying a visible-light image of a distant galaxy, and you see a dark lane cutting across the bright disk. That dark line is most likely caused by
    a. gravitational instabilities that clear the area of stars.
    b. dust in the Milky Way blocking the view of the distant galaxy.
    c. dust in the distant galaxy blocking the view of stars in the disk.
    d. a flaw in the instrumentation.

13. What causes a hydrogen atom to radiate a photon of 21-cm radio emission?
    a. The electron drops down one energy level.
    b. The formerly free electron is captured by the proton.
    c. The electron flips to an aligned spin state.
    d. The electron flips to an unaligned spin state.

14. Astronomers know that dusty accretion disks exist around protostars because
    a. a dark band often appears across the protostar.
    b. a bright band often appears across the protostar.
    c. accretion disks should be there, according to theory.
    d. planets are in the Solar System.

15. What is the single most important property of a star that will determine its evolution?
    a. mass
    b. composition
    c. temperature
    d. radius

Thinking about the Concepts

16. The interstellar medium is approximately 99 percent gas and 1 percent dust. Why does dust and not gas block a visible-light view of the galactic center?

17. Explain why observations in the infrared are necessary for astronomers to study the detailed processes of star formation.

18. How does the material in interstellar clouds and intercloud gas differ in density and distribution?

19. When a star forms inside a molecular cloud, what happens to the cloud? Can a molecular cloud remain cold and dark with one or more stars inside it? Explain your answer.

20. If you placed your hand in boiling water (100°C) for even 1 second, you would get a very serious burn. If you placed your hand in a hot oven (200°C) for a second or two, you would hardly feel the heat. Explain that difference and how it relates to million-kelvin regions of the interstellar medium.

21. How do astronomers know that the Sun is located in a "local bubble" formed by a supernova?

22. Interstellar gas atoms typically cool by colliding with other gas atoms or grains of dust; during the collision, each gas atom loses energy and hence its temperature is lowered. How does that explain why very low-density gases are generally so hot, whereas dense gases tend to be so cold?

23. Explain how the 21-cm line discussed in the Process of Science Figure supports the cosmological principle (which states that the laws of physics must be the same everywhere).

24. Molecular hydrogen is very difficult to detect from the ground, but astronomers can easily detect carbon monoxide (CO) by observing its 2.6-cm microwave emission. Describe how observations of CO might help astronomers infer the amounts and distribution of molecular hydrogen within giant molecular clouds.

25. The Milky Way contains several thousand giant molecular clouds. Describe a giant molecular cloud and its role in star formation.

26. As a cloud collapses to form a protostar, the forces of gravity experienced by all parts of the cloud (which follow an inverse square law) become stronger and stronger. One might argue that under those conditions, the cloud should keep collapsing until it becomes a single massive object. Why doesn't that happen?

27. The internal structure of a protostar maintains hydrostatic equilibrium even as more material is falling onto it. Explain how that can be.

28. What are the similarities and differences between a brown dwarf and a giant planet such as Jupiter? Would you classify a brown dwarf as a supergiant planet? Explain your answer.

29. The H^- ion acts as a thermostat in controlling the surface temperature of a protostar. Explain that process.

30. How does the composition of the molecular cloud affect the type of planets and stars that form within it?

Applying the Concepts

31. In Chapter 13, you learned that astronomers can measure the temperature of a star by comparing its brightness in blue and yellow light. Does reddening by interstellar dust affect a star's temperature measurement? If so, how?

32. When a hydrogen atom is ionized, it splits into two components.
    a. Identify the two components.
    b. If both components have the same kinetic energy, which moves faster?

33. Estimate the typical density of dust grains (grains per cubic centimeter) in the interstellar medium. A typical grain has a mass of about 10^{-17} kilogram (kg). (Hint: You know the typical density of gas and the fraction of the interstellar medium's mass that is made of dust.)

34. Referring to Figure 15.3, estimate the effective blackbody temperature of the star as shown in part (b) (without dust) and part (c) (with dust). How significant are the effects of interstellar dust when observed data are used to determine the properties of a star?

35. A typical temperature of intercloud gas is 8000 K. Using Wien's law (see Working It Out 15.1 and Chapter 5), calculate the peak wavelength at which that gas would radiate.

36. Some parts of the Orion Nebula have a blackbody peak wavelength of 0.29 μm. What is the temperature of those parts of the nebula?

37. Stellar radiation can convert atomic hydrogen (H I) to ionized hydrogen (H II).
    a. Why does a B8 main-sequence star ionize far more interstellar hydrogen in its vicinity than does a K0 giant of the same luminosity?
    b. What properties of a star are important in determining whether it can ionize large amounts of nearby interstellar hydrogen?

38. A proton has 1,850 times the mass of an electron. If a proton and an electron have the same kinetic energy ($E_K = 1/2\ mv^2$), how many times greater is the velocity of the electron than that of the proton?

39. If a typical hydrogen atom in a collapsing molecular-cloud core starts at a distance of 1.5×10^{12} km (10,000 AU) from the core's center and falls inward at an average velocity of 1.5 km/s, in how many years does it reach the newly forming protostar? Assume that a year is 3×10^7 seconds.

40. The ratio of hydrogen atoms (H) to carbon atoms (C) in the Sun's atmosphere is approximately 2,400:1 (see Table 13.1). We can reasonably assume that this ratio also applies to molecular clouds. If 2.6-cm radio observations indicate 100 M_{Sun} of carbon monoxide (CO) in a giant molecular cloud, what is the implied mass of molecular hydrogen (H_2) in the cloud? (Carbon represents 3/7 of the mass of a CO molecule.)

41. Neutral hydrogen emits radiation at a radio wavelength of 21 cm when an atom drops from a higher-energy spin state to a lower-energy spin state. On average, each atom remains in the higher-energy state for 11 million years (3.5×10^{14} seconds).
    a. What is the probability that any given atom will make the transition in 1 second?
    b. If 6×10^{59} atoms of neutral hydrogen are in a 500-M_{Sun} cloud, how many photons of 21-cm radiation will the cloud emit each second?
    c. How does that number compare with the 1.8×10^{45} photons emitted each second by a solar-type star?

42. The Sun took 30 million years to evolve from a collapsing cloud core to a star, with 10 million of those years spent on its Hayashi track. The Sun will spend 10 billion years on the main sequence. Suppose that the Sun's main-sequence lifetime were compressed into a single day.
    a. How long would the total collapse phase last?
    b. How long would the Sun spend on its Hayashi track?

43. A protostar with the mass of the Sun starts out with a temperature of about 3500 K and a luminosity about 200 times larger than the Sun's present value. Estimate that protostar's size and compare it with the size of the Sun today.

44. The star-forming region 30 Doradus is 160,000 light-years away in the nearby galaxy called the Large Magellanic Cloud and appears about one-sixth as bright as the faintest stars visible to the naked eye. If it were located at the distance of the Orion Nebula (1,300 light-years away), how much brighter than the faintest visible stars would it appear?

45. Assume that a brown dwarf has a surface temperature of 1000 K and approximately the same radius as Jupiter. What is its luminosity compared with that of the Sun? How many brown dwarfs like that one would be needed to produce the luminosity of a star like the Sun?

USING THE WEB

46. Go to the Astronomy Picture of the Day (APOD) website (https://apod.nasa.gov/apod), search for "molecular clouds," and pick out a few images. Were those pictures obtained from space or on the ground, and at what wavelengths? With which telescopes? What wavelengths do the colors in the images represent? Are they "real" or "false-color" images?

47. Go to NASA's Spitzer Space Telescope website (http://www.spitzer.caltech.edu). Click on "News" and find a recent story about star formation. What did Spitzer observe? What wavelengths do the colors in the picture represent? How does that "false color" help astronomers analyze the images? Why do astronomers study star formation in the infrared rather than in the visual part of the spectrum?

48. The Stratospheric Observatory for Infrared Astronomy (SOFIA) is a 2.5-meter telescope on a modified Boeing 747 aircraft. Go to the SOFIA website (http://sofia.usra.edu). Why have astronomers put an infrared telescope on an airplane? What has been detected with that telescope?

49. Citizen science: Go to the website for *Stardust* (http://stardustathome.ssl.berkeley.edu), a project in which volunteers use a virtual microscope to analyze digital scans of particles collected by the *Stardust* mission in 2006. The goal is to identify tiny interstellar dust grains. Follow the steps under "Get Started" (you need to create a log-in account) and help search for stardust. Click on "News." What has been learned from that project? Remember to save the images for your homework, if required.

50. Do a news search for a story about brown dwarfs. Is the story from an observatory? A NASA mission? A press release? What is new, and why is it interesting?

EXPLORATION

The Stellar Thermostat

digital.wwnorton.com/astro6

In this Exploration, you will see how the H^- thermostat works in the formation of stars. You will need about 20 coins (they do not have to be all the same type).

Place your coins on a sheet of paper and draw a circle around them—the smallest possible circle that will fit all the coins. Then divide the circle into three parts, as shown in **Figure 15.26.** This circle represents a star with a changing temperature. The coins represent H^- ions. Removing a coin from the circle means that the H^- ion has turned into a neutral hydrogen atom. Placing a coin in the circle means that the neutral hydrogen atom has become an H^- ion.

Place all the coins back on the circle.

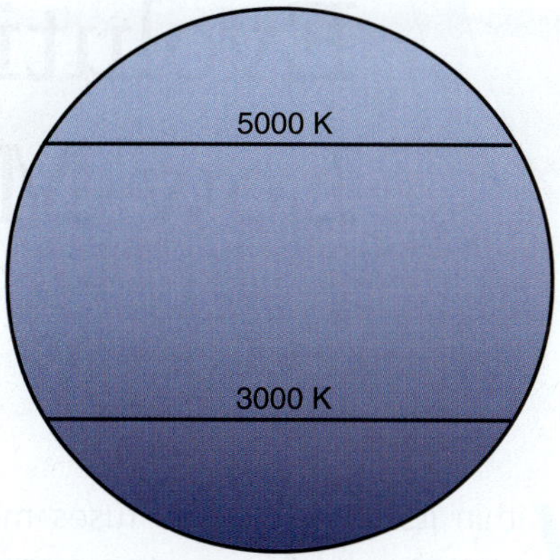

Figure 15.26 This circle represents a star with a changing temperature.

1 How many "H^- ions" are now in the star?

The "blanket" of H^- ions holds heat in the star, so the star begins to heat up until it reaches about 5000 K. At that surface temperature, the H^- ions begin to be destroyed. Now that the star is hot, begin removing coins one at a time, starting from the top of the circle and working downward. When you see the line marking 3000 K, stop removing coins.

2 How many "H^- ions" are now in the star?

3 What will happen to the surface temperature of the star now that it has fewer ions?

When the star cools off to about 3000 K, H^- ions begin to form. Place the coins back on the circle, starting from the bottom and working your way up to the line at 5000 K.

4 How many "H^- ions" are now in the star?

5 What will happen to the surface temperature of the star now that it has more ions?

Now that the star is hot, begin removing coins one at a time, starting from the top of the circle and working downward. When you see the line marking 3000 K, stop removing coins.

6 What should happen next?

7 Make a circular flowchart that includes the following steps in the proper order: the star heats up; the star cools down; H^- is formed; H^- is destroyed.

16 Evolution of Low-Mass Stars

Within its core, the Sun fuses more than 4 billion kilograms (kg) of hydrogen each second and will eventually run out of fuel. Although the Sun may seem immortal by human standards, roughly 5 billion years from now the Sun's time on the main sequence will come to an end. In this chapter, we look at how the mass of a star relates to its main-sequence lifetime. Then we examine what happens when a low-mass star like the Sun leaves the main sequence.

LEARNING GOALS

By the end of this chapter, you should be able to:

LG 1 Estimate the main-sequence lifetime of a star from its mass.

LG 2 Explain why low-mass stars grow larger and more luminous as they run out of fuel.

LG 3 Sketch post-main-sequence evolutionary tracks on a Hertzsprung-Russell (H-R) diagram and list the stages of evolution for low-mass stars.

LG 4 Describe how planetary nebulae and white dwarfs form.

LG 5 Explain how some close binary systems evolve differently from single stars.

When low-mass stars die, they leave behind an expanding nebula of gas and dust, as seen in this Hubble Space Telescope image of the Ring Nebula. ▶▶▶

Is this the future of our Sun?

16.1 The Life of a Main-Sequence Star Depends on Its Mass

A star will eventually exhaust the hydrogen fuel in its core, and when it does, its structure will begin to change dramatically and it will leave the main sequence. A star's life on the main sequence depends on the star's mass and composition, and those two qualities determine the changes that occur as the star leaves the main sequence. Just as the balance between pressure and gravity in the protostar evolves before the main sequence, this balance also evolves after the star leaves the main sequence. In this section, we explore the changes in the star that trigger its departure from the main sequence.

Observing Stellar Evolution

Suppose that you were studying the life cycle of humans, and as part of your data collection, you had 1 minute to observe all the people in a crowded stadium. You would observe people of many ages and notice some properties that would indicate that some people are young, some are old, and most are in between. It is possible, but highly unlikely, that you might observe a significant life change such as a birth or a death. More probably, you wouldn't see any individual person change during the minute because a minute is a very small fraction of a typical human lifetime.

Astronomers would have to observe a star like the Sun for several hundred years to observe the equivalent of 1 minute in a human life span. Astronomers observe only a tiny fraction of each star's lifetime, so they do not see individual stars "age." Instead, they observe many, many stars simultaneously to piece together a stellar life cycle. Sometimes, just by chance, astronomers observe a star undergoing a dramatic change. In that way, astronomers build up a picture of how stars change, more commonly known as **stellar evolution**.

While observers are collecting information about stellar properties, theorists use the most powerful computers available to model the nuclear reactions that take place inside stars. Those models predict how a star of a given mass and chemical composition will change over its lifetime. Those predictions are compared with the observations, which leads theorists to adjust and improve their models. The study of stellar evolution involves back-and-forth between observation and theory, which has led to a detailed understanding of how stars live and die. From that work, astronomers find that relatively minor differences in the masses and chemical compositions of two stars can sometimes result in significant, and possibly even dramatic, differences in their fates.

Stars can be divided into four broad categories whose members evolve in distinctly different ways. Stars of spectral types O and B are hot and luminous. Those stars have masses above 8 M_{Sun} and are called **high-mass stars**. Other B stars with masses between 3 and 8 M_{Sun} are **medium-mass stars**. Stars of spectral types A, F, G, and K, which have masses between 0.5 and 3 M_{Sun}, are considered **low-mass stars** and are typified by the Sun. M stars with masses less than 0.5 M_{Sun} are considered very low-mass stars. In this chapter, we are concerned primarily with the more common stars of mass less than 8 M_{Sun}. In the next chapter, we will consider the fate of more massive stars.

Main-Sequence Lifetime

Think about how long you can drive your car before it runs out of gas: it depends on how much gas your tank holds and on the rate at which the engine uses the gas. Both factors are important. Even if a large car or truck has a larger tank, it may run out of fuel faster than a small car if it burns gas more quickly. How long the car runs before running out of gas can be found by dividing the amount of gas in the tank by how quickly the car uses it:

$$\text{Lifetime of tank of gas (hours)} = \frac{\text{Amount of fuel (gallons)}}{\text{Rate at which fuel is used (gallons/hour)}}$$

For example, if your car has a 15-gallon tank and your engine is burning fuel at a rate of 3 gallons each hour, your car will use up all the gas in 5 hours.

The same principle applies to main-sequence stars. How much fuel is available depends on the mass of the star. More massive stars have more hydrogen available for nuclear fusion. The luminosity of the star depends on the rate at which fuel is used: in a main-sequence star, energy radiates into space from the surface as fast as energy is produced in the core. If one main-sequence star has twice the luminosity of another of the same size, it must be fusing hydrogen at twice the rate of the other.

The **main-sequence lifetime** of a star is the amount of time that it spends on the main sequence, fusing hydrogen as its primary source of energy. An expression for the main-sequence lifetime of the star looks very similar to the expression for the time your car takes to run out of fuel:

$$\text{Lifetime of star} = \frac{\text{Amount of fuel } (\propto \text{ mass of star})}{\text{Rate at which fuel is used } (\propto \text{ luminosity of star})}$$

Surprisingly, stars with higher masses live shorter lives, not longer ones. That is because both the amount of fuel (the mass) and the rate at which the fuel is used (the luminosity) depend on the star's mass. As the mass increases, the increase in luminosity is larger than the increase in the amount of fuel available. The graph in **Figure 16.1** shows that relationship. A larger mass contributes to the luminosity of a star in two ways: by increasing the core's pressure and temperature. More mass produces stronger gravity and therefore causes a higher pressure inside the star. That higher pressure pushes the atomic nuclei together into a smaller volume. Under higher pressure, nuclei are more likely to collide, and so the rate of fuel use is higher under higher pressure. Stronger gravity also increases the temperature in the core, and so the atomic nuclei move faster. Those nuclei collide more violently, increasing the chances that they will overcome the electric repulsion that pushes the positively charged nuclei apart. Therefore, increased temperature also increases the rate of fuel use. Because of the combined effects of pressure and temperature, modest increases in mass dramatically increase the amount of energy released by nuclear fusion. High-mass stars use their fuel faster, and so their main-sequence lifetimes are shorter than those of low-mass stars. Low-mass stars use their fuel more slowly than high-mass stars, so their main-sequence lifetimes are longer than those of high-mass stars. That, combined with the fact that fewer high-mass stars form initially from molecular clouds, explains why our galaxy today has far more low-mass stars than high-mass stars. **Table 16.1** shows the main-sequence lifetimes for stars of different spectral types and masses. That concept is developed further in **Working It Out 16.1**.

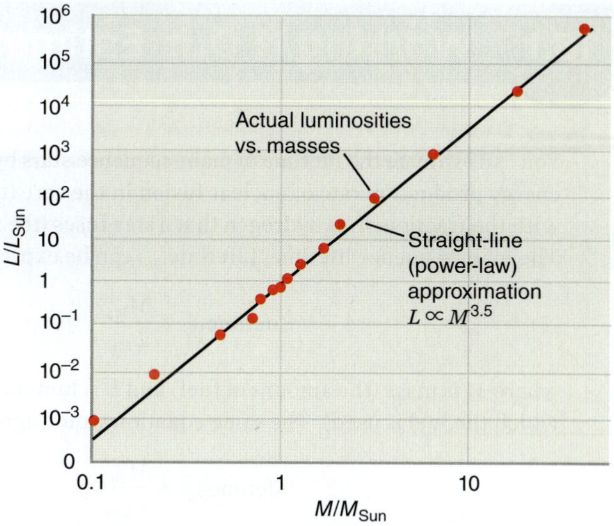

Figure 16.1 This graph plots the mass-luminosity relationship for main-sequence stars: $L \propto M^{3.5}$. The exponent can vary from 2.5 to 5.0, depending on the mass of the star. The average value, over the wide range of main-sequence star masses, is 3.5. Observational data show that how much stars deviate from the average relationship depends on their composition.

TABLE 16.1 Main-Sequence Lifetimes

Spectral Type	Mass (M_{Sun})	Luminosity (L_{Sun})	Main-Sequence Lifetime (years)
O5	60	500,000	3.6×10^5
B0	17.5	32,500	7.8×10^6
B5	5.9	480	1.2×10^8
A0	2.9	39	7×10^8
A5	2.0	12.3	1.6×10^9
F0	1.6	5.2	3.1×10^9
F5	1.4	2.6	4.3×10^9
G0	1.05	1.25	8.9×10^9
G2 (Sun)	1.0	1.0	1.0×10^{10}
G5	0.92	0.8	1.2×10^{10}
K0	0.79	0.55	1.8×10^{10}
K5	0.67	0.32	2.7×10^{10}
M0	0.51	0.08	5.4×10^{10}
M5	0.14	0.008	4.9×10^{11}
M8	~0.08	0.0003	1.1×10^{12}

16.1 Working It Out Estimating Main-Sequence Lifetimes

You can estimate the lifetime of main-sequence stars by combining the energy production rate of nuclear fusion in the core (the luminosity) with the fraction of its hydrogen that a star fuses (the mass available). The main sequence lifetime, Lifetime_{MS}, can be expressed as:

$$\text{Lifetime}_{MS} \propto \frac{M_{MS}}{L_{MS}}$$

where M is mass (the amount of fuel) and L is luminosity (the rate at which the fuel is used). The same equation would apply for the Sun:

$$\text{Lifetime}_{Sun} \propto \frac{M_{Sun}}{L_{Sun}}$$

We can express the lifetime as a ratio, adding in that the computed lifetime of a 1-M_{Sun} star like the Sun is 10 billion (1.0×10^{10}) years:

$$\frac{\text{Lifetime}_{MS}}{\text{Lifetime}_{Sun}} = \frac{\text{Lifetime}_{MS}}{10^{10}\,\text{yr}} = \frac{M_{MS}/L_{MS}}{M_{Sun}/L_{Sun}}$$

Multiplying through by 10^{10} years and rearranging the fractions yields

$$\text{Lifetime}_{MS} = 10^{10}\,\text{yr} \times \frac{M_{MS}/L_{MS}}{M_{Sun}/L_{Sun}} = 10^{10} \times \frac{M_{MS}/M_{Sun}}{L_{MS}/L_{Sun}}\,\text{yr}$$

Now let's compare the lifetime of a star with that of the Sun. The relationship between the mass and the luminosity of stars is such that relatively small differences in the masses of stars result in large differences in their main-sequence luminosities. Figure 16.1 shows the **mass-luminosity relationship**, $L \propto M^{3.5}$, for main-sequence stars. If we express that relationship in units of the Sun's mass and luminosity, we find:

$$\frac{L_{MS}}{L_{Sun}} = \left(\frac{M_{MS}}{M_{Sun}}\right)^{3.5}$$

Substituting that relationship into the lifetime equation gives

$$\text{Lifetime}_{MS} = 10^{10} \times \frac{M_{MS}/M_{Sun}}{(M_{MS}/M_{Sun})^{3.5}}\,\text{yr} = 10^{10} \times \left(\frac{M_{MS}}{M_{Sun}}\right)^{-2.5}\,\text{yr}$$

For example, let's look at a K5 main-sequence star. According to Table 16.1, a K5 star has a mass of about 0.67 M_{Sun}:

$$\text{Lifetime}_{K5} = 10^{10} \times (0.67)^{-2.5}\,\text{yr} = 2.7 \times 10^{10}\,\text{yr}$$

Instead of the 10-billion-year life span of the Sun, a K5 star has a main-sequence lifetime 2.7 times longer than the Sun's. Even though the K5 star starts out with less fuel than the Sun, it fuses that fuel so much more slowly that it lives longer.

Changes in Structure

As discussed in Chapter 14, at the end of the proton-proton chain in main-sequence stars, two ^{3}He nuclei fuse to form ^{4}He and two ^{1}H nuclei. However, at the temperatures in the centers of main-sequence stars, collisions are not energetic or frequent enough for the more massive ^{4}He nuclei to fuse.

Hydrogen fuses most rapidly at the center of a main-sequence star because the temperature and pressure are highest there. Thus, helium, the product of hydrogen fusion, accumulates most rapidly at the center of the star. **Figure 16.2** shows how the chemical composition inside a star like the Sun changes throughout its main-sequence lifetime. When the Sun formed, it had a uniform composition of about 70 percent hydrogen and 30 percent helium by mass (Figure 16.2a). Since then, the Sun converted hydrogen into helium via the proton-proton chain. As hydrogen fused into helium, the helium accumulated in the core of the Sun. Today, roughly 5 billion years later, about half the hydrogen in the core has been converted to helium; only about 35 percent of the mass in the core of the Sun is hydrogen (Figure 16.2b). Five billion years from now, as the Sun begins to leave the main sequence, the very center of the core will have no hydrogen left to fuse (Figure 16.2c).

While converting the fuel in its core from hydrogen to helium, a main-sequence star must maintain energy balance. The structure of the star must continually shift in response to the changing core composition. Between the time the Sun was born and the time it will leave the main sequence, its luminosity will roughly double, with most of that change occurring during the last billion years of its main-sequence lifetime. That main-sequence evolution is slow and modest in comparison with the events that follow.

The Lowest-Mass Stars

Brown dwarfs have masses less than 0.08 M_{Sun}. They do not become main-sequence stars, so they do not follow post-main-sequence steps. Small red dwarf stars with masses lower than about 0.4–0.5 M_{Sun} may account for most stars in our galaxy, but they are small and faint and hard to detect. Those stars have main-sequence lifetimes longer than the 13.8-billion-year age of the universe, so all those stars that have formed are still on the main sequence. Not enough time has passed for those stars to finish their lifetime on the main sequence.

CHECK YOUR UNDERSTANDING 16.1

Why does the main-sequence lifetime of a star depend on its mass? (a) Because more massive stars fuse fuel faster and therefore have shorter lives. (b) Because more massive stars have more fuel and therefore have longer lives. (c) Because more massive stars fuse different fuels and therefore have longer lives. (d) Because more massive stars have different initial compositions and therefore have shorter lives.

16.2 The Star Leaves the Main Sequence

Once the star has used up the hydrogen in the center of the core, thermal energy leaks out of the helium core into the surrounding layers of the star, but no energy is generated within the helium core to replace it. The balance that has maintained the structure of the star throughout its life is now broken. The star's life on the main sequence has come to an end, and its further evolution depends on core temperature changes, which govern fusion reactions.

Electron-Degenerate Matter in the Helium Core

Inside a star, the temperature is so high that atoms in the gas are ionized—almost all the electrons have been stripped away from their nuclei by energetic collisions. Those particles are very small, in comparison with the distances between them, so most of the space inside the star is empty, and the electrons and the atomic nuclei are all flying about freely, rarely interacting with one another. However, once a low-mass star like the Sun exhausts the hydrogen at its center, the star is no longer generating energy to keep it from collapsing. Gravity begins to "win," and the helium core collapses and becomes denser. How dense the core can get is limited because no two electrons can occupy the same state in the same place at the same time. As the core of the star collapses, it reaches that limit, in which the electrons are smashed as tightly together as possible. Matter in which electrons are packed as closely as possible is called **electron-degenerate** matter. Because the electrons resist being packed more closely together, they produce a type of pressure, known as **degeneracy pressure**, which pushes outward against gravity. Degeneracy pressure keeps the core from collapsing further. The presence of the electron-degenerate core triggers a chain of events that will dominate the evolution of a 1-M_{Sun} star for the next 50 million years, as it leaves the main sequence after the hydrogen runs out.

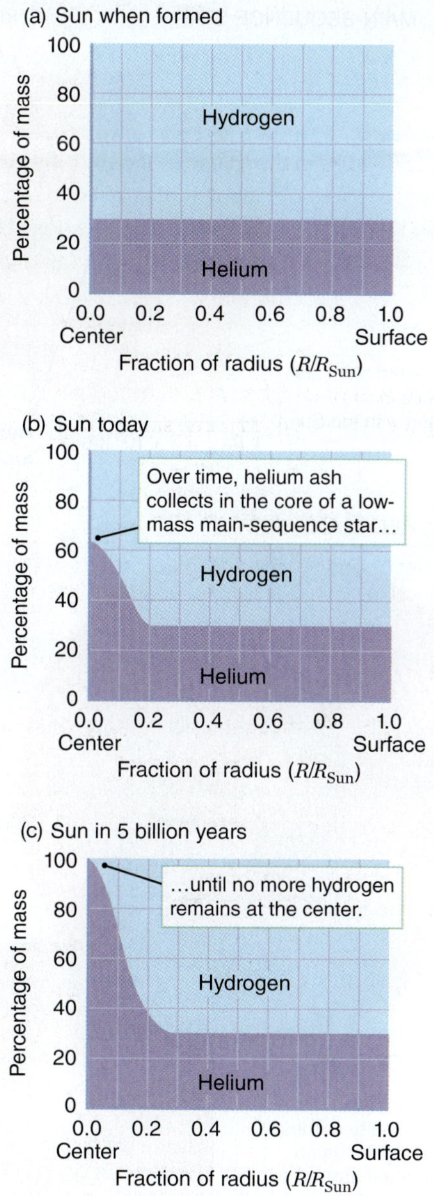

Figure 16.2 Chemical composition of the Sun is plotted here as a percentage of mass against distance from the center of the Sun. (a) When the Sun formed 5 billion years ago, the composition was uniform throughout: about 30 percent of the Sun's mass was helium and 70 percent was hydrogen. (b) Today, the material at the center of the Sun is about 65 percent helium and 35 percent hydrogen. (c) The Sun's main-sequence life will end in about 5 billion years, when all the hydrogen at its center will be exhausted.

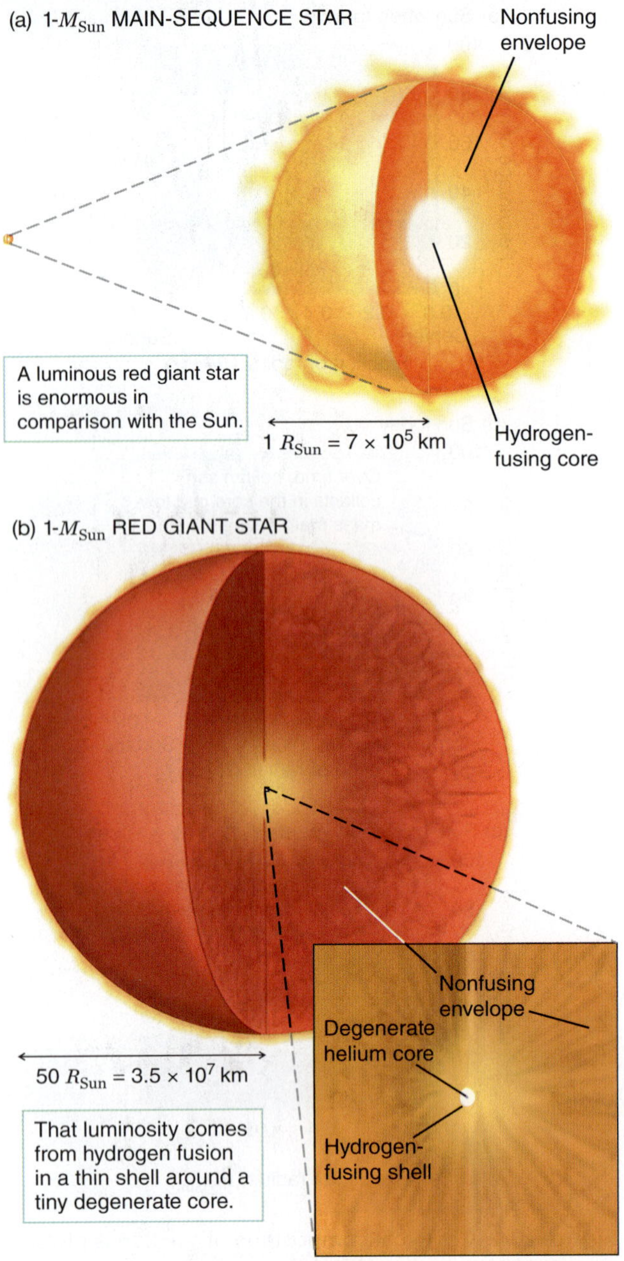

Figure 16.3 The size and structure of the Sun (a) is compared with the size of a star near the top of the red giant branch of the Hertzsprung-Russell (H-R) diagram (b). The left-hand diagrams in (a) and (b) show the relative sizes of the two stars. The magnified right-hand diagrams compare the size and structure of the Sun (a) with the size and structure of the core of the red giant (b). The diagrammatic images on the right are magnified about 50 times compared with those on the left.

Electron-degenerate matter is not like everyday, ordinary matter. Electron-degenerate matter is more than a hundred thousand times denser than lead—a block of electron-degenerate matter the size of a sugar cube has a mass of at least 1,000 kg. Oddly, a more massive electron-degenerate core *is smaller* than a less massive one. That trend is noticeably different from that of ordinary matter, such as cows: more massive cows are bigger, not smaller. More massive stellar cores are smaller because gravity is stronger, so the electrons are smashed together into a smaller volume before they stop the collapse.

Hydrogen Shell Fusion

After a low-mass star has exhausted the hydrogen at its center, nuclear fusion in the core pauses. The layers surrounding the degenerate core still contain hydrogen, however, which continues to fuse. That process is called **hydrogen shell fusion** because the hydrogen is fusing in a shell surrounding a core of inert helium. That layered structure is like that of a plum, with an internal seed (inert helium), a thin seed coat (hydrogen-fusing shell), and a large sphere of flesh (inert hydrogen).

A star with a hydrogen-fusing shell around a degenerate helium core is larger and more luminous than a star on the main sequence. A degenerate core is very compact, so it has very strong gravity and produces a higher pressure in the shell around it. Because of that higher pressure, the nuclear reaction rate in the shell is faster. Faster nuclear reactions in the shell release more energy, so the luminosity of the star increases while core fusing pauses. Over time, the mass of the degenerate helium core grows as more and more hydrogen is converted into helium in the surrounding shell. As the mass of the degenerate helium core grows, so too does its gravitational pull, further increasing the rate of energy generation in the surrounding hydrogen-fusing shell. The liberated energy drives the overlying layers of the star outward, so the star expands and grows larger.

The relation among radius, temperature, and luminosity that we discussed in Working It Out 13.2 and Chapter 15 still applies ($L = 4\pi\sigma R^2T^4$). Here, the radius and the luminosity both increase, but the surface temperature falls. The gas of the outer layers of the star expands, so it cools and therefore becomes redder. The star is red and very large, so it is called a **red giant**. A red giant fuses hydrogen in a shell around a degenerate helium core and is larger, more luminous, and redder than it was on the main sequence.

As illustrated in **Figure 16.3**, the internal structure of the main-sequence star (Figure 16.3a) changes as the star evolves to a red giant (Figure 16.3b). The red giant can grow to have a luminosity hundreds of times that of the Sun and a radius of more than 50 solar radii (50 R_{Sun}). Yet at the same time, the core of the red giant is compact: much of the star's mass becomes concentrated into a volume only a few times the size of Earth.

The Red Giant Branch

In Chapter 15, we used the Hertzsprung-Russell (H-R) diagram to show the changes in a protostar on its way to the main sequence; the diagram also is a handy device for tracking the changes in the star as it evolves away from the main sequence. As soon as the star exhausts the hydrogen in its core, it leaves the main sequence and becomes a **subgiant**: more luminous, cooler, and larger than it was on the main sequence. As it grows more luminous and cooler, its position on the H-R diagram moves upward and to the right. When the surface temperature of the subgiant star has dropped about 1000 kelvins (K) below its temperature on the main sequence, many H^- ions start to form in its atmosphere. Recall from Chapter 15 that H^- ions absorb and scatter outgoing radiation, trapping that radiation in the protostar and preventing it from cooling down. The H^- ions serve the same role in subgiants, regulating how much radiation can escape from the subgiant and preventing the temperature from dropping further.

Once that H^- "thermostat" begins working, the subgiant can cool no further. It becomes a red giant, moving almost vertically upward on the H-R diagram. The red giant grows larger and more luminous but remains about the same temperature. A red giant is both redder and larger than the star was on the main sequence. The path that a star follows on the H-R diagram as it leaves the main sequence is like a tree "branch" growing out of the "trunk" of the main sequence, as shown in **Figure 16.4**. The lower part of that track (which moves somewhat horizontally) is the **subgiant branch**. The vertical part is the **red giant branch**.

The path that a red giant follows on the H-R diagram closely parallels the path that it followed earlier as a collapsing protostar on its way toward the main sequence, except in reverse: this time, the star is moving up that path instead of coming down it. That similarity is not a coincidence. The same physical processes (such as the H^- thermostat) that give rise to the vertical Hayashi track followed by a collapsing protostar also control the relationship of luminosity, size, and surface temperature in an expanding red giant.

As the star leaves the main sequence, the changes in its structure occur slowly at first, but then the star moves up the red giant branch faster and faster. A star like the Sun takes several hundred million years to go from the main sequence to the top of the red giant branch. Roughly the first half of that time is spent on the subgiant branch as the star's luminosity increases to about 10 times the

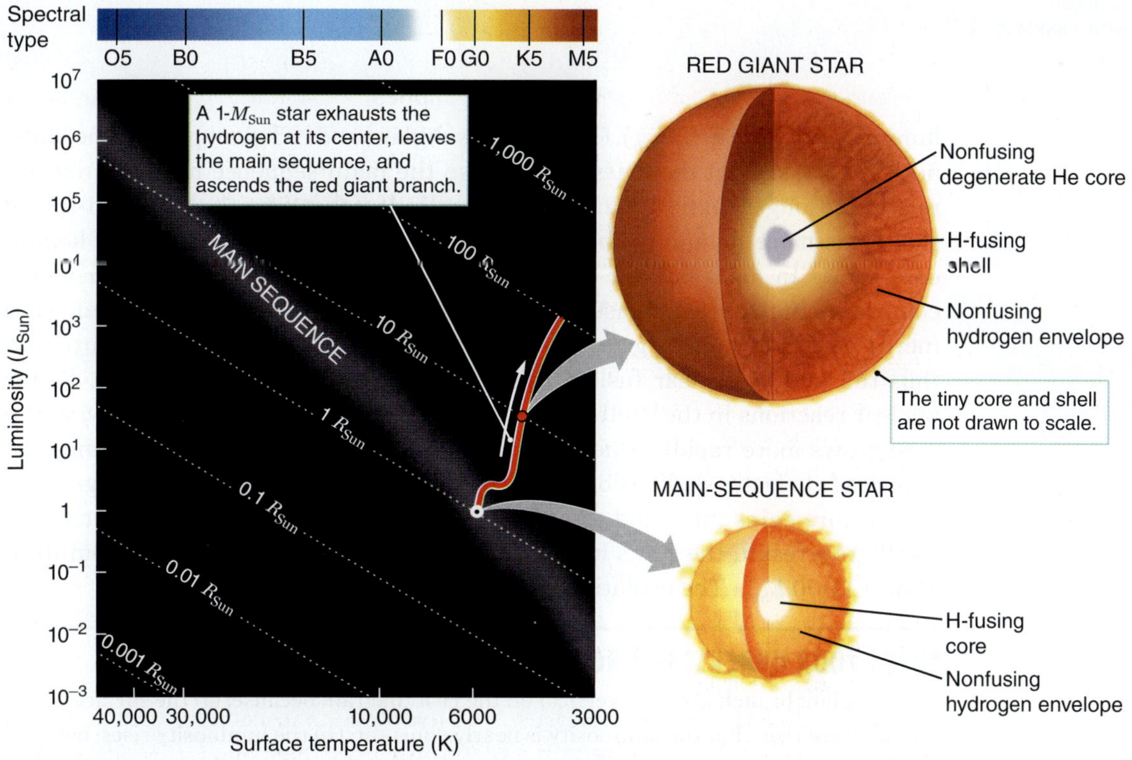

Figure 16.4 A red giant star consists of a degenerate core of helium surrounded by a hydrogen-fusing shell. Moving up the red giant branch, the star comes close to retracing the Hayashi track that it followed when it was a protostar collapsing toward the main sequence.

Figure 16.5 As a star moves up the red giant branch in the H-R diagram, the luminosity of the star grows faster and faster. The fusion of hydrogen to helium in a shell surrounding a degenerate helium core feeds on itself, creating a cycle that speeds up as time goes on.

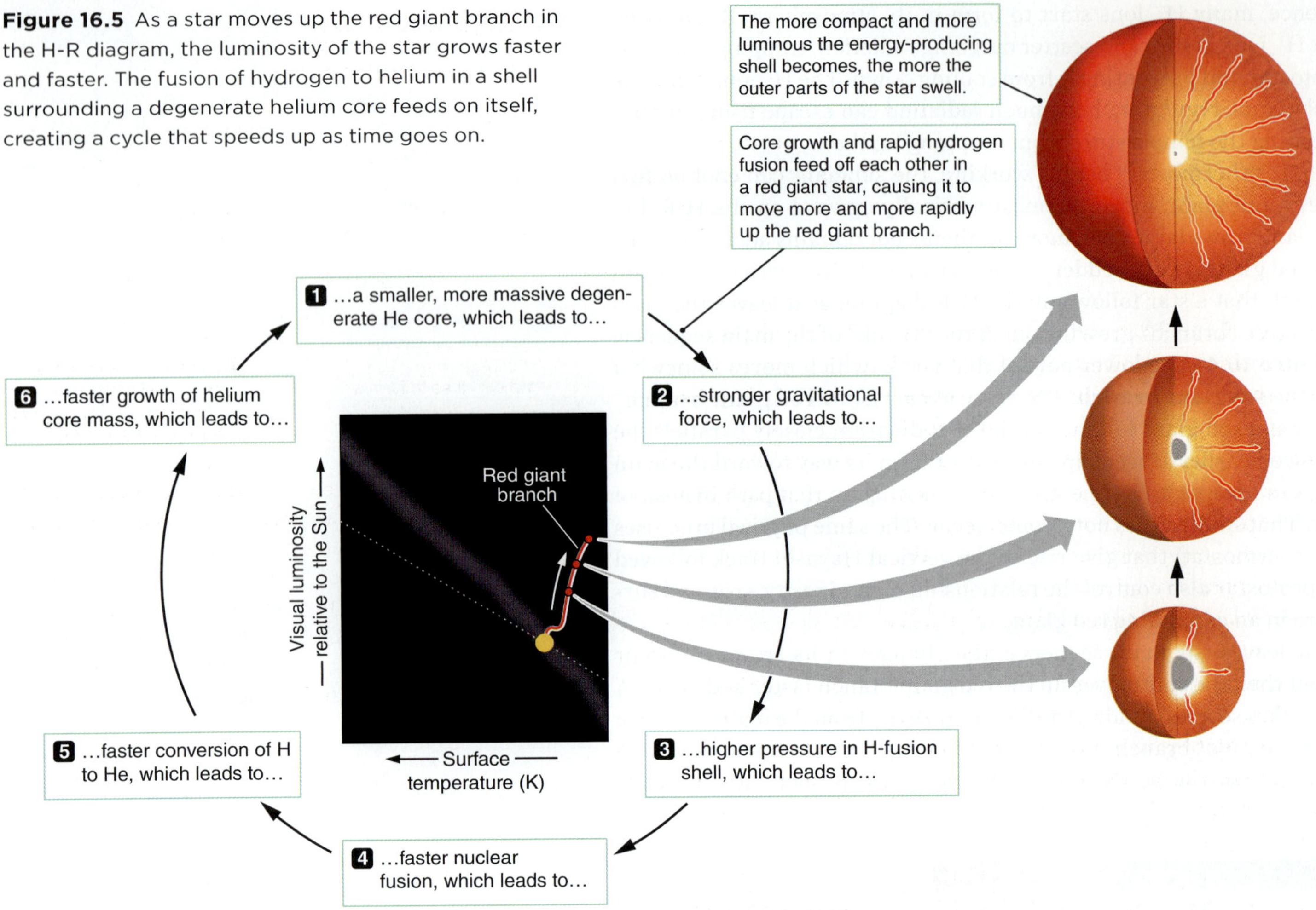

luminosity of the Sun (L_{Sun}). (During the subgiant and red giant phases, the luminosity increases but the mass does not, so the main-sequence mass–luminosity relation no longer applies.) During the second half of that time, the hydrogen-fusing shell adds more helium to the degenerate helium core of the star. That helium core grows in mass but not in radius. The increasing mass of the ever-more-compact helium core increases the force of gravity in the heart of the star. That increased gravity once again increases the pressure and the temperature and thus the rate of nuclear fusion, this time in the hydrogen-fusing shell. Faster nuclear reactions in the shell convert hydrogen into helium more quickly, so the core grows more rapidly. The star has entered a feedback loop; increasing core mass leads to faster shell fusion, and faster shell fusion leads to faster core growth. As the core gains mass and the shell becomes more luminous, the outer layers swell. As a result, the star's luminosity climbs at an ever-higher rate from 10 to almost 1,000 L_{Sun}. The evolution of the star is illustrated in **Figure 16.5**.

CHECK YOUR UNDERSTANDING 16.2

The red giant branch is nearly vertical on the H-R diagram because: (a) the surface temperature rises, but the luminosity is nearly constant; (b) the luminosity rises, but the temperature is nearly constant; (c) the luminosity and the temperature rise significantly; (d) the surface temperature and the luminosity both remain nearly constant.

16.3 Helium Fuses in the Degenerate Core

On the main sequence, a star fuses hydrogen in the core. On the red giant branch, the core consists of nonfusing helium surrounded by a shell of fusing hydrogen. Once the hydrogen in the shell is used up, the core of the red giant contracts and heats. Eventually, the temperature and pressure rise enough to start the next stage of stellar evolution: helium fusion in the core.

Helium Fusion and the Triple-Alpha Process

As the red giant evolves, its helium core grows not only smaller and more massive but also hotter. That increase in temperature is due partly to the gravitational energy released as the core shrinks and partly to the energy released by hydrogen fusion in the surrounding shell. The thermal motions of the atomic nuclei in the core become more and more energetic. Eventually, at a temperature of about 100 million (10^8) K, the ^{4}He nuclei are slammed together hard enough for the strong nuclear force to act, and helium fusion begins.

Helium fuses in a two-stage process called the **triple-alpha process**, so named because it involves the fusion of three ^{4}He nuclei, which are sometimes called **alpha particles**. The process, illustrated in **Figure 16.6**, begins when two helium-4 (^{4}He) nuclei fuse to form a beryllium-8 (^{8}Be) nucleus consisting of four protons and four neutrons. The ^{8}Be nucleus is extremely unstable and quickly decays. But if, in that short time, it collides with another ^{4}He nucleus, the two nuclei will fuse into a stable nucleus of carbon-12 (^{12}C) consisting of six protons and six neutrons. The reaction rate depends on the temperature: higher temperatures increase the number of ^{8}Be nuclei that collide with a ^{4}He nucleus and therefore increase the reaction rate.

Recall that the core of the red giant is electron-degenerate, which means that as many electrons are packed into that space as possible. Those degenerate electrons prevent the core from collapsing. The atomic nuclei, however, behave like a normal gas, moving through the sea of degenerate electrons almost as though the electrons were not there. The atomic nuclei in the core move freely about, as shown in **Figure 16.7**, just as they do throughout the rest of the star. We

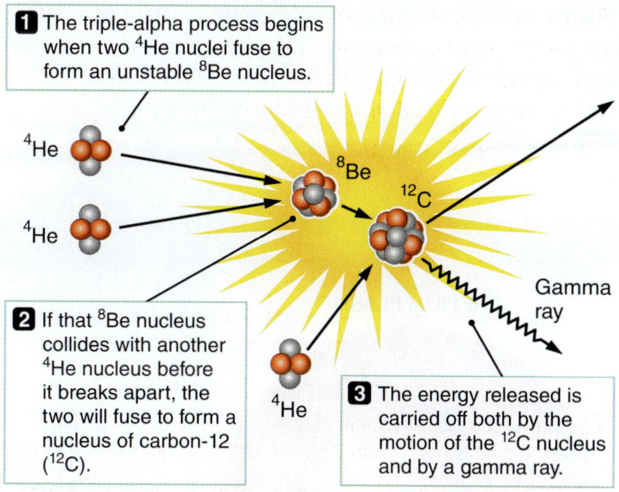

Figure 16.6 The triple-alpha process produces a stable nucleus of carbon-12. Two helium-4 (^{4}He) nuclei fuse to form an unstable beryllium-8 (^{8}Be) nucleus. If that nucleus collides with another ^{4}He nucleus before breaking apart, the two will fuse to form a stable nucleus of carbon-12 (^{12}C). The energy produced is carried off both by the motion of the ^{12}C nucleus and by a high-energy gamma ray emitted in the second step of the process.

Figure 16.7 In a red giant star, the weight of the overlying layers is supported by electron degeneracy pressure in the core arising from electrons that are packed together as tightly as quantum mechanics allows. Atomic nuclei in the core can move freely about within the sea of degenerate electrons, so they behave as a normal gas.

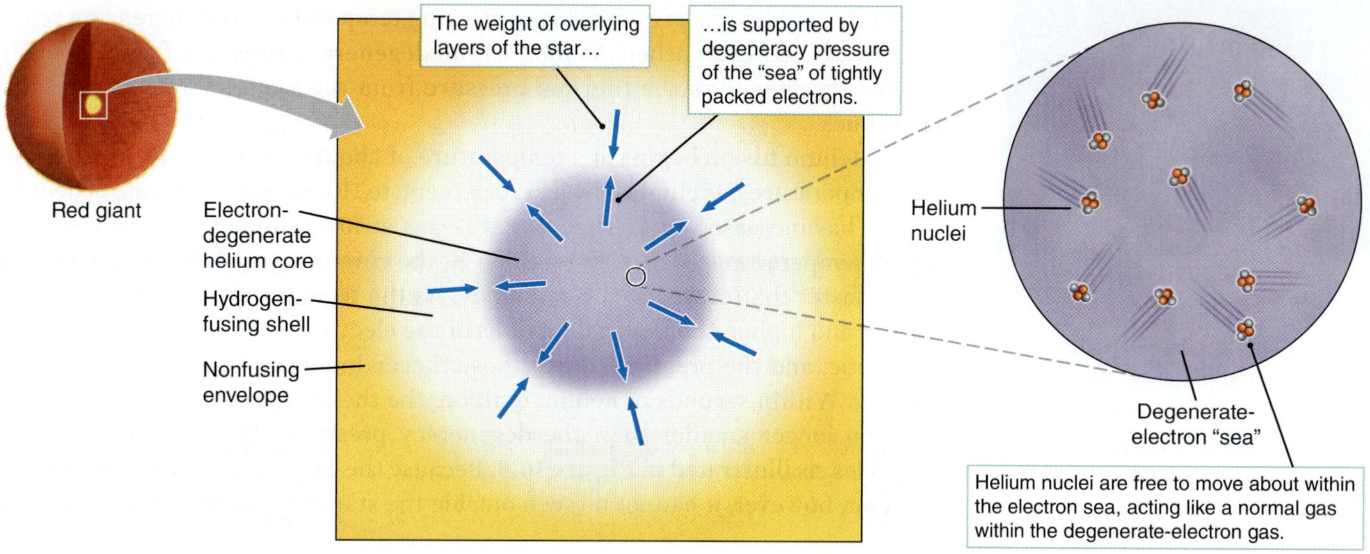

can understand the fusion of helium by treating the nuclei as matter in a normal state. Once the pressure and temperature are high enough, those nuclei begin to fuse, as the hydrogen nuclei do in main-sequence stars. However, the energy released will not affect the degenerate-electron core in the same way that it affects the core in a main-sequence star.

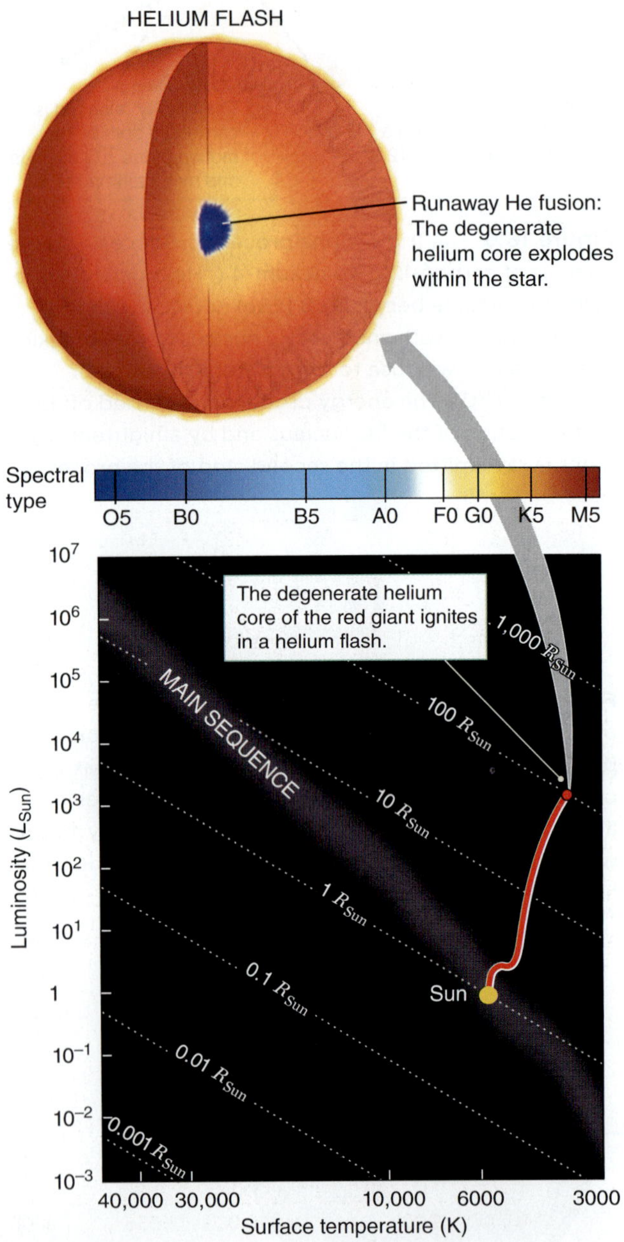

Figure 16.8 At the end of its life, a low-mass star travels a complex path on the H-R diagram. The first part of that path takes it up the red giant branch to a point where helium ignites in a helium flash. After a few hours, the core of the star begins to inflate, ending the helium flash.

The Helium Flash

Degenerate material conducts thermal energy very well, so any differences in temperature within the core rapidly disperse. As a result, when helium fusion begins at the center of the core, the energy released quickly heats the entire core. Within a few minutes, the entire core is fusing helium into carbon by the triple-alpha process. Some of the carbon will fuse with an additional ^{4}He nucleus to form stable oxygen-16 (^{16}O).

In a normal gas, such as the air around you or the core of a main-sequence star, the pressure of the gas comes from the random thermal motions of the atoms, which are faster at higher temperatures. Increasing the temperature increases the pressure of the gas. In the core of a main-sequence star, heating it causes it to expand; the temperature, density, and pressure then decrease; nuclear reactions slow; and the star settles into a new balance between gravity and pressure. Those are exactly the sorts of changes steadily occurring within the core of a main-sequence star like the Sun as the structure of the star shifts in response to the changing composition in the star's core.

However, the degenerate core of a red giant is not a normal gas. The pressure in a red giant's degenerate core comes primarily from how tightly the electrons in the core are packed together. Heating the core does not change the number of electrons that can be packed into its volume, so the core's pressure does not respond to changes in temperature. Because the pressure does not increase, the core does not expand when heated, as a normal gas would.

Although the higher temperature does not change the pressure, it does cause the helium nuclei to collide more often and with greater force, so the nuclear reactions become more vigorous. More vigorous reactions increase the temperature, and higher temperature means even more vigorous reactions. That process creates another feedback loop. Helium fusion in the degenerate core runs wildly out of control as increasing temperature and increasing reaction rates feed each other. As long as the degeneracy pressure from the electrons is greater than the thermal pressure from the nuclei, the feedback loop continues.

Helium fusion begins at a temperature of about 100 million K. By the time the temperature has climbed by just 10 percent, to 110 million K, the rate of helium fusion has increased to 40 times what it was at 100 million K. By the time the core's temperature reaches 200 million K, the core is fusing helium 460 million times faster than it was at 100 million K. As the temperature in the core grows higher and higher, the thermal motions of the electrons and nuclei become more energetic, and the pressure due to those thermal motions becomes greater and greater. Within seconds of helium ignition, the thermal pressure increases until it is no longer smaller than the degeneracy pressure. Then the helium core explodes, as illustrated in **Figure 16.8**. Because the explosion is contained within the star, however, it cannot be seen outside the star. The energy released in that runaway thermonuclear explosion lifts the intermediate layers of the star, and as the core expands, the electrons can spread out. The drama is over within a few hours because the expanded helium-fusing core is no longer degenerate, and the star is on its way toward a new equilibrium. The explosion of helium fusion in the core at the top of the red giant branch is called a **helium flash**; however, it is not visible as a "flash" from outside the star.

Energy from helium fusion in the core does not emerge from the red giant as light, so the luminosity does not change. The tremendous energy released instead goes into fighting gravity and puffing up the core. Afterward, the core (which is no longer degenerate) is much larger, so the force of gravity within it and in the surrounding shell is much smaller. Weaker gravity means less weight pushing down on the core and the shell, which means lower pressure. Lower pressure, in turn, slows the nuclear reactions. After the helium flash, core helium fusion keeps the core of the star puffed up, and the star becomes less luminous than it was as a red giant.

The Horizontal Branch

The star takes 100,000 years or so to settle into stable helium fusion. It then spends about 100 million years fusing helium into carbon in a normal, nondegenerate core while hydrogen fuses to helium in a surrounding shell. The star is now about a hundred times less luminous than it was when the helium flash occurred. The lower luminosity means that the outer layers of the star are not as puffed up as they were when the star was a red giant. Gravity becomes stronger than the outward pressure of the escaping radiation and pulls the outer layers back in. The star shrinks, and its surface temperature climbs as gravitational energy is converted to thermal energy. The star moves horizontally across the H-R diagram, remaining at the same luminosity but increasing in surface temperature. That portion of a star's path on the H-R diagram is called the horizontal branch, shown in **Figure 16.9**. At that point in their evolution, low-mass stars with chemical compositions similar to that of the Sun lie on the H-R diagram just to the left of the red giant branch. Stars that contain much less iron than the Sun tend to distribute themselves away from the red giant branch along the horizontal branch.

The structure and behavior of a star on the horizontal branch are similar to those of a main-sequence star. However, instead of fusing hydrogen into helium in the core, the horizontal branch star fuses helium into carbon in the core, with a shell of hydrogen fusion around that core.

The star's time on the horizontal branch is much shorter than its time on the main sequence. The star is more luminous than it was on the main sequence, so it is consuming fuel more rapidly. As it fuses, helium releases less energy than hydrogen, so the star has to fuse even faster to maintain equilibrium. The horizontal branch star remains stable for about 100 million years.

The temperature at the center of a horizontal branch star is not high enough for carbon to fuse, so carbon builds up in the heart of the star. When the horizontal branch star has fused all the helium at its core, gravity once again begins to overwhelm the pressure of the escaping radiation. The nonfusing carbon core is crushed by the weight of the layers of the star above it until once again the electrons in the core are packed together as tightly as possible, given its pressure. The

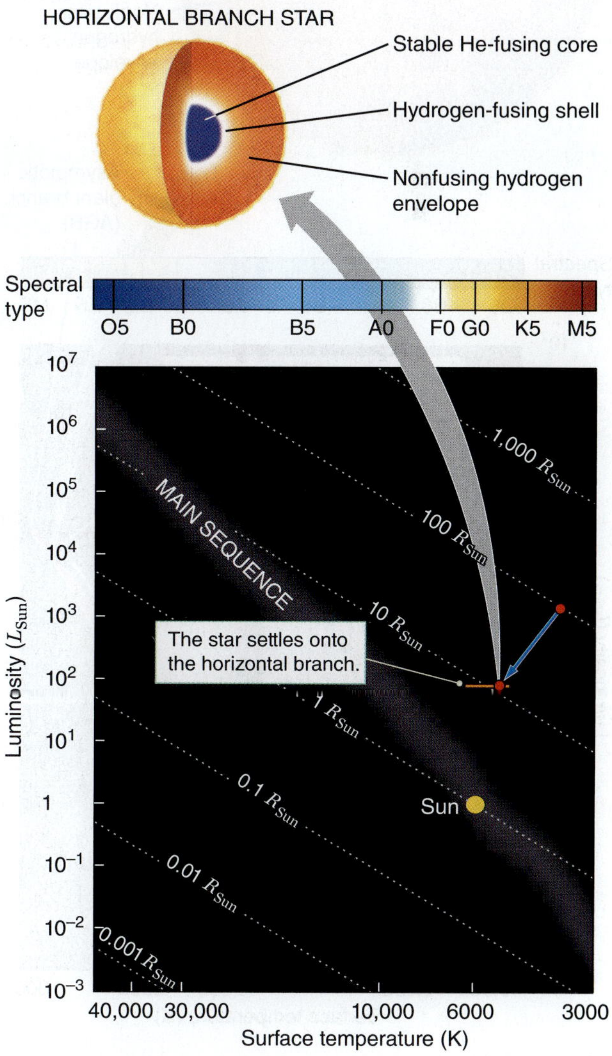

Figure 16.9 The star moves down from the red giant branch onto the horizontal branch. For about 100 million years or less, the star will remain on the horizontal branch and fuse helium in its core and hydrogen in a surrounding shell.

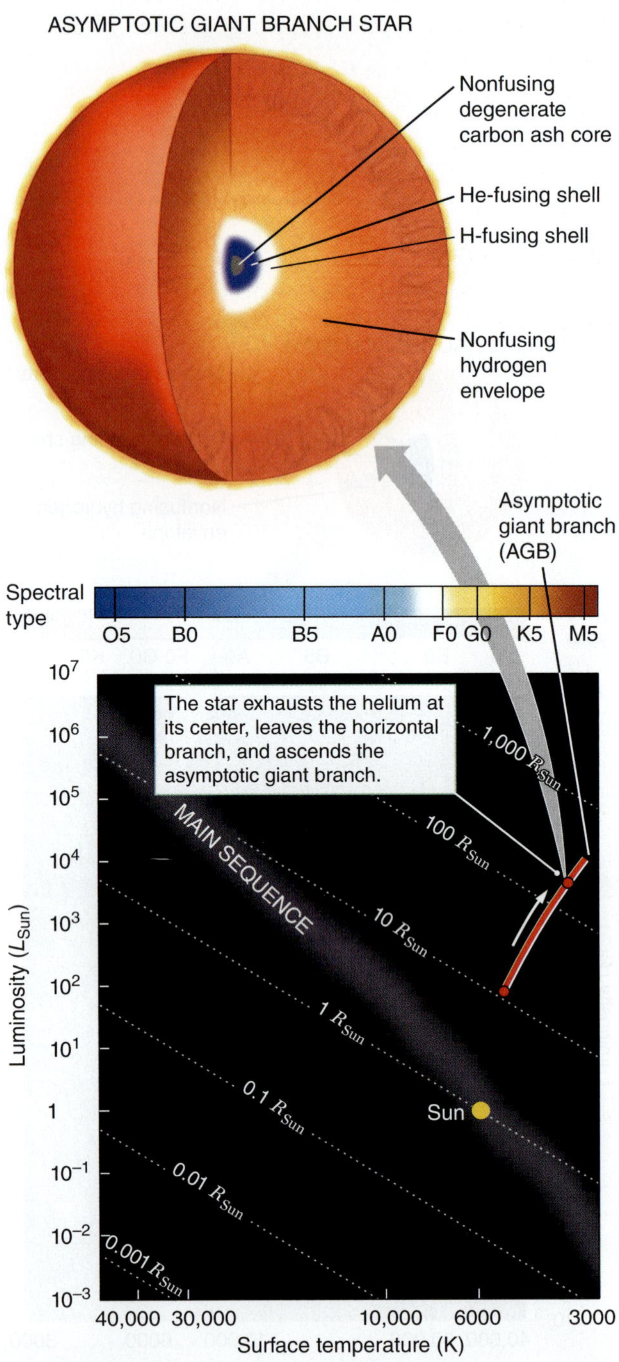

Figure 16.10 The star moves up from the horizontal branch onto the asymptotic giant branch (AGB). An AGB star consists of a degenerate carbon core surrounded by helium-fusing and hydrogen-fusing shells. As the carbon core grows, the star brightens, accelerating up the AGB just as it earlier accelerated up the red giant branch while its degenerate helium core grew.

carbon core becomes electron-degenerate, with physical properties much like those of the degenerate helium core at the center of a red giant.

CHECK YOUR UNDERSTANDING 16.3

Stars begin fusing helium to carbon when the temperature rises in the core. That temperature increase is caused by (choose all that apply): (a) gravitational collapse; (b) fusion of hydrogen into helium in the core; (c) fusion of hydrogen into helium in a shell around the core; (d) electron degeneracy pressure.

16.4 Dying Stars Shed Their Outer Layers

After the horizontal branch, small changes in the properties of a star—mass, chemical composition, strength of the magnetic field, and even the rate of rotation—can lead to noticeable differences in how the star (especially its outer envelope) evolves. In this section, we follow a 1-M_{Sun} star with solar composition as it concludes its evolutionary stages, losing its outer layers and leaving behind a cooling carbon core.

Stellar-Mass Loss and the Asymptotic Giant Branch

After the horizontal branch, a small, dense, electron-degenerate carbon core remains. That core is very compact, causing the gravity in the inner parts of the star to be very high, which in turn drives up the pressure, which speeds up the nuclear reactions, which causes the degenerate core to grow more rapidly. Those internal changes should sound familiar—they are similar to the changes that took place at the end of the star's main-sequence lifetime, and the path the star follows as it leaves the horizontal branch parallels that earlier phase of evolution. Just as the star accelerated up the red giant branch as its degenerate helium core grew, the star now leaves the horizontal branch and once again begins to grow larger, redder, and more luminous as its degenerate carbon core grows. As shown in **Figure 16.10**, the path that the star follows, called the **asymptotic giant branch** (**AGB**) of the H-R diagram, approaches the red giant branch as the star grows more luminous. An AGB star fuses helium and hydrogen in nested concentric shells around a degenerate carbon core as the star moves once again up the H-R diagram.

AGB stars are huge objects. When the Sun becomes an AGB star, its outer layers will engulf the orbits of the inner planets, possibly including Earth and maybe even Mars. When a star expands to such a size, the gravitational force at its surface is only 1/10,000 as strong as the gravity at the surface of the present-day Sun. Pushing surface material away from the star thus takes little extra energy. Before the temperature in the carbon core becomes high enough for carbon to fuse, the radiation pressure pushing outward on the outer layers of the star becomes greater than the gravitational pull inward. Those outer layers begin to drift away into space in a process called **stellar-mass loss.** That mass loss actually begins when the star is still on the red giant branch: by the time a 1-M_{Sun} main-sequence star reaches the horizontal branch, it may have lost 10–20 percent of its total mass. As the star ascends the asymptotic giant branch, it loses another 20 percent or even more of its total mass. By the time it is well up on that branch, a star that began as a 1-M_{Sun} star may have lost more

16.2 Working It Out Escaping the Surface of an Evolved Star

Why are giant stars likely to lose mass? The escape velocity from the surface of a planet or star was given in Working It Out 4.2:

$$v_{esc} = \sqrt{\frac{2GM}{R}}$$

How does v_{esc} change when a star becomes a red giant? Let's look at the Sun as an example. When the Sun is on the main sequence, the escape velocity from its surface can be calculated by using $M_{Sun} = 1.99 \times 10^{30}$ kg, $R_{Sun} = 6.96 \times 10^5$ km, and $G = 6.67 \times 10^{-20}$ km^3/(kg s^2):

$$v_{esc} = \sqrt{\frac{2 \times [6.67 \times 10^{-20}\ \text{km}^3/(\text{kg s}^2)] \times (1.99 \times 10^{30}\ \text{kg})}{6.96 \times 10^5\ \text{km}}}$$

$$v_{esc} = \sqrt{3.81 \times 10^5\ \text{km}^2/\text{s}^2} = 618\ \text{km/s}$$

What will the escape velocity be when the Sun becomes a red giant, with a radius 50 times greater than the radius it has today and a mass 0.9 times its present mass?

$$v_{esc} = \sqrt{\frac{2 \times [6.67 \times 10^{-20}\ \text{km}^3/(\text{kg s}^2)] \times 0.9 \times (1.99 \times 10^{30}\ \text{kg})}{50 \times (6.96 \times 10^5\ \text{km})}}$$

$$v_{esc} = \sqrt{6.86 \times 10^3\ \text{km}^2/\text{s}^2} = 83\ \text{km/s}$$

The escape velocity from the surface of a red giant star is only 13 percent that of a main-sequence star. That is part of the reason that red giant and AGB stars lose mass. The Sun may eventually lose half its mass.

than half its original mass. Stellar-mass loss is further explored in **Working It Out 16.2**.

Mass loss on the AGB can be spurred on by the star's unstable interior. The extreme sensitivity of the triple-alpha process to temperature in the core can lead to episodes of fast fusion and rapid energy release, which can provide the extra kick needed to expel material from the star's outer layers. Even stars that are initially similar can behave very differently when they reach that stage in their evolution.

Planetary Nebula

Toward the end of an AGB star's life, mass loss itself becomes a runaway process. When a star loses a bit of mass from its outermost layers, the weight pushing down on the underlying layers of the star is reduced. Without that weight holding them down, the remaining outer layers of the star puff up further. The post-AGB star, which is now both less massive and larger, is even less tightly bound by gravity, so less energy is needed to push its outer layers away. Mass loss leads to weaker gravity, which leads to faster mass loss, which leads to weaker gravity, and so on. By the time the last layers are lost, much of the remaining mass of the star is ejected into space, typically at speeds of 20–30 kilometers per second (km/s).

After ejection of its outer layers, all that is left of the low-mass star is a tiny, very hot, electron-degenerate carbon core surrounded by a thin envelope in which hydrogen and helium are still fusing. The star is now less luminous than when it was at the top of the AGB, but it is still much more luminous than a horizontal branch star. The remaining hydrogen and helium in the star rapidly fuse to carbon, and as more and more of the mass of the star ends up in the carbon core, the star itself shrinks and becomes hotter and hotter. Over the course of only about 30,000 years after the beginning of runaway mass loss,

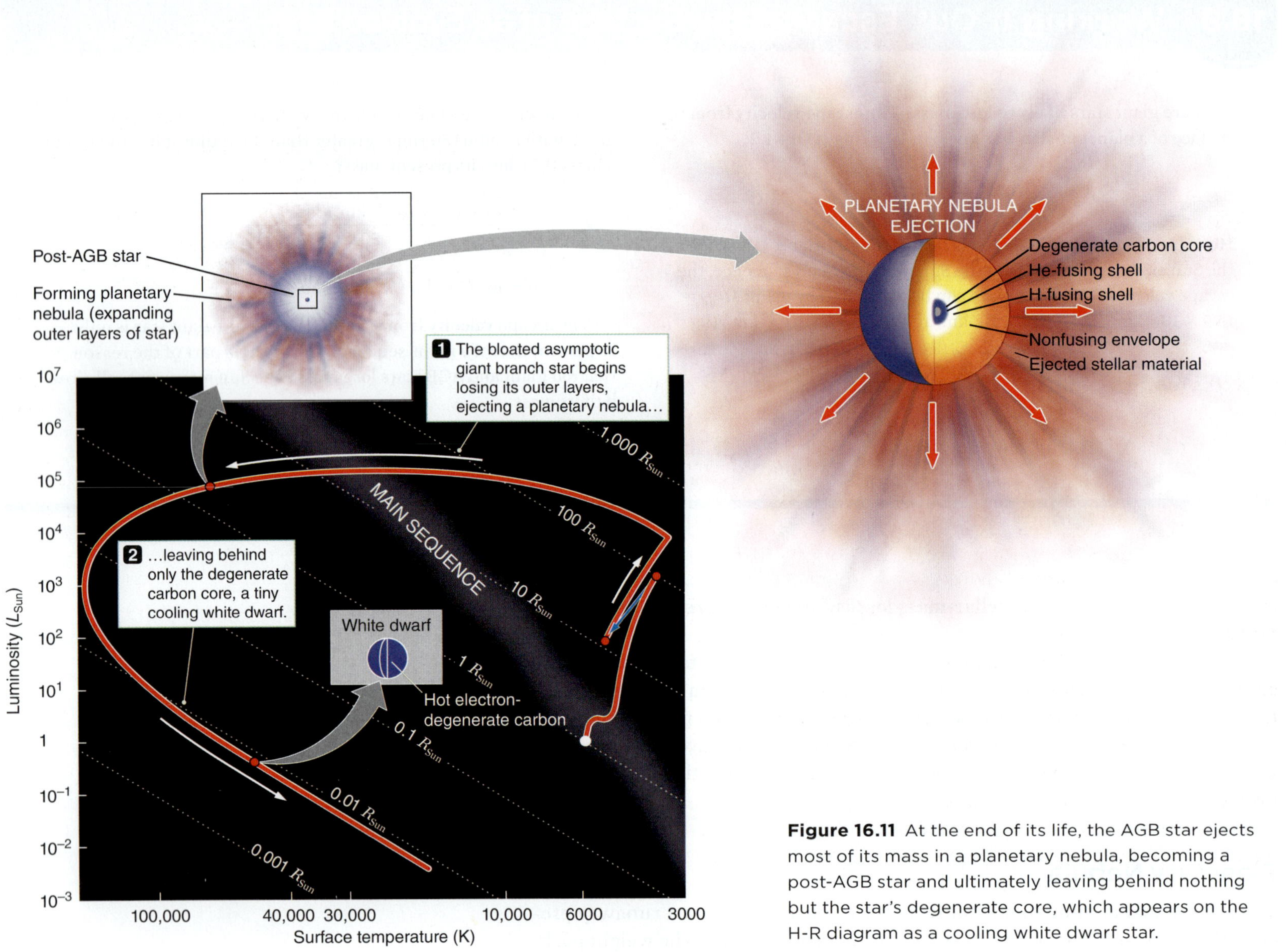

Figure 16.11 At the end of its life, the AGB star ejects most of its mass in a planetary nebula, becoming a post-AGB star and ultimately leaving behind nothing but the star's degenerate core, which appears on the H-R diagram as a cooling white dwarf star.

the star moves from right to left across the top of the H-R diagram, as shown in **Figure 16.11**.

The surface temperature of the star may eventually rise above 100,000 K. At such temperatures, the peak wavelength of the radiation is in the high-energy ultraviolet (UV) part of the spectrum, as determined by Wien's law. That intense UV light heats and ionizes the ejected, expanding shell of gas, causing it to glow in the same way that UV light from an O star causes an H II region to glow. When those glowing shells were first observed in small telescopes, they were named "planetary nebulae" because they appeared fuzzy like nebular clouds of dust and gas, but they were approximately round, like planets. Later imagery, such as the images in **Figure 16.12**, showed that those objects are not like planets at all. Rather, a **planetary nebula** is the remaining outer layers of a star, ejected into space at the end of the star's ascent of the AGB. A planetary nebula may be visible for 50,000 years or so before the gas ejected by the star disperses so far that the

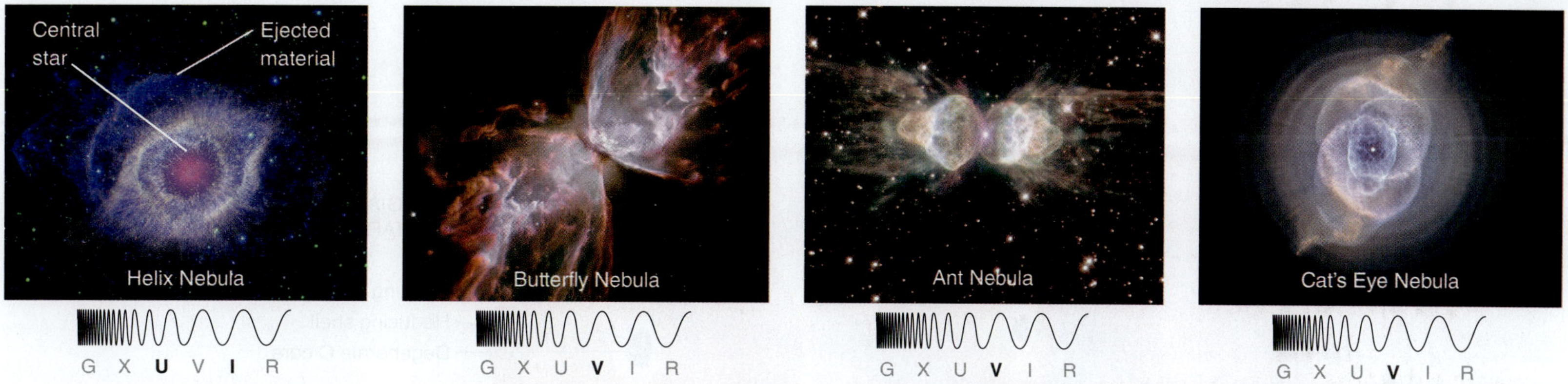

Figure 16.12 At the end of its life, a low-mass star ejects its outer layers and forms a planetary nebula consisting of an expanding shell of gas surrounding the white-hot remnant of the star. Planetary nebulae are not all simple spherical shells around their parent stars. These images of planetary nebulae from the Hubble Space Telescope and the Spitzer Space Telescope show the wealth of structures that result from the complex processes by which low-mass stars eject their outer layers.

nebula is too faint to be seen. Not all stars form planetary nebulae. Stars more massive than about 8 M_{Sun} pass through the post-AGB stage too quickly. Stars without enough mass take too long in the post-AGB stage, so their envelope evaporates before they can illuminate it. Astronomers do not know whether our own Sun will retain enough mass during its post-AGB phase to form a planetary nebula.

The detailed structure of a planetary nebula may contain concentric rings of varied density, indicating that the rate of stellar-mass loss varied—sometimes faster, sometimes slower. In the image of the Ring Nebula in the chapter-opening photograph, the colored rings are emission lines from ions located at different places in the nebula. When that nebula was forming, a lot of mass was lost nearly all at once. Then the mass loss ceased, resulting in a hollow shell around the central star. As the light makes its way out from the central star, higher-energy light gets "used up" in ionizing the inner layers. Outer layers have emission lines from atoms that require less energy to ionize. In other planetary nebulae, the material may be concentrated parallel to the equator or poles of the star, indicating that the stellar-mass loss was blocked in some directions. You can see that effect in the images of the Butterfly, Ant, and Cat's Eye nebulae in Figure 16.12.

The gas in a planetary nebula carries the chemical elements from the star's outer layers off into interstellar space. Planetary nebulae often show a greater percentage of elements such as carbon, nitrogen, and oxygen than the percentage of those elements present in the outer layers of the Sun. Those elements are by-products of nuclear fusion either from the star that produced the planetary nebula or from the stars of earlier generations. Once that chemically enriched material leaves the star, it mixes with interstellar gas, increasing the chemical diversity of the interstellar medium.

White Dwarfs

Within about 50,000 years, a post-AGB star fuses all the fuel remaining on its surface, leaving behind a nonfusing ball of carbon with a mass less than 70 percent of the mass of the original star. As that occurs, the post-AGB star becomes smaller and fainter but remains about the same temperature. Its position on the H-R diagram falls down the left side. Within a few thousand years, it shrinks to about the size of Earth, at which point it has become fully electron-degenerate and can shrink no further. That remnant of stellar evolution is called a **white dwarf**.

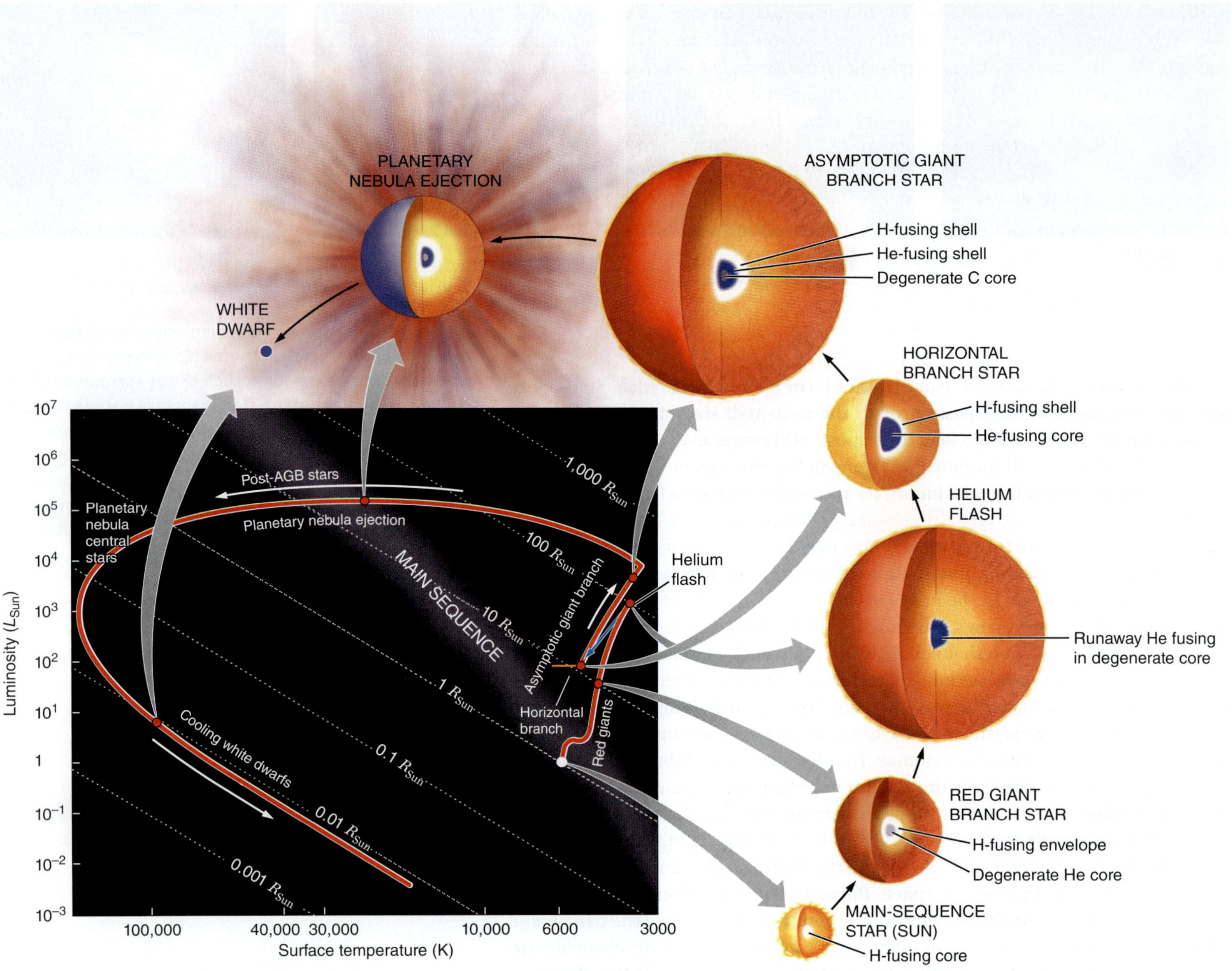

Figure 16.13 This H-R diagram summarizes the stages in the post-main-sequence evolution of a 1-M_{Sun} star.

The white dwarf, composed of nonfusing electron-degenerate carbon and maybe some oxygen, continues to radiate energy away into space. As it does so it cools, just like the heating coil on an electric stove once it is turned off. Because the white dwarf is electron-degenerate, its size remains nearly constant as it cools, so it follows a line of constant radius down and to the right on the H-R diagram. The white dwarf will remain very hot for awhile but then will slowly cool. Its tiny size means it is a thousand times less luminous than a main-sequence star like the Sun. Many white dwarfs are known, but none can be seen without a telescope. Sirius, the brightest star in Earth's sky, has a faint white dwarf as a binary companion.

Figure 16.13 summarizes the evolution of a solar-type, 1-M_{Sun} main-sequence star through its final existence as a ~0.6-M_{Sun} white dwarf. The star leaves the main sequence, climbs the red giant branch, falls to the horizontal branch, climbs back up the AGB, takes a left across the top of the diagram while ejecting a planetary nebula, and finally falls to its final resting place in the bottom left of the diagram. That process is representative of the fate of low-mass stars. Although every low-mass star forms a white dwarf at the end point of its evolution, the precise path a low-mass star follows from core hydrogen fusion on the main sequence to white dwarf depends on many details particular to the star.

Some stars less massive than the Sun may become white dwarfs composed largely of helium rather than carbon. Conversely, temperatures in the cores of evolved 2- to 3-M_{Sun} stars are high enough to allow additional nuclear reactions to occur, leading to the formation of more massive white dwarfs composed of materials such as oxygen, neon, and magnesium. Differences in chemical composition of a star can also lead to differences in its post-main-sequence evolution.

In our discussion of stellar evolution, we have focused on what happens after a star leaves the main sequence. The spectacle of a red giant or AGB star is ephemeral. Once it leaves the main sequence, the Sun will travel the path from red giant to white dwarf in less than one-tenth of the time it spent on the main sequence, steadily fusing hydrogen to helium in its core. Stars spend most of their luminous lifetimes on the main sequence, which is why most of the stars in the sky are main-sequence stars. The fainter white dwarfs constitute the final resting place for most of the stars that have been or ever will be formed.

The Fate of the Planets

What happens to the planets orbiting a low-mass star as it goes through those post-main-sequence stages? Because the first decade of extrasolar planet discoveries yielded primarily planets with orbits very close to their respective stars, astronomers assumed that any planets closer than 1–2 astronomical units (AU) would not survive the post-main-sequence expansion of the star. Therefore, they did not expect to find planets in orbit around evolved stars. However, many planets have been discovered in orbit around red giants, AGB stars, and horizontal branch stars.

Astronomers cannot be sure whether the extrasolar planets they observe have remained in the same orbital locations as when their stars were on the main sequence or whether the planets migrated to different orbits. The surface gravity of a red giant is low, and some of the mass in its outermost layers will blow away. That decrease in mass reduces the gravitational force between the star and its planets, which could lead to planetary orbits evolving outward away from the star. In some models of planetary migration, the tidal forces between planets or between a star and a planet are significant factors too, suggesting that some planetary orbits could evolve inward.

As a star loses mass during the evolutionary process, stellar or planetary companions may change their orbits. Rocky asteroids or smaller planets could migrate inward past the Roche limit (see Chapter 4) of the white dwarf and break up. Some of the material could remain in orbit around the white dwarf as a debris disk, similar to those seen in main-sequence stars with a planetary system. Those dusty debris disks have been observed around white dwarf stars with the Hubble Space Telescope and the Spitzer Space Telescope. Some of that dusty or rocky material may also fall onto the white dwarf, "polluting" its spectrum with heavy elements that were not produced in the stellar core.

What will happen to Earth? The age of the Sun as determined from radioactive dating of meteorites indicates that the Solar System formed 4.6 billion years ago. In Chapter 14, we estimated how long the Sun might last by calculating its rate of hydrogen fusion. The Sun's luminosity is about 30 percent higher now than it was early in the history of the Solar System, and it will continue to increase steadily over the rest of the Sun's main-sequence lifetime of another 5 billion years or so. The Sun's luminosity may increase enough—even while the Sun is a main-sequence star—that Earth will heat up to the point where the oceans evaporate, perhaps as soon as 1 billion to 2 billion years from now. By the time the Sun is a red giant, it might be impossible for any planet within the orbit of Jupiter to maintain liquid water on the surface.

Scientists are not certain whether the radius of the red giant Sun will extend past Earth to engulf the planet. The red giant Sun will have low surface gravity and will lose mass, which could cause Earth's orbit to expand, thereby enabling Earth to escape the encroaching solar surface. Alternatively, as the Sun expands in radius and its rotation rate slows (see Working It Out 7.1), tidal forces might pull Earth inward. The solar core will eventually become a white dwarf, perhaps with a dusty disk and a "polluted" atmosphere as the only remaining evidence of our rocky planet.

CHECK YOUR UNDERSTANDING 16.4

A planetary nebula forms from: (a) the ejection of mass from a low-mass star; (b) the collision of planets around a dying star; (c) the collapse of the magnetosphere of a high-mass star; (d) the remainders of the original star-forming nebula.

16.5 Binary Star Evolution

So far in this chapter, we have discussed the evolution of single stars in isolation. But many stars are members of binary systems. Those systems may form when a molecular-cloud core has too much angular momentum for a single star to form. Binary systems are common—as many as half of all stars may be in binary systems, and that fraction is largest among more massive stars. How is evolution different for stars in those systems? Sometimes, if the stars are close together and one star is more massive, their evolution after the main sequence may be linked. In this section, we will trace the steps from binary star system through the end point of their evolution: nova or supernova.

Mass Flows from an Evolving Star onto Its Companion

Think for a moment about what would happen if you were to travel in a spacecraft from Earth toward the Moon. When you are still near Earth, the force of Earth's gravity is far stronger than that of the Moon. As you move away from Earth and closer to the Moon, the gravitational attraction of Earth weakens, and the gravitational attraction of the Moon becomes stronger. You eventually reach an intermediate zone where neither body has the stronger pull. If you continue beyond that point, the lunar gravity begins to dominate until you find yourself firmly in the gravitational grip of the Moon. The regions surrounding the two objects—their gravitational domains—are called the **Roche lobes** of the system.

The same situation exists between two stars, as shown in **Figure 16.14**. Gas near each star clearly belongs to that star. When one star leaves the main sequence and swells up, its outer layers may cross that gravitational dividing line separating the star from its companion. Once a star expands past the boundary of its

Roche lobe, its material begins to fall onto the other star. That transfer of material from one star to the other is called **mass transfer**.

Evolution of a Close Binary System

The best way to understand how mass transfer affects the evolution of stars in a binary system is to apply what is known from studying the evolution of single low-mass stars. Figure 16.14a shows a binary system consisting of two low-mass stars: star 1 is more massive, and star 2 is less massive. That is an ordinary binary system, and each star is an ordinary main-sequence star for most of the system's lifetime.

More massive main-sequence stars evolve more rapidly than less massive main-sequence stars. Therefore, star 1 will be the first to use up the hydrogen at its center and begin to evolve off the main sequence, as shown in Figure 16.14b. If the two stars are close enough to each other, star 1 will eventually grow to overfill its Roche lobe, and material will transfer onto star 2, as shown in Figure 16.14c. The transfer of mass between the two stars results in a "drag" that causes the orbits of the two stars to shrink, bringing the stars closer together and further enhancing mass loss. In addition, as star 1 loses mass, its Roche lobe shrinks, further enhancing the mass transfer. The two stars can even reach the point at which they are effectively two cores sharing the same extended envelope of material, called a *common envelope*.

Despite those complexities, star 2 probably remains a basically normal main-sequence star throughout the process, fusing hydrogen in its core. However, over time, the mass of star 2 increases because of the accumulation of material from its companion. As it does so, the structure of star 2 must change to accommodate its new status as a higher-mass star. If we plotted star 2's position on the H-R diagram during that period, we would see it move up and to the left along the main sequence, becoming larger, hotter, and more luminous.

Star 1, because it is losing mass to star 2, never grows larger than its Roche lobe, so it does not become an isolated red giant or AGB star at the top of the H-R diagram. Yet star 1 continues to evolve, fusing helium in its core on the horizontal branch, proceeding through helium shell fusion, and finally losing its outer layers and leaving behind a white dwarf. Figure 16.14d shows the binary system after star 1 has completed its evolution. All that remains of star 1 is a white dwarf, orbiting about star 2, its bloated main-sequence companion.

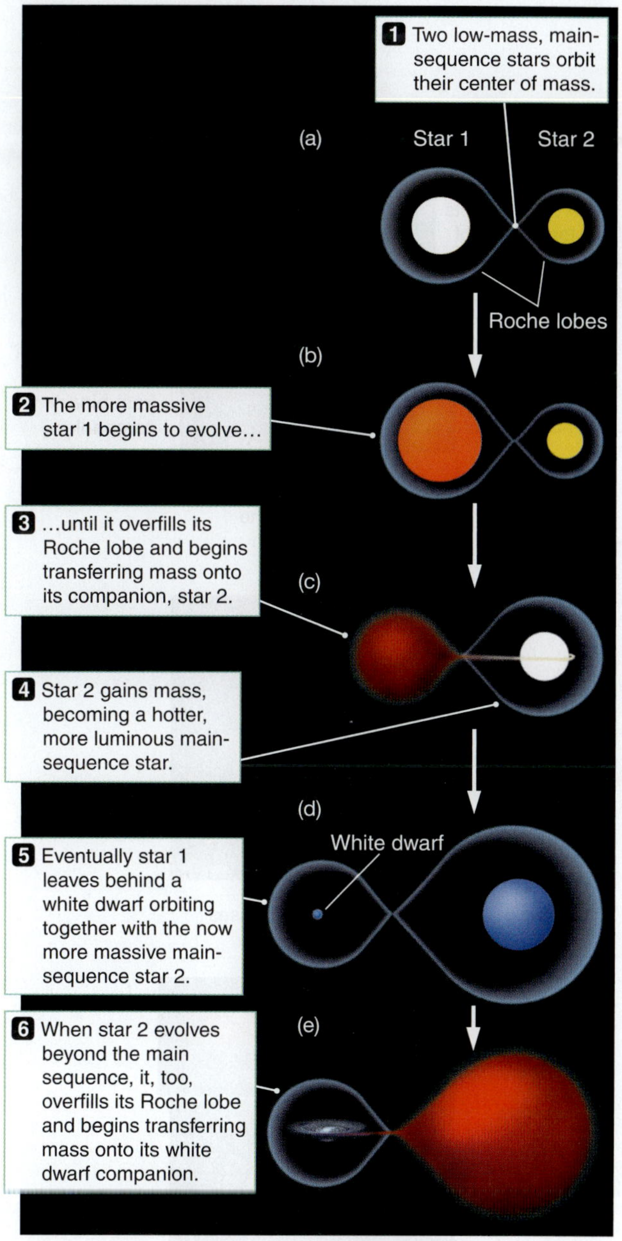

Figure 16.14 A compact binary system consisting of two low-mass stars that evolve through stages, transferring mass back and forth.

Novae

As star 2 begins to evolve off the main sequence, it expands to fill its Roche lobe, as shown in Figure 16.14e. Like star 1 before it, star 2 grows to fill its Roche lobe: material from star 2 begins to pour through the "neck" connecting the Roche lobes of the two stars. However, this time the mass is not being added to a normal star but is drawn toward the tiny white dwarf left behind by star 1. Because the system is revolving and the white dwarf is so small, the infalling material generally misses the star, instead landing on an accretion disk around the white dwarf. That disk is similar to the accretion disk that forms around a protostar. As with star formation, the accretion disk accumulates material that has too much angular momentum to hit the white dwarf directly.

A white dwarf has a mass comparable to that of the Sun but a size comparable to that of Earth. Having a large mass and a small radius means strong

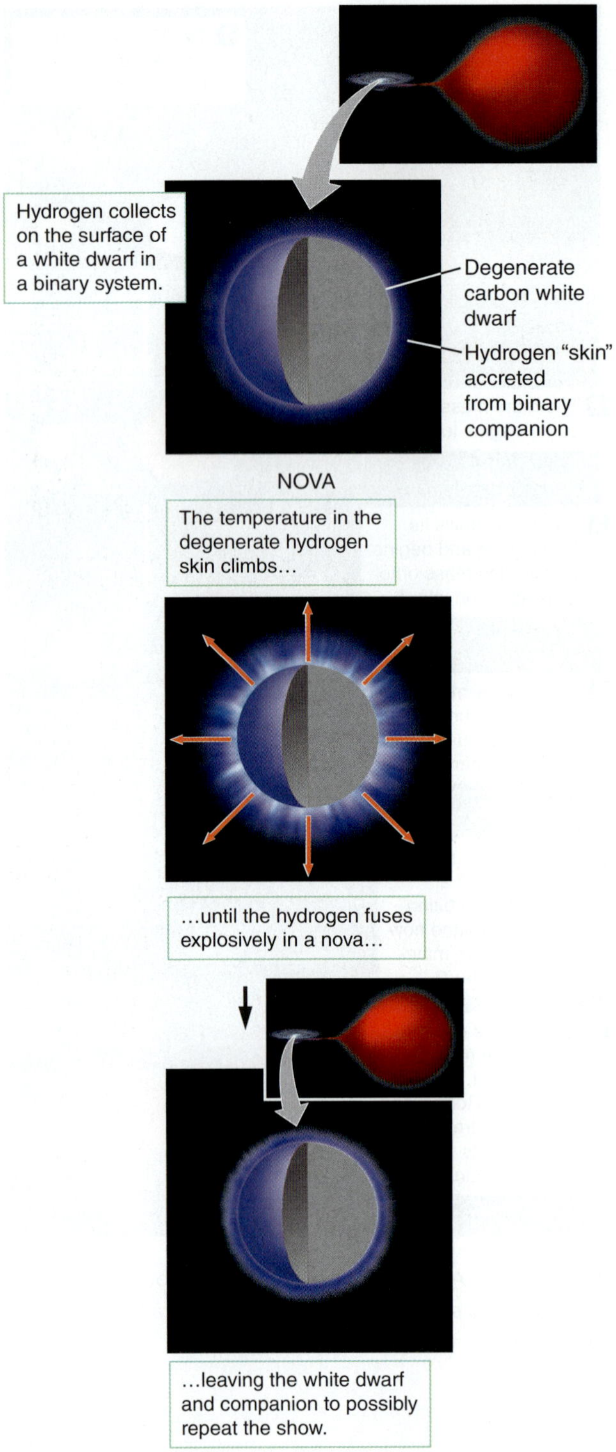

Figure 16.15 In a binary system in which mass is transferred onto a white dwarf, a layer of hydrogen builds up on the surface. That hydrogen may ignite on the surface of the white dwarf, producing a nova.

surface gravity. The material streaming toward the white dwarf in the binary system falls into an incredibly deep gravitational "well." The depth of that well affects the amount of energy with which matter impacts the white dwarf. A kilogram of material falling from space onto the surface of a white dwarf releases 100 times more energy than a kilogram of material falling from the outer Solar System onto the surface of the Sun. All that energy is turned into thermal energy. The spot where the stream of material from star 2 hits the accretion disk can be heated to millions of kelvins, where it glows in the far-UV and X-ray parts of the electromagnetic spectrum.

The infalling material accumulates on the surface of the white dwarf (**Figure 16.15**), where the enormous gravitational pull of the white dwarf compresses the material to a density close to that of the white dwarf itself. As more and more material builds up on the surface of the white dwarf, the white dwarf shrinks (just as the core of a red giant shrinks as it grows more massive). The density increases more and more while the release of gravitational energy drives the temperature of the white dwarf higher and higher. The infalling material comes from the outer, unfused layers of star 2, so it is composed mostly of hydrogen. That hydrogen is compressed to higher and higher densities and heated to higher and higher temperatures on the surface of the white dwarf.

Once the temperature at the base of the white dwarf's surface layer of hydrogen reaches about 10 million K, that hydrogen begins to fuse to helium. But that process is not the contained hydrogen fusion that takes place in the center of the Sun; it is explosive hydrogen fusion in a degenerate gas. Energy released by hydrogen fusion drives up the temperature. Because the surface is degenerate, that rising temperature does not cause an expansion, as in normal matter, but instead drives up the rate of hydrogen fusion. That runaway thermonuclear reaction is much like the runaway helium fusion that takes place during the helium flash, except now no overlying layers of a star are there to absorb the energy liberated by fusion. The result is a tremendous explosion that blows part of the layer covering the white dwarf out into space at speeds of thousands of kilometers per second, as shown in Figure 16.15. An exploding white dwarf of that kind is called a **nova**.

The explosion of a nova does not destroy the underlying white dwarf star. In fact, much of the material that had built up on the white dwarf may remain behind after the explosion. Afterward, the binary system is in much the same configuration as before: material from star 2 is still pouring onto the white dwarf. The nova can repeat many times as material builds up and ignites again and again on the surface of the white dwarf. If the underlying white dwarf is old and cooler, or the mass accumulates slowly on its surface, outbursts are separated by thousands of years, so most novae have been seen only once in historical times. But if the white dwarf is hotter, or mass is transferring quickly, such explosions can happen every few years or even every few months. Such recurring novae are called dwarf novae.

About 50 novae occur in our galaxy each year. About 10 of those can be detected with telescopes, but very few become visible to the eye. The rest are blocked from view by dust in the disk of our galaxy. Novae reach their peak brightness in only a few hours, and briefly they can be several hundred thousand times more luminous than the Sun. Although the brightness of a nova sharply drops in the weeks after the outburst, it can sometimes still be seen for years. During that time, the glow from the expanding cloud of ejected material is caused by the decay of radioactive isotopes created in the explosion. Large bursts of gamma rays have been observed to be associated with several novae.

Supernovae

A white dwarf with a mass greater than 1.4 M_{Sun} cannot remain stable. That value is referred to as the Chandrasekhar limit, named for Subrahmanyan Chandrasekhar (1910–1995), who derived it. Above that mass, even the pressure supplied by degenerate electrons is no longer enough to balance gravity, so the white dwarf will collapse. A white dwarf that is accumulating mass, however, likely does not reach the Chandrasekhar limit. As the star reaches about 1.38 M_{Sun}, core pressures and temperatures rise enough to ignite carbon and begin a simmering phase that holds off thermonuclear runaway for a while. Once the temperature reaches about 1.0×10^8 K, the runaway carbon fusion involves the entire white dwarf. Within about a second, the whole white dwarf is consumed in the resulting explosion, known as a **supernova**. In that instant, 100 times more energy is liberated than the Sun will give off over its 10-billion-year lifetime on the main sequence. Runaway fusion reactions convert a large fraction of the mass of the star into elements such as iron and nickel, and the explosion blasts the shards of the white dwarf into space at top speeds in excess of 20,000 km/s, enriching the interstellar medium with those heavier elements. The explosion is known as a **Type Ia supernova**.

How might a white dwarf gain mass and explode in that way? Three options exist, as shown in **Figure 16.16**.

First, in the binary system discussed above, star 2 eventually may simply go on to form a white dwarf, leaving behind a stable binary system consisting of two white dwarfs, as in Figure 16.16a. Those two white dwarfs may then eventually spiral together and merge. If the sum of their masses is greater than 1.4 M_{Sun}, the resulting merged star will explode. That explosion destroys both the original white dwarfs. The amount of mass involved in the explosion may range from 1.4 to 2.8 M_{Sun}. A majority of Type Ia supernovae may be of that type.

Second, star 2 may lose mass to the white dwarf while still a main-sequence star, as in Figure 16.16b. Through millions of years of mass transfer from star 2 onto the white dwarf, and possibly through countless nova outbursts, the white dwarf's mass slowly increases—until it approaches the Chandrasekhar limit and later explodes. Here, the white dwarf's mass is always almost exactly 1.4 M_{Sun}. Only one object of that type has so far been observed, from a pulse of blue light emitted when the white dwarf exploded and heated up star 2.

Third, star 2 may evolve off the main sequence to become a red giant, filling its Roche lobe. The material from that star flows onto the white dwarf over millions of years, as shown in Figure 16.16c. As in the above case, the white dwarf approaches the Chandrasekhar limit and then explodes. Here, the white dwarf's mass is almost exactly 1.4 M_{Sun} every time. A red giant companion has not yet been observed in a Type Ia supernova.

In a galaxy the size of the Milky Way, Type Ia supernovae occur about once a century. They can briefly shine with a luminosity billions of times that of our Sun, possibly outshining the galaxy itself. Those objects are particularly useful to astronomers because their luminosities can be approximately determined from a careful study of their light curves. Because that type of supernova happens during the explosion of an object with a mass between 1.4 and 2.8 M_{Sun}, the total energy involved will vary by at most a factor of 2. Therefore, the luminosity of Type Ia supernovae should vary by no more than a factor of 2. As we will see in the next few chapters, objects with known luminosity can be used to find distances to far away galaxies. Combining the luminosity of a Type Ia supernova with its

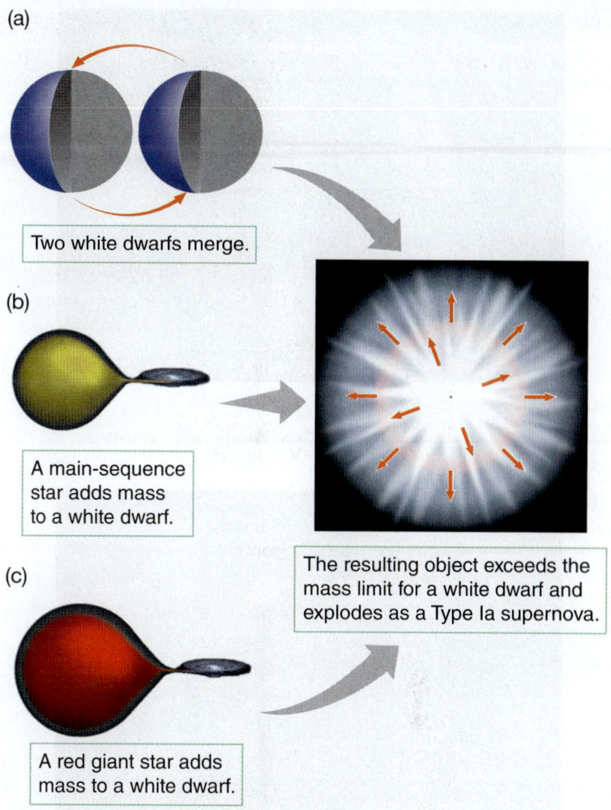

Figure 16.16 A Type Ia supernova results when a white dwarf exceeds the mass limit. That can happen because (a) two white dwarfs merge; (b) mass from a main-sequence companion falls onto the white dwarf, increasing its mass to the limit; or (c) mass from an evolved companion falls onto the white dwarf, increasing its mass to the limit.

(a)

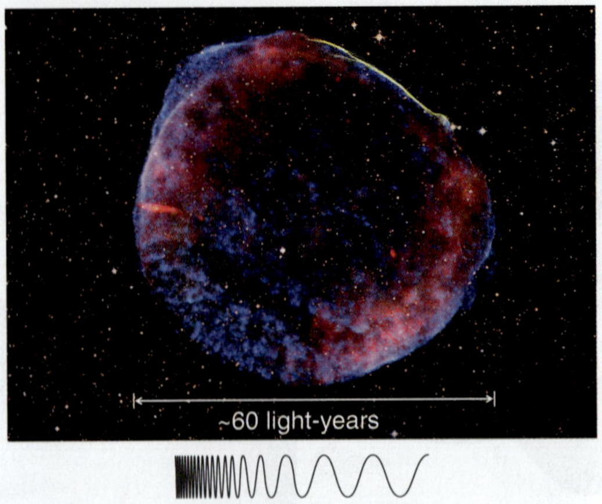

(b)

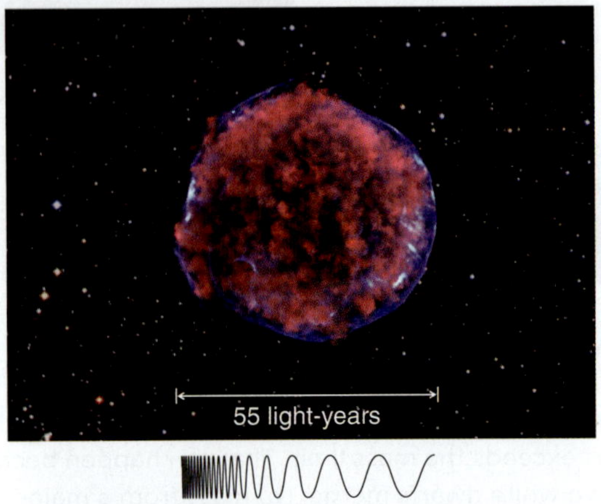

(c)

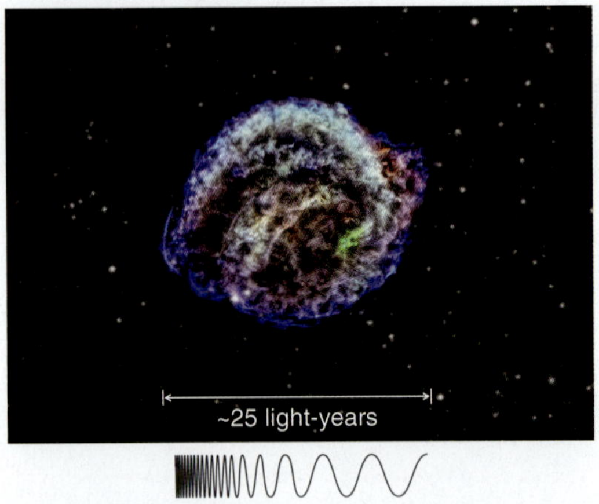

Figure 16.17 These images show the remnants of Type Ia supernovae. The material heated by the expanding blast wave from a supernova glows in the X-ray portion of the spectrum. (a) SN 1006 is the brightest recorded supernova and was observed in China, Japan, Europe, and the Middle East in 1006. X-ray data are shown in blue; radio data are shown in red. (b) Tycho's supernova was observed in 1572. Low-energy X-rays are red; high-energy X-rays are blue. (c) Kepler's supernova, observed in 1604, is shown in five X-ray wavelengths.

apparent brightness gives the distance to the host galaxy. Type Ia supernovae are still an active area of exploration in astronomy (see the **Process of Science Figure**).

Those explosions leave behind expanding shells of dust and gas called **supernova remnants**, as shown in **Figure 16.17**. The material is heated by the expanding blast wave from the supernova and glows in X-rays.

CHECK YOUR UNDERSTANDING 16.5

A white dwarf will become a supernova if: (a) the original star was more than 1.38 M_{Sun}; (b) it accretes an additional 1.38 M_{Sun} from a companion; (c) some mass falls onto it from a companion; (d) enough mass accretes from a companion to give the white dwarf a total mass of 1.38 M_{Sun}.

Origins Stellar Lifetimes and Biological Evolution

From fossil records and DNA analysis, scientists estimate that life took hold on Earth within 1 billion years after the Solar System and Earth formed 4.6 billion years ago. It took another 1.5 billion years for more complex cells to develop and another billion years to develop multicellular life. The first animals didn't appear on Earth until 600 million years ago, 4 billion years after the formation of the Sun and the Solar System.

The only known example of biology is the life on Earth. Extrapolating from one data point is always risky, and it is not known whether biology is widespread in the universe or whether Earth's biological timeline is typical. Still, reasoning from our one example is the only way to begin thinking about life in the universe. How does the preceding timeline of the evolution of life on Earth compare with the lifetimes of main-sequence stars? Table 16.1 indicates that the lifetime of an O5 star is less than half a million years; of a B5 star, about 120 million years; and of an A0 star, about 700 million years. Those stars would have run out of hydrogen in the core and started post-main-sequence evolution in less time than Earth took to settle down after its periods of heavy bombardment by debris early in the Solar System's history.

The 1-billion-year main-sequence lifetime of an A2 star corresponds to the time the simplest life-forms took to develop on Earth. The 3-billion-year lifetime for F0 stars corresponds to the time photosynthetic bacteria took to develop on Earth. Only stars cooler and less massive than F5 stars have main-sequence lifetimes longer than the 4 billion years life took to evolve into animals on Earth. Thus, searches for extrasolar planets survey stars that are F5 or cooler—because hotter and more massive stars probably don't live long enough on the main sequence for complex life to develop.

After a star leaves the main sequence, the helium-fusing red giant stage is estimated to last for about one-tenth of the main-sequence lifetime, so that doesn't help stars with short lifetimes last long enough for complex biology to evolve. Could life survive the transition of its star to a red giant? As noted earlier in the chapter, even if a planet is not destroyed, its orbit, temperature, and atmospheric conditions will drastically change, and any life might have to relocate if it is to survive.

PROCESS OF SCIENCE

Science Is Unfinished

Understanding Type Ia supernovae is at the foundation of measuring distances to the farthest galaxies and therefore for our conclusions about the universe as a whole. The observational evidence, though, remains incomplete.

Supernovae have been observed since antiquity.

As technology developed, it became clear there were **several types of supernovae**.

Type Ia supernovae are thought to come from white dwarfs that accumulate mass.

There are two possibilities:

If the white dwarf accretes mass from a red giant or other large star, and if it explodes at exactly 1.38 M_{Sun}, then the peak luminosity of the explosion should be the same for all Type Ia supernovae.

Searches for potential companions that survived the explosion, however, have only found a few candidates.

If the explosion arises from two white dwarfs merging together, then the mass of the explosion is variable and the peak luminosity may also vary.

Astronomers are studying large samples of white dwarfs to see how common binary white dwarfs are.

When observational evidence is inconclusive, each scientist adds a piece to the puzzle. Some will conduct larger observational studies, whereas others will create new theoretical models. Eventually, these efforts bring clarity to a confusing situation.

READING ASTRONOMY NEWS

ARTICLES QUESTIONS

Scientists report that the spectra of some dead stars show evidence of rocky material left over from planetary systems.

Scientists Solve Riddle of Celestial Archaeology

By University of Leicester Press Office

A decades old space mystery has been solved by an international team of astronomers led by Professor Martin Barstow of the University of Leicester and President-elect of the Royal Astronomical Society.

Scientists from the University of Leicester and University of Arizona investigated hot, young, white dwarfs—the super-dense remains of Sun-like stars that ran out of fuel and collapsed to about the size of the Earth. Their research is featured in MNRAS—the *Monthly Notices of the Royal Astronomical Society*, published by Oxford University Press.

It has been known that many hot white dwarf atmospheres, essentially of pure hydrogen or pure helium, are contaminated by other elements—like carbon, silicon, and iron. What was not known, however, was the origins of these elements, known in astronomical terms as metals.

"The precise origin of the metals has remained a mystery and extreme differences in their abundance between stars could not be explained," said Professor Barstow, a Pro-Vice-Chancellor at the University of Leicester whose research was assisted by his daughter Jo, a coauthor of the paper, during a summer work placement in Leicester. She has now gone on to be an astronomer working in Oxford—on extrasolar planets.

"It was believed that this material was 'levitated' by the intense radiation from deeper layers in the star," said Professor Barstow.

Now the researchers have discovered that many of the stars show signs of contamination by rocky material, the leftovers from a planetary system.

The researchers surveyed 89 white dwarfs, using the Far Ultraviolet Spectroscopic Explorer to obtain their spectra (dispersing the light by color) in which the "fingerprints" of carbon, silicon, phosphorus, and sulfur can be seen when these elements are present in the atmosphere.

"We found that in stars with polluted atmospheres the ratio of silicon to carbon matched that seen in rocky material, much higher than found in stars or interstellar gas.

"The new work indicates that around one-third of all hot white dwarfs are contaminated in this way, with the debris most likely in the form of rocky minor planet analogs. This implies that a similar proportion of stars like our Sun, as well as stars that are a little more massive like Vega and Fomalhaut, build systems containing terrestrial planets. This work is a form of celestial archaeology where we are studying the 'ruins' of rocky planets and/or their building blocks, following the demise of the main star.

"The mystery of the composition of these stars is a problem we have been trying to solve for more than 20 years. It is exciting to realize that they are swallowing up the leftovers from planetary systems, perhaps like our own, with the prospect that more detailed follow-up work will be able to tell us about the composition of rocky planets orbiting other stars," said Professor Barstow.

The study also points to the ultimate fate of the Earth billions of years from now—ending up as a contamination within the white dwarf Sun.

ARTICLES QUESTIONS

1. The article refers to the white dwarfs in the study as "hot, young, white dwarfs." What does "young" mean in that context?
2. The spectra described are compared to fingerprints. In what ways are white dwarf spectra like fingerprints?
3. Why were scientists surprised to find elements other than hydrogen and helium in the atmospheres of white dwarf stars? Where do they think those other elements originate?
4. Why does a telescope need to be in space to observe far-UV wavelengths?
5. How common are contaminated white dwarfs in the sample in the study? Compare that with the percentage of Sun-like stars with planets (22 percent). Does the finding in the article seem like a sensible number?

Summary

Stars with masses similar to that of the Sun form planetary nebulae like the one shown in the chapter-opening photograph. That is a likely fate for our own Sun at the end of its evolution. After leaving the main sequence, low-mass stars follow a convoluted path along the H-R diagram that includes the red giant branch, the horizontal branch, the asymptotic giant branch, and a path across the top and then down to the lower left of the diagram. Those stages of evolution are dominated by the balance between gravity and energy production from various fusion processes that "turn on and off" in the core of the star. That entire process takes much less time than the main-sequence lifetime of the star. Stars following that path have main-sequence lifetimes comparable to the evolutionary timescales of life on Earth. If life on Earth is typical, more massive stars with shorter lifetimes will not be stable long enough for complex life to evolve.

LG 1 **Estimate the main-sequence lifetime of a star from its mass.** All stars eventually exhaust their nuclear fuel as hydrogen fuses to helium in the cores of main-sequence stars. Less massive stars exhaust their fuel more slowly and have longer lifetimes than more massive stars.

LG 2 **Explain why low-mass stars grow larger and more luminous as they run out of fuel.** When a low-mass star uses up the hydrogen in its core, it begins to fuse hydrogen in a shell around the core, heating the gaseous interior. The star expands and becomes a red giant.

LG 3 **Sketch post-main-sequence evolutionary tracks on a Hertzsprung-Russell (H-R) diagram and list the stages of evolution for low-mass stars.** After exhausting its hydrogen, a low-mass star leaves the main sequence and swells to become a red giant, with a helium core made of electron-degenerate matter. The red giant fuses helium to carbon via the triple-alpha process, and quickly the core ignites in a helium flash. The star then moves onto the horizontal branch. A horizontal branch star accumulates carbon and sometimes oxygen in its core and then moves up the asymptotic giant branch. The star may become an AGB star and lose some of its mass.

LG 4 **Describe how planetary nebulae and white dwarfs form.** In their dying stages, some stars eject their outer layers to form planetary nebulae. All low-mass stars eventually become white dwarfs, which are very hot but very small.

LG 5 **Explain how some close binary systems evolve differently from single stars.** Transfer of mass within some binary systems can lead to a nuclear explosion. A nova occurs when hydrogen collects and ignites on the surface of a white dwarf in a binary system. If the mass of the white dwarf approaches 1.38 M_{Sun}, the entire star may become involved in a Type Ia supernova. That explosion can occur in several ways.

? Unanswered Questions

- Why do planetary nebulae have different shapes? Some are not simply chaotic but are well organized, with various types of symmetry. Some are spherically symmetric (like a ball), some have bipolar symmetry (like a long, hollow tube, pinched in the middle), and some are even point-symmetric (like the letter *S*). How can an essentially spherically symmetric object such as a star produce such beautifully organized outflows? Because stars are three-dimensional objects, and astronomers can view each object from only one direction, researchers can't easily determine how much of that variation is due to orientation and how much is due to actual differences in the shape of the object. For example, a bipolar nebula, viewed from one end, would appear spherically symmetric. That orientation effect, among other problems, complicates efforts to understand how those shapes are formed. No single explanation has yet satisfactorily covered all the object types.
- Could Earth be moved farther from the Sun to accommodate the Sun's inevitable changes in luminosity, temperature, and radius? One proposal suggests that Earth could capture energy from a passing asteroid and migrate outward, thus staying in the habitable zone while moving farther from the Sun as the Sun ages. Or a huge, thin "solar sail" could be constructed so that radiation pressure from the Sun would slowly push Earth into a larger orbit. Such feats of "astronomical engineering" are not feasible anytime soon, but perhaps they could be accomplished in millions of years (when they will be needed).

Questions and Problems

Test Your Understanding

1. Place the main-sequence lifetimes of the following stars in order from shortest to longest.
   a. the Sun: mass, 1 M_{Sun}; luminosity, 1 L_{Sun}
   b. Capella Aa: mass, 3 M_{Sun}; luminosity, 76 L_{Sun}
   c. Rigel: mass, 24 M_{Sun}; luminosity, 85,000 L_{Sun}
   d. Sirius A: mass, 2 M_{Sun}; luminosity, 25 L_{Sun}
   e. Canopus: mass, 8.5 M_{Sun}; luminosity, 13,600 L_{Sun}
   f. Achernar: mass, 7 M_{Sun}; luminosity, 3,150 L_{Sun}
2. Place the following steps in the evolution of a low-mass star in order.
   a. main-sequence star
   b. planetary nebula ejection
   c. horizontal branch
   d. helium flash
   e. red giant branch
   f. asymptotic giant branch
   g. white dwarf
3. If a star follows a horizontal path across the H-R diagram, the star
   a. maintains the same temperature.
   b. stays the same color.
   c. maintains the same luminosity.
   d. keeps the same spectral type.
4. Degenerate matter is different from normal matter because as the mass of degenerate material goes up,
   a. the radius goes down.
   b. the temperature goes down.
   c. the density goes down.
   d. the luminosity goes down.
5. The most massive stars have the shortest lifetimes because
   a. the temperature is higher in the core, so they fuse faster.
   b. they have less fuel in the core when the star forms.
   c. their fuel is located farther from the core.
   d. the temperatures are lower in the core, so they fuse their fuel more slowly.
6. If a main-sequence star suddenly started fusing hydrogen at a faster rate in its core, it would become
   a. larger, hotter, and more luminous.
   b. larger, cooler, and more luminous.
   c. smaller, hotter, and more luminous.
   d. smaller, cooler, and more luminous.
7. A low-mass main-sequence star's climb up the red giant branch is halted by
   a. the end of hydrogen shell burning.
   b. the beginning of helium fusion in the core.
   c. electron-degeneracy pressure in the core.
   d. instabilities in the star's expanding outer layers.
8. Post-main-sequence stars lose up to half their mass because
   a. jets from the poles release material at an increasing rate.
   b. the mass of the star drops because of mass loss from fusion.
   c. the magnetic field causes increasing numbers of coronal mass ejections.
   d. the star swells until the surface gravity is too weak to hold material.
9. A planetary nebula glows because
   a. it is hot enough to emit UV radiation.
   b. fusion is happening in the nebula.
   c. it is heating up the interstellar medium around it.
   d. light from the central star causes emission lines.
10. As an AGB star evolves into a white dwarf, it runs out of nuclear fuel, and one might guess that the star should cool off and move to the right on the H-R diagram. Why does the star move instead to the left?
    a. It becomes larger.
    b. More of the star is involved in fusion.
    c. As outer layers are lost, deeper layers are exposed.
    d. The temperature of the core rises.
11. When compressed, ordinary gas heats up but degenerate gas does not. Why, then, does a degenerate core heat up as the star continues shell fusion around it?
    a. It is heated by the radiation from fusion.
    b. It is heated by the gravitational collapse of the shell.
    c. It is heated by the weight of helium falling on it.
    d. It is insulated by the shell.
12. All Type Ia supernovae
    a. are at the same distance from Earth.
    b. always involve two stars of identical mass.
    c. are extremely luminous.
    d. always release the same amount of energy in fusion.
13. In Latin, *nova* means "new." That word is used for novae and supernovae because they are
    a. newly formed stars.
    b. newly dead stars.
    c. newly visible stars.
    d. new main-sequence stars.
14. When the Sun runs out of hydrogen in its core, it will become larger and more luminous because
    a. it will start fusing hydrogen in a shell around a helium core.
    b. it will start fusing helium in a shell and hydrogen in the core.
    c. infalling material will rebound off the core and puff up the star.
    d. energy balance will no longer hold, and the star will drift apart.

15. A white dwarf is located in the lower left of the H-R diagram. From that information alone, you can determine that the star is
    a. very massive.
    b. very dense.
    c. very hot.
    d. very bright.

Thinking about the Concepts

16. Can a star skip the main sequence and immediately begin fusing helium in its core? Explain your answer.
17. Suppose a main-sequence star suddenly started fusing hydrogen at a faster rate in its core. How would the star react? Discuss changes in size, temperature, and luminosity.
18. Describe some possible ways in which the temperature in the core of a star might increase while the density decreases.
19. Astronomers typically say that the mass of a newly formed star determines its destiny from birth to death. However, that statement is not true for one common environmental circumstance. Identify that circumstance and explain why the birth mass of a star might not fully account for the star's destiny.
20. Study the Process of Science Figure. Suppose that a new mechanism is found to explain Type Ia supernovae. In that mechanism, all Type Ia supernovae are more luminous than previously thought. Would the derived distances to galaxies be larger or smaller than we understand them to be now?
21. Do stars change structure while on the main sequence? Why or why not?
22. Suppose Jupiter is not a planet but a G5 main-sequence star with a mass of 0.8 M_{Sun}.
    a. How will life on Earth be affected, if at all?
    b. How will the Sun be affected as it comes to the end of its life?
23. Explain the similarity in the paths that a star follows along the H-R diagram as it forms from a protostar and as it leaves the main sequence to climb the red giant branch.
24. Why is a horizontal branch star (which fuses helium at a high temperature) less luminous than a red giant branch star (which fuses hydrogen at a lower temperature)?
25. Suppose the core temperature of a star is high enough for the star to begin fusing oxygen. Predict how the star will continue to evolve, including its path on the H-R diagram.
26. Why does a white dwarf move down and to the right along the H-R diagram?
27. Why does fusion in degenerate material always lead to a runaway reaction?
28. Suppose the more massive red giant star in a binary system engulfs its less massive main-sequence companion, and their nuclear cores combine. What structure will the new star have? Where will the star lie on the H-R diagram?
29. T Coronae Borealis is a well-known recurrent nova.
    a. Is it a single star or a binary system? Explain.
    b. What mechanism causes a nova to flare up?
    c. How can a nova flare-up happen more than once?
30. Why do astronomers prefer to search for planets around low-mass stars?

Applying the Concepts

31. Figure 16.1 contains the label "Straight-line (power law) approximation." What does that label tell you about the axes on the graph: are they linear or logarithmic? Explain why those data are plotted that way.
32. Use Figure 16.2 to estimate the percentage of the Sun's mass that is turned from hydrogen into helium over its lifetime.
33. Study Figure 16.13. How many times brighter is a star at the top of the giant branch than the same star (a) when it was on the main sequence and (b) when it was on the horizontal branch?
34. Study Figure 16.13. Make a graph of surface temperature versus time for the evolutionary track shown—from the time the star leaves the main sequence until it arrives at the dot showing that it is a white dwarf. Your time axis may be approximate, but it should show that the star spends different amounts of time in the different phases.
35. For most stars on the main sequence, luminosity scales with mass as $M^{3.5}$ (see Working It Out 16.1). What luminosity does that relationship predict for (a) 0.5-M_{Sun} stars, (b) 6-M_{Sun} stars, and (c) 60-M_{Sun} stars? Compare those numbers with the values in Table 16.1.
36. Calculate the main-sequence lifetimes for (a) 0.5-M_{Sun} stars, (b) 6-M_{Sun} stars, and (c) 60-M_{Sun} stars (see Working It Out 16.1). Compare those lifetimes with the values in Table 16.1.
37. What will the escape velocity be when the Sun becomes an AGB star with a radius 200 times greater and a mass only 0.7 times that of today? How will those changes in escape velocity affect mass loss from the surface of the Sun as an AGB star?

38. Each form of energy generation in stars depends on temperature.
    a. The rate of hydrogen fusion (proton-proton chain) near 10^7 K increases with temperature as T^4. If the temperature of the hydrogen-fusing core is raised by 10 percent, by how much does the hydrogen fusion energy increase?
    b. Helium fusion (the triple-alpha process) at 10^8 K increases with an increase in temperature at a rate of T^{40}. If the temperature of the helium-fusing core is raised by 10 percent, by how much does the helium fusion energy increase?

39. A planetary nebula has an expansion rate of 20 km/s and a lifetime of 50,000 years. Roughly how large will that planetary nebula grow before it disperses?

40. Suppose a companion star transferred mass onto a white dwarf at a rate of about 10^{-9} M_{Sun} per year. Roughly how long after mass transfer begins will the white dwarf explode as a Type Ia supernova? How does that length of time compare with the typical lifetime of a low-mass star? Assume that the white dwarf started with a mass of 0.6 M_{Sun}.

41. Use Kepler's third law to estimate how fast material in an accretion disk (size = 2×10^5 km) orbits around a 0.6 M_{Sun} white dwarf.

42. A white dwarf has a density of approximately 10^9 kilograms per cubic meter (kg/m^3). Earth has an average density of 5,500 kg/m^3 and a diameter of 12,700 km. If compressed to the same density as a white dwarf, what would Earth's radius be?

43. What is the density of degenerate material? Calculate how large the Sun would be if all its mass were degenerate.

44. Recall from Chapter 5 that the luminosity of a spherical object at surface temperature T is given by $L = 4\pi R^2 \sigma T^4$, where R is the object's radius. If the Sun became a white dwarf with a radius of 10^7 meters, what would its luminosity be at the following surface temperatures: (a) 10^8 K; (b) 10^6 K; (c) 10^4 K; (d) 10^2 K?

45. According to current astronomical evidence, our universe is approximately 13.8 billion years old. If that age is correct, what is the least mass that a star could possess to have already evolved into a red giant? What spectral type of star is that?

USING THE WEB

46. Go to the website for the Katzman Automatic Imaging Telescope at Lick Observatory (http://w.astro.berkeley.edu/bait/kait.html). What is that project? Why can a search for supernovae be automated? Pick a recent year. How many supernovae were discovered? Look at some of the images. How bright do the supernovae look in comparison with their galaxies?

47. Go to the website of the American Association of Variable Star Observers (AAVSO; https://aavso.org). What does that 100-year-old organization do? Read about the types of intrinsic variable stars. Click on "Getting Started." If you have access to dark skies, you can contribute to the study of variable stars. Go to the page for observers (http://aavso.org/observers) and click on each item in the "For New Observers" list, including the list of stars easy to observe. Assemble a group and observe a variable star from that list.

48. Go to the University of Washington's "Properties of Planetary Nebulae" Web page (https://sites.google.com/a/uw.edu/introductory-astronomy-clearinghouse/activities/stars/properties-of-planetary-nebulae) and download and complete the lab exercise.

49. Go to the Hubble Space Telescope's planetary nebula gallery (http://hubblesite.org/images/news/34-planetary-nebulae). For each of the three types of symmetry, find an example of a nebula that shows clearly the type of symmetry: spherical (being symmetric in every direction, like a circle), bipolar (having an axis about which they are symmetric, like a person's face), and point-symmetric (being symmetric about a point, like the letter *S*). Print each of the three images you chose, and label the type of symmetry each one represents. For all three nebulae, identify the location of the central star. For bipolar symmetry, draw a line that shows the axis about which the nebula is symmetric. For point symmetry, identify several features that are symmetric across the location of the central star.

50. In the Hubble telescope news archive, look up press releases on planetary nebulae (http://hubblesite.org/news/34-planetary-nebulae) and white dwarf stars (http://hubblesite.org/news/90-white-dwarfs). Pick a story for each. What observations were reported, and why were they important?

EXPLORATION

Low-Mass Stellar Evolution

digital.wwnorton.com/astro6

The evolution of a low-mass star, as discussed in this chapter, corresponds to many twists and turns on the H-R diagram. In this Exploration, we return to the "H-R Diagram" simulation to investigate how those twists and turns affect the appearance of the star.

Visit the Digital Resources Page and on the Student Site open the "H-R Diagram" Interactive Simulation in Chapter 16. The box labeled "Size Comparison" shows an image of both the Sun and the test star. Initially, those two stars have identical properties: the same temperature, the same luminosity, and the same size.

Examine the box labeled "Cursor Properties." That box shows the temperature, luminosity, and radius of a test star located at the "X" in the H-R diagram. Before you change anything, answer the first three questions.

1 What is the temperature of the test star?

2 What is the luminosity of the test star?

3 What is the radius of the test star?

As a star leaves the main sequence, it moves up and to the right on the H-R diagram. Move the cursor (the X on the diagram) up and to the right.

4 What changes about the image of the test star next to the Sun?

5 What is the test star's temperature? What property of the image of the test star indicates that its temperature has changed?

6 What is the test star's luminosity?

7 What is the test star's radius?

8 Ordinarily, the hotter an object is, the more luminous it is. Here, the temperature has gone down, but the luminosity has gone up. How can that be?

The star then moves around a lot in that part of the H-R diagram. Look at Figure 16.13, and then use the cursor to approximate the motion of the star as it moves up the red giant branch, back down and onto the horizontal branch, and then back to the right and up the asymptotic giant branch.

9 Are the changes you observe in the image of the star as dramatic as the ones you observed for question 4?

10 What is the most noticeable change in the star as it moves through that portion of its evolution?

Next, the star begins moving across the H-R diagram to the left, maintaining almost the same luminosity. Drag the cursor across the top of the H-R diagram to the left, and study what happens to the image of the star in the "Size Comparison" box.

11 What changed about the star as you dragged it across the H-R diagram?

12 How does the star's size now compare with that of the Sun?

Finally, the star drops to the bottom of the H-R diagram and then begins moving down and to the right. Move the cursor toward the bottom of the H-R diagram, where the star becomes a white dwarf.

13 What changed about the star as you dragged it down the H-R diagram?

14 How does its size now compare with that of the Sun?

To thoroughly cement your understanding of stellar evolution, press the "Reset" button and then move the star from main sequence to white dwarf several times. The exercise will help you remember how that part of a star's life appears on the H-R diagram.

17 Evolution of High-Mass Stars

Most stars are smaller and less massive than the Sun and live a relatively long time. O and B stars, which are much more massive than the Sun, are rarer and live a much shorter time. Those massive stars die more quickly and explosively than stars like the Sun. In this chapter, we will explore the life and death of high-mass stars.

LEARNING GOALS

By the end of this chapter, you should be able to:

LG 1 Describe how the death of high-mass stars differs from that of low-mass stars.

LG 2 List the sequence of stages for evolving high-mass stars.

LG 3 Explain the origin of chemical elements up to and heavier than iron.

LG 4 Identify how Hertzsprung-Russell (H-R) diagrams of clusters enable astronomers to measure the ages of stars and test theories of stellar evolution.

The Crab Nebula is the remains of a massive star whose explosion was observed in 1054. The center is an X-ray image of the pulsar. ▶▶▶

What caused this star to explode?

17.1 High-Mass Stars Follow Their Own Path

High-mass stars are those with masses greater than about 8 M_{Sun}. Those stars not only have several times more mass than the Sun, but they also have luminosities thousands or even millions of times greater than the Sun's. Even though they have more fuel to begin with, high-mass stars use it up much faster than low-mass stars and therefore have shorter lives. Whereas low-mass stars live for billions of years, high-mass stars live only hundreds of thousands to millions of years. In this section, we will discuss the stages in the evolution of high-mass (greater than 8 M_{Sun}) and medium-mass (3–8 M_{Sun}; also called intermediate mass) stars.

The CNO Cycle

A high-mass star has greater gravitational force pressing down on the interior than a low-mass star does. That greater force leads to higher temperature and pressure. Those two factors increase the rate of nuclear fusion, and therefore the stars are more luminous. At those higher temperatures and pressures at the center of a high-mass star, additional nuclear reactions become possible beyond the fusion of hydrogen into helium. Recall from Chapter 14 that the hydrogen nucleus has only one proton—a single positive charge—so hydrogen fuses at lower temperatures (a few million kelvins) than any other atomic nucleus. However, the probability that any two hydrogen atoms will fuse is low. The low probability of that first step in the proton-proton chain limits how rapidly the entire process can move forward.

Simulation: CNO Cycle

If the star contains elements from a previous generation of stars that produced more massive elements, carbon and other elements will be mixed with the hydrogen. In the core of a massive star, the temperature and pressure are high enough that the **carbon-nitrogen-oxygen (CNO) cycle**, a nuclear fusion process that converts hydrogen to helium in the presence of carbon, occurs. That process is illustrated in **Figure 17.1a**, which shows each step. First, a hydrogen nucleus fuses with a carbon-12 (^{12}C) nucleus to form nitrogen-13 (^{13}N). Second, a proton in that ^{13}N nucleus decays, so that the atom is once again carbon. However, that carbon nucleus has an "extra" neutron, so it is now carbon-13 (^{13}C), not carbon-12. Third and fourth, two more hydrogen nuclei then fuse with that ^{13}C nucleus, creating nitrogen-14 (^{14}N) and then oxygen-15 (^{15}O). Fifth, a proton in the oxygen nucleus decays, creating nitrogen-15 (^{15}N). One more proton enters the nucleus, causing the ejection of a helium-4 (^{4}He) nucleus and leaving behind a ^{12}C nucleus, which can participate in the cycle again.

Notice that only one of two reactions occurs at most steps in that process: either a hydrogen nucleus fuses with another nucleus to create a new element with higher atomic number or a proton spontaneously decays to a neutron to create a new element with lower atomic number. Along the way, several by-products are formed: a positron and a neutrino are ejected each time a proton decays to form a neutron. Each positron then annihilates with an electron to produce a gamma ray. Additional gamma rays are released each time fusion occurs. **Figure 17.1b** shows the net reaction: a carbon-12 nucleus and four hydrogen nuclei combine with two electrons to produce a carbon-12 nucleus, a helium-4 nucleus, two neutrinos, and seven gamma rays. Getting a hydrogen nucleus past the

Figure 17.1 (a) In high-mass stars, carbon serves as a catalyst for the fusion of hydrogen to helium. That process is the carbon-nitrogen-oxygen (CNO) cycle. (b) The CNO cycle takes carbon-12, hydrogen, and electrons as inputs and produces carbon-12, helium, neutrinos, and gamma rays.

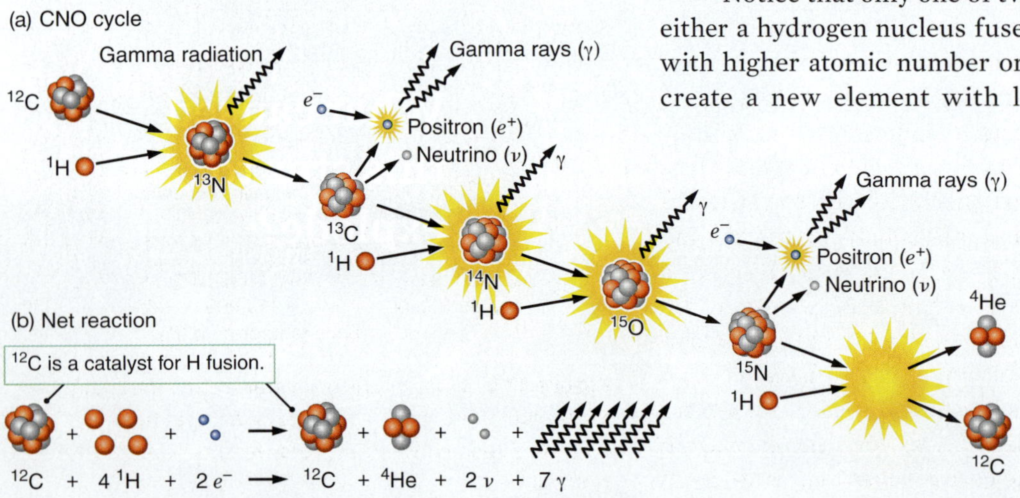

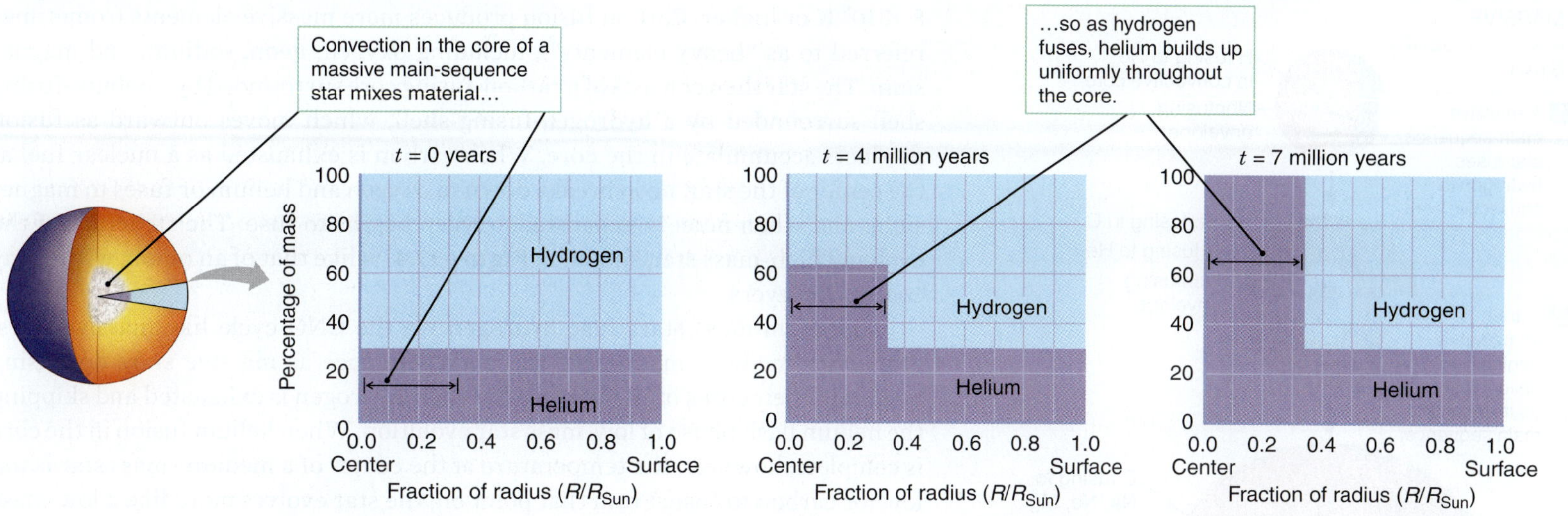

Figure 17.2 Convection keeps the core of a high-mass main-sequence star well mixed, so the composition remains uniform throughout the core as it evolves from zero age in the top graph to age 7 million years in the bottom graph. (Evolution times are for a 25-M_{Sun} star.)

electric barrier set up by a carbon nucleus, with its six protons, takes a lot of energy. But when that barrier can be overcome at high pressure and temperature, fusion is much more probable than it is in the interaction between two hydrogen nuclei. The CNO cycle is far more efficient than the proton-proton chain in stars more massive than about 1.3–1.5 M_{Sun}.

In a high-mass star, the temperature drops so quickly from the center of the star to the outside of the core that convection occurs within the core, "stirring" the core like the water in a boiling pot. As shown in **Figure 17.2**, helium residue spreads uniformly throughout the core of a high-mass star as the star consumes its hydrogen. That is unlike the case for low-mass stars, in which the helium residue builds up from the center outward (compare this figure with Figure 16.2). That difference is a result of the different processes of helium fusion in high-mass and low-mass stars.

The High-Mass Star Leaves the Main Sequence

As a high-mass star runs out of hydrogen in its core, the weight of the overlying star compresses the core, just as in a low-mass star. Yet in a high-mass star, the pressure and temperature in the core become high enough for helium fusion to begin (10^8 K) before the high-mass star forms an electron-degenerate core. The star makes a fairly smooth transition from hydrogen fusion to helium fusion as it leaves the main sequence.

Recall that when a low-mass star leaves the main sequence, the path it follows on the Hertzsprung-Russell (H-R) diagram is nearly vertical (see Figure 16.4) as the star becomes more luminous at roughly constant temperature. In contrast, as a high-mass star leaves the main sequence, the radius grows while its surface temperature falls, so its evolutionary track is nearly horizontal on the H-R diagram, as shown in **Figure 17.3**. The massive star fuses helium in its core and hydrogen in a surrounding shell. Recall that this is the same structure as a low-mass horizontal branch star. Stars more massive than 10 M_{Sun} become red supergiants during their helium-fusing phase. They have very cool surface temperatures (about 4000 K) and radii as much as 1,000 times that of the Sun.

The next stage in the evolution of a high-mass star has no analog in low-mass stars. When the high-mass star exhausts the helium in its core, the core again begins to collapse. Carbon fusion begins when the core reaches temperatures of

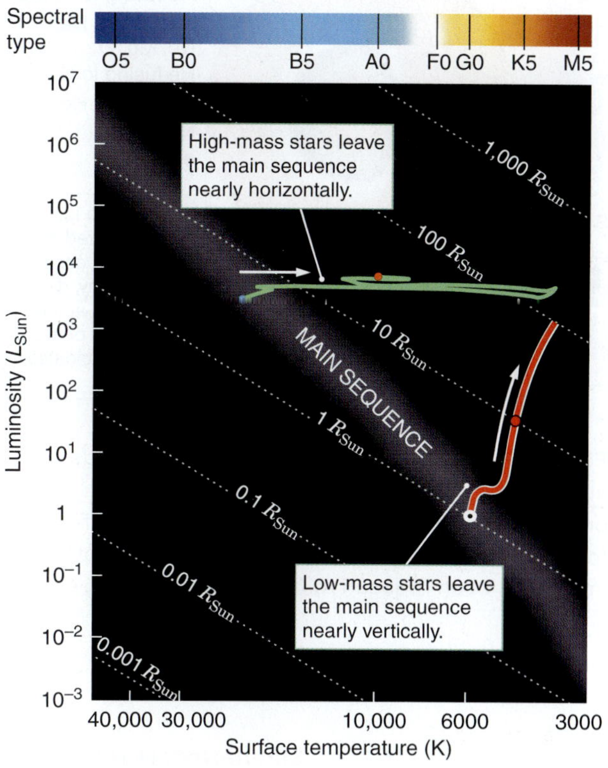

Figure 17.3 When high-mass stars leave the main sequence, they move horizontally across the H-R diagram, unlike low-mass stars, which move nearly vertically.

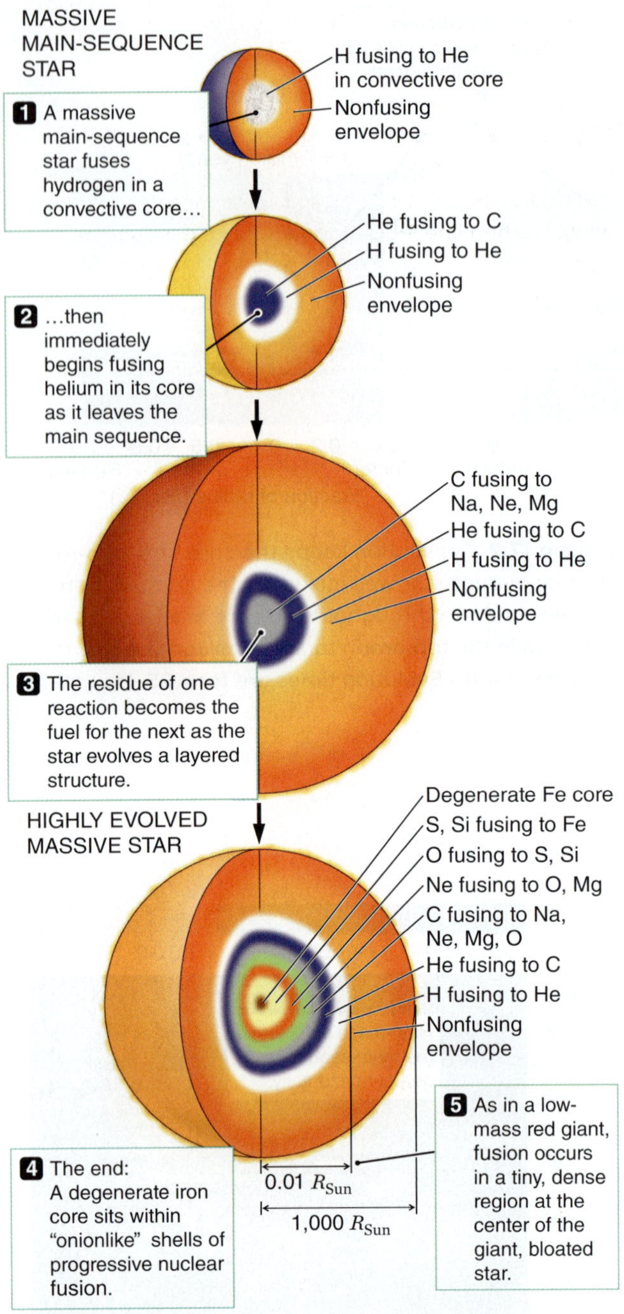

Figure 17.4 As a high-mass star evolves, it builds up a layered structure like that of an onion, with progressively more advanced stages of nuclear fusion found deeper and deeper within the star. The bottom image has been reduced to fit on the page.

Simulation: H-R Explorer

8×10^8 K or higher. Carbon fusion produces more massive elements (sometimes referred to as "heavy elements"), including oxygen, neon, sodium, and magnesium. The star then consists of a carbon-fusing core surrounded by a helium-fusing shell surrounded by a hydrogen-fusing shell, which moves outward as fusion products accumulate in the core. When carbon is exhausted as a nuclear fuel at the center of the star, neon breaks down to oxygen and helium or fuses to magnesium; and when neon is exhausted, oxygen begins to fuse. The structure of the evolving high-mass star, shown in **Figure 17.4**, is like that of an onion, with many concentric layers.

Medium-mass stars fuse hydrogen via the CNO cycle like massive stars. Stars with medium mass leave the main sequence as massive stars do, fusing helium in their cores immediately after their hydrogen is exhausted and skipping the helium flash phase of low-mass star evolution. When helium fusion in the core is complete, however, the temperature at the center of a medium-mass star is too low for carbon to fuse. From that point on, the star evolves more like a low-mass star, ascending the asymptotic giant branch (AGB), fusing helium and hydrogen in shells around a degenerate core, and then ejecting its outer layers and leaving behind a white dwarf.

Stars on the Instability Strip

While undergoing post-main-sequence evolution, a star may make one or more passes through a region of the H-R diagram known as the **instability strip**, as shown by the dashed-line region in **Figure 17.5**. **Variable stars** are stars whose brightness changes. In that region of the H-R diagram, stars grow and shrink repeatedly, so they appear to pulsate, regularly growing bright and faint and then bright again. The time for one pulsation is called the **period**. Such stars are called **pulsating variable stars**. A pulsating variable star does not achieve a steady balance between pressure and gravity; instead, it repeatedly overshoots the equilibrium radius, shrinking too far before being pushed back out by pressure or expanding too far before being pulled back in by gravity.

The most luminous pulsating variable stars are **Cepheid variables**, named after the prototype star Delta Cephei. Type I, or Classical, Cepheids are massive and luminous yellow supergiants. A Cepheid variable takes between 1 and 100 days to complete one pulsation period. The period correlates with the luminosity, creating a **period-luminosity relationship** in which stars with longer periods are more luminous, as shown in **Figure 17.6**. Henrietta Leavitt (1868–1921) experimentally determined that period-luminosity relationship for Cepheid variables in 1912. That relationship is important because it allows astronomers to use Cepheid variables as distance indicators. That process has three steps: First, astronomers observe the period. Then they use that period, combined with the period-luminosity relationship, to find the luminosity. Finally, they combine that luminosity with the observed brightness in the sky to find the distance to the star.

Thermal energy powers the pulsations of stars like Cepheid variables. An ionized gas is nearly opaque because ions can interact with light of all wavelengths, scattering many of the photons back toward the interior. In contrast, a neutral gas can absorb only light with energies that match its electron transitions. As the helium atoms in the atmosphere of the star alternate between ionized and neutral states, the atmosphere of the star alternates between opaque and transparent. That in turn alternately traps and releases light from the star. The star's atmosphere is much like a lid on a pot of boiling water. The pot builds up pressure enough to pop open the lid and let steam escape. Then gravity pulls the lid back

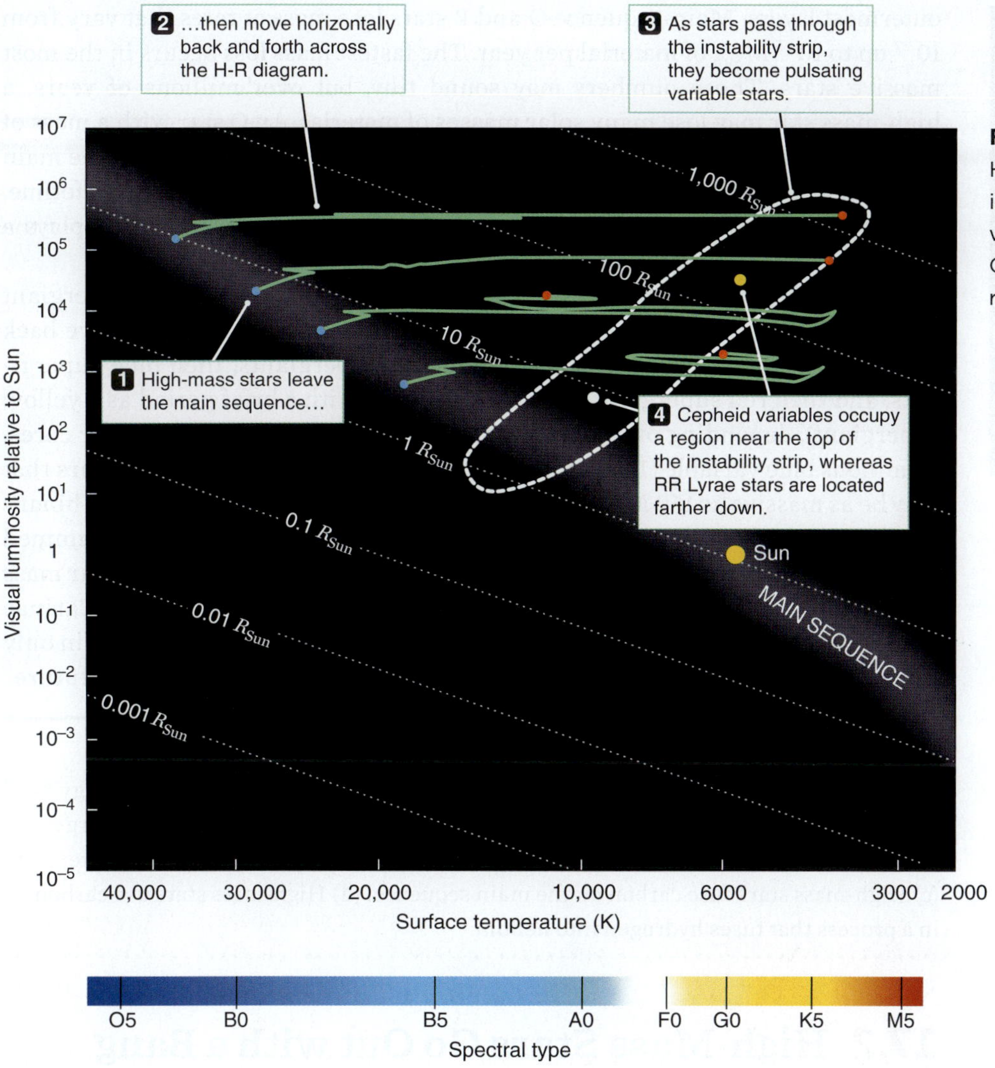

Figure 17.5 The paths of high-mass stars along the H-R diagram take them through a region known as the instability strip, shown by the region surrounded by the white dashed line. Pulsating variable stars such as Cepheid variables and RR Lyrae stars are found in that region of the H-R diagram.

down, and pressure builds again to repeat the process. Those pulsations do not affect the nuclear fusion in the star's interior; they affect just the light escaping from the star. From Chapter 13, recall that the luminosity, temperature, and radius of stars are all related, so both the luminosity and the temperature of the star change as the star expands and shrinks. The star is at its brightest and hottest (and therefore bluest) while it expands through its equilibrium size and is at its faintest and coolest (and therefore reddest) while it falls back inward.

High-mass Type I Cepheid variables are not the only type of variable star. The low-mass horizontal branch also passes through the instability strip on the H-R diagram. Low-mass ($\sim$0.8-M_{Sun}) stars can be Type II Cepheid variables or **RR Lyrae variables**, which have periods of less than a day. RR Lyrae stars pulsate by the same mechanism as Cepheid variables but are typically hundreds of times less luminous. They, too, follow a period-luminosity relationship, shown in Figure 17.6b.

High-mass stars expel a significant percentage of their mass back into space throughout their lifetimes, changing their composition. Even while on the main sequence, massive O and B stars have low-density winds with velocities as high as 3,000 kilometers per second (km/s). The pressure of the intense radiation generated in the core overcomes the star's gravity and drives away material from its

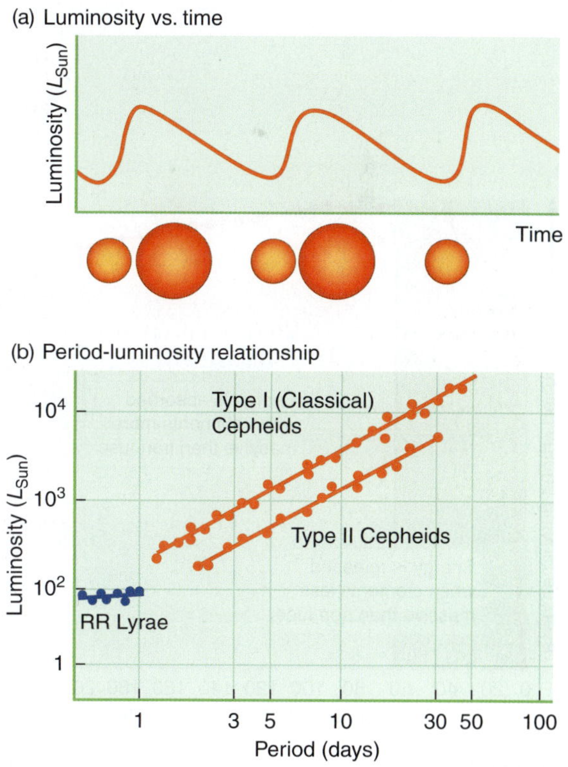

Figure 17.6 (a) Cepheid variable stars pulsate in size, and therefore their luminosity changes periodically. The shape of the resulting curve is distinctive. (b) The length of the period of pulsation is related to the star's luminosity at maximum.

outermost layers. Main-sequence O and B stars lose mass at rates that vary from 10^{-7} up to 10^{-5} M_{Sun} of material per year. The fastest mass loss occurs in the most massive stars. Those numbers may sound tiny, but over millions of years, a high-mass star may lose many solar masses of material. An O star with a mass of 20 M_{Sun} may lose about 20 percent of its mass (4 solar masses) while on the main sequence and possibly more than half its mass (10 solar masses) over its lifetime. Even an 8-M_{Sun} star may lose 5–10 percent of its mass. That mass loss plays a prominent role in the evolution of high-mass stars.

In general, stars with masses below 15 M_{Sun} go through a red supergiant stage, perhaps with a Cepheid variable phase. Stars under 30 M_{Sun} move back and forth on the H-R diagram—becoming red supergiants, then blue supergiants, and then red supergiants again (sometimes with a brief period as a yellow supergiant), depending on which fusion processes are occurring in their cores. Luminous blue variable (LBV) stars are hot, luminous, extremely rare stars that may be as massive as 150 M_{Sun}. An example is Eta Carinae (**Figure 17.7**), a binary system with at least 120 times the mass of the Sun and a luminosity (summed over all wavelengths) of 5 million L_{Sun}. Eta Carinae is now losing one solar mass every 1,000 years. However, during a 19th century eruption, when Eta Carinae became the second-brightest star in the sky, it shed ~10 M_{Sun} of material in only 20 years. Eta Carinae is expected to explode in the astronomically near future.

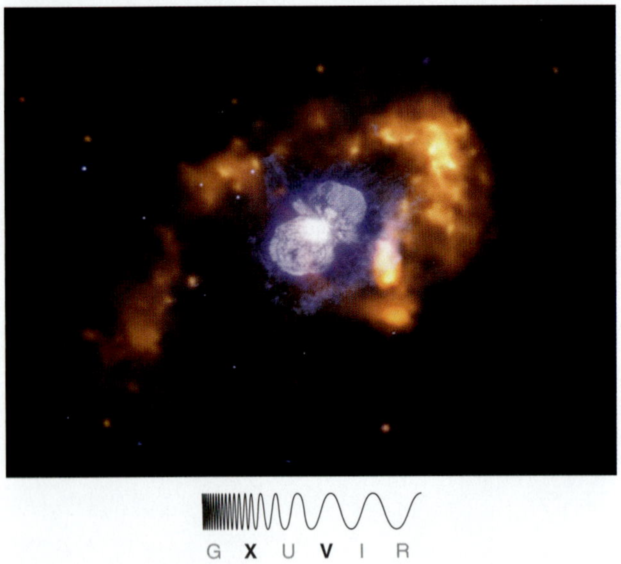

Figure 17.7 In this image of the luminous blue variable star Eta Carinae, an expanding cloud of ejected dusty material is seen in visible (blue) and X-ray (yellow) light. The star itself, largely hidden by the surrounding dust, has a luminosity of 5 million L_{Sun} and a mass probably in excess of 120 M_{Sun}. Dust is created when volatile material ejected from the star condenses.

CHECK YOUR UNDERSTANDING 17.1

How does energy production in a high-mass, main-sequence star differ from energy production in the Sun? (Choose all that apply.) (a) High-mass stars get a lot of energy through non-nuclear processes. (b) High-mass stars produce energy at a faster rate. (c) High-mass stars fuse carbon on the main sequence. (d) High-mass stars use carbon in a process that fuses hydrogen into helium.

17.2 High-Mass Stars Go Out with a Bang

A low-mass star approaches the end of its life relatively slowly and gently, ejecting its outer parts into nearby space and leaving behind a degenerate core. In contrast, a high-mass star ends its life suddenly and explosively. In this section, we explain how the energy involved in binding atomic nuclei affects nuclear fusion in the core and results in an explosive end for a high-mass star.

Binding Energy

An evolving high-mass star builds up an onionlike structure as nuclear fusion in its interior proceeds to more and more massive elements (see Figure 17.4). Hydrogen fuses to helium, helium fuses to carbon and oxygen, carbon fuses to magnesium, oxygen fuses to sulfur and silicon, and then silicon and sulfur fuse to iron. Many types of nuclear reactions occur up to that point, forming almost all the stable isotopes of elements less massive than iron. But the chain of nuclear fusion stops with iron.

The **binding energy** of an atomic nucleus is the energy required to break the nucleus into its constituent parts. A nuclear reaction releases energy if it increases the binding energy in a nucleus. Conversely, a nuclear reaction absorbs energy if it decreases the binding energy in the nucleus. **Figure 17.8** shows the binding energy per nucleon (that is, per each proton or neutron in the nucleus) for different atomic nuclei. Moving from hydrogen to helium increases the binding

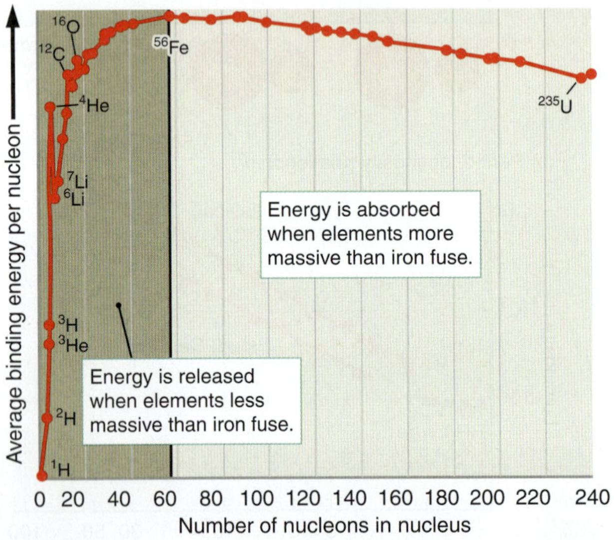

Figure 17.8 The binding energy per nucleon is plotted against the number of nucleons for each element. That is the energy required to break the atomic nucleus apart into protons and neutrons. Energy is released by nuclear fusion only if the resulting element is higher on the curve.

17.1 Working It Out Binding Energy of Atomic Nuclei

The net energy released by a nuclear reaction is the difference between the binding energy of the products and the binding energy of the reactants:

$$\text{Net energy} = \begin{pmatrix}\text{Binding energy}\\ \text{of products}\end{pmatrix} - \begin{pmatrix}\text{Binding energy}\\ \text{of reactants}\end{pmatrix}$$

In the example of the triple-alpha process, the binding energy of the initial helium nuclei is 6.824×10^{14} joules (J) per kilogram (kg) of helium, and the binding energy of the produced ^{12}C nuclei is 7.402×10^{14} J per kg of carbon. The amount of energy available from fusing 1 kg of helium nuclei into carbon is given by

$$\begin{pmatrix}\text{Net energy from}\\ \text{fusing 1 kg of He}\end{pmatrix} = \begin{pmatrix}\text{Binding energy}\\ \text{of C formed}\end{pmatrix} - \begin{pmatrix}\text{Binding energy}\\ \text{of He fused}\end{pmatrix}$$

$$= (7.402 \times 10^{14}\ \text{J}) - (6.824 \times 10^{14}\ \text{J})$$

$$= 5.780 \times 10^{13}\ \text{J}$$

That release of net energy indicates that helium is a good nuclear fuel, as Figure 17.8 shows. Fusing He to C releases energy.

What about fusing iron into more massive elements? Because iron is at the peak of the binding-energy curve, the products of iron fusion will have less binding energy than the initial reactants. Going from iron to more massive elements means moving down on the binding-energy curve in Figure 17.8, so the net energy in the reaction will be negative. Instead of producing energy, iron fusion absorbs energy.

energy, so fusing hydrogen releases energy, sustaining the nuclear reaction. Moving from helium to carbon also increases the binding energy, so that process also helps sustain the nuclear reactions in the core (**Working It Out 17.1**). Iron is at the peak of the binding-energy curve, so fusing lighter elements up to iron (Fe) also releases energy; however, fusing iron to make heavier elements decreases the binding energy. Because that process absorbs energy from the core, it lowers the temperature and the pressure. Iron fusion, therefore, cannot sustain itself. Once the star's core is filled with iron, the star will not last much longer.

The Final Days in the Life of a Massive Star

Per reaction, the helium fusion produces about one-quarter as much energy as hydrogen fusion, so when a star is fusing helium into carbon, the pressure inside the star is lower than when it was fusing hydrogen into helium on the main sequence. The star compresses, increasing the reaction rate so the helium is used up more quickly. Although hydrogen fusion can supply the energy needed to support the high-mass star on the main sequence for millions of years, helium fusion can support the star for only a few hundred thousand years.

Maintaining the balance in a star is like trying to keep a leaky balloon inflated. The larger the leak, the more rapidly air must be pumped into the balloon to keep it inflated. A star fusing hydrogen or helium is like a balloon with a slow leak. At the temperatures of hydrogen or helium fusion, energy leaks out of the interior of a star primarily by radiation and convection. Neither process is very efficient because the outer layers of the star act like a thick, warm blanket. Much of the energy is kept in the star, so nuclear fuels do not need to fuse very fast to support the weight of the outer layers of the star while keeping up with the energy escaping from the surface.

Once carbon fusion begins, that balance shifts. Energy is carried away primarily by neutrinos—referred to as neutrino cooling—rather than by radiation

TABLE 17.1 Fusion Stages in High-Mass Stars

Core Fusing Stage	Duration for a 15-M_{Sun} Star	Duration for a 25-M_{Sun} Star	Typical Core Temperature (K)
Hydrogen (H) fusion	11 million years	7 million years	$(3–10) \times 10^7$
Helium (He) fusion	2 million years	800,000 years	$(1–7.5) \times 10^8$
Carbon (C) fusion	2,000 years	500 years	$(0.8–1.4) \times 10^9$
Neon (Ne) fusion	8 months	11 months	$(1.4–1.7) \times 10^9$
Oxygen (O) fusion	2.6 years	5 months	$(1.8–2.8) \times 10^9$
Silicon (Si) fusion	18 days	0.7 day	$(2.8–4) \times 10^9$

and convection. Recall from your study of the Sun in Chapter 14 that neutrinos escape easily, carrying energy away from the core. Like air pouring out through a huge hole in the side of a balloon, neutrinos produced inside the star stream through the overlying layers of the star as though they were not even there, carrying the energy from the stellar interior out into space. As thermal energy pours out of the interior of the star, the outer layers of the star fall inward, driving up the density and temperature and increasing the rate of nuclear reactions.

Once neutrino cooling becomes significant, the star begins evolving much more rapidly. **Table 17.1** shows that the star fuses faster at each stage of nuclear fusion, so that the entire time spent fusing all the elements beyond hydrogen is very short in comparison with the main-sequence lifetime. A star that fuses silicon into iron in the core is not much more luminous than it was while fusing helium. But because of neutrino cooling, the silicon-fusing star actually releases about 200 million times more energy per second than it did while it was fusing helium—most of that energy is released in the form of neutrinos.

The Collapse of the Core and Subsequent Explosion

After silicon fuses to form an iron core in the star, the end comes suddenly and dramatically because no source of nuclear energy remains to replenish the energy emitted as neutrinos. No longer supported by thermonuclear fusion, the iron core of the massive star begins to collapse.

Figure 17.9 shows the stages a high-mass star passes through at the end of its life. The early stages of collapse of the iron core of an evolved massive star are much the same as those in the collapse of a nonfusing core in a low-mass star. As the core collapses, the force of gravity increases and the density and temperature skyrocket. The core becomes electron-degenerate when it is about the size of Earth. Unlike the electron-degenerate core of a low-mass red giant, however, the weight bearing down on the iron core is too great to be held up by electron degeneracy pressure (step 1 in Figure 17.9). As the core collapses, the core temperature climbs to 10 billion K (10^{10} K) and higher while the density exceeds 10^{10} kilograms per cubic meter (kg/m^3)—10 times the density of an electron-degenerate white dwarf.

The phenomenal temperatures and pressures trigger fundamental changes in the core. At those temperatures, the nucleus of the star is filled with thermal

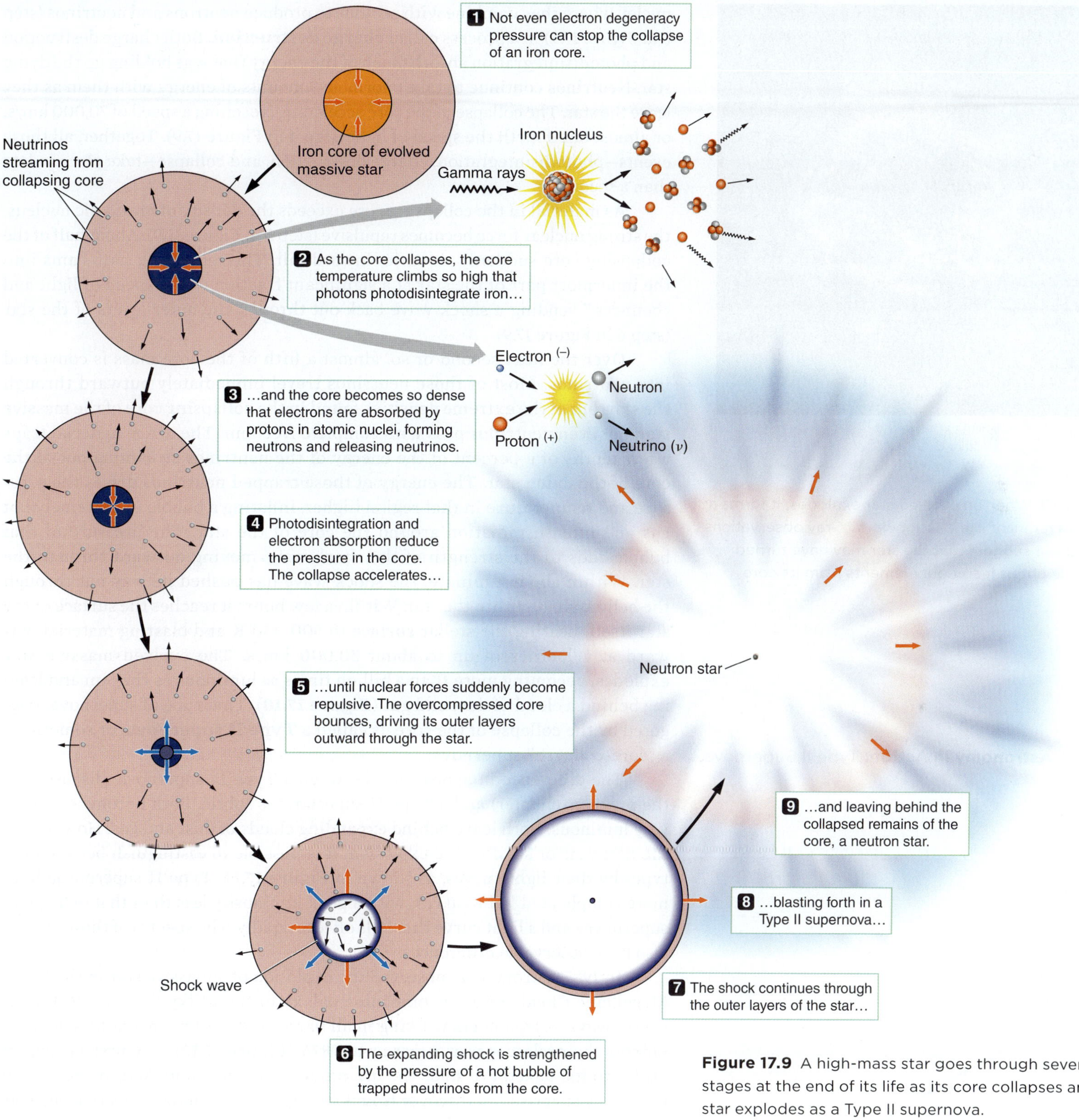

Figure 17.9 A high-mass star goes through several stages at the end of its life as its core collapses and the star explodes as a Type II supernova.

radiation so energetic that it is in the gamma-ray part of the spectrum. Those gamma-ray photons have enough energy to break iron nuclei apart into helium nuclei (step 2 in Figure 17.9) in a process called **photodisintegration**. Photodisintegration absorbs thermal energy and reverses the results of nuclear fusion. Also, the pressure in the core is now so great that electrons are forced into atomic

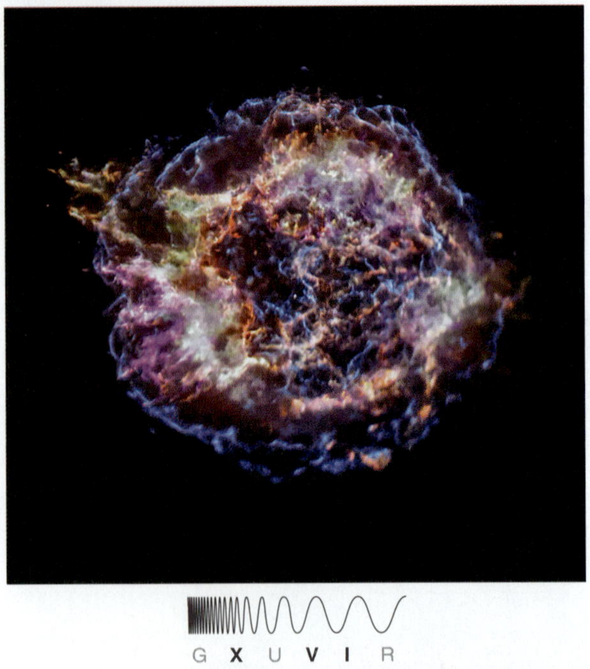

Figure 17.10 When an evolved star explodes, it forms a supernova remnant like Cassiopeia A. X-ray observations such as these suggest that the star may have turned itself inside out as it spit out elements from its core.

nuclei, where they combine with protons to produce neutrons and neutrinos (step 3 in Figure 17.9) in a process called **charge destruction**. Both charge destruction and photodisintegration absorb much of the energy that was holding up the dying star. Neutrinos continue to take enormous amounts of energy with them as they leave the star. The collapse of the core accelerates, reaching a speed of 70,000 km/s, or almost one-fourth the speed of light (step 4 in Figure 17.9). Together, all those events—photodisintegration, charge destruction, and collapse—take place in less than a second.

As material in the collapsing core exceeds the density of an atomic nucleus, the strong nuclear force becomes repulsive (step 5 in Figure 17.9). About half of the collapsing core suddenly slows its inward fall. The remaining half slams into the innermost part of the star at a significant fraction of the speed of light and "bounces," sending a shock wave back out through the outer layers of the star (step 6 in Figure 17.9).

Over the next second or so, almost a fifth of the core mass is converted into neutrinos. Most of those neutrinos travel immediately outward through the star; but at the extreme densities found in the collapsing core of the massive star, not even neutrinos pass with complete freedom. The dense material traps a few tenths of a percent of the energy of the neutrinos streaming out of the core of the dying star. The energy of those trapped neutrinos drives the pressure and temperature in that region higher, inflating a bubble of extremely hot gas and intense radiation around the core of the star. The pressure of that bubble adds to the strength of the shock wave moving outward through the star. Within about a minute, the shock wave has pushed its way out through the helium shell within the star. Within a few hours it reaches the surface of the star itself, heating the stellar surface to 500,000 K and blasting material outward at velocities of up to about 30,000 km/s. The evolved massive star explodes, becoming more than a billion times as luminous as the Sun and leaving behind a cloud of dust and gas (**Figure 17.10**). That type of supernova, triggered by the collapse of the core, is called a **Type II supernova**, or sometimes a "core-collapse" supernova.

Astronomy in Action: Type II Supernova

The difference in appearance between a Type Ia supernova (discussed in the previous chapter) and a Type II supernova is subtle. Both become suddenly very luminous. Both leave behind expanding clouds of dust and gas. However, in the first year or so after the explosion, it is possible to distinguish between the types by their light curves, as shown in **Figure 17.11**. Type II supernovae have more complicated light curves, with a peak luminosity less than that of Type Ia supernovae and a light curve that falls off less rapidly. The spectra of those objects also have observable differences.

In 1987, astronomers observed the explosion of a massive star in the Large Magellanic Cloud (LMC; a companion galaxy to the Milky Way 160,000 light-years away). Astronomers working in all parts of the electromagnetic spectrum pointed their telescopes at Supernova 1987A (**Figure 17.12**)—the first supernova visible to the naked eye since the invention of the telescope. Astronomers were ultimately surprised to discover from looking at older photographs that the star that blew up was not a red supergiant but rather a 20-M_{Sun} B3 I blue supergiant now classified as a luminous blue variable star. Neutrino telescopes recorded a burst of neutrinos passing through Earth from that tremendous stellar explosion that had occurred in the LMC. The detection of neutrinos from SN 1987A gave astronomers a rare glimpse of the very heart of a massive star at the moment of its death, confirming a fundamental prediction of theories about the collapse of the core and its effects.

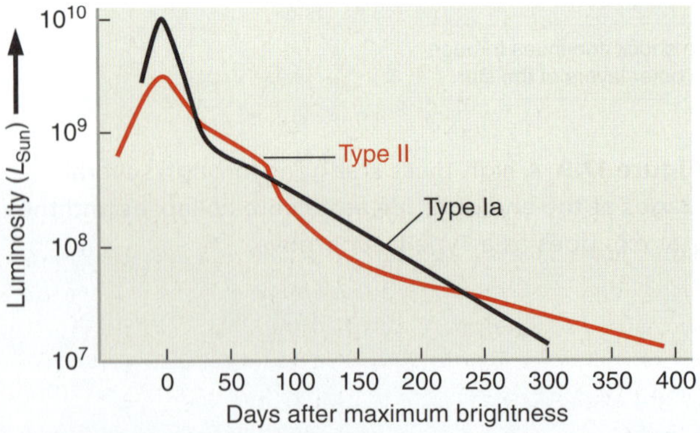

Figure 17.11 These light curves show the changes in brightness of average Type Ia and Type II supernovae.

CHECK YOUR UNDERSTANDING 17.2

What causes a high-mass star to explode as a Type II supernova? (a) The high-mass star merges with another star. (b) Iron absorbs energy when it fuses. (c) The high-mass star runs out of mass to fuse in the core. (d) The CNO cycle uses up all the carbon.

17.3 The Spectacle and Legacy of Supernovae

Supernova explosions leave a rich and varied legacy in the universe. Although only one per century or so occurs in an individual galaxy, many of them happen every second in all the galaxies of the universe. Explosions that occurred thousands of years ago have created huge expanding bubbles of million-kelvin gas that glow in ultraviolet and X-ray radiation and drive visible shock waves into the surrounding interstellar medium. The ejected energy and matter also compress nearby clouds, sometimes triggering the initial collapse that begins star formation. In this section, we discuss what happens after a massive star explodes as a supernova.

The Energetic and Chemical Legacy of Supernovae

Outflowing particles carry away approximately 10^{47} J of kinetic energy—enough energy to accelerate the entire Sun to a speed of 10,000 km/s. The kinetic energy released from both Type Ia and Type II supernovae heats the hottest phases of the interstellar medium and stirs the clouds in the interstellar medium. That kinetic energy is about 100 times as much energy as is carried away by light. Yet even that amount of energy is less than the energy carried away from the supernova explosion by neutrinos—an amount of energy at least another 100 times larger.

The chemical changes in the galaxy caused by supernova explosions may be even more important than the energy released. Only very low-mass chemical elements were present at the beginning of the universe: hydrogen, helium, and trace amounts of lithium and possibly beryllium. All the rest of the naturally occurring elements on the periodic table were released to the interstellar medium when stars died. The process of forming more massive atomic nuclei from less massive nuclei is called **nucleosynthesis**. Nucleosynthesis is responsible for the buildup of massive elements in the universe, forming those elements either during nuclear fusion in the core of a star or during the rapid nuclear reactions that occur during a supernova explosion.

Elements up to carbon, oxygen, and small amounts of neon and magnesium form from nuclear fusion in the cores of low-mass stars and are released to the interstellar medium as the star loses its outer layers. Recall that fusion up to iron creates energy, but fusion beyond iron absorbs energy (see Figure 17.8). A look at the periodic table of the elements (see Appendix 3) shows that many naturally occurring elements are more massive than iron. Elements heavier than iron fuse only under conditions in which abundant energy is available to be absorbed—as in supernova explosions. The naturally occurring elements heavier than iron were all produced in the deaths of high-mass stars.

Because like charges repel one another, electric repulsion normally keeps atomic nuclei far apart. Extreme temperatures and pressures are needed to slam nuclei together hard enough to overcome that electric repulsion. Neutrons, however, have no net electric charge, so free neutrons are not repelled by an atomic nucleus. Under normal conditions, free neutrons are rare. In the interiors of evolved stars, however, nuclear reactions occur that produce free neutrons. Under the conditions present shortly before and during a supernova, free neutrons are produced

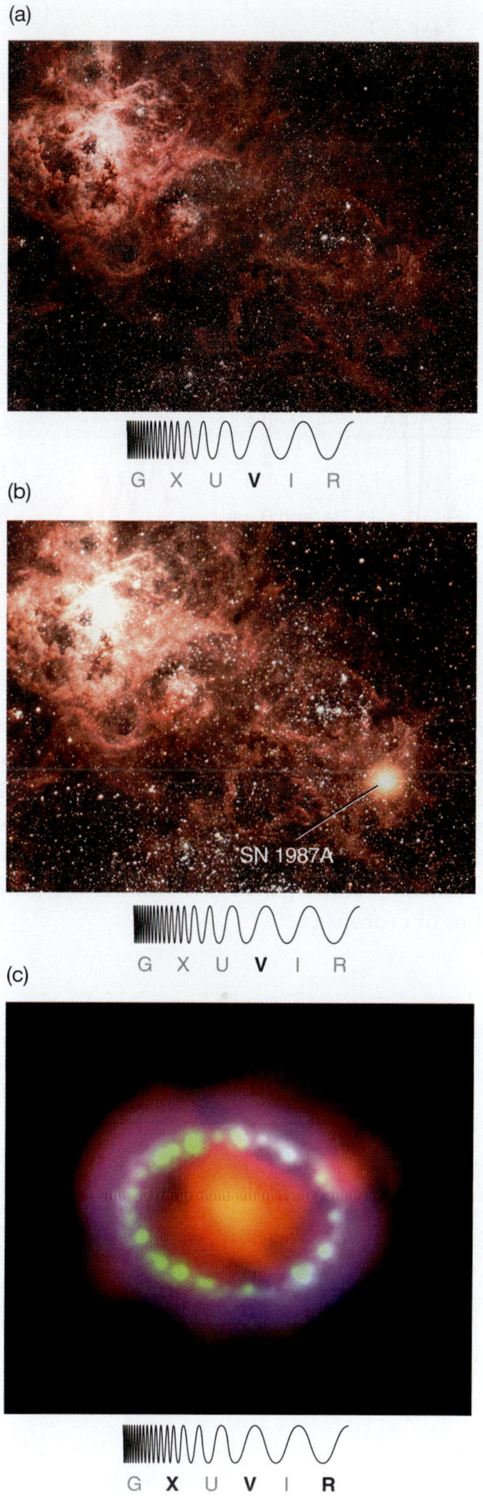

Figure 17.12 Supernova 1987A was a supernova that exploded in a small companion galaxy of the Milky Way called the Large Magellanic Cloud (LMC). These images show the LMC before the explosion (a) and while the supernova was near its peak (b). Notice the "new" bright star at lower right. (c) An X-ray image from *Chandra* is combined with Hubble Space Telescope optical and ground-based infrared images to show freshly formed dust inside the glowing rings of gas.

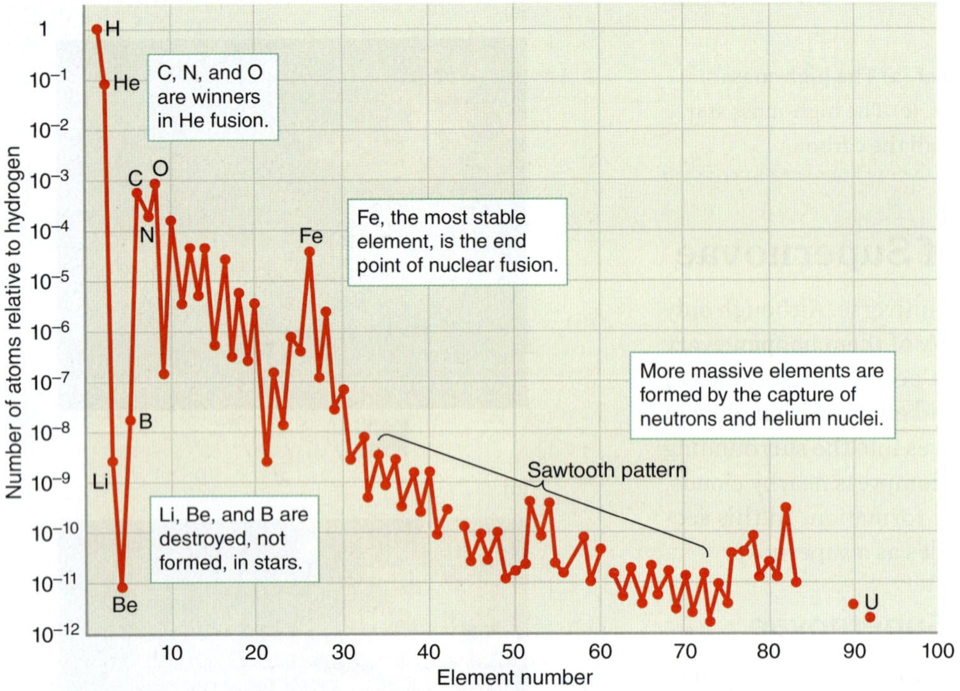

Figure 17.13 Observed relative abundances of different elements in the Solar System plotted against the element number of each element's nucleus. That pattern can be understood as a result of nucleosynthesis in stars. The periodic table of the elements in Appendix 3 identifies individual elements by their element number (the number of protons).

in very large numbers. Those free neutrons are easily captured by atomic nuclei and later decay to become protons. That process of **neutron capture** and decay forms the elements with atomic numbers and masses higher than those of iron. The abundance of helium in the outflowing material means that **helium capture**, in which a helium nucleus is captured by another nucleus, also is relatively common. Each time it occurs, helium capture increases the atomic number of the capturing atom by two protons. Proton capture also occurs, in which a free proton (a hydrogen nucleus) is captured into a larger nucleus. That type of capture is most probable if the capturing nucleus has an odd number of protons.

Nuclear physics predicts how the abundances of the elements changes as a result of those energetic explosions. Those predictions agree with abundances measured on Earth, in the Solar System (**Figure 17.13**), and in the atmospheres of stars and their remains. In general, less massive elements are more abundant than more massive elements because more massive elements are progressively built up from less massive elements. An exception to that pattern is the dip in the abundances of the light elements lithium (Li), beryllium (Be), and boron (B). Nuclear fusion easily destroys those elements, and they are not produced by the common reactions involved in fusing hydrogen (H) and helium (He). Conversely, carbon (C), nitrogen (N), and oxygen (O) are produced in quantity in the triple-alpha process, so they are abundant. The relatively high abundances of elements near iron is evidence of processes that favor those tightly bound nuclei. The sawtooth pattern observed in the figure's plot results from both how elements form in stars and how they persist in space. Atoms with odd numbers of protons are more likely to capture another proton, and atoms with even numbers of protons are more stable, so even-numbered elements are more abundant in the current universe. By comparing the predictions of nuclear physics with observations of elemental abundances, astronomers repeatedly test the theory of stellar evolution.

Neutron Stars and Pulsars

In the explosion of a Type II supernova, the outer parts of the star are blasted back into interstellar space—but what remains of the collapsing core left behind? For cores less than about 3 M_{Sun}, the collapse is halted when neutrons are packed as tightly together as the rules of quantum mechanics permit. That *neutron-degenerate* core left behind by the explosion of a Type II supernova is called a **neutron star**. It has a radius of 10–15 km, making it roughly the size of a small city, but it contains between 1.4 and around 2 M_{Sun}. (Cores greater than ~2 M_{Sun} have too much gravity to be supported by neutron degeneracy and become black holes, to be discussed in the next chapter.) Combining the mass and the size shows that the neutron star has a density of about 10^{18} kg/m^3—about a billion times denser than a white dwarf and roughly equivalent to the density of an atomic nucleus. If Earth were crushed down to the size of a football stadium, the planet would then have about the same density as a neutron star. Neutron stars are so compact that the acceleration due to gravity on the surface is more than 100 billion times the

17.2 Working It Out Gravity on a Neutron Star

Neutron stars are incredibly dense objects. As a result, the surface gravity and escape velocity of neutron stars are very high. For example, let's look at a typical neutron star with a radius of 15 km and a mass of 2 M_{Sun}.

Recall from Working It Out 4.1 that the acceleration due to gravity on the surface—here, the surface of a neutron star (NS)—is given by

$$g = \frac{GM_{NS}}{R^2_{NS}}$$

$$g = 6.67 \times 10^{-20} \frac{km^3}{kg\,s^2} \times \frac{2.0 \times (1.99 \times 10^{30}\,kg)}{(15\,km)^2}$$

$$g = 1.2 \times 10^9 \frac{km}{s^2}$$

Dividing that number by the gravitational acceleration on Earth, 9.8 m/s^2 = 0.0098 km/s^2, shows that the gravitational acceleration on a neutron star is more than 100 billion times as large as that on Earth.

What about the escape velocity from a neutron star? From Working It Out 16.2, we know that the escape velocity is given by

$$v_{esc} = \sqrt{\frac{2GM}{R}}$$

Putting in the above numbers for a typical neutron star yields

$$v_{esc} = \sqrt{\frac{2 \times [6.67 \times 10^{-20}\,km^3/(kg\,s^2)] \times 2.0 \times (1.99 \times 10^{30}\,kg)}{(15\,km)}}$$

$$v_{esc} = 190{,}000\,km/s$$

Dividing that result by the speed of light gives

$$\frac{v_{esc}}{c} = \frac{190{,}000\,km/s}{300{,}000\,km/s} = 0.63$$

The escape velocity from that neutron star is more than 60 percent of the speed of light and almost 17,000 times greater than the escape velocity from Earth (11.2 km/s). The physicist Albert Einstein showed that strange things happen at velocities near the speed of light (including modifications to Newton's equations), which we will discuss in Chapter 18.

acceleration on Earth, as shown in **Working It Out 17.2**. That extremely high surface gravity implies a very large escape velocity. A spacecraft would need to be traveling at 0.63*c* to escape from the surface of a typical neutron star.

Many of the unusual objects discussed in this and the previous chapter—such as pulsating stars, supernovae, and planetary nebulae—puzzled astronomers when they were first observed but were later understood to be associated with the end points of stellar evolution. In contrast, neutron stars were predicted in 1934, not long after neutrons themselves were discovered. Astronomers Walter Baade and Fritz Zwicky proposed that supernova explosions could lead to the formation of a neutron star. But neutron stars were not actually observed for another 30 years. Rapidly pulsing objects were first discovered in 1967 by people observing radio wavelengths (see the **Process of Science Figure**).

Those objects, which blinked like very fast, regularly ticking clocks, puzzled astronomers. Today, those rotating neutron stars are called **pulsars**. More than 2,000 pulsars are known, and more are being discovered all the time.

Recall from Chapter 15 that a collapsing molecular cloud begins to spin faster as it shrinks into a protostar because angular momentum is conserved. For the same reason, as the core of a massive star collapses, it spins faster. A massive main-sequence O star may rotate once every few days. As a neutron star, it might rotate as fast as a thousand times each second.

The collapsing star also concentrates the magnetic field to strengths trillions of times greater than the magnetic field at Earth's surface. That magnetic field does not necessarily have its poles aligned with the rotation axis of the neutron star. A neutron star has a magnetosphere just like Earth does, except that the neutron star's magnetosphere is much stronger and rotates very fast.

Astronomy in Action: Pulsar Rotation

Occam's Razor

Occam's razor is a guiding principle in science: when scientists consider two hypotheses that explain a phenomenon equally well, they should give preference to the "simpler" theory and prioritize testing it. "Simpler" does not mean that the math is easier, or even that the concept is easy to understand. It means that the fewest number of other, new assumptions need to be made.

In 1967, Jocelyn Bell, a student at Cambridge, discovered a "mystery signal" in her data. Her adviser, Anthony Hewish, half-jokingly suggested "little green men" caused the signal.

Bell and Hewish find three more such signals. It was unlikely that the same "little green men" would be sending the same signal from four separate locations in the sky.

Bell and Hewish suggested the signals were from pulsating white dwarfs or neutron stars.

Meanwhile, Franco Pacini and Thomas Gold developed a detailed explanation involving rotating neutron stars. This explanation relied entirely on previously understood physical phenomena: rotation, magnetic fields, and neutron stars. It did not require assumptions about the existence of extraterrestrials.

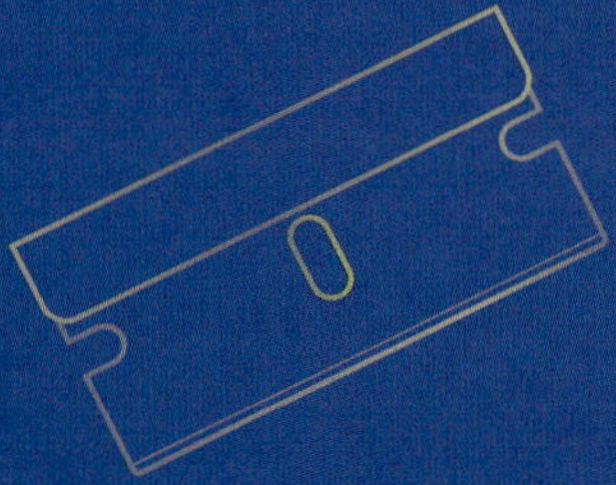

The neutron star explanation is "simpler" and has become the accepted explanation for Bell's observations.

Simpler theories are easier to rule out, so if they survive testing, then scientists are more likely to consider them more seriously.

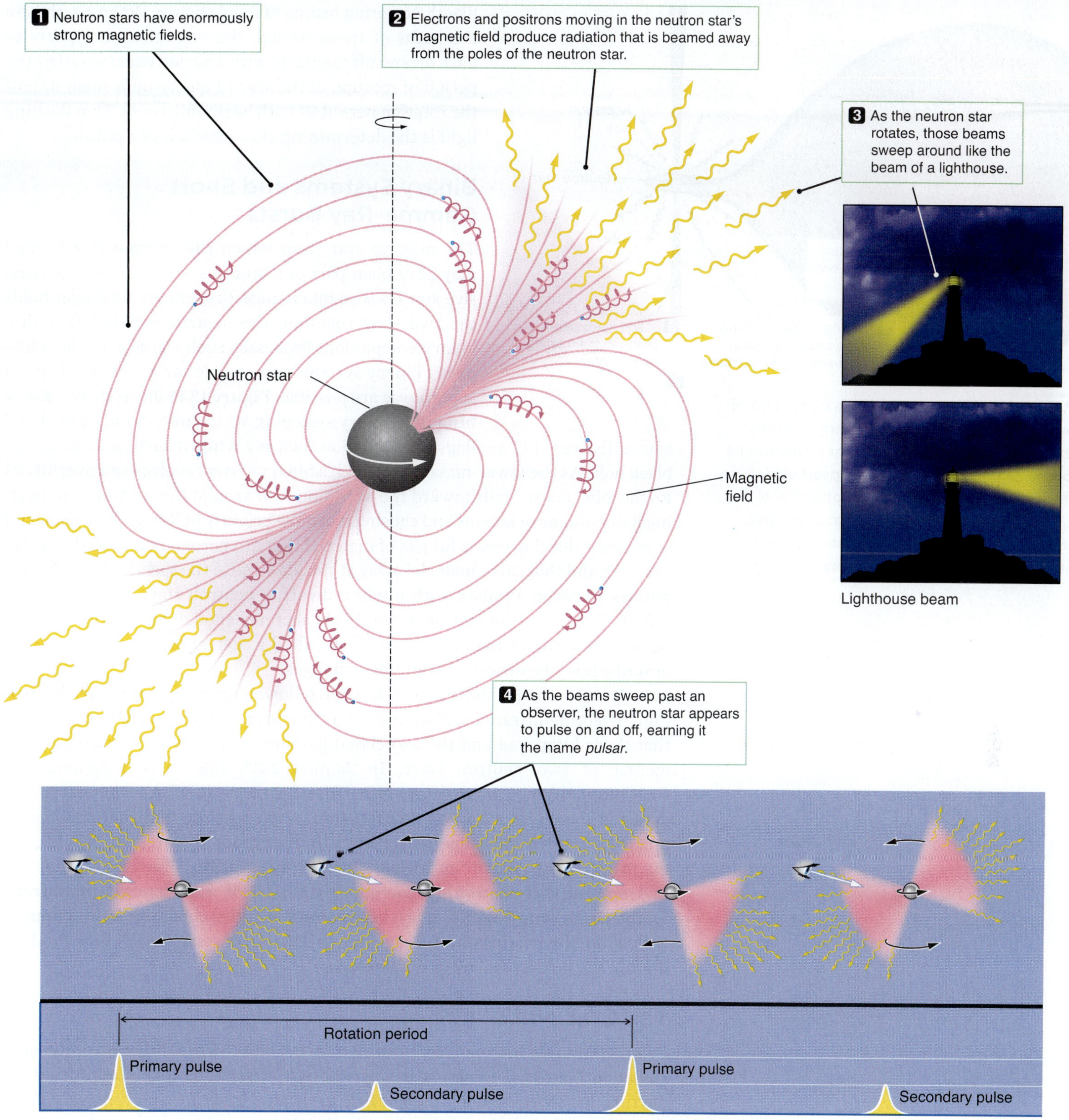

Figure 17.14 When a highly magnetized neutron star rotates rapidly, light is given off, much like the beams from a rotating lighthouse lamp. As those beams sweep past Earth, the star will appear to pulse on and off, earning it the name *pulsar*.

Electrons and positrons become trapped in the magnetic-field lines of the neutron star and are "funneled" along the field toward the magnetic poles of the system. Any accelerating charged particle produces radiation, so those particles produce beams of radiation along the magnetic poles of the neutron star, as shown in **Figure 17.14**. As the neutron star rotates, those beams sweep through space

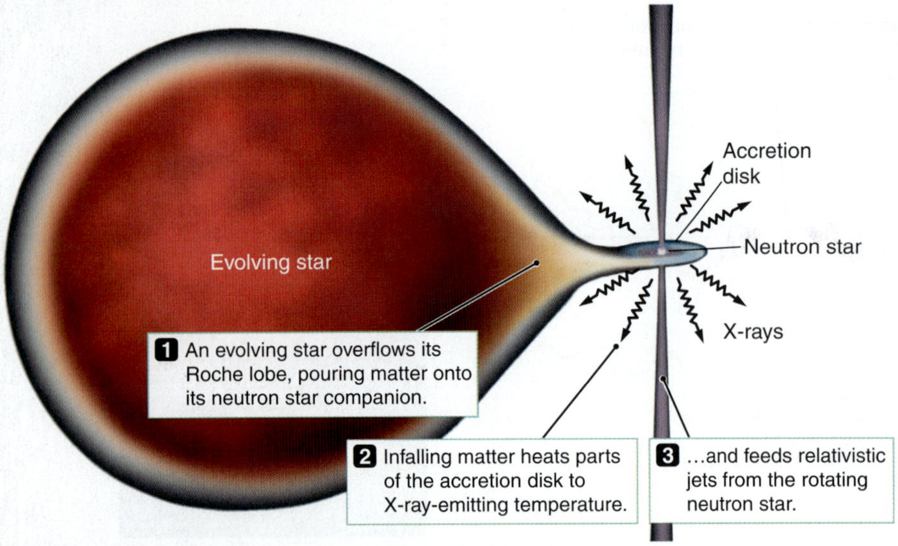

Figure 17.15 X-ray binaries are systems consisting of a normal evolving star with a white dwarf, a neutron star, or a black hole. As the evolving star overflows its Roche lobe, mass falls toward the collapsed object. The gravitational well of the collapsed object is so deep that when the material hits the accretion disk, the disk is heated to such high temperatures that it radiates away most of its energy as X-rays.

like the rotating beams of a lighthouse. When Earth is in the paths of those beams, the neutron star appears to flash on and off regularly with a period equal to either the period of rotation of the star (if one beam is seen) or half the rotation period (if both beams are seen). That flashing light is the determining characteristic of a pulsar.

Binary Systems and Short Gamma-Ray Bursts

The massive star from which the neutron star formed may have been part of a binary system. Unless the massive star loses so much mass that gravity no longer holds the two stars together, the neutron star is left with a binary companion. Processes such as those in the white dwarf binary systems responsible for novae and Type Ia supernovae are possible. **Figure 17.15** illustrates an **X-ray binary**, a binary system in which mass from an evolving star spills over onto a collapsed companion such as a white dwarf, neutron star, or black hole. As the lower-mass star in such a binary system evolves and overfills its Roche lobe, matter falls toward the accretion disk around the neutron star, heating it to millions of kelvins and causing it to glow brightly in X-rays. X-ray binaries sometimes develop powerful jets of material that are perpendicular to the accretion disk and that carry material away at speeds close to the speed of light. Binary pulsars combine a pulsar with a white dwarf, a neutron star, or even another pulsar. If the neutron star or white dwarf accretes enough matter to become a nova, once or more than once, it is often called a "cataclysmic variable star" to describe how the observed brightness changes.

In 2013, the Hubble Space Telescope followed up on a gamma-ray observation, detecting infrared radiation from a "kilonova." Astronomers hypothesized that those kilonovae and the associated gamma-ray bursts were caused by the merger of two neutron stars. In August 2017, the Laser Interferometer Gravitational-Wave Observatory (LIGO) and a plethora of other observatories detected the collision of two neutron stars. That was the first time that astronomers observed the same object both in gravitational waves and in electromagnetic waves. That "multi-messenger" observation allowed astronomers to determine that neutron star mergers are a cause of **short gamma-ray bursts**, bright gamma-ray events that last between a few milliseconds and a few minutes. Follow-up observations have suggested that neutron star mergers may produce most of the gold and silver in the universe.

The Crab Nebula

In 1054, Chinese astronomers noticed a "guest star" in the direction of the constellation Taurus. The new star was so bright that it could be seen during the daytime for 3 weeks, and it did not fade away altogether for many months. From the Chinese description of the changing brightness and color of the object, modern astronomers have concluded that the guest star of 1054 was a fairly typical Type II supernova. Today, an expanding cloud of debris from the explosion occupies that place in the sky—forming an object called the Crab Nebula (see the chapter-opening photograph).

The Crab Nebula has filaments of glowing gas expanding away from the central star at 1,500 km/s—50 times faster than the expansion rate of a planetary nebula. Those filaments contain unusually high abundances of helium and other more massive chemical elements—the result of the nucleosynthesis that took place in the supernova and its progenitor star.

The Crab pulsar at the center of the nebula flashes 60 times each second: first with a main pulse associated with one of the "lighthouse" beams and then with a fainter secondary pulse associated with the other beam. As the Crab pulsar spins, it carries its powerful magnetosphere around with it. A few thousand kilometers from the pulsar, material in its magnetosphere must move at almost the speed of light to keep up with that rotation. The rotating magnetosphere flings particles away from the neutron star in a powerful wind moving at nearly the speed of light. That wind fills the space between the pulsar and the expanding shell. The Crab Nebula is like a big balloon, but instead of being filled with hot air, it is filled with a mix of very fast particles and strong magnetic fields. The energy that accelerates those particles comes from the pulsar's rotation. The pulsar slows down as it loses energy to the particles. Images of the Crab Nebula show that bubble as a glow from synchrotron radiation—a type of beamed radiation emitted as very fast moving particles spiral around the magnetic field.

CHECK YOUR UNDERSTANDING 17.3

One reason astronomers think neutron stars were formed in supernova explosions is that: (a) all supernova remnants contain pulsars; (b) pulsars are made of heavy elements, such as those produced in supernova explosions; (c) pulsars spin very rapidly, as did the massive star just before it exploded; (d) pulsars sometimes have material around them that looks like the ejecta from supernovae.

17.4 Star Clusters Are Snapshots of Stellar Evolution

Recall from Chapter 15 that when an interstellar cloud collapses, it breaks into pieces, forming many stars of different masses. Those groups of gravitationally bound stars are called star clusters. **Globular clusters** are densely packed collections of tens of thousands to a few million stars (**Figure 17.16a**). **Open clusters** are much less tightly bound collections of a few dozen to a few thousand stars (**Figure 17.16b**). Because stars in a cluster form out of the same cloud at nearly the same time, observations of star clusters can be used to determine how stars of different masses evolve.

(a)

(b)

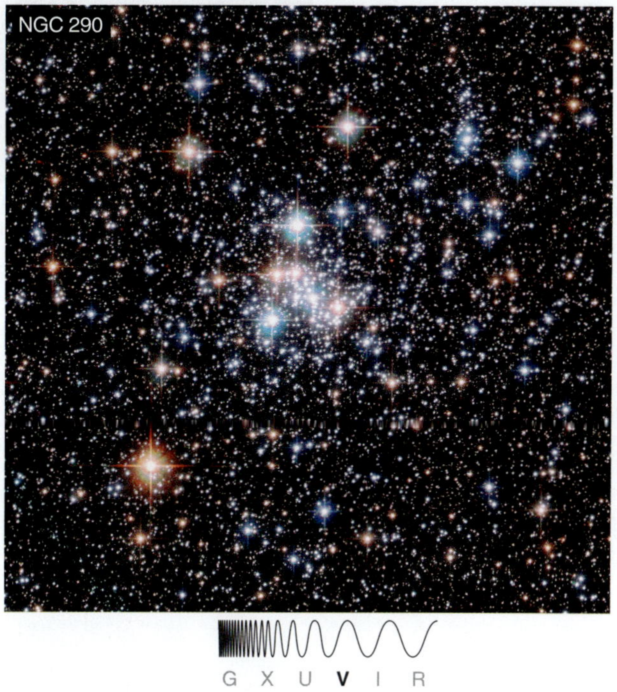

Figure 17.16 (a) Globular clusters can have tens of thousands to a few million stars. (b) Open clusters have up to hundreds or thousands of stars. These images are from the Hubble Space Telescope.

Cluster Distances and Ages

In the 1920s, astronomers plotted the observed brightness versus the spectral type for as many stars as possible in each cluster. The resulting cluster H-R diagrams showed stars of all the categories in the "textbook" H-R diagram (see Figure 13.15). Because all the stars in a given cluster are approximately the same distance from Earth, the effect of distance on the brightness of each star is the same. Shifting the main sequence of the cluster up so that it overlaps the main sequence of the "textbook" H-R diagram indicates how much the brightness of the cluster has been diminished by distance. That is one method, called "main sequence fitting," that astronomers can use to determine the distance to a cluster.

The H-R diagrams of clusters also offer snapshots of stellar evolution. All the stars in a cluster formed together at nearly the same time, so by the time the cluster is 10 million years old, all the stars in the cluster also are 10 million years

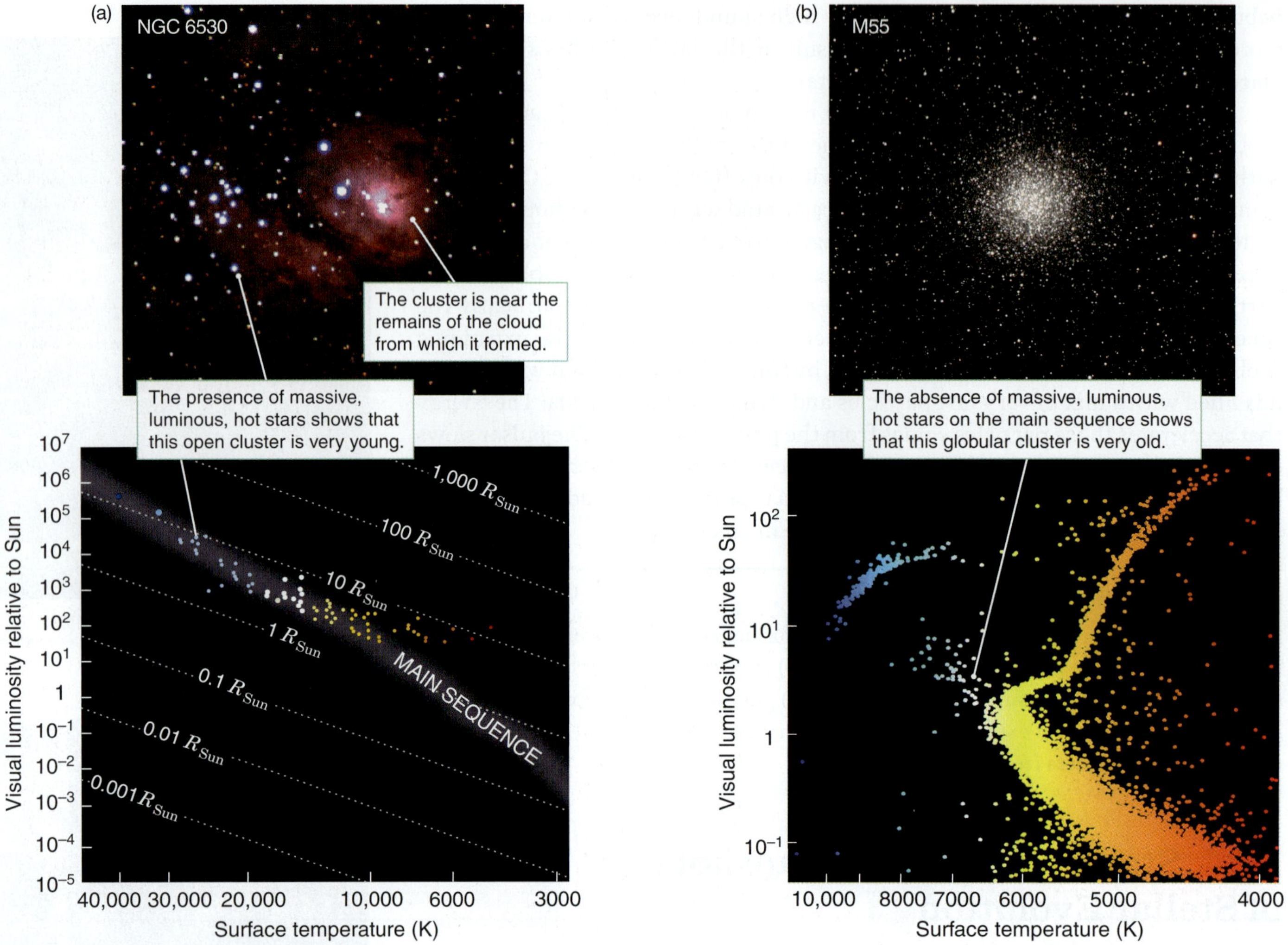

Figure 17.17 H-R diagrams of (a) a very young cluster (2 million years old), NGC 6530, and (b) a very old cluster (12 billion years old), M55. In (a), some of the stars haven't yet arrived on the main sequence. In (b), more than the top half of the main sequence has already evolved. The vertical scales are logarithmic; (b) is zoomed in compared with (a).

old. Comparing clusters of different ages shows which stars arrive on the main sequence first and which evolve off it first. **Figure 17.17a** shows an H-R diagram of a very young cluster, NGC 6530. O, B, and A stars are on the main sequence; F through M stars with lower masses are still evolving *to* the main sequence. No red giants or white dwarfs are present. In contrast, **Figure 17.17b** shows the H-R diagram of a very old cluster, M55. No high-mass (O and B) stars are on the main sequence because they have evolved off it, but stars are on the horizontal, red giant, and asymptotic giant branches and in the lower part of the main sequence. That globular cluster is about 12 billion years old.

Astronomers cannot watch an individual cluster age over millions of years, but they can observe clusters of different ages. Astronomers explore cluster evolution by examining H-R diagrams of a *simulated* cluster of 40,000 stars as it would appear at several ages and then comparing it with the observed H-R diagrams of actual clusters. Star formation in a molecular cloud is spread over several million years, and it takes considerable time for lower-mass stars to contract to reach the main sequence. The H-R diagram of a very young cluster typically shows many lower-mass stars located well above the main sequence; eventually, they move onto the main sequence. In **Figure 17.18a**, we have ignored that complication to more clearly define the main sequence itself.

The more massive a star is, the shorter its life on the main sequence will be. **Figure 17.18b** shows that after only 4 million years, all stars with masses greater

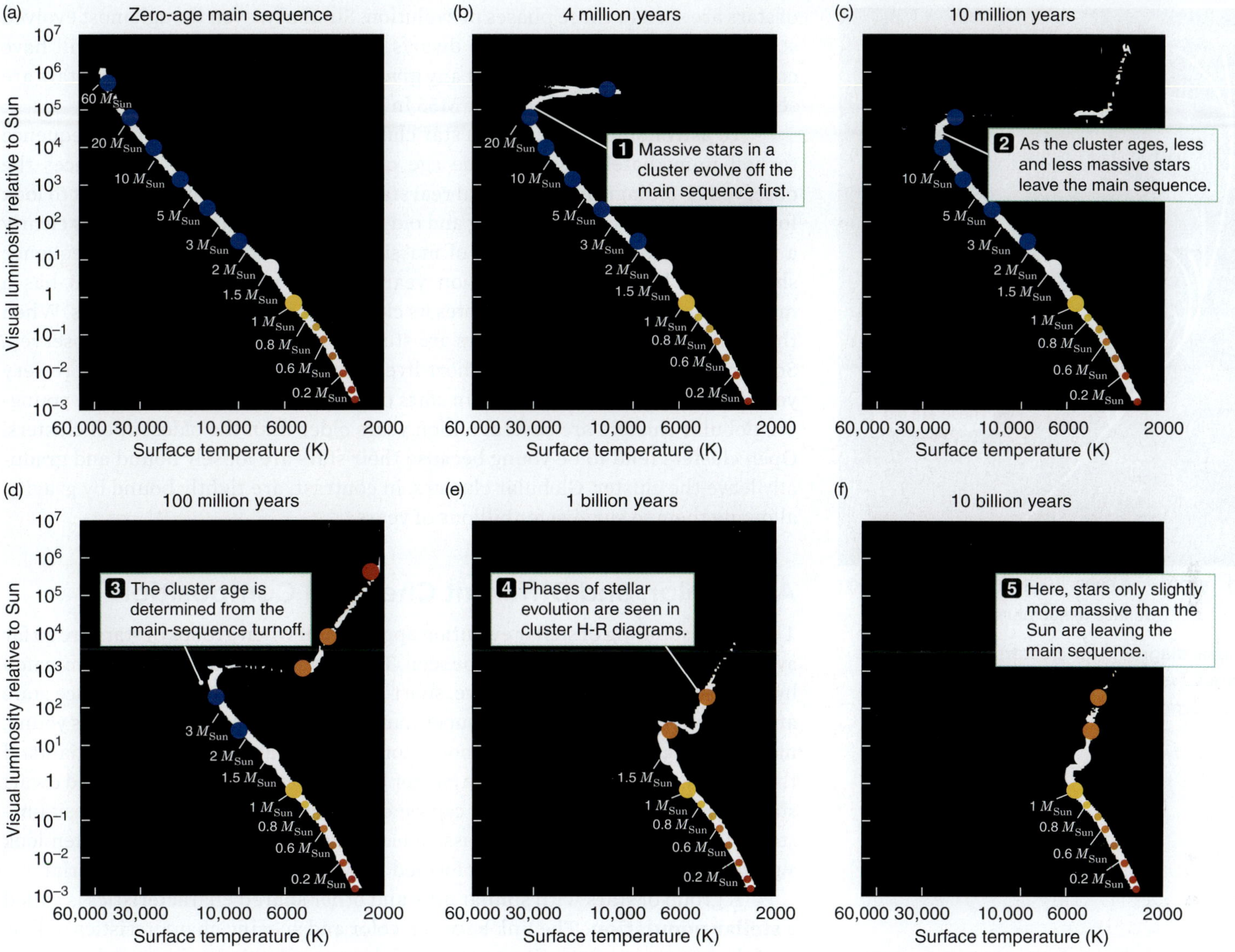

Figure 17.18 H-R diagrams of star clusters are snapshots of stellar evolution. These H-R diagrams of a simulated cluster of 40,000 stars of solar composition are shown at different times after the birth of the cluster. Note the progression of the main-sequence turnoff to lower and lower masses. In the simulation, the stars are all placed on the main sequence at zero age. In reality, however, the lowest-mass stars have not yet reached the main sequence by the time the most massive stars have left it.

than about 20 M_{Sun} have evolved off the main sequence and are now spread across the top of the H-R diagram. The most massive stars have already disappeared from the H-R diagram, having vanished in supernovae. As time goes on, stars of lower and lower mass evolve off the main sequence, and the turnoff point moves toward the bottom right in the H-R diagram. By the time the cluster is 10 million years old, illustrated in **Figure 17.18c**, only stars with masses less than about 15 M_{Sun} remain on the main sequence. The location of the most massive star that remains on the main sequence is called the **main-sequence turnoff**. As the cluster ages, the main-sequence turnoff moves farther and farther down the main sequence to stars of lower and lower mass.

As a cluster ages further, shown in **Figures 17.18d and e**, we see the details of all stages of stellar evolution. By the time the star cluster is 10 billion years old (**Figure 17.18f**), stars with masses of only 1 M_{Sun} are beginning to die. Stars slightly more massive than that are seen as giant stars of various types. Note how few supergiant and giant stars are present in any of the cluster H-R diagrams. The supergiant, giant, horizontal, and asymptotic giant branch phases in the evolution of stars pass so quickly in comparison with a star's main-sequence lifetime that even though that simulated cluster started with 40,000 stars, only a handful

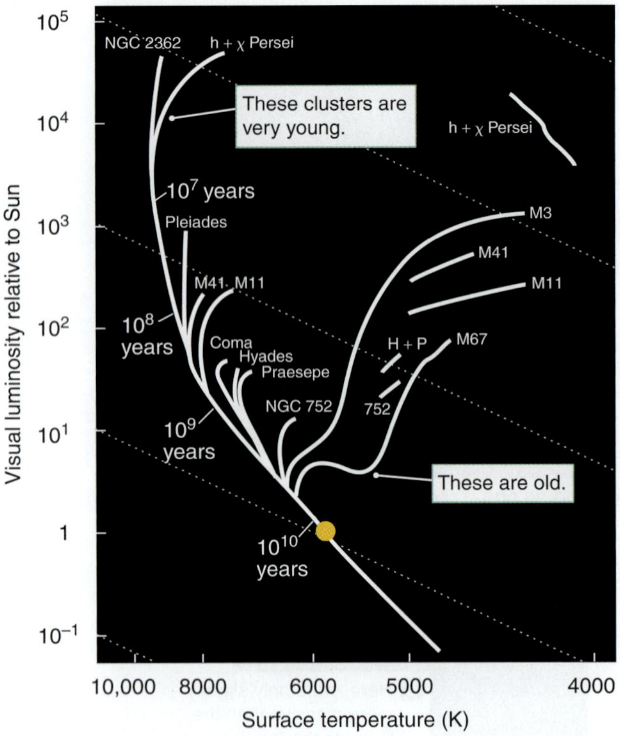

Figure 17.19 H-R diagrams for clusters having a range of ages. The ages associated with the different main-sequence turnoffs are indicated.

of stars are seen in those phases of evolution. Similarly, even though most evolved stars in an old cluster are white dwarfs, all but a few of those stars will have cooled and faded into obscurity at any given time. More of those evolved stars are seen in the larger globular cluster M55 in Figure 17.17b.

To astronomers observing a star cluster, the location of the main-sequence turnoff immediately indicates the age of the cluster. **Figure 17.19** traces the observed H-R diagrams for several real star clusters. Once you know what to look for, the difference between young and old clusters is obvious. NGC 2362 is clearly a young cluster. Its complement of massive, young stars on the main sequence shows it to be only a few million years old. In contrast, cluster M3 has a main-sequence turnoff that indicates its cluster age is about 11 billion years. When the H-R diagrams of open clusters are studied, a wide range of ages is observed. Some open clusters contain the short-lived O and B stars and are therefore very young. Other open clusters contain stars older than the Sun. But even the youngest globular clusters are several billion years older than the oldest open clusters. Open clusters tend to be young because their stars are loosely bound and gradually leave the cluster. Globular clusters, in contrast, are tightly bound by gravity, allowing them to survive for billions of years.

Age, Color, and Different Chemical Composition

That understanding of stellar evolution applies even when groups of stars are so far away that individual stars cannot be seen. The light from a star cluster is dominated by its most luminous stars: massive, short-lived blue O and B main-sequence stars and evolved post-main-sequence supergiant and giant stars. If the cluster is young, most of the light comes from luminous, hot, blue stars and some red supergiants. If the cluster is old, the light from the cluster has the color of red giants and red dwarf stars. However, that relationship can be complicated by chemical composition—stars with smaller amounts of massive elements in their atmospheres often look significantly bluer than stars that formed from chemically enriched material.

A group of stars with similar ages and other shared characteristics is called a **stellar population**. The link between color and specific characteristics will be useful when we begin to discuss the much larger collections of stars called galaxies. Astronomers usually can figure out something about the properties of a stellar population from its overall color. An especially bluish color to a galaxy or a part of a galaxy often signifies that the galaxy contains a young stellar population that still includes hot, luminous, blue stars that formed recently. In contrast, a galaxy or part of a galaxy that has a reddish color is usually composed primarily of an old stellar population.

All elements more massive than boron are formed in stars and are expelled into the interstellar medium when a star dies. In main-sequence stars, material from the core does not mix with material in the atmosphere, so the abundances of chemical elements inferred from the spectrum of a star are the same as the abundances in the interstellar gas from which the star formed. The abundance of massive elements in the interstellar medium serves as a record of the cumulative amount of star formation that has taken place so far. Gas that shows large abundances of massive elements has gone through a great deal of stellar processing—so it contains more "recycled" material. Gas with low abundances of massive elements is more pristine.

Stars in globular clusters contain only very small amounts of massive elements; some globular-cluster stars contain only 0.5 percent as much of those massive elements as the Sun does, indicating that they were among the earliest stars to form. Open clusters are younger and formed from a more enriched interstellar medium. Therefore, they have larger amounts of the more massive elements.

Even the very oldest globular-cluster stars contain massive chemical elements. Therefore, at least one generation of massive stars must have lived and died, ejecting newly synthesized massive elements into space, before even those oldest globular clusters formed. Most of the luminous matter in the universe is still composed of hydrogen and helium formed before the first stars. Even a chemically rich star like the Sun, which is made of gas processed through approximately 9 billion years of previous generations of stars, is still composed of less than 2 percent massive elements. In upcoming chapters, you will learn that those variations in the chemical content of stars indicate a lot about the chemical evolution of galaxies.

CHECK YOUR UNDERSTANDING 17.4

If the main-sequence turnoff of a globular cluster occurs near the very top of the main sequence, the cluster is: (a) very old; (b) very young; (c) very hot; (d) very dense.

Origins Seeding the Universe with New Chemical Elements

On Earth, massive elements are everywhere. The surfaces of rocky planets contain silicon, oxygen, magnesium, and sodium. The iron-and-nickel solid inner core and liquid outer core of Earth are responsible for Earth's magnetic field. The most common chemical elements in biological molecules are carbon (C), hydrogen (H), nitrogen (N), oxygen (O), phosphorus (P), and sulfur (S)—all but hydrogen created in a dying star. The fact that those elements are here means that the Sun is not a first-generation star: it formed from material in the interstellar medium enriched by material from dying massive stars.

In supergiant, giant, and AGB stars, more massive chemical elements that formed from nuclear fusion deep within their interiors are carried upward and mixed with material in the outer parts of the star. As a star ages, its core grows hotter, and the temperature gradient within the star grows steeper. Under certain circumstances, convection can spread so deep into a star that chemical elements formed by nuclear fusion within the star are dredged up and carried to the star's surface. For example, the spectra of some AGB stars show an excess of carbon and other by-products of nuclear fusion. That extra carbon originated in each star's helium-fusing shell and was carried to the surface by convection. For stars with lower masses, stellar winds and planetary nebulae carry the enriched outer layers off into interstellar space. The nuclear fusion that occurs in supergiant stars goes well beyond the formation of elements such as carbon. Supernova explosions seed the universe with much more massive atoms, from iron and nickel up to uranium.

The oxygen atoms in the air you breathe and the water you drink were created by nucleosynthesis in dying stars. The iron atoms that are a key element of hemoglobin, which makes up the red blood cells that carry oxygen from your lungs to the rest of your body, formed in the explosions of massive stars. The nickel, copper, and zinc atoms in the coins in your pocket and the rare-earth atoms in your electronics were created in exploding massive stars. The Sun, the planets (including Earth), and all life on Earth are made of recycled stars. Supernovae are in you.

ARTICLES QUESTIONS

These remaining cores of dead stars can only get so massive before they become black holes.

Astronomers Find Mass Limit for Neutron Stars Before Collapsing into Black Holes

By John Wenz, *Popular Mechanics*

The subtle difference between when a massive dying star compresses into a core and when it collapses entirely may have been found. In a study published in *Astrophysical Journal Letters*, researchers at the Goethe University in Frankfurt say they've found the dividing line between compact objects called neutron stars and black holes.

When a massive star reaches the end of its life, it goes out with an immense bang called a supernova. From there, one of two known things can happen: it either becomes a black hole, which has so much gravity not even light can escape, or a neutron star, which is a city-sized corpse of a formerly large star that's made out of incredibly dense neutron matter.

But astrophysicists have struggled to find out exactly what variations cause a large star to compress into a dense stellar remnant, a neutron star, rather than the inescapable void of matter-eating fury that is a black hole. According to the Goethe researchers, the difference is simple: 2.16 solar masses. Any leftover object after a supernova that is less than 2.16 times the mass of the sun will [become] a neutron star, while anything more than 2.16 solar masses will become a black hole.

Most neutron stars are between one and two solar masses, and most black holes discovered so far (or at least suspected so far, since we can't directly see something that gives off no light) are four solar masses or above.

So why is this important? Researchers are still studying the results of a new phenomena witnessed last year called a kilonova. It created ripples in the fabric of space-time that were detected from Earth. While it was widely reported to be a merger of two neutron stars, some researchers aren't sure if the larger object was in fact one of these dense stellar cores. The larger object was estimated to be between 1.36 and 2.26 solar masses, while the smaller was well within the mass range of an average neutron star.

If it is toward the upper end of that mass estimate, then we may have witnessed the merger of a black hole and a neutron star—which could be essential to the research taking place in the wake of the explosion, as astronomers delve deeper into what kind of object is left behind at the center of such an event. A large, unstable neutron star that swiftly became a black hole could have been left behind after the kilonova, if it was a merger of two neutron stars, or an entirely different event could have taken place where a black hole ingested the smaller neutron star. The latter type of event has been identified before, at least tentatively, in 2005.

The Goethe researchers suggest that adding even a little more mass to the object could cause it to collapse into a black hole—however, some researchers have theorized, but never proven, that another type of object exists between the mass of a neutron star and that of a black hole. But with an upper ceiling to neutron star mass determined, we can begin to figure out how a large star truly dies—and what it takes to make a star fully collapse into a light-eating inferno.

ARTICLES QUESTIONS

1. Have these astronomers found the "upper limit" on neutron star mass—the highest possible mass—or the "lower limit"—the lowest possible mass? How do you know?
2. Why is this limit important to astronomers?
3. What is the scientific name for the "ripples in the fabric of space-time" discussed in the fifth paragraph?
4. You have now learned about three types of stellar explosions: a nova, a kilonova, and a supernova. What distinguishes those three types of explosions from one another?

Summary

As high-mass stars evolve, their interiors form concentric shells of progressive nuclear fusion. Once those stars leave the main sequence, they may pass through the instability strip and become pulsating variable stars. High-mass stars eventually explode as Type II supernovae, which eject newly formed massive elements into interstellar space. The supernova explosion that ends the life of a massive star leaves behind a neutron star that contains between 1.4 and about 2 M_{Sun} of neutron-degenerate matter packed into a sphere 10–15 km in diameter. Pulsars are rapidly spinning, magnetized neutron stars. Accretion of mass onto neutron stars produces X-rays in some binary systems. Binary neutron stars sometimes merge, producing short gamma-ray bursts. The Sun, the Solar System, Earth, and all life on Earth contain heavy elements created in earlier generations of short-lived massive stars.

LG 1 **Describe how the death of high-mass stars differs from that of low-mass stars.** The larger masses of high-mass stars allow them to fuse heavier elements than those produced in low-mass stars. That process leads to a more violent death that leaves massive cores behind.

LG 2 **List the sequence of stages for evolving high-mass stars.** Evolving high-mass stars leave the main sequence as they fuse heavier elements. Once an iron core is produced, the star becomes unstable and the core collapses, heating the material to cause photodisintegration of the iron nuclei and the merging of protons and electrons into neutrons. The outer layers bounce off the dense core and produce a shock wave that travels outward. That shock wave causes neutrons to fuse with atomic nuclei and form more massive elements.

LG 3 **Explain the origin of chemical elements up to and heavier than iron.** The chain of nuclear fusion reactions consists of increasingly shorter stages, resulting in more massive elements up to iron. However, those atoms are destroyed in the set of processes that define the core collapse. Heavy elements in the universe today were formed during the rebound explosion of massive stars as high-energy neutrons penetrated atomic nuclei. Those neutrons then decay to protons, creating new elements with higher atomic numbers.

LG 4 **Identify how Hertzsprung-Russell (H-R) diagrams of clusters enable astronomers to measure the ages of stars and test theories of stellar evolution.** Clusters are groups of stars born together and are all at about the same distance from Earth. H-R diagrams of clusters show stars leaving the main sequence in a progression from the highest-mass stars to the lowest-mass stars, confirming theories of stellar evolution. The location of the main-sequence turnoff indicates the age of the cluster.

? Unanswered Questions

- Blue straggler stars, found in clusters, are bluer and brighter than the stars at the main-sequence turnoff point. How do they fit into the picture of stellar evolution? They may have resulted from mass transfer in a binary pair or from the merger of two single or two binary stars, either of which could have resulted in a more massive star than what might be expected from the age of the cluster. Astronomers study the environments of those stars by estimating the likelihood of collisions and the number of binary systems, which may be different in clusters where the density of stars is high.
- What creates magnetars? One class of pulsars is known as magnetars, which are characterized by extremely large magnetic fields. Those objects are observed to produce bursts of lower-energy gamma rays. The origin of their huge magnetic fields is not well understood. Those fields may originate from a dynamo in the interior of a superconducting region of the neutron star, but we do not know whether ordinary pulsars go through a magnetar phase.

Questions and Problems

Test Your Understanding

1. Why does the interior of an evolved high-mass star have layers like those of an onion?
   a. Heavier atoms sink to the bottom because stars are not solid.
   b. Before the star formed, heavier atoms accumulated in the centers of clouds because of gravity.
   c. Heavier atoms fuse closer to the center because the temperature and pressure are higher there.
   d. Different energy transport mechanisms occur at different densities.
2. Arrange the following elements in the order they fuse inside the nucleus of a high-mass star during the star's evolution.
   a. helium
   b. neon
   c. oxygen
   d. silicon
   e. hydrogen
   f. carbon
3. Elements heavier than iron originated
   a. in the Big Bang.
   b. in the cores of low-mass stars.
   c. in the cores of high-mass stars.
   d. in the supernova explosions of high-mass stars.
4. A pulsar pulses because
   a. its spin axis crosses Earth's line of sight.
   b. it spins.
   c. it has a strong magnetic field.
   d. its magnetic axis crosses Earth's line of sight.
5. Study Figure 17.5. If you could watch a high-mass star move to the right, along one of those post-main-sequence lines, what would you observe happening to the star's color?
   a. It would become redder.
   b. It would become bluer.
   c. It would remain the same.
6. Study Figure 17.5. If you could watch a high-mass star move to the right, along the topmost of those post-main-sequence lines, what would you observe happening to the star's size?
   a. It would become much larger.
   b. It would become much smaller.
   c. It would remain the same.
7. Study Figure 17.18. If the Sun were a member of a globular cluster, that cluster's H-R diagram would fall between
   a. (a) and (b).
   b. (b) and (c).
   c. (c) and (d).
   d. (d) and (e).
   e. (e) and (f).
8. In a high-mass star, hydrogen fusion occurs via
   a. the proton-proton chain.
   b. the CNO cycle.
   c. gravitational collapse.
   d. spin-spin interaction.
9. The layers in a high-mass star occur roughly in order of
   a. atomic number.
   b. decay rate.
   c. magnetic field strength.
   d. spin state.
10. Eta Carinae is an extreme example of
    a. a massive star.
    b. a planetary nebula.
    c. a supernova remnant.
    d. an ancient star.
11. Iron fusion cannot support a star because
    a. iron oxidizes too quickly.
    b. iron absorbs energy when it fuses.
    c. iron emits energy when it fuses.
    d. iron is not dense enough to hold up the layers.
12. The start of photodisintegration of iron in a star sets off a process that *always* results in a
    a. supernova.
    b. neutron star.
    c. supergiant.
    d. pulsar.
13. The Crab Nebula serves as a test of our ideas about supernova explosions because
    a. the system contains an X-ray binary.
    b. the nebula is slowly expanding.
    c. the supernova was observed in 1054 and now astronomers see a pulsar in the nebula.
    d. the original star was like the Sun before exploding.
14. What mechanism supplies the pressure inside a neutron star?
    a. ordinary pressure from hydrogen and helium gas
    b. degeneracy pressure from neutrons
    c. degeneracy pressure from electrons
    d. rapid rotation
15. Very young star clusters have main-sequence turnoffs
    a. that drop below the main sequence.
    b. at the top left of the main sequence.
    c. at the bottom right of the main sequence.
    d. in the middle of the main sequence.

Thinking about the Concepts

16. Explain the differences between the ways that hydrogen is converted to helium in a low-mass star (proton-proton chain) and in a high-mass star (CNO cycle). What is the catalyst in the CNO cycle, and how does it take part in the reaction?

17. How does a high-mass star begin fusing helium in its core? How is that process different from what happens in low-mass stars?

18. Why does the core of a high-mass star not become degenerate, as the core of a low-mass star does?

19. List the two reasons why each post-helium-fusion cycle for high-mass stars (carbon, neon, oxygen, silicon, and sulfur) becomes shorter than the preceding cycle.

20. Cepheids are highly luminous, variable stars in which the period of variability is directly related to luminosity. Why are Cepheids good indicators for determining stellar distances?

21. Identify and explain two important ways in which supernovae influence the formation and evolution of new stars.

22. Study the Process of Science Figure. Why is that pulsar explanation "simpler"?

23. Describe what an observer on Earth will witness when Eta Carinae explodes.

24. Recordings show that neutrinos from SN 1987A were detected on February 23, 1987. About 3 hours later the supernova was detected in optical light. What caused the delay?

25. Why can the accretion disk around a neutron star release so much more energy than the accretion disk around a white dwarf, even though the two stars have approximately the same mass?

26. In Section 17.2, you learned that Type II supernovae blast material outward at up to 30,000 km/s. The material in the Crab Nebula described in Section 17.3 is expanding at only 1,500 km/s. What explains the difference?

27. An experienced astronomer can take one look at the H-R diagram of a star cluster and immediately estimate its age. How is that possible?

28. Explain how astronomers know that an even earlier generation of stars existed before the oldest observed stars.

29. What is the binding energy of an atomic nucleus? How does that quantity help astronomers calculate the energy given off in nuclear fusion reactions?

30. Explain how Earth is made up of material from supernovae.

Applying the Concepts

31. Study Figure 17.2. What fraction of the star is helium at $t = 0$ years and at $t = 7$ million years?

32. Study Figure 17.5. How much hotter, larger, and more luminous than the Sun is the uppermost main-sequence star on that H-R diagram?

33. Study Figure 17.4. Are the radius of the core and the radius of the star represented to scale? What fraction of the star's radius is the core's radius?

34. If the Crab Nebula has been expanding at an average velocity of 1,500 km/s since the year 1054, what was its average radius in the year 2019? (Note: A year has approximately 3×10^7 seconds.)

35. Suppose you observe a Classical Cepheid variable with a period of 10 days. What is the luminosity of that star? What other piece of information would you need to find out how far away the star is?

36. Figure 17.13 shows the relative abundance of the elements. Is that graph a log or a linear plot? Explain what it means that oxygen lies on the y-axis at 10^{-3}.

37. Use Working It Out 17.2 to find the surface gravity on a neutron star with a radius of 10 km and a mass of 2.8 M_{Sun}.

38. Use Working It Out 17.2 to find the escape velocity from a neutron star with a radius of 10 km and mass of 2.8 M_{Sun}.

39. The Milky Way has about 50,000 stars of average mass (0.5 M_{Sun}) for every main-sequence star of 20 M_{Sun}. But 20-M_{Sun} stars are about 10,000 times as luminous as the Sun, and 0.5-M_{Sun} stars are only 0.08 times as luminous as the Sun.
    a. How much more luminous is a single massive star than the total luminosity of the 50,000 less massive stars?
    b. How much mass is in the lower-mass stars than the single high-mass star?
    c. Which stars—lower-mass or higher-mass stars—contain more mass in the galaxy, and which produce more light?

40. In a large outburst in 1841, the 120-M_{Sun} star Eta Carinae was losing mass at a rate of 0.1 M_{Sun} per year.
    a. The mass of the Sun is 2×10^{30} kg. How much mass (in kilograms) was Eta Carinae losing each minute?
    b. The mass of the Moon is 7.35×10^{22} kg. How does Eta Carinae's mass loss per minute compare with the mass of the Moon?

41. An O star can lose 20 percent of its mass during its main-sequence lifetime. Estimate the average mass loss rate (in solar masses per year) of a 25-M_{Sun} O star with a main sequence lifetime of 7 million years.

42. The approximate relationship between the luminosity and the period of Cepheid variables is L_{star} (in L_{Sun}) = 335 P (in days). Delta Cephei has a cycle period of 5.4 days and a parallax of 0.0033 arcsecond (arcsec). A more distant Cepheid variable appears 1/1,000 as bright as Delta Cephei and has a period of 54 days.
    a. How far away (in parsecs) is the more distant Cepheid variable?
    b. Could the distance of the more distant Cepheid variable be measured by parallax? Explain.

43. For a pulsar that rotates 30 times per second, at what radius in the pulsar's equatorial plane would a co-rotating satellite (rotating about the pulsar 30 times per second) have to be positioned to be moving at the speed of light? Compare that value with the pulsar radius of 10 km.

44. Verify the claim in Section 17.3 that Earth would be roughly the size of a football stadium if the planet were as dense as a neutron star.

45. Estimate the size of a neutron star that has the mass of the Sun.

USING THE WEB

46. Go to the Chandra X-ray Observatory's "Variable Stars" Web page (http://chandra.harvard.edu/edu/formal/variable_stars/). Do the two activities on variable stars, which ask you to estimate their changes in brightness. You might want to look at Appendix 7 to review apparent magnitudes before you do the projects.

47. Go to the International Astronomical Union's "List of 2015 Supernovae" (http://www.cbat.eps.harvard.edu/lists/RecentSupernovae.html) Pick a few of them. What type of supernova is each one? How bright was it? Are Type Ia or Type II supernovae more common?

48. Go to the website for the new Zwicky Transient Factory (ZTF) survey (https://www.ztf.caltech.edu/). What is that project? How does it improve on the Palomar Transient Factory? Click on "Science"—what do they hope to learn about supernovae? Click on "News"—has ZTF found any supernovae yet?

49. Go to the website for the Gaia mission (http://sci.esa.int/gaia/). How is that mission contributing to the study of variable stars? How is it contributing to the study of novae and supernovae? Do a news search—are any results available?

50. Go to the "Einstein@Home" website (https://einsteinathome.org). In this distributed-computing project, volunteers use their spare computer processing power to help search for gravitational waves from spinning compact objects, including pulsars. On the "Home" tab, look over the news section on the right. Have any pulsars been found lately? Join the project, create an account, download BOINC, and follow directions to look for search.

EXPLORATION

The CNO Cycle

digital.wwnorton.com/astro6

Nuclear reactions usually involve many steps. In the Exploration for Chapter 14, you investigated the proton-proton chain. In this Exploration, you will study the CNO cycle, which is even more complex. Visit the Digital Resources Page and on the Student Site open the "CNO Cycle" Interactive Simulation in Chapter 17.

First, press "Play Animation" and watch the animation all the way through. Press "Reset Animation" to clear the screen, and then press "Play Animation" again, allowing the animation to proceed past the first collision before pressing "Pause."

1 Which atomic nuclei are involved in that first collision?

____

2 What color is used to represent the proton (hydrogen nucleus)?

____

3 What does the blue squiggle represent?

____

4 What atomic nucleus is created in the collision?

____

5 The resulting nucleus is not the same type of element as either of the two that entered the collision. Why not?

____

____

Press "Play Animation" again, and then pause as soon as the yellow ball and the dashed line appear.

6 Is that a collision or a spontaneous decay?

____

7 What does the yellow ball represent?

____

8 What does the dashed line represent?

____

9 The resulting nucleus has the same number of nucleons (13), but it is a different element. What happened to the proton that was in the nitrogen nucleus but is not in the carbon nucleus?

____

____

Proceed past the next two collisions, to "^{15}O."

10 Study the pattern that is forming. When a blue ball comes in, what happens to the number of nucleons and the type of the nucleus (that is, what happens to the "12" and the "C," or the "14" and the "N")?

____

11 What is emitted in those collisions?

____

Proceed until "^{15}N" appears.

12 Is that a collision or a spontaneous decay?

____

13 Which previous reaction is that most like?

____

Now proceed to the end of the animation.

14 After the final collision, a line is drawn back to the beginning, telling you what type of nucleus the upper red ball represents. What is that nucleus?

____

15 How many nucleons are not accounted for by the upper red ball? (Hint: Don't forget the ^{1}H that came into the collision.) Those nucleons must be in the nucleus represented by the bottom red ball.

____

16 Carbon has six protons. Nitrogen has seven. How many protons are in the nucleus represented by the bottom red ball?

____

17 How many neutrons are in the nucleus represented by the bottom red ball?

____

18 What element does the bottom red ball represent?

____

19 What is the net reaction of the CNO cycle? That is, which nuclei are combined and turned into the resulting nucleus?

____

20 Why is ^{12}C not considered part of the net reaction?

____

18 Relativity and Black Holes

Some stars leave behind a black hole at the end of their lives. Black holes have conditions so extreme that the laws of Newtonian physics can't describe them. To discuss black holes, we must understand how Albert Einstein changed the way physicists thought about the nature of space and time in the early 20th century. Einstein's special theory of relativity shows that matter behaves differently when it is traveling near the speed of light. Einstein's general theory of relativity shows that space itself is warped near very massive objects. That warping of space is so extreme at a black hole, it is as though a hole exists in space. Here in Chapter 18, we move beyond Newtonian ideas of space and time to understand black holes.

LEARNING GOALS

By the end of this chapter, you should be able to:

LG 1 Describe how the motion of the observer affects the observed velocity of objects.

LG 2 Discuss the observable consequences of the relationship between space and time.

LG 3 Explain how gravity is a consequence of the way mass distorts the very shape of spacetime.

LG 4 Explain the formation of black holes from the most massive stars, and describe the key properties and observational consequences of those stellar black holes.

This artist's rendition of the black hole Cygnus X-1 shows the accretion disk and the massive star that feeds it. ▶▶▶

How are space and time changed near a black hole?

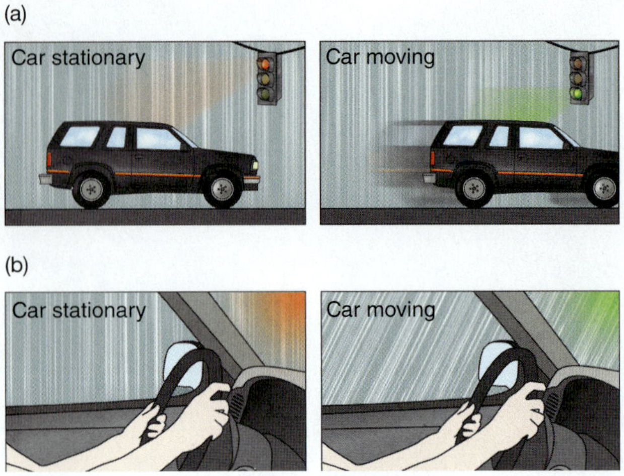

Figure 18.1 On a windless day, the direction in which rain falls depends on the frame of reference in which it is viewed. (a) From outside the car, the rain is seen to fall vertically downward whether the car is stationary or moving. (b) From inside the car, the rain is seen to fall vertically downward if the car is stationary; but if the car is moving, the rain is seen to fall at an angle determined by the speed and direction of the car's motion.

18.1 Relative Motion Affects Measured Velocities

All observers, whatever their motion, measure the same speed of light. That observed fact has profound implications for relative motion, space, and time. In this section, we will lay the groundwork for our discussion of relativity by considering how relative motion affects measurements and how those effects change at very high speeds.

Aberration of Starlight

The first direct measurement of the effect of Earth's motion around the Sun was made in the 18th century. Imagine that you are sitting in a car in a windless rainstorm (**Figure 18.1**). If the car is stationary and the rain is falling vertically, you see raindrops falling straight down when you look out your side window. When the car is moving forward, however, the situation is different. Between the time a raindrop appears at the top of your window and the time it disappears beneath the bottom of your window, the car has moved forward. The raindrop disappears beneath the window behind the point at which it appeared at the top of the window, which means the raindrop looks as though it is falling at an angle, even though in reality it is falling straight down. As you go faster, the apparent front-to-back slant of the raindrops increases, and their apparent paths become more slanted. An observer by the side of the road would say the raindrops are coming from directly overhead, but to you in the moving car they are coming from a direction in front of the car. You are observing that apparent motion of the raindrops from within your own unique frame of reference.

The light from a distant star arrives at Earth from the direction of the star (**Figure 18.2**). However, just as the raindrops appeared to be coming from in front of the moving car in Figure 18.1b, an observer on the moving Earth sees the starlight coming from a slightly different direction. Because the direction of Earth's motion around the Sun continuously changes during the year, the apparent position of a star in the sky moves in a small loop, a phenomenon known as the **aberration of starlight**. That shift in apparent position was first detected in the 1720s by two astronomers—Samuel Molyneux and James Bradley. Measurement of the aberration of starlight shows that Earth moves on its path about the Sun with an average speed of just under 30 kilometers per second (km/s). Because distance equals speed multiplied by time, astronomers used that measurement to determine the circumference, and therefore the radius, of Earth's orbit. The speed of Earth (29.8 km/s) multiplied by the number of seconds in 1 year (3.16×10^7 seconds) gives a circumference of 9.42×10^8 km. Astronomers used that circumference to estimate the radius of Earth's orbit (1.5×10^8 km). The aberration of starlight is an astronomical example of how relative motion affects a measurement.

Figure 18.2 When we observe stars, their apparent positions are deflected slightly toward the direction in which Earth is moving. As Earth orbits the Sun, stars appear to trace out small loops in the sky. That effect is called *aberration of starlight*.

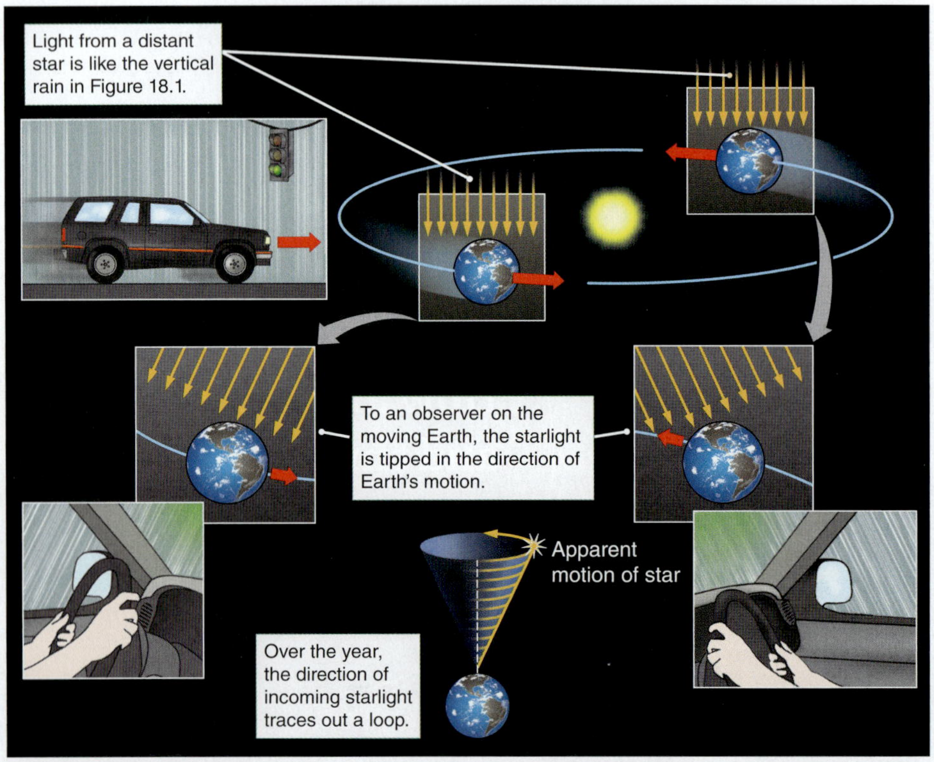

Relative Speeds Close to the Speed of Light

Every observer inhabits a reference frame, a set of coordinates in which the observer measures distances and speeds. A nonaccelerating reference frame, in which no net force is acting, is known as an **inertial reference frame**. Newton's laws of motion predict how the motion of an object will be measured by observers in different inertial reference frames. For example, the observed motion of a ball thrown from a moving car depends on the reference frame of the observer (recall the analogy for the Coriolis effect in Figure 2.11). If light behaved like other objects, the speed of light should differ from one observer to the next as a result of the observer's motion, just like the speed of a ball thrown from a moving car. However, the results of laboratory experiments with light in the late 19th century showed something very different. Scientists found instead that *all* observers measure the same value for the speed of light, regardless of the observer's frame of reference.

Imagine that you are in a red car traveling at 100 kilometers per hour (km/h) on a highway (**Figure 18.3a**), and you throw a ball at a speed of 50 km/h out the window at an oncoming green car, also traveling at 100 km/h. An observer standing by the side of the road watches the entire event. In *your* reference frame (Figure 18.3a, top panel), the red car is stationary; that is, you and the car are moving together, so the car does not move relative to you. In your reference frame, then, you measure the ball traveling at 50 km/h. To the observer standing by the road (Figure 18.3a, middle panel), however, the ball is moving at 150 km/h—50 km/h from throwing it plus 100 km/h from the motion of the car. To passengers in the oncoming green car moving at 100 km/h (Figure 18.3a, bottom panel), the speed of the ball is 250 km/h because the ball approaches them at 150 km/h, and they approach the ball at 100 km/h. That example shows that the speed of the ball depends on how the observer, the cars, and the ball are moving relative to one another. The velocities add together to yield the velocity of one object relative to another. That is Galilean relativity, which you use in everyday life.

Figure 18.3b shows how light differs from the ball in the previous example. Imagine that you are riding in the yellow spaceship at half the speed of light (0.5*c*) and you shine a beam of laser light forward (Figure 18.3b, top panel). You measure the speed of the beam of light to be *c*, or 3×10^8 meters per second (m/s)—as expected because you are holding the source of the light. But the observer on the planet also measures the speed of the passing beam of light to be *c*, not 1.5*c*, which would be the sum of the speed of light plus the speed of the spaceship (Figure 18.3b, middle panel). Even a

Figure 18.3 (a) The Newtonian rules of motion apply in daily life; however, those rules break down when speeds approach the speed of light. (b) Light itself always travels at the same speed for any observer, and that fact is the basis of special relativity, which implies that velocities don't simply add together, as they do in the Newtonian world.

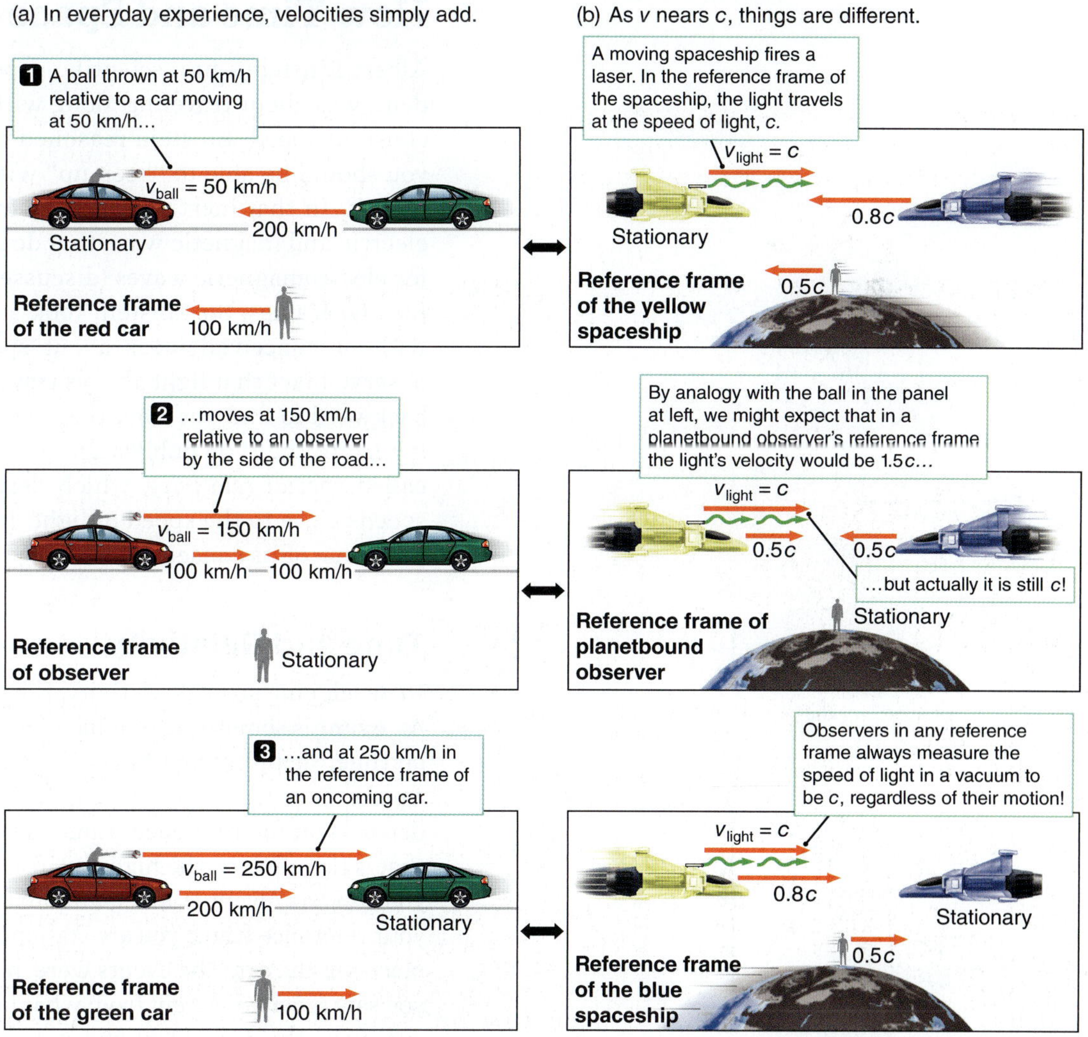

passenger in an oncoming blue spacecraft traveling at 0.5*c* (Figure 18.3b, bottom panel) finds that the beam from your light is traveling at exactly *c* in her own reference frame, and not at 2.0*c*, which would be the sum of the speeds of the spacecraft and the laser light. At speeds close to the speed of light, speeds do not add simply. That statement is true not only for light but also for all ordinary objects moving at nearly the speed of light. According to Einstein's relativistic formulas, the relative speed between the two spacecraft in the top and bottom panels of Figure 18.3b (0.5*c* + 0.5*c*) adds to 0.8*c*, not 1.0*c*! At **relativistic speeds** (speeds close to the speed of light), everyday experience no longer holds true. Every observer always finds that light in a vacuum travels at the same speed, *c*, regardless of his or her own motion or the motion of the light source. That observation directly conflicts with Newtonian theory and Galilean relativity.

CHECK YOUR UNDERSTANDING 18.1

Which beam of light is moving faster? (a) the one from the headlight of a parked car; (b) the one from the headlight of a moving car; (c) the one from the headlight of a moving spaceship; (d) all the beams are moving at *c*.

18.2 Special Relativity Explains How Time and Space Are Related

Albert Einstein's first scientific paper, written when he was a 16-year-old student, was about traveling along with a light wave, moving in a straight line at constant speed. Einstein reasoned that according to Newton's laws of motion, you should be able to "keep up" with light so that you are moving right along with it. In that inertial reference frame, the light is stationary: an oscillating electric and magnetic wave that does not move. However, Maxwell's equations for electromagnetic waves (discussed in Chapter 5) are actually the same in all inertial frames. So Einstein took a radical new approach. Instead of starting with preconceived ideas about space and time, Einstein started with the observed fact that light always travels at the same speed, and then he reasoned backward to find out what that must imply about space and time. That reasoning led to the 1905 publication of his **special theory of relativity**, sometimes called *special relativity*, which describes the effects of traveling at constant speeds close to the speed of light. In this section, we explore special relativity and some of its consequences.

Time and Relativity

In developing special relativity, Einstein focused his thinking on pairs of *events*. An **event** is something that happens at a particular location in space at a particular time. Snapping your fingers is an event because that action has both a time and a place. Everyday experience indicates that the distance between any two events depends on the reference frame of the observer. Imagine you are sitting in a car that is traveling on the highway in a straight line at a constant 60 km/h. You snap your fingers (event 1), and a minute later you snap your fingers again (event 2). In your reference frame you are stationary, and the two events happened at the same place—in the car. The events were, however, separated by a minute in time. What you saw is very different from what happens in the reference frame of an observer sitting by the road. That observer agrees that the second snap of your fingers (event 2) occurred a minute after the first snap of your fingers (event 1), but to that observer the two events were separated from each other in space by a kilometer, the distance your car traveled in the minute between snaps. In that "Newtonian" view, the distance between two events depends on the motion of the observer, but the *time* between the two events does not. Special relativity says instead that both the distance *and* the time between events vary depending on the motion of the observer.

The notion that different observers will measure time differently is a *very* counterintuitive idea, but it is central to special relativity and therefore to our scientific understanding of the universe as well. To see how Einstein arrived at the concept of relative time, consider his thought experiment known as the boxcar experiment. In that experiment, observer 1 is in a boxcar of a train moving to the right. Observer 1 has a lamp, a mirror (mounted on the roof of the boxcar), and a clock. Observer 2 is standing on the ground outside. The clock is based on a value that everyone can agree on—such as the speed of light.

Figure 18.4a shows the experimental setup as seen by observer 1, who is stationary with respect to the clock. At time t_1, event 1 happens: the lamp gives off a pulse of light. The light bounces off a mirror at a distance l meters away and then heads back toward its source. At time t_2, event 2 happens: the light arrives at the clock and is recorded by a photon detector. (Note that the light beam is shown leaving and arriving at two locations. The artist did that so that you can see both events in the figure. Both events occur at the same location.) The time between events 1 and 2 is just the distance the light travels ($2l$ meters) divided by the speed of light: $t_2 - t_1 = 2l/c$.

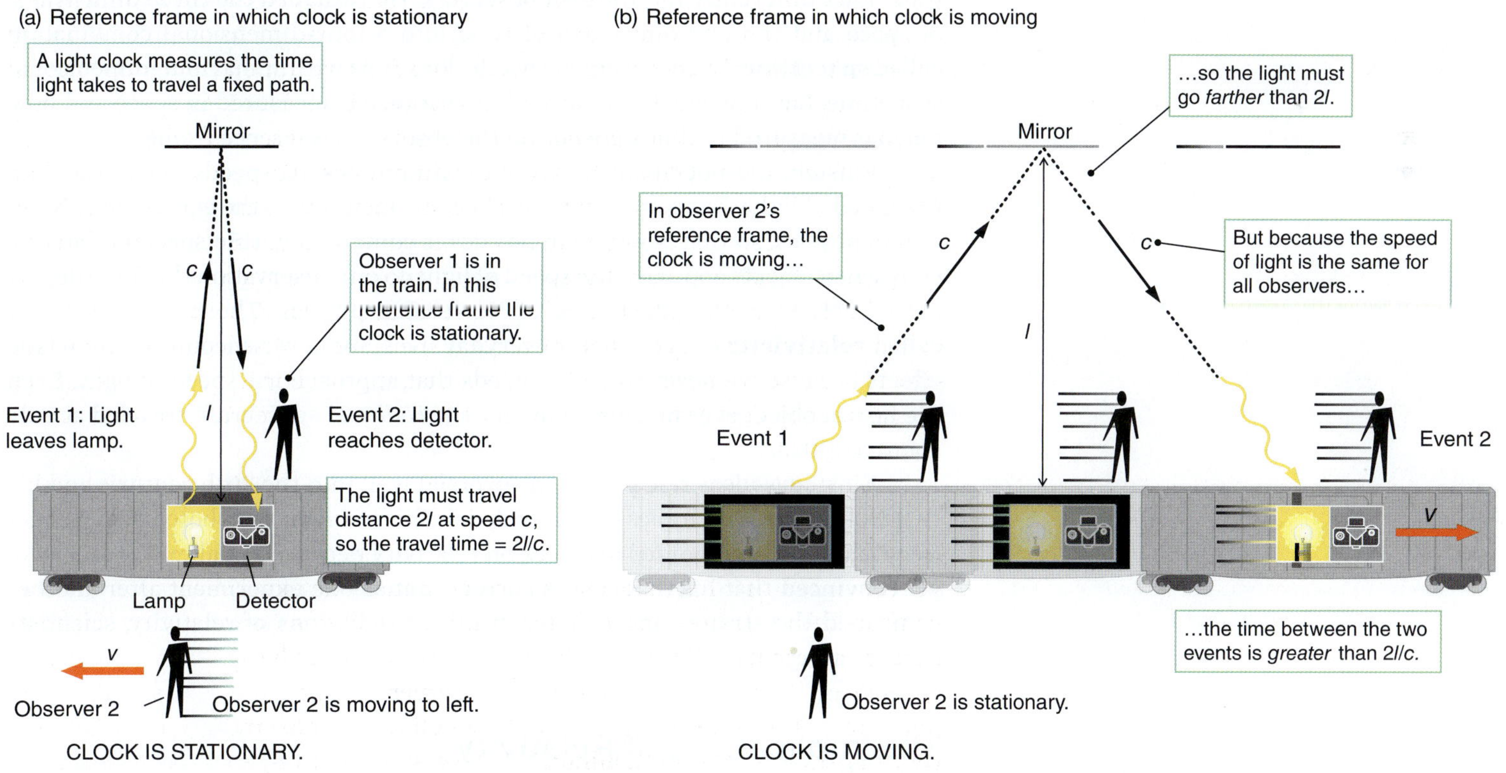

Figure 18.4 The "tick" of a light clock is different when seen in two reference frames: (a) stationary, as in the reference frame of observer 1 on the boxcar, and (b) moving, as in the reference frame of observer 2 on the tracks. As Einstein's thought experiment demonstrates, if the speed of light is the same for every observer, moving clocks *must* run more slowly than stationary clocks.

Figure 18.4b shows the experiment as seen by observer 2, who is stationary on the ground outside the train, which is moving at speed v. In observer 2's reference frame, the clock moves to the right between the two events, so the light has farther to go because of the horizontal distance. The time between the two events is still the distance traveled divided by the speed of light, but now that distance is *longer* than $2l$ meters. Because the speed of light is the same for all observers, the time between the two events must be longer as well.

The two events are the same two events, regardless of the reference frame from which they are observed. Because the speed of light is the same for all observers, more time must pass between the two events when viewed from a reference frame in which the clock is moving (observer 2). The seconds of a moving clock are stretched. That is, a moving clock takes more time than a stationary clock to complete one "tick." Therefore, the passage of time must depend on an observer's frame of reference. Because both frames of reference are equally good places to do physics, both time measurements are valid in their own frames, even though they differ from each other.

In that experiment, light travels farther between events in a moving boxcar than between events in a stationary boxcar and consequently takes longer to travel between the events in the moving boxcar. Einstein realized that the *only* way the speed of light can be the same for all observers is *if the passage of time is different from one observer to the next*. For moving observers, the time is stretched out, so that the time interval between events is longer, a phenomenon known as **time dilation**.

The Newtonian view of the world describes a three-dimensional space through which time marches steadily onward. Einstein discovered, however, that time flows differently for different observers. He reshaped the three dimensions of space and the one dimension of time into a four-dimensional combination called **spacetime**. Events occur at specific locations within that four-dimensional spacetime, but how much of a spacetime distance is measured in space and how much is measured in time depends on the observer's reference frame.

Einstein did not "disprove" Newtonian physics. At speeds much less than the speed of light, Einstein's equations become identical to the equations of Newtonian physics, so that Newtonian physics is contained within special relativity. Only when objects approach the speed of light do our observations begin to depart measurably from the predictions of Newtonian physics. Those departures are called **relativistic** effects. In our everyday lives, we never encounter relativistic effects because we never travel at speeds that approach the speed of light. Even the fastest object ever made by humans, the *Helios II* spacecraft, traveled at only about $0.00023c$.

Einstein's ideas remained controversial well into the 20th century, and his 1921 Nobel Prize in Physics was awarded for his work on the photoelectric effect (see Chapter 5) and not for his work on relativity because not all physicists were yet convinced that his theory was correct. But as one experiment after another confirmed the strange and counterintuitive predictions of relativity, scientists came to accept its validity.

The Implications of Relativity

Today, special relativity shapes our thinking about the motions of both the tiniest subatomic particles and the most distant galaxies. In this subsection, we discuss only a few of the essential insights that come from Einstein's work.

MASS AND ENERGY What we think of as "mass" and what we think of as "energy" are actually closely related. The energy of an object depends on its speed: the faster it moves, the more energy it has. But Einstein's famous equation, $E = mc^2$, says that even a stationary object has an intrinsic "rest" energy that equals the mass (m) of the object multiplied by the speed of light (c) squared. The speed of light is a very large number, so a small mass has a very large rest energy. A single tablespoon of water, for example, has a rest energy equal to the energy released in the explosion of more than 300,000 tons of TNT. We used that relationship between mass and energy in Chapter 14 when discussing the nuclear fusion that makes stars shine.

In Chapter 3, we connected the *mass* of an object to its inertia—its resistance to changes in motion. At relativistic speeds, adding to the energy of motion of an object increases its inertia. For example, a proton in a high-energy particle accelerator may travel so close to the speed of light that its total energy is 1,000 times greater than its rest energy. Such an energetic proton is harder to "push around" (in other words, it has more inertia) than a proton at rest. It also more strongly attracts other masses through gravity.

THE ULTIMATE SPEED LIMIT We already discussed Einstein's insight that if traveling at the speed of light were possible, then in that reference frame light would cease to be a traveling wave, and the laws of electromagnetism wouldn't be valid. You also can think about that limit in terms of the equivalence of mass and energy just discussed. As the speed of an object gets closer and closer to the speed of light, the energy of that object, and therefore its mass, becomes greater and greater, so it becomes increasingly resistant to further changes in its motion. Only a photon or other massless particle can travel at the speed of light. We rely on the fact that light travels at a constant speed in a vacuum whenever we use the travel time of light to describe astronomical distances.

Adding energy to an object will cause its velocity to get closer and closer to the speed of light, but the velocity will never actually reach the speed of light. To accelerate an object with a nonzero rest mass to the speed of light would take an *infinite* amount of energy. Not enough energy exists in the entire universe to accelerate even one electron to the speed of light. The electron can get arbitrarily close to that number—0.99999999999999999999 . . . $\times$ c is possible (at least in principle)—but not enough energy is available to accelerate the electron beyond that to the speed of light. **Figure 18.5** shows how a rocket ship, which experiences a constant acceleration equal to that of gravity on Earth (so that its occupants will feel "normal" gravity), moves faster and faster but never reaches the speed of light. In the Newtonian view shown by the purple curve, that limit is not present, and the speed of the spaceship would continuously increase at the same rate.

TIME Time passes more slowly in a moving reference frame: for moving objects, the seconds are stretched out by time dilation. No inertial reference frame is special. If you compared clocks with an observer moving at nine-tenths the speed of light ($0.9c$) relative to you, you would find that the other observer's clock was running 2.29 times slower than your clock. The other observer would find instead that *your* clock was running 2.29 times slower. To you, the other observer may be moving at $0.9c$, but to the other observer, *you* are the one who is moving. Both frames of reference are equally valid, so you would each find the other's clock to be slower than your own. That time dilation effect

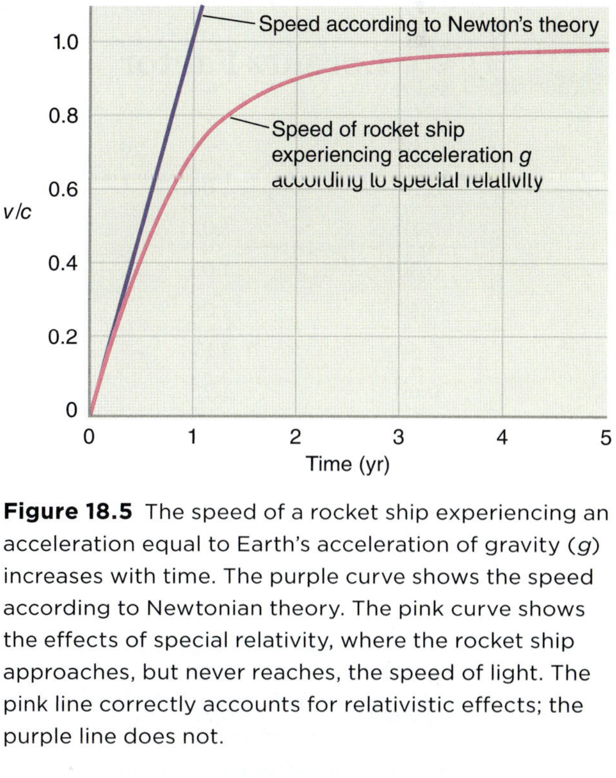

Figure 18.5 The speed of a rocket ship experiencing an acceleration equal to Earth's acceleration of gravity (g) increases with time. The purple curve shows the speed according to Newtonian theory. The pink curve shows the effects of special relativity, where the rocket ship approaches, but never reaches, the speed of light. The pink line correctly accounts for relativistic effects; the purple line does not.

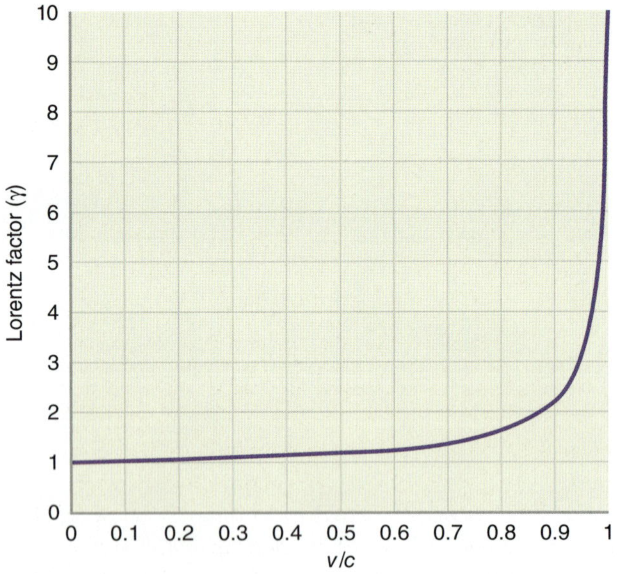

Figure 18.6 The Lorentz factor, γ, is plotted against v/c. That factor doesn't become significant until velocities are about 50 percent of the speed of light.

increases with speed, and that symmetry holds as long as neither frame accelerates. **Figure 18.6** and **Table 18.1** show that how much the time is stretched depends on the object's speed. The factor of 2.29 by which time is stretched in the preceding example is called the Lorentz factor and is usually denoted by the symbol γ.

A scientific observation demonstrates time dilation in nature. Fast particles called cosmic-ray muons (**Figure 18.7**) are produced 15 km up in Earth's atmosphere when high-energy **cosmic rays**—elementary particles moving at nearly the speed of light—strike atmospheric atoms or molecules. Muons at rest decay very rapidly into other particles. That decay happens so quickly that even if they could move at the speed of light, virtually all muons would have decayed long before traveling the 15 km to reach Earth's surface. Time dilation slows the muons' clocks, however, so the particles live longer and can travel farther and reach the ground. As muons move faster and faster, their clocks run slower and slower, and more of them can reach the ground. The same general principle is observed in particle accelerators, in which particles traveling at speeds near the speed of light live longer before decaying. **Working It Out 18.1** shows some examples of time dilation.

LENGTH An object appears shorter in motion than it is at rest. Moving objects are compressed in the direction of their motion by a factor of $1/\gamma$, where γ is the same Lorentz factor introduced in the discussion of time dilation. That phenomenon is called length contraction. A meterstick moving at $0.9c$ appears to be 0.44 meters long. That finding also explains our muon experiment from the perspective of the muons themselves. In the reference frame of the fast-moving muon produced at a height of 15 km, Earth's atmosphere is moving fast and appears to be much shorter than 15 km; indeed, it is so compressed from the muon's perspective that the muon may be able to reach the ground before decaying. That length contraction effect also increases with speed.

TABLE 18.1 Lorentz Factor

v/c	γ
0.10	1.005
0.20	1.02
0.30	1.05
0.40	1.09
0.50	1.15
0.60	1.25
0.70	1.40
0.80	1.67
0.90	2.29
0.95	3.20
0.99	7.09
0.995	10.01
0.999	22.37
0.9999	70.71

TWIN PARADOX Suppose you take a trip into space and leave your identical twin back on Earth. You accelerate to nearly the speed of light as you leave Earth. After you arrive at your destination, you return to Earth, again traveling at a speed close to c. To your twin on Earth, you were the one in a moving reference frame, so your twin measures your time as running much slower (see Working It Out 18.1), in which case you should return younger than your twin. However, from your perspective, the spaceship didn't move. Instead, Earth receded from you, stopped, and returned. Your twin is the one who moved at just under the speed of light, and your twin's time ran more slowly than yours, so your twin has aged much less. Both you and your twin cannot be correct, and that is the nature of the paradox.

One key difference between you and your twin resolves the paradox. You experienced acceleration during your trip, whereas your twin did not. Accelerated motion is not uniform motion. As a result, you changed reference frames during your trip. You changed reference frames when you left Earth, changed again when you stopped at your destination, changed a third time when you left your destination to return home, and changed reference frames one final time when you arrived back at Earth. Your twin, however, remained in Earth's reference frame. Upon your return, you would find that more time has passed for your twin on Earth than for you and that your twin has aged more than you have.

SPACE TRAVEL What is the actual reality of human travel in space? Some U.S. astronauts have been to the Moon and back, and several robotic spacecraft have been sent to explore objects throughout the Solar System. (To date, only a few robotic spacecraft are traveling at a high enough velocity to leave the Solar System.) But humans could not easily visit and explore other planetary systems in the Milky Way Galaxy because of the constraints of energy and the ultimate speed limit of light. With current technology, engineers can construct rockets that can travel at speeds of up to 20,000 m/s. At such a speed, a one-way trip to Earth's nearest neighbor star, Proxima Centauri, at a distance of 4.2 light-years, would take well over 50,000 years. Travel to more distant stars would take even longer.

In principle, travel just under the speed limit c is possible, so one could take advantage of relativistic time dilation to make such adventures well within the lifetime of a space traveler. In the example in Working It Out 18.1, an astronaut experiences a round-trip travel time to planets 20 light-years away in just 5.6 years at a speed of $0.99c$. Or an astronaut could travel to the center of the Milky Way Galaxy and back in just 2 years by traveling at $0.9999999992c$.

Although theoretically possible, travel at those speeds in practice would require an impractical amount of energy. If M is the mass of the astronauts, rocket ship, and fuel, just accelerating the rocket ship to such a high speed would take γMc^2 of energy, or $10Mc^2$ in the Proxima Centauri example and $25{,}000Mc^2$ in the

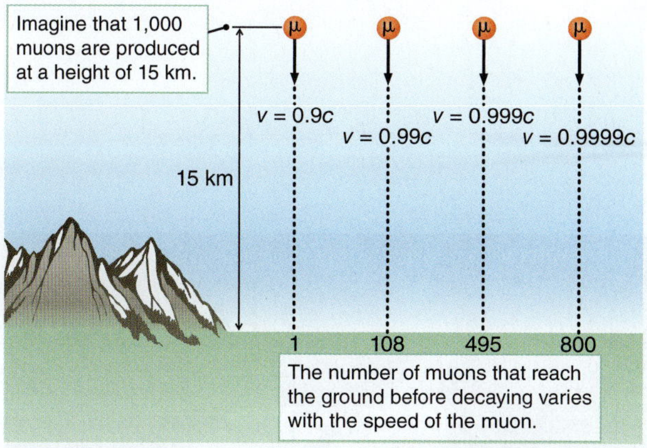

Figure 18.7 Muons created by cosmic rays high in Earth's atmosphere decay long before reaching the ground if they are not traveling at nearly the speed of light. Here, we show what happens to 1,000 muons produced at an altitude of 15 km for a variety of speeds. Faster muons have slower clocks, so more of them survive long enough to reach the ground—many more than would be expected simply because of the faster speed.

18.1 Working It Out Time Dilation

Physicist Hendrik Lorentz (1853–1928) derived the equation for how much time is dilated and how much space is contracted when something is traveling at velocities near the speed of light. That *Lorentz factor* (abbreviated γ) is given by

$$\gamma = \frac{1}{\sqrt{1 - \frac{v^2}{c^2}}}$$

Figure 18.6 shows the Lorentz factor plotted against velocity, and Table 18.1 gives the calculated value of γ for different values of velocities v/c. For something moving at half the speed of light, the Lorentz factor is 1.15. For something moving at 90 percent of c, however, the factor is 2.29, and it goes up quickly from there, becoming arbitrarily large as the velocity approaches, but never quite reaches, the speed of light.

Suppose you take a trip to a star 20 light-years away to study the "super-Earth" planets in orbit around it. You travel at $0.99c$, whereas your twin stays behind on Earth. Our common Newtonian experience on Earth would tell you how long the trip takes, using the familiar equation that time equals distance divided by speed:

$$\text{Time passed} = \frac{20\text{ light-years}}{0.99c} = 20.2\text{ years}$$

The round trip would take 40.4 years. When you return, you will find that 40.4 years passed on Earth.

To you on the spaceship moving at $0.99c$, special relativity says time will pass more slowly, by the Lorentz factor of 7.09. Your trip to the star would take $1/7 \times 20.2$ years $= 2.8$ years and another 2.8 years for the return trip. So as you traveled to that star and back, 5.6 years would have passed for you in the spaceship, but *40.4* years would have passed on Earth. Your twin on Earth would be almost 35 years older than you are!

For a more practical experiment that can be performed on Earth, we've noted that time dilation can be seen with subatomic particles. Suppose one type of particle, called a pion, can "live" for 20 nanoseconds (ns) before it decays into different particles. If pions are produced in a particle accelerator at a speed of $0.999c$, how long will physicists see the pions before they decay? Here $v = 0.999c$, so from Table 18.1, $\gamma = 22.37$. In the reference frame of the moving pions, they are still living for 20 ns: they are like the twin who traveled into space at high speed. In the reference frame of the physicists, though, special relativity predicts that the particle will last longer, 20 ns $\times$ 22.37, or 447 ns. Indeed, physicists observe that pions moving at nearly the speed of light "live" longer and travel farther before decay (like the muons in Figure 18.7).

The same factor γ applies for mass and length. If physicists measured a ruler moving at $0.999c$, the ruler would be 22.37 times shorter than its length when at rest. In the previous example, the pions traveling at $0.999c$ will behave as though their mass were 22.37 times larger: if the high-speed pions collided with other particles, the energy of the collision would be as though they had the higher mass.

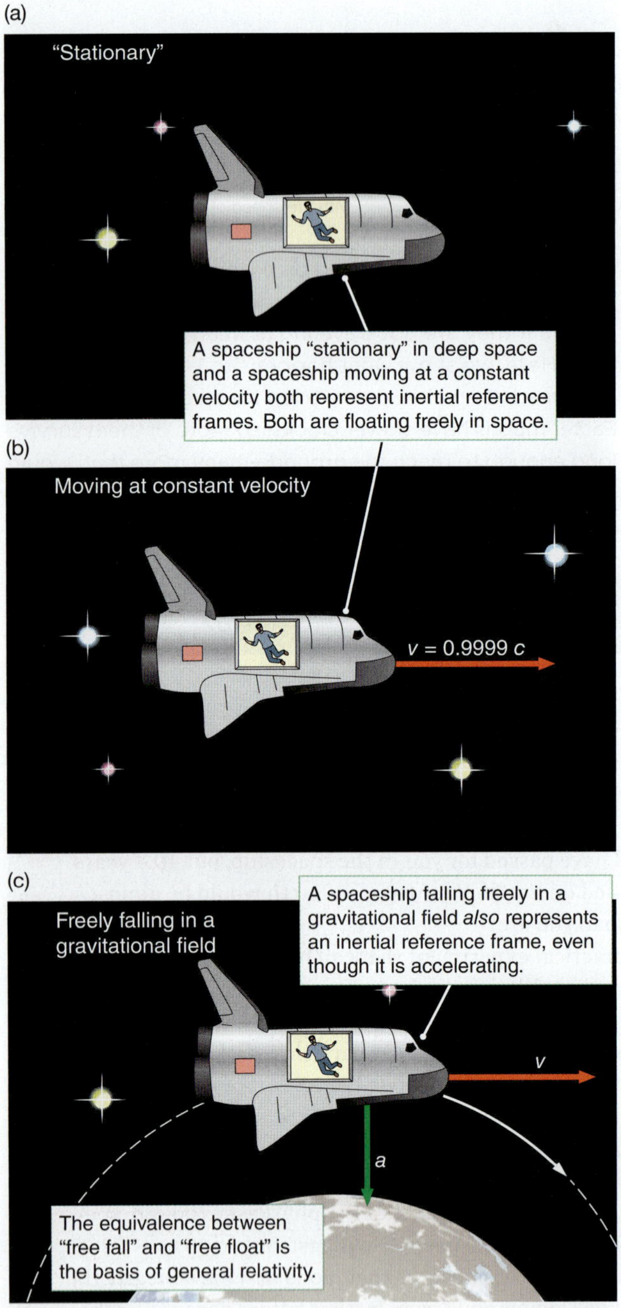

Figure 18.8 According to special relativity, no difference exists between (a) a reference frame floating stationary in space and (b) one moving through the galaxy at constant velocity. General relativity adds that no difference exists between those inertial reference frames and (c) an inertial reference frame falling freely in a gravitational field. Free fall is the same as free float, as far as the laws of physics are concerned.

example of the trip to the center of the Milky Way. For the second example, the energy to accelerate just the astronaut (not even including the spaceship) to such energies is more than that contained in 10 billion nuclear weapons. So, although not theoretically impossible, visits to other stars in our galaxy will not take place anytime soon.

CHECK YOUR UNDERSTANDING 18.2

Suppose that your friend flies past you in a spaceship she measures to be 20 m long, and both of you measure the time the spaceship takes to pass your location. Which of the following is true? (a) The time you measure is *greater* than the time your friend measures. (b) The time you measure is *less* than the time your friend measures. (c) You both measure the same amount of time.

18.3 Gravity Is a Distortion of Spacetime

Our exploration of special relativity began with the observation that the speed of light is always the same regardless of the motion of an observer or the motion of the source of the light. We have seen that three-dimensional space and time are actually just the result of a particular, limited perspective on a four-dimensional spacetime that is different for each observer. That four-dimensional spacetime is itself warped and distorted by the masses it contains. As we discuss the properties of black holes—indeed, of all massive objects in the universe—the concepts of space and time will diverge even further from the absolutes of Newtonian physics. In this section, we explore the general theory of relativity, which describes how mass affects space and time.

The Equivalence Principle

We have already discussed one fundamental connection between gravity and spacetime in Chapter 4 and showed that the *inertial mass* of an object—the mass appearing in Newton's equation $F = ma$—is *exactly* the same as the object's gravitational mass. In addition, any two objects at the same location and moving with the same velocity will follow the same path through spacetime, regardless of their masses. The astronaut in an orbiting spaceship falls around Earth, moving in lockstep with the spaceship itself. A feather dropped by an *Apollo* astronaut standing on the Moon falls toward the surface of the Moon at the same rate as a dropped hammer does. Instead of thinking of gravity as a force that acts on objects, it is more accurate to think of gravity as a consequence of the warping of spacetime in the presence of a mass. *Gravitation is the result of the shape of spacetime that objects move through.* That is one of the key insights of the **general theory of relativity**, Einstein's theory of gravity.

Special relativity tells us that any inertial reference frame is as good as any other. No experiment can distinguish between sitting in an enclosed spaceship floating "stationary" in deep space (**Figure 18.8a**) and sitting in an enclosed spaceship traveling at 0.9999c (**Figure 18.8b**). Those two situations feel the same because neither observer feels an acceleration. Both are equally valid inertial reference frames. As long as nothing accelerates either spaceship, neither observer can distinguish between the two spacecraft.

But what if an acceleration occurs? An acceleration can change either the speed or the direction of an object. Consider an astronaut inside a spaceship orbiting Earth, as shown in **Figure 18.8c**. That spaceship is accelerating because the direction of its velocity is constantly changing as it orbits Earth. The astronaut also is accelerating. Because he feels weightless, the astronaut has no way to tell the difference between being inside the spaceship as it falls around Earth and being inside a spaceship floating through interstellar space. Even though the ship's velocity is constantly changing as it falls, the inside of a spaceship orbiting Earth is an inertial frame of reference just as an object drifting along a straight line through interstellar space is an inertial frame of reference.

The idea that a freely falling reference frame is equivalent to a freely floating reference frame is called the **equivalence principle**. If you close your eyes and jump off a diving board, for the brief time that you are falling freely through Earth's gravitational field, the sensation you feel is the same as the sensation you would feel floating in interstellar space.

The natural path that an object will follow through spacetime in the absence of other forces is called the object's **geodesic**. In the absence of a gravitational field, the geodesic of an object is a straight line, in accordance with Newton's first law: an object will move at a constant speed in a constant direction unless acted on by a net external force. However, the shape of spacetime becomes distorted in the presence of mass, so an object's geodesic becomes curved.

Figure 18.9a shows two examples of inertial frames: an astronaut in a spaceship coasting through space and a person in a box falling toward the ground with an acceleration g. Both people are following their geodesics, so those two inertial reference frames are equivalent, and neither observer can distinguish between them. The equivalence principle also applies in cases of accelerations that result in a change in speed. Imagine an observer sitting in a closed box on the surface of Earth (**Figure 18.9b**). The floor of the box pushes on him to keep him from following his geodesic, and he feels that force. Now imagine the box is inside a spaceship accelerating through deep space at a rate of 9.8 meters per second per second (m/s^2) in the direction of the arrow shown in Figure 18.9b. The floor of the box pushes on the observer to overcome his inertia and cause him to accelerate at 9.8 m/s^2, so he feels as though he is being pushed into the floor of the box. In both cases, a force acts on the observer so that he does not follow the same free-falling geodesic. The observer feels as though he is being pushed into the floor of the box, so he feels the acceleration, and his frame of reference is not inertial. According to the equivalence principle, sitting in an armchair in a spaceship traveling with an acceleration of 9.8 m/s^2 is equivalent to sitting in an armchair on the surface of Earth reading this book. In each case, it is the same mass—the mass that gives an object inertia—that resists the change. Gravitational mass and inertial mass are the same thing.

The equivalence principle has an important caveat. In an accelerated reference frame such as an accelerating spaceship, the same acceleration (both magnitude and direction) is experienced everywhere.

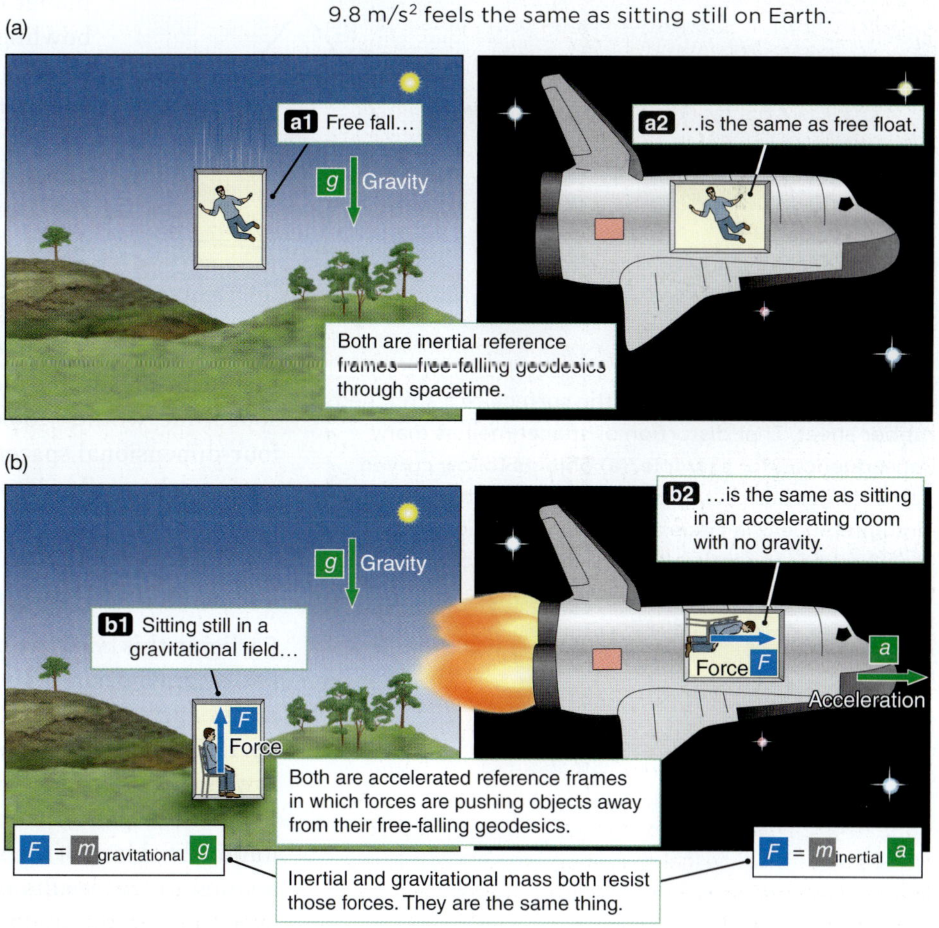

Figure 18.9 According to the equivalence principle, (a) an object falling freely in a gravitational field is in an inertial reference frame, and (b) an object at rest in a gravitational field is in an accelerated reference frame. As a result, sitting in a spaceship accelerating at 9.8 m/s^2 feels the same as sitting still on Earth.

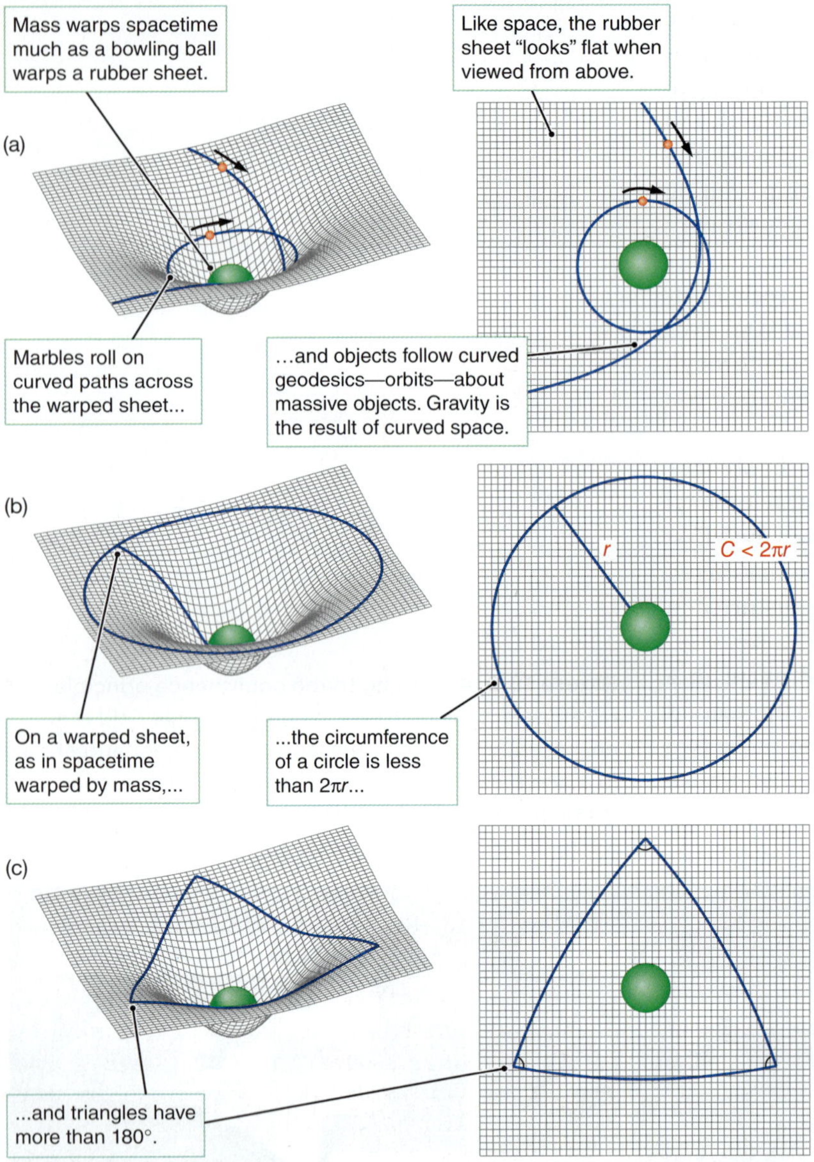

Figure 18.10 Mass warps the geometry of spacetime much as a bowling ball warps the surface of a stretched rubber sheet. That distortion of spacetime has many consequences; for example, (a) objects follow curved paths, or geodesics, through curved spacetime; (b) the circumference of a circle around a massive object is less than 2π times the radius of the circle; and (c) angles in triangles need not add to exactly 180 degrees.

In contrast, the curvature of space by a massive object is weaker farther from the object. The effects of gravity and acceleration are equivalent only locally; that is, the equivalence principle is valid only as long as attention is restricted to small enough volumes of space so that differences in gravity within that volume can be ignored.

Mass Distorts Spacetime

The general theory of relativity describes how mass distorts the geometry of spacetime. Imagine the surface of a tightly stretched, flat rubber sheet. A marble will roll in a straight line across the sheet. Euclidean geometry—the geometry of everyday life—applies on the surface of the sheet: If you draw a circle, its circumference is equal to 2π times its radius, r. If you draw a triangle, the angles add up to 180 degrees. Lines that are parallel anywhere are parallel everywhere.

Now place a bowling ball in the middle of the rubber sheet, creating a deep depression, or "well," as in **Figure 18.10**. The surface of the sheet is no longer flat, and Euclidean geometry no longer applies. If you roll a marble across the sheet (Figure 18.10a), its path dips and curves. You can roll the marble so that it moves around and around the bowling ball, like a planet orbiting about the Sun. If you draw a circle around the bowling ball (Figure 18.10b), the circumference of the circle is less than $2\pi r$. If you draw a triangle (Figure 18.10c), the angles add up to more than 180 degrees.

Mass affects the fabric of spacetime like the bowling ball affects the fabric of the rubber sheet. The bowling ball stretches the sheet, changing the distances between any two points on the surface of the sheet. Similarly, mass distorts spacetime, changing the distance between any two locations or events. Larger masses produce larger distortions in spacetime. We can see how a rubber sheet with a bowling ball on it is stretched through a third spatial dimension, but most people cannot picture what a curved four-dimensional spacetime would "look like." Yet experiments verify that the geometry of four-dimensional spacetime is distorted much like the rubber sheet, whether or not it can be easily pictured.

When One Physical Law Supplants Another

Earlier in this book, we described gravity as a force that obeys Newton's universal law of gravitation: $F = Gm_1m_2/r^2$. Here in Chapter 18, we have introduced the ideas of general relativity and asked you to change the way you think about gravity. If general relativity is correct, does that mean Newton's formulation of gravity is wrong? If so, then why does Newton's law continue to be used?

Those questions go to the heart of how science progresses. As long as a gravitational field is not too strong, Newton's law of gravitation is a very close *approximation* to the results of a calculation using general relativity. In that context, even the gravitational field near the core of a massive main-sequence star would be considered "weak." An astronomer will obtain the same results if she uses a general relativistic formulation of gravity rather than Newton's laws to calculate the structure of a main-sequence star. Similarly, even though spacetime is curved by the presence of mass, that curvature near Earth is so slight that over small regions it can be ignored entirely, and flat (Euclidean) geometry can be used. That is the same kind of approximation people use when they navigate with a flat road map, despite the curvature of Earth.

Similarly, Newton's laws of motion are *approximations* of special relativity and quantum mechanics. In fact, Newton's laws can be mathematically derived from special relativity and quantum mechanics by using the assumptions that speeds of objects are much less than the speed of light and that objects are much larger than the particles from which atoms are made. Newton's laws of motion and gravitation are used most of the time because they are far easier to apply than general relativity and because any inaccuracies introduced by using Newtonian approximations are usually far too tiny to measure. That was illustrated in Figure 18.5, in which the behavior of an accelerated object moving at a velocity much less than c is the same whether or not relativity is used. Calculations based on general relativity are required only when conditions are very different from those of everyday life (for example, the behavior of the gravitational field of a black hole) or in special cases when extremely high accuracy is needed (for example, the precise timing used by the global positioning system [GPS] satellite network).

If a new theory is to replace an earlier, highly successful scientific theory, the new theory must hold the old theory within it—that is, the new theory must be able to reproduce the successes of the earlier theory. Special relativity contains Newton's laws of motion, and general relativity holds within it the successful Newtonian description of gravity that we have relied on throughout this book (see the **Process of Science Figure**).

The Observable Consequences of General Relativity

The curved spacetime of general relativity does have observable consequences. Indeed, in our own Solar System, observations can be made that distinguish general relativity from Newtonian physics. In Newton's theory, orbits are elliptical and fixed in space. In contrast, general relativity predicts that the long axis of an elliptical orbit slowly rotates, or precesses. **Figure 18.11** illustrates the difference between an orbit in Newton's theory, which remains stationary (left panel), and one in general relativity, which precesses (right panel). In our Solar System, even after one accounts for the effects of other planets on Mercury, a very small shift in its axis remains, equal to 43 arcseconds (arcsec) per century—which cannot be explained by Newton's laws alone. General relativity predicts exactly that precession for Mercury.

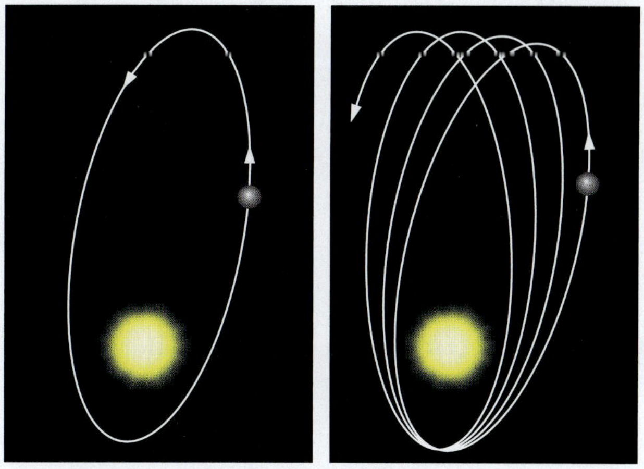

Figure 18.11 The left panel denotes an elliptical orbit about the Sun. In the Newtonian view, that elliptical orbit remains stationary. In the right panel, Mercury's orbit precesses as a result of the warped spacetime near the Sun.

General relativity has other unique implications. A beam of light moving through empty space travels in a straight line, but a beam of light moving through the distorted spacetime around a massive object is bent by gravity, just as the lines in Figure 18.10 are bent by the curvature of the sheet. That bending of the light path by curved spacetime is called **gravitational lensing** because optical lenses also bend light paths. The first measurement of gravitational lensing came during the total solar eclipse of 1919. Several months before the eclipse, astrophysicist Sir Arthur Stanley Eddington (1882–1944) measured the positions of several stars in the direction of the sky where the eclipse would occur. He then

PROCESS OF SCIENCE

New Science Can Encompass the Old

General relativity is more accurate and broadly applicable than Newton's laws. It also explains why gravity acts as it does. Still, for ordinary objects on Earth or the Solar System, Newton's and Einstein's theories agree.

One way that scientists check new theories is by considering the limits. What happens at great distances? What happens if the mass is very small? In these limits, the new and more complete theories must give the same results as old theories.

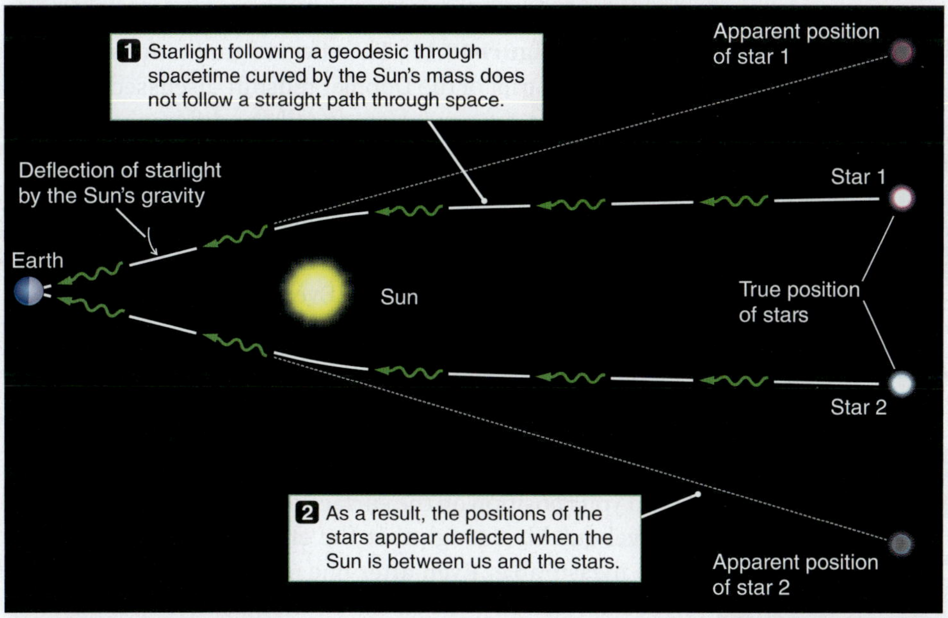

Figure 18.12 According to measurements taken during the total solar eclipse of 1919, the gravity of the Sun bends the light from distant stars by the amount predicted by Einstein's general theory of relativity. That is an example of gravitational lensing. Note that the "triangle" formed by Earth and the two stars contains more than 180 degrees, just like the triangle in Figure 18.10c.

Figure 18.13 This Hubble Space Telescope image shows four images of the same distant supernova (arrows), where the images arise from the gravitational lensing due to a massive foreground galaxy.

repeated the measurements during the eclipse. **Figure 18.12** shows how the light from distant stars curved as it passed the Sun, causing the measured positions of the stars to shift outward. The stars appeared farther apart in Eddington's second measurement than in his first by the amount predicted by general relativity. That is another example of the effects of general relativity—in curved space, parallel lines actually can intersect.

Images of galaxies are sometimes distorted by gravitational lensing by other galaxies or clusters of galaxies. In the Hubble Space Telescope image shown in **Figure 18.13**, a supernova in a distant galaxy is lensed into four images by a nearby massive galaxy. What makes that case particularly interesting is that those four images appeared at different times because the light for each image took a different path, and some paths were longer than others. In an extreme case, a lensed galaxy image can be distorted into an **Einstein ring** (**Figure 18.14**).

Mass distorts the geometry not only of space but also of time. The deeper one descends into the gravitational field of a massive object, the more slowly clocks appear to run from the perspective of a distant observer—an effect called **general relativistic time dilation**. Suppose a light is attached to a clock sitting on the surface of a neutron star. The light is timed so that it flashes once a second. Because time near the surface of the star is dilated as a result of the depth of the gravitational field, an observer far from the neutron star perceives the light to be pulsing with a lower frequency—less than once a second. Now suppose an emission line source is on the surface of the neutron star. Because time is running slowly on the surface of the neutron star, the light that reaches the distant observer will have a lower frequency than when it was emitted. It is as though photons of light have lost energy in their escape from the star's surface. A lower frequency means a longer wavelength, so the light from the source will be seen at a longer, redder wavelength than the wave-length at which it was emitted. That shift in the wavelengths of light from objects

Figure 18.14 An Einstein ring created by the gravity of a luminous red galaxy gravitationally lensing the light from a much more distant blue galaxy.

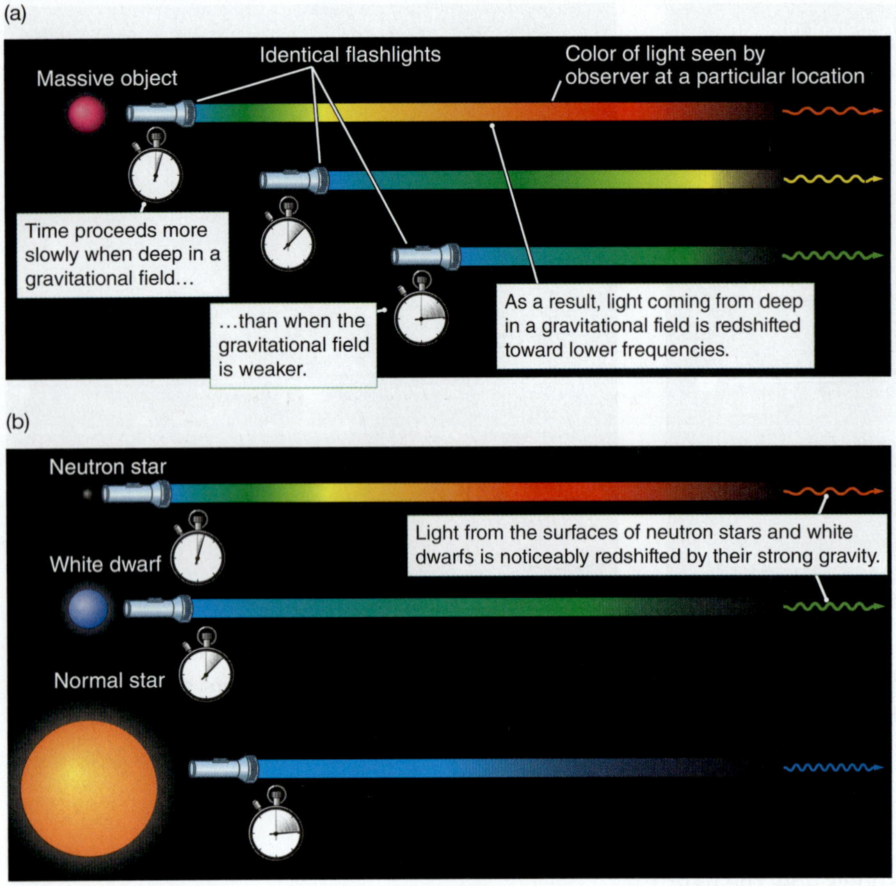

Figure 18.15 Time passes more slowly near massive objects because of the curvature of spacetime. To a distant observer, then, light from near a massive object will have a lower frequency and longer wavelength. (a) The closer the source of radiation is to the object or (b) the more massive and compact the object is, the greater the gravitational redshift will be.

deep within a gravitational well is called the **gravitational redshift** (**Figure 18.15**). The effect of gravitational redshift is similar to that of the Doppler redshift discussed in Chapter 5. In fact, no way exists to tell the difference between light that has been redshifted by gravity and light from an object moving away from you that has been Doppler shifted.

Here is an example that brings the phenomenon of general relativistic time dilation a bit closer to home. A clock on the top of Mount Everest runs faster, gaining about 80 ns a day compared with a clock at sea level. The difference between an object on the surface of Earth and an object in orbit is even greater. A GPS receiver uses the results of sophisticated calculations of the effects of general relativistic gravitational redshift to help you accurately find your position on the surface of Earth. Satellites in orbit travel quickly enough that the effects of special relativity are measurable. Even after one allows for slowing due to special relativity, the clocks on the satellites that make up the GPS run faster than clocks on the surface of Earth. If the satellite clocks *and* your GPS receiver did not correct for that and other effects of general relativity, the position your GPS receiver reported would be in error by up to half a kilometer. The fact that the GPS can be accurate to a few meters provides strong experimental confirmation of two predictions of general relativity—gravitational redshift and general relativistic time dilation.

CHECK YOUR UNDERSTANDING 18.3

In general relativity, what causes gravity? (a) It is a result of Newton's law of gravitation. (b) It is a consequence of time dilation and length contraction. (c) It is the result of masses having greater effect when they move at high speeds. (d) It is the result of the distortion in spacetime around a massive object.

18.4 Black Holes

General relativity also predicts the existence of black holes. The Newtonian world has no well-formulated theory of black holes. When placed on the surface of a rubber sheet, a mass causes a funnel-shaped distortion analogous to the distortion of spacetime by a mass. Now imagine the limit in which the funnel is *infinitely* deep—it gets narrower as it goes deeper but has no bottom. That is the rubber-sheet analog to a black hole. The mathematics describing a black hole approaches infinity, just as the mathematical expression $1/x$ behaves as x approaches zero. Such a mathematical anomaly is called a **singularity**. Black holes contain singularities in spacetime, and that mathematical complication indicates that extreme conditions exist near (and inside) a black hole.

How Black Holes Form

We have seen how stellar evolution can lead to the formation of compact stellar remnants such as white dwarfs and neutron stars. However, an even more extreme fate awaits some massive stars at the end of their evolution. Recall from

Chapter 16 that a white dwarf can have a mass of no more than the Chandrasekhar limit—about 1.4 solar masses (M_{Sun}). If the mass of the object exceeds that limit, gravity can overcome electron degeneracy pressure, and the white dwarf will begin collapsing again.

The physics of a neutron star is much like the physics of a white dwarf, except that neutrons rather than electrons are what cause a neutron star to be degenerate. Analogous to what happens with a white dwarf, if the mass of a neutron star exceeds ~2.5 M_{Sun}, gravity begins to win out over pressure once again. The neutron star grows smaller, and gravity at the star's surface becomes stronger and stronger at an ever-accelerating pace. Recall from Chapter 4 that the escape velocity from a planet or moon depends on its surface gravity. Now imagine surface gravity so strong that the escape velocity approaches the speed of light. Nothing, though, not even light, can travel faster than *c*. So when the neutron star crosses the threshold where the escape velocity from its surface exceeds the speed of light, not even light can escape its gravity. A region of space where neither matter nor radiation can escape the pull of gravity is called a **black hole**. **Figure 18.16** illustrates how stars of various masses evolve from protostars to their ultimate end, including sometimes as black holes.

A neutron star in a binary system can collapse to become a black hole if it accretes enough matter from its companion to push it over the ~2.5-M_{Sun} limit, similar to the way some Type Ia supernovae accrete matter from a companion and approach the Chandrasekhar limit of 1.4 M_{Sun}. Regardless of how it forms, any stellar remnant with a mass greater than ~2.5 M_{Sun} must be a black hole.

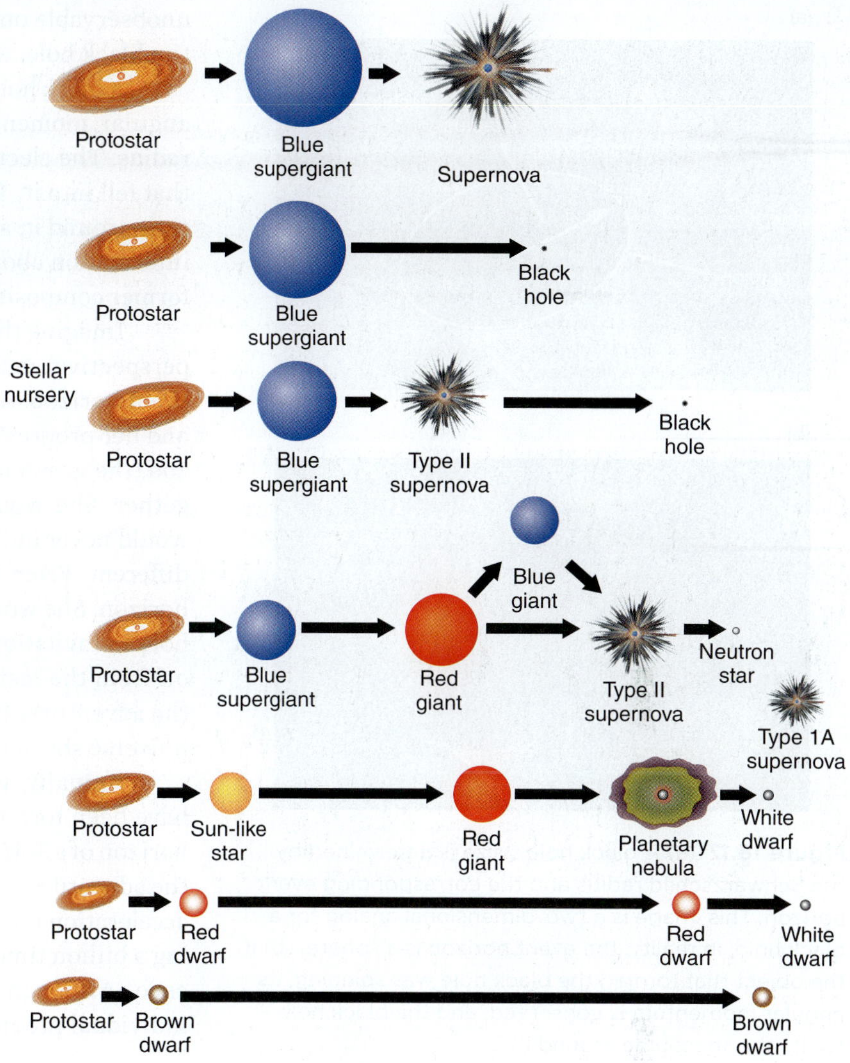

Figure 18.16 Stars of various masses evolve from left to right, going from protostars to stellar remnants, and in the process may recycle material back into the interstellar medium to produce new stars. Not to scale.

Properties of Black Holes

From the outside, you can never actually "see" a black hole. The closer an object approaches a black hole, the greater is its escape velocity (the speed that the object would need to escape from the gravity of the black hole). At a certain distance from the black hole, the escape velocity reaches the speed of light. That radius is called the **Schwarzschild radius**, named for the physicist Karl Schwarzschild (1873–1916), and it is proportional to the mass of the black hole:

$$R_S = \frac{2GM_{BH}}{c^2} = 3\,\text{km} \times \frac{M_{BH}}{M_{Sun}}$$

where R_S is the Schwarzschild radius, G is the universal gravitational constant, M_{BH} is the mass of the black hole, and c is the speed of light. The sphere around the black hole at that distance is called its **event horizon**—a surface from which nothing, not even light, can escape. A black hole with a mass of 1 M_{Sun} has a Schwarzschild radius of about 3 km. A black hole with a mass of 5 M_{Sun} has a Schwarzschild radius 5 times that, or about 15 km. If Earth were squeezed into a black hole, it would have a Schwarzschild radius of only about a centimeter. All the mass of a black hole is concentrated at its very center, but that fact is

(a)

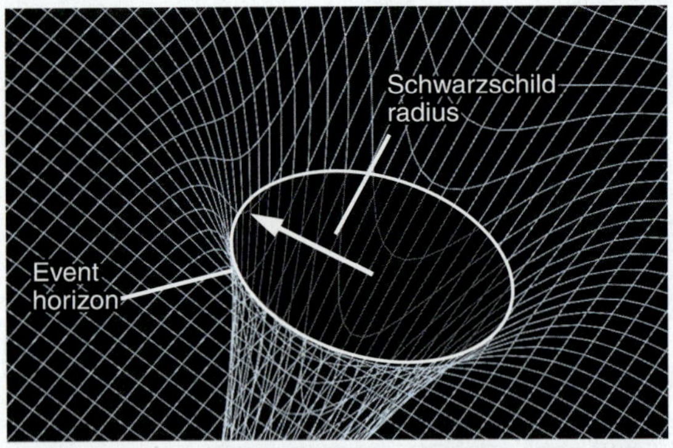

(b)

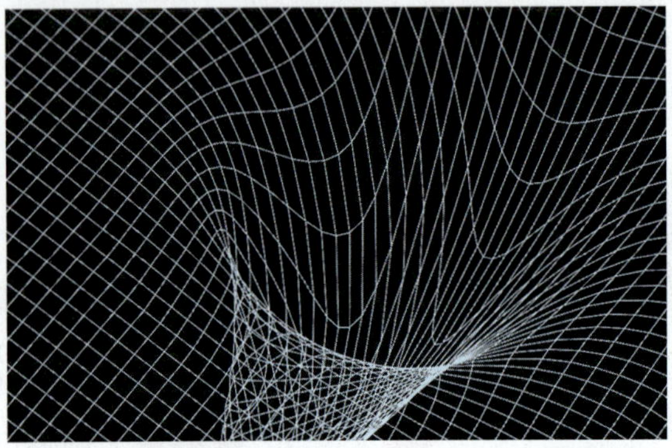

Figure 18.17 (a) A black hole's size is determined by the Schwarzschild radius and the corresponding event horizon. This image is a two-dimensional analog for a black hole. In reality, the event horizon is a sphere. (b) If the object that formed the black hole was spinning, its angular momentum is conserved, and the black hole twists the spacetime around it.

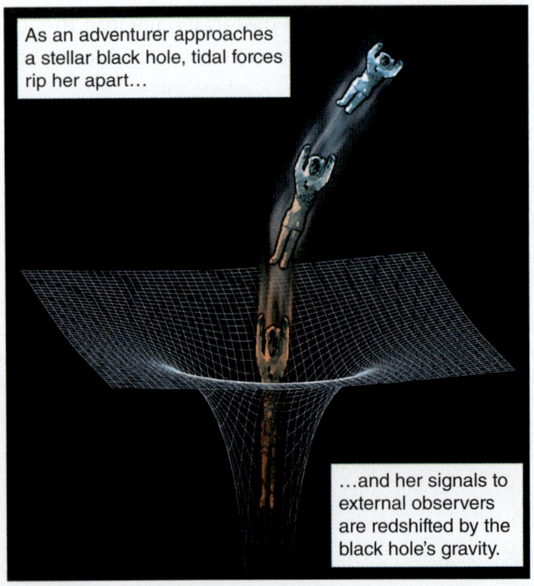

Figure 18.18 An adventurer falling into a black hole would be torn apart by the extreme tidal forces.

unobservable outside the black hole. **Figure 18.17a** shows a rubber-sheet analog to a black hole, with the Schwarzschild radius and the event horizon.

A black hole has only three observable properties: mass, electric charge, and angular momentum. The mass of a black hole determines the Schwarzschild radius. The electric charge of a black hole is the net electric charge of the matter that fell into it. The angular momentum of a rotating black hole twists the spacetime around it, as shown in **Figure 18.17b**. Apart from those three properties, all information about the material that fell into the black hole is lost. Nothing of its former composition, structure, or history survives.

Imagine that an adventurer falls into a black hole (**Figure 18.18**). From our perspective outside the black hole, the adventurer would appear to fall toward the event horizon. As she fell, her watch would appear to run more and more slowly, and her progress toward the event horizon would slow as well. At the event horizon, the gravitational redshift becomes infinite, and clocks appear to stop altogether. She would approach the event horizon, but from our perspective she would never quite make it. The adventurer's experience, however, would be very different. From her perspective, she would see nothing special about the event horizon. She would fall past the event horizon and continue deeper into the black hole's gravitational well. She would now have entered a region of spacetime cut off from the rest of the universe. The event horizon is like a one-way door: after the adventurer has passed through, she can never again pass back into the larger universe she once belonged to.

Actually, we have overlooked an important detail—the adventurer would have been torn to shreds long before she reached the black hole. Near the event horizon of a 3-M_{Sun} black hole, the difference in gravitational acceleration between the adventurer's feet and her head would be about a billion times her gravitational acceleration on the surface of Earth. In other words, her feet would be accelerating a billion times faster than her head. That would be a most unpleasant experience. Although scientific theories must produce testable predictions, not all individual predictions have to be tested directly.

"Seeing" Black Holes

In 1974, the physicist Stephen Hawking realized that black holes should be sources of radiation. In the ordinary vacuum of empty space, particles and their antiparticles spontaneously appear and then, within about 10^{-21} second, annihilate each other and disappear. Such particles are called **virtual particles** because they exist for only a very short time. If a pair of virtual particles comes into existence near the event horizon of a very small black hole, one of the particles might fall into the black hole while the other particle escapes (**Figure 18.19**). Some of the gravitational energy of the black hole will have been used up in making one of the pair of virtual particles real. Hawking showed that through that process, a black hole should emit a Planck blackbody spectrum and that the effective temperature of that spectrum would increase as the black hole became smaller through that "evaporation" process. After a very, very long time (on the order of 10^{61} years for a black hole with a mass of the Sun), the black hole would become small enough that it would become unstable and explode. Although the light that emerges, called **Hawking radiation**, is of considerable interest to physicists, the low intensity of Hawking radiation means it is not a likely way to see a black hole.

The first strong direct evidence for black holes that result from supernovae came from X-ray binary stars. The radio emission from Cygnus X-1 (an object originally identified in X-rays) flickers rapidly, changing in as little as 0.01 second.

That means the source of the X-rays must be smaller than the distance that light travels in 0.01 second, or 3,000 km—smaller than Earth. Cygnus X-1 was also identified as both a radio source and an already cataloged star called HD 226868. The spectrum of HD 226868 shows that it is a normal O9.7 I supergiant star with a mass of about 19 M_{Sun}—far too cool to produce the observed X-ray emission by itself. The wave-lengths of absorption lines in the spectrum of HD 226868 are Doppler-shifted back and forth with a period of 5.6 days, indicating that HD 226868 is part of a binary system. Using the same techniques we discussed to measure the masses of binary stars in Chapter 13, astronomers found that the mass of the unseen compact companion of HD 226868 must be about 15 M_{Sun} (**Working It Out 18.2**). The companion to HD 226868 is too compact to be a normal star, yet it is much more massive than the Chandrasekhar limit for a white dwarf or the upper mass limit of a neutron star. Such an object must be a black hole. The X-ray emission from Cygnus X-1 arises when material from the O9.7 I supergiant falls onto an accretion disk surrounding the black hole, as illustrated in the chapter-opening figure.

In some similar systems, winds have been observed blowing off the disk around the black hole. Those winds are probably caused by magnetic fields in the disk. The fastest winds observed are in the binary system IGR J17091, where the wind speeds are as high as 32 million km/h (about 3 percent of the speed of light). That wind is blowing in many directions, and it may be carrying away more mass than is being captured by the black hole (**Figure 18.20**).

Astronomers have modeled the observational data of dozens of good candidates for stellar-mass black holes in X-ray binary systems in the Milky Way. Those findings indicate that the black-hole masses are greater than 4.5–5 M_{Sun}, that is, not very close to the limit of ~2.5 M_{Sun} for a neutron star. That gap in mass between the most massive neutron stars and the least massive stellar black holes is not yet understood, and it is assumed to be a result of the mass transfer processes between the stars.

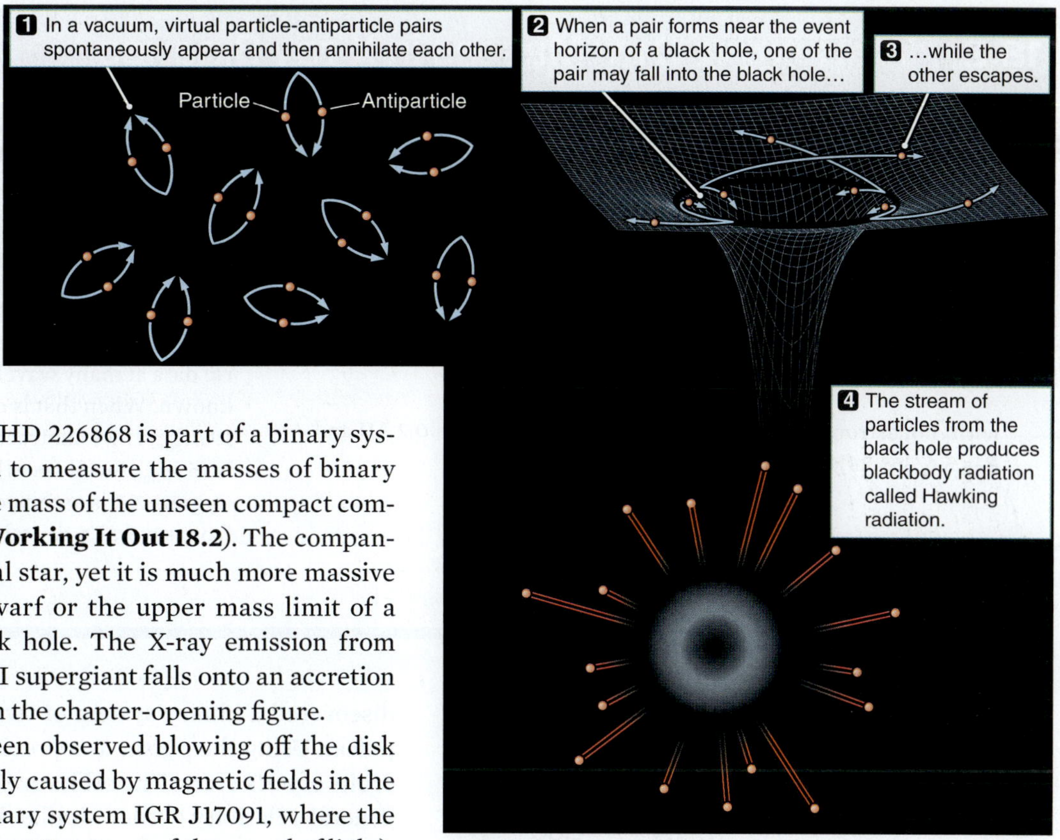

Figure 18.19 In the vacuum of empty space, particles and antiparticles are constantly being created and then annihilating each other. Near the event horizon of a black hole, however, one particle may cross the horizon and fall into the black hole before it recombines with its partner. The remaining particle leaves the vicinity of the event horizon, producing a blackbody spectrum of Hawking radiation.

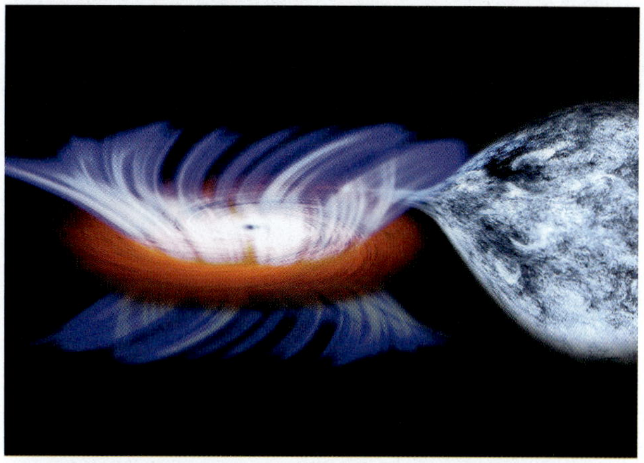

Figure 18.20 This artist's model shows strong winds being emitted from the disk around a stellar black hole. Those winds can remove more material than the amount that actually falls into the hole.

Detection of Gravitational Waves

General relativity also predicts the existence of a wholly new form of radiation. If you exert a force on the surface of a rubber sheet, accelerating it downward, waves will move away from where you struck it, like ripples spreading out over the surface of a pond. Similarly, the equations of general relativity predict that if you accelerate the fabric of spacetime (for example, through the catastrophic asymmetrical collapse of a high-mass star), ripples in spacetime, or **gravitational waves**, will move outward at the speed of light. Those gravitational waves are like electromagnetic waves in some respects. Accelerating an electrically charged particle gives rise to an electromagnetic wave. Deforming a massive object gives rise to gravitational waves. Both types of waves can travel through a vacuum, but gravitational waves do not involve electric or magnetic fields. Instead, gravitational waves are ripples in the very fabric of spacetime.

Indirect evidence has long existed for the existence of gravitational waves. General relativity predicts that the orbit of binary neutron stars should lose energy, which will be carried away as gravitational waves. In 1974, astronomers

18.2 Working It Out Masses in X-ray Binaries

Cygnus X-1 is part of a binary system with a blue supergiant star O9.7 I (very close to B0 I) and an unseen compact object located about 0.2 astronomical units (AU) away from it. The blue supergiant and the compact object orbit a common center of mass every 5.6 days. We can use the simple formula from Working It Out 13.4 to calculate the sum of the masses:

$$\frac{M_{\text{blue}}}{M_{\text{Sun}}} + \frac{M_{\text{compact}}}{M_{\text{Sun}}} = \frac{A^3}{P^2}$$

with A in astronomical units and P in years. Here, $A = 0.2$ AU and $P = 5.6/365.24$ years, so

$$\frac{M_{\text{blue}}}{M_{\text{Sun}}} + \frac{M_{\text{compact}}}{M_{\text{Sun}}} = \frac{0.2^3}{\left(\frac{5.6}{365.24}\right)^2} = 34$$

The sum of the masses of the two stars is 34 M_{Sun}.

To find the values of the two individual masses from their orbits, we need to know the velocities of the two stars or the distance of each star to the center of mass and the orbital inclination of the system. Obtaining such information is difficult when one star is compact and not observed separately. However, the mass of the blue supergiant star can be estimated from spectroscopic and photometric data at many wavelengths, as long as the distance to the system is known. When that is done, the mass of the supergiant is estimated at 19 M_{Sun}. If we subtract that from 34, the mass of the compact object is 15 M_{Sun}—well over the mass limit for a neutron star. Therefore, Cygnus X-1 is assumed to be a black hole. A recent study with data from several X-ray telescopes confirmed that mass.

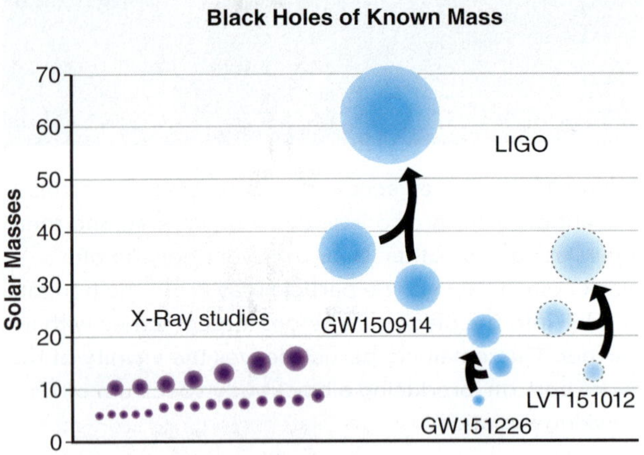

Figure 18.21 An illustration of the black hole mergers seen by LIGO and VIRGO through 2017. Also shown are the black hole masses seen in binary stars.

discovered a binary system of two neutron stars, one of which is an observable pulsar. Using the pulsar as a precise clock, astronomers accurately measured the orbits of both stars. The orbits are gradually losing energy at the rate predicted by general relativity. Other similar binary pairs have been found with an orbital energy loss consistent with the radiation of gravitational waves.

On 14 September 2015, almost exactly 100 years after the prediction of gravitational waves by Albert Einstein, the Laser Interferometer Gravitational-Wave Observatory (LIGO) detected the gravitational waves from the merger of two black holes in a distant galaxy. The signal LIGO recorded precisely agreed with the prediction of the gravitational wave signal from such a merger, and the three leaders of the LIGO team, Kip Thorne, Rainer Weiss, and Barry Barrish, won the 2017 Nobel Prize in Physics for that discovery. Since then, several other black hole mergers have been observed by LIGO and its European counterpart, VIRGO (**Figure 18.21**). They also detected the merger of two neutron stars, as discussed in Chapter 17.

The black holes we discussed here in Chapter 18 came from collapsing massive stars, but that is not the only type of black hole. In Chapters 19 and 20, you will learn that supermassive black holes can be found at the centers of galaxies, including the Milky Way.

CHECK YOUR UNDERSTANDING 18.4

If a black hole suddenly doubled in mass, the event horizon would become ________ its original radius. (a) one-quarter; (b) one-half; (c) twice; (d) 3 times; (e) 4 times

Origins Gamma-Ray Bursts

The most energetic explosions in the universe are probably related to stellar black holes or merging neutron stars. **Gamma-ray bursts (GRBs)** are intense bursts of gamma rays. The bursts are followed by a weaker "afterglow" that is observed at many wavelengths. GRBs were first observed in the 1960s by satellites designed to look for radiation from nuclear weapons being tested in space after such tests were banned on Earth. In the 1990s, gamma-ray astronomy satellites discovered that those bursts were coming from all directions in the sky and that they might be associated with supernova explosions in distant galaxies. Short-duration GRBs, which last less than 2 seconds, probably originate from the merging of two neutron stars or a neutron star and a black hole in a close binary system that collapses into a single black hole. The more common long-duration GRBs are easier to study because they have a longer afterglow. Astronomers think that those originate in the collapse of a very high-mass, rapidly spinning star to a black hole or a neutron star after a supernova explosion.

Unlike regular supernovae, which radiate equally in all directions, GRBs are beamed events, so most of their enormous energies are concentrated into two opposite jets of emission (**Figure 18.22a**). In addition to the electromagnetic radiation, relativistic jets of cosmic rays are emitted. Astronomers have not observed any GRBs in the Milky Way: a massive supernova has not occurred in our galaxy for at least a century. But the energy of GRBs is so intense that people have wondered what might happen to Earth if one went off nearby with its radiation beamed in Earth's direction, a possibility imagined in **Figure 18.22b**. A leading candidate for a future GRB in our galaxy is the massive star Eta Carinae (see Chapter 17). That star is 7,500 light-years away—a neighbor, astronomically speaking. Its rotation axis, however, is such that it is unlikely to form a GRB that beams toward Earth.

Some scientists wonder whether past supernova and GRB events could have affected the history of life on Earth. Supernova "archaeologists" may have found evidence on Earth of past supernovae. In one study, rocks deep in the Pacific Ocean were found to have amounts of a radioactive isotope of iron, iron-60, that is too short-lived to be left over from the formation of Earth. That iron-60 could have been deposited 2.8 million years ago on Earth after a supernova explosion. In another study, high concentrations of nitrates were found in some layers in Antarctic ice cores. Gamma radiation from supernovae can produce excess nitrogen oxides in the atmosphere, which then become converted to nitrates that are trapped in snowfall. Nitrate spikes correlate to 1006 and 1054 CE—two years when bright supernovae are known to have appeared in the Milky Way.

What about more drastic effects on Earth from a nearby supernova or a more distant but beamed GRB? Normally, Earth is protected from cosmic radiation and cosmic-ray particles by its ozone layer and magnetic field. Cosmic-ray particles might not be a major problem if they arose very close to Earth, but the high-energy gamma-ray radiation could have a more serious effect on Earth. The excess nitrogen oxides they produce in the atmosphere can absorb sunlight, which would cool Earth. The gamma radiation could ionize Earth's atmosphere, shrinking or destroying the ozone layer that protects life from ultraviolet radiation. Even a burst of a few seconds could lead to ozone damage lasting for decades. The gamma rays could trigger a burst of solar ultraviolet radiation at Earth's surface, which could damage the DNA of phytoplankton a few hundred meters deep in the ocean, affecting their ability to photosynthesize. Phytoplankton are the base of Earth's food chain, so a drastic reduction in phytoplankton could upset the entire biosphere. It has been hypothesized that such an event may have happened in Earth's history.

Statistically speaking, GRBs that beam to Earth may be rare, and some astronomers have argued that they are less likely to be produced in a Milky Way–type galaxy than in other types of galaxies. A lot of uncertainty surrounds any estimate of how close the supernova or GRB must be and how often those explosive events must occur to seriously affect Earth. According to one estimate, a supernova or GRB explosion close enough to alter Earth's biosphere occurs a few times every billion years, possibly leading to mass extinction events. In Chapter 17, we noted that the chemical elements that make up life were created in supernova explosions. The discussion here suggests that supernovae may have had some effect on the *evolution* of life on Earth as well.

(a)

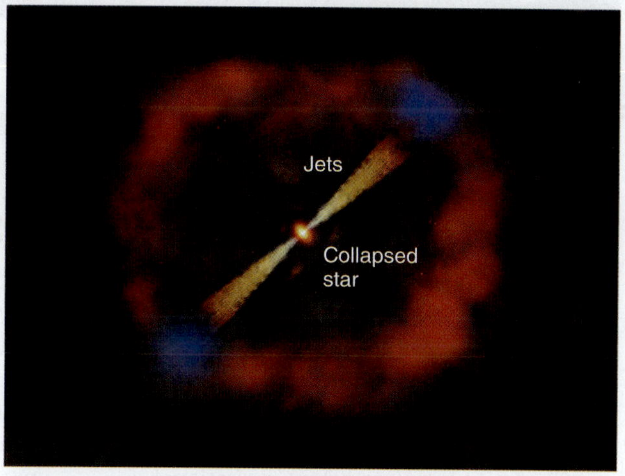

(b)

Figure 18.22 An artist's model of a gamma-ray burst (GRB). (a) Narrow beams of intense energy are sent in two opposite directions. (b) If one beam is pointed toward the observer, the GRB will appear bright.

ARTICLES QUESTIONS

Researchers have detected gravitational waves, ripples in time and space, for the third time. That discovery could offer new insight into unanswered questions surrounding dark matter and black holes.

LIGO Just Detected the Oldest Gravitational Waves Ever Discovered

By KARLA LANT

Gravitational Waves Revealing the Universe

The Laser Interferometer Gravitational-wave Observatory (LIGO) just detected gravitational waves, ripples in time and space, for the third time. Two black holes collided, forming a huge black hole 49 times more massive than our sun, and this generated the waves. This kind of collision was also the cause of the waves detected previously by LIGO, although the masses of the black holes varied. This repetition of the discovery confirms that a new area of astronomy now exists.

"We have further confirmation of the existence of stellar-mass black holes that are larger than 20 solar masses—these are objects we didn't know existed before LIGO detected them," MIT's David Shoemaker, a LIGO spokesperson, said in a press release. "It is remarkable that humans can put together a story, and test it, for such strange and extreme events that took place billions of years ago and billions of light-years distant from us. The entire LIGO and Virgo scientific collaborations worked to put all these pieces together."

In September 2015, LIGO first directly observed these gravitational waves during its first run since receiving Advanced LIGO upgrades. The second detection followed in December 2015, and this latest detection, called GW170104, followed in January of this year [2017]. In each case, both of LIGO's twin detectors perceived gravitational waves from the collisions of the black holes, but this latest observation does offer a few new pieces of information.

For example, it suggests which directions the black holes might be spinning in, and indicates that at least one of the black holes in the pair may not be aligned with the overall orbital motion. Scientists are hoping that they can learn more about how binary black holes form by making more LIGO observations.

LIGO's Future

This work is testing, and thus far providing proof for, the theories proposed by Albert Einstein. For example, the theory of relativity says that dispersion, the effect that happens as light waves in a physical medium travel at different speeds, cannot happen in gravitational waves. LIGO has not found any evidence of dispersion in gravitational waves, as predicted by relativity.

"It looks like Einstein was right—even for this new event, which is about two times farther away than our first detection," Georgia Tech's Laura Cadonati, the Deputy Spokesperson of the LIGO Scientific Collaboration (LSC), said in the press release. "We can see no deviation from the predictions of general relativity, and this greater distance helps us to make that statement with more confidence."

Moving forward, the LIGO-Virgo team will keep searching LIGO data for any hint of gravitational waves emanating from the far corners of the Universe. The sensitivity of the detector will improve during the next run starting in late 2018 after researchers apply technical upgrades, hoping to see even more. Caltech's David Reitze, the LIGO Laboratory's executive director, said in the press release, "While LIGO is uniquely suited to observing these types of events, we hope to see other types of astrophysical events soon, such as the violent collision of two neutron stars."

ARTICLES QUESTIONS

1. How do those results help confirm the theory of general relativity?
2. Why were astronomers surprised about the masses of the two black holes?
3. What was different about that particular detection?
4. Go to the website for LIGO detections (http://ligo.org/detections.php). What was found in the 2018–2019 (or later) observing runs?

Summary

The highest-mass stars leave behind black holes. In the environment surrounding black holes, relativistic effects become important. A black hole's mass determines its Schwarzschild radius: the boundary from which light cannot escape. Gamma-ray bursts are beamed high-energy explosions that result from the merger of two compact objects or the rapid collapse of a high-mass star into a black hole. The radiation from those bursts could affect life on Earth.

LG 1 **Describe how the motion of the observer affects the observed velocity of objects.** Even at relatively low relative speeds, the motion of the observer can affect the direction of the measured velocity, as in the aberration of starlight. At higher speeds, the magnitude of the measured velocity also is affected. The speed of light in a vacuum, *c*, is the ultimate speed limit. Observers in all inertial reference frames will measure the same speed of light.

LG 2 **Discuss the observable consequences of the relationship between space and time.** Special relativity connects space and time into four-dimensional spacetime. How the spacetime distance between events is divided between space and time depends on the observer's motion. As velocities approach the speed of light, observers detect that moving clocks run more slowly than their own clocks, that moving objects are contracted in length, and that moving objects behave as though they were more massive.

LG 3 **Explain how gravity is a consequence of the way mass distorts the very shape of spacetime.** Inertial mass and gravitational mass are the same, leading to the principle of equivalence, in which acceleration cannot be distinguished from gravity if the acceleration is small enough. In general relativity, mass warps the fabric of spacetime so that objects move on the shortest path in that warped geometry. Gravity is a consequence of the spacetime warping. Time runs more slowly near massive objects, and radiation from any light source near a black hole's event horizon is redshifted.

LG 4 **Explain the formation of black holes from the most massive stars, and describe the key properties of those stellar black holes.** The masses for both white dwarfs and neutron stars have an upper limit. Dense stellar remnants more massive than about $\sim 2.5\ M_{\text{Sun}}$ collapse to form black holes. The supernova explosion that ends the life of a massive star leaves behind a neutron star or a black hole. The mathematical singularity at the center of a black hole is still a mystery to science. However, scientists have now observed the mergers of many black holes through the gravitational waves emanating from such events. Black holes may, after a very, very long time, be destroyed by evaporation through Hawking radiation.

? Unanswered Questions

- What happens to the information that falls into a black hole? We said earlier that a black hole is characterized by only three properties: mass, angular momentum, and electric charge. Where did all the other information go? To a distant observer, material takes an infinitely long time to fall into a black hole, so although the observer sees less and less radiation from the material, the properties of the material seem to be the same for all time. From the perspective of the infalling material, however, crossing the event horizon takes a finite amount of time, so information other than mass, angular momentum, and charge can no longer be shared with the outside world.
- Do wormholes exist in spacetime, connecting one region with another, perhaps through black holes? Wormholes are a mathematical solution to the equations of general relativity. The idea is that when something goes into a black hole, it travels through a wormhole and emerges in a different part of the universe. In that way a wormhole acts as a shortcut through spacetime. In science fiction, wormholes are a popular means of traveling large distances by exploiting the strange geometry of spacetime. But many scientists doubt that wormholes can exist in nature, and even if they do exist, strong tidal forces would pull apart anything that falls into a black hole before it emerged from a wormhole.

Questions and Problems

Test Your Understanding

1. Rank the following (from least massive to most massive) in terms of the mass of the star that produces each.
   a. neutron star
   b. black hole
   c. white dwarf

2. A car approaches you at 50 km/h. A fly inside the car is flying toward the back of the car at 7 km/h. From your point of view by the side of the road, the fly is moving at ________ km/h.
   a. 7
   b. 28.5
   c. 43
   d. 57

3. A car approaches you at 50 km/h. The driver turns on the headlights. From your point of view, the light from the headlights is moving at
   a. $c + 50$ km/h.
   b. $c - 50$ km/h.
   c. $(c + 50$ km/h$)/2$.
   d. c.

4. Imagine that you are on a spaceship. A second spaceship rockets past yours at $0.5c$. You start a stopwatch and stop it 10 seconds later. For an astronaut in the other spaceship, the number of seconds that have ticked by during the 10 seconds on your stopwatch is
   a. more than 10 seconds.
   b. equal to 10 seconds.
   c. less than 10 seconds.

5. The International Space Station flies overhead. Using a telescope, you take a picture and measure its length to be ________ than its length as it would be measured if it were sitting on the ground.
   a. much greater
   b. slightly greater
   c. slightly less
   d. much less

6. Astronauts in the International Space Station
   a. have no mass.
   b. have no energy.
   c. are outside Earth's gravitational field.
   d. are in free fall.

7. Einstein's formulation of gravity
   a. is approximately equal to Newton's universal law of gravitation for small gravitational fields.
   b. is always used to calculate gravitational effects in modern times.
   c. explained why Newton's universal law of gravitation describes the motions of masses.
   d. both a and c

8. As the mass of a black hole increases, its Schwarzschild radius
   a. increases as the square of the mass.
   b. increases proportionately.
   c. stays the same.
   d. decreases proportionately.
   e. decreases as the square of the mass.

9. If a neutron star is more than ~2.5 times as massive as the Sun, it collapses because
   a. the force of electron degeneracy is stronger than gravity.
   b. gravity overpowers the force of electron degeneracy.
   c. gravity overpowers the force of neutron degeneracy.
   d. the force of neutron degeneracy is stronger than gravity.

10. Relative motion between two objects is apparent
    a. even at everyday speeds, such as 10 km/h.
    b. only at very large speeds, such as $0.8c$.
    c. only near very large masses.
    d. only when both objects are in the same reference frame.

11. If a spaceship approaches you at $0.5c$, and a light on the spaceship is turned on pointing in your direction, how fast will the light be traveling when it reaches you?
    a. $1.5c$
    b. between $1.0c$ and $1.5c$
    c. exactly c
    d. between $0.5c$ and $1.0c$

12. Imagine two protons traveling past each other at a distance d, with relative speed $0.9c$. Compared with two *stationary* protons a distance d apart, the gravitational force between those two protons when at their closest will be
    a. smaller because they interact for less time.
    b. smaller because the moving proton acts as though it has less mass.
    c. the same because the particles have the same mass.
    d. larger because the moving proton acts as though it has more mass.

13. If two spaceships approach each other, each traveling at $0.5c$ relative to an outside observer, spaceship 1 will measure spaceship 2 to be traveling
    a. much faster than c.
    b. slightly faster than c.
    c. at c.
    d. more slowly than c.

14. Some black holes are rapidly flickering X-ray sources. That observation indicates that
    a. X-rays are not light, because even light cannot escape from a black hole.
    b. the X-rays are coming from a very small source—so small that it could be a black hole.
    c. black holes are (at least sometimes) surrounded by hot gas.
    d. black holes have very high temperature.

15. The current model of long-duration GRBs includes jets from the collapsed star. You have seen jets like that before, when studying (choose all that apply)
    a. giant planets.
    b. white dwarfs.
    c. pulsars.
    d. star formation.
    e. supernovae.
    f. planetary nebulae.

Thinking about the Concepts

16. An astronomer sees a redshift in the spectrum of an object. Without any other information, can she determine whether it is an extremely dense object (exhibiting gravitational redshift) or one that is receding from her (exhibiting Doppler redshift)? Explain your answer.
17. Imagine you are traveling in a spacecraft at 0.9999999*c*. You point your laser pointer out the back window of the spacecraft. At what speed does the light from the laser pointer travel away from the spacecraft? What speed would be observed by someone on a planet traveling at 0.000001*c*?
18. Einstein's special theory of relativity says that no object can travel faster than, or even at, the speed of light. Recall that light is both an electromagnetic wave and a particle called a photon. If light acts as a particle, how can a photon travel at the speed of light?
19. Twin A takes a long trip in a spacecraft and returns younger than twin B, who stayed behind. Could twin A ever return before twin B was born? Explain.
20. In one frame of reference, event A occurs before event B. In another frame of reference, could the two events be reversed, so that B occurs before A? Explain.
21. Imagine you are watching someone whizzing by at very high speed in a spacecraft. Will that person's pulse rate appear to be extremely fast or extremely slow?
22. Suppose you had a density meter that could instantly measure the density of an object. You point the meter at a person in a spacecraft zipping by at very high speed. Is that person's density value larger or smaller than an average person's? Explain.
23. Imagine a future astronaut traveling in a spaceship at 0.866 times the speed of light. Special relativity says that along the direction of flight, the spaceship is only half as long as it was when at rest on Earth. The astronaut checks that prediction with a meter stick that he brought with him. Will his measurement confirm the contracted length of his spaceship? Explain your answer.
24. Within a speedy spacecraft, astronauts are playing soccer with a spherical ball. How would the shape of the ball appear to an observer watching the spacecraft speed by?
25. You observe a meter stick traveling past you at 0.9999*c*. You measure the meter stick to be 1 meter long. How is the meter stick oriented relative to you?
26. Suppose astronomers discover a 3-M_{Sun} black hole located a few light-years from Earth. Should they be concerned that its tremendous gravitational pull will lead to Earth's untimely demise?
27. If you could watch a star falling into a black hole, how would the color of the star change as it approached the event horizon?
28. Why don't people detect the effects of special and general relativity in their everyday lives here on Earth?
29. Many movies and television programs (such as *Star Wars*, *Star Trek*, and *Battlestar Galactica*) are premised on faster-than-light travel. How likely is it that such technology will be developed in the near future?
30. How could a gamma-ray burst in our galaxy potentially affect life on Earth?

Applying the Concepts

31. Does the angle of the rain falling outside the car in Figure 18.1b depend on the speed of the car? Knowing only that angle and the information on your speedometer, how could you determine the speed of the falling rain?
32. Compare Figure 18.1 with Figure 18.2. If you knew the speed of Earth in its orbit from a prior experiment, how could you determine the speed of light from the angle of the aberration of starlight?
33. According to Einstein, mass and energy are equivalent. So, which weighs more on Earth: a cup of hot coffee or a cup of iced coffee? Why? Do you think the difference is measurable?
34. As explained in the Process of Science Figure, a new theory should contain the old theory within it. Study Figure 18.5, which compares two imaginary rocket ships experiencing the same acceleration, *g*.
    a. Approximately how fast (as a fraction of the speed of light) are the two spaceships going when the effects of relativity begin to be significant?
    b. Convert that speed to kilometers per hour. How does that speed compare with the speeds at which you usually travel? Why do you not usually see relativistic effects in your life?
35. Figure 18.6 shows how the Lorentz factor depends on speed. At about what speed (in terms of *c*) does the Lorentz factor begin to differ noticeably from 1? What happens to the Lorentz factor as the speed of an object approaches the speed of light?

36. Imagine that the perihelion of Mercury advances 2 degrees per century. How many arcseconds does the perihelion advance in a year? (Recall that 60 arcseconds are in an arcminute and 60 arcminutes in a degree.) Can Mercury's position be measured well enough to measure the advance of perihelion in 1 year?

37. If the Sun were twice as massive in Figure 18.12, would the distance between the apparent positions of stars 1 and 2 increase or decrease during an eclipse?

38. Follow Working It Out 18.1 to find out how much younger than your twin you would be if you made the journey described there at 0.5*c*.

39. Use Working It Out 18.1 to predict how much younger than your twin you would be if you traveled at 0.999*c* instead of 0.5*c*, as in question 38. Then calculate the difference in ages and compare the calculated result with your prediction.

40. What is the Schwarzschild radius of a black hole that has a mass equal to the average mass of a person (~70 kilograms)?

41. What is the mass of a black hole with a Schwarzschild radius of 1.5 km?

42. The Moon has a mass equal to $3.7 \times 10^{-8}\ M_{\text{Sun}}$. Suppose the Moon suddenly collapsed into a black hole.
  a. What would be the Schwarzschild radius of the black-hole Moon?
  b. How would that collapse affect tides raised by the Moon on Earth? Explain.
  c. Would that event generate gravitational waves? Explain.

43. If a spaceship approaching Earth at 0.9*c* shines a laser beam at Earth, how fast will the photons in the beam be moving when they arrive at Earth?

44. Suppose you discover signals from an alien civilization coming from a star 25 light-years away, and you go to visit it by using the spaceship described in the discussion of the twin paradox in Working It Out 18.1.
  a. How long will you take to reach that planet, according to your clock? According to a clock on Earth? According to the aliens on the other planet?
  b. How likely is it that someone you know will be here to greet you when you return to Earth?

45. Working It Out 18.2 relates the mass of a binary pair to the period and the size of the orbit. Suppose that a spaceship orbited a black hole at a distance of 1 AU, with a period of 0.5 year. What assumptions could you make that would allow you to calculate the mass of the black hole from that information? Make those assumptions, and calculate the mass.

USING THE WEB

46. Go to NASA's Swift Gamma-Ray Burst Mission website (https://www.nasa.gov/mission_pages/swift/main). Locate a recent result related to supernovae, gamma-ray bursts, or stellar black holes. Why would two merging neutron stars be likely to form a black hole?

47. NASA's Fermi Gamma-ray Space Telescope (https://www.nasa.gov/content/fermi-gamma-ray-space-telescope and https://fermi.gsfc.nasa.gov/) is exploring the gamma-ray universe. What objectives of that mission relate to the study of black holes? What is a recent news story related to black holes?

48. Go to the LIGO website (https://ligo.org/science.php) and read about gravitational waves. Click on "Sources of Gravitational Waves" and listen to the example. What are the differences among the four listed sources of gravitational waves? Click on "Advanced LIGO flyer." What's new with the project?

49. NuSTAR (Nuclear Spectroscopic Telescopic Array) is a NASA mission to study black holes, gamma-ray bursts, and neutron stars. Go to the NuSTAR website (https://www.nustar.caltech.edu/news). What wavelengths and energies does that telescope observe? What has been observed? What new science has been learned?

50. Go to the "Inside Black Holes" website (https://jila.colorado.edu/~ajsh/insidebh/), enter, and click on "Schwarzschild." Work your way down the page, watching the videos. What does it look like when you go into a black hole? Why does gravitational lensing occur when Earth is in orbit? What happens when you fall through the event horizon: is everything black? Click on "Reissner-Nordström" to see an electrically charged black hole. What is a wormhole? Why is a warning at the top of the page? Click on "4D perspective." What does it look like if you move toward the Sun at the speed of light?

digital.wwnorton.com/astro6

Because grabbing a black hole and bringing it into the lab is not possible, and because Earth has never actually been close to one, astronomers can conduct only thought experiments to explore the properties of black holes. Following are a few thought experiments to help you think about what's happening near a black hole.

Imagine a big rubber sheet. It is very stiff and not easily stretched, but it does have some "give" to it. At the moment, it is perfectly flat. Imagine rolling some golf balls across it.

1 Describe the path of the golf balls across the sheet.

Now imagine putting a bowling ball (very much heavier than a golf ball) in the middle of the sheet, so that it makes a big, slope-sided pit. Roll some more golf balls.

2 What happens to the path of the golf balls when they are very far from the bowling ball?

3 What happens to the path of the golf balls when they come just inside the edge of the dip?

4 What happens to the path of the golf balls when they go directly toward the bowling ball?

5 How do each of the three cases in questions 2–4 change if the golf balls are moving very, very fast? What if they are moving very slowly?

6 What happens to the depth and width of the pit as the golf balls fall into the center near the bowling ball? (Imagine putting *lots* of golf balls in.)

All the preceding thought experiments relate to ordinary stuff. Stars, people, planets—everything interacts in that way because of gravity. With black holes, things are different. Here, it is more accurate to think of the bowling ball as a hole in the sheet that pulls it down rather than as an object that sits on it. But the bowling ball still affects the sheet in the same way. The hole is a good analogy for the event horizon of a black hole. Objects outside the event horizon will know that the black hole is there because the sheet is sloping, but they won't be captured unless they come within the event horizon. Think about light for a moment as though it were, say, grains of sand rolling across the sheet.

7 What happens to the light as it passes far from the pit? What happens if it reaches the hole?

Now suppose you roll another bowling ball across the sheet.

8 What happens to the sheet when the second bowling ball falls in after the first? Would that change affect your golf balls and grains of sand? How? What happens to the hole? What happens to the size of the pit?

None of those thought experiments take into account relativistic effects (length contraction and time dilation). Imagine for a moment that you are traveling close to the black hole.

9 Look out into the galaxy and describe what you see. Consider the lifetimes of stars, the distances between them, their motions in your sky, and how they die. Add anything else that occurs to you.

19 Galaxies

Less than a century has passed since astronomers realized that the universe is filled with huge collections of stars, gas, and dust called galaxies. Just as stars vary in their mass or their stage of evolution, galaxies come in many forms. Chapter 19 begins our discussion of galaxies with a survey of the types of galaxies and their basic properties to understand better the differences among them. Chapter 20 then looks in detail at our galaxy, the Milky Way, and in later chapters we describe the evolution of galaxies and of the universe itself.

LEARNING GOALS

By the end of this chapter, you should be able to:

LG 1 Determine a galaxy's type from its appearance and describe the motions of its stars.

LG 2 Explain the distance ladder and how distances to galaxies are measured.

LG 3 Describe the evidence suggesting that galaxies are composed mostly of dark matter.

LG 4 Discuss the evidence indicating that most—perhaps all—large galaxies have supermassive black holes at their centers.

The large, barred spiral galaxy NGC 1300 is about 20 million parsecs away. ▶▶▶

How do galaxies get their shapes?

19.1 Galaxies Come in Different Shapes and Sizes

A **galaxy** is a gravitationally bound collection of dust, gas, a million to hundreds of billions of stars, and dark matter. The universe contains more galaxies than the number of stars in the Milky Way Galaxy. Most of those galaxies are located at such astonishing distances that they appear too small and faint to detect with any but the most powerful telescopes. In this section, we discuss how astronomers concluded that galaxies were separate from our Milky Way, and we differentiate the types of galaxies.

The Discovery of Galaxies in the 20th Century

Observers have long known that the sky contains faint, misty patches of light. Those objects were originally called nebulae (singular: *nebula*, from Latin for "cloud") because of their nebulous (fuzzy) appearance. In 1784, astronomer Charles Messier (1730–1817) published a catalog of 103 nebulous objects. Twenty years later, because of the observations by the astronomers William Herschel and his sister, Caroline Herschel, that number jumped to 2,500. Although some of the nebulae looked diffuse and amorphous, most were round or elliptical or resembled spiraling whirlpools. Those distinctions were the basis for the original three categories—diffuse, elliptical, and spiral—used to classify nebulae.

Over the next 140 years, it was suggested that spiral nebulae might be relatively nearby planetary systems in various stages of formation. Alternatively, the influential 18th century philosopher Immanuel Kant (1724–1804) speculated that spiral nebulae were instead "island universes"—separate from the Milky Way Galaxy. The presence of interstellar dust complicated early attempts to understand the size of the Milky Way. Early astronomers did not know of the existence or consequences of the dust that blocks the passage of visible light through the Milky Way. Unable to see past that obscuring shroud, they concluded that the Milky Way is a system of stars some 1,800 parsecs (pc) across (recall from Chapter 13 that 1 parsec = 3.26 light-years). In the beginning of the 20th century, astronomer Harlow Shapley (1885–1972) used observations of globular clusters to estimate that, instead, the Milky Way is more than 50 times larger than the earlier estimates—300,000 light-years, or 92,000 pc, in size. He also showed that the Sun was not at the center of the Milky Way.

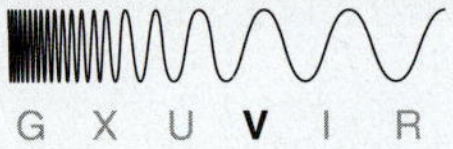

Figure 19.1 The Andromeda Galaxy, the nearest large galactic neighbor to the Milky Way, is about 2.5 million light-years (780,000 pc) away. The arrow points to the Cepheid variable star V1, a standard candle that Hubble used to estimate the distance to Andromeda. The insets show the variability in the light from V1. Hubble's measurement provided the first observational evidence of the vastness of the universe.

Shapley thought his far larger estimate for the size of the Milky Way meant that it was big enough to encompass everything in the universe, and therefore he thought that the spiral and elliptical nebulae were inside the Milky Way. Astronomer Heber D. Curtis (1872–1942) preferred the earlier, smaller model of the Milky Way. He also favored the idea that the spiral nebulae were in fact galaxies separate from the Milky Way and that therefore the whole universe was larger than the Milky Way. In 1920, Shapley and Curtis met in Washington, D.C., to debate publicly their interpretations of the nature of spiral nebulae. Historians call that meeting astronomy's *Great Debate*. Although that debate did not resolve the issue at the time, it set the stage and gave direction to the later work of Edwin P. Hubble (1889–1953), whose discoveries fundamentally changed the modern understanding of the universe.

Using the newly finished 100-inch telescope on Mount Wilson, high above the then-small city of Los Angeles, Hubble found some variable stars in the spiral nebula Andromeda (**Figure 19.1**). He recognized that those stars were very similar to but appeared fainter than other known Cepheid variable stars. Using the period-luminosity relation for Cepheid variable stars discussed in Chapter 17, Hubble turned his observations of those stars into measurements of the distances to those objects. The results showed that the distances to those nebulae are far greater than Shapley's size of the Milky Way. Kant was correct: the Milky Way is one of many island universes. Most diffuse nebulae are clouds of gas and dust near Earth in the Milky Way Galaxy, but what Messier and others thought of as elliptical and spiral nebulae are instead galaxies similar in size to the Milky Way but located at truly immense distances.

Types of Galaxies

Imagine taking a handful of coins and throwing them in the air, as shown in **Figure 19.2a**. You know that all those objects are flat and circular. When you look at the objects falling through the air, however, they do not appear all the same. Some coins appear face on and look circular. Some coins are seen edge on and look like thin lines. Most coins are seen from an angle between those two extremes and appear with various degrees of ellipticity, or flattening. Even if that image of many coins was the only information you had, you could use it to figure out that the three-dimensional shape of a coin is flat and circular.

Astronomy in Action: Galaxy Shapes

Astronomers use a similar method to discover the true three-dimensional shapes of galaxies. **Figure 19.2b** shows a set of galaxies seen from various viewing angles, from face on to edge on. You can infer from those images that, just like the coins in Figure 19.2a, some galaxies are disk-shaped and are randomly oriented on the sky.

The classifications for larger galaxies used today dates back to the 1930s, when Edwin Hubble sorted the shapes into categories. Hubble grouped galaxies according to appearance in visible light into three basic types: spirals, ellipticals, and irregulars.

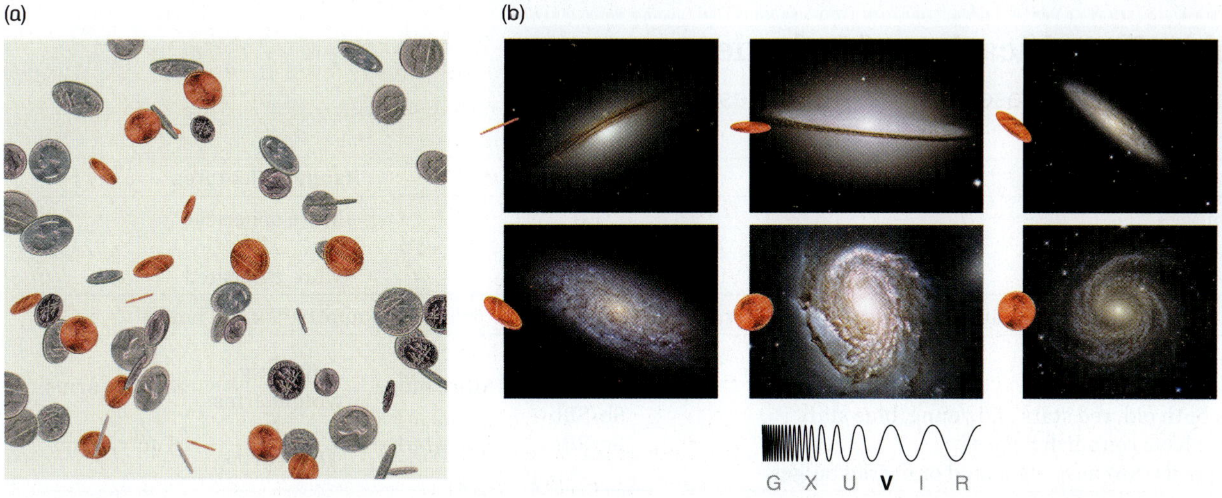

Figure 19.2 (a) A handful of coins thrown in the air illustrates the difficulties in identifying the shapes of certain types of galaxies. Like the coins in this picture, galaxies are seen in various orientations—some face on, some edge on, and most somewhere in between. (b) These disk-shaped galaxies are seen from various perspectives or angles, corresponding to the range of perspectives for the coins.

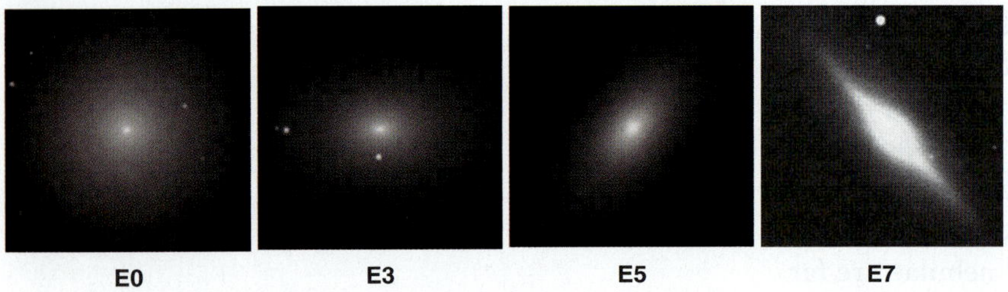

Figure 19.3 Elliptical galaxies are classified on the basis of how elliptical they appear to be from our point of view. The most circular are E0 galaxies, whereas the most elliptical are classified as E7.

Figure 19.3 shows galaxies that are generally elliptical in three dimensions. Those **elliptical galaxies** can have either spherical or ellipsoidal shapes, and they show little evidence of the flat, round disk seen in other types of galaxies. They have numbered subtypes ranging from nearly spherical (E0) to flattened (E7). As with the coins tossed in the air, the appearance of an elliptical galaxy in the sky does not necessarily tell us its true shape. For example, a galaxy might actually be shaped like a rugby ball (which has rounded ends), but if viewed end on, it looks round like a soccer ball instead.

Spiral galaxies, shown in **Figure 19.4**, are characterized by spiral arms that lie in a flattened, rotating disk. In addition to disks and arms, a spiral galaxy has a central **bulge**, which extends above and below the disk. About 2/3 of spiral galaxies have a bar-shaped bulge: those galaxies are called **barred spirals**. Both spirals and barred spirals are subdivided into types a, b, and c according to the prominence of the central bulge and how tightly the spiral arms are wound. For example, Sa and SBa galaxies have the largest bulges and display tightly wound and smooth spiral arms. Sc and SBc galaxies have small central bulges and more loosely woven spiral arms that are often very knotty in appearance. The Milky Way Galaxy is a barred spiral, classified SBbc (between SBb and SBc)

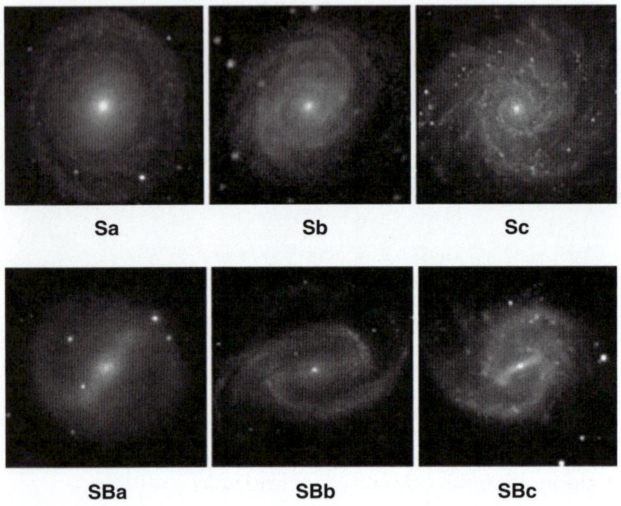

Figure 19.4 Spiral and barred spiral galaxies are classified by the openness of the arms and the prominence of the bulge.

The distinction between spiral and elliptical galaxies is not always clear. Some galaxies, known as **S0 galaxies**, are a combination of the two types. S0 galaxies have stellar disks but no spiral arms, so that the disk is smooth in appearance, like an elliptical galaxy. Hubble differentiated S0 galaxies as either barred (SB0) or unbarred (S0). Modern telescopic observations have revealed that many, if not most, elliptical galaxies contain small rotating disks at their centers, and little star formation occurs in both elliptical and S0 galaxies. Galaxies that fall into none of those classes are **irregular galaxies**. As their name implies, irregular galaxies often lack symmetry in shape or structure. Many of them once were spirals or ellipticals that became distorted by the gravity of another galaxy. **Table 19.1** summarizes the criteria that Hubble used to classify those types of galaxies.

TABLE 19.1

The Hubble Classification of Galaxies

A Classification Scheme Based on the Properties of Galaxies

Category	Criteria	Abbreviation	Range of Features		
Elliptical	Mostly bulge Old, red stellar population Smooth-appearing	E0 ↕ E7	More spherical ↕ More elongated		
S0 (unbarred/barred)	Bulge and disk with no arms and with mostly old, red stars	S0/SB0	Smooth disk and bulge		
Spiral (unbarred/barred)	Bulge and disk with arms Bulge has old, red stars Disk has both old, red stars and young, blue stars Spirals (S) have roundish bulges Barred spirals (SB) have elongated or barred bulges	Sa/SBa Sb/SBb Sc/SBc	More bulge ↕ Little bulge	Tightly wound arms ↕ Open arms	Smooth arms ↕ Knotty arms
Irregular	No arms, no bulge Some old stars, but mostly young stars, gas, and dust, giving a knotty appearance	Irr			

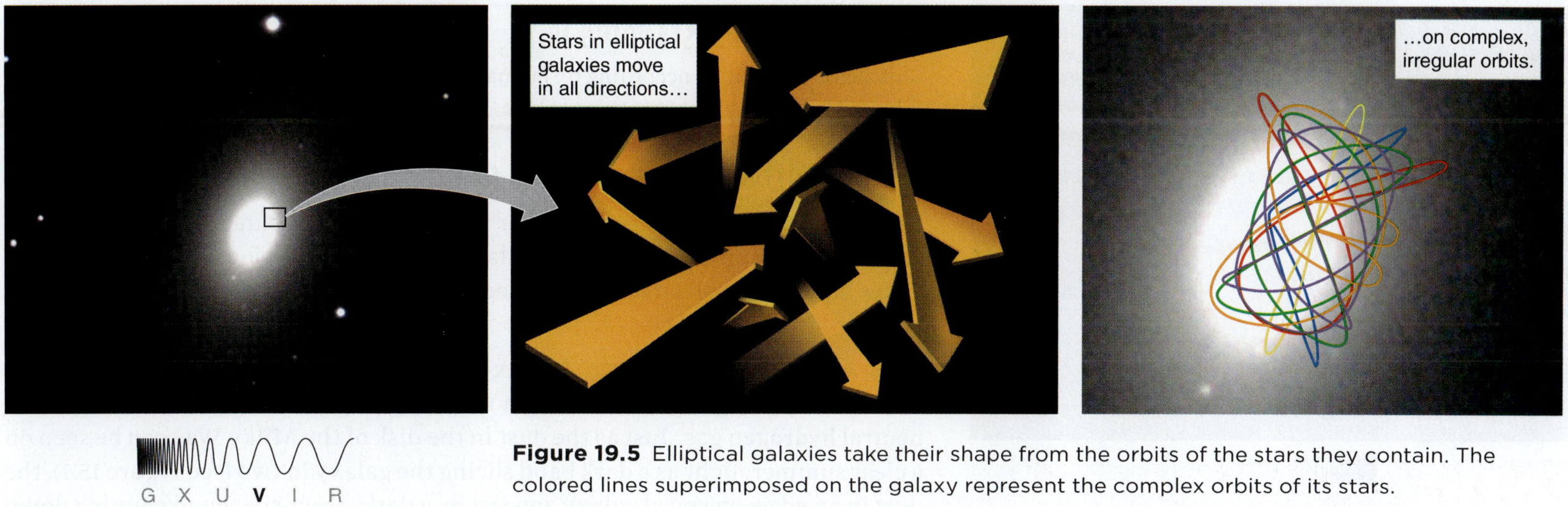

Figure 19.5 Elliptical galaxies take their shape from the orbits of the stars they contain. The colored lines superimposed on the galaxy represent the complex orbits of its stars.

More recently, using larger telescopes and more extensive deep sky surveys, astronomers have identified many **dwarf galaxies**. They are usually 1/10 or less the size of our Milky Way, with lower luminosities and lower surface brightness than those of large galaxies. Dwarf galaxies can be elliptical, spheroidal, irregular, or more rarely spiral. Some are very compact with high density and are like the cores of dwarf ellipticals; some are very compact and blue with clusters of hot, young stars. Dwarf galaxies are more difficult to detect but are probably the most common type of galaxy in the universe.

Stellar Motions and Galaxy Shape

Stellar motions determine galaxy shapes. A galaxy is not a solid object like a coin but rather a collection of stars, gas, particles, and dust. In an elliptical galaxy, stars move in all possible directions, following orbits with a wide range of shapes (**Figure 19.5**). Those orbits are more complex than the orbits of planets about a star. Taken together, all those stellar orbits give an elliptical galaxy its shape. The faster the stars are moving, the more spread out the galaxy is. If the stars in an elliptical galaxy are moving in truly random directions, the galaxy will have a spherical shape. If stars are more likely to have certain directions of motion than others at each location, however, the galaxy will be more spread out in that direction, giving it an elliptical shape.

The orbits of stars in the disks of spiral galaxies are different from those of stars in elliptical galaxies. The components of a barred spiral galaxy are shown in **Figure 19.6**. The defining feature of a spiral galaxy is that it has a flattened, rotating disk. Most stars in the disk of a spiral galaxy follow nearly circular orbits and travel in the same direction around a concentration of mass at the center of the galaxy. But the stellar orbits in a spiral galaxy's central bulge are different from those in the galaxy's disk. As with elliptical galaxies, the stars follow orbits in random directions. The bulges of unbarred spiral galaxies are thus roughly spherical.

Figure 19.6 The components of a barred spiral galaxy include a barlike bulge, a disk, and spiral arms. The orbits of stars in the rotating disk are different from the orbits of stars in the elliptically shaped bulge.

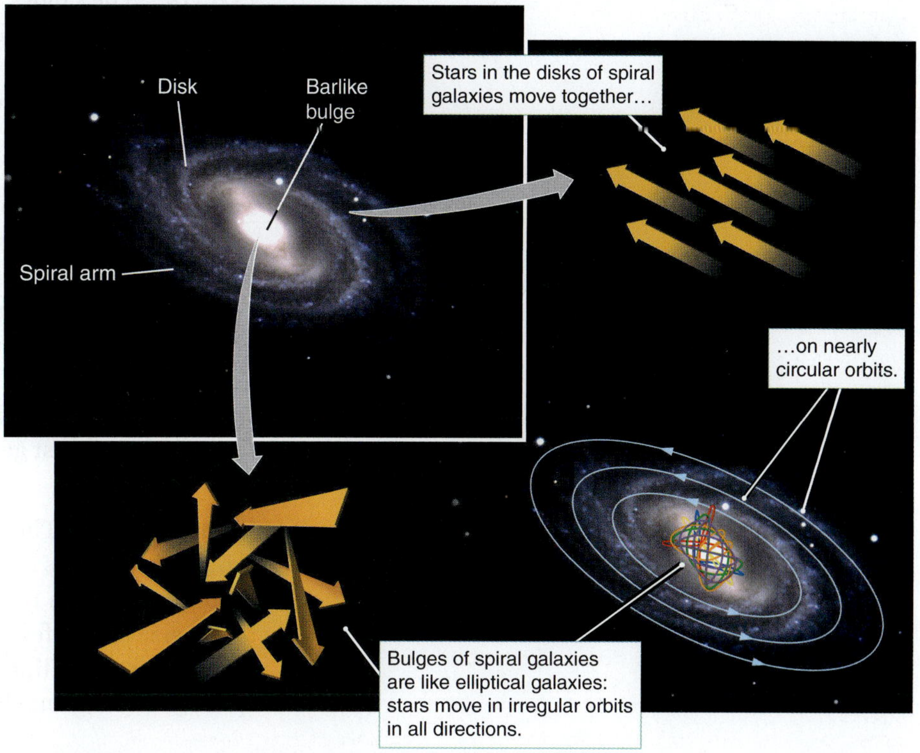

Figure 19.7 This Hubble Space Telescope (HST) image shows the nearly edge-on spiral galaxy M104 (the Sombrero Galaxy; type Sa). The dust in the plane is seen as a dark, obscuring band in the midplane of the galaxy. Note the bright halo made up of stars and globular clusters. Compare this image with Figure 15.1, which shows the dust in the plane of the Milky Way.

CHECK YOUR UNDERSTANDING 19.1A

Galaxies are classified according to: (a) mass; (b) color; (c) density; (d) shape.

Other Differences among Galaxies

In addition to the differences in their stellar orbits, other important distinctions exist between spiral and elliptical galaxies. Those distinctions carry information about how they evolved in the past and how they will evolve in the future.

GAS AND DUST Most spiral galaxies contain large amounts of dust and cold, dense molecular gas concentrated in the midplanes of their disks, in addition to neutral hydrogen gas. Just as the dust in the disk of the Milky Way can be seen on a clear summer night as a dark band slicing the galaxy in two (see Figure 15.1), the dust in an edge-on spiral galaxy appears as a dark, obscuring band running down the midplane of the disk (**Figure 19.7**). The cold molecular gas that accompanies the dust can also be seen in radio observations of spiral galaxies. In contrast, giant elliptical galaxies contain large amounts of very hot gas that astronomers see primarily by observing the X-rays emitted by the gas.

The difference in shape between elliptical and spiral galaxies offers some insight into why the gas in giant ellipticals is hot, whereas the gas in spirals is cold and dense. Conservation of angular momentum causes cold gas to settle into the disk of a spiral galaxy, just as gas settles into a disk around a forming star. In contrast, elliptical galaxies do not have a net rotation, so the gas does not settle into a disk. The only place in an elliptical galaxy where cold gas could collect is at the center. The density of stars in elliptical galaxies is so high that evolving stars and Type Ia supernovae continually reheat that gas, however, thus preventing most of it from cooling off and forming cold interstellar clouds.

COLOR The colors of spiral and elliptical galaxies reveal a great deal about their star formation histories. Recall from Chapter 15 that stars form from dense clouds of cold molecular gas. Because the gas seen in elliptical galaxies is very hot, astronomers know that active star formation is not taking place in those galaxies today. The reddish colors of elliptical and S0 galaxies confirm that little or no star formation has occurred there for some time. The stars in those galaxies are an older population of lower-mass stars. In contrast, the bluish colors of the disks of spiral galaxies indicate that massive, young, hot stars are forming in the cold molecular clouds contained within the disk. Even though most of the stars in a spiral disk are old, the massive, young stars are so luminous that their blue light dominates. Star formation in most irregular galaxies is like that in spiral galaxies. Some irregular galaxies form stars at prodigious rates, given their relatively small sizes. Irregular and disk galaxies that undergo intense bursts of star formation are called starburst galaxies.

LUMINOSITY The relationship between luminosity and radius among the types of galaxies is not straightforward. Galaxies range in luminosity from tens of thousands up to a trillion (a million million) solar luminosities (10^4 to 10^{12} L_{Sun}) and in size from a few hundred to hundreds of thousands of parsecs. Although the most luminous elliptical galaxies are more luminous than the most luminous spiral galaxies, considerable overlap is present in the range of luminosities among

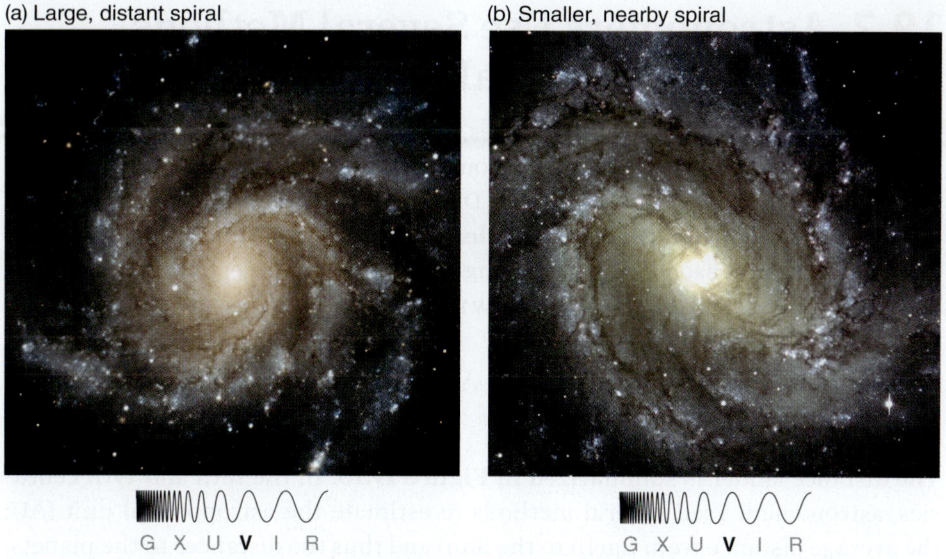

Figure 19.8 The mass or size of a spiral galaxy does not determine its appearance. Even though these galaxies *appear* to be similar in size and luminosity, the larger galaxy (a) is 4 times more distant and 10 times more luminous than the smaller galaxy (b).

all Hubble types. The least luminous galaxies known are low-luminosity dwarf spheroidals called ultrafaint dwarf galaxies.

MASS Mass is the single most important parameter in determining the properties and evolution of a star. In contrast, differences in mass and size do not lead to obvious differences among galaxies. Subtle differences in color and concentration exist between large and small galaxies, making it difficult to distinguish which are large and which are small. Even when a larger, more distant spiral galaxy (**Figure 19.8a**) is seen next to a smaller, nearby spiral galaxy (**Figure 19.8b**), telling which is which by appearance alone can be hard. Galaxies that have relatively low luminosity (less than 1 billion L_{Sun}) are the dwarf galaxies, and those that are more than 1 billion L_{Sun} are called giant galaxies, because the luminosity indicates the number of stars and therefore the total amount of stellar mass. (For comparison, the Milky Way has a luminosity of about 100 billion L_{Sun}.) The difference between a dwarf elliptical galaxy and a giant elliptical galaxy is relatively easy to see (**Figure 19.9**). Giant elliptical galaxies have a much higher density of stars, and they are more centrally concentrated than stars in dwarf ellipticals.

CHECK YOUR UNDERSTANDING 19.1B

Currently, star formation rates are highest in (a) elliptical galaxies; (b) S0 galaxies; (c) spiral galaxies; (d) dwarf galaxies.

Figure 19.9 Dwarf elliptical galaxies (a) differ in appearance from giant elliptical galaxies (b). Stars in giant elliptical galaxies are more centrally concentrated than those in dwarf elliptical galaxies.

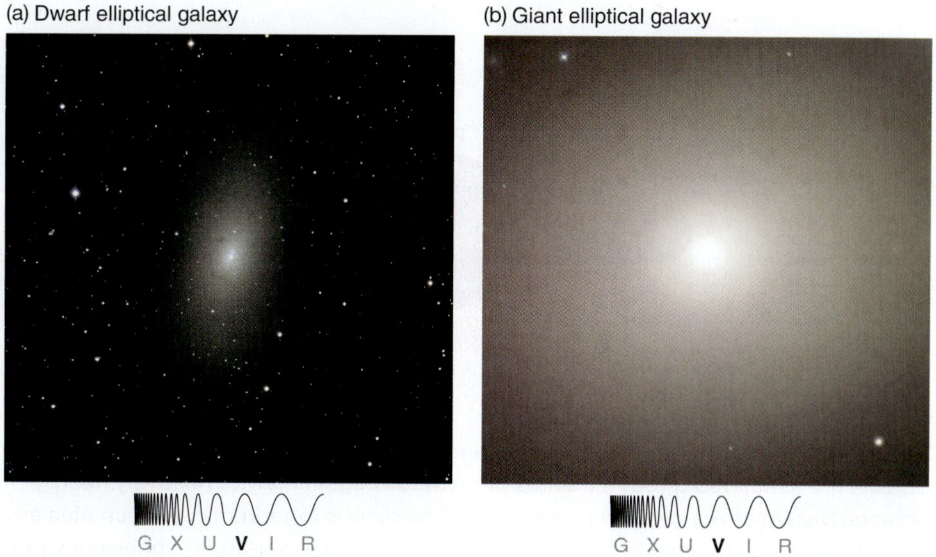

19.2 Astronomers Use Several Methods to Find Distances to Galaxies

To determine the distances to galaxies, astronomers started with the closest objects and looked for patterns that would help them find distances to the farthest objects—just as they did for stars. Distances are measured in a series of different methods called the **distance ladder**, which relates distances on a variety of overlapping scales, each method building on the previous one. In this section, we discuss those distance methods and how they led to the surprising discovery that the galaxies are moving away from us.

The Distance Ladder

The distance ladder is summarized in **Figure 19.10**. In the 18th and 19th centuries, astronomers used several methods to estimate the astronomical unit (AU; the average distance from Earth to the Sun) and thus the distances to the planets. (Since the 1960s, distances within the Solar System have been found by using radar and signals from space probes.) Once the value of the AU was known, astronomers used trigonometric parallax (see Chapter 13) to measure distances to nearby stars and thereby to build up the Hertzsprung-Russell (H-R) diagram. For more distant stars, astronomers use the spectral and luminosity classification of a star to determine its position on the H-R diagram. That position provides a star's luminosity, enabling astronomers to then estimate its distance by comparing its apparent brightness with its luminosity through using spectroscopic parallax (see Chapter 13 and Appendix 7).

Astronomers measure the distance to relatively nearby galaxies by using *standard candles*, a term borrowed from an old unit of light intensity that was based on actual candles. **Standard candles** are objects that have a known luminosity, usually because they have been observed within the Milky Way.

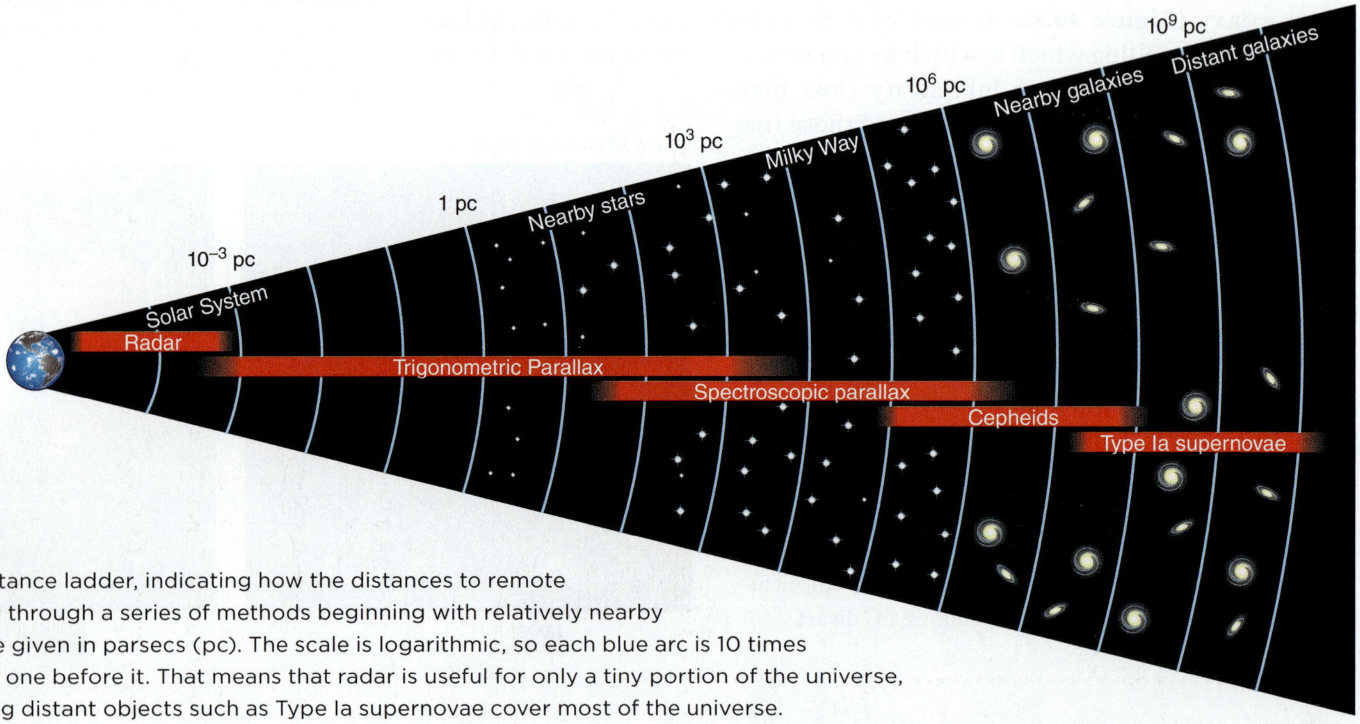

Figure 19.10 The distance ladder, indicating how the distances to remote objects are estimated through a series of methods beginning with relatively nearby objects. Distances are given in parsecs (pc). The scale is logarithmic, so each blue arc is 10 times farther away than the one before it. That means that radar is useful for only a tiny portion of the universe, and the methods using distant objects such as Type Ia supernovae cover most of the universe.

The objects must also be bright enough to be recognizable in the distant galaxy. Astronomers assume that the luminosity of each object is the same as that for a similar object of that type in the Milky Way, and then they compare the luminosity and apparent brightness of the standard candle to find its distance.

Objects that can be used as standard candles include main-sequence O stars, globular clusters, planetary nebulae, novae, Cepheid variable stars, and supernovae. For example, Hubble Space Telescope (HST) observations of Cepheid variables enable astronomers to measure distances accurately to galaxies as far away as 30 million parsecs (also called 30 megaparsecs [Mpc]). Even more luminous than the Cepheids, and thus detectable at greater distances, are Type Ia supernovae.

Recall from Chapter 16 that Type Ia supernovae can occur when gas flows from an evolved star onto its white dwarf companion, pushing the white dwarf up toward the Chandrasekhar limit for the mass of an electron-degenerate object: 1.4 M_{Sun}. When that happens, the white dwarf burns carbon, collapses, and then explodes. At first, astronomers thought that those Type Ia supernovae all occurred in white dwarfs of just below 1.4 M_{Sun}. In that situation, all such explosions would occur at the same mass and have similar luminosity, with some calibration adjustment for the rate at which the brightness declines after it peaks. But now astronomers estimate that 80 percent of Type Ia supernovae come from double-degenerate systems—for example, two white dwarfs that merge and then explode. The double white dwarf supernova could have up to twice the mass of a single white dwarf supernova. In addition, Type Ia supernovae at large distances may have different colors or compositions than nearby ones, since they formed when galaxies were younger. To test whether they all have about the same luminosity, astronomers observe nearby Type Ia supernovae in galaxies with distances determined from other methods, such as by Cepheid variables, and see that those do have similar luminosities. With a peak luminosity that can outshine a billion Suns (**Figure 19.11**), Type Ia supernovae can be seen and measured with modern telescopes at very large distances (**Working It Out 19.1**).

The motions of the stars in galaxies give rise to two "secondary" methods for estimating the distance to a galaxy. In rotating spiral galaxies, some of the light is approaching Earth and is thus blueshifted, whereas some light is moving away from Earth and is thus redshifted. Those redshifts and blueshifts together broaden a spectral line in the galaxy, and the amount of broadening indicates how fast the galaxy is rotating. Astronomers can measure the broadening from the Doppler shifts of the 21-centimeter (21-cm) radio emission line of hydrogen, which tells them the speed of rotation, which then relates to the galaxy's mass by Newton's version of Kepler's third law. The more massive galaxies have more stars and are therefore more luminous. That empirical relation between the measured width of the 21-cm line and the luminosity of the spiral galaxy is called the Tully-Fisher relation. Once the luminosity of the galaxy is known, it can be compared with the galaxy's observed apparent brightness to estimate its distance. That method is thought to work out to about 100 Mpc.

Elliptical galaxies and the bulges of S0 galaxies do not rotate, so instead astronomers look at the distribution of the surface brightness of a galaxy. Closer galaxies show more variations in the surface brightness because the distribution of stars throughout the galaxy isn't perfectly uniform. For more distant galaxies, those variations are less noticeable, and the surface brightness appears more uniform across the galaxy. That method yields less accurate results than the Tully-Fisher method for spirals, but generally it also is thought to work out to about 100 Mpc.

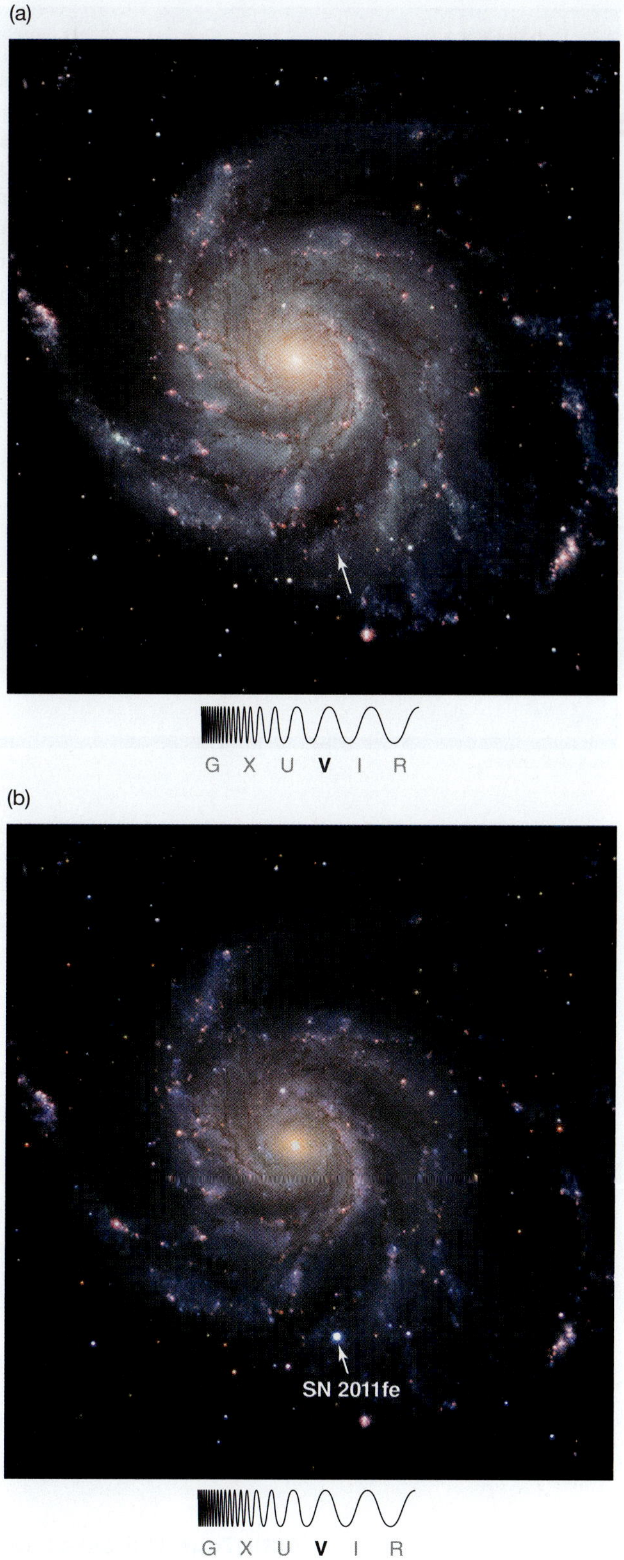

Figure 19.11 Type Ia supernovae are extremely luminous standard candles. The Pinwheel Galaxy is 6.4 Mpc away, shown before (a) and after (b) Supernova 2011fe.

19.1 Working It Out Finding the Distance from a Type Ia Supernova

How do astronomers use a standard candle to estimate distance? Figure 19.11 shows the Pinwheel Galaxy (M101), with a supernova that was observed in 2011. That supernova was detected before its brightness reached its peak. Astronomers can then compare the peak observed brightness of this supernova with the peak luminosity for that type of supernova to compute the distance.

In Section 13.1, we gave the equation relating brightness, luminosity, and distance:

$$\text{Brightness} = \frac{\text{Luminosity}}{4\pi d^2}$$

Rearranging to solve for distance gives:

$$d = \sqrt{\frac{\text{Luminosity}}{4\pi \times \text{Brightness}}}$$

The maximum observed brightness of the supernova was 7.5×10^{-12} watts per square meter (W/m^2). The graph in Figure 17.11 shows that the typical maximum luminosity L of a Type Ia supernova is 9.5×10^9 times the luminosity of the Sun:

$$L = 9.5 \times 10^9 \times L_{Sun}$$

$$L = 9.5 \times 10^9 \times (3.9 \times 10^{26}\,\text{W}) = 3.7 \times 10^{36}\,\text{W}$$

Thus, we can solve the equation:

$$d = \sqrt{\frac{3.7 \times 10^{36}\,\text{W}}{4\pi \times 7.5 \times 10^{-12}\,\text{W/m}^2}} = 2.0 \times 10^{23}\,\text{m}$$

To put that value into megaparsecs:

$$d = \frac{2.0 \times 10^{23}\,\text{m}}{3.1 \times 10^{22}\,\text{m/Mpc}} = 6.4\,\text{Mpc}$$

The distance is 6.4 Mpc. Because the Pinwheel Galaxy is relatively close, other standard candles can be observed in that galaxy to help calibrate the distance. Also note that 6.4 Mpc = 21 million light-years, so that supernova explosion took place 21 million years ago.

The Discovery of Hubble's Law

In the 1920s, Hubble and his coworkers were studying the properties of a large collection of galaxies. Another astronomer, Vesto Slipher (1875–1969), was obtaining spectra of those galaxies at Lowell Observatory in Flagstaff, Arizona. Slipher's galaxy spectra looked like the spectra of ensembles of stars with a bit of glowing interstellar gas mixed in. But he was surprised to find that the emission and absorption lines in the spectra of those galaxies were rarely seen at the same wavelengths as in the spectra of stars observed in the Milky Way Galaxy. The lines were almost always shifted to longer wavelengths, as seen in **Figure 19.12**.

Slipher characterized most of the observed shifts in galaxy spectra as redshifts because the light from those galaxies is shifted to longer (redder) wavelengths. Hubble interpreted Slipher's redshifts as Doppler shifts, concluding that almost all the galaxies in the universe are moving away from the Milky Way. Recall from Chapter 5 that objects with larger Doppler redshifts are moving away more quickly than those with smaller redshifts. When Hubble combined the measurements of galaxy velocities with his own estimates of the distances to those galaxies, he found that distant galaxies are moving away from Earth more rapidly than are nearby galaxies. Specifically, *the velocity at which a galaxy is moving away from an observer is proportional to the distance of that galaxy.* That relationship between distance and recession velocity is known as **Hubble's law**.

AstroTour: Hubble's Law

Figure 19.13 plots the measured recession velocities of galaxies against their measured distances. Because the velocity and distance are proportional to each other, the points lie along a line on the graph with a slope equal to the proportionality constant H_0, called the **Hubble constant**. Notice how well the data line up along the line. That strong correlation indicates that the nearby universe follows Hubble's law; for example, a galaxy 30 Mpc from Earth moves away twice as fast as a galaxy 15 Mpc distant. The original value for the Hubble constant was

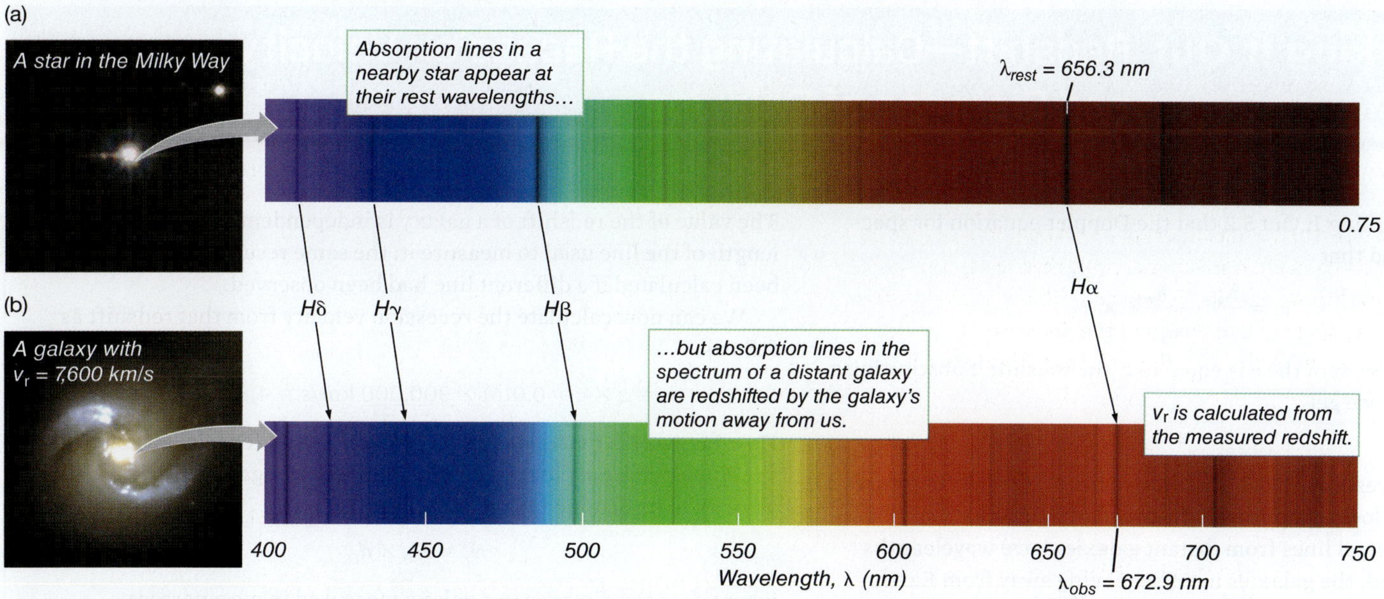

Figure 19.12 (a) The spectrum of a star in our galaxy shows absorption lines, which here lie at the rest wavelength. (b) A distant galaxy, shown with its spectrum at the same scale as that of the star, has lines redshifted to longer wavelengths. v_r is recession velocity, or radial velocity.

8 times too large, which led to inconsistencies, but the problem was resolved when astronomers realized that two types of Cepheids exist with slightly different period-luminosity relationships. Today, astronomers have measured the Hubble constant to an accuracy of a few percent by several methods, using data from several space observatories. The value is likely to be further refined in the years to come. In this text, we use a value of 70 km/s/Mpc as an approximation to the best current measured values of 67 to 74 km/s/Mpc, with uncertainties of 1–3 km/s/Mpc, depending on the measurement. The unit means that for every additional megaparsec of distance, the galaxy's motion includes an additional 70 km/s.

Using that constant, Hubble's law is written as $v_r = H_0 \times d_G$, where d_G is the distance to the galaxy, and v_r is the galaxy's recession velocity (**Working It Out 19.2**). Hubble's law gives astronomers a practical tool for measuring distances to remote galaxies. Once astronomers know the value of H_0, they can use a straightforward measurement of the redshift of a galaxy to find its distance. In other words, once H_0 is known, Hubble's law makes the once-difficult task of measuring distances in the universe relatively easy, giving astronomers a tool to map the structure of the observable universe. We will return to Hubble's law and its implications for understanding the universe as a whole in Chapter 21.

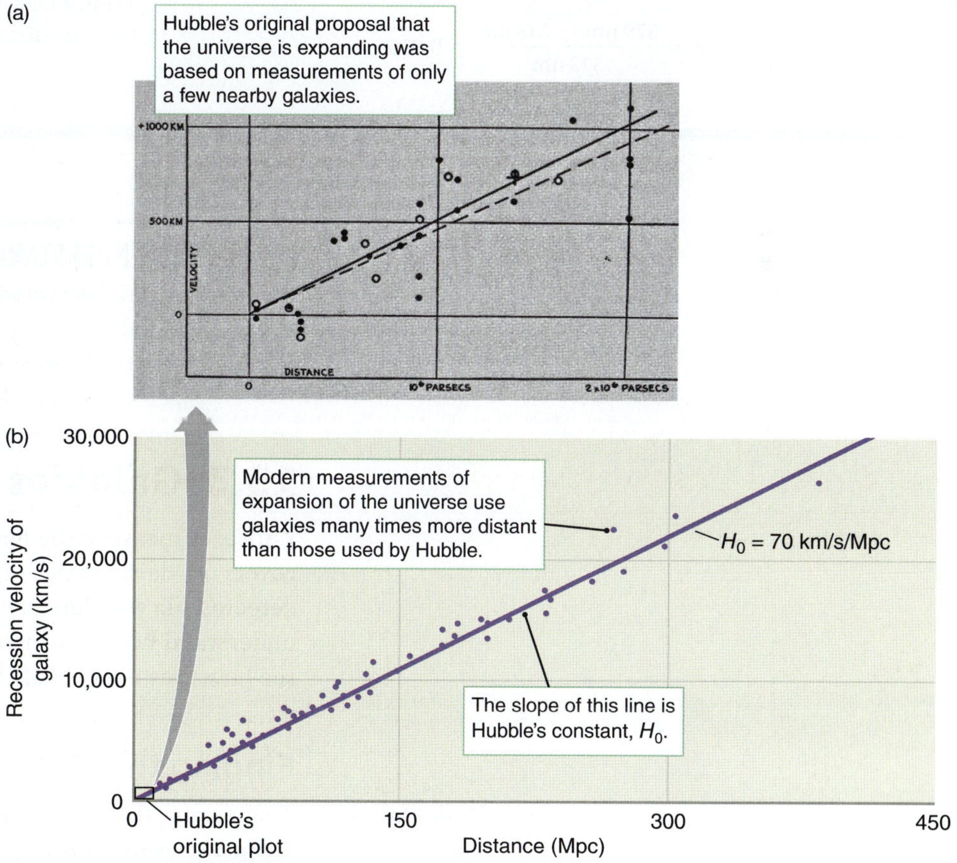

Figure 19.13 (a) Hubble's original graph shows that more distant galaxies are receding faster than less distant galaxies. (b) Modern data on galaxies many times farther away than those studied by Hubble show that recession velocity is proportional to distance.

19.2 Working It Out Redshift—Calculating the Recession Velocity and Distance of Galaxies

Recall from Working It Out 5.2 that the Doppler equation for spectral lines showed that

$$v_r = \frac{\lambda_{obs} - \lambda_{rest}}{\lambda_{rest}} \times c$$

The fraction in front of the c is equal to z, the redshift. Substituting z for the fraction, we get

$$v_r = z \times c$$

(Note: That correspondence between velocity and redshift requires a correction as velocities approach the speed of light.)

Because spectral lines from distant galaxies have wavelengths shifted to the red, the galaxies must be moving away from Earth. Suppose astronomers observe a spectral line with a rest wavelength of 373 nanometers (nm) in the spectrum of a distant galaxy. If the observed wavelength of the spectral line is 379 nm, its redshift (z) is

$$z = \frac{\lambda_{obs} - \lambda_{rest}}{\lambda_{rest}}$$

$$z = \frac{379\text{ nm} - 373\text{ nm}}{373\text{ nm}} = 0.0161$$

The value of the redshift of a galaxy is independent of the wavelength of the line used to measure it: the same result would have been calculated if a different line had been observed.

We can now calculate the recession velocity from that redshift as follows:

$$v_r = z \times c = 0.0161 \times 300{,}000\text{ km/s} = 4{,}830\text{ km/s}$$

How far away is that distant galaxy? Here is where *Hubble's law* and the *Hubble constant* apply. Hubble's law relates a galaxy's recession velocity to its distance as

$$v_r = H_0 \times d_G$$

where d_G is the distance to a galaxy measured in megaparsecs. Dividing through by $H_0 = 70$ km/s/Mpc yields

$$d_G = \frac{v_r}{H_0} = \frac{4{,}830\text{ km/s}}{70\text{ km/s/Mpc}} = 69\text{ Mpc}$$

From a measurement of the wavelength of a spectral line, we see that the distant galaxy is approximately 69 Mpc away.

CHECK YOUR UNDERSTANDING 19.2

Hubble's law was discovered by using measurements of two properties of a galaxy: ____________ and ____________. (a) size; mass (b) distance; rotation speed (c) distance; recession velocity (d) size; recession velocity

19.3 Galaxies Are Mostly Dark Matter

Efforts to measure the masses of galaxies during the 20th century led to the discovery of dark matter—mass that does not interact with light and cannot be detected via the light they emit. To understand that discovery, we first need to understand how astronomers go about measuring the mass of a galaxy and then see how they concluded that much of the mass in a galaxy is dark matter.

Finding the Mass of a Galaxy

AstroTour: Dark Matter

To measure the mass of a galaxy, astronomers add up the mass of the stars, dust, and gas that they observe. Because a galaxy's spectrum is composed primarily of starlight, once astronomers know what types of stars are in the galaxy, they can use what is known about stellar evolution to estimate the total stellar mass from the galaxy's luminosity. Astronomers then estimate the mass of the dust and gas by using the physics of radiation from interstellar gas at X-ray, infrared, and radio wavelengths. Together, the stars, gas, and dust in a galaxy are called **luminous matter** (or simply **normal matter**) because that matter emits or scatters electromagnetic radiation.

For determining mass, an independent method is available that does not involve luminosity—namely, the effect of gravity on an object's motion. Stars in disks follow orbits that are much like the orbits of planets around their parent stars and binary stars around each other (see Working It Out 13.4). To measure the mass of a spiral galaxy, astronomers apply Kepler's laws, just as they do for those other systems.

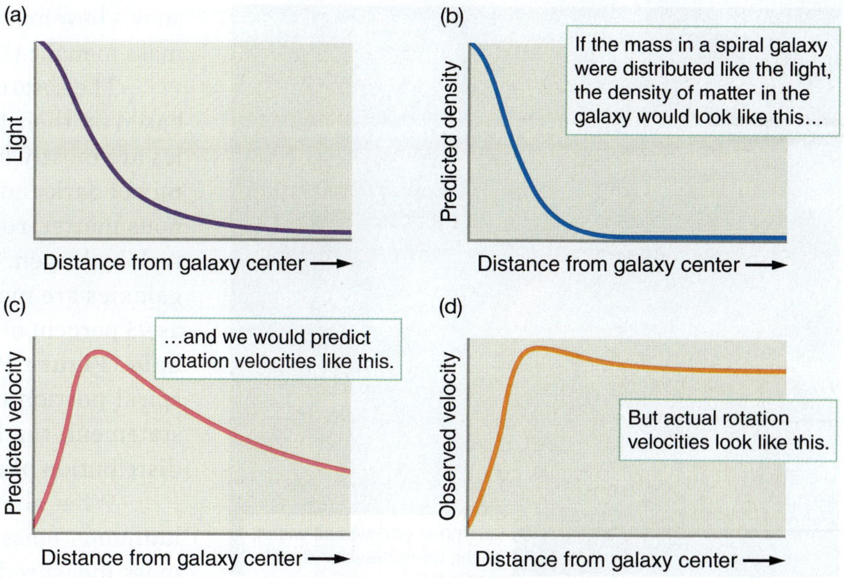

Figure 19.14 (a) The profile of visible light in a typical spiral galaxy drops off with distance from the center. (b) The predicted mass density of stars and gas located at a given distance from the galaxy's center follows the light profile. If stars and gas accounted for all of the mass of the galaxy, then the galaxy's rotation curve would be as shown in (c). However, observed galaxy rotation curves look more like the curve shown in (d).

Dark Matter

Astronomers originally hypothesized that the mass and the light in a galaxy are distributed in the same way; that is, they assumed that all the mass in a galaxy is luminous matter. They observed that the light of all galaxies, including spiral galaxies, is highly concentrated toward the center (**Figure 19.14a**). On the basis of the observed location of light, astronomers predicted that nearly all mass in a spiral galaxy is concentrated toward its center (**Figure 19.14b**). The situation is much like that in the Solar System, where nearly all the mass is in the Sun—at the center of the Solar System. Therefore, they predicted faster orbital velocities near the center of the spiral galaxy and slower orbital velocities farther out (**Figure 19.14c**).

To test that prediction, astronomers used the Doppler effect to measure orbital motions of stars, gas, and dust at various distances from a galaxy's center. The velocities of stars are obtained from observations of absorption lines in their spectra. The velocities of interstellar gas are obtained using emission lines such as those produced by hydrogen alpha (Hα) emission or 21-cm emission from neutral hydrogen (see Chapter 15). Once the velocities have been found, astronomers create a graph—called a **rotation curve**—that shows how orbital velocity in a galaxy varies with distance from the galaxy's center. The rotation curve of a spiral galaxy enables astronomers to determine directly how the mass in that galaxy is distributed.

Vera Rubin (1928–2016) pioneered work on galaxy rotation rates in the 1970s. She discovered that, contrary to earlier prediction (Figure 19.14c), the rotation velocities of spiral galaxies remain about the same out to the most distant measured parts of the galaxies (**Figure 19.14d**). Observations of 21-cm radiation from neutral hydrogen show that the rotation curves appear level, or "flat," in their outer parts even well outside the extent of the visible disks. Those observations indicated that mass and light are distributed differently.

What mass distribution would cause that unexpected rotation curve? Recall from Chapter 4 that only the mass inside a given radius contributes to the net gravitational force experienced by an orbiting object. From the rotation velocity, you can calculate the mass within the orbit of the object (**Figure 19.15**). The black line shows the measured speed of rotation of that galaxy at a particular radius. The pink line shows how much luminous mass is seen inside that radius. To produce a rotation curve like the one shown in black, that galaxy must have a second component consisting of matter that does not show up in the census of stars, gas, and dust. That material, which does not interact with light, and therefore reveals itself only by the influence of its gravity, is called **dark matter**. The purple line

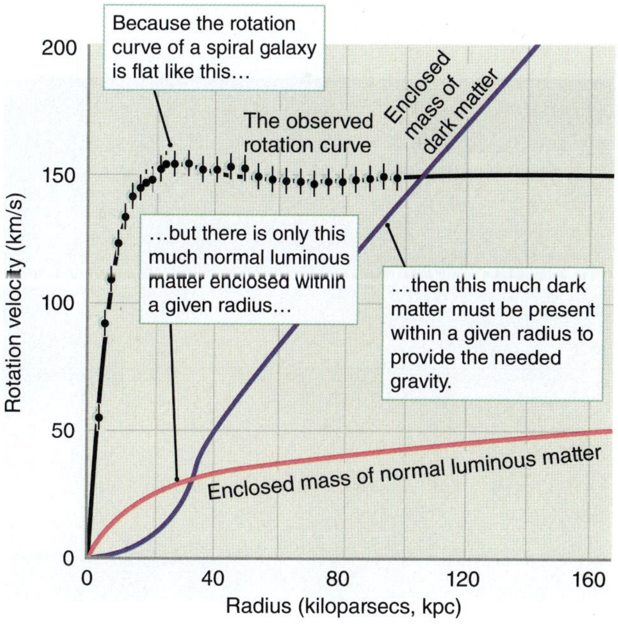

Figure 19.15 The flat rotation curve of the spiral galaxy NGC 3198 can be used to determine the total mass within a given radius. Notice that the normal mass that can be accounted for by stars and gas provides only part of the needed gravity. Extra dark matter is needed to explain the rotation curve. (1 kiloparsec [kpc] = 1,000 parsecs. The mass curves have different units.)

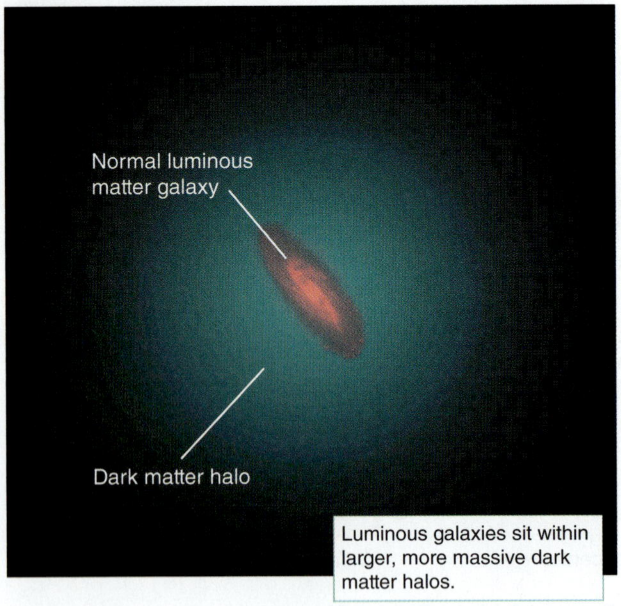

Figure 19.16 In addition to the matter that is visible, galaxies are surrounded by halos containing a large amount of dark matter.

shows how much dark matter must be inside a particular radius to supply enough mass to make the galaxy rotate as it does.

The rotation curves of the inner parts of spiral galaxies match predictions based on their luminous matter, indicating that the inner parts of spiral galaxies are mostly luminous matter. Within the entire *visual* image of a galaxy, the mix of dark and luminous matter is about half and half. However, past the luminous matter, rotation curves can be measured with 21-cm radiation from neutral hydrogen. Such measurements indicate that the outer parts of spiral galaxies are mostly dark matter. Astronomers currently estimate that as much as 95 percent of the total mass in some spiral galaxies consists of a **dark matter halo** (**Figure 19.16**), which can extend up to 10 times farther than the visible spiral portion of the galaxy located at the galaxy's center. That is a startling statement: the luminous part of a spiral galaxy is only a part of a much larger distribution of mass dominated by some type of invisible dark matter.

What about elliptical galaxies? Again, astronomers need to compare the luminous mass measured from the light they can see with the gravitational mass measured from the effects of gravity. Because elliptical galaxies do not rotate, astronomers cannot use Kepler's laws to measure the gravitational mass from the rotation of a disk. Instead, astronomers noticed that an elliptical galaxy's ability to hold on to its hot, X-ray–emitting gas depends on its mass. If the galaxy is not massive enough, the hot atoms and molecules will escape into intergalactic space. To find the mass of an elliptical galaxy, astronomers first infer the total amount of gas from X-ray images, such as the (false color) blue and purple halo seen in **Figure 19.17**. Then they calculate the total mass needed to hold on to the gas and compare that gravitational mass with the luminous mass. The amount of dark matter is the difference between what is needed to hold on to the inferred amount of gas and the observed amount of luminous matter.

Some elliptical galaxies contain up to 20 times as much mass as can be accounted for by their stars and gas alone, so they must be dominated by dark matter, just like spiral galaxies. As with spirals, the luminous matter in ellipticals is more centrally concentrated than the dark matter. The transition from the inner parts of galaxies (where luminous matter dominates) to the outer parts (where dark matter dominates) is remarkably smooth. Some galaxies may contain less dark matter than others, but about 90–95 percent of the total mass in a typical galaxy is in the form of dark matter.

The high percentage of dark matter distinguishes smaller dwarf galaxies from globular clusters, which do not have dark matter. That difference is an important observation that will need to be explained in the context of the evolution of galaxies.

Figure 19.17 In this combined visible-light and X-ray image of elliptical galaxy NGC 1132, the false-color blue and purple halo is X-ray emission from hot gas surrounding the galaxy. The hot gas extends well beyond the visible light from stars.

The Composition of Dark Matter

What is the dark matter that makes up most of a galaxy? Several candidates have been investigated, including objects such as large planets, compact stars, black holes, and exotic unknown elementary particles.

At first, astronomers considered small main-sequence M stars, Jupiter-sized planets, white dwarfs, neutron stars, or black holes as dark matter candidates. Those are collectively referred to as **MaCHOs**, which stands for *massive compact halo objects*. If the dark matter in a galaxy's halo consists of MaCHOs, a lot of those objects must be present, and they must each exert gravitational force but not emit much light. If astronomers were observing a distant star and a MaCHO passed between Earth and the star, the star's light would be deflected and, if the geometry were just right, focused by the intervening MaCHO as it passed across their line of sight. Astronomers monitored tens of millions of stars in two small companion galaxies of the Milky Way for several years, but they did not see enough of those gravitational lensing events to account for the amount of dark matter in the halo of our galaxy. Thus, MaCHOs with mass less than 100 M_{Sun} are ruled out, implying that dark matter in the Milky Way (and therefore other galaxies) must be composed of something other than MaCHOs.

Other types of candidates for dark matter are unknown elementary particles, called **WIMPs**, which stands for *weakly interacting massive particles*. WIMPs are predicted to be similar to neutrinos in that they would barely interact with ordinary matter, yet they would be more massive and would move more slowly. Many current experiments are seeking to directly or indirectly detect WIMPs. There are experiments at the Large Hadron Collider to produce and detect a WIMP. There are also multiple experiments in laboratories and in deep mines to detect halo dark matter particles as they pass through Earth. Other observations are attempting to detect the (very weak) interactions of dark matter particles in the halos of galaxies. As of this writing, there have been no confirmed direct or indirect detections of dark matter particles.

Thus, physicists and astronomers are constructing larger detectors to better catch WIMPs. Researchers are considering other types of exotic elementary particles as potential dark matter candidates. Some are exploring theoretical alternatives to dark matter, in which Newtonian dynamics or Newtonian gravity behave differently under conditions like those found in galaxies. Some of those alternative models can produce the observed rotation curves in spiral galaxies, but they cannot replace dark matter as an explanation for other aspects of galaxy evolution, as we will see in later chapters.

CHECK YOUR UNDERSTANDING 19.3

Astronomers detect dark matter: (a) by comparing luminous mass with gravitational mass; (b) because it blocks background light; (c) because more distant galaxies move away faster; (d) because it emits lots of X-rays.

19.4 Most Galaxies Have a Supermassive Black Hole at the Center

Studying the centers of galaxies is difficult because so many stars and so much dust and gas are in the way that astronomers cannot get a clear picture of the center, even for nearby galaxies. Instead, observations of the most distant objects in the universe yielded the clues to understanding what lies in the centers of massive galaxies.

The Discovery of Quasars

In the late 1950s, radio surveys detected several bright, compact objects that at first seemed to have no optical counterparts. Improved radio positions revealed that the radio sources coincided with faint, very blue, starlike objects. Unaware of

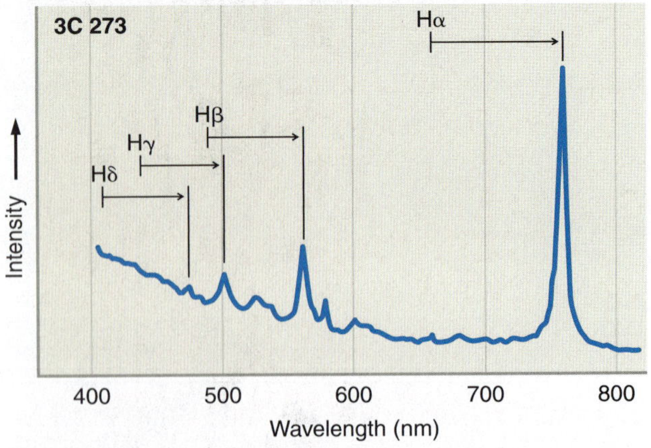

Figure 19.18 This graph shows the spectrum of quasar 3C 273, one of the closest and most luminous known quasars. The emission lines are redshifted by $z = 0.16$ from marked rest wavelengths, indicating that the quasar is at a distance of about 750 Mpc.

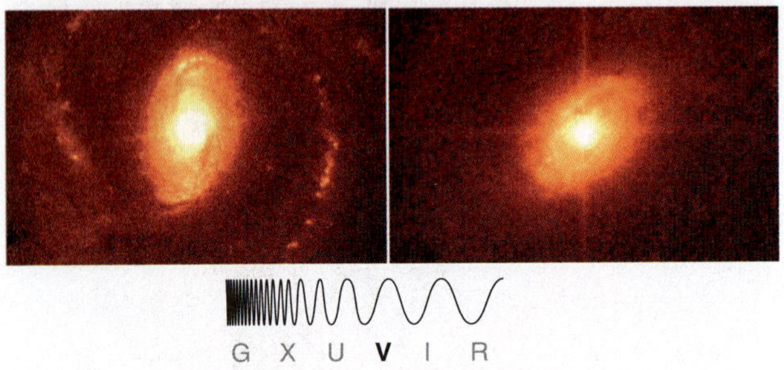

Figure 19.19 These HST images show quasars embedded in the centers of galaxies.

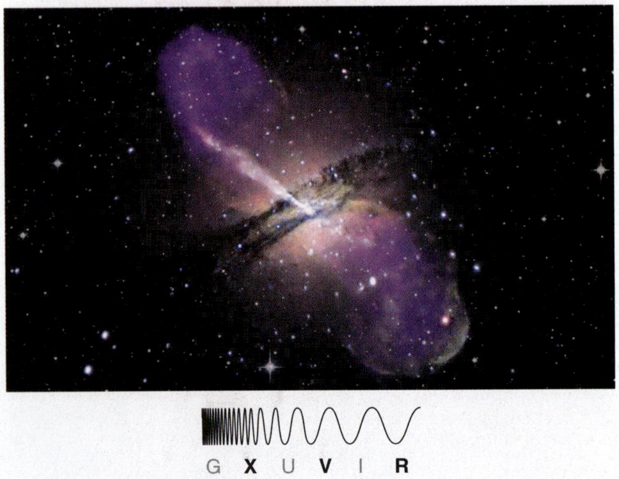

Figure 19.20 Radio galaxy Centaurus A is the closest AGN to Earth, at a distance of 3.4 Mpc. In this composite image, the visible-light image shows the galaxy, the X-ray image (pink) shows the hot gas and an energetic jet blasting from the AGN, and the radio image (purple) shows the jets and lobes.

the true nature of those objects, astronomers called them "radio stars." Obtaining spectra of the first two radio stars was a laborious task, requiring 10-hour exposures. Astronomers were greatly puzzled by the results because those spectra did not display the expected absorption lines characteristic of blue stars. Instead, the spectra showed only a single pair of emission lines that were broad—indicating very rapid motions within the objects—and that did not seem to correspond to the lines of any known substances.

For several years, astronomers believed they had discovered a new type of star until astronomer Maarten Schmidt realized that those broad spectral lines, shown in **Figure 19.18**, were the highly redshifted lines of ordinary hydrogen. The implications were surprising: those "stars" were not stars. They were extraordinarily luminous objects at enormous distances. Those "quasi-stellar radio sources" were named **quasars**. Other quasars were soon found by the same techniques, and astronomers began cataloging them.

Quasars are phenomenally powerful, shining with the luminosity of a trillion to a thousand trillion (10^{12} to 10^{15}) Suns. They also are very distant from Earth: hundreds or thousands of megaparsecs. Billions of galaxies are closer to Earth than the nearest quasar is. Recall that the distance to an object also indicates how much time has passed since the light from that object left its source. The fact that quasars are seen only at great distances implies that they are rare in the universe now but were once much more common. The discovery that quasars existed in the distant and therefore earlier universe gave astronomers one of the first pieces of evidence demonstrating that the universe has evolved.

Quasars are now recognized as the result of the most extreme form of activity that can occur in the nuclei of galaxies (**Figure 19.19**), often resulting from interactions with other galaxies. Quasars are a type of **active galactic nucleus** (or **AGN**, which can stand for either *nucleus* or *nuclei*). The distinct types of active nuclei are identified from the spectrum of the galaxy. A "normal" galaxy has an absorption spectrum that is a composite of the light from its billions of stars. A galaxy with an AGN exhibits emission lines in addition to the stellar absorption spectrum. Thus, AGN are identified by the emission lines in their spectra, which distinguishes them from normal galaxies.

AGN come in several types and can occur in spiral or elliptical galaxies. **Seyfert galaxies**, named after Carl Seyfert (1911–1960), who discovered them in 1943, are spiral galaxies whose centers contain AGN. The luminosity of a typical Seyfert nucleus can be 10 billion to 100 billion L_{Sun}, comparable to the luminosity of the rest of the host galaxy as a whole. Similarly, **radio galaxies** are elliptical galaxies whose centers contain AGN; their emission is usually most prominent in radio wavelengths. Radio galaxies and the more distant and luminous quasars are often the sources of slender jets that extend outward millions of light-years from the galaxy, powering twin lobes of radio emission (**Figure 19.20**).

Much of the light from AGN is synchrotron radiation—the same type that comes from extreme environments such as the Crab Nebula supernova remnant. Synchrotron radiation comes from relativistic charged particles spiraling around the direction of a magnetic field. The fact that AGN accelerate large amounts of material to nearly the speed of light indicates that they are very violent objects. In addition to the continuous spectrum of synchrotron emission, the spectra of many quasars and Seyfert nuclei also show emission lines that are smeared out by the Doppler effect across a wide range of wavelengths. That observation implies that gas in AGN is swirling around the centers of those galaxies at speeds of thousands or even tens of thousands of kilometers per second.

AGN Are the Size of the Solar System

The enormous radiated power and mechanical energy of AGN are made even more spectacular by the fact that all that power emerges from a region that can be no larger than a light-day or so across—comparable in size to the Solar System. Although the HST and large, ground-based telescopes show faint fuzz—light from the surrounding galaxy—around the images of some quasars and other AGN, the objects themselves remain as unresolved points of light.

To understand why astronomers think that AGN are compact objects, think about the halftime show at a football game. **Figure 19.21** illustrates a problem faced by every director of a marching band. When a band is all together in a tight formation at the center of the field, the notes you hear in the stands are clear and crisp; the band plays together beautifully. But, as the band spreads out across the field, its sound begins to get mushy. That is not because the marchers are poor musicians. Instead, it is because sound travels at a finite speed. Sound travels at a speed of about 330 meters per second (m/s). At that speed, sound takes approximately 1/3 of a second to travel from one end of the football field to the other. Even if every musician on the field plays a note at exactly the same instant in response to the director's cue, in the stands you hear the instruments

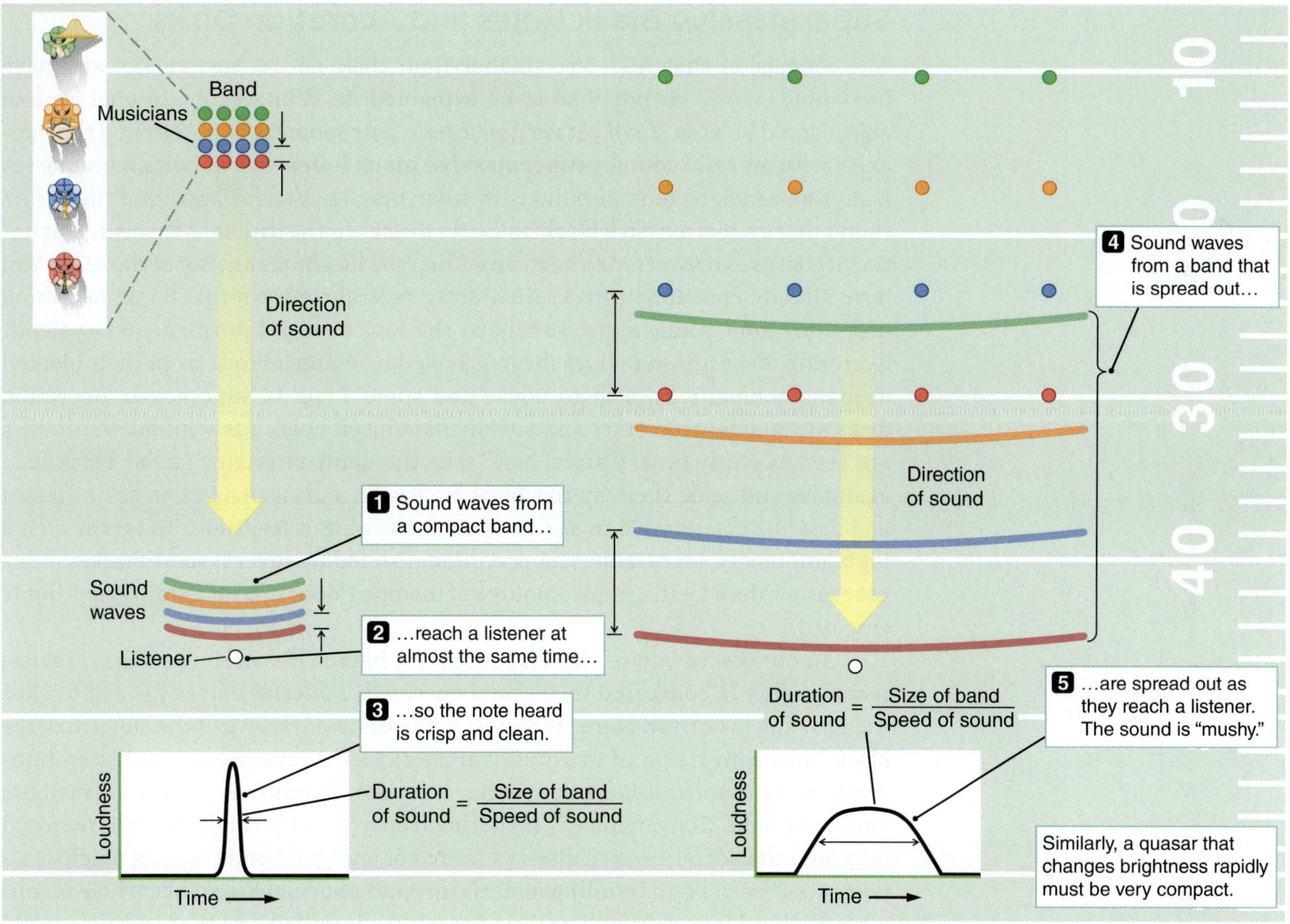

Figure 19.21 A marching band spread out across a field cannot play a clean note. Similarly, AGN must be very compact to explain their rapid variability.

close to you first but have to wait longer for the sound from the far end of the field to arrive.

AstroTour: Active Galactic Nuclei

If the band is spread from one end of the field to the other, the beginning of a note will be smeared out over about 1/3 of a second, the time sound takes to travel from one end of the field to the other. If the band were spread out over two football fields, the sound from the most distant musician would take about 2/3 of a second to arrive at your ear. If the marching band were spread out over a kilometer, it would be roughly 3 seconds—the time sound takes to travel a kilometer—before you heard a crisply played note start and stop. Even with your eyes closed, you could easily tell whether the band was in a tight group or spread out across the field.

Astronomy in Action: Size of Active Galactic Nuclei

The same principle applies to the light observed from AGN. Quasars and other AGN change their brightness dramatically over only a day or two—and sometimes as quickly as in a few hours. That rapid variability sets an upper limit on the size of the AGN, just as hearing clear music from a marching band indicates that the musicians are close together. The AGN powerhouse must therefore be no more than a light-day or so across because if it were larger, the light astronomers see could not possibly change in a day or two. An AGN has the light of up to 10,000 galaxies pouring out of a region of space that would come close to fitting within the orbit of Neptune.

Supermassive Black Holes and Accretion Disks

The conclusion that AGN are smaller than their entire host galaxy and have incredible energy outputs had to be explained. In thinking about what type of object could be very small yet very energetic, astronomers hypothesized that galaxies with an AGN contain **supermassive black holes**—black holes with masses from thousands to tens of billions of solar masses. Violent accretion disks surround those supermassive black holes. As matter in the disk falls inward, gravitational energy is converted to heat, providing the luminous energy of the AGN. You have already encountered accretion disks several times in this book. Accretion disks surround young stars, supplying the raw material for planetary systems. Accretion disks around white dwarfs, fueled by material torn from their bloated evolving companions, lead to novae and some Type Ia supernovae. Accretion disks around neutron stars and stellar-mass black holes a few kilometers across are seen as X-ray binary stars. Now take the neutron star or stellar black hole examples and scale them up to a black hole with a mass of a billion solar masses and a radius comparable in size to that of the orbit of Neptune. To attain such a high luminosity, an AGN has an accretion disk fed by several solar masses every year rather than by the small amounts of material siphoned off a star (**Working It Out 19.3**).

In our discussion of star formation in Chapter 15, we showed that gravitational energy is converted to thermal energy as material moves inward toward the growing protostar. Here, as material moves inward toward the supermassive black hole, conversion of gravitational energy heats the accretion disk to hundreds of thousands of kelvins, causing it to glow brightly in visible, ultraviolet, and X-ray light. Conversion of gravitational energy to thermal energy as material falls onto the accretion disk also is a source of energetic emission. As much as 20 percent of the mass of infalling material around a supermassive black hole is converted to luminous energy. The rest of that mass is pulled into the black hole itself, causing it to grow even more massive.

19.3 Working It Out The Size, Density, and Power of a Supermassive Black Hole

Size. What are the sizes of supermassive black holes? Recall from Chapter 18 that the Schwarzschild radii of stellar-mass black holes are kilometers in size. The formula for the Schwarzschild radius is given by

$$R_S = \frac{2GM_{BH}}{c^2}$$

where G is the gravitational constant and c is the speed of light.

The largest supermassive black holes observed have about 10 billion solar masses (M_{Sun}). For example, the black hole at the center of the galaxy M87 is 6.6 billion M_{Sun}. To compute its size, recall that $M_{Sun} = 1.99 \times 10^{30}$ kilograms (kg), $c = 3 \times 10^5$ km/s, and $G = 6.67 \times 10^{-20}\ \text{km}^3/(\text{kg s}^2)$. Then, a 6.6 billion-$M_{Sun}$ black hole has a Schwarzschild radius of

$$R_S = \frac{2 \times [6.67 \times 10^{-20}\,\text{km}^3/(\text{kg s}^2)] \times (6.6 \times 10^9 \times 1.99 \times 10^{30}\,\text{kg})}{(3 \times 10^5\,\text{km/s})^2}$$

$$R_S = 2.0 \times 10^{10}\,\text{km}$$

We can convert that value into astronomical units (AU), with 1 AU = 1.5×10^8 km. Therefore, that supermassive black hole has a radius of 130 AU—about 4 times the radius of Neptune's orbit. We know that light takes 8.3 minutes to reach Earth from the Sun at a distance of 1 AU, so 130 AU corresponds to a distance of 1,080 light-minutes, or 18 light-hours.

Density. What is the average density of that object inside of the event horizon? The mass of the black hole divided by the volume within the Schwarzschild radius is

$$\text{Density} = \frac{\text{Mass}}{\text{Volume}} = \frac{(6.69 \times 10^9) \times (1.99 \times 10^{30}\,\text{kg})}{4/3 \times \pi \times (2.0 \times 10^{10}\,\text{km})^3}$$

$$\text{Density} = 3.9 \times 10^8\,\text{kg/km}^3 = 0.39\,\text{kg/m}^3$$

That is about 1/3 of the density of air at sea level, or 1/2,500 the density of water. Supermassive black holes do not have the extremely high densities of stellar-mass black holes.

Feeding an AGN. Power for an AGN is produced when matter falls onto the accretion disk around the central supermassive black hole. Some of that high-velocity mass is radiated away according to Einstein's mass-energy equation: $E = mc^2$. About how much material has to be accreted to produce the observed luminosities? Astronomers estimate the efficiency of the accretion to be about 10–20 percent. Here, we'll assume that 15 percent of the infalling matter is radiated away as energy, or

$$E = 0.15\,mc^2$$

Astronomers can measure how much energy is produced by the infalling material and radiated to space. For a relatively weak AGN like that of the galaxy M87, $L = 5 \times 10^{35}$ joules per second (J/s), or $5 \times 10^{35}\ \text{kg m}^2/\text{s}^2$ each second. Recall that $c = 3 \times 10^8$ m/s. Dividing both sides of Einstein's equation by $0.15\,c^2$ gives us the mass consumed each second:

$$m = \frac{E}{0.15\,c^2} = \frac{5 \times 10^8\,\text{kg m}^2/\text{s}^2}{0.15 \times (3 \times 10^8\,\text{m/s})^2}$$

$$m = 3.7 \times 10^{19}\,\text{kg}$$

Multiplying that result (the mass consumed each second) by 3.2×10^7 seconds per year shows that the AGN accretes 10^{27} kg, or about half the mass of Jupiter, each year, which is then radiated away as energy.

If we consider a quasar with a luminosity (L) of 10^{39} J/s = $10^{39}\ \text{kg m}^2/\text{s}^2$ each second (= 2.5 trillion L_{Sun}), the mass accreted each second is given by

$$m = \frac{10^{39}\,\text{kg m}^2/\text{s}^2}{0.15 \times (3 \times 10^8\,\text{m/s})^2}$$

$$m = 7.4 \times 10^{22}\,\text{kg}$$

Multiplying by 3.2×10^7 seconds per year yields a mass of 2.4×10^{30} kg per year. The mass of the Sun is 1.99×10^{30} kg, so to radiate that much energy, the quasar supermassive black hole is accreting about 1.2 M_{Sun} each year. A quasar with 10 times that luminosity would be accreting 10 times the mass.

The interaction of the accretion disk with the black hole creates powerful radio jets that emerge perpendicular to the disk (as in the jet in the upper left of Figure 19.20). Throughout, twisted magnetic fields accelerate charged particles such as electrons and protons to relativistic speeds, producing synchrotron emission. Gas in the accretion disk or in nearby clouds orbiting the central black hole at high speeds produces emission lines smeared out by the Doppler effect into the broad lines seen in AGN spectra. That accretion disk surrounding a supermassive black hole is the "central engine" that powers AGN.

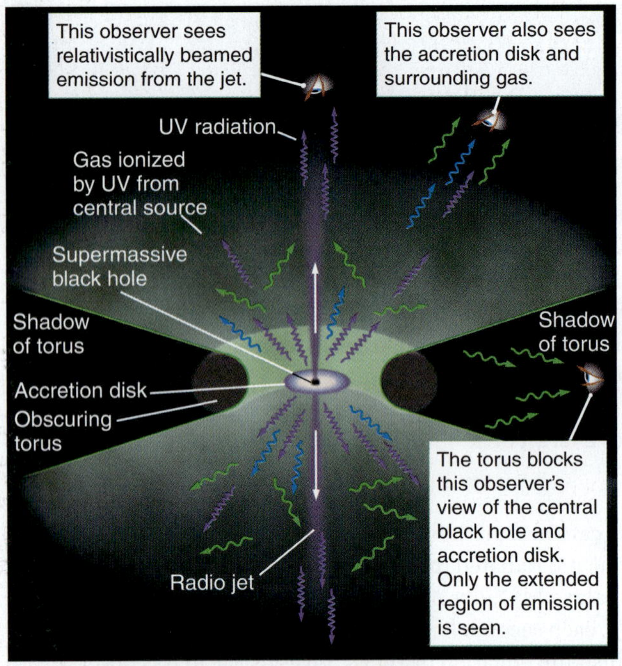

Figure 19.22 The basic model of an active galactic nucleus, with a supermassive black hole surrounded by an accretion disk at the center. A larger, dusty torus sometimes blocks the view of the black hole. The mass of the central black hole, the rate at which it is being fed, and the viewing angle determine the observational properties of an AGN.

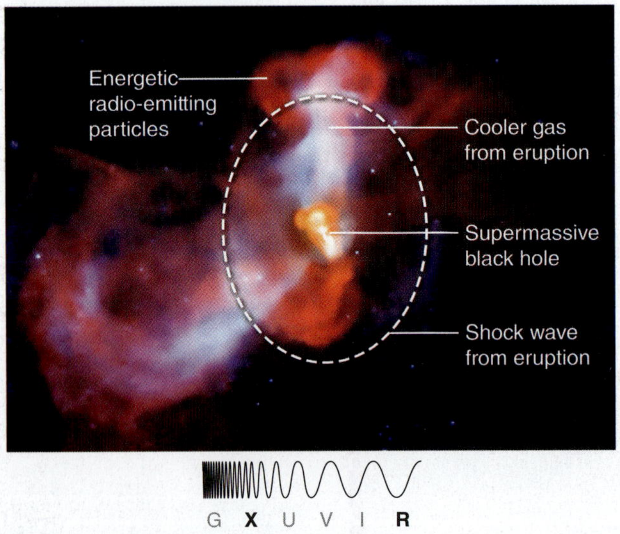

Figure 19.23 This image of M87 in radio and X-rays shows the location of the supermassive black hole.

The Unified Model of AGN

Astronomers have developed the basic picture of a supermassive black hole surrounded by an accretion disk into a more complete AGN model. The **unified model of AGN** attempts to explain all types of AGN—quasars, Seyfert galaxies, and radio galaxies. **Figure 19.22** shows the various components of that AGN model, in which an accretion disk surrounds a supermassive black hole. Much farther out from the accretion disk lies a large torus, or "doughnut" of gas and dust consisting of material feeding the central engine. Located far from the inner turmoil of the accretion disk, and far larger than the central engine, some of that torus is ionized by UV light from the AGN (see the **Process of Science Figure**).

In the unified model of AGN, the various AGN observed from Earth are partly explained by astronomers' view of the central engine. The torus of gas and dust obscures that view in different ways, depending on the viewing angle. Variation in that angle, in the mass of the black hole, and in the rate at which it is being fed accounts for a wide range of observed AGN properties. When the AGN is viewed edge on, astronomers see emission lines from the surrounding torus and other surrounding gas. Astronomers also can sometimes see the torus in absorption against the background of the galaxy. From that nearly edge-on orientation, they cannot see the accretion disk itself, so they do not see the Doppler-smeared lines that originate closer to the supermassive black hole. If jets are present in the AGN, though, they should be visible emerging from the center of the galaxy.

If astronomers observe the accretion disk more face on, they can see over the edge of the torus and thus get a more direct look at the accretion disk and the location of the black hole. Then they see more of the synchrotron emission from the region around the black hole and the Doppler-broadened lines produced in and around the accretion disk. **Figure 19.23** shows an image of one such object, the galaxy M87, at an intermediate inclination. M87 is a source of powerful jets that continue outward for 100,000 light-years but originate in the tiny engine at the heart of the galaxy. Spectra of the disk at the center of that galaxy show the rapid rotation of material around a central black hole that has a mass of 3 billion $(3 \times 10^9)\ M_{\text{Sun}}$.

The material in an AGN jet travels very close to the speed of light, so what astronomers see is strongly influenced by relativistic effects. One of those effects is called **relativistic beaming**: matter traveling at close to the speed of light concentrates any radiation it emits into a tight beam pointed in the direction in which it is moving. So, astronomers often observe only one side of the jets from AGN, even though the radio lobes of radio galaxies are usually two-sided. The jet moving away is just too faint to observe.

In rare instances when the accretion disk in a quasar or radio galaxy is viewed almost directly face on, relativistic beaming dominates the observations. In those *blazars*, emission lines and other light coming from hot gas in the accretion disk are overwhelmed by the bright glare of the jet emission beamed directly at Earth.

Normal Galaxies and AGN

The essential elements of an AGN are a central engine (an accretion disk surrounding a supermassive black hole) and a source of fuel (gas and stars flowing onto the accretion disk). Without a source of matter falling onto the black hole, an AGN would no longer be an active nucleus. Astronomers looking at such an object would observe a normal galaxy with a supermassive black hole sitting in its center.

PROCESS OF SCIENCE

Finding the Common Thread

When first discovered, active galactic nuclei (AGNs) seemed to come in many different types, with dramatically different spectra. Later, it was realized that this could be an orientation effect: the many different types of objects could all be explained with one type of object, viewed from different angles.

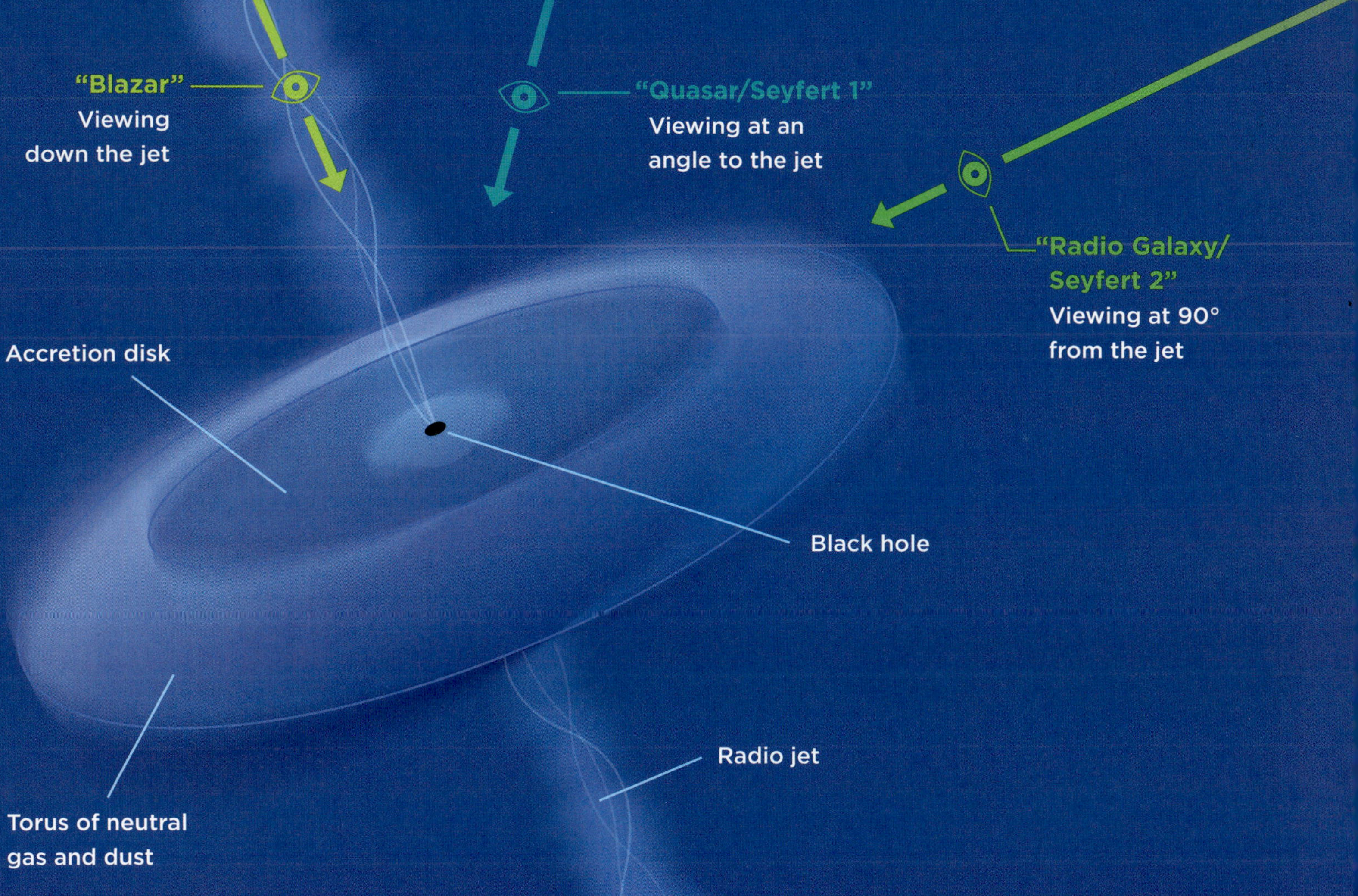

Scientists seek underlying principles that explain more than one phenomenon. In the case of AGNs, scientists developed one model to explain many types of objects. This model encompasses the idea that our classification can be affected by our viewing angle, which is true for galaxy shapes generally. A unified model is much simpler than having a different model for every type of object.

Only a few percent of present-day galaxies contain AGN as luminous as the host galaxy. But when astronomers look at more distant galaxies (and therefore look further back in time), the percentage of galaxies with AGN is much larger. These observations show that when the universe was younger, many more AGN existed than are there today. If astronomers' understanding of AGN is correct, all the supermassive black holes that powered those dead AGN should still be around. If astronomers combine what they know of the number of AGN in the past with ideas about how long a given galaxy remains in an active AGN phase, they are led to predict that many massive galaxies today contain supermassive black holes.

If supermassive black holes are present in the centers of normal galaxies, those black holes should reveal themselves in several ways. For one thing, such a concentration of mass at the center of a galaxy should draw surrounding stars close to it. The central region of such a galaxy should be much brighter than could be explained if stars alone were responsible for the gravitational field in the inner part of the galaxy. Stars experiencing the gravitational pull of a supermassive black hole in the center of a galaxy should also orbit at very high velocities and therefore show large Doppler shifts. Astronomers have found those large Doppler shifts in every normal galaxy with a substantial bulge in which a careful search has been conducted. The masses inferred for those black holes range from 10,000 to 20 billion M_{Sun}. The mass of the supermassive black hole seems to be related to the mass of the bulge in which it is found. Most large galaxies, whether elliptical or spiral, probably contain supermassive black holes. Those observations reveal something remarkable about the structure and history of normal galaxies.

Apparently, the only difference between a normal galaxy and an active galaxy is whether the supermassive black hole at its center is being fed when we see that galaxy. The rarity of present-day galaxies with very luminous AGN does not indicate which galaxies have the potential for AGN activity. Rather, it indicates which galaxy centers are being lit up at the moment. If a large amount of gas and dust were dropped directly into the center of any large galaxy, that material would fall inward toward the central black hole, forming an accretion disk and a surrounding torus. That process would change the nucleus of that galaxy into an AGN.

In Chapter 23, we will discuss galaxy evolution and note that many of the observed properties of galaxies discussed here in Chapter 19—including the type of galaxy that forms, spiral structure, star formation, and AGN—depend on the interactions and mergers between galaxies. To account for the many large galaxies visible today, interactions and mergers must have been much more prevalent in the past when the universe was younger; that is one explanation for the larger number of AGN that existed in the past. Computer models show that galaxy-galaxy interactions can cause gas located thousands of parsecs from the center of a galaxy to fall inward toward the galaxy's center, where the gas can provide fuel for an AGN. During mergers, a significant fraction of a galaxy might wind up being cannibalized. HST images of quasars often show that quasar host galaxies are tidally distorted or are surrounded by other visible matter that is probably still falling into the galaxies. Galaxies that show evidence of recent interactions with other galaxies are more likely to house AGN in their centers (**Figure 19.24**). Any large galaxy might be only an encounter away from becoming an AGN.

Figure 19.24 The Swift Gamma-Ray observatory has detected active black holes (circles) in these merging galaxies.

CHECK YOUR UNDERSTANDING 19.4

Supermassive black holes: (a) are extremely rare—only a handful exist in the universe; (b) are completely hypothetical; (c) occur in most, perhaps all, large galaxies; (d) occur only in the space between galaxies.

Origins Habitability in Galaxies

Now that we know something about the types of galaxies that have been observed, what can we say about their potential for life? No solid information is available because those galaxies are too far away for astronomers to have detected any planets around their stars. Instead, all we can do is speculate about the habitability of other galaxies. Two key requirements are the presence of heavy elements to form planets (and life) and an environment without too much radiation that might be damaging to life.

A study of the host stars of exoplanet candidates discovered by the Kepler telescope suggests that stars with a higher percentage of heavier elements may be more likely to have planets. (That finding fits with the core accretion models of planet formation discussed in Chapter 7.) The first generation of stars made from the hydrogen and helium of Big Bang nucleosynthesis do not have heavy elements. Recall from the discussion of stellar evolution that elements heavier than helium are created in the cores of dying stars and then are scattered into the galactic environment through planetary nebulae, stellar winds, and supernova explosions. So the amount of heavier elements in a star depends on the cosmic history of the material from which the star formed. Therefore, astronomers must consider the galactic environment of the star, which varies among types of galaxies and locations within the galaxies.

Spiral galaxies have had more continual star formation in their disks throughout their history. They contain more stars born from recycled material and therefore more stars with a higher fraction of heavy elements. Elliptical (and S0) galaxies have older, redder populations of stars and very little current star formation. Old, massive ellipticals have a larger percentage of lower-mass stars than that of smaller ellipticals or spirals. Astronomers had previously thought that difference meant that large elliptical galaxies would not be good environments for planet formation. But the Kepler telescope has found many planets around small, red, main-sequence stars like the ones that populate elliptical galaxies. One study of two elliptical galaxies showed that both had some fraction of stars with a heavy-element fraction similar to that of the stars hosting Kepler exoplanets in the Milky Way, so ellipticals are not ruled out.

Another issue is the presence of radiation that might be hazardous to life. That radiation would most likely come from the center of the galaxy. Galaxies in an active AGN state might have too much radiation in regions close to their centers to be conducive to life. Stars whose orbits cross spiral arms many times also might be exposed to higher-than-average levels of radiation, but that is not the case for most stars in a galaxy.

The conditions in those galaxies may also change as the galaxies evolve. Galaxy mergers can shake up stellar orbits and move stars and their planets to different locations. Mergers also may affect the growth and activity level of supermassive black holes and thus the presence of radiation. Some galactic environments just may not remain habitable for the length of time—billions of years—that life took to evolve from bacteria to intelligence on Earth.

ARTICLES QUESTIONS

Astronomers observe gravitational lensing by a cluster of galaxies

A Lot of Galaxies Need Guarding in This NASA Hubble View

By ANN JENKINS and RAY VILLARD, Space Telescope Science Institute

Much like the eclectic group of space rebels in the upcoming film *Guardians of the Galaxy Vol. 2*, NASA's Hubble Space Telescope has some amazing superpowers, specifically when it comes to observing innumerable galaxies flung across time and space.

A stunning example is a galaxy cluster called Abell 370 [**Figure 19.25**] that contains an astounding assortment of several hundred galaxies tied together by the mutual pull of gravity. That's a lot of galaxies to be guarding, and just in this one cluster!

Photographed in a combination of visible and near-infrared light, the immense cluster is a rich mix of a variety of galaxy shapes. The brightest and largest galaxies in the cluster are the yellow-white, massive, elliptical galaxies containing many hundreds of billions of stars each. Spiral galaxies—like our Milky Way—have younger populations of stars and are bluish.

Entangled among the galaxies are mysterious-looking arcs of blue light. These are actually distorted images of remote galaxies behind the cluster. These far-flung galaxies are too faint for Hubble to see directly. Instead, the cluster acts as a huge lens in space that magnifies and stretches images of background galaxies like a funhouse mirror. The massive gravitational field of the foreground cluster produces this phenomenon. The collective gravity of all the stars and other matter trapped inside the cluster warps space and affects light traveling through the cluster, toward Earth.

Nearly a hundred distant galaxies have multiple images caused by the lensing effect. The most stunning example is "the Dragon," an extended feature that is probably several duplicated images of a single background spiral galaxy stretched along an arc.

Astronomers chose Abell 370 as a target for Hubble because its gravitational lensing effects can be used for probing remote galaxies that inhabited the early universe.

Abell 370 is located approximately 4 billion light-years away in the constellation Cetus, the Sea Monster. It is the last of six galaxy clusters imaged in the recently concluded Frontier Fields project. This ambitious, community-developed collaboration among NASA's Great Observatories and other telescopes harnessed the power of massive galaxy clusters and probed the earliest stages of galaxy development. The program reveals galaxies that are 10 to 100 times fainter than any previously observed.

Figure 19.25 Galaxy cluster Abell 370 contains several hundred galaxies tied together by the mutual pull of gravity. Photographed in a combination of visible and near-infrared light, the brightest and largest galaxies are the yellow-white, massive, elliptical galaxies each containing many hundreds of billions of stars. Spiral galaxies have younger populations of stars and are bluish. Mysterious-looking arcs of blue light are distorted images of remote galaxies behind the cluster. The cluster acts as a huge lens in space that magnifies and stretches images of background galaxies like a funhouse mirror.

ARTICLES QUESTIONS

1. Which type of galaxy is the most common?
2. Why are the spiral galaxies bluer than the elliptical galaxies?
3. Explain how more remote galaxies tell us about the earlier stages of the universe.
4. What measurement tells astronomers that the blue arcs are multiple images of the same galaxy?

Summary

Galaxies are classified on the basis of their shape and the types of orbits of their stars. Most of the mass of a galaxy is dark matter, which interacts with light very weakly, if at all. The nature of that matter is not yet known. Most (non-dwarf) galaxies have a supermassive black hole at the center, which may become an AGN if gas accretes onto it. In thinking about the potential habitability of galaxies, astronomers consider the activity state of the galaxy, including the presence or absence of an AGN and mergers, and the amount of heavy elements in the stars in the galaxy, which is related to the star formation rate and galaxy type.

LG 1 **Determine a galaxy's type from its appearance and describe the motions of its stars.** Spiral galaxies are distinguished by their flat disk and spiral arms. The stars in that disk all orbit the center of the galaxy in the same direction. Elliptical galaxies are shaped like a rugby ball, and the stars orbit in all directions. Irregular galaxies are galaxies that fit neither classification, usually because they are interacting with another galaxy.

LG 2 **Explain the distance ladder and how distances to galaxies are measured.** Astronomers build a distance ladder to galaxies by observing objects of known luminosity, such as Cepheid variable stars and Type Ia supernovae, in distant galaxies. Observations of the distance and the velocity of galaxies show that the two factors are related: more distant galaxies move away from us faster. Hubble's law, $v = H_0d$, offers a method to find the distances of the most remote objects.

LG 3 **Describe the evidence suggesting that galaxies are composed mostly of dark matter.** Most of the mass in galaxies does not reside in gas, dust, or stars; instead, galaxy rotation curves indicate that about 90 percent of a galaxy's mass is in the form of dark matter, which does not emit or absorb light to any significant degree. Dark matter is so far identified by its gravitational interaction with ordinary matter. After many searches and experiments, astronomers still do not know what constitutes dark matter.

LG 4 **Discuss the evidence indicating that most—perhaps all—large galaxies have supermassive black holes at their centers.** Observations of distant quasars reveal extremely luminous, compact sources near the centers of galaxies. Those active galactic nuclei (AGN) are best explained as supermassive black holes, surrounded by an accretion disk and a torus of dust and gas. AGN can emit as much as 1,000 times the light of the whole galaxy, all coming from a region the size of the Solar System. Surveys of the centers of large galaxies show that most have black holes at their centers.

? Unanswered Questions

- How standard are "standard candles"? For example, Cepheid variable light curves are slightly different, depending on the amount of heavy elements in the stars, so that variation must be calibrated. Most Type Ia supernovae may originate from the merging of two compact objects rather than from one white dwarf accreting mass and exploding when it gets close to 1.4 M_{Sun}. Astronomers try to address those difficulties by using multiple methods to find distances to galaxies—for example, observing many Cepheids, bright O stars, and Type Ia supernovae in the same galaxy—to check that their calculated distances agree. In one recent study, astronomers used HST to observe more than 600 Cepheid variable stars in eight galaxies in which Type Ia supernova had been detected, and they reduced the uncertainty in their distances (and thus their calculated uncertainty in their value of the Hubble constant). The Type Ia supernovae calibrated in that way are consistent with earlier-calibrated maximum luminosities. However, those supernovae in galaxies that are far enough away that no other standard candles exist may have slightly different maximum luminosities, leading to a less precise distance estimate (and value of H_0).
- Where do supermassive black holes come from? To explore that question, astronomers are studying computer models in which the supermassive black hole grew from many stellar-mass black holes, or grew along with the galaxy by swallowing large amounts of central gas, or increased after the merger of two or more galaxies. We return to that question when we discuss galaxy evolution in Chapter 23.

Questions and Problems

Test Your Understanding

1. Which of the following contributes the largest percentage to the total mass of a spiral galaxy?
   a. dark matter
   b. central black hole
   c. stars
   d. dust and gas
2. In the context of spiral galaxies, Kepler's laws could be used to estimate
   a. *P*, the period of galactic rotation.
   b. *A*, the radius of the galaxy.
   c. *M*, the mass of the galaxy.
   d. *v*, the rotation speed of the galaxy.
3. Astronomers determine the radius of an AGN by measuring
   a. how much light comes from it.
   b. how hard it pulls on stars nearby.
   c. how quickly its light varies.
   d. how quickly it rotates.
4. If you observed a galaxy with an Hα emission line that had a wavelength of 756.3 nm, what would be the galaxy's redshift? Note that the rest wavelength of the Hα emission line is 656.3 nm.
   a. 0.01
   b. 0.05
   c. 0.10
   d. 0.15
5. As astronomers extend their distance ladder beyond 30 Mpc, they change their measuring standard from Cepheid variable stars to Type Ia supernovae. Why is that change necessary?
   a. Type Ia supernovae are more luminous than Cepheid variables.
   b. Type Ia supernovae are less luminous than Cepheid variables.
   c. Type Ia supernovae vary more slowly than do Cepheid variables.
   d. Type Ia supernovae vary more quickly than do Cepheid variables.
6. Which galaxy types have a spherical bulge and a well-defined disk?
   a. spiral
   b. barred spiral
   c. elliptical
   d. irregular
7. Which galaxy type is shaped like a rugby ball?
   a. Sb
   b. SBb
   c. Irregular
   d. E5
8. For a variable star to be useful as a standard candle, its luminosity must be related to its
   a. period of variation.
   b. mass.
   c. temperature.
   d. radius.
9. If all the stars in an elliptical galaxy traveled in random directions in their orbits, the elliptical galaxy would be type
   a. E0.
   b. E2.
   c. E5.
   d. E7.
10. The flat rotation curves of spiral galaxies imply that the distribution of mass
    a. is like the Solar System; most mass is concentrated in the center.
    b. is a wheel; the density remains the same as the radius increases.
    c. resembles the light distribution of the galaxy; a large concentration occurs in the middle, but significant mass exists quite far out.
    d. extends farther out than the visible galaxy.
11. Astronomers observe two galaxies, A and B. Galaxy A has a recession velocity of 2,500 km/s, whereas galaxy B has a recession velocity of 5,000 km/s. According to those data,
    a. galaxy A is 4 times as far away as galaxy B.
    b. galaxy A is twice as far away as galaxy B.
    c. galaxy B is twice as far away as galaxy A.
    d. galaxy B is 4 times as far away as galaxy A.
12. The Hubble constant is found from the
    a. slope of the line fit to the data in Hubble's law.
    b. *y*-intercept of the line fit to the data in Hubble's law.
    c. spread in the data in Hubble's law.
    d. inverse of the slope of the line fit to the data in Hubble's law.
13. Astronomers know that dark matter is present in galactic halos because the speeds of orbiting stars ________ far from the center of the galaxy.
    a. decrease
    b. increase
    c. remain about constant
    d. fluctuate dramatically
14. What accounts for the observed differences among types of AGN?
    a. the type of the host galaxy
    b. the size of the central black hole
    c. the amount of dark matter in the galaxy's halo
    d. the luminosity and our viewing angle

15. If a Seyfert galaxy's nucleus varies in brightness on the timescale of 10 hours, approximately what is the size of the emitting region?
 a. 20 AU
 b. 70 AU
 c. 90 AU
 d. 140 AU

Thinking about the Concepts

16. What was the subject of the Great Debate, and why was it important to astronomers' understanding of the scale of the universe?

17. How did observations of Cepheid variable stars finally settle the Great Debate?

18. Why is observing more than one type of standard candle in a distant galaxy better?

19. Explain what astronomers mean by *distance ladder.*

20. Why is knowing the type of progenitor of a Type Ia supernova in a distant galaxy important?

21. Some galaxies have regions that are relatively blue; other regions appear redder. What does that variation indicate about the differences between those regions?

22. Describe how elliptical galaxies and spiral bulges are similar.

23. Which is more luminous: a quasar or a galaxy with 100 billion solar-type stars? Explain your answer.

24. The nearest observed quasar is about 750 Mpc away. Why don't astronomers observe many that are closer?

25. What distinguishes a normal galaxy from one that contains an AGN?

26. Compare the size of a typical AGN with the size of the Solar System. How do astronomers know the size of an AGN?

27. Describe what astronomers think is happening at the center of a galaxy that contains an AGN.

28. Most galaxies probably contain supermassive black holes, yet many galaxies display no obvious evidence for their existence. Why do some black holes reveal their presence, whereas others do not?

29. Study Figure 19.10. The different "rungs" of the distance ladder in Figure 19.10 overlap in the distances that they measure. Why is that important?

30. Why do astronomers think that planets around stars near the centers of galaxies would not be good locations for the formation of life?

Applying the Concepts

31. In Figure 19.15, the small vertical bars (known as error bars) on the data points indicate the size of the measurement error.
 a. At a radius of 25,000 parsecs (pc), what is the approximate measurement error in the rotation velocity?
 b. What is that value as a percentage of the measured velocity?
 c. Error bars are important because they show how wrong the measurement could possibly be. One way to think about this is that the black line could be as high as the top of the error bars or as low as the bottom of the error bars. In either case, would shifting the black line change the overall conclusion about redshift and distance? Why or why not?

32. Suppose the number density of galaxies in the universe is, on average, 3×10^{-68} galaxies/m^3. If astronomers could observe all galaxies out to a distance of 10^{10} pc, how many galaxies would they find?

33. The spectrum of a distant galaxy shows the Hα line of hydrogen (λ_{rest} = 656.28 nm) at a wavelength of 750 nm. Assume that H_0 = 70 km/s/Mpc.
 a. What is the redshift (z) of that galaxy?
 b. What is its recession velocity (v_r) in kilometers per second?
 c. What is the distance of the galaxy in megaparsecs?

34. One of the closer known quasars is 3C 273. It is located in the constellation Virgo and is bright enough to be seen in a medium-sized amateur telescope. With a redshift of 0.158, what is the distance to 3C 273 in parsecs?

35. The quasar 3C 273 has a luminosity of 10^{12} L_{Sun}. Assuming that the total luminosity of a large galaxy, such as the Andromeda Galaxy, is 10 billion times that of the Sun, compare the luminosity of 3C 273 with that of the entire Andromeda Galaxy.

36. A quasar has the same brightness as a galaxy seen in the foreground 2 Mpc distant. If the quasar is 1 million times more luminous than the galaxy, what is the distance of the quasar?

37. Estimate the Schwarzschild radius for a supermassive black hole with a mass of 26 billion M_{Sun}.

38. You read on the Internet that astronomers have discovered a "new" cosmological object that appears to be flickering with a period of 83 minutes. Because you have read *21st Century Astronomy*, you can estimate quickly the maximum size of that object. How large can it be?

39. A quasar has a luminosity of 10^{41} W, or J/s, and 10^8 M_{Sun} to feed it. If you assume constant luminosity and 20 percent conversion efficiency, what is your estimate of the quasar's lifetime?

40. A solar-type star ($M = 2 \times 10^{30}$ kg) approaches a supermassive black hole. Half of its mass falls into the black hole, whereas the other half is completely converted to energy in the form of light. How much energy does that dying star send out to the rest of the universe?

41. Suppose a Type Ia supernova is found in a distant galaxy. The measured supernova brightness is 10^{-17} W/m^2. What is the distance of the galaxy?

42. Suppose that an object with the mass of Earth ($M_{Earth} = 5.97 \times 10^{24}$ kg) fell into a supermassive black hole with a 10 percent energy conversion.
    a. How much energy (in joules) would be radiated by the black hole?
    b. Compare your answer with the energy radiated by the Sun each second: 3.85×10^{26} J.

43. If a luminous quasar has a luminosity of 2×10^{41} W, or J/s, how many solar masses ($M_{Sun} = 2 \times 10^{30}$ kg) per year does that quasar consume to maintain its average energy output?

44. Material ejected from the supermassive black hole at the center of galaxy M87 extends outward from the galaxy to a distance of approximately 30,000 pc. M87 is approximately 17 Mpc away.
    a. If that material were visible to the naked eye, how large would it appear in the nighttime sky? Give your answer in degrees (1 radian = 57.3°).
    b. Compare that size with the angular size of the Moon.

45. A lobe in a visible jet from galaxy M87 is observed at a distance of 1,530 pc (5,000 light-years) from the galaxy's center moving outward at a speed of 0.99 times the speed of light (0.99*c*). If you assume constant speed, how long ago was the lobe expelled from the supermassive black hole at the galaxy's center?

USING THE WEB

46. Go to the Goddard Media Studios website and view the animation of a Cepheid variable star in a spiral galaxy (https://svs.gsfc.nasa.gov/cgi-bin/details.cgi?aid=10145). Explain how astronomers use data such as those to estimate the distance to the galaxy. What is actually observed, what is assumed, and what is calculated? (Review the discussion in Chapter 16 as needed.)

47. Go to the Astronomy Picture of the Day app or website (https://apod.nasa.gov/apod) and look at some recent pictures of galaxies. In each case, consider the following questions: Was the picture taken from a large or small telescope; from the ground or from space? Are galaxies in the image face on, edge on, or at an angle? What wavelengths were used for making the image? Are any of the colors "false colors"? If the picture is a combination of images from several telescopes, what do the different colors indicate?

48. Citizen science:
    a. Go to the website for Galaxy Zoo (https://galaxyzoo.org), the original Zooniverse citizen science project. (Log in with your Zooniverse password.) The specific project in action at any given time depends on the real data that need to be examined. One of the projects is probably a classification project. Click on and read "About," and "Classify," and then classify some galaxies. Save a copy of your classifications for your homework if necessary.
    b. Go to the website for Radio Galaxy Zoo (https://radio.galaxyzoo.org/). The project asks people to help locate supermassive black holes and their associated jets. Read the top page and "Science" to learn about the project, and then click on "Classify." Work through the given example, and then classify a few more.

49. Go to the website for the Fermi Gamma-ray Space Telescope (https://fermi.gsfc.nasa.gov). Under "Latest News," scroll down to click on "Full News Archive" and look for a story about dark matter. What has that telescope discovered about dark matter?

50. a. Go to the website for the NASA Swift Gamma-Ray observatory (https://swift.gsfc.nasa.gov), which studies gamma-ray bursts. Under "Latest Swift News," click on "Other News" and look for a story about supermassive black holes. What has been discovered?
    b. Go to the website for NuSTAR (Nuclear Spectroscopic Telescope Array—https://www.nustar.caltech.edu), a space telescope NASA launched in 2012. The mission is studying active galaxies hosting supermassive black holes. What type of telescope is this (wavelengths observed, general design)? What has been discovered?

EXPLORATION

digital.wwnorton.com/astro6

Galaxy classification sounds simple, but it can become complicated when you actually attempt it. **Figure 19.26**, taken by the Hubble Space Telescope, shows a small portion of the Coma Cluster of galaxies. The Coma Cluster contains thousands of galaxies, each containing billions of stars. Some of the objects in that image (the ones with a bright cross) are foreground stars in the Milky Way. Some of the galaxies are far behind the Coma Cluster. Working with a partner, in this Exploration you will classify the 20 or so brightest galaxies in that cluster.

First, make a map by laying a piece of paper over the image and numbering the 20 or so brightest (or largest) galaxies in the image (label them "galaxy 1," "galaxy 2," and so on). Copy the map so that you and your partner each have a list of the same galaxies.

Separately, classify each galaxy by type. If it is a spiral galaxy, is its subtype a, b, or c? If it is an elliptical, how elliptical is it? Make a table that contains the galaxy number, the type you have assigned it, and any comments that will help you remember why you made that choice. When you are done classifying, compare your list with your partner's. Now comes the fun part! Argue about the classifications until you agree—or until you agree to disagree.

Figure 19.26 This Hubble Space Telescope image of the Coma Cluster shows a diversity of shapes.

1 Which galaxy type was easiest to classify?

____

____

2 Which galaxy type was hardest to classify?

____

____

3 What makes classifying some of the galaxies hard?

____

____

4 Which galaxy type did you and your partner agree about most often?

____

____

5 Which galaxy type did you and your partner disagree about most often?

____

____

6 How might you improve your classification technique?

____

____

If you found that activity interesting and rewarding, astronomers can use your help: go to https://galaxyzoo.org to get involved in a citizen science project to classify galaxies, some of which have never been viewed before by human eyes.

20 The Milky Way—A Normal Spiral Galaxy

Of the hundreds of billions of galaxies in the universe, the Milky Way is the only one that astronomers can study at close range. Here in Chapter 20, we focus on the Milky Way and how it offers clues to understanding all galaxies.

LEARNING GOALS

By the end of this chapter, you should be able to:

LG 1 Explain how astronomers discovered the size and spiral structure of the Milky Way.

LG 2 List the clues of galaxy formation that can be found from the components of the Milky Way.

LG 3 Explain the evidence for the dark matter halo and for the supermassive black hole at the center of the Milky Way.

LG 4 Describe the Local Group of galaxies and how it offers clues about the evolution of the Milky Way.

The Milky Way in Earth's night sky has both bright patches of starlight and dark patches where stars are hidden by dust and gas. ►►►

How do we know what the Milky Way Galaxy looks like from the outside?

20.1 Astronomers Have Measured the Size and Structure of the Milky Way

The universe is full of galaxies of many sizes and types (see Chapter 19). Because Earth is embedded within the Milky Way, the details of the shape and structure are actually harder to determine for the Milky Way than for other galaxies. Comparing observations of the Milky Way with observations of more distant galaxies improves our understanding of the Milky Way. In this section, we explore how astronomers use observations from our position in the galaxy to infer the size and structure of the Milky Way

Spiral Structure in the Milky Way

Figure 20.1a shows the Milky Way Galaxy in Earth's night sky. From a dark location at night, you can see dark bands of interstellar gas and dust that obscure much of the central plane of the Milky Way. The view of the Milky Way from inside offers a different and much closer perspective of a galaxy than can be obtained by viewing external galaxies. Compare **Figure 20.1b**, which shows an edge-on spiral galaxy. The similarities between those images suggest that the Milky Way is a spiral galaxy and that we are viewing it edge on, from inside the disk.

Finding further information about the size and shape of the Milky Way requires more extensive observations in the visible, infrared, and radio regions of the electromagnetic spectrum. Recall from Chapter 15 that neutral hydrogen emits radiation at a wavelength of 21 centimeters (cm), in the radio region of the spectrum. Maps of that radiation show spiral structure in other galaxies and

(a)

Figure 20.1 (a) We see the Milky Way as a luminous band across the night sky. Prominent dark lanes caused by interstellar dust obscure the light from more distant stars. (b) The edge-on spiral galaxy NGC 891, whose disk greatly resembles that of the Milky Way.

suggest spiral structure in the Milky Way. In addition, observations of ionized hydrogen gas in visible light show two spiral arms in the Milky Way, with concentrations of young, hot O and B stars. Those observations confirm that the Milky Way is a spiral galaxy.

In 2005, Spitzer Space Telescope observations of the distribution and motions of stars in the inner part of the galaxy confirmed that the Milky Way has a substantial bar with a modest bulge at its center. **Figure 20.2** shows an artist's rendering of the major features of the Milky Way. Two major spiral arms—Scutum-Centaurus and Perseus—connect to the ends of the central bar and sweep through the galaxy's disk, just like the arms observed in external spiral galaxies. Our galaxy has several smaller arm segments, including the Orion Spur, which contains the Sun and Solar System. Astronomers conclude that the Milky Way is a giant barred spiral that is more luminous than an average spiral. Viewed from the outside, the Milky Way would look much like the barred spiral galaxy M109 (**Figure 20.3**).

Figure 20.2 Infrared and radio observations contribute to an artist's model of the Milky Way Galaxy. The galaxy's two major arms (Scutum-Centaurus and Perseus) are seen attached to the ends of a thick central bar.

Figure 20.3 From the outside, the Milky Way would look much like this barred spiral galaxy, M109.

Spiral Arms and Star Formation

In pictures of other spiral galaxies, the arms are often the most prominent feature, as in the Andromeda Galaxy (**Figure 20.4**). The spiral arms are prominent in the ultraviolet image (Figure 20.4a), and although they are less prominent in visible light (Figure 20.4b), they are still clearly defined. You might then conclude that most stars in the disk of a spiral galaxy are concentrated in the spiral arms. That turns out *not* to be the case: although stars are slightly concentrated in spiral arms, the concentration is not strong enough to account for their prominence. Structures associated with star formation, however, such as molecular clouds and associations of luminous O and B stars, are all concentrated in spiral arms. Spiral arms are prominent because star formation is occurring there, so the arms contain significant concentrations of young, massive, hot, and therefore luminous stars which emit strongly in blue light.

Stars form when dense interstellar clouds become so dense that they begin to collapse under the force of their own gravity (see Chapter 15). Because stars

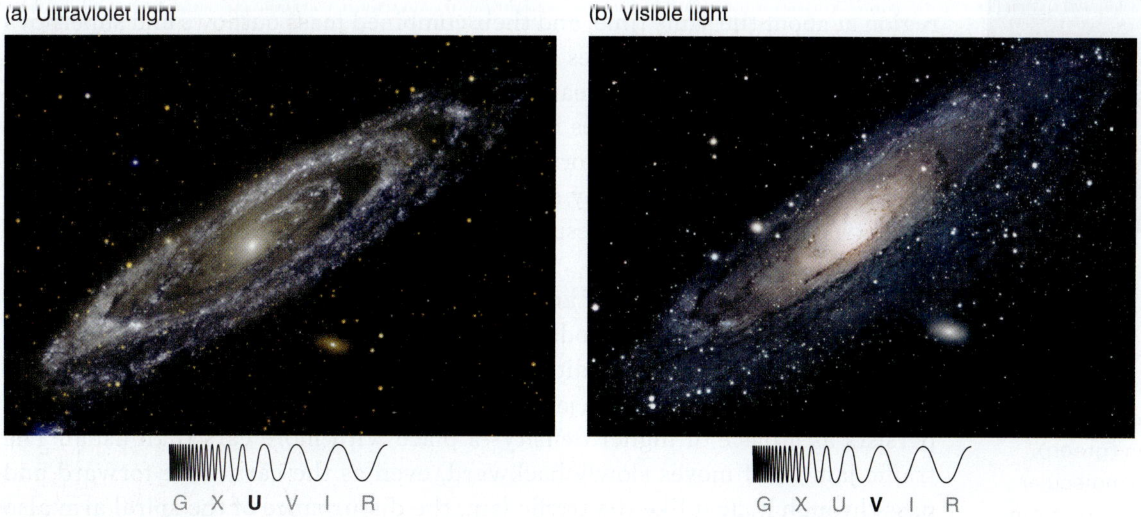

Figure 20.4 The Andromeda Galaxy in ultraviolet light (a) and visible light (b). The spiral arms, dominated by hot, young stars, are most prominent in ultraviolet light. The spiral arms are less prominent in visible light.

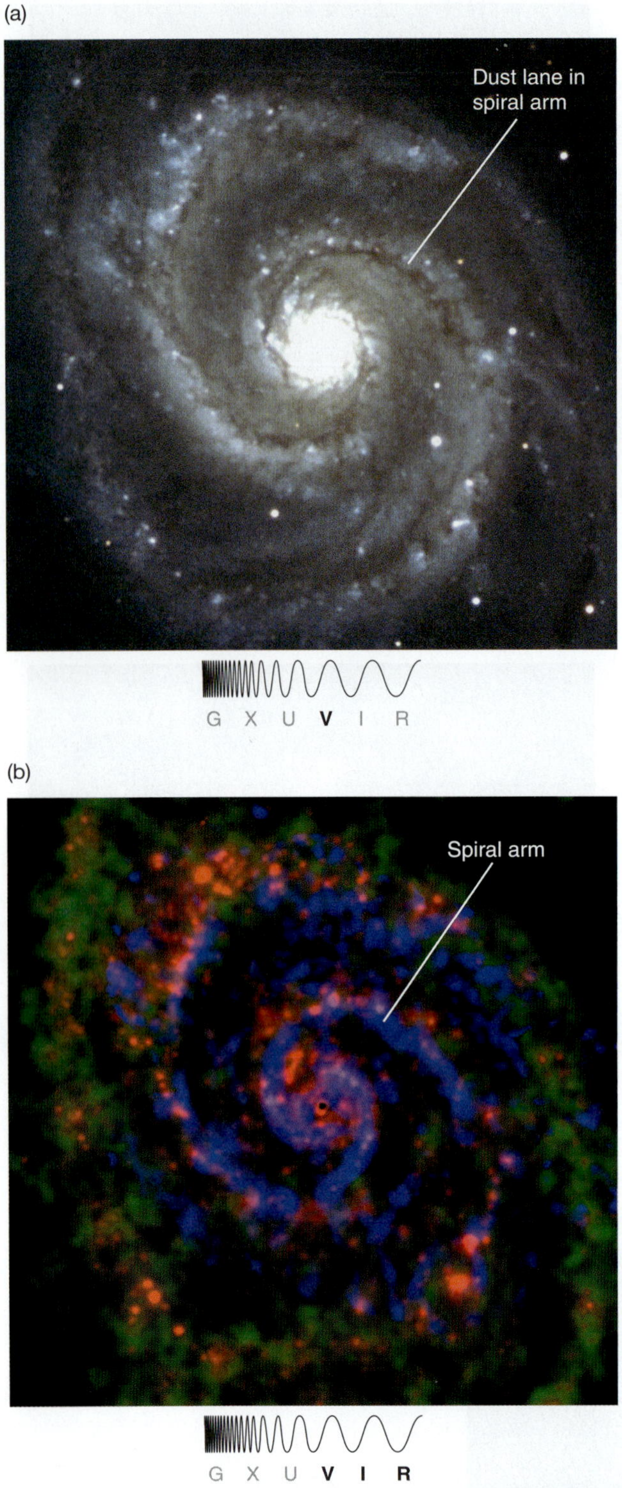

Figure 20.5 These two images of a face-on spiral galaxy show the spiral arms. (a) This visible-light image also shows dust absorption. (b) This image shows the distribution of neutral interstellar hydrogen (green), carbon monoxide (CO) emission from cold molecular clouds (blue), and hydrogen alpha (Hα) emission from ionized gas (red).

form in spiral arms, spiral arms must be places where clouds of interstellar gas and dust pile up and are compressed. Those clouds can be observed where the dust and gas block starlight (**Figure 20.5a**) or where gases such as neutral hydrogen or carbon monoxide emit at various wavelengths (**Figure 20.5b**).

Any disturbance in the disk of a spiral galaxy will cause a spiral pattern because the disk rotates. Disks of galaxies do not rotate like a solid body. Instead, material close to the center takes less time to travel around the galaxy than material farther out, and so the inner part of the disk gets ahead of the outer part. **Figure 20.6** illustrates the point: In the second frame, you can see that the outer part of the line is trailing behind the inner part. As the galaxy rotates, a straight line through the center becomes a spiral. In the time that objects in the inner part of the galaxy take to complete several rotations, objects in the outer part of the galaxy may not have completed even a single rotation.

A spiral galaxy can be disturbed, for example, by gravitational interactions with other galaxies or by a burst of star formation. However, a single disturbance will not produce a *stable* spiral-arm pattern. Spiral arms produced from a single disturbance will wind themselves up completely in two or three rotations of the disk and then disappear. Disturbances that are repetitive, however, can sustain spiral structure indefinitely. When the bulge in the center of a spiral galaxy is elongated (as seems to be the case for most spiral galaxies), the bulge gravitationally disturbs the disk. As the disk rotates through that disturbance, repeated episodes of star formation occur, and stable spiral arms form.

Many galaxies show clear evidence of a relationship between the shapes of their bulges and the structure of their spiral arms. Barred spirals, for example, have a characteristic two-armed spiral pattern that is connected to the elongated bulge, as seen in Figure 20.3. Even the bulges of galaxies that are not obviously barred may be elongated enough to contribute to the formation of a two-armed spiral structure. Smaller galaxies in orbit about larger galaxies also can give rise to a periodic gravitational disturbance, triggering the same sort of two-armed structure.

Star formation itself also can create spiral structure. Regions of star formation release considerable energy into their surroundings through UV radiation, stellar winds, and supernova explosions. That energy compresses clouds of gas and triggers more star formation. Typically, many massive stars form in the same region at about the same time, and their combined mass outflows and supernova explosions after short lifetimes occur one after another in the same region of space over only a few million years. The result can be large, expanding bubbles of hot gas that sweep out cavities in the interstellar medium and concentrate the swept-up gas into dense, star-forming clouds, much like the snow that piles up in front of a snowplow. In that way, star formation can propagate through the disk of a galaxy. Rotation bends the resulting strings of star-forming regions into spiral structures.

Stars move in and out of arms as they orbit the center of a galaxy. Consequently, the stars in an arm today are not the same stars that were in the arm 20 million years ago. The conditions are roughly analogous to a traffic jam on a busy highway. The cars in the jam are changing all the time, yet the traffic jam persists as a place of higher density—a place with more cars than usual. The traffic jam itself moves slowly backward, even as the cars move forward and pass through it. Just like the traffic jam, the disturbance of the spiral arm also moves at a different speed from that of the individual stars. Those disturbances

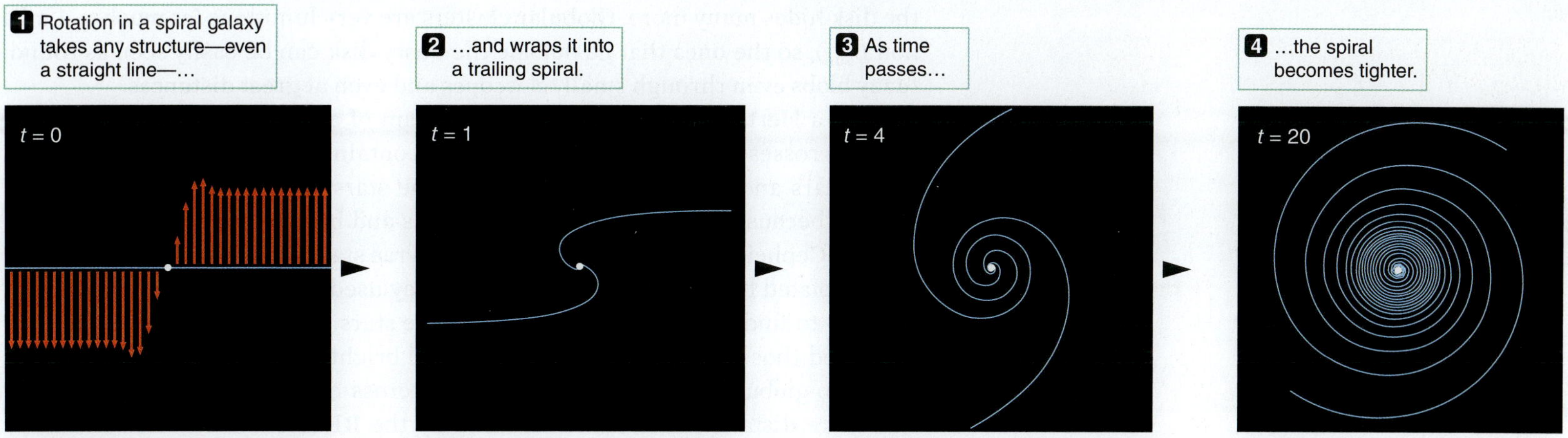

Figure 20.6 The rotation of a spiral galaxy will naturally take even an originally linear structure and wrap it into a progressively tighter spiral as time (*t*) goes by.

in the disks of spiral galaxies are called **spiral density waves** because they are waves of greater mass density and increased pressure in the galaxy's interstellar medium. Those waves move around a disk in the pattern of a two-armed spiral that does not rotate at the same rate as the stars, gas, or dust. As material in the disk orbits the center of the galaxy, it passes through those spiral density waves.

A spiral density wave has very little effect on the motions of stars as they pass through it, but it does compress the gas that flows through it. Gas flows into the spiral density wave and piles up. Stars form in the resulting compressed gas. Massive stars are concentrated in the arms because they have such short lives (typically 10 million years or so) that they never have the chance to drift far from the spiral arms where they were born. Less massive stars, however, have plenty of time to move away from their places of birth, so they form a smooth underlying disk.

The Size of the Milky Way Galaxy

If you go out on a dark night, away from any streetlights, and look toward the center of the Milky Way—located in the constellation Sagittarius—you will see the dark lane of dusty clouds shown in Figure 20.1a. Because the Solar System is inside the dusty disk of the Milky Way, the visible-light view of the galaxy itself is badly obscured. To probe the structure of the Milky Way, modern astronomers use long-wavelength infrared and radio radiation that can penetrate the dust in the disk. The most powerful tool for that work is the same 21-cm line from neutral interstellar hydrogen, which (as we described in Chapter 19) is used to measure the rotation of other galaxies. Even in those long-wavelength parts of the spectrum, the distance to the center of the galaxy cannot be measured directly.

In the 1920s, Harlow Shapley made a three-dimensional map of globular clusters in the Milky Way, which led to the first determination of the size of the Milky Way and that the Sun is offset from the center. Recall from Chapter 17 that globular clusters are large, spheroidal groups of stars held together by gravity. The Milky Way contains more than 150 cataloged globular clusters, and dust in the disk hides many more. Globular clusters are very luminous (as much as 1 million L_{Sun}), so the ones that lie outside the dusty disk can be easily seen as round, fuzzy blobs even through small telescopes and even at great distances.

In a Hertzsprung-Russell (H-R) diagram of an old cluster, the horizontal branch crosses the instability strip, which contains pulsating stars such as RR Lyrae stars and Cepheid variables. RR Lyrae stars are easy to spot in globular clusters because they are relatively luminous and have a distinctive light curve. As with Cepheid variables, the time an RR Lyrae star takes to undergo one pulsation is related to the star's luminosity. Shapley used that period-luminosity relationship to find the luminosities of RR Lyrae stars in globular clusters. He then combined those luminosities with measured brightnesses to determine the distances to globular clusters. Finally, Shapley cross-checked his results by noting that more distant clusters (as measured by the RR Lyrae stars) also tended to appear smaller in the sky, as expected.

Those globular clusters trace out the luminous part of the galactic halo of the Milky Way Galaxy, a large spherical volume of space surrounding the disk and bulge. The center of the distribution of globular clusters coincides with the gravitational center of the galaxy. Shapley realized that because he could determine the distance to the center of that distribution, he had actually determined the Sun's distance from the center of the Milky Way, as well as the size of the galaxy itself. His map showed that globular clusters occupy a roughly spherical region of space with a diameter of about 90 kiloparsecs (kpc), or 90,000 parsecs (pc). Shapley did not know about gas and dust, however, so he overestimated the distance to the globular clusters (see the **Process of Science Figure**). A modern determination indicates that the Sun is located about 8,300 pc (27,000 light-years) from the center of the galaxy, or roughly halfway out toward the edge of the disk.

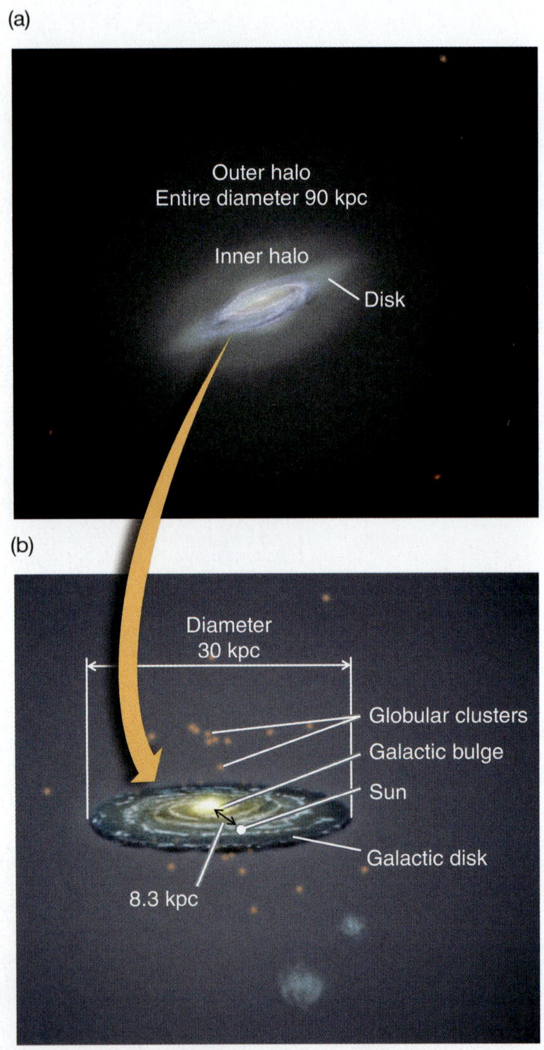

Figure 20.7 Parts of the Milky Way include the disk and the inner and outer halos (a) and the galactic bulge and disk (b). Globular clusters are located in the halo, and the Sun is located in the disk, about 8.3 kpc from the center. (1 kpc = 1,000 pc.)

CHECK YOUR UNDERSTANDING 20.1

List three pieces of evidence supporting the theory that the Milky Way is a spiral galaxy.

20.2 The Components of the Milky Way Reveal Its Evolution

Figure 20.7 illustrates the Sun's position and relationship to the rest of the Milky Way. The Sun is a middle-aged disk star located among other middle-aged stars that orbit around the galaxy within the galactic disk, as do the gas and dust in the disk. The stars in the halo move in random orbits similar to those of stars in elliptical galaxies, sometimes at high velocities. Some of those stars can be observed near the Sun as their orbits carry them swiftly through the disk. Most of those stars are much older than the Sun. The bar in the galactic bulge of the Milky Way is shaped primarily by stars and gas moving not only in highly elongated orbits up and down the long axis of the bar but also in short orbits aligned perpendicular to the bar. All those stellar orbits determine the shapes of the various parts of the galaxy, and stellar orbits are easier to measure in the Milky Way than in other galaxies. Using the ages, chemical abundances, and motions of nearby stars, astronomers can differentiate between disk and halo stars to learn more about the galaxy's structure. In this section, we describe how astronomers study the constituents of the Milky Way to find direct clues about how spiral galaxies form.

PROCESS OF SCIENCE

Unknown Unknowns

Initial efforts to measure the Milky Way did not include dark matter, because astronomers did not know about it.

Fritz Zwicky
1898–1974

Fritz Zwicky was the first to predict that dark matter existed, based on observations of clusters of galaxies.

Vera Rubin
1920–2010

Vera Rubin subsequently discovered dark matter in the Andromeda galaxy, which rotated faster than could be accounted for by the observed mass. Following up with observations of other galaxies, she confirmed that dark matter was an important component in nearly every galaxy.

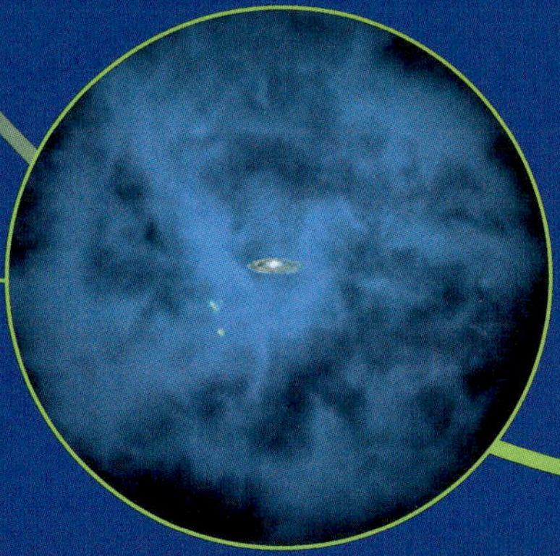

Today, astronomers routinely account for dark matter, which represents about 90% of the mass of the Milky Way.

Often, scientists don't know what is unknown until a set of observations fails to make sense. Once this "unknown unknown" is discovered, prior results must be modified to incorporate the new knowledge.

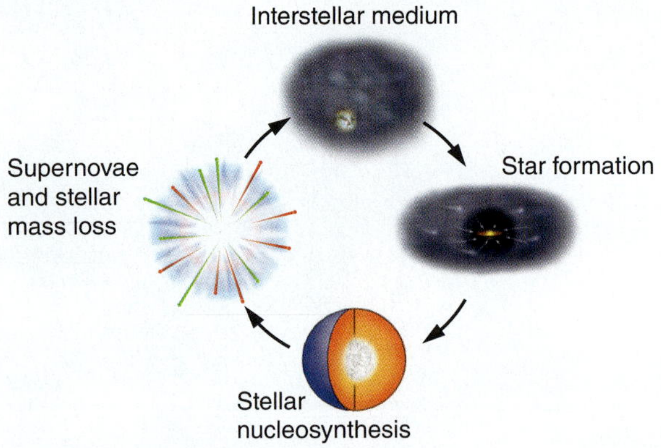

Figure 20.8 Matter moves from the interstellar medium into stars and back again in a progressive cycle that has enriched today's universe with massive elements.

Figure 20.9 As subsequent generations of stars form, live, and die, they enrich the interstellar medium with massive elements—the products of stellar nucleosynthesis. The chemical evolution of the Milky Way and other galaxies can be traced in many ways, including by the strength of interstellar emission lines and stellar absorption lines.

Age and Chemical Compositions of Stars

Over time, the chemical content of a galaxy changes as stars are born, live and die, and progressively enrich the interstellar medium (**Figure 20.8**). The interstellar medium therefore reflects all the stellar evolution that has taken place up to the present time. Gas rich in massive elements has gone through a great deal of stellar processing. That evolution is similarly reflected in the composition of stars within the galaxy. As illustrated in **Figure 20.9**, the chemical composition of a star's atmosphere reflects the cumulative amount of star formation that occurred before that star formed. Although the details are complex, several clear and important lessons can be learned from the observed patterns in the amounts of massive elements in the galaxy.

Stars in globular clusters, being among the earliest stars to form, should contain only very small amounts of massive elements. Some globular-cluster stars contain only 0.5 percent as much of those massive elements as the Sun. That relationship between age and abundances of massive elements is evident throughout much of the galaxy. Within the disk, younger stars typically have higher abundances of massive elements than those of older stars. Similarly, older stars in the outer parts of the galaxy's bulge have lower massive-element abundances than those of young stars in the disk. All the stars in the galaxy's halo have smaller amounts of heavy elements. That includes globular-cluster stars, which constitute only a minority of stars in the galactic halo.

Globular clusters orbit in the halo of the Milky Way; open clusters orbit in the disk of the Milky Way. As with globular clusters, the stars in an open cluster all formed at about the same time, so astronomers can use the H-R diagram to find the cluster's age. Some open clusters are very young, containing the very youngest stars known, whereas other open clusters contain stars older than the Sun; but all open clusters are younger than the youngest globular cluster. Because the globular clusters are old and located in the halo, astronomers infer that stars in the halo formed before stars in the disk. That epoch of halo star formation did

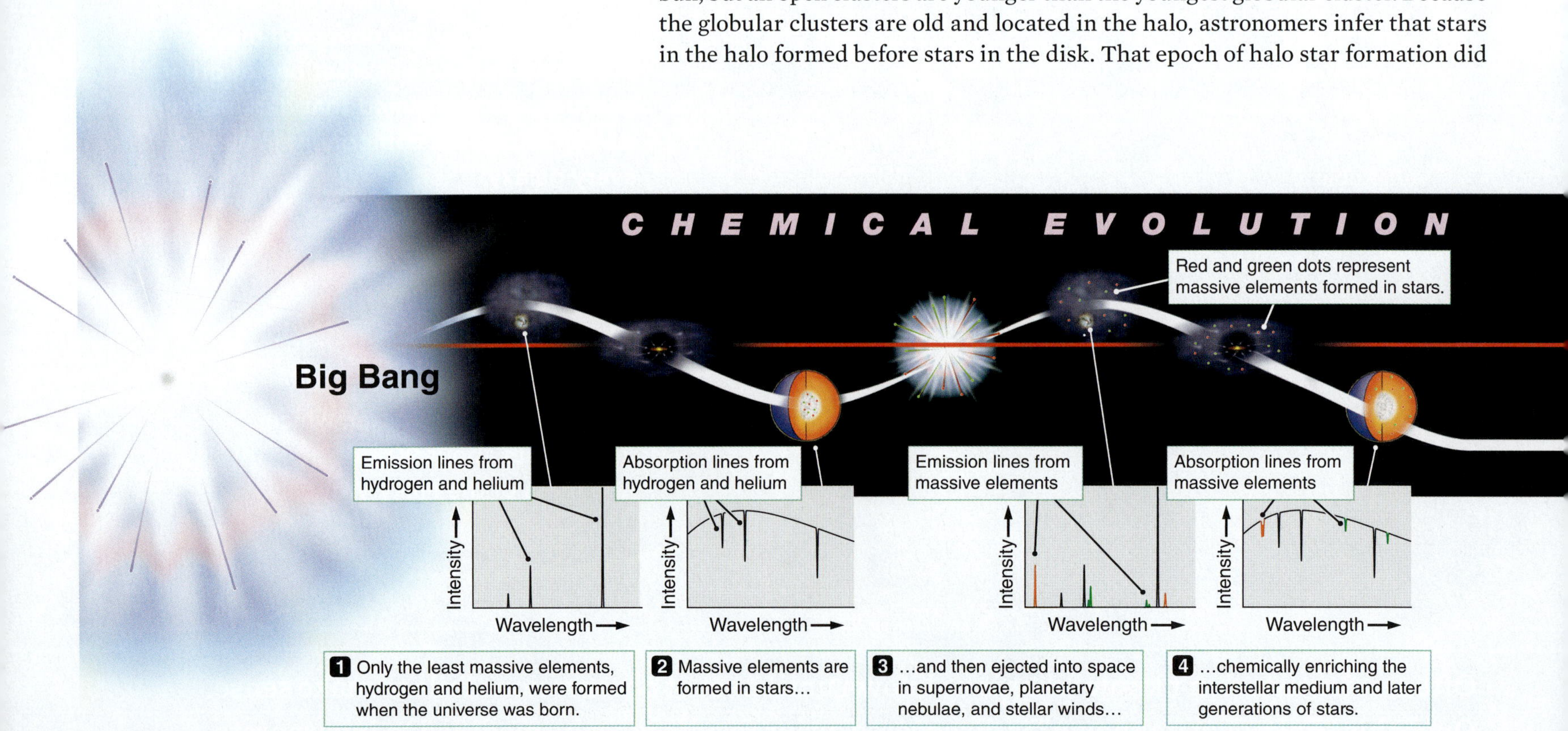

not last long. Star formation in the disk started later and has been continuing ever since.

Within the galaxy's disk, astronomers observe differences in abundances of massive elements from place to place, related to the rate of star formation in different regions. Star formation is generally more active in the denser, inner part of the Milky Way than in the outer parts. Observations of chemical abundances in the interstellar medium, based both on interstellar absorption lines in the spectra of stars and on emission lines in glowing clouds of neutral hydrogen gas known as H II regions, confirm that prediction by showing a smooth decline in abundances of massive elements from the inner to the outer parts of the disk. Astronomers have observed similar trends in other galaxies. Within a galactic disk, relatively old stars near the center of a galaxy often have greater massive-element abundances than those of young stars in the outer parts of the disk.

The basic idea that higher massive-element abundances should follow from the more prodigious star formation in the inner galaxy seems correct, but the full picture is not that simple. New material falling into the galaxy might affect the amounts of heavy elements in the interstellar medium. Chemical elements produced in the inner disk might be blasted into the halo in great "fountains" powered by the energy of massive stars, only to fall back onto the disk elsewhere. Past interactions with other galaxies might have stirred the Milky Way's interstellar medium, mixing gas from those other galaxies with gas already there. The variations of chemical abundances within the Milky Way and other galaxies—and what those variations tell us about the history of star formation and the formation of elements—remain active topics of research.

Even the very oldest globular-cluster stars contain some chemical elements fused in previous generations of more massive stars. That observation implies that globular-cluster stars and other halo stars were not the first stars in the Milky Way to form. At least one generation of massive stars lived and died, ejecting newly synthesized massive elements into space, before even the oldest globular clusters

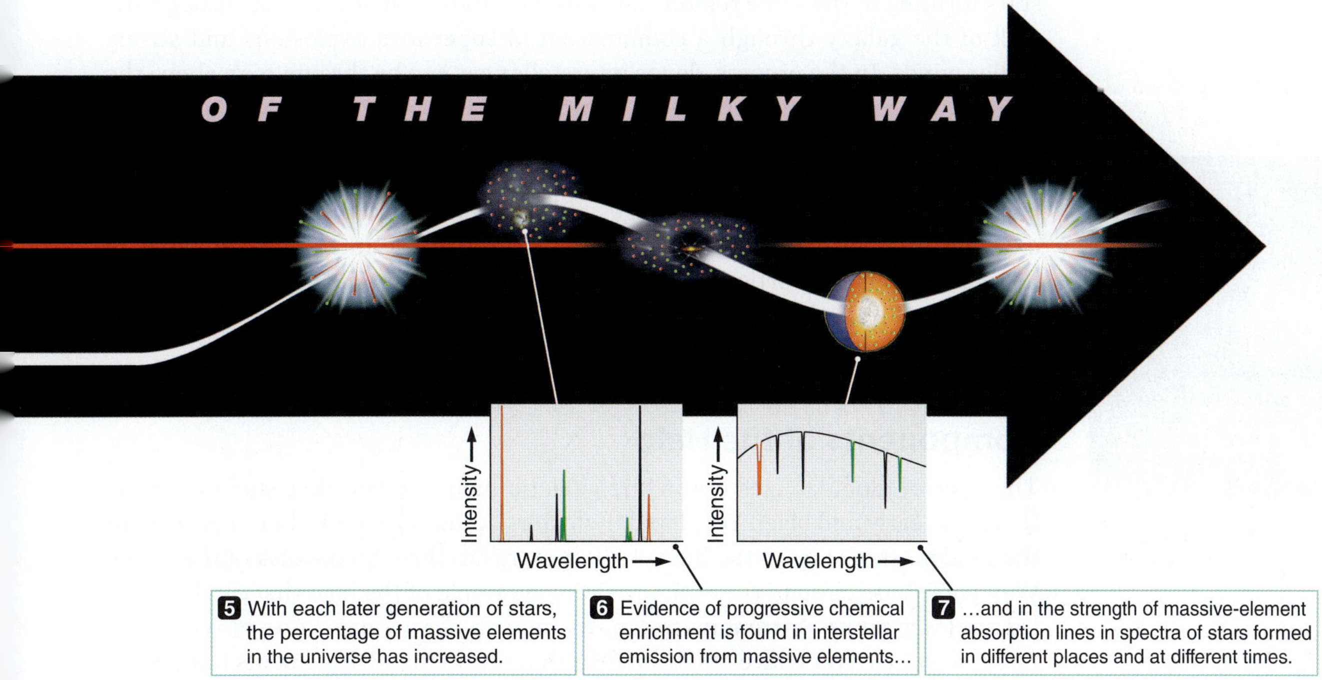

(a)

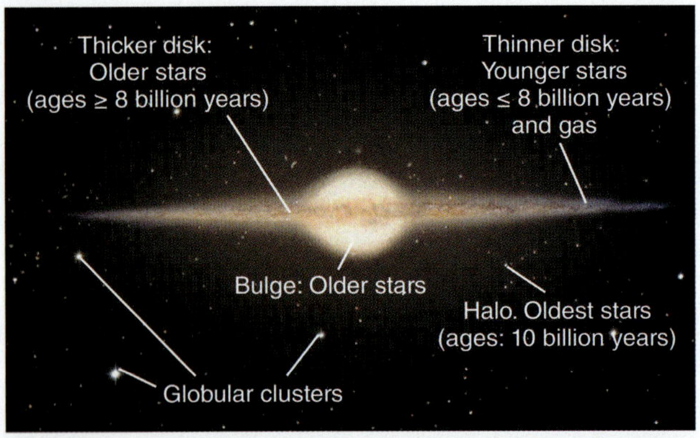

(b)

Figure 20.10 (a) This illustration shows the disks, bulge, and inner halo of the Milky Way and the location of globular clusters. (b) An all-sky view of the interstellar dust that fills the Milky Way Galaxy, from observations by ESA's *Gaia* satellite.

formed. (We return to these first stars in Chapter 23.) Every star less massive than about 0.8 M_{Sun} that ever formed is still around as a main-sequence star today. Even so, astronomers find no disk stars with exceptionally low massive-element abundances. The gas that wound up in the plane of the Milky Way must have seen a significant amount of star formation before it settled into the disk of the galaxy and made stars. Still, even a chemically "rich" star like the Sun, which is made of gas processed through approximately 9 billion years of previous generations of stars, is composed of less than 2 percent massive elements. Luminous matter in the universe is still dominated by hydrogen and helium formed just after the Big Bang, long before the first stars.

Components of the Disk

The disk of the Milky Way has a thinner component and a thicker component (**Figure 20.10a**). The youngest stars in the galaxy are most strongly concentrated in the galactic plane, defining a thinner disk about 300 pc (1,000 light-years) thick but more than 30,000 pc (100,000 light-years) across. That ratio of the diameter to the thickness of the disk is similar to that of a DVD. The older population of disk stars, distinguishable by lower abundances of massive elements, occupy the thicker disk, about 3,700 pc (12,000 light-years) thick. The youngest stars are concentrated closest to the plane of the galaxy because that is where the molecular clouds and gas are. Those stars show a decrease in heavy-element content as the distance from the galactic center increases. Older stars make up the thicker parts of the disk, with the oldest stars closer to the galactic center. Astronomers are still debating how distinct those two disks are and how they originated. Dust is also concentrated in the plane of the galaxy, as seen in a recent image from the *Gaia* mission (**Figure 20.10b**).

The interstellar medium is a dynamic place—energy from star-forming regions can shape it into impressively large structures. As mentioned, energy from regions of star formation can form interesting structures in the interstellar medium, clearing out large regions of gas in the disk of a galaxy. Many massive stars forming in the same region can blow "fountains" of hot gas out through the disk of the galaxy through a combination of supernova explosions and strong stellar winds. In the process, dense interstellar gas can be thrown high above the plane of the galaxy (see the artist's depiction in **Figure 20.11**). Once the gas is a few kiloparsecs above or below the disk, it radiates and cools, falling back to the disk. Maps of the 21-cm emission from neutral hydrogen in the galaxy, X-ray observations, and visible-light images of hydrogen emission from some edge-on external galaxies show many vertical structures in the interstellar medium of disk galaxies. Those vertical structures are often interpreted as the "walls" of the fountains. If enough massive stars are formed together, enough energy may be deposited to blast holes all the way through the plane of the galaxy.

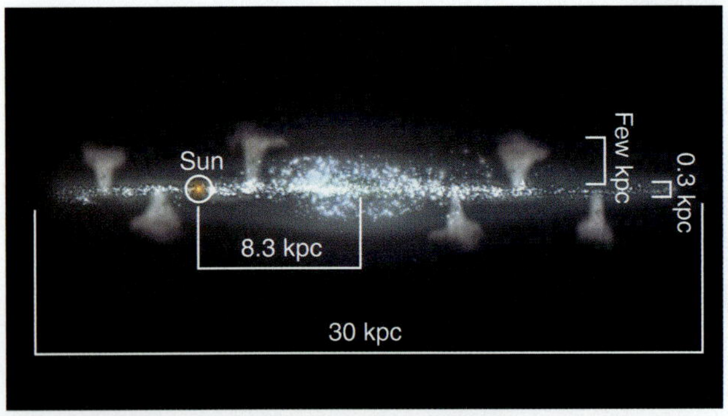

Figure 20.11 Artist's concept of a "galactic fountain," in which gas is pushed away from the plane of the galaxy by energy released by young stars and supernovae and then falls back onto the disk. The distance of the Sun from the center of the galaxy and the diameter of the disk are shown here for scale.

Components of the Halo

The ages of globular clusters in the galactic halo indicate that star formation there was early and brief. Yet globular clusters account for only about 1 percent of the total mass of stars in the halo. As halo stars fall through the disk of the Milky Way, some pass close to the Sun, providing a sample of the halo that can be studied at closer range. Astronomers can distinguish nearby halo stars in two ways. First, most halo stars have much smaller amounts of massive elements than those of disk stars. Second, halo stars appear to be moving at higher relative velocities than disk stars. Most stars near the Sun are disk stars, so they orbit the center of the galaxy at nearly the same speed, in roughly the same direction as the Sun. In contrast, halo stars orbit the center of the galaxy in random directions, so the relative velocity between the halo stars and the Sun tends to be high. Those stars are known as high-velocity stars.

By studying the orbits of high-velocity stars, astronomers have determined that the halo has two separate components: an inner halo that includes stars up to about 15 kpc (50,000 light-years) from the center and an outer halo that extends far beyond that (see Figure 20.7a). The stars in the outer halo have lower fractions of heavier elements, suggesting that those stars formed very early. Many are moving in a direction opposite to the rotation of the galaxy, suggesting that the outer halo may have its origins in a merger with a small dwarf galaxy long ago. The orbits of halo stars suggest that those stars fill a volume of space similar to that occupied by the globular clusters in the halo.

X-ray observations indicate that a halo of hot gas surrounds the Milky Way (see the artist's concept in **Figure 20.12**). That gas halo may extend for about 100–200 kpc from the galactic center, encompassing two nearby small galaxies and containing as much mass as that of all the stars in our galaxy. The halo's temperature is estimated to be about 2 million kelvins (K), so the gas particles are ionized and moving very quickly. The gas is extremely diffuse, however, so the particles are not colliding with one another and transferring energy. The gas wouldn't "feel" hot, much like the solar corona (see Chapter 14).

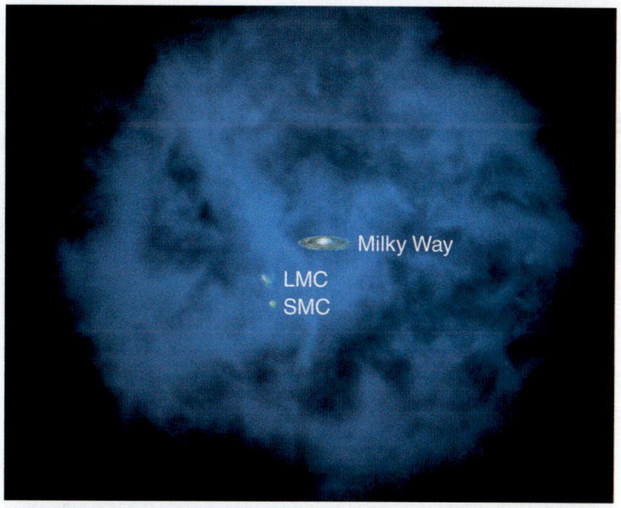

Figure 20.12 This artist's concept shows the Milky Way surrounded by a halo of hot gas, which may contain as much mass as that of all the stars in the galaxy combined. The Large Magellanic Cloud (LMC) and Small Magellanic Cloud (SMC) are two of the many nearby dwarf galaxies.

Magnetic Fields and Cosmic Rays Fill the Galaxy

The interstellar medium of the Milky Way is laced with magnetic fields that are wound up and compressed by the rotation of the galaxy's disk. The total interstellar magnetic field, however, is about a hundred thousand times weaker than Earth's magnetic field. Charged particles and magnetic fields interact strongly; the particles spiral around magnetic fields, moving along the field rather than across it. Conversely, magnetic fields cannot freely escape from a cloud of gas containing even a small number of charged particles. The dense clouds of interstellar gas in the midplane of the Milky Way (**Figure 20.13**) anchor the galaxy's magnetic field to the disk, in turn anchoring high-energy charged particles, known as cosmic rays, to the galaxy.

Cosmic rays are charged particles that originate mainly outside the Solar System and travel close to the speed of light. Despite their name, cosmic rays are not a form of electromagnetic radiation: they were named before their true nature was known. Most cosmic-ray particles are protons, but some are nuclei of helium, carbon, and other elements produced by nucleosynthesis. A few are high-energy electrons and other subatomic particles.

Cosmic rays span an enormous range in particle energy. Astronomers can observe the lowest-energy cosmic rays with interplanetary spacecraft. Those cosmic rays have energies as low as about 10^{-11} joule (J), which corresponds to the energy of a proton moving at a velocity of 1/3 the speed of light. In contrast, the most energetic cosmic rays are 10 trillion (10^{13}) times as energetic as the lowest-energy cosmic rays, and they move very close to the speed of light, 0.999999*c*. Those high-energy cosmic rays are constantly hitting Earth and are detected from the showers of elementary particles that they cause when crashing through Earth's atmosphere. Those

Figure 20.13 The mass of interstellar clouds anchors the magnetic field of the Milky Way to the disk of the galaxy. The galaxy's magnetic field in turn traps cosmic rays.

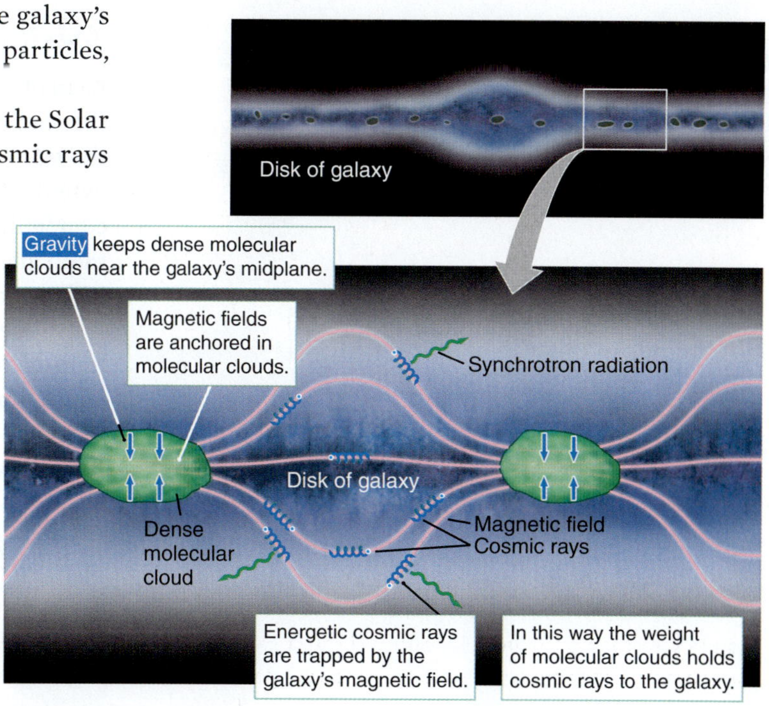

(a)

(b)

Figure 20.14 (a) Artist's sketch of the Pierre Auger Observatory in Argentina, an array of stations designed to catch the particles that shower from collisions of cosmic rays with the upper atmosphere. (b) Each station in the array is equipped with its own particle collectors, carefully protected from the elements.

particle showers are observed by special telescopes such as the High Energy Stereoscopic System (HESS) imaging telescopes in Namibia, or the Pierre Auger Observatory in Argentina (**Figure 20.14**).

Astronomers hypothesize that most of those cosmic rays are accelerated to those incredible energies by the shock waves produced in supernova explosions. The very highest-energy cosmic rays are as much as a hundred million times more energetic than any particle ever produced in a particle accelerator on Earth. Those extremely high energies make those cosmic rays much more difficult to explain than those with lower energies.

The disk of the galaxy glows from synchrotron radiation (see Chapter 17) produced by cosmic rays, mostly electrons, spiraling around the galaxy's magnetic field. Such synchrotron emission is seen in the disks of other spiral galaxies as well, indicating that they, too, have magnetic fields and populations of energetic cosmic rays. The very highest-energy cosmic rays are moving much too fast to be confined by the gravitational force of their originating galaxy. Any such cosmic rays that formed in the Milky Way would soon stream away from the galaxy into intergalactic space. Thus, some of the energetic cosmic rays reaching Earth probably originated in energetic events outside the Milky Way Galaxy.

The total energy of all the cosmic rays in the galactic disk can be estimated from the energy of the cosmic rays reaching Earth. The strength of the interstellar magnetic field can be measured by observing how the field affects the properties of radio waves passing through the interstellar medium. Those measurements indicate that in the Milky Way Galaxy, the magnetic-field energy and the cosmic-ray energy are about equal. Both are comparable to the energy present in other energetic components of the galaxy, including the motions of interstellar gas and the total energy of electromagnetic radiation within the galaxy.

CHECK YOUR UNDERSTANDING 20.2

What parts of the Milky Way contain old stars, and what parts contain young stars?

20.3 Most of the Milky Way Is Unseen

As in other galaxies, the most interesting parts of the Milky Way may be the parts that can't be seen directly but are detected only by their gravitational influence on the stars around them. Dark matter accounts for most of the mass in a galaxy and extends far beyond a galaxy's visible boundary. As in all other spiral galaxies, compelling evidence indicates that dark matter dominates the Milky Way. From radio and infrared observations, astronomers can figure out how the disk of the Milky Way moves, and from that motion, in turn, they can determine its mass.

The supermassive black hole at the center of the Milky Way poses a different kind of observation problem. That object also is detected by its gravitational effects on the stars nearby but cannot be seen directly. In this section, we explore what can be inferred about those two components of the Milky Way from how their gravity affects other objects.

Dark Matter in the Milky Way

The rotation of the disk of the Milky Way can be determined from observations of the relative velocities of interstellar hydrogen measured from 21-cm radiation. **Figure 20.15** shows how those velocities vary with viewing direction from the

Sun. Looking toward the center of the galaxy, we see that the gas is stationary relative to the Sun. When we look in the direction of the Sun's motion around the galactic center, hydrogen clouds appear to be moving toward Earth, whereas in the opposite direction, clouds are moving away from Earth. In other directions, the measured velocities are complicated by Earth's moving vantage point within the disk and so are more difficult to interpret at a glance. That is a pattern of the rotation velocity of gas in a disk like those you learned about in Chapter 19. The only difference is that instead of looking at it from outside, we see the Milky Way Galaxy's rotation curve from a vantage point located within—and rotating with—the galaxy. Even so, observed velocities of neutral hydrogen enable astronomers to measure the Milky Way Galaxy's rotation curve and even determine the structure present throughout its disk.

Recall (Chapter 19) that observations of rotation curves led astronomers to conclude that the masses of spiral galaxies consist mostly of dark matter. **Figure 20.16** shows the rotation curve of the Milky Way as inferred primarily from 21-cm observations. The orbital motion of the nearby dwarf galaxy called the Large Magellanic Cloud provides data for the outermost point in the rotation curve, at a distance of roughly 50,000 pc (160,000 light-years) from the center of the galaxy. Like other spiral galaxies, the Milky Way has a fairly flat rotation curve, which indicates the presence of large amounts of dark matter. The mass outside the Sun's orbit does not greatly affect the Sun's orbit.

The total mass of the Milky Way Galaxy is currently estimated to be about 1.0 trillion to 1.5 trillion times the mass of the Sun. The luminous mass, however, estimated by adding the masses of stars, dust, and gas, is only about one-tenth as much. Like other spiral galaxies, the Milky Way's mass consists mainly of dark

(a)

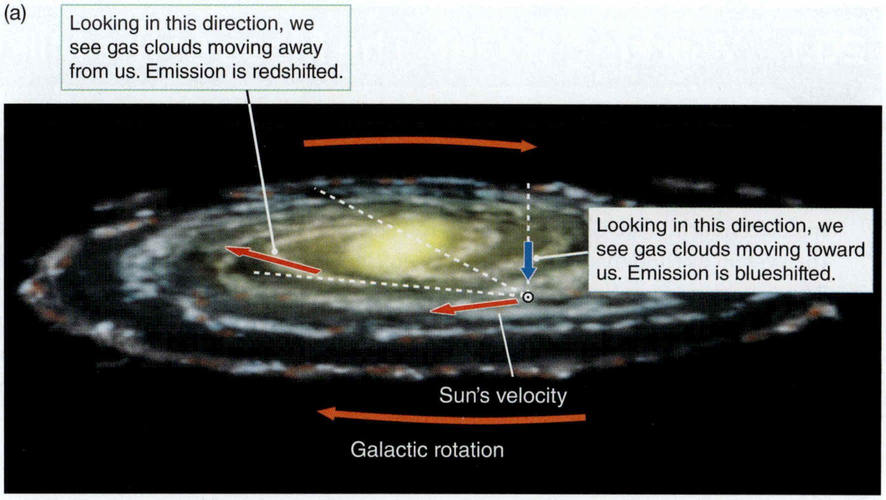

(b)

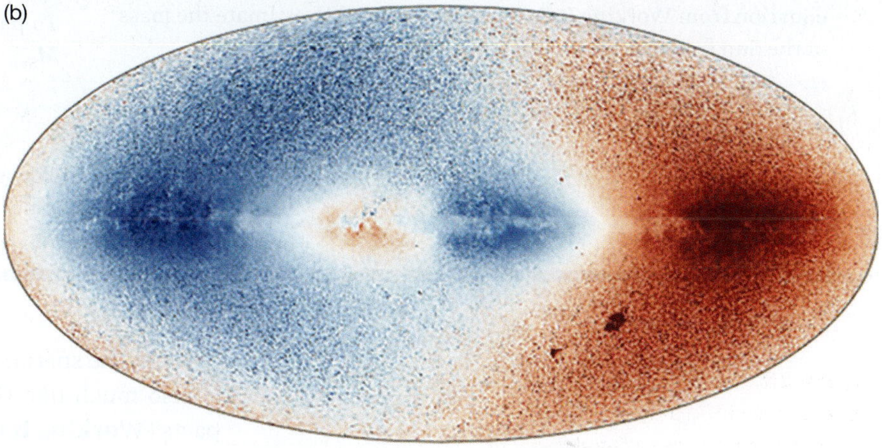

Figure 20.15 (a) From Earth's perspective within the Solar System, the signature of a rotating disk is clear from views of either side of the galactic center. (b) Those velocities vary between redshift and blueshift as an observer looks around in the plane of the Milky Way.

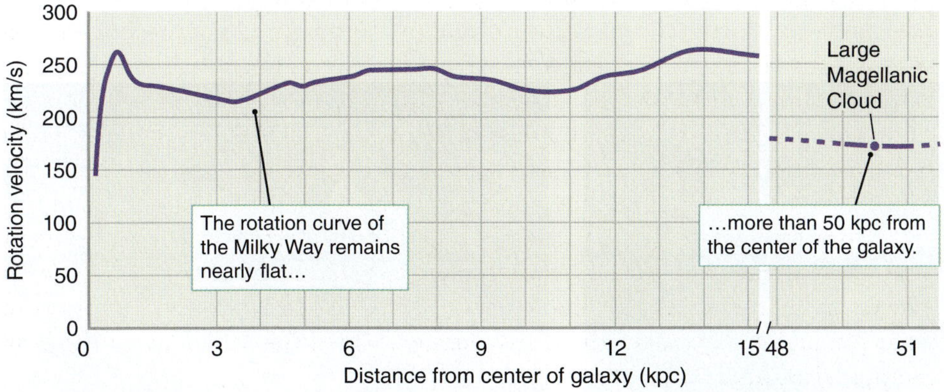

Figure 20.16 Rotation velocity is plotted against distance from the center of the Milky Way. The most distant point comes from measurements of the orbit of the Large Magellanic Cloud. The nearly flat rotation curve indicates that dark matter dominates the outer parts of the Milky Way. Part of this graph has been hidden to fit the Large Magellanic Cloud into the figure. The broken *x*-axis, marked by the diagonal hatch marks, indicates the hidden portion.

20.1 Working It Out The Mass of the Milky Way inside the Sun's Orbit

The Sun orbits about the center of the Milky Way Galaxy. Even though the gravitational pull on the Sun comes from all the material inside that orbit (see Chapter 4), Newton showed that we could treat the system as though all the mass were concentrated at the center. Thus, we can apply Newton's and Kepler's laws to calculate the mass of the Milky Way inside the Sun's orbit. Newton's version of Kepler's third law relates the period of the orbit to the orbital radius and the masses of the objects. But here, the mass of the galaxy is much larger than the mass of the Sun, so the Sun's mass is negligible by comparison. In addition, what astronomers can *measure* is the orbital speed of the Sun or other stars about the galactic center, rather than the orbital period. Thus, we can use a rearranged form of the same equation from Working It Out 4.2 that we used to estimate the mass of the Sun from the orbit of Earth:

$$M = \frac{rv^2_{\text{circ}}}{G}$$

The Sun orbits the center of the galaxy at 240 kilometers per second (km/s). The distance of the Sun from the center of the galaxy is 8,300 pc, which converts to kilometers as follows:

$$8{,}300 \text{ pc} \times (3.09 \times 10^{13} \text{ km/pc}) = 2.56 \times 10^{17} \text{ km}$$

Because we know that the gravitational constant, $G = 6.67 \times 10^{-20}$ km^3/(kg s^2), we can calculate the mass of the portion of the Milky Way inside the Sun's orbit:

$$M = \frac{(2.56 \times 10^{17} \text{ km}) \times (240 \text{ km/s})^2}{6.67 \times 10^{-20} \text{ km}^3/(\text{kg s}^2)}$$

$$M = 2.21 \times 10^{41} \text{ kg}$$

To put that result in units of the Sun's mass, we divide the answer by $M_{\text{Sun}} = 1.99 \times 10^{30}$ kg, yielding

$$M = \frac{2.21 \times 10^{41} \text{ kg}}{1.99 \times 10^{30} \text{ kg}/M_{\text{Sun}}} = 1.11 \times 10^{11} M_{\text{Sun}}$$

The mass of the Milky Way inside the Sun's orbit is about 111 billion times the mass of the Sun.

matter. The spatial distribution of dark and normal matter within the Milky Way is also much like that of other galaxies, with dark matter dominating its outer parts (**Working it Out 20.1**).

The Supermassive Black Hole

Figure 20.17 shows images of the Milky Way's center taken with the Chandra X-ray Observatory and the Spitzer Space Telescope. The X-ray view (Figure 20.17a) shows the location of a strong radio source called Sagittarius A* (abbreviated

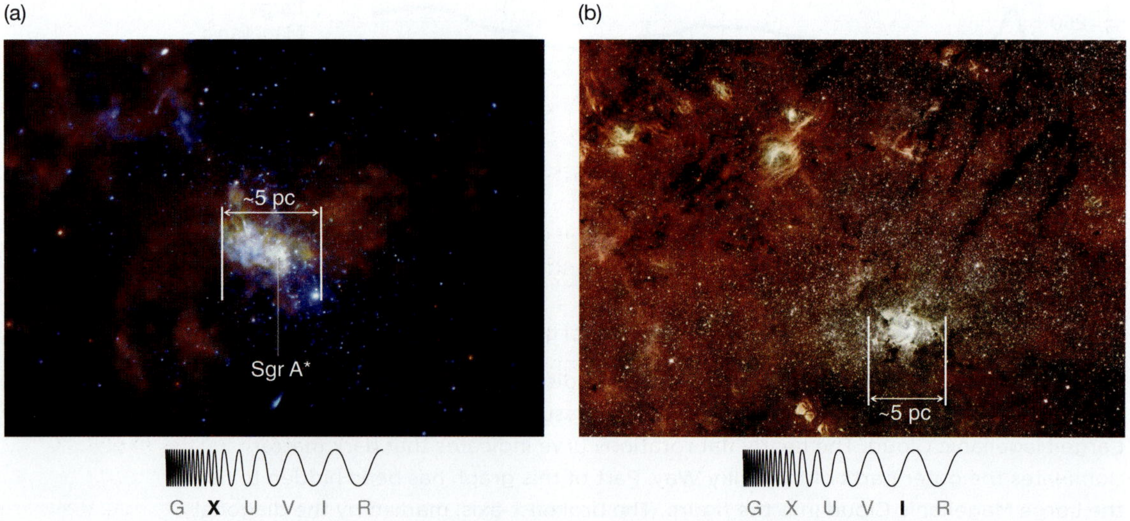

Figure 20.17 (a) This X-ray view of the Milky Way's central region shows the active source, Sagittarius A* (Sgr A*), as the brightest spot at the middle of the image. Lobes of superheated gas (shown in red) are evidence of recent, violent explosions happening near Sgr A*. (b) This infrared view of the central core of the Milky Way shows hundreds of thousands of stars. The bright white spot at the lower right marks Sgr A*, the location of the supermassive black hole.

Sgr A*), which lies at the center of the Milky Way. The infrared image (Figure 20.17b) cuts through the dust to reveal the galaxy's crowded, dense core containing hundreds of thousands of stars.

Studies of the motions of stars closest to the Sgr A* source suggest a central mass very much greater than that of the few hundred stars orbiting there. Furthermore, observations of the galaxy's rotation curve show rapid rotation velocities very close to the galactic center. Stars closer than 0.1 light-year from the galactic center follow Kepler's laws, indicating that their motion is dominated by mass within their orbit. The closest stars studied are only about 0.01 light-year from the center of the galaxy—so close that their orbital periods are only about a dozen years. The positions of those stars change noticeably over time, and astronomers can see them speed up as they whip around what can only be a supermassive black hole at the focus of their elliptical orbits (**Figure 20.18**). Using Newton's version of Kepler's third law, we can then estimate that the black hole at the center of the Milky Way Galaxy is a relative lightweight, having a mass of "only" 4 million times the mass of the Sun (**Working It Out 20.2**).

Clouds of interstellar gas at the galaxy's center are heated to millions of degrees by shock waves from supernova explosions and colliding stellar winds blown outward by young, massive stars. Superheated gas produces X-rays, and the Chandra X-ray Observatory has detected more than 9,000 X-ray sources within the central region of the galaxy. Those sources include frequent, short-lived X-ray flares near Sgr A* (see Figure 20.17a), which provide direct evidence that matter falling toward the supermassive black hole fuels the energetic activity at the galaxy's center.

The Fermi Gamma-ray Space Telescope has observed gamma-ray-emitting bubbles that extend 8 kpc (25,000 light-years) above and below the galactic plane. The bubbles may have formed after a burst of star formation a few million years ago produced massive star clusters near the center of the galaxy. If some of the gas formed stars and about 2,000 M_{Sun} of material fell into the supermassive black hole, enough energy could have been released to power the bubbles. More recently, faint gamma-ray signals were observed that look like jets coming from the center,

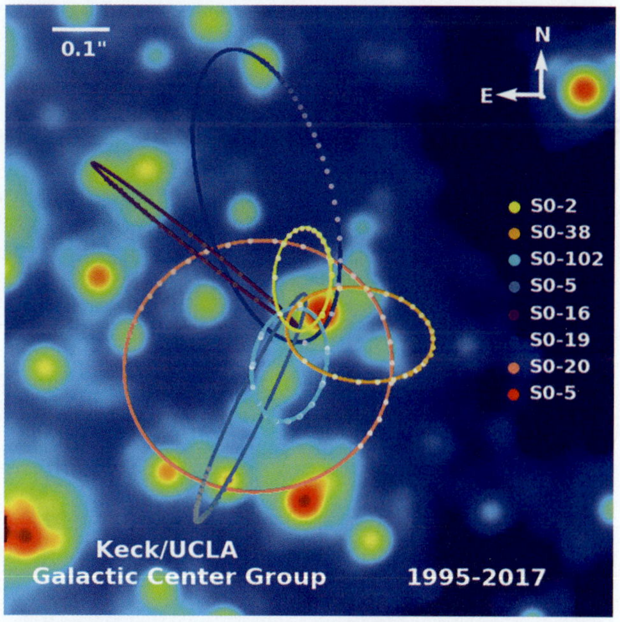

Figure 20.18 The orbits of eight stars within 0.03 pc (0.1 light-year, or about 6,000 astronomical units [AU]) of the Milky Way's center. The Keplerian motions of those stars reveal the presence of a 4-million-M_{Sun} supermassive black hole at the galaxy's center. Colored dots show the measured positions of each star over many years: the dots progress from lighter in 1995 to darker in 2016.

20.2 Working It Out The Mass of the Milky Way's Central Black Hole

Figure 20.18 illustrates data points for the stars in the central region orbiting closely to the central black hole of the Milky Way Galaxy. Those stars have highly elliptical orbits with changing speeds, but the orbital periods are short enough that they can be observed and measured. Star S0-2 in the figure has a measured orbital period of 15.8 years. The semimajor axis of its orbit is estimated to be 1.5×10^{11} km = 1,000 AU. With that information, we can use Newton's version of Kepler's third law to estimate the mass inside S0-2's orbit. Setting up the equation as we did in Working It Out 13.4:

$$\frac{m_{BH}}{M_{Sun}} + \frac{m_{S0\text{-}2}}{M_{Sun}} = \frac{A_{AU}^3}{P_{years}^2}$$

The mass of star S0-2 is much less than the mass of the black hole, so the sum of the two is very close to the mass of the black hole. Therefore, we can write

$$\frac{m_{BH}}{M_{Sun}} = \frac{A_{AU}^3}{P_{years}^2} = \frac{1{,}000^3}{15.8^2} = 4.0 \times 10^6$$

$$m_{BH} = 4.0 \times 10^6\, M_{Sun}$$

The supermassive black hole at the center of the Milky Way has a mass 4 million times that of the Sun. That value is considerably less than the billion-solar-mass black holes in some active galactic nuclei discussed in Chapter 19.

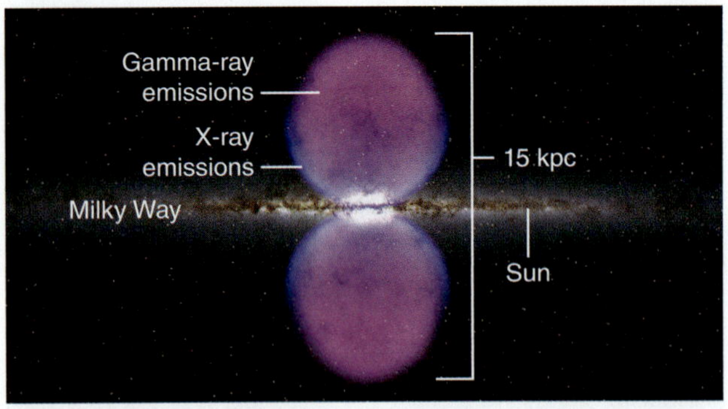

Figure 20.19 The Fermi Gamma-ray Space Telescope observed gamma-ray bubbles (purple) extending 8 kpc above and below the galactic plane. Hints of the edges of the bubbles were first observed in X-rays (blue) in the 1990s. In this artist's conceptual view from outside the galaxy, the gamma-ray jets are in magenta.

within the bubbles (see the artist's depiction in **Figure 20.19**). Using the Hubble Space Telescope, astronomers have estimated that those jets originated from material falling into the supermassive black hole about 6 million to 9 million years ago. Some astronomers predict that gas clouds are heading toward the center and will soon be accreted by the black hole. Currently, the observed activity is not as intense as that seen in active galactic nuclei with central, supermassive black holes. The inner Milky Way is a reminder that it was almost certainly "active" in the past and could become active once again.

CHECK YOUR UNDERSTANDING 20.3

Which property is detectable for both dark matter and the supermassive black hole at the center of the Milky Way? (a) luminosity; (b) temperature; (c) gravity; (d) composition

20.4 The History and Future of the Milky Way

A fundamental goal of stellar astronomy is to understand the life cycle of stars, including how they form from clouds of interstellar gas. In Chapter 15, we described that process, at least as it occurs today, and tied that story strongly to observations of Earth's galactic neighborhood. Galactic astronomy has a similar basic goal. Astronomers would like to have a complete and well-tested theory of how the Milky Way formed and to be able to make predictions about its future. The distribution of stars of different ages with different amounts of heavy elements is one clue. Additional clues come from studying other galaxies at different distances (and therefore of different ages), their supermassive black holes, and their merger history. In this section, we explore the history and the future of the Milky Way.

Figure 20.20 This graphical map shows some of the members of the Local Group of galaxies. Most are dwarf galaxies. Spiral galaxies are shown in yellow. The closest galaxies to the Milky Way (such as the Large Magellanic Cloud and Small Magellanic Cloud) are not seen on this scale. (1 Mpc = 1 megaparsec = 1 million parsecs = 3.26 million light-years.)

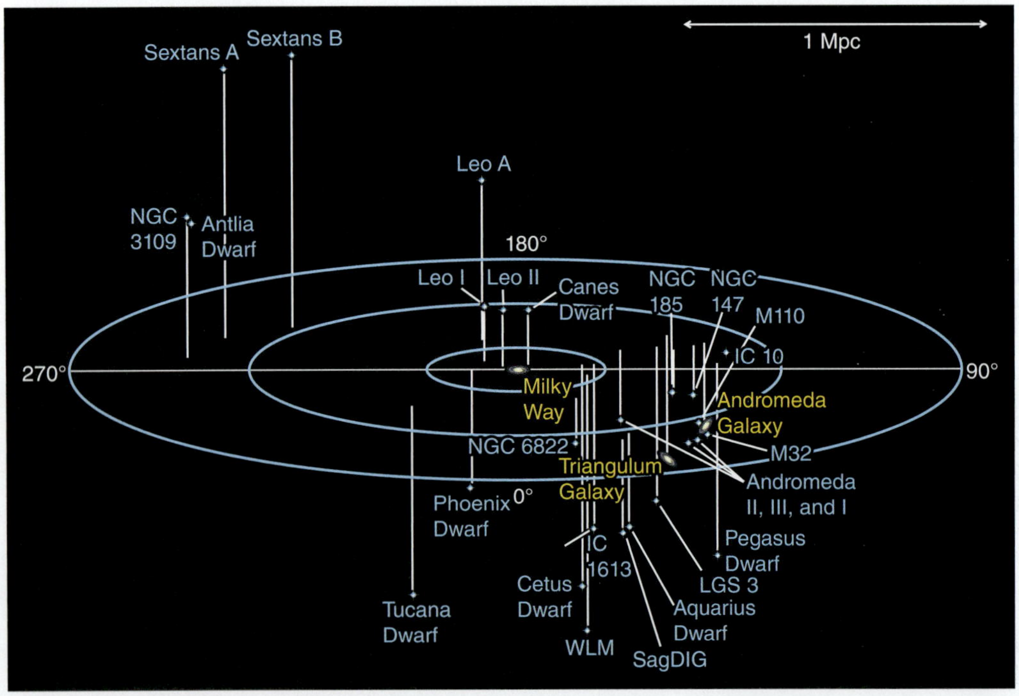

The Local Group

Galaxies do not exist in isolation. Most galaxies are parts of gravitationally bound collections of galaxies, the smallest and most common of which are called **galaxy groups**. A galaxy group contains as many as several dozen galaxies, most of them dwarf galaxies. The Milky Way is a member of the **Local Group** (see Chapter 1), first identified by Edwin Hubble in 1936. Hubble labeled 12 galaxies as part of the Local Group, but now astronomers count at least 50. The Local Group (**Figure 20.20**) includes the two giant barred spirals—the Milky Way Galaxy and the Andromeda Galaxy—along with a few ellipticals and irregulars and at least 30 smaller dwarf galaxies in a volume of space about 3 million pc (10 million light-years) in diameter. Almost 98 percent of all the galaxy mass in the Local Group resides in just those two giant galaxies. The third-largest galaxy, Triangulum, is an unbarred spiral with a few percent the mass of the Milky Way or Andromeda. Most, but not all, of the dwarf elliptical and dwarf spheroidal galaxies in the group are satellites of the Milky Way or Andromeda. The Local Group interacts with a few nearby groups, which is discussed further in Chapter 23.

More than 50 of those satellite dwarf galaxies exist, although it's not certain that all are gravitationally bound to the Milky Way. Some of the fainter dwarf galaxies were discovered only very recently because of their low luminosity. The dwarf galaxies are the lowest-mass galaxies observed, and they are dominated by an even greater percentage of invisible dark matter than are other known galaxies. They also contain stars very low in elements more massive than helium. Those old ultrafaint dwarf galaxies offer clues to the formation of the Local Group. In addition, observations of the motions and speeds of the dwarf galaxies about the Milky Way may lead to new estimates of the dark matter mass within the Milky Way itself.

Figure 20.21 The Milky Way is surrounded by more than 20 dwarf companion galaxies: the largest among them are the Large Magellanic Cloud (LMC) and Small Magellanic Cloud (SMC) (see Figure 20.12).

The Formation of the Milky Way

We have seen that globular clusters and high-velocity stars must have been among the first stars formed in the Milky Way that still exist. The fact that they are not concentrated in the disk or bulge of the galaxy indicates that they formed from clouds of gas well before those clouds settled into the galaxy's disk. That hypothesis is supported by observations that globular clusters are very old and that the youngest globular cluster is older than the oldest disk stars. The presence of extremely small amounts of massive elements in the atmospheres of halo stars also indicates that at least one generation of stars must have lived and died *before* the formation of the halo stars visible today. That line of reasoning implies that the Milky Way formed from the merger of several smaller clumps of matter, which included both stars and clouds of dust and gas.

Combining that conclusion with the presence of the central, supermassive black hole and the number of nearby dwarf galaxies, astronomers conclude that the Milky Way must have formed when the gas within a huge "clump" of dark matter collapsed into many small protogalaxies. Some of those protogalaxies then merged to form the large, barred spiral galaxies in the Local Group, but some of those smaller protogalaxies are still around today in the form of the small, satellite dwarf galaxies near the Milky Way. The largest among them are the Large Magellanic Cloud and the Small Magellanic Cloud (**Figure 20.21**), which are easily seen by the naked eye in the Southern Hemisphere. The Magellanic Clouds were named for Ferdinand Magellan (1480–1521), who headed a European expedition that ventured far enough into the Southern Hemisphere to see them.

Figure 20.22 This artist's impression shows tidal tails of stars from the Sagittarius Dwarf elliptical galaxy (reddish-orange). Those stars have been stripped from the dwarf galaxy by the much more massive Milky Way, and in billions of years the two galaxies will eventually merge.

The Future of the Milky Way

Mergers and collisions of Local Group galaxies continue today. Among the closest companions to the Milky Way is the Sagittarius Dwarf Galaxy, which is plowing through the disk of the Milky Way on the other side of the bulge. Astronomers have observed streams of stars, as sketched in **Figure 20.22**, from Sagittarius Dwarf and some of the other dwarf galaxies that are being tidally disrupted by the Milky Way. Those dwarf galaxies will be incorporated into the Milky Way—an indication that the galaxy is still growing. Computer simulations suggest that the spiral-arm structure could be the result of such mergers.

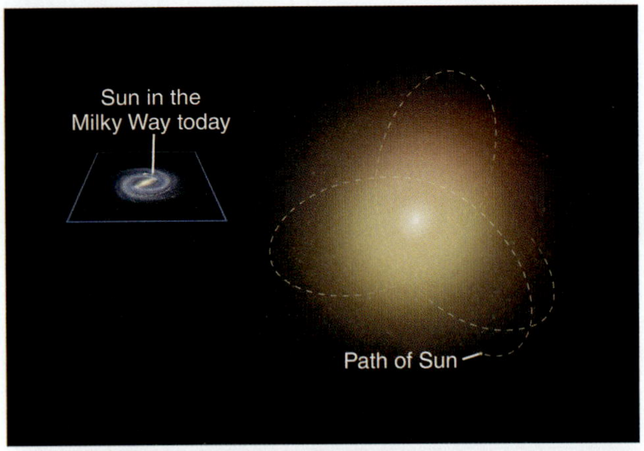

Figure 20.23 A computer simulation of the new orbit of the Sun within the Milky Way-Andromeda merger remnant 10 billion years from now.

The Andromeda Galaxy appears to violate Hubble's law because its spectrum shows blueshifts, indicating that the galaxy is moving toward, not away from, the Milky Way at 110 km/s (400,000 km/h). Andromeda and the Milky Way are the two largest galaxies in the Local Group, about 770 kpc (2.5 million light-years) apart. (If each galaxy were the size of a quarter, they would be about an arm's length apart.) They are so massive that their mutual gravitational attraction is strong and dominates over cosmological expansion, so they are moving toward each other and we see blueshifts.

Astronomers used the Hubble Space Telescope to measure the motions of Andromeda and determine whether a collision will be head-on, partial, or a total miss. They concluded that the two galaxies will "collide" head-on in about 4 billion years. Most of a galaxy is space between the stars, however, so actual collisions between the stars themselves are unlikely: although diffuse gas in the interstellar medium will collide and cause hot shocks, most of the material of one galaxy will pass through the other. The two galaxies will take another 2 billion years to merge completely and form one giant elliptical galaxy. The third spiral in the Local Group, the Triangulum Galaxy, may merge first with the Milky Way, or with Andromeda, or be ejected from the Local Group.

Note the timing here. This first close encounter with Andromeda in 4 billion years will occur *before* the Sun runs out of hydrogen in its core, although the Sun will probably have increased its luminosity enough by then that Earth's habitability will have been affected. In 6 billion years, the Sun and its planets could end up near the center of the merged galaxy, or more probably they will have a new location farther from the center with a different orbit and in a different stellar neighborhood (**Figure 20.23**). If by chance the encounter leads to a star passing close to the Sun, the orbits of the Solar System planets could be disrupted, causing them to be at different distances from the Sun. **Figure 20.24** shows how that merged galaxy might look like in Earth's sky—if anyone is still around to see it!

CHECK YOUR UNDERSTANDING 20.4

Why is Andromeda now moving toward us?

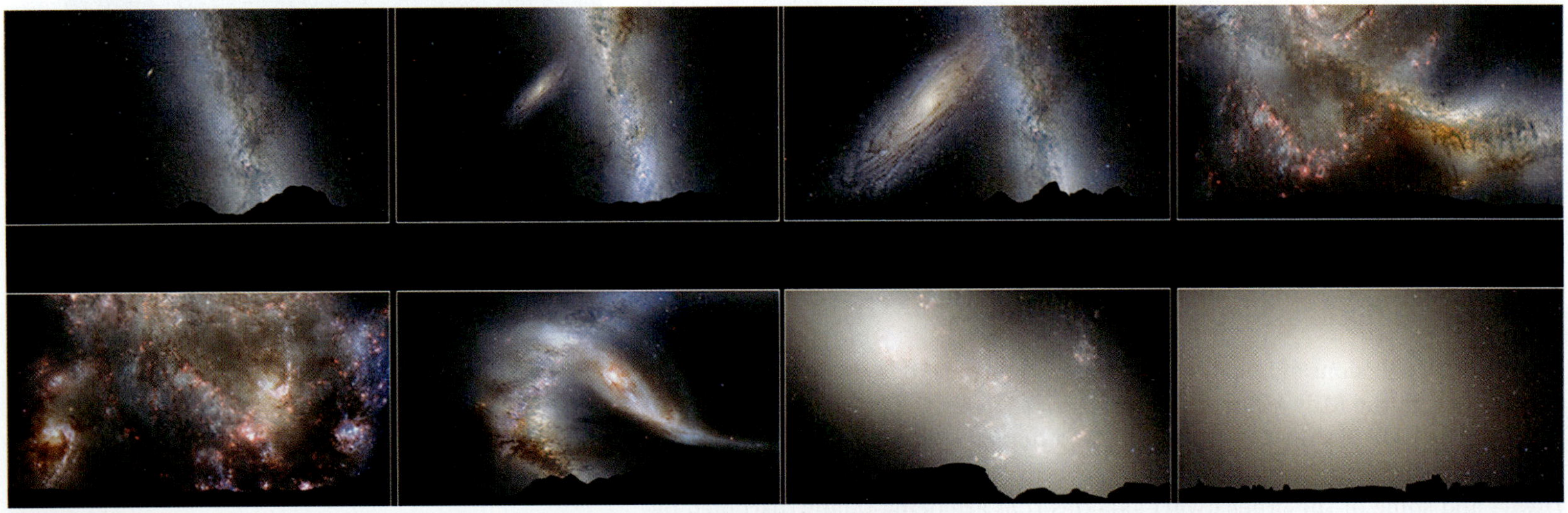

Figure 20.24 This computer simulation depicts the view in the sky on Earth when the Andromeda and Milky Way galaxies collide. The spiral Andromeda appears larger in the sky as it gets closer, and then the sky becomes brighter as the two collide.

Origins The Galactic Habitable Zone

In Chapter 19, we discussed the concept of galactic habitable zones in general; in this Origins, we focus more specifically on ideas about the habitable zone of the Milky Way. Stars situated too far from the galactic center may have protoplanetary disks with insufficient quantities of heavy elements—such as oxygen, silicon, iron, and nickel (to make up rocky planets like Earth), or carbon, nitrogen, and oxygen (to make up the molecules of life). Stars too close to the galactic center may have planets too strongly affected by its high-energy radiation environment (X-rays and gamma rays from the supermassive black hole) and by supernova explosions and gamma-ray bursts. The bulge has a higher density of stars, creating a strong radiation field, and the halo and thick disk have stars with smaller amounts of heavy elements, so perhaps only stars in the thin disk of the galaxy are candidates for residence in a galactic habitable zone.

Astronomers also must compare stellar lifetimes against the 4 billion years after the formation of Earth that life took to evolve into land animals—so only stars with masses low enough that they will live at least 4 billion years on the main sequence are considered potential hosts of more complex life. In that simple model based on the evolution times for life on Earth, the galactic habitable zone could be a doughnut-shaped region around the galactic center. In one version of the model, that zone is estimated to contain stars born 4 billion to 8 billion years ago located 7–9 kpc from the galactic center, with the Sun exactly in the middle of the doughnut. The doughnut would grow larger as heavier-element formation spread outward from the galactic center.

Some researchers have conducted additional studies and proposed more complex models. For example, one group searched for the molecule formaldehyde (H_2CO)—a key "prebiotic" molecule—by observing molecular clouds in the outer parts of the Milky Way, 12–23.5 kpc from the galactic center. Formaldehyde was detected in two-thirds of the group's sample of 69 molecular clouds, suggesting that at least one important prebiotic molecule is available far from the galactic center.

Another computer model took into account details of the evolution of individual stars within the Milky Way, including birth rates, locations, distribution within the galaxy, abundances of heavy elements, stellar masses, main-sequence lifetimes, and the likelihood that stars became or will become supernovae. The model also assumed that the development of complex life would take 4 billion years. One result of that model is that stars in the inner part of the Milky Way are more likely to be affected by supernova explosions, but those stars are even more likely to have the heavier elements for the formation of planets. In that model, then, the inner part of the galaxy, about 2.5–4 kpc from the center, in and near the midplane of the thin disk, is the most likely place for habitable planets. Here, the Sun is *not* in the middle of the most probable zone for habitable planets. As more extrasolar planets are discovered, astronomers will have a better idea of their distribution throughout the Milky Way.

Mergers with other galaxies could cause stars to migrate into or out of the galactic habitable zone. The uncertainties increase as the assumptions in those models move from the astronomical to the biological. For example, maybe life evolved faster on other planets, so stars of higher mass and shorter lifetimes should be included. Or maybe life evolved slower elsewhere, in which case the older stars would be the best candidates. Scientists do not know whether intense radiation from a supernova or a gamma-ray burst would permanently sterilize a planet or only affect evolution for a while. For example, if Earth's ozone layer were temporarily destroyed, life on land might die out, but life in the oceans would continue. Habitability in the Milky Way Galaxy is complicated, with many unanswered questions.

READING ASTRONOMY **NEWS**

ARTICLES QUESTIONS

The European spacecraft known as *Gaia* has unveiled a new view of the Milky Way.

You Are Here: Scientists Unveil Precise Map of More Than a Billion Stars

By Nell Greenfieldboyce, National Public Radio

Wednesday was the day astronomers said goodbye to the old Milky Way they had known and loved and hello to a new view of our home galaxy.

A European Space Agency mission called *Gaia* just released a long-awaited treasure trove of data: precise measurements of 1.7 billion stars (**Figure 20.25**).

It's unprecedented for scientists to know the exact brightness, distances, motions, and colors of more than a billion stars. The information will yield the best three-dimensional map of our galaxy ever.

"This is a very big deal. I've been working on trying to understand the Milky Way and the formation of the Milky Way for a large fraction of my scientific career, and the amount of information this is revealing in some sense is thousands or even hundreds of thousands of times larger than any amount of information we've had previously," said David Hogg, an astrophysicist at New York University and the Flatiron Institute. "We're really talking about an immense change to our knowledge about the Milky Way."

Figure 20.25 *Gaia*'s all-sky view of our Milky Way Galaxy and neighboring galaxies, from measurements of 1.7 billion stars. The map shows the total brightness and color of stars observed by the ESA satellite in each portion of the sky between July 2014 and May 2016.

The *Gaia* spacecraft launched in 2013 and is orbiting our sun, about a million miles away from Earth. Although it has surveyed a huge number of stars, *Gaia* is charting only about 1 percent of what is out there. The Milky Way contains around 100 billion stars.

For 7 million stars, *Gaia* even has measurements showing their velocity as they move toward or away from the spacecraft. "This is just a fantastic new addition," said Anthony Brown of the Netherlands' Leiden University, with the *Gaia* Data Processing and Analysis Consortium.

In 2016, *Gaia* released some initial results, but that was just a tease compared with Wednesday's massive data dump. Astronomers around the world have been planning how to attack it for months, knowing that discoveries are just waiting in there for the taking.

"Enjoy it," Antonella Vallenari, another member of the consortium from the Astronomical Observatory of Padua in Italy, said at a news conference as she announced that the data was becoming available.

The European team presented new images of our galaxy based on some initial analyses of the data, along with little movies that simulated what it would be like to fly through the stars. "It's not fake, it's real measurements," Brown said. "We know exactly where the stars are."

Scientists around the world who were watching the event online struggled to take it all in. "This is like massive scientific progress and we were blinked an image," said Jackie Faherty, an astronomer at the American Museum of Natural History, one of more than a dozen astronomers who woke up before dawn and gathered together at the Flatiron Institute in New York City to watch.

Then it was time to play, as the scientists bent their heads over their laptops and tried to download data. Researchers are expecting a slew of discoveries in the coming hours and days, but analyses will go on for years and even decades.

"This is the data we're going to be working on for the rest of my career. Probably no data set will rival this," Faherty said. "It's the excitement of the day that we see it. It's why we were up at 5 a.m. to get here. It's exciting to be around each other and trying to get the data all at once. It's a day we're going to remember."

ARTICLES QUESTIONS

1. What observations are needed to make a 3-dimensional map of the Milky Way?
2. How do astronomers know which direction stars are moving?
3. Go to the "360° View of Gaia's Sky" at http://sci.esa.int/gaia/60219-360-view-of-gaia-s-sky/. As you scan the image, how do you know when you are looking toward the center of the Milky Way and when you are looking away from it?
4. Do a search for "Gaia Milky Way." What has been discovered from those data?

SUMMARY

Astronomers compare images and spectra of other galaxies with observations of the Milky Way. From that, they determine that the Milky Way is a barred spiral of type SBbc. The Milky Way formed from a collection of smaller protogalaxies that collapsed out of a halo of dark matter. The idea of a galactic habitable zone is that certain parts of the Milky Way may be more suitable for the existence of habitable planets. That zone would have enough heavy elements for the formation of rocky planets and organic molecules, but not so much radiation that it would damage any life.

LG 1 **Explain how astronomers discovered the size and spiral structure of the Milky Way.** The distances to globular clusters can be found from the luminosity of variable stars within them. Because those globular clusters are symmetrically distributed around the center of the Milky Way, the center of the distribution is located at the center of the galaxy. The Sun is located about 8,300 pc (27,000 light-years) from the Milky Way's center, and the Milky Way is 30,000 pc (100,000 light-years) across. Radio observations of neutral hydrogen gas and observations of star-forming nebulae provide evidence for the Milky Way's spiral structure.

LG 2 **List the clues of galaxy formation that can be found from the components of the Milky Way.** The chemical composition of the Milky Way has evolved as material has cycled between stars and the interstellar medium. A generation of stars must have existed before the formation of the oldest halo and globular-cluster stars we see today. The Milky Way has a disk consisting of two parts, the thick disk of old stars and the thin disk of young stars, implying that gas and dust from merging galaxies settled onto the disk while the stars passed through. The galactic halo consists of an inner halo and outer halo of stars and globular clusters, as well as a large, hot gas halo. The abundance of heavy elements in the Milky Way has increased as each generation of stars has produced more of those elements during the final phases of the stars' lives.

LG 3 **Explain the evidence for the dark matter halo and for the supermassive black hole at the center of the Milky Way.** The Doppler velocities of radio spectral lines show that the rotation curve of the Milky Way is flat, like those of other galaxies. But the inferred mass cannot be accounted for by the mass that is observed directly. That finding indicates that the Milky Way's mass is mostly in the form of dark matter. Evidence for the black hole at the center of the galaxy includes rapid orbital velocities of nearby stars and symmetric X-ray and gamma-ray outflows of material.

LG 4 **Describe the Local Group of galaxies and how it offers clues about the evolution of the Milky Way.** The Milky Way is part of the Local Group of galaxies, which consists of two large, barred spirals and several dozen smaller galaxies. Collisions and mergers between those galaxies probably happened in the past, and a merger with the Andromeda Galaxy may be part of the Milky Way's future. The dwarf satellites and other neighbors in the Local Group are evidence that the Milky Way is growing through accretion.

? Unanswered Questions

- Do many more ultrafaint dwarf galaxies exist than have so far been detected? Those types of dwarf galaxies are so faint that they are hard to detect even when close to the Milky Way. As we discuss in Chapter 23, the giant spirals such as the Milky Way were built up by mergers of those small, faint galaxies, so models predict that hundreds or even thousands of them should exist. They could be so dominated by dark matter that they are not at all visible, or too small ever to have formed stars, or they might have merged with the Milky Way or other Local Group members long ago.
- Will the merger of the Milky Way and Andromeda form a quasar at the center of the new elliptical galaxy? If the dust and gas at the center of the galaxy did not block the view from Earth, such a quasar could be brighter than the full Moon. The supermassive black hole at the center of Andromeda is thought to be 25–50 times larger than the one in the Milky Way, but both black holes combined still would be on the lower end of the masses of the black holes in active galactic nuclei discussed in Chapter 19.

Questions and Problems

Test Your Understanding

1. The size of the Milky Way is determined from studying __________ stars in globular clusters.
   a. Cepheid variable
   b. blue supergiant
   c. RR Lyrae
   d. Sun-like
2. Detailed observations of the structure of the Milky Way are difficult because
   a. the Solar System is embedded in the dust and gas of the disk.
   b. the Milky Way is mostly dark matter.
   c. too many stars are in the way.
   d. the galaxy is rotating too fast (about 200 km/s).
3. In general, as time passes, the Milky Way
   a. has the same chemical composition.
   b. has more abundant hydrogen.
   c. has more abundant heavy elements.
   d. has less abundant heavy elements.
4. The magnetic field of the Milky Way has been detected by
   a. synchrotron radiation from cosmic rays.
   b. direct observation of the field.
   c. the field's interaction with Earth's magnetic field.
   d. studying molecular clouds.
5. Evidence of a supermassive black hole at the center of the Milky Way comes from
   a. direct observations of stars that orbit it.
   b. visible light from material that is falling in.
   c. strong radio emission from the black hole itself.
   d. streams of cosmic rays from the center of the galaxy.
6. The Large Magellanic Cloud and Small Magellanic Cloud will probably
   a. become part of the Milky Way.
   b. remain orbiting forever.
   c. become attached to another passing galaxy.
   d. escape from the gravity of the Milky Way.
7. Globular clusters are important to understanding the Milky Way because
   a. they are so young that they provide information about current star formation.
   b. they provide information about dwarf ellipticals from which the Milky Way formed.
   c. they reveal the size of the Milky Way and Earth's location in it.
   d. the stars in them are highly enhanced in metals.
8. A globular cluster with no variable stars would have been left out of Shapley's study because
   a. the cluster would have been too far away.
   b. the distance to the cluster could not have been determined.
   c. the cluster would have been too faint to see.
   d. the cluster would have been too young to determine its evolutionary state.
9. The best evidence for the presence of dark matter in the Milky Way comes from the observation that the rotation curve
   a. is flat at great distances from the center.
   b. rises swiftly in the interior.
   c. falls off and then rises again.
   d. has a peak at about 2,000 light-years from the center.
10. Cosmic rays are
    a. a form of electromagnetic radiation.
    b. high-energy particles.
    c. high-energy dark matter.
    d. high-energy photons.
11. What kind of galaxy is the Milky Way?
    a. elliptical
    b. spiral
    c. barred spiral
    d. irregular
12. Where are the youngest stars in the Milky Way Galaxy?
    a. in the core
    b. in the bulge
    c. in the disk
    d. in the halo
13. Halo stars are found near the Sun. What observational evidence does *not* distinguish them from disk stars?
    a. their direction of motion
    b. their speed
    c. their composition
    d. their temperature
14. Why are most of the Milky Way's satellite galaxies so difficult to detect?
    a. They are very small.
    b. They are very far away.
    c. The halo of the Milky Way blocks the view.
    d. They are very faint.
15. The concept of a galactic habitable zone does not consider
    a. the radiation field.
    b. the ages of stars.
    c. the amount of heavy elements.
    d. the distance of a planet from its central star.

Thinking about the Concepts

16. Scientists can never know everything, especially at the beginning of a set of research programs. For example, Shapley did not know about dust in the Milky Way. Yet, scientists must often set aside that problem of the unknowns they don't know and do the best they can with the knowledge they have. Explain why that is a necessary step toward scientific understanding.

17. Describe the distribution of globular clusters within the Milky Way, and explain what that distribution implies about the size of the galaxy and our distance from its center.

18. In which parts of the Milky Way do astronomers find open clusters? In which parts do they find globular clusters?

19. Old stars in the inner disk of the Milky Way have higher abundances of massive elements than those of young stars in the outer disk. Explain how that difference might have developed.

20. How do 21-cm radio observations reveal the rotation of the Milky Way Galaxy?

21. What does the rotation curve of the Milky Way indicate about the presence of dark matter in the galaxy?

22. Halo stars are found near the Sun. What observational evidence distinguishes them from disk stars?

23. What is one source of synchrotron radiation in the Milky Way, and where is it found?

24. Why must astronomers use X-ray, infrared, and 21-cm radio observations to probe the center of the galaxy?

25. What is Sgr A*, and how was it detected?

26. Explain the evidence for a supermassive black hole at the center of the Milky Way. How does the mass of the supermassive black hole at the center of our galaxy compare with that found in most other spiral galaxies?

27. To observers in Earth's Southern Hemisphere, the Large Magellanic Cloud and Small Magellanic Cloud look like detached pieces of the Milky Way. What are those "clouds," and why is it not surprising that they look so much like pieces of the Milky Way?

28. What is the origin of the Milky Way's satellite galaxies? What has been the fate of most of the Milky Way's satellite galaxies? Why are most of the Milky Way's satellite galaxies so difficult to detect?

29. Use your imagination to describe how Earth's skies might appear if the Sun and Solar System were located (a) near the center of the galaxy; (b) near the center of a large globular cluster; (c) near the center of a large, dense molecular cloud.

30. What factors do astronomers consider when thinking about a galactic habitable zone in the Milky Way?

Applying the Concepts

31. From Figure 20.7, estimate the ratio between the radius of the Milky Way's outer halo and the radius of the disk.

32. According to Figure 20.16, what is the rotation velocity of a disk star located 6,000 pc from the center of the Milky Way? If you assume a circular orbit, how long does that star take to orbit once?

33. From the data in Figure 20.16, estimate the time the Large Magellanic Cloud would take to orbit the Milky Way if the Large Magellanic Cloud were on a circular orbit.

34. The Sun completes one trip around the center of the galaxy in approximately 230 million years. How many times has the Solar System made the circuit since its formation 4.6 billion years ago?

35. The Sun is located about 8,300 pc from the center of the galaxy, and the galaxy's disk probably extends another 9,000 pc farther out from the center. Assume that the Sun's orbit takes 230 million years to complete.
    a. With a truly flat rotation curve, how long would a globular cluster located near the edge of the disk take to complete one trip around the center of the galaxy?
    b. How many times has that globular cluster made the circuit since its formation about 13 billion years ago?

36. Parallax measurements of the variable star RR Lyrae indicate that it is located 230 pc from the Sun. A similar star observed in a globular cluster located far above the galactic plane appears 160,000 times fainter than RR Lyrae.
    a. How far from the Sun is that globular cluster?
    b. What does your answer to part (a) tell you about the size of the galaxy's halo in comparison with the size of its disk?

37. How do astronomers know that the center of the Milky Way contains a black hole and not just dark matter?

38. Given what you have learned about the distribution of massive elements in the Milky Way and what you know about the terrestrial planets, where do you think such planets are most likely and least likely to form?

39. A cosmic-ray proton is traveling at nearly the speed of light (3×10^8 m/s).
    a. Using Einstein's relationship between mass and energy, $E = mc^2$, show how much energy (in joules) the cosmic-ray proton would have if m were based only on the proton's rest mass (1.7×10^{-27} kg).
    b. The actual measured energy of the cosmic-ray proton is 100 J. What, then, is the relativistic mass of the cosmic-ray proton?
    c. How much greater is the relativistic mass of that cosmic-ray proton than the mass of a proton at rest?

40. One of the fastest cosmic rays ever observed had a speed of $(1.0 - [1.0 \times 10^{-24}]) \times c$ (that is, very, very close to c). Assume that the cosmic ray and a photon left a source at the same instant. To a stationary observer, how far behind the photon would the cosmic ray be after traveling for 100 million years?

41. Consider a black hole with a mass of 5 million M_{Sun}. A star's orbit about the black hole has a semimajor axis of 0.02 light-year (1.9×10^{14} meters). Calculate the star's orbital period. (Hint: You may want to refer to Chapter 4.)

42. A star in a circular orbit about the black hole at the center of the Milky Way (whose mass $M_{BH} = 8 \times 10^{36}$ kg) has an orbital radius of 0.0131 light-year (1.24×10^{14} meters). What is the average speed of that star in its orbit? (Hint: You may want to refer to Chapter 4.)

43. What is the Schwarzschild radius of the black hole at the center of the Milky Way? What is its density? How does that value compare with the density of a stellar black hole?

44. A star is observed in a circular orbit about a black hole with an orbital radius of 1.5×10^{11} km and an average speed of 2,000 km/s. What is the mass of that black hole in solar masses?

45. One model of the galactic habitable zone contains stars in a doughnut-shaped region between 7 and 9 kpc from the center of the galaxy, in the disk. If you assume that this doughnut is as thick as the disk itself, what fraction of the disk of the Milky Way lies in that habitable zone?

USING THE WEB

46. Go to the "Night Sky" Web page of the National Park Service (NPS; https://www.nps.gov/subjects/nightskies/). Click on and read "Light Pollution" under "Science of Light" and "Measuring Lightscapes" under "Managing Lightscapes." Why is it becoming more rare for people to see the Milky Way? Why does the NPS consider viewing the Milky Way an important part of the experience for people visiting the park?

47. a. Go to the Astronomy Picture of the Day (APOD) app or website for July 2, 2012, and watch the video clip "Zooming into the Center of the Milky Way" (https://apod.nasa.gov/apod/ap120702.html). Why do the selected pictures show a shift in wavelengths? On APOD, run a search for "Milky Way" to look at some of the best photographs of it. Where were the pictures taken from? Can you see the Milky Way from *your* location on a clear night?

    b. Why does the galactic center have to be observed in infrared and X-ray wavelengths? Go to the "Milky Way Galaxy" page of the Chandra X-ray telescope website (http://chandra.harvard.edu/photo/category/milkyway.html) and to the Spitzer infrared telescope website (http://www.spitzer.caltech.edu/). Are new images of the galactic center available? What has been learned?

48. Go to the websites of the two main groups studying the supermassive black hole at the galactic center: the UCLA Galactic Center Group (http://astro.ucla.edu/~ghezgroup/gc) and the Galactic Center Research group at the Max Planck Institute for Extraterrestrial Physics (http://mpe.mpg.de/ir/GC). Watch some of the time-lapse animations of the stars orbiting something unseen. Why is it assumed that the unseen object is a black hole? What new results are those groups reporting?

49. Citizen science: Go to the Milky Way Project website (https://www.zooniverse.org/projects/povich/milky-way-project). That project examines Spitzer and WISE telescope observations of the dusty material in the galaxy. Log in with your Zooniverse account name, and read the information under "About." Participants in the project have already discovered some of these bubbles (http://spitzer.caltech.edu/images/4938-sig12-002-Finding-Bubbles-in-the-Milky-Way). Classify some images.

50. Go to the Hubble Space Telescope website and watch the video simulation of the possible collision of the Milky Way and Andromeda (http://hubblesite.org/news_release/news/2012-20; click on "Release videos". What is different when Triangulum is included (scroll down)? Why will Andromeda "eat" the Milky Way instead of the other way around?

EXPLORATION

The Center of the Milky Way

digital.wwnorton.com/astro6

Adapted from *Learning Astronomy by Doing Astronomy*, by Ana Larson

Astronomers once thought that the Sun was at the center of the Milky Way. In this Exploration, you will repeat Shapley's globular-cluster experiment that led to a more accurate picture of the size and shape of the Milky Way.

Imagine that the disk of the Milky Way is a flat, round plane, like a pizza. Globular clusters are arranged in a rough sphere around that plane. To map globular clusters on **Figure 20.26**, imagine that a line is drawn straight "down" from a globular cluster to the plane of the Milky Way. The "projected distance" in kiloparsecs is the distance from the Sun to the place where the line hits the plane. The galactic longitude indicates the direction toward that point; it is marked around the outside of the graph, along with the several constellations.

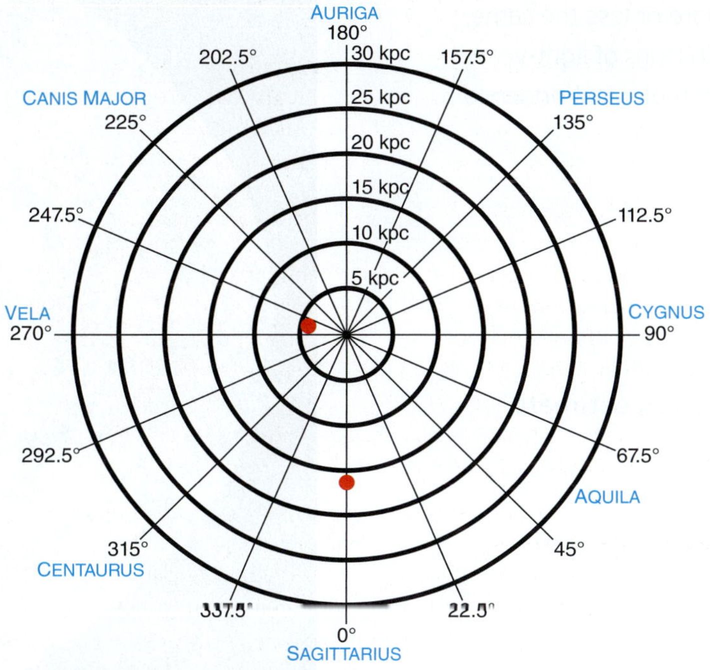

Figure 20.26 This polar graph can be used to plot distance and direction.

1 What is the approximate distance from the Sun to the center of the Milky Way?

____

2 What is the galactic longitude of the center of the Milky Way?

____

3 How do astronomers know that the Sun is not at the center of the Milky Way?

____

Make a dot at the location of each globular cluster by finding the galactic longitude indicated outside the circle and then coming in toward the center to the projected distance. The two globular clusters in boldface in **Table 20.1** have been plotted for you as examples. After plotting all the globular clusters, estimate the center of their distribution and mark it with an X. That is the center of the Milky Way.

TABLE 20.1 Globular Cluster Data

Cluster	Galactic Longitude	Projected Distance (kpc)	Cluster	Galactic Longitude	Projected Distance (kpc)
104	306	3.5	6273	357	7
362	302	6.6	**6287**	**0**	**16.6**
2808	283	8.9	6333	5	12.6
4147	**251**	**4.2**	6356	7	18.8
5024	333	3.4	6397	339	2.8
5139	309	5	6535	27	15.3
5634	342	17.6	6712	27	5.7
Pal 5	1	24.8	6723	0	7
5904	4	5.5	6760	36	8.4
6121	351	4.1	Pal 10	53	8.3
O 1276	22	25	Pal 11	32	27.2
6638	8	15.1	6864	20	31.5
6171	3	15.7	6981	35	17.7
6218	15	6.7	7089	54	9.9
6235	359	18.9	Pal 12	31	25.4
6266	353	11.6	288	147	0.3
6284	358	16.1	1904	228	14.4
6293	357	9.7	Pal 4	202	30.9
6341	68	6.5	4590	299	11.2
6366	18	16.7	5053	335	3.1
6402	21	14.1	5272	42	2.2
6656	9	3	5694	331	27.4
6717	13	14.4	5897	343	12.6
6752	337	4.8	6093	353	11.9
6779	62	10.4	6541	349	3.9
6809	9	5.5	6626	7	4.8
6838	56	2.6	6144	352	16.3
6934	52	17.3	6205	59	4.8
7078	65	9.4	6229	73	18.9
7099	27	9.1	6254	15	5.7

21 The Expanding Universe

The cosmological principle has been at the center of astronomers' conceptual understanding of the universe. The cosmological principle states that the conclusions we reach about our universe should be more or less the same, regardless of whether we live in the Milky Way or in a galaxy billions of light-years away. Here in Chapter 21, we look at the evidence for the cosmological principle and for the Big Bang.

LEARNING GOALS

By the end of this chapter, you should be able to:

LG 1 Explain in detail the cosmological principle.

LG 2 Describe how the Hubble constant can be used to estimate the age of the universe.

LG 3 Describe the observational evidence for the Big Bang.

LG 4 Explain which chemical elements were created in the early hot universe.

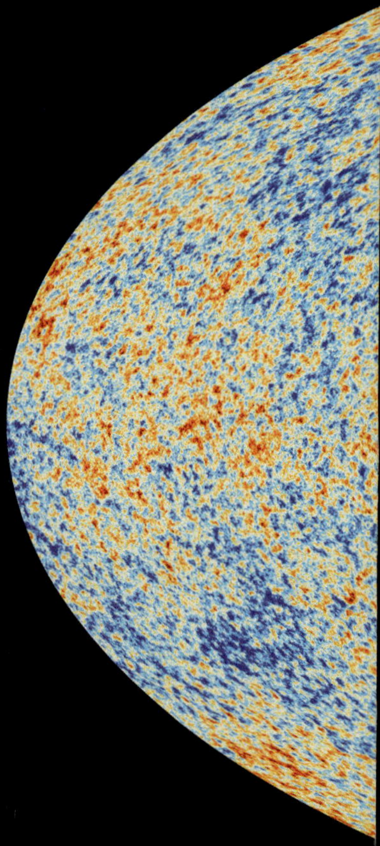

This image shows the oldest light in the universe, the cosmic microwave background radiation, detected by the Planck space observatory. ▶▶▶

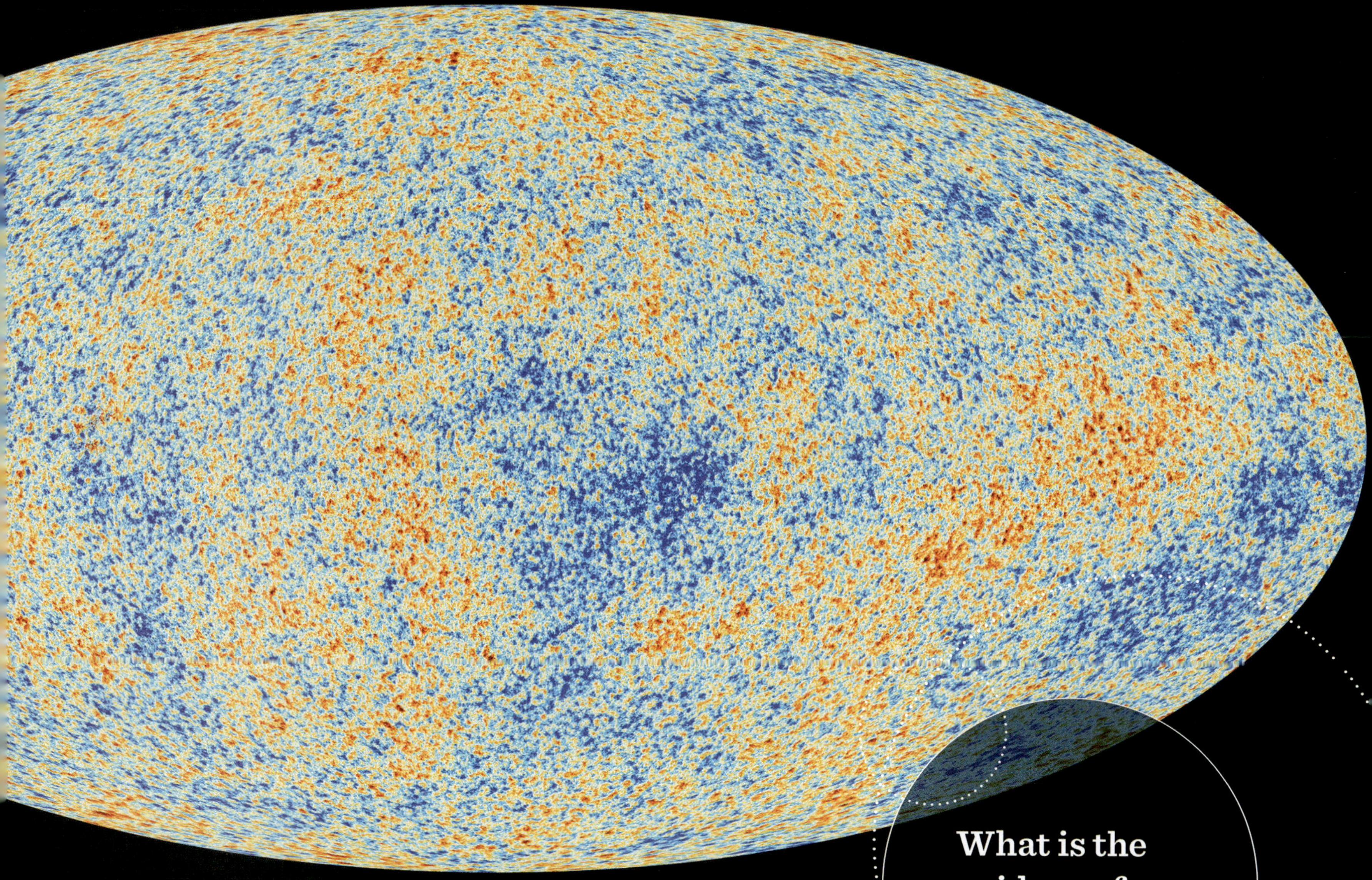

What is the evidence for the Big Bang?

21.1 The Cosmological Principle

In Chapter 19, we described observations of galaxy motions and distances. From those observations, Edwin Hubble created a graph that showed the galaxies moving apart, with the more distant ones moving faster—a result known as Hubble's law. In this section, we further explore the implications of Hubble's law.

The Homogeneous and Isotropic Universe

Cosmology is the study of space and time and the dynamics of the universe as a whole. In the 1920s, around the same time that astronomers were first measuring distances to galaxies, theoretical physicists were applying Einstein's general theory of relativity to cosmology. The cosmologist Alexander Friedmann (1888–1925) produced mathematical models of the universe that assumed the **cosmological principle**, which requires that *the physical laws and the properties of the universe are the same everywhere and in all directions.* The cosmological principle forms the basis of our study of distant galaxies and of the universe itself and is now a fundamental tenet of modern cosmology. The principle implies that we occupy no special place in the universe. Many observations have since validated that assumption.

We have used that principle throughout this book when applying laws of physics and chemistry to objects near and far in the universe. For example, according to the cosmological principle, gravity works the same way in distant galaxies as it does here on Earth. The cosmological principle is a testable scientific theory. An important prediction of the principle is that the conclusions that scientists reach about the universe should be the same whether we observe the universe from the Milky Way or from a galaxy billions of parsecs away. In other words, if the cosmological principle is correct, the universe is **homogeneous**, having the same composition and properties at all places.

In an absolute sense, the universe is *not* homogeneous because the conditions on Earth are very different from those in deep space or in the center of the Sun. In cosmology, homogeneity of the universe means that stars and galaxies in Earth's part of the universe are much the same, and behave in the same manner, as stars and galaxies in remote corners of the universe. Homogeneity also means that stars and galaxies everywhere are distributed in space in much the same way that they are distributed in Earth's cosmic neighborhood and that observers in those galaxies would see the same properties for the universe that astronomers see from here. However, homogeneity does not mean that the Universe is unchanging over time; it was different in the past and will be different in the future.

Directly verifying the prediction of homogeneity is not easy. Scientists cannot travel from the Milky Way to a galaxy in the remote universe to see whether conditions are the same. They can, however, compare light arriving from closer and farther locations in the distant universe and see the ways in which features look the same or different. For example, astronomers can look at how galaxies are distributed in distant space and see whether that distribution is similar (that is, homogeneous) to the distribution nearby.

In addition to predicting that the universe is homogeneous, the cosmological principle requires that all observers measure the same properties of the universe, regardless of the direction in which they are looking. If something is the same in all directions, it is **isotropic**. That prediction of the cosmological principle is much easier to test directly than is homogeneity. If galaxies were lined up in rows, for example, astronomers would measure very different properties of the universe, depending on the direction they looked. The universe would still be homogeneous, but not isotropic—so it would not satisfy the cosmological principle.

Usually, isotropy goes together with homogeneity, and the cosmological principle requires both. **Figure 21.1** shows examples of how the universe could have violated the cosmological principle by not being homogeneous or isotropic, as well as examples of how the universe might satisfy the cosmological principle. All observations show that the properties of the universe are basically the same, regardless of the direction in which observers are looking. When averaged over very large scales (thousands of millions of parsecs), the universe appears homogeneous as well.

The Hubble Expansion

According to Hubble's law (Chapter 19), the distance of a galaxy (d_G) is proportional to its recession velocity (v_r): $d_G = v_r/H_0$. Hubble's law helps astronomers investigate whether the universe is homogeneous and isotropic. They can confirm

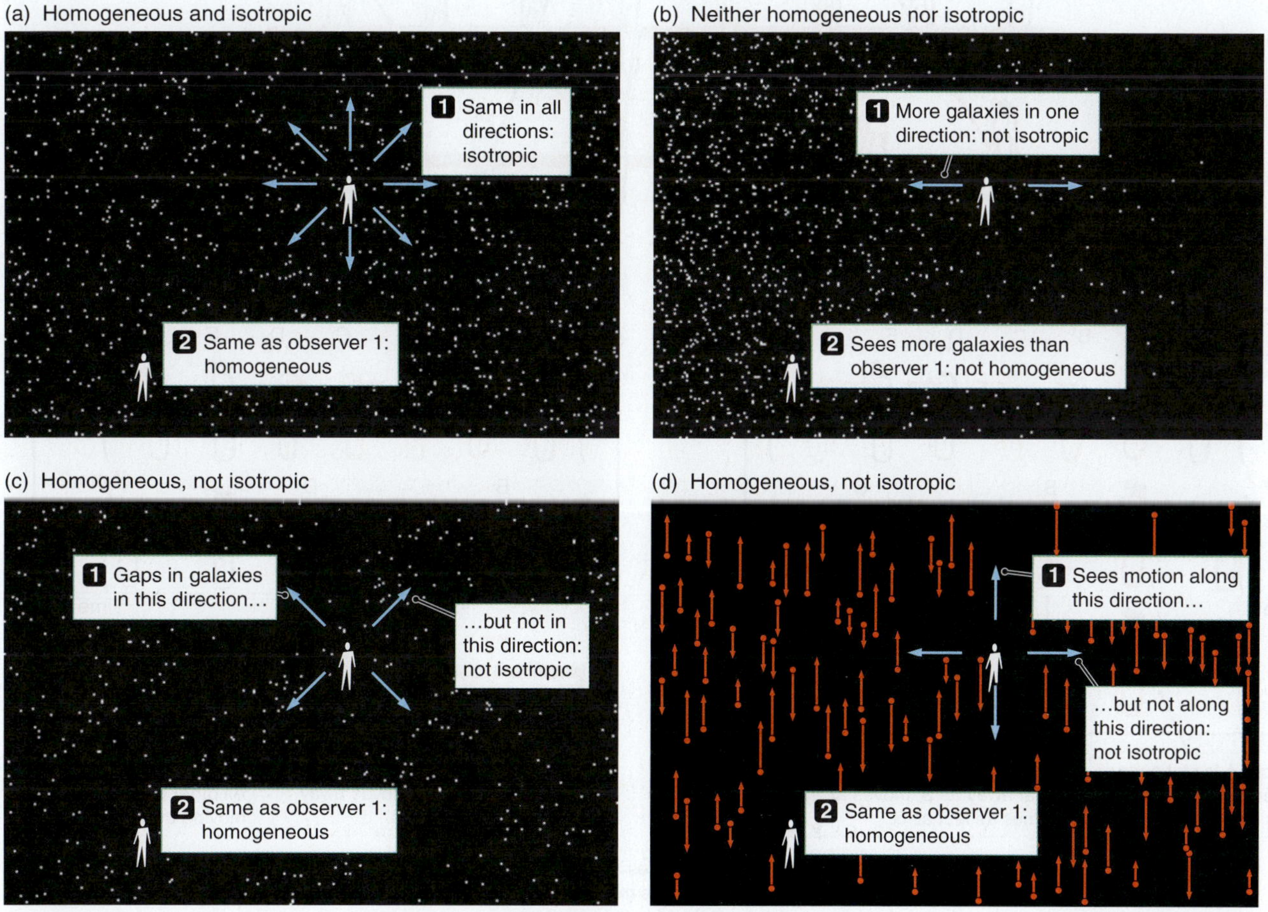

Figure 21.1 Homogeneity and isotropy in four theoretical models of a universe. Blue arrows indicate the direction of view. (a) The distribution of galaxies is uniform, so this universe is both homogeneous and isotropic. (b) The density of galaxies is decreasing in one direction, so this universe is neither homogeneous nor isotropic. (c) The bands of galaxies lie along a unique axis, making this universe not isotropic. (d) The distribution of galaxies is uniform, but galaxies move along only one direction, so this universe also is not isotropic.

its isotropy by observing that galaxies in one direction in the sky obey the same Hubble law as galaxies in other directions in the sky. Hubble's law says that Earth is located in an expanding universe and that the expansion looks the same regardless of the location of the observer. To help you visualize that uniformity, we now turn to a useful model that you can build for yourself with materials you probably have in your desk.

Figure 21.2 shows a long rubber band with paper clips attached along its length. If you stretch the rubber band, the paper clips—which represent galaxies in an expanding universe—get farther and farther apart. To get a sense of what that expansion would look like up close, imagine that you're an ant riding on paper clip A. As the rubber band is stretched, you notice that all the paper

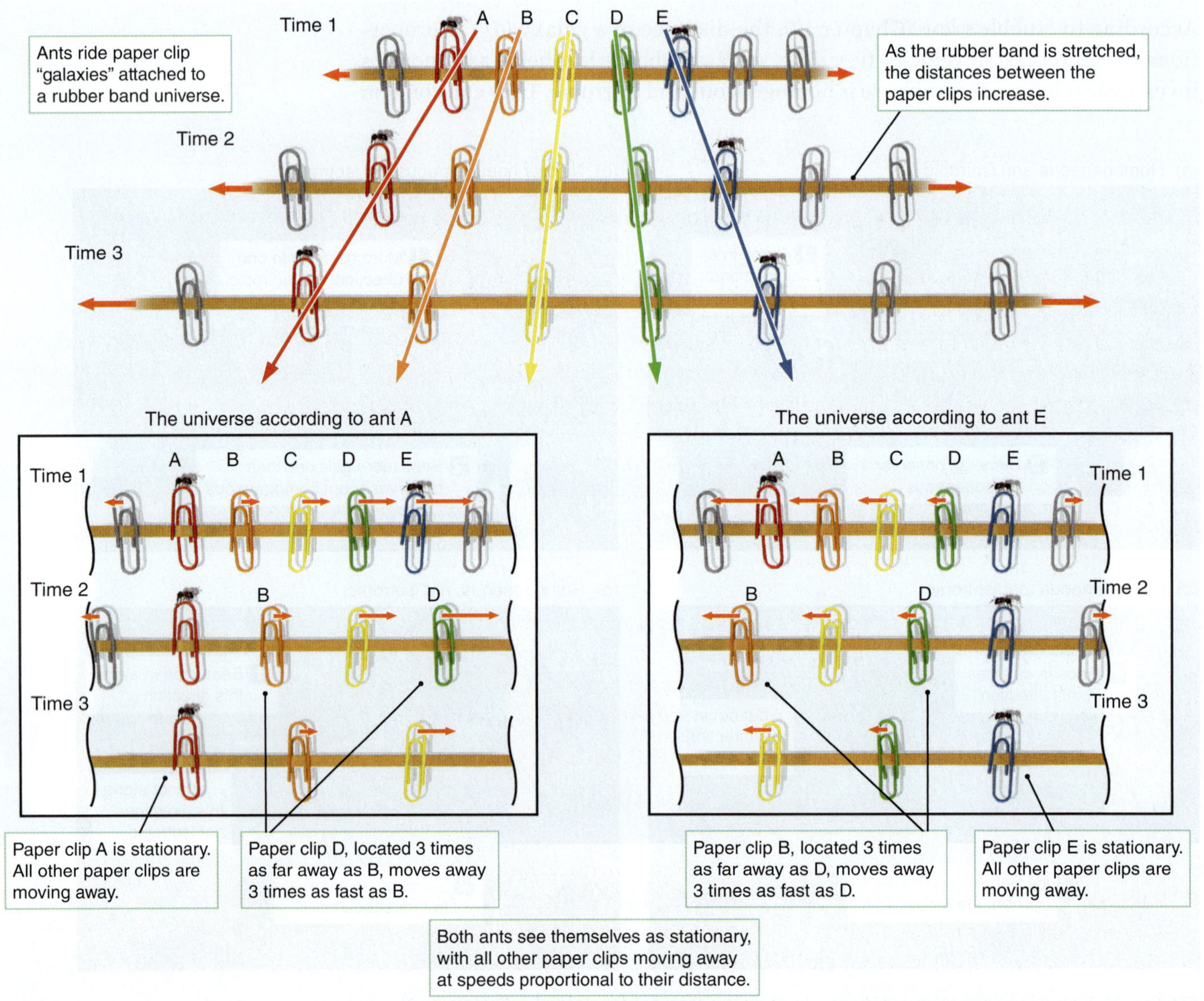

Figure 21.2 In this one-dimensional analogy of Hubble's law, a rubber band with paper clips evenly spaced along its length is stretched. As the rubber band stretches, an ant riding on clip A observes clip C moving away twice as fast as clip B. Similarly, an ant riding on clip E sees clip C moving away twice as fast as clip D. Any ant sees itself as stationary, regardless of which paper clip it is riding, and it sees the other clips moving away with speed proportional to distance.

clips are moving away from you. Clip B, the closest one, is moving away slowly. Clip C, located twice as far from you as clip B, is moving away twice as fast as clip B. Clip E, located 4 times as far away as clip B, is moving away 4 times as fast as clip B. From your perspective as an ant riding on clip A, all the other paper clips on the rubber band are moving away with a velocity proportional to their distance. The paper clips located along the rubber band obey a Hubble-like law.

The key insight to the analogy comes from realizing that nothing is special about the perspective of paper clip A. If, instead, the ant were riding on clip E, clip D would be the one moving away slowly and clip A would be moving away 4 times as fast. Repeat the experiment for any paper clip along the rubber band, and you will arrive at the same result. For an ant on any paper clip along the rubber band, the speed at which other clips are moving away from the ant is proportional to their distance. The stretching rubber band, like the universe, is "homogeneous." The same Hubble-like law applies, regardless of where the observer ant is located.

The observation that nearby paper clips move away slowly and distant paper clips move away more rapidly does not mean that the paper clip selected as a vantage point is at the center of anything. Instead, it means that the rubber band is being stretched uniformly along its length. Similarly, Hubble's law for galaxies means that nearby galaxies are carried away slowly by expanding space, and distant galaxies are carried away more rapidly (see the **Process of Science Figure**).

The expanding universe has no center; the universe is expanding uniformly. Any observer in any galaxy sees nearby galaxies moving away slowly and more distant galaxies moving away more rapidly. The same Hubble law applies from their vantage point as applies from our vantage point on Earth in our galaxy. The expansion of the universe is homogeneous. The motion of galaxies as a result of the expansion of the universe is called the **Hubble flow**.

The only exception to that rule is for galaxies that are close together, in which case gravitational attraction dominates over the expansion of space. As you saw in Chapter 20, the Andromeda Galaxy and the Milky Way are being pulled together by their gravity. The Andromeda Galaxy is approaching the Milky Way at about 110 kilometers per second (km/s), so the light from the Andromeda Galaxy is blueshifted, not redshifted. That velocity is caused by the gravitational interaction between the two nearby galaxies rather than the expansion of space. The fact that gravitational or electromagnetic forces can overwhelm the expansion of space also explains why the Solar System is not expanding, and neither are you.

The Universe in Space and Time

Hubble's law gives astronomers a practical tool for measuring distances to remote objects. Once astronomers know the value of the Hubble constant, H_0, they can use a straightforward measurement of the redshift of a galaxy to find its distance. In other words, once H_0 is known, Hubble's law makes the once-difficult task of measuring distances in the universe relatively easy, giving astronomers a tool to map the structure of the observable universe. Astronomers find H_0 from nearby galaxies by using standard candles to get the distance and Doppler shifts of spectral lines to get v. Then they use that value of H_0 to find the distances to different, more distant galaxies.

PROCESS OF SCIENCE

Data Are the Ultimate Authority

The genius of Albert Einstein is recognized by just about everyone, but even he had to change his mind in the face of new data.

1920

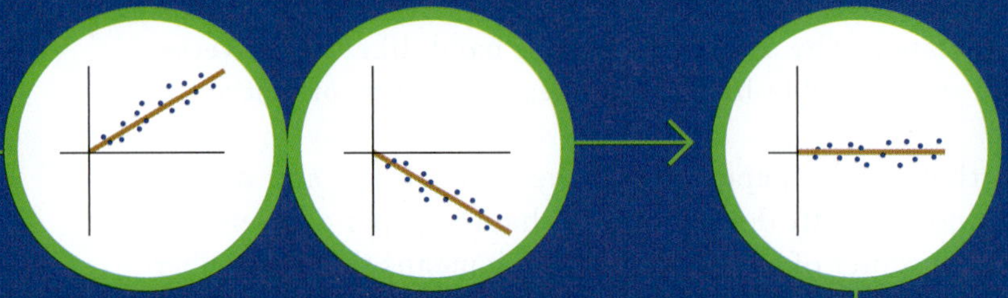

Einstein developed general relativity throughout the late 1910s. His equations predicted that the universe must be expanding or contracting.

Λ

Einstein inserted an extra term, indicated by the symbol Λ, into his equations to make the universe stationary, because he believed it should not expand or contract.

1930

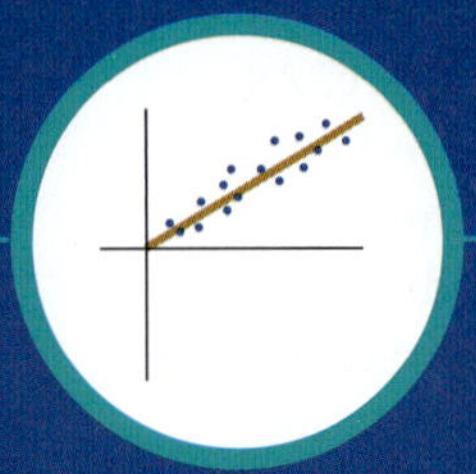

In 1929, Edwin Hubble combined observations of galaxies to show that nearly all galaxies moved away from Earth, with farther galaxies receding more quickly. These observations showed that the universe is expanding!

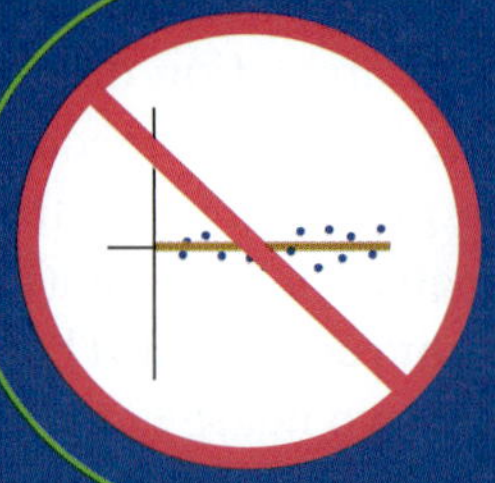

Einstein changed his mind and described his reluctance to accept the prediction of his original theory as "his biggest blunder."

Even the most brilliant scientists must adjust when the data contradict their conclusions.

Hubble's law does more than place galaxies in space; it also places galaxies in time. Light travels at a huge but finite speed. When you look at the Sun, you see it as it existed 8 minutes ago (see Chapter 1). When you look at Alpha Centauri, the nearest stellar system beyond the Sun, you see it as it existed 4.3 years ago. If you look at the center of the Milky Way, the light you see is 27,000 years old. The **look-back time** of a distant object is the time the light from that object has taken to reach a telescope on Earth. As astronomers look into the distant universe, look-back times become very large. The distance to a galaxy whose redshift of $z = 0.1$ is 1.4 billion light-years (assuming $H_0 = 70$ km/s/Mpc), so the look-back time to that galaxy is 1.4 billion years. The look-back time to a galaxy in which $z = 0.2$ is 2.7 billion years. The look-back time of the most distant galaxies observed, at about $z = 10$, is 13.2 billion years. As astronomers observe objects with greater and greater redshifts, they are seeing increasingly younger stages of the universe.

CHECK YOUR UNDERSTANDING 21.1

In astronomy, *isotropic* means that the universe is the same __________, and *homogeneous* means that the universe is the same __________. (a) in all locations; (b) in all directions; (c) at all times; (d) at all size scales.

21.2 The Universe Began in the Big Bang

Imagine watching a video of the universe, with the galaxies moving apart. Now reverse the video and run it backward in time. The galaxies become closer and closer together as the universe becomes younger and younger. In that way, the observation of the expansion of the universe leads to the idea of a beginning to the universe at a time that can be estimated from Hubble's law. In this section, we explore that implication of Hubble's law.

Expansion and the Age of the Universe

AstroTour: Hubble's Law

Astronomy in Action: Expanding Balloon Universe

Hubble's law gives an estimate of the age of the universe. For the simplest estimate, we assume that the speed of the expansion has always been constant, and the age of the universe can be estimated from the slope of the line in a graph of the velocity of galaxies plotted against their distance. The slope has units that reduce to 1/time, so its inverse has units of time. That slope is the Hubble constant, H_0, and its inverse (1 divided by H_0) is the **Hubble time**. The Hubble time is an estimate of the universe's age: 13.8 billion years (**Working It Out 21.1**). If the expansion were faster, the Hubble constant would be larger, and the universe would be younger; similarly, a slower expansion would yield a smaller Hubble constant and an older universe.

If the universe expanded uniformly, then when it was about half its current age, about 6.9 billion years ago, all the galaxies in the universe were half as far apart as they are now, and 12.4 billion years ago, all the galaxies were about a tenth as far apart. If we assume that galaxies have been moving apart all that time at the same speed as they do today, then a little less than one Hubble time ago—13.8 billion years ago—almost no space existed between the particles that constitute today's universe. All such matter, as well as energy in the universe, must then have been unimaginably dense. Because expanding gases cool down, the universe then must have been much hotter than it is today in its expanded

21.1 Working It Out Expansion and the Age of the Universe

We can use Hubble's law to estimate the age of the universe. Consider two galaxies located 30 Mpc ($d_G = 9.3 \times 10^{20}$ km) away from each other (**Figure 21.3**). If those two galaxies are moving apart, then at some time in the past they must have been together in the same place at the same time. According to Hubble's law, and on the assumption that $H_0 = 70$ km/s/Mpc, the distance (d_G) between those two galaxies is increasing at the following rate:

$$v_r = H_0 \times d_G$$

$$v_r = 70 \text{ km/s/Mpc} \times 30 \text{ Mpc}$$

$$v_r = 2{,}100 \text{ km/s}$$

Knowing the velocity (v_r) at which they are traveling, we can calculate the time the two galaxies took to become separated by 30 Mpc:

$$\text{Time} = \frac{\text{Distance}}{\text{Velocity}} = \frac{9.3 \times 10^{20} \text{ km}}{2{,}100 \text{ km/s}} = 4.4 \times 10^{17} \text{ s}$$

Dividing by the number of seconds in a year (about 3.16×10^7 s/yr) gives

$$\text{Time} = 1.4 \times 10^{10} \text{ yr} = 14 \text{ billion yr}$$

In other words, *if* expansion of the universe has been constant, two galaxies that today are 30 Mpc apart started out at the same place about 14 billion years ago.

Now let's do the same calculation with two galaxies 60 Mpc (18.6×10^{20} km) apart. Those two galaxies are twice as far apart, but the distance between them is increasing twice as rapidly:

$$v_r = H_0 \times d_G = 70 \text{ km/s/Mpc} \times 60 \text{ Mpc} = 4{,}200 \text{ km/s}$$

Therefore,

$$\text{Time} = \frac{18.6 \times 10^{20} \text{ km}}{4{,}200 \text{ km/s}} = 4.4 \times 10^{17} \text{ s} = 1.4 \times 10^{10} \text{ yr}$$

Again, we calculate time as distance divided by velocity (twice the distance divided by twice the velocity) to find that those galaxies also took about 14 billion years to reach their current locations. We can do that calculation again and again for any pair of galaxies in the universe today. The farther apart the two galaxies are, the faster they are moving. But all galaxies took the *same* amount of time to get to where they are today. (Small differences in the intermediate steps of those calculations are the result of rounding and are not significant to the argument.) The most precise measurements give a result of 13.8 billion years.

Working out the example with words instead of numbers makes it clear why the answer is always the same. Because the velocity we are calculating comes from Hubble's law, velocity equals the Hubble constant multiplied by distance. Writing that out as an equation, we get

$$\text{Time} = \frac{\text{Distance}}{\text{Velocity}} = \frac{\text{Distance}}{H_0 \times \text{Distance}}$$

Distance divides out to give

$$\text{Time} = \frac{1}{H_0}$$

where $1/H_0$ is the Hubble time. That approach is one way to estimate the age of the universe.

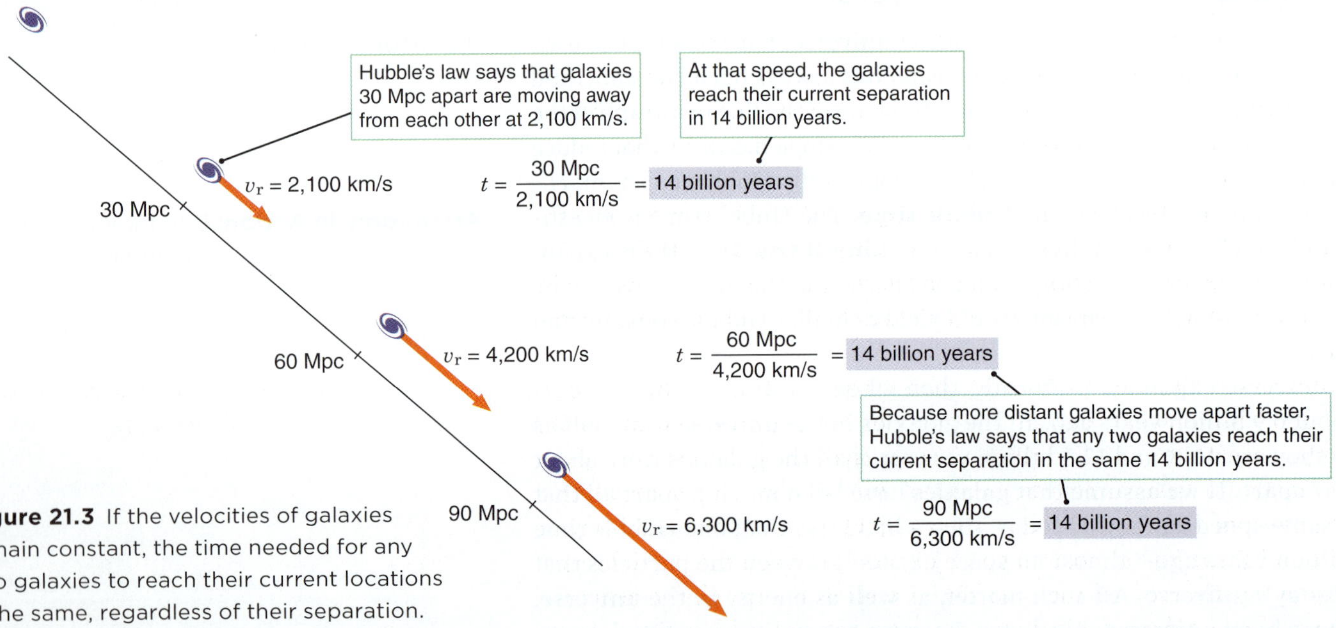

Figure 21.3 If the velocities of galaxies remain constant, the time needed for any two galaxies to reach their current locations is the same, regardless of their separation.

state. That hot, dense beginning, 13.8 billion years ago, is called the **Big Bang** (**Figure 21.4**).

Georges Lemaître (1894–1966) was the first to propose the theory of the Big Bang. The idea greatly troubled many astronomers in the early and middle years of the 20th century. Several suggestions were put forward to explain the observed fact of Hubble expansion without resorting to the idea that the universe came into existence in an extraordinarily dense "fireball" billions of years ago. As more and more distant galaxies have been observed, however, and as more discoveries about the structure of the universe have been made, the Big Bang theory has grown stronger. The major predictions of the Big Bang theory have proven to be correct.

The implications of Hubble's law forever changed the scientific concepts of the origin, history, and possible future of the universe. At the same time, Hubble's law has pointed to many new questions about the universe. To address them, we next need to consider exactly what the term *expanding universe* means.

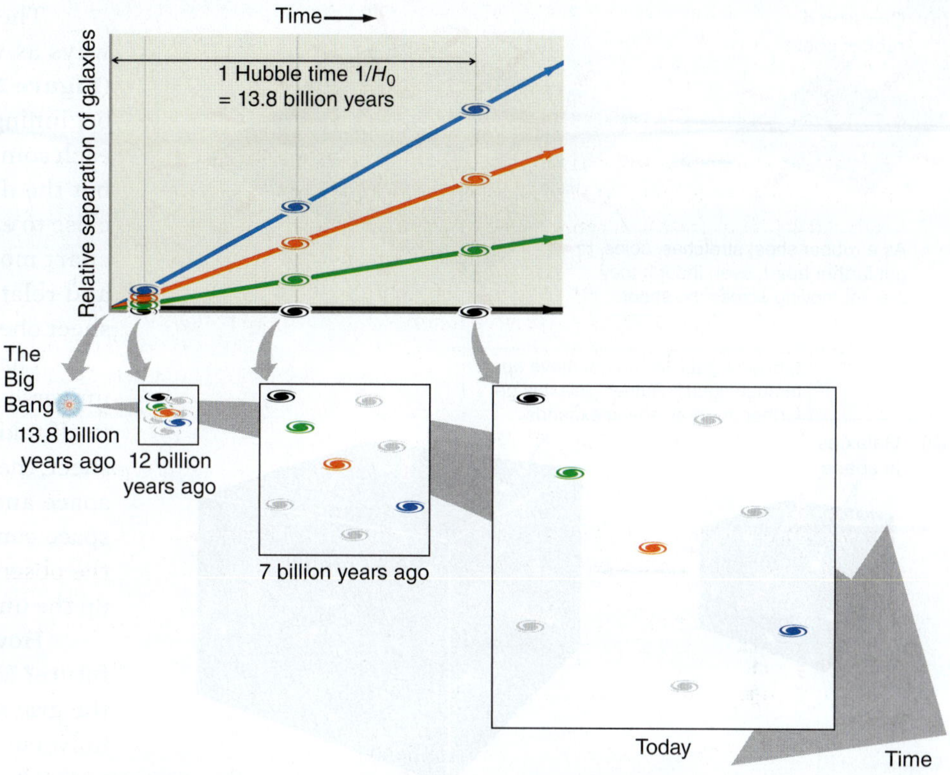

Figure 21.4 Looking backward in time, the distance between any two galaxies is smaller and smaller, until all matter in the universe is concentrated at the same point: the Big Bang.

Galaxies and the Expansion

At this point in our discussion, you may be picturing the expanding universe as a cloud of debris from an explosion flying outward through surrounding space. That is a common depiction of the Big Bang in movies and television shows, which portray a tiny bright spot that explodes to fill the screen. However, the Big Bang is not an explosion in the usual sense of the word, and in fact no surrounding space exists into which the universe is expanding.

We must draw a distinction between the universe and the observable universe. The observable universe is the part of the universe that we can see. The observable universe extends 13.8 billion light-years in every direction. That limit exists because it is the length of time the universe has been around. The light from more distant regions has not yet had time to travel to us, and so we cannot see it yet.

Where did the Big Bang take place? The Big Bang took place *everywhere*. Wherever anything is in the universe today, it is at the site of the Big Bang. The reason is that galaxies are not flying apart through space at all. Rather, space itself is expanding, carrying the stars and galaxies that populate the universe along with it.

Astronomy in Action: Observable versus Actual Universe

We have already dealt with the basic ideas that explain the expansion of space. In our discussion of black holes (Chapter 18), you encountered Einstein's general theory of relativity. General relativity says that space is distorted by the presence of mass and that the consequence of that distortion is gravity. For example, the mass of the Sun, like any other object, distorts the geometry of spacetime around it; so Earth, coasting along in its inertial frame of reference, follows a curved path around the Sun. We illustrated that phenomenon in Figure 18.10 with the analogy of a ball placed on a stretched rubber sheet, showing how the ball distorted the surface of the sheet.

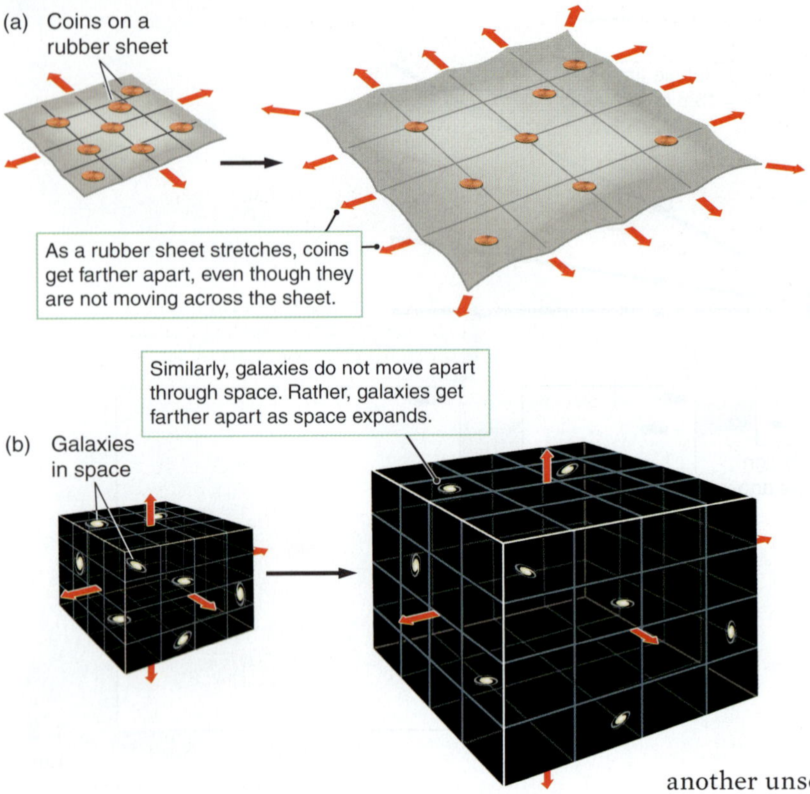

Figure 21.5 (a) As a rubber sheet is stretched, coins on its surface move farther apart, even though they are not moving with respect to the sheet itself. Any coin on the surface of the sheet observes a Hubble-like law in every direction. (b) Galaxies in an expanding universe are not flying apart through space. Rather, space itself is stretching.

The surface of a rubber sheet can be distorted in other ways as well. Imagine several coins placed on a rubber sheet (**Figure 21.5**). Then imagine grabbing the edges of the sheet and beginning to pull them outward. As the rubber sheet stretches, each coin remains at the same location on the surface of the sheet, but the distances between the coins increase. Two coins sitting close to each other move apart only slowly, whereas coins farther apart move away from each other more rapidly. The distances and relative motions of the coins on the surface of the rubber sheet obey a Hubble-like relationship as the sheet is stretched.

That movement is analogous to what is happening in the universe, with galaxies taking the place of the coins and space itself taking the place of the rubber sheet. For coins on a rubber sheet, the sheet can be stretched only so far before it breaks. With space and the real universe, no such limit exists. The fabric of space can, in principle, go on expanding forever. Hubble's law is the observational consequence of the fact that the space making up the universe is expanding.

How will that expansion of the universe behave in the future? Most of the mass in galaxies consists of dark matter, and the gravity of the dark matter slows down the expansion of the universe. In Chapter 22, you will also learn that the universe has another unseen constituent, called *dark energy*, which causes the expansion of the universe to *accelerate*. At the current stage in the expansion of the universe, the accelerating effect of dark energy dominates over the slowing effect of dark matter, suggesting that the universe will continue to expand at an ever-faster rate.

CHECK YOUR UNDERSTANDING 21.2

Where in the universe did the Big Bang take place? (a) near the Milky Way Galaxy; (b) near the center of the universe; (c) near some unknown location on the other side of the universe; (d) everywhere in the universe

21.3 Expansion Is Described with a Scale Factor

As the universe expands, the distance between any two objects increases because of the stretching of space. Astronomers find it useful to discuss that expansion in terms of the *scale factor* of the universe.

Scale Factor

Let's return to the analogy of the rubber sheet. Suppose you place a ruler on the rubber sheet and draw a tick mark every centimeter (**Figure 21.6a**). To measure the distance between two points on the sheet, you can count the marks between the two points and multiply by 1 centimeter (cm) per tick mark.

As the sheet is stretched, however, the distance between the tick marks does not remain 1 cm. When the sheet is stretched to 150 percent of the size it had when the tick marks were drawn, each tick mark is separated from its neighbors by 1 1/2 times the original distance, or 1.5 cm. The distance between two points can still come from counting the marks, but you need to scale up the distance between tick marks by 1.5 to find the distance in centimeters. If the sheet were twice the size it was when the tick marks were drawn (**Figure 21.6b**), each mark would correspond to 2 cm of actual distance. Astronomers use the term **scale factor** to indicate the size of the sheet relative to its size when the tick marks were drawn. The scale factor also indicates how much the distance between points on the sheet has changed. In the first example, the scale factor of the sheet is 1.5; in the second, the scale factor is 2.

Suppose astronomers choose today to lay out a "cosmic ruler" on the fabric of space, placing an imaginary tick mark every 10 Mpc. The scale factor of the universe at this time is defined to be 1. In the past, when the universe was smaller, distances between the points in space marked by the cosmic ruler would have been less than 10 Mpc. The scale factor of that younger, smaller universe would have been less than today's scale factor and therefore less than 1. In the future, as the universe continues to expand, the distances between the tick marks on the cosmic ruler will grow to more than 10 Mpc, and the scale factor of the universe will be greater than 1. Astronomers use the scale factor, usually written as R_U, to keep track of the changing scale of the universe.

The laws of physics are themselves unchanged by the expansion of the universe, just as stretching a rubber sheet does not change the properties of the coins on its surface. At noncosmological scales, the nuclear and electromagnetic forces within and between atoms, as well as the gravitational forces between relatively close objects, dominate over the expansion. As the universe expands, the sizes and other physical properties of atoms, stars, and galaxies also remain unchanged.

As we look back in time, the scale factor of the universe gets smaller and smaller, approaching zero as it comes closer and closer to the Big Bang. The fabric of space that today spans billions of parsecs spanned much smaller distances when the universe was young. When the universe was only a day old, all the space visible today amounted to a region only a few times the size of the Solar System. When the universe was 1/50 of a second old, the vast expanse of space that makes up today's observable universe (and all the matter in it) occupied a volume only the size of today's Earth. As we go backward in time approaching the Big Bang itself, the space that makes up today's observable universe becomes smaller and smaller—the size of a grapefruit, a marble, an atom, a proton. Every point in the fabric of space that makes up today's universe was right there at the beginning, a part of that unimaginably tiny, dense universe that emerged from the Big Bang.

The Big Bang did not occur at a specific point in space because space itself came into existence with the Big Bang. The Big Bang happened everywhere; no particular point in today's universe marked the site of the Big Bang. A Big Bang universe is homogeneous and isotropic, consistent with the cosmological principle.

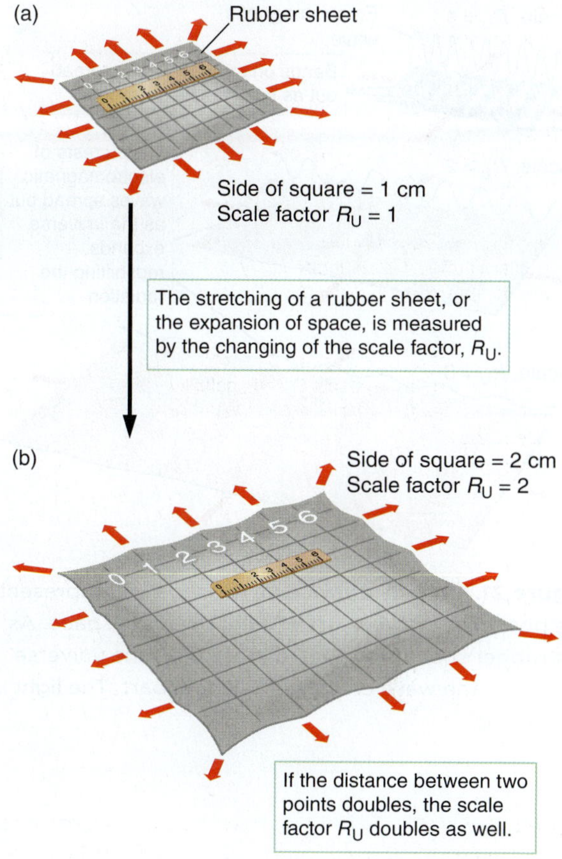

Figure 21.6 (a) On a rubber sheet, tick marks are drawn 1 cm apart. As the sheet is stretched, the tick marks move farther apart. (b) When the spacing between the tick marks is 2 cm, or twice the original value, the scale factor of the sheet, R_U, is said to have doubled. A similar scale factor, R_U, is used to describe the expansion of the universe.

CHECK YOUR UNDERSTANDING 21.3A

The scale factor keeps track of: (a) the movement of galaxies through space; (b) the current distances between many galaxies; (c) the changing distance between any two galaxies; (d) the location of the center of the universe.

Redshift Is Due to the Changing Scale Factor of the Universe

The ideas of general relativity discussed in Chapter 18 are powerful tools for interpreting Hubble's great discovery of the relationship between the velocity and distance of galaxies and of the expanding universe. Those velocities were determined from the redshift of the galaxies by using the Doppler effect. Although it is true that the distance between galaxies is increasing as a result of the expansion of the universe, and that we can use the equation for Doppler shifts to measure the

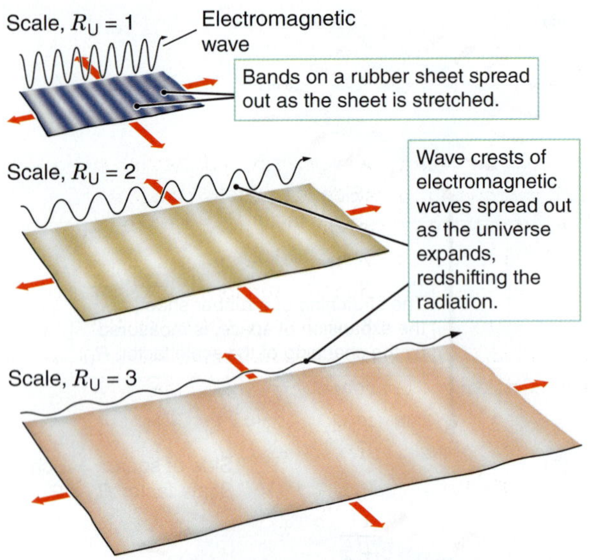

Figure 21.7 Bands drawn on a rubber sheet represent the positions of the crests of a light wave in space. As the rubber sheet is stretched—that is, as the universe expands—the wave crests get farther apart. The light is redshifted.

redshifts of galaxies, those redshifts are not due to Doppler shifts in the same way that we described for a moving star. Light that comes from very distant objects was emitted when the universe was younger and therefore smaller. As that light comes toward Earth from distant galaxies, the scale factor of the space through which the light travels is constantly increasing, and as it does, the distance between adjacent light-wave crests increases as well. The light is "stretched out" as the space it travels through expands.

Recall the rubber sheet analogy that we used when discussing black holes in Chapter 18. Then, we imagined space as a two-dimensional rubber sheet. **Figure 21.7** uses that rubber-sheet analogy to explain why the increasing distance between galaxies causes the light to be redshifted. If you draw a series of bands on the rubber sheet to represent the crests of an electromagnetic wave, you can watch what happens to the wave as the sheet is stretched out. By the time the sheet is stretched to twice its original size—that is, by the time the scale factor of the sheet is 2—the distance between wave crests has doubled. When the sheet has been stretched to 3 times its original size (a scale factor of 3), the wavelength of the wave will be 3 times what it was originally.

When light left a distant galaxy, the scale factor of the universe was smaller than it is today. As light comes toward us from distant galaxies, the space through which the light travels is stretching, and the light is also "stretched out" as the space through which it travels expands. The universe expanded while the light was in transit, and as it did so, the wavelength of the light grew longer in proportion to the increasing scale factor of the universe. The redshift of light from distant galaxies is therefore a direct measure of how much the universe has expanded since the radiation left its source. Redshift measures how much the scale factor of the universe, R_U, has changed since the light was emitted. That redshift can be greater than 1 if the galaxy is far enough away that the recession velocity is relativistic (**Working It Out 21.2**).

CHECK YOUR UNDERSTANDING 21.3B

What is the interpretation of a redshift larger than 1? (a) The object is moving faster than the speed of light. (b) The universe has more than doubled in size since the light from that object was emitted. (c) The light was shifted to longer wavelengths from gravitational radiation. (d) The rate of expansion of the universe is increasing.

21.4 Astronomers Observe Cosmic Microwave Background Radiation

What is the evidence that the Big Bang actually took place? One piece of evidence comes from observations of the early universe across the entire sky. This section introduces the cosmic microwave background radiation, one of the major confirming observations of the theory that the universe had a beginning.

Radiation from the Big Bang

In the late 1940s, cosmologists Ralph Alpher (1921–2007), Robert Herman (1914–1997), and George Gamow (1904–1968) reasoned that because a compressed gas cools as it expands, the universe also should be cooling as it expands. When the universe was very young and small, it must have consisted of an extraordinarily hot, dense gas. As with any hot, dense gas, that early universe would have been awash in blackbody radiation that exhibits a Planck blackbody spectrum (see Chapter 5).

21.2 Working It Out: When Redshift Exceeds 1

From our discussion of the Doppler shift (Chapter 5), recall that $(\lambda_{obs} - \lambda_{rest})/\lambda_{rest}$ is equal to the velocity of an object moving away (v_r) divided by the speed of light (c). Edwin Hubble used that result to interpret the observed redshifts of galaxies as evidence that galaxies throughout the universe are moving away from the Milky Way. Einstein's special theory of relativity says that nothing can move faster than the speed of light. Hubble initially assumed that redshifts are due to the Doppler effect. The resulting relation, $z = v_r/c$, would then seem to imply that no object can have a redshift (z) greater than 1. Yet that is not the case. Astronomers routinely observe redshifts significantly in excess of 1. As of this writing, the most distant objects known have redshifts as large as 9–11. How can redshifts exceed 1?

To arrive at the expression for the Doppler effect, $v_r/c = (\lambda_{obs} - \lambda_{rest})/\lambda_{rest}$, we have to *assume* that v_r is much less than c. If v_r was close to c, we would have to consider more than just the fact that the waves from an object are stretched out by the object's motion away. We also would have to consider relativistic effects, including the fact that moving clocks run slowly (see Working It Out 18.1). When combining those effects, we would find that as the speed of an object approaches the speed of light, its redshift approaches infinity, as shown in **Figure 21.8**.

Doppler's original formula is essentially correct—for objects close enough to us that their measured velocities are far less than the speed of light. When astronomers look at the motions of orbiting binary stars or the peculiar velocities of galaxies relative to the Hubble flow, that equation works just fine. But for any redshift of 0.4 or greater, relativity must be taken into account.

Another source of redshift is the gravitational redshift discussed in Chapter 18. As light escapes from deep within a gravitational well, it loses energy, so photons are shifted to longer and longer wavelengths. If the gravitational well is deep enough, the observed redshift of that radiation can be boundlessly large. In fact, the event horizon of a black hole—that is, the surface around the black hole from which not even light can escape—is where the gravitational redshift becomes infinite.

Cosmological redshift, which is most relevant to this chapter, results from the amount of "stretching" that space has undergone while the light from its original source has been en route to Earth. The amount of stretching is given by the factor $1 + z$. When astronomers observe light from a distant galaxy whose redshift of $z = 1$, the wavelength of that light is twice as long as when it left the galaxy. When the light left its source, the universe was half the size that it is today. When they see light from a galaxy with $z = 2$, the wavelength of the radiation is 3 times its original wavelength, and they are seeing the universe when it was one-third its current size. That direct relationship enables astronomers to use the observed redshift of the galaxy to calculate the size of the universe at the look-back time to that galaxy. Nearby, that means that distance and look-back time are proportional to z. As astronomers look back closer and closer to the Big Bang, however, redshift climbs more and more rapidly, approaching infinity as the look-back time approaches the age of the universe.

Written as an equation, the scale factor of the universe that astronomers see when looking at a distant galaxy is equal to 1 divided by 1 plus the redshift of the galaxy:

$$R_U = \frac{1}{1 + z}$$

For example, when astronomers report they have observed a galaxy with a redshift of 9, the scale factor when the light was emitted was

$$R_U = 1/(1 + 9) = 1/10.$$

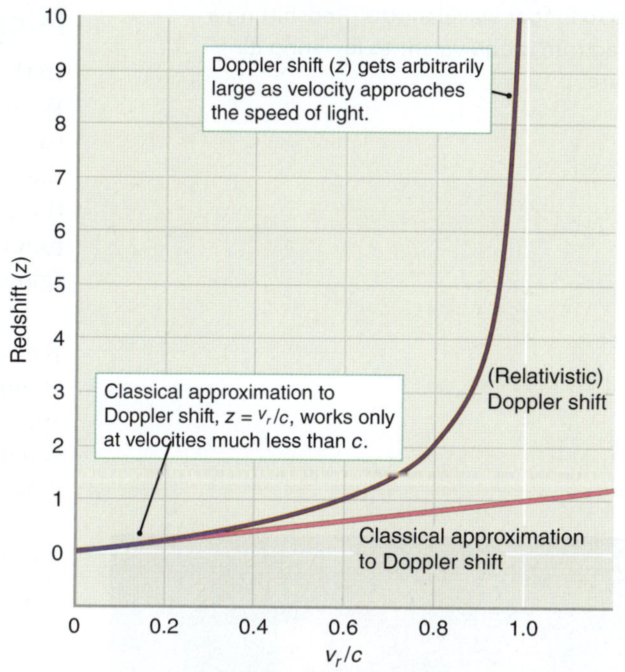

Figure 21.8 This graph shows the plot of the redshift (z) of an object versus its recession velocity (v_r) as a fraction of the speed of light. According to special relativity, the redshift becomes large without limit as v_r approaches c.

Gamow and Alpher took that idea a step further, noting that as the universe expanded, that radiation would have been redshifted to longer and longer wavelengths. According to Wien's law (Chapter 5), the temperature associated with Planck blackbody radiation is inversely proportional to the peak wavelength: $T = (2{,}900{,}000 \text{ nm K})/\lambda_{peak}$. Shifting the Planck radiation to longer and

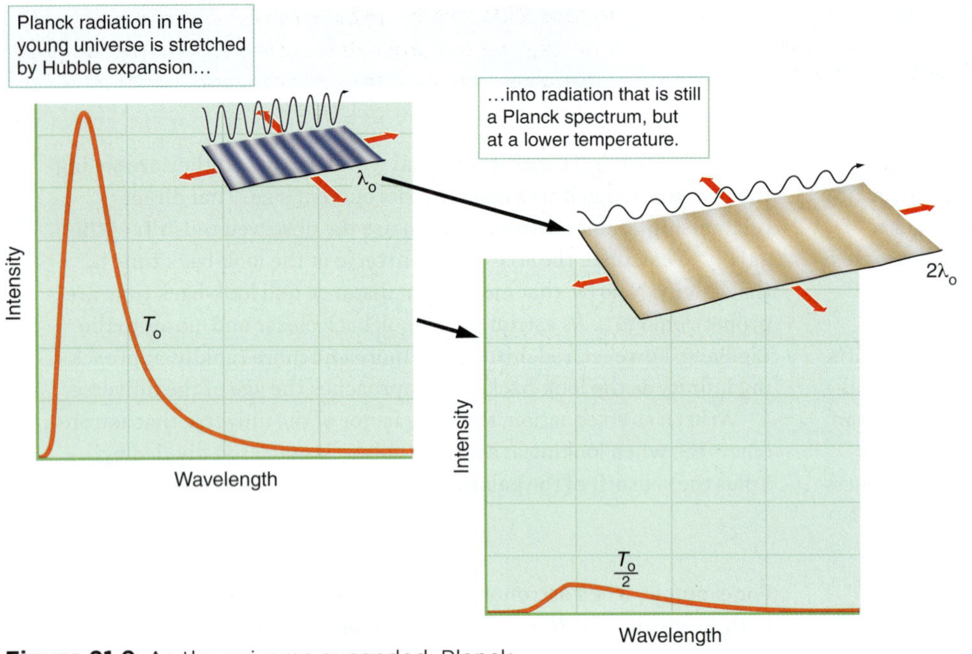

Figure 21.9 As the universe expanded, Planck radiation left over from the hot young universe was redshifted to longer wavelengths. Redshifting a Planck spectrum is equivalent to lowering its temperature.

longer wavelengths is therefore the equivalent of shifting the temperature of the radiation to lower and lower values. As illustrated in **Figure 21.9**, doubling the wavelength of the photons in a Planck blackbody spectrum by stretching space and doubling the scale factor of the universe is equivalent to cutting the temperature of the Planck spectrum in half. As a result, that radiation should still be detectable today and should have a Planck blackbody spectrum with a temperature of about 5–50 kelvins (K). Alpher searched for the signal, but the technology of the late 1940s and early 1950s was not advanced enough. A decade later, physicist Robert Dicke (1916–1997) and his colleagues at Princeton University also predicted a hot early universe, arriving independently at the same basic conclusions that Alpher and Gamow had reached earlier.

Measuring the Temperature of the Cosmic Microwave Background Radiation

In the early 1960s, two physicists at Bell Laboratories—Arno Penzias and Robert Wilson (**Figure 21.10**)—detected a faint microwave signal in all parts of the sky (see Chapter 6). Astronomers interpreted the signal as the radiation left behind by the hot early universe. The strength of the detected signal was consistent with the glow from a blackbody with a temperature of about 3 K, very close to the predicted value. Their results, published in 1965, reported the discovery of the "glow" left behind by the Big Bang.

That radiation left over from the early universe is called the **cosmic microwave background radiation** (**CMB**). When the universe was young, it was hot enough that all atoms were ionized, so that the electrons were separate from the atomic nuclei. Recall from our discussion of the structure of the Sun and stars (Chapter 14) that free electrons in such conditions interact strongly with radiation, blocking its progress, so radiation does not travel well through ionized plasma.

The conditions within the early universe were much like the conditions within a star: hot, dense, and opaque (**Figure 21.11a**). As the universe expanded, the gas filling it cooled. By the time the universe was about 380,000 years old, and about a thousandth of its current size, the temperature had dropped to a few thousand kelvins. Hydrogen and helium nuclei combined with electrons to form neutral atoms for the first time—an event called the **recombination** of the universe (**Figure 21.11b**).

Figure 21.10 Arno Penzias (left) and Robert Wilson next to the Bell Labs radio telescope antenna with which they discovered the cosmic background radiation. That antenna is now a U.S. National Historic Landmark.

Hydrogen atoms block radiation much less effectively than do free electrons, so when recombination occurred, the universe suddenly became transparent to radiation. Since that time, the radiation left behind from the Big Bang has traveled largely unimpeded throughout the universe. At the time of recombination, when the temperature of the universe was about 3000 K, the wavelength of that radiation peaked at about 1 micron (μm), according to Wien's law (see Chapter 5). As the universe expanded, that radiation was redshifted to longer and longer wavelengths—and therefore cooler temperatures, as in Figure 21.9. Today, the scale of the universe has increased a thousandfold since recombination, and the peak wavelength of the cosmic background radiation has increased by a

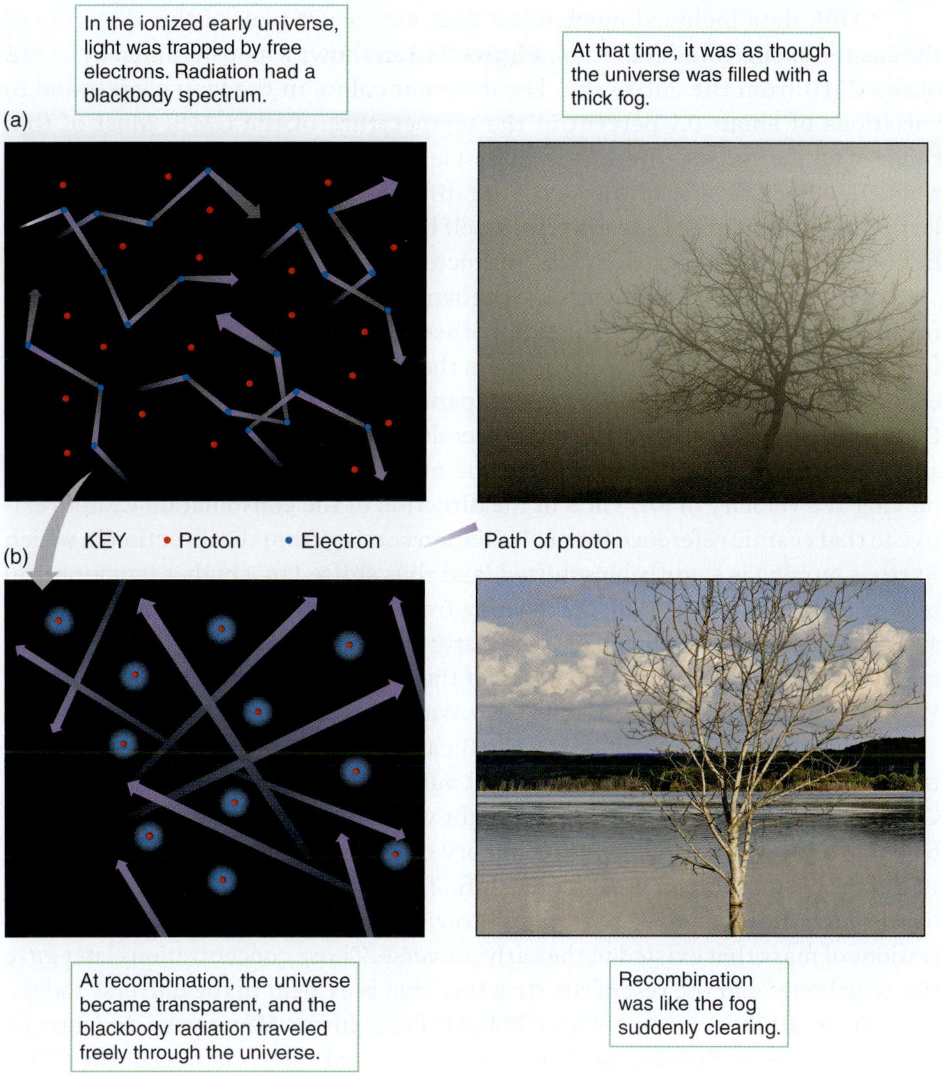

Figure 21.11 The cosmic microwave background radiation we see originated at the moment the universe became transparent. (a) Before recombination, the universe was like a foggy day, except that the "fog" was a sea of electrons and protons. Radiation interacted strongly with free electrons and so could not travel far. The trapped radiation had a Planck blackbody spectrum. The tree in the picture is analogous to a source emitting just before recombination. (b) When the constituents of the universe recombined to form neutral hydrogen atoms, the fog cleared and that radiation was free to travel unimpeded.

thousandfold as well, to a value close to 1 millimeter (mm). The spectrum of the CMB still has the shape of a Planck blackbody spectrum, but with a characteristic temperature a thousandth what it was at the time of recombination.

Variations in the CMB

The presence of cosmic background radiation with a Planck blackbody spectrum is a strong prediction of the Big Bang theory. Penzias and Wilson had confirmed that a signal with the correct strength was there, but they could not say for certain whether the signal they saw had the spectral shape of a Planck blackbody spectrum. From the late 1960s to the 1990s, experiments at different wavelengths supported those same conclusions. The Cosmic Background Explorer (COBE) satellite made extremely precise measurements of the CMB at many wavelengths, from a few microns out to 1 cm. The COBE measurements showed the CMB to be a Planck blackbody spectrum with a temperature of 2.73 K (**Figure 21.12**). The observed spectrum so perfectly matches the one predicted by Big Bang cosmology that no real doubt can exist that this is the residual radiation left behind from the primordial fireball of the early universe. Astrophysicists John Mather and George Smoot won the Nobel Prize in Physics in 2006 for their work on COBE.

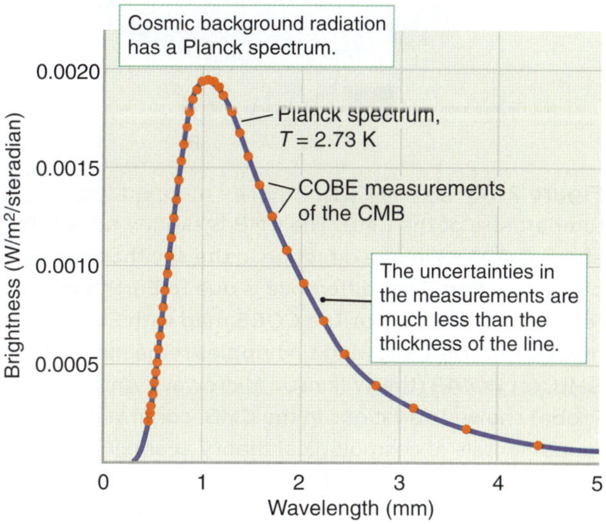

Figure 21.12 The spectrum of the CMB, as measured by the Cosmic Background Explorer (COBE) satellite, is shown by the red dots. The uncertainty in the measurement at each wavelength is much less than the size of a dot. The line running through the data is a blackbody spectrum with a temperature of 2.73 K.

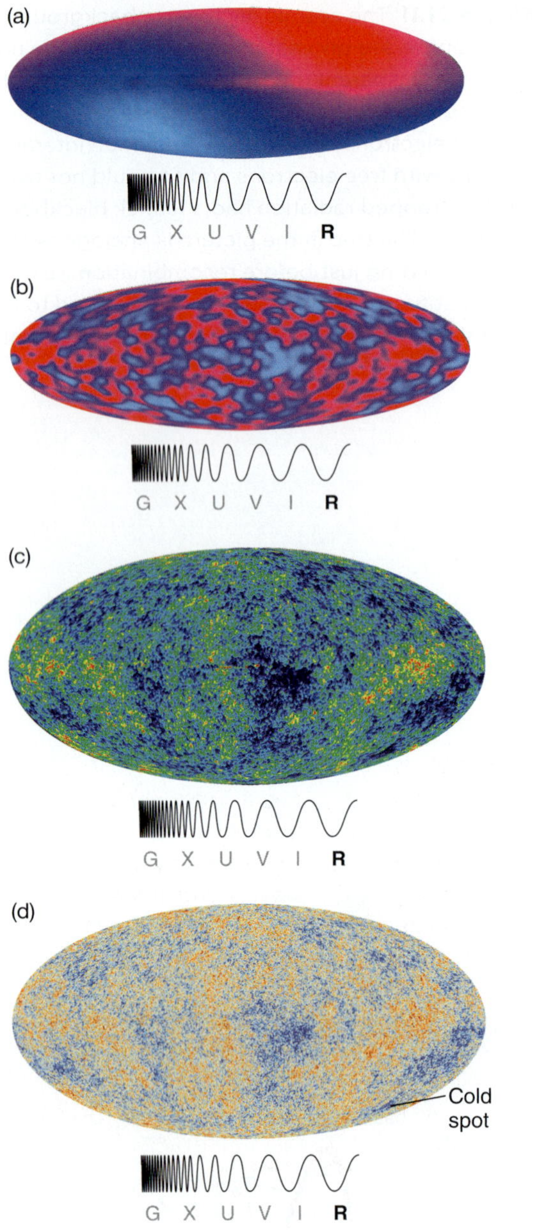

Figure 21.13 (a) The COBE satellite mapped the temperature of the CMB. The CMB is slightly hotter (by about 0.003 K) in one direction in the sky than in the other direction. That difference is due to Earth's motion relative to the CMB. (b) The COBE map with Earth's motion removed, showing tiny ripples remaining in the CMB. (c) *WMAP* (the Wilkinson Microwave Anisotropy Probe) showed variations in the CMB, confirming the fundamentals of cosmological theory at small and intermediate scales. (d) The Planck space observatory has yielded the highest resolution yet of the variations of the CMB and has detected some surprises, such as the "cold spot." The radiation seen here was emitted less than 400,000 years after the Big Bang. Blue spots are cooler, whereas red spots are warmer.

COBE data included much more than a measurement of the spectrum of the cosmic background radiation. **Figure 21.13a** shows a map obtained by COBE of the CMB from the entire sky. The different colors in the map correspond to variations of about 0.1 percent in the temperature of the CMB. Most of that range of temperature is present because one side of the sky looks slightly warmer than the opposite side of the sky. That difference has nothing to do with the large-scale structure of the universe itself but rather is the result of the Earth's motion relative to the gas that last interacted with and created the CMB.

Because the laws of physics are the same in any inertial reference frame, no frame of reference is better than any other. Yet in a certain sense a preferred frame of reference exists at every point in the universe. That is the frame of reference that is at rest with respect to the expansion of the universe and in which the CMB is isotropic, or the same in all directions. The COBE map shows that one side of the sky is slightly hotter than the other because Earth and the Sun are moving at a velocity of 370 km/s in the direction of the constellation Crater relative to that cosmic reference frame. Radiation coming from the direction in which Earth is moving is slightly blueshifted (and thus shifted to a higher temperature) by that motion, whereas radiation coming from the opposite direction is Doppler-shifted toward the red (or cooler temperatures). Earth's motion is due to a combination of factors, including the motion of the Sun around the center of the Milky Way Galaxy and the motion of the Milky Way relative to the CMB.

When that asymmetry in the CMB caused by the motion of Earth is subtracted from the COBE map, only slight variations, about 0.001 percent, in the CMB remain (**Figure 21.13b**). Those slight variations might not seem like much, but they are actually crucial in the history of the universe. Recall from Chapter 18 that gravity itself can create a redshift. Those tiny fluctuations in the cosmic background radiation are the result of gravitational redshifts caused by concentrations of mass that existed in the early universe. Those concentrations later gave rise to galaxies and the rest of the structure that is evident in the universe today.

From 2001 to 2010, NASA's *WMAP* (Wilkinson Microwave Anisotropy Probe) satellite made more precise measurements of the variations of the CMB. **Figure 21.13c** shows the ripples measured by *WMAP* with much higher resolution than could be detected by COBE. The European Space Agency's Planck space observatory collected data of even higher resolution from 2009 to 2013 (**Figure 21.13d**, and chapter opening figure). The much higher-resolution maps obtained by *WMAP* and Planck enable astronomers to refine their ideas about the development of structure in the early universe, which we discuss in Chapters 22 and 23.

CHECK YOUR UNDERSTANDING 21.4

The existence of the cosmic microwave background radiation tells us that the early universe was: (a) much hotter than it is today; (b) much colder than it is today; (c) about the same temperature as today but was much more dense.

Origins Big Bang Nucleosynthesis

The expansion of the universe and the cosmic microwave background radiation are two of the key pieces of observational evidence supporting the Big Bang theory. The third major piece of supporting evidence comes from observations of the number and types of chemical elements in the universe. For a short time after the Big Bang, the temperature and density of the universe were high enough for nuclear reactions to take place. Collisions between protons and neutrons in the early universe built up low-mass nuclei, including deuterium (heavy hydrogen) and isotopes of helium, lithium, and beryllium. That process of element creation, called **Big Bang nucleosynthesis**, determined the final chemical composition of the matter that emerged from the hot phase of the Big Bang. Because the universe was rapidly expanding and cooling, the density and temperature of the universe fell too low for fusion to heavier elements such as carbon to occur. Therefore, all elements more massive than beryllium, including most of the atoms that make up Earth and its life, must have formed in later generations of stars.

AstroTour: Big Bang Nucleosynthesis

Figure 21.14 shows the observed and calculated predictions of the amounts of deuterium, helium, and lithium from Big Bang nucleosynthesis, plotted as a function of the observed present-day density of normal (luminous) matter in the universe. Observations of current abundances are shown as horizontal bands. Theoretical predictions, which depend on the density of the universe, are shown as darker, thick lines. Big Bang nucleosynthesis predicts that about 24 percent of the mass of the normal matter formed in the early universe should have ended up in the form of the very stable isotope ^{4}He, regardless of the total density of matter in the universe. Indeed, that is what is observed. That agreement between theoretical predictions and observation serves as powerful evidence that the universe began in a Big Bang.

Unlike with helium, most other isotope abundances depend on the density of normal matter in the universe, so comparing current abundances with models of isotope formation in the Big Bang helps pin down the density of the early universe. Beginning with the abundances of isotopes such as ^{2}H (deuterium) and ^{3}He found in the universe today (shown as the horizontal, light-colored bands in Figure 21.14) and comparing them with predictions in different models of how abundant isotopes *should* be when formed at different densities (dark-colored curves in Figure 21.14), cosmologists can find the density of normal matter (the vertical yellow band in Figure 21.14). The best current measurements give a value of about 3.9×10^{-28} kg/m^3 for the average density of normal matter in the universe today (at sea level, the density of air is about 1.2 kg/m^3). That value lies well within the range predicted by the observations shown in Figure 21.14. The agreement is remarkable, and it holds for many isotopes.

Turning that process around, cosmologists can begin with an observation of the amount of normal matter in and around galaxies and then compare that with calculations of what the chemical composition emerging from the Big Bang should have been. The observations agree remarkably well with the amounts of those elements actually found in nature. The idea that light elements originated in the Big Bang is resoundingly confirmed.

That agreement also creates a powerful constraint on the nature of dark matter, which dominates the mass in the universe. Dark matter cannot consist of normal matter made up of neutrons and protons; if it did, the density of neutrons and protons in the early universe would have been much higher, and the resulting abundances of light elements in the universe would have been much different from what is actually observed.

Figure 21.14 Observed and calculated abundances of the products of Big Bang nucleosynthesis, plotted against the density of normal matter in today's universe. Big Bang nucleosynthesis correctly predicts the amounts of those isotopes found in the universe today. (Note the two scale breaks on the *y*-axis.)

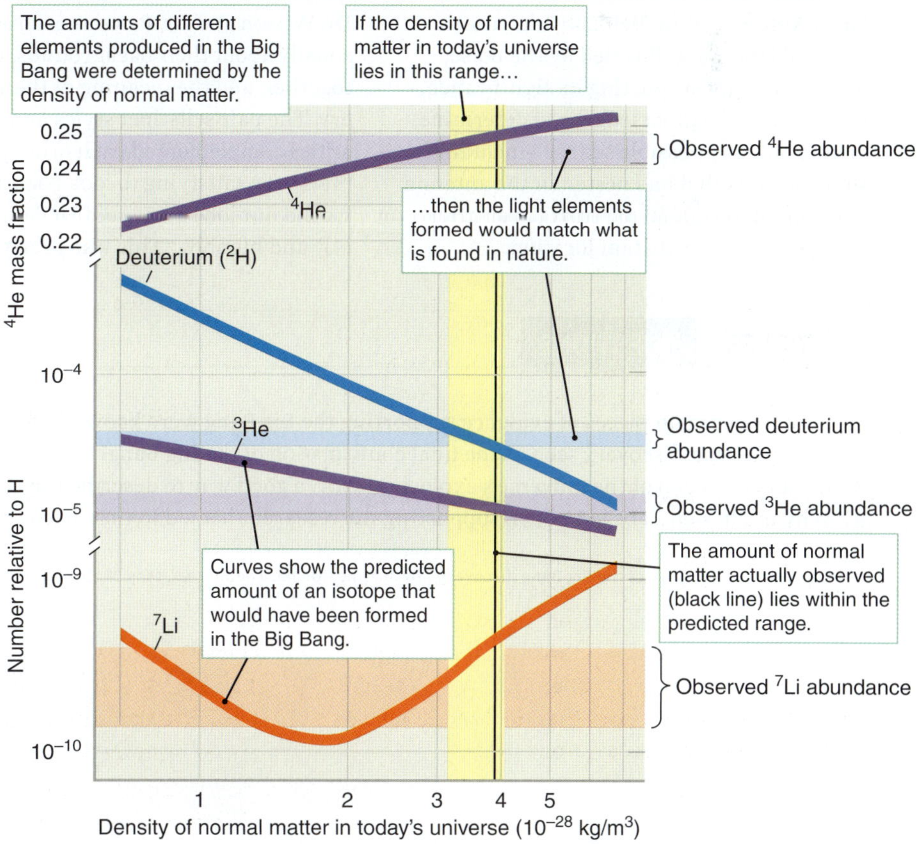

ARTICLES QUESTIONS

Scientists celebrate 50 years since the detection of the cosmic microwave background radiation.

50th Anniversary of the Big Bang Discovery

By JOANNE COLELLA, *The Journal* (NJ)

A unique gathering of some of the most brilliant minds in the world was held last month on Holmdel's Crawfords Hill on a beautiful spring day, celebrating the 50th anniversary of a truly stellar discovery: the detection of cosmic microwave background radiation (CMB), the thermal echo of the universe's explosive birth, and the evidence that proved the famed Big Bang theory, which would have taken place about 13.8 billion years ago. On May 20, 1964, American radio astronomers Robert Wilson and Arno Penzias confirmed that discovery, admittedly by accident but only after an exhaustive amount of investigative research to rule out every possible explanation for an odd buzzing sound that came from all parts of the sky at all times of day and night. The hum was detected by the enormous Horn Antenna at the Bell Labs site, now a national landmark. Puzzled by the noise, but initially not suspecting its significance, the pair went to great lengths to determine, or rule out, any possible source—including some pigeons that had nested in the antenna and were determined to return, even after being shipped to a distant location.

In 1978, Dr. Wilson and Dr. Penzias won the esteemed Nobel Prize in Physics for their work. Now 78 and 81 years old, respectively, the two came together again at the Horn Antenna site to celebrate the momentous anniversary with current and former Bell Labs colleagues. The event was headed up by Bell Labs President and Corporate CTO Marcus Weldon, who lauded their achievement and spoke about the company initiative to return "back to the future"—back to the classic model of Bell Labs, the research arm of Alcatel-Lucent, working to invent the future. Regarding the Big Bang theory, Mr. Weldon brought chuckles by stating, "In the beginning there was nothing and then it exploded . . . and then there was Arno and Bob." Subtle humor was a recurring theme by some speakers, who even poked fun at myths about Bell Labs, which has produced 12 Nobel Prize laureates . . . . Dr. Wilson and Dr. Penzias each spoke personally about their backgrounds, their work together, and the meaning of their discovery. The pair's distinct styles and personalities complemented each other perfectly. "It is very satisfying to look back and see we did our job right," said Dr. Wilson quietly and humbly. "This was pretty heady stuff," stated the talkative Dr. Penzias. "This is as close to being religious as I can be."

Throughout the presentations and celebratory luncheon, held under oversized tents on the expansive grounds, the air of excitement, pride, and mutual admiration among the attendees was palpable, as they looked back—and looked ahead—to the incredible legacy and pool of talent shared by Bell Labs personnel over the years. Robert Wilson and his wife still reside in Holmdel, as do many other Bell Labs employees, and the company has been an integral component of Holmdel and surrounding communities.

During the event, details were also announced about the establishment of the Bell Labs Prize, an annual competition to give scientists around the globe the chance to introduce their ideas in the fields of information and communications technology. The challenge offers a grand prize of $100,000, second prize of $50,000, and third prize of $25,000. Winners may also get the chance to develop their ideas at Bell Labs. The program is intended to inspire world-changing discoveries and innovations by young researchers.

ARTICLES QUESTIONS

1. What other main pieces of evidence supporting the Big Bang were known before that discovery?
2. Why was that discovery seen as the final confirmation of the Big Bang?
3. Is "thermal echo of the universe's explosive birth" a good way to describe the CMB?
4. Why might Bell Labs have been supporting the research that led to that discovery in 1964?

Source: "50th anniversary of the Big Bang discovery," by Joanne Colella. *The Journal*, May 29, 2014. Reprinted by permission of *The Holmdel Journal*, Shrewsbury, NJ.

Summary

The universe has been expanding since the Big Bang, which occurred nearly 13.8 billion years ago. The Big Bang happened everywhere: it is not an explosion spreading out from a single point. Three major pieces of evidence exist for the Big Bang. Hubble's law states that more distant galaxies have higher redshifts, and those observed redshifts result from the increasing space between them. Observations of the cosmic microwave background radiation independently confirm that the universe had a hot, dense beginning. Finally, the amounts of helium and trace amounts of other light elements measured today agree with what would be expected from nuclear reactions of normal matter in the hot early universe.

LG 1 **Explain in detail the cosmological principle.** Observations suggest that on the largest scales, the universe is homogeneous (looks the same to all observers) and isotropic (looks the same in all directions), in agreement with the cosmological principle.

LG 2 **Describe how the Hubble constant can be used to estimate the age of the universe.** Hubble found that all galaxies are moving away, with a recession velocity proportional to distance, indicating that the universe is expanding. The expansion does not affect local physics and the structure of objects. Running the Hubble expansion backward leads to the Hubble time, and the age of the universe, found from the inverse of the Hubble constant.

LG 3 **Describe the observational evidence for the Big Bang.** Hubble's law states that light from distant galaxies is redshifted; that occurs because space itself is expanding. Hubble's law suggests that the universe was once very hot and very dense, a beginning known as the Big Bang. Big Bang theory predicts that we should be able to observe the radiation from a few hundred thousand years after the Big Bang. That radiation, called the cosmic microwave background radiation, has the same spectrum in every direction.

LG 4 **Explain which chemical elements were created in the early hot universe.** During the first few minutes after the Big Bang, the universe was hot enough for nucleosynthesis to take place through fusion. Deuterium, helium, lithium, and beryllium, but not the heavier elements, were created before the universe cooled too quickly for that fusion to take place further.

? Unanswered Questions

- What existed before the Big Bang? The usual (and less than satisfying) answer is that the Big Bang was the beginning of space and *time*, so time could not exist before it happened. A more recent answer is that this universe may be just one of many universes, and the Big Bang was the beginning of this universe only. We return to that topic in Chapter 22.
- If evidence existed that the cosmological principle were not true, then could we deduce anything about the evolution of the universe?

Questions and Problems

Test Your Understanding

1. What do astronomers mean when they say that the universe is homogeneous?
   a. The universe looks the same from every perspective.
   b. Galaxies are generally distributed evenly throughout the universe.
   c. All stars in all galaxies have planetary systems just like ours.
   d. The universe has looked the same at all times in its history.
2. What do astronomers mean when they say that the universe is isotropic?
   a. More distant parts of the universe look just like nearby parts.
   b. Intergalactic gas has the same density everywhere in the universe.
   c. The laws of physics apply everywhere in the universe.
   d. The universe looks the same in every direction.

3. Cosmological redshifts are calculated from observations of spectral lines from
   a. individual stars in distant galaxies.
   b. clouds of dust and gas in distant galaxies.
   c. spectra of entire galaxies.
   d. rotations of the disks of distant galaxies.
4. When they look into the universe, astronomers observe that nearly all galaxies are moving away from the Milky Way. That observation suggests that
   a. the Milky Way is at the center of the universe.
   b. the Milky Way must be at the center of the expansion.
   c. the Big Bang occurred at the current location of the Milky Way.
   d. an observer in a distant galaxy would make the same observation.
5. Some galaxies have redshifts z that if equated to v_r/c correspond to velocities greater than the speed of light. Special relativity is not violated then
   a. because of relativistic beaming.
   b. because of superluminal motion.
   c. because redshifts carry no information.
   d. because those velocities do not measure motion through space.
6. The Big Bang theory predicted (select all that apply)
   a. the Hubble law.
   b. the cosmic microwave background radiation.
   c. the cosmological principle.
   d. the abundance of helium.
   e. the period-luminosity relationship of Cepheid variables.
7. The simplest way to estimate the age of the universe is from
   a. using the slope of Hubble's law.
   b. the age of Moon rocks.
   c. models of stellar evolution.
   d. measurements of the abundances of elements.
8. The CMB includes information about (select all that apply)
   a. the age of the universe.
   b. the temperature of the early universe.
   c. the density of the early universe.
   d. density fluctuations in the early universe.
   e. the motion of Earth around the center of the Milky Way.
9. Repeated measurements showing that the current helium abundance is much less than the value predicted by the Big Bang would imply that
   a. some part of the Big Bang theory is incorrect or incomplete.
   b. the current helium abundance is wrong.
   c. scientists don't know how to measure helium abundances.
10. The cosmological principle says that
    a. the universe is expanding.
    b. the universe began in the Big Bang.
    c. the rules that govern the universe are the same everywhere.
    d. the early universe was 1,000 times hotter than the characteristic temperature of the CMB.
11. Why is the Milky Way Galaxy not expanding together with the rest of the universe?
    a. Because it is at the center of the expansion.
    b. It is expanding, but the expansion is too small to measure.
    c. The Milky Way is a special location in the universe.
    d. Local gravity dominates over the expansion of the universe.
12. The scale factor keeps track of
    a. the movement of galaxies through space.
    b. the current distances between many galaxies.
    c. the changing distance between any two galaxies.
    d. the location of the center of the universe.
13. The Big Bang is
    a. the giant supernova explosion that triggered the formation of the Solar System.
    b. the explosion of a supermassive black hole.
    c. the eventual demise of the Sun.
    d. the beginning of space and time.
14. The CMB is essentially uniform in all directions in the sky. That fact is an example of
    a. anisotropy.
    b. isotropy.
    c. thermal fluctuations.
    d. Wien's law.
15. Which of the following was *not* created as a result of Big Bang nucleosynthesis?
    a. helium
    b. lithium
    c. hydrogen
    d. deuterium
    e. carbon

Thinking about the Concepts

16. Imagine that you are standing in the middle of a dense fog.
    a. Would you describe your environment as isotropic? Why or why not?
    b. Would you describe it as homogeneous? Why or why not?
17. As the universe expands from the Big Bang, galaxies are not actually flying apart from one another. What is really happening?
18. We see the universe around us expanding, which gives distant galaxies an apparent velocity of 70 km/s/Mpc. If you were an astronomer living today in a galaxy located 1 billion light-years away from us, at what rate would you see the galaxies moving away from you?

19. Does Hubble's law imply that our galaxy is sitting at the center of the universe? Explain.

20. What does the value of R_U, the scale factor of the universe, tell us?

21. Does the expansion of the universe make the Sun bigger? What about the Milky Way? Why or why not?

22. Science's greatest strength is the self-correcting nature of scientific inquiry: minds can be changed in the face of new data, as described in the Process of Science Figure. Consider a modern paradigm of science such as the Big Bang, climate change, or the power source of stars. In a short paragraph, describe your current viewpoint, and give a piece of evidence that would make you change your mind if it were true. For example, Einstein believed the universe was static. Measurements of the movement of galaxies caused him to change his mind.

23. Name two predictions of the standard Big Bang theory that have been verified by observations.

24. Knowing that you are studying astronomy, a curious friend asks where the center of the universe is located. You answer, "Right here and everywhere." Explain in detail why you give this answer.

25. The general relationship between recession velocity (v_r) and redshift (z) is $v_r = cz$. That simple relationship fails, however, for very distant galaxies with large redshifts. Explain why.

26. Why is it significant that the CMB displays a Planck blackbody spectrum?

27. What is the significance of the tiny brightness variations observed in the CMB?

28. What important characteristics of the early universe are revealed by today's observed abundances of various isotopes, such as ^2H and ^3He?

29. Why were only a few of the chemical elements created in the Big Bang?

30. How do astronomers know that some of the observed helium is left over from the Big Bang?

Applying the Concepts

31. In Figure 21.6, a rubber sheet is shown as an analogy to help you think about the scale factor. Between the moments shown in parts (a) and (b), each square doubles in size on every edge. How does the area of a square change? Imagine that the sheet is now a block of rubber, expanding in three dimensions instead of two. How would the volume of a cube change between the moment shown in part (a) and the moment shown in part (b)?

32. Error bars have not been plotted in Figure 21.12. Why not? Was that a very precise measurement or a very imprecise measurement? How does the precision of the measurement affect your confidence in the conclusions drawn from it?

33. Figure 21.14 includes both predictions and observations.
    a. What do the vertical yellow bar and the slanted lines and curves represent: theory or observation?
    b. What do the pastel horizontal lines and the vertical black line represent: predictions or observations?
    c. Do the predictions and observations match? Choose one example, and explain how you know.

34. The Hubble time ($1/H_0$) represents the age of a universe that has been expanding at a constant rate since the Big Bang. Calculate the age of the universe in years if H_0 equals 80 km/s/Mpc. (Note: 1 year $= 3.16 \times 10^7$ seconds, and 1 Mpc $= 3.09 \times 10^{19}$ km.)

35. Throughout the latter half of the 20th century, estimates of H_0 ranged from 50 to 100 km/s/Mpc. Calculate the age of the universe in years for each of those estimated values of H_0. Do any answers contradict the ages obtained from stellar evolution or geology on Earth?

36. Suppose a galaxy is observed with a redshift of $z = 2$. How much has the universe expanded since that light was emitted from those galaxies?

37. How much has the universe expanded since light was emitted from a galaxy with a redshift of $z = 8$?

38. You observe a distant quasar in which a spectral line of hydrogen with rest wavelength $\lambda_{rest} = 121.6$ nm is found at a wavelength of 547.2 nm. What is its redshift? When the light from that quasar was emitted, how large was the universe in comparison with its current size?

39. A distant galaxy has a redshift of $z = 5.82$ and a recession velocity $v_r = 287{,}000$ km/s (about 96 percent of the speed of light).
    a. If $H_0 = 70$ km/s/Mpc and if Hubble's law remains valid out to such a large distance, how far away is that galaxy?
    b. Assuming a Hubble time of 13.8 billion years, how old was the universe at the look-back time of that galaxy?
    c. What was the scale factor of the universe at that time?

40. The spectrum of the CMB is shown as the red dots in Figure 21.12, along with a blackbody spectrum for a blackbody at a temperature of 2.73 K. From the graph, determine the peak wavelength of the CMB spectrum. Use Wien's law to find the temperature of the CMB. How does that rough measurement that you just made compare with the accepted temperature of the CMB?

41. COBE observations show that the Solar System is moving in the direction of the constellation Crater at a speed of 368 km/s relative to the cosmic reference frame. What is the blueshift (negative value of z) associated with that motion?

42. The average density of normal matter in the universe is 4×10^{-28} kg/m^3. The mass of a hydrogen atom is 1.66×10^{-27} kg. On average, how many hydrogen atoms are present in each cubic meter in the universe?

43. To get a feeling for the emptiness of the universe, compare its density (4×10^{-28} kg/m^3) with that of Earth's atmosphere at sea level (1.2 kg/m^3). How much denser is Earth's atmosphere? Write that ratio in standard notation.

44. Assume that the most distant galaxies have a redshift of $z = 10$. The average density of normal matter in the universe today is 4×10^{-28} kg/m^3. What was its density when light was leaving those distant galaxies? (Hint: Keep in mind that volume is proportional to the *cube* of the scale factor.)

45. What was the size of the universe (compared with the present) when the CMB was emitted, at $z = 1{,}000$?

USING THE WEB

46. For more details on the history of the discovery of the expanding universe, go to the American Institute of Physics' "Cosmic Journey: A History of Scientific Cosmology" website (https://history.aip.org/exhibits/cosmology/index.htm). Read through the sections titled "Island Universes," "The Expanding Universe," and "Big Bang or Steady State?" Why was Albert Einstein "irritated" by the idea of an expanding universe? What was the contribution of Belgian astrophysicist (and Catholic priest) Georges Lemaître? What is the steady-state theory, and what was the main piece of evidence against it?

47. Go to the American Institute of Physics page on the age of the universe, https://history.aip.org/exhibits/curie/age-of-earth.htm. How does the age of the universe from Hubble's law compare with the age of the Solar System? How does that age compare with the age of the oldest globular clusters? Do a search for "age of the universe"—has the value been updated?

48. A series of public lectures at Harvard commemorating the 50th anniversary of the discovery of the CMB can be found here: https://astronomy.fas.harvard.edu/news/public-talk-50-year-anniversary-discovery-cosmic-micrrowave-background. Robert Wilson's talk starts at the 27-minute mark. Why does he say this is the most important thing he ever measured? How did he identify that radiation?

49. Updated results from observations of the CMB from the Planck space observatory are reported on the European Space Agency's Planck website (http://www.esa.int/Our_Activities/Space_Science/Planck). What has been learned from that mission? What is their value of the Hubble constant?

50. Go to the University of Washington Astronomy Department's "Hubble's Law: An Introductory Astronomy Lab" Web page (https://sites.google.com/a/uw.edu/introductory-astronomy-clearinghouse/activities/galaxies-and-cosmology/hubbles-law) and download the lab exercise, which uses real data from galaxies to calculate Hubble's constant. Your instructor will indicate whether you should use the "current" or the "shortened" version.

EXPLORATION

Hubble's Law for Balloons

digital.wwnorton.com/astro6

The expansion of the universe is extremely difficult to visualize, even for professional astronomers. In this Exploration, you will use the surface of a balloon to get a feel for how an "expansion" changes distances between objects. Throughout this Exploration, remember to think of the surface of the balloon as a two-dimensional object, much as the surface of Earth is a two-dimensional object for most people. The average person can move east or west, or north or south, but moving into Earth and out to space are not options. For this Exploration you will need a balloon, 11 small stickers, a piece of string, and a ruler. A partner is helpful as well. **Figure 21.15** shows some of the steps involved.

Blow up the balloon partially and hold it closed, but *do not tie it shut*. Stick the 11 stickers on the balloon (those represent galaxies) and number them. Galaxy 1 is the reference galaxy.

Measure the distance between the reference galaxy and each of the galaxies numbered 2–10. The easiest way to do that is to use your piece of string. Lay it along the balloon between the two galaxies and then measure the length of the string. Record those data in the "Distance 1" column of a table like the one shown later.

Simulate the expansion of your balloon universe by *slowly* blowing up the balloon the rest of the way. Have your partner count the seconds you take to do that, and record that number in the "Time Elapsed" column of the table (each row has the same time elapsed because the expansion occurred for the same amount of time for each galaxy). Tie the balloon shut. Measure the distance between the reference galaxy and each numbered galaxy again. Record those data under "Distance 2."

Subtract the first measurement from the second. Record the difference in the table.

Divide that difference, which represents the distance traveled by the galaxy, by the time that blowing up the balloon took. Distance divided by time gives an average speed.

Make a graph with velocity on the y-axis and distance 2 on the x-axis to get "Hubble's law for balloons." You may wish to roughly fit a line to those data to clarify the trend.

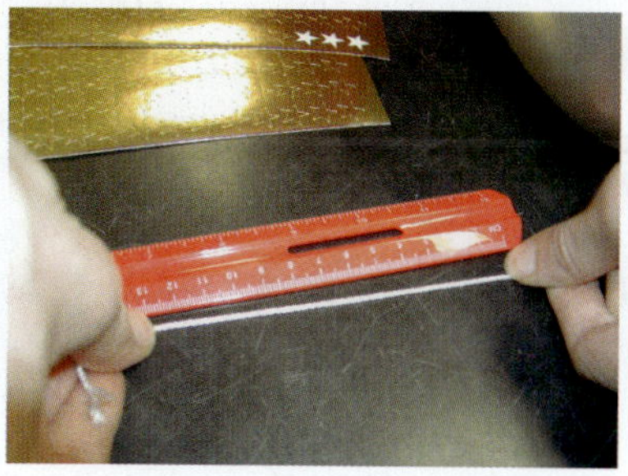

Figure 21.15 Measure the distance around a curved balloon by using a string.

Galaxy Number	Distance 1	Distance 2	Difference	Time Elapsed	Velocity
1 (reference)	0	0	0		0
2					
3					
4					
5					
6					
7					
8					
9					
10					
11					

1 Describe your data. If you fit a line to them, is it horizontal or does it trend upward or downward?

2 Is anything special about your reference galaxy? Is it different in any way from the others?

3 If you had picked a different reference galaxy, would the trend of your line be different? If you are not sure of the answer, get another balloon and try it.

4 The expansion of the universe behaves similarly to the movement of the galaxies on the balloon. We don't want to carry the analogy too far, but let's think about one more thing. In your balloon, some areas probably expanded less than others because the material was thicker; more "balloon stuff" was holding it together. How is that analogy similar to some places in the actual universe?

22 Cosmology

Cosmology is the study of the large-scale universe, including its nature, origin, evolution, and ultimate destiny. Here in Chapter 22, we take a closer look at the nature of the universe, how it has evolved, and its ultimate fate. We also discuss the physics of the smallest particles, which is necessary for describing the earliest moments of the universe.

LEARNING GOALS

By the end of this chapter, you should be able to:

LG 1 Explain how mass within the universe and the gravitational force it produces affect the history, shape, and fate of the universe.

LG 2 Describe the evidence for the accelerating expansion of the universe.

LG 3 Describe the early period of rapid expansion of the universe known as inflation.

LG 4 Explain how the events that occurred in the earliest moments of the universe are related to the forces that operate in the modern universe.

The arrows point to a Type Ia supernova observed in the galaxy NGC 1365. Such supernovae provide evidence about the fate of the universe. ▶▶▶

What will be the fate of the universe?

22.1 Gravity and the Expansion of the Universe

What is the fate of the universe? That is a central question of modern cosmology, and the simplest answer depends in part on the average mass in the universe within a fixed volume. Several factors determine the way mass is distributed on large scales across the universe. In this section, we look at one of those factors: gravity.

Mass Distribution

To see how gravity affects the expansion of the universe, recall gravity's effects on the motion of projectiles and the discussion of escape velocity (Chapter 4). The fate of a projectile fired straight up from the surface of a planet depends on its speed. As long as the projectile speed is *less* than the planet's escape velocity, gravity will eventually stop the rise of the projectile and pull it back to the planet's surface. But if the speed of the projectile is *greater* than the planet's escape velocity, gravity will lose. In that case, the projectile will slow down, but it will never stop, and it will escape from the planet entirely.

The escape velocity from a planet's surface depends on both the mass and radius of the planet. If the planet is massive enough, it will slow the projectile, stop it, and pull it back down. However, if the planet is *not* massive enough, the projectile will slow down but still escape to space. The size of the planet also is important. A smaller planet's mass is more densely packed: if the projectile is fired from the surface, which is closer to the planet's center, the gravitational pull is stronger, and the planet can pull the projectile down. However, if the planet has lower density but the same mass, the surface from which the projectile starts is farther from the center of the planet, the gravitational pull is weaker, and the projectile will escape. Whether the projectile escapes depends on both the planet's mass and the planet's radius.

Just as the mass of the planet gravitationally pulls on a projectile to slow its climb, the mass distributed across the universe, as characterized by its average density, gravitationally slows the expansion of the universe. If enough mass is in the universe, gravity will be strong enough to stop the expansion. However, if the radius of the universe is large, the escape velocity will be small. Thus, the density is critical in determining the fate of the universe. If the mass is packed closely together, so that the average density is high, the expansion will slow, stop, and reverse. If the mass is very spread out, so that the average density is low, the universe will expand forever.

Critical Density

The "escape velocity" of the universe also is determined by its average density. A faster expansion requires a higher density to stop the expansion. Astronomers define a **critical density** for which the mass in the universe would cause it to just barely stop expanding after a very long time. That critical density determines the dividing line between two possible fates of the universe: expanding forever or collapsing. If the universe is less dense than the critical density, gravity will be too weak, and the universe will expand forever. If the universe is denser on average than the critical density, gravity will be strong enough to stop and reverse the expansion eventually.

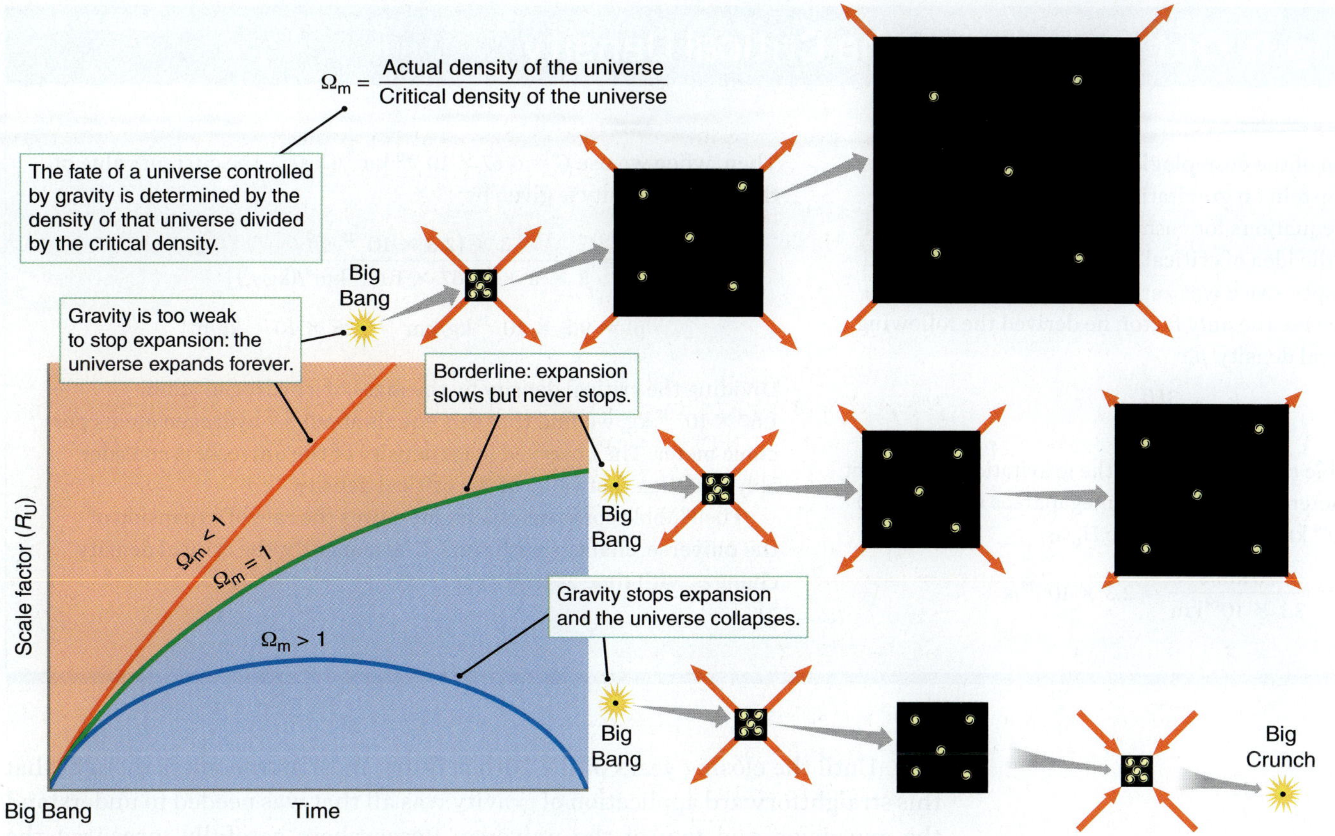

Figure 22.1 The universe has three possible fates if gravity is the only important factor.

Astronomers use omega (Ω), the last letter in the Greek alphabet, to characterize densities and the ultimate fate of the universe. The ratio of actual density to critical density is called the *matter density parameter*, written as Ω_m (pronounced "omega matter"). It is a ratio of two densities, so it has no units.

Figure 22.1 plots the expansion of the universe versus time for various values of Ω_m for a universe in which the expansion is controlled solely by the gravitational attraction of its mass. The colored lines show possible values of Ω_m. If the universe has a density higher than the critical density, Ω_m is greater than 1, and gravity is strong enough to turn the expansion around. The expansion will slow and eventually stop, and the universe will then fall back in on itself. Conversely, if the universe has a density lower than the critical density, Ω_m is less than 1. The expansion of the universe will be slowed by gravity, but it will still expand forever. The dividing line, where Ω_m equals 1, corresponds to a universe that expands more and more slowly—continuing forever, but never quite stopping. Such a universe is expanding at exactly the "escape velocity." If gravity were not present, the Hubble expansion would not be changing, and the plots in Figure 22.1 would be straight lines instead of curved ones. Note that the Hubble parameter H_0 and the density parameter Ω_m can depend on time, although in this book, we will always refer to today's values when discussing those parameters.

The faster the universe is expanding, the more mass is needed to turn that expansion around. Therefore, the critical density today depends on the value of several parameters, including the gravitational constant, *G*, and the Hubble constant, H_0. If gravity is the only factor affecting the expansion, the critical density of the universe is less than the mass of six hydrogen atoms in every cubic meter (**Working It Out 22.1**).

22.1 Working It Out Calculating the Critical Density

Using the assumption of the cosmological principle and using a set of equations from Einstein on gravitation, Alexander Friedmann (1888–1925) derived equations for cosmology in an expanding universe and discussed the idea of critical density. When he assumed that the energy of empty space was zero, making gravity from the total mass of the universe the only factor, he derived the following equation for the critical density, ρ_c:

$$\rho_c = \frac{3H_0^2}{8\pi G}$$

where H_0 is the Hubble constant and G is the gravitational constant. Using $H_0 = 70$ kilometers per second per megaparsec (km/s/Mpc) and 1 Mpc $= 3.1 \times 10^{19}$ km, we can rewrite H_0 as

$$H_0 = \frac{70\text{ km/s}}{3.1 \times 10^{19}\text{ km}} = 2.3 \times 10^{-18}/\text{s}$$

Then, when we use $G = 6.67 \times 10^{-20}\text{ km}^3/(\text{kg s}^2)$, the current value of the critical density is given by

$$\rho_c = \frac{3 \times (2.3 \times 10^{-18}/\text{s})^2}{8 \times \pi \times [6.67 \times 10^{-20}\text{ km}^3/(\text{kg s}^2)]}$$

$$\rho_c = 9.5 \times 10^{-18}\text{ kg/km}^3 = 9.5 \times 10^{-27}\text{ kg/m}^3$$

Dividing the critical density by the mass of a hydrogen atom, 1.67×10^{-27} kg, we find that this equals about 5.7 hydrogen atoms per cubic meter. The observed mass density of the universe is considerably less than that value of the critical density.

The Hubble constant, which measures the rate of expansion of the universe, changes with time. Consequently, the critical density changes with time as well.

Until the closing years of the 20th century, most astronomers thought that this straightforward application of gravity was all that was needed to understand the expansion and fate of the universe. Researchers carefully measured the masses of galaxies and assemblages of galaxies in the expectation that those data would reveal the density and therefore the fate of the universe. The luminous matter seen in galaxies and groups of galaxies gives an Ω_m value of about 0.02. But galaxies contain about 10 times as much dark matter as normal luminous matter, so adding in the dark matter in galaxies pushes the value of Ω_m up to about 0.2. When the mass of dark matter *between* galaxies is included (a subject we return to in Chapter 23), Ω_m could increase to 0.3 or higher. By that accounting, the universe has at most only a third as much mass as is needed to stop the expansion. Many astronomers were convinced that the expansion of the universe would continue but that it would slow down because of gravity. They were wrong.

CHECK YOUR UNDERSTANDING 22.1

If the density of the universe is higher than the critical density, the universe will: (a) expand forever; (b) expand, but gradually slow down; (c) eventually collapse; (d) neither expand nor contract.

22.2 The Accelerating Universe

Edwin Hubble showed that the universe is expanding. If the expansion of the universe is slowing down because of gravity, the expansion must have been faster in the past. If so, objects very far away—so that we see them as they were long ago—are moving faster. They should therefore have larger redshifts than Hubble's law derived from local galaxies would suggest. However, if objects very far away have *smaller* redshifts than Hubble's law derived from local galaxies indicates, the expansion of the universe is speeding up. In this section, we examine the evidence that the expansion of the universe is accelerating.

The Cosmological Constant

During the 1990s, astronomers measured the brightness of Type Ia supernovae in very distant galaxies and compared the brightness of those supernovae with their expected brightness, which was based on the redshift distances of those galaxies. (Recall from Chapter 19 that Type Ia supernovae have a very high peak luminosity that can be calibrated, so they are used as "standardizable" candles for measuring distances to galaxies.) The findings of those studies sent a wave of excitement through the astronomical community. The observational data of the Type Ia supernovae at different distances are plotted in **Figure 22.2**. Instead of showing that the expansion of the universe has slowed down, the data indicated that the expansion is *speeding up*—accelerating. An "accelerating universe" does *not* mean that the universe is zooming through space faster and faster like a car along a road. It means instead that the expansion is happening faster and faster. For that to be true, a previously unknown force stronger than gravity must be pushing the *entire universe* apart in opposition to gravity. Results from the *WMAP* (Wilkinson Microwave Anisotropy Probe) spacecraft early in the 21st century confirmed the result independently, and in 2011, the Nobel Prize in Physics was awarded to Saul Perlmutter, Brian P. Schmidt, and Adam G. Riess for their observations of Type Ia supernovae and the discovery of the accelerating universe.

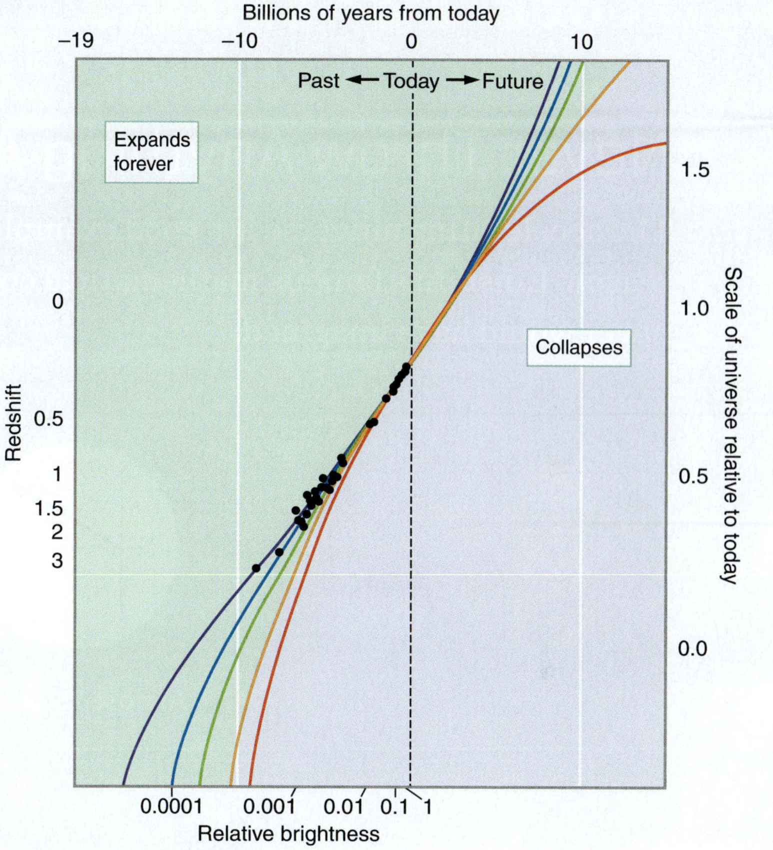

Figure 22.2 This graph plots the observed brightness of Type Ia supernovae as a function of their redshift. The different-colored lines represent different values of the cosmological constant. The observations (data points) indicate that the redshifts are too small for their distances: they best fit the line for an accelerating universe—a universe expanding faster today than it did in the past. The colored lines are more separate at higher redshifts, so astronomers look for higher-redshift supernovae to differentiate the data better.

The idea of a repulsive force that opposes the attractive force of gravity is not new. In the early 20th century, Einstein used his newly formulated equations of general relativity to calculate the structure of spacetime in the universe. The equations clearly indicated that any universe containing mass could not be static, any more than a ball can hang motionless in the air. He found the same result that Figure 22.1 illustrates—namely, that gravity always makes the universe move toward slower expansion or even collapse. However, Einstein's formulation of spacetime came more than a decade before Hubble discovered the expansion of the universe, and the conventional wisdom then was that the universe was static—that it neither expands nor collapses.

To force his new general theory of relativity to allow for a static universe, Einstein inserted a "fudge factor" called the **cosmological constant** into his equations. Einstein's cosmological constant, which is constant in both space and time, acts as a repulsive force that opposes gravity. If it has just the right value, the cosmological constant can lead to a static universe in which galaxies remain stationary despite their mutual gravitational attraction.

When Hubble announced his discovery that the universe is expanding, Einstein realized his mistake. Einstein could have predicted that the universe must be either expanding or contracting with time, but instead he forced his equations to comply with conventional wisdom. He called the introduction of the cosmological constant the "biggest blunder" of his scientific career. The much more recent discovery of an accelerating universe, however, restores the cosmological constant to general relativity. With the results on the brightness of Type Ia supernovae, Einstein's cosmological constant turned out to be critical to understanding the expansion of the universe. A repulsive force, like the force associated with the cosmological constant in Einstein's equations, is just what is needed to describe a universe expanding at an accelerating rate (see the **Process of Science Figure**).

Today, we write the cosmological constant as Λ (uppercase lambda). The fraction of the critical density provided by the cosmological constant is written as

Never Throw Anything Away

The accelerating universe is a good example of a scientist being right for the wrong reason. Einstein's famous mathematical constant has become useful in a way he never would have imagined.

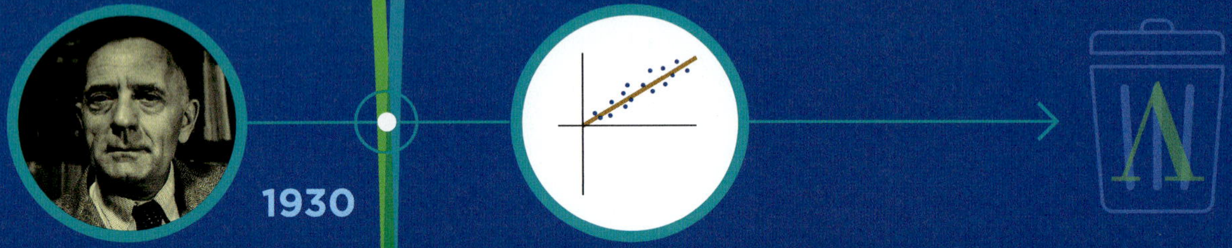

Almost 75 years after Hubble proved Einstein wrong . . .

. . . astronomers were working to measure the slowing down of the expansion of the universe, and to their surprise, they found that the expansion rate was accelerating.

Einstein's "fudge factor" was revived as the cosmological constant, which accounts for the acceleration.

Science has a memory, so old ideas can be revisited in the light of new data, often saving a great deal of effort, even when the interpretation has changed.

Ω_Λ (pronounced "omega lambda"). In a universe with an accelerating expansion, Ω_Λ is not zero. If Ω_Λ is not zero, the fate of the universe is no longer controlled exclusively by Ω_m but rather is affected by *both* Ω_Λ and Ω_m. If something is effectively pushing outward from within the universe, adding to its expansion, gravity will have a harder time turning the expansion around. Then the mass density needed to halt the expansion of the universe will be greater than the critical mass density. **Figure 22.3** shows plots of size of the universe versus time that are similar to those in Figure 22.1, but now we have included the effects of Ω_Λ. The fate of the universe depends on the balance between Ω_Λ and Ω_m. If the mass density (or Ω_m) is large enough, gravity will stop the expansion, and the universe will collapse back on itself, regardless of the value of Ω_Λ. However, if Ω_m is less than 1, and therefore not large enough to stop the expansion, the universe expands forever, and whether the expansion of the universe will accelerate or decelerate depends on the value of Ω_Λ.

When the universe was young and compact, the effect of gravity was strong enough to dominate the effect of the cosmological constant. As the universe expanded, gravity grew weaker and weaker relative to the cosmological constant because the mass was more and more spread out. Unlike gravity, however, the effect of the cosmological constant has become increasingly *greater* as the universe expands. Unless gravity is strong enough to counter the cosmological constant, the cosmological constant will win in the end, causing the expansion to continue accelerating forever. Indeed, even if the density of the universe is greater than the critical density (that is, Ω_m is greater than 1), so that otherwise it would have collapsed back on itself because of gravity, a large enough cosmological constant could overwhelm gravity and make the universe expand forever (dashed blue line in Figure 22.3).

When Einstein added the cosmological constant to his equations of general relativity, he considered it a new fundamental constant, similar to Newton's universal gravitational constant *G*. Today, physicists call "empty space" the **vacuum** and recognize that it has some distinct physical properties. For example, the vacuum can have nonzero energy even in the total absence of matter. That energy of empty space produces exactly the kind of repulsive force that could accelerate the expansion of the universe. That repulsive force is called **dark energy**. Einstein's cosmological constant is an example of dark energy that accelerates the expansion of the universe.

Although the cosmological constant does not change, other versions of dark energy can evolve. Dark energy is a very active area of study, as scientists work to figure out where that energy comes from, its form and properties, and whether the amount changes. One major observational effort to both measure the amount of dark energy and whether it changes with time is an international collaboration called the Dark Energy Survey. That effort uses multiple approaches to measuring dark energy, including finding many Type Ia supernovae, measuring the observed distortions of galaxies from dark matter fluctuations, and using other phenomena sensitive to dark energy.

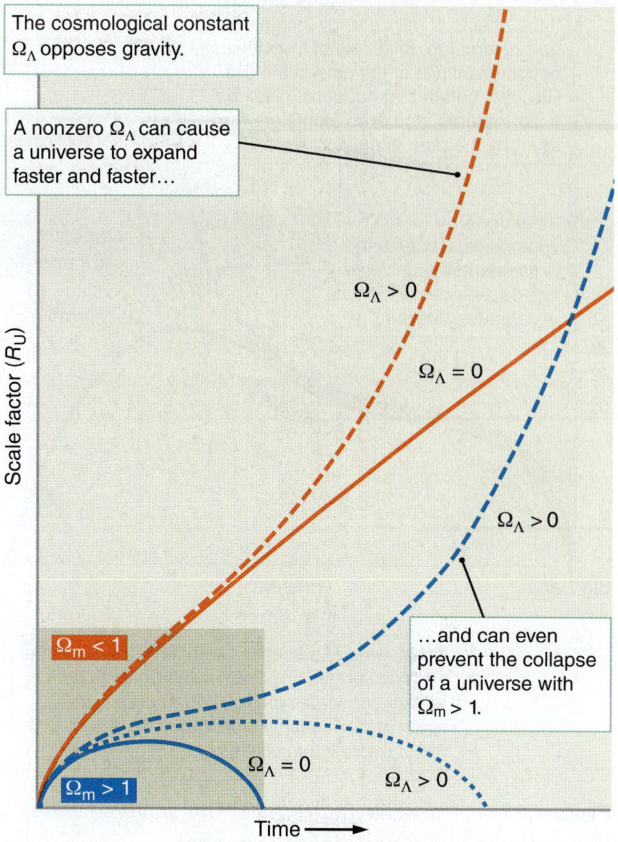

Figure 22.3 This graph plots scale factor R_U versus time for a possible universe with or without a cosmological constant, Ω_Λ. The solid curves for $\Omega_\Lambda = 0$ are the same as in Figure 22.1. The dashed curves show universes with $\Omega_\Lambda > 0$ for different Ω_m. If enough mass is in a universe, gravity could still overcome the cosmological constant and cause that universe to collapse. In the presence of a cosmological constant, any universe without enough mass to collapse will instead eventually end up expanding at an ever-increasing rate.

Dark Energy and the Fate of the Universe

What fate, then, actually awaits the universe? Astronomers have data from several experiments showing the range of possible values for Ω_m and Ω_Λ. The data from Type Ia supernovae—from *WMAP*, from *Planck*, and from clusters of galaxies (discussed in Chapter 23)—are all consistent with values for Ω_m and Ω_Λ of

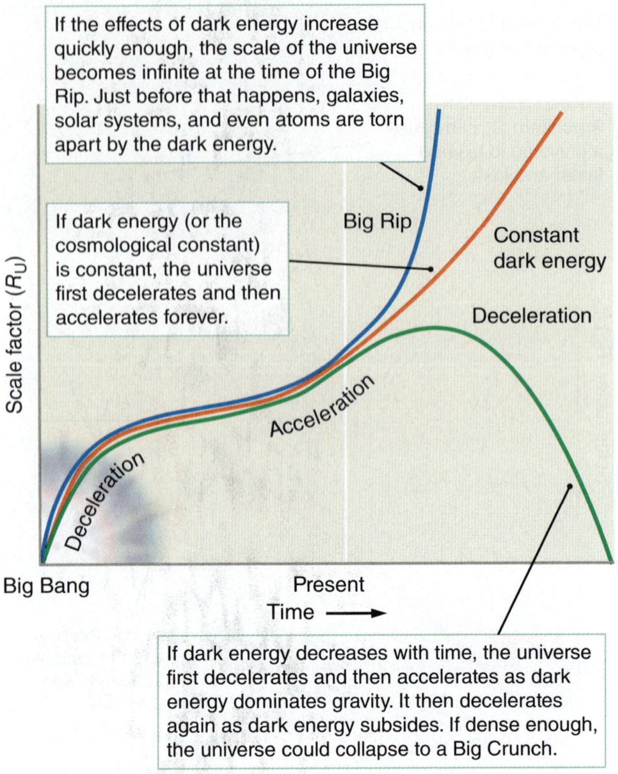

Figure 22.4 The scale factor R_U of the universe varies depending on how dark energy changes.

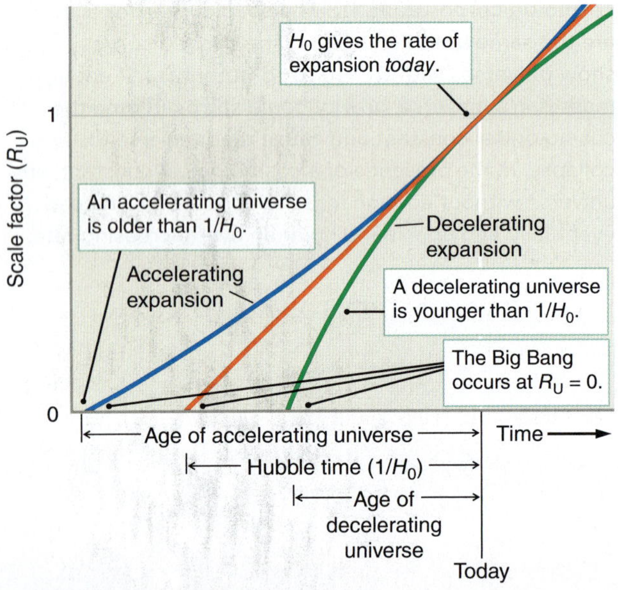

Figure 22.5 This graph shows plots of the scale factor R_U versus time for three possible universes. If the universe has expanded at a constant rate, its age is equal to the Hubble time, $1/H_0$. If the expansion of the universe has slowed, the universe is younger than the Hubble time. If the expansion has sped up, the universe is older than $1/H_0$.

about 0.3 and 0.7, respectively. Thus, the expansion of the universe is apparently accelerating under the dominant effect of dark energy and has been doing so for 5 billion to 6 billion years.

Because scientists do not yet understand the origin of dark energy, it is possible that it is not really a constant of nature and instead could be either increasing or decreasing. A changing cosmological constant would significantly change the future of the universe (**Figure 22.4**). For example, if dark energy were to decrease rapidly enough with time, the accelerating expansion of the universe that is observed now would change to a *deceleration* as the mass once again dominates over dark energy. In fact, if the universe were much denser than measured values indicate, the expansion could reverse, and the universe could collapse to what astronomers call the **Big Crunch**. If the effect of dark energy were to increase with time, by contrast, the universe would accelerate its expansion at an ever-increasing rate. Ultimately, expansion could be so rapid that the scale factor would become infinite within a finite period—a phenomenon called the **Big Rip**. In the Big Rip, the repulsive force of dark energy would become so dominant that the entire universe would come apart. First, gravity would no longer keep groups of galaxies together; then, gravity would no longer be able to hold individual galaxies together; and so on. Just before the end, the Solar System would come apart, and even atoms would be ripped into their constituent components. Don't worry too much about the Big Crunch or the Big Rip, though: the best observational data seem consistent with constant dark energy. In that case, the universe continues expanding, and after 100 trillion years or so, the universe is cold and dark, filled with black holes, dead stars, and dead planets (sometimes called the *Big Chill*).

The Age of the Universe

The values for Ω_m and Ω_Λ not only affect predictions for the future of the universe but also influence how astronomers interpret the past. **Figure 22.5** shows plots of the scale factor of the universe versus time. Measurement of the Hubble constant (H_0) indicates how fast the universe is expanding *today*. That is, the Hubble constant indicates the slope of the curves in Figure 22.5 at the current time. If the expansion of the universe has not changed, the plot of the scale factor versus time is the straight red line in Figure 22.5. The age of the universe in that case is equal to the Hubble time: $1/H_0$. If the expansion of the universe has been slowing down (green line in Figure 22.5), the universe is actually younger than the Hubble time. If the expansion of the universe has been speeding up (blue line), the true age of the universe is greater than the Hubble time. That scenario is also illustrated in **Figure 22.6**, where looking down the blue time line from the present to the past shows that the different models take different amounts of time to go from the Big Bang to the present.

Recall that a Hubble constant of $H_0 = 70$ km/s/Mpc corresponds to a Hubble time ($1/H_0$) of about 13.8 billion years. If the expansion of the universe has slowed, the universe is actually younger than 13.8 billion years. Having a younger universe is a problem if the measured ages of globular clusters—13 billion years—is correct because globular clusters cannot be older than the universe that contains them. But if the expansion of the universe has sped up, as suggested by the observations of Type Ia supernovae and of the 2.7-K cosmic background radiation, the universe is at least 13.8 billion years old—comfortably older than globular clusters. The effects from gravity and dark energy have nearly canceled out now.

The Shape of the Universe

We have already discussed such properties of the universe as density, dark energy, and age. The universe also has another key property: its shape in spacetime. Recall the concept of spacetime as described by general relativity. Space is a "rubber sheet" that has stretched outward from the Big Bang. In Chapter 18, you saw that the rubber sheet of space is also curved by the presence of mass. You saw how the shape of space around a massive object is detected through changes in geometric relationships, such as the ratio of the circumference of a circle to its radius or the sum of the angles in a triangle. Recall that the mass of a star, planet, or black hole causes a distortion in the shape of space; similarly, the mass of everything in the universe—including galaxies, dark matter, and dark energy—distorts the shape of the universe *as a whole*.

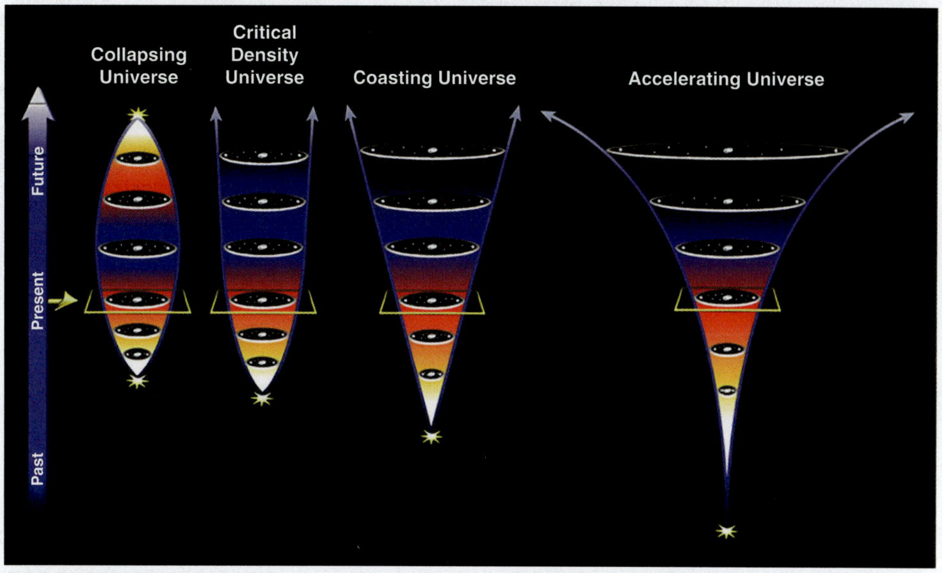

Figure 22.6 Past, present, and future for different scenarios. All models start in the past with a Big Bang (yellow star), expand to the present, and then have different possible futures. The present age of the universe is different in each model, as in Figure 22.5.

Three basic shapes are possible for the universe (**Figure 22.7**). Which shape actually describes the universe is determined by the total amount of mass and energy—in other words, the sum of Ω_m and Ω_Λ. As we continue with the rubber-sheet analogy, the first possibility is a **flat universe** (Figure 22.7a), corresponding to $\Omega_m + \Omega_\Lambda = 1$. A flat universe is described overall by the rules of the basic Euclidean geometry. That is, circles in a flat universe have a circumference of 2π times their radius ($2\pi r$), and triangles contain angles whose sum is 180 degrees. A flat universe stretches on forever.

The second possibility, an **open universe** (Figure 22.7b), corresponds to $\Omega_m + \Omega_\Lambda < 1$ and is shaped something like a saddle. In an open universe, the circumference of a circle is greater than $2\pi r$, and triangles contain less than 180 degrees. An open universe, like a flat universe, is infinite.

The third possibility, a **closed universe** (Figure 22.7c) corresponds to $\Omega_m + \Omega_\Lambda > 1$ and is like the surface of a sphere. The geometric relationships on a sphere are similar to those near a massive object (see Chapter 18). Thus the circumference of a circle on a sphere is less than $2\pi r$, and triangles contain more than 180 degrees. Space in a closed universe is finite and closes back on itself. Imagine a universe with two spatial dimensions on the surface of a sphere. Then the surface (and universe) is finite with no boundary. The cosmological principle is satisfied. Space is locally flat, and the expansion of the universe maintains both the cosmological principle and the local flatness of space.

Figure 22.7 These two-dimensional representations show possible geometries that space can have in a universe. In a flat universe (a), Euclidean geometry holds: triangles have angles that sum to 180 degrees, and the circumference of a circle equals 2π times the radius. In an open universe (b) or a closed universe (c), those relationships are no longer correct over very large distances.

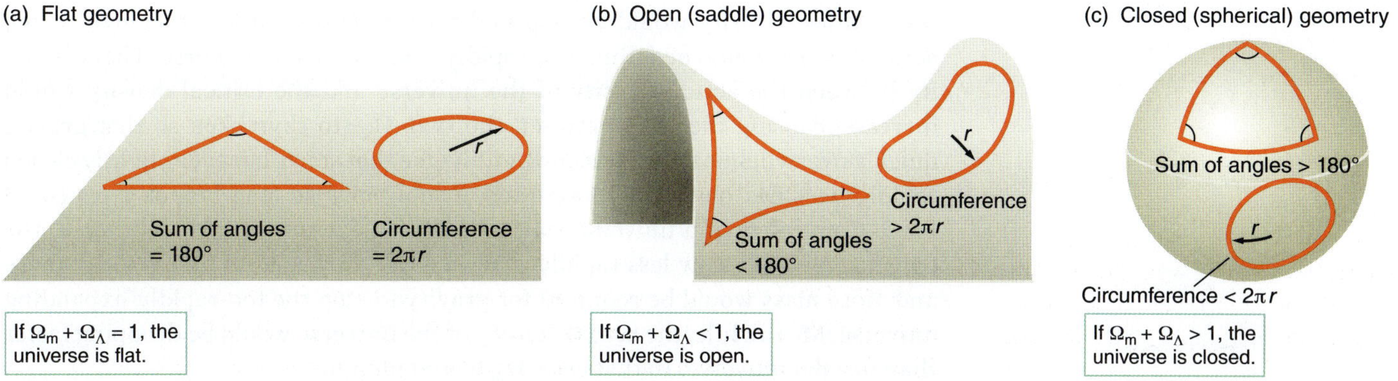

The measurements to estimate directly which of those shapes describes the universe are difficult. As noted earlier, the data suggest that $\Omega_m = 0.3$ and $\Omega_\Lambda = 0.7$, so $\Omega_m + \Omega_\Lambda$ is close to 1, in which case the universe is very nearly flat.

CHECK YOUR UNDERSTANDING 22.2

Dark energy has been hypothesized to solve which problem? (a) The universe is expanding. (b) The cosmic microwave background radiation is too smooth. (c) The expansion of the universe is accelerating. (d) Stars orbit the centers of galaxies too fast.

22.3 Inflation Solves Several Problems in Cosmology

A century ago, astronomers were struggling to understand the size of the universe. Today scientists have a comprehensive theory that ties together many diverse facts about nature: the constancy of the speed of light, the properties of gravity, the motions of galaxies, and even the origins of the atoms that make up planets and life. The case for the Big Bang is compelling. Even so, improved observations of the cosmic background radiation and measurements of the expansion of the universe have raised some questions about how the universe expanded when it was very young. In this section, we look at those questions and some potential answers.

The Flatness Problem

From the cosmic microwave background radiation (CMB), astronomers have found that the universe is flat—too close to being exactly flat for that to have happened by chance. Any deviation from flatness would grow, so if the universe originally had a value of $\Omega_m + \Omega_\Lambda$ even slightly different from 1, the value would by now be drastically different and easily detectable. For the present-day value of $\Omega_m + \Omega_\Lambda$ to be as close to 1 as it is, when the universe was 1 second old, $\Omega_m + \Omega_\Lambda$ must have been equal to 1 all the way out to at least the tenth decimal place. At even earlier times, it had to be much flatter still. That situation is too special to be the result of chance—a fact referred to in cosmology as the flatness problem: the universe is so flat that something about the early universe must have caused $\Omega_m + \Omega_\Lambda$ to have a value incredibly close to 1.

A universe that contains mass and does not start out perfectly flat has a very different fate. If a universe started out with Ω_m even slightly greater than 1, its expansion would slow more rapidly than that of the flat universe, meaning that less and less density would be required to stop the expansion. Meanwhile, the actual density would be falling less rapidly than in the flat universe. That disparity between the actual density of the universe and the critical density would increase, causing the ratio between the two, Ω_m, to skyrocket, so that gravity quickly overwhelms the expansion. A universe that starts out even slightly closed rapidly becomes obviously closed and would collapse long before stars could form. Conversely, if a universe started with Ω_m even a tiny bit less than 1, the expansion would slow less rapidly than in a flat universe. As time passed, more and more mass would be required for gravity to stop the too-rapidly-expanding universe. Meanwhile, the actual density of the universe would be dropping faster than in a flat universe. In that case, Ω_m would plummet.

Adding Ω_Λ to the picture makes the math more complex but does not change the basic results. Try balancing a razor blade on its edge. If the blade is tipped just a tiny bit in one direction, it quickly falls that way. If the blade is tipped just a tiny bit in the other direction, it quickly falls in the other direction instead. It would seem that the actual universe should be obviously open or obviously closed—analogous to the tipped razor blade. Instead, the universe has $\Omega_m + \Omega_\Lambda$ so close to 1 that telling which way the razor blade is tipped is difficult—if it is tipped at all. Discovering that $\Omega_m + \Omega_\Lambda$ is extremely close to 1 after more than 13 billion years is like balancing a razor blade on its edge and coming back 10 years later to find that it still has not tipped over.

The Horizon Problem

Another problem faced by cosmological models is that the CMB is surprisingly smooth. After its discovery in the 1960s, many observational cosmologists turned their attention to mapping that background radiation. At first, result after result showed that the temperature of the CMB is remarkably constant, with variations of less than one part in 3,000, regardless of where one looks in the sky. Over time, though, that strong confirmation of Big Bang cosmology challenged cosmologists' view of the early universe. Once Earth's motion relative to the CMB is removed from the picture, the CMB is not just smooth—it is *too* smooth.

In Chapter 5, we discussed the bizarre world of quantum mechanics that shapes the world of atoms, light, and elementary particles. When the universe was extremely young, it was so small that quantum mechanical effects played a role in shaping the structure of the universe as a whole. The early universe was subject to the quantum mechanical **uncertainty principle**, which says that as a system is studied at extremely small scales, the properties of that system become less and less well determined. That principle applies to the properties of an electron in an atomic orbital or to the entire universe when it was very young and would have fit within the size of an atom.

Consider a simple analogy of how the uncertainty principle applies to the universe. Imagine sitting on the beach looking out across the ocean. Off in the distance, you see more total ocean and average the surface over larger scales; therefore, the surface of the ocean appears smooth and flat. The horizon looks almost like a geometric straight line. Yet the apparent smoothness of the ocean as a whole hides the tumultuous structure present at smaller scales, where waves and ripples fluctuate dramatically from place to place. Similarly, while the universe on large scales seems steady and smooth, quantum mechanics says that conditions must fluctuate unpredictably at smaller and smaller scales in the universe. In particular, quantum mechanics says that those fluctuations are more dramatic earlier and earlier in the history of the universe. When the universe was young, it could not have been smooth. Dramatic variations ("ripples") must have existed in the density and temperature of the universe from place to place.

If the universe had expanded slowly, those ripples would have smoothed themselves out, but the universe expanded much too rapidly for such smoothing to be possible. After the Big Bang, not enough time was available for a smoothing signal to travel from one region to the other. So when cosmologists look at the universe today, they should see the fingerprint of those early ripples imprinted on the cosmic background radiation—but they do not. The fact that the CMB is so smooth is called the *horizon problem* in cosmology. The **horizon problem**

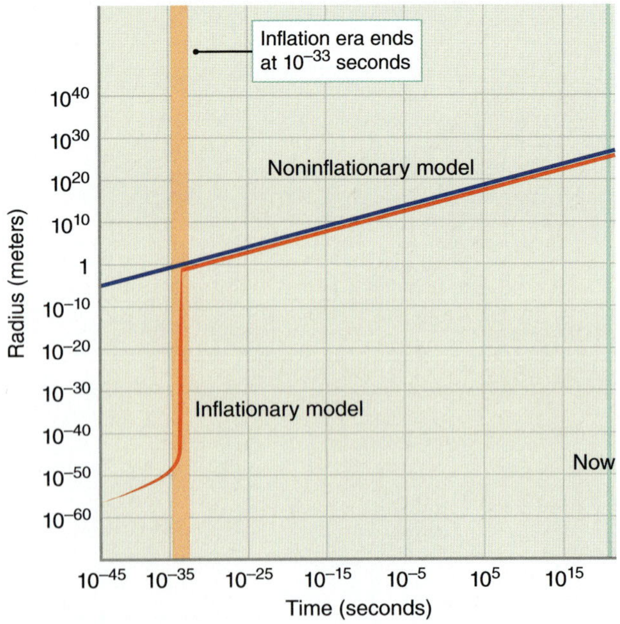

Figure 22.8 This graph illustrates expansion of the observable universe. In an inflationary model, between 10^{-35} and 10^{-33} second after the Big Bang, the universe expanded greatly.

states that different parts of the universe are too much like other parts of the universe that should have been "over their horizon" and beyond the reach of any signals that might have smoothed out the early quantum fluctuations. In essence, the horizon problem is as follows: How can different parts of the universe that underwent different fluctuations and were never able to communicate with one another still show the same temperature in the cosmic background radiation to an accuracy of better than one part in 100,000?

Inflation: Early, Rapid Expansion

In the early 1980s, physicist Alan Guth (1947–) offered a solution to the flatness and horizon problems of cosmology. Guth suggested that the universe has not expanded at a steady pace but that instead it started out much more compact than steady expansion would predict. Then, briefly, the young universe expanded at a rate *far* in excess of the speed of light. That rapid expansion of the universe is called **inflation**. In the first 10^{-33} second of the universe, the distance between points in space increased by a factor of at least 10^{30} and perhaps very much more (**Figure 22.8**). In that incomprehensibly brief instant, the size of the observable universe grew from ten-trillionths the size of the nucleus of an atom to a region about 1 meter across. That is like a grain of very fine sand growing to the size of today's observable universe—all in a billionth of the time that light takes to cross the nucleus of an atom. During inflation, space itself expanded so rapidly that the distances between points in space increased faster than the speed of light. Inflation does *not* violate the rule that no signal can travel through space faster than the speed of light because the space itself was expanding.

To understand how inflation solves the flatness and horizon problems of cosmology, imagine that you are an ant living in the two-dimensional universe defined by the surface of a golf ball (**Figure 22.9**). This universe is positively curved, like the surface of Earth. If you were to walk around the circumference of a circle in your two-dimensional universe and then measure the radius of the circle, you would find the circumference to be less than $2\pi r$. If you were to draw a triangle in your universe, the sum of its angles would be greater than 180 degrees. Another obvious characteristic would be the dimples, approximately a half-millimeter deep, on the surface of the golf ball.

Now imagine that the golf-ball universe suddenly grew to the size of Earth. The curvature of the universe would no longer be apparent. An ant (or person) walking along the surface of the golf-ball universe would think the universe is flat. The circumference of a circle would be $2\pi r$, and a triangle would have 180 degrees. (In fact, it took most of human history for people to realize that Earth is a sphere.) For inflationary cosmology, the universe after inflation would be extraordinarily flat (that is, having $\Omega_m + \Omega_\Lambda$ extraordinarily close to 1) *regardless* of what the geometry of the universe was before inflation. Because the universe was inflated by a factor of at least 10^{30}, $\Omega_m + \Omega_\Lambda$ immediately after inflation must have been equal to 1 within one part in 10^{60}, which is flat enough for $\Omega_m + \Omega_\Lambda$ to remain close to 1 today. If inflation occurred, today's universe is not flat by chance. It is flat because any universe that underwent inflation would become flat.

What about the horizon problem? When the golf-ball universe inflates to the size of Earth, the dimples that covered the surface of the golf ball stretch out as well. Instead of being a half millimeter or so deep and a few millimeters across, those dimples now are only an atom deep but are hundreds of kilometers across. The ant would not detect any dimples at all. For our universe, inflation took the

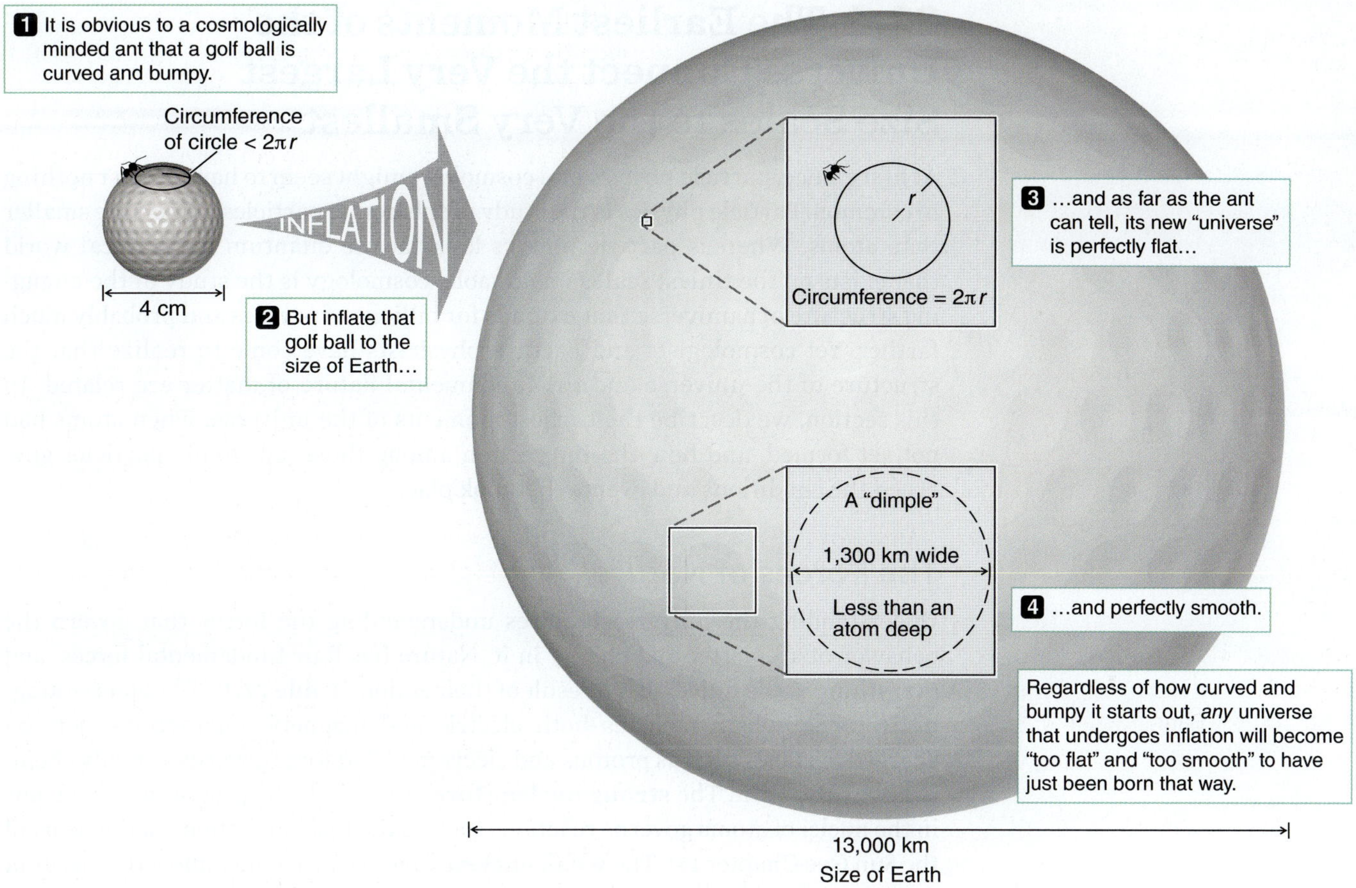

Figure 22.9 If a round, lumpy golf ball were suddenly inflated to the size of Earth, the ball would seem flat and smooth to an ant on its surface. Similarly, after inflation, any universe would seem both extremely flat and extremely smooth, regardless of the exact geometry and irregularities it started with.

large fluctuations in conditions caused by quantum uncertainty in the preinflationary universe and stretched them out so much that they are not measurable in today's postinflationary local universe. The slight irregularities observed in the CMB are the faint ghosts of quantum fluctuations that occurred as the universe inflated.

An early era of inflation in the history of the universe offers a way to solve the horizon and flatness problems, but that the universe should have undergone a period during which it expanded at such a high rate seems remarkable. The cause of inflation must lie in the fundamental physics that governed the behavior of matter and energy at the earliest moments of the universe, but current physical theories have not yet been able to predict the details of an early inflationary phase of the universe. Although the existence of an inflationary epoch is difficult to test, it is not impossible, and astronomers are devising ways to test whether inflation occurred in the early universe. Note that one prediction of inflation is that the universe will be filled with a background of gravitational radiation, just as it is filled with the cosmic background radiation.

CHECK YOUR UNDERSTANDING 22.3

Identify the two problems of cosmology that inflation solves.

22.4 The Earliest Moments of the Universe Connect the Very Largest Size Scales to the Very Smallest

At first glance, particle physics and cosmology might seem to have almost nothing in common. Particle physics is the study of subatomic particles, which are smaller than atoms. Whereas particle physics looks at the quantum mechanical world that exists on the tiniest scales imaginable, cosmology is the study of the changing structure of a universe that extends for billions of parsecs and probably much farther. Yet cosmologists and particle physicists have come to realize that the structure of the universe and the fundamental nature of matter are related. In this section, we describe the earliest moments of the universe, when atoms had not yet formed, and how the interaction among those subatomic particles governed the conditions and events that took place.

The Forces of Nature

Understanding the universe requires understanding the forces that govern the behavior of all matter and energy in it. Nature has four fundamental forces, and everything in the universe is a result of their action (**Table 22.1**). The **electromagnetic force**, which includes both electric and magnetic interactions, acts on charged particles such as protons and electrons. That force governs not only chemistry but also light. The **strong nuclear force** that binds the protons and neutrons in the nuclei of atoms governs reactions such as the fusion reactions in the heart of the Sun (see Chapter 14). The **weak nuclear force** governs the radioactive decay of unstable nuclei. Finally, *gravity*, which plays a major role throughout astronomy, governs how matter affects the geometry of spacetime. In models of particle physics, the first three forces combine in a single force, leaving only gravity to stand alone. To understand how those four forces formed and came to govern the universe today, we must explore backward in time, toward the Big Bang itself.

The underlying concept of the standard model is that forces between particles are caused by the exchange of carrier particles. Light can be described either as an electromagnetic wave resulting from electric and magnetic fields or as a stream of particles called photons (see Chapter 5). Those two descriptions of electromagnetism coexist. The branch of physics that deals with a quantum description of radiation is called **quantum electrodynamics (QED)**.

QED treats charged particles almost as though they were baseball players engaged in an endless game of catch. As the players throw and catch baseballs, they

TABLE 22.1 The Four Fundamental Forces of Nature

Force	Relative Strength	Range of Force	Particles That Can Carry the Force	Example of What the Force Does
Strong nuclear	1	10^{-15} m	Gluons	Holds protons and neutrons together in atomic nuclei.
Electromagnetic	10^{-2}	Infinite	Photons	Binds the electrons in an atom to the nucleus.
Weak nuclear	10^{-4}	10^{-16} m	W^+, W^-, and Z^0	Responsible for beta decay.
Gravitational	10^{-38}	Infinite		Holds you to Earth; binds planetary systems, stars, galaxies, clusters of galaxies, and so forth.

experience forces. Similarly, in QED, charged particles "throw" and "catch" an endless stream of "virtual photons" (**Figure 22.10**). Quantum mechanics is a science of probabilities rather than certainties. The QED description of the electromagnetic interaction between two charged particles is an average of all the possible ways that the particles could throw photons back and forth. The resulting force acts, over large scales, like the classical electromagnetic force that we observe in chemistry. Physicists describe that effect as the electromagnetic force being mediated by the exchange of photons.

The central idea of QED—forces mediated by the exchange of carrier particles—serves as a template for understanding two of the other three fundamental forces in nature. The electromagnetic and weak nuclear forces have been combined into a single theory called **electroweak theory**. That theory predicts the existence of three particles—labeled W^+, W^-, and Z^0—that mediate the weak nuclear force. Sheldon Glashow, Abdus Salam, and Steven Weinberg received the 1979 Nobel Prize in Physics for their work on the theory of the unified weak and electromagnetic forces. In the 1980s, physicists identified those particles in laboratory experiments and confirmed the essential predictions of electroweak theory.

The strong nuclear force is described by a distinct theory called **quantum chromodynamics (QCD)**. QCD states that particles such as protons and neutrons are composed of more fundamental building blocks called **quarks**, which are bound together by the exchange of another type of carrier particle, dubbed **gluons**. Together, electroweak theory and QCD compose the **standard model** of particle physics. Excluding gravity, the standard model explains all the currently observed interactions of matter and has made many predictions that laboratory experiments later confirmed. Despite that great success, the standard model leaves many questions unanswered, such as whether neutrinos have mass or why strong interactions are so much stronger than weak interactions.

At high temperatures, the different forces are indistinguishable because the different carrier particles have such high energy. Therefore, our universe started out with all the forces unified, as described by one (as yet unknown) theory of everything. As the universe expanded and cooled, symmetry between the particles was broken, and carrier particles for different forces became distinguishable. The Higgs field, which is responsible for breaking the symmetry between different kinds of carrier particles, is manifested as an elementary particle called the Higgs boson.

The Higgs boson is the particle that all other particles must interact with to gain their masses. In the standard model, all particles are created without mass. When the electroweak symmetry breaks (as expansion and cooling continue), that "special" particle is created throughout the universe. All existing particles interact with the Higgs field and gain their mass in the process. The existence of that particle was predicted in 1964, and it was finally detected at the Large Hadron Collider in Europe in 2012. In 2013, Peter Higgs and François Englert shared the Nobel Prize for their work predicting that particle. As the universe cooled, the carrier particles (and other particles, such as electrons) gained mass due to the Higgs field, and so the forces also began to be distinguishable.

A Universe of Particles and Antiparticles

In the standard model, every particle in nature has an **antiparticle** that is identical in mass but opposite in charge. For example, the positron emitted in the proton-proton chain (see Chapter 14) is the antiparticle of an electron. A positron is identical to an electron except that it has a positive charge instead of a negative

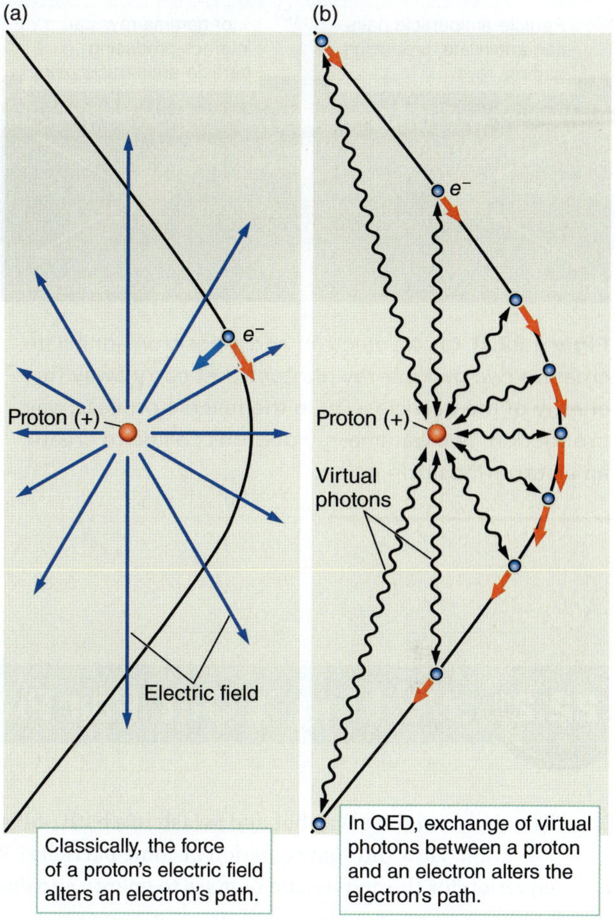

Figure 22.10 (a) This figure shows the classical view of an electron being deflected from its course (black line) by the electric field from a proton. (b) According to quantum electrodynamics models, the interaction is viewed as an ongoing exchange of virtual photons between the two particles.

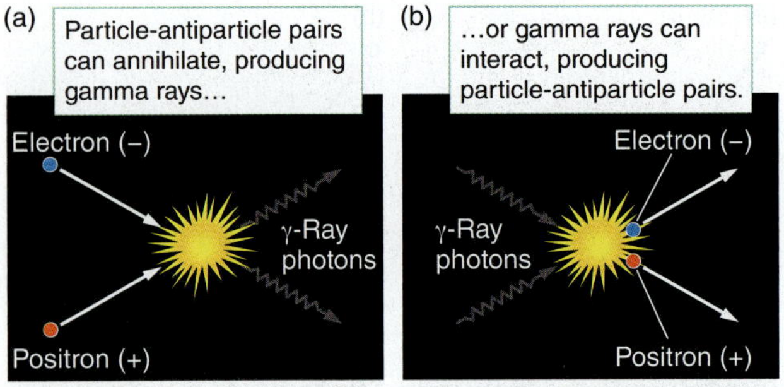

Figure 22.11 (a) An electron and a positron annihilate, creating two gamma-ray photons that carry away the energy of the particles. (b) In the reverse process, pair production, two gamma-ray photons collide to create an electron-positron pair.

charge. For the proton there is the antiproton; for the neutron, the antineutron; and so on down the list. Collectively, those antiparticles are called **antimatter**.

One property of particle-antiparticle pairs is that if you bring such a pair together, the two particles annihilate each other. When a particle-antiparticle pair annihilates, the mass of the two particles is converted into energy in accordance with Einstein's equation $E = mc^2$. In **Figure 22.11a**, for example, an electron and a positron annihilate each other, and the energy is carried away by a pair of gamma-ray photons. Particle-antiparticle pairs were produced when two high-energy photons collided with each other (**Figure 22.11b**), creating in their place an electron-positron pair. **Pair production**—when an energetic collision creates a particle and its corresponding antiparticle—has been observed in particle accelerators.

In principle, *any* type of particle and its antiparticle can be created by pair production. The only limitation comes when not enough energy is available to supply the mass of the particles being created (**Working It Out 22.2**). If two gamma-ray photons with a combined energy greater than the rest mass energy of an electron-positron pair collide, the two photons may disappear and leave an

22.2 Working It Out Pair Production in the Early Universe

The early universe was hot and awash in a bath of Planck blackbody radiation. How did that radiation become particles? Recall Einstein's equation for the conversion of mass to energy and energy to mass:

$$E = mc^2$$

To produce a pair consisting of a particle plus an antiparticle of a certain rest mass, a minimum amount of energy is required. The formula shows that producing a higher-mass particle plus antiparticle requires more energy than does producing a lower-mass pair. For example, a proton has 1,836 times the mass of an electron, so it will require 1,836 times as much energy to produce a proton-antiproton pair of particles than to produce an electron-positron pair.

We can relate that amount of energy to the average energy of particles at a given temperature through the following equation:

$$E = 3/2\, kT$$

where k is Boltzmann's constant: 1.38×10^{-23} kg m^2/s^2/K (which is joules per kelvin; J/K). Equating the two energies yields

$$mc^2 = 3/2\, kT$$

Rearranging,

$$T = \frac{2mc^2}{3k}$$

The proton and antiproton each have a mass of 1.67×10^{-27} kg. What temperature would the radiation have to be to produce that proton-antiproton pair?

$$T = \frac{2 \times (2 \times 1.67 \times 10^{-27}\ \text{kg}) \times (3.0 \times 10^8\ \text{m/s})^2}{3 \times (1.38 \times 10^{-23}\ \text{kg m}^2/\text{s}^2/\text{K})}$$

$$T = 1.45 \times 10^{13}\ \text{K}$$

Such high temperatures are thought to have existed only during the first few seconds after the Big Bang.

Similarly, to produce an electron-positron pair, each of which has a mass of 9.11×10^{-31} kg, the temperature of the CMB would have to be

$$T = \frac{2 \times (2 \times 9.11 \times 10^{-31}\ \text{kg}) \times (3.0 \times 10^8\ \text{m/s})^2}{3 \times (1.38 \times 10^{-23}\ \text{kg m}^2/\text{s}^2/\text{K})}$$

$$T = 7.9 \times 10^9\ \text{K}$$

That is still very hot, but less than the temperature required to form a proton-antiproton pair, so electron-positron production lasted several minutes, a longer time after the Big Bang than did proton-antiproton production.

We also can think of that in terms of the energy of the photons involved in creating those particles. From Chapter 5, recall that the energy of a photon is related to its wavelength by

$$E = \frac{hc}{\lambda} \quad \text{or} \quad \lambda = \frac{hc}{E}$$

where h is Planck's constant ($h = 6.63 \times 10^{-34}$ kg m^2/s). We then use $E = 3/2\, kT$ with our value of T for the electron-positron production above, yielding

$$\lambda = \frac{2hc}{3kT} = \frac{2 \times (6.63 \times 10^{-34}\ \text{kg m}^2/\text{s}) \times (3.0 \times 10^8\ \text{m/s})}{3 \times (1.38 \times 10^{-23}\ \text{kg m}^2/\text{s}^2/\text{K}) \times (7.92 \times 10^9\ \text{K})}$$

$$\lambda = 1.21 \times 10^{-12}\ \text{m}$$

According to the electromagnetic spectrum shown in Figure 5.6, photons with that wavelength are high-energy gamma-rays.

electron-positron pair behind in their place. If the photons have more than the necessary energy, the extra energy goes into the kinetic energy of the two newly formed particles.

Now we apply that idea to a hot universe awash in a bath of blackbody radiation. When the universe was less than about 100 seconds old and had a temperature greater than a billion kelvins, it was filled with energetic photons constantly colliding, creating electron-positron pairs, and those electron-positron pairs were constantly annihilating each other, creating pairs of gamma-ray photons. Those two processes reached equilibrium, determined strictly by temperature, in which pair creation and pair annihilation exactly balanced each other. Instead of being filled only with a swarm of photons, the universe then was filled with a swarm of photons, electrons, and positrons. Earlier, when the universe was even hotter, photons would have produced a swarm of protons and antiprotons. Still earlier, a swarm of quarks/antiquarks and gluons existed, called a "quark-gluon plasma," as has been observed in some heavy-nucleus accelerators on Earth.

Grand Unified Theories

In pair production, a symmetry exists between matter and antimatter: for every particle created, its antiparticle is created as well. As the universe cooled, there was no longer enough energy to create particle pairs, so the swarm of particles and antiparticles that filled the early universe annihilated each other and were not replaced. When that cooling happened, first every proton should have been annihilated by an antiproton. Then at still cooler temperatures, every electron should have been annihilated by a positron. That was almost the case, but not quite. For every proton and electron in the universe today, 10 billion and one protons and electrons existed in the early universe, but only 10 billion antiprotons and positrons. That one-part-in-10-billion excess of electrons over positrons meant that when electron-positron pairs finished annihilating each other, some electrons were left over—enough to account for all the electrons in all the atoms in the universe today (**Figure 22.12**). Similarly, the early universe had an excess of protons over antiprotons, and the protons observed today are all that is left from the annihilation of proton-antiproton pairs.

If the standard model of particle physics were a complete description of nature, the imbalance of one part in 10 billion between matter and antimatter would not have been present in the early universe. The symmetry between matter and antimatter would have been complete. No matter at all would have survived into today's universe, and galaxies, stars, and planets would not exist. The fact that you are reading this page demonstrates that something more needs to be added to the model.

Several competing ideas seek to explain why the amounts of matter and antimatter were unequal in the early universe. One set of ideas is called **grand unified theories (GUTs)** because they join the electromagnetic force, weak nuclear force, and strong nuclear force together into a single force. Such theories do not include gravity, however. GUTs explain why the universe is composed of matter rather than antimatter. When the universe was very young (younger than about 10^{-35} second) and very hot (hotter than about 10^{27} K), enough energy was available for particles associated with a GUT to be freely created. During that time, the distinction among the electromagnetic, weak nuclear, and strong nuclear forces had not yet taken place. Only the one unified force existed. During that era of GUTs, the apparent size of the entire observable universe was less than a trillionth the size of a single proton.

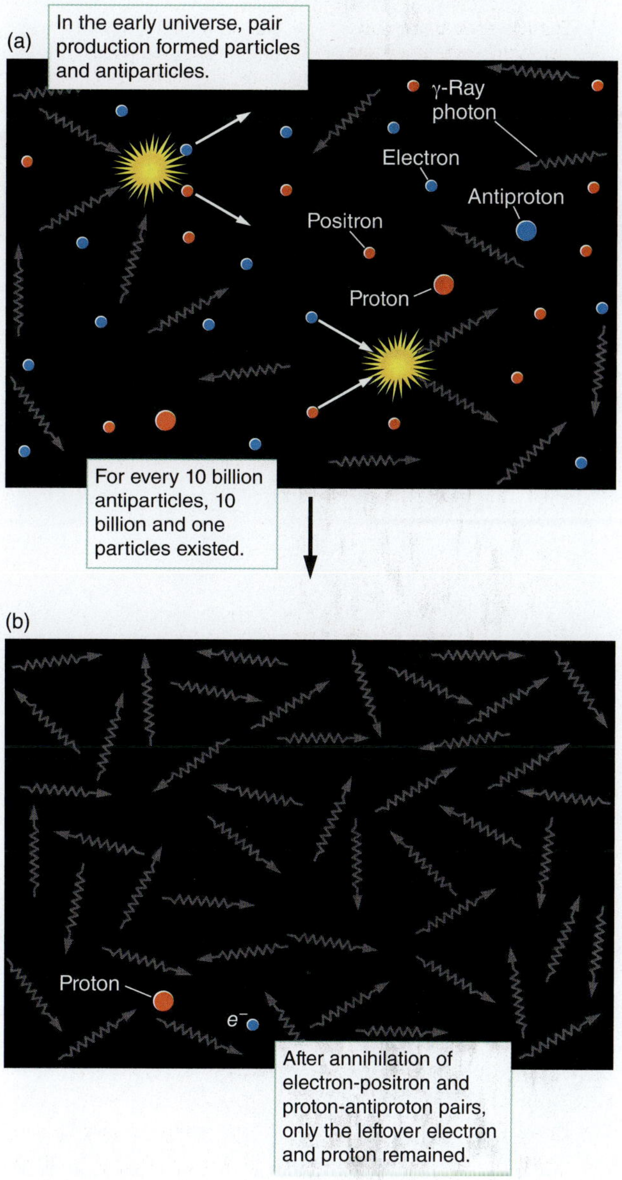

Figure 22.12 (a) For every 10 billion antiparticles in the early universe, 10 billion and one particles existed. (b) After those particles annihilated, only one electron and one proton remained.

Many possible GUTs exist, and they make many predictions about the universe. Unfortunately, most of those predictions are impossible to test with even the largest of today's particle colliders. The problem is that the particles carrying the grand unification forces are so massive that creating them takes enormous amounts of energy—roughly a trillion times as much energy as can be achieved in today's particle accelerators. Even so, some predictions of GUTs can be tested with current technology. For example, GUTs predict that protons should be unstable particles that, given enough time, will decay into other types of elementary particles. That process is *very* slow. Over 100 years, GUTs predict that as much as a 1 percent chance may exist that *one* of the 10^{28} or so protons in your body will decay. Experiments have given a lower limit on the lifetime of a proton of 10^{34} years, but as of this writing, proton decay has yet to be observed.

How does gravity fit into that scheme? General relativity provides a description of gravity that correctly predicts the orbits of planets, describes the ultimate collapse of stars, and enables astronomers to calculate the structure of the universe. Yet general relativity's description of gravity is very different from the theories of the other three forces. Instead of talking about the exchange of photons or gluons or other mediating particles, general relativity talks about the large, smooth, continuous canvas of spacetime that events are painted on. The era of GUTs is described perfectly if gravity is treated as a force separate from the other three forces. However, a time even closer to the Big Bang exists when a description of the universe requires that gravity be treated as a quantum phenomenon.

Toward a Theory of Everything

Even earlier in the universe than when the three forces were unified, when the universe was younger than about 10^{-43} second, its density was incomprehensibly high. The observable universe was so small that 10^{60} universes would have fit into the volume of a single proton. Under those extreme conditions, general relativity can no longer describe spacetime: quantum physics is needed to describe not only particles but also spacetime itself. Rather than a smooth sheet, spacetime was a quantum mechanical "foam." The failure of general relativity to describe that early universe is much like the failure of Newtonian mechanics to describe the structure of atoms. An electron in an atom must be thought of in terms of probabilities rather than certainties. Similarly, no deterministic history is available for the earliest moments after the Big Bang. That era in the history of the universe is called the **Planck era**, signifying that physicists can understand the structure of the universe during that period only by using the ideas of quantum mechanics.

The conflict between general relativity and quantum mechanics is at the current limits of human knowledge. Known physics can explain things back to when the universe was a ten-millionth of a trillionth of a trillionth of a trillionth of a second old, but to push back any earlier, something new is needed. To understand the earliest moments of the universe, physicists need a theory that combines general relativity and quantum mechanics into a single theoretical framework unifying all *four* fundamental forces—a **theory of everything** (**TOE**).

A successful TOE would do more than unify general relativity with quantum mechanics. It would suggest which of the possible GUTs is correct and would tell us the nature of dark matter. A successful TOE would also explain the how, when, and why of inflation and the underlying physics of the dark energy accelerating the expansion of the universe. Physicists are grappling with what a TOE might look like. One leading contender is superstring theory (see Section 22.5).

The Forces Separated in the Cooling Universe

To understand the very earliest moments in the history of the universe, physicists look backward to earlier and earlier times and, consequently, to higher and higher energies. Now let's organize the events the other way, beginning at the beginning.

In the early universe, the TOE could unite all four forces, and as the universe expanded and cooled, the various forces emerged separately. **Figure 22.13** illustrates how the four fundamental forces emerged in the evolving universe. In the first 10^{-43} second after the Big Bang, as described by the TOE, the physics of elementary particles and the physics of spacetime were one and the same. As the universe expanded and cooled, gravity separated from the forces described by the GUT. Spacetime took on the properties described by general relativity. Inflation also may have been taking place then.

As the universe continued to expand and its temperature fell further, less and less energy was available to create particle-antiparticle pairs. When the particles responsible for mediating GUT interactions could no longer form, the strong force split off from the others. As the unity of the original TOE was lost, the symmetry between matter and antimatter was broken. As a result, the universe ended up with more matter than antimatter.

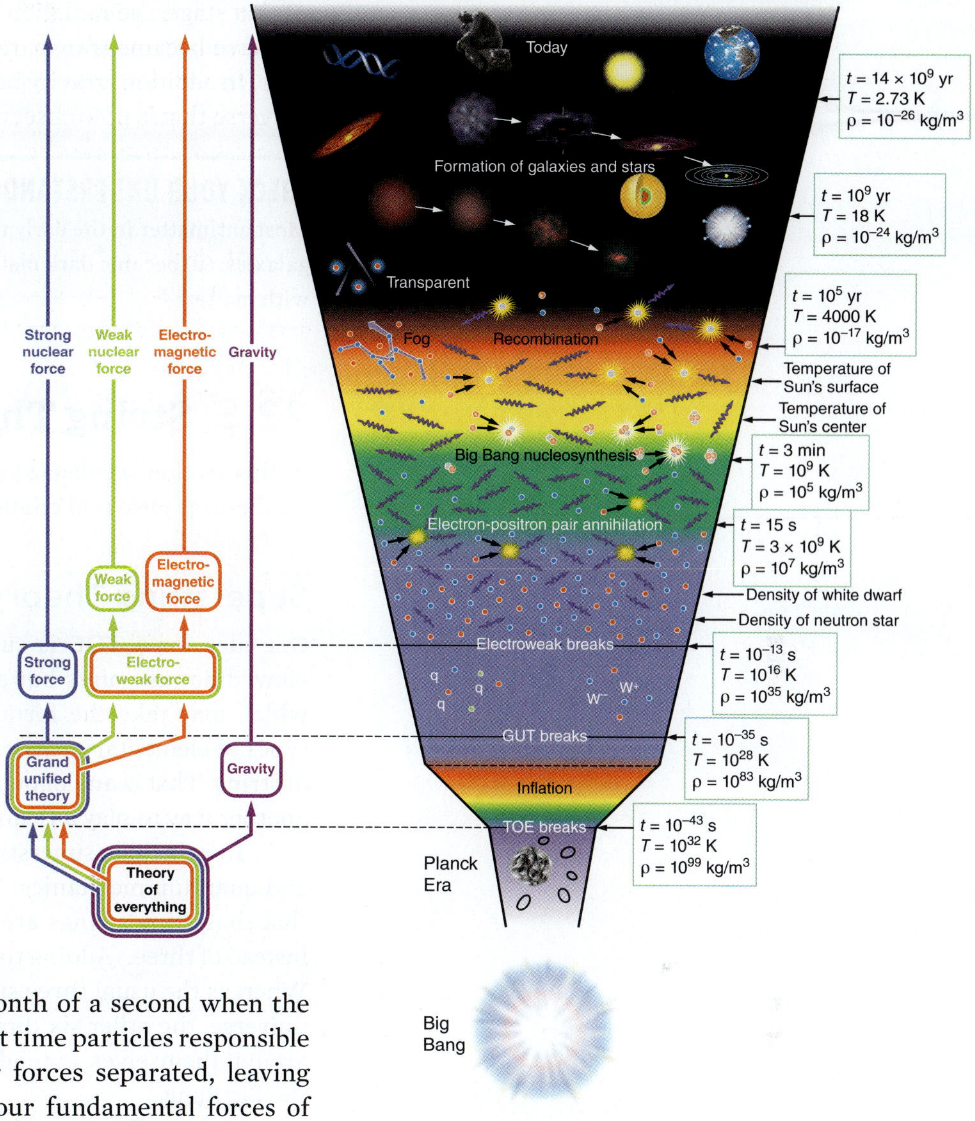

Figure 22.13 This figure conceptualizes eras in the evolution of the universe. The left side shows the four forces separating in stages after the Big Bang. The right side shows the temperature and density of the universe at different key times. As it expanded and cooled after the Big Bang, the universe went through phases determined by what types of particles could be created freely at that temperature. Later, the structure of the universe was set by the gravitational collapse of material to form galaxies and stars and by the chemistry made possible by elements formed in stars.

The next big change took place at one 10-trillionth of a second when the temperature of the universe had fallen to 10^{16} K. At that time particles responsible for unifying the electromagnetic and weak nuclear forces separated, leaving those two forces independent of one another. All four fundamental forces of nature that govern today's universe were then separate. It was a full minute or two later before the universe cooled to the billion-kelvin mark, below which not even pairs of electrons and positrons could form.

The universe then was too cool to form additional particles and their antiparticles. It was still hot enough, though, for the fast-moving protons to overcome the electric barriers between them, allowing nuclear reactions to take place. Those reactions formed the least massive elements and isotopes, including deuterium, helium, and lithium. Recall from stellar evolution that increasingly high temperatures are needed for the nucleosynthesis of increasingly heavy elements. So as the universe continued to expand, it soon became too cool for the nucleosynthesis of more massive elements.

The Big Bang nucleosynthesis discussed at the end of Chapter 21 ended by the time the universe was about 15–20 minutes old and the temperature of the universe had dropped below about 800 million K. By then the density of the universe had fallen to only about a tenth that of water. Normal matter consisted of atomic nuclei and electrons, but the universe was still dominated by the radiation from the Big Bang. After several hundred thousand years, the temperature dropped so low that electrons could combine with atomic nuclei to form neutral atoms. That was the era of recombination, which is seen directly when astronomers observe the CMB.

At that stage, the radiation background could no longer dominate over matter: the universe became transparent, and light could move freely through it for the first time. In addition, gravity began playing its role in forming the vast structure of the universe that is now observed.

CHECK YOUR UNDERSTANDING 22.4

Most antimatter in the early universe: (a) is still around today, filling the space between galaxies; (b) became dark matter; (c) formed antimatter galaxies and stars; (d) annihilated with matter.

22.5 String Theory and Multiverses

In this section, we discuss some speculative ideas in theoretical cosmology whose predictions pose real challenges to test.

Superstring Theory

One idea for a TOE is superstring theory, in which elementary particles are viewed not as points but as excitations or waves traveling along tiny "strings," which may take the form of loops. According to superstring theory, different types of elementary particles are like different "notes" played by vibrating loops of string. That is analogous to how a guitar string vibrates in one way to play an F, another way to play a G, and yet another way to play an A.

In principle, superstring theory offers a way to reconcile general relativity and quantum mechanics. To make superstring theory work, physicists imagine that those tiny strings are vibrating in a universe with nine spatial dimensions instead of three. (Adding time to the list would make the universe 10-dimensional.) Whereas the usual three spatial dimensions spread out across the vastness of the universe, the other six dimensions predicted by superstring theory wrap tightly around themselves, extending no further today than they did a brief instant after the Big Bang.

To visualize that configuration, imagine living in a three-dimensional universe (like the one you experience) in which one of those dimensions extended for only a tiny distance. Living in such a universe would be like living within a thin sheet of paper that extended billions of parsecs in two directions but was far smaller than an atom in the third. In such a universe, you would easily be aware of length and width: you could move in those directions at will. However, you would have no freedom to move in the third dimension at all, and you might not even realize that the third dimension existed. Perhaps your only inkling of the true nature of space would come from the fact that in order to explain the results of particle physics experiments, you would have to assume that particles extended into a third, unseen dimension. If superstring theory is correct, everyone now sees three spatial dimensions, extending possibly forever, but doesn't notice that each point in that three-dimensional space actually has a tiny but finite extent in the six other dimensions at the same time.

Although called a "theory," superstring theory is not like the well-tested theories that have been discussed throughout this book: it is no more than a promising idea that gave direction to theorists searching for a TOE. Therefore, calling superstring theory a "string hypothesis" or a "string idea" would be more consistent with the definitions of idea, hypothesis, and theory given in Chapter 1. Physicists will probably never be able to build particle accelerators that enable them to search directly for the most fundamental particles predicted by a TOE: the energies required are simply too high. Other ways to test the theory may be available, however, and some progress may be made by studying the ultimate particle accelerator: the Big Bang itself.

Multiverses

Is our universe the only one? Because we've defined the universe as "everything," what does it mean to say "multiple universes"? Do parallel universes—either separated in space or even occupying exactly the same space as "our" universe—exist? Can experiments or observations test the idea of multiple universes? Those ideas are speculative, but some cosmologists think seriously about the idea of multiple universes, or **multiverses**—collections of parallel universes.

The simplest example of such parallel universes is illustrated in **Figure 22.14**. The age of our universe—that is, the time that has passed since the Big Bang—is 13.8 billion years. The universe has been expanding since then. Imagine freezing the expansion of the universe today. Then the distance between us and a source of light emitted at the time of the Big Bang that we see today is the comoving size of the universe. That observable universe—everything astronomers can possibly observe today—is within a sphere of radius ~47 billion light-years. Anything farther than that is outside the observable universe and cannot ever be seen. As we discussed earlier in the chapter, the observational evidence suggests that the geometry of space is flat. A flat universe is infinite in size and must therefore contain an infinite number of similar spheres. As dark energy causes the universe to expand faster and faster, the separate observable universes move farther apart and will never overlap. Those parallel universes are simply too far away to ever be observed from Earth, and they are moving farther away all the time.

What are these other parallel universes like? Physicists suggest several possibilities from what has been learned about the observable universe. First, if the cosmological principle holds, on large scales each of those observable universes should look pretty much like our own, although the details could be very different. In a truly infinite universe, an infinite number of observable universes exactly like this one must exist, with exact copies of you reading identical versions of *21st Century Astronomy*. The argument is that if our own observable universe is cooler than about 10^8 K everywhere, no more than 10^{118} particles can be in the observable universe, and those particles can be distributed in only so many ways. If you then ask how far you must go before you will find an observable universe *identical to* our own, the answer is about $10^{10^{118}}$ m. (Yes, that's 10 raised to the power 10^{118}.) So, in an infinite universe—as enormous as it might be—the nearest identical universe must be at about that distance.

Other types of multiverses include those in which a universe undergoes **eternal inflation**, with no beginning or end to the inflation. If such a universe exists, small quantum fluctuations may cause some regions to expand more slowly than the rest of the universe. As a result, such a region may form a bubble whose inflating phase ends quickly. In that scenario, Earth is inside such a region, and "our" Big Bang would just be the condensation of our bubble within the eternally inflating universe. That type of multiverse neatly answers the question of what existed before the Big Bang. Because the universe has been inflating and will continue to inflate forever, it has no beginning or end. "Our" own bubble or

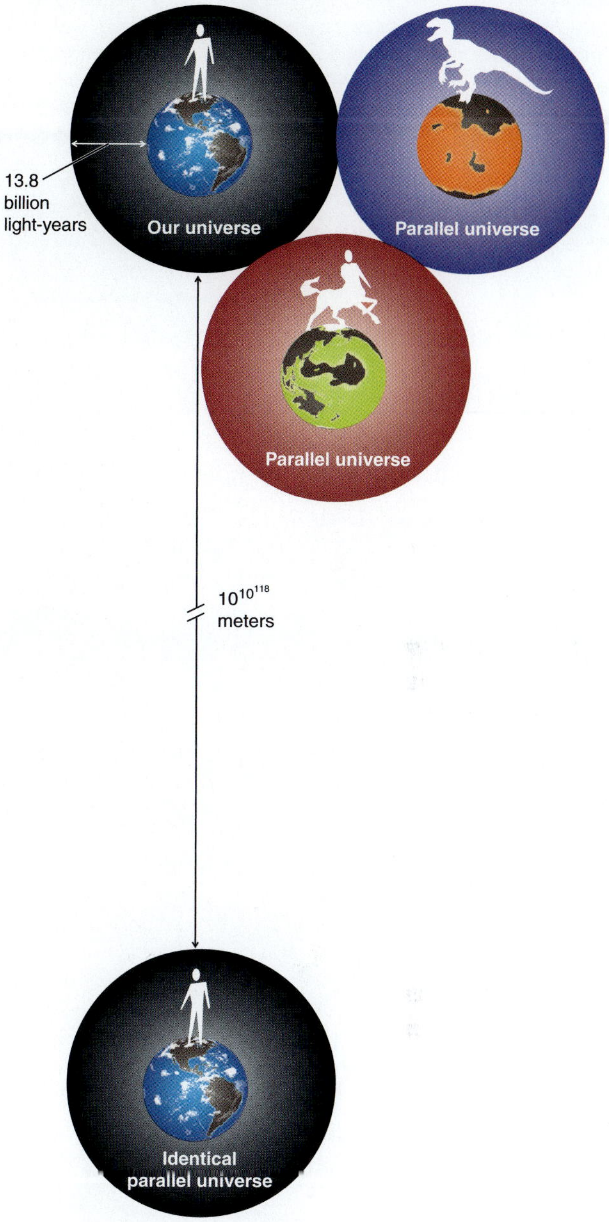

Figure 22.14 The observable universe is a sphere with a radius equal to the distance light has traveled since the Big Bang (13.8 billion light-years). Because the universe is infinite, an infinite number of similar spheres must exist. The rules of probability dictate that some of those spheres are *exactly* like our own.

parallel universe separated from the rest of the universe at a time called the Big Bang, but other bubbles are constantly separating and becoming their own parallel-universe big bangs.

In another type of multiverse derived from quantum physics, which describes a probabilistic universe, each event in the universe spawns multiple universes in which each possible outcome of the event exists. Although cosmologists consider that phenomenon at the level of particle interactions, it is more simply explained at the human scale: you made a decision about having breakfast this morning. That event caused two universes: one in which you did have breakfast and one in which you didn't. Which one are you in while you read this book? Yet another type of multiverse is characterized by parallel universes that have different mathematical structure to describe the different physics within those universes. In that all-encompassing case, almost any behavior for the universe is possible.

Multiverses are common in science fiction and in popular science books and form a type of common mythology about alternative realities that is explored with enthusiasm in popular culture. Many astronomers, though, ask whether the idea of parallel universes, or multiverses, is really science. Throughout this book, we have emphasized that any legitimate scientific theory must be testable and ultimately falsifiable. Do tests exist that can prove those multiverse ideas to be wrong? Possibly. For example, the first multiverse we described, involving distinct observable bubbles, is tested when astronomers measure the isotropy of the CMB, the flatness of the universe, or the large-scale distribution of galaxies. The eternal inflation model is difficult to test because observing parallel universes directly is impossible, though such a theory could produce a feature in the cosmic background radiation. But if physicists obtain a TOE that predicts eternal inflation, and if that theory is itself falsifiable, a connection exists between eternal inflation and observation.

Considerable debate occurs within the scientific community as to whether the multiverse hypotheses, except for the first type, can be tested and falsified. Even if some tests are possible, the concern is that they are not truly meaningful. For example, you might test Newton's theory of gravity by releasing an apple and watching whether it falls upward or downward. But that test is not very discerning: it might not always distinguish between two or more sensible theories. Similarly, for the multiverse hypothesis, whether the tests that seem possible would be meaningful tests of the theory is still debatable.

CHECK YOUR UNDERSTANDING 22.5

Which is a false statement about multiverses? (a) They represent a scientific theory only if they can be tested and falsified. (b) They might explain what happened before the Big Bang. (c) They are an inevitable consequence of modern observational cosmology. (d) Two identical multiverses can exist.

Origins Our Own Universe Must Support Life

In some models of an inflationary universe, each bubble could contain different values of the fundamental constants of physics. In some bubble universes, for example, the strength of the nuclear force might be larger, the electric charge might be smaller, and the gravitational constant G might be much smaller than in others. To address what are the properties of our universe, scientists invoke the **anthropic principle**, which states that our universe (or our bubble in the universe) must have physical properties that allow intelligent life to develop. Because humans exist, are intelligent, and can observe the surrounding universe, our universe must have the properties that would allow intelligent life to evolve. That is, this universe must have had the right physical properties and existed long enough for atoms, stars, galaxies, planets, and life to have formed; otherwise, we wouldn't be here to observe it.

For a multiverse that contains bubbles with different physical constants in each of them, the anthropic principle provides information about the values of those physical constants. Consider a few examples. In a bubble universe where the gravitational constant G was much bigger than G as measured in our universe, stellar evolution would occur much faster, and intelligent life might not have time to evolve on a planet before its star burned itself out. Similarly, the anthropic principle provides a relatively narrow range for the strength of the strong nuclear force that holds nuclei together. If that force were much weaker than what physicists now measure, nuclei would not be able to overcome their electric repulsion in order to fuse. Without nuclear fusion, stars could not shine, heavy elements would not form, and planets and life as we know it could not evolve. Alternatively, if the nuclear force were stronger, two protons would more easily fuse in the early universe; thus, most hydrogen will fuse to helium, and less hydrogen would exist to form water and organic molecules that are necessary for life as we know it.

As another example of the anthropic principle applied to multiverses, recall the fraction of the critical density provided by the cosmological constant, Ω_Λ. If Ω_Λ were 20 times larger than what is observed now, the universe would have begun accelerating much earlier, and galaxies would not have had time to form. Without galaxies, little star formation would occur. Therefore, stellar nucleosynthesis would not have taken place, and rocky planets and life could not have evolved. Alternatively, if the cosmological constant were negative, this entire bubble universe would have reached a maximum size and begun to collapse even before galaxies and stars could have evolved. A universe with an intermediate value might last only a few billion years—enough time for stars and galaxies to be established, but perhaps not enough time for sufficient amounts of heavy elements to form or for life to evolve to intelligence. As those examples illustrate, the cosmological constant must be within a particular range of values to allow the intelligent life that exists on Earth to evolve.

ARTICLES QUESTIONS

Scientists produce indirect evidence of gaseous filaments and sheets known as Whims linking clusters of galaxies in the cosmic web.

Astronomers Find Half of the Missing Matter in the Universe

By HANNAH DEVLIN, *The Guardian*

It is one of cosmology's more perplexing problems: that up to 90% of the ordinary matter in the universe appears to have gone missing.

Now astronomers have detected about half of this missing content for the first time, in a discovery that could resolve a longstanding paradox.

The conundrum first arose from measurements of radiation left over from the Big Bang, which allowed scientists to calculate how much matter there is in the universe and what form it takes. This showed that about 5% of the mass in the universe comes in the form of ordinary matter, with the rest being accounted for by dark matter and dark energy.

Dark matter has never been directly observed and the nature of dark energy is almost completely mysterious, but even tracking down the 5% of ordinary stuff has proved more complicated than expected. When scientists have counted up all the observable objects in the sky—stars, planets, galaxies and so on—this only seems to account for between a 10th and a fifth of what ought to be out there.

The deficit is known as the "missing baryon problem," baryons being ordinary subatomic particles like protons and neutrons.

Richard Ellis, a professor of astrophysics at the University College London, said: "People agree that there's a lot missing, raising the question where is it?"

The distribution of galaxies in the universe follows a web-like pattern and scientists have speculated that the missing baryons could be floating in diffuse gaseous filaments and sheets linking the galaxy clusters in the cosmic web.

Theoretical calculations suggest these gaseous threads, known as the warm–hot intergalactic medium, or the Whim, ought to be around a million degrees celsius. A mist of gas at this temperature is too cold to emit X-rays that could be spotted by ordinary telescopes from the Earth—but not cold enough to absorb significant amounts of light passing through it.

"The trouble is, it's in this unusual temperature regime where we can't see it," said Ellis.

Now two separate teams of scientists, one at the University of Edinburgh, the other at the Institute of Space Astrophysics in Orsay, France, have produced compelling indirect evidence for the Whim. Both teams relied on the fact that when radiation travels through a hot gas, it is scattered, meaning that the Whim ought to appear as a dim outline in the cosmic microwave background.

The scientists overlaid observations of the cosmic background radiation, made by the Planck space observatory, and the most detailed three-dimensional map of the cosmic web, created by the Sloane Digital Sky Survey (SDSS). They hypothesized that if there were gas threads linking galaxy clusters, these should show up in the Planck data.

The Edinburgh team found the regions between galaxies appeared to be about six times as dense as the surrounding bits of space and when summed up, these gaseous threads could amount to about 30% of the ordinary matter in the universe. The French teams' calculation came out at slightly less than this, but the numbers are consistent.

Ellis, who was not involved in either project, describes the findings as "inspirational." "These two papers have been very prominently discussed and people are excited," he added. "The Whim is out there."

The initial measurements still do not account for all the ordinary matter, and some believe the remaining portion could be made up by exotic unobserved objects such as black holes or dark stars. Cosmologists are also still yet to discover the nature of dark matter, which makes up even more of the universe.

ARTICLES QUESTIONS

1. What is "missing"?
2. What is a Whim? How is it detected?
3. Why can't Whims be seen with X-ray telescopes?
4. Why is it important that two separate teams are reporting that result?

Summary

Both gravity and the cosmological constant (or dark energy) determine the fate of the universe. Observations indicate that instead of slowing down, the expansion of the universe is accelerating. The best explanation for that phenomenon is dark energy. The very early universe may have gone through a brief but dramatic period of exceptionally rapid expansion, known as inflation. If so, inflation would explain both the flatness and the uniformity of the universe we see today. During the very earliest moments in the universe, the four fundamental forces of nature were unified. Ideas about multiple universes have been proposed, but none are yet directly testable. Our observable universe must be one in which physics can support the formation of life.

LG 1 **Explain how mass within the universe and the gravitational force it produces affect the history, shape, and fate of the universe.** The rate of expansion of the universe is affected by its density (mass/volume). Mass can slow and even stop the expansion of the universe. The density also affects the shape of the universe: if density is high, the universe is shaped more like a sphere; if density is low, the shape is more like a saddle. In between, at the critical density, the universe is flat.

LG 2 **Describe the evidence for the accelerating expansion of the universe.** Recent observations of distant Type Ia supernovae suggest that the expansion of the universe was slower in the past. That means the expansion is speeding up, implying that a force must be acting to increase the expansion rate. That force may come from dark energy: the energy of empty space. Because the rate of expansion of the universe is increasing, the universe will probably expand forever.

LG 3 **Describe the early period of rapid expansion of the universe known as inflation.** Inflationary models, which state that the universe expanded very rapidly in a very short time soon after the Big Bang, were proposed to solve several problems in cosmology. Astronomers are looking for an observational signature of inflation.

LG 4 **Explain how the events that occurred in the earliest moments of the universe are related to the forces that operate in the modern universe.** During the earliest moments of the universe, the four forces (gravity, electromagnetism, strong nuclear, and weak nuclear) split off, each becoming separate at a different time. Modern physicists search for the combined theory of all four forces: the theory of everything.

? Unanswered Questions

- What is the origin of dark energy, and why is $\Omega_\Lambda + \Omega_m$ so close to 1 now? Is that a coincidence? We have already said that dark energy is a form of vacuum energy, and explaining how big Ω_Λ really is is one of the grand challenges of any successful theory of everything. The simplest estimates for the size of Ω_Λ yield results much larger than the observed value—by a factor of about 10^{120}. So an as yet undetermined mechanism must exist that affects the size and evolution of Ω_Λ, and that is one of the bigger questions in modern cosmology.
- What mechanism leads to an extra electron for every 10 billion electron-positron pairs in the early universe? Here in Chapter 22, we mentioned that grand unified theories predict an asymmetry between particles and antiparticles, as well as proton decay. But physicists do not know which GUT is the correct one, and so far they have not actually observed proton decay. Measuring the actual lifetime of a proton would enable physicists to determine the correct GUT and therefore the mechanism leading to particle-antiparticle asymmetry. In addition, if the correct TOE was really understood, that theory could predict which GUT describes the universe and therefore the mechanism and amount of asymmetry.

Questions and Problems

Test Your Understanding

1. If astronomers ignored any cosmological constant (or dark energy), the future of the universe could be determined solely from
   a. the mass of the universe.
   b. the volume of the universe.
   c. the amount of light in the universe.
   d. the density of the universe.
2. The cosmological constant accounts for the effects of
   a. dark matter.
   b. the Big Bang.
   c. dark energy.
   d. gravity.
3. The cosmic microwave background radiation (CMB) indicates that the early universe
   a. was uniform.
   b. varied greatly in density from one place to another.
   c. varied greatly in temperature from one place to another.
   d. was shaped differently from the modern universe.
4. According to the definitions of these terms in Chapter 1, superstring theory is
   a. a hypothesis.
   b. a theory.
   c. a law.
   d. a principle.
5. Place in order the following events in the history of the universe.
   a. Planck era
   b. grand unified theory breaks
   c. today
   d. Big Bang nucleosynthesis
   e. electroweak breaks
   f. theory of everything breaks
   g. electron-positron pair annihilation
   h. formation of galaxies and stars
   i. recombination
   j. inflation
6. Astronomers will never directly observe the first few minutes of the universe because
   a. the universe was opaque then.
   b. the universe is too large now.
   c. no particles or other matter existed for astronomers to see.
   d. no photons existed.
7. As applied to the universe, what is the meaning of *critical density*?
   a. Above that density, nebulae collapse to form stars.
   b. Above that density, dark matter becomes important.
   c. Above that density, the universe will eventually collapse.
   d. Above that density, matter becomes degenerate.
8. Of the four fundamental forces in nature, which two become unified at the lowest energy?
   a. gravitational force
   b. electromagnetic force
   c. strong nuclear force
   d. weak nuclear force
9. Suppose you measure the angles of a triangle and find that they add to 185 degrees. From that you can determine that the space the triangle occupies is
   a. flat.
   b. positively curved.
   c. negatively curved.
   d. filled with dark matter.
10. Which of the following is the correct order for these objects to form after the Big Bang?
    a. neutral atoms, protons, nuclei
    b. protons, nuclei, neutral atoms
    c. nuclei, neutral atoms, protons
    d. protons, neutral atoms, nuclei
11. Quarks are
    a. virtual particles.
    b. massless particles.
    c. candidates for dark matter.
    d. building blocks of larger particles.
12. Current understanding indicates that the universe (choose all that apply)
    a. is closed.
    b. is flat.
    c. is open.
    d. is inflating.
    e. has accelerating expansion.
13. When a particle and an antiparticle come together, they
    a. annihilate each other, releasing photons.
    b. create a black hole.
    c. release enormous amounts of energy.
    d. create new particles.
14. Observations of Type I supernovae in distant galaxies have shown that
    a. the rate of star formation in galaxies decreases with increasing redshift.
    b. the expansion rate of the universe is increasing.
    c. the cosmological constant is zero.
    d. dark energy is negligible now.

15. The anthropic principle states that
    a. the universe was created so that life exists.
    b. life exists, so the universe must be such that life can exist.
    c. if the universe were otherwise, life would not exist.
    d. life has made the universe the way it is.

Thinking about the Concepts

16. What set of circumstances would cause an expanding universe to reverse its expansion and end up in a "Big Crunch"?
17. Describe the observational evidence suggesting that Einstein's cosmological constant (a repulsive force) may be needed to explain the historical expansion of the universe. Explain how Einstein was "right for the wrong reason."
18. What do astronomers mean by *dark energy*?
19. If the universe is being forced apart by dark energy, why isn't the Milky Way Galaxy, the Solar System, or the planet Earth being torn apart?
20. In Chapter 21, we said we could estimate the age of the universe with Hubble time ($1/H_0$). Why does that method not give the best answer?
21. What is the flatness problem, and why has it created difficulties for cosmologists?
22. During the period of inflation, the universe may have briefly expanded at 10^{30} (a million trillion trillion) or more times the speed of light. Why did that ultra-rapid expansion not violate Einstein's special theory of relativity, which says that neither matter nor communication can travel faster than the speed of light?
23. Why is particle physics important for understanding the early universe?
24. The fundamental forces of the universe are generally assumed not to change.
    a. How would the fate of the universe be affected if Newton's gravitational constant changed with time?
    b. What if, instead, the electric force between charged particles changed with time?
25. The standard model cannot explain why neutrinos have mass or why electron-positron asymmetry existed in the early universe. Do those failings make it an incomplete theory? Should all its predictions be ignored until the theory can resolve those remaining issues?
26. Explain pair production.
27. Describe the Planck era.
28. What are the basic differences between a grand unified theory (GUT) and a theory of everything (TOE)?
29. Consider the term *superstring theory* in light of the discussion of scientific theory in Chapter 1. Some scientists object to using the word *theory* to describe superstring theory. Why?
30. Suggest another example of how a different value of a physical constant would affect conditions in a "parallel" universe.

Applying the Concepts

31. Consider Figure 22.2.
    a. Is the vertical axis linear or logarithmic?
    b. The horizontal axes have two labels. The top label is measured in billions of years. Is that axis linear or logarithmic?
    c. The bottom label for the horizontal axis is measured in relative brightness. Is that axis linear or logarithmic?
    d. What is the relationship between billions of years and relative brightness?
32. The blue curve in Figure 22.3 indicates a model in which the universe first decelerated and then accelerated, whereas the red curve indicates continual deceleration.
    a. How are the two curves different?
    b. What is it about one of those curves that indicates deceleration? What indicates acceleration?
    c. If a straight line were plotted on that graph, what would the model that the new line represents indicate about the expansion of the universe?
33. On the graph in Figure 22.2, the colored lines represent various models, and the black dots represent data taken in the actual universe.
    a. Why are no data points on the right-hand side of the graph?
    b. Which models are excluded by the data?
    c. Roughly how far back in time do the data go?
    d. What fraction of the age of the universe is the answer to part (c) (if you assume an age of 13.8 billion years)?
34. Is the plot in Figure 22.8 linear or logarithmic? How much did the universe increase in size during the time of inflation? How much smaller was the universe at the beginning in an inflationary model than in the noninflationary model?
35. Suppose that new data coming in from a new instrument give the value 0.1 for both Ω_Λ and Ω_m. How would astronomers probably respond to those new data?
36. Consider Figure 22.13.
    a. Is the time axis (the vertical dimension of the figure) approximately linear or approximately logarithmic?
    b. By how many orders of magnitude (factors of 10) has the density ρ of the universe dropped since earliest time?
    c. By how many orders of magnitude has the temperature dropped since earliest time?

37. Currently, the Hubble constant has an uncertainty of about 4 percent. What are the corresponding maximum and minimum ages allowed for the universe?

38. Estimate the average density of the Milky Way. How does it compare to the critical density?

39. The universe today has an average density $\rho_0 = 3 \times 10^{-28}$ kg/m^3. If you assume that the average density depends on the scale factor, as $\rho = \rho_0/R_U{}^3$, what was the scale factor of the universe when its average density was about the same as Earth's atmosphere at sea level ($\rho = 1.2$ kg/m^3)?

40. The proton and antiproton each have the same mass, $m_p = 1.67 \times 10^{-27}$ kg. What is the energy (in joules) of each of the two gamma rays created in a proton-antiproton annihilation?

41. About 500 million CMB photons are in the universe for every hydrogen atom. Using $E = mc^2$, what is the equivalent mass of those photons? Is it large enough to factor into the overall density of the universe?

42. Suppose you brought together a gram of ordinary-matter hydrogen atoms (each composed of a proton and an electron) and a gram of antimatter hydrogen atoms (each composed of an antiproton and a positron). Keeping in mind that 2 grams is less than the mass of a dime:
    a. Calculate how much energy (in joules) would be released as the ordinary-matter and antimatter hydrogen atoms annihilated one another.
    b. Compare that amount of energy with the energy released by a 1-megaton hydrogen bomb (4.2×10^{15} J).

43. One GUT predicts that a proton will decay in about 10^{31} years, which means if you have 10^{31} protons, you should see one decay per year. The Super-Kamiokande observatory in Japan holds about 20 million kg of water in its main detector, and it did not see any decays in 5 years of continuous operation. What limit does that observation place on proton decay and on the GUT described here?

44. Assume a planet's orbit is perfectly circular as it travels in the gravitational well of its star. If that were true, would the orbit's circumference be greater than, less than, or equal to 2π times the radius of the orbit? Explain.

45. To get a feeling for the emptiness of the universe, compare its density (about 9.9×10^{-27} kg/m^3) with that of Earth's atmosphere at sea level (1.2 kg/m^3). How much denser is our atmosphere? Write that ratio in standard notation.

USING THE WEB

46. Go to the website for the Dark Energy Survey, an international project that began in 2013 (https://www.darkenergysurvey.org). What observations will be made for that project? What will it tell scientists about dark energy? Click on "News." What is the status of the project? What are some recent results?

47. Go to the website of the European Organization for Nuclear Research (CERN—https://home.cern/about/physics/early-universe) and read through the pages indexed on the right. What was the role of the Higgs boson after the Big Bang? (Note: The World Wide Web was invented at CERN.)

48. Citizen science: Go to Higgs Hunters (https://www.higgshunters.org/) and log in with your Zooniverse account. Click on "Science" and watch the brief videos; keep going until "How you can help." Why do they expect that human eyes are more likely than computers to find exotic decays? Click on "Classify" and "Restart the Tutorial" to see examples of how to mark the images with "Off-center vertex" or "Something weird." Mark up a few images, keeping a record for your homework.

49. Scientists debate whether scientific evidence for a multiverse can ever exist. A good example is a discussion in the journal *Scientific American*. Read the article "Does the Multiverse Really Exist?" in the August 2011 issue (it is probably accessible online through your school library) and the response at http://scientificamerican.com/article/multiverse-the-case-for-parallel-universe. What are some of the arguments for and against multiverses?

50. Clips and full episodes from a four-part series of the television show *NOVA* called *The Fabric of the Cosmos* can be accessed on PBS's website (http://pbs.org/wgbh/nova/physics/fabric-of-cosmos.html). Watch at least one episode. Are the arguments made in those programs compelling? Is the science explained in a way that makes sense to a general audience?

EXPLORATION

Studying Particles

digital.wwnorton.com/astro6

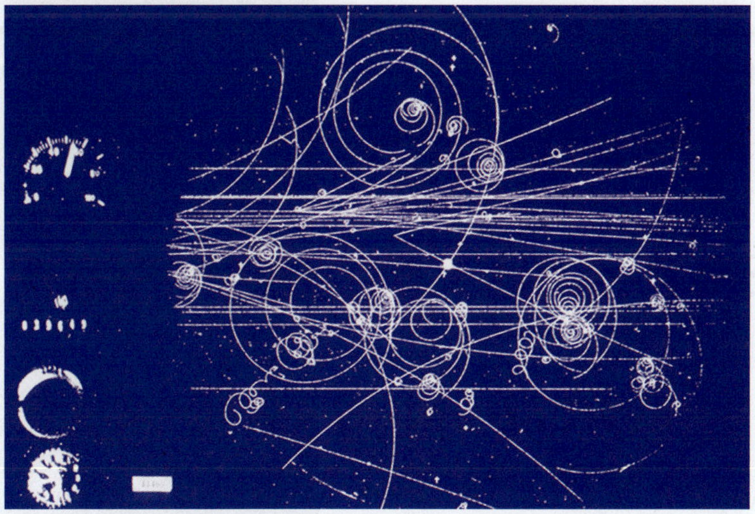

Figure 22.15 The decay of a particle in a hydrogen bubble chamber.

Just as astronomers use images to study distant clouds of dust and gas, so too do particle physicists use images to study the tiny particles. Those images capture the tracks the particles make. As the particle passes through vapor in a chamber, it ionizes some of the molecules of the vapor. The vapor then condenses around the ionized molecules, leaving a thin trail of cloud. A strong magnetic field fills the chamber, so that the charged particles will turn as they travel through it.

Figure 22.15 shows tracks of particles moving through a bubble chamber. Studying that figure in more detail will give you a feel for how scientists have gained an understanding of those tiny particles.

1 Estimate the number of particle tracks that you can see in Figure 22.15.

____

2 Recall that charged particles spiral around magnetic field lines. From the image of Figure 22.15, determine which way the magnetic field points: perpendicular or parallel to the image.

____

3 Identify each of the following types of tracks in Figure 22.15:

- **a.** a track that looks straight
- **b.** a track that spirals tightly (a small spiral) and a track that spirals loosely (a large spiral)
- **c.** a bright track and a fainter track
- **d.** two tracks that form a "V" shape
- **e.** a track with an obvious beginning and ending
- **f.** a track with a kink, or abrupt change of direction

Oppositely charged particles spiral in opposite directions. For example, if the positive particles spiral clockwise, the negative ones spiral counterclockwise. More massive particles are harder to turn than less massive particles. Faster-moving particles also are harder to turn. The fact that both the mass and the velocity of a particle affect how easily it turns makes separating the two properties difficult.

4 Only charged particles interact with the vapor in the cloud chamber to produce tracks, so all the tracks should be curved in the magnetic field. But some of those tracks are very nearly straight. What can you say about the velocity of particles that leave straight tracks?

____

5 Consider the tight spiral and the loose spiral tracks.

- **a.** If both tracks were made by electrons, which particle was traveling faster?
- **b.** If both tracks were *not* made by electrons, what can you say about the total energy of each particle, including its mass?

6 The brightness of the track tells you how long the particle spent in each position: a brighter track is produced when the particle stays near the same spot longer.

- **a.** Consider the bright track and the faint track. Which track shows the faster particle?
- **b.** Now apply that reasoning to the tight and loose spiral tracks. Can you tell which particle was moving faster?
- **c.** If the looser track is also brighter, were both the loose and the tight tracks made by the same type of particle?

7 A track that is shaped like a *V* indicates that two charged particles were produced at the base of the *V* and then proceeded onward. That can occur, for example, if a neutral particle decays to produce a positive and a negative charge. If the two particles were a proton and an electron, but both arms of the *V* curved the same (possibly very small) amount, what can you say about their relative velocities?

____

8 A track with an obvious beginning and endpoint shows an event initiated by a neutral particle. Because neutral particles do not leave tracks in the chamber, that type of track must indicate a charged particle that suddenly begins moving. How might a neutral particle initiate one of those tracks?

____

9 A kink occurs when a charged particle decays, emitting a particle and in the process becoming a different charged particle.

- **a.** Is the particle emitted in the decay neutral or charged? How do you know?
- **b.** Does the remaining charged particle have more or less mass than before it decayed?
- **c.** Sketch the kink that you are studying, and indicate the direction the particle was traveling along the track. (Hint: Consider the tightness of the spiral!)

To learn more, you would have to know more about the experimental setup—the strength of the magnetic field, the speed of the particles as they enter the chamber, and so on. By studying images like those, and "crunching the numbers," scientists determine the properties and interactions of the fundamental building blocks of the universe.

23 Large-Scale Structure in the Universe

The universe that emerged from the Big Bang was incredibly uniform—wholly unlike today's universe of galaxies, stars, and planets. Here in Chapter 23, we investigate the origin of the current structure of the universe and find that complex structure is a natural consequence of the action of physical laws in an evolving universe.

LEARNING GOALS

By the end of this chapter, you should be able to:

LG 1 Describe the distribution of galaxies in the universe.

LG 2 Explain how the large-scale structure of today's universe evolved from the structure that began to form shortly after the Big Bang.

LG 3 Describe the formation of the first stars and the first galaxies.

LG 4 Explain how the observations of galaxies at different redshifts illustrate the evolution of the large-scale structure of the universe.

The Hubble Ultra Deep Field 2014 image shows thousands of galaxies in a small section of space in ultraviolet, visible, and infrared light. ▶▶▶

What does this image tell us about galaxy evolution?

Figure 23.1 The Virgo Cluster of galaxies, at a distance of about 16.5 Mpc (54 million light-years), is the closest cluster to the Milky Way.

23.1 Galaxies Form Groups, Clusters, and Larger Structures

Just as stars and clouds of glowing gas reveal the structure of the Milky Way, the distribution of galaxies indicates the structure of the universe. And just as gravity holds galaxies together, giving them their shape, gravity is what produces structures larger than galaxies. Most galaxies are parts of gravitationally bound collections of galaxies. In this section, we look at how those clusters of galaxies form.

Types of Galaxy Structures

Galaxy groups are the smallest and most common collection of galaxies and contain up to several dozen galaxies, most of which are dwarf galaxies. The Milky Way is a member of the Local Group (see Chapter 20), which consists of two giant spiral galaxies—the Milky Way Galaxy and the Andromeda Galaxy—along with more than 50 smaller dwarf elliptical or dwarf spheroidal galaxies in a volume of space roughly 3 megaparsecs (Mpc) in diameter. Most of the mass in the Local Group, dark and luminous, resides in the two giant galaxies.

Larger gravitationally bound systems of galaxies, called **galaxy clusters**, can consist of thousands of galaxies. Galaxy clusters are larger than groups, typically occupying a volume of space 2–10 Mpc across. Galaxy clusters also contain far more dwarf galaxies than giant galaxies. Most of the galaxy mass in galaxy clusters resides, however, in the giant galaxies. Most clusters contain spiral galaxies, though the more massive clusters have a much higher fraction of giant elliptical galaxies. The Virgo Cluster (**Figure 23.1**), located 16.5 Mpc from the Local Group, is an example of a cluster that contains mostly spiral galaxies. Giant elliptical and S0 galaxies dominate the more distant Coma Cluster.

Galaxy clusters and groups of galaxies bunch together to form enormous **superclusters**, which contain tens of thousands or even hundreds of thousands of galaxies and span regions of space typically larger than 30 Mpc. Our Local Group is part of the Laniakea Supercluster, which also includes the Virgo Cluster (**Figure 23.2**; see also the Reading Astronomy News article).

Figure 23.2 This computer-generated visualization shows a slice of the Laniakea Supercluster. Individual galaxies are white dots, blue areas are voids, red dots indicate the Virgo Cluster, and green areas contain many galaxies. White lines indicate the movement of galaxies toward the center of the supercluster. The orange contour encloses the outer limits of those galaxies streaming to the center. The dark blue dot shows the location of the Milky Way.

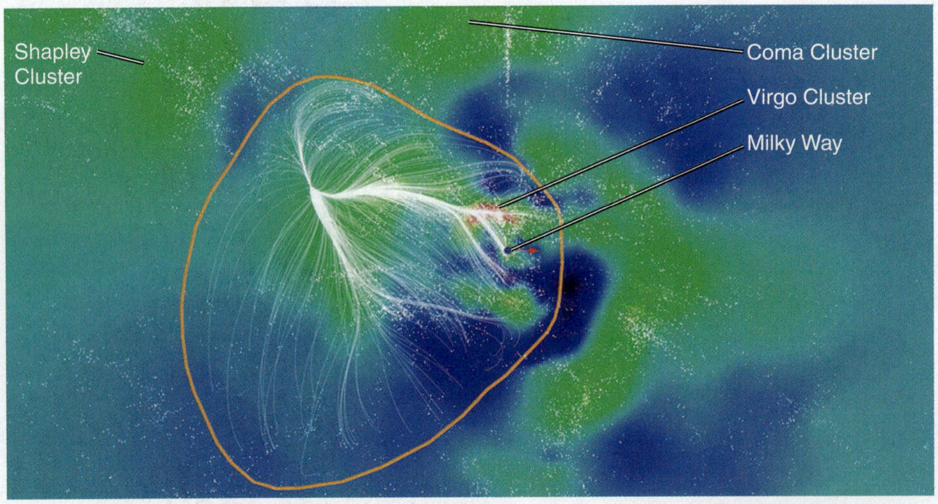

Mapping the Universe

Hubble's law for the expansion of the universe ($v_r = H_0 \times d_G$) is a powerful tool for mapping the distribution of galaxies, groups, clusters, and superclusters in space. Using that law, astronomers estimate the distance to a galaxy (d_G) by measuring the redshift in the galaxy's spectrum. The first redshifts were measured from spectra recorded on photographic plates, which required exposures of several hours to capture the faint signal. Now astronomers can measure the redshifts of galaxies by using larger telescopes equipped with electronic detectors and spectrographs that can observe many galaxies at once. We now know the redshifts of several million galaxies, so we now know the approximate distances to those galaxies. From that information, we can develop a map of the structure of the universe on the largest scales.

The first large redshift survey looked only nearby, but later surveys extended much farther. Those surveys were used to illustrate a "slice of the universe" (**Figure 23.3a**). Observations show that clusters and superclusters of galaxies are not scattered randomly through space but instead are linked in an intricate network of relatively thin structures known as filaments and walls. Those concentrations of galaxies surround large regions of space with very few galaxies, known as **voids**. Voids are some of the largest structures seen in the universe. Though the voids may seem empty, we do not know that they are devoid of matter—only that they have very few observable galaxies.

Clusters and superclusters are located within the walls and filaments. Later surveys have looked at much larger volumes of space. The Sloan Digital Sky Survey (SDSS) includes the Sloan Great Wall (**Figure 23.3b**), a string of galaxies 400 Mpc long. For as far out as current observations can measure, the universe has a porous structure much like that of a sponge or a pile of soap bubbles. Together, galaxies and the larger groupings in which they are found are called **large-scale structure**.

The **peculiar velocity** of a galaxy is its motion relative to the frame in which the cosmic microwave background radiation (CMB; see Chapter 21) is at rest. Observations of the peculiar velocity yield information about the distribution of mass near that galaxy. For example, the peculiar velocity of the Local Group was originally attributed to the Great Attractor, which has a mass of several thousand times the mass of the Milky Way and is located about 75 Mpc away. More recently, it was found that the Great Attractor is actually the center of mass of the larger supercluster called Laniakea, with the Local Group and the Milky Way located at the outer edge (Figure 23.2).

Those observations of the large-scale structure of the universe provide direct observational evidence for the cosmological principle—that the universe is homogeneous (the same everywhere) and isotropic (the same in every direction)—because they show that on that very largest scale, the structure of the universe is the same everywhere, in every direction. If, for example, the universe in the top half of the multicolored image of Figure 23.3b showed a uniform distribution of galaxies, whereas the universe in the bottom half showed walls and voids, the universe would not be homogeneous or isotropic. All conclusions based on the cosmological principle would have been called into doubt. As it is, however, observations support that underlying principle of cosmology.

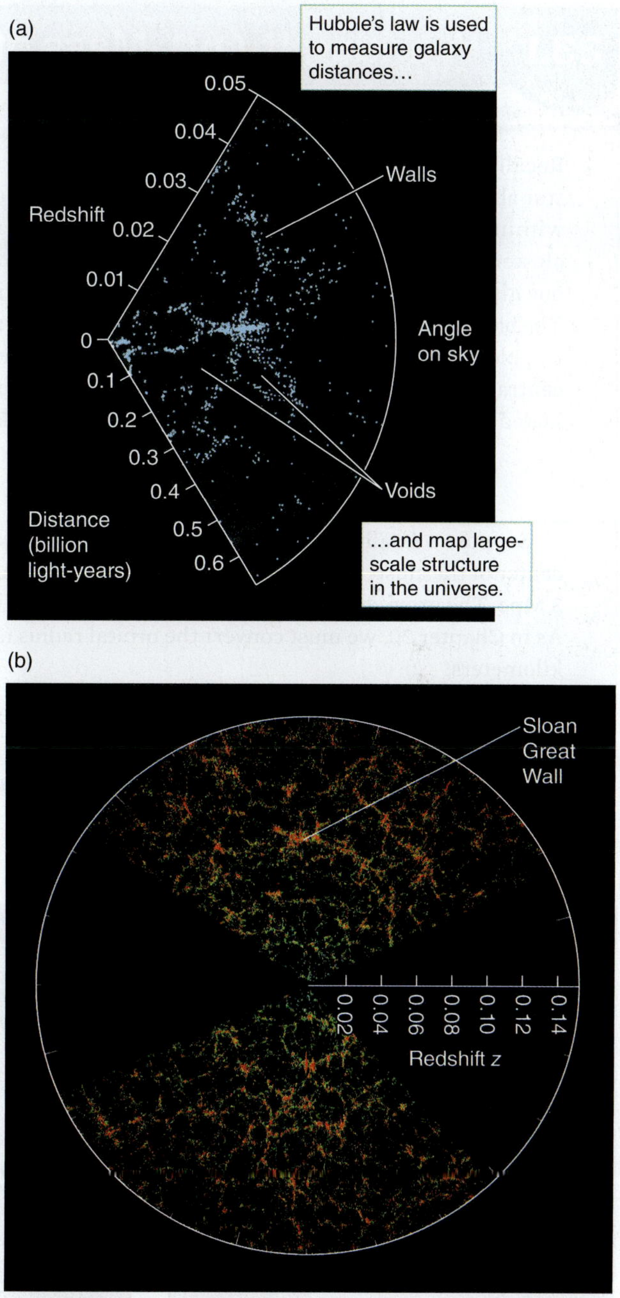

Figure 23.3 Redshift surveys use Hubble's law to map the universe. (a) In 1986, the Harvard-Smithsonian Center for Astrophysics redshift survey, called "A Slice of the Universe," first showed that clusters and superclusters of galaxies are part of even larger-scale structures. (b) The 2008 Sloan Digital Sky Survey map of the universe extends outward to a distance of about 600 Mpc. Shown here is a sample of 67,000 galaxies colored according to the ages of their stars, with the redder, more strongly clustered points being galaxies made of older stars.

Dark Matter in Galaxy Groups and Clusters

Just as dark matter dominates galaxies, it also dominates galaxy groups and clusters. Astronomers can make that inference in several ways. They can look at the motion of a small satellite galaxy orbiting the central dominant galaxy of the cluster and then estimate the mass of the cluster inside that orbit—similar to the way they measure dark matter in spiral galaxies. Or they can look at the motions of all of a cluster's galaxies and calculate how strong gravity must be to hold the cluster together. And again, the conclusion is that the total mass of the clusters, including dark matter, must be about 8–10 times greater than the normal matter they contain (**Working It Out 23.1**).

Another piece of evidence for the presence of dark matter is that the space between galaxies in a cluster is filled with extremely hot gas that is 10 million to

23.1 Working It Out Mass of a Cluster of Galaxies

Recall from Working It Out 20.1 that astronomers use the orbit of a star about the center of a galaxy to estimate the mass of the galaxy within the star's orbit. A similar estimation is made with groups or clusters of galaxies. Those groups or clusters are often dominated by one giant galaxy at the center, with smaller ones orbiting around it. The orbital velocities of the smaller galaxies are measured from the Doppler shifts of the lines in their spectra. The distance between the central and orbiting galaxies (the radius of a circular orbit) is estimated. The equation then looks like the one from Working It Out 20.1:

$$M = \frac{rv_{\text{circ}}^2}{G}$$

Suppose a smaller galaxy is orbiting the large galaxy at the center of the cluster at a speed of 1,000 km/s at a distance of about 3 Mpc. The gravitational constant is $G = 6.67 \times 10^{-20}$ km^3/(kg s^2). As in Chapter 20, we must convert the orbital radius (3 Mpc) to kilometers:

$$(3 \times 10^6 \text{ pc}) \times (3.09 \times 10^{13} \text{ km/pc}) = 9.3 \times 10^{19} \text{ km}$$

The mass of the cluster core is given by

$$M = \frac{(9.3 \times 10^{19} \text{ km}) \times (10^3 \text{ km/s})^2}{6.67 \times 10^{-20} \text{ km}^3/(\text{kg s}^2)}$$

$$M = 1.4 \times 10^{45} \text{ kg}$$

We can divide that result by the mass of the Sun, 2.0×10^{30} kg, to get a cluster mass of 7.0×10^{14} M_{Sun}. If we divide that core cluster mass by the mass of the Milky Way Galaxy, 10^{12} M_{Sun}, we see that the cluster core has a mass of about 700 Milky Way galaxies. That value includes the dark matter inside the orbit.

In 1933, astronomer Fritz Zwicky (1898–1974) measured the velocities of many galaxies within a collection of clusters, and in each case he found the entire cluster contained more mass than could reside in the stars that produce the cluster's visible light. He concluded there was a considerable amount of extra and yet unseen mass in the clusters. In the 1970s, Vera Rubin (1928–2016) and colleagues measured rotation curves of individual galaxies and discovered evidence of dark matter in those galaxies, too.

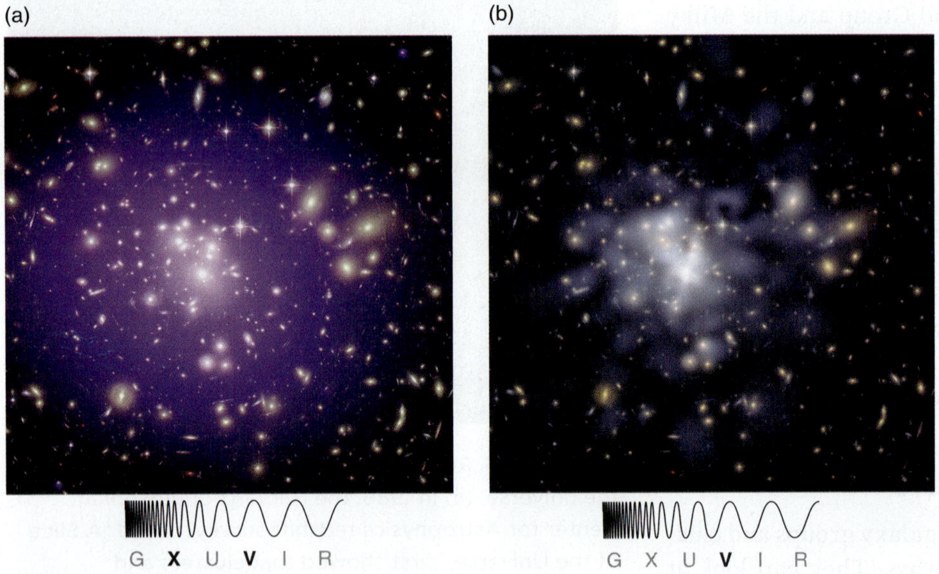

Figure 23.4 (a) Galaxy clusters, such as Abell 1689 shown here, are rich in hot, X-ray-emitting gas (shown in purple overlaying the image taken in the visible part of the spectrum). (b) Abell 1689, with the inferred dark matter distribution shown in blue-white.

100 million kelvins (K), making it bright in X-rays (**Figure 23.4**). Even though that gas is of extremely low density, the volume of space that it occupies is enormous: the mass of that hot gas can be up to 5 times the mass of all the stars in that cluster. X-ray spectra show that the gas contains significant amounts of massive elements that must have formed in stars. That chemically enriched gas has been either blown out of galaxies in winds driven by the energy of massive stars or stripped from galaxies during encounters with neighboring galaxies. The amount of luminous mass is not enough to keep that hot gas from escaping, so it would have dispersed long ago if not for the gravity of the dark matter filling the volume of the cluster.

Another way to look for dark matter relies on the predictions of Einstein's general theory of relativity (see Chapter 18), which states that mass distorts the geometry of spacetime, causing even light to bend near a massive object. In particular, light from a distant object is bent by the gravity of a galaxy or cluster of galaxies, so that images of the distant object can be seen magnified on both sides of the intervening galaxy or cluster. The result is a gravitational lens (**Figure 23.5a**). Recall from the discussion of massive compact halo objects (MaCHOs) in Chapter 19 that lenses can make background objects appear brighter. Lenses also can show multiple images of background objects, and those magnified images are often drawn out into arcs. The amount of gravitational lensing increases as the amount mass in the cluster increases. **Figure 23.5b** shows a galaxy cluster acting as a gravitational lens for several background galaxies. Analysis of such images reveals the mass of the lensing cluster.

Regardless of how astronomers measure the masses of galaxy clusters—by looking at the motions of their galaxies, by measuring their hot gas, or by using them as gravitational lenses—the results are the same. Dark matter dominates the mass of galaxy clusters and superclusters.

CHECK YOUR UNDERSTANDING 23.1

Place the following types of galaxy collections in order of increasing size: (a) wall; (b) cluster; (c) group; (d) supercluster

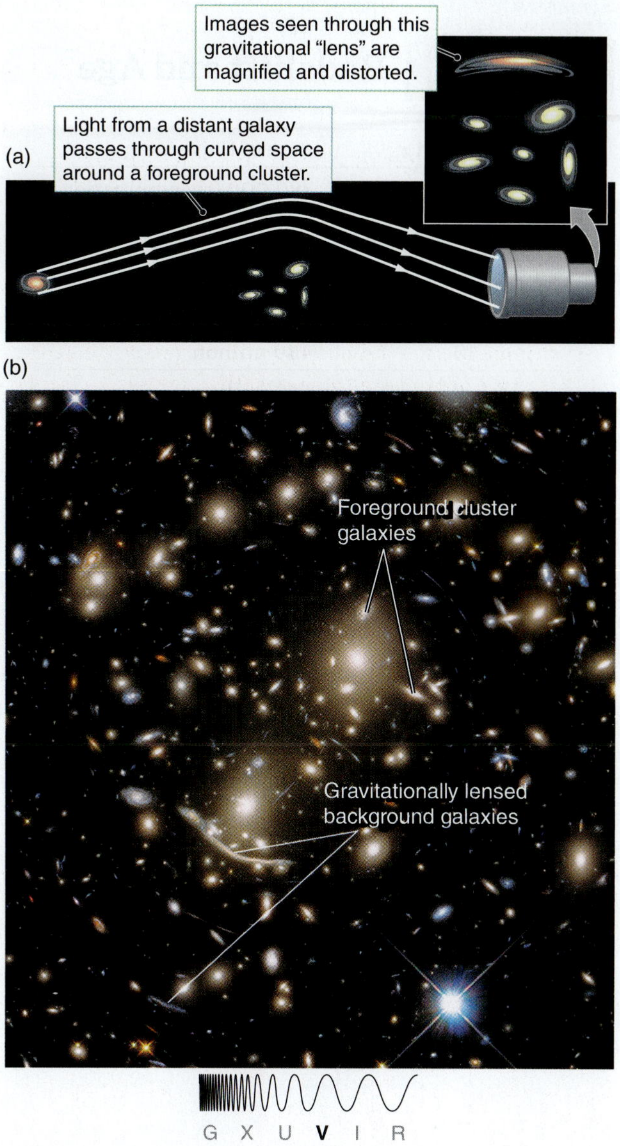

Figure 23.5 (a) This illustration shows the geometry of a gravitational lens. A mass can gravitationally focus the light from a distant object, thereby magnifying and distorting the image. (b) A Hubble Space Telescope image of the cluster Abell 370 shows many gravitationally lensed galaxies, seen as arcs.

23.2 Gravity Forms Large-Scale Structure

How did the universe evolve from a very smooth distribution of radiation and matter after the Big Bang to the large-scale structure of filaments and voids seen today? Astronomers approach that question observationally and theoretically: Observers use telescopes to study the most distant objects and the CMB, whereas theorists use the largest supercomputers to simulate the growth of small- and large-scale structure. In this section, we examine the role of gravity and dark matter in the formation of structure.

Gravitational Instabilities

In Chapter 15, we discussed gravitational instabilities involved in star formation. Star formation can begin in a molecular cloud with clumps inside it. Gravity causes those clumps to collapse faster than their surroundings; thus, gravity can turn density variations of clouds into stars. The same gravitational instability can turn density variations of the universe into galaxies.

As discussed in Chapter 21, space missions such as the Cosmic Background Explorer (COBE), Wilkinson Microwave Anisotropy Probe (*WMAP*), and the Planck space observatory have revealed variations in the CMB. Those tiny variations reflect the imprinted structure of the early universe at the time of inflation, providing the "clumps" or "seeds" from which galaxies and collections of galaxies grew. Over time, gravity amplified those clumps. Smaller structures such as dwarf galaxies formed first, whereas larger structures such as clusters, superclusters, filaments, and voids formed later. That process is called **hierarchical clustering** because the structure forms in a "bottom up" hierarchy. Hierarchical clustering is supported by observations and is fundamental to how structure formed in the universe.

If the universe is not totally uniform but instead has small variations in density from place to place, gravity will amplify those variations over time. On the scale of a galaxy, for example, the early universe may have regions denser than average by perhaps 1 percent. Over time, gravity can amplify that difference in density to more than twice the average, and a galaxy can form. The seeds are essential because gravity cannot produce structure in a perfectly uniform universe, which would have no difference in density for gravity to amplify. Those seeds are thought to have formed from quantum fluctuations during inflation in the early universe. That model has a profound implication: the seeds leading to galaxies, clusters, and superclusters (the largest structures in the universe) arose from the same quantum physics that describes the smallest structures in the universe (atoms, nuclei, and elementary particles).

TABLE 23.1 **Redshift and Age**

Observed z	Age of Universe (years)
1,100	380,000 (recombination)
30	100 million
20	200 million
15	270 million
10	480 million
9	560 million
8	650 million
7	750 million
6	900 million
5	1.2 billion
4	1.6 billion
3	2.2 billion
2	3.3 billion
1	5.9 billion
0.5	8.6 billion
0.25	10.5 billion
0	13.8 billion

Galaxies Formed because of Dark Matter

Whereas cosmology theorists think about gravity, inflation, and dark energy and consider the balance among radiation, normal matter, and dark matter in the early universe, observational astronomers measure the constants of cosmology, such as the Hubble constant (H_0), the cosmological constant (Ω_Λ), and matter density (Ω_m). The available observations and measured constants of cosmology become inputs to theoretical models and supercomputer simulations of how the large-scale structure formed. By the beginning of the 21st century, a "standard model" of Big Bang cosmology had broad support for explaining the large-scale structure and accelerated expansion of the universe, as well as the CMB and the amounts of the light elements from Big Bang nucleosynthesis (discussed in Chapters 21 and 22). That model is called **Lambda-CDM**—from *lambda* for the cosmological constant Ω_Λ and *CDM* for the cold dark matter that played a vital role in structure formation.

Astronomers measure the ratio of normal and dark matter and look for galaxies with the highest redshifts. More distant galaxies have higher measured values of redshift (z), so the observed light was emitted when the universe was *younger*, closer to the time of the Big Bang (**Table 23.1**). Observations of galaxies with different redshifts show how the universe has changed over time. However, a gap of several hundred million years occurs between the CMB maps of the universe at age 380,000 years and the highest-redshift galaxies observed. The international Atacama Large Millimeter/submillimeter Array (ALMA) in Chile has observed even younger galaxies. The future James Webb Space Telescope (JWST) may detect galaxies at even higher redshifts.

The details of the very early universe affected the growth of the large-scale structure seen today. The values of Ω_Λ and Ω_m are important in part because they determine how rapidly the universe expanded and therefore how difficult it is for gravity to overcome that expansion in a particular region. The more rapidly the universe is expanding, or the less mass it contains, the more difficult it will be for gravity to pull material together into galaxies and larger-scale structures.

Variations in the CMB are found at a level of about one part in 100,000. The theoretical models clearly show that such tiny variations at the time of recombination (when the universe was about 380,000 years old) are far too small to explain the structure observed in today's universe. Gravity is not strong enough for galaxies and clusters of galaxies to have grown from such small clumps. Those models indicate that for today's galaxies to have formed, the density of those clumps must have been a few tenths of a percent greater than the average density of the universe at the time of recombination. But if normal luminous matter in the early universe had clumps with that higher density, the variations in the CMB today would be at least 30 times larger than they are. How do astronomers reconcile that problem? Dark matter holds the key.

Dark matter is an essential ingredient in the formation of the structure we observe. The comparison of Big Bang nucleosynthesis models with observed amounts of light elements predicts a current density of normal matter in the universe very close to what we observe. That agreement in the observed density of normal matter powerfully constrains the nature of dark matter, which dominates the mass in the universe. Dark matter cannot consist of normal matter made up of neutrons and protons. If it did, the density of neutrons and protons in the early universe would have been much higher, and the resulting amounts of light elements in the universe from Big Bang nucleosynthesis would have been very different from what we actually observe (see Chapter 21).

Dark matter must be something that interacts only weakly with normal matter and has no electric charge and therefore does not interact with electromagnetic radiation. Dark matter and normal matter behaved differently in the early universe. Because clumps of such dark matter in the early universe did not interact with radiation or normal matter, astronomers would not see them directly when looking at the CMB. Dark matter clumpiness can be large enough to form galaxies without producing too much variation in the CMB as long as the clumpiness in the ordinary matter is much smaller. That unseen dark matter solves the problems of modeling the formation of galaxies and clusters of galaxies.

After the Big Bang, radiation pressure smoothed out ripples in the distribution of normal matter. Feebly interacting dark matter is immune to those processes, so whereas normal matter smoothed out, dark matter remained clumpy. **Figure 23.6** shows how the distributions of normal and dark matter differ as a result of the difference in how they interact with radiation. In addition, the dark matter in those clumps does not glow, so astronomers do not see it in the CMB. Although those clumps of dark matter do cause slight gravitational redshifts in the light coming from normal matter, the resulting variations fit well with current observations of the CMB. The mass of dark matter controlled the growth of gravitational instabilities in the early universe, analogous to how dark matter dominates the gravitational fields of today's galaxies and clusters. Dark matter solves the problems of the formation of galaxies and clusters of galaxies by producing a stronger gravitational attraction without being smoothed out by radiation in the early universe.

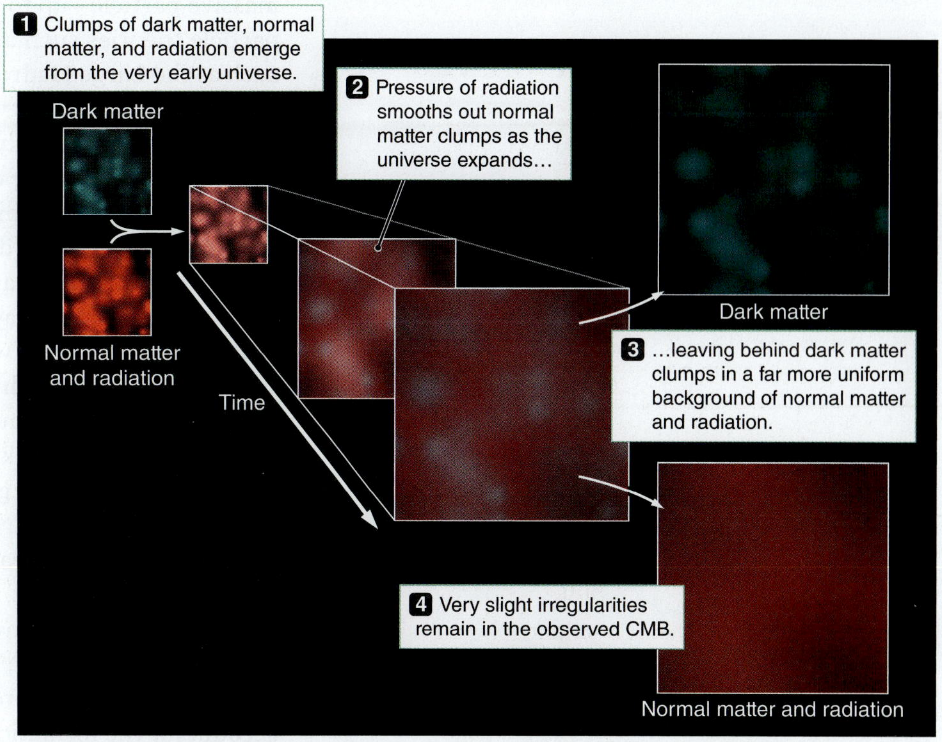

Figure 23.6 Radiation pressure and other processes in the early universe smoothed out variations in normal matter, but clumps in the dark matter survived and later pulled in normal matter to form galaxies.

Hot and Cold Dark Matter

Dark matter in the early universe was much more strongly clumped than normal matter. Within a few million years after recombination, those dark matter clumps pulled in the surrounding normal matter. Later, gravitational instabilities caused those clumps to collapse. The normal matter in the clumps went on to form visible galaxies. The details of how that happened depend greatly on the properties of the dark matter itself. Recall (see Chapter 20) that MaCHOs and weakly interacting massive particles (WIMPs) were considered possible explanations of dark matter, but it turned out that MaCHOs could *not* account for the unseen mass. Even though we do not yet know exactly what dark matter is made of, we can distinguish two broad classes, called cold dark matter and hot dark matter, based on how dark matter behaves.

Cold dark matter consists of feebly interacting particles moving about relatively slowly, like the slow-moving atoms and molecules in a cold gas. Several candidates are possible for the composition of cold dark matter. Most likely, cold dark matter consists of an unknown **elementary particle**. One candidate is the **axion**, a hypothetical particle first proposed to explain some observed properties of neutrons. Axions should have very low mass, and they would have been produced in great abundance in the Big Bang. Another candidate is the photino, a massive particle—up to 1,000 times the mass of the proton—predicted in certain particle theories such as string theory. Physicists are using particle accelerators such as the Large Hadron Collider to look for those types of particles, and several experiments are under way to search for axions and photinos trapped in the dark matter halo of the Milky Way.

Hot dark matter consists of particles moving so rapidly that gravity cannot confine them to the same region as the luminous matter in the galaxy. Neutrinos are one example of hot dark matter. Neutrinos interact with matter so weakly (see Chapter 14) that they can flow freely outward from the center of the Sun, passing through the overlying layers of matter as though they were not there. The universe is filled with neutrinos, which might account for a few percent of its mass. Although that percentage is not high enough to account for most of the dark matter in the universe, it may still have noticeably affected the formation of structure.

Cold and hot dark matter affect structure formation differently because of how they respond to a gravitational field. Slow-moving particles are more easily held by gravity than are fast-moving particles, so particles of cold dark matter clump together more easily into galaxy-sized structures than do particles of hot dark matter. As a result, theoretical models show that on the largest scales of massive superclusters, both hot dark matter and cold dark matter can form the kinds of structures observed; on much smaller scales, however, only cold dark matter can clump enough to produce structures like the galaxies filling the universe. To account for the formation of today's galaxies, we need cold dark matter.

We can best see how models of galaxy formation work by following the events predicted by the models step by step. Consider a universe made up primarily of cold dark matter, clumped together with normal matter in a manner consistent with observations of the CMB. On the scale of an individual galaxy, the effect of the cosmological constant is negligible.

Figure 23.7 A spiral galaxy passes through roughly six stages as it forms from the collapse of a clump of cold dark matter.

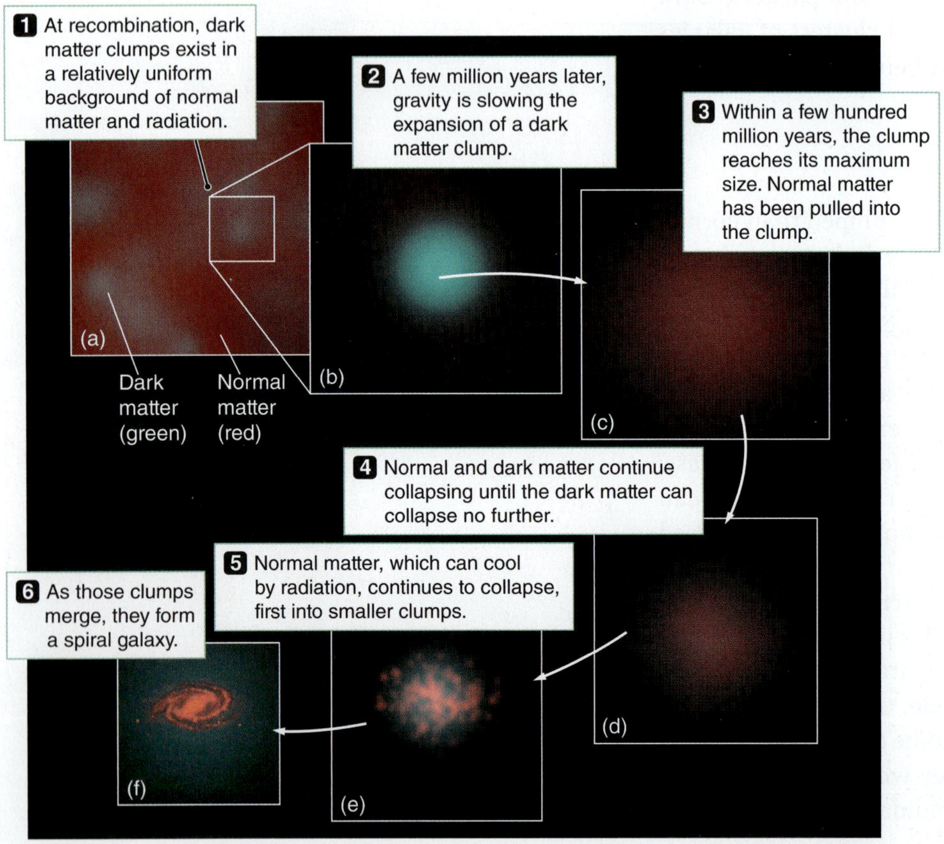

Figure 23.7a shows the first step of a model simulation of one clump of dark matter at the time of recombination. According to the model, the dark matter was slightly clumpier than normal matter, but overall the distribution of matter was remarkably uniform. **Figure 23.7b** shows that by a few million years after recombination, the universe in the simulation had expanded severalfold; spacetime is expanding, so the clumps of dark matter also are expanding. However, the clumps of dark matter did not expand as rapidly as their surroundings because their self-gravity slowed down their expansion. The clumps of dark matter stood out more with respect to their surroundings. The gravity of the dark matter clumps began to pull in normal matter. Eventually, the clumps of dark matter stopped expanding when their own self-gravity slowed and then stopped their initial expansion. By the stage shown in **Figure 23.7c**, normal matter clumped in much the same way as dark matter. Unlike dark matter, which cannot emit radiation, the normal matter in the clumps radiated away energy and cooled, collapsing toward the center of the dark matter clumps.

Those clumps did not exist in isolation: they were tugged on by the gravity of neighboring clumps and were pushed around by the pressure waves that ran through the young universe, smoothing out its structure. As a result, each clump may have been rotating slightly when it began to collapse. As normal matter fell inward toward the center of a rotating dark matter clump, that rotation forced much of the gas to settle into a rotating disk, just as the collapsing clouds of protostars settled first into an accretion disk. Later, that rotating disk became the disk of a spiral galaxy (**Figures 23.7d–f**). That model has succeeded because it yields the right masses for observed galaxies. For example, a system too low in mass would not cool fast enough to separate the visible matter from the dark matter. The dark matter we have been talking about as a major component of galaxies turns out to be responsible for their formation as well (see the **Process of Science Figure**).

CHECK YOUR UNDERSTANDING 23.2

The dominant factor in the formation of galaxies is the distribution of ________ in the early universe. (a) ordinary matter; (b) dark matter; (c) energy; (d) dark energy

23.3 First Light of Stars and Galaxies

Recombination occurred approximately 380,000 years after the Big Bang (see Chapter 21), when the universe cooled enough for electrons and protons to combine to form hydrogen and helium atoms. Before that time the universe was opaque; after recombination it became transparent. The CMB was emitted at that moment, and we can observe it today. That epoch in the history of the universe is called the **Dark Ages** because no visible "light" came from astronomical objects. The only light available then was from the cooling—and darkening—CMB, and some amount of 21-centimeter (21-cm) radio radiation from hydrogen. Several projects are ongoing to detect that 21-cm radio radiation, which has been redshifted to much longer radio wavelengths today.

The Dark Ages lasted from about 380,000 to several hundred million years after the Big Bang. By several hundred million years after the Big Bang, the first stars were forming from the elements created in the Big Bang. As those stars formed, they heated up until they emitted ultraviolet (UV) photons with enough energy to reionize neutral hydrogen in interstellar space. During that **reionization** stage, starting about 500 million years after the Big Bang, the UV light from the first stars stripped electrons from hydrogen atoms, and the ionized hydrogen became transparent to the light generated by the first stars. Reionization continued with star formation in the first low-luminosity galaxies and with radiation from the first supermassive black holes inside the first quasars. Reionization was completed by about 900 million to 1 billion years after the Big Bang (**Figure 23.8**).

Only within the past decade have astronomers been detecting objects representing light from the first billion years of the universe: those objects have redshifts greater than 6. (Table 23.1 shows how the times in the history of the universe compare with observed redshifts.) Many astronomers were surprised by the identification of galaxies, quasars, and gamma-ray bursts (GRBs) at such high redshifts because those objects had been thought to not form until after the universe was at least a billion years old. GRBs, for example, are extremely luminous and result from the explosive deaths of massive stars (see Chapter 18). For GRBs to be detected at $z = 8$, massive stars must have existed that had already died by 650 million years after the Big Bang. Similarly, the detection of quasars at $z = 7$ indicates that supermassive black holes must have formed in less than 750 million years after the Big Bang. The most distant galaxies observed have redshifts of 10–11, corresponding

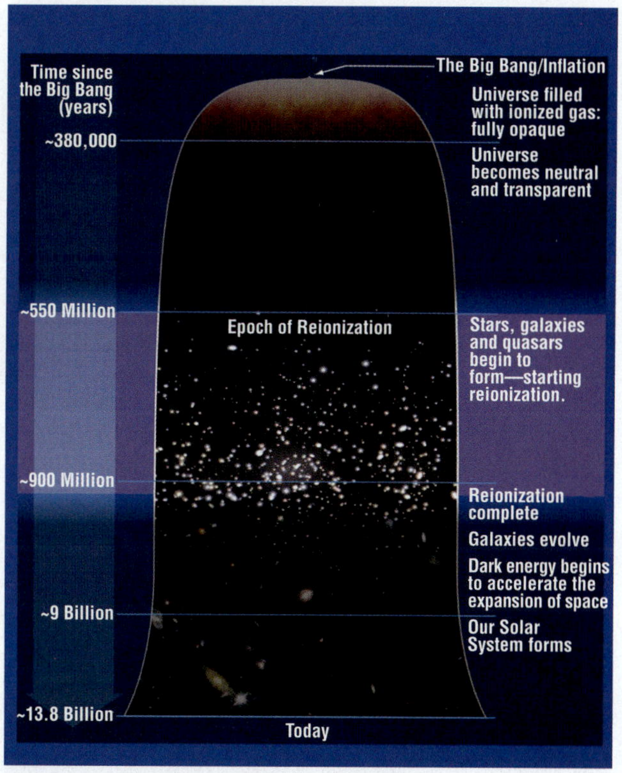

Figure 23.8 This timeline illustrates the formation of the first stars and galaxies and the epoch of reionization.

PROCESS OF SCIENCE

Multiple Streams of Evidence

Roughly 85 percent of the mass in the universe is an unknown form or composition called dark matter. So many observations, models, and experiments provide supporting evidence for dark matter that scientists have concluded that it exists.

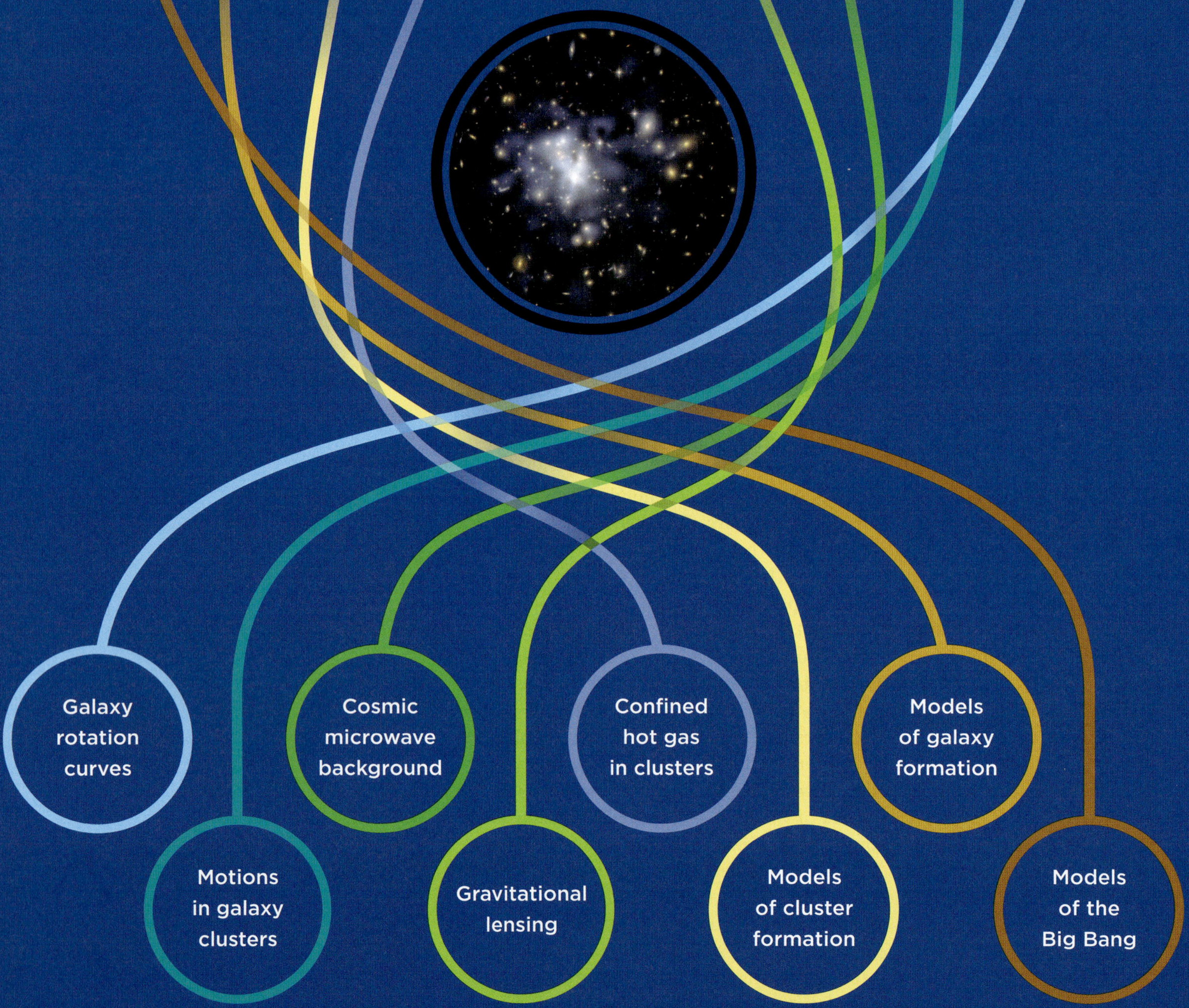

Scientists do not have the luxury of ignoring evidence that does not fit current theories. They must see "being wrong" as an opportunity to learn and as a challenge to try harder.

to 400 million to 500 million years after the Big Bang. The study of those highest-redshift objects and what they tell astronomers about the early universe is one of the most dynamic topics in astronomy today. New telescopes and new instruments are regularly detecting objects with higher and higher redshifts. By the time you read this book, the highest known redshifts will probably be higher.

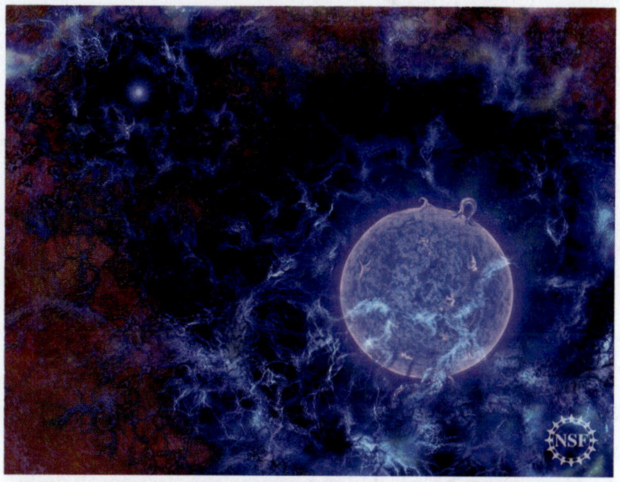

Figure 23.9 This supercomputer simulation shows the formation of the first stars from primordial gas in a dark matter minihalo a few hundred million years after the Big Bang. Here, two massive stars are forming a few hundred astronomical units apart. The brighter yellow regions are denser than the purple regions.

The First Stars

Astronomers use what has been learned in stellar evolution theory to study galaxy evolution and cosmology. The very first stars must have formed from the elements created in Big Bang nucleosynthesis: hydrogen, helium, and a very small amount of lithium. Observational astronomers look for stars that have only those elements, but so far none have been detected. Old stars in the halo of the Milky Way contain very low amounts of heavy elements, but not as low as zero. Astronomers use computer simulations that combine data from those old stars with the conditions in the early universe to figure out what happened before those old stars formed.

The formation of the first stars was different from the processes we discussed in Chapter 15. Because heavy elements did not exist, the universe had no dust or molecular clouds filled with cold, dense gas for the stars to form in. Instead, those first stars formed inside dark matter minihalos, which were about 0.5 million to 1.0 million solar masses (M_{Sun}) and 100 parsecs (pc) across. Those minihalos formed a few hundred million years after the Big Bang ($z \approx 20$ to 30). Primordial gas clouds within those minihalos contained neutral hydrogen, and over time some small amounts of molecular hydrogen (H_2) formed. Radiation emitted by molecular hydrogen cooled the gas, and as it cooled, the gas pressure dropped and the gas collapsed to the center of the minihalo. A protostar grew in the gas cloud, accreting more gas to become a star. Some theoretical models and computer simulations predict that those first stars were likely to be hot and massive. The illustration in **Figure 23.9** shows how such stars may have formed. Estimates of the masses of the first stars range from 10 to more than 100 M_{Sun} for single stars and 10–40 M_{Sun} for double stars. Those stars had high luminosity, peaking in the ultraviolet, which ionized the gas near the star. Those stars were singles, doubles, or small multiples. Large star clusters probably didn't form.

Those first stars were probably much more massive than the average star observed today. Stars with those high masses would have very short lifetimes (see Table 16.1 and Chapter 17), burning hydrogen in their cores for 10 million years or less. Today, massive stars use the carbon-nitrogen-oxygen (CNO) cycle for more efficient hydrogen burning, but carbon, nitrogen, and oxygen were not available to those first stars. They ended their brief lives in supernova explosions, scattering some heavy elements into nearby space. If the core of such a star had rapid rotation during the supernova explosion, it might have emitted a GRB of extremely high luminosity. Astronomers have observed a few GRBs that might be from redshift 8 or 9.

Some of those stars were massive enough to have become black holes after exploding. Black holes with companions can become energetic X-ray binary systems as mass falls onto the accretion disk of the black hole as the companion evolves. Because those stars were all so massive, both stars in a binary system (or all stars in a multiple system) would probably become black holes, which could then merge. (Recall from Chapter 18 that gravitational waves emitted during black-hole mergers have been detected.) Some theorists think that those merged black holes might have become the seeds for the supermassive black holes found

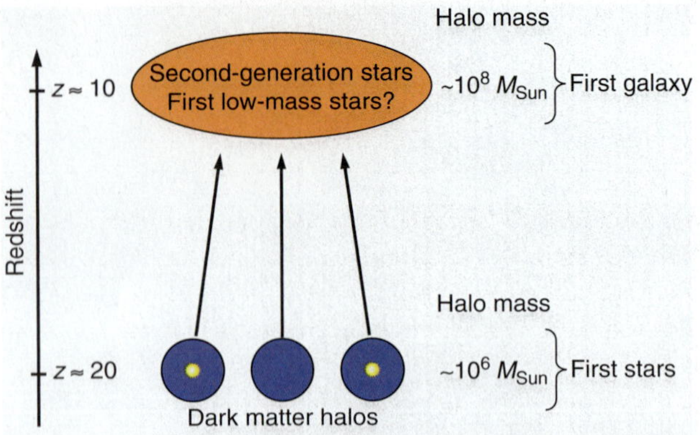

Figure 23.10 In "bottom up" (hierarchical) growth, the first stars formed in minihalos and then came together to form larger halos, which became the first small dwarf galaxies.

in galaxies, but other models suggest that those stellar black holes would take too long to have built up to a mass of 1 million to 1 billion M_{Sun}. When those first-generation stars died, supernova explosions expelled the heavy elements they formed through nucleosynthesis in their interiors.

Carbon, oxygen, and other elements might have mixed in and cooled nearby star-forming gas clouds. Some of those elements condensed into dust grains, which further cooled the clouds, so that the next generation of stars formed similarly to how stars form in today's cold molecular clouds. Those "second-generation stars" had very low abundances of heavy elements but measurably more than in the first stars. Because they formed in a cooler environment than the first stars, those second-generation stars could have had less mass. Any stars less massive than 0.8–0.9 M_{Sun} have such long lifetimes that they are still burning hydrogen on the main sequence today. Those stars are not very luminous, but a few of them have been found in the halo of the Milky Way. Such stars have very small amounts of heavy elements, but their spectra show small amounts of many elements in the periodic table—including uranium. Astronomers are interested in studying those small second-generation stars because they offer clues about the nature of the first stars and the conditions of the very young Milky Way.

The First Galaxies

How did the first galaxies form? From a theoretical perspective, the minihalos just discussed are not considered galaxies. A galaxy is supposed to have stellar systems that live a long time and can retain gas heated by the UV light of stars or by supernova explosions. After the first stars died, the energy from supernovae, GRBs, or X-ray binaries either may have heated any remaining gas in the minihalo too much for further star formation or such gas may have escaped the minihalo because of its relatively low gravitational pull. Thus, one generation of short-lived massive stars may have been produced from the minihalos, which with their small mass may not have been able to hang on to heated or blasted gas.

One theory is that the first galaxies were made up of the first systems of stars gravitationally bound in a dark matter halo. Those stars may have been first-generation stars or chemically enriched second-generation stars. The properties of the first galaxies were shaped by the first stars: their radiation, their production of some heavier elements, and the black holes resulting from the deaths of those stars. The masses of those galaxies are thought to have been about 100 million M_{Sun}, and they were built up hierarchically from the merging of minihalos (**Figure 23.10**).

One piece of evidence for that theory comes from infrared observations (**Figure 23.11**). Figure 23.11a includes the usual nearby stars and galaxies, but when those are all subtracted, a glow remains. That remaining structure (Figure 23.11b) probably arose from the first stars and galaxies several hundred million years after the Big Bang.

Another piece of evidence comes from the discoveries of galaxies and quasars at higher and higher redshifts, so we see them when they were very young. Those observations constrain the time line by indicating how soon the first galaxies—and first supermassive black holes—formed after the Big Bang. The peak of the spectrum of those early galaxies has been cosmologically redshifted into the infrared (**Working It Out 23.2**). The near-infrared instruments installed on the Hubble Space Telescope (HST) and the infrared instruments on the Spitzer Space Telescope and the Herschel Space Observatory have provided images of those very young, highly redshifted objects (**Figure 23.12**). The images of the highest-

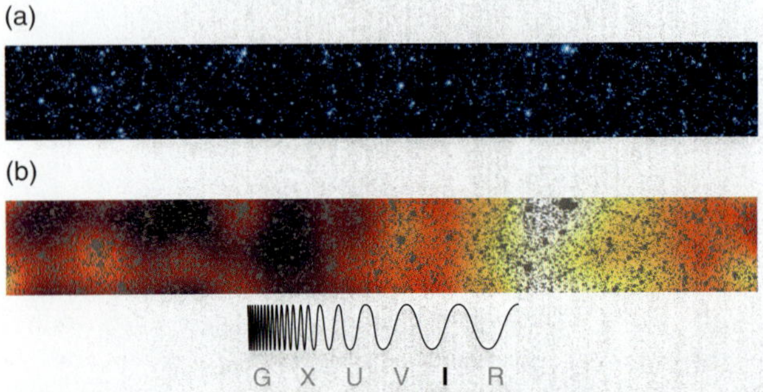

Figure 23.11 (a) A standard infrared image from the Spitzer Space Telescope shows stars and some galaxies in this strip of sky. (b) Here, the nearby stars and galaxies have been subtracted out (gray) and the remaining glow enhanced, showing some structure from the time when the earliest stars and galaxies were forming.

23.2 Working It Out Observing High-Redshift Objects

New instruments and new telescopes constructed in the past few years enable astronomers to detect galaxies at higher and higher redshifts. At high redshifts, the observed wavelengths of radiation are very different from the wavelengths emitted (see Working It Out 21.2). The equation is as follows:

$$1 + z = \frac{\lambda_{\text{observed}}}{\lambda_{\text{emitted}}}$$

where z is the redshift. For $z = 1$, the observed wavelength is twice that emitted; for $z = 2$, it is 3 times that emitted; and so on. For high-redshift galaxies with $z = 8$ to 11, the observed wavelengths are 9–12 times that emitted.

Neutral hydrogen in gas clouds along the line of sight to a distant galaxy scatters light at a wavelength of 121.6 nanometers (nm). As a result of that scattering by many clouds, a galaxy's spectrum is observed to be brighter at emitted wavelengths longer than 121.6 nm and is fainter, or "drops out," at emitted wavelengths shorter than 121.6 nm. That dropout is noticeable even in a weak spectrum from a very faint galaxy and even if other spectral details cannot be discerned. For a nearby galaxy at $z = 0$, that dropout occurs in the far-ultraviolet, but for distant galaxies the dropout is redshifted.

For example, at $z = 9$:

$$1 + 9 = \frac{\lambda_{\text{observed}}}{121.6\ \text{nm}}$$

Solving for the observed wavelength of the dropout: $\lambda_{\text{observed}} = 10 \times 121.6\ \text{nm} = 1{,}216\ \text{nm} = 1.2\ \text{microns}\ (\mu\text{m})$.

Galaxies at the highest redshifts are detected in observations at wavelengths longer than 1.2 μm in the near-infrared, but they do not show up in optical images. The infrared James Webb Space Telescope (the successor to HST) should observe many high-redshift galaxies.

redshift objects are small and faint, without the detail seen in the pictures of closer galaxies. Astronomers are excited by those images because just detecting those objects contributes to an understanding of when and how galaxies formed.

The first galaxies are thought to have formed by about $z \approx 15$ to 10, or by several hundred million years after the Big Bang. For the second generation of stars, the physics of formation was more complicated than for the first, so astronomers need to account for heavy elements and dust from the first generation mixed into the halo, magnetic fields, and turbulence. The heavier elements carbon, oxygen, and iron cooled the gas, which then collapsed to the center of that larger dark matter halo, probably to a disk, and stars formed in the dark matter halo. Those highest-redshift—that is, youngest—galaxies appear to be small (1/20 the size of the Milky Way), which adds support for the bottom-up models of galaxy formation. Only in the past decade have astronomers identified galaxies at

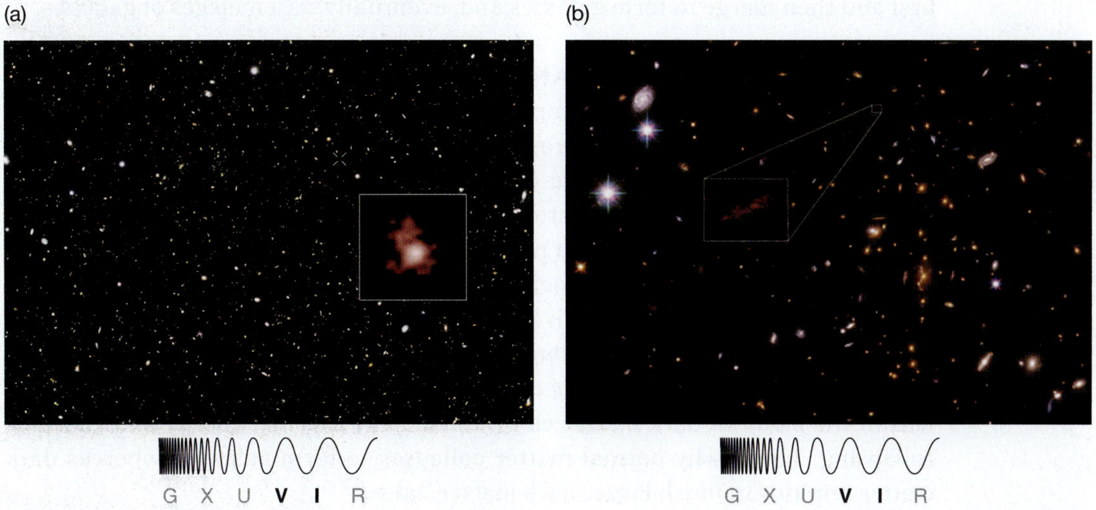

Figure 23.12 (a) This image from the Hubble Space Telescope shows a high-redshift galaxy, GN-z11 (at the X; enlarged in the inset). Spectroscopy indicates the galaxy is at redshift $z = 11.1$, seen as it was 13.4 billion years ago, when the universe was only 400 million years old. (b) Hubble and Spitzer image of a $z = 10$ galaxy that has been gravitationally lensed. This galaxy is estimated to be only 2,500 light-years in diameter.

$z = 7$ to 11, so as yet there are fewer confirmed galaxies from that time than older galaxies at $z = 2$ to 6.

The Local Group has small, faint dwarf galaxies orbiting the Milky Way (see Chapter 20). Streams of material from the dwarfs are falling onto the Milky Way, indicating that the Milky Way is still accreting mass. Recently, scientists have discovered many more of those dwarf galaxies. About a few dozen of the faint dwarf galaxies are called **ultrafaint dwarf galaxies** because they are dim—only 1,000–100,000 times the Sun's luminosity. They contain mostly old, faint stars with small amounts of heavy elements. One such galaxy is Segue 1. It contains only a few hundred stars and fewer heavy elements than any other observed galaxy, suggesting it may not have had much more star formation after its first stars formed (unlike the additional bursts of star formation that occurred in regular dwarf galaxies). The ultrafaint dwarfs may have contributed to building the Milky Way's halo, and they may be the oldest galaxies around that have been involved in galaxy mergers. Ultrafaint dwarfs may even be the fossil remains of the first galaxies or of the first minihalos. Because those very old galaxies are more massive than their luminosity suggests, dark matter dominates even them.

Parallels between Galaxy and Star Formation

Both star formation (see Chapter 15) and galaxy formation involve the gravitational collapse of vast clouds to form denser, more concentrated structures. To help you see some similarities and differences, this section compares the formation of galaxies and the formation of stars after the first generation.

GRAVITATIONAL INSTABILITY In both star and galaxy formation, the collapse begins with a gravitational instability. Regions only slightly denser than their surroundings are pulled together by their own self-gravity. As the matter in those regions becomes more compact, gravity becomes stronger, and the collapse process snowballs. One key difference between galaxy and star formation is that for a galaxy to form, the dark matter clump must collapse rapidly enough to counteract the overall expansion of the universe itself.

FRAGMENTATION The order of fragmentation differs between star and galaxy formation. In molecular clouds, first large regions begin to collapse, and then they fragment further to form individual stars. In contrast to that "top down" star formation process, galaxy formation is "bottom up": smaller structures collapse first and then merge to form galaxies and, eventually, assemblages of galaxies.

COMPRESSION, HEATING, AND THERMAL SUPPORT As an interstellar molecular cloud collapses, its temperature increases, causing the pressure in the cloud to increase. The higher pressure would eventually be enough to prevent further collapse, except that the cloud core can radiate away thermal energy. That energy is the bright infrared radiation that enables astronomers to see star-forming cores. Compare that process with galaxy formation: as a dark matter clump collapses, the random velocities of its particles increase, and it too quickly reaches a point at which gravity is balanced with the random motions of the dark matter particles. Dark matter *cannot* radiate away energy, however, so once that balance is reached, the collapse of the dark matter is over. Only the normal matter within the cloud of dark matter can radiate away thermal energy and continue collapsing. That's why normal matter collapses to form galaxies, whereas dark matter remains in much larger dark matter halos.

As galaxies form, dark matter remains in extended halos. Dark matter may be the dominant form of matter in the universe, and it may determine the structure of galaxies, but dark matter can never collapse enough to play a role in the processes that shape stars, planets, or the interstellar medium.

ANGULAR MOMENTUM AND THE FORMATION OF DISKS Conservation of angular momentum is responsible for the formation of disk galaxies, just as it is responsible for the formation of the accretion disks around young stars and for the flatness of both the Milky Way and the Solar System. The origin of the angular momentum is different, though. Whereas turbulent motions within star-forming molecular clouds produce the net angular momentum for stellar disks, gravitational interactions with nearby clumps are responsible for the angular momentum of the Milky Way.

TIMESCALE The time a star spends as a protostar is a small fraction of the time it will spend on the main sequence. The formation of a galaxy is a slower process, taking from redshift 10–20 down to redshift 1–2, or about half the current age of the galaxy.

THE END PRODUCT Once a stellar accretion disk forms, most of the matter moves inward and is collected into a star. In contrast, much of the matter in a spiral galaxy remains in the disk (see Chapter 19).

CHECK YOUR UNDERSTANDING 23.3

The first stars formed in the universe had ________ than the stars formed today. (a) more heavy elements and higher mass; (b) more heavy elements and lower mass; (c) fewer heavy elements and higher mass; (d) fewer heavy elements and lower mass

23.4 Galaxies Evolve

Galaxies continued to evolve hierarchically in the young universe, with smaller "protogalactic" fragments merging to form larger ones. The young universe was smaller and denser. The light from a galaxy now at redshift z was emitted when the universe was $z + 1$ times smaller than it is today (see Chapter 21), with a volume $(1 + z)^3$ times smaller than its volume today. Those early fragments and galaxies were closer together because the universe was smaller, and therefore mergers were more common. Computer simulations indicate that small concentrations of normal matter within the dark matter would have clumped and collapsed under their own gravity as they radiated and cooled, forming clumps of normal matter that ranged from the size of globular clusters to the size of dwarf galaxies.

In a large spiral galaxy such as the Milky Way, faint dwarf spheroidal galaxies (with dark matter) and the oldest globular clusters (without much dark matter) may be leftover protogalactic fragments. The gas collapsed to form a rotating disk as it cooled. A supercomputer simulation that included dark matter, gravity, star formation, and supernova explosions reproduced a Milky Way–like galaxy with a large disk and a small bulge (**Figure 23.13**). Observationally, astronomers conduct "stellar archaeology" on the oldest parts of the Milky Way to better understand how the components are all assembled into a galaxy. For example, the oldest globular clusters may be 1 billion to 2 billion years older than the halo. Most stars in the Milky Way formed between 11 billion and 7 billion years ago. The disk and bulge formed at about that same time.

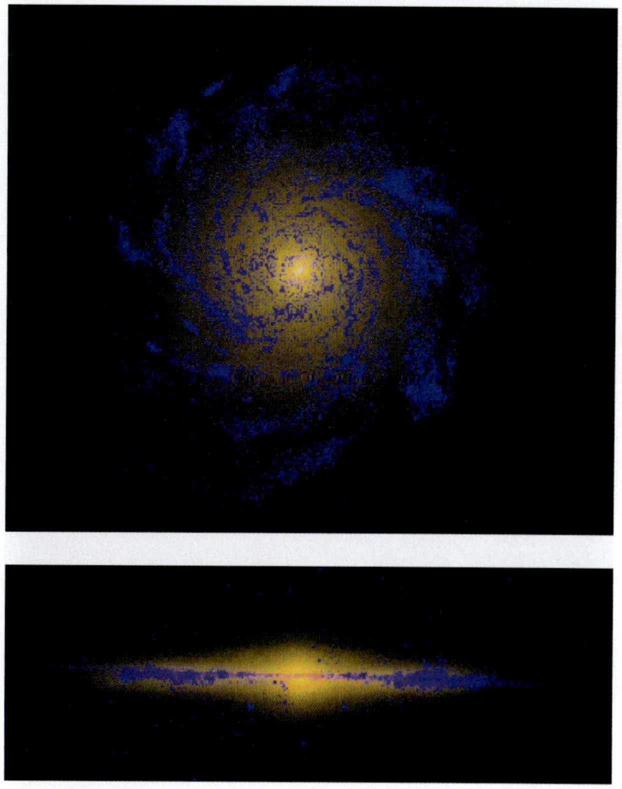

Figure 23.13 This supercomputer simulation of the formation of a Milky Way–sized spiral galaxy has reproduced the small bulge and big disk by using the Lambda-CDM model. At the bottom is an edge-on view of the galaxy. Blue colors indicate recent star formation, whereas older stars are redder.

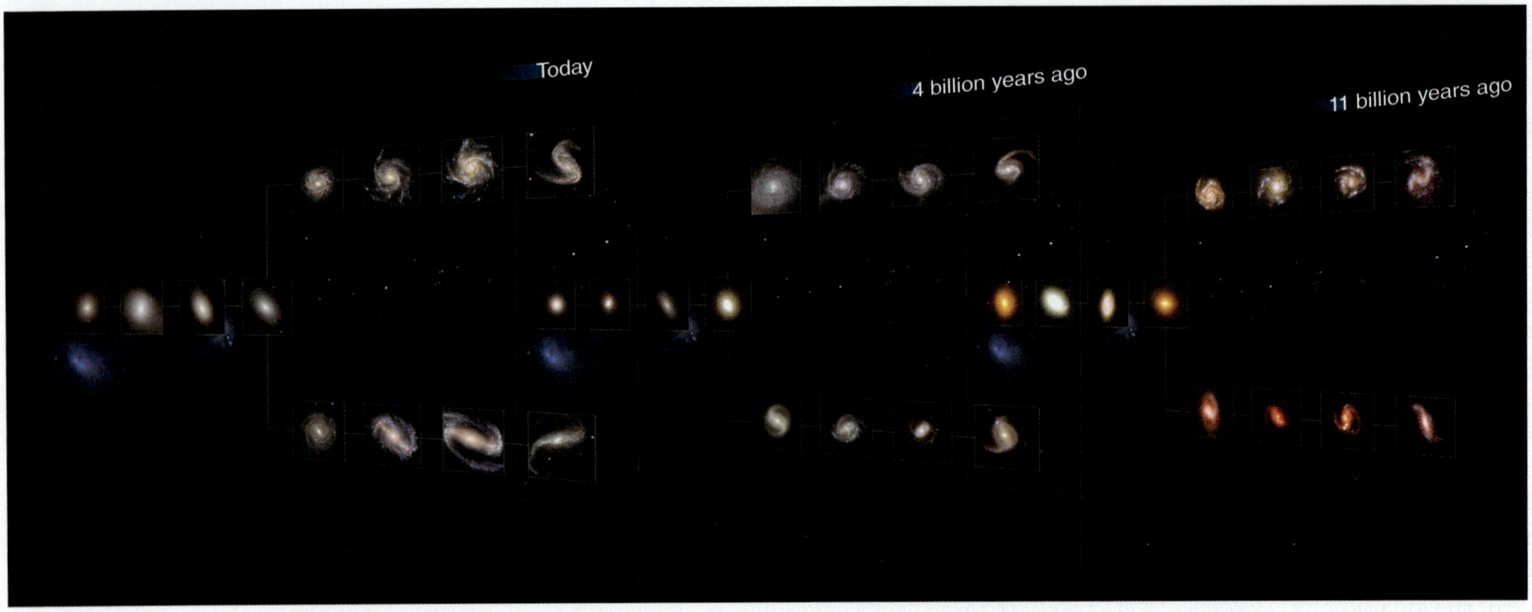

Figure 23.14 A comparison of the Hubble classification of galaxies today (spirals on the top, barred spirals on the bottom, elliptical galaxies in the middle) with galaxies throughout the history of the universe. More irregular galaxies existed in the past, indicating that spirals took some time to form.

The Most Distant Galaxies

Figure 23.14 shows images of galaxies throughout the history of the universe. The galaxies observed in the very early universe, before about 11 billion years ago, are so faint that no structure is visible. Observations of galaxies at about 11 billion years ago have shown visible structure much less regular than that of galaxies today. Even at 4 billion years ago, galaxies were much more irregular. Early irregular galaxies are merging galaxies, which fits with our ideas that in the early universe, when galaxies were closer together, more mergers were taking place. Most of today's galaxies conform to the Hubble classification, and only about 10 percent are irregular. Four billion years ago, however, more than half the galaxies were irregular, the number of elliptical and S0 galaxies was about the same, and many fewer spirals existed. That difference in galaxy types at different times suggests that spirals took time to form. Those later mergers probably produced spiral galaxies over time.

The bottom-up hierarchical merging also may have triggered the formation of the supermassive black holes at the centers of galaxies. The first supermassive black holes, which power the distant quasars seen at $z = 7$ with masses of $10^9\ M_{\text{Sun}}$, could have grown from the merging of minihalos with stellar black holes left after the first stars. Or they could have formed through the accretion of gas from the material between the galaxies during mergers of the first galaxies or through rapid collapse from hot, dense gas at the center of the first galaxies. In nearby galaxies, the mass of the supermassive black hole and the bulge properties are related, suggesting that the growth of the black hole and the bulge might have been linked when they were younger. Supermassive black holes could have grown even more massive from the mergers of large galaxies, too. **Figure 23.15** shows a nearby galaxy with two supermassive black holes about 900 pc (3,000 light-years) apart that are merging.

Figure 23.15 Chandra X-ray observations (red, orange, and yellow), combined with Hubble Space Telescope observations (blue and white), show that NGC 6240 has two black holes less than 1,000 parsecs apart. The black holes at the center of the white features will probably merge in about 100 million years.

The hierarchical merging and growth of the supermassive black holes also affected the rates of star formation in the evolving galaxies. The tidal interactions between the galaxies and the collisions between gas clouds in the galaxies probably triggered many regions of star formation throughout the combined

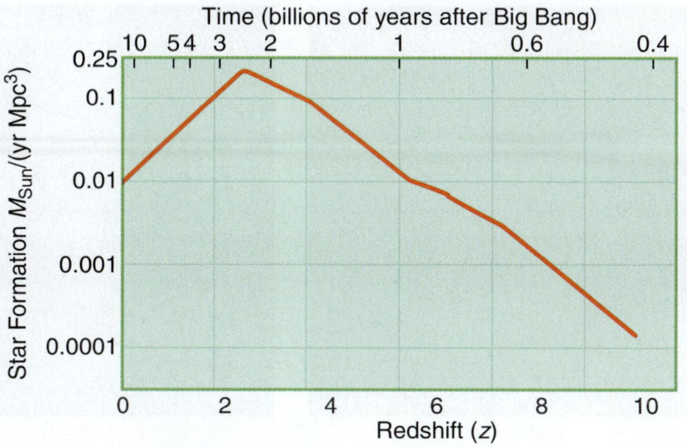

Figure 23.16 The rate of star formation per volume has changed in the universe, peaking at about 2 billion to 3 billion years after the Big Bang.

system. Star formation generally increased sporadically, including a rapid increase in the 200 million years between $z = 10$ and 8. The graph in **Figure 23.16** shows that the star formation rate seems to have peaked around $z = 3$ (2.5 billion to 3 billion years after the Big Bang) before decreasing again to the current rate.

Young galaxies that are merging are seen in HST images (**Figure 23.17a**). By observing galaxy mergers at different distances, astronomers can see how mergers differ at various times in the history of the universe. Large ellipticals are now thought to result from the merger of two or more spiral galaxies. The dark matter halos of the galaxies merge, and the stars eventually settle down into the blob-like shape of an elliptical galaxy. Elliptical galaxies are more common in dense clusters, where mergers are likely to have been more frequent. Compare those young mergers with those of closer, older galaxies (**Figure 23.17b**—a computer simulation of such a merger is shown in Figure 6.32).

Just as galaxies merge, so do clusters of galaxies. **Figure 23.18a** shows the high-speed collision and merging of two galaxy clusters in the Bullet Cluster. Images in visible light show the individual galaxies. Ordinary matter, mostly hot gas, is seen in X-rays (shown in red), and the distribution of the total mass can be found from the gravitational lensing produced by the clusters. The ordinary matter slowed down in the collision, but the dark matter (shown in blue) did not. That

(a)

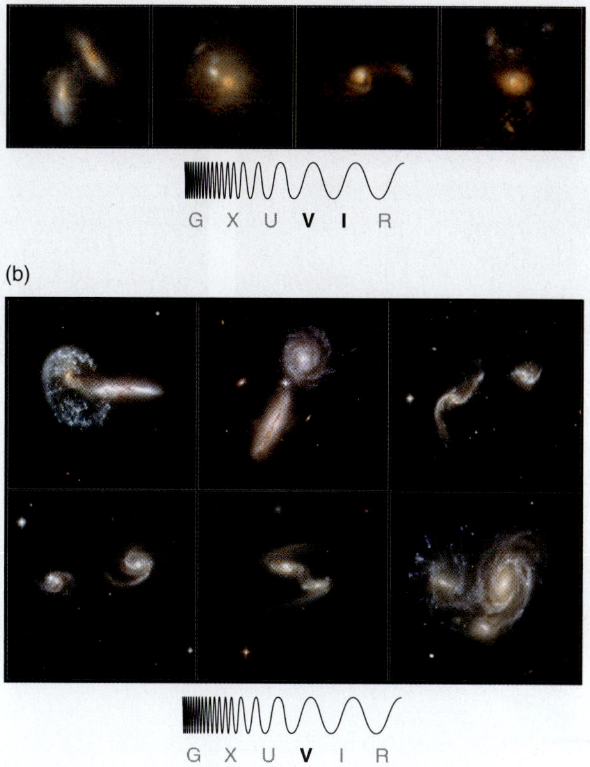

Figure 23.17 (a) These HST images show young, merging galaxies. From left to right, galaxies at 2.4 billion to 6.2 billion light-years from Earth merging at 11 billion to 7.5 billion years after the Big Bang. (b) These tidally interacting galaxies in the more nearby universe show severe distortions, including stars and gas drawn into long tidal tails.

(a) (b)

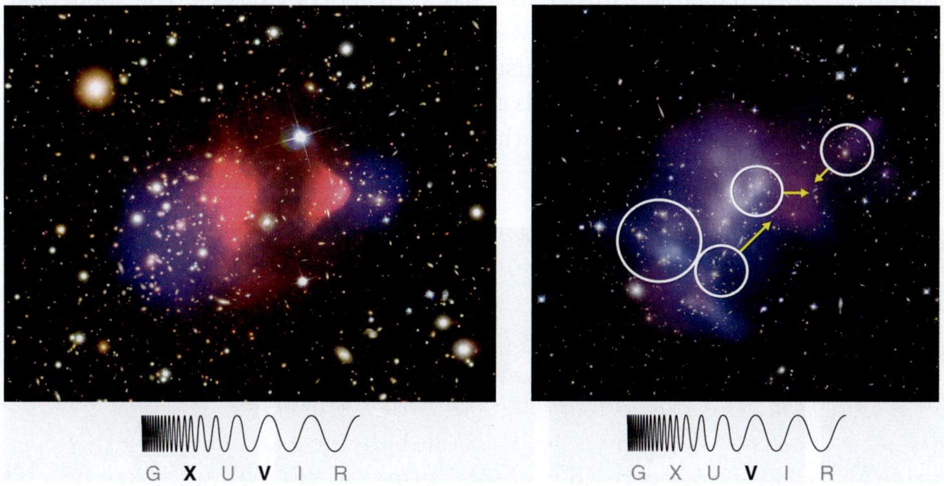

Figure 23.18 (a) The Bullet Cluster of galaxies, at a redshift of $z = 0.3$, represents a later stage in the merging of two giant clusters of galaxies. The smaller cluster on the right seems to have moved through the larger cluster like a bullet. (b) Four galaxy clusters (circled) are merging in the direction of the yellow arrows. In the Chandra X-ray Observatory image, the cooler gas is magenta and the hotter gas is blue. This is one of the most complex galaxy clusters observed.

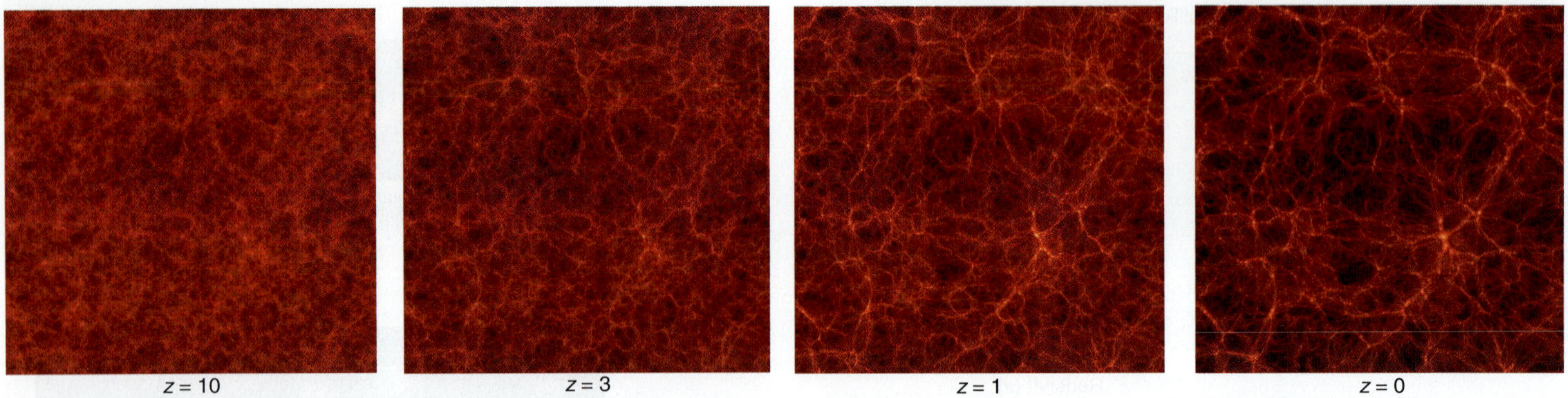

Figure 23.19 The Bolshoi supercomputer simulation of the formation of very large-scale structure in a universe filled with cold dark matter. These images show the growth of filaments and voids.

separation serves as evidence for dark matter in galaxy clusters. The collision of four galaxy clusters is shown in **Figure 23.18b**. Clusters of galaxies also evolve hierarchically, growing from smaller structures to larger ones. As with galaxies themselves, younger, distant clusters are messier than older, nearby ones, yielding more evidence that the formation of structure in the universe was hierarchical.

Simulating Structure

Astronomers use the most powerful supercomputers available to simulate the universe. Those simulations start with billions of particles of dark matter and use the most recent observations of the CMB. The simulations model the formation and evolution of dark matter clumps and halos, filaments and voids, small and large galaxies, and galaxy groups and clusters. Those computations also simulate the flow of ordinary gas within those structures as stars form, and researchers use those simulations to create images of what the universe should have looked like at different times (and different redshifts). Those images are then compared with images of the actual universe. That comparison sets limits on the parameters of the universe: the amount of mass, for example, or the type of dark matter. If the inputs to the simulation are correct, the two sets of images will look very similar. If not, the two sets of images will look very different.

In one example, a Lambda-CDM simulation called the Bolshoi was run on NASA supercomputers. The simulation shows that slight variations in density after inflation led to higher-density regions that became the seeds for the growth of structure (**Figure 23.19**). During the first few billion years, dark matter fell together into structures comparable in size to today's clusters of galaxies. The spongelike filaments, walls, and voids became well defined later. Zooming in on some simulated filaments and voids shows a cluster of galaxies (**Figure 23.20**). The similarities between the results of the models and observations of large-scale structure are remarkable. **Figure 23.21** compares the simulated view with the observed slice of

Figure 23.20 Simulations enable astronomers to model structure at different-sized scales. Each image zooms in more to show smaller structures. Dark matter halos are seen as dense blobs. The smallest blob in the last image could become a giant spiral galaxy like the Milky Way.

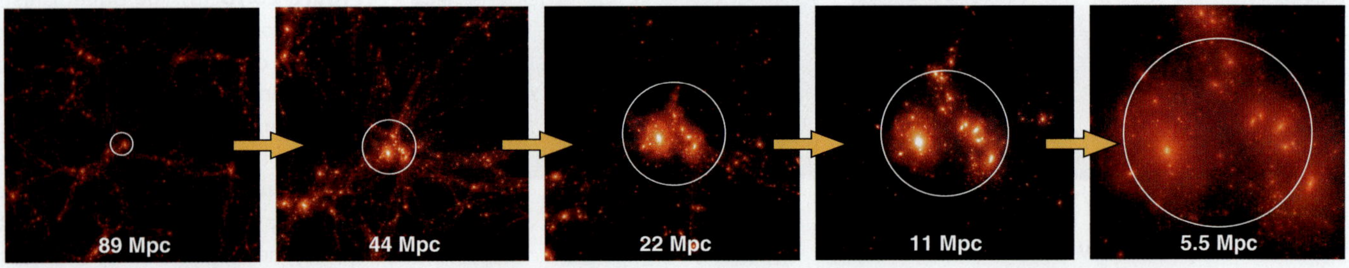

the universe from the Sloan Digital Sky Survey. Only simulations with certain combinations of mass, CMB variations, types of dark matter, and values for the cosmological constant will produce structure similar to what is actually observed. That result is very important. Models incorporate assumptions consistent with observational and theoretical knowledge of the early universe, and they predict the formation of large-scale structure similar to what is actually seen in today's universe.

The precision cosmology developed over the past two decades has given astronomers a detailed model of the universe in space and time, so that galaxy evolution can now be reliably sequenced. **Figure 23.22** neatly summarizes the galaxy formation process, in which smaller objects form first and merge into ever-larger structures, leading ultimately to the Hubble Ultra Deep Field 2014 image shown as the opening figure of this chapter.

Predictions about the Deep Future

What does the future hold for the universe and its structure? Scientists use well-established physics and our current cosmological understanding to speculate how the existing structures in the universe will evolve over a very long time. Of course, any speculation might change with newer discoveries.

In the distant future, studying galaxies and cosmology will be much more difficult, if not impossible. In 90 billion years or so, the only visible galaxy will be the one that resulted from the merger of the Milky Way and Andromeda (see Chapter 20). Because of the acceleration of the expansion of the universe, the other galaxies will be too far away to be detectable from here. In a trillion years, even the wavelength of the CMB will be greater than the size of the observable universe, making the CMB invisible too. In 100 trillion (10^{14}) years from now (about 10,000 times as long as the current age of the universe), the last molecular cloud will collapse to form stars, and a mere 10 trillion years later the least massive of those stars will evolve to form white dwarfs.

At that time, most of the normal matter in the universe will be locked up in degenerate stellar objects: brown dwarfs, white dwarfs, and neutron stars. Unless a Big Rip (where everything is torn apart by expansion) occurs, every condensed object, including white dwarfs and neutron stars, will ultimately decay via

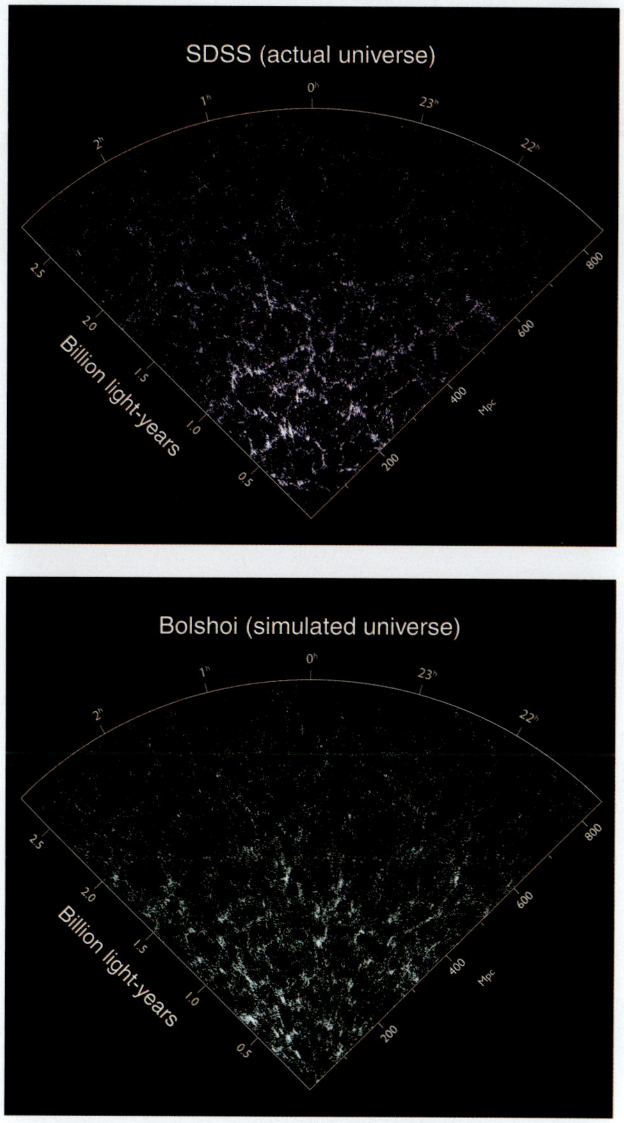

Figure 23.21 The large-scale structure of dark matter halos produced by the Bolshoi simulation (bottom) is remarkably similar to the distribution of galaxies observed in the Sloan Digital Sky Survey (SDSS; top).

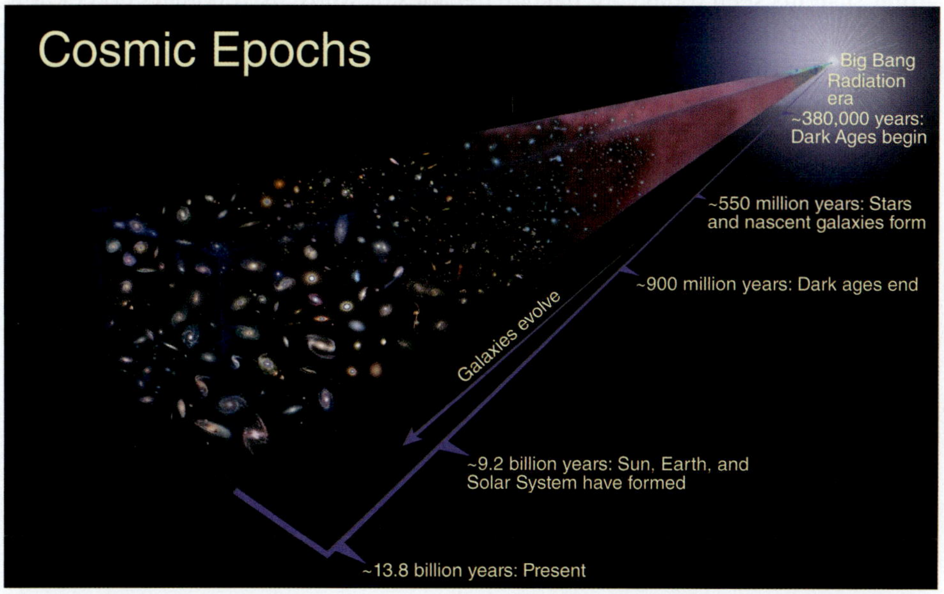

Figure 23.22 A schematic view of how structure formed in the universe, from smaller systems to larger ones.

quantum effects. Eventually, the only significant remaining concentrations of mass will be black holes, ranging from small ones with the masses of single stars to supermassive black holes that grew to the size of galaxy clusters. Those black holes will slowly evaporate into elementary particles through the emission of Hawking radiation (see Chapter 18). A black hole with a mass of a few solar masses will evaporate into elementary particles in 10^{65} years, and galaxy-sized black holes will evaporate in about 10^{98} years. By the time the universe reaches an age of 10^{100} years, even the largest black holes will be gone. A universe vastly larger than our current one will contain little but photons with colossal wavelengths, neutrinos, electrons, positrons, and other waste products of black-hole evaporation. The universe will continue to expand forever—into the long, cold, dark night of eternity. Ultimately, the universe will be structureless and empty except for any residual uniform dark energy.

CHECK YOUR UNDERSTANDING 23.4

We expect the kinds of galaxies that we see at a redshift of $z = 4$ to be: (a) much like what we see today; (b) smaller and much more irregular looking than today; (c) far more numerous but with more spiral galaxies; (d) larger versions of what we see today.

Origins We Are the 4 or 5 Percent

The very first stars initiated post–Big Bang nucleosynthesis, which chemically enriched the universe with elements heavier than lithium, including the elements necessary to form planets and life on Earth. The first stars and galaxies "lit up" the universe and brought it out of the Dark Ages, thanks to the glow of their ordinary matter. But that ordinary matter is not the primary constituent of the universe. In their study of the largest structures in the universe—the galaxies and groups and clusters of galaxies—astronomers have concluded that those objects are dominated by dark matter, not the stuff of stars and planets.

Several types of independent observations suggest that dark matter accounts for about 25 percent of the universe (see Chapters 19–22). Dark matter dominated over normal matter when the first stars and galaxies formed. Dark matter dominated the evolution of the galaxies as they went through mergers to become the larger systems seen today. But despite decades of study, astronomers don't know exactly what dark matter is.

As the universe evolved, dark energy became more important to the structure of the universe. When it was younger and galaxies were forming, the universe was dominated by matter. The expansion of the universe was slowing because of the pull of gravity from all that matter. About 5 billion to 7 billion years ago, though, dark energy began to dominate over matter in the universe, and the expansion of the universe accelerated. Dark energy now makes up about 70 percent of the mass and energy of the universe—nearly 3 times as much as the dark matter—and astronomers don't know exactly what it is, either.

Ordinary matter is only about 5 percent of the mass and energy of the present-day universe. Most of what astronomers have studied since people first looked at the sky is that 5 percent. The parts of the local universe important to life on Earth—the Sun, its planets and their moons, and the local environment in the Milky Way—are composed of that 5 percent. The matter that constitutes us is a surprisingly small part of the universe.

ARTICLES QUESTIONS

Astronomers report on the discovery that the Milky Way is part of a huge supercluster of galaxies.

Welcome to Laniakea, Your Galactic Supercluster Home

By AMINA KHAN, *Los Angeles Times*

Home, supercluster home. Astronomers studying the skies have found that our galaxy, the Milky Way, is part of an enormous supercluster of galaxies that has been named Laniakea, which comes from the Hawaiian words for "immeasurable heaven."

Laniakea, described in the journal *Nature*, stretches some 500 million light-years across and holds 100,000 galaxies with the mass of a hundred million billion Suns. This supercluster, scientists said, provides a whole new lens on the mysterious internal dynamics of such giant clusters, the biggest structures in the universe (see Figure 23.3).

The universe has a certain order, defined by gravity: Planets circle stars, which spin around in galaxies, which hang out together in large clusters, which are nestled in gargantuan superclusters. These superclusters give the cosmos its large-scale structure—matter concentrates in massive, dense nodes and thin tendrils connect those nodes across the empty spaces in between.

Even though superclusters define the nodes of this cosmic web, scientists know very little about how they work. Until now, they didn't even know where exactly our own Milky Way lives—just that we're in a local group of at least 54 galaxies, which is itself part of a larger conglomeration long called the Virgo (or Local) Supercluster.

It's a little like knowing your street address without knowing which city you live in—which sounds like a pretty basic problem, but has proved deeply challenging to solve.

"This is what I've been doing for a long time, trying to understand where we live," said lead author R. Brent Tully, an astronomer at the University of Hawaii in Honolulu.

The giveaway is gravity. Gravity holds superclusters together, so watching the movements of galaxies speeding around their neighborhoods could help define a supercluster's borders.

Here's the problem. The universe is expanding faster and faster, due to a mysterious force called dark energy. This mucks up the speed readings: Scientists don't know how much of a particular galaxy's movement is due to the inward pull from the gravity of a cluster, and how much is due to the outward pull of the universe expanding faster and faster.

So a team of cosmic cartographers used the Superflows-2 database, a catalog of the motions of galaxies, to tease apart these movements. For each of more than 8,000 galaxies, they subtracted the expansion-related movement from the overall motions leaving only the local, gravitationally driven movement.

The researchers found that the galaxies were flowing along these long, beautiful lines, many of them toward a gravitational dense basin of galaxies known as the Great Attractor. The boundary between those flowing toward this spot and those flowing away marked the edges of the galaxy supercluster.

It's like watching a watershed where the flows divide, Tully said.

"If you're standing at a certain place at the divide, water might be going in two different directions—one going into the Mississippi basin, one going off to the Great Lakes," Tully said.

The Milky Way sits in a backwater among these flowing galaxies, near a far edge of the supercluster. In fact, the whole Virgo Supercluster is just a small fraction of the Laniakea Supercluster.

"The Local Supercluster, we now see quite clearly, is just an appendage on Laniakea," Tully said. "It's like a suburb in a city of a metropolitan area."

Elmo Tempel, an astronomer at Tartu Observatory in Estonia who was not involved in the paper, praised the findings.

"This is the first clear definition of a supercluster," Tempel wrote in a commentary. "The downside of it is that it requires dynamical information that is available only for the nearby universe."

Understanding the dynamics of our own supercluster could help shed light on the large-scale forces at work in the universe, Tully said, particularly dark matter, the mysterious unseen mass whose gravity holds such enormous structures together, and dark energy, the strange force that is tearing them inevitably apart.

ARTICLES QUESTIONS

1. How large is Laniakea in megaparsecs?
2. Why was figuring out the shape and extent of the Milky Way's supercluster hard?
3. Are the Milky Way and the Local Group in the center of that supercluster?
4. Why will using a similar method to map out more distant superclusters be difficult?
5. Watch the video at https://www.nature.com/nature/videoarchive/laniakea/ for a 3-D view of Laniakea. How do astronomers set the boundaries of the different clusters of galaxies?

Summary

Galaxies are clumped into groups, clusters, superclusters, and walls, not distributed uniformly. Galaxies develop because of gravitational instabilities in the presence of cold dark matter. The first stars formed in minihalos of dark matter, whereas the first galaxies formed later, in larger dark matter halos. Over time, smaller galaxy fragments merged to form larger galaxies. Mergers still happen today. Dark matter and dark energy are responsible for the formation of structure and the future of the universe. Ordinary matter, which makes up stars, planets, and us, is only about 20 percent of the total mass in the universe and, once dark energy is included, only 4 or 5 percent of all the universe. Dark energy is playing an increasingly dominant role in the universe and will shape its future.

LG 1 **Describe the distribution of galaxies in the universe.** Galaxies are hierarchically gathered into groups, clusters, superclusters, and larger structures. The walls surround voids in which very few galaxies are present. Dark matter dominates the mass of galaxy groups and clusters.

LG 2 **Explain how the large-scale structure of today's universe evolved from the structure that began to form shortly after the Big Bang.** Structure formed in the universe as slight variations grew in the density of the dark matter emerging from the Big Bang. Those "seeds" then collapsed under the force of gravity, pulling in normal matter as well. Structure formed from the bottom up.

LG 3 **Describe the formation of the first stars and the first galaxies.** The first stars formed in dark matter minihalos rather than in clouds of dust and gas. Because of the absence of heavy elements in those dark matter halos, star formation and evolution were different from that in the nearby universe. Gravitationally bound groupings of stars formed, which were the earliest galaxies. Radiation from the first stars and their supernovae affected the growth of the first galaxies. Radiation from the first stars, galaxies, and black holes ended the Dark Ages.

LG 4 **Explain how the observations of galaxies at different redshifts illustrate the evolution of the large-scale structure of the universe.** Larger structures form from smaller structures. The earliest galaxies merged to become larger galaxies, which in turn accumulated into clusters. The clusters grew hierarchically, through mergers, to become larger clusters, superclusters, and walls. Distant young galaxies look very different from the nearby galaxies in the present-day universe: they are smaller, fainter, and more likely to be merging.

? Unanswered Questions

- How do supermassive black holes form? The detection of quasars at very high redshifts indicates that supermassive black holes formed early in the Dark Ages, but exactly how or how much they may have contributed to reionization is unknown.
- How can astrophysicists better understand dark matter? Will a way be developed to detect dark matter by means other than gravity? If dark matter is a particle, is it stable against decay? Experiments in particle accelerators on the ground and observations of Local Group halos in space have put some limits on the type of dark matter particles that might be out there, but an answer still is not apparent.

Questions and Problems

Test Your Understanding

1. Place the following in order of size, from smallest to largest.
   a. a galaxy
   b. star clusters
   c. the Local Group
   d. a wall
   e. Virgo Cluster
   f. Laniakea
   g. a star
2. The dominant force in the formation of galaxies is
   a. gravity.
   b. angular momentum.
   c. the electromagnetic force.
   d. the strong nuclear force.
3. Larger galaxies form from the merging of small protogalaxies. That process is similar to the formation of
   a. stars.
   b. planets.
   c. molecular clouds.
   d. asteroids.
4. Which is *not* a characteristic difference between cold and hot dark matter?
   a. temperature
   b. ability to emit radiation
   c. the way they clump under the influence of gravity
   d. mass density

5. Gravitational lenses can be used to find
   a. dwarf galaxies near the Milky Way.
   b. dust and gas in the voids.
   c. the masses of galaxy clusters.
   d. the structure of Laniakea.

6. If dark energy is constant, the universe in the far distant future
   a. will be cold and dark.
   b. will be bright and hot.
   c. will collapse and re-form.
   d. will be the same as it is now, on large scales.

7. What is the primary difference between galaxy groups and galaxy clusters?
   a. how tightly they are bound by gravity
   b. the size of the largest galaxy
   c. the total mass of the galaxies
   d. dark matter does not exist in galaxy groups

8. Once the redshift of a galaxy has been found, its __________ is also known.
   a. mass
   b. velocity
   c. distance
   d. both b and c

9. Galaxy formation is similar to star formation because both
   a. are the result of gravitational instabilities.
   b. are dominated by the influence of dark matter.
   c. end with the release of energy through fusion.
   d. result in the formation of a disk.

10. Dark matter clumps stop collapsing because
    a. angular momentum must be conserved.
    b. they are not affected by normal gravity.
    c. fusion begins, and radiation pressure stops the collapse.
    d. the particles are moving too fast to collapse any further.

11. Elliptical galaxies come from
    a. the gravitational collapse of clouds of normal and dark matter.
    b. the collision of smaller elliptical galaxies.
    c. the fragmentation of large clouds of normal and dark matter.
    d. the merging of two or more spiral galaxies.

12. Which of the following statements about the Dark Ages of the universe is *false*?
    a. The first stars began forming during the Dark Ages.
    b. The end of the Dark Ages coincided with reionization.
    c. The Dark Ages lasted from 200 million to 600 million years after the Big Bang.
    d. During the Dark Ages, photons could travel freely through the universe.

13. Astronomers have never observed a star that has no elements heavier than lithium. What does that imply about the first stars?
    a. They must have died before galaxies were fully formed.
    b. The first stars did not form until after galaxies formed.
    c. The first stars must have had very low masses.
    d. The first stars must have been enriched in heavy elements.

14. Reionization of the neutral gas in the universe occurred because of the
    a. decay of dark matter particles.
    b. emission of neutrinos by the first stars that formed.
    c. release of jets of charged particles from supermassive black holes.
    d. radiation from the first stars, supernovae, and black holes that formed.

15. Place the following in increasing order of size.
    a. the fraction of the universe that is stars, planets, dust, and gas
    b. the fraction of the universe that is dark energy
    c. the fraction of the universe that is dark matter

Thinking about the Concepts

16. Suppose you could view the early universe when galaxies were first forming. How would it be different from today's universe?
17. Are voids likely to be filled with dark matter? Why or why not?
18. Imagine that the universe has galaxies composed mostly of dark matter, with relatively few stars or other luminous normal matter. If that were true, how might you learn of the existence of such galaxies?
19. How are star formation and galaxy formation similar? How do they differ?
20. What is the origin of large-scale structure?
21. Why is dark matter essential to galaxy formation?
22. Which of the following is the correct evolutionary sequence? (a) Small star clusters formed first, which were bound together into galaxies, which were later bound together in clusters and superclusters; or (b) supercluster-sized regions collapsed to form clusters, which then later collapsed to form galaxies, which formed small clusters of stars. Justify your answer.
23. Why does the current model of large-scale structure require dark matter?
24. What is the difference between a galaxy cluster and a supercluster? Is our galaxy part of either? How do we know?
25. How does a roughly spherical cloud of gas collapse to form a disk-like, rotating spiral galaxy?
26. What are some observational signs that dark matter exists? Explain why that evidence challenged earlier theories and forced astronomers to change their minds about the existence of matter they could not see.
27. Using the current model of galaxy formation, describe how galaxies should appear as you look further back in time. Are the features you described observed?
28. Why do astronomers think that the dark matter in the universe must be mostly cold, rather than hot?
29. Describe structure formation in the universe, starting at recombination and ending today.
30. Why do scientists think that gravity, and not the other fundamental forces, is responsible for large-scale structure?

Applying the Concepts

31. Figure 23.3a shows the redshifts and velocities for many galaxies. Find the average recession velocity for the galaxies in the wall indicated by the line labeled "Walls."

32. The theory of cosmology assumes that on large scales, the structure in the universe is uniform no matter where you look. Maps of structure, like the ones shown in Figure 23.3, support that assumption. Does the presence of large masses such as Laniakea violate that principle? Explain your answer.

33. As clumps containing cold dark matter and normal matter collapse, they heat up. When a clump collapses to about half its maximum size, the increased thermal motion of particles tends to inhibit further collapse. Whereas normal matter can overcome that effect and continue to collapse, dark matter cannot. Explain the reason for that difference.

34. In previous chapters, we painted a fairly comprehensive picture of how and why stars form. Why, then, is the star formation history of a young galaxy hard to model? Is that difficulty a failure of scientific theories?

35. If 300 million neutrinos fill each cubic meter of space, and if neutrinos account for only 5 percent of the mass density (including dark matter) of the universe, estimate the mass of a neutrino.

36. What are the approximate masses of (a) an average group of galaxies, (b) an average cluster, and (c) an average supercluster?

37. The lifetime of a black hole varies in direct proportion with the cube of the black hole's mass. How much longer does a supermassive black hole of 3 million M_{Sun} take to decay than a stellar black hole of 3 M_{Sun}?

38. Knowing what elliptical galaxies are made of, estimate how old they must be. Knowing that ellipticals form via mergers of spirals, and knowing when galaxies first formed, estimate how long the merging events that formed the elliptical galaxies seen today took to complete.

39. The chapter-opening figure is the Hubble Ultra Deep Field 2014.
    a. Explain how you might identify which galaxies might be at higher redshifts.
    b. The image shows about one 10-millionth of the entire sky. Estimate the number of galaxies in the entire sky. Explain how you used the cosmological principle in the calculation.

40. The Bullet Cluster image in Figure 23.18a shows the collision of two galaxy clusters. Estimate the number of galaxies you can see in each cluster.

41. The initial fluctuations leading to large-scale structure probably arose from quantum fluctuations in the early universe. How would the universe look different today if those fluctuations had been 10 times bigger? What about 10 times smaller?

42. Is the early universe on the left or the right in the graph in Figure 23.16? (Alternatively, you could wonder whether "now" was on the left or right in the graph.) Compare the star formation rate today with the star formation rate at the peak. How much more star formation occurred during the peak than occurs now?

43. Figure 23.21 shows real data in the top panel and simulated data in the bottom panel. Those two panels are not in exact agreement. Do those differences indicate a significant problem in the simulation's ability to represent reality? Why or why not?

44. Compare galaxies with redshifts of $z = 0.5$, 4, and 8. About how old was the universe when the light was emitted from each galaxy? At what spectral wavelengths would you see the "dropout" of the spectrum? Can you observe those dropouts from the ground, or do you need a telescope in space?

45. Suppose a dwarf galaxy is orbiting a giant elliptical galaxy at the center of a cluster at a distance of 4 Mpc and a speed of 800 km/s. Estimate the core mass of the cluster.

USING THE WEB

46. Go to the website for the Sloan Digital Sky Survey, which has made a three-dimensional map of the sky. A video fly-through can be accessed at http://www.sdss.org/science/, and new ones may be posted as the project acquires more data. Why did the SDSS scientists make that map? What are the goals of the SDSS project? What do astronomers learn from that fly-through? What are the new results from that and from the BOSS projects?

47. Go to the website for the Dark Energy Survey (https://www.darkenergysurvey.org). What are the science goals of the project? What are the observations? Are any results available yet?

48. Use your favorite search engine to find the highest-redshift galaxy, quasar, or GRB observed to date. Why are astronomers interested in finding objects at higher and higher redshifts? Why is estimating the relative frequency of such objects in comparison with their frequency at $z = 6$ or 2 also important for astronomers?

49. Go to the website for the ALMA telescope (http://almaobservatory.org/en). What is unique about that telescope? How does it study the Dark Ages? Look at the items under "News" and "Press Releases." Are any reports about galaxy formation available?

50. Go to the website for the James Webb Space Telescope (https://www.jwst.nasa.gov/). What is its status? Are any results available on the earliest stars and galaxies?

EXPLORATION

The Story of a Proton

digital.wwnorton.com/astro6

Now that you have surveyed the current astronomical understanding of the universe, you are prepared to put the pieces together to make a story of how you came to be sitting in your chair and reading these pages. Taking a moment to work your way backward through the book is valuable: start at the Big Bang and review all the intervening steps that had to occur, back to the beginning of the book, which began with looking at the sky.

1 In the Big Bang, how did a proton form?

2 How might that proton have become part of one of the first stars?

3 Suppose that proton later became part of a carbon atom in a 4-M_{Sun} star. Through what type of nebula would it have passed before returning to the interstellar medium?

4 Suppose that carbon atom then became part of the molecular-cloud core forming the Sun and the Solar System. What two physical processes dominated the core's collapse as the Solar System formed and that carbon atom became part of a planet?

5 Beginning with the Big Bang, create a time line that traces the full history of a proton that becomes a part of the nucleus of a carbon atom in you.

24 Life

Throughout history, many people have wondered whether we are alone in the universe. But only in recent times has science has been able to address the underlying questions that must be answered, such as: How common are planets? What is the range of conditions under which life can thrive? How does life begin and evolve? Even answering the question "What is life?" is surprisingly complicated. Answering those questions is part of a science called **astrobiology**—the study of the origin, evolution, distribution, and future of life in the universe. Throughout this book, the "Origins" section in each chapter has discussed how astronomers think about those questions. In this chapter, we expand on some of those topics and provide a more systematic overview of how astronomers think about life in the universe and how they search for signs of it.

LEARNING GOALS

By the end of this chapter, you should be able to:

LG 1 Explain our current understanding of how and when life began on Earth and how it has evolved.

LG 2 Explain how life is a structure that has evolved through the action of the physical and chemical processes that shape the universe.

LG 3 Describe the locations in our Solar System and around other stars where astronomers think life might be possible.

LG 4 Describe some methods used to search for intelligent extraterrestrial life.

The Allen Telescope Array searches for signs of intelligent life. ▶▶▶

Are we alone in the universe?

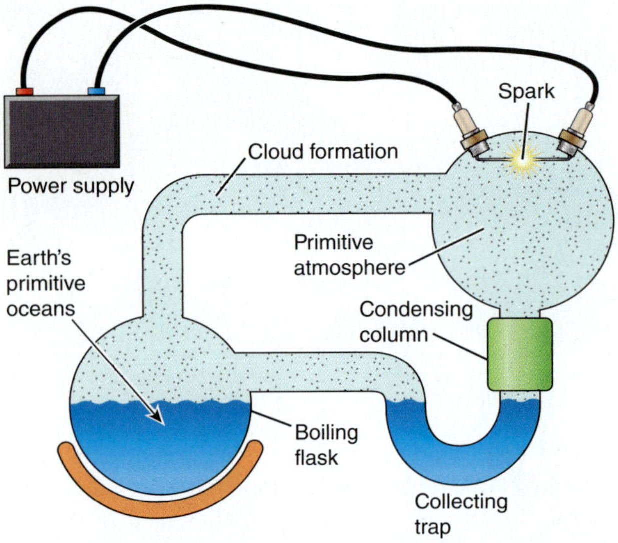

Figure 24.1 The Urey-Miller experiment was designed to simulate conditions in an early-Earth atmosphere.

(a)

(b)

Figure 24.2 (a) Life on Earth may have arisen near ocean hydrothermal vents such as this one. Similar environments might exist elsewhere in the Solar System. (b) Living organisms around hydrothermal vents, such as the giant tube worms shown here, rely on hydrothermal rather than solar energy for their survival.

24.1 Life Evolves on Earth

What is **life**? Many scientists suggest that no single definition of life would encompass all the forms of life that may exist in the universe. To date, we have discovered only a single example of life: that found here on Earth. Life on Earth may be very different from life found in other places in the universe. Indeed, even comparing varied organisms on Earth leads to complications in the definition of life. Viruses, for example, meet some criteria for life but not others. A complete definition of life may also have to take into account life-forms not yet discovered. In general, life draws energy from the environment to survive and reproduce. From studies of known life, we conclude that like planets, stars, and galaxies, life is a structure that has evolved in the universe. On Earth, all life involves carbon-based chemistry and uses liquid water as its biochemical solvent, whereas specific biological molecules such as ribonucleic acid (RNA) and deoxyribonucleic acid (DNA) enable life to reproduce and evolve. In this section, we will briefly review what is known about the origin and evolution of life on Earth.

The Origin of Life on Earth

How did life begin on Earth? Recall from Chapter 7 that Earth's secondary atmosphere was formed in part by carbon dioxide and water vapor emitted by volcanoes. Comets and asteroids probably added large quantities of water, methane, and ammonia to the mix. Liquid water is considered essential for any terrestrial-type life to get its start and evolve because water is an effective solvent that can move other atoms and molecules around, making them more accessible to cells. Early Earth had abundant sources of energy, such as lightning and ultraviolet solar radiation, that fragmented those simple molecules. Those fragments later reassembled into molecules of greater mass and complexity. Some of those were organic molecules—that is, molecules that contained hydrogen and carbon. Rain carried the heavier molecules out of the atmosphere into Earth's oceans, forming a primordial soup.

In 1952, chemists Harold Urey (1893–1981) and Stanley Miller (1930–2007) attempted to create conditions similar to those early-Earth conditions. Using equipment illustrated in **Figure 24.1**, they placed water in a sterilized laboratory jar to represent the ocean and then added methane, ammonia, and hydrogen as a primitive atmosphere; electric sparks simulated lightning as a source of energy. Within a week, the Urey-Miller experiment yielded molecules associated with life: amino acids and components of nucleic acids. Proteins, the structural molecules of life, are made of 20 amino acids. Eleven of those acids were synthesized in the Urey-Miller experiment. Nucleic acids are the precursors of RNA and DNA.

Additional sealed samples from that old experiment were examined 50 years later. In those samples, hydrogen sulfide had been added to the "primitive atmosphere." When the samples were analyzed, 23 amino acids were found. That finding suggests that hydrogen sulfide, which would have come from volcanic plumes in the early Earth, was important. More recent experiments with carbon dioxide and nitrogen as the primitive atmosphere have produced results similar to those of Urey and Miller. A plausible atmospheric composition with an energy source can produce significant quantities of amino acids and other substances important to life.

From laboratory experiments such as those, scientists have developed various models to explain how life might have begun in an early-Earth environment. However, the details of how those precursor molecules evolved into the molecules of life are not yet clear. Some biologists think life began in the ocean depths, where volcanic vents provided the localized heat needed to create the highly organized molecules responsible for biochemistry (**Figure 24.2**). Other researchers think that life originated in tide pools, where lightning and ultraviolet radiation supplied the energy (**Figure 24.3**). In either case, short strands of molecules that could replicate themselves may have formed first, later evolving into RNA, and finally into DNA, the huge molecule that serves as the biological "blueprint" for self-replicating organisms.

A few scientists have suggested that life on Earth may have been "seeded" from space in the form of microorganisms brought here by meteoroids or comets. However, no scientific evidence exists to support the seeding hypothesis. In addition, although that hypothesis might explain how life came to Earth, it does not explain how life itself began.

Figure 24.3 Life may have begun in tide pools, where lightning and ultraviolet light supply the energy for chemical processes.

When Life Began

If life did indeed get its start in Earth's oceans, when did it happen? Recall from Chapter 7 that Earth was bombarded by Solar System debris for several hundred million years after it formed roughly 4.6 billion years ago. Those conditions might have been too harsh for life to form and evolve on Earth. Once the bombardment abated and oceans formed, the opportunities for life to begin greatly improved. Terrestrial life seems to have quickly taken advantage of that more favorable environment. In the Nuvvuagittuq belt in Quebec, Canada, scientists recently found what they think are fossilized microorganisms that are 3.77 billion to 4.28 billion years old and seem similar to microorganisms from modern hydrothermal vents. Scientists debate whether carbonized material in Greenland rocks dating back 3.65 billion to 3.85 billion years provides indirect evidence of early life. Stronger and more direct evidence for early life appears in the form of fossilized **stromatolites** (masses of simple microorganisms) that date back about 3.5 billion years. Fossilized stromatolites have been found in western Australia and southern Africa, and living examples still exist today (**Figure 24.4**). An analysis of eleven 3.465-billion-year-old microfossils from western Australia found five species—two of which carried out primitive photosynthesis and the other three produced or consumed methane. That evidence suggests that the earliest life formed less than a billion years after the Solar System formed and within a few hundred million years of the end of the late heavy-bombardment period.

Figure 24.4 These modern-day stromatolites are growing in colonies along an Australian shore.

All life on Earth shares a similar genetic code that originated from a common ancestor. Close comparison of DNA of different species enables biologists to trace backward to the time when different types of life first appeared on Earth and to identify the species from which those life-forms evolved. The earliest organisms were **extremophiles**—life-forms that not just survive but actually thrive under extreme environmental conditions. Extremophiles include organisms such as thermophiles, which flourish in water temperatures as high as 120°C that occur near deep-ocean hydrothermal vents, such as the one shown in Figure 24.2a. Other extremophiles thrive in conditions of extraordinary cold, low oxygen, salinity, pressure, dryness, acidity, or alkalinity. Scientists today study extremophiles in boiling-hot sulfur springs in Yellowstone National Park, in salt crystals beneath the Atacama Desert in Chile, at the bottoms of glaciers, in ice fields in the Arctic, and in other extreme environments (**Figure 24.5**).

Figure 24.5 These thermophiles in the Grand Prismatic Spring in Yellowstone National Park live in temperatures of 70°C. The colors result from different amounts of chlorophyll.

Among the early life-forms was an ancestral form of **cyanobacteria**, single-celled organisms otherwise known as *blue-green algae*. Those microorganisms

(a)

(b)

Figure 24.6 (a) Cyanobacteria today form sheets on lakes and other bodies of water. (b) Under a microscope, the individual microorganisms are visible.

form extensive sheets on the surface of bodies of water, as shown in **Figure 24.6a**. Under a microscope, it becomes clear that those sheets are colonies of individual microorganisms, as seen in **Figure 24.6b**. Cyanobacteria **photosynthesize**, using sunlight and carbon dioxide as food and generating oxygen as a waste product. Initially, the highly reactive oxygen that cyanobacteria produced was quickly removed from Earth's atmosphere by oxidation, or rusting, of surface minerals. Once most of the exposed minerals were oxidized and could no longer absorb oxygen, atmospheric levels of oxygen began to rise. Oxygenation of Earth's atmosphere and oceans began about 2 billion years ago, and the current level was reached only about 250 million years ago, as shown in **Figure 24.7**. Without cyanobacteria and other photosynthesizing organisms, Earth's atmosphere would be as oxygen-free as the atmospheres of Venus and Mars.

Biologists comparing DNA sequences find that terrestrial life is divided into two types: prokaryotes and eukaryotes. Prokaryotes, which include Bacteria and Archaea, are simple organisms that consist of free-floating DNA inside a cell wall; as shown in **Figure 24.8a**, they lack both cell structure and a nucleus. Eukaryotes, which form the cells in animals, plants, and fungi, have a more complex form of DNA contained within the cell's membrane-enclosed nucleus, illustrated in **Figure 24.8b**. The first eukaryote fossils date from about 2 billion years ago, coincident with the rise of free oxygen in the oceans and atmosphere, although the first multicellular eukaryotes did not appear until a billion years later.

Increasing Complexity

Scientists have used DNA sequencing to establish what is known as the "phylogenetic tree of life," shown in **Figure 24.9**. That complex tree describes the evolutionary interconnectivity of all species of Bacteria, Archaea, and Eukarya and has revealed some interesting relationships. For example, Archaea were initially thought to be the same as Bacteria, but genetic studies show they diverged long ago, and the Archaea have genes and metabolic pathways more similar to those of Eukarya than to those of Bacteria. On the macroscopic scale, the phylogenetic tree places animals closest to fungi, which branched off the evolutionary tree after slime molds and plants.

Living creatures in Earth's oceans remained much the same—a mixture of single-celled and relatively primitive multicellular organisms—for more than 3 billion years after terrestrial life appeared. Between 540 million and 500 million years ago, the number and diversity of biological species increased spectacularly. Biologists call that event the **Cambrian explosion**. The trigger of that sudden surge in biodiversity remains unknown, but possibilities include rising oxygen levels, an increase in genetic complexity, major climate change, or some combination of those factors. The "Snowball Earth" hypothesis suggests that before the Cambrian explosion, Earth was in a period of extreme cold between about 750 million and 550 million years ago and was covered almost entirely by ice. During that period of extreme cold, predatory animals died out, making it easier for new species to adapt and thrive. Another possibility is that the marked increase in atmospheric oxygen (O_2) would have been accompanied by a corresponding increase in stratospheric ozone (O_3), which shields Earth's surface from deadly solar ultraviolet radiation. With that protective ozone layer in place, life was free to leave the oceans and

Figure 24.7 The amount of oxygen in Earth's atmosphere has built up as a result of cyanobacteria and plant life on the planet.

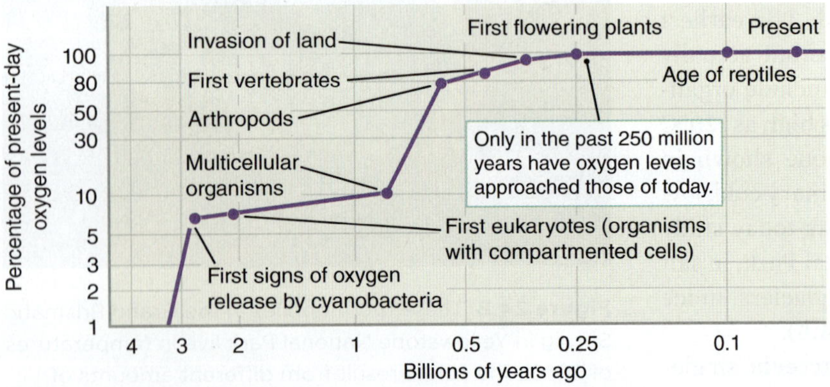

move to land. *Tiktaalik,* a fish with limblike fins and ribs, was an animal in a midevolutionary step of leaving the water for dry land, as shown in the artist's illustration in **Figure 24.10**.

The first plants appeared on land about 475 million years ago. Large forests and insects go back 360 million years. The age of dinosaurs began 230 million years ago and ended abruptly 65 million years ago, when an asteroid or comet collided with Earth. The collision threw so much dust into the atmosphere that the sunlight was dimmed for months, causing the extinction of more than 70 percent of all existing plant and animal species. Mammals were the big winners in the aftermath. Primates evolved from the last ancestor common with other mammals about 70 million years ago. The great apes (gorillas, chimpanzees, bonobos, and orangutans) split off from the lesser apes about 20 million years ago (Figure 24.9, inset). DNA tests show that humans and chimpanzees share about 98 percent of their DNA, indicating that they evolved from a common ancestor about 6 million years ago. By comparison, all humans share 99.9 percent of their DNA. The earliest human ancestors appeared a few million years ago, and the first civilizations occurred a mere 10,000 years ago. Present-day industrial society, barely more than two centuries old, is but a moment in the history of life on Earth.

Humans are here today because of a series of events that occurred throughout the history of the universe. Some of those events are common in the universe, such as the formation of heavy elements in earlier generations of stars and the formation of planets. Other events in Earth's history may have been less likely to happen elsewhere, such as the formation of a planet with life-supporting conditions like Earth or the development of self-replicating molecules that led to Earth's earliest life. A few events stand out, such as major extinctions that allowed the evolution of mammalian life and, ultimately, human beings.

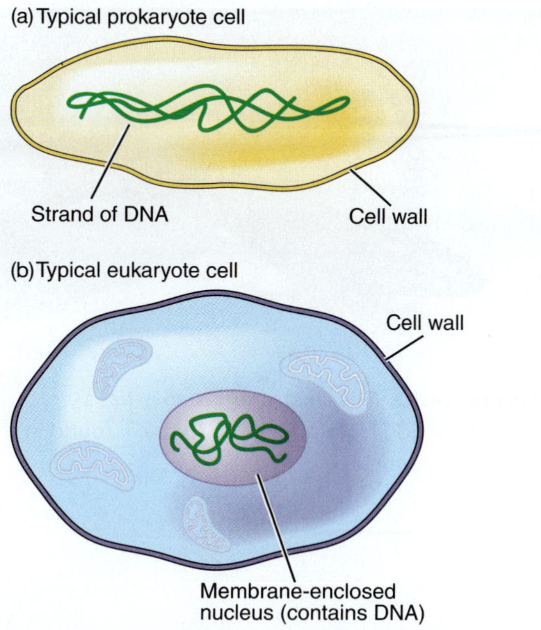

Figure 24.8 (a) A simple prokaryote cell contains little more than the cell's genetic material. (b) A eukaryote cell contains several membrane-enclosed structures, including a nucleus that houses the cell's genetic material.

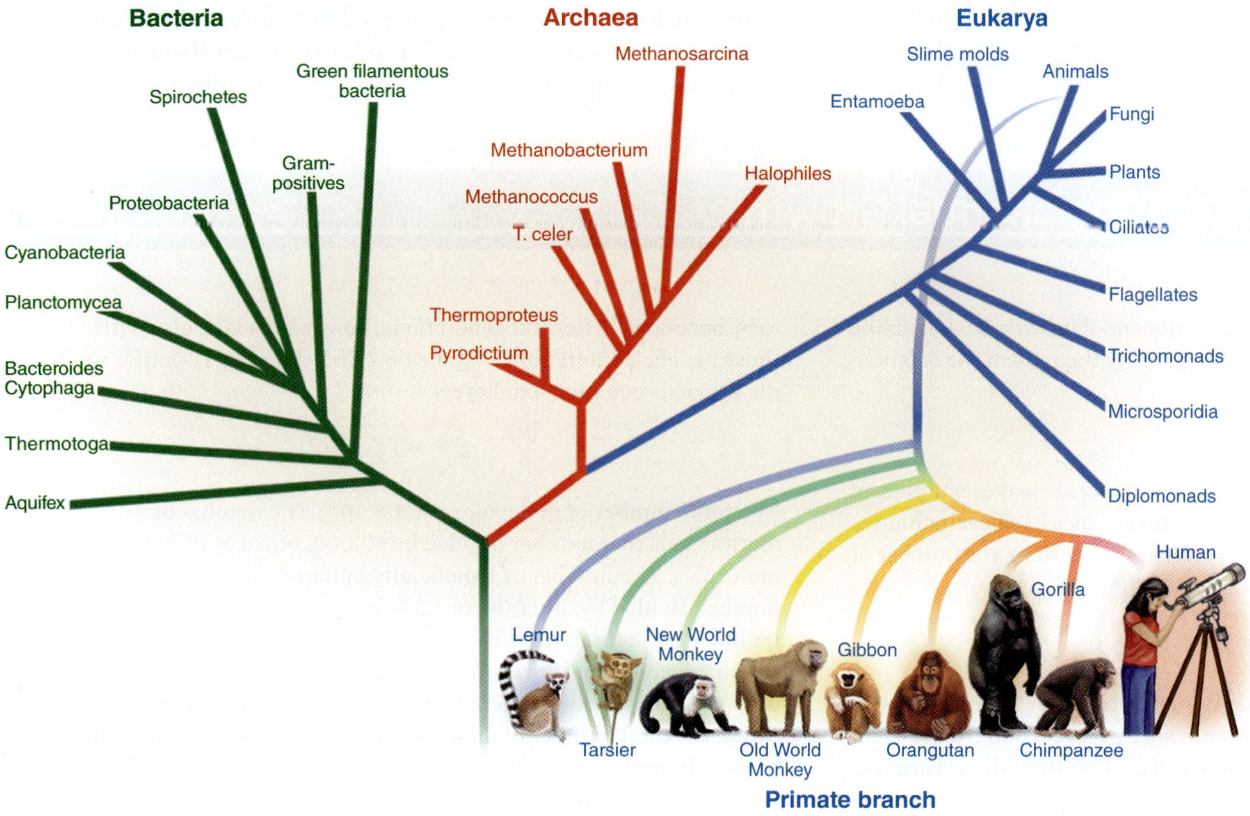

Figure 24.9 This simplified version of the phylogenetic tree has been constructed from analysis of the DNA strands of different life-forms. Humans are included in the "Animals" twig on the Eukarya branch. The primate branch, which includes humans, is shown in the inset on the bottom. By tracing those common ancestors through DNA and other means, scientists can reconstruct the evolutionary history of a species.

Figure 24.10 This illustration is an artist's reconstruction from a fossil of *Tiktaalik* found in the Canadian Arctic.

Evolution as a Mechanism of Change

Imagine that just once during the first few hundred million years after Earth formed, a single molecule formed somewhere in Earth's oceans. That molecule had a very special property: chemical reactions between that molecule and other molecules in the surrounding water caused the molecule to make a copy of itself. The molecule became *self-replicating*. Chemical reactions then produced copies of each of those two molecules, making four molecules. Four molecules became eight, eight became 16, 16 became 32, and so on. By the time the original molecule had copied itself just 100 times, more than a million trillion trillion (10^{30}) of those molecules existed. That is about 100 million times more of those molecules than the number of stars in the observable universe. The molecules of DNA that make up the chromosomes in the nuclei of the cells of all advanced life today are direct descendants of those early self-replicating molecules that flourished in the oceans of the young Earth.

Over time, not all replications are exact. The likelihood that a copying variation will occur while a molecule is replicating increases significantly with the number of copies being made. For DNA, which contains the genetic code for an entire organism, a change in the genetic code is called a **mutation**. Sometimes a mutation has no effect. Other times, a mutation can prevent an organism from flourishing. In still other cases, a mutation can make an organism better suited to its environment. Organisms with those advantageous mutations will survive to reproduce successfully. Even if mutations are rare and only a small fraction turn out to be beneficial, after just 100 generations trillions of mutations occur that, by luck, might improve on the original (**Working It Out 24.1**). **Heredity**—the ability of one generation to pass on its genetic code to future generations—allow beneficial mutations to persist and be incorporated into a species' genetic code.

As the organisms of the early Earth continued to interact with their surroundings and make copies of themselves, mutations caused them to diversify into many species. Sometimes the resources they needed to reproduce became scarce. In the face of that scarcity, varieties that were more successful reproducers became more numerous. Competition for resources, predation by one species on

24.1 Working It Out Exponential Growth

Self-replication is an example of exponential growth. The doubling time, n, for exponential growth is given by the ratio of the original and final amounts:

$$\frac{P_F}{P_O} = 2^n$$

Assume a hypothetical self-replicating molecule makes one copy of itself each minute, and each copy in turn copies itself each minute. How many molecules will exist after an hour? Here, the number of generations is given by $n = 60$, for 60 minutes in an hour:

$$\frac{P_F}{P_O} = 2^{60} = 1.2 \times 10^{18}$$

A billion billion of those molecules will exist after 1 hour.

Now suppose a mutation occurs once every 50,000 times that a molecule reproduces itself, and one of 200,000 mutations turns out to be beneficial. After 100 generations, how many molecules with those beneficial mutations might exist? That equation is similar to the previous equation, but here $n = 100$:

$$\frac{P_F}{P_O} = 2^{100} = 1.3 \times 10^{30}$$

The total number of molecules is 1.3×10^{30}. The number of mutations is that number divided by 50,000, or 2.6×10^{25} mutated molecules. The number of beneficially mutated molecules is that number divided by 200,000, or 1.3×10^{20} molecules. So 100 million trillion (10^{20}) mutations will occur that, by chance, might improve the survivability of the original molecule. Because that number does not count earlier beneficial changes that themselves replicated, the total number of molecules with beneficial changes will be even larger!

another, and cooperation between organisms became important to the survival of different varieties. Some varieties were more successful and reproduced to become more numerous, whereas less successful varieties became less and less common. That process, in which better-adapted organisms reproduce and thrive, whereas less well-adapted organisms become extinct, is called **natural selection**.

Life has existed on Earth for about 4 billion years, which is a very long time—long enough for the combined effects of heredity and natural selection to shape the descendants of that early self-copying molecule into a huge variety of complex, competitive, successful structures. Geological processes on Earth have preserved a fossil record of the history of some of those structures (**Figure 24.11**). Among those descendants are human "structures" that can think about their own existence and unravel the mysteries of the stars.

Figure 24.11 Fossils, such as this *Parasaurolophus* ("near crested lizard"), record the history of the evolution of life on Earth. This 10 meter long plant-eating dinosaur lived in North America about 75 million years ago.

CHECK YOUR UNDERSTANDING 24.1

Extremophiles are organisms that: (a) are extremely reactive; (b) are extremely rare; (c) have an extreme quality, such as mass or size; (d) live in extreme conditions.

24.2 Life Involves Complex Chemical Processes

The evolution of life on Earth cannot be separated from the narrative of astronomy: it is one of many examples of the emergence of structure in an evolving universe (see the **Process of Science Figure**). That emergence leads naturally to a profound question: Has life arisen elsewhere? Unlike the study of planets, stars, and galaxies, only one known case exists for the study of life—Earth—and scientists do not know how much can be generalized to other places. To explore that question, we need to take a closer look at the processes that have led to life on Earth. In this section, we explore the chemical and physical properties of life on Earth.

The infant universe was composed basically of hydrogen and helium and very little else. After 9 billion years of stellar nucleosynthesis, all the heavier chemical elements essential to life were present and available in the molecular cloud that gave birth to the Solar System. Those heavier elements were formed by nuclear fusion in the cores of earlier generations of stars and were then dispersed into space. At times, that dispersal was passive. For example, low-mass stars such as the Sun, when they become puffed-up, dying red giants, may shed their extended atmospheres, sending some newly created carbon into space. Other dispersals were more violent. High-mass stars produce even heavier elements through nucleosynthesis in their cores—up to and including iron. But some of the trace elements essential to biology on Earth are even more massive than iron. They are produced within a matter of minutes during the violent supernova explosions that mark the death of high-mass stars and then are thrown into the chemical mix found in molecular clouds. Or they are produced in the collisions of neutron stars, the dense remains of prior supernova explosions (**Figure 24.12**).

Figure 24.12 This version of the periodic table shows the astronomical origin of the naturally occurring elements (up to uranium) in the Solar System.

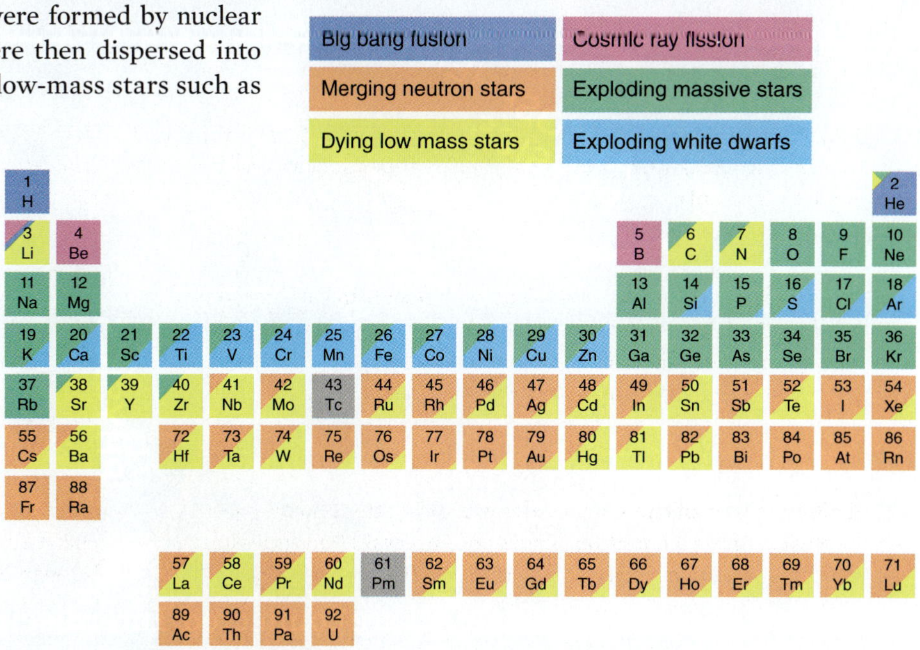

All known living organisms on Earth are composed of a more or less common suite of complex chemicals. Approximately two-thirds of the atoms in the

PROCESS OF SCIENCE

All of Science Is Interconnected

More than most other subjects, astrobiology relies on concepts from a number of scientific fields—showing that all of science is interconnected.

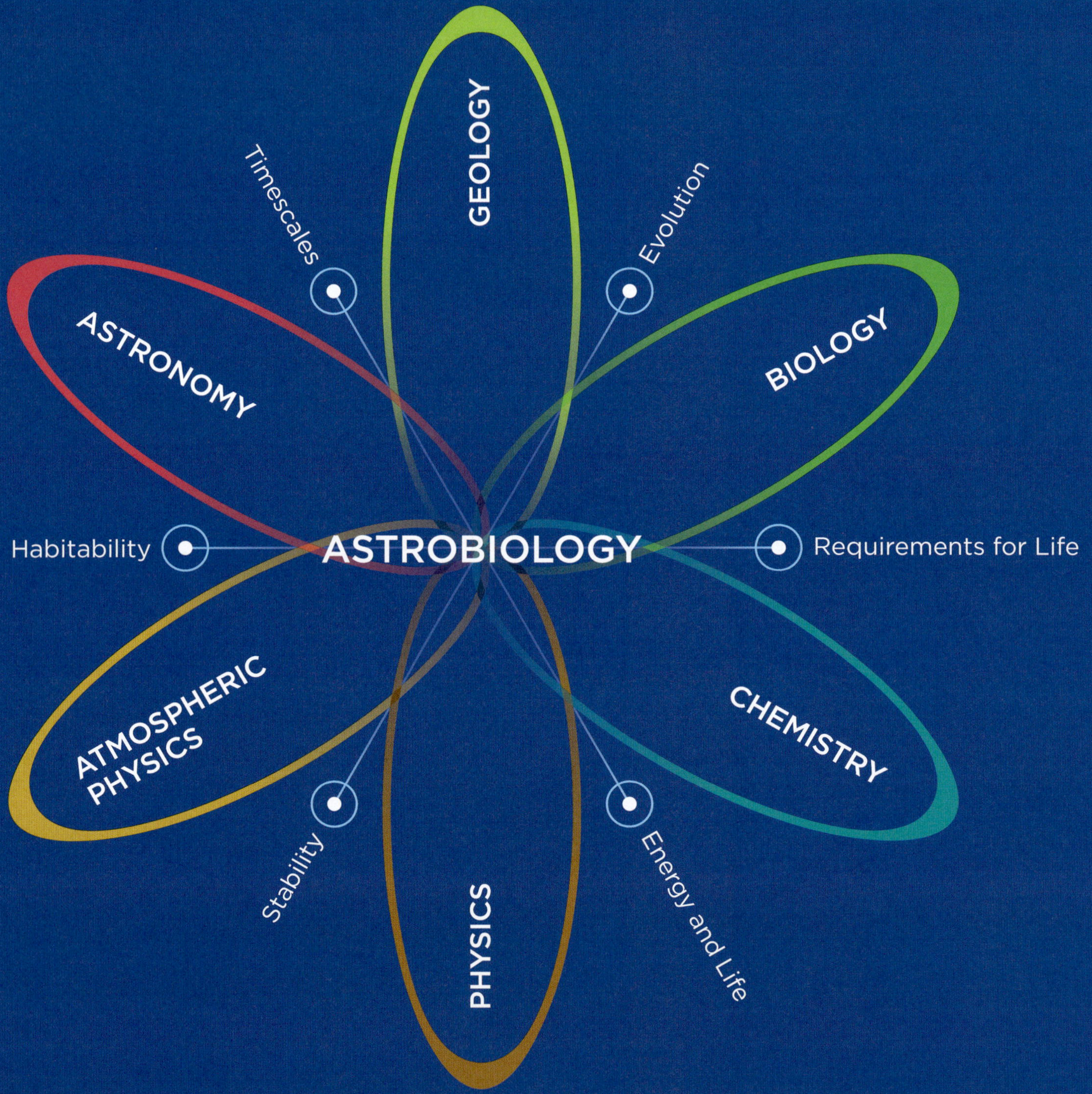

No science stands alone. All are connected. Interdisciplinary fields of study like astrobiology provide opportunities for new tests for theories of many fields.

human body are hydrogen (H), about one-fourth are oxygen (O), a tenth are carbon (C), and a few hundredths are nitrogen (N). Carbon, nitrogen, and oxygen are the three most abundant products of stellar nucleosynthesis after helium (see the "astronomer's periodic table" in Figure 5.15). The several dozen remaining atomic elements in the human body make up only 0.2 percent of the total. All known living creatures are assemblages of molecules composed almost entirely of those four elements, sometimes called CHON (carbon, hydrogen, oxygen, nitrogen), along with small amounts of phosphorus and sulfur. Some of those molecules are enormous. Consider DNA, which is responsible for genetic codes, illustrated in **Figure 24.13**. DNA is made up entirely of only five atomic elements: CHON and phosphorus. But the DNA in each cell of the human body is composed of combinations of *tens of billions* of atoms of those same five elements. Proteins, the huge molecules responsible for the structure and function of living organisms, are long chains of smaller molecules called amino acids. Terrestrial life uses 20 specific amino acids, which also consist of no more than five atomic elements—CHON plus sulfur.

Although life primarily uses the half-dozen atoms of CHON plus sulphur and phosphorus, some other elements, present in smaller amounts, are essential to the chemical processes that living organisms carry out. Those elements include sodium, chlorine, potassium, calcium, magnesium, iron, manganese, and iodine. Trace elements such as copper, zinc, selenium, and cobalt also play a crucial role in biochemistry but are needed in only tiny amounts.

Notice that carbon, which can bond to four other atoms or molecules, forms the backbone of the DNA molecule shown in Figure 24.13. That is why carbon is so important to life on Earth. Forms of extraterrestrial life could exist that are also carbon-based but have chemistries different from that of life on Earth. For example, countless varieties of amino acids exist in addition to the 20 used by terrestrial life. Most other atoms are more limited than carbon in the number of bonds they can make, but silicon, like carbon, can bond to four other atoms, so that many combinations are possible. As a potential life-enabling atom, silicon has both advantages and disadvantages in comparison with carbon. Silicon-based molecules remain stable at much higher temperatures than carbon-based molecules, perhaps enabling possible silicon-based life to thrive in high-temperature environments, such as on planets that orbit close to their parent star. But silicon is also a larger and more massive atom than carbon: it cannot form molecules as complex as those based on carbon. Any silicon-based life probably would be simpler than life-forms here on Earth, but it might exist in high-temperature niches somewhere within the universe. Although carbon's unique properties make it readily adaptable to the chemistry of life on Earth, other types of life might be found elsewhere.

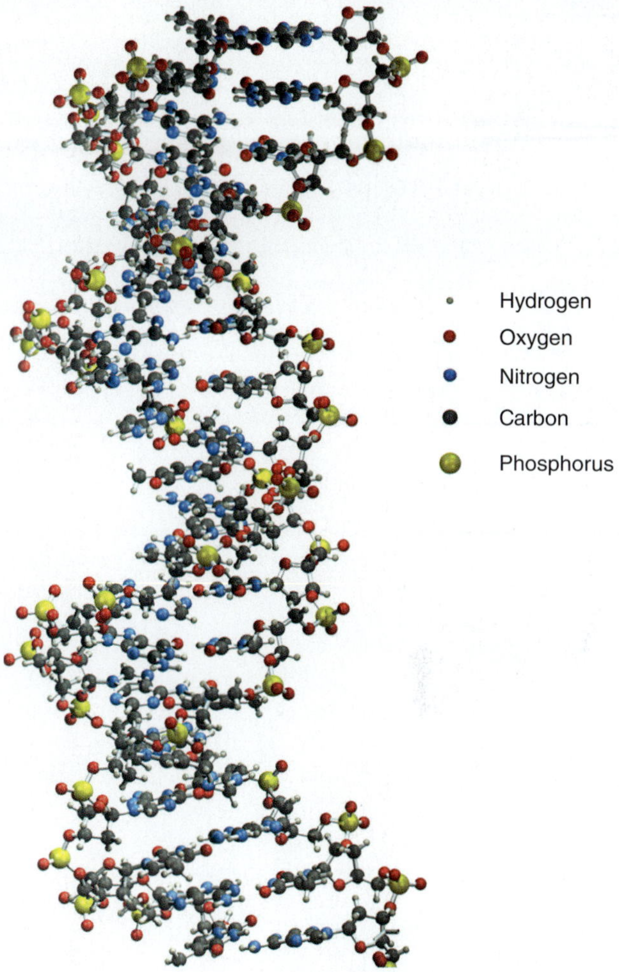

Figure 24.13 DNA, the heritable molecule that forms the basis for life on Earth, contains the atoms of only five elements. Even so, those atoms are combined in billions of ways, giving rise to the diversity of life on Earth.

CHECK YOUR UNDERSTANDING 24.2

Carbon is a favorable base for life because: (a) it can bond to many other atoms in long chains; (b) it is nonreactive; (c) it forms weak bonds that can be readily reorganized as needed; (d) it is organic.

24.3 Where Do Astronomers Look for Life?

One approach to the scientific search for extraterrestrial life is to use robotic spacecraft to explore the planets and moons of the Solar System (see Chapter 6). Spacecraft have visited all the planets and some moons and sent back at least some information about the conditions on those worlds. Another approach is to use telescopes to detect planets outside our Solar System (see Chapter 7). Telescopes on the ground and in space have detected several thousand planets orbiting other stars. In this section, we will survey the locations where life might be found, both in our own Solar System and around other stars.

Life within Our Solar System

Scientists start the search for evidence of extraterrestrial life here in our own Solar System. Early conjectures about life in our Solar System seem naïve, considering what we now know. Two centuries ago, the eminent astronomer Sir William Herschel, discoverer of Uranus, proclaimed, "We need not hesitate to admit that the Sun is richly stored with inhabitants." In 1877, astronomer Giovanni Schiaparelli (1835–1910) observed what appeared to be linear features on Mars and dubbed them *canali* ("channels" in Italian). The famous observer of Mars, Percival Lowell (1855–1916), misinterpreted Schiaparelli's *canali* as "canals," suggesting that they were constructed by intelligent beings.

Because Mercury and the Moon lacked atmospheres, astronomers determined that those worlds were not conducive to life. The giant planets and their moons were thought to be too remote and too cold to sustain life. By the 1960s it was understood that the surface of Venus is too hot to permit life. At the same time, astronomers using ground-based telescopes had discovered that Mars has a thin atmosphere, water ice, and carbon dioxide ice. During the 1960s, the United States and the Soviet Union sent reconnaissance spacecraft to the Moon, Venus, and Mars, but the instruments on those spacecraft probed the physical and geological properties of those astronomical bodies instead of searching for life. Serious efforts to look for signs of life—past or present—require more advanced spacecraft with specialized instrumentation.

In the mid-1970s, two American *Viking* spacecraft were sent to Mars with detachable landers containing a suite of instruments designed to find evidence of a terrestrial type of life. When the *Viking* landers failed to find convincing evidence of life on Mars, hopes faded for finding life on any other body orbiting the Sun. Since then, however, further exploration of the Solar System has generated renewed optimism. A better understanding of the history of Mars indicates the planet's climate has changed. Mars was once wetter and warmer than it is today.

In the 1990s, Mars missions began to map the planet's surface from the ground and from space. In 2008, NASA's *Phoenix* spacecraft landed at a far-northern latitude, inside the planet's arctic circle, where specialized instruments dug into and analyzed the martian water-ice permafrost. *Phoenix* found that the martian arctic soil has a chemistry similar to that of the Antarctic dry valleys on Earth, where life exists deep below the surface at the ice-soil boundary. Minerals that form in water, such as calcium carbonate, have been detected. That finding suggests that oceans existed in the past on Mars. However, *Phoenix* found no evidence of life.

Figure 24.14 The Mars *Curiosity* rover detected evidence that Mars had a watery past. The rounded gravel surrounding the bedrock suggests that an ancient, flowing stream once existed.

The *Curiosity* rover (originally known as the *Mars Science Laboratory*) landed in Gale Crater on Mars in 2012. That rover studies the rocks and soil of Mars to acquire data for a better understanding of the history of the planet's climate and geology. Shortly after landing, *Curiosity* found evidence that a stream of liquid water had once flowed in the crater. The rover observed rounded, gravelly pebbles stuck together, which have been interpreted as coming from a stream that varied at times from ankle-deep to hip-deep, and moved at about 1 meter per second (**Figure 24.14**). The rover also found sedimentary rocks containing clay, which suggested the presence of a freshwater lake bed at one time. Later observations found that the surface soil contained up to 2 percent water by weight—or about 1 quart per cubic foot of martian dirt. Those conditions are too dry to support Earthlike plant life, which permanently wilts in soil that is less than about 10% water by weight.

Curiosity also detected evidence of organic molecules in 3-billion-year-old rocks located in that crater (**Figure 24.15**). Those organic molecules could have come from space on asteroids and comets, from geological processes on Mars, or from ancient life. *Curiosity* also detected seasonal variation in the amount of methane gas within the crater; methane can come from water-rock chemistry or from life. Thus, a water lake inside Gale Crater could have been hospitable to life billions of years ago, but astrobiologists do not know whether any life existed.

The *Mars Atmosphere and Volatile EvolutioN* (*MAVEN*) mission, which arrived at Mars in September 2014, is studying the upper atmosphere to learn more about the escape of carbon dioxide, hydrogen, and nitrogen from the planet's atmosphere and how losing those gases affected surface pressure and the existence of liquid water. Recently the *Mars Reconnaissance Orbiter* (*MRO*) found water ice on eight steep, eroded slopes located at midlatitudes on Mars (**Figure 24.16**). *Mars Express* may have found evidence of a lake of liquid water under the south polar ice cap. Future experiments will look for liquid water—and fossil or living microorganisms—below the martian surface. Eventually other missions will return samples of Martian rock and soil to Earth for more advanced analysis.

NASA's instrumented robotic spacecraft reached the outer Solar System starting in the 1970s, and many astrobiologists were surprised by the findings. Although the outer planets themselves did not appear to be habitats for life, some of their moons became objects of special interest. Jupiter's moon Europa is covered with a layer of water ice that appears to overlie a great ocean of briny liquid water (**Figure 24.17**). The water remains liquid because of high pressure and tidal heating by Jupiter. Impacts by comet nuclei may have added a mix of organic material, another essential ingredient for life. Once thought to be a frozen, inhospitable world, Europa is now a candidate for biological exploration. Recently, scientists using the Hubble Space Telescope observed water geysers taller than Mount Everest erupting from the icy surface of Europa. Ejected material from those geysers may make searching for life on Europa possible without drilling through the ice. NASA's *Europa Clipper* mission, scheduled for launch in the 2020s, will fly by Europa a few dozen times. Jupiter's moons Ganymede and Calisto also have icy surfaces and possibly subsurface salty oceans.

Figure 24.15 A "selfie" of the *Curiosity* Mars rover at the site where it drilled into some rocks and identified organic compounds.

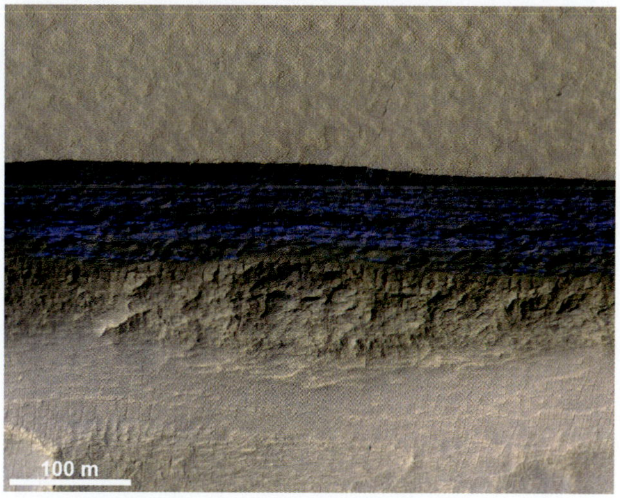

Figure 24.16 This image from the *MRO* shows 80 meters of exposed water ice on a steep slope. The color is exaggerated. Similar steep slopes with ice at other midlatitude locations also were seen.

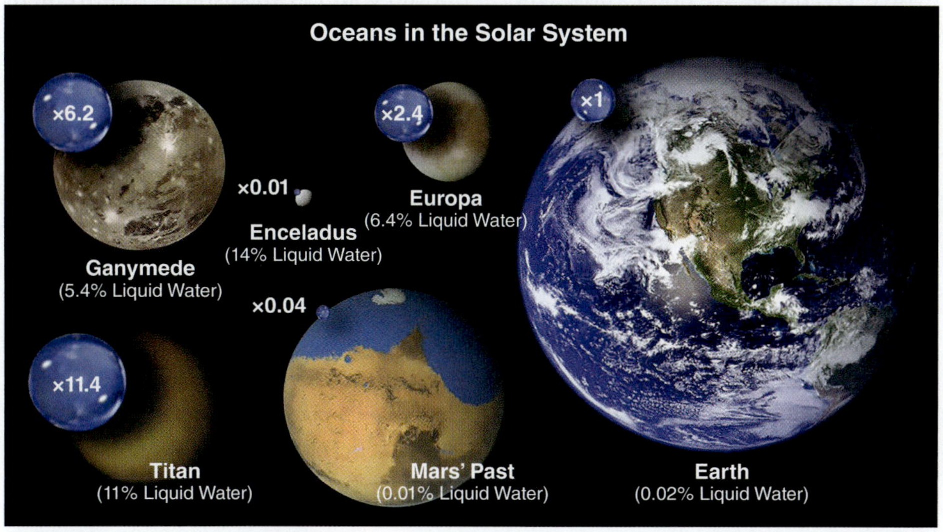

Figure 24.17 The total amount of liquid water (blue balls) on Jupiter's moons and Saturn's moon Titan compared with the amount on Earth. All figures are drawn to scale and assume average ocean depths of 4 km (Earth), 100 km (Europa), and 200 km (Titan).

Saturn's moon Titan has an atmosphere rich in organic chemicals, many of which are thought to be precursor molecules of a type that existed on Earth before life appeared here. A probe from the *Cassini* spacecraft in orbit around Saturn descended through Titan's atmosphere and found additional evidence for a variety of molecules that might be necessary for life, as well as liquid lakes of methane on the surface and probably a liquid-water ocean under the surface. The *Cassini* spacecraft also detected water-ice crystals spouting from cryovolcanoes (which erupt ice crystals instead of rocks) near the south pole of Saturn's tiny moon Enceladus. Liquid water must lie beneath its icy surface, and Enceladus therefore is also a possible habitat of extremophile life, perhaps life similar to that found near hydrothermal vents deep within Earth's oceans.

The discovery of life on even one Solar System body beyond Earth would be exciting. If scientists discover that life arose independently *twice* in the same planetary system, that finding could suggest that the spontaneous appearance of life is not rare at all.

CHECK YOUR UNDERSTANDING 24.3A

Which of the following Solar System objects is *not* a good candidate for future searches for life? (a) Mars; (b) Jupiter's moon Europa; (c) Saturn's moon Titan; (d) Uranus

Habitable Zones

Recall from Chapter 7 that thousands of confirmed and thousands of candidate exoplanets exist within the Milky Way Galaxy. To decide which planets to focus on for further study, astronomers are narrowing the possibilities by searching for planets with environments conducive to the formation and evolution of life, as we understand it, while eliminating clearly unsuitable planets. Astronomers consider issues such as each planet's orbit, inferred temperature, distance from its star, and location in the galaxy. One criterion astrobiologists look for is stability of planetary systems. As noted in Chapters 2, 3, and 4, astronomers think about the effects of a planet's rotation and orbit. Planets in stable systems have nearly circular orbits that preserve relatively uniform climatological environments. Planets in very elliptical orbits or planets with a large axial tilt can experience more intense temperature swings that could be detrimental to the survival of life. A stable temperature that maintains the existence of water in a liquid state might be important. We know that liquid water was essential for life on Earth to form and evolve. Of course, we don't know whether liquid water is an absolute requirement for life elsewhere, but it's a reasonable starting assumption.

In the "Origins" sections of Chapters 7 and 13, we discussed the idea of the habitable zone, the location of a planet relative to its parent star that provides a range of temperatures in which liquid water can exist on the surface. On planets too close to their parent stars, water would exist only as a vapor—if at all. On planets too far from their stars, water would be permanently frozen as ice. Planet size is another consideration: Large gas giants retain most of their light gases during formation and have no solid surface. Small planets may be rocky or a mix of water, rock, and ice: measuring the mass and radius enables scientists to estimate the density. Very small planets may not have enough surface gravity to retain their atmospheric gases and so end up like our Moon. Calculating whether any particular planet is in the habitable zone is complicated. Recall from Chapter 7 that even if a planet is located in the habitable zone, that means only that liquid water *could* exist on the surface: it does not mean that astronomers have confirmed the presence of liquid water or that the planet has inhabitants.

In our own Solar System, Venus, which orbits at 0.7 times Earth's distance from the Sun, has become an inferno because of its runaway greenhouse effect. Any liquid water that might once have existed on Venus has long since evaporated and been lost to space. Mars orbits about 1.5 times farther from the Sun than the orbit of Earth, and the water that we see on Mars today is nearly always frozen. But the orbit of Mars is more elliptical and variable than Earth's, giving the planet a greater variety of climate, including long-term cycles that might occasionally permit liquid water to exist. Most astrobiologists put the habitable zone of our Solar System at about 0.9–1.4 astronomical units (AU), which includes Earth but just misses Venus and Mars. However, that range ignores the possibility of liquid water under ice, as occurs on the moons of the outer Solar System.

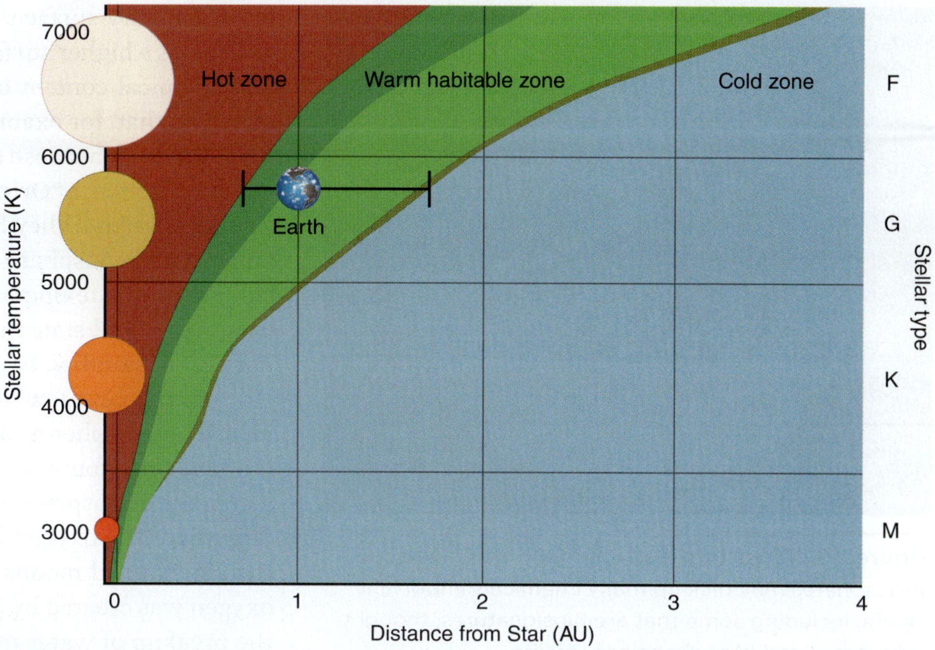

Figure 24.18 The habitable zone changes with the mass and temperature of a star. Habitable zones around hot, high-mass stars are larger and more distant than the zones around cooler, lower-mass stars.

Astronomers also must think about the type of star they are observing in their search for planets that could have liquid water. Stars less massive than the Sun and thus cooler will have narrower habitable zones. A planet in the habitable zone of a cool star is close in to its star. As a result, the planet is more likely to be tidally locked to the star so that no day/night cycle exists. Stars more massive than the Sun are hotter and will have a larger and more distant habitable zone (**Figure 24.18**). However, massive stars have shorter main-sequence lifetimes and might not last long enough for evolution to take place. For example, a star of 3 M_{Sun} has a lifetime of only a few hundred million years. Here on Earth, a billion years was long enough for bacterial life to form and cover the planet, but insufficient for anything more advanced to evolve. It took 3.5 billion years of evolution on Earth to reach the period of the Cambrian explosion. Even though evolution might happen at a different pace elsewhere, stellar lifetime is still a sufficiently strong consideration that astronomers focus their efforts on stars with longer lifetimes—specifically, stars of 0.6–1.4 M_{Sun}, which corresponds to spectral types F, G, K, and M. Most stars in our galaxy are low-mass M main-sequence stars, and many of those seem to have small, rocky planets. One example is the Trappist-1 system of seven planets in close orbits to their cool M dwarf star (**Figure 24.19**). Some of those planets could be tidally locked (see the Chapter 4 "Origins") to their star. Two or three of the planets are in the star's habitable zone.

Another factor to consider is a planet's atmosphere. A planet's ability to keep an atmosphere depends on the planet's mass and radius (and therefore its escape velocity) and its temperature. Very small planets may not have enough surface gravity to retain their atmospheric gases. In the inner Solar System, the Moon and Mercury were too small to keep any atmosphere. Mars lost most of its atmosphere, but the larger Earth and Venus kept thick atmospheres. Another important consideration is the greenhouse effect, which traps heat underneath an atmosphere and raises the temperature

Figure 24.19 The Trappist-1 system. The Trappist planets are very close to their star: their distances are more comparable to those of Jupiter and its large moons than to our Solar System planets. (The star is not to scale.)

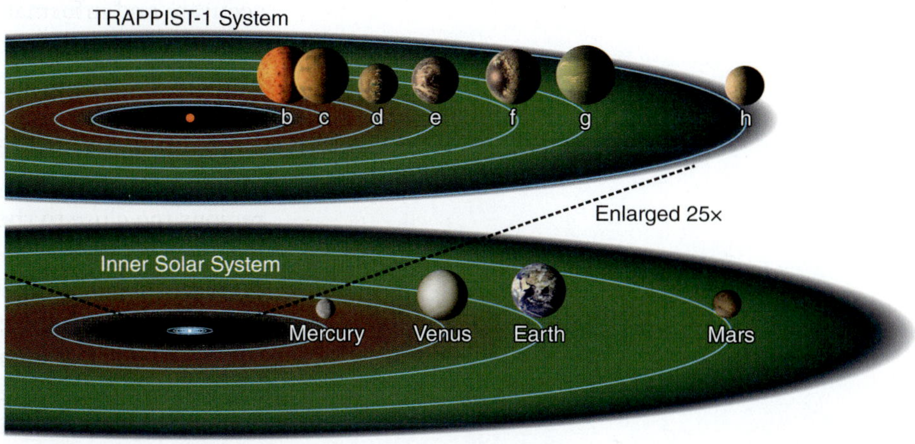

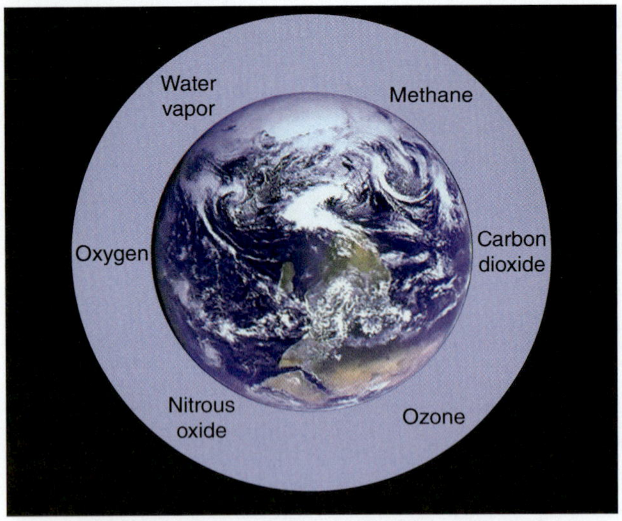

Figure 24.20 Spectroscopy of rocky exoplanet atmospheres may pick up many chemicals important for life, including some that are biosignatures, thought to be produced overwhelmingly by life.

on a planetary surface. That has happened on Venus, Earth, and Mars, each of which has a higher surface temperature because of its atmosphere. The thickness and chemical content of the atmosphere affect the strength of the greenhouse effect, so that, for example, Venus is much hotter than its distance from the Sun would suggest because of its thick atmosphere of carbon dioxide (see Figure 5.24 and the Chapter 5 "Origins"). Some of those exoplanets might be more like Venus than like Earth if they have atmospheres filled with greenhouse gases. The total amount of atmosphere affects the atmospheric pressure at the surface, which, along with the temperature, determines whether water (or other molecules) can exist in a liquid state on the surface. The current thin atmosphere of Mars does not permit standing liquid water on its surface.

In the next few years, with new telescopes, many more observations will identify atmospheres on exoplanets. Water vapor, hydrogen, and carbon monoxide have been found on a few exoplanets already. In particular, discovering oxygen in the atmosphere of an exoplanetary atmosphere would be exciting, but not definitive. The oxygen in Earth's atmosphere makes it stand out from the rest of the planets and moons in the Solar System, and we know that most terrestrial oxygen was created by photosynthetic life. However, oxygen also can come from the breakup of water molecules, so the presence of oxygen alone doesn't necessarily mean life exists there. Aside from oxygen, astrobiologists are thinking about other possible **biosignatures**—chemical compounds that would show up in the spectra of an exoplanet atmosphere that would suggest life may be present (**Figure 24.20**).

Astrobiologists continue to develop ways to classify planets more clearly as they try to narrow the possibilities about where life might exist. One such classification uses the currently available data on an exoplanet to estimate how much it is like Earth. Factors include the radius, density, escape velocity, and surface temperature. That is an Earth-centric approach based on the experience of life on Earth. Another measure aims to be less Earth-centric and to broaden the options for habitability, but it depends on factors not yet measured or measurable for most exoplanets. That measure depends on whether the planet has a surface on which organisms can grow, as well as the right kind of chemistry, a source of energy, and the ability to hold a liquid solvent. Saturn's moon Titan or Jupiter's moon Europa might satisfy those conditions, and Mars might have done so in the past. Over the next decade, improvements in observations will probably lead to enough information that at least some exoplanets can be classified in that way. As observations of exoplanets become more complete, astrobiologists will undoubtedly develop new classification schemes that are more accurate and informative.

Astronomers also consider the *galactic habitable zone*—the idea that the Milky Way Galaxy may have some locations where planets might have a higher probability of hosting life. Stars situated too far from the galactic center may be without enough heavy elements—such as oxygen, silicon (silicates), iron, and nickel—in their protoplanetary disks to form rocky planets like Earth. Conversely, regions too close to the galactic center experience less star formation and therefore fewer opportunities to gather heavy elements into planetary environments. Stars too close to the galactic center may be affected by the high-energy radiation environment (X-rays and gamma rays from supermassive black holes or gamma-ray bursts), which can damage RNA and DNA. Stars that migrate within the galaxy and change their distance from the galactic center may move in and out of any galactic habitable zone.

CHECK YOUR UNDERSTANDING 24.3B

The habitable zone around a star depends most on the star's: (a) mass and age; (b) radius and distance; (c) age and radius; (d) color and distance; (e) luminosity and velocity.

24.4 Scientists Are Searching for Signs of Intelligent Life

Are we alone? Scientists approach that question from many directions. Biologists consider the origin and evolution of life and the definition of intelligence. Astronomers send messages and search for alien signals in the vast array of astronomical data. In this section, you will learn about the search for intelligent life and how scientists think about the probability of finding it.

Sending Messages

During the 1970s, messages were sent from Earth to space. The *Pioneer 11* spacecraft, which will probably spend eternity drifting through interstellar space, carries the plaque shown in **Figure 24.21**. It pictures humans and the location of Earth for any future interstellar traveler who might happen to find it and understand its content. Another message to the cosmos accompanied the two *Voyager* spacecraft on identical phonograph records that contained greetings from planet Earth in 60 languages, samples of music, animal sounds, and a message from then-President Jimmy Carter. Some politicians were concerned that scientists were dangerously advertising our location in the galaxy, even though radio signals had already been broadcast into space for nearly 80 years. Some philosophers also worried that those messages contained anthropomorphic assumptions about aliens being enough like us to decode the messages. However, sending messages on spacecraft is not an efficient way to make contact with extraterrestrial life. The probability that an alien species will actually find any of those messages is very, very small.

In 1974, astronomers used the 300-meter-wide dish of the Arecibo radio telescope to beam a message in binary code (**Figure 24.22**) toward the star cluster M13, located 25,000 light-years away. That distance is far enough that by the time the message arrives, the core of M13 will have moved, and the radio signal will not actually arrive there. However, the intention of the experiment was to show that such a message could be sent, not to make contact, because 50,000 years would pass before any reply could come back. In 2008, a radio telescope in Ukraine sent a message to the exoplanet Gliese 581c. The message was composed of 501 digitized images and text messages selected by users on a social networking site and will arrive at Gliese 581c in 2029.

The Drake Equation

The first serious effort to quantify the probability of the existence of intelligent extraterrestrial life was made by astronomer Frank Drake in 1960. He developed the **Drake equation**, which estimates the likely number (N) of intelligent civilizations currently existing in the Milky Way Galaxy. The Drake equation is different from the other equations in this book because the values for many of the variables are uncertain. However, it is a useful way to categorize some of the

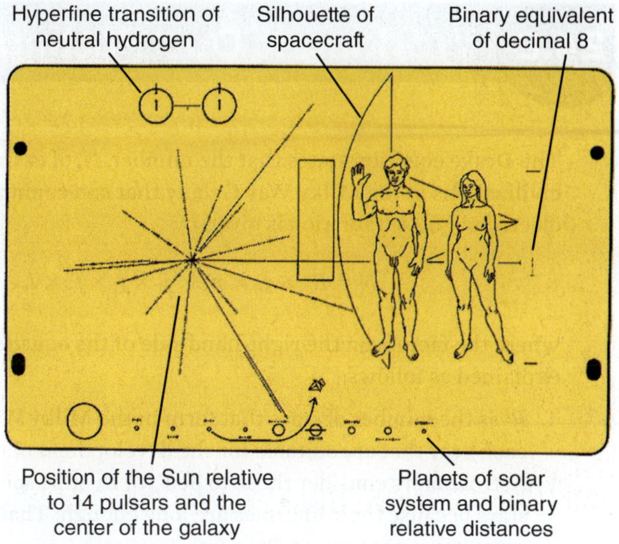

Figure 24.21 This plaque is carried by the *Pioneer 11* probe, which launched in 1973 and will eventually leave the Solar System to travel in interstellar space.

Figure 24.22 This message was beamed toward the star cluster M13 in 1974. Reading right to left, this binary-encoded message contains the numbers 1–10, hydrogen and carbon atoms, some interesting molecules, DNA, a human figure and its size, the basics of the Solar System, and a depiction of the Arecibo telescope.

24.2 Working It Out Putting Numbers into the Drake Equation

The Drake equation states that the number, N, of extraterrestrial civilizations in the Milky Way Galaxy that can communicate by electromagnetic radiation is given by

$$N = R^* \times f_p \times n_e \times f_l \times f_i \times f_c \times L$$

where the factors on the right-hand side of the equation are explained as follows:

1. R^* is the number of stars that form in the Milky Way Galaxy each year that are suitable for the development of intelligent life. Astronomers consider those to be F, G, K, or M spectral-type stars because their lifetimes are long enough. That is about five to seven stars per year, so $R^* = 5$ to 7.
2. f_p is the fraction of stars that form planetary systems. The discoveries of exoplanets over the past two decades have shown that planets form as a natural by-product of star formation and that many—perhaps most—stars have planets. For that factor, astronomers assume that f_p is between 0.5 and 1.
3. n_e is the number of planets and moons in each planetary system with an environment suitable for life. In the Solar System, that number is at least 1 (for Earth), but it could be more if Mars or an outer-planet moon or two has suitability for life. Only recently have stars with multiple planets been discovered, so astronomers are just starting to get data on that factor. Generally they estimate that $n_e = 0$ to 3.
4. f_l is the fraction of suitable planets and moons on which life actually arises. Remember that just a single self-replicating molecule may be enough to start the process. Some biochemists think that life *will* develop if the right chemical and environmental conditions are present, but others disagree. Values of f_l range from 100 percent (life always develops) to 1 percent (life is more rare). Astronomers use a range of 0.01–1 for f_l.
5. f_i is the fraction of those planets harboring life that eventually develop intelligent life. Intelligence is certainly the kind of survival trait that might often be strongly favored by natural selection. Yet on Earth, tool-building intelligence took about 4 billion years—nearly half the expected lifetime of the Sun—to evolve. The correct value for f_i might be close to 0.01 or it might be closer to 1. No one knows.
6. f_c is the fraction of intelligent life-forms that develop technologically advanced civilizations—that is, civilizations that send communications into space. With only one example of a technological civilization to work with, f_c also is unknown. Astronomers estimate f_c to be between 0.1 and 1.
7. L is the number of years that technologically advanced civilizations exist. That factor is certainly difficult to estimate because it depends on the long-term stability of those civilizations. On Earth, the longest-lived civilizations have existed for, at most, thousands of years. Those civilizations, however, were not at the level of technology that allows interstellar communication—thus far, the first "technologically advanced civilization" on Earth is less than 100 years old. We do not know whether life intelligent enough to manipulate its environment technologically can maintain its planetary resources for any extended length of time. Astronomers usually put a value of L between 1,000 years and 1 million years in their estimates.

factors that relate to the conditions that must be met for a civilization to exist. The equation is discussed further in **Working It Out 24.2**.

As illustrated in **Figure 24.23**, the conclusions we draw using the Drake equation depend a great deal on the assumptions we make and therefore on the numbers used in the equation. For the most pessimistic estimates, the Drake equation sets the number of technological civilizations in our galaxy at about 1, in which case we are the *only* technological civilization in the Milky Way at this time. Such a universe could still be full of intelligent life. With 100 billion galaxies in the observable universe, even those pessimistic assumptions mean that 100 billion technological civilizations could be out there somewhere. However, if the nearest neighbors are in another galaxy, they are *very* far away—millions of parsecs, on average.

At the other extreme—with the most optimistic numbers, which assume that intelligent life arises and survives everywhere it gets the chance—the Milky Way alone could have tens of millions of technological civilizations! If so, the nearest neighbors may be "only" 40 or 50 light-years away.

If humans did meet a technologically advanced civilization, what would it be like? The Drake equation suggests that neighbors nearby are highly unlikely, unless civilizations typically live for many thousands or even millions of years (see Working It Out 24.2). If so, any civilization we encountered would probably have been around for much longer than we have.

Technologically Advanced Civilizations

During lunch with colleagues, the physicist Enrico Fermi (1901–1954), a firm believer in extraterrestrial life, is reported to have asked, "If the universe is teeming with aliens . . . where is everybody?" Fermi's question—first posed in 1950 and sometimes called the *Fermi paradox*—remains unanswered. If intelligent life-forms are common but interstellar travel is difficult or impossible, would the aliens send out messages—perhaps by electromagnetic waves instead? And if they did, why haven't astronomers detected their signals?

Drake used what was then astronomy's most powerful radio telescope to listen for signals from intelligent life around two nearby stars, but he found nothing unusual. His original project has grown over the years into a much more elaborate program called the Search for Extraterrestrial Intelligence, or **SETI**. Scientists from around the world have thought carefully about what strategies might be useful for finding life in the universe. Most of those endeavors use radio telescopes to listen for signals from space that bear an unambiguous signature of an intelligent source. Some have focused on significant parts of the spectrum, assuming that a civilization will broadcast on a channel that astronomers throughout the galaxy should find interesting—for example, the 21-centimeter (21-cm) line from hydrogen gas. More recent searches have used advances in technology to record as broad a range of radio signals from space as possible. Analysts search those databases for regular signals that might be intelligent in origin.

The SETI Institute's Allen Telescope Array (ATA) received much of its initial financing from Microsoft cofounder Paul Allen. The ATA consists of a "farm" of small, inexpensive radio dishes like those used to capture signals from orbiting television-broadcasting satellites (see the chapter-opening photograph). One key project of the ATA is to observe the planets discovered by the Kepler Mission. Each dish has a diameter of 6.1 meters, but all the telescopes working together have a total signal-receiving area greater than that of a 100-meter radio telescope. Just as your brain can sort out sounds coming from different directions, that array of radio telescopes can determine the direction a signal is coming from, allowing it to listen to many stars at the same time. Over several years' time, astronomers using the ATA are expected to survey as many as a million stars, hoping to find a civilization that has sent a signal toward Earth.

As stated earlier, finding even one nearby civilization in the Milky Way Galaxy—that is, a *second* technological civilization in Earth's small corner of the universe—will make scientists optimistic that the universe as a whole is teeming with intelligent life. The likelihood of SETI's success is difficult to predict, but its potential payoff is enormous. Few discoveries would have a more profound impact than the certain knowledge that we on Earth are not alone.

Science fiction is filled with tales of humans who leave Earth to "seek out new life and new civilizations." Unfortunately, those scenarios are not scientifically realistic. The distances to the stars and their planets are enormous: to explore a significant sample of stars would require extending the physical search over tens or hundreds of light-years. Special relativity limits how fast one can travel. The speed of light is the limit, and even at that rate reaching the *nearest*

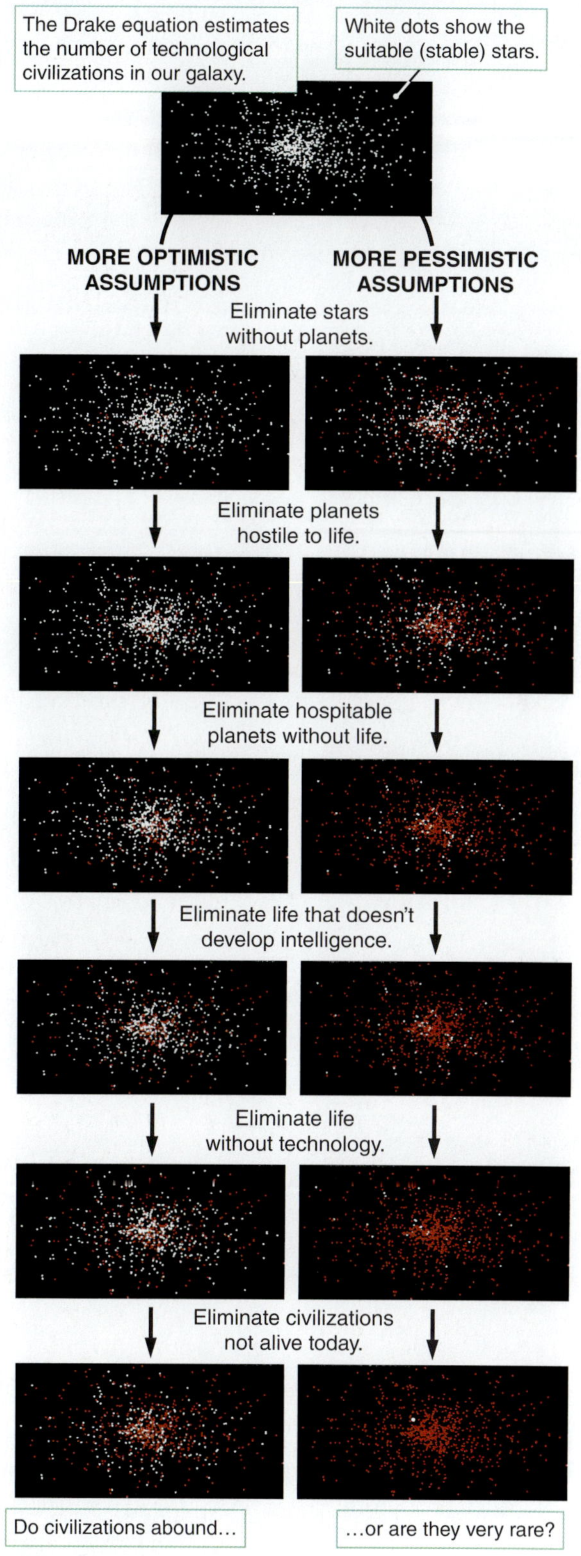

Figure 24.23 The two columns show estimates of the existence in the Milky Way Galaxy of intelligent, communicative civilizations as made using the Drake equation. White dots represent stars with possible civilizations. Notice how widely the estimates vary, given optimistic and pessimistic assumptions about the seven factors in the equation.

star would take more than 4 years. The relativistic effect of time dilation means that time would pass slower for astronauts traveling at very high speeds, and they would return to Earth younger than if they had stayed at home. For example, suppose astronauts visited a star 15 light-years distant. Even if they traveled at speeds close to the speed of light, by the time they returned to Earth, 30 years would have passed at home.

Some science fiction writers get around that problem by invoking "warp speed" or "hyperdrive," which enables travel faster than the speed of light, or by using wormholes as shortcuts across the galaxy—but absolutely no evidence exists that any of those options are possible. And most of those imaginative stories ignore the vast number of other complications that accompany human space travel. Humans are just beginning to learn to live in space even for short periods. Much work remains to be done before we can realistically contemplate even voyages within the Solar System.

Some people claim that aliens have already visited Earth: tabloid newspapers, books, and websites are filled with tales of sightings of unidentified flying objects (UFOs), government conspiracies and cover-ups, alleged alien abductions, and UFO religious cults. However, none of those reports meet the basic standards of science. They are not falsifiable—they lack verifiable evidence and repeatability—and we must conclude that no scientific evidence exists for any alien visitations.

CHECK YOUR UNDERSTANDING 24.4

The Drake equation enables astronomers to: (a) calculate precisely the number of alien civilizations; (b) organize their thoughts about probabilities for life; (c) locate the stars they should study to find life; (d) find new kinds of life.

Origins The Fate of Life on Earth

Astronomers have used their understanding of physics and cosmology to look back through time and watch as structure formed throughout the universe and to look forward to the future of our Sun, our galaxy, and the ultimate fate of the universe. About 5 billion years from now, the Sun will end its long period of relative stability. The Sun will expand to become a red giant star, swelling to hundreds of times larger than it is now and thousands of times more luminous. The giant planets, orbiting outside the extended red giant atmosphere, will probably survive. But at least some of the planets of the inner Solar System will not. Just as an artificial satellite is slowed by drag in Earth's tenuous outer atmosphere and eventually falls to the ground, so, too, will a planet caught in the Sun's atmosphere be engulfed by the expanding Sun. If that happens to Earth, no trace of this planet will remain other than a slight increase in the amount of massive elements in the Sun's atmosphere.

Another possibility is that the red giant Sun will lose mass in a powerful wind, its gravitational pull on the planets will weaken, and the orbits of both the inner and outer planets will spiral outward. If Earth moves out far enough, it may survive as a seared cinder, orbiting the small, hot, white dwarf star that the Sun will become. Barely larger than Earth and with its nuclear fuel exhausted, the white dwarf Sun will slowly cool, eventually becoming a cold sphere of densely packed carbon, orbited by what remains of its planets. The ultimate outcome for Earth—consumed in the Sun or left behind as a cold, burned rock orbiting a long-dead white dwarf—is not yet known.

Or the threat to the future of Earth can come from within, as humanity itself changes conditions on Earth that are necessary for our habitability (for example, by releasing chemicals into the atmosphere that alter the ozone layer, or are greenhouse gases that alter the trapping of radiation).

In either case, however, Earth's status as a garden spot in the habitable zone will be at an end. If the Sun does not expand too far or Mars also migrates outward, Mars could become the habitable planet in the Solar System, at least for a while. As the dying Sun loses more and more of its atmosphere in a stellar wind, Earth's atoms might be expelled back into the reaches of interstellar space from which they came, perhaps to be recycled into new generations of stars, planets, and even life itself.

But even before the Sun's change into a red giant star, the Sun's luminosity will begin to rise. As solar luminosity increases, so will temperatures on all the planets. The inner edge of the Sun's habitable zone will slowly move out past the orbit of Earth. Eventually, Earth's temperatures will climb so high that all animal and plant life will perish. Even the extremophiles that inhabit the oceanic depths will die as the oceans themselves boil away. Models of the Sun's evolution are still not precise enough for astronomers to predict with certainty when that fatal event will occur, but the end of all terrestrial life may be 1 billion to 4 billion years away. In addition, the Milky Way Galaxy will collide with the Andromeda Galaxy in 4 billion to 5 billion years' time. Galaxies are mostly empty space, so the Sun is not likely to collide with another star, but one effect of the collision is that our Solar System may be gravitationally flung to a different part of our galaxy.

However, that the descendants of today's humanity will even be around a billion years from now is far from certain. Some threats to life come from beyond Earth. For the rest of the Sun's life, the terrestrial planets, including Earth, will continue to be bombarded by asteroids and comets. Perhaps a hundred or more of those impacts will involve kilometer-sized objects, capable of causing the kind of devastation that eradicated the dinosaurs. Those impacts may create new surface scars but they will have little effect on the integrity of Earth itself. Earth's geological record is filled with such events, and each time they happen, life manages to recover and reorganize. But we do not know if human life or civilization could survive.

Humans might protect themselves from the fate of the dinosaurs, but in the long run humanity will either leave this world or die out. Planetary systems are common to other stars, and many other Earth-like planets may well exist throughout the Milky Way Galaxy. Colonizing other planets is currently the stuff of science fiction, but if the descendants of modern-day humans are ultimately to survive the death of the home planet, off-Earth colonization must at some point become science fact.

By studying astronomy, you have learned where you come from. You have learned about the self-correcting nature of science—how it continually adapts to new information to give us the ability to make better and better predictions about the behavior of the physical world. That predictive ability makes science an extremely powerful tool. No other species in the history of Earth has been able to understand its position, predict what will happen next, and therefore adapt its behavior to seek the best possible future for its members.

Figure 24.24 shows Earth as seen by the *Cassini* spacecraft, looking near Saturn's rings. That tiny dot, which is Earth, is the only place in the entire universe where we know that life exists. Compare the size of that dot with the size of the universe. Compare the history of life on Earth with the history of the universe. Compare Earth's future with the fate of the universe. Astronomy is humbling. We occupy a tiny part of space and time. Yet we are unique, as far as we know. Think for a moment about what that means to you. That may be the most important lesson the universe has to offer.

Figure 24.24 This image from the *Cassini* spacecraft shows Earth as seen from Saturn. The pale dot in the lower right (arrow) is Earth.

ARTICLES QUESTIONS

Astronomers struggle with how to handle announcements (or nonannouncements) about extraterrestrial contact.

When Reporting News about Aliens, Caution Is Advised

By ELIZABETH HOWELL, airspacemag.com

Speculation about extraterrestrials seems to be everywhere these days. Last week it was "Tabby's Star" (more officially known as KIC 8462852), whose mysterious dimming and brightening, according to the latest analysis, is likely due to dust blocking different wavelengths of light rather than "alien megastructures." Before that came reports of an interstellar asteroid—not a spacecraft—entering our solar system and a UFO monitoring program conducted by the Department of Defense.

The attention given to such stories has some scientists worried, especially as social media amplifies claims of alien contact over other, more prosaic explanations.

"Currently, most SETI-related news seems to be interfering with conventional scientific discoveries, stealing the limelight—without following basic rules of science," wrote Dutch exoplanet researcher Ignas Snellen of Leiden Observatory, on a Facebook exoplanets discussion group for professional astronomers.

Although he has "great respect for SETI scientists," Leiden wrote, "there is no place for alien civilizations in a scientific discussion on new astrophysical phenomena, in the same way as there is no place for divine intervention as a possible solution. One may view it as harmless fun, but I see parallels in athletes taking banned substances. It may lead to short-term fame and medals, but in the long run it harms the sport. Same for astronomy: we should be very careful not to be ridiculed. I really hope we can stop mentioning SETI for every unexplained phenomenon."

Such worries aren't new. Nearly 20 years ago researchers in the field came up with the Rio Scale to guide them in reporting the significance of any candidate SETI signal. Modeled after the Richter Scale for earthquakes, it assigns a value between 0—no significance—and 10—extraordinary significance—to any detected signal. The method was first proposed by Ivan Almar and Jill Tarter at the 51st International Astronautical Congress in Rio de Janeiro, in 2000. Although the International Academy of Astronautics' SETI Permanent Committee adopted the scale two years later, it hasn't seen widespread use.

Meanwhile, the news business has changed drastically since 2000. Today much of the public doesn't get its information from TV and newspapers, but from Facebook updates and Twitter posts, which move—and change—at a much faster pace. In recognition of the altered media landscape, a new paper submitted to the *International Journal of Astrobiology* proposes streamlining the Rio Scale and having scientists take a short quiz to answer a few questions about their discovery. Journalists and other news providers could use the answers in their stories.

"You can go through the steps and ask yourself, do I believe the instrumentation is working properly? Do I believe I am looking at it objectively? Do I have enough people who looked at the signal with different instruments? Has there been a lot of scientific discussion? Have there been alternative explanations? Is it a hoax?" says the paper's lead author, Duncan Forgan, a postdoctoral scholar at the University of St. Andrews' School of Physics and Astronomy in Scotland, whose work includes problems related to astrobiology and SETI.

Forgan says that although it can be difficult to convey uncertainty about a particular result to the public, the revised Rio Scale would make that job easier. Some scientific speculation can be "a bit wibbly-wobbly," he says, "and those wibbly-wobbly bits end up in the press." He agrees that spurious SETI claims can sometimes distract from more legitimate science.

In recognition of the sensitivity around alien signal detection, SETI has a voluntary list of protocols to follow when something interesting is found. The first principle urges researchers to "verify that the most plausible explanation for the evidence is the existence of extraterrestrial intelligence, rather than some other natural phenomenon or anthropogenic phenomenon, before making any public announcement."

Morris Jones, an Australian space observer with both scientific and journalistic training, says we should label "fringe SETI" claims for what they are. "The media is under pressure to deliver attention-grabbing news, but it's hard to expect them to judge fringe SETI as spurious when it comes from reputable institutions and qualified researchers. The best way to reduce these reports is to stop the production of questionable scientific papers in the first place."

ARTICLES QUESTIONS

1. Why are astrobiologists concerned about media coverage of SETI?
2. How do the suggestions for journalists fit with what you have learned about the process of science?
3. Think of an article you have seen recently on evidence for alien life. Was it overly speculative?
4. How do you think people on Earth would react to an actual discovery of alien life?

Summary

Astrobiology seeks to answer the question, Are we alone? All the sciences have a part to play in answering that question, from astronomy to zoology. Theories of how life began and evolved on Earth help limit the number of targets astronomers study in their search for life elsewhere. Even before the Sun expands and evolves to a red giant star, its luminosity will increase enough to alter the location of its habitable zone. When that happens, Earth may no longer be the planet at the right location for maintaining liquid water.

LG 1 **Explain our current understanding of how and when life began on Earth and how it has evolved.** Life on Earth is a form of complex carbon-based chemistry, made possible by self-replicating molecules. Life probably formed in Earth's oceans and then evolved chemically from simple molecules into self-replicating organisms through a combination of mutation and heredity.

LG 2 **Explain how life is a structure that has evolved through the action of the physical and chemical processes that shape the universe.** All terrestrial life is composed primarily of only six elements: carbon, hydrogen, oxygen, nitrogen, sulfur, and phosphorus. However, life-forms very different from those on Earth, including those based on silicon chemistry, cannot be ruled out.

LG 3 **Describe the locations in our Solar System and around other stars where astronomers think life might be possible.** Within the Solar System, Mars and some moons of Jupiter and Saturn are the most promising candidates for life. A habitable zone around a star is a location within which the temperature of a planet will support liquid water on its surface. Thus, astronomers look for exoplanets that orbit in habitable zones surrounding solar-type stars. That concept of habitable zone can be extended by special circumstances, such as tidal heating from a giant planet or atmospheres containing greenhouse gases.

LG 4 **Describe some methods used to search for intelligent extraterrestrial life.** The Drake equation includes factors that astronomers consider when thinking about the possibility of life in the universe. Astronomers use radio telescopes to search for signals from extraterrestrial life, particularly in astronomically important regions of the spectrum. In recent times, that effort has been expanded to a search through as broad a region of the radio spectrum as possible. No intelligent extraterrestrial life has yet been detected.

? Unanswered Questions

- Will humans spread life into space? Some scientists have suggested that seeding from Earth may have already happened as Earth microorganisms scattered into space after giant impacts. Humans also may have unintentionally sent microorganisms to space aboard our spacecraft. More intentional methods of seeding include sending microorganisms from Earth to other planets or moons to try to jump-start evolution, "terraforming" Mars or a moon to change conditions on it to make it more habitable for humanity, or sending humans in spaceships to colonize the galaxy.
- Will humans themselves spread into space? At some point, humans must leave planet Earth if the species is to survive. But space is a dangerous place, and many of the problems that humans encounter in space have not yet been solved. Those problems are as diverse as purely physical issues, such as the loss of bone density in low gravity, and the societal problems that occur when a few people are confined together for long periods. We do not yet know whether humanity can overcome those problems and journey to other planets.

Questions and Problems

Test Your Understanding

1. The study of life and the study of astronomy are connected because (select all that apply)
   a. life may be commonplace in the universe.
   b. studying other planets may help explain why life exists on Earth.
   c. explorations of extreme environments on Earth suggest where to look for life elsewhere.
   d. life is a structure that evolved through physical processes, and life on Earth may not be unique.
   e. life elsewhere is most likely to be found by astronomers.
2. Scientists look for water to indicate places where life might exist because
   a. water is a common molecule in interstellar space.
   b. life on Earth depends on it.
   c. no other molecules are solvents.
   d. the spectrum of water is very complex.
3. The Urey-Miller experiment produced ________ in a laboratory jar.
   a. life
   b. RNA and DNA
   c. amino acids
   d. proteins
4. A mutation is
   a. always a deadly change to DNA.
   b. always a beneficial change to DNA.
   c. a change to DNA that is sometimes beneficial and sometimes not.
   d. a change to DNA that cannot be inherited.
5. Scientists think that terrestrial life probably originated in Earth's oceans, rather than on land, because (select all that apply)
   a. all the chemical pieces were in the ocean.
   b. energy was available in the ocean.
   c. the earliest evidence for life on Earth is from fossils of ocean-dwelling organisms.
   d. the deepest parts of the ocean have hydrothermal vents.
6. Any system with the processes of heredity, mutation, and natural selection will, over time (choose all that apply),
   a. change.
   b. become larger.
   c. become more complex.
   d. develop intelligence.
7. The fact that no alien civilizations have yet been detected indicates that
   a. they are not there.
   b. they are rare.
   c. Earth is in a "blackout," and they are not talking to us.
   d. we don't know enough yet to draw any conclusions.
8. During the Cambrian explosion
   a. the dinosaurs were killed.
   b. all the carbon that is now here on Earth was produced.
   c. biodiversity increased significantly.
   d. a lot of carbon dioxide was released into the atmosphere.
9. The habitable zone is the place around a star where
   a. life has been found.
   b. atmospheres can contain oxygen.
   c. liquid water has been confirmed.
   d. liquid water can exist on the surface of a planet.
10. The difference between a prokaryote and a eukaryote is that prokaryotes
    a. have no DNA.
    b. have no cell wall.
    c. have no nucleus.
    d. do not exist today.
11. A thermophile is an organism that lives in extremely ________ environments.
    a. salty
    b. hot
    c. cold
    d. dry
12. The search for life elsewhere in the Solar System is carried out primarily by
    a. astronauts.
    b. robotic spacecraft.
    c. astronomers using optical telescopes.
    d. astronomers using radio telescopes.
13. Astronomers think that intelligent life is more likely to be found around stars of types F, G, K, and M because
    a. those stars are hot enough to have planets and moons with liquid water.
    b. those stars are cool enough to have planets and moons with liquid water.
    c. those stars live long enough for life to begin and evolve.
    d. those stars produce no ultraviolet radiation or X-rays.
14. Life first appeared on Earth
    a. billions of years ago.
    b. millions of years ago.
    c. hundreds of thousands of years ago.
    d. thousands of years ago.
15. In the phrase "theory of evolution," the word *theory* means that evolution
    a. is an idea that can't be tested scientifically.
    b. is an educated guess to explain natural phenomena.
    c. probably doesn't happen anymore.
    d. is a tested and corroborated scientific explanation of natural phenomena.

Thinking about the Concepts

16. How does the evolution of life on Earth depend on RNA and DNA?
17. How do scientists think that amino acids first formed on Earth?
18. Today, most organisms on Earth enjoy relatively moderate climates and temperatures. Compare that environment with some of the conditions in which early life developed.
19. The Process of Science Figure represents many areas of science that all inform the science of astrobiology. Choose any two of those, and explain how one depends on the other.
20. How was Earth's carbon dioxide atmosphere changed into today's oxygen-rich atmosphere? How long did that transformation take?
21. What was the Cambrian explosion, and what might have caused it?
22. Why did plants and forests appear in large numbers before large animals did?
23. What is a habitable zone? What defines its boundaries?
24. Which general conditions were needed on Earth for life to arise?
25. Is evolution under way on Earth today? If so, how might humans continue to evolve?
26. Where did all the atoms in your body come from?
27. The two *Viking* spacecraft found no convincing evidence of life on Mars when they visited that planet in the late 1970s, nor did the *Phoenix* lander when it examined the martian soil in 2009. Do those results imply that life never existed on the planet? Why or why not?
28. Some scientists believe that humans may be the only advanced life in the galaxy today. If so, which factors in the Drake equation must be extremely small?
29. The second law of thermodynamics says that the entropy (a measure of disorder) of the universe is always increasing. Yet complex living organisms exist. Why does that fact not violate the second law of thermodynamics?
30. Why is life on Earth as we know it likely to end long before the Sun runs out of nuclear fuel?

Applying the Concepts

31. Several missions are searching for planets in the habitable zones of stars. Explain which factors in the Drake equation are affected by that search and how the final number N will be affected if these missions find that most stars have planets in their habitable zones.
32. Study Figure 24.22. The white blocks at the top represent the numbers 1–10. The bottom row of the set of four is a placeholder, and the top three rows of the four represent the numbers, in order from right to left. Explain the "rule" for the kind of counting shown here. (For example, how do three white blocks represent the number 7?)
33. Suppose that an organism replicates itself each second. If you start with a single specimen, what will the final population be after 10 seconds?
34. Suppose that an organism replicates itself each second. After how many seconds will the population increase by a factor of 1,024?
35. In general, you can find the doubling time for exponential growth by dividing 70 by the rate of increase. So, if the population increases by 7 percent per year, the doubling time is 10 years ($70/7 = 10$). Suppose Earth's human population continues to grow by 1 percent annually. What is the doubling time? How much time will pass before Earth has 4 times as many humans?
36. As discussed in Working It Out 24.1, the doubling time for exponential growth is given by $P_F/P^n = 2n$. Assume a self-replicating molecule that makes one copy of itself each second. Make a graph of the number of molecules versus time for the first 60 seconds after the molecule begins replicating.
37. The doubling time for *Escherichia coli* is 20 minutes, and you start getting sick when just 10 bacteria enter your system. How many bacteria are in your body after 12 hours?
38. If the chance that a given molecule will mutate is one in 100,000, how many generations are needed before, on average, at least one mutation has occurred?
39. Study the Drake equation in Working It Out 24.2. Make your own most optimistic and most pessimistic assumptions for each variable in the equation. What values do you find for N?
40. Study the Drake equation in Working It Out 24.2. Keep the other variables as is, but put in different values for the lifetime of a technological civilization. How does doing so affect your value of N? How does that affect the possibility of making contact with extraterrestrials?
41. Look back at Figure 5.24 in Chapter 5. Trace (or photocopy) that graph, and then add horizontal lines for the temperatures at which water freezes and boils. Which planets in the Solar System have measured temperatures that fall within those lines? Which planets have predicted temperatures (based on the equilibrium model) that fall within those lines? What does that tell you about assumptions about the habitable zone?
42. Figure 5.24 shows that Mercury's measured range of temperatures overlaps the temperatures at which water is a liquid. Is Mercury in the habitable zone? Why or why not?

43. Suppose astronomers announce the discovery of a new planet around a star with a mass equal to the Sun's. The planet has an orbital period of 87 days. Is that planet in the habitable zone for a Sun-like star?

44. Why do you think astronomers sent a coded radio signal to the globular cluster M13 in 1974—rather than, say, to a nearby star?

45. As noted in Section 24.1, some scientists suspect the early Earth was "seeded" with primitive life stored in comets and meteoroids. Now that you know when and how our Solar System, galaxy, and universe formed, what time line is required for such seeding to be possible?

USING THE WEB

46. a. Go to the online *Astrobiology Magazine* (https://www.astrobio.net), which covers many topics included in this chapter. Under "Origins," click on "Extreme Life." What are some new findings about extremophiles? Why is that important for astrobiologists? Under "Deep Space" click on "New Planets." What is a recent discovery?
    b. Go to the "Life, Unbounded" blog (https://blogs.scientificamerican.com/life-unbounded). What is a recent topic of interest? Is the discussion based on some new data? A new theory?

47. Solar System space missions:
    a. Go to the website for the *Curiosity Mars Science Laboratory* (https://mars.nasa.gov/msl/mission). Are there any results from the *Mars Insight* mission (https://mars.nasa.gov/insight/)? What is the status of the *Mars 2020* lander, scheduled for launch in 2020?
    b. The *MAVEN* Mars mission (https://mars.nasa.gov/maven/) arrived at Mars in 2014. What are the mission's science goals? Why are astrobiologists interested in the history of the climate and atmosphere of Mars? Are any results available?

48. Go to the "Habitable Exoplanets Catalog" website (http://phl.upr.edu/projects/habitable-exoplanets-catalog). Click on "Methods" to read about the criteria for evaluating a planet's habitability. How many confirmed habitable exoplanets exist? How many candidates? How might the number of confirmed exoplanets affect the terms in the Drake equation?

49. a. Go to the website for the European Space Agency's *Gaia* satellite (http://sci.esa.int/gaia). What are the mission's science objectives? Under "Mission Science" on the left, click on "Exoplanets." How does *Gaia* identify new planets? What properties of the planet will it be able to measure? What are some recent results?
    b. Go to the website for the NASA TESS (Transiting Exoplanet Survey Satellite; https://tess.gsfc.nasa.gov/). What is the mission's goal? What are some of its discoveries?
    c. Go to the website for the NASA JWST (James Web Space Telescope; https://www.jwst.nasa.gov/). On the left, click on "Science" and then "Planets & Life." How will the telescope study exoplanet atmospheres? Click on "News"—are any results available?

50. Go to the website for the Breakthrough Initiative (https://breakthroughinitiatives.org/), a privately funded series projects started in 2015 to address the question, Are we alone? What are those projects and what is their status?

EXPLORATION Fermi Problems and the Drake Equation

digital.wwnorton.com/astro5

The Drake equation is a way to organize ideas about other intelligent, communicating civilizations in the galaxy. That type of thinking is useful for estimating a value, particularly when analyzing systems for which counting is not possible. The types of problems that can be solved in that way are often called Fermi problems after Enrico Fermi, who was mentioned in this chapter. For example, we might ask, "What is the circumference of Earth?"

You could Google that question, or you could already "know" the answer, or you might look it up in this textbook. Alternatively, you could very carefully measure the shadow of a stick in two locations at the same time on the same day. Or you could drive around the planet.

Or you could reason this way:

How many time zones are between New York and Los Angeles?
Three time zones. You know that from traveling or from television.

How many miles is it from New York to Los Angeles?
Three thousand miles. You know that from traveling.

So, how many miles per time zone?
3,000/3 = 1,000

How many time zones are in the world?
Twenty-four, because 24 hours are in a day, and each time zone marks an hour.

So, what is the circumference of Earth?
Twenty-four thousand miles, because Earth has 24 time zones, each 1,000 miles wide.

The measured circumference is 24,900 miles, which agrees with our estimate to within 4 percent.

Here we list several Fermi problems. Time yourself for an hour, and work as many of them as possible. (You don't have to do them in order!)

1 How much has the mass of the human population on Earth increased in the past year?

2 How much energy does a horse consume in its lifetime?

3 How many pounds of potatoes are consumed in the United States annually?

4 How many cells are in the human body?

5 If your lifetime earnings were given to you by the hour, how much is your time worth per hour?

6 What is the weight of solid garbage that U.S. families throw out each year?

7 How fast does human hair grow (in feet per hour)?

8 If all the people on Earth were crowded together, how much area would be covered?

9 How many people could fit on Earth if every person occupied 1 square meter of land?

10 How much carbon dioxide (CO_2) does an automobile emit each year?

11 What is the mass of Earth?

12 What is the average annual cost of an automobile, including overhead (maintenance, looking for parking, insurance, cleaning, and so on)?

APPENDIX 1: Mathematical Tools

Mathematics helps scientists understand the patterns they see and then communicate that understanding to others. Appendix 1 presents some tools that will be useful in our study of astronomy.

Powers of 10 and Scientific Notation

Astronomy is a science of both the very large and the very small. The mass of an electron, for example, is

0.0000000000000000000000000000009109 kilograms (kg)

whereas the distance to a galaxy far, far away might be about

100,000,000,000,000,000,000,000,000 meters

A quick glance at those two numbers shows why astronomers need a more convenient way to express numbers.

Our number system is based on **powers of 10**. Going to the left of the decimal place,

$$10 = 10 \times 1$$
$$100 = 10 \times 10 \times 1$$
$$1{,}000 = 10 \times 10 \times 10 \times 1$$

and so on. Going to the right of the decimal place,

$$0.1 = \frac{1}{10} \times 1$$

$$0.01 = \frac{1}{10} \times \frac{1}{10} \times 1$$

$$0.001 = \frac{1}{10} \times \frac{1}{10} \times \frac{1}{10} \times 1$$

and so on. In other words, each place to the right or left of the decimal place in a number represents a power of 10. For example, 1 million can be written as

$$1 \text{ million} = 1{,}000{,}000 = 1 \times 10 \times 10 \times 10 \times 10 \times 10 \times 10$$

That is, 1 million is "1 multiplied by six factors of 10." **Scientific notation** combines those factors of 10 in convenient shorthand. Rather than all being written out, the six factors of 10 are expressed using an exponent:

$$1 \text{ million} = 1 \times 10^6$$

which also means "1 multiplied by six factors of 10."

Moving to the right of the decimal place, each step *removes* a power of 10 from the number. One-millionth can be written as

$$1 \text{ millionth} = 1 \times \frac{1}{10} \times \frac{1}{10} \times \frac{1}{10} \times \frac{1}{10} \times \frac{1}{10} \times \frac{1}{10}$$

That divides by powers of 10, so that number can be written by use of a negative exponent

$$\frac{1}{10} = 10^{-1}$$

as

$$\begin{aligned} 1 \text{ millionth} &= 1 \times 10^{-1} \times 10^{-1} \times 10^{-1} \times 10^{-1} \times 10^{-1} \times 10^{-1} \\ &= 1 \times 10^{-6} \end{aligned}$$

Let's return to our earlier examples. The mass of an electron is 9.109×10^{-31} kg, and the distant galaxy is located 1×10^{26} meters away. Scientific notation is a much more convenient ways of writing those values. Notice that *the exponent in scientific notation gives you a feel for the size of a number at a glance.* The exponent of 10 in the electron mass is −31, which quickly indicates that it is a very small number. The exponent of 10 in the distance to a remote galaxy, +26, quickly indicates that it is a very large number. The exponent is often called the *order of magnitude* of a number. When you see a number written in scientific notation while reading *21st Century Astronomy* (or elsewhere), just remember to look at the exponent to understand the size of the number.

Scientific notation also is convenient because it makes multiplying and dividing numbers easier. For example, 2 billion multiplied by eight-thousandths can be written as

$$2{,}000{,}000{,}000 \times 0.008$$

but writing those two numbers in scientific notation is:

$$(2 \times 10^9) \times (8 \times 10^{-3})$$

We can regroup those expressions in the following form:

$$(2 \times 8) \times (10^9 \times 10^{-3})$$

The first part of the problem is just $2 \times 8 = 16$. The more interesting part of the problem is the multiplication in the right-hand parentheses. The first number, 10^9, is just shorthand for $10 \times 10 \times 10 \ldots$ nine times. That is, it represents nine factors of 10. The second number stands for three factors of 1/10—or removing three factors of 10 if you prefer to think of it that way. Altogether, that makes $9 - 3 = 6$ factors of 10. In other words,

$$10^9 \times 10^{-3} = 10^{(9-3)} = 10^6$$

Putting the problem together, we get

$$\begin{aligned}(2 \times 10^9) \times (8 \times 10^{-3}) &= (2 \times 8) \times (10^9 \times 10^{-3}) \\ &= 16 \times 10^6\end{aligned}$$

By convention, when a number is written in scientific notation, only one digit is placed to the left of the decimal point. Here, though, we have two. However, 16 is 1.6×10, so we can add that additional factor of 10 to the exponent at right, making the final answer

$$1.6 \times 10^7$$

Dividing is just the inverse of multiplication. Dividing by 10^3 means removing three factors of 10 from a number. Using the previous number,

$$\begin{aligned}(1.6 \times 10^7) \div (2 \times 10^3) &= (1.6 \div 2) \times (10^7 \div 10^3) \\ &= 0.8 \times 10^{(7-3)} \\ &= 0.8 \times 10^4\end{aligned}$$

This time we have only a zero to the left of the decimal point. To get the number into proper form, we can substitute 8×10^{-1} for 0.8, giving

$$0.8 \times 10^4 = (8 \times 10^{-1}) \times 10^4 = 8 \times 10^3$$

Adding and subtracting numbers in scientific notation is harder because all numbers must be written as values multiplied by the *same* power of 10 before they can be added or subtracted. Therefore, you will need to use a calculator that has scientific notation. Most scientific calculators have a button that says EXP or EE. Those abbreviations mean "times 10 to the ____." So for 4×10^{12}, you would type [4][EXP][1][2] or [4][EE][1][2] into your calculator. Usually, that number shows up in the window on your calculator either just as you see it written in this book or as a 4 with a smaller 12 all the way over in the right side of the window.

Significant Figures

In the previous example, we actually broke some rules in the interest of explaining how powers of 10 are treated in scientific notation. The rules we broke involve the *precision* of the numbers. When expressing quantities in science, it is extremely important to know not only the value of a number but also how precise that value is.

The most complete way to keep track of the precision of numbers is to actually write down the uncertainty in the number. For example, suppose you know that the distance to a store (call it *d*) is between 0.8 and 1.2 kilometers (km); you can then write

$$d = 1.0 \pm 0.2 \text{ km}$$

where the symbol "±" is pronounced "plus or minus." Here, *d* is between $1.0 - 0.2 = 0.8$ km and $1.0 + 0.2 = 1.2$ km. That is an unambiguous statement about the limitations on knowing the value of *d*, but carrying along the formal errors with every number written would be cumbersome at best. Instead, you keep track of the approximate precision of a number by using *significant figures*.

The convention for significant figures is this: Assume the written number has been rounded from a number that had one additional digit to the right of the decimal point. If a quantity *d*, which might represent the distance to the store, is "1.", then *d* is close to 1. It is probably not as small as "0.", and it is probably not as large as "2.". If instead it is written as

$$d = 1.0$$

then *d* is probably not 0.9 and is probably not 1.1. It is roughly 1.0 to the nearest tenth. The greater the number of significant figures, the more precisely the number is being specified. For example, 1.00000 is not the same number as 1.00. The first number, 1.00000, represents a value that is probably not as small as 0.99999 and probably not as large as 1.00001. The second number, 1.00, represents a value that is probably not as small as 0.99 nor as large as 1.01. The number 1.00000 is much more precise than the number 1.00.

In mathematical operations, significant figures are important. For example, consider $2.0 \times 1.6 = 3.2$. It does *not* equal 3.20000000000. *The product of two numbers cannot be known to any greater precision than the numbers themselves.* In general, when you multiply and divide, the answer should have the same number of significant figures as the less precise of the numbers being multiplied or divided. In other words, $2.0 \times 1.602583475 = 3.2$. Because all you know is that the first factor is probably closer to 2.0 than to 1.9 or 2.1, all you know about the product is that it is between about 3.0 and 3.4. It is 3.2. It is not 3.205166950 (*even if that is the answer on your calculator*). The rest of the digits to the right of 3.2 just do not mean anything.

When two numbers are added or subtracted, if one number has a significant figure with a particular place value but another number does not, their sum or difference cannot have a significant figure in that place value. For example,

$$\begin{array}{r} 1{,}045. \\ +1.34567 \\ \hline 1{,}046. \end{array}$$

The answer is "1,046." *not* "1,046.34567". Again, the extra digits to the right of the decimal place have no meaning because "1,045." is not known to that precision.

What is the precision of the number 1,000,000? As it is written, the answer is unclear. Are all those zeros really significant, or are they merely placeholders? If the number is written in scientific notation, however, you never have to wonder. Instead of 1,000,000, you write 1.0×10^6 for a number that is known to the nearest hundred thousand or so, or you write 1.00000×10^6 for a number that is known to the nearest 10.

So the earlier example would have been more correct if written as

$$(2.0 \times 10^9) \times (8.0 \times 10^{-3}) = 1.6 \times 10^7$$

Algebra

Mathematics has many branches. The branch that focuses on the relationships between quantities is called **algebra**. Basically, algebra begins by using symbols to represent quantities.

For example, you could write the distance you travel in a day as *d*. As it stands, *d* has no value. It might be 10,000 miles. It might be 30 feet. It does, however, have **units**—here, the units of distance. The average speed at which you travel is equal to the distance you travel divided by the time you take. By using the symbol v to represent your average speed and the symbol t to represent the time you take, instead of writing out "Your average speed is equal to the distance you travel divided by the time you take" you can write

$$v = \frac{d}{t}$$

The meaning of that algebraic expression is the same as the sentence quoted before it, but it is much more concise. As it stands, v, d, and t still have no specific values. No numbers are assigned to them yet. However, the expression indicates what the relationship between those numbers will be when you look at a specific example. For example, if you go 500 km ($d = 500$ km) in 10 hours ($t = 10$ hours), that expression tells you that your average speed is

$$v = \frac{d}{t} = \frac{500\,\text{km}}{10\,\text{h}} = 50\,\text{km/h}$$

Notice that the units in this expression act exactly like the numerical values. Dividing the two shows that the units of v are kilometers divided by hours, or km/h (pronounced "kilometers per hour").

We introduced algebra as shorthand for expressing relations between quantities, but it is far more powerful than that. Algebra provides rules for manipulating the symbols used to represent quantities. We begin with a bit of notation for *powers* and *roots*. Similar to powers of 10, raising a quantity to a power means multiplying the quantity by itself some number of times. For example, if S is a symbol for something (anything), S^2 (pronounced "S-squared" or "S to the second power") means $S \times S$, and S^3 (pronounced "S-cubed" or "S to the third power") means $S \times S \times S$. Suppose S represents the length of the side of a square. The area of the square is given by

$$\text{Area} = S \times S = S^2$$

If $S = 3$ meters (m), the area of the square is

$$S^2 = 3\text{ m} \times 3\text{ m} = 9\text{ m}^2$$

(pronounced "9 square meters"). That is why raising a quantity to the second power is called *squaring* the quantity. We could have done the same thing for the sides of a cube and found that the volume of the cube is

$$\text{Volume} = S \times S \times S = S^3$$

If $S = 3$ meters, the volume of the cube is

$$S^3 = 3\text{ m} \times 3\text{ m} \times 3\text{ m} = 27\text{ m}^3$$

(pronounced "27 cubic meters"). That is why raising a quantity to the third power is called *cubing* the quantity.

Roots are the reverse of that process. The square root of a number is the value that, when squared, gives the original quantity. The square root of 4 is 2, which means that $2 \times 2 = 4$. The square root of 9 is 3, which means that $3 \times 3 = 9$. Similarly, the cube root of a quantity is the value that, when cubed, gives the original quantity. The cube root of 8 is 2, which means that $2 \times 2 \times 2 = 8$. Roots are written with the symbol $\sqrt{\ }$. For example, the square root of 9 is written as

$$\sqrt{9} = 3$$

and the cube root of 8 is written as

$$\sqrt[3]{8} = 2$$

If the volume of a cube is $V = S^3$, then

$$S = \sqrt[3]{V} = \sqrt[3]{S^3}$$

Roots can also be written as powers. Powers and roots behave like the exponents of 10 in our discussion of scientific notation. (The exponents used in scientific notation are just powers of 10.) For example, if a, n, and m are all algebraic quantities, then

$$a^n \times a^m = a^{n+m} \quad \text{and} \quad \frac{a^n}{a^m} = a^{n-m}$$

(The square root of a can also be written $a^{1/2}$ and the cube root of a can be written $a^{1/3}$.)

The rules of arithmetic can be applied to the symbolic quantities of algebra. As long as the rules of algebra are applied properly, the relationships among symbols arrived at through algebraic manipulation remain true for the physical quantities that those symbols represent.

Here we summarize a few algebraic rules and relationships. Here, a, b, c, m, n, r, x, and y are all algebraic quantities.

Associative rule:

$$a \times b \times c = (a \times b) \times c = a \times (b \times c)$$

Commutative rule:

$$a \times b = b \times a$$

Distributive rule:

$$a \times (b + c) = (a \times b) + (a \times c)$$

Cross-multiplication:

$$\text{If } \frac{a}{b} = \frac{c}{d}, \text{ then } ad = bc$$

Working with exponents:

$$\frac{1}{a^n} = a^{-n} \qquad a^n a^m = a^{n+m}$$

$$\frac{a^n}{a^m} = a^{n-m} \qquad (a^n)^m = a^{n\times m} \qquad \left(\frac{a}{b}\right)^n = \frac{a^n}{b^n}$$

Equation of a line with slope m and y-intercept b:

$$y = mx + b$$

Equation of a circle with radius r centered at $x = 0, y = 0$:

$$x^2 + y^2 = r^2$$

Angles and Distances

The farther away something is, the smaller it appears. That is common sense and everyday experience. Because astronomers cannot walk up to the object they are studying and measure it with a meterstick, their knowledge about the sizes of things usually depends on relating the size of an object, its distance, and the angle it covers in the sky.

One way to measure angles is to use a unit called the **radian**. As shown in **Figure A1.1a**, the size of an angle in radians is the length of the arc subtending the angle, divided by the radius of the circle. In the figure, the angle $x = S/r$ radians.

Because the circumference of a circle is 2π multiplied by the radius ($C = 2\pi r$), a complete circle has an angular measure of $(2\pi r)/r = 2\pi$ radians. In more conventional angular measure, a complete circle is 360°, so

$$360° = 2\pi \text{ radians}$$

or

$$1\,\text{radian} = \frac{360°}{2\pi} = 57.2958°$$

Often, seconds of arc (**arcseconds**) are used to measure angles for stars and galaxies. A degree is divided into 60 minutes of arc (**arcminutes**), each of which is divided into 60 seconds of arc—so 3,600 seconds of arc are in a degree. Therefore,

$$3{,}600\,\frac{\text{arcseconds}}{\text{degree}} \times 57.2958\,\frac{\text{degree}}{\text{radian}} = 206{,}265\,\frac{\text{arcseconds}}{\text{radian}}$$

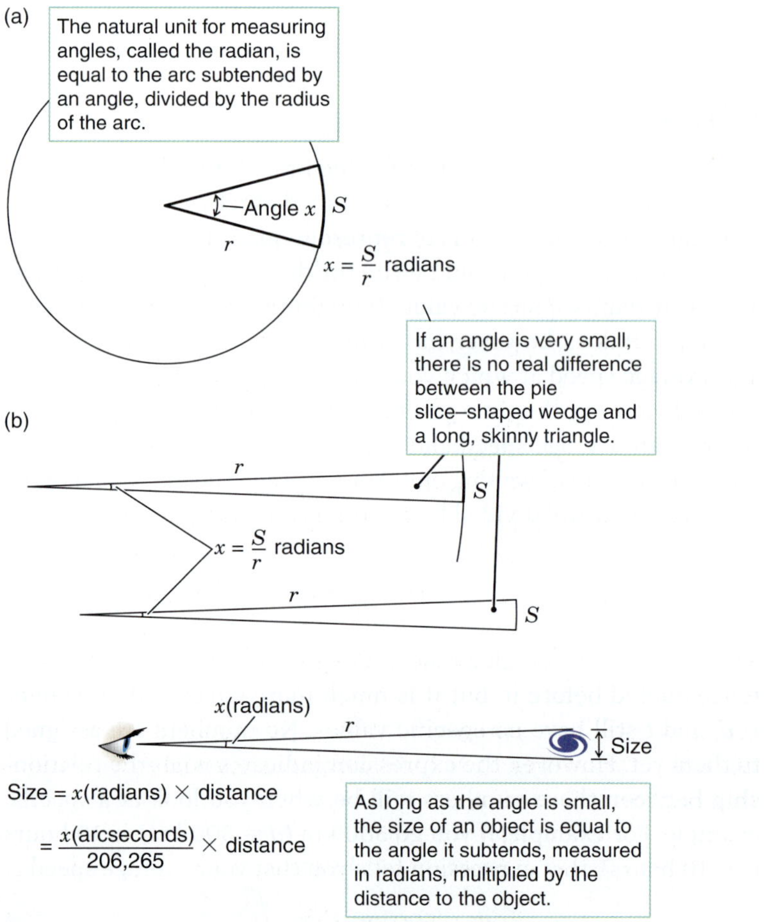

Figure A1.1 Measuring angles.

If the angle is small enough (which it usually is in astronomy), very little difference exists between the pie slice just described and a long skinny triangle with a short side of length *S*, as Figure A1.1b illustrates. So, if you know the distance *d* to an object and you can measure the angular size *x* of the object, the size of the object is given by

$$S = x\,(\text{in radians}) \times d = \frac{x\,(\text{in degrees})}{57.2958\ \text{degrees/radian}} \times d$$

$$S = \frac{x\,(\text{in arcseconds})}{206{,}265\ \text{arcseconds/radian}} \times d$$

That is how astronomers relate an object's angular size, distance, and physical size.

Circles and Spheres

To round out those mathematical tools, here are a few useful formulas for circles and spheres. The circle or sphere in each case has a radius *r*.

$$\text{Circumference}_{\text{circle}} = 2\pi r$$

$$\text{Area}_{\text{circle}} = \pi r^2$$

$$\text{Surface area}_{\text{sphere}} = 4\pi r^2$$

$$\text{Volume}_{\text{sphere}} = \frac{4}{3}\pi r^3$$

Working with Proportionalities

Most of the mathematics in *21st Century Astronomy* involves **proportionalities**—statements about how one physical quantity changes when another quantity changes. We began a discussion of proportionality in Working It Out 1.1; here, we offer a few examples of working with proportionalities.

To use a statement of proportionality to compare two objects, begin by turning the proportionality into a ratio. For example, the price of a bag of apples is **proportional** to the weight of the bag:

$$\text{Price} \propto \text{Weight}$$

Here, the symbol ∝ is pronounced "is proportional to." That means the ratio of the prices of two bags of apples is equal to the ratio of the weights of the two bags:

$$\text{Price} \propto \text{Weight means}$$

$$\frac{\text{Price of A}}{\text{Price of B}} = \frac{\text{Weight of A}}{\text{Weight of B}}$$

Let's work a specific example. Suppose bag A weighs 2 pounds and bag B weighs 1 pound. That means bag A will cost twice as much as bag B. We can turn that proportionality into the following equation:

$$\frac{\text{Price of A}}{\text{Price of B}} = \frac{\text{Weight of A}}{\text{Weight of B}} = \frac{2\ \text{lb}}{1\ \text{lb}} = 2$$

In other words, the price of bag A is 2 times the price of bag B. The price per pound is an example of a **constant of proportionality**.

Now let's work another, more complicated example. In Chapter 13, we discuss how the luminosity, brightness, and distance of stars are related. The luminosity of a star—the total energy that the star radiates each second—is proportional to the star's brightness multiplied by the square of its distance:

$$\text{Luminosity} \propto \text{Brightness} \times \text{Distance}^2$$

We can turn that proportionality into a ratio for two stars, A and B:

$$\frac{\text{Luminosity of A}}{\text{Luminosity of B}} = \frac{\text{Brightness of A}}{\text{Brightness of B}} \times \left(\frac{\text{Distance of A}}{\text{Distance of B}}\right)^2$$

If we use the symbols *L*, *b*, and *d* to represent luminosity, brightness, and distance, respectively, that equation becomes

$$\frac{L_A}{L_B} = \frac{b_A}{b_B} \times \left(\frac{d_A}{d_B}\right)^2$$

As an example, suppose that star A appears twice as bright in the sky as star B, but star A is located 10 times as far away as star B. How do the luminosities of the two stars compare? We know that

$$\text{Luminosity} \propto \text{Brightness} \times \text{Distance}^2$$

we write

$$\frac{\text{Luminosity of A}}{\text{Luminosity of B}} = \frac{\text{Brightness of A}}{\text{Brightness of B}} \times \left(\frac{\text{Distance of A}}{\text{Distance of B}}\right)^2$$

$$\frac{\text{Luminosity of A}}{\text{Luminosity of B}} = \frac{2}{1} \times \left(\frac{10}{1}\right)^2 = 200$$

In other words, star A is 200 times as luminous as star B.

Two quantities may also be inversely proportional, such that making one of them smaller makes the other larger. For example, when you are driving to another town, if you drive twice as fast, getting there takes half the time. The travel time is inversely proportional to the travel speed. We write that relationship as

$$\text{Time} \propto \frac{1}{\text{Speed}}$$

Proportionalities are used to compare one object with another. Constants of proportionality are used to calculate actual values. In *21st Century Astronomy*, the proportionality is usually what is important.

APPENDIX 2:

Physical Constants and Units

Fundamental Physical Constants

Constant	Symbol	Value
Speed of light in a vacuum	c	2.99792×10^8 m/s
Universal gravitational constant	G	6.6741×10^{-11} m^3/(kg s^2) 6.6741×10^{-20} km^3/(kg s^2)
Planck constant	h	6.62607×10^{-34} J-s
Boltzmann constant	k	1.38065×10^{-23} J/K
Stefan-Boltzmann constant	σ	5.67037×10^{-8} W/(m^2 K^4)
Mass of electron	m_e	9.10938×10^{-31} kg
Mass of proton	m_p	1.67262×10^{-27} kg
Mass of neutron	m_n	1.67493×10^{-27} kg
Electric charge of electron or proton	e	1.60218×10^{-19} C

SOURCE: Data from the Particle Data Group (http://pdg.lbl.gov).

Unit Prefixes

Prefix*	Name	Factor†
n	nano-	10^{-9}
μ	micro-	10^{-6}
m	milli-	10^{-3}
k	kilo-	10^3
M	mega-	10^6
G	giga-	10^9
T	tera-	10^{12}

*When appended to a unit, these prefixes change the size of the unit by the factor (†) given. For example, 1 km is 10^3 meters.

Units and Values

Quantity	Fundamental Unit	Values
Length	meters (m)	radius of Sun (R_{Sun}) = 6.957×10^8 m astronomical unit (AU) = 1.49598×10^{11} m 1 AU = 149,598,000 km light-year (ly) = 9.4605×10^{15} m 1 ly = 6.324×10^4 AU 1 parsec (pc) = 3.262 *ly* = 3.0857×10^{16} m 1 m = 3.281 feet
Volume	cubic meters (m^3)	1 m^3 = 1,000 liters = 264.2 gallons
Mass	kilograms (kg)	1 kg = 1,000 grams mass of Earth (M_{Earth}) = 5.9724×10^{24} kg mass of Sun (M_{Sun}) = 1.9885×10^{30} kg
Time	seconds (s)	1 hour (h) = 60 minutes (min) = 3,600 s solar day (noon to noon) = 86,400 s sidereal day (Earth rotation period) = 86,164.1 s tropical year (equinox to equinox) = 365.24219 days = 3.15569×10^7 s sidereal year (Earth orbital period) = 365.25636 days = 3.15581×10^7 s
Speed	meters/second (m/s)	1 m/s = 2.237 miles/h 1 km/s = 1,000 m/s = 3,600 km/h $c = 2.99792 \times 10^8$ m/s = 299,792 km/s
Acceleration	meters/second^2 (m/s^2)	g = gravitational acceleration on Earth = 9.81 m/s^2
Energy	joules (J)	1 J = 1 kg m^2/s^2 1 megaton = 4.18×10^{15} J
Power	watts (W)	1 W = 1 J/s solar luminosity (L_{Sun}) = 3.828×10^{26} W
Force	newtons (N)	1 N = 1 kg m/s^2 1 pound (lb) = 4.448 N 1 N = 0.22481 lb
Pressure	newtons/meter^2 (N/m^2)	atmospheric pressure at sea level = 1.013×10^5 N/m^2 = 1.013 bar
Temperature	kelvins (K)	absolute zero = 0 K = −273.15°C = −459.67°F

SOURCE: Data from the Particle Data Group (http://pdg.lbl.gov).

APPENDIX 3:

Periodic Table of the Elements

Key	
1	Atomic number
H	Symbol
Hydrogen	Name
1.00794	Average atomic mass

Metals

Metalloids

Nonmetals

Period	1 1A	2 2A	3 3B	4 4B	5 5B	6 6B	7 7B	8 8B	9 8B	10 8B	11 1B	12 2B	13 3A	14 4A	15 5A	16 6A	17 7A	18 8A
1	1 **H** Hydrogen 1.00794																	2 **He** Helium 4.002602
2	3 **Li** Lithium 6.941	4 **Be** Beryllium 9.012182											5 **B** Boron 10.811	6 **C** Carbon 12.0107	7 **N** Nitrogen 14.0067	8 **O** Oxygen 15.9994	9 **F** Fluorine 18.9984032	10 **Ne** Neon 20.1797
3	11 **Na** Sodium 22.98976928	12 **Mg** Magnesium 24.3050											13 **Al** Aluminum 26.9815386	14 **Si** Silicon 28.0855	15 **P** Phosphorus 30.973762	16 **S** Sulfur 32.065	17 **Cl** Chlorine 35.453	18 **Ar** Argon 39.948
4	19 **K** Potassium 39.0983	20 **Ca** Calcium 40.078	21 **Sc** Scandium 44.955912	22 **Ti** Titanium 47.867	23 **V** Vanadium 50.9415	24 **Cr** Chromium 51.9961	25 **Mn** Manganese 54.938045	26 **Fe** Iron 55.845	27 **Co** Cobalt 58.933195	28 **Ni** Nickel 58.6934	29 **Cu** Copper 63.546	30 **Zn** Zinc 65.38	31 **Ga** Gallium 69.723	32 **Ge** Germanium 72.64	33 **As** Arsenic 74.92160	34 **Se** Selenium 78.96	35 **Br** Bromine 79.904	36 **Kr** Krypton 83.798
5	37 **Rb** Rubidium 85.4678	38 **Sr** Strontium 87.62	39 **Y** Yttrium 88.90585	40 **Zr** Zirconium 91.224	41 **Nb** Niobium 92.90638	42 **Mo** Molybdenum 95.96	43 **Tc** Technetium [98]	44 **Ru** Ruthenium 101.07	45 **Rh** Rhodium 102.90550	46 **Pd** Palladium 106.42	47 **Ag** Silver 107.8682	48 **Cd** Cadmium 112.411	49 **In** Indium 114.818	50 **Sn** Tin 118.710	51 **Sb** Antimony 121.760	52 **Te** Tellurium 127.60	53 **I** Iodine 126.90447	54 **Xe** Xenon 131.293
6	55 **Cs** Cesium 132.9054519	56 **Ba** Barium 137.327	57 **La** Lanthanum 138.90547	72 **Hf** Hafnium 178.49	73 **Ta** Tantalum 180.94788	74 **W** Tungsten 183.84	75 **Re** Rhenium 186.207	76 **Os** Osmium 190.23	77 **Ir** Iridium 192.217	78 **Pt** Platinum 195.084	79 **Au** Gold 196.966569	80 **Hg** Mercury 200.59	81 **Tl** Thallium 204.3833	82 **Pb** Lead 207.2	83 **Bi** Bismuth 208.98040	84 **Po** Polonium [209]	85 **At** Astatine [210]	86 **Rn** Radon [222]
7	87 **Fr** Francium [223]	88 **Ra** Radium [226]	89 **Ac** Actinium [227]	104 **Rf** Rutherfordium [261]	105 **Db** Dubnium [262]	106 **Sg** Seaborgium [266]	107 **Bh** Bohrium [264]	108 **Hs** Hassium [277]	109 **Mt** Meitnerium [268]	110 **Ds** Darmstadtium [271]	111 **Rg** Roentgenium [272]	112 **Cn** Copernicium [285]	113 **Uut** Ununtrium [284]	114 **Fl** Flerovium [289]	115 **Uup** Ununpentium [288]	116 **Lv** Livermorium [292]	117 **Uus** Ununseptium [294]	118 **Uuo** Ununoctium [294]

6 Lanthanides	58 **Ce** Cerium 140.116	59 **Pr** Praseodymium 140.90765	60 **Nd** Neodymium 144.242	61 **Pm** Promethium [145]	62 **Sm** Samarium 150.36	63 **Eu** Europium 151.964	64 **Gd** Gadolinium 157.25	65 **Tb** Terbium 158.92535	66 **Dy** Dysprosium 162.500	67 **Ho** Holmium 164.93032	68 **Er** Erbium 167.259	69 **Tm** Thulium 168.93421	70 **Yb** Ytterbium 173.05	71 **Lu** Lutetium 174.967
7 Actinides	90 **Th** Thorium 232.03806	91 **Pa** Protactinium 231.03588	92 **U** Uranium 238.02891	93 **Np** Neptunium [237]	94 **Pu** Plutonium [244]	95 **Am** Americium [243]	96 **Cm** Curium [247]	97 **Bk** Berkelium [247]	98 **Cf** Californium [251]	99 **Es** Einsteinium [252]	100 **Fm** Fermium [257]	101 **Md** Mendelevium [258]	102 **No** Nobelium [259]	103 **Lr** Lawrencium [262]

We have used the U.S. system as well as the system recommended by the International Union of Pure and Applied Chemistry (IUPAC) to label the groups in this periodic table. The system used in the United States includes a letter and a number (1A, 2A, 3B, 4B, etc.), which is close to the system developed by Mendeleev. The IUPAC system uses numbers 1–18 and has been recommended by the American Chemical Society (ACS). While we show both numbering systems here, we use the IUPAC system exclusively in the book. Elements with atomic numbers higher than 112 have been reported but not yet fully authenticated.

APPENDIX 4:
Properties of Planets, Dwarf Planets, and Moons

Physical Data for Planets and Dwarf Planets

Planet	EQUATORIAL RADIUS (km)	EQUATORIAL RADIUS (R/R_{Earth})	MASS (kg)	MASS (M/M_{Earth})	Average Density (relative to water*)	Rotation Period (days)	Tilt of Rotation Axis (degrees, relative to orbit)	Equatorial Surface Gravity (relative to Earth†)	Escape Velocity (km/s)	Average Surface Temperature (K)§
Mercury	2,440	0.383	3.30×10^{23}	0.055	5.427	58.65	0.03	0.378	4.3	340 (100, 700)
Venus	6,052	0.950	4.87×10^{24}	0.815	5.243	243.02‡	177.4	0.905	10.36	737
Earth	6,378	1.000	5.97×10^{24}	1.000	5.514	1.000	23.44	1.000	11.19	288 (185, 331)
Mars	3,396	0.532	6.42×10^{23}	0.107	3.934	1.026	25.19	0.377	5.03	210 (120, 293)
Ceres	473	0.074	9.39×10^{20}	0.0002	2.09	0.378	4	0.26	0.5	168
Jupiter	71,492	11.209	1.90×10^{27}	317.8	1.326	0.4135	3.13	2.530	59.5	165
Saturn	60,268	9.449	5.68×10^{26}	95.16	0.687	0.444	26.73	1.065	35.5	134
Uranus	25,559	4.007	8.68×10^{25}	14.54	1.271	0.7183‡	97.77	0.905	21.3	76
Neptune	24,764	3.883	1.02×10^{26}	17.15	1.638	0.6713	28.32	1.14	23.5	72
Pluto	1,187	0.186	1.30×10^{22}	0.0022	1.860	6.387‡	122.53	0.063	1.21	50
Haumea	~800	0.13	4.0×10^{21}	0.0007	1.9–2.6	0.163	?	0.045	0.81	<50
Makemake	715	0.11	4.0×10^{21}	0.0007	1.4–3.2	0.324	3	~ 0.5	~ 0.8	~ 30
Eris	1,163	0.182	1.67×10^{22}	0.0028	2.5	1.08	?	0.083	1.38	43

*The density of water is 1,000 kg/m^3.

†The surface gravity of Earth is 9.81 m/s^2.

‡Venus, Uranus, and Pluto rotate opposite to the directions of their orbits. Their north poles are south of their orbital planes.

§Values in parentheses give extremes of recorded temperatures.

Orbital Data for Planets and Dwarf Planets

	MEAN DISTANCE FROM SUN (A*)					
Planet	(10^6 km)	(AU)	Orbital Period (P) (sidereal years)	Eccentricity	Inclination (degrees, relative to ecliptic)	Average Speed (km/s)
Mercury	57.9	0.387	0.241	0.2056	7.0	47.36
Venus	108.2	0.723	0.615	0.0067	3.39	35.02
Earth	149.6	1.000	1.000	0.0167	0.000	29.78
Mars	227.9	1.524	1.881	0.0935	1.85	24.07
Ceres	414.0	2.768	4.603	0.079	10.59	17.90
Jupiter	778.6	5.204	11.862	0.0484	1.30	13.06
Saturn	1,433.53	9.582	29.457	0.0565	2.49	9.68
Uranus	2,872.5	19.20	84.011	0.0457	0.77	6.80
Neptune	4,495.1	30.047	164.79	0.0113	1.769	5.43
Pluto	5,906.4	39.48	247.94	0.2488	17.16	4.67
Haumea	6,432.0	43.2	284.12	0.198	28.19	4.53
Makemake	6,783.3	45.7	309.1	0.164	29.00	4.42
Eris	10,180	67.8	558	0.434	43.8	3.43

*A is the semimajor axis of the planet's elliptical orbit.

Properties of Selected Moons*

		ORBITAL PROPERTIES		PHYSICAL PROPERTIES		Relative Density
Planet	Moon	*P* (days)	*A* (10^3 km)	*R* (km)	*M* (10^{20} kg)	(g/cm^3) (water = 1.00)
Earth (1 moon)	Moon	27.32	384.4	1,738.1	735	3.34
Mars (2 moons)	Phobos	0.32	9.38	13.0 × 11.4 × 9.1	0.0001	1.9
	Deimos	1.26	23.46	7.8 × 6.0 × 5.1	0.00002	1.75
Jupiter (79 known moons)	Metis	0.30	128	30 × 20 × 17	0.001	3
	Amalthea	0.50	181.4	125 × 73 × 64	0.075	3.1
	Io	1.77	421.8	1,822	893	3.53
	Europa	3.55	671.1	1,561	480	3.01
	Ganymede	7.15	1,070	2,631	1,482	1.94
	Callisto	16.69	1,883	2,410	1,076	1.83
	Himalia	250.6	11,460	85	0.095	2.6
	Pasiphae	744†	23,620	18	0.0030	2.6
	Callirrhoe	759†	24,100	4.3	0.00001	2.6
Saturn (62 known moons)	Pan	0.58	133.58	17 × 16 × 10	0.00005	0.42
	Prometheus	0.61	139.35	68 × 40 × 30	0.0016	0.48
	Pandora	0.63	141.70	52 × 41 × 32	0.0014	0.49
	Mimas	0.94	185.52	198	0.38	1.15
	Enceladus	1.37	238.02	252	1.08	1.6
	Tethys	1.89	294.66	531	6.18	0.99
	Dione	2.73	377.40	561	11.0	1.48
	Rhea	4.52	527.04	763	23.1	1.24
	Titan	15.95	1,222	2,575	1,346	1.88
	Hyperion	21.28	1,481	180 × 133 × 103	0.056	0.55
	Iapetus	79.33	3,561	735	18.1	1.09
	Phoebe	548†	12,944	107	0.08	1.64
	Paaliaq	687.5	15,200	11	0.0007	2.3
Uranus (27 known moons)	Cordelia	0.34	49.8	20	0.0004	1.3
	Miranda	1.41	129.9	236	0.66	1.20
	Ariel	2.52	199.9	579	12.9	1.59
	Umbriel	4.14	266.0	585	12.2	1.46
	Titania	8.71	436.3	789	34.2	1.66
	Oberon	13.46	583.5	761	28.8	1.56
	Setebos	2,235	17,418	24	0.0008	1.5

(continued)

Properties of Selected Moons*

(continued)

Planet	Moon	ORBITAL PROPERTIES P (days)	A (10^3 km)	PHYSICAL PROPERTIES R (km)	M (10^{20} kg)	Relative Density (g/cm^3) (water = 1.00)
Neptune (14 known moons)	Naiad	0.29	48.2	48 × 30 × 26	0.002	1.3
	Larissa	0.55	73.5	108 × 102 × 84	0.05	1.3
	Proteus	1.12	117.6	220 × 208 × 202	0.5	1.3
	Triton	5.88†	354.8	1,353	214	2.05
	Nereid	360.14	5,513.4	170	0.3	1.5
Pluto (5 moons)	Charon	6.39	19.60	606	15.9	1.70
Haumea (2 moons)	Namaka	18	25.66	85	0.018	~1
	Hi'iaka	49	49.88	170	0.179	~1
Eris	Dysnomia	15.8	37.4	350	?	?

*Innermost, outermost, largest, and/or a few other moons for each planet.

†Irregular moon (has retrograde orbit).

APPENDIX 5:
Space Missions

Selected Recent and Current Solar System Missions

Spacecraft	Sponsoring Nation(s)*	Destination	Launch Year	Type	Status (mid-2018)
Voyager 1 and *2*	USA	Jupiter, Saturn, Uranus (2), Neptune (2)	1977	Flyby	Actively exploring outer edge of Solar System
Galileo	USA	Jupiter	1989	Orbiter/probe	Ended 2003
Ulysses	USA, Europe	Sun	1990	Solar polar orbiter	Ended 2008
SOHO	USA, Europe	Sun	1995	Orbiter	Active
Mars Global Surveyor	USA	Mars	1996	Orbiter	Ended 2006
Cassini-Huygens	USA, Europe, Italy	Saturn, Titan	1997	Saturn orbiter, Titan probe/lander	Ended 2017
Stardust	USA	Comets	1999	Sample return/flyby	Ended 2011
Mars Odyssey	USA	Mars	2001	Orbiter	Active
Mars Exploration Rover	USA	Mars	2003	Two landers	One rover active
Hayabusa	Japan	Asteroid	2003	Sample return	Ended 2010
Mars Express	Europe	Mars	2003	Orbiter	Active
Messenger	USA	Mercury (2011)	2004	Orbiter	Ended 2015
Rosetta	Europe	Comet 67P/Churyumov–Gerasimenko (2014)		Orbiter and lander	Ended 2016
Venus Express	Europe	Venus	2005	Orbiter	Ended 2014
Mars Reconnaissance Orbiter (MRO)	USA	Mars	2005	Orbiter	Active
Deep Impact/EPOXI	USA	Comet Hartley (2010)	2005	Impactor/flyby	Ended 2010
STEREO	USA	Sun	2006	Two orbiters	Active
New Horizons	USA	Pluto (2015)	2006	Flyby	Active
Chang'e 1	China	Moon	2007	Orbiter	Ended 2009
Kayuga	Japan	Moon	2007	Orbiter	Ended 2009
Artemis	USA	Moon, solar wind	2007	Orbiter	Active
Dawn	USA	Vesta (2011), Ceres (2015)	2007	Orbiter	Ended 2018
Chandrayaan	India	Moon	2008	Orbiter/impactor	Ended 2009
Lunar Reconnaissance Orbiter (LRO)	USA	Moon	2009	Orbiter	Active
Lunar Crater Observation and Sensing Satellite (LCROSS)	USA	Moon	2009	Impactor	Ended 2009

(continued)

Selected Recent and Current Solar System Missions

(continued)

Spacecraft	Sponsoring Nation(s)*	Destination	Launch Year	Type	Status (mid-2018)
Chang'e 2	China	Moon	2010	Orbiter	Ended 2011
Akatsuki	Japan	Venus (2015)	2010	Orbiter	Active
Juno	USA	Jupiter (2016)	2011	Orbiter	Active
Gravity Recovery and Interior Laboratory (GRAIL)	USA	Moon	2011	Two orbiters	Ended 2012
Mars Science Laboratory (Curiosity rover)	USA	Mars	2011	Lander	Active
Mars Atmosphere and Volatile EvolutioN (MAVEN) mission	USA	Mars	2013	Orbiter	Active
Chang'e 3	China	Moon	2013	Lander	On lunar surface
Mangalyaan	India	Mars	2013	Orbiter	Active
Hayabasu 2	Japan	Asteroid 162173 Ryugu	2014	Orbiter, scheduled sample return	Active
Osiris-REx (Origins, Spectral Interpretation, Resource Identification, Security-Regolith Explorer)	USA	Asteroid 101955 Bennu	2016	Orbiter, scheduled sample return	En route
ExoMars Trace Gas Orbiter (TGO)	Europe, Russia	Mars (2016)	2016	Orbiter (lander crashed)	Active
InSight	USA	Mars	2018	Lander	En route
Parker Solar Probe	USA	Sun	2018	Orbiter	En route

*Countries are represented by the following agencies: China = CNSA (China National Space Administration); Europe = ESA (European Space Agency); India = ISRO (Indian Space Research Organisation); Italy = Italian Space Agency; Japan = JAXA (Japan Aerospace Exploration Agency); Russia = ROSCOSMOS (State Corporation for Space Activities); USA = NASA (National Aeronautics and Space Administration).

APPENDIX 6:
Brightest Stars

The 25 Brightest Stars in the Sky

Name	Common Name	Distance (ly)	Spectral Type	Relative Visual Luminosity* (Sun = 1.000)	Apparent Visual Magnitude	Absolute Visual Magnitude
Sun	Sun	1.58×10^{-5}	G2V	1.000	−26.74	4.83
Alpha Canis Majoris	Sirius	8.60	A1V	22.9	−1.46	1.43
Alpha Carinae	Canopus	309	F0II	14,900	−0.72	−5.60
Alpha^1 Centauri	Rigil Kentaurus A	4.36	G2V	1.51	−0.01	4.38
Alpha^2 Centauri	Rigil Kentaurus B	4.36	K1V	0.44	1.33	5.71
Alpha Bootis	Arcturus	37	K1.5III	113	−0.04	−0.30
Alpha Lyrae	Vega	25	A0Va	49.2	0.03	0.60
Alpha Aurigae	Capella	43	G5IIIe+G0III	137	0.08	−0.51
Beta Orionis	Rigel	860	B8Iab	54,000	0.12	−7.0
Alpha Canis Minoris	Procyon	11.5	F5IV-V	7.73	0.34	2.61
Alpha Eridani	Achernar	140	B3Vpe	1,030	0.46	−2.70
Beta Centauri	Hadar	392	B1III	7,180	0.61	−4.81
Alpha Orionis	Betelgeuse	570	M2Iab	13,600	0.7	−5.5
Alpha Aquilae	Altair	16.7	A7V	11.1	0.77	2.22
Alpha Crucis	Acrux	325	B0.5IV+B1V	3,100	1.3	−3.9
Alpha Tauri	Aldebaran	67	K5III	163	0.85	−0.70
Alpha Scorpii	Antares	550	M1.5Ib	16,300	0.96	−5.7
Alpha Virginis	Spica	250	B1IV+B4V	1,920	1.04	−3.38
Beta Geminorum	Pollux	34	K0III	32.2	1.14	1.06
Alpha Piscis	Fomalhaut	25	A3V	17.4	1.16	1.73
Beta Crucis	Mimosa	280	B0.5III	1,980	1.25	−3.41
Alpha Cygni	Deneb	1,425	A2Ia	58,600	1.25	−7.09
Alpha Leonis	Regulus	79	B7V	146	1.35	−0.58
Epsilon Canis Majoris	Adhara	405	B2II	3,400	1.50	−4.0
Alpha Gemini	Castor	51	A1V+A5Vm	49	1.58	0.61
Gamma Crucis	Gacrux	88	M3.5III	138	1.63	−0.52

SOURCES: Data from Jim Kaler's *STARS* page (http://stars.astro.illinois.edu/sow/bright.html); SIMBAD Astronomical Database (http://simbad.u-strasbg.fr/simbad).

*Luminosity in this table refers only to radiation in "visual" light.

APPENDIX 7:
Observing the Sky

This appendix gives you enough information to make sense of a star chart or list of astronomical objects and find a few objects in the sky.

Celestial Coordinates

In Chapter 2, we discuss the **celestial sphere**—the imaginary sphere with Earth at its center on which celestial objects appear to lie. Several coordinate systems are used to specify the positions of objects on the celestial sphere. The simplest of these is the *altitude-azimuth coordinate system*. The altitude-azimuth coordinate system is based on the "map" direction to an object (the object's azimuth, with north = 0°, east = 90°, south = 180°, and west = 270°) combined with how high the object is above the horizon (the object's altitude, with the horizon at 0° and the zenith at 90°). For example, an object 10° above the eastern horizon has an altitude of 10° and an azimuth of 90°. An object 45° above the horizon in the southwest is at altitude 45°, azimuth 225°.

The altitude-azimuth coordinate system is the simplest way to tell someone where in the sky to look at the moment, but it is not a good coordinate system for cataloging the positions of objects. The altitude and azimuth of an object are different for each observer, depending on the observer's position on Earth, and they are constantly changing as Earth rotates on its axis. To specify the direction to an object in a way that is the same for everyone requires a coordinate system that is fixed relative to the celestial sphere. The most common such coordinates are called *celestial coordinates.*

Celestial coordinates are illustrated in **Figure A7.1**. Celestial coordinates are much like the traditional system of latitude and longitude used on the surface of Earth. On Earth, latitude specifies how far you are from Earth's equator, as discussed in Chapter 2. If you are on Earth's equator, your latitude is 0°. If you are at Earth's North Pole, your latitude is 90° north. If you are at Earth's South Pole, your latitude is 90° south.

The latitude-like coordinate on the celestial sphere is called **declination**, often signified with the lowercase Greek letter δ (delta). The celestial equator has $\delta = 0°$. The north celestial pole has $\delta = +90°$. The south celestial pole has $\delta = -90°$. (See Chapter 2 if you need to refresh your memory about the celestial equator or celestial poles.) Declination is usually expressed in degrees, minutes of arc, and seconds of arc. For example, Sirius, the brightest star in the sky, has $\delta = -16°42'58''$ meaning that it is located not quite 17° south of the celestial equator.

On Earth, east–west position is specified by longitude. Lines of constant longitude run north–south from one pole to the other. Unlike latitude, for which the equator provides a natural place to call "zero," longitude has no natural starting point, so one was invented. By arbitrary convention, the Royal Observatory in Greenwich, England, is defined to lie at a longitude of 0°. On the celestial sphere, the longitude-like coordinate is called **right ascension**, often signified with the lowercase Greek letter α (alpha). Unlike longitude, right ascension *does* have a natural starting point on the celestial sphere: the vernal equinox, or the point at which the ecliptic crosses the celestial equator with the Sun moving from the southern sky into the northern sky. The (Northern Hemisphere) vernal equinox defines the line of right ascension at which $\delta = 0°$. The (Northern Hemisphere) autumnal equinox, located on the opposite side of the sky, is at $\delta = 180°$.

Normally, right ascension is measured in units of time rather than degrees. Earth takes 24 hours (of sidereal time) to rotate on its axis, so the celestial sphere is divided into 24 hours of right ascension, with each hour of right ascension corresponding to 15°. Hours of right ascension are then subdivided into minutes and seconds of time. Right ascension increases going to the east. The right ascension of Sirius, for example, is $\alpha = 06^h45^m08.9^s$, meaning that Sirius is about 101° (that is, 06^h45^m) east of the vernal equinox. Time is a natural unit for measuring right ascension because time naturally tracks the motion of objects due to Earth's rotation on its axis. If stars on the meridian at a certain time have $\alpha = 06^h$, then an hour later the stars on the meridian will have $\alpha = 07^h$, and an hour after that they will have $\alpha = 08^h$. The *local sidereal time*, or *star time*, at your location right now is equal to the right ascension of the stars on your meridian at the moment. Because of Earth's motion around the Sun, a sidereal day is about 4 minutes shorter than a solar day, so local sidereal time constantly gains on solar time. At midnight on September 22, the local sidereal time is 0^h. By midnight on December 21, local sidereal time has advanced to 06^h. On March 20, local sidereal time at midnight is 12^h. And at midnight on June 20, local sidereal time is 18^h.

Putting all that together, right ascension and declination provide a convenient way to specify the location of any object on the celestial sphere. Sirius is located at $\alpha = 06^h45^m08.9^s$, $\delta = -16°42'58''$ which means that at midnight on December 21

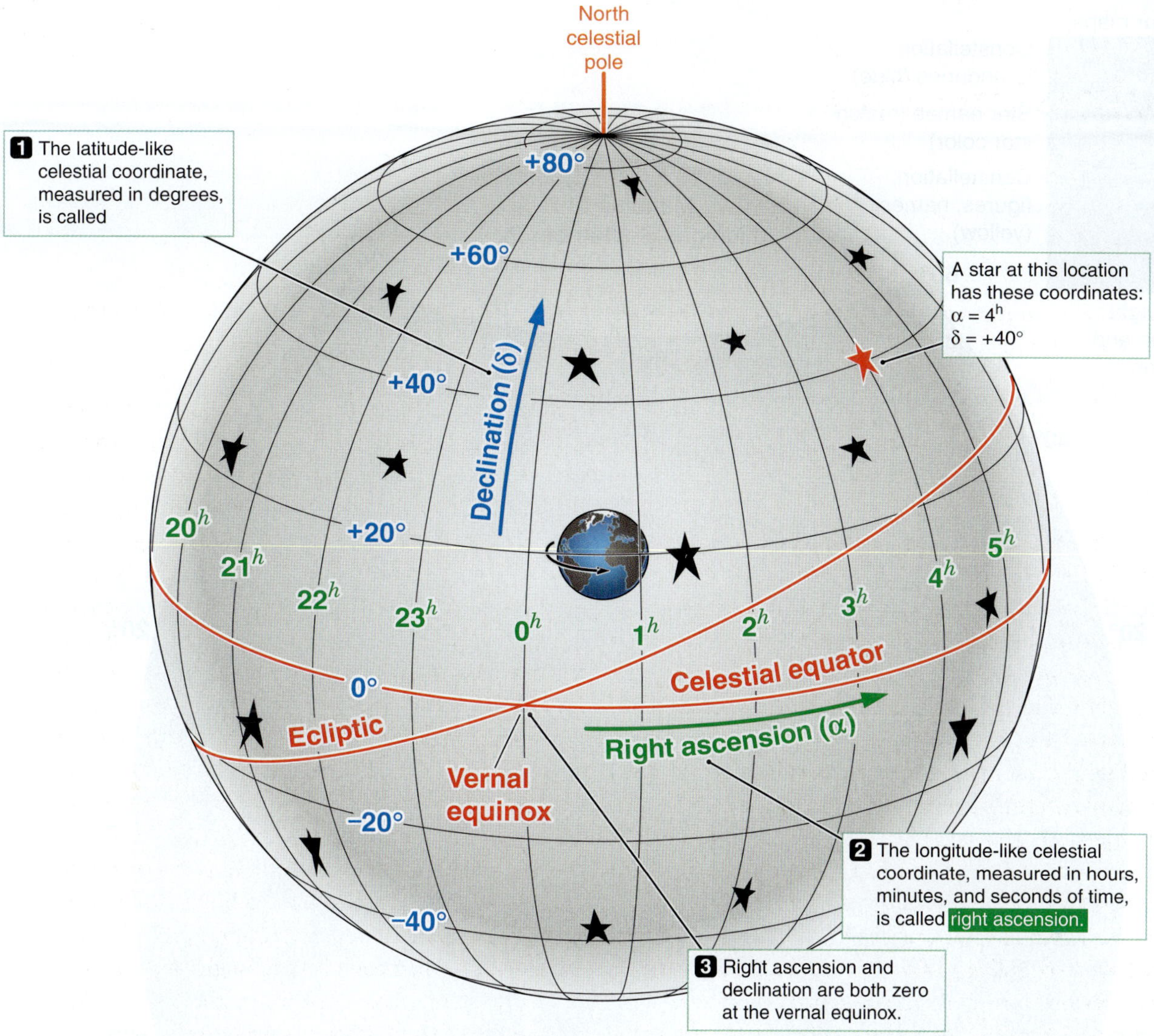

Figure A7.1 Celestial coordinates.

(local sidereal time = 06^h), you will find Sirius about 45^m east of the meridian, not quite 17° south of the celestial equator.

Just one final caveat remains. As we discussed in Chapter 2, the directions of the celestial equator, celestial poles, and vernal equinox are constantly changing as Earth's axis wobbles like the axis of a spinning top. In Chapter 2, we called that 26,000-year wobble the **precession of the equinoxes**, meaning that the location of the equinoxes is slowly advancing along the ecliptic. So when we specify the celestial coordinates of an object, we need to specify the date at which the positions of the vernal equinox and celestial poles were measured. By convention, coordinates are usually referred to with the position of the vernal equinox on January 1, 2000. A complete, formal specification of the coordinates of Sirius would then be $\alpha(2000) = 06^h45^m08.9^s$, $\delta(2000) = -16°42'58''$, where the "2000" in parentheses refers to the equinox of the coordinates.

Constellations and Names

Although it is certainly possible to specify any location on the surface of Earth exactly by giving its latitude and longitude, using a more descriptive address is usually more convenient. We might say, for example, that one of the coauthors of this book works near latitude 37° north, longitude 122° west; but it would probably mean a lot more to you if we said that George Blumenthal works in Santa Cruz, California.

Just as the surface of Earth is divided into nations and states, the celestial sphere is divided into 88 **constellations**, the names of which are often used to refer to objects within their boundaries (see the star charts in **Figure A7.2**). The brightest stars within the boundaries of a constellation are named using a Greek letter combined with the name of the constellation. For example, the star Sirius is the brightest star in the constellation Canis Major

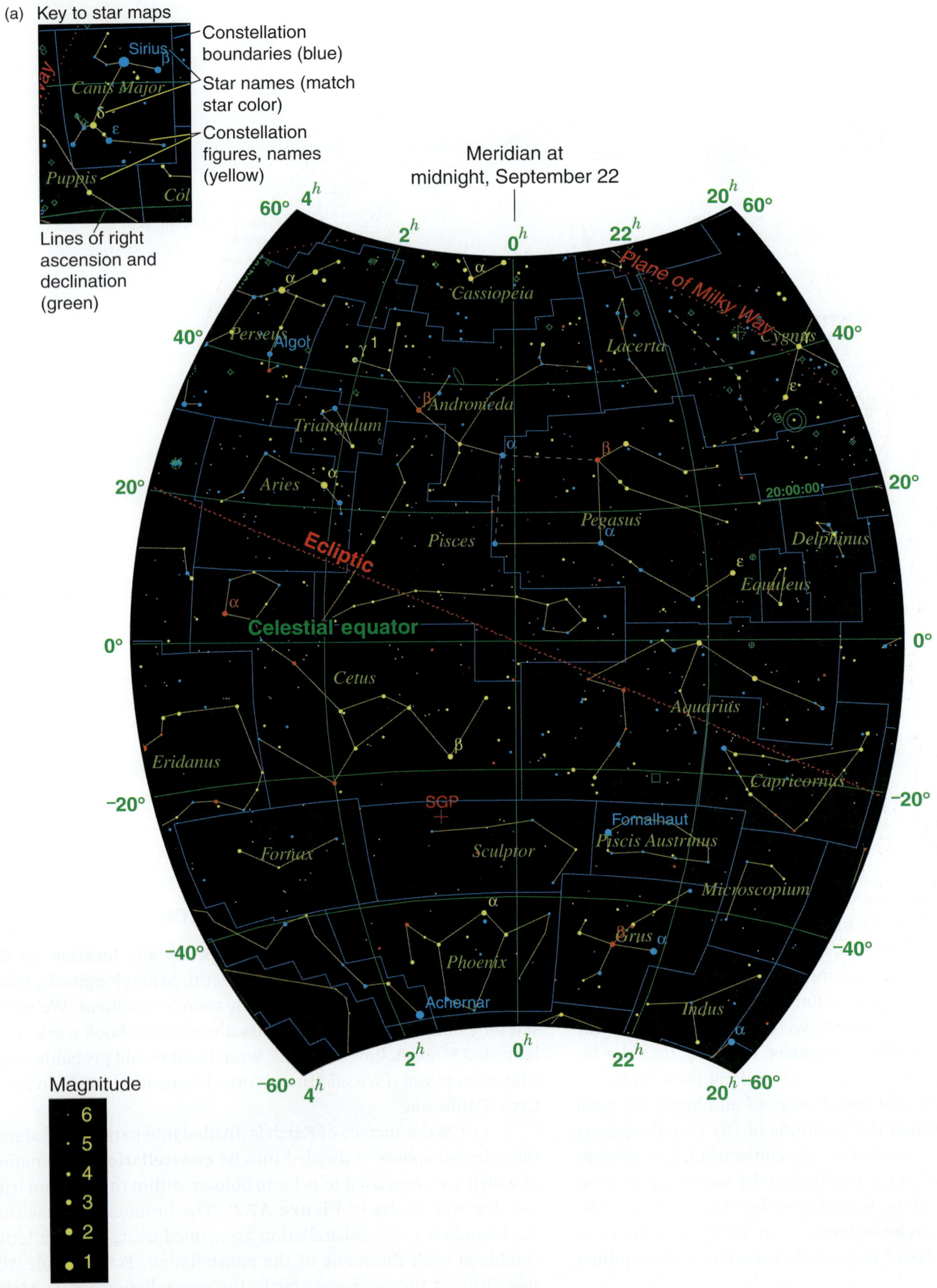

Figure A7.2A The sky from right ascension 20h to 04h and declination −60° to +60°.

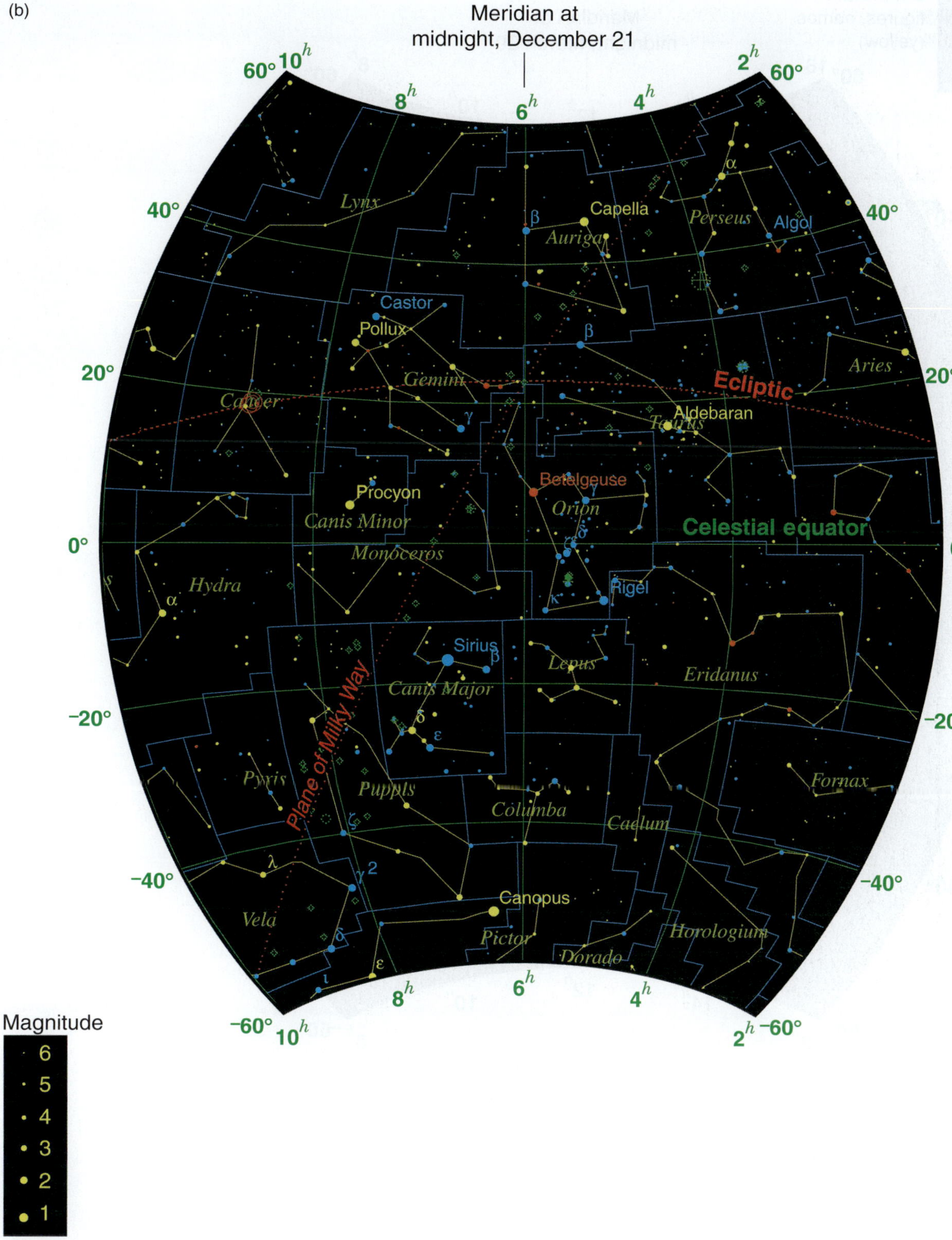

Figure A7.2B The sky from right ascension 02h to 10h and declination −60° to +60°.

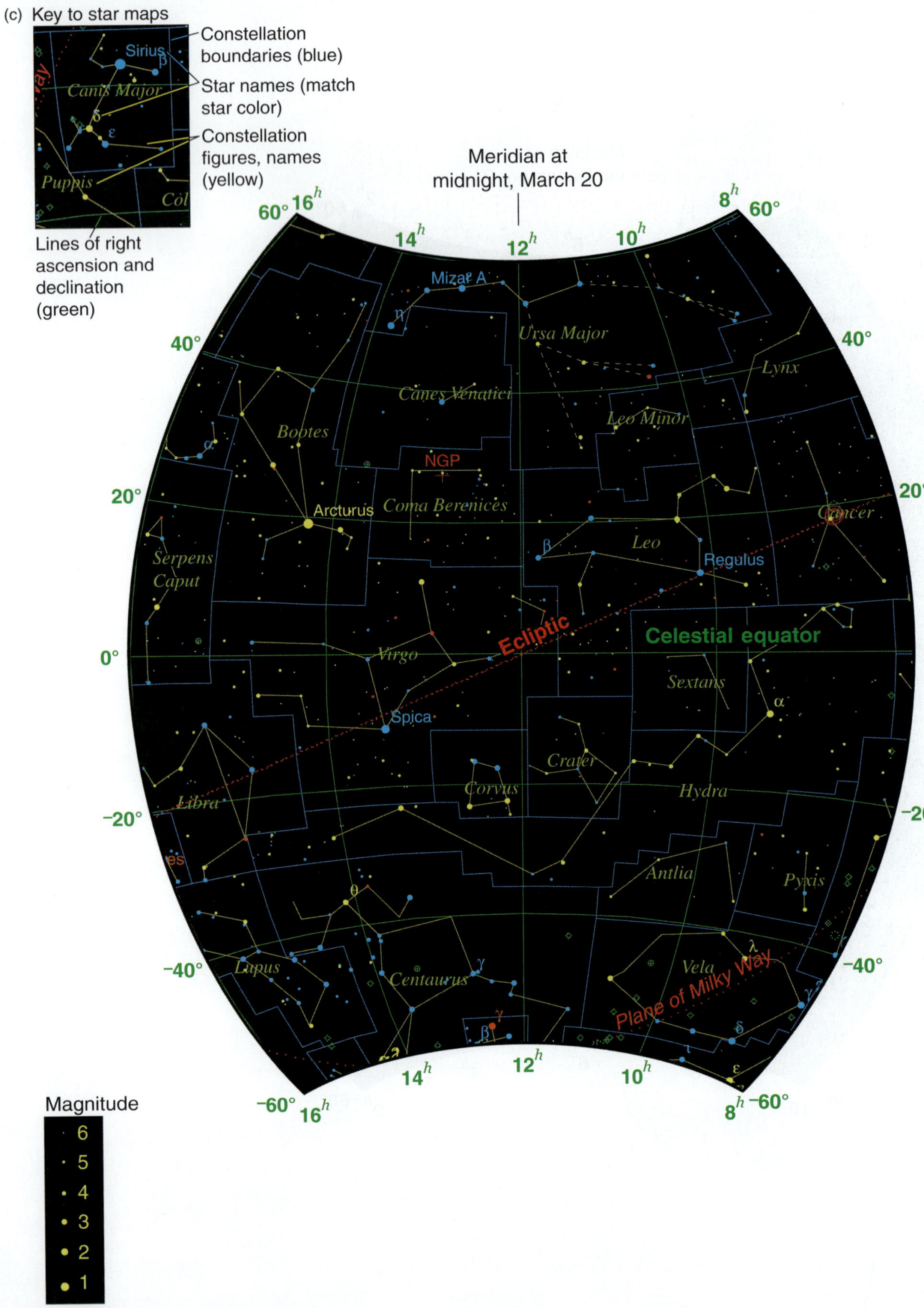

Figure A7.2C The sky from right ascension 08^h to 16^h and declination −60° to +60°.

(d)

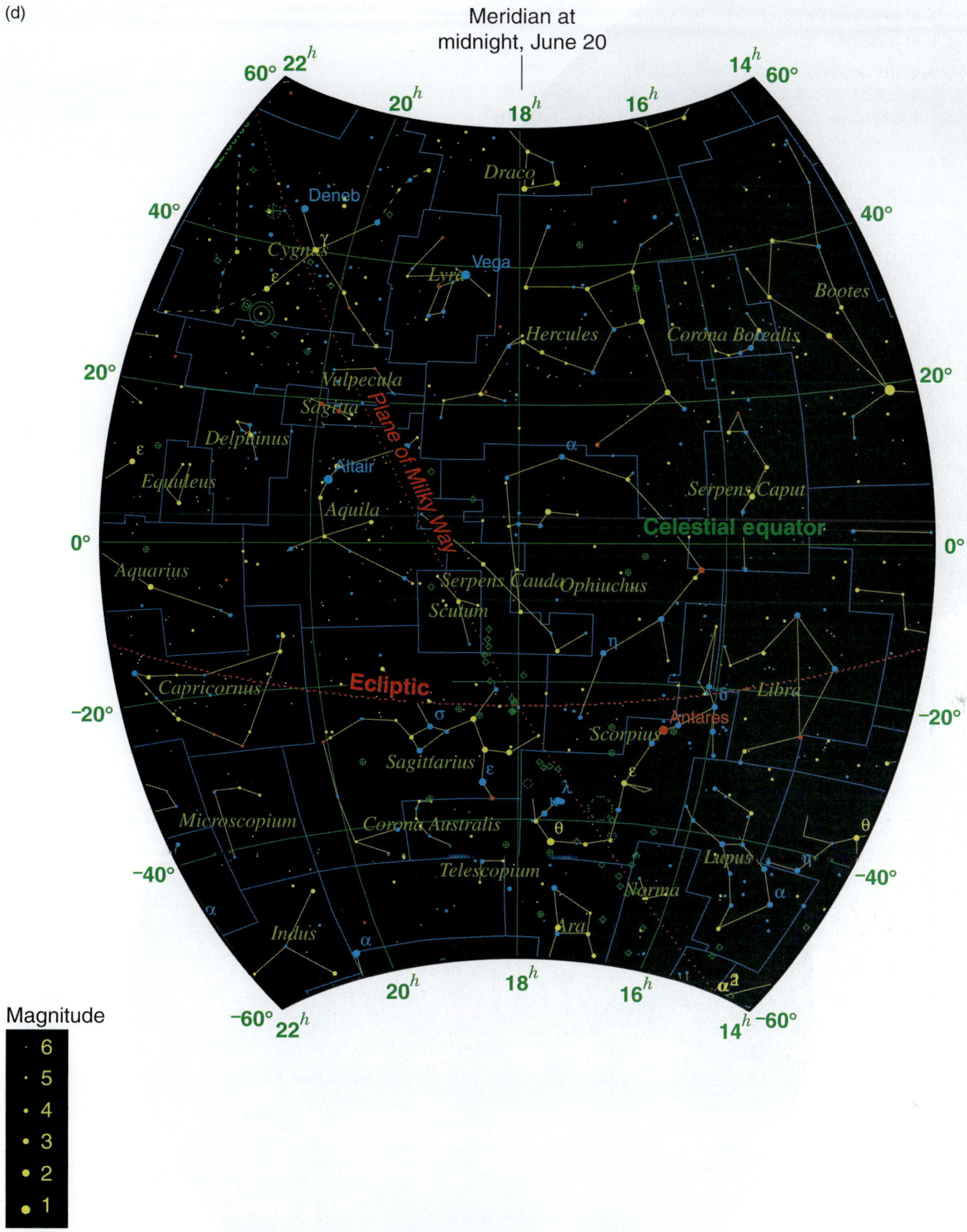

Figure A7.2D The sky from right ascension 14^h to 22^h and declination −60° to +60°.

(e)

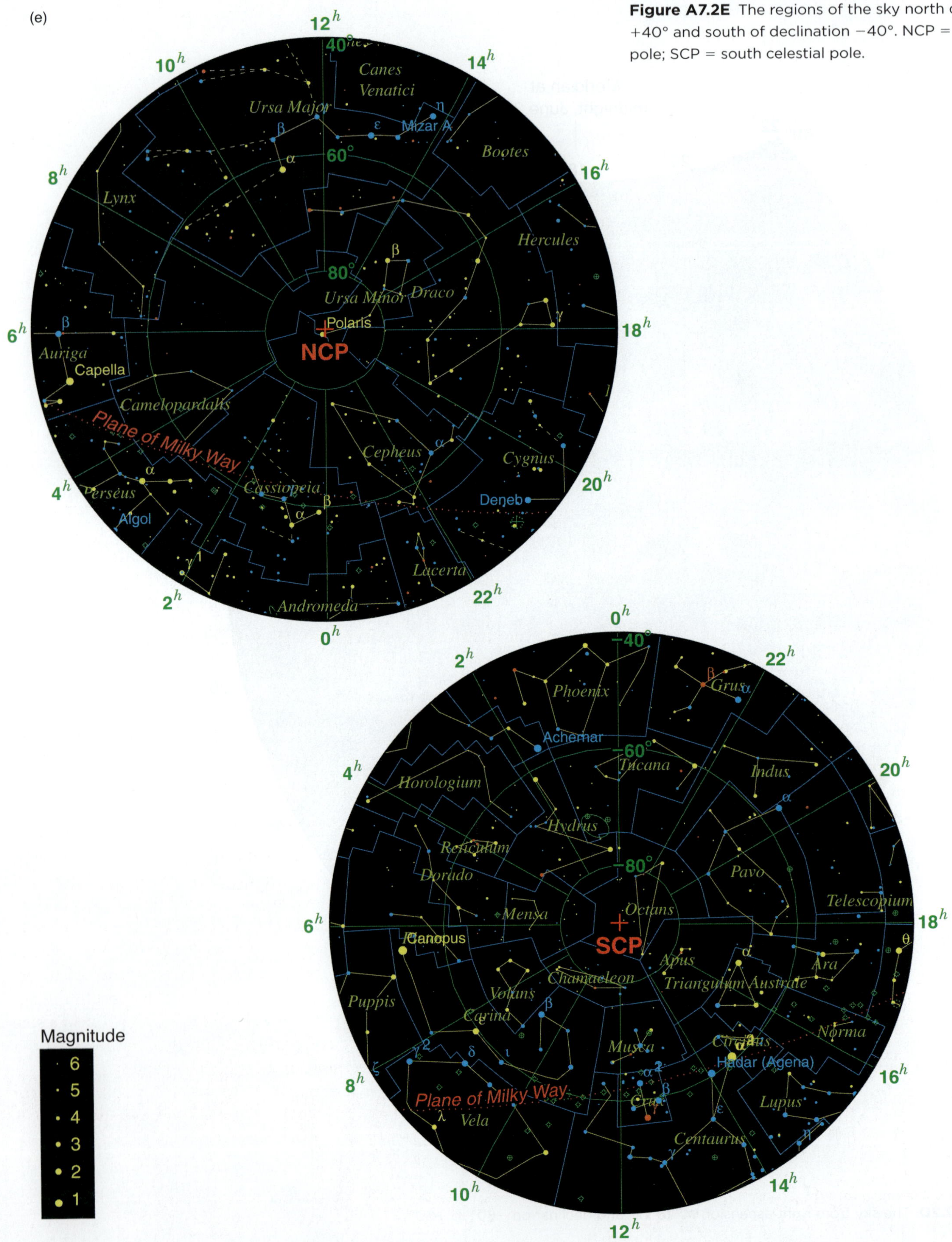

Figure A7.2E The regions of the sky north of declination +40° and south of declination −40°. NCP = north celestial pole; SCP = south celestial pole.

(literally, the "great dog"), so it is called "Alpha Canis Majoris." The bright red star in the northeastern corner of the constellation Orion is called "Alpha Orionis," also known as Betelgeuse. Rigel, the bright blue star in the southwest corner of Orion, is also called "Beta Orionis."

Astronomical objects can take on a bewildering range of names. For example, the bright southern star Canopus, also known as "Alpha Carinae" (the brightest star in the constellation Carina), has no fewer than 34 names, most of which are about as memorable as "SAO 234480" (number 234,480 in the Smithsonian Astrophysical Observatory catalog of stars).

A constellation is spelled slightly differently when it becomes part of a star's name. For example, Sirius is called "Alpha Canis Majoris," not "Alpha Canis Major"; Rigel is referred to as "Beta Orionis," not "Beta Orion"; and Canopus becomes "Alpha Carinae," not "Alpha Carina." That is because the Latin genitive, or possessive, case is used with star names; for example, *Orionis* means "of Orion."

Astronomical Magnitudes

Apparent Magnitudes

We first introduced magnitudes in Working It Out 13.2; here, we provide more information. You are most likely to see that system if you take a lab course in astronomy or if you use a star catalog. Astronomers use the logarithmic system of **apparent magnitudes** to compare the apparent brightness of objects in the sky. Other common systems of logarithmic measurements that you may have encountered include decibels for measuring sound levels and the Richter scale for measuring the strength of earthquakes. For example, an earthquake of magnitude 6 is not just a little stronger than an earthquake of magnitude 5; it is, in fact, 10 times stronger.

As discussed in Working It Out 13.2, a difference of five magnitudes between the apparent brightness of two stars (say, a star with $m = 6$ and a star with $m = 1$) corresponds to 100 times difference in brightness, and *the greater the magnitude, the fainter the object*. If five steps in magnitude corresponds to a factor of 100 in brightness, one step in magnitude must correspond to the fifth root of 100; that is, a factor of $100^{1/5}$ = approximately 2.512 in brightness ($100^{1/5} \times 100^{1/5} \times 100^{1/5} \times 100^{1/5} \times 100^{1/5} = 100$).

If star 1 has a brightness of b_1 and star 2 has a brightness of b_2, the ratio of the brightness of the stars is given by

$$\frac{b_1}{b_2} = (2.512)^{m_2 - m_1} = 100^{\frac{(m_2 - m_1)}{5}}$$

We can put that into the more common base 10 by noting that $100 = 10^2$, so the expression becomes

$$\frac{b_1}{b_2} = 10^{2 \times \frac{(m_2 - m_1)}{5}} = 10^{0.4(m_2 - m_1)}$$

After we take the log of both sides and divide by 0.4, the difference in magnitude $(m_2 - m_1)$ between the two stars is given by

$$m_2 - m_1 = 2.5 \log_{10}\frac{b_1}{b_2}$$

The following table shows some examples using the preceding equations.

Apparent Magnitude Difference $(m_2 - m_1)$	Ratio of Apparent Brightness (b_1/b_2)
1	2.512
2	$2.512^2 = 6.3$
3	$2.512^3 = 15.8$
4	$2.512^4 = 39.8$
5	$2.512^5 = 100$
10	$2.512^{10} = 100^2 = 10{,}000$
15	$2.512^{15} = 100^3 = 1{,}000{,}000$
20	$2.512^{20} = 100^4 = 10^8$
25	$2.512^{25} = 100^5 = 10^{10}$

Absolute Magnitudes

Recall that stars differ in their brightness for two reasons: the amount of light they are actually emitting and their distance from Earth. The magnitude system also is used for **luminosity**, with the same scale as for brightness: a difference of five magnitudes corresponds to 100 times difference in luminosity. Astronomers call those **absolute magnitudes** (M_{abs}), and the idea is to imagine how bright the star would be if it were at a distance of 10 parsecs (pc). Absolute magnitudes enable comparison of how luminous two stars really are, without the factor of distance. The Sun is very bright because it is so close (apparent visual magnitude = −27), but if the Sun were at a distance of 10 pc, its magnitude would be only about 5.[1] Thus, the absolute magnitude of the Sun is $M_{abs} = 5$. Recall that the luminosity of a star is usually expressed by comparing it with the luminosity of the Sun. As with apparent magnitudes, higher magnitude numbers correspond to lower luminosity. Thus, a star 1/100 as luminous as the Sun will be 5 absolute magnitudes fainter, or $M_{abs} = 10$. A

[1]The apparent and absolute magnitudes of the Sun are −26.74 and +4.83, respectively. We use +5 for the Sun's absolute magnitude as an approximation.

star 10,000 times more luminous than the Sun will be 10 absolute magnitudes brighter, or $M_{abs} = -5$.

Absolute magnitudes and luminosities follow the same equations as those we provided already, using L instead of b and M_{abs} instead of m:

$$\frac{L_1}{L_2} = 10^{2 \times \frac{M_{abs(2)} - M_{abs(1)}}{5}} = 10^{0.4(M_{abs(2)} - M_{abs(1)})}$$

and

$$M_{abs(2)} - M_{abs(1)} = 2.5 \log_{10}\frac{L_1}{L_2}$$

Most often, astronomers think about the luminosity of a star in comparison with the luminosity of the Sun. Here $L_1 = L_{star}$ and $L_2 = L_{Sun}$. The following table compares luminosity (where $L_{Sun} = 1$) with absolute magnitude of a star.

L_{Star}/L_{Sun}	M_{abs}
1,000,000	−10
10,000	−5
100	0
1	5
1/100	10
1/10,000	15

Distance Modulus

The difference between the apparent magnitude and the absolute magnitude depends on the star's distance. By definition, a star at a distance of exactly 10 pc will have an apparent magnitude equal to its absolute magnitude. Astronomers can always measure the brightness of a star and thus its apparent magnitude and can estimate the luminosity of a star and thus its absolute magnitude by using the Hertzsprung-Russell (H-R) diagram. That is how the distances to most stars are found.

Using the preceding equations and the definition of absolute magnitude, we can get to the following relatively simple expression:

$$m - M_{abs} = 5 \log_{10} d - 5$$

where distance d is in parsecs.

We can rewrite that equation to solve for distance as follows:

$$d = 10^{\frac{(m - M_{abs} + 5)}{5}}$$

The following table shows how the difference between an object's apparent and absolute magnitudes leads to its distance in parsecs.

$m - M_{abs}$	Distance (pc)
−3	2.5
−2	4.0
−1	6.3
0	10
1	16
2	25
3	40
4	63
5	100
10	1,000
15	10,000
20	100,000

Although the system of astronomical magnitudes is convenient in many ways—which is why astronomers continue to use it—it can also be confusing to new students. Just remember three things and you will probably get by:

1. The greater the magnitude, the fainter the object.
2. One magnitude *smaller* means about two and a half times *brighter*.
3. The brightest stars in the sky have magnitudes of less than 1, and the faintest stars visible to the naked eye on a dark night have magnitudes of about 6.

A final note: Astronomers sometimes use "colors" based on the ratio of the brightness of a star as seen in two parts of the spectrum. The "b_B/b_V color," for example, is the ratio of the brightness of a star seen through a blue filter, divided by the brightness of a star seen through a yellow-green (visual) filter. Normally, astronomers instead discuss the "$B - V$ color" of a star, which is equal to the difference between a star's blue magnitude and its visual magnitude. We can use the previous expression for a magnitude difference to write

$$B - V \text{ color} = m_B - m_V = -2.5 \log_{10} (b_B/b_V)$$

Thus, a star with a b_B/b_V color of 1.0 has a $B - V$ color of 0.0, and a star with a b_B/b_V color of 1.4 has a $B - V$ color of −0.37. Notice that, as with magnitudes, $B - V$ colors are "backward": the bluer a star, the greater its b_B/b_V color but the less its $B - V$ color.

APPENDIX 8:
Uniform Circular Motion and Circular Orbits

Uniform Circular Motion

In Chapter 4 (see Section 4.2 and Figure 4.7), we discuss the motion of an object moving in a circle at a constant speed. That motion, called **uniform circular motion**, is the result of the fact that centripetal force always acts toward the center of the circle. The key question when thinking about uniform circular motion is, How hard does something have to pull to keep the object moving in a circle? Part of the answer to that question is obvious: the more massive an object is, the harder it will be to keep it moving on its circular path. According to Newton's second law of motion, $F = ma$, or here, the centripetal force equals the mass multiplied by the centripetal acceleration. The larger the mass, the greater the force required to keep it moving in its circle.

The centripetal force needed to keep an object moving in constant circular motion also depends on two other quantities: the speed of the object and the size of the circle. The faster an object is moving, the more rapidly it has to change direction to stay on a circle of a given size. The second quantity that influences the needed acceleration is the radius of the circle. The smaller the circle, the greater the pull needed to keep it on track. You can understand that relationship by looking at the motion. A small circle requires a continuous "hard" turn, whereas a larger circle requires a more gentle change in direction. It takes more force to keep an object moving faster in a smaller circle than it does to keep the same object moving more slowly in a larger circle. (To get a better feel for how that works, think about the difference between riding in a car taking a tight curve at high speed and a car moving slowly around a gentle curve.)

To arrive at the circular velocity and other results discussed in Chapter 4, those intuitive ideas about uniform circular motion are turned into a quantitative expression of exactly how much centripetal acceleration is needed to keep an object moving in a circle with radius r at speed v. **Figure A8.1** shows a ball moving around a circle of radius r at a constant speed v at two times. The centripetal acceleration keeping the ball on the circle is a. Remember that the acceleration is always directed toward the center of the circle, whereas the velocity of the ball is always perpendicular to the acceleration. The ball's velocity and its acceleration are always at right angles to each other. As the object moves around the circle, the direction of motion and the direction of the acceleration change together in lockstep.

Figure A8.1 contains two triangles. Triangle 1 shows the velocity (speed and direction) at both times. The arrow labeled "Δv" connecting the heads of the two velocity arrows shows how much the velocity changed between time 1 (t_1) and time 2 (t_2). That change is the effect of the centripetal acceleration. If you imagine that points 1 and 2 are very close together—so close that the direction of the centripetal acceleration does not change by much between the two—the centripetal acceleration equals the change in the velocity divided by the time between the two, $\Delta t = t_2 - t_1$. So, $\Delta v = a\Delta t$.

Triangle 2 shows something similar. Here, the arrow labeled "Δr" indicates the change in the position of the ball between time 1 and time 2. Again, if you imagine that the time between the two points is very short, Δr is equal to the velocity multiplied by the time, or $\Delta r = v\Delta t$.

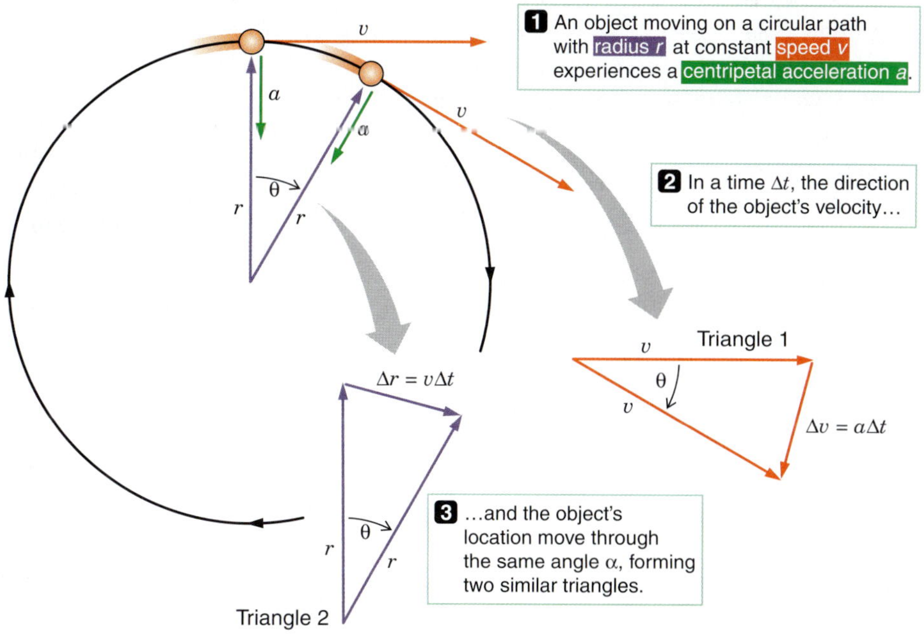

Figure A8.1 Similar triangles are used to find the centripetal force needed to keep an object moving at a constant speed on a circular path.

The line between the center of the circle and the ball is always perpendicular to the velocity of the ball. So if the direction of the ball's velocity changes by an angle θ, the direction of the line between the ball and the center of the circle must also change by the same angle α. In other words, triangles 1 and 2 are "similar triangles." They have the same *shape*. If the triangles are the same shape, the ratio of two sides of triangle 1 must equal the ratio of the two corresponding sides of triangle 2. Then,

$$\frac{a\Delta t}{v} = \frac{v\Delta t}{r}$$

If we divide by Δt on both sides of the equation and then cross-multiply, we obtain

$$ar = v^2$$

which, after we divide both sides of the equation by r, becomes

$$a_{\text{centripetal}} = \frac{v^2}{r}$$

The subscript "centripetal" is added to a to signify that this is the centripetal acceleration needed to keep the object moving in a circle of radius r at speed v. The centripetal force required to keep an object of mass m moving on such a circle is then

$$F_{\text{centripetal}} = ma_{\text{centripetal}} = \frac{mv^2}{r}$$

Circular Orbits

For an object moving in a circular orbit, no string is available to hold the ball on its circular path. Instead, that force is provided by **gravity**.

Think about an object with mass m in orbit about a much larger object with mass M. The orbit is circular, and the distance between the two objects is given by r. The force needed to keep the smaller object moving at speed v in a circle with radius r is given by the previous expression for $F_{\text{centripetal}}$. The force actually provided by gravity (see Chapter 4) is

$$F_{\text{grav}} = G\frac{Mm}{r^2}$$

If gravity is responsible for holding the mass in its circular motion, it should be true that $F_{\text{grav}} = F_{\text{centripetal}}$. That is, if mass m is moving in a circle under the force of gravity, the force provided by gravity must equal the centripetal force needed to explain that circular motion. Setting the two expressions for $F_{\text{centripetal}}$ and F_{grav} equal to each other gives

$$\frac{mv^2}{r} = G\frac{Mm}{r^2}$$

All that remains is a bit of algebra. Dividing by m on both sides of the equation and multiplying both sides by r gives

$$v^2 = G\frac{M}{r}$$

Taking the square root of both sides then yields the desired result:

$$v_{\text{circular}} = \sqrt{\frac{GM}{r}}$$

That is the **circular velocity** we presented in Chapter 4. It is the velocity at which an object in a circular orbit must be moving. If the object were not moving at that velocity, gravity would not be providing the needed centripetal force, and the object would not move in a circle.

Glossary

A

aberration of starlight A star's apparent displacement in position due to the finite speed of light and Earth's orbital motion around the Sun. (Ch. 18)

absolute magnitude A measure of the intrinsic brightness, or luminosity, of a celestial object, generally a star. Specifically, the apparent magnitude an object would have if it were located at a standard distance of 10 parsecs (pc). Compare *apparent magnitude*. (Ch. 13)

absolute zero The temperature at which thermal motions cease. The lowest possible temperature. Zero on the Kelvin temperature scale. (Ch. 5)

absorption The process by which an atom captures energy from a passing photon. Compare *emission*. (Ch. 5)

absorption line A minimum in the intensity of a spectrum that is due to the absorption of electromagnetic radiation at a specific wavelength determined by the energy levels of an atom or molecule. Compare *emission line*. (Ch. 5)

acceleration The rate at which the speed and/or direction of an object's motion is changing. (Ch. 3)

accretion disk A flat, rotating disk of gas and dust surrounding an object, such as a young stellar object, a forming planet, a collapsed star in a binary system, or a black hole. (Ch. 7)

achondrite A stony meteorite that does not contain chondrules. Compare *chondrite*. (Ch. 12)

active comet A comet nucleus that approaches close enough to the Sun to show signs of activity, such as the production of a coma and tail. (Ch. 12)

active galactic nucleus (AGN) A highly luminous, compact galactic nucleus whose luminosity may exceed that of the rest of the galaxy. (Ch. 19)

active region An area of the Sun's chromosphere anchoring bursts of intense magnetic activity. (Ch. 14)

adaptive optics Electro-optical systems that largely compensate for image distortion caused by Earth's atmosphere. (Ch. 6)

AGB See *asymptotic giant branch*. (Ch. 16)

AGN See *active galactic nucleus*. (Ch. 19)

albedo The fraction of electromagnetic radiation striking a surface that is reflected by that surface. (Ch. 5)

algebra A branch of mathematics in which letters represent numeric variables. (Appendix 1)

alpha particle A ^{4}He nucleus, consisting of two protons and two neutrons. Alpha particles are given off in the type of radioactive decay referred to as *alpha decay*. (Ch. 16)

altitude The location of an object above the horizon, measured by the angle formed between an imaginary line from an observer to the object and a second line from the observer to the point on the horizon directly below the object. (Ch. 2)

Amors A group of asteroids whose orbits cross the orbit of Mars but not the orbit of Earth. Compare *Apollos* and *Atens*. (Ch. 12)

amplitude In a wave, the maximum deviation from its undisturbed or relaxed position. For example, in a water wave the amplitude is the vertical distance from the wave's crest to the undisturbed water level. (Ch. 5)

angular momentum In a rotating or revolving system, a conserved property whose value depends on the velocity and distribution of the system's mass. (Ch. 7)

angular resolution The ability of an imaging device such as a telescope (or the eye) to separate two objects that appear close together. (Ch. 6)

annular solar eclipse The type of solar eclipse that occurs when the apparent diameter of the Moon is less than that of the Sun, leaving a visible ring of light ("annulus") surrounding the dark disk of the Moon. Compare *partial solar eclipse* and *total solar eclipse*. (Ch. 2)

Antarctic Circle The circle on Earth with latitude 66.5° south, marking the northern limit where at least one day per year is in 24-hour daylight. Compare *Arctic Circle*. (Ch. 2)

anthropic principle The idea that this universe (or this bubble in the universe) must have physical properties that allow intelligent life to develop. (Ch. 22)

anthropogenic climate change The release of greenhouse gases into the atmosphere from human activities, such as the burning of fossil fuels. (Ch. 9)

anticyclonic motion The rotation of a weather system resulting from the Coriolis effect as air moves outward from a region of high atmospheric pressure. Compare *cyclonic motion*. (Ch. 9)

antimatter Matter made up of antiparticles. (Ch. 14, 22)

antiparticle An elementary particle of antimatter identical in mass but opposite in charge and all other properties to its corresponding ordinary matter particle. (Ch. 22)

aperture The clear diameter of a telescope's objective lens or primary mirror. (Ch. 6)

aphelion (pl. aphelia) The point in a solar orbit that is farthest from the Sun. Compare *perihelion*. (Ch. 12)

Apollos A group of asteroids whose orbits cross the orbits of both Earth and Mars. Compare *Amors* and *Atens*. (Ch. 12)

apparent daily motion As seen from Earth's surface, the path along which each object seems to move across the sky. (Ch. 2)

apparent magnitude A measure of the apparent brightness of a celestial object, generally a star. Compare *absolute magnitude*. (Ch. 13)

arcminute (arcmin) A minute of arc (′), a unit used to measure angles. An arcminute is 1/60 of a degree of arc. (Ch. 6, 13)

arcsecond (arcsec) A second of arc (″), a unit used to measure very small angles. An arcsecond is 1/60 of an arcminute, or 1/3,600 of a degree of arc. (Ch. 6, 13)

Arctic Circle The circle on Earth with latitude 66.5° north, marking the southern limit where at least one day per year is in 24-hour daylight. Compare *Antarctic Circle*. (Ch. 2)

asteroid Also called *minor planet*. A primitive rocky or metallic body (planetesimal) that has survived planetary accretion. Asteroids are parent bodies of meteoroids. (Ch. 12)

asteroid belt The region between the orbits of Mars and Jupiter that contains most of the asteroids in our Solar System. (Ch. 12)

astrobiology An interdisciplinary science combining astronomy, biology, chemistry, geology, and physics to study life in the cosmos. (Ch. 1, 24)

astrology The belief that the positions and aspects of stars and planets influence human affairs and characteristics, as well as terrestrial events. (Ch. 1)

astronomical seeing A measurement of the degree to which Earth's atmosphere degrades the resolution of a telescope's view of astronomical objects. (Ch. 6)

astronomical unit (AU) The average distance from the Sun to Earth: approximately 150 million kilometers (km). (Ch. 1, 3)

astronomy The scientific study of planets, stars, galaxies, and the universe as a whole. (Ch. 1)

astrophysics The application of physical laws to the understanding of planets, stars, galaxies, and the universe as a whole. (Ch. 13)

asymptotic giant branch (AGB) The path on the H-R diagram that goes from the horizontal branch toward higher luminosities and lower temperatures, asymptotically approaching and then rising above the red giant branch. (Ch. 16)

Atens A group of asteroids whose orbits cross the orbit of Earth but not the orbit of Mars. Compare *Amors* and *Apollos*. (Ch. 12)

atmosphere The gravitationally bound outer gaseous envelope surrounding a planet, moon, or star. (Ch. 9)

atmospheric greenhouse effect A warming of planetary surfaces produced by atmospheric gases that transmit optical solar radiation but partially trap infrared radiation. Compare *greenhouse effect*. (Ch. 9)

atmospheric probe An instrumented package designed to provide on-site measurements of the chemical and/or physical properties of a planetary atmosphere. (Ch. 6)

atmospheric window A region of the electromagnetic spectrum in which radiation can penetrate a planet's atmosphere. (Ch. 6)

atom The smallest unit of a chemical element that retains the properties of that element. Each atom is composed of a nucleus (neutrons and protons) surrounded by a cloud of electrons.

AU See *astronomical unit*. (Ch. 1, 3)

aurora Emission in the upper atmosphere of a planet from atoms that have been excited by collisions with energetic particles from the planet's magnetosphere. (Ch. 9)

autumnal equinox 1. One of two points where the Sun crosses the celestial equator. 2. The day on which the Sun appears at that location, marking the first day of autumn (about September 22 in the Northern Hemisphere and March 20 in the Southern Hemisphere). Compare *vernal equinox*. See also *summer solstice* and *winter solstice*. (Ch. 2)

axion A hypothetical elementary particle first proposed to explain certain properties of the neutron and now considered a candidate for cold dark matter. (Ch. 23)

B

backlighting Illumination from behind a subject as seen by an observer. Fine material such as human hair and dust in planetary rings stands out best when viewed under backlighting conditions. (Ch. 11)

bar A unit of pressure. One bar is equivalent to 10^5 newtons per square meter—approximately equal to Earth's atmospheric pressure at sea level. (Ch. 9)

barred spiral galaxy A spiral galaxy with a bulge having an elongated, barlike shape. Compare *elliptical galaxy, irregular galaxy, S0 galaxy*, and *spiral galaxy*. (Ch. 19)

basalt Gray to black volcanic rock, rich in iron and magnesium. (Ch. 8)

beta decay 1. The decay of a neutron into a proton by emission of an electron (beta ray) and an antineutrino. 2. The decay of a proton into a neutron by emission of a positron and a neutrino. (Ch. 22)

Big Bang The event that occurred 13.8 billion years ago that marks the beginning of time and the universe. (Ch. 1, 21)

Big Bang nucleosynthesis The formation of low-mass nuclei (H, He, Li, Be) during the first few minutes after the Big Bang. (Ch. 21)

Big Crunch A hypothetical cosmic future in which the expansion of the universe reverses and the universe collapses onto itself. (Ch. 22)

Big Rip A hypothetical cosmic future in which all matter in the universe, from stars to subatomic particles, is progressively torn apart by expansion of the universe. (Ch. 22)

binary star A system in which two stars are in gravitationally bound orbits about their common center of mass. (Ch. 13)

binding energy The minimum energy required to separate an atomic nucleus into its component protons and neutrons. (Ch. 17)

biosignature Chemical compounds detectable in the spectra of an exoplanet's atmosphere that might indicate the presence of life. (Ch. 24)

biosphere The global sum of all living organisms on Earth (or any planet or moon). Compare *hydrosphere* and *lithosphere*. (Ch. 18)

bipolar outflow Material streaming away in opposite directions from both sides of the accretion disk of a young star. (Ch. 15)

black hole An object so dense that its escape velocity exceeds the speed of light; a *singularity* in spacetime. (Ch. 18)

blackbody An object that absorbs and can reemit all electromagnetic energy it receives. (Ch. 5)

blackbody spectrum See *Planck spectrum*. (Ch. 5)

blue-green algae See *cyanobacteria*. (Ch. 24)

blueshift The Doppler shift toward shorter (bluer) wavelengths of light from an approaching object. Compare *redshift*. (Ch. 5)

Bohr model A model of the atom, proposed by Niels Bohr in 1913, in which a small positively charged nucleus is surrounded by orbiting electrons, similar to a miniature solar system. (Ch. 5)

bound orbit An orbit in which an object is gravitationally bound to the body it is orbiting. A bound orbit's velocity is less than the escape velocity. Compare *unbound orbit*. (Ch. 4)

bow shock The boundary at which the speed of the solar wind abruptly drops from supersonic to subsonic in its approach to a planet's magnetosphere; the boundary between the region dominated by the solar wind and the region dominated by a planet's magnetosphere. (Ch. 9)

brightness The apparent intensity of light from a luminous object. Brightness depends on both the *luminosity* of a source and its distance. Units at the detector: watts per square meter (W/m^2). (Ch. 5)

brown dwarf A "failed star" without enough mass to fuse hydrogen in its core. An object whose mass is intermediate between that of the least massive stars and that of supermassive planets. (Ch. 7, 15)

bulge The central region of a spiral galaxy that is similar in appearance to a small elliptical galaxy. (Ch. 19)

C

C See *Celsius*. (Ch. 5)

C-type asteroid An asteroid made of material that has remained mostly unmodified since the formation of the Solar System; the most primitive type of asteroid. Compare *M-type asteroid* and *S-type asteroid*. (Ch. 12)

caldera The summit crater of a volcano. (Ch. 11)

Cambrian explosion The spectacular rise in the number and diversity of biological species that occurred between 540 million and 500 million years ago. (Ch. 24)

carbon-nitrogen-oxygen (CNO) cycle One way in which hydrogen is converted to helium (hydrogen fusion in the interiors of main-sequence stars. Compare *proton-proton chain*. (Ch. 17)

carbon star A cool red giant or asymptotic giant branch star that has an excess of carbon in its atmosphere. (Ch. 16)

carbonaceous chondrite A primitive stony meteorite that contains chondrules and is rich in carbon and volatile materials. (Ch. 12)

Cassini Division The largest gap in Saturn's rings, discovered by Jean-Dominique Cassini in 1675. (Ch. 11)

catalyst An atomic and molecular structure that permits or encourages chemical and nuclear reactions but does not change its own chemical or nuclear properties. (Ch. 9, 14)

CCD See *charge-coupled device*. (Ch. 6)

celestial equator The imaginary great circle that is the projection of Earth's equator onto the celestial sphere. (Ch. 2)

celestial sphere An imaginary sphere with celestial objects on its inner surface and Earth at its center. The celestial sphere has no physical existence but is a convenient tool for picturing the directions in which celestial objects are seen from Earth's surface. (Ch. 2)

Celsius (C) Also called *centigrade scale*. The arbitrary temperature scale, formulated by Anders Celsius (1701–1744), that defines 0°C as the freezing point of water and 100°C as the boiling point of water at sea level. Unit: degrees Celsius (°C). Compare *Fahrenheit* and *Kelvin temperature scale*. (Ch. 5)

center of mass 1. The weighted average location of all the mass in a system of objects. The point in any isolated system that moves according to Newton's first law of motion. 2. In a binary star system, the point between the two stars that is the focus of both their elliptical orbits. (Ch. 4, 13)

centigrade scale See *Celsius*. (Ch. 5)

centripetal force A force directed toward the center of curvature of an object's curved path. (Ch. 4)

Cepheid variable An evolved high-mass star with an atmosphere that is pulsating, leading to variability in the star's luminosity and color. (Ch. 17)

Chandrasekhar limit The upper limit on the mass of an object supported by electron degeneracy pressure; approximately 1.4 solar masses (M_{Sun}). (Ch. 16)

chaotic Behavior in complex systems in which a small change in the initial state of a system can lead to a large change in the final state of the system. (Ch. 7)

charge-coupled device (CCD) A common type of solid-state detector of electromagnetic radiation that transforms the intensity of light directly into electric signals. (Ch. 6)

charge destruction Electrons are absorbed by protons in atomic nuclei, forming neutrons and releasing neutrinos. (Ch. 17)

chondrite A stony meteorite that contains chondrules. Compare *achondrite*. (Ch. 12)

chondrule A small, crystallized, spherical inclusion of rapidly cooled molten droplets found inside some meteorites. (Ch. 12)

chromatic aberration A detrimental property of a lens in which rays of different wavelengths are brought to different focal distances from the lens. (Ch. 6)

chromosphere The region of the Sun's atmosphere located between the *photosphere* and the *corona*. (Ch. 14)

circular velocity The orbital velocity needed to keep an object moving in a circular orbit. (Ch. 4)

circumpolar Describing the part of the sky, near either celestial pole, that can always be seen above the horizon from a specific location on Earth. (Ch. 2)

circumstellar disk See *protoplanetary disk*. (Ch. 7)

climate The state of an atmosphere averaged over an extended time. Compare *weather*. (Ch. 9)

closed universe A finite universe with a curved spatial structure such that the sum of the angles of a triangle always exceeds 180 degrees. Compare *flat universe* and *open universe*. (Ch. 22)

CMB See *cosmic microwave background radiation*. (Ch. 6, 21)

CNO cycle See *carbon-nitrogen-oxygen cycle*. (Ch. 17)

cold dark matter Particles of dark matter that move slowly enough to be gravitationally bound even in the smallest galaxies. Compare *hot dark matter*. (Ch. 23)

color index The color of a celestial object, generally a star, based on the ratio of its brightness in blue light to its brightness in "visual" (yellow-green) light. The difference between an object's blue (B) magnitude and visual (V) magnitude, $B - V$. (Ch. 13, Appendix 7)

coma (pl. **comae**) The nearly spherical cloud of gas and dust surrounding the nucleus of an active comet. (Ch. 12)

comet A complex object consisting of a small, solid, icy nucleus; an atmospheric halo; and a tail of gas and dust. (Ch. 7)

comet nucleus A primitive planetesimal composed of ices and refractory materials that has survived planetary accretion. The "heart" of a comet, containing nearly the entire mass of the comet. A "dirty snowball." (Ch. 7)

comparative planetology The study of planets by comparing their chemical and physical properties. (Ch. 8)

composite volcano A large, cone-shaped volcano formed by viscous, pasty lava flows alternating with pyroclastic (explosively generated) rock deposits. Compare *shield volcano*. (Ch. 8)

compound lens A lens made up of two or more elements with different refractive indices, the purpose of which is to minimize chromatic aberration. (Ch. 6)

concave mirror A telescope mirror with a surface that curves inward toward the incoming light. (Ch. 6)

conduction The transfer of energy in which the thermal energy of particles is transferred to adjacent particles by collisions or other interactions. Conduction is the most important way that thermal energy is transported in solid matter. Compare *convection*. (Ch. 14)

conservation law A physical law stating that the amount of a particular physical quantity (such as energy or angular momentum) of an isolated system does not change. (Ch. 7)

conservation of angular momentum The physical law stating that the amount of angular momentum of an isolated system does not change. (Ch. 7)

conservation of energy The physical law stating that the amount of energy of an isolated, closed system does not change. (Ch. 7)

constant of proportionality The number by which one quantity is multiplied to get another number. (Ch. 1, 4)

constellation An imaginary image formed by patterns of stars; any of 88 defined areas on the celestial sphere that astronomers use to locate celestial objects. (Ch. 2)

continental drift The slow motion (centimeters per year) of Earth's continents relative to one another and to Earth's mantle. See also *plate tectonics*. (Ch. 8)

continuous radiation Electromagnetic radiation with intensity that varies smoothly over a wide range of wavelengths. (Ch. 5)

continuous spectrum A spectrum containing all wavelengths, without specific spectral lines. (Ch. 5)

convection The transport of thermal energy from the lower (hotter) to the higher (cooler) layers of a fluid by motions within the fluid driven by variations in buoyancy. Compare *conduction*. (Ch. 8, 14)

convective zone A region within a star where energy is transported outward by convection. Compare *radiative zone*. (Ch. 13)

core 1. The innermost region of a planetary interior. Compare *crust* and *mantle*. 2. The innermost part of a star. (Ch. 8, 14)

core accretion–gas capture A process for forming giant planets, in which large amounts of surrounding hydrogen and helium gas are gravitationally captured onto a massive rocky core. (Ch. 7)

Coriolis effect The apparent displacement of objects in a direction perpendicular to their true motion as viewed from a rotating frame of reference. On a rotating planet, that effect arises from different latitudes' rotating at different speeds. (Ch. 2)

corona The hot, outermost part of the Sun's atmosphere. Compare *chromosphere* and *photosphere*. (Ch. 14)

coronal hole A low-density region in the solar corona containing "open" magnetic-field lines along which coronal material is free to stream into interplanetary space. (Ch. 14)

coronal mass ejection (CME) An eruption on the Sun that ejects hot gas and energetic particles at much higher speeds than are typical in the solar wind. (Ch. 14)

cosmic microwave background radiation (CMB) Also called simply *cosmic background radiation*. Isotropic microwave radiation from every direction in the sky having a 2.73-kelvin (K) blackbody spectrum. The CMB is residual radiation from the Big Bang. (Ch. 6, 21)

cosmic ray A very fast-moving particle (usually protons or another atomic nucleus) that originated in outer space; cosmic rays fill the disk of the Milky Way. (Ch. 18, 20)

cosmological constant A constant, introduced into general relativity by Einstein, that characterizes an extra, repulsive force in the universe due to the vacuum of space itself. (Ch. 22)

cosmological principle The (testable) assumption that the same physical laws that apply here and now also apply everywhere and at all times and that the universe has no special locations or directions. (Ch. 1, 21)

cosmological redshift The redshift that results from the expansion of the universe rather than from the motions of galaxies or gravity. Compare *gravitational redshift*. (Ch. 21)

cosmology The study of the large-scale structure and evolution of the universe as a whole. (Ch. 21)

crescent Any phase of the Moon, Mercury, or Venus in which the object appears less than half illuminated by the Sun. Compare *gibbous*. (Ch. 2)

Cretaceous-Paleogene (K-Pg) boundary The boundary between the Cretaceous and Paleogene periods in Earth's history. That boundary corresponds to the time of the impact of an asteroid or comet and the extinction of the dinosaurs. (Ch. 8)

critical density The value of mass density of the universe that, ignoring any cosmological constant, can just barely halt expansion of the universe. (Ch. 22)

crust The relatively thin, outermost, hard layer of a planet, which is chemically distinct from the interior. Compare *core* and *mantle*. (Ch. 8)

cryovolcanism Low-temperature volcanism in which the magmas are composed of molten ices rather than rocky material. (Ch. 11)

cyanobacteria Also called *blue-green algae*. Single-celled organisms that created oxygen in Earth's atmosphere by photosynthesizing carbon dioxide and releasing oxygen as a waste product. (Ch. 24)

cyclonic motion The rotation of a weather system resulting from the Coriolis effect as air moves toward a region of low atmospheric pressure. Compare *anticyclonic motion*. (Ch. 9)

D

Dark Ages The epoch in the history of the universe during which no visible "light" came from astronomical objects. (Ch. 23)

dark energy A form of energy that permeates all space (including the vacuum), producing a repulsive force that accelerates the expansion of the universe. (Ch. 22)

dark matter Matter in galaxies that does not emit or absorb electromagnetic radiation. Dark matter is thought to constitute most of the mass in the universe. Compare *luminous matter.* (Ch. 19)

dark matter halo The centrally condensed, greatly extended dark matter component of a galaxy that accounts for up to 95 percent of the galaxy's mass. (Ch. 19)

daughter product An element resulting from radioactive decay of a more massive *parent element.* (Ch. 8)

decay 1. The process of a radioactive nucleus changing into its daughter product. 2. The process of an atom or molecule dropping from a higher energy state to a lower energy state. 3. The process of a satellite's orbit losing energy. (Ch. 5)

declination A measure, analogous to *latitude*, that tells you the angular distance of a celestial body north or south of the celestial equator (from 0° to ±90°). Compare *right ascension.* (Ch. 2)

degeneracy pressure Pressure exerted by closely packed electrons in the collapsing core of a star. Such pressure pushes outward against gravity and keeps the star from collapsing further. (Ch. 16)

density The measure of an object's mass per unit of volume. Possible units include kilograms per cubic meter (kg/m^3). (Ch. 4, 7)

differential rotation Rotation of different parts of a system at different rates. (Ch. 14)

differentiation The process by which materials of higher density sink toward the center of a molten or fluid planetary interior. (Ch. 8)

diffraction The spreading of a wave after it passes through an opening or beyond the edge of an object. (Ch. 6)

diffraction grating An optical component with many narrow parallel lines that separate the wavelengths of light to produce a spectrum. (Ch. 6)

diffraction limit The limit of a telescope's angular resolution caused by diffraction. (Ch. 6)

diffuse ring A sparsely populated planetary ring spread out both horizontally and vertically.

dispersion The separation of rays of light into their component wavelengths. (Ch. 6)

distance ladder A sequence of techniques for measuring cosmic distances: each method is calibrated using the results from other methods that have been applied to closer objects. (Ch. 19)

Doppler effect The change in wavelength of sound or light as a result of the relative motion of the source toward or away from the observer. (Ch. 5)

Doppler shift The amount by which the Doppler effect shifts the wavelength of light. (Ch. 5)

Drake equation A prescription for estimating the number of intelligent civilizations existing in the Milky Way Galaxy. (Ch. 24)

dust devil A small tornado-like column of air containing dust or sand. (Ch. 9)

dust tail A type of comet tail consisting of dust particles pushed away from the comet's head by radiation pressure from the Sun. Compare *ion tail.* (Ch. 12)

dwarf galaxy A small galaxy with a luminosity ranging from 1 million to 1 billion solar luminosities (L_{Sun}). Compare *giant galaxy.* (Ch. 19)

dwarf planet A body with characteristics similar to those of a planet except that it has not cleared smaller bodies from the neighboring regions around its orbit. Compare *planet* (definition 2). (Ch. 12)

dynamic equilibrium A state in which a system is constantly changing but its configuration remains the same because one source of change is exactly balanced by another source of change. Compare *static equilibrium.* (Ch. 5)

dynamo theory A theory postulating that Earth's magnetic field (and those of other planets) is generated from a rotating and electrically conducting liquid core. (Ch. 8)

E

eccentricity (*e*) The ratio of the distance between the two foci of an ellipse to the length of its major axis, which measures how noncircular the ellipse is. (Ch. 3)

eclipse 1. The total or partial obscuration of one celestial body by another. 2. The total or partial obscuration of light from one celestial body as it passes through the shadow of another celestial body. (Ch. 2)

eclipse season Any time during the year when the Moon's line of nodes points towards the Sun and eclipses can occur. (Ch. 2)

eclipsing binary A binary system in which the orbital plane is oriented such that the two stars appear to pass in front of each other as seen from Earth. Compare *spectroscopic binary* and *visual binary.* (Ch. 13)

ecliptic 1. The apparent annual path of the Sun against the background of stars. 2. The projection of Earth's orbital plane onto the celestial sphere. (Ch. 2)

ecliptic plane The plane of Earth's orbit around the Sun. The ecliptic is the projection of that plane onto the celestial sphere. (Ch. 2)

effective temperature The temperature at which a blackbody, such as a star, appears to radiate. (Ch. 14)

Einstein ring Light bent by gravitational lensing into a ring. (Ch. 18)

ejecta 1. Material thrown outward by the impact of an asteroid or comet on a planetary surface, leaving a crater behind. 2. Material thrown outward by a stellar explosion. (Ch. 8)

electric field A field that can exert a force on a charged object, whether at rest or moving. Compare *magnetic field.* (Ch. 5)

electric force The force exerted on electrically charged particles such as protons and electrons, arising from their electric charges. Compare *magnetic force.* See also *electromagnetic force.* (Ch. 5)

electromagnetic force The force, including both electric and magnetic forces, that acts on electrically charged particles. One of four fundamental forces of nature, along with the *strong nuclear force, weak nuclear force,* and *gravity* (definition 1). The force is mediated by the exchange of photons. (Ch. 22)

electromagnetic radiation A traveling disturbance in the electric and magnetic fields caused by accelerating electric charges. In quantum mechanics, a stream of photons. Light. (Ch. 5)

electromagnetic spectrum The spectrum made up of all possible frequencies or wavelengths of electromagnetic radiation, ranging from gamma rays through radio waves and including the portion our eyes can use. (Ch. 5)

electromagnetic wave A wave consisting of oscillations in the electric-field strength and the magnetic-field strength. (Ch. 5)

electron (e^-) A subatomic particle having a negative electric charge of 1.6×10^{-19} coulomb (C), a rest mass of 9.1×10^{-31} kilogram (kg), and a rest energy of 8×10^{-14} joule (J). The antiparticle of the *positron.* Compare *proton* and *neutron.* (Ch. 5)

electron-degenerate Describing matter, compressed to the point at which electron density reaches the limit imposed by the rules of quantum mechanics. (Ch. 16)

electroweak theory The quantum theory that combines descriptions of both the electromagnetic force and the weak nuclear force. (Ch. 22)

element One of 92 naturally occurring substances (such as hydrogen, oxygen, and uranium) and more than 20 human-made ones (such as plutonium). Each element is chemically defined by the specific number of protons in the nuclei of its atoms. (Ch. 5)

elementary particle One of the basic building blocks of nature that is not known to have substructure, such as the *electron* or the *quark.* (Ch. 23)

ellipse A conic section produced by the intersection of a plane with a cone when the plane is passed through the cone at an angle to the axis other than 0° or 90°. The shape that results when you attach the two ends of a piece of string to a piece of paper, stretch the string tight with the tip of a pencil, and then draw around those two points while keeping the string taut. (Ch. 3)

elliptical galaxy A galaxy of Hubble type "E" class, with a circular to elliptical outline on the sky, and containing almost no disk and a population of old stars. Compare *barred spiral galaxy, irregular galaxy, S0 galaxy,* and *spiral galaxy.* (Ch. 19)

emission The production of a photon when an atom decays to a lower energy state. Compare *absorption.* (Ch. 5)

emission line A peak in the intensity of a spectrum that is due to the emission of electromagnetic radiation at a specific wavelength determined by the energy levels of an atom or molecule. Compare *absorption line.* (Ch. 5)

empirical science Descriptive scientific investigation based primarily on observations and experimental data rather than on theoretical inference. (Ch. 3)

energy The conserved quantity that gives objects and systems the ability to do work. Possible units include joules (J). (Ch. 5)

energy transport The transfer of energy from one location to another. In stars, energy transport is carried out mainly by radiation or convection. (Ch. 14)

entropy A measure of the disorder of a system related to the number of ways a system can be rearranged without its appearance being affected. (Ch. 24)

equator The imaginary great circle on the surface of a body midway between its poles that divides the body into northern and southern hemispheres. The equatorial plane passes through the center of the body and is perpendicular to its rotation axis. Compare *meridian*. (Ch. 2)

equilibrium The state of an object in which physical processes balance each other so that its properties or conditions remain constant. (Ch. 5)

equinox Literally, "equal night." 1. One of two positions on the ecliptic where it intersects the celestial equator. 2. Either of the two times of year (the *autumnal equinox* and *vernal equinox*) when the Sun is at one of these two positions. At this time, night and day are of the same length everywhere on Earth. Compare *solstice*. (Ch. 2)

equivalence principle The principle stating that no difference exists between a frame of reference freely floating through space and one freely falling within a gravitational field. (Ch. 18)

erosion The degradation of a planet's surface topography by the mechanical action of wind, water, or living organisms. (Ch. 8)

escape velocity The minimum velocity needed for an object to achieve a parabolic trajectory and thus permanently leave the gravitational grasp of another mass. (Ch. 4)

eternal inflation The idea that a universe might inflate forever. In such a universe, quantum effects could randomly cause regions to slow their expansion, eventually stop inflating, and experience an explosion resembling the Big Bang. (Ch. 22)

event Something that happens at a particular location in spacetime. (Ch. 18)

event horizon The effective "surface" of a black hole. Nothing inside that surface—not even light—can escape from a black hole. (Ch. 18)

evolutionary track The path that a star follows across the H-R diagram as it evolves through its lifetime. (Ch. 15)

excited state Any energy level of a system or part of a system, such as an atom, molecule, or particle, that is higher than its ground state. Compare *ground state*. (Ch. 5)

exoplanet See *extrasolar planet*. (Ch. 7)

exosphere A very thin atmosphere or layer of atmosphere, where the molecules are bound by gravity to the moon or planet but their density is too low to behave like a gas of colliding particles. (Ch. 9)

extrasolar planet Also called *exoplanet*. A planet orbiting a star other than the Sun. (Ch. 7)

extremophile A life-form that thrives under extreme environmental conditions. (Ch. 24)

eyepiece A lens that is closest to the eye in a telescope. Changing the eyepiece will change the magnification of the image in the telescope. (Ch. 6)

F

F See *Fahrenheit*. (Ch. 5)

Fahrenheit (F) The arbitrary temperature scale, formulated by Daniel Gabriel Fahrenheit (1686–1736), that defines 32°F as the melting point of water and 212°F as the boiling point of water at sea level. Unit: degrees Fahrenheit (°F). Compare *Celsius* and *Kelvin temperature scale*. (Ch. 5)

falsified A hypothesis shown to be false. (Ch. 1)

fault A fracture in the crust of a planet or moon along which blocks of material can slide. (Ch. 8)

filter An instrument element that transmits a limited wavelength range of electromagnetic radiation. For the optical range, such elements are typically made of different kinds of glass and take on the hue of the light they transmit. (Ch. 13)

first quarter Moon The phase of the Moon in which only the western half of the Moon, as viewed from Earth, is illuminated by the Sun. It occurs about a week after a new Moon. Compare *third quarter Moon*. See also *full Moon* and *new Moon*. (Ch. 2)

fissure A fracture in the planetary lithosphere from which magma emerges. (Ch. 8)

flat rotation curve A rotation curve of a spiral galaxy in which rotation rates do not decline in the outer part of the galaxy but remain relatively constant to the outermost points. (Ch. 19)

flat universe An infinite universe whose spatial structure obeys Euclidean geometry, such that the sum of the angles of a triangle always equals 180 degrees. Compare *closed universe* and *open universe*. (Ch. 22)

flatness problem The surprising result that the sum of Ω_m plus Ω_Λ is extremely close to 1 in the present-day universe; equivalent to saying that it is surprising the universe is so close to being exactly flat. (Ch. 22)

flux The total amount of energy passing through each square meter of a surface each second. Unit: watts per square meter (W/m^2). (Ch. 5)

flux tube A strong magnetic field contained within a tubelike structure. Flux tubes are found in the solar atmosphere and connecting the space between Jupiter and its moon Io. (Ch. 10)

flyby A spacecraft that first approaches and then continues flying past a planet or moon. Flybys can visit multiple objects, but they remain near their targets only briefly. Compare *orbiter*. (Ch. 6)

focal length The optical distance between a telescope's objective lens or primary mirror and the plane (called the focal plane) on which the light from a distant object is focused. (Ch. 6)

focal plane The plane, perpendicular to the optical axis of a lens or mirror, on which an image is formed. (Ch. 6)

focus (pl. **foci**) 1. One of two points that define an ellipse. 2. A point in the focal plane of a telescope. (Ch. 3)

force A push or a pull on an object. (Ch. 3)

frame of reference A coordinate system within which an observer measures positions and motions. (Ch. 2)

free fall The motion of an object when the only force acting on it is gravity. (Ch. 4)

frequency The number of times per second that a periodic process occurs. Unit: hertz (Hz), or cycles per second (1/s). (Ch. 5)

full Moon The phase of the Moon in which the near side of the Moon, as viewed from Earth, is fully illuminated by the Sun. It occurs about two weeks after a *new Moon*. See also *first quarter Moon* and *third quarter Moon*. (Ch. 2)

G

galaxy A gravitationally bound system that consists of stars and star clusters, gas, dust, and dark matter; typically greater than 1,000 light-years across and recognizable as a discrete, single object. (Ch. 19)

galaxy cluster A large, gravitationally bound collection of galaxies containing hundreds to thousands of members; typically 3–5 megaparsecs (Mpc) across. Compare *galaxy group* and *supercluster*. (Ch. 23)

galaxy group A small, gravitationally bound collection of galaxies containing from several to a hundred members; typically 1–2 megaparsecs (Mpc) across. Compare *galaxy cluster* and *supercluster*. (Ch. 20)

gamma ray Also called *gamma radiation*. Electromagnetic radiation with higher frequency, higher photon energy, and shorter wavelength than all other types of electromagnetic radiation. (Ch. 5)

gamma-ray burst (GRB) A brief, intense burst of gamma rays from a distant energetic explosion. (Ch. 18)

gas giant A giant planet formed mostly of hydrogen and helium. In the Solar System, Jupiter and Saturn are the gas giants. Compare *ice giant*. (Ch. 10)

general relativistic time dilation The verified prediction that time passes more slowly in a gravitational field than in the absence of a gravitational field. Compare *time dilation*. (Ch. 18)

general relativity See *general theory of relativity*. (Ch. 18)

general theory of relativity Sometimes referred to as simply *general relativity*. Einstein's theory explaining gravity as the distortion of spacetime by massive objects, such that particles travel on the shortest path between two events in spacetime. This theory deals with all types of motion. Compare *special theory of relativity*. (Ch. 18)

geocentric model A historical cosmological model with Earth at its center, and all the other objects in the universe in orbit around Earth. Compare *heliocentric model*. (Ch. 3)

geodesic The path an object will follow through spacetime in the absence of external forces. (Ch. 18)

giant galaxy A galaxy with luminosity greater than about 1 billion solar luminosities (L_{Sun}). Compare *dwarf galaxy*. (Ch. 19)

giant molecular cloud An interstellar cloud composed primarily of molecular gas and dust and having hundreds of thousands of solar masses. (Ch. 15)

giant planet Also called *Jovian planet*. One of the largest planets in the Solar System (Saturn, Jupiter, Uranus, or Neptune), typically 10 times the size and many times the mass of any *terrestrial planet* and lacking a solid surface. (Ch. 7)

gibbous Any phase of the Moon, Mercury, or Venus in which the object appears more than half illuminated by the Sun. Compare *crescent*. (Ch. 2)

global circulation The overall, planetwide circulation pattern of a planet's atmosphere. (Ch. 9)

globular cluster A spherically symmetric, highly condensed group of stars, containing tens of thousands to a million members. Compare *open cluster*. (Ch. 17)

gluon The particle that carries (or, equivalently, mediates) interactions due to the strong nuclear force. (Ch. 22)

grand unified theory (GUT) A unified quantum theory that combines the strong nuclear, weak nuclear, and electromagnetic forces but does not include gravity. (Ch. 22)

granite Rock that is cooled from magma and is relatively rich in silicon and oxygen. (Ch. 8)

grating An optical surface containing many narrow, closely and equally spaced parallel grooves or slits that spectrally disperse reflected or transmitted light. (Ch. 6)

gravitational lens A massive object that gravitationally focuses the light of a more distant object to produce multiple brighter, magnified, possibly distorted images. (Ch. 7, 18)

gravitational lensing The bending of light by gravity. (Ch. 18)

gravitational potential energy The stored energy in an object that is due solely to its position within a gravitational field. (Ch. 7)

gravitational redshift The shifting to longer wavelengths of radiation from an object deep within a gravitational well. Compare *cosmological redshift*. (Ch. 18)

gravitational wave A wave in the fabric of spacetime emitted by accelerating masses. (Ch. 6, 18)

gravity 1. The mutually attractive force between massive objects. One of four fundamental forces of nature, along with the *electromagnetic force*, the *strong nuclear force*, and the *weak nuclear force*. 2. An effect arising from the bending of spacetime by massive objects. (Ch. 4)

GRB See *gamma-ray burst*. (Ch. 18)

Great Red Spot The giant, oval, brick red anticyclone seen in Jupiter's southern hemisphere. (Ch. 10)

greenhouse effect The solar heating of air in an enclosed space, such as a closed building or car, resulting primarily from the inability of the hot air to escape. Compare *atmospheric greenhouse effect*. (Ch. 9)

greenhouse gas One of a group of atmospheric gases such as carbon dioxide that are transparent to visible radiation but absorb infrared radiation. (Ch. 9)

greenhouse molecule A molecule such as water vapor or carbon dioxide that transmits visible radiation but absorbs infrared radiation. (Ch. 9)

Gregorian calendar The modern calendar. A modification of the Julian calendar decreed by Pope Gregory XIII in 1582. By then, the less accurate Julian calendar had developed an error of 10 days over the 13 centuries since its inception. (Ch. 2)

ground state The lowest possible energy state for a system or part of a system, such as an atom, molecule, or particle. Compare *excited state*. (Ch. 5)

GUT See *grand unified theory*. (Ch. 22)

H

H II region A region of interstellar gas that has been ionized by ultraviolet radiation from nearby hot, massive stars. (Ch. 15)

H-R diagram The Hertzsprung-Russell diagram, a plot of the luminosities versus the surface temperatures of stars. The evolving properties of stars are plotted as tracks across the H-R diagram. (Ch. 13)

habitable zone The distance from its star at which a planet must be located to have a temperature suitable for water to exist in a liquid state. (Ch. 7, 13)

Hadley circulation A simplified, and therefore uncommon, atmospheric global circulation that carries thermal energy directly from the equator to the polar regions of a planet. (Ch. 9)

half-life The time that half a sample of a particular radioactive parent element takes to decay into a daughter product. (Ch. 8)

halo The spherically symmetric, low-density distribution of stars and dark matter that defines the outermost regions of a galaxy. (Ch. 19)

harmonic law See *Kepler's third law*. (Ch. 3)

Hawking radiation Radiation from a black hole. (Ch. 18)

Hayashi track The path that a protostar follows on the H-R diagram as it contracts toward the main sequence. (Ch. 15)

head The part of a comet that includes both the nucleus and the inner part of the coma. (Ch. 12)

heavy element Also called *massive element*. Any element more massive than helium. (Ch. 10, 13)

Heisenberg uncertainty principle The physical limitation that the product of the position and the momentum of a particle cannot be smaller than a well-defined value, Planck's constant (h). (Ch. 22)

heliocentric model A model of the Solar System, with the Sun at its center, and the planets, including Earth, in orbit around the Sun. Compare *geocentric model*. (Ch. 3)

helioseismology The use of solar oscillations to study the interior of the Sun. (Ch. 14)

heliosphere A region surrounding the Solar System in which the solar wind blows against the interstellar medium and clears out an area like the inside of a bubble. The heliosphere protects the Solar System from cosmic rays. (Ch. 14)

helium capture A helium nucleus is captured by another nucleus during nucleosynthesis (Ch. 17)

helium flash The runaway explosive fusion of helium in the degenerate helium core of a red giant star. (Ch. 16)

Herbig-Haro (HH) object A glowing, rapidly moving knot of gas and dust that is excited by bipolar outflows in very young stars. (Ch. 15)

heredity The process by which one generation passes on its characteristics to future generations. (Ch. 24)

hertz (Hz) A unit of frequency equivalent to cycles per second. (Ch. 5)

Hertzsprung-Russell diagram See *H-R diagram*. (Ch. 13)

HH object See *Herbig-Haro object*. (Ch. 15)

hierarchical clustering The "bottom up" process of forming large-scale structure. Small-scale structure first produces groups of galaxies, which in turn form clusters, which then form superclusters. (Ch. 23)

high-mass star A star with a main-sequence mass of greater than about 8 solar masses (M_{Sun}). Compare *low-mass star* and *medium-mass star*. (Ch. 16)

homogeneous In cosmology, describing a universe in which observers in any location would observe the same properties. Compare *isotropic*. (Ch. 21)

horizon The boundary that separates the sky from the ground. (Ch. 2)

horizon problem The puzzling observation that the cosmic background radiation is so uniform in all directions, even though widely separated regions should have been "over the horizon" from each other in the early universe. (Ch. 22)

horizontal branch A region on the H-R diagram defined by stars fusing helium to carbon in a stable core. (Ch. 16)

hot dark matter Particles of dark matter that move so fast that gravity cannot confine them to the volume occupied by a galaxy's normal luminous matter. Compare *cold dark matter*. (Ch. 23)

hot Jupiter A large, Jupiter-type extrasolar planet located very close to its parent star. (Ch. 7)

hot spot A place where hot plumes of mantle material rise near the surface of a planet. (Ch. 8)

Hubble constant (H_0) The constant of proportionality relating the recession velocities of galaxies to their distances. Compare *Hubble time*. (Ch. 19)

Hubble flow The motion of galaxies as a result of the expanding universe. (Ch. 21)

Hubble's law The law stating that the speed at which a galaxy is moving away from Earth is proportional to the distance of that galaxy. (Ch. 19)

Hubble time An estimate of the age of the universe from the inverse of the *Hubble constant*, $1/H_0$. (Ch. 21)

hurricane A large tropical cyclonic system circulating counterclockwise in the Northern Hemisphere and clockwise in the Southern Hemisphere. Hurricanes can extend outward from their center to more than 600 kilometers (km) and generate winds in excess of 300 kilometers per hour (km/h). (Ch. 8)

hydrogen fusion The release of energy from the nuclear fusion of four hydrogen atoms into a single helium atom. (Ch. 14)
hydrogen shell fusion The fusion of hydrogen in a shell surrounding a stellar core that may be either degenerate or fusing more massive elements. (Ch. 16)
hydrosphere The portion of Earth that is largely liquid water. Compare *biosphere* and *lithosphere.* (Ch. 8)
hydrostatic equilibrium The condition in which the weight bearing down at a particular point within an object is balanced by the pressure within the object. (Ch. 8, 14)
hypernova (pl. **hypernovae**) A very energetic supernova from a very high-mass star. (Ch. 18)
hypothesis A well-considered idea, based on scientific principles and knowledge, that leads to testable predictions. Compare *theory.* (Ch. 1)
Hz See *hertz.* (Ch. 5)

I

ice The solid form of a volatile material; sometimes the *volatile material* itself, in any form. (Ch. 7)
ice giant A giant planet formed mostly of the liquid form of volatile substances (ices). In the Solar System, Uranus and Neptune are the ice giants. Compare *gas giant.* (Ch. 10)
impact crater The scar of the impact left on a solid planetary or moon surface by collision with another object. Compare *secondary crater.* (Ch. 8)
impact cratering The process in which solid planetary objects collide with each other, leaving distinctive scars. (Ch. 8)
index of refraction (***n***) The ratio of the speed of light in a vacuum (c) to the speed of light in an optical medium (v). (Ch. 6)
inert gas A gaseous element that combines with other elements only under conditions of extreme temperature and pressure. Examples include helium, neon, and argon. (Ch. 16)
inertia The tendency for objects to retain their state of motion. (Ch. 3)
inertial frame of reference 1. A frame of reference moving in a straight line at constant speed, that is, not accelerating. 2. In general relativity, a frame of reference falling freely in a gravitational field. (Ch. 3, 18)
inertial reference frame See *inertial frame of reference.* (Ch. 18)
inferior planet A Solar System planet that orbits closer to the Sun than Earth does. Compare *superior planet.* (Ch. 3)
inflation An extremely brief phase of ultra-rapid expansion of the very early universe. After inflation, the standard Big Bang models of expansion apply. (Ch. 22)
infrared (IR) radiation Electromagnetic radiation with frequencies, photon energies, and wavelengths between those of visible light and microwaves. (Ch. 5)
instability strip A region of the H-R diagram containing stars that pulsate with a periodic variation in luminosity. (Ch. 17)
integration time The time interval during which photons are collected and added up in a detecting device. (Ch. 6)
intensity Of light, the amount of radiant energy emitted per second per unit area. Units for electromagnetic radiation: watts per square meter (W/m^2). (Ch. 5)
intercloud gas A low-density region of the interstellar medium that fills the space between interstellar clouds. (Ch. 15)
interference The interaction of two sets of waves producing high and low intensity, depending on whether their amplitudes reinforce (*constructive interference*) or cancel (*destructive interference*). (Ch. 6)
interferometer Linked optical or radio telescopes whose overall separation determines the angular resolution of the system. (Ch. 6)
interferometric array An interferometer made up of several telescopes arranged in an array. (Ch. 6)
interstellar cloud A discrete, high-density region of the interstellar medium made up mostly of atomic or molecular hydrogen and dust. (Ch. 15)
interstellar dust Small particles or grains (0.01–10 microns [μm] in diameter) of matter, primarily carbon and silicates, distributed throughout interstellar space. (Ch. 15)
interstellar extinction The dimming of visible and ultraviolet light by interstellar dust. (Ch. 15)
interstellar gas The tenuous gas, far less dense than air, composing 99 percent of the matter in the interstellar medium. (Ch. 7, 15)
interstellar medium The gas and dust that fill the space between the stars within a galaxy. (Ch. 14, 15)
inverse square law The rule stating that a quantity or effect diminishes with the square of the distance from the source. (Ch. 4)
ion An atom or molecule that has lost or gained one or more electrons. (Ch. 9, 15)
ionize see *ionization.* (Ch. 9, 15)
ionization The process by which electrons are stripped free from an atom or molecule, resulting in free electrons and a positively charged atom or molecule. (Ch. 9, 15)
ionosphere A layer high in Earth's atmosphere in which most atoms are ionized by solar radiation. (Ch. 9)
ion tail A type of comet tail consisting of ionized gas. Particles in the ion tail are pushed directly away from the comet's head in the antisolar direction at high speeds by the solar wind. Compare *dust tail.* (Ch. 12)
IR Infrared. See *infrared radiation.* (Ch. 5)
iron meteorite A metallic meteorite composed mostly of iron-nickel alloys. Compare *stony-iron meteorite* and *stony meteorite.* (Ch. 12)
irregular galaxy A galaxy without regular or symmetric appearance. Compare *barred spiral galaxy, elliptical galaxy, S0 galaxy,* and *spiral galaxy.* (Ch. 19)
irregular moon A moon that has been captured by a planet instead of having formed along with that planet. Some irregular moons revolve in a direction opposite to the rotation of the planet, and many are in distant, unstable orbits. Compare *regular moon.* (Ch. 11)
isotope A forms of the same chemical element that has the same number of protons but a different number of neutrons. (Ch. 5)
isotropic In cosmology, having the same appearance to an observer in all directions. Compare *homogeneous.* (Ch. 21)

J

J See *joule.* (Ch. 5)
jansky (Jy) The basic unit of flux density. Unit: watts per square meter per hertz ($W/m^2/Hz$). (Ch. 6)
jet 1. A stream of gas and dust ejected from a comet nucleus by solar heating. 2. A stream of material that moves away from a protostar or active galactic nucleus at hundreds of kilometers per second. (Ch. 15)
joule (J) A unit of energy or work. 1 J = 1 newton meter. (Ch. 5)
Jovian planet See *giant planet.* (Ch. 7)
Jy See *jansky.* (Ch. 6)

K

K See *kelvin.* (Ch. 5)
K-Pg boundary See *Cretaceous-Paleogene boundary.* (Ch. 8)
KBO See *Kuiper Belt object.* (Ch. 12)
kelvin (K) The basic unit of the Kelvin scale of temperature. (Ch. 5)
Kelvin temperature scale The temperature scale, formulated by William Thomson, better known as Lord Kelvin (1824–1907), that uses Celsius-sized degrees but defines 0 K as absolute zero instead of as the melting point of water. Unit: kelvins (K). Compare *Celsius* and *Fahrenheit.* (Ch. 5)
Kepler's first law A rule of planetary motion that Johannes Kepler inferred, stating that planets move in elliptical orbits with the Sun at one focus. (Ch. 3)
Kepler's laws The three rules of planetary motion that Johannes Kepler inferred from data collected by Tycho Brahe. (Ch. 3)
Kepler's second law Also called *law of equal areas.* A rule of planetary motion that Johannes Kepler inferred, stating that a line drawn from the Sun to a planet sweeps out equal areas in equal times as the planet orbits the Sun. (Ch. 3)
Kepler's third law Also called *harmonic law.* A rule of planetary motion that Johannes Kepler inferred, describing the relationship between the period of a planet's orbit and its distance from the Sun. The law states that the square of the period of a planet's orbit, measured in years, is equal to the cube of the semimajor axis of the planet's orbit, measured in astronomical units: $(P_{years})^2 = (A_{AU})^3$. (Ch. 3)
kiloparsec A unit of distance equal to 1,000 parsecs, or 3,260 light-years. (Ch. 19)
kinetic energy (E_K) The energy of an object resulting from its motions. $E_K = \frac{1}{2}mv^2$. Possible units include joules (J). (Ch. 5, 7)

Kirkwood gap A gap in the main asteroid belt related to orbital resonances with Jupiter. (Ch. 12)

Kuiper Belt A disk-shaped population of comet nuclei extending from Neptune's orbit to perhaps several thousand astronomical units (AU) from the Sun. The highly populated innermost part of the Kuiper Belt has an outer edge approximately 50 AU from the Sun. (Ch. 12)

Kuiper Belt object (KBO) Also called *trans-Neptunian object*. An icy planetesimal (comet nucleus) that orbits within the Kuiper Belt beyond the orbit of Neptune. (Ch. 12)

L

Lambda-CDM The standard model of the Big Bang universe in which most of the energy density of the universe is dark energy (similar to Einstein's cosmological constant), and most of the mass in the universe is cold dark matter. (Ch. 23)

lander An instrumented spacecraft designed to land on a planet or moon. Compare *rover.* (Ch. 6)

large-scale structure Observable aggregates on the largest scales in the universe, including galaxy groups, clusters, and superclusters. (Ch. 23)

latitude The angular distance north (+) or south (–) from the equatorial plane of a nearly spherical body. Compare *longitude.* (Ch. 2)

lava Molten rock flowing out of a volcano during an eruption; also the rock that solidifies and cools from that liquid. (Ch. 8)

law of equal areas See *Kepler's second law.* (Ch. 3)

law of gravitation See *universal law of gravitation.* (Ch. 4)

leap year A year that contains 366 days. Leap years occur every 4 years when the year is divisible by 4, correcting for the accumulated excess time in a normal year, which is approximately 365 1/4 days long. (Ch. 2)

length contraction The relativistic compression of moving objects in the direction of their motion. (Ch. 18)

Leonids A November meteor shower associated with the dust debris left by comet Tempel-Tuttle. (Ch. 12)

life A biochemical process in which living organisms can reproduce, evolve, and sustain themselves by drawing energy from their environment. All terrestrial life involves carbon-based chemistry, assisted by the self-replicating molecules ribonucleic acid (RNA) and deoxyribonucleic acid (DNA). (Ch. 24)

light All electromagnetic radiation, which composes the entire electromagnetic spectrum. (Ch. 5)

light-year (ly) The distance that light travels in 1 year—about 9.5 trillion kilometers (km). (Ch. 1)

limb The outer edge of the visible disk of a planet, moon, or the Sun. (Ch. 11,14)

limb darkening The darker appearance caused by increased atmospheric absorption near the limb of a planet or star. (Ch. 14)

limestone A common sedimentary rock composed of calcium carbonate. (Ch. 9)

line of nodes 1. A line defined by the intersection of two orbital planes. 2. The line defined by the intersection of Earth's equatorial plane and the plane of the ecliptic. (Ch. 2)

lithosphere The solid, brittle part of Earth (or any planet or moon), including the crust and the upper part of the mantle. Compare *biosphere* and *hydrosphere.* (Ch. 8)

lithospheric plate A separate piece of Earth's lithosphere that can move independently. See also *continental drift* and *plate tectonics.* (Ch. 8)

Local Group The group of galaxies that includes the Milky Way and Andromeda galaxies. (Ch. 1)

long-period comet A comet with an orbital period of greater than 200 years. Compare *short-period comet.* (Ch. 12)

longitude The angular distance east (+) or west (–) from the prime meridian at Greenwich, England. Compare *latitude.* (Ch. 2)

longitudinal wave A wave that oscillates parallel to the direction of the wave's propagation. Compare *transverse wave.* (Ch. 8)

look-back time The amount of time that the light from an astronomical object has taken to reach Earth. (Ch. 21)

low-mass star A star with a main-sequence mass of less than about 3 solar masses (M_{Sun}). Compare *high-mass star* and *medium-mass star.* (Ch. 16)

luminosity The total amount of light emitted by an object. Unit: watts (W). Compare *brightness.* (Ch. 5)

luminosity class A spectral classification based on stellar size, ranging from supergiants at the large end to white dwarfs at the small end. (Ch. 13)

luminosity-temperature-radius relationship A relationship among those three properties of stars indicating that if any two are known, the third can be calculated. (Ch. 13)

luminous matter Also called *normal matter.* Matter in galaxies—including stars, gas, and dust—that emits electromagnetic radiation. Compare *dark matter.* (Ch. 19)

lunar eclipse An eclipse that occurs when the Moon is partially or entirely in Earth's shadow. Compare *solar eclipse.* (Ch. 2)

lunar tide A tide on Earth caused by the differential gravitational pull of the Moon. Compare *solar tide.* (Ch. 4)

lunisolar calendar Calendar created by the Babylonians in which a month began with the first sighting of the lunar crescent, and a 13th month was added when needed to catch up to the solar year. (Ch. 2)

ly See *light-year.* (Ch. 1)

M

μm See *micron.* (Ch. 5)

M-type asteroid An asteroid made of material that was once part of the metallic core of a larger, differentiated body that has since broken into pieces; made mostly of iron and nickel. Compare *C-type asteroid* and *S-type asteroid.* (Ch. 12)

MaCHO Short for *massive compact halo object.* MaCHOs include brown dwarfs, white dwarfs, and black holes and are candidates for dark matter. Compare *WIMP.* (Ch. 19)

magma Molten rock, often containing dissolved gases and solid minerals. (Ch. 8)

magnetic field A field that can exert a force on a moving electric charge. Compare *electric field.* (Ch. 5)

magnetic force The force exerted on electrically charged particles such as protons and electrons, arising from their motion. Compare *electric force.* See also *electromagnetic force.* (Ch. 5)

magnetosphere The region surrounding a planet that is filled with relatively intense magnetic fields and plasmas. (Ch. 8)

magnitude A system used by astronomers to describe the brightness or luminosity of stars. The brighter the star, the lower its magnitude. (Ch. 13)

major axis The long axis of an ellipse. (Ch. 3)

main asteroid belt See *asteroid belt.* (Ch. 12)

main sequence The strip on the H-R diagram where most stars are found. Main-sequence stars are fusing hydrogen to helium in their cores. (Ch. 13)

main-sequence lifetime The amount of time a star spends on the main sequence, fusing hydrogen into helium in its core. (Ch. 16)

main-sequence turnoff The location on the main sequence of an H-R diagram made from a population of stars of the same age (such as a star cluster) where stars are just evolving off the main sequence. That location is determined by the age of the population of stars. (Ch. 17)

mantle The solid portion of a rocky planet that lies between the *crust* and the *core.* (Ch. 8)

mare (pl. **maria**) A dark region on the Moon composed of basaltic lava flows. (Ch. 8)

mass 1. Inertial mass: the property of matter that determines its resistance to changes in motion. Compare *weight.* 2. Gravitational mass: the property of matter defined by its attractive force on other objects. According to general relativity, the two are equivalent. (Ch. 3, 4)

mass-luminosity relationship An empirical relationship between the luminosity (L) and mass (M) of main-sequence stars; for example, $L \propto M^{3.5}$. (Ch. 13, 16)

mass transfer The transfer of mass from one member of a binary star system to its companion. Mass transfer occurs when one of the stars evolves to the point that it overfills its Roche lobe, so that its outer layers are pulled toward its binary companion. (Ch. 16)

massive element Also called *heavy element.* Any element more massive than helium. (Ch. 13)

matter 1. Objects made of particles that have mass, such as protons, neutrons, and electrons. 2. Anything that occupies space and has mass. (Ch. 5)

Maunder Minimum The period from 1645 to 1715 during which very few sunspots were observed. (Ch. 14)

medium The substance that a wave, such as light, travels through—for example, air or glass. Compare *vacuum.* (Ch. 5)

medium-mass star A star with a main-sequence mass between 3 and 8 solar masses (M_{Sun}). Compare *high-mass star* and *low-mass star.* (Ch. 16)

megaparsec (Mpc) A unit of distance equal to 1 million parsecs, or 3.26 million light-years. (Ch. 19)

meridian The imaginary arc in the sky running from the horizon at due north through the zenith to the horizon at due south. The meridian divides the observer's sky into eastern and western hemispheres. Compare *equator.* (Ch. 2)

mesosphere The layer of Earth's atmosphere immediately above the stratosphere, extending from an altitude of 50 kilometers (km) to about 90 km. Compare *troposphere, stratosphere,* and *thermosphere.* (Ch. 9)

meteor A meteoroid that enters and burns up in a planetary atmosphere, often leaving an incandescent trail. Compare *meteorite* and *meteoroid.* (Ch. 8)

meteor shower A larger-than-normal display of meteors, occurring when Earth passes through the orbit of a disintegrating comet, sweeping up its debris. Compare *sporadic meteor.* (Ch. 12)

meteorite A piece of rock or other fragment of material (a meteoroid) that survives to reach a planet's surface. Compare *meteor* and *meteoroid.* (Ch. 7, 8)

meteoroid A small cometary or asteroidal fragment, ranging in size from 100 microns (μm) to 100 meters. When entering a planetary atmosphere, the meteoroid creates a *meteor.* Compare *meteor* and *meteorite*; also *planetesimal* and *zodiacal dust.* (Ch. 8)

micrometer (μm) See *micron.* (Ch. 5)

micron (μm) One-millionth (10^{-6}) of a meter; a unit of length used for the wavelength of infrared light. (Ch. 5)

microwave radiation Electromagnetic radiation with frequencies, photon energies, and wavelengths between those of infrared radiation and radio waves. (Ch. 5)

Milky Way Galaxy The galaxy in which the Sun and Solar System reside. (Ch. 1)

minor planet See *asteroid.* (Ch. 16)

minute of arc See *arcminute.* (Ch. 13)

molecular cloud An interstellar cloud composed primarily of molecular hydrogen. (Ch. 15)

molecular-cloud core A dense clump within a molecular cloud that forms as the cloud collapses and fragments. Protostars form from molecular-cloud cores. (Ch. 15)

molecule Generally, the smallest particle of a substance that retains its chemical properties and is composed of two or more atoms. (Ch. 5)

momentum The product of the mass and velocity of a particle. Possible units include kilograms times meters per second (kg m/s). (Ch. 7)

moon A less massive satellite orbiting a more massive object. Moons are found around planets, dwarf planets, asteroids, and Kuiper Belt objects. The term is usually capitalized when referring to Earth's Moon. (Ch. 7)

Mpc See *megaparsec.* (Ch. 19)

multiverse A collection of parallel universes that together make up all that exists. (Ch. 22)

mutation In biology, an imperfect reproduction of self-replicating material. (Ch. 24)

N

N See *newton.* (Ch. 3)

nadir The point on the celestial sphere located directly below an observer, opposite the *zenith.* (Ch. 2)

nanometer (nm) One-billionth (10^{-9}) of a meter; a unit of length used for the wavelength of visible light. (Ch. 5)

natural selection The process by which forms of structure, ranging from molecules to whole organisms, that are best adapted to their environment become more common than less well-adapted forms. (Ch. 24)

NCP See *north celestial pole.* (Ch. 2)

neap tide An especially weak tide that occurs around the time of the first or third quarter Moon, when the gravitational forces of the Moon and the Sun on Earth are at right angles to each other. Compare *spring tide.* (Ch. 4)

near-Earth asteroid An asteroid whose orbit brings it close to the orbit of Earth. See also *near-Earth object.* (Ch. 12)

near-Earth object (NEO) An asteroid, comet, or large meteoroid whose orbit intersects Earth's orbit. (Ch. 12)

nebula (pl. **nebulae**) A cloud of interstellar gas and dust, either illuminated by stars (for bright nebulae) or seen in silhouette against a brighter background (for dark nebulae). (Ch. 7)

nebular hypothesis The first plausible theory of the formation of the Solar System, proposed by Immanuel Kant in 1755, which stated that the Solar System formed from the collapse of an interstellar cloud of rotating gas. (Ch. 7)

NEO See *near-Earth object.* (Ch. 12)

neutrino A very low-mass, electrically neutral particle emitted during beta decay. Neutrinos interact with matter only very feebly and so can penetrate great quantities of matter. (Ch. 6, 14)

neutron capture The process in which an atomic nucleus forms a heavier nucleus after colliding with a neutron (Ch. 17)

neutrino cooling The process in which thermal energy is carried out of the center of a star by neutrinos rather than by electromagnetic radiation or convection. (Ch. 17)

neutron A subatomic particle having no net electric charge and a rest mass and rest energy nearly equal to that of the proton. Compare *electron* and *proton.* (Ch. 5)

neutron star The neutron-degenerate stellar core left behind by a Type II supernova. (Ch. 17)

new Moon The phase of the Moon in which the Moon is between Earth and the Sun; from Earth, we see only the side of the Moon not being illuminated by the Sun. Compare *full Moon.* See also *first quarter Moon* and *third quarter Moon.* (Ch. 2)

newton (N) The force required to accelerate a 1-kilogram (kg) mass at a rate of 1 meter per second per second (m/s^2). Unit: kilograms multiplied by meters per second squared (kg m/s^2). (Ch. 3)

Newton's first law of motion The law, formulated by Isaac Newton, stating that an object will remain at rest or will continue moving along a straight line at a constant speed until an unbalanced force acts on it. (Ch. 3)

Newton's laws The three physical laws of motion that Isaac Newton formulated. (Ch. 3)

Newton's second law of motion The law that Isaac Newton formulated, stating that if an unbalanced force acts on a body, the body will accelerate in proportion to the unbalanced force and in inverse proportion to the object's mass: $a = F/m$. The acceleration will be in the direction of the unbalanced force. (Ch. 3)

Newton's third law of motion The law that Isaac Newton formulated, stating that for every force an equal force in the opposite direction exists. (Ch. 3)

nm See *nanometer.* (Ch. 5)

normal matter See *luminous matter.* (Ch. 19)

north celestial pole (NCP) The northward projection of Earth's rotation axis onto the celestial sphere. Compare *south celestial pole.* (Ch. 2)

North Pole The location in the Northern Hemisphere where Earth's rotation axis intersects Earth's surface. Compare *South Pole.* (Ch. 2)

nova (pl. **novae**) A stellar explosion that results from runaway nuclear fusion in a layer of material on the surface of a white dwarf in a binary system. (Ch. 16)

nuclear burning See nuclear fusion. (Ch. 14)

nuclear fusion The combination of two less massive atomic nuclei into a single more massive atomic nucleus. (Ch. 14)

nucleosynthesis The formation of more massive atomic nuclei from less massive nuclei, either in the Big Bang (Big Bang nucleosynthesis) or in the interiors of stars (stellar nucleosynthesis). (Ch. 17)

nucleus (pl. **nuclei**) 1. The dense, central part of an atom. 2. The central core of a galaxy, comet, or other diffuse object. (Ch. 5)

O

objective lens The primary optical element in a telescope or camera that produces an image of an object. (Ch. 6)

oblateness The flattening of an otherwise spherical planet or star caused by its rapid rotation. (Ch. 10)

obliquity The inclination of a celestial body's equator to its orbital plane. (Ch. 2)

observational uncertainty The fact that real measurements are never perfect; all observations are uncertain by some amount. (Ch. 13)

Occam's razor The principle that the simplest hypothesis is the most likely, named after William of Occam (circa 1285–1349), the medieval English cleric to whom the idea is attributed. (Ch. 1)

Oort Cloud A spherical distribution of comet nuclei stretching from beyond the Kuiper Belt to more than 50,000 astronomical units (AU) from the Sun. (Ch. 12)

opacity A measure of how effectively a material blocks the radiation going through it. (Ch. 14)

open cluster A loosely bound group of a few dozen to a few thousand stars that formed together in the disk of a spiral galaxy. Compare *globular cluster.* (Ch. 17)

open universe An infinite universe with a negatively curved spatial structure (much like the surface of a saddle) such that the sum of the angles of a triangle is always less than 180 degrees. Compare *closed universe* and *flat universe.* (Ch. 22)

orbit The path taken by one object moving around another object under the influence of their mutual gravitational or electric attraction. (Ch. 4)

orbital resonance A situation in which the orbital periods of two objects are related by a ratio of small integers. (Ch. 11)

orbiter A spacecraft placed in orbit around a planet or moon. Compare *flyby.* (Ch. 6)

organic Containing the element carbon. (Ch. 7)

P

P wave See *primary wave.* (Ch. 8)

pair production The creation of a particle-antiparticle pair from a source of electromagnetic energy. (Ch. 22)

paleoclimatology The study of changes in Earth's climate throughout its history. (Ch. 9)

parallax Also called *parallactic angle.* The displacement in the apparent position of a nearby star caused by the changing location of Earth in its orbit. (Ch. 13)

parent element A radioactive element that decays to form more-stable *daughter products.* (Ch. 8)

parsec (pc) Short for *parallax second.* The distance to a star with a parallax of 1 arcsecond (arcsec) using a base of 1 astronomical unit (AU). One parsec is approximately 3.26 light-years. (Ch. 13)

partial lunar eclipse An eclipse that occurs when the Moon is partially in Earth's shadow. (Ch. 2)

partial solar eclipse The type of eclipse that occurs when Earth passes through the penumbra of the Moon's shadow, so that the Moon blocks only a portion of the Sun's disk. Compare *annular solar eclipse* and *total solar eclipse.* (Ch. 2)

pc See *parsec.* (Ch. 13)

peculiar velocity The motion of a galaxy relative to the overall expansion of the universe. (Ch. 23)

penumbra (pl. **penumbrae**) 1. The outer part of a shadow, where the source of light is only partially blocked. Compare *umbra* (definition 1). 2. The region surrounding the umbra of a sunspot. The penumbra is cooler and darker than the surrounding surface of the Sun but is not as cool or dark as the umbra. Compare *umbra* (definition 2). (Ch. 2, 14)

penumbral lunar eclipse A lunar eclipse in which the Moon passes through the penumbra of Earth's shadow. Compare *total lunar eclipse.* (Ch. 2)

perihelion (pl. **perihelia**) The point in a solar orbit that is closest to the Sun. Compare *aphelion.* (Ch. 12)

period The time that a regularly repetitive process takes to complete one cycle. (Ch. 17)

period-luminosity relationship The relationship between the period of variability of a pulsating variable star, such as a Cepheid or RR Lyrae variable, and the luminosity of the star. Longer-period pulsating variable stars are more luminous than shorter-period ones. (Ch. 17)

Perseids A prominent August meteor shower associated with the dust debris left by comet Swift-Tuttle. (Ch. 12)

phase One of the various appearances of the sunlit surface of the Moon or a planet caused by the change in viewing location of Earth relative to both the Sun and the object. Examples include crescent phase and gibbous phase. (Ch. 2)

photino A hypothetical subatomic particle related to the photon. One of the candidates for cold dark matter. (Ch. 23)

photochemical Resulting from light acting on chemical systems. (Ch. 11)

photodisintegration An atomic nucleus absorbs an energetic gamma ray and emits some particles (Ch. 17)

photodissociation The breaking apart of molecules into smaller fragments or individual atoms by the action of photons. Compare *recombination* (definition 1). (Ch. 11)

photoelectric effect The emission of electrons from a substance illuminated by electromagnetic radiation greater than a certain critical frequency. (Ch. 18)

photon Also called *quantum of light.* A discrete unit or particle of electromagnetic radiation. The energy of a photon is equal to Planck's constant (h) multiplied by the frequency (f) of its electromagnetic radiation: $E_{photon} = h \times f$. The photon is the carrier of the electromagnetic force. (Ch. 5)

photosphere The apparent surface of the Sun as seen in visible light. Compare *chromosphere* and *corona.* (Ch. 14)

photosynthesize The process by which plants and algae absorb sunlight, water, and carbon dioxide, and release oxygen. (Ch. 9, 24)

physical law A broad statement that predicts a particular aspect of how the physical universe behaves and that is supported by many empirical tests. See also *theory.* (Ch. 1)

pixel The smallest picture element in a digital image array. (Ch. 6)

Planck era The early time, just after the Big Bang, for which the universe as a whole must be described with quantum mechanics. (Ch. 22)

Planck spectrum Also called *blackbody spectrum.* The spectrum of electromagnetic energy emitted by a blackbody per unit area per second, which is determined only by the temperature of the object. (Ch. 5)

Planck's constant (h) The constant of proportionality between the energy and the frequency of a photon. This constant defines how much energy a single photon of a given frequency or wavelength has. Value: 6.63×10^{-34} joule-second. (Ch. 5)

planet 1. A large body that orbits the Sun or other star that shines only by light reflected from the Sun or star. 2. In the Solar System, a body that orbits the Sun, has enough mass for self-gravity to overcome rigid body forces so that it assumes a spherical shape, and has cleared smaller bodies from the neighborhood around its orbit. Compare *dwarf planet.* (Ch. 7)

planet migration The theory that a planet can move to a location away from where it formed, through gravitational interactions with other bodies or loss of orbital energy from interaction with gas in the protoplanetary disk. (Ch. 7)

planetary nebula The expanding shell of material ejected by a dying asymptotic giant branch star. A planetary nebula glows from fluorescence caused by intense ultraviolet light coming from the hot, stellar remnant at its center. (Ch. 16)

planetary system A system of planets and other smaller objects in orbit around a star. (Ch. 7)

planetesimal A primitive body of rock and ice, 100 meters or more in diameter, that combines with others to form a planet. Compare *meteoroid* and *zodiacal dust.* (Ch. 7)

plasma A gas composed largely of charged particles but that also may include some neutral atoms. (Ch. 10, 14)

plate tectonics The geological theory concerning the motions of lithospheric plates, which in turn serves as the theoretical basis for *continental drift.* (Ch. 8)

positron A positively charged subatomic particle; the antiparticle of the *electron.* (Ch. 14)

power Energy per unit time. Possible units include watts (W) and joules per second (J/s). (Ch. 15)

precession of the equinoxes The slow change in orientation between the ecliptic plane and the celestial equator caused by the wobbling of Earth's axis. (Ch. 2)

pressure Force per unit area. Possible units include newtons per square meter (N/m^2) and bars.

primary atmosphere An atmosphere, composed mostly of hydrogen and helium, that forms at the same time as its host planet. Compare *secondary atmosphere.* (Ch. 7)

primary mirror The principal optical mirror in a reflecting telescope. The primary mirror determines the telescope's light-gathering power and resolution. Compare *secondary mirror.* (Ch. 6)

primary wave Also called *P wave.* A longitudinal seismic wave, in which the oscillations involve compression and decompression parallel to the direction of travel. Compare *secondary wave.* (Ch. 8)

principle A general idea or sense about the universe that guides us in constructing new scientific theories. Principles can be testable theories. (Ch. 1)

prograde motion 1. Rotational or orbital motion of a moon that is in the same direction as the planet it orbits. 2. The counterclockwise orbital motion of Solar System objects as seen from above Earth's orbital plane. Compare *retrograde motion.* (Ch. 3)

prominence An archlike projection above the solar photosphere often associated with a sunspot. (Ch. 14)

proportional See *proportionality.* (Ch. 1, Appendix 1)

proportionality A relationship between two things whose ratio is a constant. (Ch. 1, Appendix 1)

proton (***p*** or ***p***$^+$) A subatomic particle having a positive electric charge of 1.6×10^{-19} coulomb (C), a rest mass of 1.67×10^{-27} kilogram (kg), and a rest energy of 1.5×10^{-10} joule (J). Compare *electron* and *neutron.* (Ch. 5)

proton-proton chain One way in which hydrogen fusion can take place. That path is the most important for hydrogen fusion in low-mass stars such as the Sun. Compare *carbon-nitrogen-oxygen cycle.* (Ch. 14)

protoplanetary disk The remains of the accretion disk around a young star from which a planetary system may form. Sometimes called *circumstellar disk.* (Ch. 7)

protostar A young stellar object that derives its luminosity from converting gravitational energy to thermal energy rather than from nuclear reactions in its core. (Ch. 7, 15)

pulsar A rapidly rotating neutron star that beams radiation into space in two searchlight-like beams. To a distant observer, the star appears to flash on and off. (Ch. 17)

pulsating variable star A variable star that undergoes periodic radial pulsations. (Ch. 17)

Q

QCD See *quantum chromodynamics.* (Ch. 22)

QED See *quantum electrodynamics.* (Ch. 22)

quantized Existing as discrete, irreducible units. (Ch. 5)

quantum chromodynamics (QCD) The quantum theory describing the strong nuclear force and its mediation by gluons. Compare *quantum electrodynamics.* (Ch. 22)

quantum efficiency The likelihood that a particular photon falling on a detector will actually produce a response in the detector. (Ch. 6)

quantum electrodynamics (QED) The quantum theory describing the electromagnetic force and its mediation by photons. Compare *quantum chromodynamics.* (Ch. 22)

quantum mechanics The branch of physics that deals with the quantized and probabilistic behavior of atoms and subatomic particles. (Ch. 5)

quantum of light See *photon.* (Ch. 5)

quark The building block of protons and neutrons. (Ch. 22)

quasar Short for *quasi-stellar radio source.* The most luminous of the active galactic nuclei, seen only at great distances from the Milky Way. (Ch. 19)

R

radial velocity The component of velocity directed toward or away from the observer. (Ch. 5)

radian The angle at the center of a circle subtended by an arc equal to the length of the circle's radius; 2π radians equals 360°, and 1 radian equals approximately 57.3°. (Appendix 1)

radiant The point in the sky from which the meteors in a meteor shower appear to come. (Ch. 12)

radiation Waves or particles of energy traveling through space or a medium. (Ch. 14)

radiation belt A toroidal ring of high-energy particles surrounding a planet. (Ch. 9)

radiative transfer The transport of energy from one location to another by electromagnetic radiation. (Ch. 14)

radiative zone A region within a star where energy is transported outward by radiation. Compare *convective zone.* (Ch. 14)

radio galaxy A type of elliptical galaxy that has an active galactic nucleus at its center and very strong emission (10^{35} to 10^{38} watts [W]) in the radio part of the electromagnetic spectrum. Compare *Seyfert galaxy.* (Ch. 19)

radio telescope An instrument for detecting and measuring radio frequency emissions from celestial sources. (Ch. 6)

radio wave Electromagnetic radiation in the extreme long-wavelength region of the spectrum, beyond the region of microwaves. (Ch. 5)

radioisotope A radioactive element. (Ch. 8)

radiometric dating Use of the radioactive decay of elements to measure the ages of materials such as minerals. (Ch. 8)

ratio The relationship in quantity or size between two or more things. (Ch. 1, Appendix 1)

ray 1. A beam of electromagnetic radiation. 2. A bright streak emanating from a young impact crater. (Ch. 6; 8)

recombination 1. The combining of ions and electrons to form neutral atoms. Compare *photodissociation.* 2. An event early in the evolution of the universe in which hydrogen and helium nuclei combined with electrons to form neutral atoms. The removal of electrons caused the universe to become transparent to electromagnetic radiation. (Ch. 11, 21)

red giant A low-mass star that has evolved beyond the main sequence and is now fusing hydrogen in a shell surrounding a degenerate helium core. (Ch. 16)

red giant branch A region on the H-R diagram defined by low-mass stars evolving from the main sequence toward the horizontal branch. (Ch. 16)

reddening The effect by which stars and other objects, when viewed through interstellar dust, appear redder than they actually are. Reddening occurs because blue light is more strongly absorbed and scattered than red light. (Ch. 15)

redshift The Doppler shift toward longer (redder) wavelengths of light from an approaching object. Compare *blueshift.* (Ch. 5)

reflecting telescope A telescope that uses mirrors to collect and focus incoming electromagnetic radiation to form an image in their focal planes. The size of a reflecting telescope is defined by the diameter of the primary mirror. Compare *refracting telescope.* (Ch. 6)

reflection The redirection of a beam of light that strikes, but does not cross, the surface between two media having different refractive indices. If the surface is flat and smooth, the angle of incidence equals the angle of reflection. Compare *refraction.* (Ch. 6)

refracting telescope A telescope that uses objective lenses to collect and focus incoming electromagnetic radiation to form an image. Compare *reflecting telescope.* (Ch. 6)

refraction The redirection or bending of a beam of light when it crosses the boundary between two media having different refractive indices. Compare *reflection.* (Ch. 6)

refractory material Material that remains solid at high temperatures. Compare *volatile material.* (Ch. 7)

regular moon A moon that formed together with the planet it orbits. Compare *irregular moon.* (Ch. 11)

reionization A period after the Dark Ages during which objects formed that radiated enough energy to ionize neutral hydrogen, at redshift $6 < z < 20$. (Ch. 23)

relative humidity The amount of water vapor held by a volume of air at a given temperature compared (stated as a percentage) to the total amount of water that could be held by the same volume of air at the same temperature. (Ch. 9)

relative motion The difference in motion between two individual frames of reference. (Ch. 2)

relativistic Describing systems that travel at nearly the speed of light or are located near very strong gravitational fields. (Ch. 18)

relativistic beaming The effect created when material moving at nearly the speed of light beams the radiation it emits in the direction of its motion. (Ch. 19)

relativistic speed A speed high enough that special relativity, rather than Newtonian physics, is needed to describe the motion. Speeds greater than about 10% the speed of light are relativistic. (Ch. 18)

remote sensing The use of images, spectra, radar, or other techniques to measure the properties of an object from a distance. (Ch. 6)

resolution The ability of a telescope to separate two point sources of light. Resolution is determined by the telescope's aperture and the wavelength of light it receives. (Ch. 6)

rest wavelength The wavelength of light that is seen coming from an object at rest with respect to the observer. (Ch. 5)

retrograde motion 1. Rotation or orbital motion of a moon that is in the opposite direction to the rotation of the planet it orbits. 2. The clockwise orbital motion of Solar System objects as seen from above Earth's orbital plane. Compare *prograde motion.* 3. Apparent retrograde motion is a motion of the planets with respect to the "fixed stars," in which the planets appear to move westward for a time before resuming their normal eastward motion. (Ch. 3)

revolve Motion of one object in orbit around another. (Ch. 2)
right ascension A measure, analogous to *longitude*, that tells you the angular distance of a celestial body eastward along the celestial equator from the vernal equinox. Compare *declination*. (Ch. 2)
ring An aggregation of small particles orbiting a planet or star. The rings of the four giant planets of the Solar System are composed variously of silicates, organic materials, and ices. (Ch. 11)
ring arc A discontinuous, higher-density region within an otherwise continuous, narrow ring. (Ch. 11)
ringlet A narrowly confined concentration of ring particles. (Ch. 11)
Roche limit The distance at which a planet's tidal forces exceed the self-gravity of a smaller object—such as a moon, asteroid, or comet—causing the object to break apart. (Ch. 4)
Roche lobes The hourglass-shaped or figure eight–shaped volume of space surrounding two stars, which constrains material that is gravitationally bound by one or the other. (Ch. 16)
rotation curve A plot showing how the orbital velocity of stars and gas in a galaxy changes with radial distance from the galaxy's center. (Ch. 19)
rover A remotely controlled instrumented vehicle designed to move and explore the surface of a terrestrial planet or moon. Compare *lander*. (Ch. 6)
RR Lyrae variable A variable giant star whose regularly timed pulsations are good predictors of its luminosity. RR Lyrae variables are used in measuring the distance to globular clusters. (Ch. 17)

S

S-type asteroid An asteroid made of material that was once part of the outer layer of a larger, differentiated body that has since broken into pieces. Compare *C-type asteroid* and *M-type asteroid*. (Ch. 12)
S wave See *secondary wave*. (Ch. 8)
S0 galaxy A galaxy with a bulge and a disk-like spiral, but smooth in appearance like ellipticals. Compare *barred spiral galaxy*, *elliptical galaxy*, *irregular galaxy*, and *spiral galaxy*. (Ch. 19)
satellite An object in orbit around a more massive body, for example, a moon of any planet or an artificial satellite. (Ch. 4)
scale factor (R_U) A dimensionless number proportional to the distance between two points in space. The scale factor increases as the universe expands. (Ch. 21)
scattering The random change in the direction of travel of photons, caused by their interactions with molecules or dust particles. (Ch. 9, 16)
Schwarzschild radius The distance from the center of a nonrotating, spherical black hole at which the escape velocity equals the speed of light. (Ch. 18)
scientific method The formal procedure—including hypothesis, prediction, and experiment or observation—used to test (that is, to attempt to falsify) the validity of scientific hypotheses and theories. (Ch. 1)
scientific notation The standard expression of numbers with one digit (which can be zero) to the left of the decimal point and multiplied by 10 to the exponent required to give the number its correct value. Example: $2.99 \times 10^8 = 299{,}000{,}000$. (1, Appendix 1)
SCP See *south celestial pole*. (Ch. 2)
second law of thermodynamics The law stating that the entropy or disorder of an isolated system always increases as the system evolves. (Ch. 24)
second of arc See *arcsecond*. (Ch. 6, 13)
secondary atmosphere An atmosphere that forms—as a result of volcanism, comet impacts, or another process—sometime after its host planet has formed. Compare *primary atmosphere*. (Ch. 7)
secondary crater A crater formed from ejecta thrown from an *impact crater*. (Ch. 8)
secondary mirror A small mirror placed on the optical axis of a reflecting telescope that returns the beam back through a small hole in the *primary mirror*, thereby shortening the mechanical length of the telescope. (Ch. 6)
secondary wave Also called *S wave*. A transverse seismic wave, which involves the sideways motion of material. Compare *primary wave*. (Ch. 8)
sediment Solid material such as rock or sand that has been affected by erosion and weather, and transported by water, wind, or ice. (Ch. 8)
seismic wave A vibration due to an earthquake, a large explosion, or an impact on the surface that travels through a planet's interior. (Ch. 8)
seismometer An instrument that measures the amplitude and frequency of seismic waves. (Ch. 8)
self-gravity The gravitational attraction among all parts of the same object. (Ch. 4)
semimajor axis Half of the longer axis of an ellipse. (Ch. 3)
SETI The Search for Extraterrestrial Intelligence project, which uses advanced technology combined with radio telescopes to search for evidence of intelligent life elsewhere in the universe. (Ch. 24)
Seyfert galaxy A type of spiral galaxy with an active galactic nucleus at its center; first discovered in 1943 by Carl Seyfert. Compare *radio galaxy*. (Ch. 19)
shepherd moon A moon that orbits close to rings and gravitationally confines the orbits of the ring particles. (Ch. 11)
shield volcano A volcano formed by very fluid lava flowing from a single source and spreading out from that source. Compare *composite volcano*. (Ch. 8)
short gamma-ray burst Bursts of gamma rays lasting a few milliseconds to a few minutes, thought to originate in neutron star mergers. (Ch. 17)
short-period comet A comet with an orbital period of less than 200 years. Compare *long-period comet*. (Ch. 12)
sidereal day Earth's period of rotation with respect to the stars—about 23 hours 56 minutes—which is the time Earth takes to make one rotation and face the same star on the meridian. It differs from the *solar day* because of Earth's motion around the Sun. (Ch. 2)
sidereal period An object's orbital or rotational period measured with respect to the stars. Compare *synodic period*. (Ch. 2)
silicate One of the family of minerals composed of silicon and oxygen in combination with other elements. (Ch. 7)
singularity The point at which a mathematical expression or equation becomes meaningless, such as a fraction whose denominator approaches zero See also *black hole*. (Ch. 18)
solar abundance The relative amount of an element detected in the atmosphere of the Sun, expressed as the ratio of the number of atoms of that element to the number of hydrogen atoms. (Ch. 5)
solar day The day in common use—24 hours, which is Earth's period of rotation that brings the Sun back to the same local meridian where the rotation started. Compare *sidereal day*. (Ch. 2)
solar eclipse An eclipse that occurs when the Moon partially or entirely blocks the Sun. Compare *lunar eclipse*. (Ch. 2)
solar flare An explosion on the Sun's surface associated with a complex sunspot group and a strong magnetic field. (Ch. 14)
solar maximum (pl. **maxima**) The time, occurring about every 11 years, when the Sun is at its peak activity, meaning that sunspot activity and related phenomena (such as prominences, flares, and coronal mass ejections) are at their peak. (Ch. 14)
solar neutrino problem The historical observation that only about a third as many neutrinos as predicted by theory seemed to be coming from the Sun. (Ch. 14)
Solar System The gravitationally bound system made up of the Sun, planets, dwarf planets, moons, asteroids, comets, and Kuiper Belt objects, along with their associated gas and dust. (Ch. 1)
solar tide A tide on Earth caused by the differential gravitational pull of the Sun. Compare *lunar tide*. (Ch. 4)
solar wind The stream of charged particles emitted by the Sun and that flows at high speeds through interplanetary space. (Ch. 9)
solstice Literally, "Sun standing still." 1. One of the two most northerly and southerly points on the ecliptic. 2. Either of the two times of year (the *summer solstice* and *winter solstice*) when the Sun is at one of these two positions. Compare *equinox*. (Ch. 2)
south celestial pole (**SCP**) The southward projection of Earth's rotation axis onto the celestial sphere. Compare *north celestial pole*. (Ch. 2)
South Pole The location in the Southern Hemisphere where Earth's rotation axis intersects Earth's surface. Compare *North Pole*. (Ch. 2)
spacetime A concept that combines space and time into a four-dimensional continuum with three spatial dimensions plus one dimension of time. (Ch. 18)
special relativity See *special theory of relativity*. (Ch. 18)
special theory of relativity Sometimes referred to as simply *special relativity*. Einstein's theory explaining how the fact that the speed of light is a constant affects nonaccelerating frames of reference. Compare *general theory of relativity*. (Ch. 18)

spectral type A classification system for stars that is based on the presence and relative strength of absorption lines in their spectra. Spectral type is related to the surface temperature of a star. (Ch. 13)

spectrograph Also called *spectrometer.* A device that spreads out the light from an object into its component wavelengths. (Ch. 6)

spectrometer See *spectrograph.* (Ch. 6)

spectroscopic binary A binary star system whose existence and properties are revealed to astronomers only by the Doppler shift of its spectral lines. Most spectroscopic binaries are close pairs. Compare *eclipsing binary* and *visual binary.* (Ch. 13)

spectroscopic parallax Use of the spectroscopically determined luminosity and the observed brightness of a star to determine the star's distance. (Ch. 13)

spectroscopy The study of an object's electromagnetic radiation in terms of its component wavelengths. (Ch. 6)

spectrum (pl. **spectra**) Waves sorted by wavelength. See also *electromagnetic spectrum.* (Ch. 5)

speed The rate of change of an object's position with time, without regard to the direction of movement. Possible units include meters per second (m/s) and kilometers per hour (km/h). Compare *velocity.* (Ch. 3)

spherically symmetric Describing an object whose properties depend only on distance from the object's center, so that the object has the same form viewed from any direction. (Ch. 4)

spin-orbit resonance A relationship between the orbital and rotation periods of an object such that the ratio of their periods can be expressed with simple integers. (Ch. 11)

spiral density wave A stable, spiral-shaped change in the local gravity of a galactic disk that can be produced by periodic gravitational kicks from neighboring galaxies or from nonspherical bulges and bars in spiral galaxies. (Ch. 20)

spiral galaxy A galaxy of Hubble type "S" class, with a discernible disk in which large spiral patterns exist. Compare *barred spiral galaxy, elliptical galaxy, irregular galaxy,* and *S0 galaxy.* (Ch. 19)

spoke One of several narrow radial features seen occasionally in Saturn's B Ring. Spokes appear dark in backscattered light and bright in forward, scattering light, indicating that they are composed of tiny particles. Their origin is not well understood. (Ch. 11)

spreading center A zone from which two tectonic plates diverge. (Ch. 8)

spring tide An especially strong tide that occurs around the time of a new or full Moon, when lunar tides and solar tides reinforce each other. Compare *neap tide.* (Ch. 4)

stable equilibrium An equilibrium state in which the system returns to its former condition after a small disturbance. Compare *unstable equilibrium.* (Ch. 5)

standard candle An object whose luminosity either is known or can be predicted in a distance-independent way, so its brightness can be used to determine its distance via the inverse square law of radiation. (Ch. 19)

standard model The theory of particle physics that combines electroweak theory with quantum chromodynamics to describe the structure of known forms of matter. (Ch. 22)

star A luminous ball of gas held together by gravity. A normal star is powered by nuclear reactions in its interior. (Ch. 1)

star cluster A group of stars that all formed at the same time and in the same general location. (Ch. 15)

static equilibrium A state in which the forces within a system are all in balance so that the system does not change. Compare *dynamic equilibrium.* (Ch. 5)

Stefan-Boltzmann constant (σ) The constant of proportionality that relates the flux emitted by an object to the fourth power of its absolute temperature. Value: 5.67×10^{-8} W/(m^2 K^4) (W = watts, m = meters, K = kelvins). (Ch. 5)

Stefan-Boltzmann law The law, formulated by Josef Stefan and Ludwig Boltzmann, stating that the amount of electromagnetic energy emitted from the surface of a body (flux), summed over the energies of all photons of all wavelengths emitted, is proportional to the fourth power of the temperature of the body: $\mathcal{F} = \sigma T^4$. (Ch. 5)

stellar evolution The stages of development that a star goes through over its life cycle. Astronomers observe many stars at the same time to build a picture of how stars change. (Ch. 16)

stellar-mass loss The loss of mass from the outermost parts of a star's atmosphere during the star's evolution. (Ch. 16)

stellar occultation An event in which a planet or other Solar System body moves between the observer and a star, eclipsing the light emitted by that star. (Ch. 10)

stellar population A group of stars with similar ages, chemical compositions, and dynamic properties. (Ch. 17)

stereoscopic vision The way an animal's brain combines the different information from its two eyes to perceive the distances to objects around it. (Ch. 13)

stony-iron meteorite A meteorite consisting of a mixture of silicate minerals and iron-nickel alloys. Compare *iron meteorite* and *stony meteorite.* (Ch. 12)

stony meteorite A meteorite composed primarily of silicate minerals, similar to those found on Earth. Compare *iron meteorite* and *stony-iron meteorite.* (Ch. 12)

stratosphere The atmospheric layer immediately above the troposphere. On Earth, it extends upward to an altitude of 50 kilometers (km). Compare *troposphere, mesosphere,* and *thermosphere.* (Ch. 9)

string theory See *superstring theory.* (Ch. 22)

stromatolite A structure created by living or fossilized cyanobacteria. (Ch. 24)

strong nuclear force The attractive short-range force between protons and neutrons that holds atomic nuclei together. One of the four fundamental forces of nature, along with the *electromagnetic force,* the *weak nuclear force,* and *gravity* (definition 1). The force is mediated by the exchange of gluons. (Ch. 14, 22)

subduction zone A region where two tectonic plates converge, with one plate sliding under the other and being drawn downward into the interior. (Ch. 8)

subgiant A giant star that is smaller and lower in luminosity than normal giant stars of the same spectral type. Subgiants evolve to become giants. (Ch. 16)

subgiant branch A region of the H-R diagram defined by stars that have left the main sequence but have not yet reached the red giant branch. (Ch. 16)

sublimation The process by which a solid becomes a gas without first becoming a liquid. (Ch. 12)

summer solstice 1. One of two points where the Sun is at its greatest distance from the celestial equator. 2. The day on which the Sun appears at that location, marking the first day of summer (about June 20 in the Northern Hemisphere and December 21 in the Southern Hemisphere). Compare *winter solstice.* See also *autumnal equinox* and *vernal equinox.* (Ch. 2)

Sun The star at the center of the Solar System. (Ch. 1)

sungrazer A comet whose perihelion is within a few solar diameters of the surface of the Sun. (Ch. 12)

sunspot A cooler, transitory region on the solar surface produced when loops of magnetic flux break through the surface of the Sun. (Ch. 14)

sunspot cycle The approximate 11-year cycle during which sunspot activity increases and then decreases. This is one-half of a full 22-year cycle, in which the magnetic polarity of the Sun first reverses and then returns to its original configuration. (Ch. 14)

supercluster A large conglomeration of galaxy clusters and galaxy groups, typically more than 100–300 megaparsecs (Mpc) in size and containing tens of thousands to hundreds of thousands of galaxies. Compare *galaxy cluster* and *galaxy group.* (Ch. 1, 23)

super-Earth An extrasolar planet with about 2–10 times the mass of Earth. (Ch. 7, 24)

superior planet A Solar System planet that orbits farther from the Sun than Earth does. Compare *inferior planet.* (Ch. 3)

superluminal motion The appearance (though not the reality) that a jet is moving faster than the speed of light. (Ch. 21)

supermassive black hole A black hole of 1,000 solar masses (M_{Sun}) or more that resides in the center of a galaxy and whose gravity powers active galactic nuclei. (Ch. 19)

supernova (pl. **supernovae**) A stellar explosion resulting in the release of tremendous amounts of energy, including the high-speed ejection of matter into the interstellar medium. See also *Type Ia supernova* and *Type II supernova.* (Ch. 16)

supernova remnant The material ejected from the outer layers of a star after a supernova explosion. (Ch. 16)

supersonic Moving within a medium at a speed faster than the speed of sound in that medium. Compare *subsonic.* (Ch. 15)

superstring theory The theory that conceives of particles as strings in 10 dimensions of space and time; the current contender for a theory of everything. (Ch. 22)

surface brightness The amount of electromagnetic radiation emitted or reflected per unit area. (Ch. 6)

surface wave A seismic wave that travels on the surface of a planet or moon. (Ch. 8)

symmetry In theoretical physics, the properties of physical laws that remain constant when certain things change, such as the symmetry between matter and antimatter even though their charges may be different. (Ch. 22)

synchronous rotation The case that occurs when a body's rotation period equals its orbital period around another body. A special type of spin-orbit resonance. (Ch. 2)

synchrotron radiation Radiation from electrons moving at close to the speed of light as they spiral in a strong magnetic field, so named because that kind of radiation was first identified on Earth in particle accelerators called synchrotrons. (Ch. 10)

synodic period An object's orbital or rotational period measured with respect to the Sun. Compare *sidereal period.* (Ch. 2)

T

T Tauri star A young stellar object that has dispersed enough of the material surrounding it to be seen in visible light. (Ch. 15)

tail A stream of gas and dust swept away from the coma of a comet by the solar wind and by radiation pressure from the Sun. See also *ion tail* and *dust tail.* (Ch. 12)

tectonism Deformation of the lithosphere of a planet. (Ch. 8)

telescope The basic tool of astronomers. Working over the entire range from gamma rays to radio waves, astronomical telescopes collect and concentrate electromagnetic radiation from celestial objects. (Ch. 6)

temperature A measure of the average kinetic energy of the atoms or molecules in a gas, solid, or liquid. (Ch. 5)

terrestrial planet An Earth-like planet, made of rock and metal and having a solid surface. In the Solar System, the terrestrial planets are Mercury, Venus, Earth, and Mars. Compare *giant planet.* (Ch. 7)

theoretical model A description of the properties of a particular object or system in terms of known physical laws or theories; often, a computer calculation of predicted properties based on such a description. (Ch. 1)

theory A well-developed idea or group of ideas that are tied solidly to known physical laws and make testable predictions about the world. A very well-tested theory may be called a *physical law*, or simply a fact. Compare *hypothesis.* (Ch. 1)

theory of everything (TOE) A theory that unifies all four fundamental forces of nature: strong nuclear, weak nuclear, electromagnetic, and gravitational forces. (Ch. 22)

thermal conduction See *conduction.* (Ch. 14)

thermal energy The energy that resides in the random motion of atoms, molecules, and particles, by which we measure their temperature. (Ch. 5)

thermal equilibrium The state in which the rate of thermal-energy emission by an object is equal to the rate of thermal-energy absorption. (Ch. 5)

thermal motion The random motion of atoms, molecules, and particles that gives rise to thermal radiation. (Ch. 5)

thermal radiation Electromagnetic radiation resulting from the random motion of the charged particles in every substance. (Ch. 5)

thermosphere The layer of Earth's atmosphere at altitudes greater than 90 kilometers (km), above the mesosphere. Near its top, at an altitude of 600 km, the temperature can reach 1000 K. Compare *troposphere, stratosphere,* and *mesosphere.* (Ch. 9)

third quarter Moon The phase of the Moon in which only the eastern half of the Moon, as viewed from Earth, is illuminated by the Sun. It occurs about one week after the full Moon. Compare *first quarter Moon.* See also *full Moon* and *new Moon.* (Ch. 2)

tidal bulge Distortion of a body resulting from tidal stresses. (Ch. 4)

tidal force A force caused by the change in the strength of gravity across an object. (Ch. 4)

tidal locking Synchronous rotation of an object caused by internal friction as the object rotates through its tidal bulge. (Ch. 4)

tide On Earth, the rise and fall of the oceans as Earth rotates through a tidal bulge caused by the Moon and the Sun. See also *lunar tide, neap tide, solar tide,* and *spring tide.* (Ch. 4)

time dilation The relativistic "stretching" of time. Compare *general relativistic time dilation.* (Ch. 18)

TOE See *theory of everything.* (Ch. 22)

Topography The study of the natural physical features of the surface of a planet or moon (Ch. 8)

tornado A violent rotating column of air, typically 75 meters across with 200-kilometer-per-hour (km/h) winds. Some tornadoes can be more than 3 km across, and winds up to 500 km/h have been observed. (Ch. 9)

torus (pl. **tori**) A three-dimensional, doughnut-shaped ring. (Ch. 10, 19)

total lunar eclipse A lunar eclipse in which the Moon passes through the umbra of Earth's shadow. Compare *penumbral lunar eclipse.* (Ch. 2)

total solar eclipse The type of eclipse that occurs when Earth passes through the umbra of the Moon's shadow, so that the Moon blocks the disk of the Sun. Compare *annular solar eclipse* and *partial solar eclipse.* (Ch. 2)

transit method A method of detecting extrasolar planets by measuring the decrease in light from a star as its orbiting planet passes in front of the star as viewed from Earth. (Ch. 7)

trans-Neptunian object See *Kuiper Belt object.* (Ch. 12)

transverse wave A wave that oscillates perpendicular to the direction of the wave's propagation. Compare *longitudinal wave.* (Ch. 8)

triple-alpha process The nuclear fusion reaction that combines three helium nuclei (alpha particles) into a single nucleus of carbon. (Ch. 16)

Trojans A group of asteroids orbiting in the L_4 and L_5 Lagrangian points of Jupiter's orbit. (Ch. 12)

tropical year The time between two crossings of the vernal equinox. Because of the precession of the equinoxes, a tropical year is slightly shorter than the time Earth takes to orbit once around the Sun. Compare *year.* (Ch. 2)

Tropics The region on Earth between latitudes 23.5° south and 23.5° north, where the Sun appears directly overhead twice during the year. (Ch. 2)

tropopause The top of a planet's troposphere. (Ch. 9)

troposphere The convection-dominated layer of a planet's atmosphere. On Earth, the atmospheric region closest to the ground within which most weather phenomena take place. Compare *stratosphere, mesosphere,* and *thermosphere.* (Ch. 9)

turbulence The random motion of blobs of gas within a larger cloud of gas. (Ch. 15)

Type Ia supernova A supernova explosion with a calibrated peak luminosity that occurs as a result of runaway carbon fusion in a white dwarf star that accretes mass from a companion and approaches the Chandrasekhar mass limit of 1.4 M_{Sun}. (Ch. 16)

Type II supernova A supernova explosion in which the degenerate core of an evolved massive star suddenly collapses and rebounds. (Ch. 17)

U

ultrafaint dwarf galaxy A dim dwarf galaxy with only 1,000–100,000 times the Sun's luminosity. Ultrafaint dwarf galaxies differ from globular clusters in that they are composed of large amounts of dark matter. (Ch. 23)

ultraviolet (UV) radiation Electromagnetic radiation having frequencies and photon energies greater than those of visible light but less than those of X-rays and having wavelengths shorter than those of visible light but longer than those of X-rays. (Ch. 5)

umbra (pl. **umbrae**) 1. The darkest part of a shadow, where the source of light is blocked. Compare *penumbra* (definition 1). 2. The darkest, innermost part of a sunspot. Compare *penumbra* (definition 2). (Ch. 2, 14)

unbound orbit An orbit in which an object is no longer gravitationally bound to the body it was orbiting. An unbound orbit's velocity is greater than the escape velocity. Compare *bound orbit.* (Ch. 4)

uncertainty principle See *Heisenberg uncertainty principle.* (Ch. 22)

unified model of AGN A model in which many types of activity in the nuclei of galaxies are all explained by accretion of matter around a supermassive black hole. (Ch. 19)

uniform circular motion Motion in a circular path at a constant speed. (Ch. 4)

unit A fundamental quantity of measurement. The meter is an example of a metric unit; the foot is an example of an English unit. (Ch. 1, Appendix 1)

universal gravitational constant (*G*) The constant of proportionality in the universal law of gravitation. Value: 6.67×10^{-11} meters cubed per kilogram second squared [$m^3/kg\,s^2 = N\,m^2/kg^2$]. (Ch. 4)

universal law of gravitation The law, formulated by Isaac Newton, stating that the gravitational force between any two objects is proportional to the product of their masses and inversely proportional to the square of the distance between them: $F_{grav} = G \times \frac{m_1 \times m_2}{r^2}$ (Ch. 4)

universe 1. All of space and everything contained therein. 2. Our own universe in a collection of parallel universes that together make up all that exists. (Ch. 1)

unstable equilibrium An equilibrium state in which a small disturbance will cause a system to move away from equilibrium. Compare *stable equilibrium*. (Ch. 5)

UV Ultraviolet. See *ultraviolet radiation*. (Ch. 5)

V

vacuum A region of space devoid of matter. In quantum mechanics and general relativity, however, even a perfect vacuum has physical properties. (Ch. 5, 22)

variable star A star with varying luminosity. Many periodic variables are found within the instability strip on the H-R diagram. (Ch. 17)

velocity The rate and direction of change of an object's position with time. Possible units include meters per second (m/s) and kilometers per hour (km/h). Compare *speed*. (Ch. 3)

vernal equinox 1. One of two points where the Sun crosses the celestial equator. 2. The day on which the Sun appears at that location, marking the first day of spring (about March 20 in the Northern Hemisphere and September 22 in the Southern Hemisphere). Compare *autumnal equinox*. See also *summer solstice* and *winter solstice*. (Ch. 2)

virtual particle A particle that, according to quantum mechanics, comes into existence only momentarily. According to theory, fundamental forces are mediated by the exchange of virtual particles. (Ch. 18)

visual binary A binary system in which the two stars can be seen individually from Earth. Compare *eclipsing binary* and *spectroscopic binary*. (Ch. 13)

void A region in space containing little or no matter. Examples include regions in cosmological space that are largely empty of galaxies. (Ch. 23)

volatile material Generally called *ice* in its solid form. Material that remains gaseous at moderate temperature. Compare *refractory material*. (Ch. 7)

volcanism A form of geological activity on a planet or moon in which molten rock (magma) erupts onto the surface. (Ch. 8)

W

W See *watt*. (Ch. 5)

waning Describing the changing phases of the Moon as it becomes less fully illuminated between full Moon and new Moon as seen from Earth. Compare *waxing*. (Ch. 2)

waning crescent Moon The phases of the Moon between third quarter and new Moon (Ch. 2)

waning gibbous Moon The phases of the moon between full Moon and third quarter. (Ch. 2)

water cycle The flow of water on, above, and through Earth's surface. (Ch. 9)

watt (W) A measure of *power*. Unit: joules per second (J/s). (Ch. 5)

wave A disturbance moving along a surface or passing through a space or a medium. (Ch. 5)

wavefront The imaginary surface of an electromagnetic wave, either plane or spherical, oriented perpendicular to the direction of travel. (Ch. 6)

wavelength The distance on a wave between two adjacent points having identical characteristics. The distance a wave travels in one period. Possible units include meters (m). (Ch. 5)

waxing Describing the changing phases of the Moon as it becomes more fully illuminated between new Moon and full Moon as seen from Earth. Compare *waning*. (Ch. 2)

waxing crescent Moon The phases of the Moon between new and first quarter. (Ch. 2)

waxing gibbous Moon The phases of the Moon between first quarter and full. (Ch. 2)

weak nuclear force The force underlying some forms of radioactivity and certain interactions between subatomic particles. It is responsible for radioactive beta decay and for the initial proton-proton interactions that lead to nuclear fusion in the Sun and other stars. One of the four fundamental forces of nature, along with the *electromagnetic force*, the *strong nuclear force*, and *gravity* (definition 1). The force is mediated by the exchange of *W* and *Z* particles. (Ch. 22)

weather The state of an atmosphere at any given time and place. Compare *climate*. (Ch. 9)

weight The gravitational force acting on an object, that is, the force equal to the mass of an object multiplied by the local acceleration due to gravity. In general relativity, the force equal to the mass of an object multiplied by the acceleration of the frame of reference in which the object is observed. Compare *mass*. (Ch. 4)

white dwarf The stellar remnant left at the end of the evolution of a low-mass star. A typical white dwarf has a mass of 0.6 solar mass (M_{Sun}) and a size about equal to that of Earth; it is made of non-fusing, electron-degenerate carbon. (Ch. 16)

Wien's law A law, named for Wilhelm Wien, stating that location of the peak wavelength in the electromagnetic spectrum of an object is inversely proportional to the temperature of the object. (Ch. 5)

WIMP Short for *weakly interacting massive particle*. A hypothetical massive particle that interacts through the weak nuclear force and gravity but not with electromagnetic radiation. WIMPs are candidates for dark matter. Compare *MaCHO*. (Ch. 19)

winter solstice 1. One of two points where the Sun is at its greatest distance from the celestial equator. 2. The day on which the Sun appears at that location, marking the first day of winter (about December 21 in the Northern Hemisphere and June 20 in the Southern Hemisphere). Compare *summer solstice*. See also *autumnal equinox* and *vernal equinox*. (Ch. 2)

X

X-ray Electromagnetic radiation having frequencies and photon energies greater than those of ultraviolet (UV) light but less than those of gamma rays and having wavelengths shorter than those of UV light but longer than those of gamma rays. (Ch. 5)

X-ray binary A binary system in which mass from an evolving star spills over onto a collapsed companion, such as a neutron star or black hole. The material falling in is heated to such high temperatures that it glows brightly in X-rays. (Ch. 17)

Y

year The time Earth takes to make one revolution around the Sun. A solar year is measured from equinox to equinox. A sidereal year, Earth's true orbital period, is measured relative to the stars. Compare *tropical year*. (Ch. 2)

Z

zenith The point on the celestial sphere located directly overhead of an observer. Compare *nadir*. (Ch. 2)

zero-age main sequence The strip on the H-R diagram plotting where stars of all masses in a cluster begin their lives. (Ch. 15)

zodiac The 12 constellations lying along the plane of the ecliptic. (Ch. 2)

zodiacal dust Particles of cometary and asteroidal debris smaller than 100 microns (μm) that orbit the inner Solar System close to the plane of the ecliptic. Compare *meteoroid* and *planetesimal*. (Ch. 12)

zodiacal light A band of light in the night sky caused by sunlight reflected by zodiacal dust. (Ch. 12)

zonal wind The east–west component of a wind. (Ch. 9)

Selected Answers

These pages contain selected answers for all 24 chapters in the complete volume of *21st Century Astronomy*, Sixth Edition. The Solar System Edition does not include Chapters 15–23. The Stars and Galaxies Edition does not include Chapters 8–12.

Chapter 1

CHECK YOUR UNDERSTANDING

1.1 (d, a, e, c, f, b)
1.2 (b)
1.3 (c)

TEST YOUR UNDERSTANDING

4. (a)
7. (c)
10. (d)
11. (c)
14. (b)

THINKING ABOUT THE CONCEPTS

18. 8.3 minutes (see Figure 1.3).
23. A *theory* is generally understood to mean an idea, regardless of any proof, evidence, or way to test it. A *scientific theory* is an explanation for an occurrence in nature that must be based on observations and data and make testable predictions.
29. Some scientific fields rely heavily on math, but not all. Following the scientific method is the important part, not use of mathematics.

APPLYING THE CONCEPTS

31. (a) 7×10^9 (b) 3.46×10^{-3} (c) 1.238×10^3
36. 5,333 km/hr. 6.7 times faster than a jet plane.

Chapter 2

CHECK YOUR UNDERSTANDING

2.1A (b)
2.1B (d)
2.2 (d)
2.3 A full Moon is rising as the Sun sets. At noon, a third-quarter Moon will be setting.
2.4 If the synodic period of the Moon phases were precisely 30 days, and Earth took precisely 12 months to orbit the Sun, the current "wandering" holidays would always fall on the same days from year to year.
2.5 (b)

TEST YOUR UNDERSTANDING

1. (c)
4. (b)
6. (e)
9. (d)
14. (b)

THINKING ABOUT THE CONCEPTS

16. Magellan could not use the North Star (Polaris) for navigation because he was in the Southern Hemisphere; thus, Polaris was never above the horizon.
18. If Gemini is high in the night sky in winter, it is high in the daytime sky in summer, and the Sun's light blots out all light from the stars.
23. (a) The Earth takes 24 hours to complete one rotation (about its axis) with respect to the Sun—23 h 56 m with respect to the stars. (b) The Earth takes 26,000 years to complete one "wobble."

APPLYING THE CONCEPTS

31. 12,795 km.
34. 25° N latitude. Hawaii, southern Florida.

Chapter 3

CHECK YOUR UNDERSTANDING

3.1A Ancient civilizations knew planets were different from stars because planets wandered through the sky—hence the name planet, from the Greek *planetes,* meaning "wanderer."
3.1B (c)
3.2 (c, b, d, a)
3.3 (c, d)
3.4 (b)

TEST YOUR UNDERSTANDING

1. (d)
5. (a)
9. (a)
12. (c)
14. (b)

THINKING ABOUT THE CONCEPTS

16. During normal motion, Mars moves toward the east. Retrograde motion for Mars is motion toward the west. In this image, Mars was in retrograde motion for about 2.5 months, or about 75 days. Mars observers would have seen Earth in retrograde motion as well.
18. Saturn takes 29.5 years to orbit the Sun, so after 1 Earth orbit it has not moved far in its orbit. Earth catches up to and passes Saturn is just over one Earth year.
27. We wear seat belts to protect us from acceleration. We travel at the same speed as the vehicle, but because we have inertia of our own, we will keep moving forward (Newton's first law) until something stops us (that is, the seat belt) if the vehicle suddenly decelerates.

APPLYING THE CONCEPTS

33. 354 years.
42. 11 days.

Chapter 4

CHECK YOUR UNDERSTANDING

4.1A (d)
4.1B (c), (a)
4.2 (c, a, b)
4.3 (d, b, a, c)
4.4 (a)

TEST YOUR UNDERSTANDING

3. (c)
5. (c)
8. (a)
10. (b)
13. (b)

THINKING ABOUT THE CONCEPTS

17. Circular velocity is the speed needed to orbit a body in a circular path. Escape velocity is the speed needed to escape the gravitational force of a body. Escape velocity will always be larger because more energy is needed to escape the gravitational force of an object.
20. The comet in an elliptical orbit will eventually return to the immediate vicinity of the Sun, depending on the size of its orbit. The comet in an unbound orbit will escape the Sun entirely and head out into interstellar space, never to return.
28. Solid bodies can be tidally distorted, and tidal distortion occurs on Earth. That distortion of the planet's landmasses is dissipated in frictional heating of rock.

APPLYING THE CONCEPTS

32. 5.4 km/s.
37 9.81 m/s^2

Chapter 5

CHECK YOUR UNDERSTANDING

5.1A (e, c, b, d, a)
5.1B (c)
5.2 The spectral features correspond to the composition of elements present in the stars' atmospheres. Each element—even its ions and isotopes—has a unique set of lines.
5.3 (d)
5.4 (d, b, a, c, e)
5.5 (c)

TEST YOUR UNDERSTANDING

2. (b)
6. (d)
8. (a)
11. (c)
15. (b)

THINKING ABOUT THE CONCEPTS

16. Light can travel at a speed slower than 3×10^5 km/s when it is traveling in a medium. Light, like sound, travels at a certain speed in each kind of medium.
19. A beam of red light can have as much energy as a blue beam if the red beam has many more photons.
23. Redshifted objects are moving away from us; blueshifted objects are moving toward us.

APPLYING THE CONCEPTS

33. 472.4 Hz.
36. 0.005 percent of its brightness as seen from Earth.

Chapter 6

CHECK YOUR UNDERSTANDING

6.1A (a, b, c)
6.1B (a)
6.2 (c)
6.3 (d)
6.4 (c)
6.5 (a, b, d)

TEST YOUR UNDERSTANDING

4. (b)
7. (a)
10. (b)
13. (d)
15. (b)

THINKING ABOUT THE CONCEPTS

16. Simple lenses have chromatic aberration; they bend different colors by different amounts. The resulting image is not nearly as crisp as in telescopes with multiple lenses.
19. Different types of astronomical objects are at different temperatures and thus emit light at different peak wavelengths. Many astronomical sources that look normal at visible wavelengths may be much brighter or more complex in other regions of the spectrum.
25. The stars are so far away that a telescope in space is only infinitesimally closer to the stars than a telescope on Earth. Telescopes in space avoid atmospheric effects and can observe at wavelengths or energies that Earth's atmosphere blocks.

APPLYING THE CONCEPTS

33. 16 times more light. Stars 1/4 as bright, or stars of 1 to 2 magnitudes dimmer.
40. (a) 0.35 milliarcseconds
(b) 1.1 milliarcseconds

Chapter 7

CHECK YOUR UNDERSTANDING

7.1 (a, b, d)
7.2 (a)
7.3 (b)
7.4 (a)
7.5 (c)

TEST YOUR UNDERSTANDING

1. (a, b, h, c, e, d, f, g)
3. (b)
8. (b)
14. (c)
15. (d)

THINKING ABOUT THE CONCEPTS

16. The hydrogen and most helium atoms that make up the Sun and the rest of the Solar System all came from the Big Bang. Everything else was formed in previous generations of stars.
20. Scientists will have to propose a formation mechanism consistent with all other evidence for planet formation. Scientists will consider all the relevant physics, propose the most likely scenarios, and then propose tests of those scenarios and search to validate or invalidate them.
29. Many of the initial discoveries of exoplanets were made using the radial velocity method, which at the time could do no better than resolve massive planets in orbits close to their stars with periods of about a week.

APPLYING THE CONCEPTS

31. 30 m/s, about 70 mph. 0.1 percent
34. 0.003

Chapter 8

CHECK YOUR UNDERSTANDING

8.1 (b)
8.2 (d)
8.3A (d)
8.3B (c)
8.4 (c)
8.5 (b)
8.6 (a, c, d, e)

TEST YOUR UNDERSTANDING

3. (e)
6. (c)
7. (c)
12. (a)
15. (a)

THINKING ABOUT THE CONCEPTS

18. Rock ages can be measured with radiometric dating. Rocks get older as one goes deeper into the canyon.
20. The Moon is smaller than Earth, so the Moon contained less thermal energy in its core to begin with, and a smaller core cools faster than a larger one. The core of Earth also contains more radioactive material than the core of the Moon.
23. The impactor theory of the Moon is supported by the compositional similarities of Earth and the Moon, the lack of volatiles on the Moon, and the inclination of Earth's spin axis. However, evidence of the impacting object's own composition has yet to be observed.

APPLYING THE CONCEPTS

34. Figure 8.18 shows a relatively smooth surface with lots of small craters. Figure 8.23 also shows a very smooth surface with fewer small craters, suggesting that this region is younger.
41. (a) 6 half-lives old.
    (b) 34,200 years old.

Chapter 9

CHECK YOUR UNDERSTANDING

9.1 (a, b, c, d)
9.2 (c)
9.3A (c, a, e, b, d)
9.3B (a)
9.4 (a, d, c, b)
9.5 (a, b, c)

TEST YOUR UNDERSTANDING

1. (d, b, c, f, a, e)
4. (c)
7. (b)
10. (a)
11. (c)

THINKING ABOUT THE CONCEPTS

17. After the primary atmospheres were stripped from the terrestrial planets, the planets outgassed volatile material from their interiors. The gas expelled from planetary interiors combined with those materials delivered by comet impacts to produce secondary atmospheres.
22. The troposphere is the lowest layer of the atmosphere; that is where ozone is a pollutant. In the higher layers (mainly the stratosphere), ozone absorbs UV photons, protecting the planet from that harmful radiation.
27. The slow rotation of Venus means that strong wind currents do not occur across the planet's longitude, only in latitude, so hot gas at the equator is constantly mixed with cooler gas at the poles to make a uniform temperature. Venus has a thick atmosphere which traps heat efficiently.

APPLYING THE CONCEPTS

31. (a) The axes are linear.
    (b) The variations of CO_2 and temperature are correlated.
    (c) The highest CO_2 level was about 300 ppm, but it is 410 ppm today.
36. 1.8×10^5 N.
    (b) $\approx$ 20,700 kg.
    (c) $\approx$ 350 kangaroos on a person's head.

Chapter 10

CHECK YOUR UNDERSTANDING

10.1 (a)
10.2 (b)
10.3 (b)
10.4 (c)
10.5 (a, d, b, c)

TEST YOUR UNDERSTANDING

2. (b)
6. (b)
8. (a)
13. (d)
15. (d)

THINKING ABOUT THE CONCEPTS

20. Uranus is discovered → Hypothesis: We can predict its orbit → Prediction: On the basis of its distance from the Sun, we predict Uranus's orbit → Test: Uranus's orbit does not match predictions → FAIL → New hypothesis: Something might be pulling on Uranus → New prediction: Another planet might exist → New test: Find the planet → Success: We found Neptune.
25. They measure the positions of clouds as they move across the surface of the planet, and see how they vary in latitude. They also compare these speeds with the rotation speed of the interior of the planet, according to its magnetic fields.
30. Migration of the outer giant planets could leave inner ones alone completely as they eject objects orbiting nearby, or they could completely disrupt their orbits and send them flying out into space.

APPLYING THE CONCEPTS

33. The interiors of Jupiter and Saturn reach very high pressures and temperatures over 10,000 K, violent and hot enough to strip the electrons off the hydrogen atoms.
44. 0.09 atm/km, 0.04 atm/km, 1.33 atm/km, 4 atm/km.

Chapter 11

CHECK YOUR UNDERSTANDING

11.1 (a, b, c)
11.2A (a)
11.2B (a, d, b, c)
11.3 (a)
11.4 (b)

TEST YOUR UNDERSTANDING

1. (a)
4. (d)
8. (a)
11. (a)
13. (d)

THINKING ABOUT THE CONCEPTS

17. Cryovolcanism is volcanic activity with eruptions of materials that are liquid at very low temperatures, such as water, ammonia, methane. Seen on Enceladus, Titan, Ganymede and others.
20. Titan resembles Earth with the nitrogen composition of its atmosphere and surface features similar to rivers and lakes. On Titan, methane appears to play the same role that water plays in the hydrologic cycle on Earth.
27. The discovery of kinks and knots in Saturn's rings meant that either our understanding of rings was completely wrong or that we had an incomplete theory. Because scientific understanding is provisional, our understanding of rings *was* correct *until* we learned more about them and required a theory that incorporated more physics. Thus, our understanding grew with more observations to become a more complete explanation.

APPLYING THE CONCEPTS

31. (a) 2.5 kilometers per second (km/s).
    (b) 2.5 times faster than the typical speed of gas escaping from Io's vents. That means most of the vented material will fall back on the moon rather than escape into space.
40. 14.3 hours

Chapter 12

CHECK YOUR UNDERSTANDING

12.1 According to the IAU definitions, those objects are dwarf planets and not moons because they orbit the Sun, not a planet.
12.2 (c)
12.3A (c)
12.3B (c)
12.4 (a, b, c, e)
12.5 The origins of a meteorite come from its precollision orbit as measured by multiple observations, as well as its structure, and its chemical makeup.

TEST YOUR UNDERSTANDING

2. (b)
5. (c)
8. (a)
11. (b)
15. (d)

THINKING ABOUT THE CONCEPTS

16. Pluto does not have a nearly circular orbit and does not orbit in the plane of the other planets, which suggests that its formation was different from that of the eight "classical" planets. Pluto also is not massive enough to have gravitationally swept out the space around it.
19. *Meteoroids* are particles of space debris smaller than about 100 meters in diameter. *Meteors* are ionized trails left by the passage of meteoroids and interplanetary dust through the upper atmosphere. *Meteorites* are chunks of meteoroids that survive their atmospheric passage to land on the surface.
28. Ion tails are pushed directly away from the comet by the charged particles of the solar wind, whereas the dust tails are pushed away both by solar wind and solar radiation. The two tails need not point in the same direction because the shape of the dust tail depends on the particle makeup of the dust and how it is accelerated away from the comet by the Sun's light.

APPLYING THE CONCEPTS

32. 340 km/hr
41. (a) 29 orbits.
    (b) 9×10^{12} kg lost.
    (c) 4 percent of its total mass.

Chapter 13

CHECK YOUR UNDERSTANDING

13.1 (b)
13.2 (d)
13.3 (a, b, d)
13.4 (b, d)

TEST YOUR UNDERSTANDING

1. (b)
5. (b)
10. (d)
12. (a)
15. (a)

THINKING ABOUT THE CONCEPTS

19. Every blackbody emits light at all wavelengths. A star with a 2500 K blackbody emits mostly in the red and infrared but still gives off a measurable amount of blue light.
20. Betelgeuse is cooler than Rigel. Nothing can be said about their luminosity or size without knowing their distance.
24. Binary stars can transit each other just like planets, so any mission looking for regular variability in the light from the stars is going to discover binary stars (and variable stars).

APPLYING THE CONCEPTS

33. The range in luminosities is about a hundred-billion (10^{11}); extremely hard to display in a linear plot. Logarithmic.
41. Betelgeuse; Achenar.

Chapter 14

CHECK YOUR UNDERSTANDING

14.1A (c)
14.1B (b)
14.2 (c)
14.3 (b)
14.4 (d)

TEST YOUR UNDERSTANDING

1. (a)
9. (d)
10. (c)
13. (c)
14. (d)

THINKING ABOUT THE CONCEPTS

17. To fuse two nuclei, very high temperatures (millions of degrees), pressures, and densities are required so the two nuclei can overcome their mutual repulsion and fuse. The oceans do not satisfy those conditions.
24. The Sun is made of gas and plasma only, so it has no actual surface at all. The photosphere is as deep into the Sun as we can see because it becomes opaque at the base of that layer.
30. The solar wind blows against the interstellar medium and clears out an area like the inside of a bubble. This area, the heliosphere, protects us from galactic high-energy cosmic rays as the Sun and Solar System move through our galaxy,

APPLYING THE CONCEPTS

31. (a) The height axis increases by adding 1,000 km to each tick mark; thus, it is linear.
    (b) The density axis increases by a multiplicative factor of 100 for each tick mark; thus, it is logarithmic.
40. 4.3 days.

Chapter 15

CHECK YOUR UNDERSTANDING

15.1 (a, d)
15.2 (c)
15.3 (a)
15.4 (b)

TEST YOUR UNDERSTANDING

4. (c)
5. (d)
8. (c)
10. (c)
15. (a)

THINKING ABOUT THE CONCEPTS

17. Star formation occurs in the dense, cold, and dusty interstellar medium, where dust blocks visible light. Infrared radiation is less affected by interstellar dust. Also protostars emit significantly in the infrared.
26. Smaller regions (clumps) can be denser and contract into individual stars. As the cloud collapses, the central regions of the protostars in those clumps become more massive, more compressed, and hotter. Eventually, the center of each protostar gets hot and dense enough for fusion to start, and a single star is born.
28. Some brown dwarfs in the ~10–80 $M_{Jupiter}$ mass range can temporarily have low levels of nuclear fusion in their cores; others cannot. Anything larger than about 10–20 $M_{Jupiter}$ is usually considered a planet.

APPLYING THE CONCEPTS

34. 6300 K, 4030 K. Dust has a significant effect on dimming and reddening light from stars.
45. ~1/112,000 L_{Sun}, so 112,000 brown dwarfs would produce the luminosity of the Sun.

Chapter 16

CHECK YOUR UNDERSTANDING

16.1 (a)
16.2 (b)
16.3 (a)
16.4 (a)
16.5 (d)

TEST YOUR UNDERSTANDING

2. (a, e, d, c, f, b, g)
5. (a)
8. (d)
9. (d)
12. (c)

THINKING ABOUT THE CONCEPTS

17. The energy generated would increase, which would be used to lift the atmosphere of the star, making it grow larger and cool down. That would drop the core temperature, returning the star to its original size and luminosity.
20. The brightness we measure won't change, so if all Type Ia supernovae are shown to be more luminous than previously thought, they must be farther away.
26. A white dwarf is just a cooling blackbody of constant size. As the white dwarf emits radiation and cools, its temperature will decrease, and hence its luminosity drops as well.

APPLYING THE CONCEPTS

37. 37 km/s.
44. $1.8 \times 10^{13} L_{Sun}$, $1.8 \times 10^{5} L_{Sun}$, $1.8 \times 10^{-3} L_{Sun}$, $1.8 \times 10^{-11} L_{Sun}$

Chapter 17

CHECK YOUR UNDERSTANDING

17.1 (b, d)
17.2 (b)
17.3 (c)
17.4 (b)

TEST YOUR UNDERSTANDING

2. (e, a, f, b, c, d)
3. (d)
7. (e)
12. (a)
15. (b)

THINKING ABOUT THE CONCEPTS

19. In each successive fusion cycle, fewer and fewer nuclei exist to be fused. Also, fusion of heavier nuclei requires much higher energy, so the collisions happen more often.
21. New generations of stars and their planets incorporate the heavy elements fused in supernova and scattered into the interstellar medium. Also the shock wave from the expanding supernova bubble can compress interstellar clouds, triggering their collapse into new stellar systems.
30. Although AGB stars create some C and O, most of the heavy elements in the universe were made in supernovae and distributed into space in their ejecta. Later generations of stars like the Sun (and its planets) formed from this enriched material.

APPLYING THE CONCEPTS

40. (a) 3.8×10^{23} kg; (b) 5.25 M_{Moon}/minute
43. 160 times the radius of a pulsar (10 km)

Chapter 18

CHECK YOUR UNDERSTANDING

18.1 (d)
18.2 (b)
18.3 (d)
18.4 (c)

TEST YOUR UNDERSTANDING

2. (c)
10. (a)
11. (c)
12. (d)
15. (c, d, f)

THINKING ABOUT THE CONCEPTS

18. A photon is a massless particle, and all massless particles travel at exactly the speed of light in a vacuum.
21. Moving clocks run slowly, so the person's pulse rate will appear extremely slow to you.
23. The astronaut is moving with the ship and is in the same reference frame; thus, he will see no length contraction. An outside viewer who is *not* moving at that high speed would observe length contraction.

APPLYING THE CONCEPTS

40. 10^{-25} meters
43. 3×10^{8} meters per second

Chapter 19

CHECK YOUR UNDERSTANDING

19.1A (d)
19.1B (c)
19.2 (c)
19.3 (a)
19.4 (c)

TEST YOUR UNDERSTANDING

2. (c)
7. (d)
10. (d)
11. (c)
15. (b)

THINKING ABOUT THE CONCEPTS

17. Cepheids that Hubble found in the Andromeda Galaxy (M31) showed that M31 was too far away to be part of our own Milky Way.
20. A Type Ia supernova is a standard candle only if we know exactly how luminous each one is. For that, we have to know the type of its progenitor.
30. In the crowded center of galaxies, several tidal encounters among stars could disrupt planetary orbits. Even if the center of a galaxy is not an active AGN, an overabundance of radiation harmful to life may still be present.

APPLYING THE CONCEPTS

32. 10^{12} galaxies
33. (a) 0.143
    (b) 4.28×10^7 m/s
    (c) 612 Mpc

Chapter 20

CHECK YOUR UNDERSTANDING

20.1 Answers will vary; may include comparisons, content, observations, motions, star formation.
20.2 The disk (and bulge) contains old and young stars; the halo contains only old stars.
20.3 (c)
20.4 Andromeda and the Milky Way are moving toward each other because of their mutual gravitational attraction.

TEST YOUR UNDERSTANDING

2. (a)
5. (a)
7. (c)
10. (b)
15. (d)

THINKING ABOUT THE CONCEPTS

18. Open clusters are young and therefore will generally be found near spiral arms and regions of recent star formation. Globular clusters are found primarily in the galactic halo.
23. Synchrotron radiation causes the disk of the galaxy to glow and it comes from cosmic rays, mostly electrons, spiraling through the galaxy's magnetic field.
30. Astronomers exclude regions of rapid star formation (both because of the UV light and frequent supernovae), the galactic center (high radiation), close stellar encounters, and regions that do not have enough heavy elements to form rocky planets.

APPLYING THE CONCEPTS

33. 1.8×10^9 yr
42. 2.1×10^3 km/s

Chapter 21

CHECK YOUR UNDERSTANDING

21.1 (b, a)
21.2 (d)
21.3A (c)
21.3B (b)
21.4 (a)

TEST YOUR UNDERSTANDING

1. (b)
2. (d)
8. all
11. (d)
12. (c)

THINKING ABOUT THE CONCEPTS

17. Space itself is expanding. Imagine space filled with a grid: the space between gridlines gets larger. See Figure 21.5.
21. Objects held together by their own self-gravity or the gravitational attraction of its contents do not expand. The expansion happens only on the largest scales, in which gravitational attraction is weaker than the expansion rate.
24. Regardless of your position in the universe, all distant galaxies recede from you as the universe expands. If you observe a distant galaxy moving rapidly away, an observer in that distant galaxy would say the Milky Way is rapidly moving away from them. The Big Bang took place everywhere; all observers, regardless of location, think they are at the center.

APPLYING THE CONCEPTS

34. ~12 billion years
39. (a) 4,100 Mpc; (c) 0.147

Chapter 22

CHECK YOUR UNDERSTANDING

22.1 (c)
22.2 (c)
22.3 Inflation solves the horizon and flatness problems.
22.4 (d)
22.5 (a)

TEST YOUR UNDERSTANDING

2. (c)
3. (a)
5. (a, f, j, b, e, g, d, i, h, c)
7. (c)
14. (b)

THINKING ABOUT THE CONCEPTS

17. When we look at distant galaxies and observe Type Ia supernovae, we see that their recessional velocity increases with distance faster than the linear relationship of Hubble's law, suggesting a nonzero cosmological constant. Einstein brought in the cosmological constant to explain why the universe didn't collapse in on itself.
20. Hubble time does not take into account the changing dynamics that result from having different cosmological densities and a cosmological constant (dark energy).
23. The early universe was *hot* and *dense*, which are situations that might be replicated in particle accelerators. At high temperatures and densities, all sorts of exotic forms of matter can exist, which may have substantially influenced how the early universe evolved.

APPLYING THE CONCEPTS

39. 6.1×10^{-10}
45. 1.2×10^{26}

Chapter 23

CHECK YOUR UNDERSTANDING

23.1 (c, b, d, a)
23.2 (b)
23.3 (c)
23.4 (b)

TEST YOUR UNDERSTANDING

3. (b)
5. (c)
8. (d)
11. (d)
14. (d)

THINKING ABOUT THE CONCEPTS

17. If voids were filled with considerable dark matter, we would expect to see gravitational lensing of galaxies behind those voids. We also would have expected galaxies to have formed inside the dark matter voids. So it is much more likely that they are fairly empty of any mass.
28. Hot dark matter could not be gravitationally bound enough to form the halos needed to contain galaxies and to explain galactic rotation curves.
30. The nuclear forces are too short ranged to work over cosmic distances, and electromagnetic forces work only for charged particles; most matter in the universe is neutral. That leaves only gravity.

APPLYING THE CONCEPTS

31. 1.05×10^7 m/s
36. (a) $10^{13}\ M_{Sun}$; (b) 10^{14} to $10^{15}\ M_{Sun}$; (c) 10^{16} to $10^{18}\ M_{Sun}$

Chapter 24

CHECK YOUR UNDERSTANDING

24.1 (d)
24.2 (a)
24.3A (d)
24.3B (a)
24.4 (b)

TEST YOUR UNDERSTANDING

4. (c)
7. (d)
12. (b)
14. (a)
15. (d)

THINKING ABOUT THE CONCEPTS

21. A burst of biodiversity around 500 million years ago. Possibly caused by a decrease in predators, a warming of Earth, or oxygen reaching high enough levels and with an adequate ozone layer for modern life to evolve.
26. All the hydrogen and most of the helium are from the Big Bang. All the other atoms in our bodies are from dying stars, especially supernovae.
30. Main-sequence stars slowly become more luminous. In a few billion years, the increase in the Sun's luminosity will cause the oceans to evaporate and generally scorch the Earth.

APPLYING THE CONCEPTS

33. 1,024 organisms
38. About 17 generations

Credits

Photo

Frontmatter

(in chronological order): NASA/JPL-Caltech/MSSS; © Damian Peach; NASA; robertharding/Alamy Stock Photo; Caltech/NASA/JPL; Paul Fearn/Alamy Stock Photo; NASA/JPL-Caltech/Univ. of Arizona; Stockli, Nelson, Hasler, Goddard Space Flight Center/NASA; NASA/JPL/Space Science Institute; ESA/Rosetta/NavCam, CC BY-SA IGO 3.0; Visuals Unlimited/Getty Images; NASA/SDO/AIA; NASA and ESA; NASA/CXC/SAO; NASA/ESA; Brad Goldpaint/Aurora Open/Alamy; NASA/JPL/Space Science Institute; Seth Shostak/SETI Institute; Caltech/NASA/JPL; Courtesy of Stacy Palen.

Author bio photos

Zac Williams; Courtesy of Laura Kay; Courtesy of George Blumenthal.

Chapter 1

Pages 2–3 NASA **5 both** NASA **6 second to bottom right** Ryan Somma, CC BY-SA **6 bottom right** NASA **11** Courtesy of the Archives, California Institute of Technology **12** Matheisl/Getty Images **15** ESA/NASA.

Chapter 2

Pages 22–23 Science Photo Library/Science Source **24 left** Roger Ressmeyer/Getty Images **24 center** J Gerard Sidaner/Getty Images **24 right** Henry Westheim Photography/Getty Images **28 right** Pekka Parviainen/Science Source **28 left** David Nunuk/Science Source **38** Arnulf Husmo/Getty Images **44** akg-images/Rabatti - Domingie/The Image Works **46** Laura Kay **47 top** Heinz-Peter Bader/Reuters/Newscom **47 bottom** Laura Kay and Lisa Rand **49 left** knickohr/Getty Images **49 right** © Wang Letian.

Chapter 3

Pages 58–59 © Damian Peach **60 top** Image Select/Art Resource, NY **60 bottom** © Tunc Tezel **61 center** Forum/UIG/AgeFotostock **61 bottom** Bettmann/Getty Images **64** Kean Collection/Getty Images **65 top** De Mundi Aetherei Recentioribus Phaenomenis, 1588. The Tycho Brahe Museum, Sweden. www.tychobrahe.com **65 center** Science & Society Picture Library/Getty Images **67 top** Universal Images/Superstock **67 top center** Iberfoto/Superstock **67 bottom center** Iberfoto/Superstock **67 bottom** Heinz-Dieter Falkenstein/age fotostock/Superstock **70 top** Leemage/Getty Images **70 bottom** The Granger Collection, NYC **71 top** Jamie Cooper/Getty Images **71 bottom** DEA/G. Nimatallah/Getty Images.

Chapter 4

Pages 82–83 NASA **91 top right** NASA **91 bottom right** NASA **91 bottom left** JSC/NASA **91 center left** JSC/NASA **91 top left** Heinz-Dieter Falkenstein/age fotostock/Superstock **97 top** Eric Carr/Alamy Stock Photo **97 bottom** Education Images/UIG via Getty Images **101** NASA.

Chapter 5

Pages 108–109 robertharding/Alamy Stock Photo **115 top left** Paul Fearn/Alamy Stock Photo **115 top right** Heinz-Dieter Falkenstein/age fotostock/Superstock **115 bottom left** Iberfoto/Superstock **115 bottom right** Underwood Photo Archives/Superstock **133 all** NASA **134** ESO/Y. Beletsky.

Chapter 6

Pages 140–141 Rick Peterson/W.M. Keck Observatory **142** ASU Physics Instructional Resource Team **143** Yerkes Observatory photographs: Richard Dreiser **145 top** ASU Physics Instructional Resource Team **145 bottom** Jim Sugar/Getty Images **147 top** Enrico Sacchetti/Science Source **147 center (both)** The Space Telescope Science Institute **147 bottom** ASU Physics Instructional Resource Team **149 bottom (both)** NASA, ESA, HEIC, and The Hubble Heritage Team (STScI/AURA) **150 top** ESO/L. Calada/ACe Consortium **150 bottom** NASA/NOAA **151 top** Hulton-Deutsch Collection/CORBIS/Corbis via Getty Images **151 bottom** Jean-Charles Cuillandre (CFHT) **153 top** Rick Peterson/W.M. Keck Observatory **153b** NASA **153c** NASA **153d** NASA **153e** NASA **153f** Joint Astronomy Center in Hilo, Hawaii **153g** Photo by Dave Finley, courtesy National Radio Astronomy Observatory and Associated Universities, Inc. **153h** NRAO/AUI, James J. Condon, John J. Broderick, and George A. Seielstad **153i** NASA/Courtesy of the Arecibo Observatory, a facility of the NSF **154 top** Roger Ressmeyer/Getty Images **154 center** NASA/Courtesy of the Arecibo Observatory, a facility of the NSF **154 bottom** Photo by Dave Finley, courtesy National Radio Astronomy Observatory and Associated Universities, Inc. **155 top** National Geographic Creative/Alamy Stock Photo **155 center** NASA Photo / Carla Thomas **bottom right** NASA/Tom Tschida **155 bottom** NASA Photo/Carla Thomas **bottom right** NASA/Tom Tschida **157 top** NASA/JPL-Caltech/MSSS **157 bottom** NASA/CXC/NGST **158 top** NASA/JPL-Caltech/MSSS **158 bottom** NASA/JPL-Caltech/MSSS **159 top** NASA/JPL/ESA, D. Ducros **159 bottom** NASA **160 top** CERN **160 bottom** Jim Haugen/NSF **161 center** LIGO Laboratory/CalTech/U.S. National Science Foundation **161 left** Volker Steger/Science Source **161 right** EADS Astrium/ESA **162 all** Patrik Jonsson, Greg Novak & Joel Primack, UC Santa Cruz, 2008.

Chapter 7

Pages 170–171 Caltech/NASA/JPL **173 top left** Science Source **173 top right** S. Andrews (Harvard-Smithsonian CfA); B. Saxton (NRAO/AUI/NSF); ALMA (ESO/NAOJ/NRAO) **173 bottom** Massimo Brega/Science Source **top** NASA/JPL-Caltech/T. Pyle **175 bottom** John Schults/ Reuters/Newscom **178** NASA **182** NASA/Johns Hopkins University Applied Physics Laboratory/Carnegie Institution of Washington **187** NASA **189** NASA Ames/SETI Institute/JPL-Caltech.

Chapter 8

Pages 196–197 NASA/JPL-Caltech/Univ. of Arizona **198** NASA **199 all** RGB Ventures/SuperStock/Alamy Stock Photo **200 top** D.J. Roddy and K.A. Zeller, USGS, Flagstaff, AZ **200 bottom** NASA/Goddard/MIT/Brown **201** NASA/JPL/Caltech **202** NASA Goddard **205 both** NASA **209** Spencer Grant/Science Source **210** Francois Gohier/Science Source **214 top** NASA **214 bottom left** NASA/JSC **214 bottom right** NSSDC/NASA **215** NASA/Magellan Image/JPL **216 top** Tomas Janececk **216 bottom** Jukurae/Shutterstock **217 top** NASA/JSC **217 center** NASA/JSC/Arizona State University **217 bottom** NASA/Johns Hopkins University Applied Physics Laboratory/Arizona State University/Carnegie Institution of Washington. Image reproduced courtesy of Science/AAAS. **218 top** Corbis via Getty Images **218 center and bottom** NASA/JPL/ESA, D. Ducros **219** NASA **220 top** NASA/JPL/University of Arizona **220 bottom** NASA **221 both** NASA **223** Joe Tucciarone/Science Source.

Chapter 9

Pages 230–231 Stockli, Nelson **240** NOAA **244 top** NASA **244 bottom** All Canada Photos/Alamy Stock Photo **247 top** JAXA/ISAS/DARTS/Damia Bouic **247 bottom** JAXA/ISAS/DARTS/Damia Bouic **248 top** NASA/JPL-Caltech/MSSS **248 bottom** NASA/JPL/Caltech **249 top** NASA/STScI **249 bottom** NASA/JPL-Caltech/UA **256 left** JAXA/ISAS/DARTS/Damia Bouic **256 center and right** NASA **257** NASA's Goddard Space Flight Center.

Chapter 10

Pages 264–265 NASA/SwRI/MSSS/Gerald Eichstadt/Sean Doran **266 all** NASA **267 all** NASA **268 top left** JPL/NASA **268 bottom right** JPL/NASA **269 top** NASA/JPL/Space Science Institute **269 bottom** NASA/JPL **270** NASA, ESA, and the Hubble Heritage Team (STScI/AURA) **271 top** NASA/SwRI/MSSS/Gerald Eichstadt/Sean Doran **271 bottom** NASA/JPL-Caltech/SwRI/ASI/INAF/JIRAM **272 top left** NASA **272 top right** NASA, ESA **272 bottom left** NASA/JPL/Caltech. **272 center right** NASA, ESA, and M.H. Wong and J. Tollefson (UC Berkeley) **272 bottom right** NASA **276** NASA/JPL/Caltech. **277** NASA, ESA, L. Sromovsky and P. Fry (University of Wisconsin), H. Hammel (Space Science Institute), and K. Rages (SETI Institute) **281** freenaturepictures.com/Wikimedia Commons **282 bottom left** NASA Goddard Photo and Video **282 both insets** NASA Goddard Photo and Video **282 bottom right** NASA Goddard Photo and Video **284** NASA's Goddard Space Flight Center/S. Wiessinger **286** NASA/JPL-Caltech/SwRI/MSSS/Jason Major **291 top** NASA/JPL **291 bottom (both)** NASA/JPL.

Chapter 11

Pages 292–293 NASA/JPL/Space Science Institute **294** NASA/JPL/Caltech **295 right** ESA/DLR/FU Berlin (G. Neukum) **295 left** NASA/JPL/University of Arizona **297 top** NASA **297 bottom** NASA/JPL/Caltech **297 bottom insets** NASA/JPL/Caltech **298 all** NASA/JPL-CalTech **299 bottom** NASA/JPL Caltech/UCLA/ MPS/DLR/IDA/ **299 top** NASA/JPL/Space Science Institute **300 top** NASA/JPL/Space Science Institute **300 bottom** NASA/JPL/Space Science Institute **300 center** NASA/JPL/University of Arizona/University of Idaho **301 top** NASA **301 bottom** NASA/JPL/USGS **302** NASA, ESA and A. Feild/STScI **303 top** NASA/JPL/Space Science Institute **303 bottom** NASA/JPL/Ted Stryk **304** NASA/JPL/Space Science Institute **306** NASA/JPL/Space Science Institute **307** NASA/JPL/Space Science Institute **308 top** NASA/JPL/Space Science Institute **308 bottom** NASA/JPL/Space Science Institute **309 bottom** NASA **309 center** NASA/JPL/Space Science Institute **309 top** Stephanie Swartz / Dreamstime **310 top** NASA **310 center** NASA/JPL/Space Science Institute **310 bottom** NASA/JPL/Space Science Institute **311 left** NASA/JPL-Caltech **311 center** NASA/JPL/Space Science Institute **311 right** NASA/JPL/Space Science Institute **312 bottom** NASA/JPL/Caltech. **312 top** NASA/JPL/Space Science Institute **313** NASA **319 all** NASA/JPL/Caltech.

Chapter 12

Pages 320–321 ESA/Rosetta/NavCam, CC BY-SA IGO 3.0 **322 center** NASA/JHUAPL/SwRI **322 bottom** NASA/JHUAPL/SwRI **322 top** NASA/JPL **323 center** JPL/NASA **323 top** NASA/JPL-Caltech/UCLA/MPS/DLR/IDA **323 bottom** NASA/JPL-Caltech/UCLA/MPS/DLR/IDA/ **324 right** JPL/NASA/Johns Hopkins University Applied Physics Laboratory/Southwest Research Institute **324 center** JPL/NASA/Johns Hopkins University Applied Physics Laboratory/Southwest Research Institute **324 left** NASA/JPL-Caltech **325** NASA **330 bottom** NASA/JPL-Caltech/UCLA/MPS/DLR/IDA **330 top** NASA/JPL/USGS **331 top** NASA/JPL-Caltech/UCLA/MPS/DLR/IDA **331 bottom** NASA/JPL-Caltech/UCLA/MPS/DLR/IDA **336 bottom** Courtesy of Terry Acomb **336 top** Science Source **337 top** NASA/JPL/Caltech **337 bottom left** NASA/JPL/Caltech/UMD **337 bottom right** NASA/JPL/Caltech/UMD. **338** NASA/JPL-Caltech/UMD **340 top** Alexandar/Shutterstock **340 bottom** Barrie Rokeach/Getty Images **341 top left** Courtesy of Ron Greeley **341 top right** Courtesy of Ron Greeley **341 center left** Courtesy of Ron Greeley **341 center right** Courtesy of Ron Greeley **341 bottom** NASA/JPL Caltech/LANL/CNES/IRAP/ LPGNantes/CNRS/IAS/MSSS **342** ESO/Y. Beletsky **343 bottom** Lunin Gleb/ITAR-TASS Photo/Avalon/Photoshot **343 top** NASA **343 center** Science Photo Library/Science Source **351 all** Larry Marschall and Christy Zuidema, Gettysburg College.

Chapter 13

Pages 352–353 Visuals Unlimited/Getty Images **367** ESA/Hubble **371** Norman Lockyer Observatory, Sidmouth **372 left** GL Archive/Alamy Stock Photo **372 top right** Science History Images/Alamy Stock Photo **372 center right** Paul Fearn/Alamy Stock Photo **372 top center** GL Archive/Alamy Stock Photo **372 center right** Interfoto/Alamy Stock Photo **375** Norman Lockyer Observatory, Sidmouth.

Chapter 14

Pages 382–383 NASA/SDO/AIA **390** Hinode JAXA/NASA/PPARC **394 both** Brocken Inaglory, 2006; http://creativecommons.org/licenses/by-sa/3.0/deed.en **395 top** Nigel Sharp, NOAO/NSO/Kitt Peak FTS/AURA/NSF **395 right** Westend61 GmbH/Alamy Stock Photo **395 center bottom** John Chumack/Science Source **395 center left** Hinode JAXA/NASA/NASA **396** NASA **398 left** NASA/SDO/Solar Dynamics Observatory **398 right** NASA/SDO/Solar Dynamics Observatory **399** SOHO/ESA/NASA **400** NASA/SDO **401 top left** SOHO/ESA and NASA **401 top center** NASA SDO **401 top right** SOHO/ESA and NASA **401 bottom** SOHO/ESA and NASA **403** JPL/NASA.

Chapter 15

Pages 410–411 ESO **412 top** Axel Mellinger **412 both** NASA/JPL-Caltech/IRAS/2MASS/COBE **413** U.S. Coastguard **415 bottom** IRAS/COBE (NASA) **415 top** NASA/JPL-Caltech/UCLA **416 bottom** Dr. Douglas Finkbeiner **416 top** Masashi Kimura/MAXI/SSC **417 top center** Courtesy of Emmanuel Mallart www.astrophoto-mallart.com & www.axisinstruments.com, **417 top right** ESO **417 top center** Jean-Charles Cuillandre (CFHT), Hawaiian Starlight, CFHT. **417 bottom** NRAO/AUI **418 top** ESO **418 bottom** ESO **420** ESA/Planck Collaboration **423 left** ESA/Herschel/PACS/SPIRE/Hill, Motte, HOBYS Key Programme Consortium **423 center** MPG/ESO **423 right** VLT/ISAAC/McCaughrean & Andersen/AIP/ESO **426** NASA/JPL-Caltech/UCB **429 bottom** ESO/M. McCaughrean **429 top** NASA/JPL-Caltech/UIUC **431 bottom** NASA/JPL-Caltech/UIUC **431 top** Science Photo Library/Alamy Stock Photo.

Chapter 16

Pages 438–439 NASA and ESA **453 all** NASA, ESA and The Hubble Heritage Team **460 all** NASA.

Chapter 17

Pages 468–469 NASA/CXC/SAO **474** NASA/CXC/GSFC/M.Corcoran et al.; Optical: NASA/STScI **478** NASA/CXC/SAO **479 center** Alexandra Angelich (NRAO/AUI/NSF); NASA Hubble; NASA Chandra **479 bottom** NASA/CXC/SAO/PSU **479 top** NASA/CXC/SAO/PSU **482 bottom center** Emilio Segré Visual Archives/American Institute of Physics/Science Source **482 top center** PA Images/Alamy Stock Photo **482 bottom left** Romolo Paradisi/ESO **482 top left** Science and Society/Superstock **482 bottom right** Science Photo Library/Superstock **482 top right** Science Photo Library/Superstock **485 bottom** Hillary Mathis/REU Program/NOAO/AURA/NSF/Science Source **485 top** NASA **486 right** Barbara Mochejska Hedberg, J. Kaluzny (Copernicus Astronomical Center) **486 left** European Space Agency & NASA.

Chapter 18

Pages 496–497 NASA/ESA **510 right** Heinz-Dieter Falkenstein/age fotostock/Superstock **510 left** Underwood Photo Archives/Superstock **511 bottom**

ESA/Hubble & NASA **511 top** NASA/ESA **515** NASA/CXC/M.Weiss **517 top** NASA/CXC/Cal; Chandra X-ray Observatory Center **517 bottom** NASA/CXC/Cal; Chandra X-ray Observatory Center.

Chapter 19

Pages 524–525 NASA, ESA, and The Hubble Heritage Team (STScI/AURA) **526** NASA, ESA, and the Hubble Heritage Team (STScI/AURA) **527 top left** ESA/Hubble & NASA **527 bottom left** ESA/Hubble & NASA **527 bottom right** ESA/Hubble & NASA; Acknowledgement: Judy Schmidt **527 top right** ESO/INAF-VST/A. Grado/L. Limatola/INAF-Capodimonte Observatory **527 bottom center** NASA **527 top center** NASA/Hubble Heritage Team **528 top left** NOAO/AURA/NSF **528 top center left** NASA/JPL-Caltech **528 top center right** NASA/JPL-Caltech **528 top right** NASA/JPL-Caltech **528 center (all)** NASA/JPL-Caltech **528 bottom (all)** NASA/JPL-Caltech **529 top** NOAO/AURA/NSF **529 bottom** NOAO/AURA/NSF **530** NASA and The Hubble Heritage Team (STScI/AURA) **531 top both** European Southern Observatory **531 bottom right** NASA, ESA, Hubble Heritage Team (STScI/AURA) **531 bottom left** NOAO/AURA/NSF/Science Source **533 top** BJ Fulton/Bryne Observatory at Sedgewick Reserve and the Palomar Transient Factory **533 bottom** NASA/Swift/Peter Brown, Univ. of Utah **535 both** NASA **538 top** NASA/CXC/Penn State/G. Garmire; Optical: NASA/ESA/STScI/M. West **538 bottom** NASA/CXC/Penn State/G. Garmire; Optical: NASA/ESA/STScI/M. West **540 top** HST **540 bottom** NASA/CXC/CfA/R.Kraft et al; Radio: NSF/VLA/Univ.Hertfordshire/M.Hardcastle; Optical: ESO/WFI/M.Rejkuba et al. **544** X-ray (NASA/CXC/KIPAC/N. Werner, E. Million et al); Radio (NRAO/AUI/NSF/F. Owen **546** NASA/CXC/CfA/R.Kraft et al; Radio: NSF/VLA/Univ.Hertfordshire/M.Hardcastle; Optical: ESO/WFI/M.Rejkuba et al. **548** NASA, ESA, and J. Lotz and the HFF Team (STScI) **553** NASA, ESA, and the Hubble Heritage Team (STScI/AURA).

Chapter 20

Pages 554–555 Brad Goldpaint/Aurora Open/Alamy **556 bottom** C. Howk and B. Savage (University of Wisconsin); N. Sharp (NOAO)/WIYN, Inc. **556 top** NG Images/Alamy Stock Photo **557 top** NASA/JPL-Caltech/ESO/R. Hurt **557 center** NOAO/AURA/NSF **557 bottom left** NASA/JPL-Caltech **557 bottom right** Robert Gendler/Corbis/Getty Images **558 bottom** Richard Rand, University of New Mexico **558 top** Todd Boroson/ NOAO/AURA/NSF **561 top left** Emilio Segré Visual Archives/American Institute of Physics/Science Source **561 bottom left** John Irwin Collection/AIP/Science Source **561 center right** NASA Goddard **561 top right** NASA/JPL-Caltech/ESA/Institute of Astrophysics of Andalusia, University of Basque **561 bottom right** NASA/JPL-Caltech/ESA/Institute of Astrophysics of Andalusia, University of Basque **564 center** ESA/Gaia/DPAC **564 bottom** ESA/Gaia/DPAC **565** NASA **566 top** ASPERA/G.Toma/A.Saftoiu **566 bottom** Courtesy of Pierre Auger Observatory **567** DPAC/ESA **568 both** NASA **569** UCLA Galactic Center Group - W.M. Keck Observatory Laser Team **570** NASA's Goddard Space Flight Center Image of Fermi all-sky map, with bubbles outlined courtesy of NASA/GSFC/DOE/Fermi LAT/D.Finkbeiner et al. **571 bottom** Amanda Smith/University of Cambridge **571 center** ESA/Gaia/DPAC **571 top** Royal Observatory, Edinburgh/Science Source **572 top** NASA **572 bottom (all)** NASA, ESA, Z. Levay and R. van der Marel (STScI), T. Hallas, and A. Mellinger **574** ESA/Gaia/DPAC.

Chapter 21

Pages 580–581 ESA and the Planck Collaboration **586 bottom** The LIFE Images Collection./Getty Images **586 top** Underwood Photo Archives/Superstock **594** Lucent Technologies, Bell Labs **595 top** Emil Oprisa/Alamy Stock Photo **595 bottom** Sasho Bogoev/Alamy Stock Photo **596 bottom** ESA and the Planck Collaboration **596 top** NASA COBE Science Team **596 top center** NASA COBE Science Team **596 bottom center** NASA COBE Science Team **603 all** Courtesy of Stacy Palen.

Chapter 22

Pages 604–605 NASA/JPL/Space Science Institute **610 bottom** The LIFE Images Collection./Getty Images **610 top** Underwood Photo Archives/Superstock **633** CERN PhotoLab.

Chapter 23

Pages 634–635 NASA, ESA, H. Teplitz and M. Rafelski (IPAC/Caltech), A. Koekemoer (STScI), R. Windhorst (Arizona State University), and Z. Levay (STScI) **636 top** NASA/ESA/ESO/NAOJ/G. Paglioli **636 bottom** R. Brent Tully (U. Hawaii) et al., SDvision, DP, CEA/Saclay **637 both** Max Tegmark/SDSS Collaboration **638 both** ASA/CXC/MIT/E.-H Peng et al; Optical: NASA/STScI **638** NASA, ESA, Jennifer Lotz and the HFF Team (STScI) **639 both** NASA, A. Fruchter and the ERO team (STScI, ST-ECF). NASA, ESA **643** NASA/WMAP Science Team **644** NASA/JPL-Caltech/ESA/Institute of Astrophysics of Andalusia, University of Basque **645** Nicolle R. Fuller, National Science Foundation **646 both** NASA/JPL-Caltech/GSFC **647 right** NASA , ESA, and B. Salmon (STScI) **647 left** NASA, ESA, and P. Oesch (Yale University) **649 both** Guedes, Javiera, et al. The Astrophysical Journal, Volume 742, Issue 2, article id. 76 (2011). **650 bottom** NASA/CXC/SAO/P **650 top** NASA/Hubble Space Telescope Cosmic Assembly Near-infrared Deep Extragalactic Legacy Survey **651 top, both** NASA, ESA, and N. Pirzkal (STScI/ESA); NASA, ESA, J. Lotz (STScI), M. Davis (University of California, Berkeley), and A. Koekemoer (STScI) **651 bottom left** NASA/CXC/CfA/M.Markevitch et aL. **651 bottom right** NASA/STScI; Magellan/U.Arizona/D. Clowe et al. **652 top, all** Courtesy Joel Primack and George Blumenthal **652 bottom, all** Courtesy Joel Primack and George Blumenthal **653 top, both** Courtesy Joel Primack and George Blumenthal **653 bottom** NASA, ESA, and A. Feild (STScI).

Chapter 24

Pages 660–661 Seth Shostak/SETI Institute **662 top** Dr. Michael Perfit, University of Florida, Robert Embley/NOAA **662 bottom** Woods Hole Oceanographic Institution, Deep Submergence Operations Group, Dan Fornari **663 center** Chris Boydell/Getty Images **663 top** Gaertner/Alamy **663 bottom** National Park Service Photo by Jim Peaco **664 bottom** Dr. Robert Calentine/Getty Images **664 top** Peter Essick/Aurora Photos **666** Courtesy of National Science Foundation **667** Lucas Oleniuk/Getty Images **670** NASA/JPL-Caltech/MSSS **671 top** NASA/JPL-Caltech/MSSS **671 center** NASA/JPL-Caltech/UA/USGS **671 bottom** PHL@UPR Arecibo, NASA **674** NASA/Wikimedia Commons/Joshua Krissansen-Totton **675 bottom** F. Drake (UCSC) et al., Arecibo Observatory (Cornell, NAIC) **675 top** NASA **677** F. Drake (UCSC) et al., Arecibo Observatory (Cornell, NAIC) **679** NASA/JPL-Caltech/Space Science Institute.

Line Art

Figure 7.17: Graph: "Radial Velocity/ Year" from Exoplantets.org. Reprinted by permission of the Department of Terrestrial Magnetism, Carnegie Institution of Washington.
Figure 10.18: Figure from "First Earth-Sized Planets Found," Harvard-Smithsonian Center for Astrophysics/ David A. Aguilar (CfA), December 20, 2011. Reprinted with permission.
Figure 18.21: "Black Holes with Known Masses." Courtesy Caltech/MIT/LIGO Laboratory.
Figure 23.10: Figure from "The First Galaxies," by V. Bromm & N. Yoshida. Reproduced with permission of Annual Review of Astronomy & Astrophysics, Volume 49 (373-407) © by Annual Reviews, http://www.annualreviews.org
Figure 24.12: "The Origin of the Solar System Elements," Graphic created by Jennifer Johnson. Astronomical Image Credits: ESA/NASA/AASNova. Licensed under a Creative Commons Attribution-ShareAlike 4.0 International License.
Figure 24.17: "Oceans of the Solar System," from PHL @UPR Arecibo. Reprinted by permission of PHL@UPR Arecibo.
Figure 24.18: Figure 1 from "Summary of the Limits of the New Habitable Zone." PHL.UPR.edu, March 23, 2013. Reprinted by permission of PHL@UPR Arecibo.

Index

Page numbers in *italics* refer to illustrations and tables.